T0145094

Lecture Notes in Artificial Intelligence   11052

Subseries of Lecture Notes in Computer Science

LNAI Series Editors

Randy Goebel
University of Alberta, Edmonton, Canada
Yuzuru Tanaka
Hokkaido University, Sapporo, Japan
Wolfgang Wahlster
DFKI and Saarland University, Saarbrücken, Germany

LNAI Founding Series Editor

Joerg Siekmann
DFKI and Saarland University, Saarbrücken, Germany

More information about this series at http://www.springer.com/series/1244

Michele Berlingerio · Francesco Bonchi
Thomas Gärtner · Neil Hurley
Georgiana Ifrim (Eds.)

Machine Learning and Knowledge Discovery in Databases

European Conference, ECML PKDD 2018
Dublin, Ireland, September 10–14, 2018
Proceedings, Part II

 Springer

Editors
Michele Berlingerio
IBM Research - Ireland
Dublin, Ireland

Neil Hurley ⓘ
University College Dublin
Dublin, Ireland

Francesco Bonchi
Institute for Scientific Interchange
Turin, Italy

Georgiana Ifrim ⓘ
University College Dublin
Dublin, Ireland

Thomas Gärtner
University of Nottingham
Nottingham, UK

ISSN 0302-9743   ISSN 1611-3349   (electronic)
Lecture Notes in Artificial Intelligence
ISBN 978-3-030-10927-1   ISBN 978-3-030-10928-8   (eBook)
https://doi.org/10.1007/978-3-030-10928-8

Library of Congress Control Number: 2018965756

LNCS Sublibrary: SL7 – Artificial Intelligence

This Springer imprint is published by the registered company Springer Nature Switzerland AG
The registered company address is: Gewerbestrasse 11, 6330 Cham, Switzerland

Preface

We are delighted to introduce the proceedings of the 2018 edition of the European Conference on Machine Learning and Principles and Practice of Knowledge Discovery in Databases (ECML-PKDD 2018). The conference was held in Dublin, Ireland, during September 10–14, 2018. ECML-PKDD is an annual conference that provides an international forum for the discussion of the latest high-quality research results in all areas related to machine learning and knowledge discovery in databases, including innovative applications. This event is the premier European machine learning and data mining conference and builds upon a very successful series of ECML-PKDD conferences.

The scientific program was of high quality and consisted of technical presentations of accepted papers, plenary talks by distinguished keynote speakers, workshops, and tutorials. Accepted papers were organized in five different tracks:

- The Conference Track, which featured research contributions, presented as part of the main conference program
- The Journal Track, which featured papers that were reviewed and published separately in special issues of the Springer journals *Machine Learning* and *Data Mining and Knowledge Discovery*, and that were selected as suitable for presentation at the conference
- The Applied Data Science Track, which focused on the application of data science to practical real-world scenarios, including contributions from academia, industry, and non-governmental organizations
- The Demo Track, which presented working demonstrations of prototypes or fully operational systems that exploit data science techniques
- The Nectar Track, which presented an overviews of recent scientific advances at the frontier of machine learning and data mining in conjunction with other disciplines, as published in related conferences and journals

In addition to these tracks, the conference also included a PhD Forum in which PhD students received constructive feedback on their research progress and interacted with their peers. Co-located with the conference, this year there were 17 workshops on related research topics and six tutorial presentations.

In total, 95% of the accepted papers in the conference have accompanying software and/or data and are flagged as Reproducible Research papers in the proceedings. This speaks to the growing importance of reproducible research for the ECML-PKDD community. In the online proceedings each Reproducible Research paper has a link to the code and data made available with the paper. We believe this is a tremendous resource for the community and hope to see this trend maintained over the coming years.

We are very happy with the continued interest from the research community in our conference. We received 353 papers for the main conference track, of which 94 were accepted, yielding an acceptance rate of about 26%. This allowed us to define a very rich program with 94 presentations in the main conference track. Moreover, there was a 26% acceptance rate (143 submissions, 37 accepted) to the Applied Data Science Track and 15% to the Journal Track (151 submissions, 23 accepted). Including the Nectar Track papers and some Journal Track papers from the 2017 special issues that were held over for presentation until 2018, we had in total 166 parallel scientific talks during the three main conference days.

The program also included five plenary keynotes by invited speakers: Misha Bilenko (Head of Machine Intelligence and Research Yandex, Moscow, Russia), Corinna Cortes (Head of Google Research New York, USA), Aristides Gionis (Professor, Department of Computer Science, Aalto University, Helsinki, Finland), Cynthia Rudin (Associate Professor of Computer Science and Electrical and Computer Engineering, Duke University, Durham, North Carolina, USA), and Naftali Tishby (Professor, School of Engineering and Computer Science, Hebrew University of Jerusalem, Israel).

This year, ECML-PKDD attracted over 630 participants from 42 countries. It attracted substantial attention from industry both through sponsorship and submission/participation at the conference and workshops. Moreover, ECML-PKDD hosted a very popular Nokia Women in Science Luncheon to discuss the importance of awareness of equal opportunity and support for women in science and technology.

The Awards Committee selected research papers considered to be of exceptional quality and worthy of special recognition:

- ML Student Best Paper Award: "Hyperparameter Learning for Conditional Mean Embeddings with Rademacher Complexity Bounds," by Kelvin Hsu, Richard Nock and Fabio Ramos
- KDD Student Best Paper Award: "Anytime Subgroup Discovery in Numerical Domains with Guarantees," by Aimene Belfodil, Adnene Belfodil and Mehdi Kaytoue.

We would like to thank all participants, authors, reviewers, and organizers of the conference for their contribution to making ECML-PKDD 2018 a great scientific event.

We would also like to thank the Croke Park Conference Centre and the student volunteers. Thanks to Springer for their continuous support and Microsoft for allowing us to use their CMT software for conference management and providing support throughout. Special thanks to our many sponsors and to the ECML-PKDD Steering Committee for their support and advice. Finally, we would like to thank the organizing institutions: the Insight Centre for Data Analytics, University College Dublin, Ireland, and IBM Research, Ireland.

September 2018

Michele Berlingerio
Francesco Bonchi
Thomas Gärtner
Georgiana Ifrim
Neil Hurley

Organization

ECML PKDD 2018 Organization

General Chairs

Michele Berlingerio	IBM Research, Ireland
Neil Hurley	University College Dublin, Ireland

Program Chairs

Michele Berlingerio	IBM Research, Ireland
Francesco Bonchi	ISI Foundation, Italy
Thomas Gärtner	University of Nottingham, UK
Georgiana Ifrim	University College Dublin, Ireland

Journal Track Chairs

Björn Bringmann	McKinsey & Company, Germany
Jesse Davis	Katholieke Universiteit Leuven, Belgium
Elisa Fromont	IRISA, Rennes 1 University, France
Derek Greene	University College Dublin, Ireland

Applied Data Science Track Chairs

Edward Curry	National University of Ireland Galway, Ireland
Alice Marascu	Nokia Bell Labs, Ireland

Workshop and Tutorial Chairs

Carlos Alzate	IBM Research, Ireland
Anna Monreale	University of Pisa, Italy

Nectar Track Chairs

Ulf Brefeld	Leuphana University of Lüneburg, Germany
Fabio Pinelli	Vodafone, Italy

Demo Track Chairs

| Elizabeth Daly | IBM Research, Ireland |
| Brian Mac Namee | University College Dublin, Ireland |

PhD Forum Chairs

| Bart Goethals | University of Antwerp, Belgium |
| Dafna Shahaf | Hebrew University of Jerusalem, Israel |

Discovery Challenge Chairs

| Martin Atzmüller | Tilburg University, The Netherlands |
| Francesco Calabrese | Vodafone, Italy |

Awards Committee

Tijl De Bie	Ghent University, Belgium
Arno Siebes	Uthrecht University, The Netherlands
Bart Goethals	University of Antwerp, Belgium
Walter Daelemans	University of Antwerp, Belgium
Katharina Morik	TU Dortmund, Germany

Professional Conference Organizer

| Keynote PCO | Dublin, Ireland |

www.keynotepco.ie

ECML PKDD Steering Committee

Michele Sebag	Université Paris Sud, France
Francesco Bonchi	ISI Foundation, Italy
Albert Bifet	Télécom ParisTech, France
Hendrik Blockeel	KU Leuven, Belgium and Leiden University, The Netherlands
Katharina Morik	University of Dortmund, Germany
Arno Siebes	Utrecht University, The Netherlands
Siegfried Nijssen	LIACS, Leiden University, The Netherlands
Chedy Raïssi	Inria Nancy Grand-Est, France
João Gama	FCUP, University of Porto/LIAAD, INESC Porto L.A., Portugal
Annalisa Appice	University of Bari Aldo Moro, Italy
Indré Žliobaité	University of Helsinki, Finland
Andrea Passerini	University of Trento, Italy
Paolo Frasconi	University of Florence, Italy
Céline Robardet	National Institute of Applied Science Lyon, France

Jilles Vreeken   Saarland University, Max Planck Institute for Informatics,
                        Germany
Sašo Džeroski   Jožef Stefan Institute, Slovenia
Michelangelo Ceci   University of Bari Aldo Moro, Italy
Myra Spiliopoulu   Magdeburg University, Germany
Jaakko Hollmén   Aalto University, Finland

Area Chairs

Michael Berthold   Universität Konstanz, Germany
Hendrik Blockeel   KU Leuven, Belgium and Leiden University, The Netherlands
Ulf Brefeld   Leuphana University of Lüneburg, Germany
Toon Calders   University of Antwerp, Belgium
Michelangelo Ceci   University of Bari Aldo Moro, Italy
Bruno Cremilleux   Université de Caen Normandie, France
Tapio Elomaa   Tampere University of Technology, Finland
Johannes Fürnkranz   TU Darmstadt, Germany
Peter Flach   University of Bristol, UK
Paolo Frasconi   University of Florence, Italy
João Gama   FCUP, University of Porto/LIAAD, INESC Porto L.A.,
                        Portugal
Jaakko Hollmén   Aalto University, Finland
Alipio Jorge   FCUP, University of Porto/LIAAD, INESC Porto L.A.,
                        Portugal
Stefan Kramer   Johannes Gutenberg University Mainz, Germany
Giuseppe Manco   ICAR-CNR, Italy
Siegfried Nijssen   LIACS, Leiden University, The Netherlands
Andrea Passerini   University of Trento, Italy
Arno Siebes   Utrecht University, The Netherlands
Myra Spiliopoulu   Magdeburg University, Germany
Luis Torgo   Dalhousie University, Canada
Celine Vens   KU Leuven, Belgium
Jilles Vreeken   Saarland University, Max Planck Institute for Informatics,
                        Germany

Conference Track Program Committee

Carlos Alzate   Roberto Bayardo   Indrajit Bhattacharya
Aijun An   Martin Becker   Marenglen Biba
Fabrizio Angiulli   Srikanta Bedathur   Silvio Bicciato
Annalisa Appice   Jessa Bekker   Mario Boley
Ira Assent   Vaishak Belle   Gianluca Bontempi
Martin Atzmueller   Andras Benczur   Henrik Bostrom
Antonio Bahamonde   Daniel Bengs   Tassadit Bouadi
Jose Balcazar   Petr Berka   Pavel Brazdil

Dariusz Brzezinski	Cesar Ferri	Latifur Khan
Rui Camacho	Răzvan Florian	Frank Klawonn
Longbing Cao	Eibe Frank	Jiri Klema
Francisco Casacuberta	Elisa Fromont	Tomas Kliegr
Peggy Cellier	Fabio Fumarola	Marius Kloft
Loic Cerf	Esther Galbrun	Dragi Kocev
Tania Cerquitelli	Patrick Gallinari	Levente Kocsis
Edward Chang	Dragan Gamberger	Yun Sing Koh
Keke Chen	Byron Gao	Alek Kolcz
Weiwei Cheng	Paolo Garza	Irena Koprinska
Silvia Chiusano	Konstantinos Georgatzis	Frederic Koriche
Arthur Choi	Pierre Geurts	Walter Kosters
Frans Coenen	Dorota Glowacka	Lars Kotthoff
Mário Cordeiro	Nico Goernitz	Danai Koutra
Roberto Corizzo	Elsa Gomes	Georg Krempl
Vitor Santos Costa	Mehmet Gönen	Tomas Krilavicius
Bertrand Cuissart	James Goulding	Yamuna Krishnamurthy
Boris Cule	Michael Granitzer	Matjaz Kukar
Tomaž Curk	Caglar Gulcehre	Meelis Kull
James Cussens	Francesco Gullo	Prashanth L. A.
Alfredo Cuzzocrea	Stephan Günnemann	Jorma Laaksonen
Claudia d'Amato	Tias Guns	Nicolas Lachiche
Maria Damiani	Sara Hajian	Leo Lahti
Tijl De Bie	Maria Halkidi	Helge Langseth
Martine De Cock	Jiawei Han	Thomas Lansdall-Welfare
Juan Jose del Coz	Mohammad Hasan	Christine Largeron
Anne Denton	Xiao He	Pedro Larranaga
Christian Desrosiers	Denis Helic	Silvio Lattanzi
Nicola Di Mauro	Daniel Hernandez-Lobato	Niklas Lavesson
Claudia Diamantini	Jose Hernandez-Orallo	Binh Le
Uwe Dick	Thanh Lam Hoang	Freddy Lecue
Tom Diethe	Frank Hoeppner	Florian Lemmerich
Ivica Dimitrovski	Arjen Hommersom	Jiuyong Li
Wei Ding	Tamas Horvath	Limin Li
Ying Ding	Andreas Hotho	Jefrey Lijffijt
Stephan Doerfel	Yuanhua Huang	Tony Lindgren
Carlotta Domeniconi	Eyke Hüllermeier	Corrado Loglisci
Frank Dondelinger	Dino Ienco	Peter Lucas
Madalina Drugan	Szymon Jaroszewicz	Brian Mac Namee
Wouter Duivesteijn	Giuseppe Jurman	Gjorgji Madjarov
Inŝ Dutra	Toshihiro Kamishima	Sebastian Mair
Dora Erdos	Michael Kamp	Donato Malerba
Fabio Fassetti	Bo Kang	Luca Martino
Ad Feelders	Andreas Karwath	Elio Masciari
Stefano Ferilli	George Karypis	Andres Masegosa
Carlos Ferreira	Mehdi Kaytoue	Florent Masseglia

Ernestina Menasalvas
Corrado Mencar
Rosa Meo
Pauli Miettinen
Dunja Mladenic
Karthika Mohan
Anna Monreale
Joao Moreira
Mohamed Nadif
Ndapa Nakashole
Jinseok Nam
Mirco Nanni
Amedeo Napoli
Nicolo Navarin
Benjamin Negrevergne
Benjamin Nguyen
Xia Ning
Kjetil Norvag
Eirini Ntoutsi
Andreas Nurnberger
Barry O'Sullivan
Dino Oglic
Francesco Orsini
Nikunj Oza
Pance Panov
Apostolos Papadopoulos
Panagiotis Papapetrou
Ioannis Partalas
Gabriella Pasi
Dino Pedreschi
Jaakko Peltonen
Ruggero Pensa
Iker Perez
Nico Piatkowski
Andrea Pietracaprina
Gianvito Pio
Susanna Pirttikangas
Marc Plantevit
Pascal Poncelet
Miguel Prada
Philippe Preux
Buyue Qian
Chedy Raissi
Jan Ramon
Huzefa Rangwala
Zbigniew Ras

Chotirat Ratanamahatana
Jan Rauch
Chiara Renso
Achim Rettinger
Fabrizio Riguzzi
Matteo Riondato
Celine Robardet
Juan Rodriguez
Fabrice Rossi
Celine Rouveirol
Stefan Rueping
Salvatore Ruggieri
Yvan Saeys
Alan Said
Lorenza Saitta
Tomoya Sakai
Alessandra Sala
Ansaf Salleb-Aouissi
Claudio Sartori
Pierre Schaus
Lars Schmidt-Thieme
Christoph Schommer
Matthias Schubert
Konstantinos Sechidis
Sohan Seth
Vinay Setty
Junming Shao
Nikola Simidjievski
Sameer Singh
Andrzej Skowron
Dominik Slezak
Kevin Small
Gavin Smith
Tomislav Smuc
Yangqiu Song
Arnaud Soulet
Wesllen Sousa
Alessandro Sperduti
Jerzy Stefanowski
Giovanni Stilo
Gerd Stumme
Mahito Sugiyama
Mika Sulkava
Einoshin Suzuki
Stephen Swift
Andrea Tagarelli

Domenico Talia
Letizia Tanca
Jovan Tanevski
Nikolaj Tatti
Maguelonne Teisseire
Georgios Theocharous
Ljupco Todorovski
Roberto Trasarti
Volker Tresp
Isaac Triguero
Panayiotis Tsaparas
Vincent S. Tseng
Karl Tuyls
Niall Twomey
Nikolaos Tziortziotis
Theodoros Tzouramanis
Antti Ukkonen
Toon Van Craenendonck
Martijn Van Otterlo
Iraklis Varlamis
Julien Velcin
Shankar Vembu
Deepak Venugopal
Vassilios S. Verykios
Ricardo Vigario
Herna Viktor
Christel Vrain
Willem Waegeman
Jianyong Wang
Joerg Wicker
Marco Wiering
Martin Wistuba
Philip Yu
Bianca Zadrozny
Gerson Zaverucha
Bernard Zenko
Junping Zhang
Min-Ling Zhang
Shichao Zhang
Ying Zhao
Mingjun Zhong
Albrecht Zimmermann
Marinka Zitnik
Indré Žliobaité

Applied Data Science Track Program Committee

Oznur Alkan	Sidath Handurukande	Nikunj Oza
Carlos Alzate	Souleiman Hasan	Ioannis Partalas
Nicola Barberi	Georges Hebrail	Milan Petkovic
Gianni Barlacchi	Thanh Lam	Fabio Pinelli
Enda Barrett	Neil Hurley	Yongrui Qin
Roberto Bayardo	Hongxia Jin	Rene Quiniou
Srikanta Bedathur	Anup Kalia	Ambrish Rawat
Daniel Bengs	Pinar Karagoz	Fergal Reid
Cuissart Bertrand	Mehdi Kaytoue	Achim Rettinger
Urvesh Bhowan	Alek Kolcz	Stefan Rueping
Tassadit Bouadi	Deguang Kong	Elizeu Santos-Neto
Thomas Brovelli	Lars Kotthoff	Manali Sharma
Teodora Sandra	Nick Koudas	Alkis Simitsis
Berkant Barla	Hardy Kremer	Kevin Small
Michelangelo Ceci	Helge Langseth	Alessandro Sperduti
Edward Chang	Freddy Lecue	Siqi Sun
Soumyadeep Chatterjee	Zhenhui Li	Pal Sundsoy
Abon Chaudhuri	Lin Liu	Ingo Thon
Javier Cuenca	Jiebo Luo	Marko Tkalcic
Mahashweta Das	Arun Maiya	Luis Torgo
Viktoriya Degeler	Silviu Maniu	Radu Tudoran
Wei Ding	Elio Masciari	Umair Ul
Yuxiao Dong	Luis Matias	Jan Van
Carlos Ferreira	Dimitrios Mavroeidis	Ranga Vatsavai
Andre Freitas	Charalampos	Fei Wang
Feng Gao	Mavroforakis	Xiang Wang
Dinesh Garg	James McDermott	Wang Wei
Guillermo Garrido	Daniil Mirylenka	Martin Wistuba
Rumi Ghosh	Elena Mocanu	Erik Wittern
Martin Gleize	Raul Moreno	Milena Yankova
Slawek Goryczka	Bogdan Nicolae	Daniela Zaharie
Riccardo Guidotti	Maria-Irina Nicolae	Chongsheng Zhang
Francesco Gullo	Xia Ning	Yanchang Zhao
Thomas Guyet	Sean O'Riain	Albrecht Zimmermann
Allan Hanbury	Adegboyega Ojo	

Nectar Track Program Committee

Annalisa Appice	Peter Flach	Ernestina Menasalvas
Martin Atzmüller	Johannes Fürnkranz	Mirco Musolesi
Hendrik Blockeel	Joao Gama	Franco Maria
Ulf Brefeld	Andreas Hotho	Maryam Tavakol
Tijl De	Kristian Kersting	Gabriele Tolomei
Kurt Driessens	Sebastian Mair	Salvatore Trani

Demo Track Program Committee

Gustavo Carneiro	Brian Mac Namee	Konstantinos Skiannis
Derek Greene	Susan McKeever	Jerzy Stefanowski
Mark Last	Joao Papa	Luis Teixeira
Vincent Lemaire	Niladri Sett	Grigorios Tsoumakas

Contents – Part II

Online and Active Learning

Pattern and Sequence Mining

Probabilistic Models and Statistical Methods

Recommender Systems

Transfer Learning

Contents – Part III

ADS Engineering and Design

ADS Financial/Security

ADS Health

ADS Sensing/Positioning

Demo Track

Contents – Part I

Deep Learning

Ensemble Methods

Evaluation

Graphs

Temporally Evolving Community Detection and Prediction in Content-Centric Networks

Ana Paula Appel[1]([⊠]), Renato L. F. Cunha[1], Charu C. Aggarwal[2],
and Marcela Megumi Terakado[3]

[1] IBM Research, São Paulo, Brazil
{apappel,renatoc}@br.ibm.com
[2] IBM Research, Yorktown, NY, USA
charu@us.ibm.com
[3] University of São Paulo, São Paulo, Brazil
terakado@ime.usp.br

Abstract. In this work, we consider the problem of combining link, content and temporal analysis for community detection and prediction in evolving networks. Such temporal and content-rich networks occur in many real-life settings, such as bibliographic networks and question answering forums. Most of the work in the literature (that uses both content and structure) deals with static snapshots of networks, and they do not reflect the dynamic changes occurring over multiple snapshots. Incorporating dynamic changes in the communities into the analysis can also provide useful insights about the changes in the network such as the migration of authors across communities. In this work, we propose *Chimera* (https://github.com/renatolfc/chimera-stf), a shared factorization model that can simultaneously account for graph links, content, and temporal analysis. This approach works by extracting the latent semantic structure of the network in multidimensional form, but in a way that takes into account the temporal continuity of these embeddings. Such an approach simplifies temporal analysis of the underlying network by using the embedding as a surrogate. A consequence of this simplification is that it is also possible to use this temporal sequence of embeddings to predict future communities. We present experimental results illustrating the effectiveness of the approach. Code related to this paper is available at: https://github.com/renatolfc/chimera-stf.

1 Introduction

Structural representations of data are ubiquitous in different domains such as biological networks, online social networks, information networks, co-authorship networks, and so on. The problem of community detection or graph clustering aims to identify densely connected groups of nodes in the network [8], one of

M. M. Terakado—Work done while at IBM Research.

the central tasks in network analysis. Examples of useful applications include that of finding clusters in protein-protein interaction networks [24] or groups of people with similar interests in social networks [31]. Recently, it has become easier to collect content-centric networks in a time-sensitive way, enabling the possibility of using tightly-integrated analysis across different factors that affect network structure. Aggregate topological and content information can enable more informative community detection, in which cues from different sources are integrated into more powerful models.

Another important aspect of complex networks is that such networks evolve, meaning that nodes may move from one community to another, making some communities grow, and others shrink. For example, authors that usually publish in the data mining community could move to the machine learning community. Furthermore, the temporal aspects of changes in community structure could interact with the content in unusual ways. For example, it is possible for an author in a bibliographic network to change their topic of work, preceding a corresponding change in community structure. The converse is also possible, with a change in community structure affecting content-centric attributes.

Matrix factorization methods are traditional techniques that allow us to reduce the dimensional space of network adjacency representations. Such methods have broad applicability in various tasks such as clustering, dimensionality reduction, latent semantic analysis, and recommender systems. The main point of matrix factorization methods is that they embed matrices in a *latent space* where the clustering characteristics of the data are often amplified. A useful variant of matrix factorization methods is *shared* matrix factorization, which factors two or more different matrices simultaneously. Shared matrix factorization is not new, and is used in various settings where different matrices define different parts of the data (e.g., links and content). This method could be used to embed link and content in a shared feature space, which is convenient because it allows the use of traditional clustering techniques, such as k-means. However, incorporating a temporal aspect to the shared factorization process adds some challenges concerning the adjustment of the shared factorization as data evolves.

Related to the problem of community detection is that of community prediction, in which one attempts to predict future communities from previous snapshots of the network. It is notoriously difficult to predict future clustering structures from a complex combination of data types such as links and content. However, the matrix factorization methodology provides a nice abstraction, because one can now use the multidimensional representations created by sequences of matrices over different snapshots. The basic idea is that we can consider each of the entries in the latent space representation as a stream of evolving entries, which implicitly creates a time-series.

In this work, we present *Chimera*, a method that uses link and content from networks over time to detect and predict community structure. To the best of our knowledge, there is no work addressing these three aspects *simultaneously* for both detection and prediction of communities. The main contributions of this paper are:

- An efficient algorithm based on shared matrix factorization that uses link and content over time; the uniform nature of the embedding allows the use of any traditional clustering algorithm on the corresponding representation.
- A method for predicting future communities from embeddings over snapshots.

2 Related Work

In this section, we review the existing work for community detection using link analysis, content analysis, temporal analysis and their combination. Since these methods are often proposed in different contexts (sometimes even by different communities), we will organize these methods into separate sections.

Topological Community Detection: These methods are based mainly on links among nodes. The idea is to minimize the number of edges across nodes belonging to different communities. Thus, the nodes inside the community should have a higher density (number of edges) with other nodes inside the community than with nodes outside the community. There are several ways of defining and quantifying communities based on their topology, modularity [4], conductance [16], betweeness [9], and spectral partition [1]. More information can be found in Fortunato [8].

Content-Centric Community Detection: Topic modeling is a common approach for content analysis and is often used for clustering, in addition to dimensionality reduction. PLSA-PHITS [12] and LDA [6] are the most traditional methods for content analysis, but they are susceptible to words that appear very few times. Extended methods that are more reliable are Link-PLSA-LDA [20] and Community-User-Topic model [35]. In most cases, the combination of link and content provides insights that are missing with the use of a single modality.

Link and Temporal Community Detection: A few authors address the problem of temporal community detection that aims to identify how communities emerge, grow, combine, and decay over time [15,17]. Tang and Yang [30] use temporal Dirichlet processes to detect communities and track their evolution. Chen, Kawadia, and Urgaonkar [5] tackle the problem of overlapping temporal communities. Bazzi et al. [2] propose the detection of communities in temporal networks represented as multilayer networks. Pietilänen and Diot [23] identify clusters of nodes that are frequently connected for long periods of time, and such sets of nodes are referred to as temporal communities. He and Chen [11] propose an algorithm for dynamic community detection in temporal networks, which takes advantage of community information at previous time steps. Yu, Aggarwal and Wang [34] present a model-based matrix factorization for link prediction and also for community prediction. However, their work uses only links for the prediction process.

Link and Content-Centric Community Detection: In recent years, some approaches were developed to use link and content information for community detection [18,26,32,33]. Among them, probabilistic models have been applied to fuse content analysis and link analysis in a unified framework. Examples include generative models that combine a generative linkage model with a generative content-centric model through some shared hidden variables [7,21]. A discriminative model is proposed by Yang et al. [33], where a conditional model for link analysis and a discriminative model for content analysis are unified. In addition to probabilistic models, some approaches integrate the two aspects from other directions. For instance, a similarity-based method [36] adds virtual attribute nodes and edges to a network, and computes the similarity based on the augmented network. Gupta et al. [10] use matrix factorization to combine sources to improving tagging. It is evident that none of the aforementioned works combine all the three factors of link, content, and temporal information within a unified framework; caused in part by the fact that these modalities interact with one another in complex ways. Therefore, the use of latent factors is a particularly convenient way to achieve this goal.

Community Prediction: There has been a growing interest in the dynamics of communities in evolving social networks, with recent studies addressing the problem of building a predictive model for community detection. Most of the community prediction techniques described in these works are about community evolution prediction that aim to predict events such as growth, survival, shrinkage, splits and merges [27,29]. İlhan and Öğüdücü [13] use ARIMA models to predict community events in a network without using any previous community detection method. İlhan and Öğüdücü [14] propose to use a small number of features to predict community events. Pavlopoulou [22] employ several structural and temporal features to represent communities and improve community evolution prediction.

The community prediction addressed in our work can predict not only community evolution but also a more accurate prediction about each node of the network, in which community the node will be and if its community will change or not. We do so by using topological characteristics and also content associated with nodes.

3   Problem Definition

We assume we have T graphs $G_1 \ldots G_T$ that form a time-series. The graphs are defined over a fixed set of nodes $\mathcal{N}$ of cardinality n. In each timestamp, a different set of edges may exist over time. For example, in the case of a co-authorship network, the node set may correspond to the authors in the network, and the graph G_t might correspond to the co-author relations among them in the tth year. These co-author relations are denoted by the $n \times n$ adjacency matrix A_t. Note that the entries in A_t need not be binary, but might contain arbitrary weights. For example, in a co-authorship network, the entries might correspond to the number of publications between a pair of authors. For undirected graphs, the adjacency matrix A_t is symmetric, while in directed graphs the adjacency matrix is asymmetric. Our approach can handle both settings. Hence, the graph G_t is denoted by the pair $G_t = (\mathcal{N}, A_t)$.

We assume that for each timestamp t, we have an $n \times d$ content matrix C_t. C_t contains one row for each node, and each row contains d attribute values representing the content for that node at the tth timestamp. For example, in the case of the co-authorship network, d might correspond to the lexicon size, and each row might contain the word frequencies of various keywords in the titles. Therefore, one can fully represent the content and structural pair at the tth timestamp with the triplet $(\mathcal{N}, A_t, C_t)$.

In this paper, we study the problem of content-centric community detection in networks. We study two problems: temporal community *detection*, and community *prediction*. While the problem of temporal community detection has been studied in the literature, as presented in Sect. 2, the problem of community prediction, as defined in this work, has not been studied to any significant extent. We define these problems as follows.

Definition 1 (Temporal Community Detection). *Given a sequence of snapshots of graphs $G_1 \dots G_T$, with $n \times n$ adjacency matrices $A_1 \dots A_T$, and $n \times d$ content matrices $C_1 \dots C_T$, create a clustering of the nodes into k partitions at each timestamp $t \leq T$.*

The clustering of the nodes at each timestamp t may use only the graph snapshots up to and including time t. Furthermore, the clusters in successive timestamps should be temporally related to one another. Such a clustering provides better insights about the evolution of the graph. In this sense, the clustering of the nodes for each timestamp will be somewhat different from what is obtained using an independent clustering of the nodes at each timestamp.

Definition 2 (Temporal Community Prediction). *Given a sequence of snapshots of graphs $G_1 \dots G_T$ with $n \times n$ adjacency matrices $A_1 \dots A_T$, and $n \times d$ content matrices $C_1 \dots C_T$, **predict** the clustering of the nodes into k partitions at **future** timestamp $T + r$.*

The community prediction problem attempts to predict the communities at a *future timestamp*, before the structure of the network is known. To the best of our knowledge, this problem is new, and it has not been investigated elsewhere in the literature. Note that the temporal community prediction problem is more challenging than temporal community detection, because it requires us to predict the community structure of the nodes *without any knowledge of the adjacency matrix* at that timestamp.

Temporal prediction is generally a much harder problem in the structural domain of networks as compared to the multidimensional setting. In the multidimensional domain, one can use numerous time-series models such as the auto-regressive (AR) model to predict future trends. However, in the structural domain, it is far more challenging to make such predictions.

4   Mathematical Model

In this section, we discuss the optimization model for converting the temporal sequences of graphs and content to a multidimensional time-series. To achieve this goal, we use a non-negative matrix factorization framework. Although the non-negativity is not essential, one advantage is that it leads to a more interpretable analysis. Consider a setting in which the rank of the factorization is denoted by k. The basic idea is to use three sets of latent factor matrices in a *shared* factorization process, which is able to combine content and structure in a holistic way:

1. The matrix U_t is an $n \times k$ matrix, which is specific to each timestamp t. Each row of the matrix U_t describes the k-dimensional latent factors of the corresponding node at time stamp t, while taking into account *both* the structural and content information.
2. The matrix V is an $n \times k$ matrix, which is global to all timestamps. Each row of the matrix V describes the k-dimensional latent factors of the corresponding node over all time stamps, based on *only* the structural information.
3. The matrix W is an $d \times k$ matrix, which is global to all timestamps. Each row of the matrix W describes the k-dimensional latent factors of one of the d keywords over all time stamps, based on *only* the content information.

The matrices $U_1 \ldots U_T$ are more informative than the other matrices, because they contain latent information specific to the content and structure, and they are also specific to each timestamp. However, the matrices V and W are global, and they contain *only* information corresponding to the structure and the content in the nodes, respectively. This is a setting that is particularly suitable to *shared* matrix factorization, where the matrices $U_1 \ldots U_T$ are shared between the factorization of the adjacency and content matrices.

Therefore, we would like to approximately factorize the adjacency matrices $A_1 \ldots A_T$ as $A_t \approx U_t V^T$, for all $t \in \{1 \ldots T\}$. Similarly, we would like to approximately factorize the content matrices $C_1 \ldots C_T$ as $C_t \approx U_t W^T$. With this setting, we propose the following optimization problem:

$$\text{Minimize } J = \sum_{t=1}^{T} \|A_t - U_t V^T\|^2 + \beta \sum_{t=1}^{T} \|C_t - U_t W^T\|^2 + \lambda_1 \Omega(U_t, V, W). \quad (1)$$

where β is a balancing parameter, λ_1 is the regularization parameter, and $\Omega(U_t, V, W)$ is a regularization term to avoid overfitting. The notation $\| \cdot \|^2$ denotes the Frobenius norm, which is the sum of the squares of the entries in the matrix. The regularization term is defined as

$$\Omega(U_t, V, W) = \|V\|^2 + \|W\|^2 + \sum_{t=1}^{T} \|U_t\|^2. \quad (2)$$

We would also like to ensure that the embeddings between successive timestamps do not change suddenly because of random variations. For example, an author might publish together with a pair of authors every year, but might not be publishing in a particular year because of random variations. To ensure that the predicted values do not change suddenly, we add a temporal regularization term:

$$\Omega_2(U_1 \ldots U_T) = \sum_{t=1}^{T-1} \|U_{t+1} - U_t\|^2 \tag{3}$$

This additional regularization term ensures the variables in any pair of successive years do not change suddenly. The additional regularization term is added to the objective function, after multiplying it with λ_2. The enhanced objective function is defined as

$$J = \sum_{t=1}^{T} \|A_t - U_t V^T\|^2 + \beta \sum_{t=1}^{T} \|C_t - U_t W^T\|^2 +$$
$$+ \lambda_1 \left(\|V\|^2 + \|W\|^2 + \sum_{t=1}^{T} \|U_t\|^2 \right) + \lambda_2 \sum_{t=1}^{T-1} \|U_{t+1} - U_t\|^2 . \tag{4}$$

In order to ensure a more interpretable solution, we impose non-negativity constraints on the factor matrices

$$U_t \geq 0, V \geq 0, W \geq 0 . \tag{5}$$

One challenge with this optimization model is that it can become very large. The main size of the optimization model is a result of the adjacency matrix. The content matrix is often manageable, because one can often reduce the keyword-lexicon in many real settings. However, the adjacency matrix scales with the square of the number of nodes, which can be onerous in real settings. An important observation here is that the adjacency matrix is sparse, and most of its values are zeros. Therefore, one can often use sampling on the zero entries of the adjacency matrix in order to reduce the complexity of the problem. This also has a beneficial effect of ensuring that the solution is not dominated by the zeros in the matrix.

4.1 Solving the Optimization Model

In this section, we discuss a gradient-descent approach for solving the optimization model. The basic idea is to compute the gradient of J with respect to the various parameters. Note that $U_t V^T$ can be seen as the "prediction" of the value of A_t. Obviously, this predicted value may not be the same as the observed entries in the adjacency matrices. Similarly, while the product $U_t W^T$ predicts C_t, the predicted values may be different from the observed values. The gradient descent steps are dependent on the errors of the prediction. Therefore, we define the error for the structural and content-centric entries as $\Delta_t^A = A_t - U_t V^T$ and $\Delta_t^C = C_t - U_t W^T$. Also let $\Delta_t^U = U_t - U_{t+1}$, with $\Delta_T^U = 0$, since the difference is not defined at this boundary value.

Our goal is to compute the partial derivative of J with respect to the various optimization variables, and then use it to construct the gradient-descent steps. By computing the partial derivatives of (4) with respect to each of the decision variables, we obtain

$$\frac{\partial J}{\partial U_t} = 2\lambda_1 U_t - 2\left(\Delta_t^A V + \beta \Delta_t^C W\right) + 2\lambda_2 \Delta_t^U, \tag{6}$$

$$\frac{\partial J}{\partial V_t} = 2\lambda_1 V_t - 2\sum_{t=1}^{T}\left[\Delta_t^A\right] U_t \tag{7}$$

$$\frac{\partial J}{\partial W_t} = 2\lambda_1 W_t - 2\beta\sum_{t=1}^{T}\left[\Delta_t^C\right] U_t. \tag{8}$$

The gradient-descent steps use these partial derivatives for the updates. The gradient-descent steps may be written as

$$U_t \leftarrow U_t - \alpha\frac{\partial J}{\partial U_t}\ \forall t, \tag{9}$$

$$V \leftarrow V_t - \alpha\frac{\partial J}{\partial V}, \tag{10}$$

$$W \leftarrow W_t - \alpha\frac{\partial J}{\partial W}. \tag{11}$$

Here, $\alpha > 0$ is the step-size, which is a small value, such as 0.01. The matrices U_t, V, and W are initialized to non-negative values in $(0, 1)$, and the updates (9–11) are performed until convergence or until a pre-specified number of iterations is performed. Non-negativity constraints are enforced by setting an entry in these matrices to zero whenever it becomes negative due to the updates.

Δ_t^A is a sparse matrix, and should be stored using sparse data structures. As a practical matter, it makes sense to first compute those entries in Δ_t^A that correspond to non-zero entries in A, and then store those entries using a sparse matrix data structure. This is because a $n \times n$ matrix may be too large to hold using a non-sparse representation.

Combining Eqs. (6–11), we obtain the following update rule:

$$U_t \leftarrow U_t(1 - 2\alpha\lambda_1) + 2\alpha\Delta_t^A V + 2\alpha\beta\Delta_t^C W + 2\lambda_2\Delta_t^U$$

$$V \leftarrow V(1 - 2\alpha\lambda_1) + 2\alpha\sum_{t=1}^{T}[\Delta_t^A]^T U_t \tag{12}$$

$$W \leftarrow W(1 - 2\alpha\lambda_1) + 2\alpha\beta\sum_{t=1}^{T}[\Delta_t^C]^T U_t.$$

The set of updates above are typically performed "simultaneously" so that the entries in U_t, V and W (on the right-hand side) are fixed to their values in the previous iteration during a particular block of updates. Only after the new values of U_t, V, and W have been computed (using temporary variables), can they be used in the right-hand side in the next iteration.

4.2   Complexity Analysis

With the algorithm fully specified, we can now analyze its asymptotic complexity. Per gradient descent iteration, the computational cost of the algorithm is the sum of (i) the complexity of evaluating the objective function (4) and (ii) the complexity of the update step (12). Recall from Sect. 4 that A_t, C_t, U_t, V, and W have dimensions $n \times n$, $n \times d$, $n \times k$, $n \times k$, and $d \times k$, respectively. Since matrix factorization reduces the dimensions of the data, we can safely assume $n \gg k$ and that $d \gg k$.

Assuming the basic matrix multiplication algorithm is used, the complexity of multiplying matrices of dimensions $m \times p$ and $p \times n$ is $O(mnp)$. Therefore, the complexity of computing $\|U_t V^T\|^2 = O(n^2 k) + O(n^2)$, since the norm can be computed by iterating over all elements of the matrix, squaring and summing them. Hence, the complexity of evaluating the objective function (4) is

$$J = T \left[O(n^2 k) + O(n^2) + O(dkn) + O(dn) + O(kn) + O(kn) \right] + O(kn) + O(dk)$$

$$= O(\max(n^2 k, dkn)).$$

To obtain the asymptotic complexity of the updates, note that $\Delta_t^A = A_t - U_t V^T$, and $\Delta_t^C = C_t - U_t W^T$. Hence, $\Delta_t^A V = O(n^2 k)$, $[\Delta_t^A]^T U_t = O(n^2 k)$, $\Delta_t^C W = O(dkn)$, and $[\Delta_t^C]^T U_t = O(dkn)$. Therefore, the asymptotic complexity of the gradient descent update is $T[O(kn) + O(n^2 k) + O(dkn)] = O(\max(n^2 k, dkn))$.

5   Applications to Clustering

5.1   Temporal Community Detection

The learned factor matrices can be used for temporal community detection. In this context, the matrix U_t is very helpful in determining the communities at time t, because it accounts for structure, content, and smoothness constraints. The overall approach is:

1. Extract the $n \cdot T$ rows from $U_1 \ldots U_T$, so that each of the $n \cdot T$ rows is associated with a timestamp from $\{1 \ldots T\}$. This timestamp will be used in step 3 of the algorithm.
2. Cluster the $n \cdot T$ rows into k clusters $\mathcal{C}_1 \ldots \mathcal{C}_k$ using a k-means clustering algorithm.
3. Partition each $\mathcal{C}_i$ into its T different timestamped clusters $\mathcal{C}_i^1 \ldots \mathcal{C}_i^T$, depending on the timestamp of the corresponding rows.

In most cases, the clusters will be such that the T different avatars of the ith row in $U_1 \ldots U_T$ will belong to the same cluster. However, in some cases, rows may drift from one cluster to the other. Furthermore, some clusters may shrink with time, whereas others may increase with time. All these aspects provide interesting insights about the community structure in the network. Even though the data is clustered into k groups, it is often possible for one or more timestamps to contain clusters without any members. This is likely when the number of clusters expands or shrinks with time.

5.2   Temporal Community Prediction

This approach can also be naturally used for community prediction. The basic idea here is to treat $U_1 \ldots U_T$ as a time-series of matrices, and predict how the weights evolve with time. The overall approach is as follows:

1. For each (i, j) of the non-zero entries of matrix A, represent the time series $\mathcal{T}_{ij}$.
2. Use an autoregressive model on $\mathcal{T}_{ij}$ to predict u_{ij}^{t+r} for each (i, j) of the non-zero entries. Set all other entries in U_{t+r} to 0.
3. Perform node clustering on the rows of U_{t+r} to create the predicted node clusters at time $(t + r)$. This provides the predicted communities at a future timestamp.

Thus, *Chimera* can provide not only the communities in the current timestamp, but also the communities in a future timestamp.

6   Experiments

This section describes the experimental results of the approach. We describe the datasets, evaluation methodology, and the results obtained.

A key point in choosing a dataset to evaluate algorithms such as *Chimera* is that there must be co-evolving interactions between network and content. In order to check our model's consistency, and to have a fair comparison with other state-of-the-art algorithms, we generated a couple of synthetic dataset.

Synthetic dataset: The synthetic dataset was generated in the following way: first, we create the matrix A_1 with 5 groups. Then, we follow a randomized approach to rewire edges. According to some probability, we connect edges from one group to another. In this dataset, all link matrices (A) have 5,000 nodes and 20,000 edges. For the content matrices (C), we generate five groups of five words. As in the link case, we have a probability of a word being in more than one group. Due to the nature of its construction, all content matrices have 25 words. For transitioning between timestamps, we have another probability that defines whether a node changes group or not. The transitions are constrained to be at most 10% of the nodes. We generated 3 timestamps for each synthetic dataset. The rewire probabilities $1 - p$ used in each synthetic dataset were $p = 0.75$ (Synthetic 1) and $p = 0.55$ (Synthetic 2).

Real Dataset: We used the arXiv API[1] to download information about preprints submitted to the arXiv system. We extracted information about 7107 authors during a period of five years (from 2013 to 2017). We used the papers' titles and abstracts to build the author-content network with 10256 words, and we selected words with more than 25 occurrences after removal of stop words and stemming.

---

[1] https://arxiv.org/help/api/index.

Since every preprint submitted to the arXiv has a category, we used the category information as a group label. We selected 10 classes: cs.IT, cs.LG, cs.DS, cs.CV, cs.SI, cs.AI, cs.NI, cs, math, and stat. Authors were added to the set of authors if they published for at least three years in the five-year period we consider. In years without publications, we assume authors belong to the temporally-closest category.

There are several metrics for evaluating cluster quality. We use two well-known supervised metrics: the Jaccard index and cluster purity. Cluster purity [19] measures the quality of the communities by examining the dominant class in a given cluster. It ranges from 0 to 1, with higher purity values indicating better clustering performance.

We compared our approach with state-of-the-art algorithms in four categories: Content-only, Link-only, Temporal-Link-only and Link-Content-only. By following this approach, we are also able to isolate the specific effects of using data in different modalities.

Content-Only Method. We use GibbsLDA++ as a baseline for the content-only method. As input for this method, we considered that a document consists of the words used in the title and abstract of a paper.

Link-Only Method. For link we use the Louvain [4] method for community detection.

Temporal-Link-Only Method. For temporal link-only method we used the work presented by He and Chen [11], which we refer to as DCTN.

Combination of Link and Content[2]. For link and content combination, we used the work presented by Liu et al. [18], with algorithms CPRW-PI, CPIP-PI, CPRW-SI, CPIP-SI. Since all them perform very similarly and we have a space constraint we will report only the results obtained with CPIP-PI.

6.1   Evaluation Results

In this section, we present the results of our experiments.

The Louvain and DCTN methods are based on link structure and do not allow fixed numbers of clusters. They use topological structure to find the number of communities. All methods in the baseline were used in their default configuration.

First, we present the results with synthetic data we generated (Synthetic 1 and Synthetic 2) in Table 1. In synthetic datasets we use $\alpha = 0.00001$, $\beta = 1000$, $\lambda = 0.1$ and $\lambda_2 = 0.0001$ with $k = 5$ and 1000 steps.

The only methods that are able to find the clusters in all datasets are CPIP-PI and *Chimera*, both using content and link information. In the synthetic data the changes between timestamps were small. Thus, CPIP-PI and *Chimera* performed similarly. However, *Chimera* displayed almost perfect performance in all datasets and timestamps. Louvain and DCTN, which use only link information, were not

---

[2] Code from authors obtained from https://github.com/LiyuanLucasLiu/Content-Propagation.

able to find the clusters. Despite the purity of 1, they cluster all the data into only one cluster. DCTN finds clusters only for the two first timestamps of synthetic 2, obtaining 3 and 4 clusters respectively. Louvain found 3 clusters in timestamps 1 and 3 of synthetic 2.

Table 1. Jaccard (J) and Purity (P) of the *Synthetic 1* and *Synthetic 2* dataset from all timestamps and methods. *Chimera* outperforms baseline methods in almost every year.

Algorithm	Synthetic 1						Synthetic 2					
	1		2		3		1		2		3	
	J	P	J	P	J	P	J	P	J	P	J	P
Louvain	0.4	1	0.2	1	0.4	1	0.2	1	0.2	1	0.2	1
GibbsLDA++	0.542	0.714	0.267	0.463	0.399	0.611	0.279	0.473	0.533	0.657	0.326	0.575
CPIP-PI	0.909	1	1	1	0.866	0.917	1	1	0.999	0.997	0.999	0.995
DCTN	0.4	1	0.2	1	0.2	1	0.2	1	0.2	1	0.2	1
Chimera	1	1	1	1	1	1	0.999	0.994	0.998	0.992	0.997	0.990

Table 2. Purity (P) and Jaccard (J) Index obtained in the *arXiv* dataset for all years and methods. *Chimera* outperforms baseline methods in almost every year.

Algorithm	2013		2014		2015		2016		2017	
	J	P	J	P	J	P	J	P	J	P
Louvain	0	0.041	0	0.062	0	0.073	0	0.100	0	0.086
GibbsLDA++	0.087	0.373	0.080	0.394	0.182	0.387	**0.166**	0.399	0.168	0.389
CPIP-PI	**0.096**	**0.523**	0.097	0.518	0.149	0.412	0.090	0.365	0.105	0.361
DCTN	0	0.039	0	0.052	0	0.069	0	0.077	0	0.085
Chimera	0.078	0.456	**0.261**	**0.601**	**0.281**	**0.610**	0.105	**0.573**	**0.291**	**0.628**

Table 2 presents the Jaccard and Purity metrics over all methods for the real dataset *arXiv*. In *arXiv*, the Louvain method found 3636, 2679, 2006, 1800 and 2190 communities respectively for each year. CDTN, which is based on Louvain has a very similar result with 3636, 2656, 1829, 1500 and 1791 communities respectively for each year. Since they are methods based on link, they consider specially disconnected nodes as isolated communities. Methods that combine link and content use content to aggregate such nodes in a community. Also, as we can note in Table 2, our method can learn with time and improve its results in the following years. GibbsLDA++ presents a nice performance because the content was much more stable and had more quality over the years than the link information. This is another reason to combine various sources to achieve better performance.

To tune the hyperparameters of *Chimera*, we used Bayesian Optimization [3,28] to perform a search in the hyperparameter space. Bayesian Optimization is the appropriate technique in this setting, because minimizing the model loss (4) does not necessarily translate into better performance. We defined an objective function that minimizes the mean silhouette coefficient [25] of the labels assigned by *Chimera*, as described in Sect. 5.1. We used Bayesian Optimization to determine the number of clusters as well. With this approach, the optimization process is completely unsupervised and, although we have access to the true labels, they were not used during optimization, a situation closer to reality. With Bayesian Optimization, our model was able to learn that the actual number of clusters was in the order of 10. The full set of hyperparameters and their ranges are shown in Table 3, with best results shown in bold face.

Table 3. Hyperparameters used for tuning *Chimera* with Bayesian Optimization. Elements in bold indicate the best parameter for that hyperparameter. The set of all elements in bold defines the hyperparameters used for training the model.

Hyperparameter	Values
α	$\{0.01, \mathbf{0.1}\}$
β	$\{0.1, 0.25, 0.5, 0.75, \mathbf{0.9}\}$
λ_1	$\{\mathbf{1 \times 10^{-5}}, 1 \times 10^{-6}\}$
λ_2	$\{\mathbf{1 \times 10^{-4}}, 1 \times 10^{-5}\}$
K	$\{\mathbf{10}, 20, 30, 40, 50\}$
Clusters	$\{2, 4, 8, \mathbf{10}, 16, 18, 32\}$

Table 4. The Jaccard index and Purity of *arXiv* for prediction. In the "Original U's" row, we used the original matrices to make the prediction in 2015, 2016 and 2017. Whereas in the "Predicted U's" row, we used the output of *Chimera* to make the predictions. Hence, for 2016 we used the prediction for 2015, and for 2017 we used the predictions of both 2015 and 2016.

	2015		2016		2017	
	Jaccard	Purity	Jaccard	Purity	Jaccard	Purity
Original U's	0.0709	0.5180	0.2273	0.4395	0.1145	0.3766
Predicted U's			0.0589	0.5177	0.0765	0.4981

In Table 4 we show our results for prediction. Here, we will not compare our results with other methods that estimate or evaluate the size of each community. The idea here is to predict in which community an author will be in the future. One advantage of our method is that we can augment our time series with our predictions. Clearly, doing so will add noise to further predictions, but the results presented are very similar to the ones present in the original dataset. *Chimera* is the only one that allows us to do that kind of analysis in an easy way, since the embeddings create multidimensional representations of the nodes in the graph.

7   Conclusions

In this work, we presented *Chimera* a novel shared factorization overtime model that can simultaneously take the link, content, and temporal information of networks into account improving over the state-of-the-art approaches for community detection. Our approach model and solve in efficient time the problem of combining link, content and temporal analysis for community detection and prediction in network data. Our method extracts the latent semantic structure of the network in multidimensional form, but in a way that takes into account the temporal continuity of the embeddings. Such approach greatly simplifies temporal analysis of the underlying network by using the embedding as a surrogate. A consequence of this simplification is that it is also possible to use this temporal sequence of embeddings to predict future communities with good results. The experimental results illustrate the effectiveness of *Chimera*, since it outperforms the baseline methods. Our experiments also show that the prediction is efficient in using embeddings to predict near future communities, which opens a vast array of new possibilities for exploration.

Acknowledgments. Charu C. Aggarwal's research was sponsored by the Army Research Laboratory and was accomplished under Cooperative Agreement Number W911NF-09-2-0053. The views and conclusions contained in this document are those of the authors and should not be interpreted as representing the official policies, either expressed or implied, of the Army Research Laboratory or the U.S. Government. The U.S. Government is authorized to reproduce and distribute reprints for Government purposes notwithstanding any copyright notation here on.

References

1. Barnes, E.R.: An algorithm for partitioning the nodes of a graph. SIAM J. Algebr. Discret. Methods **3**(4), 541–550 (1982). https://doi.org/10.1137/0603056
2. Bazzi, M., Porter, M.A., Williams, S., McDonald, M., Fenn, D.J., Howison, S.D.: Community detection in temporal multilayer networks, with an application to correlation networks. Multiscale Model. Simul. **14**(1), 1–41 (2016)
3. Bergstra, J., Yamins, D., Cox, D.: Making a science of model search: Hyperparameter optimization in hundreds of dimensions for vision architectures. In: International Conference on Machine Learning, pp. 115–123 (2013)
4. Blondel, V.D., Guillaume, J.L., Lambiotte, R., Lefebvre, E.: Fast unfolding of communities in large networks. J. Stat. Mech.: Theory Exp. **2008**(10), P10008 (2008)
5. Chen, Y., Kawadia, V., Urgaonkar, R.: Detecting overlapping temporal community structure in time-evolving networks. arXiv preprint arXiv:1303.7226 (2013)
6. Cohn, D., Hofmann, T.: The missing link: a probabilistic model of document content and hypertext connectivity. In: Proceedings of the 13th International Conference on Neural Information Processing Systems, NIPS 2000, pp. 409–415. MIT Press (2000)
7. Cohn, D., Hofmann, T.: The missing link-a probabilistic model of document content and hypertext connectivity. In: Advances in Neural Information Processing Systems, pp. 430–436 (2001)

8. Fortunato, S.: Community detection in graphs. Phys. Rep. **486**(3), 75–174 (2010)
9. Girvan, M., Newman, M.E.: Community structure in social and biological networks. Proc. Natl. Acad. Sci. **99**(12), 7821–7826 (2002)
10. Gupta, S.K., Phung, D., Adams, B., Tran, T., Venkatesh, S.: Nonnegative shared subspace learning and its application to social media retrieval. In: Proceedings of the 16th ACM SIGKDD International Conference on Knowledge Discovery and Data Mining, pp. 1169–1178. ACM (2010)
11. He, J., Chen, D.: A fast algorithm for community detection in temporal network. Phys. A: Stat. Mech. Appl. **429**, 87–94 (2015). https://doi.org/10.1016/j.physa.2015.02.069
12. Hofman, J.M., Wiggins, C.H.: Bayesian approach to network modularity. Phys. Rev. Lett. **100**(25), 258701 (2008)
13. İlhan, N., Öğüdücü, Ş.G.: Predicting community evolution based on time series modeling. In: Proceedings of the 2015 IEEE/ACM International Conference on Advances in Social Networks Analysis and Mining 2015, ASONAM 2015, pp. 1509–1516. ACM (2015). https://doi.org/10.1145/2808797.2808913
14. İlhan, N., Öğüdücü, Ş.G.: Feature identification for predicting community evolution in dynamic social networks. Eng. Appl. Artif. Intell. **55**, 202–218 (2016). https://doi.org/10.1016/j.engappai.2016.06.003
15. Kawadia, V., Sreenivasan, S.: Sequential detection of temporal communities by estrangement confinement. Sci. Rep. **2**, 794 (2012)
16. Leskovec, J., Lang, K.J., Dasgupta, A., Mahoney, M.W.: Statistical properties of community structure in large social and information networks. In: Proceedings of the 17th International Conference on World Wide Web, WWW 2008, pp. 695–704. ACM (2008). https://doi.org/10.1145/1367497.1367591
17. Lin, Y.R., Chi, Y., Zhu, S., Sundaram, H., Tseng, B.L.: FacetNet: a framework for analyzing communities and their evolutions in dynamic networks. In: Proceedings of the 17th International Conference on World Wide Web, pp. 685–694. ACM (2008)
18. Liu, L., Xu, L., Wangy, Z., Chen, E.: Community detection based on structure and content: a content propagation perspective. In: 2015 IEEE International Conference on Data Mining (ICDM), pp. 271–280. IEEE (2015)
19. Manning, C.D., Raghavan, P., Schütze, H.: Introduction to Information Retrieval. Cambridge University Press, New York (2008)
20. Nallapati, R.M., Ahmed, A., Xing, E.P., Cohen, W.W.: Joint latent topic models for text and citations. In: Proceedings of the 14th ACM SIGKDD International Conference on Knowledge Discovery and Data Mining, KDD 2008, pp. 542–550. ACM (2008). https://doi.org/10.1145/1401890.1401957
21. Nallapati, R.M., Ahmed, A., Xing, E.P., Cohen, W.W.: Joint latent topic models for text and citations. In: Proceedings of the 14th ACM SIGKDD International Conference on Knowledge Discovery and Data Mining, pp. 542–550. ACM (2008)
22. Pavlopoulou, M.E.G., Tzortzis, G., Vogiatzis, D., Paliouras, G.: Predicting the evolution of communities in social networks using structural and temporal features. In: 2017 12th International Workshop on Semantic and Social Media Adaptation and Personalization (SMAP), pp. 40–45 (2017). https://doi.org/10.1109/SMAP.2017.8022665
23. Pietilänen, A.K., Diot, C.: Dissemination in opportunistic social networks: the role of temporal communities. In: Proceedings of the Thirteenth ACM International Symposium on Mobile Ad Hoc Networking and Computing, MobiHoc 2012, pp. 165–174. ACM (2012). https://doi.org/10.1145/2248371.2248396

24. Ravasz, E., Somera, A.L., Mongru, D.A., Oltvai, Z.N., Barabási, A.L.: Hierarchical organization of modularity in metabolic networks. Science **297**(5586), 1551–1555 (2002)
25. Rousseeuw, P.J.: Silhouettes: a graphical aid to the interpretation and validation of cluster analysis. J. Comput. Appl. Math. **20**, 53–65 (1987)
26. Ruan, Y., Fuhry, D., Parthasarathy, S.: Efficient community detection in large networks using content and links. In: Proceedings of the 22nd International Conference on World Wide Web, WWW 2013, pp. 1089–1098. ACM (2013). https://doi.org/10.1145/2488388.2488483
27. Saganowski, S.: Predicting community evolution in social networks. In: 2015 IEEE/ACM International Conference on Advances in Social Networks Analysis and Mining (ASONAM), pp. 924–925 (2015). https://doi.org/10.1145/2808797.2809353
28. Shahriari, B., Swersky, K., Wang, Z., Adams, R.P., De Freitas, N.: Taking the human out of the loop: a review of Bayesian optimization. Proc. IEEE **104**(1), 148–175 (2016)
29. Takaffoli, M., Rabbany, R., Zaïane, O.R.: Community evolution prediction in dynamic social networks. In: 2014 IEEE/ACM International Conference on Advances in Social Networks Analysis and Mining (ASONAM 2014), pp. 9–16 (2014). https://doi.org/10.1109/ASONAM.2014.6921553
30. Tang, X., Yang, C.C.: Dynamic community detection with temporal Dirichlet process. In: 2011 IEEE Third International Conference on Privacy, Security, Risk and Trust (PASSAT) and 2011 IEEE Third International Conference on Social Computing (SocialCom), pp. 603–608. IEEE (2011)
31. Watts, D.J., Dodds, P.S., Newman, M.E.: Identity and search in social networks. Science **296**(5571), 1302–1305 (2002)
32. Xu, H., Martin, E., Mahidadia, A.: Exploiting paper contents and citation links to identify and characterise specialisations. In: 2014 IEEE International Conference on Data Mining Workshop, pp. 613–620. IEEE (2014)
33. Yang, T., Jin, R., Chi, Y., Zhu, S.: Combining link and content for community detection: a discriminative approach. In: Proceedings of the 15th ACM SIGKDD International Conference on Knowledge Discovery and Data Mining, KDD 2009, pp. 927–936. ACM (2009). https://doi.org/10.1145/1557019.1557120
34. Yu, W., Aggarwal, C.C., Wang, W.: Temporally factorized network modeling for evolutionary network analysis. In: Proceedings of the Tenth ACM International Conference on Web Search and Data Mining, WSDM 2017, pp. 455–464. ACM (2017). https://doi.org/10.1145/3018661.3018669
35. Zhou, D., Manavoglu, E., Li, J., Giles, C.L., Zha, H.: Probabilistic models for discovering e-communities. In: Proceedings of the 15th International Conference on World Wide Web, WWW 2006, pp. 173–182. ACM (2006). https://doi.org/10.1145/1135777.1135807
36. Zhou, Y., Cheng, H., Yu, J.X.: Graph clustering based on structural/attribute similarities. Proc. VLDB Endow. **2**(1), 718–729 (2009)

Local Topological Data Analysis to Uncover the Global Structure of Data Approaching Graph-Structured Topologies

Robin Vandaele[1,2]([✉]), Tijl De Bie[1], and Yvan Saeys[2]

[1] IDLab, Department of Electronics and Information Systems, Ghent University,
Technologiepark-Zwijnaarde 19, 9052 Gent, Belgium
{robin.vandaele,tijl.debie}@ugent.be
[2] Data Mining and Modelling for Biomedicine (DaMBi),
VIB Inflammation Research Center,
Technologiepark-Zwijnaarde 927, 9052 Gent, Belgium
yvan.saeys@irc.vib-ugent.be

Abstract. Gene expression data of differentiating cells, galaxies distributed in space, and earthquake locations, all share a common property: they lie close to a graph-structured topology in their respective spaces [1,4,9,10,20], referred to as *one-dimensional stratified spaces* in mathematics. Often, the uncovering of such topologies offers great insight into these data sets. However, methods for dimensionality reduction are clearly inappropriate for this purpose, and also methods from the relatively new field of *Topological Data Analysis (TDA)* are inappropriate, due to noise sensitivity, computational complexity, or other limitations. In this paper we introduce a new method, termed *Local TDA (LTDA)*, which resolves the issues of pre-existing methods by unveiling (*global*) graph-structured topologies in data by means of robust and computationally cheap *local* analyses. Our method rests on a simple graph-theoretic result that enables one to identify isolated, end-, edge- and multifurcation points in the topology underlying the data. It then uses this information to piece together a graph that is homeomorphic to the unknown one-dimensional stratified space underlying the point cloud data. We evaluate our method on a number of artificial and real-life data sets, demonstrating its superior effectiveness, robustness against noise, and scalability. Code related to this paper is available at: https://bitbucket.org/ghentdatascience/gltda-public.

Keywords: Topological Data Analysis · Persistent homology
Metric spaces · Graph theory · Stratified spaces

© Springer Nature Switzerland AG 2019
M. Berlingerio et al. (Eds.): ECML PKDD 2018, LNAI 11052, pp. 19–36, 2019.
https://doi.org/10.1007/978-3-030-10928-8_2

1 Introduction

Motivation. Identifying and visualizing graph-structured topologies underlying point cloud data sets is a non-trivial and active topic of research, with known applications in many fields of science, such as biology, physics, geology, geography, and computer science [1,4,10,19,20].

E.g., consider data of differentiating cells in a high-dimensional expression space. The way in which different cell stages are interconnected during cell differentiation can be represented by means of a graph (which may contain cycles) in the expression space, such that each of the differentiating cells lie close to it. More formally, the point cloud data approaches a topological structure *homeomorphic* to (i.e., obtainable from by 'bending' and 'stretching') the embedding of a corresponding graph in the expression space. In the mathematical literature, such an embedding is know as a *one-dimensional stratified space* (in this paper referred to as a graph-structured topology), composed of 0-D *strata* (here called the vertices) and 1-D linear strata (here called the edges or loops), glued together in a particular way.

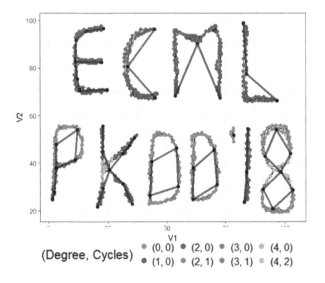

Fig. 1. When the underlying graph-structured topology of D is well-modeled by a proximity graph, counting connected components in induced subgraphs suffices to learn topological structures locally, as well as the presence of cycles (see Algorithm 1 in Sect. 2, $|D| = 873$, $\epsilon = 3.5$, $r = 3$, comp. time: 0.43 s). By using these identified local topologies, we are able to reconstruct a graph homeomorphic to the underlying space (see Algorithm 2 in Sect. 3, $r' = 4$, comp. time: 8.04 s).

A toy data set D is shown in Fig. 1 for illustration. Here the colored dots represent data points, and the black dots and lines represent vertices and edges of the graph-structured topology. The different colors express both *local and global topological information*, which we simply refer to as *local topologies*. E.g., near the center of the '8 component', points are marked by a $(4, 2)$ local topology, meaning four branches emerge from this location, and induce two cycles by convergence. We will formally explain this below. As this data set is 2-dimensional, its graph-structured topology is readily noticed. However, it is clear that such topologies, in high-dimensional data, are hard to uncover, and standard dimensionality reduction techniques will fail in all but the most trivial cases.

The emergent area of Topological Data Analysis (TDA) [5], which aims to understand the *shape of data* [23], seems to be the obvious approach to handle this problem. Its power for uncovering the underlying topology of data sets has been demonstrated in several recent works [3,7,13,19–21]. However, TDA methods designed for this problem, such as *Mapper* [19,20], *local persistent (co)homology* [11,21], *functional persistence* [6], and *metric graph reconstruction* [1], are either computationally inefficient, restricted to specific graph-structured topologies, vulnerable to noise, or simply do not consider reconstructing the topology.

In this paper, we develop a novel method to fill this gap, under the name of *Local Topological Data Analysis* (LTDA). Investigating structures locally allows one to detect the degree, denoted δ_0, i.e., the number of branches emerging from a point, as well as the number of cycles, denoted δ_1, induced by the convergence of the same branches away from this point. LTDA provides methods for classifying data points according to their local topology (δ_0, δ_1), identifying isolated, end-, edge- and multifurcation points, as well as cycles, by only tracking the number of connected components in graphs [2,15] (Algorithm 1 in Sect. 2). Note that the discovery of cycles in such data using state-of-the-art TDA techniques requires the computation of the first order Betti number, the computation of which is challenging [24]. Combining the information retrieved by LTDA with clustering techniques allows for a fast reconstruction of the underlying graph-structured topology (Algorithm 2 in Sect. 3). These concepts are illustrated on Fig. 1.

Contributions

- We develop a method, under the name of *Local Topological Data Analysis* (LTDA). This method allows us to detect isolated, end-, edge- and multifurcation points, as well as cycles, underlying data approaching graph-structured topologies, by merely counting the number of connected components in proximity graphs (Algorithm 1 in Subsect. 2.3).
- We develop a framework that combines the information retrieved from LTDA with clustering techniques to reconstruct and visualize the unknown underlying topology of such data sets (Algorithm 2 in Sect. 3).

– We clarify and empirically validate the usefulness of our methods on a variety of simulated and real data sets (Sects. 2, 3 and 4). We show that our methods are competitive with current state-of-the-art approaches in terms of results and computational efficiency.
– We discuss how future research on the potential of LTDA may open up new possibilities to the set of TDA methods (Sect. 5).

2   LTDA of Graph-Structured Topologies

Given a Euclidean point cloud data set $D \subseteq \mathbb{R}^n$ with an unknown underlying topological structure, we wish to investigate the *global topology*, i.e., the complete and unknown topological structure, by applying TDA to small patches of data, indicating (unknown) properties of the *local topology*. We start by showing how knowing both the local topological structures, as well as how these affect the global structure, may unravel graph-structured topologies. This leads to an algorithm proposed in this paper for identifying and locating multifurcation points and cycles in point cloud data approaching such topologies (Algorithm 1, Subsect. 2.3).

2.1   Overview: Illustrating the Idea Behind LTDA on a Toy Example

Here we first introduce LTDA in an intuitive and constructive way. We will do this by means of a simple two-dimensional toy data set. The used underlying topological structure of the toy data will show to be quite useful to understand the intuition behind Theorem 1 (Subsect. 2.2), which forms the foundation for the proposed approach of LTDA for graph-structured topologies (Subsect. 2.3).

A Toy Data Set. The toy data set $D \subseteq \mathbb{R}^2$ we consider is the subset of the data illustrated in Fig. 1, that has the underlying topological structure of 'the number 8', illustrated in Fig. 2. Without going much into detail, an *n-manifold* is a *topological space*[1] locally resembling the Euclidean space of dimension n near every point on the space. There are essentially two (non-homeomorphic) connected 1-manifolds: the circle $\mathcal{S}^1$ and the real line $\mathbb{R}$. The underlying topology τ of D is that of (homeomorphic to) two circles $\mathcal{S}_1^1$ and $\mathcal{S}_2^1$, intersecting in one singular point $x \in \mathcal{S}_1^1 \cap \mathcal{S}_2^1$.

---

[1] Formally, a *topological space* is an ordered pair (X, τ), where X is a set and τ is a collection of subsets of X, satisfying particular axioms. The elements of τ are called *open sets* and the collection τ is called a *topology on* X. In this paper, we abuse notation for simplicity, and use τ to refer to the set $X = \bigcup \tau$.

The Idea Behind LTDA. One may assign a
point $y \in \tau$ to two classes: either $y \neq x$ or $y = x$.
If $y \neq x$, then y inherits its local topology from
exactly one of the circles S_1^1 or S_2^1. As these are
1-manifolds, y has a neighborhood homeomorphic to $\mathbb{R}$, or equivalently, to $]0, 1[$. Removing
any point c from $]0, 1[$ breaks the interval into
two disjoint connected components, as one can
either move left or right from c in $]0, 1[$. The
same behavior occurs at y: starting from y, we
can move into two directions, i.e., two *branches*
emerge from y. If we would remove y from a
neighborhood of y homeomorphic to $]0, 1[$, then
this neighborhood would break into two disjoint
connected components as well. If $y = x$, then
four branches emerge from y, and removing y
from a small neighborhood of y in τ breaks the
neighborhood into four components.

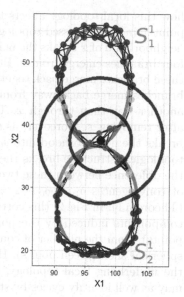

Fig. 2. The idea behind LTDA
for data that approaches a
graph-structured topology $\tau =
S_1^1 \cup S_2^1$. For appropriate prox-
imity graphs, one finds the
underlying degree of a data
point z (black) by counting the
connected components in the
graph induced by the intersection of a spherical shell and the
data (green points), representing branches emerging from z.
Convergence of these branches
away from z indicates cycles
through z, which may be identified by comparing the obtained
degree with the number of connected components in the graph
induced by the points away
from z (blue and green points).
(Color figure online)

When a point cloud data set approaches
a graph-structured topology, it reflects similar
properties as that underlying topology. Consider
the centered black data point $z \in D$ in Fig. 2,
representing the singular point x in the underlying topology τ. A neighborhood of x in τ
now corresponds to the points contained in a
small open ball centered at z. Removing x from
this neighborhood in τ corresponds to removing points in an even smaller ball centered at z,
strictly contained within the original ball. The
points remaining in the *spherical shell* determined by these two concentric circles, or in general, hyperspheres, now represent the four components that result from removing x from a
small neighborhood of x in τ (green points in
Fig. 2). Moreover, for an appropriate proximity
graph constructed from D (see below and Fig. 2),
the remaining points induce exactly four connected components in this graph. Hence, by only
tracking the number of connected components
in graphs [2, 15], we deduce the underlying degree δ_0, denoting the number of
branches emerging from a data point.

While classical approaches for TDA of data approaching graph-structured
topologies stop at this point [1, 11], our concept of LTDA goes one step beyond.
Not only are we interested in the local topology underlying a data point, i.e.,
the number of branches emerging from this point, but we are also interested in

how this local topology affects the global topology. Consider again the singular point x in our discussed topology τ. As stated before, removing x from a small neighborhood of x breaks the neighborhood into four connected components, i.e., four branches emerge from x. However, moving further from x, two times two of these branches merge back together, and form cycles passing through x. As these branches merge back away from x, this implies that they must be connected in another way than through x. They are connected in the global topology even after removing x. Moreover, as removing x from a small neighborhood of x in τ breaks the neighborhood into $\delta_0 = 4$ components, but removing x from the full topological structure breaks the structure only into two connected components, the difference between these two denotes a practical lower bound on the number of convergences $\delta_1 = \delta_0 - 2 = 2$ induced by the branches emerging from x (Theorem 1). In Fig. 2, this corresponds to subtracting the number of connected components induced by the points outside the smallest circle (green and blue points), from the number of connected components induced by the points in the spherical shell (green points). Hence, we may not only apply LTDA to identify the underlying local topology, i.e., the number of emerging branches, but we may as well identify cycles by studying how the local topology affects the global topology.

The Vietoris-Rips Complex. As D is a point cloud data set, it does not make much sense to talk exactly about the local topology of some point $x \in D$ within the topological (normed vector) space $(D, \|\cdot\|)$, as this would be just a set of isolated points. However, for appropriate distance parameters $\epsilon \in \mathbb{R}^+$, which may be found by means of *persistent homology* (Appendix A), the *Vietoris-Rips complex*

$$\mathcal{V}_\epsilon(D) := \left\{ S \in 2^D : (|S| \leq \dim(D) + 1) \wedge (\forall v, w \in S)(\|v - w\| < \epsilon) \right\},$$

'well-models' topological behavior of the underlying topology τ of D (Appendix A), and it makes more sense to talk about the local topology of a point $\{x\} \in \mathcal{V}_\epsilon(D)$. The complex corresponds to the hypergraph induced by the

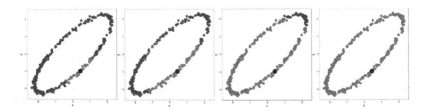

Fig. 3. Investigating the local topology of $z \in D$ (black) by studying the underlying topology of $B_{\mathbb{R}^2}(z, r) \cap D$ for increasing values of r. Points in $B_{\mathbb{R}^2}(z, r) \cap D$ are marked in red ($r = 1, 2, 3, 4$), remaining points in blue. This method starts off well, but quickly becomes susceptible to the restrictions imposed by the underlying topology on paths. (Color figure online)

cliques up to size $\dim(D) + 1$ of its graph 'skeleton', i.e., the graph consisting of all nodes from D and all edges $\{v, w\} \in 2^D$, where $0 < \|v - w\| < \epsilon$ (Fig. 2, $\epsilon = 3.5$). We will also talk about the (Vietoris-Rips) graph $\mathcal{V}_\epsilon(D)$ when referring to the skeleton of the complex, as we only consider *simplicial 1-complexes*, i.e., graphs in this paper.

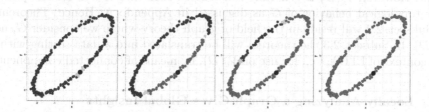

Fig. 4. Investigating the local topology of $z \in D$ (black) by studying topological properties of $\mathcal{V}_{0.3}(B_{\mathcal{V}_{0.3}(D)}(z, h + 1))$ for increasing values of h. Vertices from $\mathcal{V}_{0.3}(B_{\mathcal{V}_{0.3}(D)}(z, h))$ are marked in red, from $\mathcal{V}_{0.3}(B_{\mathcal{V}_{0.3}(D)}(z, h + 1)) \backslash B_{\mathcal{V}_{0.3}(D)}(z, h))$ in green ($h = 1, 10, 20, 27$), and remaining points in blue. The underlying linear structure is preserved until all points are included at $h = 27$. (Color figure online)

A Metric for LTDA Derived from the Vietoris-Rips Graph. The open balls in Fig. 2 are drawn using the Euclidean metric, i.e., the balls denote sets

$$B_{\mathbb{R}^2}(z, r) := \{y \in \mathbb{R}^2 : \|z - y\| < r\},$$

for some $r > 0$. Using this 'original' metric to investigate local topologies in $\mathcal{V}_\epsilon(D)$ seems like a natural approach. However, in the general case, we may not be able to reach one point from another by following a straight line within the topological structure itself. In general, we are restricted to follow *paths*, corresponding to new distances defined by integrating over these when possible. Following this intuition, we 'redefine' the metric on D by defining the distance between two points as the distance within the graph $\mathcal{V}_\epsilon(D)$. These *geodesic distances*, i.e., lengths of the shortest paths between nodes in the graph, are used to approximate the lengths of the shortest paths between the nodes' projections on the underlying topology. This metric corresponds to new open balls in D, containing finitely many data points, and defined as

$$B_{\mathcal{V}_\epsilon(D)}(z, h) := \{y \in D : d_{\mathcal{V}_\epsilon(D)}(z, y) < h\}.$$

Figures 3 and 4 illustrate this for a point cloud data set approaching an ellipse.

Remark. We emphasize the difference between the (embedding of a) graph G underlying a point cloud data set D, and the Vietoris-Rips graph $\mathcal{V}_\epsilon(D)$ constructed from D. These are generally non-homeomorphic in a graph-theoretical sense [16]. The unknown structure of G is often simple, with only a few multifurcation points and cycles. The known graph topology of $\mathcal{V}_\epsilon(D)$ itself is often complex, with many multifurcation points and cycles present in the graph. This may be seen on the toy data set in Fig. 2. As a graph itself, $\mathcal{V}_\epsilon(D)$ is quite complex, with many cycles and multifurcation points, i.e., nodes with degree at least equal to 3, whereas the underlying 8-structured topology of D is homeomorphic to the planar embedding of a graph with only two cycles and one multifurcation point. However, $\mathcal{V}_\epsilon(D)$ is generally constructed such that it well-models particular topological behavior of G as discussed in Appendix A. Hence, Theorem 1 in Subsect. 2.2 will reside in the field of graph theory where we consider G, not $\mathcal{V}_\epsilon(D)$. In Subsect. 2.3 the theorem will be translated into a data setting within the context of LTDA, i.e., for use on $\mathcal{V}_\epsilon(D)$, by means of connected components.

2.2   Locally Analyzing a Graph Gives Global Insights

We now formalize the insights obtained from the discussion above in a graph-theoretical theorem. While this theorem applies to general graphs, in Subsect. 2.3, we show how it can be applied to proximity graphs representing the underlying topology of point cloud data. We assume graphs to be simple[2], finite, and undirected, and that the reader is familiar with basic concepts of graph theory.

Notations. For a graph $G = (V, E)$, we denote the number of connected components by[3] $\beta_0(G)$, and the degree of a node $v \in V$ by $\delta_0(v)$. The degree of any edge $e \in E$ is by definition $\delta_0(e) := 2$. If $\alpha \in V \cup E$, we denote by $G \backslash \alpha$ the graph that results from removing α from G, as well as all edges incident to α if $\alpha \in V$.

Theorem for LTDA of Data Approaching Graph-Structured Topologies. The following theorem illustrates how the local topology of a node or an edge α in a graph G, expressed by its degree $\delta_0(\alpha)$, and how this local topology affects the connectedness of the global topology, expressed by the term $\beta_0(G) - \beta_0(G \backslash \alpha)$, may be used to learn a practical lower bound on the number of cycles passing through α. Moreover, the theorem allows us to exactly determine whether a cycle passes through a node or an edge in a graph or not.

Theorem 1. *Let $G = (V, E)$ be a graph. Then for each $\alpha \in V \cup E$, the number of cycles $C \subseteq E$ passing through α is bounded from below by*

$$\delta_1(\alpha) := \delta_0(\alpha) + \beta_0(G) - (\beta_0(G \backslash \alpha) + 1) \geq 0.$$

Moreover, for each $\alpha \in V \cup E$, a cycle passes through α iff $\delta_1(\alpha) > 0$.

Proof. The statements easily follows by induction from the well-known fact that inserting an edge into a graph either merges two connected components, or adds a cycle through that edge. Details are omitted for conciseness. □

---

[2] Loops and parallel edges may be subdivided without changing the graph topology.
[3] We maintain the terminology of *homology*, where β_0 refers to the *zeroth Betti number*.

2.3   LTDA of Data Approaching Graph-Structured Topologies

To be applicable for LTDA of point cloud data approaching graph-structured topologies, we show how to translate Theorem 1 into a data setting. This will allow us to construct an algorithm identifying multifurcation points and cycles present in the underlying topology by merely counting the number of connected components in a proximity graph constructed from such data (Algorithm 1).

We again emphasize the difference between the (embedding of a) graph G underlying a point cloud data set D, and the simplicial complex $\mathcal{V}_\epsilon(D)$ constructed from D. As remarked in Subsect. 2.1: these are generally non-homeomorphic in the graph-theoretical meaning. However, they approximate each other in terms of topological behavior as discussed in Appendix A.

Graph-Structured Topologies in a Data Setting. When a point cloud data set D approaches (the embedding of) a graph $G = (V, E)$ in $\mathbb{R}^n$ that is well-modeled by $\mathcal{V}_\epsilon(D)$ for some $\epsilon \in \mathbb{R}^+$, we may study the topology near $x \in D$, represented by $\alpha_x \in V \cup E$, by letting

- $\beta_0(G)$ correspond to $\beta_0(\mathcal{V}_\epsilon(D))$,
- $\beta_0(G \backslash \alpha_x)$ correspond to $\beta_0(\mathcal{V}_\epsilon(D \backslash B_{\mathcal{V}_\epsilon(D)}(x, r)))$,
- $\delta_0(\alpha_x)$ correspond to $\beta_0(\mathcal{V}_\epsilon(B_{\mathcal{V}_\epsilon(D)}(x, r') \backslash B_{\mathcal{V}_\epsilon(D)}(x, r)))$,

for some $0 \leq r < r'$ (see the discussion in Subsect. 2.1 and Fig. 2). All results in this paper were obtained by taking $r' - 1 = r \in \{2, 3\}$.

Hence, we may provide a mapping $D \to \mathbb{N} \times \mathbb{N} : x \mapsto (\delta_0(x), \delta_1(x))$, expressing the underlying local topology at α_x, as well as a lower bound on the number of cycles through α_x, furthermore indicating whether or not a cycle passes through α_x (Algorithm 1). We illustrate the use of this algorithm on an artificially constructed data set D based on the conference acronym, see Fig. 1.

E.g., the 'ends' of the four homeomorphic C, M, L and I-structured topologies are truthfully marked as (1,0) local topologies, i.e., structures resembling half-lines. The quotation mark is completely marked as having a (0,0) local topology, meaning this structure represents an isolated point. This shows that our algorithm may as well identify outlying points or areas, if the used proximity graphs models the underlying topology well. The (4,2) local topology in the 8-structured component marks an area with a local star-like topology with four legs, through which, in this case exactly, two cycles pass within the global topology.

Algorithm. For computational efficiency, the proposed algorithm marks neighbors of a node with a particular local topology with the same local topology. We implemented the algorithm such that nodes at a particular distance from another node are determined by a breadth-first search construction [2]. Hence, the total number of connected components in G is not needed to compute δ_1. If the inputted graph G has n vertices and m edges, where $m = \mathcal{O}(\delta n)$ for some 'average' degree δ, the while loop will be executed $\mathcal{O}(n/\delta)$ times. As each step

input : Prox. graph G & dist. par. r
output: (δ_0, δ_1)-classification of the nodes
que $\leftarrow$ G.nodes();
LG $\leftarrow$ matrix(length(G.nodes()), 2);
while *que* **do**
$\quad$ $\delta_0 \leftarrow \beta_0(G(\{v \in V(G) : d_G(\text{que}[0], v) = r\}));$
$\quad$ $\delta_1 \leftarrow \delta_0 - \beta_0(G(\{v \in V(G) : r \leq d_G(\text{que}[0], v) < \infty\}));$
$\quad$ **for** *v* in *(que[0] — G.neighbors[que[0]])* **do**
$\quad\quad$ LG[v] $\leftarrow$ (δ_0, δ_1);
$\quad\quad$ que.remove(v);
$\quad$ **end**
end
return *LG*

Algorithm 1. Pseudocode for (δ_0, δ_1)-classification

in the loop can be executed in linear, i.e., $\mathcal{O}(n + m) = \mathcal{O}(n + \delta n)$ time [2], the total complexity is $\mathcal{O}(n^2)$.

Tuning ϵ and r. The distance parameters ϵ and r may usually be tuned by manual investigation. For all results in this paper, it was sufficient to investigate the use of either $r = 2$ or $r = 3$. Tuning ϵ is more data dependent, and may be done by *persistent homology* as well (Figs. 13 and 14 in Appendix A). One may also integrate over different parameter ranges, which are bounded by the maximal pairwise distance for ϵ, and by the radius of the graph for r. Consequently, one inspects how well the reconstructed graph (Sect. 3) approximates the original graph, checking for a balance between reducing the Hausdorff distance, MSE, or metric distortion, (e.g., one may redefine distances as their projected distances on the reconstruction,) and reduction of the graph size, as also discussed in [1].

3   LTDA for Reconstructing Graph-Structured Topologies

In this section, we show the importance of LTDA for reconstructing the underlying topology. More concretely, we illustrate why the information retrieved by LTDA needs to be both *stored* and *used*, and why a simple *'edge or no-edge'* classification as used in the *metric graph reconstruction* algorithm [1] may not always lead to optimal results for noisy samples. The latter method uses, similar to our approach, spherical shell clustering in a Vietoris-Rips graph to identify branching structures, but only classifies points according to $\delta_0 = 2$ (edge) or $\delta_0 \neq 2$ (branch). The graph reconstruction is based on placing an edge between connected components of branch points, if they are both near one connected component of edge points. For further details on this method, we refer to [1].

Consider the simulated noisy two-dimensional data set D approaching a Y-structured topology with nonuniform density in Fig. 5. Our method of subgraph clustering (Algorithm 1) correctly infers the location of the (1,0) and (3,0) local topologies. However, due to the high amount of noise relative to the length of the branches, no (2,0) local topologies are detected. In this case, an *'edge or no-edge'*

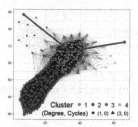

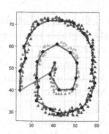

Fig. 5. Classifying the local topologies ($\epsilon = 15$, $r = 3$, comp. time: 0.17 s), and using these to reconstruct the underlying graph topology (comp. time: 0.34 s) for a noisy sample of 395 points approaching a Y-structured topology with nonuniform density.

Fig. 6. By a breadth-first traversal of the (2,0)-cluster, one may construct even better approximations of the underlying structure (black) than the original reconstructed graph (red). (Color figure online)

classification as in [1] would lead to one connected component of branch points, of which the reconstructed graph [1] would be a single vertex.

Nevertheless, the (3,0) local topology 'hints' the presence of three surrounding branches. Simply clustering the (1,0) local topologies in their induced subgraph would not lead to three connected components, as two of the branches would not be separated (cluster 2 & 3 in Fig. 5). This is a straightforward consequence of the underlying topology: even when we remove the bifurcation point, the branches are still at distance 0 from each other. Inseparability of the branches may even occur for less noisy data with uniform density, when the distance parameter ϵ was tuned too high. However, in this particular example there does not even exist a single distance value ϵ for which the three clusters would be pairwise separable in their induced subgraph of $\mathcal{V}_\epsilon(D)$, due to the nonuniform density.

Algorithm. A different clustering algorithm exploiting the information of the (3,0) local topology is needed. Applying hierarchical clustering (we use complete-linkage clustering unless stated otherwise), allows us to separate the points neighboring the (3,0) local topologies in three clusters (Fig. 5), leading to Algorithm 2 for reconstructing general underlying graph-structured topology. The pseudocode assumes the used graph G and distance object d stored in the output of Algorithm 1.

The pseudocode of Algorithm 2 allows for many variants in its implementation. E.g., many steps implicitly assume most pairwise distances defined by d to be unique, and we use the original Euclidean metric used to construct our proximity graph for Algorithm 1. We define the center of a set $X \subseteq D$ as the data point $c_X := \arg\min_{x \in X}(\max_{y \in Y} d(x, y))$, which leads to better results than the point closest to the mean in the case of nonuniform density. Representing the center in our current way works well for short patches of the underlying topology, but is less efficient for patches representing long and curvy trajectories (red graph in Fig. 6). Using a new metric defined by distances in the weighted

input : Output LG of Alg. 1 & dist. par. $\tilde{r}$
output: A graph representing the underlying topology
Cluster $\{x \in D = V(G) : \delta_0(x) \geq 3\}$ by LG-group in G;
Let N_1 be the collection of obtained clusters;
$\forall C \in N_1$, Use d to obtain a representative center $x_C \in D$;
$\forall C \in N_1$, use d to cluster $\{x \in D\backslash C : d_G(x_C, x) \leq \tilde{r}\}$ in $\delta_0(x_C)$ components;
Let N_2 be the collection of obtained clusters;
$\forall C \in N_2$, Use d to obtain a representative center $x_C \in D$;
If for $C_1, C_2 \in N_2$, $C_1 \cap C_2 \neq \emptyset$, split $C_1 \cup C_2$ into two equally sized disjoint sets
   by ordering the distances to x_{C_1}, according to d, of the included points;
Connect $C_1 \in N_1$ and $C_2 \in N_2$ by an edge if C_2 merged from C_1 in Step 5 or 8;
Cluster $D\backslash(\bigcup N_1 \cup \bigcup N_2)$ by LG-group in G;
Let N_3 be the collection of obtained clusters;
Split each $C \in N_3$ with uniform (2,1) local topology and disconnected from N_2
   in at least three consecutive connected components (this is an isolated cycle);
Connect $C_1 \in N_2 \cup N_3$ and $C_2 \in N_3$ by an edge if they are connected in G;
Connect $C_1, C_2 \in N_2$ by an edge if they are connected in G, unless this
   contradicts $\delta_0(x_{C_1})$ or $\delta_0(x_{C_2})$ in the current construction (this reduces
   ϵ-sensitivity);
return *A graph with (centers of)* $\bigcup_{i=1}^3 N_i$ *as vertices and the obtained edges*
  Algorithm 2. Pseudocode for reconstructing the graph topology

graph $\mathcal{V}_\epsilon(D)$, with the Euclidean lengths of the edges as weights, may lead to even better results for computing centers of long and curvy patches and (hierarchical) clustering into a given number of clusters, at the cost of computational efficiency. An alternative method is to use a breadth-first traversal to decompose long clusters representing edges into short and consecutive patches (black graph in Fig. 6, note that both graphs are nevertheless homeomorphic), or one may connect different centers by shortest paths as well. Isolated circles are separated into four components by starting a breadth-first traversal at a random point, dividing points according to low, medium, or high distance from the root, and dividing the points at medium distance into two separate components. Finally, we replace the representative point of a (1,0) component such that it is furthest from its adjacent center.

Tuning $\tilde{r}$. The distance parameter $\tilde{r}$ may be either tuned manually (all results in this paper were obtained by using either $\tilde{r} = r$ or $\tilde{r} = r + 1$, r being the distance parameter used to obtain the output of Algorithm 1), or tuned in an integration scheme as discussed in Subsect. 2.3. However, a new distance parameter $\tilde{r}$ is not needed for components resembling isolated points, edges, cycles or multifurcating trees. This last observations follows from

$$\begin{cases} |E| = \frac{1}{2}\sum_{v \in V} \delta_0(v) = \frac{1}{2}|\{v \in V : \delta_0(v) = 1\}| + \frac{1}{2}\sum_{\substack{v \in V \\ \delta_0(v) \geq 3}} \delta_0(v), \\ |E| = |V| - 1 = |\{v \in V : \delta_0(v) = 1\}| + |\{v \in V : \delta_0(v) \geq 3\}| - 1, \end{cases}$$

for a tree $T = (V, E)$ with $|E| \geq 1$ and no vertices of degree 2 (these are irrelevant for representing the underlying topology). This implies that the union

of points having either $(1,0)$ or $(2,0)$ local topologies must be clustered into
$|E| = \sum_{\substack{v \in V \\ \delta_0(v) \geq 3}} \delta_0(v) - |\{v \in V : \delta_0(v) \geq 3\}| + 1$ components, where this number
is computed with respect to the connected components with $\delta_0 \geq 3$. If the tree
has at least one multifurcation point, all such obtained clusters of edges will
be incident to at least one multifurcation point and represented by at least two
nodes in the reconstructed graph topology. This allows for another variant of
Algorithm 2 for tree-structured topologies: cluster the union of $(1,0)$ and $(2,0)$
local topologies in the obtained number of clusters, and connect each component
with $\delta_0 \geq 3$ to all adjacent clusters of edges.

4   Experimental Results

Our method is validated on two more real point cloud data sets approaching
graph-structured topologies. All our results were obtained using non-optimized
R code on a basic laptop.

Earthquake Data. We considered a geological data set D of 1479 strong to
great earthquakes (*Richter magnitude* $M_L > 6.5$), scattered across the world in
the rectangular domain $[140, 315] \times [-75, 65]$ of (longitude, latitude)-coordinates
($180°$ were added to negative longitudes to obtain a continuous structure). The
raw data is freely accessible from USGS Earthquake Search. A *distance to measure* [8] from the R-package **TDA** was used to remove most outliers ($m0 = 0.1$),
keeping 1440 observations with DTM < 30. The local topologies were classified
in 0.90 s ($\epsilon = 10, r = 2$), after which the underlying graph was reconstructed
in 4.16 s ($\tilde{r} = r = 2$). Two clusters representing long edges were decomposed
into respectively 15 and 5 consecutive patches, resulting in the graph depicted
in black in Fig. 7, approximating the underlying graph-structured topology well.

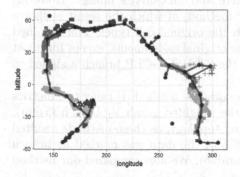
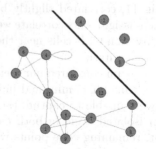

Fig. 7. LTDA and underlying graph reconstruction of earthquake data. Separating
long trajectories in consecutive patches
allows for a smooth reconstruction.

Fig. 8. Reconstructed graphs of the
earthquake data set by the method
discussed in [1].

We compared our method with the original underlying graph reconstruction method as discussed in [1], where parameters were tuned to capture the single self-loop present in the underlying topology. We used both the original Euclidean metric (Fig. 8, bottom left, 4 min 11 s), as well as the metric induced by the weighted graph $\mathcal{V}_{10}(D)$ (Fig. 8, top right, 2 min 41 s), but were unable to retrieve the full underlying topology with either of the metrics.

Cell Trajectory Data. We considered a normalized expression data set D of 4647 manually analyzed bone marrow cells containing measurements of five surface markers (CD34, CD1632, CD117, CD127 & Sca1). These cells are known to differentiate from long-term hematopoietic stem cells (LT-HSC) into short-term hematopoietic stem cells (ST-HSC), which can in turn differentiate into either common myeloid progenitor cells (CMP) or common lymphoid progenitor cells (CLP) [10]. I.e., the topology underlying this data set is that of an embedding in $\mathbb{R}^5$ of the graph depicted in Fig. 9. No data preprocessing was applied, and the Euclidean distance was used as the original metric. A PCA plot of the data is shown in Fig. 10. Comparing Figs. 9 and 10, we indeed note the presence of the Y-structured topology. However, it is clear that identifying this topology

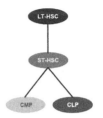

Fig. 9. The 4647 analyzed bone marrow cells consist of four cell types that are interconnected by means of cell differentiation.

would be a crucial problem in absence of the cell labeling. Hence, our method may serve as a first step in the context of cell trajectory inference [4,10], identifying the branching structure and different stages within a cell differentiation process. Our method classified local topologies in 15.55 s ($\epsilon = r = 2$), and used these to reconstruct the underlying topology in 5.46 s. Note that the local topology classes ((1,0) and (3,0)) imply an underlying tree-structured topology, and no new distance parameter $\tilde{r}$ is needed for the graph-reconstruction. We inferred the exact same graph using both complete and McQuitty's linkage. However, the labeling induced by using the latter method, of which the result is shown in Fig. 11, correlated slightly better with the original cell types. The obtained branch-assignments correlate well with the original assignments, except for, most notably, non-CLP cells near the base of the ST-HSC→CLP branch assigned to the branch itself.

We again compared our method to the original method [1] using two metrics (Euclidean: 1 h 17 min, and induced by the weighted graph $\mathcal{V}_2(D)$: 1 h 35 min), but were unable to capture the underlying topology, as these methods resulted in an isolated cycle in both cases (> 98% of the data was marked as branch point, remaining edge points were inseparable). We also compared our method

with Mapper[4] [19,20], using the freely accessible tool from the R package
TDAmapper. Experimenting with different filter functions, only the projection onto the first principal component allowed us to correctly infer the underlying topology in 11.85 s. However, this was a matter of luck, as the assignments
induced by the Mapper graph correlate badly with the original assignments
(Fig. 12).

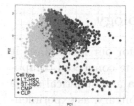
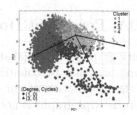
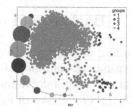

Fig. 10. PCA plot of the expression data.   **Fig. 11.** LTDA of the expression data.   **Fig. 12.** Mapper graph and its induced assignments.

5   Conclusion and Further Work

Applying clustering techniques to study local topologies, and how these affect
the global topology, introduces new possibilities for learning graph-structured
topologies underlying point cloud data sets, as one may even detect cycles without the need of 1-dimensional homology. Current state-of-the-art approaches for
investigating local topological structures either do not bother with reconstruction techniques, are vulnerable to noise, or miss out on the fact that knowledge
of the local topologies is crucial for reconstructing underlying graph-structured
topologies. We combined both LTDA and reconstruction techniques in a simple and intuitive way, leading to a framework for reconstructing the underlying
graph in many practical examples, improving both on the computational level
as well as the obtained results compared to current state-of-the-art approaches.

Contrary to [1], we prioritized explaining and validating our method by means
of empirical results on simulated and real data sets, rather than providing theoretical results guaranteeing the correctness of the reconstructed graph topology. Real data will most often violate the stated assumptions, and the 'one-for-
all' parameter approach posed by these may not be suitable when extending
our method to even more complex and high-dimensional data sets approaching graph-structured topologies with nonuniform noise. For this, one needs local

---

[4] Mapper uses a *filter* $f : D \to \mathbb{R}^d$ that maps the data to a lower-dimensional space
$\mathbb{R}^d$ (usually $d \in \{1, 2\}$), builds a grid of overlapping bins (intervals for $d = 1$, squares
for $d = 2, \ldots$) on top of $\mathbb{R}^d$, clusters the preimage $f^{-1}(B)$ for each bin B, and
connects clusters based on the overlap of the data and bins. This method results in
the construction of a graph meant to resemble the unknown underlying topology,
and has shown it may reveal a Y-structured topology in expression data before. For
such an example and further details on the Mapper algorithm, we refer to [19].

parameter integration schemes, combining results from the the fields of TDA (e.g., persistent local homology [11]), statistics, and machine learning. This provides new research both on the mathematical and experimental level.

Acknowledgments. This work was funded by the ERC under the European Union's Seventh Framework Programme (FP7/2007-2013) / ERC Grant Agreement no. 615517, and the FWO (G091017N, G0F9816N).

A   Background on TDA: Persistent Homology

Finding an appropriate proximity graph is a crucial step for our method, as it identifies the number of emerging branches by counting the connected components in subgraphs induced by the intersection of our data and spherical shells (Fig. 2). Our choice of using Vietoris-Rips graphs is not arbitrary, as experimental results have shown that these are far more useful for our method than a wide variety of other proximity graphs, such as, e.g., k-nearest neighbor graphs. Moreover, Rips-graphs are well studied within the field of TDA, more concretely *persistent homology*, allowing to appropriately tune the distance parameter ϵ.

Persistent Homology [12,22] tracks the (dis)appearance of distinct shape features across a *filtration*, i.e., a sequence of *simplicial complexes* [14]

$$\sigma_{\epsilon_1}(D) \subseteq \sigma_{\epsilon_2}(D) \subseteq \ldots \subseteq \sigma_{\epsilon_n}(D),$$

constructed from a point cloud data set D embedded in a metric space, for an increasing sequence of parameters $\epsilon_1, \ldots, \epsilon_n$. By evaluating how long certain

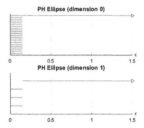

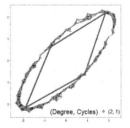

Fig. 13. Persistent homology of a point cloud data set approaching an ellipse. (Top) Each bar represents a connected component in $\mathcal{V}_\epsilon(D)$ for varying ϵ. The long persisting bar indicates that there is one connected component present in the underlying topological structure. (Bottom) Each bar represents one of the non-equivalent cycles in $\mathcal{V}_\epsilon(D)$ for varying ϵ. The long persisting bar indicates that there is one cycle present in the underlying topological structure.

Fig. 14. The resulting graph (skeleton of) $\mathcal{V}_\epsilon(D)$ for one of the distance parameters $\epsilon = 0.3$ occurring at both persisting bars in Fig. 13 (edges in black). The uniform (2,1) local topology indicates a cycle (see Subsect. 2.3, $r = 2$, comp. time: 0.14 s), and allows us to reconstruct the underlying topology (edges in red, see Sect. 3, comp. time: 0.17 s). (Color figure online)

features exist, one is able to deduce topological invariants of its underlying topological structure [17]. The evolution of these (dis)appearing shape features may be modelled by means of *barcodes*, computed by methods of linear algebra [18,24]. The number of bars occuring at a fixed value of ϵ denotes the k-th *Betti number*, i.e., the number of k-dimensional holes, at the point ϵ in the filtration. Long bars resemble topological features that 'persist' for many consecutive values $\epsilon_i, \epsilon_{i+1}, \ldots, \epsilon_j$, and indicate features of the underlying topology of point cloud data. See Fig. 13, where the used filtration consists of Vietoris-Rips complexes.

References

1. Aanjaneya, M., Chazal, F., Chen, D., GLisse, M., Guibas, L., Morozov, D.: Metric graph reconstruction from noisy data. Int. J. Comput. Geom. Appl. **22**(04), 305–325 (2012)
2. Bernhard, K., Vygen, J.: Combinatorial Optimization: Theory and Algorithms. Springer, Heidelberg (2012). https://doi.org/10.1007/3-540-29297-7
3. Cámara, P.G., Rosenbloom, D.I.S., Emmett, K.J., Levine, A.J., Rabadán, R.: Topological data analysis generates high-resolution, genome-wide maps of human recombination. Cell Syst. **3**(1), 83–94 (2016)
4. Cannoodt, R., Saelens, W., Saeys, Y.: Computational methods for trajectory inference from single-cell transcriptomics. Eur. J. Immunol. **46**(11), 2496–2506 (2016)
5. Carlsson, G.: Topology and data. Bull. Am. Math. Soc. **46**(2), 255–308 (2009). https://doi.org/10.1090/S0273-0979-09-01249-X
6. Carlsson, G.: Topological pattern recognition for point cloud data (2013)
7. Carlsson, G., Ishkhanov, T., de Silva, V., Zomorodian, A.: On the local behavior of spaces of natural images. Int. J. Comput. Vis. **76**(1), 1–12 (2008)
8. Chazal, F., Cohen-Steiner, D., Mérigot, Q.: Geometric inference for measures based on distance functions (2009)
9. Choi, E., Bond, N.A., Strauss, M.A., Coil, A.L., Davis, M., Willmer, C.N.A.: Tracing the filamentary structure of the galaxy distribution at z ~ 0.8. Mon. Not. R. Astron. Soc. **406**(1), 320–328 (2010)
10. De Baets, L., Van Gassen, S., Dhaene, T., Saeys, Y.: Unsupervised trajectory inference using graph mining. In: Angelini, C., Rancoita, P.M.V., Rovetta, S. (eds.) CIBB 2015. LNCS, vol. 9874, pp. 84–97. Springer, Cham (2016). https://doi.org/10.1007/978-3-319-44332-4_7
11. Fasy, B.T., Wang, B.: Exploring persistent local homology in topological data analysis. In: 2016 IEEE International Conference on Acoustics, Speech and Signal Processing (ICASSP), pp. 6430–6434 (2016)
12. Ghrist, R.: Barcodes: the persistent topology of data. Bull. (New Ser.) Am. Math. Soc. **45**(107), 61–75 (2008)
13. Giusti, C., Ghrist, R., Bassett, D.S.: Two's company, three (or more) is a simplex. J. Comput. Neurosci. **41**(1), 1–14 (2016)
14. Hatcher, A.: Algebraic Topology. Cambridge University Press, Cambridge (2002)
15. Hopcroft, J.E., Ullman, J.D.: Set merging algorithms. SIAM J. Comput. **2**(4), 294–303 (1973)
16. Lapaugh, A.S., Rivest, R.L.: The subgraph homeomorphism problem. J. Comput. Syst. Sci. **20**(2), 133–149 (1980)

17. Medina, P., Doerge, R.: Statistical methods in topological data analysis for complex, high-dimensional data. In: Annual Conference on Applied Statistics in Agriculture (2015)
18. Nanda, V., Sazdanović, R.: Simplicial models and topological inference in biological systems. In: Jonoska, N., Saito, M. (eds.) Discrete and Topological Models in Molecular Biology. NCS, pp. 109–141. Springer, Heidelberg (2014). https://doi.org/10.1007/978-3-642-40193-0_6
19. Nicolau, M., Levine, A.J., Carlsson, G.: Topology based data analysis identifies a subgroup of breast cancers with a unique mutational profile and excellent survival. Proc. Nat. Acad. Sci. **108**(17), 7265–7270 (2011)
20. Rizvi, A.H., et al.: Single-cell topological RNA-seq analysis reveals insights into cellular differentiation and development. Nat. Biotechnol. **35**, 551–560 (2017)
21. Wang, B., Summa, B., Pascucci, V., Vejdemo-Johansson, M.: Branching and circular features in high dimensional data. IEEE Trans. Visual. Comput. Graph. **17**, 1902–1911 (2011)
22. Wang, K.: The basic theory of persistent homology (2012)
23. Wasserman, L.: Topological data analysis. Ann. Rev. Stat. Appl. **5**(1), 501–532 (2018)
24. Zomorodian, A., Carlsson, G.: Computing persistent homology. Discrete Comput. Geom. **33**(2), 249–274 (2005)

Similarity Modeling on Heterogeneous Networks via Automatic Path Discovery

Carl Yang[✉], Mengxiong Liu, Frank He, Xikun Zhang,
Jian Peng, and Jiawei Han

University of Illinois at Urbana-Champaign, Urbana, IL 61801, USA
{jiyang3,mliu60,shibihe,xikunz2,jianpeng,hanj}@illinois.edu

Abstract. Heterogeneous networks are widely used to model real-world semi-structured data. The key challenge of learning over such networks is the modeling of node similarity under both network structures and contents. To deal with network structures, most existing works assume a given or enumerable set of meta-paths and then leverage them for the computation of meta-path-based proximities or network embeddings. However, expert knowledge for given meta-paths is not always available, and as the length of considered meta-paths increases, the number of possible paths grows exponentially, which makes the path searching process very costly. On the other hand, while there are often rich contents around network nodes, they have hardly been leveraged to further improve similarity modeling. In this work, to properly model node similarity in content-rich heterogeneous networks, we propose to automatically discover useful paths for pairs of nodes under both structural and content information. To this end, we combine continuous reinforcement learning and deep content embedding into a novel semi-supervised joint learning framework. Specifically, the supervised reinforcement learning component explores useful paths between a small set of example similar pairs of nodes, while the unsupervised deep embedding component captures node contents and enables inductive learning on the whole network. The two components are jointly trained in a closed loop to mutually enhance each other. Extensive experiments on three real-world heterogeneous networks demonstrate the supreme advantages of our algorithm. Code related to this paper is available at: https://github.com/yangji9181/AutoPath.

Keywords: Similarity modeling · Heterogeneous networks
Deep embedding

1 Introduction

Networks are commonly used to model relational data such as people with social relations and proteins with biochemical interactions. Recently, increasing research attention has been paid to heterogeneous networks, highlighting

Electronic supplementary material The online version of this chapter (https://doi.org/10.1007/978-3-030-10928-8_3) contains supplementary material, which is available to authorized users.

ⓒ Springer Nature Switzerland AG 2019
M. Berlingerio et al. (Eds.): ECML PKDD 2018, LNAI 11052, pp. 37–54, 2019.
https://doi.org/10.1007/978-3-030-10928-8_3

multi-typed nodes and connections. Their modeling of rich semantics in terms of both node contents and typed links enables the integration of real-world data from various sources and facilitates wide applications [13, 22, 30, 31, 33].

The key challenge of learning with heterogeneous networks is the modeling of node *similarities* (also known as *proximities*) [21]. To deal with this, meta-paths have been introduced to constrain the counting of path instances [22, 28] or guide meaningful network embedding [3, 18]. However, we summarize the drawbacks of most existing heterogeneous network learning algorithms into the following two aspects and explain them in details with our toy example in Fig. 1.

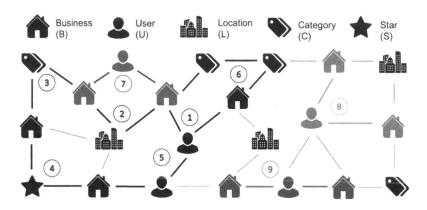

Fig. 1. A toy example of modeling the Yelp data with heterogeneous networks.

Drawback 1: Assumption of Given or Enumerable Sets of Meta-paths.
Most existing methods for heterogeneous network modeling assume a known set of useful meta-paths, either given by domain experts or exhaustively enumerable. Then they combine the information of multiple meta-paths through uniform addition [3, 8, 18, 22], or importance weighing [4, 5, 14, 28, 33]. However, given any arbitrary heterogeneous network, the process of composing meta-paths according to domain knowledge is ad hoc. Moreover, it is not always efficient or even feasible to enumerate or search for all potentially useful paths, since the number of paths grows exponentially as we consider longer paths, and it is notoriously costly to instantiate the paths on the network.

Consider our toy example in Fig. 1, which is a simple heterogeneous network constructed with the Yelp data similarly as done in [33]. We only consider five node types: businesses (B), users (U), locations (L), categories (C) and stars (S). As for links, we only consider users reviewing businesses (U – B), businesses residing in locations (B – L), businesses belonging to categories (B – C), businesses having stars (B – S), users being friends with users (U – U) and categories belonging to categories (C – C), while other links such as those between adjacent star levels and pairs of geographically nearby locations are ignored for the simplicity of the example.

On this simple heterogeneous network, if we only consider meta-paths between pairs of businesses with length no longer than 4, we already have 6 paths (1–6). Once we increase the length to 5, since meta-paths of length 5 can be composed by two meta-paths of length 3 or one meta-path of length 4 with an additional node, the number of meta-paths of length 5 alone is around 20 (4 × 4 for the combination of two paths of length 3 plus 2 for paths with 3 categories or users in a row between two businesses).

Note that this is a simplified heterogeneous network with a few node types and link types, and we are only considering meta-paths with lengths no longer than 5, while each meta-path can have millions of instances. Real-world heterogeneous networks can be much more complex. Also, while existing works argue that longer paths are less useful, there exists no solid support for this argument, nor a good way of setting the maximum length of paths to consider.

Drawback 2: No Leverage of Rich Contents Around Network Nodes. Furthermore, networks can have various contents [32], but no existing algorithm on heterogeneous networks has considered the integration of such rich information. For instance, in the example in Fig. 1, users have attributes like number of reviews, time since joining Yelp, number of fans, average rating of all reviews. Such contents well characterize user properties like *preference* and *expertise*.

In this paper, we argue that even instances of the same meta-path can carry rather different semantic meanings. To give a few examples, suppose the user on path 7 enjoys high-end restaurants while that on path 8 prefers cheap ones. The two pairs of businesses on the ends of the paths are then close in different ways. Likewise, if the two users on path 7 and 9 have been to very different numbers of places, they may choose the places to go based on quite different criteria, and thus again lead to different path semantics. Besides users, categories can be differentiated based on the generality, while locations cover different ranges. Stars also correspond to different similarities, as 1-star means equally bad whereas 5-star means comparably fantastic. Due to such observations, existing heterogeneous network learning algorithms are incompetent, because they do not consider the node contents and, as a consequence, model every instance of a meta-path as the same. It is urgent that we develop a powerful framework to incorporate such semantics and better model node similarity on heterogeneous networks.

Insight: Semi-supervised Learning with Limited Labeled Examples. In this work, we propose to leverage SSL to capture both structural and content information that is important for measuring the similarities among nodes on heterogeneous networks. Given an arbitrary network, unlike existing methods, we do not require a known set of useful meta-paths, nor do we try to enumerate all of them up to a heuristic length limit. Instead, we depend on a small number of example pairs of similar nodes which can be easily composed. Then we design an efficient algorithm to automatically explore useful paths on the network under the supervision of these labeled node pairs. In this way, the structural information on the heterogeneous networks can be fully leveraged.

Moreover, to incorporate content information such as node attributes, we combine an unsupervised objective of content embedding with the supervised path discovery into an SSL framework. By modeling the unlabeled node contents in an unsupervised way, it allows our algorithm to induce the similarity among unlabeled nodes on the whole network, as well as unseen nodes that might be added to the network in the future. It also avoids the requirements for large amounts of training data that cover the whole network.

Approach: Reinforcement Learning with Deep Content Embedding. In this work, we propose AUTOPATH, to solve the problem of similarity modeling on content-rich heterogeneous networks.

As we discussed before, the number of paths between nodes is exponential to the length. Moreover, searching for paths on networks is notoriously expensive. To deal with such challenges, we leverage reinforcement learning, which has been found efficient in sequential decision making and successfully applied for path exploration on knowledge bases [2,29]. However, to the best of our knowledge, there is no previous work on employing reinforcement learning to model heterogeneous networks, which have quite a few unique properties, such as the large action spaces at each node when growing the paths and the large numbers of valid paths between each pair of nodes. Such properties make the direct application of existing algorithms on knowledge bases to heterogeneous networks impossible. Another major distinction between heterogeneous networks and knowledge bases is the prevalence of rich node contents, which has hardly been explored before by existing algorithms. The existence of such node contents that potentially differentiate the semantics on instances of the same meta-paths further increases the difficulty of similarity modeling over heterogeneous networks. Such situations, as we will discuss more in Sect. 2, urge the development of a specifically designed reinforcement learning framework.

To overcome the challenges of large action spaces and node contents simultaneously, we leverage continuous reinforcement learning and incorporate deep content embedding to learn the state representations. Specifically, continuous policy gradient effectively estimates similar actions and avoids the explicit search over all discrete actions. Moreover, we devise conjugate deep autoencoders to capture node types and contents, and jointly train them with the policy and value networks of the reinforcement learning agent in a closed loop, so as to allow the mutual enhancement between embedding and learning. More details of our models are discussed in Sect. 3.

As we will demonstrate in Sect. 4, our proposed AUTOPATH algorithm is able to break free the requirements of known sets of meta-paths, leverage node contents, and achieve state-of-the-art performance on the task of similarity search with very limited supervision. Extensive quantitative experiments and qualitative analysis on three real-world heterogeneous networks demonstrate the advantages of AUTOPATH over various state-of-the-art heterogeneous network modeling algorithms.

2   Preliminaries

In this section, we briefly introduce the key concepts and relevant techniques of heterogeneous network modeling and reinforcement learning. Due to space limit, a broader discussion of related works is placed into our Supplementary Materials.

2.1   Heterogeneous Network Modeling

Heterogeneous network has been intensively studied due to its power of accommodating multi-typed interconnected data [3,21,22,30]. In this work, we stress that rich contents are prevalently available on nodes in the networks, and we define content-rich heterogeneous networks as follows.

Definition 1. *Content-Rich Heterogeneous Network. A content-rich heterogeneous network is defined as a directed graph $\mathcal{N} = \{\mathcal{V}, \mathcal{E}, \mathcal{A}\}$. For each node $v \in \mathcal{V}$ and its corresponding node type $\phi(v) = T$, a content vector $A_v^T \in \mathcal{A}$ is associated with v. Depending on the node type T and available data, A^T can be categorical, numerical, textual, visual, etc., or any mixture of them.*

To properly model heterogeneous networks, [22] introduces the concept of meta-path, which has been the golden measure of similarity among nodes on heterogeneous networks [4,14,19,22,27,28], and recently have also enabled various heterogeneous network embedding algorithms [3,5,8,18,20,26]. However, most existing heterogeneous network modeling algorithms assume a given or enumerable set of useful meta-paths up to a certain empirically decided length, which is not always practical. Moreover, they do not consider contents in the networks, and thus regard all instances of the same meta-paths as the same.

2.2   Reinforcement Learning

The main challenge of heterogeneous network modeling without a known set of meta-paths is to automatically explore and find the useful ones, which is naturally a combinatorial problem. For automatic path discovery on heterogeneous networks, as we consider K types of nodes and meta-paths of length L, the number of all possible meta-paths can be at the same scale as K^L. Moreover, we stress that on content-rich heterogeneous networks, instances of the same meta-paths can carry different semantics, and the search space is further enlarged to approximately ρ^L, where ρ is the average out-degree of nodes on the network and is often much larger than K.

Reinforcement learning has been intensively studied for solving complex planning problems with consecutive decision makings, such as robot control and human-computer games [15,23]. Recently, there are several approaches based on reinforcement learning to tackle the combinatorial optimization problems over network data [1,9], as well as reasoning over knowledge bases [2,29], which are shown to be effective. Motivated by their success, we aim to leverage reinforcement learning to efficiently solve the combinatorial problem of automatic path discovery on heterogeneous networks.

Different from knowledge bases, although content-rich heterogeneous networks have fewer node types, each type has much larger number of nodes. Categorical actor networks used in [2,29] have poor convergence property in our heterogeneous network setting. To address this issue, continuous reinforcement learning serves as an appropriate paradigm. Our action is applied in the deep embedding space which is trained together with conjugate autoencoders to represent node types and contents. Unlike DDPG [12] or Q-learning [15] which learn a deterministic policy, our algorithm is designed to learn a probability distribution over actions as a policy. By sampling from the learned policy, our framework assigns large probabilities to high-quality paths. To briefly summarize, our algorithm leverages both structural and content information and automatically discover meaningful paths, under the guidance of limited labeled data.

3   AutoPath

In this section, we describe our AUTOPATH algorithm, which combines reinforcement learning and deep embedding over content-rich heterogeneous networks into a semi-supervised learning framework.

3.1   Overall Semi-supervised Learning Framework

We start with a formal definition of our problem.

Definition 2. *Similarity Modeling. Consider a content-rich heterogeneous network $\mathcal{G} = \{\mathcal{V}, \mathcal{E}, \mathcal{A}\}$ with a corresponding type function ϕ. The problem of similarity modeling is to measure the similarity between any pair of nodes, under the consideration of various meta-paths and rich node contents on the path instances.*

We stress that similarity modeling is the key challenge of learning with content-rich heterogeneous networks, as its solution naturally enables various subsequent tasks like link prediction, node classification, community detection and so on.

In this work, we aim to automatically learn the important meta-paths and node contents by leveraging limited labeled data. Therefore, besides a graph $\mathcal{G} = \{\mathcal{V}, \mathcal{E}, \mathcal{A}\}$, we consider the basic input as a set of example similar pairs of nodes $\mathcal{P}$, upon which we build a supervised learning module using reinforcement learning to explore their prominent connecting paths characterized by network links $\mathcal{E}$. To make the learning algorithm efficient and aware of node contents $\mathcal{A}$, we further build an unsupervised learning module with deep content embedding, which also enables inductive learning on the whole network $\mathcal{G}$ not necessarily covered by $\mathcal{P}$. Figure 2 shows the overall framework of AUTOPATH, and in what follows, we describe the two major components of this framework in details.

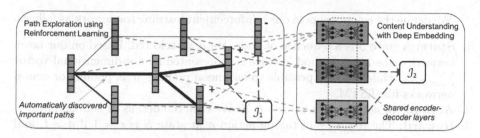

Fig. 2. Overall framework of AUTOPATH: Nodes are encoded by their embedding vectors based on both structure and content information on the network, where different colors denote different node types. The black solid lines denote actual network links, and the line weights denote the importance of the paths discovered by the algorithm. The colored dash lines indicate the connections between node embeddings and their content embedding models *w.r.t.* the corresponding node types. The supervised module with an objective $\mathcal{J}_1$ is trained *w.r.t.* the labeled node pairs in the given example set, and the unsupervised module with an objective $\mathcal{J}_2$ is trained over the whole network.

3.2   Path Exploration Using Reinforcement Learning

Learning Paradigm. As we discussed before, automatic path exploration is essentially a combinatorial problem over enormous search spaces, which cannot be well solved by exhaustive enumeration or searching with greedy pruning. Motivated by the recent success of reinforcement learning on sequential decision making, we propose to leverage the following paradigm for efficient path exploration.

For an example pair of similar nodes $p = \{s, t\} \in \mathcal{P}$, we call s a *start node* and t a *target node*. From each start node, we repeatedly train the reinforcement learning agent by looking for the next node to go on the network. A partial solution is represented as a sequence $S = (s, v_1, v_2, \ldots)$. At each step, based on the model parameters and the current state, the agent will either choose a neighboring node to go to (v_k) or return to the start node (s). Every time the agent reaches the target node (t), it gets a positive reward and returns to the start node (s). We will depict the details of $\mathcal{P}$ on different datasets in Sect. 4.

Framework Representation. To deal with the aforementioned large action space challenge, we propose to leverage a novel network embedding method, which captures the node types and contents into a low-dimensional latent space. The details of the embedding method are deferred to the next subsection. Similar to [1], our neural network architecture models a stochastic policy $\pi(a \mid S, \mathcal{G})$, where a is the action of selecting the next node from the network $\mathcal{G}$ and S is the current partial solution.

We define the components in our reinforcement learning framework as follows.

1. **State:** A state S is a sequence of nodes we have selected. Based on our novel network embedding method, a state is represented by a κ-dimensional vector $\sum_{v \in S} \mathbf{x}_v$, while it is also possible to use mean pooling, max pooling or neural networks like LSTM.
2. **Action:** An action a is a node $v \in \mathcal{V}$. We cast the details of actions later.
3. **Reward:** The reward r of taking action a at state S is $r = 1$ if $a = t$, and $r = 0$ otherwise.
4. **Transition:** The transition is deterministic by simply adding the node v we have selected according to action a to the current state S. Thus, the next state $S' := (S, v)$.

Our actor network (policy network) $\mu_\Theta(S)$ and critic network (value function) $\nu_\Theta(S)$ are both fully connected feedforward neural networks, each containing four layers including two hidden layers of size H, as well as the input and output layers. Rectified linear unit (ReLU) is used as the activation function for each layer, and the first hidden layer is shared between two networks. Both networks' inputs are κ-dimensional node embeddings. The output of the actor network $\mu_\Theta(S)$ are κ-dimensional vectors μ and σ^2, whereas the output of critic network $\nu_\Theta(S)$ is a real number.

Learning Algorithm. To overcome the large action space problem, we adopt continuous policy gradient as our learning algorithm. Our policy selects actions in node embedding space [12,17]. At each time step, we select a continuous vector and then retrieve the closest node from the current neighborhood plus the start node by comparing the action vector with the node embeddings.

Consider our policy $\pi(a \mid S)$, unlike in the discrete action domain where the action output is a *softmax* function, here the two outputs of the policy network are two real number vectors which we treat as the mean μ and variance σ^2 of a multi-dimensional normal distribution with a spherical covariance $\Sigma = \sigma^2 I$. To act, the input is passed through the model to the output layer where a Gaussian exploration is determined by μ and σ^2 as

$$\pi(a \mid S, \{\mu, \Sigma\}) = \frac{1}{\sqrt{2\pi |\Sigma|}} \exp\left(-\frac{1}{2}(a - \mu)^T \Sigma^{-1}(a - \mu) \right). \tag{1}$$

Since our goal is to find the important path S, our training loss is

$$\mathcal{J}_p(\Theta \mid \mathcal{G}) = -\mathbb{E}_{\tau \sim p_\Theta(S|\mathcal{G})} R(\tau), \tag{2}$$

where τ denotes an episode of the state-action trajectory, Θ is the set of parameters, and R is the reward. $\mathcal{J}_p$ is called the surrogate loss in reinforcement learning which evaluates the quality of the entire path S constructed by τ. To derive the gradient of $\mathcal{J}_p$, we use the policy gradient theorem [23] which gives

$$\nabla_\Theta \mathcal{J}_p = -\frac{1}{\alpha} \sum_{i=1}^{\alpha} \sum_{t=0}^{T-1} \nabla_\Theta \pi_\Theta(a_t^{(i)} \mid S_t^{(i)}) \hat{A}_t, \tag{3}$$

$$\hat{A}_t = \left(\sum_{k=t}^{T-1} r(S_k^{(i)}, a_k^{(i)}) - b(S_k^{(i)}) \right), \tag{4}$$

where α is the number of trajectories, T is the trajectory length, $\hat{A}_t$ is advantage and b is the baseline for variance reduction. By exploiting the fact that

$$\nabla_\Theta \pi_\Theta(a \mid S) = \pi_\Theta(a \mid S) \frac{\nabla_\Theta \pi_\Theta(a \mid S)}{\pi_\Theta(a \mid S)} = \pi_\Theta(a \mid S) \nabla_\Theta \log \pi_\Theta(a \mid S), \tag{5}$$

we have the approximate gradient estimator as

$$g = \mathbb{E}_t[\nabla_\Theta \log \pi_\Theta(a_t \mid S_t) \hat{A}_t], \tag{6}$$

where $\mathbb{E}_t$ denotes the empirical average over a mini-batch of samples in the algorithm that alternates between sampling and optimization using policy gradient.

In order to reduce the variance, we choose the value function $\mathcal{V}_\Theta$ as the baseline. $\mathcal{V}_\Theta$ is learned by using Monte Carlo method to minimize the loss

$$\mathcal{J}_v = \| \mathcal{V}_\Theta(S_t) - \sum_{k=t}^{T-1} r(S_k, a_k) \|_2^2. \tag{7}$$

Subsequently, we define our policy gradient loss as the sum of surrogate loss and value function loss, i.e., $\mathcal{J}_1 = \mathcal{J}_p + \mathcal{J}_v$, which can be regarded as a supervised loss under the example similar pairs of nodes.

3.3   Content Understanding with Deep Embedding

Conjugate Autoencoders. In order to make AUTOPATH aware of node contents and able to perform inductive learning on the whole network, we design a novel unsupervised node embedding method. Unlike existing network embedding methods designed to capture link structures, we aim to represent node types and contents in a shared low-dimensional space. To this end, we get inspired by recent success in deep learning for feature composition [11], which has been proven advantageous in capturing intrinsic features within complex contents in an unsupervised learning fashion.

To be specific, we propose conjugate autoencoders, which is a novel variant of deep denoise autoencoder. It consists of two non-linear feedforward neural network layers, i.e., two encoder layers and two decoder layers. The first encoder layers and the last decoder layers have individual embedding weights for each node type, while the other two layers are shared across different node types, as demonstrated in Fig. 2. Therefore, the embedding $\mathbf{x}_i$ for node v_i of type k (*i.e.*, $\phi(v_i) = k$) is computed as

$$\mathbf{x}_i = \mathbf{f}_e^o(\mathbf{f}_e^k(\mathbf{a}_i)), \text{ where } \mathbf{f}_e^j(\mathbf{x}) = ReLU(\mathbf{W}_e^j Dropout(\mathbf{x}) + \mathbf{b}_e^j). \tag{8}$$

Similarly, the reconstructed feature $\tilde{\mathbf{a}}_i$ of node v_i is computed as

$$\tilde{\mathbf{a}}_i = \mathbf{f}_d^k(\mathbf{f}_d^o(\mathbf{x}_i)), \text{ where } \mathbf{f}_d^j(\mathbf{x}) = ReLU(\mathbf{W}_d^j Dropout(\mathbf{x}) + \mathbf{b}_d^j). \tag{9}$$

The parameters in $\mathbf{f}_e^o$ and $\mathbf{f}_d^o$ are shared across all node types, while the parameters in $\{\mathbf{f}_e^k, \mathbf{f}_d^k\}_{k=1}^K$ are different for each node type.

Content Reconstruction Loss. To learn the intrinsic node features in an unsupervised fashion, a *content reconstruction loss* is computed over the whole network as

$$\mathcal{J}_r = \sum_{i=1}^n l(\mathbf{a}_i, \tilde{\mathbf{a}}_i). \tag{10}$$

Depending on the contents in the datasets, l can be implemented either as a cross entropy (for binary features, such as user attributes) or a mean squared error (for continuous features, such as TF-IDF scores of words).

Type Discrimination Loss. While $\mathcal{J}_r$ enforces the capture of node contents, node embeddings computed in this way does not necessarily discriminate different types of nodes in the shared embedding space, which weakens the ability of the algorithm to differentiate various meta-paths. To deal with this, we further impose a *type discrimination loss* over the whole network as

$$\mathcal{J}_d = -\sum_{i=1}^n log(p(i)), \text{ where } p(i) = \frac{exp(\mathbf{W}_c^{\phi(v_i)}\mathbf{x}_i)}{\sum_k exp(\mathbf{W}_c^k \mathbf{x}_i)}. \tag{11}$$

It is basically a softmax classifier towards node types with cross-entropy loss, which acts as adversarial to the shared reconstruction loss to make sure different types of nodes do not mingle too much in the shared embedding space.

The two losses can be combined with a tunable weighting parameter λ as $\mathcal{J}_2 = \mathcal{J}_r + \lambda \mathcal{J}_d$. We use Φ to denote all parameters related to these two losses.

3.4   Joint Training of Reinforcement Learning and Deep Embedding

Training Pipeline. To realize our SSL framework, we integrate the training of reinforcement learning and deep embedding into a joint learning pipeline, with the overall loss $\mathcal{J} = \mathcal{J}_1 + \mathcal{J}_2$. We firstly pre-train the content embedding with all parameters in Φ until $\mathcal{J}_2$ is sufficiently small, which captures the intrinsic distribution of node contents in a low-dimensional space. Then we detach the encoder layers and learn the rest of the model through co-training. Such detachment and separation of pre-training and co-training are necessary for allowing the node embeddings to become different for nodes even with the same contents to respect the network structures. Specifically, during co-training, we iteratively train the actor and critic networks by updating the parameters in Θ, and the embedding networks by updating the parameters in Φ except for those in the encoder. Note that, in both processes, the node embeddings $\mathcal{X}$ will also get updated, to reflect both important network structures and node contents. In each epoch, when updating Θ and $\mathcal{X}$, we sample a set Ω of α trajectories of length m using the current policy $\pi_\Theta(a \mid \mathcal{S})$, with each trajectory starting from a random start node in the set of example node pairs $\mathcal{P}$, and construct the surrogate loss and value function loss in $\mathcal{J}_1$; when updating Φ, we sample a set Ψ of β nodes from all nodes $\mathcal{V}$ in the whole network $\mathcal{G}$, and compute the reconstruction loss and discrimination loss in $\mathcal{J}_2$. Mini-batch SGD is then used to optimize the objectives iteratively for γ epochs, where all model parameters in $\{\Theta, \Phi, \mathcal{X}\}$ are updated by Adam [10]. We released our code with a demo function on Github[1] and also included it in our Supplementary Materials.

Computational Complexity. We theoretically analyze the complexity of AUTOPATH. For the reinforcement learning component, during each step of training, AUTOPATH generates a target mean μ_Θ in constant time and then selects a node from $\mathcal{G}$ that is the closest to μ_Θ. Note that, to grow a path, we only need to compare nodes in the direct neighborhood of the current node plus the start node, the size of which is much smaller than n and can be regarded as a constant number ρ. Since computing the quality function and updating the neural network model based on particular trajectories take constant time, the overall complexity of training and planning with the reinforcement learning agent is $O(\alpha\rho m)$ in each epoch. For the deep embedding component, AUTOPATH uniformly samples the nodes in $O(\beta)$ time, and then compute the losses and update the models in $O(1)$ time. Therefore, the overall training time of AUTOPATH is $O((\alpha\rho m + \beta)\gamma)$. The time of model inference for particular nodes is ignorable compared with model training.

4 Experimental Evaluations

In this section, we evaluate the performance of our proposed AUTOPATH algorithm on three real-world content-rich heterogeneous networks in different domains, *i.e.*, IMDb from a movie rating platform[2], DBLP from an academic publication collection[3], and Yelp from a business review service[4]. Through extensive quantitative experiments and qualitative analysis in comparison with various baselines, we show that AUTOPATH can efficiently leverage both structural and

---

[1] https://github.com/yangji9181/AutoPath.
[2] http://www.imdb.com/.
[3] https://dblp.uni-trier.de/.
[4] https://www.yelp.com/.

content information on heterogeneous networks, which leads to supreme performance on the key task of similarity modeling.

4.1 Experimental Settings

Datasets. We describe the datasets we use as follows and the statistics are summarized in Table 1.

1. **IMDb:** We use the MovieLens-100K dataset[5] made public by [7]. There are four types of nodes in the network, *i.e.*, users (U), movies (M), actors (A), and directors (D). The edge types include users reviewing movies, actors featuring in movies, and director making movies. The contents we use for users include simple demographics like age, gender, occupation, zipcode. For movies, actors and directors, we collect the first textual paragraph of the main content in their corresponding Wikipedia[6] page if available.
2. **DBLP:** We use the Arnetminer dataset V8[7] collected by [25]. It contains four types of nodes, *i.e.*, authors (A), papers (P), venues (V), and years (Y). The edge types include authors writing papers, papers citing papers, papers published in venues, and papers published in years. As for contents, we use titles and abstracts for papers, full names for venues, and also the first textual paragraph of the main content in Wikipedia for authors if available.
3. **Yelp:** We use the public dataset from the Yelp Challenge Round 11[8]. Following an existing work that models Yelp data with heterogeneous networks [33], we extract five types of nodes, *i.e.*, businesses (B), users (U), locations (L), *categories* (C), and stars (S). The edge types include users reviewing businesses, businesses belonging to categories, businesses residing in locations, businesses having average stars, category related to categories and users being friends with users. We further extract contents for businesses like latitudes, longitudes, review counts, *etc.*, and for users like review counts, time since joining Yelp, number of fans, average stars, *etc.* For nodes with no additional contents but a name like categories (*e.g.*, Mexican, Burgers, Gastropubs) and locations (*e.g.*, San Francisco, Chicago, London), we use the pre-trained word embeddings[9] provided by [16] as initial contents.

Table 1. Statistics of the three experimented public datasets.

Dataset	Size	#Types	#Nodes	#Links	#Classes	#Pairs
IMDb	16.1 MB	4	45,913	153,645	23	4,000
DBLP	4.33 GB	4	335,185	2,704,655	4	10,000
Yelp	6.52 GB	5	1,123,649	8,912,736	6	20,000

---

[5] https://grouplens.org/datasets/movielens/100k/.
[6] https://en.wikipedia.org/wiki/Main_Page.
[7] https://aminer.org/citation.
[8] https://www.yelp.com/dataset.
[9] https://nlp.stanford.edu/projects/glove/.

As we can see, the structures and sizes of networks are quite different across the experimented datasets, and the network contents are of various types including categorical, numerical, textual and mixtures of them. In this work, we model all textual contents simply as bag-of-words.

Baselines. We compare with both path matching and network embedding based heterogeneous network modeling algorithms to comprehensively evaluate the performance of AUTOPATH.

- **PathSim** [22]: Normalized meta-path constrained path counts for measuring node similarity on heterogeneous networks.
- **RelSim** [28]: Exhaustive meta-path enumeration up to a given length and supervised weighting for combining the normalized counts of multiple meta-paths.
- **FSPG** [14]: Greedy meta-path search to a given length and similarity computation through a linear combination of biased path constrained random walks.
- **PTE** [24]: Heterogeneous network embedding by decomposing the network into a set of bipartite networks and capturing first and second order proximities.
- **Metapath2vec** [3]: Heterogeneous network embedding through heterogeneous random walks and negative sampling.
- **ESim** [18]: Heterogeneous network embedding through meta-path guided path sampling and noise-contrastive estimation.

Evaluation Protocols. We study the efficacy of all algorithms on similarity modeling, which can be naturally evaluated under the setting of standard link prediction. The links are generated from additional labels of semantic classes not directly captured by the networks. For IMDb, we use all 23 available genres such as drama, comedy, romance, thriller, crime and action. For DBLP, we use the manual labels of authors from four research areas, *i.e.*, database, data mining, machine learning and information retrieval provided by [22]. For Yelp, we extract six sets of businesses based on some available attributes, *i.e.*, good for kids, take out, outdoor seating, good for groups, delivery and reservation. For each dataset, we assume that movies (businesses, authors) within each semantic class are similar in certain ways, and generate pairwise links among them.

Following the common practice in [4,14], we firstly sample certain amounts of linked pairs of nodes, the numbers of which are listed in Table 1. We use them as training data, *i.e.*, example pairs of similar nodes. Since all pairs are positive, we also randomly generate an equal amount of negative pairs, each consisting of two entities not in the same semantic class. *PathSim* needs no training, while *RelSim* and *FSPG* are both trained on the training data in a supervised way. For embedding algorithms, we compute the embeddings in an unsupervised way on the whole network, and train a standard SVM[10] on the training data. For

---

[10] http://scikit-learn.org/stable/modules/svm.html.

AUTOPATH, we train the reinforcement learning agent with the training data and deep embedding on the whole network. After training, similarity scores can be computed by starting from any particular node, planning with the agent for multiple times, and taking the empirical probabilities of reaching the target nodes. For testing, we randomly select 10% start nodes disjointly with the training pairs, and retrieve all target nodes from the same semantic class for each of them to form the ground-truth lists. Each baseline ranks all nodes on the network *w.r.t.* each start node, and we compute the average *precision at K*, *recall at K* and *AUC* over all selected start nodes, which are the standard evaluation metrics for link prediction [6]. We also record the runtimes of all algorithms.

Parameter Settings. When comparing AutoPath with the baseline methods, we slightly tune the parameters via cross-validation. For the IMDb dataset, the parameters are empirically set to the following values: For reinforcement learning, we set the length of trajectories m to 10, the sample size α to 400; for deep embedding, we set the sample size β to 2000 and the weighting factor λ to 0.1; for both components, we set the size of hidden layers to 64, and the number of epochs γ to 200. The parameters on other datasets are slightly different due to different data sizes. During cross-validation, we find AutoPath to be quite robust across different parameter settings. All parameters of the compared baselines are either set as given in the original work on the same datasets, or tuned to the best through standard five-fold cross validation on each dataset.

4.2   Quantitative Evaluation

As we can observe from Fig. 3 and Table 2: (1) the compared algorithms have varying results, while AUTOPATH is able to constantly outperform all of them with significant margins on all experimented datasets, demonstrating its general and robust advantages; (2) the performance improvements of AUTOPATH are more significant on DBLP and Yelp datasets where rich node contents are available, indicating the advantage of content embedding; (3) *FSPG* and *Rel-Sim* perform much better than *PathSim*, and even better than the advanced

Table 2. Quantitative evaluation results: *AUC* and *runtime* of compared algorithms.

Algorithm	AUC			Runtime		
	IMDb	DBLP	Yelp	IMDb	DBLP	Yelp
PathSim	0.584 ± 0.018	0.692 ± 0.022	0.541 ± 0.006	119 s	241 s	468 s
RelSim	0.602 ± 0.023	0.788 ± 0.028	0.595 ± 0.011	325 s	1498 s	4394 s
FSPG	0.568 ± 0.011	0.759 ± 0.024	0.612 ± 0.013	186 s	1062	3186 s
PTE	0.544 ± 0.008	0.707 ± 0.018	0.608 ± 0.015	46 s	238 s	424 s
Metapath2vec	0.539 ± 0.010	0.726 ± 0.021	0.622 ± 0.015	127 s	1170 s	2824 s
ESim	0.573 ± 0.012	0.715 ± 0.016	0.636 ± 0.018	256 s	312 s	684 s
AutoPath	$\mathbf{0.635 \pm 0.015}$	$\mathbf{0.840 \pm 0.018}$	$\mathbf{0.713 \pm 0.016}$	163 s	466 s	1620 s

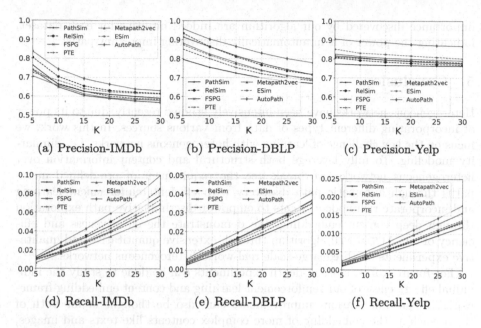

Fig. 3. Quantitative evaluation results: *Precision* and *recall* of compared algorithms.

network embedding algorithms, especially on DBLP, probably because they consider different weights of meta-paths. AUTOPATH also performs well on DBLP, indicating the advantage of reinforcement learning in automatically discovering important paths; (4) the runtimes of AUTOPATH are shorter than *FSPG* and *RelSim*, which try to enumerate or search for all useful meta-paths, especially on large networks like DBLP and Yelp, indicating its efficiency and scalability. Due to space limit, we put more discussions into our Supplementary Materials and defer more detailed experimental studies into the future work.

4.3 Qualitative Analysis

As we stress in this work, a unique advantage of AUTOPATH is the automatic discovery of useful meta-paths from enormous search spaces without a pre-defined maximum length. To demonstrate such utility, after training our model, we plan on random nodes for 10,000 times and summarize the most frequently traveled meta-paths in Table 3. As we can see, the meta-paths with variable lengths and

Table 3. Top 3 meta-paths automatically found and deemed important by AUTOPATH.

IMDb	DBLP	Yelp
M – A – M *(0.372)*	A – P – V – P – A *(0.781)*	B – L – B *(0.414)*
M – D – M *(0.315)*	A – P – A *(0.132)*	B – U – B *(0.277)*
M – U – M *(0.298)*	A – P – A – P – A *(0.046)*	B – S – B *(0.236)*

importance discovered by our algorithm are indeed intuitive for each dataset, indicating the power of it in automatically discovering important paths.

5 Conclusions

Heterogeneous networks have been intensively studied recently, due to its power of incorporating different types of data from various sources. In this work, we focus on the key challenge of learning with heterogeneous networks, *i.e.*, similarity modeling. To fully leverage both structural and content information over heterogeneous networks, we break free the requirement of pre-defined meta-paths through automatic path discovery with efficient reinforcement learning and incorporate rich node contents to empower discriminative path exploration through deep content embedding. We demonstrate the effectiveness and efficiency of our AUTOPATH algorithm through extensive quantitative and qualitative experiments on three large-scale real-world heterogeneous networks.

For future works, more in-depth experiments can be done to study the individual effectiveness of our reinforcement learning and content embedding frameworks. Meanwhile, various improvements can also be thought of for both of them, such as the embedding of more complex contents like texts and images, the interpretation of discovered paths, and the generation of heterogeneous network embedding for various other downstream applications.

Acknowledgement. Research was sponsored in part by U.S. Army Research Lab. under Cooperative Agreement No. W911NF-09-2-0053 (NSCTA), DARPA under Agreement No. W911NF-17-C-0099, National Science Foundation IIS 16-18481, IIS 17-04532, and IIS-17-41317, DTRA HDTRA11810026, and grant 1U54GM114838 awarded by NIGMS through funds provided by the trans-NIH Big Data to Knowledge (BD2K) initiative (www.bd2k.nih.gov).

References

1. Bello, I., Pham, H., Le, Q.V., Norouzi, M., Bengio, S.: Neural combinatorial optimization with reinforcement learning. In: ICLR (2017)
2. Das, R., et al.: Go for a walk and arrive at the answer: reasoning over paths in knowledge bases using reinforcement learning. In: ICLR (2018)
3. Dong, Y., Chawla, N.V., Swami, A.: metapath2vec: scalable representation learning for heterogeneous networks. In: KDD, pp. 135–144 (2017)
4. Fang, Y., Lin, W., Zheng, V.W., Wu, M., Chang, K., Li, X.L.: Semantic proximity search on graphs with metagraph-based learning. In: ICDE, pp. 277–288 (2016)
5. Fu, T.Y., Lee, W.C., Lei, Z.: HIN2Vec: explore meta-paths in heterogeneous information networks for representation learning. In: CIKM, pp. 1797–1806 (2017)
6. Han, J., Pei, J., Kamber, M.: Data Mining: Concepts and Techniques. Elsevier, Amsterdam (2011)
7. Harper, F.M., Konstan, J.A.: The movielens datasets: history and context. TIIS 5(4), 19 (2016)
8. Huang, Z., Mamoulis, N.: Heterogeneous information network embedding for meta path based proximity. arXiv preprint arXiv:1701.05291 (2017)

9. Khalil, E., Dai, H., Zhang, Y., Dilkina, B., Song, L.: Learning combinatorial optimization algorithms over graphs. In: NIPS, pp. 6351–6361 (2017)
10. Kingma, D.P., Ba, J.: Adam: a method for stochastic optimization. In: ICLR (2015)
11. Le, Q.V.: Building high-level features using large scale unsupervised learning. In: ICASSP, pp. 8595–8598 (2013)
12. Lillicrap, T.P., et al.: Continuous control with deep reinforcement learning. arXiv:1509.02971 (2015)
13. Liu, Z., et al.: Semantic proximity search on heterogeneous graph by proximity embedding. In: AAAI, pp. 154–160 (2017)
14. Meng, C., Cheng, R., Maniu, S., Senellart, P., Zhang, W.: Discovering meta-paths in large heterogeneous information networks. In: WWW, pp. 754–764 (2015)
15. Mnih, V., et al.: Human-level control through deep reinforcement learning. Nature 518(7540), 529 (2015)
16. Pennington, J., Socher, R., Manning, C.D.: Glove: global vectors for word representation. In: EMNLP, pp. 1532–1543 (2014)
17. Schulman, J., Levine, S., Abbeel, P., Jordan, M., Moritz, P.: Trust region policy optimization. In: ICML, pp. 1889–1897 (2015)
18. Shang, J., Qu, M., Liu, J., Kaplan, L.M., Han, J., Peng, J.: Meta-path guided embedding for similarity search in large-scale heterogeneous information networks. arXiv preprint arXiv:1610.09769 (2016)
19. Shi, Y., Chan, P.W., Zhuang, H., Gui, H., Han, J.: Prep: path-based relevance from a probabilistic perspective in heterogeneous information networks. In: KDD, pp. 425–434 (2017)
20. Shi, Y., Gui, H., Zhu, Q., Kaplan, L., Han, J.: AspEm: embedding learning by aspects in heterogeneous information networks. In: SDM (2018)
21. Sun, Y., Han, J.: Mining heterogeneous information networks: principles and methodologies. Synth. Lect. Data Min. Knowl. Discov. 3(2), 1–159 (2012)
22. Sun, Y., Han, J., Yan, X., Yu, P.S., Wu, T.: Pathsim: meta path-based top-k similarity search in heterogeneous information networks. VLDB 4(11), 992–1003 (2011)
23. Sutton, R.S., McAllester, D.A., Singh, S.P., Mansour, Y.: Policy gradient methods for reinforcement learning with function approximation. In: NIPS, pp. 1057–1063 (2000)
24. Tang, J., Qu, M., Mei, Q.: PTE: predictive text embedding through large-scale heterogeneous text networks. In: KDD, pp. 1165–1174 (2015)
25. Tang, J., Zhang, J., Yao, L., Li, J., Zhang, L., Su, Z.: ArnetMiner: extraction and mining of academic social networks. In: KDD, pp. 990–998 (2008)
26. Wan, M., Ouyang, Y., Kaplan, L., Han, J.: Graph regularized meta-path based transductive regression in heterogeneous information network. In: SDM, pp. 918–926 (2015)
27. Wang, C., Song, Y., Li, H., Zhang, M., Han, J.: KnowSim: a document similarity measure on structured heterogeneous information networks. In: ICDM, pp. 1015–1020 (2015)
28. Wang, C., et al.: RelSim: relation similarity search in schema-rich heterogeneous information networks. In: SDM, pp. 621–629 (2016)
29. Xiong, W., Hoang, T., Wang, W.Y.: DeepPath: a reinforcement learning method for knowledge graph reasoning. In: EMNLP (2017)
30. Yang, C., Bai, L., Zhang, C., Yuan, Q., Han, J.: Bridging collaborative filtering and semi-supervised learning: a neural approach for poi recommendation. In: KDD, pp. 1245–1254 (2017)

31. Yang, C., Zhang, C., Chen, X., Ye, J., Han, J.: Did you enjoy the ride: under-standing passenger experience via heterogeneous network embedding. In: ICDE (2018)
32. Yang, C., Zhong, L., Li, L.J., Jie, L.: Bi-directional joint inference for user links and attributes on large social graphs. In: WWW, pp. 564–573 (2017)
33. Zhao, H., Yao, Q., Li, J., Song, Y., Lee, D.L.: Meta-graph based recommendation fusion over heterogeneous information networks. In: KDD, pp. 635–644 (2017)

Dynamic Hierarchies in Temporal Directed Networks

Nikolaj Tatti[1,2]([✉])

[1] F-Secure, Helsinki, Finland
[2] Aalto University, Espoo, Finland
nikolaj.tatti@aalto.fi

Abstract. The outcome of interactions in many real-world systems can be often explained by a hierarchy between the participants. Discovering hierarchy from a given directed network can be formulated as follows: partition vertices into levels such that, ideally, there are only forward edges, that is, edges from upper levels to lower levels. In practice, the ideal case is impossible, so instead we minimize some penalty function on the backward edges. One practical option for such a penalty is agony, where the penalty depends on the severity of the violation. In this paper we extend the definition of agony to temporal networks. In this setup we are given a directed network with time stamped edges, and we allow the rank assignment to vary over time. We propose 2 strategies for controlling the variation of individual ranks. In our first variant, we penalize the fluctuation of the rankings over time by adding a penalty directly to the optimization function. In our second variant we allow the rank change at most once. We show that the first variant can be solved exactly in polynomial time while the second variant is **NP**-hard, and in fact inapproximable. However, we develop an iterative method, where we first fix the change point and optimize the ranks, and then fix the ranks and optimize the change points, and reiterate until convergence. We show empirically that the algorithms are reasonably fast in practice, and that the obtained rankings are sensible. Code related to this paper is available at: https://bitbucket.org/orlyanalytics/temporalagony/.

1 Introduction

The outcome of interactions in many real-world systems can be often explained by a hierarchy between the participants. Such rankings occur in diverse domains, such as, hierarchies among athletes [3], animals [8,14], social network behaviour [11], and browsing behaviour [10].

Discovering a hierarchy in a directed network can be defined as follows: given a directed graph $G = (V, E)$, find an integer $r(v)$, representing a rank of v, for each vertex $v \in V$, such that ideally $r(u) < r(v)$ for each edge $(u, v) \in E$. This is

Electronic supplementary material The online version of this chapter (https://doi.org/10.1007/978-3-030-10928-8_4) contains supplementary material, which is available to authorized users.

© Springer Nature Switzerland AG 2019
M. Berlingerio et al. (Eds.): ECML PKDD 2018, LNAI 11052, pp. 55–70, 2019.
https://doi.org/10.1007/978-3-030-10928-8_4

possible only if G is a DAG, so in practice, we penalize each edge with a penalty $q(r(u), r(v))$, and minimize the total penalty. One practical choice for a penalty is *agony* [6,15,16], $q(r(u), r(v)) = \max(r(u) - r(v) + 1, 0)$. If $r(u) < r(v)$, an ideal case, then the agony is 0. On the other hand, if $r(u) = r(v)$, then we penalize the edge by 1, and the penalty increases as the edge becomes more 'backward'. The major benefit of computing agony is that we can solve it in polynomial time [6,15,16].

In this paper we extend the definition of agony to temporal networks: we are given a directed network with time stamped edges[1] and the idea is to allow the rank assignment to vary over time; in such a case, the penalty of an edge with a time stamp t depends only on the ranks of the adjacent vertices at time t.

We need to penalize or constrain the variation of the ranks, as otherwise the optimization problem of discovering dynamic agony reduces to computing the ranks over individual snapshots. In order to do so, we consider 2 variants. In our first variant, we compute the fluctuation of the rankings over time, and this fluctuation is added directly to the optimization function, multiplied by a parameter λ. In our second variant we allow the rank to change at most once, essentially dividing the time line of a single vertex into 2 segments.

We show that the first variant can be solved exactly in $\mathcal{O}(m^2 \log m)$ time. On the other hand, we show that the second variant is **NP**-hard, and in fact inapproximable. However, we develop a simple iterative method, where we first fix the change points and optimize the ranks, and then fix the ranks and optimize the change points, and reiterate until convergence. We show that the resulting two subproblems can be solved exactly in $\mathcal{O}(m^2 \log m)$ time.

We show empirically that, despite the pessimistic theoretical running times, the algorithms are reasonably fast in practice: we are able to compute the rankings for a graph with over 350 000 edges in 5 min.

The remainder of the paper is organized as follows. We introduce the notation and formalize the problem in Sect. 2. In Sect. 3 we review the technique for solving static agony, and in Sect. 4 we will use this technique to solve the first two variants of the dynamic agony. In Sect. 5, we present the iterative solution for the last variant. Related work is given in Sect. 6. Section 7 is devoted to experimental evaluation, and we conclude the paper with remarks in Sect. 8. The proofs for non-trivial theorems are given in Appendix in supplementary material.

2   Preliminaries and Problem Definition

We begin with establishing preliminary notation, and then continue by defining the main problem.

The main input to our problem is a *weighted temporal directed graph* which we will denote by $G = (V, E)$, where V is the set of vertices and E is a set of tuples of form $e = (u, v, w, t)$, meaning an edge e from u to v at time t with a

---

[1] An edge may have several time stamps.

weight w. We allow multiple edges to have the same time stamp, and we also allow two vertices u and v to have multiple edges. If w is not provided we assume that an edge has a weight of 1. To simplify the notation we will often write $w(e)$ to mean the weight of an edge e. Let T be the set of all time stamps.

A *rank assignment* $r : V \times T \to \mathbb{N}$ is a function mapping a vertex and a time stamp to an integer; the value $r(u; t)$ represents the rank of a vertex u at a time point t.

Our next step is to penalize backward edges in a ranking r. In order to do so, consider an edge $e = (u, v, w, t)$. We define the penalty as

$$p(e; r) = w \times \max(0, r(u; t) - r(v; t) + 1).$$

This penalty is equal to 0 whenever $r(v; t) > r(u; t)$, if $r(v; t) = r(u; t)$, then the $p(e; r) = w$, and the penalty increases as the difference $r(u; t) - r(v; t)$ increases.

We are now ready to define the cost of a ranking.

Definition 1. *Assume an input graph $G = (V, E)$ and a rank assignment r. We define a score for r to be*

$$q(r, G) = \sum_{e \in E} p(e; r).$$

Static Ranking: Before defining the main optimization problems, let us first consider the optimization problem where we do *not* allow the ranking to vary over time.

Problem 1 (AGONY). Given a graph $G = (V, E)$, an integer k, find a ranking r minimizing $q(r, G)$, such that $0 \le r(v; t) \le k - 1$ and $r(v; t) = r(v; s)$, for every $v \in V$ and $t, s \in T$.

Note that AGONY does not use any temporal information, in fact, the exact optimization problem can be defined on a graph where we have stripped the edges of their time stamps. This problem can be solved exactly in polynomial time, as demonstrated by Tatti [16]. We should also point out that k is an optional parameter, and the optimization problem makes sense even if we set $k = \infty$.

Dynamic Ranking: We are now ready to define our main problems. The main idea here is to allow the rank assignment to *vary* over time. However, we should penalize or constrain the variation of a ranking. Here, we consider 2 variants for imposing such a penalty.

In order to define the first variant, we need a concept of fluctuation, which is the sum of differences between the consecutive ranks of a given vertex.

Definition 2. *Let r be a rank assignment. Assume that T, the set of all time stamps, is ordered, $T = t_1, \ldots, t_\ell$. The* fluctuation *of a rank for a single vertex u is defined as*

$$fluc(u; r) = \sum_{i=1}^{\ell-1} |r(u, t_{i+1}) - r(u, t_i)|.$$

Note that if $r(u, t)$ is a constant for a fixed u, then $fluc(u; r) = 0$. We can now define our first optimization problem.

*Problem 2 (*FLUC-AGONY*).* Given a graph $G = (V, E)$, an integer k, and a penalty parameter λ, find a rank assignment r minimizing

$$q(r, G) + \lambda \sum_{v \in V} fluc(v; r),$$

such that $0 \leq r(v; t) \leq k - 1$ for every $v \in V$ and $t \in T$.

The parameter λ controls how much emphasis we would like to put in constraining *fluc*: If we set $\lambda = 0$, then the *fluc* term is completely ignored, and we allow the rank to vary freely as a function of time. In fact, solving FLUC-AGONY reduces to taking snapshots of G at each time stamp in T, and applying AGONY to these snapshots individually. On the other hand, if we set λ to be a very large number, then this forces $fluc(v; r) = 0$, that is the ranking is constant over time. This reduces FLUC-AGONY to the static ranking problem, AGONY.

In our second variant, we limit how many *times* we allow the rank to change. More specifically, we allow the rank to change only once.

Definition 3. *We say that a rank assignment r is a* rank segmentation *if each u changes its rank $r(u; t)$ at most once. That is, there are functions $r_1(u)$, $r_2(u)$ and $\tau(v)$ such that*

$$r(u; t) = \begin{cases} r_1(u), & t < \tau(u), \\ r_2(u), & t \geq \tau(u). \end{cases}$$

This leads to the following optimization problem.

*Problem 3 (*SEG-AGONY*).* Given a graph $G = (V, E)$ and an integer k, find a *rank segmentation* r minimizing $q(r; G)$ such that $0 \leq r(v; t) \leq k - 1$ for every $v \in V$ and $t \in T$.

Note that the obvious extension of this problem is to allow rank to change ℓ times, where $\ell > 1$. However, in this paper we focus specifically on the $\ell = 1$ case as this problem yields an intriguing algorithmic approach, given in Sect. 5.

3   Generalized Static Agony

In order to solve the dynamic ranking problems, we need to consider a minor extension of the static ranking problem.

To that end, we define a *static graph* $H = (W, A)$ to be the graph, where W is a set of vertices and A is a collection of directed edges (u, v, c, b), where $u, v \in V$, c is a positive—possibly infinite—weight, and b is an integer, negative or positive.

Problem 4 (GEN-AGONY). Given a static graph $H = (W, A)$ find a function $r : W \rightarrow \mathbb{Z}$ minimizing

$$\sum_{(u,v,c,b) \in A} \max(c \times (r(u) - r(v) + b), 0).$$

Note that c in (u, v, c, b) may be infinite. This implies that if the solution has a finite score, then $r(u) + b \leq r(v).$[2]

We can formulate the static ranking problem, AGONY, as an instance of GEN-AGONY: Assume a graph $G = (V, E)$, and a(n optional) cardinality constraint k. Define a graph $H = (W, A)$ as follows. The vertex set W consists of the vertices V and two additional vertices α and ω. For each edge $(u, v, w, t) \in E$, add an edge $(u, v, c = w, b = 1)$ to A. If there are multiple edges from u to v, then we can group them and combine the weights. This guarantees that the sum in GEN-AGONY corresponds exactly to the cost function in AGONY. If k is given, then add edges $(\alpha, u, c = \infty, b = 0)$ and $(u, \omega, c = \infty, b = 0)$ for each $u \in V$. Finally, add $(\omega, \alpha, c = \infty, b = 1 - k)$. This guarantees that the for the optimal solution we must have $r(\alpha) \leq r(u) \leq r(\omega) \leq r(\alpha) + k - 1$, so now the ranking defined $r(u; t) = r(u) - r(\alpha)$ satisfies the constraints by AGONY.

Example 1. Consider a temporal network given in Fig. 1a. The corresponding graph H is given in Fig. 1b.

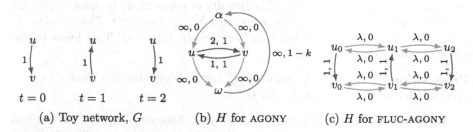

(a) Toy network, G (b) H for AGONY   (c) H for FLUC-AGONY

Fig. 1. Graph G, and the corresponding graphs H used in AGONY and FLUC-AGONY. In (b), the edges with omitted parameters have $c = \infty$ and $b = 0$. In (c), vertices α and ω, and the adjacent edges, are omitted.

As argued by Tatti [16], GEN-AGONY is a dual problem of capacitated circulation, a classic variant of a max-flow optimization problem. This problem can be solved using an algorithm by Orlin [13] in $\mathcal{O}\left(|A|^2 \log |W|\right)$ time. In practice, the running time is faster.

---

[2] Here we adopt $0 \times \infty = 0$, when dealing with the case $r(u) - r(v) + b = 0$.

4   Solving fluc-agony

In this section we provide a polynomial solution for FLUC-AGONY by mapping the problem to an instance of GEN-AGONY.

Assume that we are given a temporal graph $G = (V, E)$, a parameter λ and a(n optional) constraint on the number of levels, k.

We will create a static graph $H = (W, A)$ for which solving GEN-AGONY is equivalent of solving FLUC-AGONY for G. First we define W: for each vertex $v \in V$ and a time stamp $t \in T$ such that there is an edge adjacent to v at time t, add a vertex v_t to W. Add also two vertices α and ω. The edges A consists of three groups A_1, A_2 and A_3:

(i) For each edge $e = (u, v, w, t) \in E$, add an edge $(u_t, v_t, c = w, b = 1)$.
(ii) Let $v_t, v_s \in W$ such that $s > t$ and there is no $v_o \in W$ with $t < o < s$, that is v_t and v_s are 'consecutive' vertices corresponding to v. Add an edge $(v_t, v_s, c = \lambda, b = 0)$, also add an edge $(v_s, v_t, c = \lambda, b = 0)$.
(iii) Assume that k is given. Connect each vertex u_t to ω with $b = 0$ and weight $c = \infty$. Connect α to each vertex u_t with $b = 0$ and weight $c = \infty$. Connect ω to α with $b = 1 - k$ and $c = \infty$. This essentially forces $r(\alpha) \leq r(u_t) \leq r(\omega) \leq r(\alpha) + k - 1$.

Example 2. Consider a temporal graph in Fig. 1a. The corresponding graph, without α and ω, is given in Fig. 1c.

Let r be the rank assignment for H with a finite cost, and define a rank assignment for G, $r'(v; t) = r(v_t)$. The penalty of edges in A_1 is equal to $q(r', G)$ while the penalty of edges in A_2 is equal to $\lambda \sum_{v \in V} fluc(v, r')$. The edges in A_3 force r' to honor the constraint k, otherwise $q(r, H) = \infty$. This leads to the following proposition.

Proposition 1. *Let r be the solution of* GEN-AGONY *for H. Then $r'(v; t) = r(v_t) - r(\alpha)$ solves* FLUC-AGONY *for G.*

We conclude with the running time analysis. Assume G with n vertices and m edges. A vertex $v_t \in W$ implies that there is an edge $(u, v, w, t) \in E$. Thus, $|W| \in \mathcal{O}(m)$. Similarly, $|A_1| + |A_2| + |A_3| \in \mathcal{O}(m)$. Thus, solving GEN-AGONY for H can be done in $\mathcal{O}(m^2 \log m)$ time.

5   Computing seg-agony

In this section we focus on SEG-AGONY. Unlike the previous problem, SEG-AGONY is very hard to solve (see Appendix for the proof).

Proposition 2. *Discovering whether there is a rank segmentation with a 0 score is an **NP**-complete problem.*

This result not only states that the problem is hard to solve exactly but it is also very hard to approximate: there is no polynomial-time algorithm with a multiplicative approximation guarantee, unless $\mathbf{NP} = \mathbf{P}$.

5.1   Iterative Approach

Since we cannot solve the problem exactly, we have to consider a heuristic app-roach. Note that the rank assignment of a single vertex is characterized by 3 values: a change point, the rank before the change point, and the rank after the change point. This leads to the following iterative algorithm: (i) fix a change point for each vertex, and find the optimal ranks before and after the change point, (ii) fix the ranks for each vertex, and find the optimal change point. Repeat until convergence.

More formally, we need to solve the following two sub-problems iteratively.

*Problem 5 (*CHANGE2RANKS*).* Given a graph $G = (V, E)$ and a function τ map-ping a vertex to a time stamp, find $r_1 : V \to N$ and $r_2 : V \to N$ mapping a vertex to an integer, such that the rank assignment r defined as

$$r(v; t) = \begin{cases} r_1(v), & t < \tau(v), \\ r_2(v), & t \geq \tau(v) \end{cases}$$

minimizes $q(r; G)$.

*Problem 6 (*RANKS2CHANGE*).* Given a graph $G = (V, E)$ and two functions $r_1 : V \to N$ and $r_2 : V \to N$ mapping a vertex to an integer, find a rank segmentation r minimizing $q(r; G)$ such that there is a function τ such that

$$r(v; t) = \begin{cases} r_1(v), & t < \tau(v), \\ r_2(v), & t \geq \tau(v). \end{cases}$$

Surprisingly, we can solve both sub-problems exactly as we see in the next two subsections. This implies that during the iteration the score will always decrease. We still need a starting point for our iteration. Here, we initialize the change point of a vertex v as the median time stamp of v.

5.2   Solving CHANGE2RANKS

We begin by solving the easier of the two sub-problems.

Assume that we are given a temporal network $G = (V, E)$ and a function $\tau : V \to T$. We will map CHANGE2RANKS to GEN-AGONY. In order to do so, we define a graph $H = (W, A)$. The vertex set W consists of two copies of V; for each vertex $v \in V$, we create two vertices v^1 and v^2, we also add vertices α and ω to enforce the constraint k. For each edge $e = (u, v, w, t) \in E$, we introduce an edge $(u^i, v^j, c = w, b = 1)$ to A, where

$$i = \begin{cases} 1 & \text{if } t < \tau(u), \\ 2 & \text{if } t \geq \tau(u), \end{cases} \quad \text{and} \quad j = \begin{cases} 1 & \text{if } t < \tau(v), \\ 2 & \text{if } t \geq \tau(v). \end{cases}$$

Finally, like before, we add $(\alpha, v, c = \infty, b = 0)$, $(v, \omega, c = \infty, b = 0)$ and $(\omega, \alpha, c = \infty, b = 1 - k)$ to enforce the constraint k.

We will denote this graph by $G(\tau)$.

Example 3. Consider the toy graph given in Fig. 1a. Assume $\tau(u) = 1$ and $\tau(v) = 2$. The resulting graph $G(\tau)$ is given in Fig. 2a.

The following proposition shows that optimizing agony for H is equivalent of solving CHANGE2RANKS. We omit the proof as it is trivial.

Proposition 3. *Let r be a ranking for H. Define r' as*

$$r'(v;t) = \begin{cases} r(v^1) - r(\alpha), & t < \tau(v), \\ r(v^2) - r(\alpha), & t \geq \tau(v). \end{cases}$$

Then $q(r', G) = q(r, H)$. Reversely, given a ranking r' satisfying conditions of CHANGE2RANKS, define a ranking r for G by setting $r(v^i) = r_i(v)$. Then $q(r', G) = q(r, H)$.

We conclude with the running time analysis. Assume G with n vertices and m edges. We have at most $2n + 2$ vertices in W and $|A| \in \mathcal{O}(m)$. Thus, solving CHANGE2RANKS for H can be done in $\mathcal{O}(m^2 \log n)$ time.

5.3   Solving RANKS2CHANGE

Our next step is to solve the opposite problem, where we are given the two alternative ranks for each vertex, and we need to find the change points. Luckily, we can solve this problem in polynomial time. To solve the problem we map it to GEN-AGONY, however unlike in previous problems, the construction will be quite different.

Assume that we are given a graph $G = (V, E)$, and the two functions r_1 and r_2. To simplify the following definitions, let us first define

$$r_{min}(v) = \min(r_1(v), r_2(v)) \quad \text{and} \quad r_{max}(v) = \max(r_1(v), r_2(v)).$$

Assume an edge $e = (u, v, w, t) \in E$. A solution to RANKS2CHANGE must use ranks given by r_1 and r_2, that is the rank of u is either $r_{min}(u)$ or $r_{max}(u)$, and the rank of v is either $r_{min}(v)$ or $r_{max}(v)$, depending where we mark the change point for u and v. This means that there are only 4 possible values for the penalty of e. They are

$$p_{00}(e) = w \times \max(0, r_{min}(u) - r_{min}(v) + 1),$$
$$p_{10}(e) = w \times \max(0, r_{max}(u) - r_{min}(v) + 1),$$
$$p_{01}(e) = w \times \max(0, r_{min}(u) - r_{max}(v) + 1),$$
$$p_{11}(e) = w \times \max(0, r_{max}(u) - r_{max}(v) + 1).$$

Among these penalties, $p_{01}(e)$ is the smallest, and ideally we would pay only $p_{01}(e)$ for each edge. This is rarely possible, so we need to design a method that takes other penalties into account.

Next we define a static graph $H = (W, A)$ that will eventually solve RANKS2CHANGE. For each vertex $v \in V$ and a time stamp $t \in T$ such that there is an edge adjacent to v at time t, add a vertex v_t to W. Add also two additional vertices α and ω. We will define the edges A in groups. The first two sets of edges in A essentially force $r(u_t) = 0, 1$, and that the ranking is monotonic as a function of t. Consequently, there will be at most only one time stamp for each vertex u, where the ranking changes. This will be the eventual change point for u. The edges are:

(i) Connect each vertex u_t to ω with $b = 0$ and weight $c = \infty$. Connect α to each vertex u_t with $b = 0$ and weight $c = \infty$. Connect ω to α with $b = -1$ and $c = \infty$. Connect α to ω with $b = 1$ and $c = \infty$. This forces $r(\alpha) \leq r(u_t) \leq r(\omega) = r(\alpha) + 1$.

(ii) Let $v_t, v_s \in W$ such that $s > t$ and there is no $v_o \in W$ with $t < o < s$. If $r_2(v) \geq r_1(v)$, then connect v_t to v_s with $b = 0$ and $c = \infty$. This forces $r(v_s) \geq r(v_t)$. If $r_2(v) < r_1(v)$, then connect v_s to v_t with $b = 0$ and $c = \infty$. This forces $r(v_s) \leq r(v_t)$.

For notational simplicity, let us assume that $r(\alpha) = 0$. The idea is then that once we have obtained the ranking for H, we can define the ranking for G as

$$r'(v; t) = r_{min}(v) + (r_{max}(v) - r_{min}(v))r(v_t).$$

Our next step is to define the edges that correspond to the penalties in the original graph. We will show later in Appendix that the agony of r' is equal to $P_1 + P_2 + P_3 + const$, where

$$P_1 = \sum_{v_t | r(v_t)=0} \sum_{e=(u,v,w,t)\in E} p_{00}(e) - p_{01}(e),$$

$$P_2 = \sum_{u_t | r(u_t)=1} \sum_{e=(u,v,w,t)\in E} p_{11}(e) - p_{01}(e),$$

$$P_3 = \sum_{\substack{e=(u,v,w,t)\in E \\ r(v_t)=0, r(u_t)=1}} p_{10}(e) - p_{00}(e) - p_{11}(e) + p_{01}(e).$$

Let us first define the edges that lead to these penalties.

(i) Connect ω to each vertex v_t with $b = 0$ and weight

$$c = \sum_{e=(u,v,w,t)\in E} p_{00}(e) - p_{01}(e).$$

In the sum v and t are fixed, and correspond to v_t. This edge penalizes vertices with $r(v_t) = 0$ with a weight of c. Summing these penalties yields P_1.

(ii) Connect each vertex u_t to α with $b = 0$ and weight

$$c = \sum_{e=(u,v,w,t)\in E} p_{11}(e) - p_{01}(e).$$

In the sum u and t are fixed, and correspond to u_t. This edge penalizes vertices with $r(u_t) = 1$ with a weight of c. Summing these penalties yields P_2.

(iii) For each edge $e = (u, v, w, t) \in E$, connect u_t and v_t with $b = 0$ and

$$c = p_{10}(e) - p_{00}(e) - p_{11}(e) + p_{01}(e).$$

This edge penalizes cases when $r(u_t) = 1$ and $r(v_t) = 0$, and constitute P_3.

We will denote the resulting H by $G(r_1, r_2)$.

Example 4. Consider the toy graph given in Fig. 1a. Assume that the rank assignments are $r_1(u) = 0$, $r_1(v) = 1$, $r_2(u) = 2$, $r_2(v) = 3$. The resulting graph $G(r_1, r_2)$ is given in Fig. 2b. The optimal ranking for $G(r_1, r_2)$ assigns 0 to α, u_0, v_0, and v_1; the rank for the remaining vertices is 1.

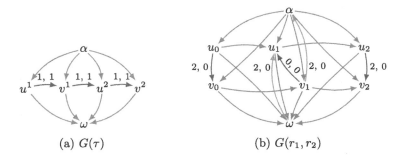

(a) $G(\tau)$ (b) $G(r_1, r_2)$

Fig. 2. Graphs used for solving SEG-AGONY. In both figures, the edges with omitted parameters have $c = \infty$ and $b = 0$. For clarity, we omit edges between α and ω in both figures, in addition, in (b) we omit parameters for the edges (x, α) and (ω, x) with $c = 0$.

Before we show the connection between the ranks in G and $H = G(r_1, r_2)$, we first need to show that the edge weights are non-negative. This is needed to guarantee that we can find the optimal ranking of H using GEN-AGONY.

Proposition 4. *The weights of edges in H are non-negative.*

The proof is given in Appendix.

We will state our main result: we can obtain the solution for RANKS2CHANGE using the optimal ranking for H; see Appendix for the proof.

Proposition 5. *Let r be the optimal ranking for H. Then*

$$r'(v; t) = r_{min}(v) + (r_{max}(v) - r_{min}(v))(r(v_t) - r(\alpha))$$

solves RANKS2CHANGE.

We conclude this section with the running time analysis. Assume G with n vertices and m edges. A vertex $v_t \in W$ implies that there is an edge $(u, v, w, t) \in E$. Thus, $|W| \in \mathcal{O}(m)$. Similarly, $|A| \in \mathcal{O}(m)$. Thus, solving RANKS2CHANGE for H can be done in $\mathcal{O}(m^2 \log m)$ time.

6 Related Work

Perhaps the most classic way of ranking objects based on pair-wise interactions is Elo rating proposed by Elo [3], used to rank chess players. A similar approach was proposed by Jameson et al. [8] to model animal dominance.

Maiya and Berger-Wolf [11] proposed discovering directed trees from weighted graphs such that parent vertices tend to dominate the children. A hierarchy is evaluated by a statistical model where the probability of an edge is high between a parent and a child. A good hierarchy is then found by a greedy heuristic.

Penalizing edges using agony was first considered by Gupte et al. [6], and a faster algorithm was proposed by Tatti [15]. The setup was further extended to handle the weighted edges, which was not possible with the existing methods, by Tatti [16], as well to be able to limit the number of distinct ranks (parameter k in the problem definitions).

An alternative to agony is a penalty that penalizes an edge (u, v) with $r(u) \geq r(v)$ with a constant penalty. In such a case, optimizing the cost is equal to FEEDBACK ARC SET (FAS), an **APX**-hard problem with a coefficient of $c = 1.3606$ [2]. Moreover, there is no known constant-ratio approximation algorithm for FAS, and the best known approximation algorithm has ratio $O(\log n \log \log n)$ [4]. In addition, Tatti [16] demonstrated that minimizing agony is **NP**-hard for any concave penalties while remains polynomial for any convex penalty function.

An interesting direction for future work is to study whether the rank obtained from minimizing agony can be applied as a feature in role mining tasks, where the goal is to cluster vertices based on similar features [7,12].

SEG-AGONY essentially tries to detect a change point for each vertex. Change point detection in general is a classic problem and has been studied extensively, see excellent survey by Gama et al. [5]. However, these techniques cannot be applied directly for solving SEG-AGONY since we would need to have the ranks for individual time points.

The difficulty of solving SEG-AGONY stems from the fact that we allow vertices to have different change points. If we require that the change point must be the equal for all vertices, then the problem is polynomial. Moreover, we can easily extend such a setup for having ℓ segments. Discovering change points then becomes an instance of a classic segmentation problem which can be optimized by a dynamic program [1].

7 Experiments

In this section we present our experimental evaluation.

Datasets and Setup: We considered 5 datasets. The first 3 datasets, *Mention*, *Retweet*, and *Reply*, obtained from SNAP repository [9], are the twitter interaction networks related to Higgs boson discovery. The 4th dataset, *Enron* consists of the email interactions between the core members of Enron. In addition, for illustrative purposes, we used a small dataset: *NHL*, consisting of National

Hockey League teams during the 2015–2016 regular season. We created an edge (x, y) if team x has scored more goals against team y in a single game during the 2014 regular season. We assign the weight to be the difference between the points and the time stamp to be the date the game was played. We used hours as time stamps for Higgs datasets, days for *Enron*. The sizes of the graphs are given in Table 1.

Table 1. Basic characteristics of the datasets and the experiments. The third data column, $|T|$, represents the number of unique time stamps, while the last column is the number of unique (v, t) pairs such that the vertex v is adjacent to an edge at time t, $\left| \bigcup_{(u,v,w,t) \in E} \{(v,t), (u,t)\} \right|$.

Name	$|V|$	$|E|$	$|T|$	$|\{(v,t)\}|$
Enron	146	105 522	964	24 921
Reply	38 683	36 395	168	54 892
Retweet	256 491	354 930	168	390 583
Mention	115 684	164 156	168	183 693
NHL	30	1 230	178	2 460

For each dataset we applied FLUC-AGONY, SEG-AGONY, and the static variant, AGONY. For FLUC-AGONY we set $\lambda = 1$ for the Higgs datasets, $\lambda = 2$ for *NHL* and *Enron*.

We implemented the algorithms in C++, and performed experiments using a Linux-desktop equipped with a Opteron 2220 SE processor.[3]

Computational Complexity: First, we consider the running times, reported in Table 2. We see that even though the theoretical running time is $\mathcal{O}(m^2 \log n)$ for FLUC-AGONY and for a single iteration of SEG-AGONY, the algorithms perform well in practice. We are able to process graphs with 300 000 edges in 5 min. Naturally, SEG-AGONY is the slowest as it requires multiple iterations—in our experiments 3–5 rounds—to converge.

Table 2. Agony, running time, and number of unique ranks in the ranking.

Name	Score			Number of ranks			Time		
	AGONY	FLUC	SEG	AGONY	FLUC	SEG	AGONY	FLUC	SEG
Enron	57 054	21 434	50 393	6	9	7	3 s	4 s	26 s
Reply	6 017	5 401	4 147	13	12	16	0.4 s	10 s	15 s
Retweet	2 629	1 384	1 070	23	21	18	8 s	4 m	5 m
Mention	12 756	10 082	8 219	20	19	18	4 s	1 m	2 m
NHL	2 090	1 414	1 883	2	4	4	0.6 s	0.3 s	1 s

---

[3] See https://bitbucket.org/orlyanalytics/temporalagony for the code.

Table 3. Statistics measuring fluctuation of the resulting rankings: *fluc* is equal to the fluctuation $fluc(u; r)$ averaged over u, *maxdiff* is the maximum difference between the ranks of a single vertex u, averaged over u, *change* is the number of times rank is changed for a single vertex u, averaged over u. Note that *fluc* = *maxdiff* for SEG-AGONY as the assignment is allowed to change only once.

Name	*fluc*		*maxdiff*		*change*	
	FLUC	SEG	FLUC	SEG	FLUC	SEG
Enron	28.2	1	3.2	1	21.8	0.66
Reply	0.013	0.43	0.012	0.43	0.01	0.36
Retweet	0.003	0.17	0.003	0.17	0.002	0.13
Mention	0.016	0.3	0.014	0.3	0.011	0.2
NHL	2.7	0.73	1.5	0.73	2.6	0.5

Statistics of Obtained Rankings: Next, we look at the statistics of the obtained rankings, given in Table 2. We first observe that the agony of the dynamic variants is always lower than the static agony, as expected.

Let us compare the constraint statistics, given in Table 3. First, we see that FLUC-AGONY yields the smallest *fluc* in Higgs databases. SEG-AGONY produces smaller *fluc* in the other two datasets but it also produces a higher agony.

Interestingly enough, FLUC-AGONY yields a surprisingly low average number of change points for Higgs datasets. The low average is mainly due to most resulting ranks being constant, and only a minority of vertices changing ranks over time. However, this minority changes its rank more often than just once.

Agony vs Fluctuation: The parameter λ of FLUC-AGONY provides a flexible way of controlling the fluctuation: smaller values of λ leads to smaller agony but larger fluctuation while larger values of λ leads to larger agony but smaller fluctuation. This can be seen in Table 2, where relatively large λ forces small fluctuation for the Higgs datasets, while relatively small λ allows variation and a low agony for *Enron* dataset. This flexibility comes at a cost: we need to have a sensible way of selecting λ. One approach to select this value is to study the joint behavior of the agony and the fluctuation as we vary λ. This is demonstrated in Fig. 3 for *Enron* data, where we scatter plot the agony versus the average fluctuation, and vary λ. We see that agony decreases steeply as we allow some fluctuation over time but the obtained benefits decrease as we allow more variation.

Use Case: Finally, let us look on the rankings by SEG-AGONY of *NHL* given in Fig. 4. We limit the number of possible rank levels to $k = 3$.

The results are sensible: the top teams are playoff teams while the bottom teams have a significant losing record. Let us highlight some change points that reflect significant changes in teams: for example, the collapse of *Montreal Canadiens* (MTL) from the top rank to the bottom rank coincides with the injury of their star goaltender. Similarly, the rise of the *Pittsburgh Penguins* (PIT) from the middle rank to the top rank reflects firing of the head coach as well as retooling their strategy, *Penguins* eventually won the Stanley Cup.

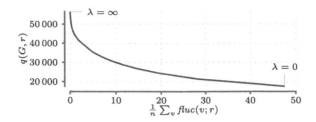

Fig. 3. Agony plotted against *fluc* of the optimal ranking for FLUC-AGONY by varying the parameter λ (*Enron*).

Ranking before the change:

1. MTL WIN BOS MIN DAL FLO WSH
2. VAN NYR CHI SJ TOR COL PIT NSH STL TBL DET NJ NYI
3. CAL LAK OTT BUF CAR EDM PHI CBJ ARI ANA

Ranking after the change:

1. NYR SJ LAK PIT NSH STL TBL PHI ANA WSH
2. CAL CHI BOS OTT BUF MIN DAL CAR CBJ DET ARI NJ NYI FLO
3. VAN MTL TOR WIN COL EDM

Change points:

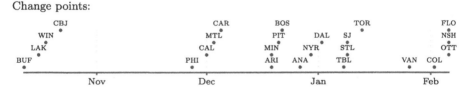

Fig. 4. Rank segmentations for *NHL* with $k = 3$. The bottom figure shows only the teams whose rank changed. The y-axis is used only to reduce the clutter.

8   Concluding Remarks

In this paper we propose a problem of discovering a dynamic hierarchy in a directed temporal network. To that end, we propose two different optimization problems: FLUC-AGONY and SEG-AGONY. These problems vary in the way we control the variation of the rank of single vertices. We show that FLUC-AGONY can be solved in polynomial time while SEG-AGONY is **NP**-hard. We also developed an iterative heuristic for SEG-AGONY. Our experimental validation showed that the algorithms are practical, and the obtained rankings are sensible.

FLUC-AGONY is the more flexible of the two methods as the parameter λ allows user to smoothly control how much rank is allowed to vary. This comes at a price as the user is required to select an appropriate λ. One way to select λ is to vary the parameter and monitor the trade-off between the agony and the fluctuation. An interesting variant of FLUC-AGONY—and potential future line of work—is to minimize agony while requiring that the fluctuation should not increase over some given threshold.

The relation between SEG-AGONY and the sub-problems RANKS2CHANGE and CHANGE2RANKS is intriguing: while the joint problem SEG-AGONY is **NP**-hard not only the sub-problems are solvable in polynomial time, they are solved with the same mechanism.

A straightforward extension for SEG-AGONY is to allow more than just one change point, that is, in such a case we are asked to partition the time line of each vertex into ℓ segments. However, we can no longer apply the same iterative algorithm. More specifically, the solver for RANKS2CHANGE relies on the fact that we need to make only one change. Developing a solver that can handle the more general case is an interesting direction for future work.

References

1. Bellman, R.: On the approximation of curves by line segments using dynamic programming. Commun. ACM **4**(6), 284 (1961)
2. Dinur, I., Safra, S.: On the hardness of approximating minimum vertex cover. Ann. Math. **162**(1), 439–485 (2005)
3. Elo, A.E.: The Rating of Chessplayers, Past and Present. Arco Publisher, La Palma (1978)
4. Even, G., (Seffi) Naor, J., Schieber, B., Sudan, M.: Approximating minimum feedback sets and multicuts in directed graphs. Algorithmica **20**(2), 151–174 (1998)
5. Gama, J., Zliobaite, I., Bifet, A., Pechenizkiy, M., Bouchachia, A.: A survey on concept drift adaptation. ACM Comput. Surv. **46**(4), 44:1–44:37 (2014)
6. Gupte, M., Shankar, P., Li, J., Muthukrishnan, S., Iftode, L.: Finding hierarchy in directed online social networks. In: WWW, pp. 557–566 (2011)
7. Henderson, K., et al.: RolX: structural role extraction & mining in large graphs. In: KDD, pp. 1231–1239 (2012)
8. Jameson, K.A., Appleby, M.C., Freeman, L.C.: Finding an appropriate order for a hierarchy based on probabilistic dominance. Anim. Behav. **57**, 991–998 (1999)
9. Leskovec, J., Krevl, A.: SNAP Datasets: stanford large network dataset collection, January 2005. http://snap.stanford.edu/data
10. Macchia, L., Bonchi, F., Gullo, F., Chiarandini, L.: Mining summaries of propagations. In: ICDM, pp. 498–507 (2013)
11. Maiya, A.S., Berger-Wolf, T.Y.: Inferring the maximum likelihood hierarchy in social networks. In: ICSE, pp. 245–250 (2009)
12. McCallum, A., Wang, X., Corrada-Emmanuel, A.: Topic and role discovery in social networks with experiments on enron and academic email. J. Artif. Int. Res. **30**(1), 249–272 (2007)
13. Orlin, J.B.: A faster strongly polynomial minimum cost flow algorithm. Oper. Res. **41**(2), 338–350 (1993)

14. Roopnarine, P.D., Hertog, R.: Detailed food web networks of three Greater Antillean coral reef systems: the Cayman Islands, Cuba, and Jamaica. Dataset Pap. Ecol. **2013**, 9 (2013)
15. Tatti, N.: Faster way to agony–discovering hierarchies in directed graphs. In: ECML PKDD, pp. 163–178 (2014)
16. Tatti, N.: Hierarchies in directed networks. In: ICDM, pp. 991–996 (2015)

Risk-Averse Matchings over Uncertain Graph Databases

Charalampos E. Tsourakakis[1], Shreyas Sekar[2(✉)], Johnson Lam[1],
and Liu Yang[3]

[1] Boston University, Boston, USA
{ctsourak,jlam17}@bu.edu
[2] Harvard University, Cambridge, USA
ssekar@hbs.edu
[3] Yale University, New Haven, USA
liu.yang@yale.edu

Abstract. In this work we study a problem that naturally arises in the context of several important applications, such as online dating, kidney exchanges, and team formation.

Given an uncertain, weighted (hyper)graph, how can we efficiently find a (hyper)matching with high expected reward, and low risk?

We introduce a novel formulation for finding matchings with maximum expected reward and bounded risk under a general model of uncertain weighted (hyper)graphs. Given that our optimization problem is NP-hard, we turn our attention to designing efficient approximation algorithms. For the case of uncertain weighted graphs, we provide a $\frac{1}{3}$-approximation algorithm, and a $\frac{1}{5}$-approximation algorithm with near optimal run time. For the case of uncertain weighted hypergraphs, we provide a $\Omega(\frac{1}{k})$-approximation algorithm, where k is the rank of the hypergraph (i.e., any hyperedge includes at most k nodes), that runs in almost (modulo log factors) linear time.

We complement our theoretical results by testing our algorithms on a wide variety of synthetic experiments, where we observe in a controlled setting interesting findings on the trade-off between reward, and risk. We also provide an application of our formulation for providing recommendations of teams that are likely to collaborate, and have high impact. Code related to this paper is available at: https://github.com/tsourolampis/risk-averse-graph-matchings.

1 Introduction

Graphs model a wide variety of datasets that consist of a set of entities, and pairwise relationships among them. In several real-world applications, these relationships are inherently uncertain. For example, protein-protein interaction (PPI) networks are associated with uncertainty since protein interactions are obtained via noisy, error-prone measurements [4]. In privacy applications, deterministic

© Springer Nature Switzerland AG 2019
M. Berlingerio et al. (Eds.): ECML PKDD 2018, LNAI 11052, pp. 71–87, 2019.
https://doi.org/10.1007/978-3-030-10928-8_5

edge weights become appropriately defined random variables [7,23], in dating applications each recommended link is associated with the probability that a date will be successful [11], in viral marketing the extent to which an idea propagates through a network depends on the 'influence probability' of each social interaction [24], in link prediction possible interactions are assigned probabilities [30,39], and in entity resolution a classifier outputs for each pair of entities a probability that they refer to the same object.

Mining uncertain graphs poses significant challenges. Simple queries—such as distance queries—on deterministic graphs become #**P**-complete ([42]) problems on uncertain graphs [19]. Furthermore, approaches that maximize the expected value of a given objective typically involve high risk solutions, e.g., solutions where there is an unacceptably large probability that the realized value of the objective is much smaller than its expected value. On the other hand, risk-averse methods are based on obtaining several graphs samples, a procedure that is computationally expensive, or even prohibitive for large-scale uncertain graphs.

Two remarks about the uncertain graph models used in prior work are worth making before we discuss the main focus of this work. The datasets used in the majority of prior work are *uncertain, unweighted graphs*. There appears to be less work related to *uncertain, weighted hypergraphs* that are able to model a wider variety of datasets, specifically those containing more than just pairwise relationships (i.e., hyperedges). Secondly, the model of uncertain graphs used in prior work [9,18,25,27,31,34–36] are homogeneous random graphs [8]. More formally, let $\mathcal{G} = (V, E, p)$ be an uncertain graph where $p : E \rightarrow (0, 1]$, is the function that assigns a probability of success to each edge independently from the other edges. According to the possible-world semantics [8,13] that interprets $\mathcal{G}$ as a set $\{G : (V, E_G)\}_{E_G \subseteq E}$ of $2^{|E|}$ possible deterministic graphs (worlds), each defined by a subset of E. The probability of observing any possible world $G(V, E_G) \in 2^E$ is

$$\mathbf{Pr}\,[G] = \prod_{e \in E_G} p(e) \prod_{e \in E \setminus E_G} (1 - p(e)).$$

Such a model restricts the distribution of each edge to be a Bernoulli distribution, and does not capture various important applications such as privacy applications where noise (say Gaussian) is injected on the weight of each edge [7,23].

In this work, we focus on *risk-averse matchings over uncertain (hyper)graphs*. To motivate our problem consider Fig. 1 that shows a probabilistic graph (i.e., a 2-regular hypergraph) with two perfect matchings, $M_1 = \{(A, B), (C, D)\}$ and $M_2 = \{(A, C), (B, D)\}$. Each edge e follows a Bernoulli distribution with success probability $p(e)$, and is associated with a reward $w(e)$ that is obtained only when the edge is successfully realized. These two parameters $(p(e), w(e))$ annotate each edge e in Fig. 1. The maximum weight matching *in expectation* is M_1 with expected reward $100 \times \frac{1}{2} \times 2 = 100$. However, with probability $(1 - \frac{1}{2}) \times (1 - \frac{1}{2}) = \frac{1}{4}$ the reward we receive from M_1 equals zero. However, the second matching M_2 has expected reward equal to 80 with probability 1. In other words, matching M_1 offers potentially higher reward but entails *higher risk* than M_2. Indeed, in many situations with asymmetric rewards, one observes that high reward solutions are accompanied by higher risks and that such solutions may be shunned by agents in favor of safer options [26].

Another way to observe that matching M_1 entails greater risk is to draw graph samples from this probabilistic graph multiple times, and observe that around 25% of the realizations of M_1 result in zero reward. However, sampling is computationally expensive on large-scale uncertain graphs. Furthermore, in order to obtain statistical guarantees, a large number of samples may be needed [35] which makes the approach computationally intensive or infeasible even for medium-scale graphs. Finally, it is challenging and sometimes not always clear how to aggregate different samples [35]. These two drawbacks are well-known to the database community, and recently Parchas et al. [35] suggested a heuristic to extract representative instances of uncertain graphs. While

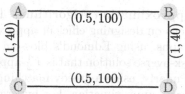

Fig. 1. Probabilistic graph, each edge e is annotated with $(p(e), w(e))$, its probability and its reward/weight. The matching $(A, B), (C, D)$ has higher expected weight than $(A, C), (B, D)$. However, the reward of the former matching is 0 with probability $\frac{1}{4}$, but the reward of the latter matching is 80 with probability 1. For details, see Sect. 1

their work makes an important practical contribution, their method is an intuitive heuristic whose theoretical guarantees and worst-case running time are not well understood [35].

Motivated by these concerns, we focus on the following central question:

> How can we design *efficient, risk-averse algorithms* with *solid theoretical guarantees* for finding maximum weight matchings in uncertain weighted graphs and hypergraphs?

This question is well-motivated, as it naturally arises in several important applications. In online dating applications a classifier may output a probability distribution for the probability of matching two humans successfully [41]. In kidney exchange markets, a kidney exchange is successful according to some probability distribution that is determined by a series of medical tests. Typically, this distribution is unknown but its parameters such as the mean and the variance can be empirically estimated [11]. Finally, the success of any large organization that employs skilled human resources crucially depends on the choice of teams that will work on its various projects. Basic team formation algorithms output a set of teams (i.e., hyperedges) that combine a certain set of desired skills [3,17,20,29,33]. A classifier can leverage features that relate to crowd psychology, conformity, group-decision making, valued diversity, mutual trust, effective and participative leadership [22] to estimate the probability of success of a team.

In detail, our contributions are summarized as follows.

Novel Model and Formulation. We propose a general model for weighted uncertain (hyper)graphs, and a novel formulation for risk-averse maximum matchings. Our goal is to select (hyper)edges that have *high expected reward, but also bounded risk of failure.* Our problem is a novel variation of the well-studied stochastic matching problem [5,11].

Approximation Algorithms. The problem that we study is NP-Hard so we focus on designing efficient approximation algorithms. For the case of uncertain graphs, using Edmond's blossom algorithm [15] as a black-box, we provide a risk-averse solution that is a $\frac{1}{3}$-approximation to the optimal risk-averse solution. Similarly, using a greedy matching algorithm as a black box we obtain a $\frac{1}{5}$-risk-averse approximation. For hypergraphs of rank k (i.e., any hyperedge contains at most k nodes) we obtain a risk-averse $\Omega(\frac{1}{k})$-approximation guarantee. Our algorithms are risk-averse, do not need to draw graph samples, and come with solid theoretical guarantees. Perhaps more importantly, the proposed algorithms that are based on greedy matchings have a running time of $O(m \log^2 m + n \log m)$, where n, m represent the number of nodes, and (hyper)edges in the uncertain (hyper)graph respectively—this makes the algorithm easy to deploy on large-scale real-world networks such as the one considered in our experiments (see Sect. 4).

Experimental Evaluation. We evaluate our proposed algorithm on a wide variety of synthetic experiments, where we observe interesting findings on the trade-offs between reward and risk. There appears to be little (or even no) empirical work on *uncertain, weighted hypergraphs.* We use the Digital Bibliography and Library Project (DBLP) dataset to create a hypergraph where each node is an author, each hyperedge represents a team of co-authors for a paper, the probability of a hyperedge is the probability of collaboration estimated from historical data, and the weight of a hyperedge is its citation count. This uncertain hypergraph is particularly interesting as there exist edges with high reward (citations) but whose authors have low probability to collaborate. On the other hand, there exist papers with a decent number of citations whose co-authors consistently collaborate. Intuitively, the more risk-averse we are, the more we should prefer the latter hyperedges. We evaluate our proposed method on this real dataset, where we observe several interesting findings. The code and the datasets are publicly available at https://github.com/tsourolampis/risk-averse-graph-matchings.

2   Related Work

Uncertain Graphs. Uncertain graphs naturally model various datasets including protein-protein interactions [4,28], kidney exchanges [37], dating applications [11], sensor networks whose connectivity links are uncertain due to various kinds of failures [38], entity resolution [34], viral marketing [24], and privacy-applications [7].

Given the increasing number of applications that involve uncertain graphs, researchers have put a lot of effort in developing algorithmic tools that tackle several important graph mining problems, see [9,18,25,27,31,34–36]. However, with a few exceptions these methods suffer from a critical drawback; either they are not risk-averse, or they rely on obtaining many graphs samples. Risk-aversion has been implicitly discussed by Lin et al. in their work on reliable clustering [31], where the authors show that interpreting probabilities as weights does not result in good clusterings. Jin et al. provide a risk-averse algorithm for distance queries on uncertain graphs [19]. Parchas et al. have proposed a heuristic to extract a good possible world in order to combine risk-aversion with efficiency [35]. However, their work comes with no guarantees.

Graph Matching is a major topic in combinatorial optimization. The interested reader should confer the works of Lovász and Plummer [32] for a solid exposition. Finding maximum matchings in weighted graphs is solvable in polynomial time [15,16]. A faster algorithm sorts the edges by decreasing weight, and adds them to a matching greedily. This algorithm is a $\frac{1}{2}$-approximation to the optimum matching. Finding a maximum weight hypergraph matching is NP-hard, even in unweighted 3-uniform hypergraphs (a.k.a 3-dimensional matching) [21]. The greedy algorithm provides a $\frac{1}{k}$-approximation (intuitively for each hyperedge we greedily add to the matching, we lose at most k hyperedges) where k is the maximum cardinality of an edge.

Stochastic Matchings. Various stochastic versions of graph matchings have been studied in the literature. We discuss two papers that lie close to our work [5,11]. Both of these works consider a random graph model with a Bernoulli distribution on each edge. In contrast to our work, these models allow the central designer to *probe* each edge to verify its realization: if the edge exists, it gets irrevocably added to the matching. While Chen et al. [11] provide a constant factor approximation on unweighted graphs based on a simple greedy approach, Bansal et al. [5] obtain a $O(1)$-factor for even weighted graphs using an LP-rounding algorithm. On the other hand, our work focuses on designing fast algorithms that achieve good matchings with bounded risk on weighted graphs without probing the edges. Finally, since the hypergraph matching is also known as set packing, the above problems are special cases of stochastic set packing problem [14].

3 Model and Proposed Method

Uncertain Weighted Bernoulli hypergraphs. Before we define a general model for uncertain weighted hypergraphs that allows for both continuous and discrete probability distributions, we introduce a simple probabilistic model for weighted uncertain hypergraphs that generalizes the existing model for random graphs. Each edge e is distributed as a weighted Bernoulli variable independently from the rest: with probability $p(e)$ it exists, and its weight/reward is equal to $w(e)$, and with the remaining probability $1 - p(e)$ it does not exist, i.e., its weight is zero. More formally, let $\mathcal{H} = ([n], E, p, w)$ be an uncertain hypergraph on n nodes with $|E| = m$ potential hyperedges, where $p : E \to (0, 1]$, is the function that assigns a probability of existence to each hyperedge independently from the other hyperedges, and $w : E \to \mathbb{R}^+$. The value $w(e)$ is the reward we receive from hyperedge e if it exists. Let $r_e \overset{\text{def}}{=} p(e)w(e)$ be the expected reward from edge e. According to the possible-world semantics [8,13], the probability of observing any possible world $H(V, E_H) \in 2^E$ is $\mathbf{Pr}[H] = \prod_{e \in E_H} p(e) \prod_{e \notin E_H} (1 - p(e))$.

Uncertain Weighted Hypergraphs. More generally, let $\mathcal{H}([n], E, \{f_e(\theta_e)\}_{e \in E})$ be an uncertain hypergraph on n nodes, with hyperedge set E. The reward $w(e)$ of each hyperedge $e \in E$ is drawn according to some probability distribution f_e with parameters θ_e, i.e., $w(e) \sim f_e(x; \theta_e)$. We assume that the reward for each hyperedge is drawn independently from the rest; each probability distribution is assumed to have finite mean, and finite variance. Given this model, we define the probability of a given hypergraph H with weights $w(e)$ on the hyperedges as:

$$\mathbf{Pr}[H; \{w(e)\}_{e \in E}] = \prod_{e \in E} f_e(w(e); \theta_e).$$

Our model allows for both discrete and continuous distributions, as well as mixed discrete and continuous distributions. In our experiments (Sect. 4) we focus on the weighted Bernoulli, and Gaussian cases.

Problem Definition. Our goal is to output a matching M with high expected reward and low variance. A crucial assumption that we make is that for any given edge e, the algorithm designer does not have access to the complete distribution $f_e(\cdot)$ but only simple statistics such as its mean and standard deviation (s.t.d). Let $\mathcal{M}$ be the set of all matchings from the hyperedge set E. The total associated reward with a matching $M \in \mathcal{M}$ is the expected reward, i.e.,

$$R(M) \overset{\text{def}}{=} \sum_{e \in M} r_e = \sum_{e \in M} E_{f_e}[w(e)].$$

Similarly, the associated risk in terms of the standard deviation is defined as

$$risk(M) \overset{\text{def}}{=} \sum_{e \in M} \sigma_e,$$

where σ_e denotes the standard deviation of the distribution $f_e(x; \theta_e)$.

Given an uncertain weighted hypergraph, and a risk upper-bound B, our goal is to maximize the expected reward over all matchings with risk at most B. We refer to this problem as the Bounded Risk Maximum Weighted Matching (BR-MWM) problem. Specifically,

$$
\begin{aligned}
\max_{M \in \mathcal{M}} \quad & R(M) \qquad \text{[BR-MWM problem]} \qquad\qquad (1)\\
\text{s.t} \quad & risk(M) \le B
\end{aligned}
$$

For example, in the case of a weighted Bernoulli hypergraph where each hyperedge $e \in E$ exists with probability $p(e)$ and has weight $b(e)$ when it exists, formulation (1) becomes

$$\max_{M \in \mathcal{M}} \sum_{e \in M} p(e)b(e)$$
$$\text{s.t} \quad \sum_{e \in M} b(e)\sqrt{p(e)(1-p(e))} \leq B \tag{2}$$

Similar formulations can be obtained for other specific distributions such as Gaussian. Finally, we remark that the BR-MWM problem is NP-Hard even on graphs via a simple reduction from Knapsack.

Other Measures of Risk. It is worth outlining that our model and proposed method adapts easily to other risk measures. For example, if we define the risk of a matching M in terms of its variance, i.e.,

$$risk(M) \overset{\text{def}}{=} \sum_{e \in M} \sigma_e^2, \tag{3}$$

then all of our theoretical guarantees and the insights gained via our experiments still hold with minor changes in the algorithm. At the end of this section, we discuss in detail the required changes. For the sake of convenience and concreteness, we present our results in terms of the standard deviation.

An LP-approximation Algorithm. The BR-MWM problem is a special case of the *Hypermatching Assignment Problem* (HAP) introduced in [12]: given a k-uniform hypergraph $H(V, E)$, a budget B, a profit and a cost $w_e, c_e \geq 0$ for hyperedge e respectively, the goal of HAP is to compute a matching M so that the total profit $\sum_{e \in M} w_e$ is maximized and the budget constraint $\sum_{e \in M} c_e \leq B$ is satisfied. Therefore, we can invoke the randomized $\frac{1}{k+1+\epsilon}$-approximation algorithm for HAP [12] to solve our problem, here $\epsilon > 0$ is constant. However, this approach—at least for the moment—is unlikely to scale well: it requires solving a linear program with an exponential number of variables in terms of $\frac{1}{\epsilon}$, and then strengthen this LP by one round of the Lasserre's lift-and-project method. This motivates the design of scalable approximation algorithms.

3.1 Proposed Algorithm and Guarantees

Our algorithm is described in pseudocode 1. It takes as input a hypergraph matching algorithm MATCH-ALG as a black-box: the black-box takes a weighted hypergraph and returns a hypergraph matching. First, our algorithm removes all hyperedges e that have negative reward and for which $\sigma_e > B$ as they are not part of any optimal solution. For any given edge $e \in E$, define $\alpha_e \overset{\text{def}}{=} \frac{r_e}{\sigma_e}$. Now, we label the edges in E as $e_1, e_2, \ldots, e_m$ such that $\alpha_{e_1} \geq \alpha_{e_2} \geq \ldots \geq \alpha_{e_m}$, breaking ties arbitrarily. Sorting the α values requires $O(m \log m)$ time. Next, we consider the nested sequence of hypergraphs $\emptyset = H^{(0)} \subset H^{(1)} \subset \ldots \subset H^{(m)} = H$, where $H^{(i)}$ contains the i hyperedges $(e_1, e_2, \ldots, e_i)$, and each edge e is weighted by the expected reward r_e.

Let $M^{(i)}$ be the matching returned by MATCH-ALG on $H^{(i)}$ with weights $(r_e)_{e \in H^{(i)}}$. We first compute the maximum weight matching on $H^{(m)}$. If the quantity $risk(M^{(m)})$ is less than or equal to B, then we output $M^{(m)}$. Otherwise, we binary search the nested sequence of hypergraphs to find *any* index ℓ^* for which

$$risk(M^{(\ell^*)}) \leq B < risk(M^{(\ell^*+1)}).$$

The final output matching M_{OUT} is either $M^{(\ell^*)}$ or e_{ℓ^*+1}, depending on which one achieves greater expected reward. Intuitively, the latter case is required when there exists a single high-reward hyperedge whose risk is comparable to the upper bound B. In general, there may be more than one index ℓ^* that satisfies the above condition since the variance of $M^{(i)}$ is *not* monotonically increasing with i. Figure 2 provides such an example that shows that increasing the set of allowed edges can actually decrease the overall risk of the optimum matching.

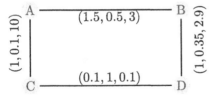

Fig. 2. The risk $risk(M^{(i)})$ of the optimum matching $M^{(i)}$ is *not* monotonically increasing with i. For details, see Sect. 3

Specifically, Fig. 2 shows an uncertain graph, each edge e is annotated with $(r_e, \sigma_e, \alpha_e)$. One can always find distributions that satisfy these parameters. We consider Algorithm 1 with the black-box matching algorithm MATCH-ALG as the optimum matching algorithm on weighted graphs. As our algorithm considers edges in decreasing order of their α-value, we get that $M^{(1)} = \{(A, C)\}, M^{(2)} = \{(A, B)\}, M^{(2)} = \{(A, B)\}, M^{(3)} = \{(A, C), (B, D)\}$. The risk of the above three matchings are $0.1, 0.5$, and 0.45 respectively. Thus, the quantity $risk(M^{(i)})$ is *not* monotonically increasing with i.

While it is not hard to see how a binary search would work, we provide the details for completeness. We know that $risk(M^{(1)}) = \sigma(e_1) \leq B$, and $risk(M^{(m)}) > B$. Let $low = 1, high = m$. We search the middle position mid between low and high, and $mid + 1$. If $risk(M^{(mid)}) \leq B < risk(M^{(mid+1)})$, then we set ℓ^* equal to mid and return. If not, then if $risk(M^{(mid)}) \leq B$, we repeat the same procedure with $low = mid + 1, high = m$. Otherwise, if $risk(M^{(mid)}) > B$ we repeat with $low = 1, high = mid$. This requires $O(\log m)$ iterations, and each iteration requires the computation of at most two matchings using the black-box MATCH-ALG.

Our proposed algorithm uses the notion of a black-box reduction: wherein, we take an arbitrary c-approximation algorithm for computing a maximum-weight hypermatching (MATCH-ALG, $c \leq 1$) and leverage its properties to derive an algorithm that in addition to maximizing the expected weight also has low risk. This black-box approach has a significant side-effect: organizations may have already invested in graph processing software for deterministic graphs can continue to use the same methods (as a black-box) regardless of the uncertainty inherent in the data. Our search takes time $O(\log m \times T(n,m))$ where $T(n,m)$ is the running time of maximum weighted matching algorithm MATCH-ALG.

$\frac{1}{3}$-**approximation for Uncertain Weighted Graphs.** First we analyze our algorithm for the important case of uncertain weighted graphs. Unlike general hypergraphs, we can find a maximum weight graph matching in polynomial time using Edmond's algorithm [16]. Our main result is stated below.

Theorem 1. *Assuming an exact maximum weight matching algorithm* MATCH-ALG, *Algorithm 1 returns a matching M_{OUT} whose risk is less than or equal to B, and whose expected reward is at least $\frac{1}{3}$ of the optimal solution to the Bounded Risk Maximum Weighted Matching problem on uncertain weighted graphs.*

Before we prove Theorem 1, it is worth reiterating that our proposed algorithm provides a better approximation than the factor guaranteed in [12], i.e., $\frac{1}{3} > \frac{1}{3+\epsilon}$ for any constant $\epsilon > 0$. Additionally, our approach is orders of magnitude faster than the algorithm from [12] as the latter uses an LP-rounding technique, whereas our approach is simple and combinatorial.

Proof. Let M^{OPT} denote an optimum matching whose risk is at most B. Since it is immediately clear by the description of our algorithm that $risk(M_{OUT}) \leq B$, our goal is to prove that the matching returned by our algorithm has reward at least one-third as good as the reward of the optimum matching, i.e., $R(M_{OUT}) = \sum_{e \in M_{OUT}} r_e \geq \frac{R(M^{OPT})}{3}$.

In order to show this bound, we prove a series of inequalities. By definition, $H^{(\ell^*+1)}$ differs from $H^{(\ell^*)}$ in exactly one edge, that is e_{ℓ^*+1}. We also know that the maximum weight matching in $H^{(\ell^*+1)}$ (i.e., $M^{(\ell^*+1)}$) is different from the maximum weight matching in $H^{(\ell^*)}$ ($M^{(\ell^*)}$) since the former entails risk that exceeds the budget B. We conclude that $M^{(\ell^*+1)}$ contains the edge e_{ℓ^*+1}.

Therefore, we have that $R(M^{(\ell^*+1)}) = R(M^{(\ell^*+1)} \setminus e_{\ell^*+1}) + r_{e_{\ell^*+1}} \leq R(M^{(\ell^*)}) + r_{e_{\ell^*+1}}$. This is true because $M^{(\ell^*)}$ is the maximum weight matching in $H^{(\ell^*)}$ and so its weight is larger than or equal to that of $M^{(\ell^*+1)} \setminus e_{\ell^*+1}$. In conclusion, our first non-trivial inequality is:

$$R(M^{(\ell^*)}) + r_{e_{\ell^*+1}} \geq R(M^{(\ell^*+1)}) \tag{4}$$

Algorithm 1. Algorithm for computing a $\frac{c}{2+c}$-approximate matching for the BR-MWM problem on uncertain weighted hypergraphs

---

input : $\mathcal{H}([n], E), (r_e = E_{f_e}[w_e])_{e \in E}, (\sigma_e = \sqrt{E_{f_e}[(w_e - r_e)^2]})_{e \in E}$, MATCH-ALG

Remove all hyperedges e that have either $r_e \leq 0$ or $\sigma_e > B$;

Sort E in decreasing order of $\alpha_e = \frac{r_e}{\sigma_e}$, let $\alpha_{e_1} \geq \ldots \geq \alpha_{e_m} \geq 0$.;

$M^{(m)} \leftarrow$ MATCH-ALG$(H^{(m)})$;

if $risk(M^{(m)}) \leq B$ **then**
   | $\ell^* \leftarrow m$; Return $\ell^*, M^{(\ell^*)}$;
end

$low \leftarrow 1, high \leftarrow m$;

while *True* **do**
   | $mid \leftarrow \lfloor \frac{low+high}{2} \rfloor$; Compute $M^{(mid)}, M^{(mid+1)}$;
   | **if** $risk(M^{(mid)}) \leq B < risk(M^{(mid+1)})$ **then**
   | | $\ell^* \leftarrow mid$; Return $\ell^*, M^{(\ell^*)}$
   | **else if** $risk(M^{(mid)}) \leq B$ **then**
   | | $low \leftarrow mid + 1$;
   | **else**
   | | $high \leftarrow mid$;
   | **end**
end

Next, we lower-bound $M^{(\ell^*+1)}$ by using the facts that $\alpha_e \geq \alpha_{e_{\ell^*+1}}$ for all $e \in M^{(\ell^*+1)}$, and that the total risk of $M^{(\ell^*+1)}$ is at least B by definition. Specifically,

$$R(M^{(\ell^*+1)}) = \sum_{e \in M^{(\ell^*+1)}} r_e = \sum_{e \in M^{(\ell^*+1)}} \alpha_e \sigma_e \tag{5}$$

$$\geq \sum_{e \in M^{(\ell^*+1)}} \alpha_{e_{\ell^*+1}} \sigma_e$$

$$= \alpha_{e_{\ell^*+1}} \sum_{e \in M^{(\ell^*+1)}} \sigma_e > \alpha_{e_{\ell^*+1}} B. \tag{6}$$

Now we show upper bounds on the optimum solution to the BR-MWM problem M^{OPT}. We divide M^{OPT} into two parts: M_1^{OPT} and M_2^{OPT}, where the first part is the set of edges in $M^{OPT} \cap H^{(\ell^*)}$ and the second part is the edges not present in $H^{(\ell^*)}$. We present separate upper bounds on M_1^{OPT} and M_2^{OPT}. By definition, M_1^{OPT} is a matching on the set of edges $H^{(\ell^*)}$. Therefore, its reward is smaller than or equal to that of the optimum matching on $H^{(\ell^*)}$, which happens to be $M^{(\ell^*)}$. Hence,

$$R(M_1^{OPT}) \leq R(M^{(\ell^*)}). \tag{7}$$

Next, consider M_2^{OPT}. To upper-bound $R(M_2^{OPT})$ we also use inequalities 4, 6:

$$R(M_2^{OPT}) = \sum_{e \in M_2^{OPT}} r_e = \sum_{e \in M_2^{OPT}} \alpha_e \sigma_e$$

$$\leq \sum_{e \in M_2^{OPT}} \alpha_{e_{\ell^*+1}} \sigma_e - \alpha_{e_{\ell^*+1}} \sum_{e \in M_2^{OPT}} \sigma_e$$

$$\leq \alpha_{e_{\ell^*+1}} B < R(M^{(\ell^*+1)})$$
$$\leq R(M^{(\ell^*)}) + r_{e_{\ell^*+1}}.$$

Now, we are ready to complete the proof. Recall that the output of the algorithm M_{OUT} satisfies $R(M_{OUT}) = \max(R(M^{(\ell^*)}), r_{e_{\ell^*+1}})$. Combining the upper bounds for M_1^{OPT} and M_2^{OPT} yields the desired approximation factor of $\frac{1}{3}$:

$$R(M^{OPT}) \leq R(M^{(\ell)}) + R(M^{(\ell)}) + r(e_{\ell+1})$$
$$= 2R(M^{(\ell)}) + r(e_{\ell+1}) \leq 3R(M_{OUT}). \blacksquare$$

Running time: Assuming that the $O(mn + n^2 \log n)$ [16] implementation of Edmond's algorithm is used as a black-box, we remark that the run time of Algorithm 1 is $O(mn \log m + n^2 \log m \log n)$.

Fast $\frac{1}{5}$-approximation for Uncertain Weighted Graphs. Since the running time using Edmond's algorithm is somewhat expensive, we show how the approximation guarantee changes when we use the (much faster) greedy algorithm for maximum weighted matchings as MATCH-ALG. Recall, the greedy matching algorithm runs in $O(m \log m + n)$ time.

Theorem 2. *If the black-box MATCH-ALG is set to be the greedy matching algorithm, then Algorithm 1 computes a $\frac{1}{5}$-approximation to the optimal solution of the BR-MWM problem in $O(m \log^2 m + n \log m)$-time.*

The proof is omitted as it is essentially identical to the proof of Theorem 1, with the only change that the greedy matching algorithm provides a $\frac{1}{2}$-approximation to the maximum weighted matching problem.

Fast $\frac{c}{2+c}$-approximation for Uncertain Weighted Hypergraphs. Recall that finding a maximum weight hypergraph matching is NP-hard even for unweighted, 3-regular hypergraphs [21]. However, there exist various algorithms, that achieve different approximation factors $c < 1$. For example, the greedy algorithm provides a $\frac{1}{k}$-approximation guarantee, where k is the rank of the hypergraph (i.e., any hyperedge contains at most k nodes). Our main theoretical result follows.

Theorem 3. *Given any polynomial-time c-approximation algorithm MATCH-ALG (c $\leq$ 1) for the maximum weighted hypergraph matching problem, we can compute in polynomial time a hypermatching M_{OUT} such that its risk is at most B and its expected weight is a $\frac{c}{2+c}$-approximation to the expected weight of the optimal hypermatching that has risk at most B.*

Again the proof proceeds step by step as the proof of Theorem 1, and is omitted. In what follows, we restrict our attention to using the greedy hypermatching algorithm as a black-box. Our focus on greedy matchings stems from the fact that its approximation factor ($\frac{1}{k}$) is asymptotically optimal [6,10], that it is easy to implement, and runs in $O(m \log m + n)$ time using appropriate data structures. Since we will be using the greedy algorithm in our experiments (Sect. 4), we provide the following corollary.

Corollary 4. *For any hypergraph of rank k, we can compute in polynomial time a hypergraph-matching whose risk is at most B and whose weight is a $\Omega(\frac{1}{k})$ approximation to the optimum bounded-risk hypergraph matching.*

Algorithm 1 using the greedy hypermatching algorithm in lieu of MATCH-ALG runs in $O(m \log^2 m + n \log m)$ time.

Remark. We reiterate the point that our algorithm can be used to compute risk-averse matchings for other notions of risk such as variance. For instance, if we define risk as in Eq. (3), then the only thing that changes in our algorithm is the definition of the α_e, namely that α_e is set equal to $\frac{r_e}{\sigma_e^2}$ for each (hyper)edge $e \in E$. The rest, including the theoretical guarantees remain identical.

4   Experimental Results

Experimental Setup and Normalization. We test our proposed algorithm on a diverse range of datasets, where the orders of magnitude of risk (e.g., standard deviation) can vary greatly across datasets. In order to have a consistent interpretation of the trade-off between expected reward and risk across datasets, we normalize the allowed risk B relative to the maximum possible standard deviation of a benchmark matching, $B_{\max}$. For the purpose of computing or more precisely approximating $B_{\max}$, we run the greedy matching algorithm on the (hyper)graph G (H) where the weight on edge e is σ_e, and set $B_{\max}$ to be the aggregate risk of the computed matching. While in theory one may observe a matching with greater risk than the obtained value $B_{\max}$, this does not occur in any of our simulations. We range B according to the rule $B = B_n \times B_{\max}$, where $B_n \in [0,1]$ and is incremented in steps of 0.05. We refer to B_n as the *normalized risk* from now on. Due to space constraints we have not included a wide variety of synthetic experiments that can be found in an extended version of our work [40]. In the following we show our results on a real-world uncertain, weighted hypergraph. We implement our proposed fast approximation algorithm for uncertain weighted hypergraphs in Python. The code is available at Github [2]. All experiments were performed on a laptop with 1.7 GHz Intel Core i7 processor and 8 GB of main memory.

Recommending Impactful but Probable Collaborations. In many ways, academic collaboration is an ideal playground to explore the effect of risk-averse team formation for research projects as there exist teams of researchers that have the potential for high impact but may also collaborate less often. To explore this further, we use our proposed algorithm for uncertain weighted hypergraphs as a tool for identifying a set of disjoint collaborations that are both impactful and likely to take place. For this purpose, we use the Digital Bibliography and Library Project (DBLP) database. From each paper, we obtain a team that corresponds to the set of authors of that paper. As a proxy for the impact of the paper we use the citation count. Unfortunately, we could not obtain the citation counts from Google Scholar for the whole DBLP dataset as we would get rate limited by

Google after making too many requests. Therefore, we used the AMiner citation network dataset [1] that contains citation counts, but unfortunately is not as up-to-date as Google Scholar is.

We preprocessed the dataset by removing all single-author papers since the corresponding hyperedge probabilities are one. Furthermore, multiple hyperedges are treated as one, with citation count equal to the sum of the citation counts of the multiple hyperedges. To give an example, if there exist three papers in the dataset that have been co-authored by authors A_1, A_2 with citation counts w_1, w_2, w_3 we create one hyperedge on the nodes that correspond to A_1, A_2 with weight equal $w_1 + w_2 + w_3$. If there exists another paper co-authored by A_1, A_2, A_3, this yields a different hyperedge/team $\{A_1, A_2, A_3\}$, and we do not include its citations in the impact of team $\{A_1, A_2\}$.

For hyperedge $e = (u_1, \ldots, u_\ell)$ we find the set of papers $\{P_1, \ldots, P_\ell\}$ authored by authors $u_1, \ldots, u_\ell$ respectively. We set the probability of hyperedge e as

$$p_e = \frac{|P_1 \cap P_2 \cap \ldots \cap P_\ell|}{|P_1 \cup P_2 \cup \ldots \cup P_\ell|}.$$

Intuitively, this is the empirical probability of collaboration between the specific set of authors.

To sum up, we create an uncertain weighted hypergraph using the DBLP dataset, where each node corresponds to an author, each hyperedge represents a paper whose reward follows a Bernoulli distribution with weight equal to the number of its citations, and probability p_e is the likelihood of collaboration. The final hypergraph consists of $n = 1,752,443$ nodes and $m = 3,227,380$ edges, and will be made publicly available on the first author's website. The largest collaboration involves a paper co-authored by 27 people, i.e., the rank k of the hypergraph is 27.

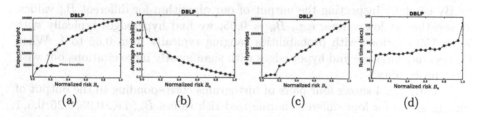

| (a) | (b) | (c) | (d) |

Fig. 3. (a) Expected reward, (b) average probability (over hypermatching's edges), (c) number of edges in the hypermatching, and (d) running time in seconds versus normalized risk B_n. For details, see Sect. 4

Figure 3 shows our findings when we vary the normalized risk bound B_n and obtain a hypermatching for each value of this parameter, using our algorithm. For the record, when $B_n = 1$, then $B = B_{\max} = 454\,392.0$. Figure 3(a) plots the expected weight of the hypermatching versus B_n. We observe an interesting phase transition when B_n changes from 0.15 to 0.2. This is because after

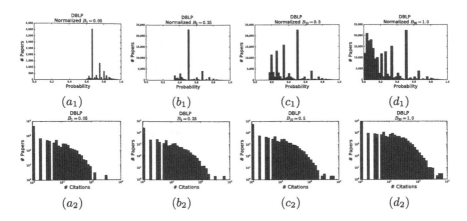

Fig. 4. Figures in first row $(a_1), (b_1), (c_1), (d_1)$ (second row $(a_2), (b_2), (c_2), (d_2)$): histograms showing the hyperedge probabilities (citations) in the hypermatching returned by our algorithm for normalized risk values B_n equal to $0.05, 0.25, 0.5, 1$ respectively. For details, see Sect. 4

$B_n = 0.15$ the average probability of the hyper-matching drops from ~ 0.7 to ~ 0.5. This is shown in Fig. 3(b) that plots the average probability of the edges in each hypermatching computed by our algorithm vs. B_n. Figure 3(a),(b) strongly indicate what we verified by inspecting the output: up to $B_n = 0.15$, our algorithm picks teams of co-authors that tend to collaborate frequently. This finding illustrates that our tool may be used for certain anomaly detection tasks. Figure 3(c),(d) plot the number of hyperedges returned by our algorithm, and its running time in seconds vs B_n. We observe that a positive side-effect of using small risk bounds is speed: for small B_n values, the algorithm computes fewer maximum matchings.

By carefully inspecting the output of our algorithm for different B_n values, we see that at low values, e.g., $B_n = 0.05$, we find hyperedges typically with 50 to 150 citations with probabilities ranging typically from 0.66 to 1. When B_n becomes large we find hyper-edges with significantly more citations but with lower probability.

Finally, Fig. 4 shows four pairs of histograms corresponding to the output of our algorithm for four different normalized risk values B_n, i.e., $0.05, 0.25, 0.5, 1$ respectively. Each pair ($\{(a_1), (a_2)\}$, $\{(b_1), (b_2)\}$, $\{(c_1), (c_2)\}$, and $\{(d_1), (d_2)\}$) plots the histogram of the probabilities, and the number of citations of the hyperedges selected by our algorithm for $B_n \in \{0.05, 0.25, 0.5, 1\}$ respectively. The histograms provide a view of how the probabilities decrease and citations increase as we increase B_n, i.e., as we allow higher risk.

5   Conclusion

In this work we study the problem of finding matchings with high expected reward and bounded risk on large-scale uncertain hypergraphs. We introduce a general model for uncertain weighted hypergraphs that allows for both continuous and discrete probability distributions, we provide a novel stochastic matching formulation that is NP-hard, and develop fast approximation algorithms. We verify the efficiency of our proposed methods on several synthetic and real-world datasets.

In contrast to the majority of prior work on uncertain graph databases, we show that it is possible to combine risk aversion, time efficiency, and theoretical guarantees simultaneously. Moving forward, a natural research direction is to design risk-averse algorithms for other graph mining tasks such as motif clustering, the k-clique densest subgraph problem, and k-core decompositions?

References

1. Aminer citation network dataset, August 2017. https://aminer.org/citation
2. Risk-averse matchings over uncertain graph databases, January 2018. https://github.com/tsourolampis/risk-averse-graph-matchings
3. Anagnostopoulos, A., Becchetti, L., Castillo, C., Gionis, A., Leonardi, S.: Online team formation in social networks. In: Proceedings of WWW, vol. 2012, pp. 839–848 (2012)
4. Asthana, S., King, O.D., Gibbons, F.D., Roth, F.P.: Predicting protein complex membership using probabilistic network reliability. Genome Res. 14(6), 1170–1175 (2004)
5. Bansal, N., Gupta, A., Li, J., Mestre, J., Nagarajan, V., Rudra, A.: When LP is the cure for your matching woes: improved bounds for stochastic matchings. Algorithmica 63(4), 733–762 (2012)
6. Berman, P.: A d/2 approximation for maximum weight independent set in d-claw free graphs. SWAT 2000. LNCS, vol. 1851, pp. 214–219. Springer, Heidelberg (2000). https://doi.org/10.1007/3-540-44985-X_19
7. Boldi, P., Bonchi, F., Gionis, A., Tassa, T.: Injecting uncertainty in graphs for identity obfuscation. Proc. VLDB Endow. 5(11), 1376–1387 (2012)
8. Bollobás, B., Janson, S., Riordan, O.: The phase transition in inhomogeneous random graphs. Random Struct. Algorithms 31(1), 3–122 (2007)
9. Bonchi, F., Gullo, F., Kaltenbrunner, A., Volkovich, Y.: Core decomposition of uncertain graphs. In: Proceedings of the KDD 2014, pp. 1316–1325 (2014)
10. Chan, Y.H., Lau, L.C.: On linear and semidefinite programming relaxations for hypergraph matching. Math. Program. 135(1–2), 123–148 (2012)
11. Chen, N., Immorlica, N., Karlin, A.R., Mahdian, M., Rudra, A.: Approximating matches made in heaven. In: Albers, S., Marchetti-Spaccamela, A., Matias, Y., Nikoletseas, S., Thomas, W. (eds.) ICALP 2009. LNCS, vol. 5555, pp. 266–278. Springer, Heidelberg (2009). https://doi.org/10.1007/978-3-642-02927-1_23
12. Cygan, M., Grandoni, F., Mastrolilli, M.: How to sell hyperedges: the hypermatching assignment problem. In: Proceedings of SODA 2013, pp. 342–351 (2013)
13. Dalvi, N.N., Suciu, D.: Efficient query evaluation on probabilistic databases. VLDB J. 16(4), 523–544 (2007)
14. Dean, B.C., Goemans, M.X., Vondrák, J.: Adaptivity and approximation for stochastic packing problems. In: Proceedings of SODA 2005, pp. 395–404 (2005)
15. Edmonds, J.: Paths, trees, and flowers. Can. J. Math. 17(3), 449–467 (1965)
16. Gabow, H.N.: Data structures for weighted matching and nearest common ancestors with linking. In: Proceedings of SODA 1990, pp. 434–443 (1990)

17. Gajewar, A., Das Sarma, A.: Multi-skill collaborative teams based on densest subgraphs. In: Proceedings of ICDM 2012, pp. 165–176 (2012)
18. Huang, X., Lu, W., Lakshmanan, L.V.: Truss decomposition of probabilistic graphs: semantics andalgorithms. In: Proceedings of SIGMOD 2016, pp. 77–90 (2016)
19. Jin, R., Liu, L., Aggarwal, C.C.: Discovering highly reliable subgraphs in uncertain graphs. In: Proceedings of KDD 2011, pp. 992–1000 (2011)
20. Kargar, M., An, A., Zihayat, M.: Efficient bi-objective team formation in social networks. In: Flach, P.A., De Bie, T., Cristianini, N. (eds.) ECML PKDD 2012. LNCS (LNAI), vol. 7524, pp. 483–498. Springer, Heidelberg (2012). https://doi.org/10.1007/978-3-642-33486-3_31
21. Karp, R.M.: Reducibility among combinatorial problems. In: Miller, R.E., Thatcher, J.W., Bohlinger, J.D. (eds.) Complexity of Computer Computations, pp. 85–103. Springer, Boston (1972). https://doi.org/10.1007/978-1-4684-2001-2_9
22. Katzenbach, J.R.: Peak Performance: Aligning the Hearts and Minds of Your Employees. Harvard Business Press (2000)
23. Kearns, M., Roth, A., Wu, Z.S., Yaroslavtsev, G.: Private algorithms for the protected in social network search. Proc. Natl. Acad. Sci. 113(4), 913–918 (2016)
24. Kempe, D., Kleinberg, J., Tardos, É: Maximizing the spread of influence through a social network. In: Proceedings of KDD 2003, pp. 137–146. ACM (2003)
25. Khan, A., Chen, L.: On uncertain graphs modeling and queries. Proc. VLDB Endow. 8(12), 2042–2043 (2015)
26. Kolata, G.: Grant system leads cancer researchers to play it safe. New York Times, vol. 24 (2009)
27. Kollios, G., Potamias, M., Terzi, E.: Clustering large probabilistic graphs. IEEE Trans. Knowl. Data Eng. 25(2), 325–336 (2013)
28. Krogan, N.J., et al.: Global landscape of protein complexes in the yeast saccharomyces cerevisiae. Nature 440(7084), 637 (2006)
29. Lappas, T., Liu, K., Terzi, E.: Finding a team of experts in social networks. In: Proceedings of KDD 2009, pp. 467–476. ACM (2009)
30. Liben-Nowell, D., Kleinberg, J.: The link-prediction problem for social networks. J. Assoc. Inf. Sci. Technol. 58(7), 1019–1031 (2007)
31. Liu, L., Jin, R., Aggarwal, C., Shen, Y.: Reliable clustering on uncertain graphs. In: Proceedings of ICDM 2012, pp. 459–468. IEEE (2012)
32. Lovász, L., Plummer, M.D.: Matching Theory, vol. 367. American Mathematical Society (2009)
33. Majumder, A., Datta, S., Naidu, K.V.M.: Capacitated team formation problem on social networks. In: Proceedings of KDD 2012, pp. 1005–1013 (2012)
34. Moustafa, W.E., Kimmig, A., Deshpande, A., Getoor, L.: Subgraph pattern matching over uncertain graphs with identity linkage uncertainty. In: Proceedings of ICDE 2014, pp. 904–915. IEEE (2014)
35. Parchas, P., Gullo, F., Papadias, D., Bonchi, F.: The pursuit of a good possible world: extracting representative instances of uncertain graphs. In: Proceedings SIGMOD 2014, pp. 967–978 (2014)
36. Potamias, M., Bonchi, F., Gionis, A., Kollios, G.: K-nearest neighbors in uncertain graphs. Proc. VLDB Endow. 3(1–2), 997–1008 (2010)
37. Roth, A.E., Sönmez, T., Ünver, M.U.: Kidney exchange. Q. J. Econ. 119(2), 457–488 (2004)
38. Saha, A.K., Johnson, D.B.: Modeling mobility for vehicular ad-hoc networks. In: Proceedings of the 1st ACM International Workshop on Vehicular Ad Hoc Networks, pp. 91–92. ACM (2004)

39. Tsourakakis, C.E., Mitzenmacher, M., Błasiok, J., Lawson, B., Nakkiran, P., Nakos, V.: Predicting positive and negative links with noisy queries: theory & practice. arXiv preprint arXiv:1709.07308 (2017)
40. Tsourakakis, C.E., Sekar, S., Lam, J., Yang, L.: Risk-averse matchings over uncertain graph databases. arXiv preprint arXiv:1801.03190 (2018)
41. Tu, K., et al.: Online dating recommendations: matching markets and learning preferences. In: Proceedings of WWW 2014, pp. 787–792 (2014)
42. Valiant, L.G.: The complexity of computing the permanent. Theor. Comput. Sci. 8(2), 189–201 (1979)

Discovering Urban Travel Demands Through Dynamic Zone Correlation in Location-Based Social Networks

Wangsu Hu[1], Zijun Yao[2], Sen Yang[1], Shuhong Chen[1], and Peter J. Jin[1(✉)]

[1] Rutgers University, New Brunswick, USA
{wh251,sy358,sc1624,peter.j.jin}@rutgers.edu
[2] IBM Thomas J. Watson Research Center, Yorktown, USA
zijun.yao@ibm.com

Abstract. Location-Based Social Networks (LBSN), which enable mobile users to announce their locations by checking-in to Points-of-Interests (POI), has accumulated a huge amount of user-POI interaction data. Compared to traditional sensor data, check-in data provides the much-needed information about trip purpose, which is critical to motivate human mobility but was not available for travel demand studies. In this paper, we aim to exploit the rich check-in data to model dynamic travel demands in urban areas, which can support a wide variety of mobile business solutions. Specifically, we first profile the functionality of city zones using the categorical density of POIs. Second, we use a Hawkes Process-based State-Space formulation to model the dynamic trip arrival patterns based on check-in arrival patterns. Third, we developed a joint model that integrates Pearson Product-Moment Correlation (PPMC) analysis into zone gravity modeling to perform dynamic Origin-Destination (OD) prediction. Last, we validated our methods using real-world LBSN and transportation data of New York City. The experimental results demonstrate the effectiveness of the proposed method for modeling dynamic urban travel demands. Our method achieves a significant improvement on OD prediction compared to baselines. Code related to this paper is available at: https://github.com/nicholasadam/PKDD2018-dynamic-zone-correlation.

Keywords: Origin-Destination (OD) analysis
Travel demand prediction · Location-Based Social Networks

1 Introduction

Due to the ubiquity of smartphone and the pervasiveness of social media, there have been rapid developments in Location-Based Social Networks (LBSN) research. Using LBSN, mobile users are able to check-in to Points-of-Interest (POI) for sharing their experiences and enjoying a variety of location-based services such as POI recommendation. This type of check-in data shows the presence

© Springer Nature Switzerland AG 2019
M. Berlingerio et al. (Eds.): ECML PKDD 2018, LNAI 11052, pp. 88–104, 2019.
https://doi.org/10.1007/978-3-030-10928-8_6

of users at different locations, and so can reveal large-scale human mobility in urban areas over time. With proper analysis, check-in data can be a rich source of intelligence for supporting real-time decision making in smart city applications.

While literature has shown the promising effectiveness of analyzing urban area check-in data in many applications (such as POI recommendation, business demand forecasting, human mobility analysis, and users' behavior prediction [1–4]), there are limited studies exploiting check-in data for modeling urban travel demands across city zones. Traditional transportation studies mainly use sensor-based approaches for travel demand analysis [5]. However, sensor data usually does not contain trip purpose information which is the fundamental motivation for traveling. Recent studies have shown that check-in data can be an alternative source to sense both static and time-of-day Origin-Destination (OD) flow patterns [6]. This motivates investigation into exploiting check-in data to model dynamic travel demands in urban areas.

The work presented in this study is firstly motivated by the need to fully utilize the trip purpose information revealed by check-in data. Trip purpose greatly influences a traveler's decisions, such as whether to travel, the departure time, and the destination choice [7]. Since mobile users check-in to POIs of different tags (multiple-level categorical information - e.g., nightlife spot), their check-ins can consistently expose their trip purposes (e.g., recreation). By aggregating all the individual-level check-ins at the level of zones and correlating trip purpose portfolios, we can discover the OD flow patterns between zones. Second, the proposed method is motivated by the temporal effects of travel demands. OD flow patterns between zones changes across time, but most of the existing work only provides static results. We instead propose to learn a time-aware correlation between zones by modeling dynamic OD flow patterns. To achieve this, we incorporate transportation models (i.e., zone gravity modeling and radiation modeling) into our dynamic OD analysis framework. Lastly, this work is motivated by the uniqueness of zonal analysis applications. Zonal OD flow patterns study the destination choices at an aggregated level (such as census tracts, neighborhoods, or regions) for supporting urban planning, traffic operation, and transportation management [5]. However, most check-in modeling approaches focus on individual-level prediction; therefore, they are difficult to directly apply to zonal analysis applications. To overcome this challenge, we propose to use the dynamic variations of check-in activities at different zones and different times to study zonal OD flow patterns. For example, we may see a decrease of check-in activities at residential areas followed by an increase at workplaces area in the morning period. From this observation, we may infer the AM Peak commuting trip from residential zones to commercial zones.

Alone this line, in this paper, we develop a dynamic framework to predict the time-aware travel demands across city zones through dynamic zone correlation modeling using massive mobile check-ins. Depart from prior works [3] which applied check-in data to estimate zonal trip arrivals, the proposed models move one-step forward to generate the dynamic OD flow patterns among zonal check-in activities. Specifically, we first introduce a profiling method to infer the functionality of city zones based on the POI categorical distribution and density. Second, we adopt a Hawkes Process-based State-Space (HPSS) formulation to infer the dynamic trip arrivals using check-in arrivals. Moreover, in order to capture the travel demands for any pair of zones, we develop a dynamic OD flow predictive framework called Pearson Product-Moment Correlation Gravity Model (PPMC-GM) which consists of dynamic zone correlation identification and zone gravity optimization. Last, we conduct experiments to validate the performance of PPMC-GM using a real-world dataset collected from Foursquare. The experimental results indicate that PPMC-GM outperforms other state-of-the-art baselines at predicting OD flow patterns. In addition, we reported qualitative results which study the pairwise time-of-day OD flow patterns between zones of different functionality.

2   Related Work

There are two areas related to this study: OD estimation models and check-in data-based spatial-temporal correlation models.

2.1   OD Estimation Model

OD estimation models are crucial for understanding the mechanisms of human mobility. There are three major approaches for predicting the OD flow patterns at an aggregated zonal level: the gravity model, the intervening opportunities model, and the regression model. The gravity model, inspired by the Newton's law of gravitation, assumes that the amount of trips between two locations is proportional to their populations and decays by a function of the distance [8]. The gravity model was originally applied to model large-scale migration patterns, then has been widely used for predicting destination choices of network users [6,9,10]. These models were able to replicate the OD patterns on static scenarios to reveal insightful temporal and spatial patterns. Jin *et al.* [6] predicted the OD flows over a target time period to reveal a static urban travel demand pattern. The model was developed for offline planning purposes and the prediction required recalibration to satisfy different scenarios. In contrast to the gravity model, Stouffer [11] proposed the intervening opportunities model, which assumes that individuals are more attracted to locations with higher interest (i.e. employment opportunities, venue capacities) rather than closer distance. Since then, extensive studies have been conducted to model human mobility patterns [12–14]. These models were applied to either model static zonal OD flow patterns or recommend individual-level locations. The third approach deploys regression. Regression requires survey-based partial OD data [15–17], which includes spatial-temporal information of participating travelers (i.e., GPS survey data, cellular data, and location-based services data). The real-time requirements are met by an auto-regressive process [17].

2.2   Spatial-Temporal Correlation Model

The spatial-temporal correlation model has been extensively studied from three approaches in check-in data-based research. Firstly, studies applied collaborative filtering techniques on check-in data [18–20]. These studies focused on measuring similarities between locations, such as the visit popularity of a geographic region, and the hierarchical properties of geographic spaces. The user-based collaborative filtering techniques have been extensively applied to support individual location recommendation applications. Secondly, the spatial influence modeling has been widely utilized to improve spatial-temporal correlation analysis. These studies [21–24] consider spatial information of current locations and the travel distance of visited locations to determine the travelers' potential destination choice. Meanwhile, temporal influence modeling has been widely used to identify the temporal periodic patterns of check-in behaviors. Some research efforts [25–27] proposed discrete time slots, then separately modeled the temporal influence for each slot based on collaborative filtering techniques. Some research dynamically integrated both spatial and temporal influence models. Cho *et al.* [28] proposed a time-aware Gaussian Mixture model combining periodic short-range movements and sporadic long-distance travels. Wang *et al.* [29] provided a Regularity Conformity Heterogeneous (RCH) model to predict user location at specific times, considering both regularity and conformity. Lian *et al.* [30] incorporated temporal regularity into a Hidden Markov framework to predict regular user locations. Finally, taking advantage of sequential patterns in human movement [28], various sequential mining techniques [4,29,31] have been developed for location predictions based on the sequential pattern of individual's visit. Chong *et al.* explored Latent Dirichlet Allocation (LDA) topic models for venue prediction given users' history of other visited venues.

3   Methodology Overview

In this section, we first define some basic concepts in our method. Then we formulate the problem of zonal travel demand modeling. Finally we show the overview of the proposed dynamic OD flow pattern matching based framework.

3.1   Preliminary Definitions

Definition 1 *(Zone correlation). Given an origin zone and a destination zone, a zone correlation is a quantity measuring the extent of interdependence between the origin's outflow and destination's inflow.*

Definition 2 *(Trip arrival). For a zone at a time slot, trip arrivals are the number of trip counts that arrived in this zone; aggregated arrivals can describe general human mobility patterns.*

Definition 3 *(Check-in arrival). For a zone at a time slot, check-in arrivals represent the number of mobile users who visited a POI at this zone reported by check-in data.*

3.2 Problem Statement

Let $X = \{x_1, x_2, ..., x_n\}$ be a set of check-ins where each check-in has a location (e.g., latitude and longitude), a timestamp and a POI category. Let $Z = \{z_1, z_2, ..., z_m\}$ be a set of zones. We aggregate check-ins by each zones and store them as a 3-dimensional tensor V indexed by zone, category, and timeslot. Each entry of V indicates the number of observed check-ins. Our goal is to predict the 3-dimensional OD flow tensor F indexed by origin zone, destination zone, and timeslot. Each value in F indicates the number of trips from a origin zone to a destination zone during a timeslot.

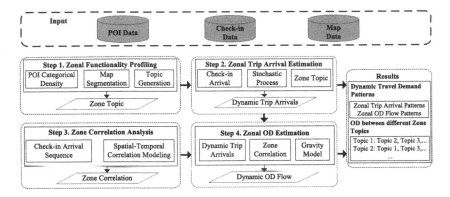

Fig. 1. An overview of our dynamic travel demand estimation framework.

3.3 General Framework

We propose a procedure for estimating the dynamic OD flow patterns (Fig. 1), including four major parts: zonal functionality profiling, zonal trip arrival estimation, zone correlation analysis, and zonal OD flow pattern estimation.

For zonal functionality profiling, we treat zonal functionality as a latent "topic" variable to discover from POI categorical density. By classifying zones by these zone topics, we can now analyze interactions between zones of different functionality.

For zonal trip arrival estimation, we applied HPSS formulation to model the trip arrival patterns based on the check-in arrival patterns and the discovered zone topics. The predicted trip arrivals are a form of dynamic urban travel demand patterns. The result is then applied as the input to the dynamic zonal OD estimation model.

For zone correlation analysis, we calculated the PPMC matrix to represent the pairwise zone correlation based on check-in arrival sequence. The result is then used in zonal OD estimation.

Finally, given the gravity model framework, we proposed a joint PPMC-GM model to predict the dynamic OD flow patterns. The zone topics will be combined with the predicted OD flow patterns to discover time-of-day variations between pairwise zones of different zonal functionality.

4   A Dynamic Travel Demand Model

In this section, we describe the four components of the proposed dynamic travel demand estimation framework.

4.1   Zonal Functionality Profiling

To infer the zonal functionality, we introduced M zone topics determined by analyzing the POI distributions $D_i = \{d_{i,c}\}$. First, we calculated the POI density $d_{i,c}$ of each POI category c at zone i:

$$d_{i,c} = \frac{numbers\ of\ POI_{c\in C}}{area\ of\ zone\ i}. \tag{1}$$

Using Latent Dirichlet allocation (LDA) method, we treat the zone functionalities $m \in \{1, ..., M\}$ as the document topics, the zones as documents (each POI distribution D_i as one documentation), and POI categorical density $d_{i,c}$ as words in the documentation. Then the zone functionality can be uncovered from the POIs. We construct a hierarchical topic model as following:

$$\theta_i \sim Dir(\eta_1); Multinomial(z_{i,d_{i,c}}, \theta_i); \varphi_m \sim Dir(\eta_2); Multinomial(\omega_{i,d_{i,c}}, \varphi_m), \tag{2}$$

where η_1 and η_2 are the prior Dirichlet parameters on the per-document topic distribution and word distribution, θ_i represent the topic distribution for zone i, φ_m is the word distribution for topic m, $z_{i,d_{i,c}}$ and $\omega_{i,d_{i,c}}$ are the chosen topic and word for zone i. Then the probability of POIs within one zone being covered by zone topic m is:

$$P(m|D_i) = \prod_m P(\varphi_m|\eta_2) \prod_i P(\theta_i|\eta_1) \prod_c P(z_{i,d_{i,c}}|\theta_i)P(\omega_{i,d_{i,c}}|\varphi_{1:M}, z_{i,d_{i,c}}). \tag{3}$$

4.2   Zonal Trip Arrival Estimation

Efforts have been made to model zonal trip arrival based on check-in data [3]. Here, we leverage the HPSS formulation to predict zonal trip arrival patterns based on the discovered zone topics and the check-in arrival patterns. The method includes a state equation and an observation equation. The state equation represented in Eq. 4 introduces every check-in arrival as an event occurrence following the Hawkes point process. Then an observation equation represented in Eq. 5 generates zonal trip arrivals, which are compared with the observed check-in arrivals. Through the state-space model framework, check-in arrivals can then be coupled with the Hawkes process model to estimate dynamic zonal trip arrivals. The zonal trip arrival modeling can be represented as follows:

$$\lambda_{i,m}^t = \frac{\tilde{A}_i^{t-\Delta t}}{\Delta t} + \alpha_{i,m}^t \sum_{t:t_1<t_2} exp(-\beta_{i,m}^t(t_1 - t_2)) \tag{4}$$

$$\tilde{A}_i^t = \delta_{i,m}^t (\int_t^{t+\Delta t} \lambda_{i,m}^t dn) + (1 - \delta_{i,m}^t)\gamma_{i,m}^t x_i^t, \tag{5}$$

where $\lambda_{i,m}^t$ represents a counting process for trip arrivals at zone i at time slot t for zone topic m, $\tilde{A}_i^{t-\Delta t}$ is the predicted zonal trip arrivals at zone i at previous time slot $t - \Delta t$, $\tilde{A}_i^t$ is the predicted zonal trip arrivals at zone i at time slot t, x_i^t is check-in arrivals at zone i at time slot t, n is the predicted sequence of trip arrival occurrence, and the set of parameters $(\alpha, \beta, \gamma, \delta)$ to be calibrated. Given the ground truth data of zonal trip arrival A_i and the time-of-day arrival A^t, we applied the following objective function for parameter estimation:

$$\min_{\alpha,\beta,\gamma,\delta} (\sum_i abs((\sum_t \tilde{A}_i^t) - A_i)) + \sum_t abs((\sum_i \tilde{A}_i^t) - A^t). \tag{6}$$

4.3   Zone Correlation Analysis

We used PPMC to measure the zone correlation. Let s_i^t represent check-in arrival sequences that consist of a set of check-in arrivals $\{x_i^t, x_i^{t+1}, \ldots, x_i^{t+w}\}$. Then we normalized s_i^t as follows:

$$f_i^t(w) = \frac{s_i^t(w) - \mu}{\sigma}, \tag{7}$$

where w is the selected sequence length to be calibrated, μ and σ is the mean and standard deviation of sequence, respectively. The zone correlation was quantified as follows:

$$RCC_{ij}^{td} = f_i^t(w)f_j^d(w), \tag{8}$$

where RCC_{ij}^{td} is a vector that contains $(2w - 1)$ elements and $d = t + \tau$ indicates the time slot after a POI-category-dependent time delay τ. Each element of RCC_{ij}^{td} has a value between $+1$ and -1 inclusive, where 1 indicates linear correlation, and 0 indicates no correlation. We selected element as the correlation coefficient rcc_{ij}^{td} based on:

$$rcc_{ij}^{td} = \max_{k \in 2w-1} abs(\{RCC_{ij}^{td}\}) \ s.t. RCC_{ij}^{td}(k) >= r_{threshold}, \tag{9}$$

where $r_{threshold}$ is the threshold to be calibrated, $I_{[clause]}$ is an indicator function for a logic clause. The PPMC procedure's purpose is to generate the rcc_{ij}^{td} followed Algorithm 1.

Algorithm 1. Zone Correlation Identification Learning

Require: Identification of zone spatial-temporal correlation
Input: $x_{i,c}^t$ # Check-in arrivals at zone i at time slot t.
Output: rcc_{ij}^{td} # PPMC coefficient for zone correlation between zone i at time slot t and zone j at time slot d.
1: **Initialization** $\forall rcc_{ij}^{td} = 0$
2: **for** $t \in T$ **do** # Let T be a set of time slots.
3:    **for** $i \in N$ **do** # Let N be a set of zones
4:      $s_i^t = \{x_i^t, x_i^{t+1}, \ldots, x_i^{t+w}\}$
5:      **for** $j \in N$ **do**
6:        $d = t + \tau$
7:        $s_j^d = \{x_j^d, x_j^{d+1}, \ldots, x_j^{d+w}\}$
8:        Generate correlation coefficient vector RCC_{ij}^{td} with the normalized s_i^t, s_j^d
9:        Select one element of RCC_{ij}^{td} as rcc_{ij}^{td}
10:     **end for**
11:   **end for**
12: **end for**
13: return all rcc_{ij}^{td}

4.4   Zonal OD Estimation

We propose a joint PPMC-Gravity Model (PPMC-GM) to predict the dynamic OD flows. The gravity model assumes knowledge of travel cost when calculating relative zone attractiveness. Normally, such travel cost is a function of travel distance or travel time. When integrated with the inflow and outflow of travelers, the GM model predicts the traffic flows from one zone to another. Considering both the HPSS formulation and PPMC coefficients, we jointly predicted dynamic OD flow patterns by incorporating the predicted dynamic trip arrivals A_i^t, A_j^d and PPMC coefficient rcc_{ij}^{td} as follows:

$$P_{ij}^{td} = \frac{A_i^t A_j^d g(rcc_{ij}^{td})}{\sum_j A_j^d g(rcc_{ij}^{td})}, \qquad (10)$$

where P_{ij}^{td} stands for the probability of a trip from zone i at time slot t to zone j at time slot d, A_i^t are predicted trip arrivals of zone i at time t, and $g(rcc_{ij}^{td})$ is the travel cost function using the PPMC coefficient. Given the ground truth OD flow matrix $F = \{f_{ij}\}$, we sample $N = \sum_{i,j} f_{ij}$ trips following the predicted probability p_{ij}^{td} and additional constraints to generate the OD flow matrix $\tilde{F} = \{\tilde{f}_{ij}\}$. We consider four different types of constraints on the proposed joint PPMC-GM model:

Unconstrained model (UM). The only constraint on UM is that the total number of predicted trips $\tilde{N} = \sum_{i,j,t,d} \tilde{f}_{ij}^{td}$ is equal to the total number of trips N in the ground truth data. The N trips are randomly sampled from the multinomial distribution:

$$Multinomial(N, (P_{ij}^{td})). \qquad (11)$$

Singly-production-constrained model (PCM). PCM ensures preservation of the total number of predicted origin zone's trips $O_i = \sum_j f_{ij}$. For each origin zone i, the O_i trips are randomly sampled from the multinomial distribution:

$$Multinomial(O_i, \frac{\sum_{t,d} P_{ij}^{td}}{\sum_{j,t,d} P_{ij}^{td}}). \tag{12}$$

Singly-attraction-constrained model (ACM). ACM ensures preservation of the total number of predicted destination zone's trips $d_j = \sum_i f_{ij}$. For each destination zone j, the D_j trips are randomly sampled from the multinomial distribution:

$$Multinomial(D_j, \frac{\sum_{t,d} P_{ij}^{td}}{\sum_{i,t,d} P_{ij}^{td}}). \tag{13}$$

Doubly-constrained model (DCM). DCM ensures preservation of the number of both origin's and destination zone's trips. For each origin zone i and destination zone j, the N trips are randomly sampled from the multinomial distribution:

$$\tilde{f}_{ij}^{td} = B_i B_j P_{ij}^{td}; \sum_{j,t,d} \tilde{f}_{ij}^{td} = O_i; \sum_{i,t,d} \tilde{f}_{ij}^{td} = D_j; Multinomial(N, \frac{\sum_{t,d} \tilde{f}_{ij}^{td}}{\sum_{i,j,t,d} \tilde{f}_{ij}^{td}}), \tag{14}$$

where B are balancing factors from Iterative Proportional Fitting procedure [32].

5 Evaluation

5.1 Experiment Setting

Experimental Datasets. Manhattan Island of New York City was selected as the study area. We used the following datasets to evaluate our approach.

POI data. The POI dataset obatined from Foursquare covers 96,263 POIs of Manhattan. Each POI was recorded with location information (i.e., latitude and longitude) and category. The POI category is given in a comprehensive multiple-level classification provided by Foursquare [33]. We chose the first level, as the second level had more than 100 categories, making each category too trivial for our analysis. The selected 9 POI categories were: Nightlife Spot, Food, Shop & Service, College & University, Arts & Entertainment, Travel & Transport, Professional & Other Places, Outdoors & Recreation, and Residence.

Check-in data. We extracted check-in arrivals and their sequences from 1,168, 073 anonymous records collected in Manhanttan between August 1st, 2016 and March 31st, 2017 to extract check-in arrivals and check-in arrival sequences. Each check-in data contains spatial-temporal information demonstrating where and when it was generated.

Map segmentation. The study area can be partitioned into zones with different methods, e.g., grid-based or road network-based [18]. Based on the spatial resolution of the ground truth OD data, we selected census tracts as our zone partitions. As a result, we obtained 318 zones in the study area.

Travel demand observation data. The ground truth data includes weekday zonal OD matrices and time-of-day arrival of year 2017 provided by the New York Metropolitan Transportation Council.

We seperate the check-in data and ground truth data into two datasets, one for model training and the other for testing. The training set contains a random 50 out of the 318 zones in Manhattan. The learned parameters were then evaluated on the testing set. All 318 zones are included to ensure complete visualization and analysis of the full mobility pattern in the study area.

Baseline Methods. We compared our proposed approach with the following baseline methods.

Normalized Gravity Model with exponential distance decay (NGravExp). In this popular form of the gravity model [34], the probability of a trip between two zones is proportional to the outflow of the origin zone O_i and the inflow of the destination zone D_j, and is inversely proportional to the travel cost $cost_{ij}$, which is modeled with an exponential distance decay function:

$$P_{ij} = \frac{O_i D_j g(cost_{ij})}{\sum_j D_j g(cost_{ij})}; g(cost_{ij}) = exp(-\beta distance_{ij}). \tag{15}$$

Normalized Gravity Model with power distance decay (NGravPow). Unlike the NGravExp model, the NGravPow considers travel cost modeled with a power distance decay function:

$$P_{ij} = \frac{O_i D_j g(cost_{ij})}{\sum_j D_j g(cost_{ij})}; g(cost_{ij}) = (distance_{ij})^{-\beta}. \tag{16}$$

Schneider Intervening Opportunity Model (Schneider). In this model, the probability of a trip between two zones is proportional to the conditional probability that a traveler departure from zone i with outflow O_i is attracted to zone j, given that there are S_{ij} populations in between [12]:

$$P_{ij} = O_i \frac{exp(-\beta S_{ij}) - exp(-\beta(S_{ij} + O_i))}{\sum_j exp(-\beta S_{ij}) - exp(-\beta(S_{ij} + O_i))}. \tag{17}$$

Radiation Model (Rad). Simini *et al.* [35] reformulated the intervening opportunities model in terms of radiation and absorption processes:

$$P(1|O_i D_j, S_{ij}) = \frac{O_i D_j}{(O_i + S_{ij})(O_i + D_j + S_{ij})}. \tag{18}$$

Evaluation Measurement. We used Mean Absolute Error (MAE), Normalized Root Mean Square Error (NRMSE), and Coincidence Ratios (CR) [36] as metrics to evaluate the performance of zonal OD estimation:

$$MAE = \frac{\sum_{i,j} abs(f_{ij} - \tilde{f}_{ij})}{\sum_{i,j} 1}; NRMSE = \frac{\sum_{i,j} abs(f_{ij} - \tilde{f}_{ij})^2}{\sum_{i,j} f_{ij}} \tag{19}$$

$$CR = \frac{\sum_k min(\tilde{tl}_{distance_k}, tl_{distance_k})}{\sum_k min(\tilde{tl}_{distance_k}, tl_{distance_k})}, \tag{20}$$

where $tl_{distance_k}$ represents the percentage of trips in interval k of trip length distance, CR measures the common area of the trip length distribution for the predicted and ground truth OD matrices. The result takes the value in $[0, 1]$. When $CR = 0$, two distributions are completely different; while $CR = 1$, two distributions are identical.

5.2   Experimental Results

Zonal Functionality Profiling. To uncover the zonal functionality, we generated 5 latent zone topics based on the POI distribution under 9 POI categories shown in Table 1. We found that the POIs such as "Shop & Service", "Food", and "Art & Entertainment" had a relatively high rank within different topics compared to other POI categories. This reflects the fact that most trips reported by POI check-ins are discretionary trips such as social/recreational activities. Therefore, we considered the POI categories containing not only the highest but also the 2nd or 3rd probability to determine the zonal functionality. Five land use types guided by the New York City zoning and land use data [37] were selected as zone topic labels: "Commercial-Retail", "Commercial-Work", "Residence", "Transportation", and "Open Space". We mapped out in Fig. 2 the distribution of POIs and zone topics. We observed that POIs were most densely distributed in the Midtown and lower Manhattan area, while few were observed in the ring area of Central Park and Upper Manhattan. The zone topics discovered from the POI data indeed resemble the functional diversity of Manhattan's census tracts: commercial-work area for "Financial District", open space area "Central Park", and residential area in "Upper West Side".

Table 1. Zonal topic profiling.

Topic 1	Prob.	Topic 2	Prob.	Topic 3	Prob.	Topic 4	Prob.	Topic 5	Prob.
S	0.395	S	0.298	S	0.197	F	0.191	S	0.167
N	0.170	T	0.184	F	0.154	A	0.145	P	0.161
T	0.119	P	0.157	R	0.092	O	0.133	F	0.139
F	0.102	S	0.146	T	0.067	N	0.108	T	0.113
P	0.098	O	0.125	O	0.063	C	0.071	A	0.109
R	0.094	N	0.124	P	0.027	T	0.066	O	0.100
Commercial-retail		Transportation hub		Residence		Open space		Commercial-work	

S-Shop & Service; N-Nightlife Spot; T-Travel & Transport; F-Food; R-Residence; A-Art & Entertainment; P-Professional & Other Building; O-Outdoor & Recreation; C-College & University.

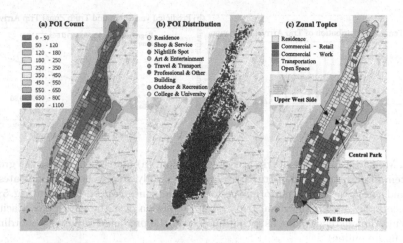

Fig. 2. Spatial distribution of the collected POIs and the generated zonal topics. (a) A heat map representing the POI categorical density. (b) We colored each POI based on the POI categories. (c) The zonal topics distribution generated by LDA. (Color figure online)

Zonal Trip Arrival Estimation. Regarding the mimicking of the trip arrival patterns in the ground truth data, the predicted trip arrivals from HPSS formulation were aggregated hourly to generate the trip arrival patterns in Fig. 3. The calibrated results show that the predicted trip arrival patterns from the proposed model match well with the ground truth time-of-day trip arrival under the aggregation of 318 total zones. There are two distinct peaks during the AM/PM periods and relatively few trips during the midday and nighttime. Meanwhile, an average distribution can be found during the lunch break. Furthermore, given the high variations of trips among 318 total zones, we plot the modeled result versus the ground truth data to visualize estimation accuracy. The regression line has a slope of 0.66, and the $R^2 = 0.80$ under the statistically significant level $p - value < 0.01$.

Zonal OD Estimation. We evaluated the generated daily OD flow patterns against four baseline models using three indicators: MAE, NRMSE, and CR. The MAE and NRMSE metrics indicate the zonal trip count differences between the ground truth and predicted OD matrices, while CR measures the similarity of trip length distribution curve between the ground truth and predicted OD flow patterns. The performance of five OD estimation models is presented under four different constraints as shown in Fig. 4. A total 20 model-constraint combinations were explored. Since the constraint models contain a sampling step from the multinomial distribution, we consider the average metrics over 100 runs of the OD estimation. We observed that the OD estimation model with a singly-constrained model (ACM/PCM) is better for estimating the zonal OD trip counts, while the doubly-constrained model (DCM) better predicts trip length

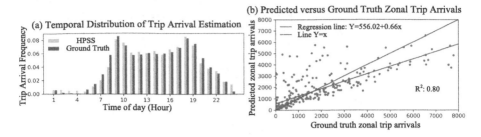

Fig. 3. Model Performance of trip arrival estimation. (a) blue bar indicates the ground truth temporal distribution of trip arrival over the study area; yellow bar indicates the predicted one. The model performance for specific zones can be found in Fig. 5; (b) Regression analysis for the zonal trip arrival prediction over the entire day. Each dot represents the reference value as x coordinate and the predicted value as y coordinate. (Color figure online)

distribution. Globally, the result obtained with the proposed PPMC-GM model achieves the lowest MAE/NRMSE value and the highest CR.

OD Flow Patterns Under Zone Topics. Since the travel demand observation data only has uniform sets of time-of-day arrival for each trip purpose, it cannot provide the time-of-day pattern for different land use type. The proposed model can report the time-of-day OD flow patterns under different zone topics. There are a total of 5 discovered zone topics shown in Fig. 5 including residential area (R), transportation hub area (TH), open space area (OS), commercial-retail area (CR), and commercial-workplace area (CW). Given the six selected zones representing five different zone topics, we evaluate different outflow patterns.

Residential area. Zone 81 is selected as one typical residential (R) area to explore its outflow pattern to other zones during the weekday. A morning peak period can be observed in the outflow of the commercial-workplace (CW) and the transportation hub (TH) area; this may reflect morning commuting activities. We notice that there is a time delay between the AM peak of the R-CW trips compared with that of the R-TH trips. This is consistent with the transit stop during commuters' trip from their homes to the office. Meanwhile, the outflows to CR area and OS area increase into the day starting from the late morning period. This is consistent with typical starting times of trips attempting to avoid rush hours. Finally, for R-CR trips, another two fluctuations can be observed in mid-afternoon and evening for late afternoon shoppers and dinnertime activities.

Commercial-Retail area. Zone 110 contains the major landmark "Time Square" in a commercial area. Trip patterns originating from this zone are analyzed across different destinations. For CR-R trips, the model did capture peaks indicating home-returning activities before and after dinnertime. Meanwhile, travelers leaving the nightlife spots and other recreational attractions within the targeted zone also generated significant late-night outflow trips to the residential area. For CR-TH trips, the afternoon peak to the transportation hub area

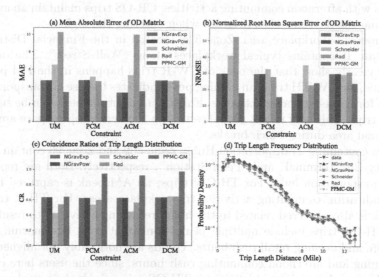

Fig. 4. Performance of the unconstrained model (UM), the production constrained model (PCM), the attraction constrained model (ACM) and the doubly constrained model (DCM) according to 4 baseline models and the proposed model. (a) Average MAE. (b) Average NRMSE. (c) Average CR. (d) Average trip length distribution curve. As the different performance indicator gives the different best combination of the OD estimation model and constraint model, we refer to the best constrained OD estimation model when mentioning the model in trip length distribution curve in (d).

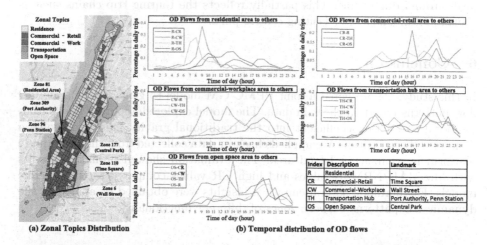

Fig. 5. Sample distribution of time-of-day OD flow patterns. (a) the selected zones and landmark involved to represent different zone topics. (b) the outflow from one zone under one zone topic to others under different zone topics.

coincides with afternoon commuting activities. CR-OS trips maintain an average rate during the daytime, then decrease when entering the night.

Commercial-Workplace area. Zone 5 is located in the Financial District of Manhattan. It contains typical workplaces along "Wall Street". The outflow patterns clearly show that the peak of CW-R trips happens in the PM period. Meanwhile, for CW-TH trips, an evening peak indicates the use of transportation facilities for home-returning activities. Finally, a lunchtime peak can be seen for CW-OS trips. This may reflect people resting in the nearby open space area and recreational area during lunch breaks.

Zones containing Transportation Hubs. Zone 309 and Zone 96 contain "Port Authority Bus Terminal" and "Penn Station", respectively. Both are representative transportation hubs. For TH-CW trips, an AM peak is captured by the model indicating commuting activities to workplace areas. For TH-R trips, a PM peak is also observed related to the home-returning activities to residential areas. TH-CR trips include multiple peaks consistent with late morning, late afternoon, and evening retail rush hours. Trends are noticeably not aligned with the morning and afternoon commuting rush hours, since the users here consist of mostly casual travelers and tourists. TH-OS trips reach their peaks during the late afternoon and evening periods consistent with touring and recreational activities at e.g. Central Park.

Open Space area. Zone 177 is fully occupied by "Central Park", and classified as open space. The outflow patterns indicate an early PM peak. This peak may be explained as the office-returning activities caused by the CW-OS trips during the lunch break. Meanwhile, the model captures the evening peaks for both OS-TH and OS-R trips reflecting home-returning activities. The OS-CR trips exhibit high flow during most of the morning and early afternoon and another peak around dinnertime. This partially reflects the touring trip chains such as visiting retail shops after visiting central park.

6   Conclusion

In this study, we presented a joint PPMC-GM model to generate dynamic OD flow patterns based on check-in data. The model explored adopting the spatial-temporal correlation coefficient into the traditional gravity model to evolve a dynamic OD estimation. We applied check-in data collected from the Foursquare platform in Manhattan Island area. The evaluation showed promising results with low MAE/NRMSE values and high CR values compared to the baseline models. Furthermore, several empirical insights were obtained by analyzing the dynamic OD patterns between zones of different functionality.

References

1. Liu, B., Xiong, H.: Point-of-interest recommendation in location based social networks with topic and location awareness. In: SDM, pp. 396–404. SIAM (2013)
2. Cheng, Z., Caverlee, J., Lee, K., Sui, D.Z.: Exploring millions of footprints in location sharing services. In: ICWSM, pp. 81–88 (2011)
3. Hu, W., Jin, P.J.: An adaptive hawkes process formulation for estimating zonal trip arrivals with LBSN data. TRP-C: Emerg. Technol. 79, 136–155 (2017)
4. Chong, W.-H., Dai, B.-T., Lim, E.-P.: Prediction of venues in foursquare using flipped topic models. In: Hanbury, A., Kazai, G., Rauber, A., Fuhr, N. (eds.) ECIR 2015. LNCS, vol. 9022, pp. 623–634. Springer, Cham (2015). https://doi.org/10.1007/978-3-319-16354-3_69
5. de Grange, L., Fernández, E., de Cea, J.: A consolidated model of trip distribution. TRP Part E 46(1), 61–75 (2010)
6. Jin, P., Cebelak, M., Yang, F., Zhang, J., Walton, C., Ran, B.: Exploration into use of doubly constrained gravity model for OD estimation. TRR 2430(1), 72–82 (2014)
7. Elldér, E.: Residential location and daily travel distances: the influence of trip purpose. J. Transp. Geogr. 34, 121–130 (2014)
8. Carey, H.C.: Principles of Social Science, vol. 3. JB Lippincott & Company, Philadelphia (1867)
9. Medina, A., Taft, N., Salamatian, K., Bhattacharyya, S., Diot, C.: Traffic matrix estimation: existing techniques and new directions. In: SIGCOMM, vol. 32, no. 4, pp. 161–174 (2002)
10. Zhang, J.D., Chow, C.Y.: Spatiotemporal sequential influence modeling for location recommendations: a gravity-based approach. ACM TIST 7(1), 11 (2015)
11. Stouffer, S.A.: Intervening opportunities: a theory relating mobility and distance. Am. Sociol. Rev. 5(6), 845–867 (1940)
12. Schneider, M.: Gravity models and trip distribution theory. Reg. Sci. 5(1), 51–56 (1959)
13. McArdle, G., Lawlor, A., Furey, E., Pozdnoukhov, A.: City-scale traffic simulation from digital footprints. In: SIGKDD UrbComp, pp. 47–54. ACM (2012)
14. Tarasov, A., Kling, F., Pozdnoukhov, A.: Prediction of user location using the radiation model and social check-ins. In: SIGKDD UrbComp, p. 8. ACM (2013)
15. Lee, J.H., Gao, S., Goulias, K.G.: Can Twitter data be used to validate travel demand models. In: IATBR (2015)
16. Bierlaire, M., Crittin, F.: An efficient algorithm for real-time estimation and prediction of dynamic OD tables. Oper. Res. 52(1), 116–127 (2004)
17. Calabrese, F., Colonna, M., Lovisolo, P., Parata, D., Ratti, C.: Real-time urban monitoring using cell phones: a case study in Rome. IEEE Trans. Intell. Transp. Syst. 12(1), 141–151 (2011)
18. Liu, B., Fu, Y., Yao, Z., Xiong, H.: Learning geographical preferences for point-of-interest recommendation. In: SIGKDD, pp. 1043–1051. ACM (2013)
19. Yuan, J., Zheng, Y., Xie, X.: Discovering regions of different functions in a city using human mobility and POIs. In: SIGKDD, pp. 186–194. ACM (2012)
20. Shi, Y., Serdyukov, P., Hanjalic, A., Larson, M.: Nontrivial landmark recommendation using geotagged photos. ACM TIST 4(3), 47 (2013)
21. Bao, J., Zheng, Y., Mokbel, M.F.: Location-based and preference-aware recommendation using geosocial network data. In: SIGSPATIAL, pp. 199–208. ACM (2012)

22. Ference, G., Ye, M., Lee, W.C.: Location recommendation for out-of-town users in location-based social networks. In: CIKM, pp. 721–726, ACM (2013)
23. Wang, H., Terrovitis, M., Mamoulis, N.: Location recommendation in location-based social networks using user check-in data. In: SIGSPATIAL, pp. 374–383. ACM (2013)
24. Ying, J.J.C., Kuo, W.N., Tseng, V.S., Lu, E.H.C.: Mining user check-in behavior with a random walk for urban POI recommendations. ACM TIST 5(3), 40 (2014)
25. Gao, H., Tang, J., Hu, X., Liu, H.: Exploring temporal effects for location recommendation on location-based social networks. In: RecSys, pp. 93–100. ACM (2013)
26. Yuan, Q., Cong, G., Ma, Z., Sun, A., Thalmann, N.M.: Time-aware point-of-interest recommendation. In: SIGIR, pp. 363–372. ACM (2013)
27. Yuan, Q., Cong, G., Sun, A.: Graph-based point-of-interest recommendation with geographical and temporal influences. In: CIKM, pp. 659–668. ACM (2014)
28. Cho, E., Myers, S.A., Leskovec, J.: Friendship and mobility: user movement in location-based social networks. In: SIGKDD, pp. 1082–1090. ACM (2011)
29. Wang, Y., et al.: Location prediction using heterogeneous mobility data. In: SIGKDD, pp. 1275–1284. ACM (2015)
30. Lian, D., Xie, X., Zheng, V.W., Yuan, N.J., Zhang, F., Chen, E.: A collaborative exploration and returning model for location prediction. ACM TIST 6(1), 8 (2015)
31. Li, X., Lian, D., Xie, X., Sun, G.: Lifting the predictability of human mobility on activity trajectories. In: ICDMW, pp. 1063–1069. IEEE (2015)
32. Deming, W.E., Stephan, F.F.: On a least squares adjustment of a sampled frequency table when the expected marginal totals are known. Ann. Math. Stat. 11(4), 427–444 (1940)
33. Foursquare. https://developer.foursquare.com/docs/resources/categories
34. Barthélemy, M.: Spatial networks. Phys. Rep. 499(1–3), 1–101 (2011)
35. Simini, F., González, M.C., Maritan, A., Barabási, A.L.: A universal model for mobility and migration patterns. Nature 484(7392), 96 (2012)
36. Martin, W.A., McGuckin, N.A.: Travel Estimation Techniques for Urban Planning, vol. 365. National Academy Press, Washington (1998)
37. Zola. http://maps.nyc.gov/doitt/nycitymap/template?applicationName=ZOLA

Social-Affiliation Networks: Patterns and the SOAR Model

Dhivya Eswaran[1]([⊠]), Reihaneh Rabbany[2], Artur W. Dubrawski[1], and Christos Faloutsos[1]

[1] School of Computer Science, Carnegie Mellon University, Pittsburgh, USA
{deswaran,awd,christos}@cs.cmu.edu
[2] School of Computer Science, McGill University, Montreal, Canada
reihaneh.rabbany@mcgill.ca

Abstract. Given a social-affiliation network – a friendship graph where users have many, binary attributes e.g., check-ins, page likes or group memberships – what *rules* do its structural properties such as edge or triangle counts follow, *in relation to its attributes*? More challengingly, how can we synthetically generate networks which *provably* satisfy those rules or patterns? Our work attempts to answer these closely-related questions in the context of the increasingly prevalent social-affiliation graphs. Our contributions are two-fold: (a) **Patterns:** we discover three new rules (power laws) in the properties of *attribute-induced subgraphs*, substructures which connect the friendship structure to affiliations; (b) **Model:** we propose SOAR– short for SOcial-Affiliation graphs via Recursion– a stochastic model based on recursion and self-similarity, to *provably* generate graphs obeying the observed patterns. Experiments show that: (i) the discovered rules are useful in detecting deviations as anomalies and (ii) SOAR is fast and scales linearly with network size, producing graphs with millions of edges and attributes in only a few seconds. Code related to this paper is available at: www.github.com/dhivyaeswaran/soar.

Keywords: Graph mining · Attributes · Patterns · Anomalies · Generator

1 Introduction

With the proliferation of the web and online social networks, social-affiliation networks – social/friendship networks where users have many, *binary* attributes or affiliations – have become increasingly common. Examples include social networking sites such as Facebook and Google+ which record user engagement, e.g., pages liked (attributes are pages – yes if liked, no if not), media-sharing social platforms such as Flickr and Youtube where users can form groups based on their interests (attributes are groups – yes if member, no if not), location-based social networks like GOWALLA where users can check-in at a location they physically visit (attributes are locations – yes if visited).

R. Rabbany—Work performed while at Carnegie Mellon University.

© Springer Nature Switzerland AG 2019
M. Berlingerio et al. (Eds.): ECML PKDD 2018, LNAI 11052, pp. 105–121, 2019.
https://doi.org/10.1007/978-3-030-10928-8_7

We consider two closely-related research questions concerning these networks: **[RQ1]** What *rules* (patterns) do the various structural properties of social-affiliation graphs – e.g., edge or triangle count – follow, *in relation to its attributes*? **[RQ2]** How can we synthetically generate realistic networks which *provably* satisfy these patterns? These questions fall under the umbrella of pattern analysis and modeling, a well-explored research area and a standard practice in understanding real-world graphs [6,16,17,19]. Our interest in considering these research questions stems in part from the scientific and practical impact that the works on pattern analysis and modeling have had in the past. The discoveries of the scale-free property (skewed degree distributions [10]) and the small world property (small graph diameters [28]) and respectively their preferential-attachment [4] and forest-fire [19] models, for instance, have had numerous applications in graph algorithm design, anomaly detection, graph sampling and more [3,18].

While works on patterns and models for non-attributed graphs abound in the literature, studies dealing with social-affiliation networks are somewhat limited [14,29] (see Sect. 2). Our work complements these by discovering rules which the structural properties of social-affiliation graphs follow *in relation to their attributes*. Specifically, we study *"attribute-induced subgraphs"* (AIS, in short) – each of which is a subgraph induced by the nodes affiliated to a given attribute – substructures which connect the structure of friendship graph to the distribution of attribute values. See Sect. 3 for more details and Fig. 1 for an example. Studying the patterns exhibited by the structural properties of AIS allows us to understand homophily effects ('birds of the same feather flock together') and consider questions of form 'If the number of users affiliated to attribute a doubles, what happens to the number of friendships between them?' As we show later, the patterns discovered based on AIS and the associated capability to answer 'what-if' questions are subsequently useful in (i) detecting anomalies and (ii) developing and testing a realistic model for social-affiliation graphs.

Our contributions are two-fold: **(a) Patterns:** We study four large real-world social-affiliation graphs and discover three new consistent patterns concerning the structural properties of attribute-induced subgraphs. With the help of a case study, we illustrate how the findings can be leveraged for anomaly detection. **(b) Model:** We propose the SOAR model to produce synthetic social-affiliation graphs *provably matching all observed patterns*. SOAR is based on self-similarity, implicitly incorporates attribute correlations, scales linearly with graph size and is up to $50\times$ *faster* than the prior models for social-affiliation graphs.

Reproducibility. We use publicly-available datasets and open-source our code at www.github.com/dhivyaeswaran/soar.

2 Related Work

We group related work into three categories: models for social networks with no attributes [A] and those for social-affiliation graphs when attributes are given [B] and not given [C].

Table 1. Comparison with other models for social-affiliation graphs

Properties	MAG [15]	AGM [24]	Zhel [29]	SAN [14]	SOAR
Generates edges and attributes *simultaneously*			✔	✔	✔
Scalable with increasing number of edges and attributes			✔	✔	✔
Provably obeys all observed patterns					✔

[A] Social graphs with no attributes. Several outstanding network models have been proposed to explain the observed structural characteristics of real-world non-attributed networks. Notably, the Barabási-Albert model for heavy-tail degree distributions [4], Forest Fire model for shrinking diameter [19], Butterfly model for the evolution of giant connected component [20], Kronecker model for community structure [18] and Random Typing Graph Model for self-similar temporal evolution [2]. Excellent surveys are given in [6,13,22]. As such, it is not clear how these models could be extended to produce attributes, given the complex interplay between attributes and friendship structure [9,11,25].

[B] Social-affiliation graphs when attributes are given. The problem of modeling network structure in the presence of *known* nodal attributes has been studied. Notably, Multiplicative Attribute Graph (MAG) model [15] connects nodes according to user-specified attribute-based link affinities. Attributed Graph Model (AGM) [24] presents a generic approach using an accept-reject sampling framework to augment a given non-attributed network model with correlated attributes. Both MAG and AGM apply to settings with categorical (not just binary) nodal attributes; however, they scale poorly with the number of attributes: each edge is sampled proportional to roughly the dot product of nodal attribute vectors, which is an expensive operation, considering that the social-affiliation graph datasets we study have around $30K$ to $1.28M$ affiliations.

[C] Social-affiliation graphs when attributes are not given. The *simultaneous* generation of attributes and friendships, in the context of social-affiliation graphs (i.e., with many binary attributes), has received some attention. The pioneering work by [29] discovers several patterns in social-affiliation graphs (e.g., power law relation between number of friends and average count of affiliations). It proposes ZHEL model by adapting the non-attributed microscopic graph evolutionary model [17] for this setting. [14] studies the evolution of directed social network of Google+ and its affiliations, focusing on the density, diameter, degrees and clustering coefficients of users and affiliations. It proposes SAN model augmenting [17] with attribute-augmented preferential attachment and triangle-closing mechanisms to replicate the observations on Google+. The patterns we discover in this paper are complementary to the above discoveries. Further, both ZHEL and SAN model the *evolution* of social-affiliation graphs, by generating attributes and edges of one node at a time, while in contrast, we investigate a *one-shot* approach to graph generation (i.e., without modeling its evolution) which leads to *input parsimony and* ∼50× *speed-up* (see Sect. 5).

A qualitative comparison of social-affiliation graph models is given in Table 1.

Table 2. Frequently used symbols and their meanings

Symbol	Term	Description
$\mathcal{G}$	Social-affiliation graph	Undirected unweighted graph with many binary nodal attributes
n		Number of nodes in $\mathcal{G}$
k		Number of attributes in $\mathcal{G}$
$\mathbf{A}$	Adjacency matrix	$n \times n$ binary matrix showing edge existence
$\mathbf{F}$	Membership matrix	$n \times k$ binary matrix showing attribute possession
$\mathcal{G}_a$	Attribute-induced subgraph	Subgraph induced by nodes affiliated to attribute a
n_a	Node count	Number of nodes in $\mathcal{G}_a$
m_a	Edge count	Number of edges in $\mathcal{G}_a$
Δ_a	Triangle count	Number of triangles in $\mathcal{G}_a$
σ_a	Spectral norm	Highest singular value of the adjacency matrix of $\mathcal{G}_a$

3 Preliminaries

Notation. Let $\mathcal{G} = (\mathcal{V}, \mathcal{E}, \mathcal{A}, \mathcal{M})$ be a social-affiliation graph, where $\mathcal{V}$ is the set of nodes (users), $\mathcal{A}$ is the set of binary attributes (affiliations[1]), $\mathcal{E}$ is the set of unweighted undirected *who-is-friends-with-whom* edges among nodes and $\mathcal{M}$ is the set of *who-is-affiliated-to-what* attribute memberships between nodes and attributes. That is, if node u is connected to node u', then, $\mathcal{E}$ includes edges (u, u') and (u', u); similarly, $(u, a) \in \mathcal{M}$ iff node u is affiliated with attribute a. $\mathcal{G}$ is equivalently expressed as a tuple $(\mathbf{A}, \mathbf{F})$ of the $n \times n$ symmetric adjacency matrix $\mathbf{A}$ and the $n \times k$ membership matrix $\mathbf{F}$, where

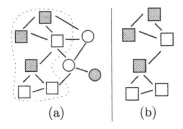

(a)   (b)

Fig. 1. (a) A social-affiliation graph with *isSquare*, *isStriped* attributes and (b) the subgraph induced by *isSquare* attribute

$n = |\mathcal{V}|$ and $k = |\mathcal{A}|$ denote the number of nodes and attributes respectively. The matrices are binary with 1 indicating the presence of an edge (in $\mathbf{A}$) or an attribute membership (in $\mathbf{F}$). Table 2 gives the frequently used notation.

Attribute-Induced Subgraph (AIS). Given a social-affiliation graph $\mathcal{G} = (\mathcal{V}, \mathcal{E}, \mathcal{A}, \mathcal{M})$, the attribute-induced subgraph $\mathcal{G}_a$ corresponding to a given attribute $a \in \mathcal{A}$ is obtained by selecting the nodes affiliated to attribute a and the edges which link two such nodes. Formally, $\mathcal{G}_a = (\mathcal{V}_a, \mathcal{E}_a)$ where $\mathcal{V}_a = \{u \in \mathcal{V} \mid (u, a) \in \mathcal{M}\}$ and $\mathcal{E}_a = \{(u, u') \in \mathcal{E} \mid u, u' \in \mathcal{V}_a\}$. Let $n_a = |\mathcal{V}_a|$ and $m_a = |\mathcal{E}_a|$ denote its number of nodes and edges respectively. *Triangle count* Δ_a is the number of triangles in $\mathcal{G}_a$ while *spectral radius* σ_a is the largest eigenvalue of its adjacency matrix. An example of an AIS is given in Fig. 1.

---

[1] We use the following pairs of terms interchangeably throughout the paper: (graph, network), (node, user), (attribute, affiliation).

Datasets. We study four large publicly-available datasets, each of which contains a social network formed by friendship (or family) relations and also side-information regarding affiliations of users. Based on the nature of affiliations, we describe the datasets in two categories: *(i) Online-affiliation networks:* In FLICKR [21] and YOUTUBE [23], online photo-sharing and video-sharing websites respectively, users are allowed to form groups based on their common interests. We consider each group as a binary attribute, i.e., a user u has a group g if she participates in it. The friendship networks in these datasets are directed, but still, they have a high link symmetry or edge reciprocity [21]. Hence, for simplicity, we drop the direction of edges and retain a single copy of each resulting edge to get an undirected graph without multi-edges. *(ii) Offline-affiliation networks:* BRIGHTKITE and GOWALLA datasets [8] contain undirected friendship network along with user check-in information, i.e., who visited where and when. We use each location as a binary attribute; a user u has a location attribute l if she has visited l at least once. For a detailed description of these datasets, we refer readers to the papers cited above. Some useful statistics are provided in Table 3. The next section details our pattern discoveries on these datasets.

Table 3. Social-affiliation graph datasets studied

Dataset	Reference	$|\mathcal{V}|$	$|\mathcal{E}|$	$|\mathcal{A}|$	$|\mathcal{M}|$
YOUTUBE	[23]	$77K$	$0.4M$	$30K$	$0.3M$
FLICKR	[21]	$1.8M$	$16M$	$0.1M$	$8.5M$
BRIGHTKITE	[8]	$58K$	$0.2M$	$0.8M$	$1M$
GOWALLA	[8]	$0.2M$	$1M$	$1.28M$	$4M$

4   Pattern Discoveries

Given an attribute-induced subgraph $\mathcal{G}_a = (\mathcal{V}_a, \mathcal{E}_a)$, there is an infinite set of graph properties that one could investigate to look for patterns (number of nodes/edges, degree distributions, one or more eigenvalues, core number, etc.). Which ones should we focus on? Intuitively, we want to study properties that are (i) *fundamental,* easy to understand and interpret, (ii) *fast to compute,* exactly or approximately, in near-linear time in the number of edges and (iii) *lead to prevalent patterns* that AISs obey consistently across different datasets. After extensive experiments, we shortlist the following four properties of attribute-induced subgraphs: (i) $n_a = |\mathcal{V}_a|$: number of nodes in $\mathcal{G}_a$, i.e., number of users affiliated with attribute a. (ii) $m_a = |\mathcal{E}_a|$: number of edges in $\mathcal{G}_a$, i.e., number of friendships among users affiliated with attribute a. (iii) Δ_a: number of triangles in $\mathcal{G}_a$, typically indicative of the extent to which nodes in $\mathcal{G}_a$ tend to cluster together (e.g., via clustering coefficient). (iv) σ_a: spectral radius, or the principal eigenvalue of adjacency matrix of $\mathcal{G}_a$, roughly indicative of how large and how dense the giant connected component in $\mathcal{G}_a$ is. We list our observations regarding these properties in Sect. 4.1 and postpone explanations to Sect. 4.2.

4.1   Observations

Following standard terminology, we say that variables x and y obey a power law with exponent c, if $y \propto x^c$ [1]. Our pattern discoveries are all power laws with non-negative (and usually non-integer) exponents, as stated below.

Observation 1 ([P1] Edge count vs. node count). *Edge count m_a and node count n_a of AISs obey a power law: $m_a \propto n_a^\alpha$, $0 \le \alpha \le 2$.*

In the datasets we studied, $\alpha \in [1.17, 1.51]$. That is, double the nodes in an AIS, over double (roughly, triple) its edges.

Observation 2 ([P2] Triangle count vs. node count). *Triangle count Δ_a and node count n_a of AISs obey a power law: $\Delta_a \propto n_a^\beta$, $0 \le \beta \le 3$.*

In the datasets we studied, $\beta \in [1.24, 1.96]$. That is, as the number of nodes in an AIS doubles, its triangle count becomes about 3–4 times larger.

Observation 3 ([P3] Spectral radius vs. triangle count). *Spectral radius σ_a and triangle count Δ_a of AISs obey a power law: $\sigma_a \propto \Delta_a^\gamma$, $\gamma \ge 0$.*

In the datasets we studied, $\gamma \in [0.31, 0.33]$. That is, doubling the spectral radius of an AIS leads to an eight-fold increase in its number of triangles.

Figure 2, which plots the relevant quantities (m_a vs. n_a, Δ_a vs. n_a and σ_a vs. Δ_a), illustrates these observations. The cloud of gray points in these figures show values corresponding to various AISs and darker areas signify regions of higher density. The relevant exponents α, β, γ are computed following standard practice (e.g., as in [16]). We bucketize x-axis logarithmically and compute per-bucket y averages (black triangles). The slope of the black line, which is the least-squares fit to the black triangles, gives the exponent. In addition, we report the *Pearson correlation coefficient* ρ of the per-bucket averages as a proxy for the goodness-of-fit of the power law relation. This value lies in $[0, 1]$ and intuitively, the higher the value is, the better is the fit. In our experiments, ρ was consistently above 0.95, suggesting a near-perfect fit.

4.2   Explanations, Use in Anomaly Detection, and Discussion

Here, we attempt to explain our observations in terms of known/expected properties of social-affiliation networks and hypothesize the nature of anomalies deviation from each pattern above would give rise to.

[P1] Edge count vs. node count. As the number of nodes in an AIS doubles, the number of edges remains the same ($\alpha = 0$) for empty social-affiliation graphs having no edges and quadruples ($\alpha = 2$) for complete graphs. As real-world social-affiliation networks tend to be sparse ($|\mathcal{E}| = \mathcal{O}(|\mathcal{V}|)$), one might expect the exponent α to be roughly 1. However, in experiments, α was *much higher*, e.g., ~1.5 for FLICKR dataset. This suggests *homophily*, i.e., more friend-ships among people sharing the same attributes, which causes the number of edges to more than double (in fact, triple) when the number of nodes is doubled.

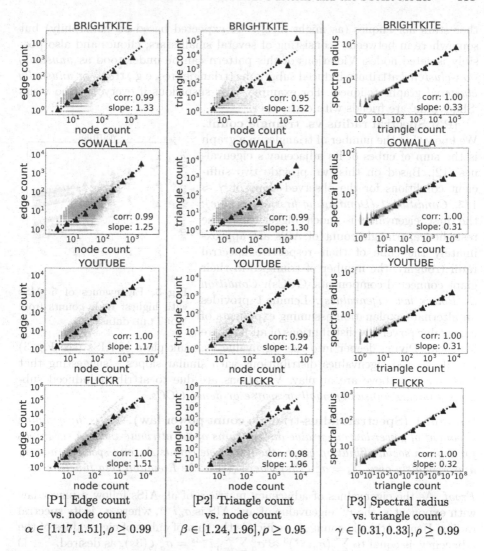

Fig. 2. Patterns exhibited by attribute-induced subgraphs (each point is an AIS)

Attribute-induced graphs violating this pattern can be understood as *unusually sparse or dense* having too few/many friendships between users sharing an attribute, e.g., when no two people who go to Starbucks are friends with each other.

[P2] Triangle count vs. node count. As the number of nodes in an AIS doubles, triangle count remains the same ($\beta = 0$) for empty or tree/star-like graphs with no triangles and becomes eight times ($\beta = 3$) for fully connected graphs. In experiments, β was been 1 and 2; that is, the triangle count becomes 2–4 times when the node count doubles. This suggests that the AISs are neither stars nor cliques (as might ideally be expected based on homophily) but somewhere in between – consisting of several small stars, cliques and also possibly isolated nodes. Violations of this pattern can be understood as *unusually non-clustered* attribute-induced subgraphs (triangle-free, e.g., trees) or *unusually clustered* graphs (cliques). For example, it is suspicious if everyone who visits 'ShadySide' are friends with each other.

[P3] Spectral radius vs. triangle count.
We know that the number of triangles in a graph is the sum of cubes of its adjacency's eigenvalues [12]. Based on this, we provide two sufficient conditions for the observed slope of $\gamma \approx 1/3$. *Condition 1 (Dominating first eigenvalue):* the first eigenvalue is much bigger than the rest; hence, triangle count of AISs are approximately the cube of their respective spectral radii (roughly, the number of triangles in their giant connected components, GCCs). *Condition 2 (Power law eigenvalues):* Lemma 1 provides an alternate explanation assuming exponents of eigenvalue power law distributions of all AISs are

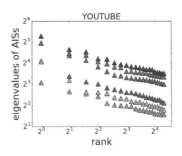

Fig. 3. Eigenvalues of 5 AISs with highest node counts from YOUTUBE dataset

identical. Diving deeper into the eigenvalue vs. rank plots of AISs (see Fig. 3) reveals skewed eigenvalues distributions with similar slopes – suggesting that both reasons above are at play. Violations are due to attribute-induced subgraphs having *unusually small or sparse or dense GCCs.*

Lemma 1 (Spectral radius-triangle count power law). *If s is the common exponent of power law eigenvalue distributions of the attribute-induced subgraphs for a given social-affiliation graph, their triangle counts Δ_a and spectral radii σ_a approximately obey $\Delta_a = \sigma_a^3 \, \zeta(3s)$ where $\zeta(\cdot)$ is the Riemann zeta function [27].*

Proof. As the eigenvalues of adjacency matrices of all AISs follow a power law with exponent s, the ith eigenvalue of any AIS is $\sigma_a i^{-s}$, where σ_a is its spectral radius. Hence, triangle count Δ_a, which is the sum of cubes of eigenvalues of the adjacency, is equal to $\sum_i (\sigma_a i^{-s})^3 \approx \sigma_a^3 \sum_{i=0}^{\infty} i^{-3s} = \sigma_a^3 \, \zeta(3s)$, as desired.   □

Anomaly Detection. Our pattern discoveries represent normal behavior of attributes in a social-affiliation graph, deviations from which can be flagged as anomalies. For example, the spectral radius vs. triangle count plot for YOUTUBE yields a dense cloud of points mostly distributed along a straight line in log-log scales (Fig. 4a); the red triangle marks an exception due to an anomalous attribute. It turns out that, as expected, the deviation was due to its *unusually sparse GCC*, which consisted of a giant star plus a few triangles (see Fig. 4b for its GEPHI visualization [5]). In contrast, a typical AIS with a comparable triangle count (green triangle in Fig. 4a) has a denser GCC (Fig. 4c).

Discussion. It is natural to suppose that the data scraping methodology (sampling size/strategy) would have a considerable impact on the pattern discoveries.

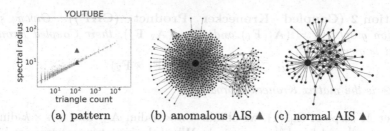

(a) pattern   (b) anomalous AIS ▲   (c) normal AIS ▲

Fig. 4. Anomaly detection using pattern [P3] reveals an attribute-induced subgraph (AIS) with an unusually sparse giant connected component (GCC) (Color figure online)

However, the consistency of our observations across datasets sampled in various ways – multiple sizes (GOWALLA and BRIGHTKITE – almost whole public data; FLICKR, YOUTUBE – large fraction of the giant weakly connected component [8,21]) and strategies (no sampling, snowball sampling using forward and/or reverse links depending on the public API) – suggest that the patterns are indeed generalize across many reasonable data scraping mechanisms. Also, note that our study is limited to the case of binary attributes; similar explorations of categorical and real-valued attributes are possible but left to future work.

5 SOAR Model

In this section, we show how to generate graphs which *provably* obey the discovered patterns using a *coupled* version of the matrix Kronecker product [26]. The resulting model, called SOAR– short for SOcial-Affiliation graphs via Recursion– has two steps: (i) an *initiator graph* $\mathcal{G}_1$, consisting of carefully coupled *initiator matrices* $\mathbf{A}_1$ for adjacency and $\mathbf{F}_1$ for membership, is chosen; (ii) the initiator graph is *recursively* multiplied with itself via *Coupled Kronecker Product* (Definition 2) for a desired number of steps to obtain the final social-affiliation graph. Sect. 5.1 presents SOAR model in detail. Our important contribution here is the proof that Coupled Kronecker Product is a *pattern-preserving* operation, i.e., if the initiator graph obeys patterns P1–P3, so does the final graph (see Sect. 5.2).

5.1 Proposed SOAR Model

Recall from Sect. 3 that $\mathcal{G}$ is a tuple $(\mathbf{A}, \mathbf{F})$ of the $n \times n$ symmetric adjacency matrix $\mathbf{A}$ and the $n \times k$ membership matrix $\mathbf{F}$, where $n = |\mathcal{V}|$ and $k = |\mathcal{A}|$ denote the number of nodes and attributes respectively. Given an initiator social-affiliation graph $\mathcal{G}_1 = (\mathbf{A}_1, \mathbf{F}_1)$, where $\mathbf{A}_1$ is the $n_1 \times n_1$ symmetric *initiator matrix* for adjacency and $\mathbf{F}_1$ is the $n_1 \times k_1$ initiator matrix for membership, we propose to derive the final social-affiliation graph $\mathcal{G} = (\mathbf{A}, \mathbf{F})$ via the recursive equation:

$$\mathcal{G}_{t+1} = \mathcal{G}_t \bar{\otimes} \mathcal{G}_1 \tag{1}$$

where $\bar{\otimes}$ is the Coupled Kronecker Product, as defined below:

Definition 2 (Coupled Kronecker Product (CKP)). *Given social-affiliation graphs* $\mathcal{G}_1 = (\mathbf{A}_1, \mathbf{F}_1)$ *and* $\mathcal{G}_2 = (\mathbf{A}_2, \mathbf{F}_2)$, *their Coupled Kronecker Product is given by*

$$\mathcal{G}_1 \bar{\otimes} \mathcal{G}_2 = (\mathbf{A}_1 \otimes \mathbf{A}_2, \mathbf{F}_1 \otimes \mathbf{F}_2) \tag{2}$$

where $\otimes$ *is the matrix Kronecker product.*

After M steps of Eq. (1), we obtain a $n \times n$-dim $\mathbf{A}_M$ and a $n \times k$-dim $\mathbf{F}_M$ where $n = n_1^M$ and $k = k_1^M$ respectively. When the initiator matrices are binary, so are the final matrices and thus can be directly used as the adjacency $\mathbf{A}$ and membership $\mathbf{F}$ matrices, respectively. It turns out that the above process captures the required power laws but has several discrete jumps (fluctuations). Hence, we use the stochastic version below.

The main idea is to produce at every recursive step, matrices of edge/membership occurrence probabilities instead of discrete (binary) edges/memberships. Thus, we begin with initiator matrices having real number entries in $[0, 1]$ (they do not need to sum to 1) and add a small relative noise η to the initiator matrices independently at every recursive step t. This process results in the final dense probability matrices $\mathbf{A}_M$ and $\mathbf{F}_M$, from which we recover $\mathbf{A}$ and $\mathbf{F}$ by sampling each entry proportional to its final value. A scalable implementation of the above approach by sampling one edge or membership at a time is given in Algorithm 1. The Hadamard product $\odot$ in lines 6 and 8 performs an element-wise matrix multiplication to add the desired noise to the initiators.

Running Time Analysis. Initialization (*ln 1–11*) contributes a fixed overhead of $\mathcal{O}(M(n_1^2 + n_1 k_1))$. The generation of edges (*ln 12–20*) and memberships (*ln 21–29*) take $\mathcal{O}(n_1^2 M)$ per edge and $\mathcal{O}(n_1 k_1 M)$ per membership respectively. As n_1, k_1 and M are small in practice (<10), Algorithm 1 is *linear in the number of edges and attribute memberships.*

5.2   Theoretical Properties

The structural properties of graphs generated using Kronecker product are well-studied and a number of desirable properties have been proved, e.g., multinomial distribution of degrees and singular values, etc. [18]. These properties directly carry over to the proposed model. More surprisingly, for careful coupling of initiators, SOAR graphs provably obey all the discovered power laws from Sect. 4. This is due to the *pattern-preserving* property of the Coupled Kronecker Product operation. That is, if graphs $\mathcal{G}_1$ and $\mathcal{G}_2$ obey the patterns P1–P3 with the same exponent, then, so does their Coupled Kronecker Product $\mathcal{G}_1 \bar{\otimes} \mathcal{G}_2$. This is stated in Lemmas 3–5 (proofs in appendix).

Lemma 3 (CKP preserves [P1]). *If* $\mathcal{G}_1$ *and* $\mathcal{G}_2$ *obey the edge count vs. node count power law with exponent* α, *i.e.,* $m_a \propto n_a^\alpha$, *so does* $\mathcal{G}_1 \bar{\otimes} \mathcal{G}_2$.

Lemma 4 (CKP preserves [P2]). *If* $\mathcal{G}_1$ *and* $\mathcal{G}_2$ *obey the triangle count vs. node count power law with exponent* β, *i.e.,* $\Delta_a \propto n_a^\beta$, *so does* $\mathcal{G}_1 \bar{\otimes} \mathcal{G}_2$.

Algorithm 1. SOAR model

input   : $\mathbf{A}_1 \in [0,1]^{n_1 \times n_1}$, $\mathbf{F}_1 \in [0,1]^{n_1 \times k_1}$, $M \in \mathbb{N}$, $\eta \in [0,1]$
output: $(\mathbf{A}, \mathbf{F}) = \mathrm{SOAR}(\mathbf{A}_1, \mathbf{F}_1, M, \eta)$

1 num_edges $\leftarrow \lfloor$(sum of entries in $\mathbf{A}_1)^M \rfloor$
2 num_memberships $\leftarrow \lfloor$(sum of entries in $\mathbf{F}_1)^M \rfloor$
   /* create M noisy copies of initiators $(\mathbf{A}_1, \mathbf{F}_1), \ldots, (\mathbf{A}_M, \mathbf{F}_M)$ */
3 $\mathbf{A}_0, \mathbf{F}_0 \leftarrow \mathbf{A}_1, \mathbf{F}_1$
4 for $t = 1, 2, \ldots, M$ do
5 | Sample $\mathbf{N}_{\mathbf{A},t} \sim [-0.5, 0.5]^{n_1 \times n_1}$ // i.i.d, uniform
6 | $\mathbf{A}_t \leftarrow \mathbf{A}_0 + \eta \mathbf{A}_0 \odot \mathbf{N}_{\mathbf{A},t}$ // $\mathbf{A}_t \in [0,1]^{n_1 \times n_1}$
7 | Sample $\mathbf{N}_{\mathbf{F},t} \sim [-0.5, 0.5]^{n_1 \times k_1}$ // i.i.d, uniform
8 | $\mathbf{F}_t \leftarrow \mathbf{F}_0 + \eta \mathbf{F}_0 \odot \mathbf{N}_{\mathbf{F},t}$ // $\mathbf{F}_t \in [0,1]^{n_1 \times k_1}$
9 end
   /* generate edges   */
10 $\mathbf{A} \leftarrow \mathbf{0}^{n_1^M \times n_1^M}$ // zero matrix in sparse format
11 for $i = 1, \ldots,$ num_edges do
12 | for $t = 1, \ldots, M$ do $r_t, c_t \leftarrow$ Sample a position in $\mathbf{A}_t$ prop. to its value ;
13 | $r \leftarrow \sum_{t=1}^M r_t \times n_1^{t-1}$ and $c \leftarrow \sum_{t=1}^M c_t \times n_1^{t-1}$
14 | $\mathbf{A}_{rc} \leftarrow 1$ and $\mathbf{A}_{cr} \leftarrow 1$ // add an undirected unweighted edge
15 end
   /* generate attribute memberships   */
16 $\mathbf{F} \leftarrow \mathbf{0}^{n_1^M \times k_1^M}$ // zero matrix in sparse format
17 for $i = 1, \ldots,$ num_memberships do
18 | for $t = 1, \ldots, M$ do $r_t, c_t \leftarrow$ Sample a position in $\mathbf{F}_t$ prop. to its value ;
19 | $r \leftarrow \sum_{t=1}^M r_t \times n_1^{t-1}$ and $c \leftarrow \sum_{t=1}^M c_t \times k_1^{t-1}$
20 | $\mathbf{F}_{rc} \leftarrow 1$
21 end

Lemma 5 (CKP preserves [P3]). *If $\mathcal{G}_1$ and $\mathcal{G}_2$ obey the spectral radius vs. triangle count power law with exponent γ, i.e., $\sigma_a \propto \Delta_a^\gamma$, so does $\mathcal{G}_1 \bar{\otimes} \mathcal{G}_2$.*

The proofs, given in appendix, use the properties of matrix Kronecker product [26] and two key observations: (1) edge count, node count, triangle count and spectral radius of AIS for an attribute a are *explicit* algebraic functions of the adjacency matrix $\mathbf{A}$ and the column in $\mathbf{F}$ which corresponds to a; (2) each column in $\mathbf{F}_1 \otimes \mathbf{F}_2$ is the Kronecker product of a column in $\mathbf{F}_1$ and a column in $\mathbf{F}_2$. Given this, our main result is:

Theorem 6 (SOAR graphs provably obey patterns P1–P3). *If $\mathcal{G}_1 = (\mathbf{A}_1, \mathbf{F}_1)$ obeys patterns P1–P3 with exponents α, β and γ respectively, then $\mathcal{G} = \mathrm{SOAR}(\mathbf{A}_1, \mathbf{F}_1, M, \eta = 0)$ also obeys P1–P3, with the same exponents α, β and γ.*

Proof. We prove this using induction on the number of steps $t = 1, \ldots, M$. It is given that $\mathcal{G}_1$ follows P1–P3, hence the base case for $t = 1$ is true. Now suppose for $1 \leq t < M$, $\mathcal{G}_t$ follows P1–P3. Then, using Lemmas 3, 4 and 5, $\mathcal{G}_t \bar{\otimes} \mathcal{G}_1 = \mathcal{G}_{t+1}$ follows P1–P3. Thus, by induction, $\mathcal{G} = \mathcal{G}_M$ obeys P1–P3.   □

Although Theorem 6 assumes no noise, it can be easily extended to the stochastic version of the SOAR generator to give similar guarantees in expectation. Our simulation studies, presented in Sect. 5.3, confirm our theoretical results.

Discussion. We elaborate on various aspects of the proposed SOAR model. *(a) Input parsimony:* SOAR, belonging to the paradigm of *one-shot graph generation*, has only four knobs to set: two (small) initiator matrices $(\mathbf{A}_1, \mathbf{F}_1)$, number of recursive steps M and noise level η. In contrast, *evolutionary* models typically need knobs for node-arrival, lifetime, sleep-time and linking processes (e.g., [29]). *(b) Attribute correlations:* SOAR implicitly incorporates attribute correlations, as Kronecker product naturally leads to recursive community structure [18]. Contrast this with [24] which explicitly models attribute correlations. *(c) Parameter fitting:* Given a social-affiliation network $\mathcal{G} = (\mathbf{A}, \mathbf{F})$, its parameters for SOAR model can be learned by employing KronFit [18] for $\mathbf{A}$ and $\mathbf{F}$ separately. *(d) Parameter selection:* To create social-affiliation graphs with homophily, we recommend choosing initiators such that the entries of $\mathbf{F}_1 \mathbf{F}_1^T$ are correlated with those of $\mathbf{A}_1$. Intuitively, this ensures that nodes with similar attributes are linked in the initiator and the self-similarity of Kronecker product passes this property on to the final graph.

5.3   Simulation Studies

We compare SOAR to two representative baselines – AGM [24] and SAN [14] – which were the most recent works in categories [B] and [C] from Sect. 2. Quantitative experiments compare the time taken by the models to generate graphs of comparable sizes, while qualitative experiments verify whether the models are able to generate graphs obeying the three discovered patterns – **[P1]** Edge count vs. node count power law relation, **[P2]** Triangle count vs. node count power law relation, **[P3]** Spectral radius vs. triangle count power law relation – as well as the following well-known properties: **[P4]** Skewed distributions[2] of #friends per node (node degree), #attributes per node (attribute degree of node) and #nodes per attribute (AIS node count) [29], **[P5]** Skewed distribution of eigenvalues of adjacency matrix [7].

We use the open-sourced code for SAN as is, but adapted AGM to get a skewed distribution of #nodes per attribute (i.e., group size [29]) and subsequently generated edges using the default Fast Chung Lu model. For SOAR, we use initiators from Fig. 6a–b (observe the correlation between $\mathbf{F}_1 \mathbf{F}_1^T$ and $\mathbf{A}_1$) replacing $1 \to 0.6, 0 \to 10^{-4}$ for stochasticity (and scaling the remaining entries appropriately), recursive steps $M = 8$ and noise level $\eta = 0.5$. This yields a graph with $0.4M$ nodes, $5.6M$ edges, $65K$ attributes and $2M$ attribute memberships.

---

[2] Distributions having an asymmetric long or heavy tail, e.g., log-normal, log-logistic.

Quantitative Evaluation. Figure 5 compares generation time of SOAR vs. SAN for five different graph sizes (AGM, due to the explicit enforcing of attribute correlations, scaled poorly with #attributes). Running times are averaged over 10 runs and experiments were performed on Mac OSX Yosemite with 2.7 GHz Intel i5 core and 16 GB main memory. We find that SOAR scales linearly, i.e., slope $\approx$1 in log-log scale. SAN also shows the desired linear scalability, but was 50× slower for ~1M edges plus memberships.

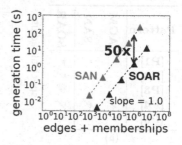

Fig. 5. Speed and scalability.

Qualitative Evaluation. From Figs. 6 and 7, we observe that only the proposed SOAR model is able to generate graphs obeying all these five patterns (Fig. 6), whereas the baselines fail at least one of them (Fig. 7a). In the interest of space, we show only one failed pattern per baseline: AGM leads to very low triangle count for AIS, perhaps due to its undesirably high importance to attribute correlation and homophily, which leads to few edges between nodes sharing attributes when the number of attributes is large (Fig. 7b); SAN produces an almost flat eigenvalue distribution (excluding first three values), likely due to the underlying preferential attachment model (Fig. 7c).

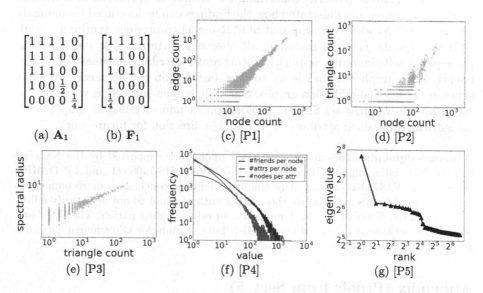

Fig. 6. SOAR generates realistic graphs: initiators in (a–b) lead to the discovered patterns P1–P3 (c–e) and skewed degree and eigenvalue distributions P4–P5 (f–g).

In sum, our simulations demonstrate that SOAR is able to generate social-affiliation graphs obeying all observed patterns in a fast and scalable manner.

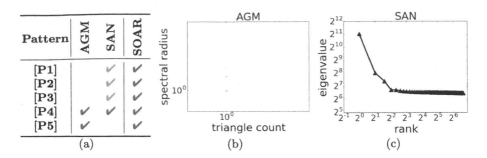

Fig. 7. (a) Graphs generated by baselines (AGM, SAN) disobey at least one pattern, e.g., (b) [P3] of AGM and (c) [P5] of SAN. Here, ✓ denotes empirical adherence based on a few parameters, while ✔ indicates *theoretical* adherence as well. (Color figure online)

6   Conclusion

We investigated the problem of pattern analysis and modeling of social-affiliation graphs – a friendship graph where users have many, binary attributes e.g., check-ins, page likes or group memberships – with the help of four large publicly-available real-world datasets. Our contributions are: (i) **Patterns:** We discovered three new consistent patterns concerning the structural properties of attribute-induced subgraphs and illustrated how the findings can be leveraged for anomaly detection. (ii) **Model:** We proposed SOAR model to produce synthetic social-affiliation graphs *provably* matching all observed patterns. It is based on the principle of self-similarity, implicitly incorporates attribute correlations, scales linearly with graph size and is up to 50× faster than the currently available generators for social-affiliation graphs. Our code is open-sourced at www.github. com/dhivyaeswaran/soar. Similar exploration of node-attributed graphs with categorical/real-valued attributes is a valuable direction for future work.

Acknowledgments. This material is based upon work supported by the National Science Foundation under Grants No. CNS-1314632, IIS-1408924 and by DARPA under award FA8750-17-2-0130. Any opinions, findings, conclusions or recommendations expressed in this material are those of the author(s) and do not necessarily reflect the views of the National Science Foundation, or other funding parties. The U.S. Government is authorized to reproduce and distribute reprints for Government purposes notwithstanding any copyright notation here on.

Appendix (Proofs from Sect. 5)

First, recall the following properties of the Kronecker product [26] for any four suitably sized matrices A, B, C and D: $(A \otimes B)^T = A^T \otimes B^T$; $(A \otimes B)(C \otimes D) = AB \otimes CD$; $\mathrm{Tr}[A \otimes B] = \mathrm{Tr}[A]\mathrm{Tr}[B]$; $\sigma(A \otimes B) = \sigma(A)\sigma(B)$ where $\sigma(\cdot)$ is the spectral radius.

Next, observe that edge count, node count, triangle count and spectral radius of AIS for an attribute a can be explicitly expressed as a function of adjacency matrix $\mathbf{A}$ and the ath column in $\mathbf{F}$ (call it f_a) as follows: (i) Node count of AIS, $n_a = f_a^T f_a$; (ii) Edge count of AIS, $m_a = \frac{1}{2} f_a^T \mathbf{A} f_a$; (iii) Triangle count of AIS, $\Delta_a = \frac{1}{6} \mathrm{Tr}[(\mathcal{D}(f_a)\mathbf{A}\mathcal{D}(f_a))^3]$, assuming no self loops – here, $\mathcal{D}(f_a)$ denotes the diagonalization of vector f_a; (iv) Spectral radius of AIS, $\sigma_a = \sigma(\mathcal{D}(f_a)\mathbf{A}\mathcal{D}(f_a))$.

Let the compact notation, $\bigotimes_{j=1}^n A_j$ denote $A_1 \otimes A_2 \ldots \otimes A_n$. Accordingly, every column of $\bigotimes_{j=1}^n A_j$ can be expressed as the Kronecker product of a column from each A_j, $j \in \{1, \ldots, n\}$. We are now ready to state our proofs.

Proof (Lemma 3). Any column f_a in $\mathbf{F}_1 \otimes \mathbf{F}_2$ is a Kronecker product of columns $f_{i,1}$ in $\mathbf{F}_1$ and $f_{j,2}$ in $\mathbf{F}_2$ for some i, j. The node count of AIS of a is $f_a^T f_a = (f_{i,1} \otimes f_{j,2})^T (f_{i,1} \otimes f_{j,2})$ which simplifies to $(f_{i,1}^T f_{i,1})(f_{j,2}^T f_{j,2})$ i.e., $n_a = n_{i,1} n_{j,2}$. Similarly, the edge count of AIS of a is $m_a = \frac{1}{2}(f_{i,1} \otimes f_{j,2})^T (\mathbf{A}_1 \otimes \mathbf{A}_2)(f_{i,1} \otimes f_{j,2})$ which can be written as $2(\frac{1}{2} f_{i,1}^T \mathbf{A}_1 f_{i,1})(\frac{1}{2} f_{j,2}^T \mathbf{A}_2 f_{j,2}) \propto m_{i,1} m_{j,2}$. Now, as $\mathcal{G}_1, \mathcal{G}_2$ follow [P1] with exponent α (given), $m_{i,1} \propto n_{i,1}^\alpha$ and $m_{j,2} \propto n_{j,2}^\alpha$. Hence, $m_a \propto n_a^\alpha$.

Proof (Lemma 4). Again, let $f_a = f_{i,1} \otimes f_{j,2}$ for attributes i, j, a in $\mathcal{G}_1, \mathcal{G}_2$ and $\mathcal{G}_1 \otimes \mathcal{G}_2$ respectively. The node count of AIS of a, again, is $n_a \propto n_{i,1} n_{j,2}$. The triangle count of AIS of a is $\Delta_a = \frac{1}{6} \mathrm{Tr}[(\mathcal{D}(f_a)(\mathbf{A}_1 \otimes \mathbf{A}_2)\mathcal{D}(f_a))^3]$ which can be simplified as $\frac{1}{6} \left(\mathrm{Tr}[(\mathcal{D}(f_{i,1})\mathbf{A}_1\mathcal{D}(f_{i,1}))^3] \right) \left(\mathrm{Tr}[(\mathcal{D}(f_{j,2})\mathbf{A}_2\mathcal{D}(f_{j,2}))^3] \right) \propto \Delta_{i,1} \Delta_{j,2}$ using the first, second and third Kronecker properties stated above. Now, as i and j follow [P2] with exponent β (given), $\Delta_{i,1} \propto n_{i,1}^\beta$, and $\Delta_{j,2} \propto n_{j,2}^\beta$. This results in $\Delta_a \propto n_a^\beta$.

Proof (Lemma 5). Once again, let $f_a = f_{i,1} \otimes f_{j,2}$ for attributes i, j, a in $\mathcal{G}_1, \mathcal{G}_2$ and $\mathcal{G}_1 \otimes \mathcal{G}_2$ respectively. We know from the previous proof that triangle count of AIS of a follows $\Delta_a \propto \Delta_{i,1} \Delta_{j,2}$. Now, spectral radius of AIS of a is $\sigma_a = \sigma(\mathcal{D}(f_a)(\mathbf{A}_1 \otimes \mathbf{A}_2)\mathcal{D}(f_a))$ which is $\sigma(\mathcal{D}(f_{i,1})\mathbf{A}_1\mathcal{D}(f_{i,1}))\sigma(\mathcal{D}(f_{j,2})\mathbf{A}_2\mathcal{D}(f_{j,2})) = \sigma_{i,1}\sigma_{j,2}$ due to the second and fourth Kronecker properties stated above. As graphs $\mathcal{G}_1, \mathcal{G}_2$ follow [P3] with exponent γ (given), i.e., $\sigma_{i,1} \propto \Delta_{i,1}^\gamma$ and $\sigma_{j,2} \propto \Delta_{j,2}^\gamma$. Therefore, $\sigma_a \propto \Delta_a^\gamma$.

References

1. Power law. https://en.wikipedia.org/wiki/Power_law
2. Akoglu, L., Faloutsos, C.: RTG: a recursive realistic graph generator using random typing. Data Min. Knowl. Discov. **19**(2), 194–209 (2009)
3. Akoglu, L., Tong, H., Koutra, D.: Graph based anomaly detection and description: a survey. Data Min. Knowl. Discov. **29**(3), 626–688 (2015)
4. Barabási, A.L., Albert, R.: Emergence of scaling in random networks. Science **286**(5439), 509–512 (1999)
5. Bastian, M., Heymann, S., Jacomy, M.: Gephi: an open source software for exploring and manipulating networks. In: ICWSM. The AAAI Press (2009)
6. Chakrabarti, D., Faloutsos, C.: Graph mining: laws, generators, and algorithms. ACM Comput. Surv. **38**(1), 2 (2006)

7. Chakrabarti, D., Zhan, Y., Faloutsos, C.: R-MAT: a recursive model for graph mining. In: SDM, pp. 442–446. SIAM (2004)
8. Cho, E., Myers, S.A., Leskovec, J.: Friendship and mobility: user movement in location-based social networks. In: KDD, pp. 1082–1090. ACM (2011)
9. Crandall, D.J., Cosley, D., Huttenlocher, D.P., Kleinberg, J.M., Suri, S.: Feedback effects between similarity and social influence in online communities. In: KDD, pp. 160–168. ACM (2008)
10. Faloutsos, M., Faloutsos, P., Faloutsos, C.: On power-law relationships of the internet topology. In: SIGCOMM, pp. 251–262 (1999)
11. Fond, T.L., Neville, J.: Randomization tests for distinguishing social influence and homophily effects. In: WWW, pp. 601–610. ACM (2010)
12. Goh, K.I., Kahng, B., Kim, D.: Spectra and eigenvectors of scale-free networks. Phys. Rev. E **64**(5), 051903 (2001)
13. Goldenberg, A., Zheng, A.X., Fienberg, S.E., Airoldi, E.M.: A survey of statistical network models. Found. Trends Mach. Learn. **2**(2), 129–233 (2009)
14. Gong, N.Z., Xu, W., Huang, L., Mittal, P., Stefanov, E., Sekar, V., Song, D.: Evolution of social-attribute networks: measurements, modeling, and implications using Google+. In: IMC, pp. 131–144. ACM (2012)
15. Kim, M., Leskovec, J.: Multiplicative attribute graph model of real-world networks. Internet Math. **8**(1–2), 113–160 (2012)
16. Koutra, D., Koutras, V., Prakash, B.A., Faloutsos, C.: Patterns amongst competing task frequencies: super-linearities, and the ALMOND-DG model. In: Pei, J., Tseng, V.S., Cao, L., Motoda, H., Xu, G. (eds.) PAKDD 2013. LNCS (LNAI), vol. 7818, pp. 201–212. Springer, Heidelberg (2013). https://doi.org/10.1007/978-3-642-37453-1_17
17. Leskovec, J., Backstrom, L., Kumar, R., Tomkins, A.: Microscopic evolution of social networks. In: KDD, pp. 462–470. ACM (2008)
18. Leskovec, J., Chakrabarti, D., Kleinberg, J.M., Faloutsos, C., Ghahramani, Z.: Kronecker graphs: an approach to modeling networks. JMLR **11**, 985–1042 (2010)
19. Leskovec, J., Kleinberg, J.M., Faloutsos, C.: Graph evolution: densification and shrinking diameters. TKDD **1**(1), 2 (2007)
20. McGlohon, M., Akoglu, L., Faloutsos, C.: Weighted graphs and disconnected components: patterns and a generator. In: KDD, pp. 524–532. ACM (2008)
21. Mislove, A., Marcon, M., Gummadi, P.K., Druschel, P., Bhattacharjee, B.: Measurement and analysis of online social networks. In: IMC, pp. 29–42. ACM (2007)
22. Newman, M.E.J., Watts, D.J., Strogatz, S.H.: Random graph models of social networks. Proc. Natl. Acad. Sci. **99**(Suppl. 1), 2566–2572 (2002)
23. Perozzi, B., Akoglu, L., Sánchez, P.I., Müller, E.: Focused clustering and outlier detection in large attributed graphs. In: KDD, pp. 1346–1355. ACM (2014)
24. Pfeiffer III, J.J., Moreno, S., La Fond, T., Neville, J., Gallagher, B.: Attributed graph models: modeling network structure with correlated attributes. In: WWW, pp. 831–842. ACM (2014)
25. Rabbany, R., Eswaran, D., Dubrawski, A.W., Faloutsos, C.: Beyond assortativity: proclivity index for attributed networks (ProNe). In: Kim, J., Shim, K., Cao, L., Lee, J.-G., Lin, X., Moon, Y.-S. (eds.) PAKDD 2017. LNCS (LNAI), vol. 10234, pp. 225–237. Springer, Cham (2017). https://doi.org/10.1007/978-3-319-57454-7_18
26. Schacke, K.: On the Kronecker product. Master's thesis, University of Waterloo (2004)

27. Titchmarsh, E.C., Heath-Brown, D.R.: The Theory of the Riemann Zeta-Function. Oxford University Press, Oxford (1986)
28. Watts, D.J., Strogatz, S.H.: Collective dynamics of 'small-world' networks. Nature **393**(6684), 440–442 (1998)
29. Zheleva, E., Sharara, H., Getoor, L.: Co-evolution of social and affiliation networks. In: KDD, pp. 1007–1016. ACM (2009)

ONE-M: Modeling the Co-evolution of Opinions and Network Connections

Aastha Nigam[1], Kijung Shin[2], Ashwin Bahulkar[3], Bryan Hooi[2],
David Hachen[1], Boleslaw K. Szymanski[3], Christos Faloutsos[2],
and Nitesh V. Chawla[1(✉)]

[1] University of Notre Dame, Notre Dame, IN, USA
{anigam,dhachen,nchawla}@nd.edu
[2] Carnegie Mellon University, Pittsburgh, PA, USA
{kijungs,christos}@cs.cmu.edu, bhooi@andrew.cmu.edu
[3] Rensselaer Polytechnic Institute, Troy, NY, USA
{bahula,szymab}@rpi.edu

Abstract. How do opinions of individuals on controversial issues such
as marijuana and gay marriage and their underlying social network connections evolve over time? Do people alter their network to have more
like-minded friends or do they change their own opinions? Does the society eventually develop echo chambers? In this paper, we study dynamically evolving networks and changing user opinions to answer these
questions. Our contributions are as follows: (a) *Discovering Evolution of
Polarization in Networks:* We present evidence of growing divide among
users based on their opinions who eventually form homophilic groups (b)
Studying Opinion and Network Co-Evolution: We present observations of
how individuals change opinions and position themselves in dynamically
changing networks (c) *Forecasting Persistence and Change in Opinions
and Network:* We propose ONE-M to forecast individual beliefs and persistence or dissolution of social ties. Using a unique real-world network
dataset including periodic user surveys, we show that ONE-M performs
with high accuracy, while outperforming the baseline approaches. Code
related to this paper is available at: https://github.com/anigam/ONE-M
and Data related to this paper is available at: http://netsense.nd.edu/.

1 Introduction

How do opinions of individuals on topics, specially controversial topics, and
their network structures co-evolve? That is, we are interested in understanding
and modeling how opinions develop and diversify, and their intricacies with the
underlying network structure. We posit that it is not just the network effect that
parlays diffusion or preponderance or change of an ego's opinion, but rather
the opinions also dictate how the network of connections changes around an
ego. When egos and alter-egos begin to evoke similar opinion, they form echo
chambers, which are clique-like patterns.

Prior research [7–9,13,15,17,24,25] has looked at polarization and opinion
dynamics, but there is a paucity of work in studying how opinion and network

© Springer Nature Switzerland AG 2019
M. Berlingerio et al. (Eds.): ECML PKDD 2018, LNAI 11052, pp. 122–140, 2019.
https://doi.org/10.1007/978-3-030-10928-8_8

co-evolve over time, and how it translates into persistence or change in opinions and network structure both. As such, the literature also lacks a principled model that brings together these characteristics to be able to forecast both spread of an opinion and change in the network structure (through persistence or dissolution of links).

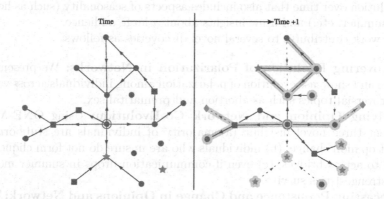

Fig. 1. Effectiveness of ONE-M on real data set: A real communication ego-network (central node marked by a boundary) snapshot from the NetSense study and the results obtained by using ONE-M. The nodes are color-coded by their opinions where red (○), yellow (⋆) and blue (□) indicate conservative, unsure and liberal opinion about political alignment, respectively. A grey (◇) node identifies an inactive node whereas the dotted edges indicate dissolved edges at $t + 1$. Using opinion and network topology at t (left), ONE-M is able to correctly forecast the opinion and the persistence/dissolution of edges (as highlighted by green color) at $t + 1$. (Color figure online)

Therefore, in this paper, we ask the following question:

Informal Problem: *Given information about evolving user network interactions either derived from communication or collocation patterns, user opinions on diverse beliefs (e.g. gay marriage, marijuana, etc.) and external user attributes (e.g. age, gender, etc.), how can we model and leverage the two-way effect that opinions and networks have on each other, to forecast persistence and change in opinions and social ties?*

To address this question, we propose the ONE-M model (**O**pinion and **N**etwork co-**E**volution-**M**odel) to study and model how opinions and networks co-evolve. To jointly model the co-evolution of opinions and networks, ONE-M incorporates several data-driven features that capture opinions on beliefs at different time segments such as opinion pre-disposition and persistence at the ego and population level, as well as the network characteristics including strength of ties, reciprocity, and triads.

To evaluate our model, we use a longitudinal dataset that we have collected at the University of Notre Dame as part of the NetSense study. About 200 incoming freshmen in 2011 were enrolled in the NetSense study. Enrolling students on their arrival at the University gives us a unique opportunity to study the formation and evolution of their network, and also whether their opinions about beliefs changed or remained the same over time. We acknowledge that we used only this one dataset in the paper, but given the longitudinality and impressionable age of the participants (18 and older) it provides us with a novel opportunity to study the behavior as well as be able to forecast both opinion and link persistence or dissolution over time that also includes aspects of seasonality (such as holiday break, summer, etc.) and draw insights about a large audience.

Our work contributes to several novel discoveries as follows:

- **Discovering Evolution of Polarization in Networks:** We present evidence and showcase evolution of polarization among individuals across various controversial topics such as abortion and premarital sex.
- **Studying Opinion and Network Co-Evolution:** Using ONE-M, we present three novel insights: (a) majority of individuals are stubborn and resist opinion change; (b) individuals who are unsure do not form cliques and tend to act as bridges; (c) even if communication drops in summer months, the strongest ties survive.
- **Forecasting Persistence and Change in Opinions and Network:** Using ONE-M, we report high performance for both forecasting opinion and network change. Figure 1 highlights the results of ONE-M demonstrating both the co-evolution of opinion and network, and also the forecasting of opinion and persistence or dissolution of links on real-world data.

Table 1. ONE-M is comprehensive compared to existing models of opinion and network dynamics.

Considerations	Das et al. [8]	Kim and Leskovec [21]	Durrett et al. [10]	Badev [2]	Bilò et al. [6]	ONE-M (proposed)
Node opinions	✓		✓		✓	✓
Node attributes		✓		✓		✓
Edge presence	✓	✓	✓	✓	✓	✓
Edge weights	✓				✓	✓
Network characteristics*		✓		✓		✓
Multiple networks**						✓
Multiple topics***						✓
Multiple time steps	✓	✓	✓	✓		✓

* reciprocity, transitivity, etc.   ** co-location, communication, etc.   *** euthanasia, death penalty, etc.

Reproducibility: Our code is available at https://github.com/anigam/ONE-M. The data used in the study is available from the NetSense Team at University of Notre Dame (http://netsense.nd.edu/).

2   Background and Related Work

In this section, we review existing models of opinion evolution and network evolution.

Evolution of Opinions. The interaction of individuals' opinions plays a role in nearly every social, economic, and political process. Understanding this interaction is particularly useful for political voting, public health campaigns, viral marketing, and information dissemination. A rich line of work in theoretical economics has studied mathematical models of opinion exchange; see [26] for a survey. The models are roughly divided into heuristic models and Bayesian models. In heuristic models, individuals follow simple update rules when interacting with each other. Notable examples of heuristic models are (a) averaging models [9,13], where each individual's opinion is updated to the average opinion of its neighbors, (b) voter models [7,17], where each individual follows a randomly-chosen neighbor's opinion or the majority opinion in its neighbors, and (c) flocking models [15], where individuals give more weight to opinions that conform to theirs, and (d) the combination of these models [8]. On the other hand, in Bayesian models, rational individuals maximize their expected utilities that depend on their actions as well as randomness [24,25]. In addition to this theoretical work, a rich body of empirical work in social and developmental psychology has studied how individuals update their opinions based on the opinions of those around them [1,12].

Evolution of Networks. Social networks change over time by nature. Consequently, there has been great interest in dynamics of social networks, especially in underlying micro-mechanisms that result in macro-level evolution of networks. Models of how networks evolve are roughly divided into stochastic models and game-theoretic models. Stochastic models view each edge as a random variable whose probability of presence depends on many different effects; see [27] for a survey. Notable examples of such effects are (a) transitivity [16]: friends of friends tend to be connected, (b) popularity [4]: high-degree nodes tend to be connected with new nodes, (c) assortativity [23]: nodes with similar degrees tend to be connected, and (d) homophily [21]: nodes with similar sociodemographic characteristics tend to be connected. In game-theoretic models, nodes are rational individuals whose utilities depend on social network structures, and edges are formed at the discretion of the nodes [20,30]; see [19] for a survey.

Co-evolution of Opinions and Networks. Relatively little attention has been given to the co-evolution of opinions and networks despite the considerable interest in the evolution of each of them. Recently, several models of opinion dynamics were extended so that nodes update their edges before updating their opinions [5,6,10,14,18]. In the extend models, nodes either follow simple heuristic rules (e.g., following a randomly chosen neighbor's opinion) [10,18] or optimize simple utility functions that are solely based on the agreement of opinion between neighboring nodes [5,6]. In our model, however, diverse user attributes (age, gender, etc.) and network characteristics (reciprocity, transitivity, etc.) are taken into consideration as well as the agreement of opinion between neighboring nodes. Our work is also closely related to models of mutual interaction between networks and behaviors (e.g., drinking [29] and smoking [2]). In Table 1, we compare our model with several other models of opinion and network dynamics.

3   Proposed Method

In this section, we motivate the problem and provide intuition behind the proposed approach. Next, we define the notation used in the paper for easier comprehension followed by the details of the proposed model ONE-M.

3.1   Intuition and Problem

In human societies, opinions are guided by social interactions, and conversely, network connections are influenced by the opinions. For instance, all individuals supporting the conservative party might opt to maintain strong relationships and subsequently minimize communication with individuals who might be liberal—leading to echo chambers and polarization. Similarly, an individual might be against the use of marijuana, but the majority of his/her friends may support the usage—leading to the individual changing his/her opinion. On the contrary, some views might be more ingrained in an individual which he/she refuses to change irrespective of the network and would rather lose friends than change opinion.

While studying this two way effect might be challenging, other factors can also govern opinions and network connections. Personal factors such as age, gender, ethnicity and hometown could also impact the underlying network characteristics. For instance, an individual may choose to form same-gender friendships whereas another individual may select to conform his friendship among individuals of the same ethnic background.

Therefore, to truly understand and build a model around opinion and network co-evolution it is crucial to incorporate varied network interactions based on who calls whom or who spends time with whom, opinions on a diverse set of topics and personal attributes about the individuals. Intuitively, each action a user takes whether it be changing (or persisting) with their opinion and (or) altering their social connections, would be influenced by their surroundings and inherent characteristics. With this motivation, we build the ONE-M as described in the subsequent sections.

3.2   Notation

For a population of N individuals, we have a time-evolving collection of T networks across C modalities (represented by $\mu \in \{c, b, wb\}$ when $C = 3$): communication (c), collocation (b) and collocation over the weekends (wb). For each timestep, we record user networks G^t for each modality in μ where $G^t = \{g_c^t, g_b^t, g_{wb}^t\}$.

Every network for each time step in each modality (g_μ^t) captures relationships between N individuals, with each node represented as u_i, is of size $N \times N$. Edges are directed and weighted by the strength of the connection represented as $w_c^t(i,j), w_b^t(i,j)$ and $w_{wb}^t(i,j)$ respectively. In the communication network, an edge $u_i \to u_j$ denotes if u_i chooses to call/message u_j. Further, in the collocation network, an edge $u_i \to u_j$ indicates if u_i selects to be physically present near u_j. Similarly, for collocation network over the weekend, an edge $u_i \to u_j$ denotes if u_i meets u_j over the weekends. Therefore, collectively, we have $G = \{G^1, G^2, \ldots, G^T\}$ which is a tensor of size $N \times N \times C \times T$.

In addition to the network information, for each u_i, we record their evolving opinions for a set of K beliefs. Typically, at each time t, for a belief a_k, u_i reports their opinion $a_{i,k}^t \in \{1, 2, 3\}$. Therefore, A of size $N \times K \times T$ captures opinion information for N users, K beliefs across T timesteps. Further, u_i is also associated with M external user attributes such as age, gender and ethnicity denoted by $X_i = \{X_{i1}, \ldots, X_{iM}\}$ which stay constant throughout the study for each individual. Thus, X capturing the user attributes is of size $N \times M$. Table 2 lists the symbols and their definitions.

In summary, the inputs to our setting are the following:

- Evolving network information: G, a tensor of size $N \times N \times C \times T$,
- Evolving user opinions: A, a tensor of size $N \times K \times T$,
- User attribute information: X, a matrix of size $N \times M$.

Recall that N is the number of users, C is the number of communication modalities (phonecall, bluetooth, etc.), T is the count of time-ticks, K is the number of topics (marijuana, abortion, etc.) and M is the number of demographic attributes (gender, etc.).

3.3   Proposed Model: ONE-M

General Problem Definition. *Given the networks* (G^t), *opinions* (A^t) *and user attributes* (X) *for individuals at time* t, *how can we model and forecast persistence and change in opinions* (A^{t+1}) *and network connections* (G^{t+1}) *at time* $t+1$?

To that end, we propose a model aimed at jointly capturing opinion and network co-evolution called ONE-M (**O**pinion and **N**etwork co-**E**volution-**M**odel). We define a function for each individual over a set of 8 derived factors. Using information about G, A and X, we learn the importance/weights (β) for each of the factors. For each network-related factor, we learn weights corresponding to the type of the network as denoted by $\mu \in \{c, b, wb\}$. To build a more general system and have fewer parameters, we assume that the weight for each factor, stays constant over time and users. We propose that using 8 derived factors and their relative weights, we can capture the interplay between an individual's opinion and the corresponding network topology. Next, we present the definition and intuitive description for each of the factors.

Table 2. Symbols and definitions

Symbols	Definitions
$\mathcal{N}$	Set of all individuals
G	Set of networks across time $\{G^1, G^2, ..., G^T\}$
G^t	Collection of networks $\{g_c^t, g_b^t, g_{wb}^t\}$ at time t
g_μ^t	General representation of the desired network at time t where $\mu \in \{c, b, wb\}$
$w_\mu^t(i, j)$	Tie strength between u_i and u_j at time t in either networks where $\mu \in \{c, b, wb\}$
$e_\mu^t(i, j)$	0/1 if the tie is present u_i and u_j at time t in either networks where $\mu \in \{c, b, wb\}$
$\eta(i)$	Friends of i, who have a directed edge $i \to j$ and $j \to i$
A	Tensor of size $N \times K \times T$ capturing opinions on K beliefs for N individuals across T time-steps
A_i^t	Individual i's opinion for all beliefs at time t
$a_{i,k}^t$	Individual i's opinion for belief a_k at time t
$v(X_i)$	Probability of opinion a_i given the user attributes for individual i
X	Matrix of size $N \times M$ capturing M user attributes for N individuals
X_i	A vector of M user attributes (such as age, hometown, ethnicity and gender) for individual i

1. Opinion Persistence: Given the controversial nature of many beliefs, this feature encapsulates the consistency of individuals in their opinions. For many, staying true to their opinions would be of priority over changing their opinions to conform to their surroundings. In order to incorporate for such persistence, we include opinion of user i at time t as shown below:

$$f_{stubborn} = a_{i,k}^t \tag{1}$$

2. Attribute Predisposition: Captures the phenomena that an individual can be predisposed to have a certain opinion based on their external attributes such as age, gender, hometown and ethnicity. For example, women might be more likely to support abortion or participants from a conservative leaning state would have a higher tendency to be conservative. In order to compute the dependency based on the individual's external attributes (X_i):

$$f_{predispose} = v(X_i) \sim P(a_{i,k}^t | X_i) \tag{2}$$

3. Population Belief: Measures the impact the global population can have on an individual's opinion. For instance, if all students around a participant i would indulge in marijuana usage, i is more likely to get influenced and conform to the surroundings. This phenomena can be measured as follows:

$$f_{population} = \frac{\sum_{j \in \mathcal{N} \setminus \{i\}} a_{j,k}^t}{N - 1} \tag{3}$$

4. Attribute Similarity: Are we more likely to communicate or spend time with another student because they come from the same city or have the same gender as us? This feature captures the strength of communication between user i and their friends based on how similar they are in terms of these external attributes such as ethnicity, hometown, age and gender. To account for this, we include the following feature:

$$f_{friend-sim} = sim(X_i, X_j) \tag{4}$$

5. Reciprocity Effect: While a student would like to be in harmony with their surroundings, they would be most influenced by their immediate friends whom they talk or spend time with. The strongest friendships would have the most impact on their opinion which can be measured as follows:

$$f_{friend-bias} = \frac{\sum_{j\in\eta(i)}(log(w_\mu^t(i,j)+1) + log(w_\mu^t(j,i)+1))a_{j,k}^t}{\sum_{j\in\eta(i)}(log(w_\mu^t(i,j)+1) + log(w_\mu^t(j,i)+1))} \quad (5)$$

6. Triads: Network and opinions could be impacted by the emergence of triads/ cliques. If u_i communicates or socializes with u_j and similarly u_j and u_k are friends, then u_i and u_k are more likely to communicate or influence each other beliefs either by agreeing or disagreeing. In order to capture the clique effect, we include the following:

$$f_{triads} = \sum_{q\neq j\in\eta(i)} (log(w_\mu^t(i,j)+1) + log(w_\mu^t(j,q)+1) + log(w_\mu^t(q,i)+1)) \quad (6)$$

7. Social Tie Persistence: Barring external shocks (such as disagreement or enrolling in same classes), typically, two friends tend to be consistent with their level of communication. In order, to account for such persistence between ties we include the strength of communication between two individuals:

$$f_{tie-persist} = log(w_\mu^t(i,j)+1) + log(w_\mu^t(j,i)+1) \quad (7)$$

8. Belief Similarity: In principle, for a friendship to survive the two students should have similar beliefs to minimize conflict. We include this similarity between u_i and u_j based on their opinions across all beliefs. In this research, we use cosine similarity.

$$f_{same-beliefs} = sim(A_i^t, A_j^t) \quad (8)$$

We break down the problem definition into following sub-problems:

- **Sub-problem 1 (Opinion Forecasting):** *Given the networks (G^t), opinions (A^t) and user attributes (X) for individuals at time t, how can we model and forecast persistence and change in opinions (A^{t+1}) at time $t+1$?*
- **Sub-problem 2 (Edge Forecasting):** *Given the networks (G^t), opinions (A^t) and user attributes (X) for individuals at time t, how can we model and forecast*
   - **Sub-problem 2.1 (Tie Strength)** *the strengths of ties and*
   - **Sub-problem 2.2 (Persistence/Dissolution of Ties):** *the persistence and dissolution of ties in networks (G^{t+1}) at time $t+1$?*

Using ONE-M, in Sect. 5 we explore how the model can be used to study the opinion and network co-evolution and present novel insights about time-evolving networks. Further, we show its effectiveness to forecast opinions and changes of links.

4   Data

We use the NetSense data that we have collected at the University of Notre Dame. The dataset comprises of 199 freshmen students who joined in Fall 2011 and captures various facets of student opinions and interactions through periodic surveys and mobile phone monitoring for a 2 year period [28]. The participants were carefully selected to represent the general population. For our study, we leverage the survey, communication and collocation information as explained below:

1. Survey Data: Students were presented a survey at the beginning of each semester (summer included) beginning Fall 2011 to Summer 2013. A total of 6 surveys were conducted in the two year period where the students were requested to report their opinions on various beliefs such as premarital sex, euthanasia, death penalty, gay marriage, marijuana, political alignment towards the liberal party, abortion and homosexuality. In general, for each belief, the student could nominate as being against, unsure or in support of the belief. In addition to their views on such controversial topics, the students reported personal attributes such as age, gender, hometown and ethnicity. From among the 199 students, only 108 students participated in all 6 surveys and have been for the purpose of this study.

2. Communication Network (CN): During the two year period, as a part of the NetSense study, we also recorded the communication patterns among the students. Primarily, for all students in the study, we recorded call and message events. Using this information, we constructed a directed and weighted communication network, where each node is a student and the weight signifies the number of calls and messages sent from u_i to u_j aggregated across the semester.

3. Collocation Network (BN, WBN): In addition, to the call and message patterns, we also recorded bluetooth interactions between mobile devices of the participants. Each bluetooth interaction is associated with a signal strength, called the Received Signal Strength Indicator (RSSI). Higher RSSI value signify that the mobile devices were closer to each other which can safely be used as a proxy of in-person interactions [22]. To build a collocation network, we extracted the interactions which occurred more than 50 times to reduce the noise and ensure the quality of an edge [3]. The network was directed and weighted based on the number of times the two mobile devices were collocated. This network is referred to as the BN network. Next, to obtain a more representative and strong network of student interactions, we extracted the connections that occurred outside of school hours, typically outside of 9 AM and 5 PM on weekdays, and constructed a weekend-based network (WBN). We argue that WBN would be more representative of friendships over BN. BN could have edges that occur only due to students taking classes together, however a connection outside class hours indicates the students choosing to spend time together.

For all three networks, among the 108 students, we observe that only a subset of them are active in each semester. However, we observe a drop in the number of interactions for the summer semesters. We comment about the nature of these ties in Sect. 5.

5   Experiments

In this section we discuss ONE-M and its applicability for studying opinion and network co-evolution. We are primarily interested in answering the following questions:

Q1. **Discovering Evolution of Polarization in Networks:** How does polarization evolve? Are people changing their opinions or are they dropping unlike-minded friends?

Q2. **Studying Opinion and Network Co-Evolution:** Are people resistant to change in opinion? How do the unsure individuals differ from polarized population? Are the strong ties able to survive? or do they become in-active?

Q3. **Forecasting Persistence and Change in Opinions and Network:** How well does ONE-M work on real data for opinion forecasting and edge forecasting?

5.1   Discovering Evolution of Polarization in Networks

In order to study the presence/absence of polarization in dynamic networks, we study the communication patterns for the polarized users who are in-support and against a belief. The unsure/neutral individuals are dropped from this network as they are not polarized yet. A connection between individuals having the same opinion is called a homophilic edge. Conversely, a connection between students with different opinions is referred to as a cross edge.

We then conduct a test for polarization on each communication network snapshot (g_c^t). The test is based on the standard test for homophily in networks [11]. Let, p and q be the fraction of users against and for a belief, respectively. Then, the probability of a cross edge can be given by $\mu_0 = 2pq$ for a directed network. In order for polarization to exist, we expect to see the number of cross edges to be significantly lower than $2pq$ in the observed network (μ). Therefore, we test, $H_0 : \mu \geq \mu_0$ against $H_1 : \mu < \mu_0$. We present the results for polarization in Table 3.

Evidence of Polarization: In general, we witness *growing polarization* from Fall 2011 to Summer 2013 (as shown in Table 3). We also observe that topics such as abortion and premarital sex are most dividing across time. This makes sense because discussions about these topics are fervent at the Notre Dame campus which is a catholic university. Moreover, views on such topics can be governed by the cultural practices as observed in one's surrounding. For instance, an individual could more likely be in favor of marijuana usage if everyone around him/her are in favor of it. Further, we do not observe significant evidence of polarization for topics such as euthanasia and death penalty which could be defined as the "non-wedge" topics. Given the demographics of our participants, who are mostly aged between 17–19 years, we do not expect them to be discussing or have firm opinions on such topics.

Evolution of Polarization: While we do witness polarization in our data, we ask the next most natural question. Are people changing their belief over time causing the polarity of edges to flip (denoted as flipped cross edges) or are people simply dropping connections with people who do not share the same opinion as them (termed as dropped cross edges)? In order to answer this question, we present a breakdown of the evolving communication network in case of marijuana as shown in Table 4. On analyzing the evolution, we observe that the number of nodes in support of marijuana over time get much more active than their counterparts who consider marijuana not legal. Further, as communication transpires, we notice that at each time step we do not observe a lot of edges flipping the polarity, however we observe that most of the cross edges keep dropping in subsequent time intervals. We observe a similar pattern for the other beliefs too however due to space constraint we do not add those tables. Overall, we discover that at the beginning of the study individuals start with casual friends but over time they drop ties (cross edges) and form more like-minded groups (or echo chambers).

Table 3. Evidence of Polarization: results for each belief across 6 semesters using the communication network. ✓ indicates the presence of homophily at varying confidence levels: 99% (***), 95% (**), and 90% (*). We have color-coded based on significance for easy comprehension (darker shade indicate higher the confidence). × marks (or the absence of color) indicate that we do not have enough evidence to reject the null hypothesis (H_0).

	Fall 2011	Spring 2012	Summer 2012	Fall 2012	Spring 2013	Summer 2013
Premarital Sex	✓**	✓**	✓**	✓*	✓***	✓**
Gay Marriage	×	×	×	✓**	×	✓**
Marijuana	×	✓**	✓***	×	✓*	✓**
Political	✓*	✓*	×	✓**	✓**	✓***
Abortion	✓***	✓***	✓*	✓*	✓***	✓*
Homosexual	✓*	✓***	×	✓***	✓**	✓*
Euthanasia	×	×	×	×	×	×
Death Penalty	×	×	×	×	×	×

Table 4. Evolution of Polarization: using marijuana usage as an example we break-down the evolution of the communication network among the students. In this illustration, we only consider the polarized individuals (Against Group: Not legal and For Group: Legal). Symbols mean the following: N: #nodes, E: #edges, N+/E+: #nodes/edges added at $t + 1$, N−/E−: #nodes/edges dropped at $t + 1$,

Time	N	E	N+	N−	E+	E−	Against nodes	For nodes	Homo-philic edges	Cross edges	Flipped cross edges	Dropped cross edges
Fall 2011	79	166	-	-	-	-	51	28	83	83	-	-
Spring 2012**	69	151	9	19	70	85	38	31	86	65	13	52
Summer 2012***	57	89	12	24	30	92	33	24	61	28	8	46
Fall 2012	70	122	21	8	60	27	37	33	56	66	0	13
Spring 2013*	67	124	12	15	49	47	31	36	69	55	12	28
Summer 2013**	54	100	6	19	34	58	19	35	64	36	0	34

5.2   Studying Opinion and Network Co-Evolution

We leverage ONE-M and explore opinion and network characteristics of individuals across the communication network and present the following key observations:

– **Stubbornness:** Given the controversial nature of the topics, we explore how resistant people are to changing their opinions. To that end, we leverage the Opinion Persistence factor of ONE-M. We present the persistent (and/or changing) opinions of individuals across the different beliefs in Fig. 2. A user with same color coding for all semesters indicates that they did not change their opinion across the two year period. We observe that the polarized individuals (who are either against or for the belief) are more consistent with their opinions. Conversely, the unsure/neutral individuals frequently sway between the two extremes. Overall, we conclude that many individuals are stubborn about changing their opinions—being consistent with their own opinions is of more value than altering their opinion or network.

– **Bridges Between Cliques:** We characterize the participants belonging in different opinion groups based on their tendency to form triads/cliques. In order to understand this phenomenon, we leverage the Triads component of ONE-M. We compute a metric called *cliquishness* which captures the conditional probability of an individual forming a clique given their opinion of being against, unsure or in support of the belief. As shown in Fig. 3, we observe that the tendency of forming cliques for both the extremes stay relatively constant across time. However, the unsure individuals not only have a lower probability of forming cliques but over time they are less likely to participate in cliques. We conclude that people with strong beliefs (i.e. individuals who are either for or against) tend to form cliques and follow their cliques over time. However, the unsure/neutral individuals have a lower tendency of participating in cliques and tend to act as bridges between the cliques.

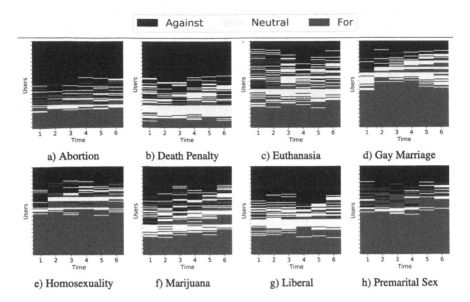

Fig. 2. Stubbornness: figures (a)–(h) capture persistence and change in user opinions for 8 beliefs: abortion, death penalty, euthanasia, gay marriage, homosexuality, marijuana use, political alignment towards liberal group and premarital sex. For each figure, the 6 semesters are marked on the x-axis and each row represents a user. Red, yellow and blue indicate against, unsure/neutral and favor of the belief. (Color figure online)

– **Strong Ties Persist:** As reported earlier in Sect. 4, we observe that a lot of connections disappear in the summer months. While this is understandable as most students might not be present on campus, it is interesting to study the links that do persist. To that end, we study the Social Tie Persistence factor of ONE-M. As shown in Fig. 4, we compare the strength of edges in the communication network between two consecutive semesters. As expected, we observe many edges to die in Fig. 4b and e, however we observe that the links that do persist maintain the same strength of communication (falling close to the 45° line). We conclude that even though we observe many ties breaking between consecutive semesters, the ones that do survive are indeed the strong ties.

5.3   Forecasting Persistence and Change in Opinions and Network

Sub-problem 1: Opinion Forecasting. We leverage ONE-M to forecast evolving opinions of individuals as described in Sub-Problem 1 (Sect. 3)—using networks (G^t), opinions (A^t) and user attributes (X) for individuals at time t, we forecast opinions (A^{t+1}) at time $t+1$. We address the problem as a classification task, where we apply ONE-M to extract features and predict $a_{k,i}^{t+1} \in \{1, 2, 3\}$. For opinion forecasting, we derive the following features: Opinion Persistence,

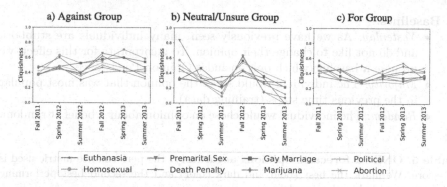

Fig. 3. Bridges between Cliques: people with strong beliefs have clique-ish ego networks ($\sim$ "echo chambers") whereas unsure people do not.

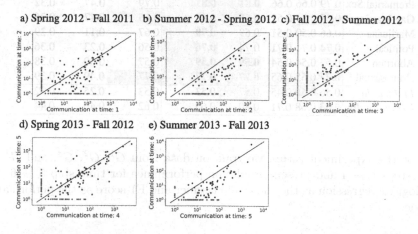

Fig. 4. Strong Ties Persist: communication strength between $t + 1$ (y-axis) and t (x-axis). Blue points: edges that persisted; Green points: new active edges; Red points: Dropped edges. Black line: 45°. Observations: Communication drops in summer months, stays consistent between academic semesters. (Color figure online)

Attribute Predisposition, Population Belief and Reciprocity Effect and learn their weights for the task. The weights for other features are set to 0.

We use ONE-M, to combine information across different network modalities, opinions and attributes, and study the following variations:

- **CN:** Extract features from the communication network (g_c^t) only.
- **BN:** Leverage the collocation network (g_b^t) only.
- **WBN:** Derive features from weekend-based collocation network (g_{wb}^t) only.
- **CN+BN:** Derive features from CN (g_c^t) and BN (g_b^t).
- **CN + WBN:** Combine features from CN (g_c^t) and WBN (g_{wb}^t) networks.

- **Baselines:**
  - *Yesterday:* As we have previously seen, many individuals are stubborn and do not like to change their opinion. To incorporate for this effect, we propose a baseline which only includes Opinion Persistence.
  - *Majority:* The individual would select the opinion that was most popular in the previous time steps (training data).
  - *Random:* The individual would chose an opinion about a belief at random.

Table 5. ONE-M forecasts opinions accurately: the performance metric used is F1-score. We mark the best (bold and dark color) and the second best performing (underline and light color) model.

	ONE-M					Baselines		
	CN	BN	WBN	CN+BN	CN+WBN	Yesterday	Majority	Random
Premarital Sex	0.79	0.66	0.66	0.81	0.81	0.79	0.47	0.32
Gay Marriage	0.65	0.54	0.52	0.65	0.65	0.65	0.47	0.36
Marijuana	0.66	0.50	0.51	0.69	0.70	0.67	0.11	0.29
Political	0.74	0.71	0.71	0.74	0.76	0.56	0.23	0.36
Abortion	0.57	0.54	0.54	0.59	0.59	0.57	0.30	0.38
Homosexual	0.77	0.65	0.65	0.77	0.77	0.77	0.49	0.39
Euthanasia	0.56	0.66	0.45	0.57	0.53	0.53	0.25	0.32
Death penalty	0.83	0.68	0.71	0.80	0.65	0.62	0.46	0.44

For the experiment setup, we train on data from G^1, G^2, G^3, and G^4 for time steps $1 \rightarrow 4$ and present results on performance for time step $5 \rightarrow 6$. We use logistic regression as the choice of classifier and F1-score as the performance metric.

Result: Based on the results in Table 5, we observe that for majority of the beliefs ONE-M achieves the best performance. We also observe that between the collocation (BN) and weekend-based collocation network (WBN), adding the latter provides a stronger signal and contributes more towards forecasting an individual's opinion. This makes sense because if two students are choosing to meet outside classes over the weekends they are more likely to influence each others opinions. We would also like to note the difference in the perception of some beliefs: while opinions about beliefs such as premarital sex might be more ingrained in an individual based on their religious value (explaining why the 'Yesterday' model based on opinion persistence performs better), opinions about marijuana usage and political views might evolve based on the cultural norms of the environment an individual is. Additionally, given that our population consists of students aged between 17–19 years, we do not expect them to discuss or have a firm opinion about the "non-wedge" issues such as death penalty and euthanasia (as also seen in Table 3). Overall, based on the performance of top two performing models (bold and underlined in Table 5), we conclude that

ONE-M is able to forecast opinions and adding network features is of value for understanding evolving opinions. Moreover, we recommend using ONE-M with communication and collocation networks.

Sub-problem 2: Edge Forecasting. We evaluate the performance of ONE-M at forecasting information about the network connections as defined in sub-problem 2. To that end, we define experiments for both sub-problems:

1. **Sub-problem 2.1: Tie Strength:** Using features from G^t, A^t and X^t, forecast the strength of the communication between u_i and u_j at $t + 1$, that is $log(w_c^{t+1}(i, j) + 1)$. This can be treated as a standard regression problem.
2. **Sub-problem 2.2: Persistence/Dissolution:** Using features from G^t, A^t and X^t, forecast whether an edge that exists between u_i and u_j at time t would persist or dissolve at time $t + 1$ in the communication network (G^{t+1}). This can be considered as a classification problem with the target variable as 0/1 class.

For both sub-problems, we derive the following features: Social Tie Persistence, Belief Similarity, Attribute Similarity and Triads and learn their weights for the task. The weights for other features are set to 0. For both experiments, we employ a series of ONE-M variations as described below:

- **CN:** Extract features from only the communication network (g_c^t).
- **BN:** Leverage only the collocation network (g_b^t).
- **WBN:** Derive features from weekend-based collocation network (g_{wb}^t) only.
- **CN+BN:** Derive features from CN (g_c^t) and BN (g_b^t).
- **CN+WBN:** Combine features from CN (g_c^t) and WBN (g_{wb}^t) networks.
- **Baseline:**
  - 'Yesterday': incorporates the persistence of the communication strength and edge by including only the Social Tie Persistence feature and excludes any network or opinion effects.
  - *For Tie Strength Forecasting:* we include two baselines. 'Mean' predicts the average tie strength, whereas 'Median' forecasts the median tie strength observed in the training data as the predicted strength.
  - *For Persistence/Dissolution Forecasting:* we include a 'Random' baseline for the edge persistence/dissolution experiment, which randomly assigns if a link persists or dissolves in the consecutive time step.

For both experiments, we train on data from G^1, G^2, G^3, and G^4 for time steps $1 \rightarrow 4$ and present results on performance for time step $5 \rightarrow 6$. For tie strength, we employ linear regression with mean squared error (MSE) as the performance measure. For persistence/dissolution, we use logistic regression with F1-score as the performance metric. The results for the both the experiments are listed in Table 6.

Table 6. ONE-M forecasts edges accurately: for forecasting (1) tie strength (2) persistence/dissolution of a tie, we use mean-squared error (MSE) in experiment 1 (lower the better) and F1-score in experiment 2 (higher the better). We mark the best (bold and dark color) and the second-best (underline and light color) model.

| | ONE-M | | | | | Baselines | | |
	CN	BN	WBN	CN+BN	CN+WBN	Yesterday	Mean	Median
Tie Strength (MSE)	2.50	3.59	3.34	2.48	2.35	2.36	4.21	4.00

| | ONE-M | | | | | Baselines | |
	CN	BN	WBN	CN+BN	CN+WBN	Yesterday	Random
Persistence/Dissolution (F1-score)	0.76	0.61	0.65	0.72	0.75	0.75	0.52

Result. Based on results in Table 6, we observe that ONE-M performs better for tie strength and persistence/dissolution experiment as compared to baselines. We observe that for predicting tie strength between two nodes, ONE-M using communication and collocation network based on weekends combined obtains the lowest error. For persistence/dissolution, we observe that CN wins. Between the two collocation networks, we again observe a strong signal from the WBN (as also seen from opinion forecasting experiment). Our ONE-M is able to leverage the connections between friends who are spending time with each other outside classes rather than random meetings for communication strength forecasting. We also notice that using simple social tie persistence ('Yesterday' model) is not enough to understand how connections persist and dissolve in a polarizing environment. Again, based on the results, we recommend using ONE-M with the communication and strong collocation networks that capture friendships for edge forecasting.

6   Conclusions

In summary, the contributions of our work are as follows:

- **Discovering Evolution of Polarization in Networks:** We discover (a) polarization indeed happens, for several topics and (b) people prefer to severe ties with dis-agreeing contacts, than change opinion.
- **Studying Opinion and Network Co-Evolution:** With our proposed ONE-M model, we made additional discoveries: "stubborness" (people keep their opinions); "bridges between cliques" (strong-opinion people tend to belong to clique-like groups); "strong ties persist" (despite summer breaks).
- **Forecasting Persistence and Change in Opinions and Network:** Our proposed model, ONE-M, outperforms baselines on forecasting accuracy, for beliefs as well as network changes.

Acknowledgements. This work is supported by the Army Research Laboratory under Cooperative Agreement Number W911NF-09-2-0053 and by the National Science Foundation (NSF) Grant IIS-1447795.

References

1. Allen, J.P., Porter, M.R., McFarland, F.C.: Leaders and followers in adolescent close friendships: susceptibility to peer influence as a predictor of risky behavior, friendship instability, and depression. Dev. Psychopathol. **18**(1), 155–172 (2006)
2. Badev, A.: Discrete games in endogenous networks: equilibria and policy. arXiv preprint arXiv:1705.03137 (2017)
3. Bahulkar, A., et al.: Coevolution of a multilayer node-aligned network whose layers represent different social relations. Comput. Soc. Netw. **4**(1), 11 (2017)
4. Barabási, A.L., Albert, R.: Emergence of scaling in random networks. Science **286**(5439), 509–512 (1999)
5. Bhawalkar, K., Gollapudi, S., Munagala, K.: Coevolutionary opinion formation games. In: STOC, pp. 41–50. ACM (2013)
6. Bilò, V., Fanelli, A., Moscardelli, L.: Opinion formation games with dynamic social influences. In: Cai, Y., Vetta, A. (eds.) WINE 2016. LNCS, vol. 10123, pp. 444–458. Springer, Heidelberg (2016). https://doi.org/10.1007/978-3-662-54110-4_31
7. Clifford, P., Sudbury, A.: A model for spatial conflict. Biometrika **60**(3), 581–588 (1973)
8. Das, A., Gollapudi, S., Munagala, K.: Modeling opinion dynamics in social networks. In: WSDM, pp. 403–412. ACM (2014)
9. DeGroot, M.H.: Reaching a consensus. J. Am. Stat. Assoc. **69**(345), 118–121 (1974)
10. Durrett, R., et al.: Graph fission in an evolving voter model. PNAS **109**(10), 3682–3687 (2012)
11. Easley, D., Kleinberg, J.: Networks, Crowds, and Markets: Reasoning about a Highly Connected World. Cambridge University Press, Cambridge (2010)
12. Evans, W.N., Oates, W.E., Schwab, R.M.: Measuring peer group effects: a study of teenage behavior. J. Polit. Econ. **100**(5), 966–991 (1992)
13. Friedkin, N.E., Johnsen, E.C.: Social positions in influence networks. Soc. Netw. **19**(3), 209–222 (1997)
14. Gu, Y., Sun, Y., Gao, J.: The co-evolution model for social network evolving and opinion migration. In: Proceedings of the 23rd ACM SIGKDD International Conference on Knowledge Discovery and Data Mining, pp. 175–184. ACM (2017)
15. Hegselmann, R., Krause, U., et al.: Opinion dynamics and bounded confidence models, analysis, and simulation. J. Artif. Soc. Soc. Simul. **5**(3) (2002)
16. Holland, P.W., Leinhardt, S.: A dynamic model for social networks. J. Math. Sociol. **5**(1), 5–20 (1977)
17. Holley, R.A., Liggett, T.M.: Ergodic theorems for weakly interacting infinite systems and the voter model. Ann. Probab. **3**(4), 643–663 (1975)
18. Holme, P., Newman, M.E.: Nonequilibrium phase transition in the coevolution of networks and opinions. Phys. Rev. E **74**(5), 056108 (2006)
19. Jackson, M.O.: A survey of network formation models: stability and efficiency. In: Group Formation in Economics: Networks, Clubs, and Coalitions, pp. 11–49 (2005)
20. Jackson, M.O., Wolinsky, A.: A strategic model of social and economic networks. J. Econ. Theory **71**(1), 44–74 (1996)
21. Kim, M., Leskovec, J.: Multiplicative attribute graph model of real-world networks. Internet Math. **8**(1–2), 113–160 (2012)
22. Liu, S., Jiang, Y., Striegel, A.: Face-to-face proximity estimation using bluetooth on smartphones. TMC **13**(4), 811–823 (2014)
23. Morris, M., Kretzschmar, M.: Concurrent partnerships and transmission dynamics in networks. Soc. Netw. **17**(3–4), 299–318 (1995)

24. Mossel, E., Sly, A., Tamuz, O.: Asymptotic learning on Bayesian social networks. Probab. Theory Relat. Fields **158**(1–2), 127–157 (2014)
25. Mossel, E., Sly, A., Tamuz, O.: Strategic learning and the topology of social networks. Econometrica **83**(5), 1755–1794 (2015)
26. Mossel, E., Tamuz, O.: Opinion exchange dynamics. Probab. Surv. **14**, 155–204 (2017)
27. Snijders, T.A., Van de Bunt, G.G., Steglich, C.E.: Introduction to stochastic actor-based models for network dynamics. Soc. Netw. **32**(1), 44–60 (2010)
28. Striegel, A., et al.: Lessons learned from the netsense smartphone study. SIGCOMM Comput. Commun. Rev. **43**(4), 51–56 (2013)
29. Wang, C., Hachen, D.S., Lizardo, O.: The co-evolution of communication networks and drinking behaviors. In: Proceedings of AAAI Fall Symposium Series (2013)
30. Watts, A.: A dynamic model of network formation. Games Econ. Behav. **34**(2), 331–341 (2001)

Think Before You Discard: Accurate Triangle Counting in Graph Streams with Deletions

Kijung Shin[⊠], Jisu Kim, Bryan Hooi, and Christos Faloutsos

School of Computer Science, Carnegie Mellon University, Pittsburgh, PA, USA
{kijungs,christos}@cs.cmu.edu, {jisuk1,bhooi}@andrew.cmu.edu

Abstract. Given a stream of edge additions and deletions, how can we estimate the count of triangles in it? If we can store only a subset of the edges, how can we obtain unbiased estimates with small variances?

Counting triangles (i.e., cliques of size three) in a graph is a classical problem with applications in a wide range of research areas, including social network analysis, data mining, and databases. Recently, streaming algorithms for triangle counting have been extensively studied since they can naturally be used for large dynamic graphs. However, existing algorithms cannot handle edge deletions or suffer from low accuracy.

Can we handle edge deletions while achieving high accuracy? We propose THINKD, which accurately estimates the counts of global triangles (i.e., all triangles) and local triangles associated with each node in a fully dynamic graph stream with edge additions and deletions. Compared to its best competitors, THINKD is **(a) Accurate:** up to *4.3× more accurate* within the same memory budget, **(b) Fast:** up to *2.2× faster* for the same accuracy requirements, and **(c) Theoretically sound:** always maintaining *unbiased* estimates with small variances. Code related to this paper is available at: https://github.com/kijungs/thinkd.

Keywords: Triangle counting · Local triangles
Streaming algorithms · Fully dynamic graph streams · Edge deletions

1 Introduction

Given a fully dynamic graph stream with edge additions and deletions, how can we accurately estimate the count of triangles in it with fixed memory size?

The count of triangles (i.e., cliques of size three) is a key primitive in graph analysis with a wide range of applications, including spam/anomaly detection [5,14], link recommendation [8,22], community detection [6], degeneracy estimation [18], and query optimization [3]. In particular, many important metrics in

Electronic supplementary material The online version of this chapter (https://doi.org/10.1007/978-3-030-10928-8_9) contains supplementary material, which is available to authorized users.

Table 1. Comparison of streaming algorithms for triangle counting. Notice that ThinkD is accurate while satisfying all the criteria.

	TRIEST$_{FD}$ [7]	ESD [10]	Other Local [14]	[7,17]*	Other Global [2,11,16,20]	THINKD$_{FAST}$	THINKD$_{ACC}$ (Proposed)
Local triangles	∨		∨	∨		∨	∨
Large graphs**	∨		∨	∨	∨	∨	∨
Edge deletions	∨	∨				∨	∨
Accuracy***	-	-	∨	Ṽ	?	∨	Ṽ

*TRIEST$_{IMPR}$ [7], **Graphs that do not fit in memory, ***Ṽ: highest, ∨: high, ?: highest-low. -: low

social network analysis, including the clustering coefficient [24], the transitivity ratio [15], and the triangle connectivity [4], are based on the count of triangles.

Many real graphs are best represented as a sequence of edge additions and deletions, and they often need to be processed in real time. For example, many social networking service companies aim to detect fraud or spam as quickly as possible in their online social networks, which evolve indefinitely with both edge additions and deletions. Another example is to examine graphs of data traffic and improve the network performance in real time.

As a result, there has been great interest in graph stream algorithms, which gradually update their outputs as each edge insertion or deletion is received rather than operating on the entire graph at once. However, existing streaming algorithms for triangle counting focus on insertion-only streams [2,11,14,16,17, 19] or greatly sacrifice accuracy to support edge deletions [7,10,13].

In this work, we propose THINKD (**Think** before you **D**iscard), an accurate streaming algorithm for triangle counting in a fully dynamic graph stream with both edge additions and deletions. THINKD maintains and updates estimates of the counts of global triangles (i.e., all triangles) and local triangles incident to each node. THINKD is named after the fact that, upon receiving each edge addition or deletion, THINKD uses it to improve its estimates even if the edge is about to be discarded without being stored. This allows THINKD to achieve higher accuracy than if it were to only use edges in memory for estimation. As a result, our proposed algorithm THINKD has the following strengths:

- **Accurate:** THINKD gives up to *4×* and *4.3× smaller estimation errors* for global and local triangle counts, respectively, than its best competitors within the same memory budget (Fig. 2).
- **Fast:** THINKD *scales linearly* with the size of the input stream (Fig. 1, Corollary 1, and Theorem 4). Especially, THINKD is up to *2.2× faster* than its best competitors with similar accuracies (Fig. 3).
- **Theoretically Sound:** We prove the formulas for the bias and variance of the estimates provided by THINKD (Theorems 1 and 2). In particular, we show that THINKD always maintains *unbiased* estimates (Fig. 1).

Reproducibility: The source code and datasets used in the paper are available at http://www.cs.cmu.edu/~kijungs/codes/thinkd/.

In Sect. 2, we review related work. In Sect. 3, we present notations and the problem definition. In Sect. 4, we describe our proposed algorithm THINKD. After providing experimental results in Sect. 5, we conclude in Sect. 6.

2   Related Work

See Table 1 for a comparison of streaming algorithms for triangle counting. Streaming algorithms for triangle counting in insertion-only graph streams have been studied extensively, including multi-pass [12,21] or single-pass [2,11,16,20] algorithms for the count of global triangles, and multi-pass [5] or single-pass [7,14,17,19] algorithms for the counts of both global and local triangles.

The first algorithm for triangle counting in fully dynamic graph streams with edge deletions was proposed in [13]. The algorithm estimates the count of global triangles by making a single pass over the input stream. However, the algorithm is inapplicable to real-time applications since it expensively computes an estimate once at the end of the stream instead of always maintaining an estimate. Although ESD [10] maintains and updates an estimate of the global triangle count, its scalability is limited since it requires the entire input graph to be stored in memory. TRIEST$_{FD}$ [7], which maintains and updates estimates of both global and local triangle counts, scales better than ESD since TRIEST$_{FD}$ samples edges within a given memory budget and discards the other edges. However, TRIEST$_{FD}$, which simply discards those unsampled edges, is significantly less accurate than our proposed algorithm THINKD, which utilizes those unsampled edges to update estimates before discarding them. Although the idea of using unsampled edges has been considered for insertion-only streams [7,14,17,19], applying the idea to fully dynamic graph streams has remained unexplored.

Table 2. Table of frequently-used symbols.

	Symbol	Definition
Notations for fully dynamic graph streams (Sect. 3)	$e^{(t)} = (\{u, v\}, \delta)$	Change in the input graph $\mathcal{G}$ at time t
	$\mathcal{G}^{(t)} = (\mathcal{V}^{(t)}, \mathcal{E}^{(t)})$	Graph $\mathcal{G}$ at time t
	$\{u, v\}$	Edge between nodes u and v
	$\{u, v, w\}$	Triangle with nodes u, v, and w
	$\mathcal{T}^{(t)}$	Set of global triangles in $\mathcal{G}^{(t)}$
	$\mathcal{T}^{(t)}[u]$	Set of local triangles of node u in $\mathcal{G}^{(t)}$
Notations for algorithms and analyses (Sect. 4)	$\mathcal{S}$	Set of sampled edges
	$\mathcal{N}[u]$	Set of neighbors of node u in $\mathcal{S}$
	$\bar{c}$	Estimate of the count of global triangles
	$c[u]$	Estimate of the count of local triangles of node u
	r	Sampling probability in THINKD$_{FAST}$
	k	Maximum number of sampled edges in THINKD$_{ACC}$
	$\mathcal{A}^{(t)}$	Set of added triangles at time t
	$\mathcal{D}^{(t)}$	Set of deleted triangles at time t

3   Notations and Problem Definition

Notations: Table 2 lists the symbols frequently used in the paper. Consider an undirected graph $\mathcal{G} = (\mathcal{V}, \mathcal{E})$ with nodes $\mathcal{V}$ and edges $\mathcal{E}$. Each edge $\{u, v\} \in \mathcal{E}$ connects two distinct nodes $u \neq v \in \mathcal{V}$. We say a subset $\{u, v, w\} \subset \mathcal{V}$ of size 3 is a *triangle* if every pair of distinct nodes u, v, and w is connected by an edge in $\mathcal{E}$. We denote the set of *global triangles* (i.e., all triangles) in $\mathcal{G}$ by $\mathcal{T}$ and the set of *local triangles* of each node $u \in \mathcal{V}$ (i.e., all triangles containing u) by $\mathcal{T}[u] \subset \mathcal{T}$.

Assume the graph $\mathcal{G}$ evolves from the empty graph. We consider the *fully dynamic graph stream* representing the sequence of changes in $\mathcal{G}$, and denote the stream by $(e^{(1)}, e^{(2)}, ...)$. For each $t \in \{1, 2, ...\}$, the pair $e^{(t)} = (\{u, v\}, \delta)$ of an edge $\{u, v\}$ and a sign $\delta \in \{+, -\}$ denotes the change in $\mathcal{G}$ at time t. Specifically, $(\{u, v\}, +)$ indicates the addition of a new edge $\{u, v\} \notin \mathcal{E}$, and $(\{u, v\}, -)$ indicates the deletion of an existing edge $\{u, v\} \in \mathcal{E}$. We use $\mathcal{G}^{(t)} = (\mathcal{V}^{(t)}, \mathcal{E}^{(t)})$ to indicate $\mathcal{G}$ at time t. That is,

$$\mathcal{E}^{(0)} = \emptyset \quad \text{and} \quad \mathcal{E}^{(t)} = \begin{cases} \mathcal{E}^{(t-1)} \cup \{\{u, v\}\}, & \text{if } e^{(t)} = (\{u, v\}, +), \\ \mathcal{E}^{(t-1)} \setminus \{\{u, v\}\}, & \text{if } e^{(t)} = (\{u, v\}, -). \end{cases}$$

Lastly, we let $\mathcal{T}^{(t)}$ denote the set of global triangles in $\mathcal{G}^{(t)}$ and $\mathcal{T}^{(t)}[u] \subset \mathcal{T}^{(t)}$ denote the set of local triangles of each node $u \in \mathcal{V}^{(t)}$ in $\mathcal{G}^{(t)}$.

Problem Definition (Problem 1): In this work, we address the problem of estimating the counts of global and local triangles in a fully dynamic graph stream. We assume the standard data stream model where the elements in the input stream, which may not fit in memory, can be accessed once in the given order unless they are explicitly stored in memory.

Problem 1 (Global and Local Triangle Counting in a Fully Dynamic Graph Stream).

- **Given:** a fully dynamic graph stream $(e^{(1)}, e^{(2)}, ...)$
    (i.e., sequence of edge additions and deletions in graph $\mathcal{G}$)
- **Maintain:** estimates of global triangle count $|\mathcal{T}^{(t)}|$ and local triangle counts $\{(u, |\mathcal{T}^{(t)}[u]|)\}_{u \in \mathcal{V}^{(t)}}$ of graph $\mathcal{G}^{(t)}$ for current $t \in \{1, 2, ...\}$
- **to Minimize:** the estimation errors.

We follow a general approach of reducing the biases and variances of estimates simultaneously rather than minimizing a specific measure of estimation error.

4   Proposed Method: Think Before You Discard (ThinkD)

We propose THINKD (**Think** before you **D**iscard), which estimates the counts of global and local triangles in a fully dynamic graph stream. For estimation with limited memory, THINKD samples edges and maintains those sampled edges, while discarding the other edges. The main idea of THINKD is to fully utilize unsampled edges before they are discarded. Specifically, whenever each element in the input stream arrives, THINKD first updates its estimates using the element. After that, if the element is an addition of an edge, THINKD decides whether to sample the edge or not.

We present two versions of THINKD and theoretically analyze their accuracies and complexities. To this end, we use $\bar{c}$ to denote the maintained estimate of the count of global triangles. Likewise, for each node u, we use $c[u]$ to denote the maintained estimate of the count of local triangles of node u. In addition, we let S be the set of currently sampled edges, and for each node u, we let $\hat{\mathcal{N}}[u]$ be the set of neighbors of u in the graph composed of the edges in S.

Algorithm 1. THINKD$_{FAST}$: Simple and Fast Version of THINKD

    Inputs   : fully dynamic graph stream: $(e^{(1)}, e^{(2)}, ...)$, sampling probability: r
    Outputs: estimate of the global triangle count: $\bar{c}$
                     estimates of the local triangle counts: $c[u]$ for each node u

1   $S \leftarrow \emptyset$
2   **for each** element $e^{(t)} = (\{u, v\}, \delta)$ in the input stream **do**
3   UPDATE($\{u, v\}, \delta$)
4   **if** $\delta = +$ **then** INSERT($\{u, v\}$)
5   **else if** $\delta = -$ **then** DELETE($\{u, v\}$)

6   **Procedure** UPDATE($\{u, v\}, \delta$):
7   **for each** common neighbor $w \in \hat{\mathcal{N}}[u] \cap \hat{\mathcal{N}}[v]$ **do**
8   **if** $\delta = +$ **then** increase $\bar{c}$, $c[u]$, $c[v]$, and $c[w]$ by $1/r^2$
9   **else if** $\delta = -$ **then** decrease $\bar{c}$, $c[u]$, $c[v]$, and $c[w]$ by $1/r^2$

10 **Procedure** INSERT($\{u, v\}$):
11   **if** a random number in Bernoulli(r) is 1 **then** $S \leftarrow S \cup \{\{u, v\}\}$

12 **Procedure** DELETE($\{u, v\}$):
13   **if** $\{u, v\} \in S$ **then** $S \leftarrow S \setminus \{\{u, v\}\}$

4.1   ThinkD$_{fast}$: Simple and Fast Version of ThinkD

THINKD$_{FAST}$, which is a simple and fast version of THINKD, is described in Algorithm 1. THINKD$_{FAST}$ initially has no sampled edges (line 1). Whenever each element $(\{u, v\}, \delta)$ of the input stream arrives (line 2), THINKD$_{FAST}$ first updates its estimates by calling the procedure UPDATE (line 3). Then, if the element is an addition (i.e., $\delta = +$), THINKD$_{FAST}$ samples the edge $\{u, v\}$ with a given sampling probability r (line 11) by calling the procedure INSERT (line 4). If the element is a deletion (i.e., $\delta = -$), THINKD$_{FAST}$ removes the edge $\{u, v\}$ from the existing samples (line 13) by calling the procedure DELETE (line 5).

In the procedure UPDATE, THINKD$_{FAST}$ finds the triangles connected by the arrived edge $\{u, v\}$ and two edges from the existing samples S (line 7). To this end, THINKD uses the fact that each common neighbor w of the nodes u and v in the graph composed of the sampled edges in S indicates the existence of such a triangle $\{u, v, w\}$. In the case of additions (i.e., $\delta = +$), since such triangles are new triangles added to the input stream, THINKD$_{\text{FAST}}$ increases the estimates of the global count and the corresponding local counts (line 8). In the case of deletions (i.e., $\delta = -$), since such triangles are those removed from the input stream, THINKD$_{\text{FAST}}$ decreases the estimates of the global count and the corresponding local counts (line 9). Notice that the amount of change per triangle is $1/r^2$, which is the reciprocal of the probability that each added or deleted triangle is discovered by THINKD$_{\text{FAST}}$. Note that each such triangle $\{u, v, w\}$ is discovered if and only if $\{w, u\}$ and $\{v, w\}$ are in $\mathcal{S}$, whose probability is r^2, as formalized in Lemma 1. This makes the expected amount of changes in the corresponding estimates for each such triangle be exactly one and thus makes THINKD$_{\text{FAST}}$ give unbiased estimates, as explained in detail in Sect. 4.3.

Lemma 1 (Discovery Probability of Triangles in ThinkD$_{\text{fast}}$). *In* THINKD$_{\text{FAST}}$, *any two distinct edges in graph* $\mathcal{G}^{(t)} = (\mathcal{V}^{(t)}, \mathcal{E}^{(t)})$ *are sampled with probability* r^2. *That is, if we let* $\mathcal{S}^{(t)}$ *be* $\mathcal{S}$ *in Algorithm 1 after the* t*-th element* $e^{(t)}$ *is processed, then*

$$Pr[\{u, v\} \in \mathcal{S}^{(t)} \cap \{w, x\} \in \mathcal{S}^{(t)}] = r^2, \; \forall t \geq 1, \; \forall \{u, v\} \neq \{w, x\} \in \mathcal{E}^{(t)}. \quad (1)$$

Proof. Equation (1) holds since each edge is sampled independently with probability r. See Sect. A.1 of the supplementary document [1] for a formal proof. ∎

(Dis)advantages of ThinkD$_{\text{fast}}$: Due to its simplicity, THINKD$_{\text{FAST}}$ is faster than its competitors, as shown empirically in Sect. 5.4. However, it is less accurate than THINKD$_{\text{ACC}}$, described in the following subsection, since it may discard edges even when memory is not full, leading to avoidable loss of information.

4.2   ThinkD$_{\text{acc}}$: Accurate Version of ThinkD

THINKD$_{\text{ACC}}$, which is an accurate version of THINKD, is described in Algorithm 2. Unlike THINKD$_{\text{FAST}}$, which may discard edges even when memory is not full, THINKD$_{\text{ACC}}$ maintains as many samples as possible within a given memory budget k (≥ 2) to minimize information loss.

To this end, THINKD$_{\text{ACC}}$ uses a sampling method called Random Pairing (RP) [9]. Given a fully dynamic stream with deletions, and a memory budget k, RP maintains at most k samples while satisfying the uniformity of the samples. That is, if we let $\mathcal{E}$ be the set of edges remaining (without being deleted) in the input stream so far and $\mathcal{S} \subset \mathcal{E}$ be the set of samples being maintained by RP, then the following equations hold:

$$|\mathcal{S}| \leq k \quad \text{and} \quad Pr[\mathcal{S} = \mathcal{A}] = Pr[\mathcal{S} = \mathcal{B}], \; \forall \mathcal{A} \neq \mathcal{B} \subset \mathcal{E} \text{ s.t. } |\mathcal{A}| = |\mathcal{B}|.$$

Updating the set $\mathcal{S}$ of samples using RP is described in lines 10–23. Whenever a deletion of an edge arrives, RP increases n_b or n_g depending on whether the edge is in $\mathcal{S}$ or not (lines 22 and 23). Roughly speaking, n_b and n_g denote the number of deletions that need to be "compensated" by additions (lines 16–18). If there is no deletion to compensate, RP processes each addition of an edge as in Reservoir Sampling [23]. That is, if memory is not full (i.e., $|\mathcal{S}| < k$), RP adds the new edge to $\mathcal{S}$ (line 13), while otherwise, RP replaces a random edge in $\mathcal{S}$ with the new edge with a certain probability (lines 14–15). We refer to [9] for the intuition behind the compensation and the details of RP; and we focus on how to use RP for triangle counting in the rest of this section.

Updating the estimates in THINKD$_{\text{ACC}}$ is the same as that in THINKD$_{\text{FAST}}$ except for the amount of change per triangle (lines 8 and 9), which is the reciprocal of the probability that each added or deleted triangle is discovered. When each element $e^{(t)} = (\{u, v\}, \delta)$ arrives, each added or deleted triangle $\{u, v, w\}$ is discovered if and only if $\{w, u\}$ and $\{v, w\}$ are in $\mathcal{S}$. As shown in Lemma 2, if we let $y = \min(k, |\mathcal{E}| + n_b + n_g)$, then the probability of such an event is

$$p(|\mathcal{E}|, n_b, n_g) := \frac{y}{|\mathcal{E}| + n_b + n_g} \times \frac{y - 1}{|\mathcal{E}| + n_b + n_g - 1}. \tag{2}$$

Algorithm 2. THINKD$_{\text{ACC}}$: Accurate Version of THINKD

Inputs : fully dynamic graph stream: $(e^{(1)}, e^{(2)}, ...)$, memory budget: k (≥ 2)
Outputs: estimate of the global triangle count: $\bar{c}$
        estimates of the local triangle counts: $c[u]$ for each node u

1 $\mathcal{S} \leftarrow \emptyset$, $|\mathcal{E}| \leftarrow 0$, $n_b \leftarrow 0$, $n_g \leftarrow 0$
2 **for each** element $e^{(t)} = (\{u, v\}, \delta)$ in the input stream **do**
3   UPDATE($\{u, v\}, \delta$)
4   **if** $\delta = +$ **then** INSERT($\{u, v\}$)
5   **else if** $\delta = -$ **then** DELETE($\{u, v\}$)

6 **Procedure** UPDATE($\{u, v\}, \delta$):
7   **for each** common neighbor $w \in \hat{\mathcal{N}}[u] \cap \hat{\mathcal{N}}[v]$ **do**
8   **if** $\delta = +$ **then** increase $\bar{c}$, $c[u]$, $c[v]$, and $c[w]$ by $1/p(|\mathcal{E}|, n_b, n_g)$
9   **else if** $\delta = -$ **then** decrease $\bar{c}$, $c[u]$, $c[v]$, and $c[w]$ by $1/p(|\mathcal{E}|, n_b, n_g)$

10 **Procedure** INSERT($\{u, v\}$):
11      $|\mathcal{E}| \leftarrow |\mathcal{E}| + 1$
12   **if** $n_b + n_g = 0$ **then**
13   **if** $|\mathcal{S}| < k$ **then** $\mathcal{S} \leftarrow \mathcal{S} \cup \{\{u, v\}\}$
14   **else if** *a random number in* Bernoulli($k/|\mathcal{E}|$) is 1 **then**
15   replace a random edge in $\mathcal{S}$ with $\{u, v\}$
16   **else if** *a random number in* Bernoulli($n_b/(n_b + n_g)$) is 1 **then**
17         $\mathcal{S} \leftarrow \mathcal{S} \cup \{\{u, v\}\}$, $n_b \leftarrow n_b - 1$
18   **else** $n_g \leftarrow n_g - 1$

19 **Procedure** DELETE($\{u, v\}$):
20      $|\mathcal{E}| \leftarrow |\mathcal{E}| - 1$
21   **if** $\{u, v\} \in \mathcal{S}$ **then**
22         $\mathcal{S} \leftarrow \mathcal{S} \setminus \{\{u, v\}\}$, $n_b \leftarrow n_b + 1$
23   **else** $n_g \leftarrow n_g + 1$

Lemma 2 (Discovery Probability of Triangles in ThinkD$_{acc}$). *In* THINKD$_{ACC}$, *any two distinct edges in graph* $\mathcal{G}^{(t)} = (\mathcal{V}^{(t)}, \mathcal{E}^{(t)})$ *are sampled with probability as in Eq.* (2). *That is, if we let* $p^{(t)}$ *and* $\mathcal{S}^{(t)}$ *be the values of Eq.* (2) *and* $\mathcal{S}$, *resp., in Algorithm 2 after the t-th element* $e^{(t)}$ *is processed, then*

$$Pr[\{u,v\} \in \mathcal{S}^{(t)} \cap \{w,x\} \in \mathcal{S}^{(t)}] = p^{(t)}, \; \forall t \geq 1, \; \forall \{u,v\} \neq \{w,x\} \in \mathcal{E}^{(t)}. \quad (3)$$

Proof. See Sect. A.2 of the supplementary document [1] for a proof. ∎

(Dis)advantages of ThinkD$_{acc}$: Within the same memory budget, THINKD$_{ACC}$ is slower than THINKD$_{FAST}$ since THINKD$_{ACC}$ maintains and processes more samples on average. However, THINKD$_{ACC}$ is more accurate than THINKD$_{FAST}$ by utilizing more samples. These are shown empirically in Sects. 5.3 and 5.4.

Reducing Estimation Errors by Sacrificing Unbiasedness: The estimates (i.e., $\bar{c}$ and $c[u]$ for each node u) in Algorithms 1 and 2 can have negative values. Since true triangle counts are always non-negative, lower bounding the estimates by zero always reduces the estimation errors. However, the estimates become biased, and Theorem 1 in the following section does not hold anymore.

4.3   Accuracy Analyses

We prove that THINKD$_{FAST}$ and THINKD$_{ACC}$ maintain unbiased estimates with the expected values equal to the true global and local triangle counts. Then, we analyze the variances of the estimates that THINKD$_{FAST}$ maintains. To this end, for each variable (e.g., $\bar{c}$) in Algorithms 1 and 2, we use superscript (t) (e.g., $\bar{c}^{(t)}$) to denote the value of the variable after the t-th element $e^{(t)}$ is processed.

We first define *added triangles* and *deleted triangles* in Definitions 1 and 2.

Definition 1 (Added Triangles). *Let* $\mathcal{A}^{(t)}$ *be the set of triangles that have been added to graph* $\mathcal{G}$ *at time t or earlier. Formally,*

$$\mathcal{A}^{(t)} := \{(\{u,v,w\}, s) : 1 \leq s \leq t \text{ and } \{u,v,w\} \notin T^{(s-1)} \text{ and } \{u,v,w\} \in T^{(s)}\},$$

where addition time s is for distinguishing triangles composed of the same nodes but added at different times.[1]

Definition 2 (Deleted Triangles). *Let* $\mathcal{D}^{(t)}$ *be the set of triangles that have been removed from graph* $\mathcal{G}$ *at time t or earlier. Formally,*

$$\mathcal{D}^{(t)} := \{(\{u,v,w\}, s) : 1 \leq s \leq t \text{ and } \{u,v,w\} \in T^{(s-1)} \text{ and } \{u,v,w\} \notin T^{(s)}\},$$

where deletion time s is for distinguishing triangles composed of the same nodes but deleted at different times (see footnote 1).

---

[1] Note that triangles composed of the same nodes can be added multiple times (and thus can be removed multiple times) only if deleted edges are added again.

Similarly, for each node $u \in \mathcal{V}^{(t)}$, we use $\mathcal{A}^{(t)}[u] \subset \mathcal{A}^{(t)}$ and $\mathcal{D}^{(t)}[u] \subset \mathcal{D}^{(t)}$ to denote the added and deleted triangles with node u, respectively. Lemma 3 formalizes the relationship between these concepts and the number of triangles.

Lemma 3 (Count of Triangles in the Current Graph). *The count of triangles in the current graph equals to the count of added triangles subtracted by the count of deleted triangles. Formally,*

$$|\mathcal{T}^{(t)}| = |\mathcal{A}^{(t)}| - |\mathcal{D}^{(t)}|, \ \forall t \geq 1, \tag{4}$$

$$|\mathcal{T}^{(t)}[u]| = |\mathcal{A}^{(t)}[u]| - |\mathcal{D}^{(t)}[u]|, \ \forall t \geq 1, \ \forall u \in \mathcal{V}^{(t)}. \tag{5}$$

Proof. Equations (4) and (5) follow from Definitions 1 and 2. See Sect. A.3 of the supplementary document [1] for a formal proof. ∎

Based on these concepts, we prove that THINKD$_{\text{FAST}}$ and THINKD$_{\text{ACC}}$ maintain unbiased estimates in Theorem 1. For the unbiasedness of the estimate $\bar{c}$ of the global count, we show that the expected amount of change in $\bar{c}$ for each added triangle is $+1$, while that for each deleted triangle is -1. Then, by Lemma 3, the expected value of $\bar{c}$ equals to the true global count. Likewise, we show the unbiasedness of the estimate of the local triangle count of each node by considering only the added and deleted triangles incident to the node.

Theorem 1 ('Any Time' Unbiasedness of ThinkD). THINKD *gives unbiased estimates at any time. Formally, in Algorithms 1 and 2,*

$$\mathbb{E}[\bar{c}^{(t)}] = |\mathcal{T}^{(t)}|, \ \forall t \geq 1, \tag{6}$$

$$\mathbb{E}[c^{(t)}[u]] = |\mathcal{T}^{(t)}[u]|, \ \forall t \geq 1, \ \forall u \in \mathcal{V}^{(t)}. \tag{7}$$

Proof. Consider a triangle $(\{u, v, w\}, s) \in \mathcal{A}^{(t)}$, and let $e^{(s)} = (\{u, v\}, +)$ without loss of generality. The amount $\alpha_{uvw}^{(s)}$ of change in each of $\bar{c}$, $c[u]$, $c[v]$, and $c[w]$ due to the discovery of $(\{u, v, w\}, s)$ in line 8 of Algorithm 1 or Algorithm 2 is

$$\alpha_{uvw}^{(s)} = \begin{cases} 1/r^2 & if\{v, w\} \in \mathcal{S}^{(s-1)} \text{ and } \{w, u\} \in \mathcal{S}^{(s-1)} \text{ in Algorithm 1} \\ 1/p^{(s-1)} & if\{v, w\} \in \mathcal{S}^{(s-1)} \text{ and } \{w, u\} \in \mathcal{S}^{(s-1)} \text{ in Algorithm 2} \\ 0 & otherwise. \end{cases}$$

Then, from Eqs. (1) and (3), the following equation holds:

$$\alpha_{uvw}^{(s)} = \begin{cases} \frac{1}{Pr[\{v,w\} \in \mathcal{S}^{(s-1)} \cap \{w,u\} \in \mathcal{S}^{(s-1)}]} & if\{v, w\} \in \mathcal{S}^{(s-1)} \text{ and } \{w, u\} \in \mathcal{S}^{(s-1)} \\ 0 & otherwise. \end{cases}$$

Hence,

$$\mathbb{E}[\alpha_{uvw}^{(s)}] = 1. \tag{8}$$

Consider a triangle $(\{u, v, w\}, s) \in \mathcal{D}^{(t)}$, and let $e^{(s)} = (\{u, v\}, -)$ without loss of generality. The amount $\beta_{uvw}^{(s)}$ of change in each of $\bar{c}$, $c[u]$, $c[v]$, and $c[w]$

due to the discovery of $(\{u, v, w\}, s)$ in line 9 of Algorithm 1 or Algorithm 2 is

$$\beta_{uvw}^{(s)} = \begin{cases} -1/r^2 & if\{v, w\} \in \mathcal{S}^{(s-1)} \text{ and } \{w, u\} \in \mathcal{S}^{(s-1)} \text{ in Algorithm 1} \\ -1/p^{(s-1)} & if\{v, w\} \in \mathcal{S}^{(s-1)} \text{ and } \{w, u\} \in \mathcal{S}^{(s-1)} \text{ in Algorithm 2} \\ 0 & otherwise. \end{cases}$$

Then, from Eqs. (1) and (3), the following equation holds:

$$\beta_{uvw}^{(s)} = \begin{cases} \frac{-1}{Pr[\{v,w\}\in\mathcal{S}^{(s-1)}\cap\{w,u\}\in\mathcal{S}^{(s-1)}]} & if\{v, w\} \in \mathcal{S}^{(s-1)} \text{ and } \{w, u\} \in \mathcal{S}^{(s-1)} \\ 0 & otherwise. \end{cases}$$

Hence,

$$\mathbb{E}[\beta_{uvw}^{(s)}] = -1. \tag{9}$$

By definition, the following holds:

$$\bar{c}^{(t)} = \sum_{(\{u,v,w\},s)\in\mathcal{A}^{(t)}} \alpha_{uvw}^{(s)} + \sum_{(\{u,v,w\},s)\in\mathcal{D}^{(t)}} \beta_{uvw}^{(s)}.$$

By linearity of expectation, Eqs. (8), (9), and Lemma 3, the following holds:

$$\mathbb{E}[\bar{c}^{(t)}] = \sum_{(\{u,v,w\},s)\in\mathcal{A}^{(t)}} \mathbb{E}[\alpha_{uvw}^{(s)}] + \sum_{(\{u,v,w\},s)\in\mathcal{D}^{(t)}} \mathbb{E}[\beta_{uvw}^{(s)}]$$

$$= \sum_{(\{u,v,w\},s)\in\mathcal{A}^{(t)}} 1 + \sum_{(\{u,v,w\},s)\in\mathcal{D}^{(t)}} (-1) = |\mathcal{A}^{(t)}| - |\mathcal{D}^{(t)}| = |\mathcal{T}^{(t)}|.$$

Likewise, for each node $u \in \mathcal{V}^{(t)}$, the following holds:

$$c^{(t)}[u] = \sum_{(\{u,v,w\},s)\in\mathcal{A}^{(t)}[u]} \alpha_{uvw}^{(s)} + \sum_{(\{u,v,w\},s)\in\mathcal{D}^{(t)}[u]} \beta_{uvw}^{(s)}.$$

By linearity of expectation, Eqs. (8), (9), and Lemma 3, the following holds:

$$\mathbb{E}[c^{(t)}[u]] = \sum_{(\{u,v,w\},s)\in\mathcal{A}^{(t)}[u]} \mathbb{E}[\alpha_{uvw}^{(s)}] + \sum_{(\{u,v,w\},s)\in\mathcal{D}^{(t)}[u]} \mathbb{E}[\beta_{uvw}^{(s)}]$$

$$= \sum_{(\{u,v,w\},s)\in\mathcal{A}^{(t)}[u]} 1 + \sum_{(\{u,v,w\},s)\subset\mathcal{D}^{(t)}[u]} (-1) = |\mathcal{A}^{(t)}[u]| - |\mathcal{D}^{(t)}[u]| = |\mathcal{T}^{(t)}[u]|.$$

∎

In Sect. B of the supplementary document [1], we prove the formulas for the variances of estimates given by THINKD$_{\text{FAST}}$. Theorem 2 is implied by them.

Theorem 2 (Variance of ThinkD$_{\text{fast}}$). *Given an input graph stream, the variances of estimates maintained by* THINKD$_{\text{FAST}}$ *with the sampling probability* r *is proportional to* $1/r^2$. *Formally, in Algorithm 1,*

$$Var[\bar{c}^{(t)}] = O(1/r^2), \quad \forall t \geq 1, \quad and \quad Var[c^{(t)}[u]] = O(1/r^2), \quad \forall t \geq 1, \forall u \in \mathcal{V}^{(t)}.$$

Proof. See Theorem 5 in Sect. B of the supplementary document [1]. ∎

4.4 Complexity Analyses

We analyze the time and space complexities of $\textsc{ThinkD}_{\text{FAST}}$ and $\textsc{ThinkD}_{\text{ACC}}$. In our analyses, we use $\bar{\mathcal{V}}^{(t)} := \bigcup_{s=1}^{t} \mathcal{V}^{(s)}$ to denote the set of nodes that appear in the t-th or earlier elements in the input stream.

Space Complexity: To process the first t elements in the input graph stream, $\textsc{ThinkD}_{\text{FAST}}$ and $\textsc{ThinkD}_{\text{ACC}}$ maintain one estimate for the global triangle count and at most $|\bar{\mathcal{V}}^{(t)}|$ estimates for the local triangle counts. In addition, $\textsc{ThinkD}_{\text{FAST}}$ maintains $|\mathcal{E}^{(t)}| \cdot r$ edges on average, while $\textsc{ThinkD}_{\text{ACC}}$ maintains up to k edges. Thus, the average space complexities of $\textsc{ThinkD}_{\text{FAST}}$ and $\textsc{ThinkD}_{\text{ACC}}$ are $O(|\mathcal{E}^{(t)}| \cdot r + |\bar{\mathcal{V}}^{(t)}|)$ and $O(k + |\bar{\mathcal{V}}^{(t)}|)$, respectively. The complexities become $O(|\mathcal{E}^{(t)}| \cdot r)$ and $O(k)$ when only the global triangle count needs to be estimated.

Time Complexity: We prove the average time complexity of $\textsc{ThinkD}_{\text{FAST}}$ in Theorem 3, which implies Corollary 1, and the worst-case time complexity of $\textsc{ThinkD}_{\text{ACC}}$ in Theorem 4. Corollary 1 and Theorem 4 state that, given a fixed memory budget k, $\textsc{ThinkD}_{\text{FAST}}$ and $\textsc{ThinkD}_{\text{ACC}}$ scale linearly with the number of elements in the input stream.

Theorem 3 (Time Complexity of ThinkD$_{\text{fast}}$). *Algorithm 1 takes $O(t + t^2 r)$ on average to process the first t elements in the input stream.*

Table 3. Summary of the real-world and synthetic graph streams used in our experiments. B: billion, M: million, K: thousand.

Name	#Nodes	#Edges	Type	Name	#Nodes	#Edges	Type
Friendster	65.6 M	1.81 B	Friendship	Youtube	3.22 M	9.38 M	Friendship
Orkut	3.07 M	117 M	Friendship	BerkStan	685 K	6.65 M	Web
Flickr	2.30 M	22.8 M	Friendship	Facebook	63.7 K	817 K	Friendship
Patent	3.77 M	16.5 M	Citation	Epinion	132 K	711 K	Trust
Random (800 GB)	1 M	0.1 B–100 B	Synthetic				

Proof. In Algorithm 1, the most expensive step in processing each element $e^{(s)} = (\{u, v\}, \delta)$ is to intersect $\hat{\mathcal{N}}[u]$ and $\hat{\mathcal{N}}[v]$ (line 7), which takes $O(1 + \mathbb{E}[|\hat{\mathcal{N}}[u]| + |\hat{\mathcal{N}}[v]|]) = O(1 + \mathbb{E}[|\mathcal{S}|]) = O(1 + sr)$ on average. Hence, processing the first t elements takes $\sum_{s=1}^{t} O(1 + sr) = O(t + t^2 r)$ on average. ∎

Corollary 1 (Time Complexity of ThinkD$_{\text{fast}}$ with Fixed Memory k). *If $r = O(k/t)$ for a constant k (≥ 1), then Algorithm 1 takes $O(tk)$ on average to process the first t elements in the input stream.*

Theorem 4 (Time Complexity of ThinkD$_{\text{acc}}$). *Algorithm 2 takes $O(tk)$ to process the first t elements in the input stream.*

Proof. In Algorithm 2, the most expensive step in processing each element $e^{(s)} = (\{u, v\}, \delta)$ is to intersect $\hat{\mathcal{N}}[u]$ and $\hat{\mathcal{N}}[v]$ (line 7), which takes $O(1 + |\hat{\mathcal{N}}[u]| + |\hat{\mathcal{N}}[v]|) = O(k)$. Thus, processing the first t elements takes $O(tk)$. ∎

5   Experiments

In this section, we review our experiments for answering the following questions:

- **Q1. Illustration of Theorems:** Does THINKD give unbiased estimates? Does THINKD scale linearly with the size of the input stream?
- **Q2. Accuracy:** Is THINKD more accurate than its best competitors?
- **Q3. Speed:** Is THINKD faster than its best competitors?
- **Q4. Effects of Deletions:** Is THINKD consistently accurate regardless of the ratio of deleted edges?

In addition, in Sect. D of the supplementary document [1], we describe how THINKD can be used to detect the sudden emergence of dense subgraphs, and we experimentally show that it outperforms state-of-the-art competitors.

5.1   Experimental Settings

Machines: We used a machine with a 3.60 GHz CPU and 32 GB RAM unless otherwise stated.

Datasets: We created fully dynamic graph streams with deletions using the real-world graphs listed in Table 3 as follows: (a) create the additions of the edges in the input graph and shuffle them, (b) choose α% of the edges and create the deletions of them, (c) locate each deletion in a random position after the corresponding addition. We set α to 20% unless otherwise stated (see Sect. 5.5 for its effect on accuracy). The created streams were streamed from the disk.

Implementations: We implemented THINKD$_{FAST}$, THINKD$_{ACC}$, TRIEST$_{FD}$ [7], TRIEST$_{IMPR}$ [7], ESD [10], and MASCOT [14] in Java 1.7. In all of them, sampled edges are stored in the adjacency list format, and as described in the last paragraph of Sect. 4.2, estimates are lower bounded by zero.

Evaluation Metrics: Let x and $\{(u, x[u])\}_{u \in \mathcal{V}}$ be the true counts of global triangles and local triangles at the end of the input stream. Let $\hat{x}$ and $\{(u, \hat{x}[u])\}_{u \in \mathcal{V}}$

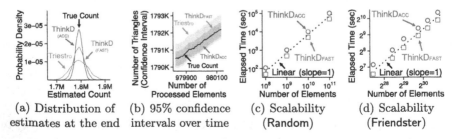

| (a) Distribution of estimates at the end | (b) 95% confidence intervals over time | (c) Scalability (Random) | (d) Scalability (Friendster) |

Fig. 1. ThinkD is provably accurate and scalable. (a) THINKD gives unbiased estimates with smaller variances than its best competitor. (b) THINKD maintains more accurate estimates with smaller confidence intervals than its best competitor. (c–d) THINKD scales linearly with the size of the input stream.

be the corresponding estimates obtained by the evaluated algorithm. We used *global error*, defined as $\frac{|x-\hat{x}|}{1+x}$, and *RMSE*, defined as $\sqrt{\frac{1}{|\mathcal{V}|}\sum_{u\in\mathcal{V}}(x[u]-\hat{x}[u])^2}$, to evaluate the accuracy of global and local triangle counting, respectively.

5.2 Q1. Illustration of Theorems

ThinkD Gives Unbiased Estimates (Theorem 1). We compared 10,000 estimates of the global triangle count obtained by $\text{THINKD}_{\text{FAST}}$, $\text{THINKD}_{\text{ACC}}$, and $\text{TRIEST}_{\text{FD}}$, whose parameters were set so that on average 10% of the edges are stored at the end of each graph stream. Figure 1(a) shows the distributions of the estimates at the end of the Facebook dataset. The means of the estimates were close to the true triangle count, consistently with Theorem 1 (i.e., unbiasedness of THINKD). Moreover, $\text{THINKD}_{\text{ACC}}$ and $\text{THINKD}_{\text{FAST}}$ gave estimates with smaller variances than $\text{TRIEST}_{\text{FD}}$. Figure 1(b) shows how the 95% confidence intervals change over time in the Facebook dataset. $\text{THINKD}_{\text{FAST}}$ and $\text{THINKD}_{\text{ACC}}$ maintained more accurate estimates with smaller confidence intervals than $\text{TRIEST}_{\text{FD}}$. Between $\text{THINKD}_{\text{FAST}}$ and $\text{THINKD}_{\text{ACC}}$, $\text{THINKD}_{\text{ACC}}$ was more accurate.

ThinkD Scales Linearly (Corollary 1 and Theorem 4). We measured the elapsed times taken by $\text{THINKD}_{\text{FAST}}$ and $\text{THINKD}_{\text{ACC}}$ to process all elements in graph streams with different numbers of elements. To measure their speeds independently of the speed of the input stream, we ignored time taken to wait for the arrival of elements. In both algorithms, we set k and r so that on average 10^7 edges are stored at the end of each input stream. Figure 1(c) shows the results in the Random datasets, which were created by the Erdös-Rényi model. Both $\text{THINKD}_{\text{FAST}}$ and $\text{THINKD}_{\text{ACC}}$ scaled linearly with the number of elements, as expected in Corollary 1 and Theorem 4. Notice that the largest dataset is **800** GB **with 100 billion elements**. As seen in Fig. 1(d), $\text{THINKD}_{\text{FAST}}$ and $\text{THINKD}_{\text{ACC}}$ showed linear scalability also in a graph stream with realistic structure, which we created by sampling different numbers of elements from the Friendster dataset.

5.3 Q2. Accuracy (ThinkD Is More Accurate Than Its Competitors)

We compared the accuracies of four algorithms that support edge deletions. As we changed the ratio of stored edges at the end of each input stream from 5% to 40%, we measured the accuracies of $\text{THINKD}_{\text{FAST}}$, $\text{THINKD}_{\text{ACC}}$, and $\text{TRIEST}_{\text{FD}}$. ESD always stores the entire input stream in memory, and we set its parameter to 1.0 to maximize its accuracy. Each evaluation metric was averaged over 100 trials in the Friendster and Orkut datasets and 1,000 trials in the others.[2] As seen in Fig. 2, $\text{THINKD}_{\text{FAST}}$ and $\text{THINKD}_{\text{ACC}}$ consistently gave the best trade-off between space and accuracy. Specifically, within the same memory budget, $\text{THINKD}_{\text{ACC}}$ was up to **4×** and **4.3× more accurate** than $\text{TRIEST}_{\text{FD}}$ in terms of global error and RMSE, respectively. Between our algorithms, $\text{THINKD}_{\text{ACC}}$ consistently outperformed $\text{THINKD}_{\text{FAST}}$. We observed the same trend in the other datasets (see Fig. 5 in the supplementary document [1]).

---

[2] We used a machine with 2.67 GHz CPUs and 1 TB memory for the Friendster dataset.

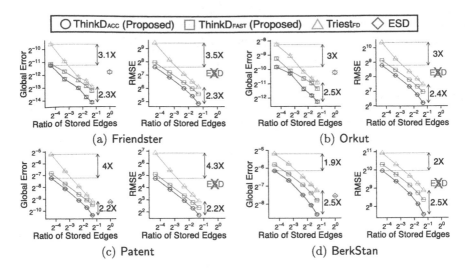

Fig. 2. ThinkD is accurate. ThinkD gives the best trade-off between space and accuracy. In particular, ThinkD$_{ACC}$ is up to **4.3× more accurate** than Triest$_{FD}$ within the same memory budget. Error bars denote ±1 standard error. ESD is inapplicable to local triangle counting.

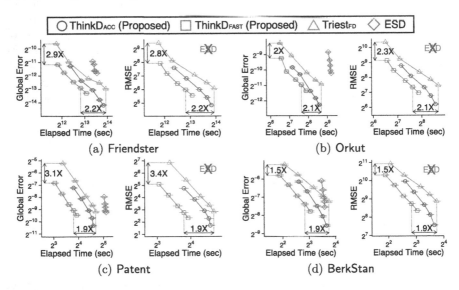

Fig. 3. ThinkD is fast. ThinkD gives the best trade-off between speed and accuracy. In particular, ThinkD$_{FAST}$ is up to **2.2× faster** than Triest$_{FD}$ when they are similarly accurate. Error bars denote ±1 standard error. ESD is inapplicable to local triangle counting.

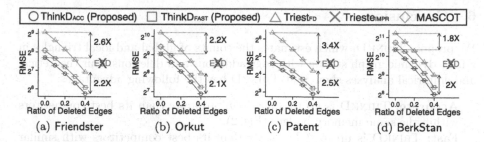

Fig. 4. ThinkD is consistently accurate regardless of the ratio of deleted edges. Error bars denote ±1 standard error. TRIEST$_{\text{IMPR}}$ and MASCOT are inapplicable when there are deletions. ESD is inapplicable to local triangle counting.

5.4   Q3. Speed (ThinkD Is Faster Than Its Competitors)

We compared the speeds and accuracies of four algorithms that support edge deletions. The detailed settings were the same as those in Sect. 5.3 except that we measured the performance of ESD as we changed its parameter from 0.2 to 1.0. To measure the speeds of the algorithms independently of the speed of the input stream, we ignored time taken to wait for the arrival of elements. As seen in Fig. 3, THINKD$_{\text{FAST}}$ and THINKD$_{\text{ACC}}$ consistently gave the best trade-off between speed and accuracy. Specifically, for the same global error and RMSE, THINKD$_{\text{FAST}}$ was up to **2.2× faster** than TRIEST$_{\text{FD}}$. Between our algorithms, THINKD$_{\text{FAST}}$ consistently outperformed THINKD$_{\text{ACC}}$. We observed the same trend in the other datasets (see Fig. 6 in the supplementary document [1]).

5.5   Q4. Effects of Deletions (ThinkD Is Consistently Accurate)

We measured how the ratio of deleted edges (i.e., α in Sect. 5.1) in input graph streams affects the accuracies of the considered algorithms. In every algorithm, we set the ratio of stored edges at the end of each input stream to 10%. As seen in Fig. 4, all algorithms that support edge deletions became more accurate as input graphs became smaller with more deletions. THINKD$_{\text{FAST}}$ and THINKD$_{\text{ACC}}$ were similarly accurate with MASCOT and TRIEST$_{\text{IMPR}}$, respectively, in the streams without deletions. In the streams with deletions, which MASCOT and TRIEST$_{\text{IMPR}}$ cannot handle, THINKD$_{\text{FAST}}$ and THINKD$_{\text{ACC}}$ were **1.8 − 3.4× more accurate** than TRIEST$_{\text{FD}}$ regardless of the ratio of deleted edges. We observed the same trend in the other datasets (see Fig. 7 in the supplementary document [1]).

6 Conclusion

We propose THINKD, which estimates the counts of global and local triangles in a fully dynamic graph stream with edge additions and deletions. Our theoretical and empirical analyses show that THINKD has the following advantages:

- **Accurate:** THINKD is up to *4.3× more accurate* than its best competitors within the same memory budget (Fig. 2).
- **Fast:** THINKD is up to *2.2× faster* than its best competitors with similar accuracies (Fig. 3). THINKD processes *terabyte*-scale graph streams with linear scalability (Fig. 1, Corollary 1, and Theorem 4).
- **Theoretically Sound:** THINKD maintains *unbiased* estimates (Theorem 1) with small variances (Theorem 2) *at any time* while the input graph evolves.

Reproducibility: The source code and datasets used in the paper are available at http://www.cs.cmu.edu/~kijungs/codes/thinkd/.

Acknowledgements. This material is based upon work supported by the National Science Foundation under Grants No. CNS-1314632 and IIS-1408924. Research was sponsored by the Army Research Laboratory and was accomplished under Cooperative Agreement Number W911NF-09-2-0053. Shin was supported by the KFAS Scholarship, and Kim was supported by the Samsung Scholarship. Any opinions, findings, and conclusions or recommendations expressed in this material are those of the author(s) and do not necessarily reflect the views of the National Science Foundation, or other funding parties. The U.S. Government is authorized to reproduce and distribute reprints for Government purposes notwithstanding any copyright notation here on.

References

1. Supplementary   document   (2018).   http://www.cs.cmu.edu/~kijungs/codes/thinkd/supple.pdf
2. Ahmed, N.K., Duffield, N., Willke, T.L., Rossi, R.A.: On sampling from massive graph streams. PVLDB **10**(11), 1430–1441 (2017)
3. Bar-Yossef, Z., Kumar, R., Sivakumar, D.: Reductions in streaming algorithms, with an application to counting triangles in graphs. In: SODA (2002)
4. Batagelj, V., Zaveršnik, M.: Short cycle connectivity. Discret. Math. **307**(3), 310–318 (2007)
5. Becchetti, L., Boldi, P., Castillo, C., Gionis, A.: Efficient algorithms for large-scale local triangle counting. TKDD **4**(3), 13 (2010)
6. Berry, J.W., Hendrickson, B., LaViolette, R.A., Phillips, C.A.: Tolerating the community detection resolution limit with edge weighting. Phys. Rev. E **83**(5), 056119 (2011)
7. De Stefani, L., Epasto, A., Riondato, M., Upfal, E.: Trièst: counting local and global triangles in fully-dynamic streams with fixed memory size. In: KDD (2016)
8. Epasto, A., Lattanzi, S., Mirrokni, V., Sebe, I.O., Taei, A., Verma, S.: Ego-net community mining applied to friend suggestion. PVLDB **9**(4), 324–335 (2015)
9. Gemulla, R., Lehner, W., Haas, P.J.: Maintaining bounded-size sample synopses of evolving datasets. VLDB J. **17**(2), 173–201 (2008)

10. Han, G., Sethu, H.: Edge sample and discard: a new algorithm for counting triangles in large dynamic graphs. In: ASONAM (2017)
11. Jha, M., Seshadhri, C., Pinar, A.: A space efficient streaming algorithm for triangle counting using the birthday paradox. In: KDD (2013)
12. Kolountzakis, M.N., Miller, G.L., Peng, R., Tsourakakis, C.E.: Efficient triangle counting in large graphs via degree-based vertex partitioning. In: Kumar, R., Sivakumar, D. (eds.) WAW 2010. LNCS, vol. 6516, pp. 15–24. Springer, Heidelberg (2010). https://doi.org/10.1007/978-3-642-18009-5_3
13. Kutzkov, K., Pagh, R.: Triangle counting in dynamic graph streams. In: Ravi, R., Gørtz, I.L. (eds.) SWAT 2014. LNCS, vol. 8503, pp. 306–318. Springer, Cham (2014). https://doi.org/10.1007/978-3-319-08404-6_27
14. Lim, Y., Kang, U.: MASCOT: memory-efficient and accurate sampling for counting local triangles in graph streams. In: KDD (2015)
15. Newman, M.E.: The structure and function of complex networks. SIAM Rev. **45**(2), 167–256 (2003)
16. Pavan, A., Tangwongsan, K., Tirthapura, S., Wu, K.L.: Counting and sampling triangles from a graph stream. PVLDB **6**(14), 1870–1881 (2013)
17. Shin, K.: WRS: waiting room sampling for accurate triangle counting in real graph streams. In: ICDM (2017)
18. Shin, K., Eliassi-Rad, T., Faloutsos, C.: Patterns and anomalies in k-cores of real-world graphs with applications. Knowl. Inf. Syst. **54**(3), 677–710 (2018)
19. Shin, K., Hammoud, M., Lee, E., Oh, J., Faloutsos, C.: Tri-Fly: distributed estimation of global and local triangle counts in graph streams. In: Phung, D., Tseng, V.S., Webb, G.I., Ho, B., Ganji, M., Rashidi, L. (eds.) PAKDD 2018. LNCS (LNAI), vol. 10939, pp. 651–663. Springer, Cham (2018). https://doi.org/10.1007/978-3-319-93040-4_51
20. Tangwongsan, K., Pavan, A., Tirthapura, S.: Parallel triangle counting in massive streaming graphs. In: CIKM (2013)
21. Tsourakakis, C.E.: Fast counting of triangles in large real networks without counting: algorithms and laws. In: ICDM (2008)
22. Tsourakakis, C.E., Drineas, P., Michelakis, E., Koutis, I., Faloutsos, C.: Spectral counting of triangles via element-wise sparsification and triangle-based link recommendation. Soc. Netw. Anal. Min. **1**(2), 75–81 (2011)
23. Vitter, J.S.: Random sampling with a reservoir. TOMS **11**(1), 37–57 (1985)
24. Watts, D.J., Strogatz, S.H.: Collective dynamics of 'small-world' networks. Nature **393**(6684), 440–442 (1998)

Semi-supervised Blockmodelling
with Pairwise Guidance

Mohadeseh Ganji[1(✉)], Jeffrey Chan[2], Peter J. Stuckey[1], James Bailey[1],
Christopher Leckie[1], Kotagiri Ramamohanarao[1], and Laurence Park[3]

[1] School of Computing and Information Systems, University of Melbourne,
Melbourne, Australia
{mohadeseh.ganji,pstuckey,baileyj,caleckie,kotagiri}@unimelb.edu.au
[2] School of Computer Science and Software Engineering, RMIT University,
Melbourne, Australia
jeffrey.chan@rmit.edu.au
[3] Computer Science Department, Western Sydney University, Penrith, Australia
l.park@westernsydney.edu.au

Abstract. Blockmodelling is an important technique for detecting
underlying patterns in graphs. Existing blockmodelling algorithms are
unsupervised and cannot take advantage of the existing information that
might be available about objects that are known to be similar. This background information can help finding complex patterns, such as hierarchical or ring blockmodel structures, which are difficult for traditional
blockmodelling algorithms to detect. In this paper, we propose a new
semi-supervised framework for blockmodelling, which allows background
information to be incorporated in the form of pairwise membership information. Our proposed framework is based on the use of Lagrange multipliers and can be incorporated into existing iterative blockmodelling algorithms, enabling them to find complex blockmodel patterns in graphs.
We demonstrate the utility of our framework for discovering complex
patterns, via experiments over a range of synthetic and real data sets.
Code related to this paper is available at: https://people.eng.unimelb.
edu.au/mganji/.

Keywords: Blockmodelling · Pairwise information
Lagrange multipliers

1 Introduction

Understanding the latent structure beneath real world complex interactions
allows us to gain a deeper insight into the underlying reason and purpose of these
interactions. Representing these interactions as a graph allows us to discover

Electronic supplementary material The online version of this chapter (https://
doi.org/10.1007/978-3-030-10928-8_10) contains supplementary material, which is
available to authorized users.

these informative structures. Community structure has been extensively studied in the context of graph mining and many algorithms have been introduced to locate communities in graphs [2,12]. Commonly, communities are defined as a group of vertices that are densely connected among themselves but have sparse connections to the rest of the graph.

However, graphs may contain other inherent and latent structures as well. For example, consider a network of interactions in a question answering forum where members ask or answer questions. In such forums often there is a group of novices who ask many questions but rarely reply to questions, and a group of experts who answer questions but may not ask many questions. If you consider the graph representing such a question answering relationship, community detection fails to find the underlying groups as it mixes all the members (novices and experts) together due to the many edges between them. Hence, a more general approach is needed that not only detects communities but is also able to reveal deeper patterns in the network.

Blockmodelling is a powerful approach that partitions graphs into groups of equivalent vertices (also called blocks or positions) that play a similar role in the graph [22]. Equivalent vertices have connections to similar vertices that may or may not be in the same group. For example, vertices in a community structure mainly link to their own community, while for the question answering forum example, the expert group (answerers) have similar connections (e.g., they are mainly connected to novices), and vice versa. Hence, blockmodelling is a general approach to discover deeper graph structures, not only communities. In blockmodelling, the inherent structure of the graph can be identified by visualizing the interactions within and between blocks, which is captured in the so-called image matrix. Given k, the number of blocks, the image matrix is a $k \times k$ non-negative real-valued matrix whose elements represent the probability of interaction between and within the blocks. For instance, the image matrix of a graph with a community structure, would have high values along the main diagonal showing the highly probability of edges appearing inside the community and low off-diagonal values showing the sparse connections between communities.

Blockmodelling so far has been studied as an unsupervised task that just relies on the network topology to discover latent patterns and structures, ignoring any potential existing background information about the latent groups.

However, rather than relying completely on unsupervised structure discovery, there may be pre-existing knowledge available about expected patterns in the graph. For example, in the question answering forum, a subset of participants may be known as experts, or some indicators such as the number of upvotes for forum participants may highlight participants as experts. As another example, side-information may be derived using an expensive/invasive medical test that discovers ground truth (block membership) for a small sample of objects. Background information is typically represented as known labels (such as "expert" or "novice") or pairwise information, which can be seen as constraints that show whether two vertices should be in same group (*must-link* constraint) or they should belong to different groups (*cannot-link* constraint). An example of pairwise side-information is a must-link between two sets of proteins with known same functionality in a protein-protein interaction network.

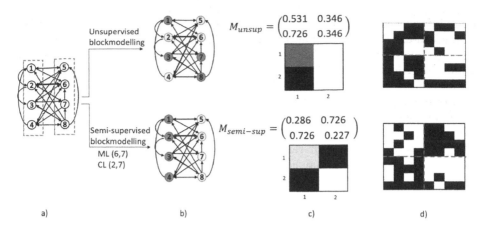

Fig. 1. An example of comparision between unsupervised blockmodeling (top row) and our semi-supervised blockmodelling (bottom row, ML = must-link, CL = cannot-link) on an almost-bipartite graph. (a) The original almost-bipartite graph and the ground truth. (b) Resulting membership assignments, represented by colors. (c) Resulting image matrices and their visualization. (d) The reordered adjacency matrices with dash lines representing the block borders.

Incorporating such background knowledge in the blockmodelling process can improve the performance and result in a more accurate role discovery of vertices. In addition, it can enable existing blockmodelling algorithms to discover complex structures that they would not be able to find otherwise. For instance, consider an almost bipartite graph shown in Fig. 1, similar to the question answering forum example in which the two groups of askers and answerers (left and right groups) communicate densely to each other but the interactions within the groups are sparse. However, as it is shown in part (b) of Fig. 1, the unsupervised blockmodelling algorithm (top-row) cannot find the true block assignments and mixes the two groups together while the bottom row of the figure shows that by just incorporating one must-link and one cannot-link supervision, our proposed semi-supervised blockmodelling is able to correctly group the vertices. Parts (c) and (d) in Fig. 1 compare how accurately the structure of the graph is captured in the image matrix and the reordered adjacency matrix respectively. As explained earlier, one would expect low diagonal values and high anti-diagonal values in the image matrix of an almost bipartite graph. This has been accurately captured by our semi-supervised method where the higher probabilities in anti-diagonal elements represent the bipartite nature of the graph while the image matrix from the unsupervised blockmodelling does not reflect such a pattern. Similarly, no bipartite pattern is captured in the reordered adjacency matrix of the unsupervised blockmodelling method. However, the higher density of links in the top-right and bottom-left sections of the reordered adjacency matrix of our semi-supervised method illustrates the bipartite characteristic of the graph.

Incorporating side-information and background knowledge has been shown to improve the results for other machine learning tasks. For instance, the effect of must-link and cannot-link constraints has been studied on clustering [8,9] and community detection [10,14,15], and it has been shown that they can improve the quality of the solutions and robustness to noise [8,10,13,14].

However, to the best of our knowledge, none of the existing blockmodelling algorithms offers the flexibility to incorporate side information in the form of pairwise constraints. Hence, the side information is ignored by these methods even if the constraints are required or highly desirable to be satisfied.

Klein et al. [18] showed that partitioning in the presence of cannot-link constraints is NP-complete because it can be represented as a reduction from the Graph K-Colorability problem (K-Color). We extended this NP-Completeness proof for blockmodelling in Appendix A (see supplementary material). It has been shown that the CL-feasibility problem is NP-complete even when the number of constraints is linear in the number of points [7]. Note that applying an unsupervised partitioning and then fixing the constraint violations afterwards (by flipping the 0/1 assignments in the output result) is also shown to be NP-complete [7]. In addition, in the case of blockmodelling, this does not naturally reflect the constraints in the structure captured by the image matrix, causing inconsistencies in the results.

In this paper, we propose a semi-supervised blockmodelling framework based on the method of Lagrange multipliers, which can be coupled with several state of the art algorithms in blockmodelling to benefit from pairwise supervision in the form of must-link and cannot-link constraints. The method of Lagrange multipliers is a powerful constrained optimization technique, which has performed very well in difficult NP-hard graph coloring problems and has been used successfully for constrained community detection and clustering problems [13,15]. In this paper we focus on nonnegative matrix tri-factorization based blockmodelling [5,19,21] to discover the membership assignments and graph structure at the same time. Our Lagrange multipliers method encourages satisfaction of the supervision constraints throughout the blockmodelling task by introducing and increasing the penalty for violated constraints from one iteration to the next. The main contributions of this paper are as follows:

- We present a flexible method of blockmodelling that allows background and expert knowledge to be incorporated in the form of pairwise constraints, which results in improved ability to find complex patterns in graphs;
- We demonstrate the high accuracy and noise resistance of our framework using synthetic and real-life data sets.

2   Background

Consider a graph $G(V, E)$ where V is a set of vertices and E is a set of edges. The graph can be represented by its adjacency matrix, A, where for each pair i and j, A_{ij} indicates whether or not an edge exists between i and j (or from i to j in a directed graph).

Blockmodelling aims to decompose the graph into groups of equivalent vertices, which are called positions. Relations between positions are captured by an image matrix whose entries show the probability of communication (edges) between positions. The positions and the image matrix together form a blockmodel. Blockmodelling has been modeled as a nonnegative matrix tri-factorization problem [5,19,21] in which positions are represented by a membership matrix $C \in [0,1]^{n \times k}$ and $M \in [0,1]^{k \times k}$ represents the image matrix where n is the number of graph vertices and k is the number of blocks or positions. The membership matrix shows the assignment of vertices to blocks and the image matrix summarizes the communications within and between positions. The decomposition aims to approximate the graph adjacency matrix A as CMC^T. Blockmodelling is then the task of finding $M \geqslant 0$ and $C \geqslant 0$ that minimizes a pre-defined approximation error function, for instance, the sum of squared differences (Eq. (1)) where $||.||_F$ is the Frobenius norm.

$$\arg\min_{M,C} ||A - CMC^T||_F^2 \tag{1}$$

It is known that blockmodelling of three or more positions is NP-hard [11]. Hence, most blockmodelling algorithms try to find locally optimal M and C. One common approach is to iteratively solve the optimization problem for C given that M is fixed, and then solve the optimization problem for M given that C is fixed. The iteration of these alternating steps continues until the algorithm converges. The optimization is done using update rules. In optimizing/updating the membership (image) matrix, the entries can be forced to accept binary or real values, which are called hard or soft membership (image), respectively. It is also possible to turn soft membership to hard membership assignments by assigning each vertex i to the position k with maximum C_{ik} value.

There are several types of equivalence for blockmodelling. In this paper, we focus on structural equivalence [22], according to which, two vertices are in the same position if they have similar sets of in and out neighbours. For instance, if two students have the same supervisor, the students are structurally equivalent. According to structural equivalence, the elements of the image matrix have densities ideally close to 0 or 1.

3   Related Work

It has been shown that even unsupervised blockmodelling is an NP-hard problem [11]. Therefore, blockmodelling algorithms try to find a good solution which has a (locally) minimal approximation error. Blockmodelling has been formulated as a nonnegative matrix tri-factorization problem [5,19,21]. Zhang et al. [23] introduced a coordinate descent optimization approach for overlapping blockmodelling. Karrer and Newman [17] considered the heterogeneity in vertex degrees and proposed a stochastic blockmodelling approach. Chan et al. [5] proposed a framework focusing on sparse and noisy graphs. They also introduced objective functions and proposed an incremental approach for optimizing the membership matrix, which only updates the necessary entries of the C matrix each time. The same authors later [4] introduced a soft membership formulation that ensured the vertex to cluster memberships sum to 1, and showed that this made a significant difference to the discovered blockmodels. Reichardt and White [20] used a simulated annealing method to optimize an objective function based on the difference between the adjacency matrix and its blockmodel approximation. However, none of the existing blockmodelling techniques, to the best of our knowledge, can incorporate pairwise background information to find complex structures in graphs.

In some existing algorithms [5,20], an instance of the desired structure can be applied as an initialization of the image matrix. In another approach [3], new forms of equivalence were introduced and a blockmodel could consist of different block types (each corresponding to a different type of equivalence). These blockmodels can incorporate supervision in the form of desired block types. But for all these algorithms, there is no guarantee that they will eventually find the specified structure. In addition, none of the algorithms can incorporate pairwise instance level constraints. Recently, constraint programming has been coupled with a non-negative matrix tri-factorization method to force some structural patterns in image-constrained blockmodelling [16]. While, in contrast to the earlier approaches, the constraint programming framework can incorporate block(image)-level supervision in blockmodelling, it is not able to incorporate pairwise instance-level information, e.g., must-links and cannot-links. Hence, in this paper we close this gap in the literature by proposing semi-supervised blockmodelling incorporating pairwise instance-level constraints. Our method also has the advantage that it can be coupled with the existing approaches above to enable them to benefit from instance-level information.

The effect of must-link and cannot-link constraints has been studied for clustering tasks using a SAT formulation [8], constraint programming [9] and Lagrange multiplier methods [13]. Community detection has also been shown to benefit from pairwise supervision [10,15]. Ganji et al. [14] used constraint programming to model several constraint types for community detection, including community size, distribution and pairwise instance level constraints. However, scalability is an issue for this method. In addition, even by incorporating constraints, none of the existing semi-supervised community detection and clustering methods can find other structural patterns in the graph such as the groups in the question answering forum example.

To the best of our knowledge, incorporating pairwise background knowledge has not been studied for blockmodelling and none of the existing methods are able to incorporate pairwise constraints. This paper addresses this gap and proposes a semi-supervised framework for blockmodeling based on non-negative matrix tri-factorization and Lagrange multipliers.

4 Proposed Semi-supervised Framework

In this section we elaborate on our proposed framework and provide more details on modeling and optimization of our semi-supervised blockmodelling method.

4.1 Method for Modelling Semi-supervised Blockmodelling Using Lagrange Multipliers

Pairwise information for blockmodelling is given in two ways. A *must-link* (ML) constraint between two vertices (i, j) requires that the two vertices are mapped to the same position, i.e., they are known to take the same role in the graph. A *cannot-link* (CL) constraint between two vertices (i, j) requires that they are not mapped to the same position, i.e., they are known to take different roles in the graph. Let ML be the set of must-link constraints and CL the set of cannot-link constraints.

We enforce these constraints using the Lagrange multipliers method which requires mapping them to a penalty term for each violation and adding the penalties to the original objective function to guide the optimization algorithm towards satisfying the constraints.

In order to define the penalty terms, we introduce matrices Q_{ML} and Q_{CL} which represent the cost coefficients for each pairwise constraint. Q_{ML} and Q_{CL} are $n \times n$ non-negative real valued matrices quantifying the cost of violating each of the must-link and cannot-link constraints respectively. If no must-link and cannot-link are imposed on the pair (i, j), then the corresponding element in Q_{ML} and Q_{CL} is equal to zero, meaning assignments of the pair (i, j) to same or different blocks will not have any supervision violation cost. If the pair (i, j) is involved in a must-link constraint ($ML(i, j)$ or $ML(j, i)$), then $Q_{ML}(i, j)$ and $Q_{ML}(j, i)$ are equal to the corresponding Lagrange multiplier for that constraint, denoted by $\lambda_{[i,j]}$. Similarly $Q_{CL}(i, j) = Q_{CL}(j, i) = \mu_{[i,j]}$ if and only if there exist a cannot-link constraint between i and j. Note that λ and μ are vectors (of Lagrange multipliers corresponding to each constraint type) whose lengths are equal to the number of must-link and cannot-link constraints respectively; the notation $[i, j]$ is only for indexing these vectors and the order is not important.

Given the cost matrix Q_{ML} and a partition represented by C, the total cost associated with must-link constraint violation can be obtained using $\frac{1}{2}(1 - C) \otimes (Q_{ML} \times C)$ where 1 is the matrix of the same size as C whose elements are equal to 1 and $\otimes$ denotes element-wise multiplication (and then sum). $R = (Q_{ML} \times C)$ is a $n \times k$ matrix which has a straightforward interpretation. Considering the binary membership case, R_{ik} represents the (must-link) cost of not assigning vertex i to position k. Hence, the total cost for all vertices can be captured by element-wise multiplication of $1/2(1 - C)$ to R. The constant coefficient is because each violated constraint has been penalized twice in our matrix representation. Note that one could prevent this by defining the Q_{ML} as an upper triangular matrix. However, R would no longer have the explained interpretation.

Similarly, given the cost matrix Q_{CL} and a partition represented by C, the total cost associated with cannot-link constraint violation can be obtained using $\frac{1}{2}C \otimes (Q_{CL} \times C)$. The interpretation of $R' = (Q_{CL} \times C)$ in this case is R'_{ik} represents the (cannot-link) cost of assigning vertex i to position k.

These penalty functions for must-link and cannot-link constraint violations naturally generalize to the soft membership version of the problem, where C may not have a unique non-zero (1-valued) entry for each row.

The Lagrangian objective function shown in Eq. (2) can be calculated by the addition of the penalty terms corresponding to must-link and cannot-link constraint violations to the original blockmodelling objective of Eq. (1). The problem of constrained blockmodelling is then equivalent to finding a good M and C that minimize the Lagrangian function of Eq. (2).

$$L(M, C, Q_{ML}, Q_{CL}) = ||A - CMC^T||_F^2 + \frac{1}{2}(1-C) \otimes (Q_{ML} \times C) + \frac{1}{2} C \otimes (Q_{CL} \times C)$$
(2)

Example: We illustrate the calculations of the violation penalty terms in Eq. (2) using the following small example. Note that we consider a hard (or binary) membership in this example. However, our method naturally works for soft memberships as well.

Let $n = 3$ and in our first scenario, suppose we have two must-link constraints, $ML(1, 2)$ and $ML(1, 3)$. Given the following membership matrix C, we aim to calculate the ML penalty term as follows:

$$Q_{ML} = \begin{bmatrix} 0 & \lambda_{[1,2]} & \lambda_{[1,3]} \\ \lambda_{[1,2]} & 0 & 0 \\ \lambda_{[1,3]} & 0 & 0 \end{bmatrix} \quad C = \begin{bmatrix} 1 & 0 \\ 1 & 0 \\ 0 & 1 \end{bmatrix} \quad R = Q_{ML} \times C = \begin{bmatrix} \lambda_{[1,2]} & \lambda_{[1,3]} \\ \lambda_{[1,2]} & 0 \\ \lambda_{[1,3]} & 0 \end{bmatrix}$$
(3)

And finally, the must-link penalty value for the partition C is equal to $1/2(1-C) \otimes (Q_{ML} \times C) = \lambda_{[1,3]}$. This occurs because $ML(1, 3)$ is the only violation within partition C.

In the second scenario, consider the same membership matrix C but this time suppose we only have the cannot-link constraints $CL(1, 3)$.

$$Q_{CL} = \begin{bmatrix} 0 & 0 & \mu_{[1,3]} \\ 0 & 0 & 0 \\ \mu_{[1,3]} & 0 & 0 \end{bmatrix} \quad R' = Q_{CL} \times C = \begin{bmatrix} 0 & \mu_{[1,3]} \\ 0 & 0 \\ \mu_{[1,3]} & 0 \end{bmatrix} \quad \frac{1}{2} C \otimes (Q_{CL} \times C) = 0$$
(4)

We find that the cannot-link penalty is zero since no cannot-link constraints were violated in C.

4.2   Optimization Procedure

In this section we elaborate upon the optimization procedure to minimize the Lagrangian function of Eq. (2). Recall that blockmodelling algorithms attempt to find good M and C matrices that satisfy Eq. (1). Coordinate descent [23] and projected gradient descent [1] are the two main approaches for optimizing image and membership matrices in a soft manner. If hard membership is desired, the incremental approach of [5] can be used, which only updates the necessary entries of the C matrix and hence is more efficient than recomputing the entire objective value for each single vertex and position.

For our semi-supervised blockmedelling approach, we use a similar iterative approach where we iteratively optimize for C given fixed M, and M given
fixed C, to minimize the Lagrangian objective function. In the optimization
of the image matrix, M, we can directly use the existing unsupervised coordinate descent [23] or projected gradient descent [1] or other image optimization
approaches. This is because the image matrix, M, does not participate in any of
the penalty terms in our Lagrangian objective function. Hence, the M matrix is
only responsible in minimizing the first part of the Lagrangian function, which
is the original unsupervised blockmodelling objective function.

However, in optimizing the membership matrix C, we adapt the hard membership algorithm of [5] to update the C matrix in order to minimize the
Lagrangian function (2). After each iterative optimization for both M and C
is done, our algorithm updates (increases) the Lagrange multipliers λ and μ corresponding to the violated must-link and cannot-link constraints based on the
update rule in Eqs. (5) and (6). This is the main mechanism of the Lagrange
multipliers method to encourage satisfying the constraints from one iteration
to the next. The corresponding cost matrices Q_{ML} and Q_{CL} are then updated
based on the updated Lagrange multipliers λ and μ.

$$\lambda_{i,j} = \lambda_{i,j} + \alpha(1 - (C_{i.} \times C_{j.})) \qquad \forall (i,j) \in ML \qquad (5)$$
$$\mu_{i,j} = \mu_{i,j} + \alpha(C_{i.} \times C_{j.}) \qquad \forall (i,j) \in CL \qquad (6)$$

In the update shown in Eqs. (5) and (6), α is the learning rate determining how
fast each penalty increases if a constraint remains violated from one iteration to
the next and $C_{i.}$ refers to the ith row of the membership matrix C.

Pseudo-code of our proposed semi-supervised blockmodelling framework is
shown in Fig. 2. As shown in Fig. 2, after random initialization of the image and
membership matrices, the Lagrange multipliers are initialized with the value 1
and the corresponding cost matrices are built. The algorithm then iterates until
the improvement on the objective value $L(M, C, Q_{ML}, Q_{CL})$ is smaller than a
threshold (ϵ), which indicates convergence of the algorithm. In each iteration,
the image and membership matrices are updated according to a given optimization approach. Note that, in optimizing the M matrix assuming C is fixed, the
Lagrange penalty terms for must-link and cannot-link constraints are constant
values because they are only dependant to the membership matrix C, which is
fixed throughout the image optimization. Hence, the optimization of the image
matrix (the *Image_Optimizer* in Fig. 2) can be done using existing unsupervised
approaches such as projected gradient descent [1] or coordinate descent [23] and
the objective function of Eq. (1).

In optimizing the membership matrix, we use the incremental hard membership algorithm of Chan et al. [5], where each vertex is assigned to a position
such that the objective value is locally minimized, and it iterates until no vertex
changes position. Chan et al. suggested some precomputed elements to calculate
the change in the objective value by flipping a vertex to a different position rather
than recomputing the whole objective. We also follow a similar idea for calculating the change in the penalty values, rather than checking all the constraints.

Hence, in membership optimization, a vertex is flipped to another position only if this results in a drop in the $L(M, C, Q_{ML}, Q_{CL})$ value, which means a drop in the original approximation error and/or a decrease in the constraint violation penalty.

After optimizing the membership matrix, the Lagrange multipliers are updated (lines 9 and 10). If a constraint remains violated, its corresponding Lagrange multiplier is increased by a factor $\alpha > 1$, while it remains the same as in the previous iteration if the constraint is satisfied. This increase in the Lagrange multipliers imposes a higher penalty on violating the same constraint in the next iteration, driving the algorithm to satisfy that constraint eventually.

Procedure *SemiSupBlock(A,k, ML,CL)*
▷ Initializing image, membership, Lagrange multipliers and violation cost matrices:
1. Random initialization of membership (C^0) and image (M^0)
2. $\lambda^0_{[i,j]} = 1 \quad \forall (i,j) \in ML$
3. $\mu^0_{[i,j]} = 1 \quad \forall (i,j) \in CL$
4. Initialize Q^0_{ML} and Q^0_{CL} based on λ^0, μ^0, ML and CL.
5. $t \leftarrow 1$
▷ Update image, membership, Lagrange multipliers and violation cost matrices until convergence:
6. **Repeat until** $L^t(M^t, C^t, Q^t_{ML}, Q^t_{CL}) - L^{t-1}(M^{t-1}, C^{t-1}, Q^{t-1}_{ML}, Q^{t-1}_{CL}) < \epsilon$
7. $M^t \leftarrow Image\_Optimizer(M^{t-1})$
8. $C^t \leftarrow Membership\_Optimizer(C^{t-1}, Q^{t-1}_{ML}, Q^{t-1}_{CL})$
9. $\lambda^t_{[i,j]} = \lambda^{t-1}_{[i,j]} + \alpha(1 - \overline{(C^{t-1}_{i.} \times C^{t-1}_{j.})}) \quad \forall (i,j) \in ML$
10. $\mu^t_{i,j} = \mu^{t-1}_{i,j} + \alpha(C^{t-1}_{i.} \times C^{t-1}_{j.}) \quad \forall (i,j) \in CL$
11.   Update Q^t_{ML} and Q^t_{CL} based on λ^t, μ^t, ML and CL
12. $t \leftarrow t + 1$
13. **end**

Fig. 2. Semi-supervised blockmodelling algorithm

5 Experiments and Discussion

In this section we evaluate our proposed semi-supervised blockmodelling method and compare it with other state of the art blockmodelling algorithms.

We compare with the best algorithms proposed by Chan et al. [5]: hard incremental membership algorithm with coordinate descent (*Coord*) and gradient descent (*Grad*) image optimization algorithms. These two algorithms have been shown to outperform other algorithms proposed in [5] and also the blockmodelling algorithm of Reichardt and White [20].

Our semi-supervised framework has the advantage that it can be integrated with different unsupervised blockmodelings. Our two semi-supervised blockmodelling algorithms in this section are semi-supervised gradient descent based (called *S-Grad*) and semi-supervised coordinate descent based (called *S-Coord*) blockmodelling algorithms. The learning rate α in our experiments is set to 1.5. We observed that our proposed methods are mostly robust to the choice of α.

To evaluate the quality and accuracy of the discovered blockmodels, we use Normalized Mutual Information (NMI) [6], which is an information theoretic measure to evaluate the quality of each solution in comparison to the ground truth. All the experiments in this section are performed on a 8 GB RAM, 2.7 GHz Core i5 Mac. The source code and data sets used in our experiments are available at http://people.eng.unimelb.edu.au/mganji.

5.1   Experiments on Real Benchmarks

In this experiment, we compare the performance of the unsupervised and our semi-supervised blockmodelling algorithms on a set of real benchmark data sets that are commonly used for finding community structure. We randomly generated pairwise constraints (equally divided into must-link and cannot-link constraints) from the ground truth. To generate the must-link (cannot-link) constraints, we pick a vertex randomly and pair it with another randomly selected vertex of the same (different) block based on the ground truth (the true labels).

The real data sets used in this experiment are benchmarks from Mark Newman's homepage[1] and Pajek repository[2]. The statistics of the data sets and constraints are shown in Table 1 where n, k and #const refer to the number of vertices, blocks and pairwise constraints respectively.

The results of this experiment are shown in Table 2 where the average quality of the solutions (NMI) and number of violated constraints are reported. As shown in this table, our semi-supervised blockmodelling techniques (*S-Grad* and *S-Coord*) improve the performance of both baseline methods significantly. Note that the Lagrange methods are not guaranteed to satisfy every constraint, since they are simply treated as penalties. However, there are very few constraints left unsatisfied compared to the original methods that ignore them. A pairwise Friedman statistical test is performed between each pair of the unsupervised method and our corresponding semi-supervised method. The hypothesis of the statistical tests is that the ranking of the two sets of average results are not different. The consistently small p-values indicate that the difference is highly unlikely due to random chance which confirms the statistically significant effect of incorporating side-information into blockmodelling.

Table 1. Real data sets and number of generated constraints

Data	Karate	Strike	Sampson	Mexican	Dolphin	Polbooks	Adjword	Polblogs
n	34	24	25	35	62	105	112	1490
k	2	3	2	2	2	3	2	2
#Const	34	24	24	34	62	104	112	1490

---

[1] http://www-personal.umich.edu/~mejn/.

[2] http://vlado.fmf.uni-lj.si/pub/networks/pajek/.

Table 2. Solution quality (NMI, higher is better) and number of violated constraints on real benchmarks

Data	NMI				Violation count			
	Grad	S-Grad	Coord	S-Coord	Grad	S-Grad	Coord	S-Coord
Karate	0.08	**0.54**	0.09	**0.61**	7.91	**0.00**	8.35	**0.00**
Strike	0.29	**0.40**	0.32	**0.57**	4.45	**0.00**	4.21	**0.00**
Sampson	0.12	**0.54**	0.11	**0.77**	5.51	**0.00**	5.08	**0.01**
Mexican	0.07	**0.58**	0.07	**0.66**	6.95	**0.03**	7.09	**0.00**
Dolphin	0.07	**0.55**	0.06	**0.81**	13.73	**0.00**	13.72	**0.00**
Political books	0.16	**0.39**	0.26	**0.68**	21.86	**0.01**	18.80	**0.00**
Adjacent word	0.06	**0.59**	0.06	**0.71**	29.33	**0.00**	28.78	**0.00**
Political blogs	0.01	**0.19**	0.01	**0.17**	365.63	**73.91**	358.06	**82.04**
P-value	0.0047		0.0047		0.0047		0.0047	

Table 3. Solution quality (NMI, higher is better) and number of violations on synthetic data

Data	k	NMI				Violation count			
		Grad	S-Grad	Coord	S-Coord	Grad	S-Grad	Coord	S-Coord
ring	5	0.76	**0.82**	0.94	**1.00**	26.88	**5.34**	7.05	**0.17**
star	5	0.44	**0.61**	0.46	**0.68**	63.45	**10.48**	62.46	**5.24**
chain	5	0.67	**0.79**	0.88	**0.99**	37.17	**5.83**	13.43	**0.68**
hierarchy	5	0.68	**0.76**	0.86	**0.91**	35.67	**6.76**	15.26	**1.72**
bipartite	2	0.53	**0.91**	0.80	**1.00**	23.32	**0.08**	10.38	**0.00**
core-periphery	3	0.58	**0.85**	0.65	**0.95**	24.01	**0.18**	20.24	**0.07**
P-value		0.014		0.014		0.014		0.014	

5.2 Experiments on Synthetic Benchmarks

In this experiment, we evaluate the performance of our method on different graph structures other than community structures. For this purpose, we generated synthetic datasets with ring, star, hierarchy, chain, bipartite and core-periphery structures according to the method described in [5]. Given random memberships (C) and the image matrix (M), the adjacency matrix (A) is generated using $A = CMC^T$. To generate C, first the block sizes are drawn from a uniform distribution and then the position memberships of vertices are determined by drawing from a multivariate hyper-geometric distribution according to which the probability of each position is relative to its size. Different graph structures such as ring and star are also replicated in M. Uniform random background noise is also added to M, with an specific noise ratio in each experiment.

In this experiment, for each graph structure, we generated 10 different graphs of 100 vertices and perturbed the structure with 20% uniformly distributed noise. We generated three sets of constraints (n must-link and n cannot-link constraints) based on the ground truth of each of the generated graphs. The structures and number of positions, k, as well as the average results based on NMI and the number of constraint violations are shown in Table 3. The results show a substantial improvement in accuracy of the semi-supervised methods over the unsupervised ones, and a substantial reduction in the number of violated constraints.

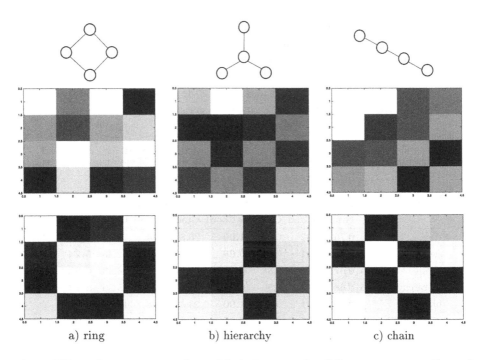

a) ring   b) hierarchy   c) chain

Fig. 3. Effect of supervision on latent block discovery for different structures. Ground truth Image diagram (top row), discovered image matrix by unsupervised (middle row) and proposed semi-supervised (bottom row) methods.

5.3 Effect of Supervision on Latent Structure Discovery

So far we investigated the effect of adding supervision constraints (and the performance of our semi-supervised blockmodelling) on finding accurate vertex assignments to positions. Apart from that, in blockmodelling, the latent structure of the graph is discovered by the image matrix. In this section we evaluate the effect of adding supervision constraints, and the performance of our semi-supervised framework, on structure discovery of the graph.

We generated synthetic data sets of different structures containing 4 blocks and 40% background noise, according to the procedure described in Sect. 3. Figure 3 shows the visualized ground truth block interactions using image diagrams and also visualized image matrices found by the coordinate descent blockmodelling method and our corresponding semi-supervised *S Coord* algorithm (incorporating n must-link and n cannot-link constraints) on a sample graph of different structures[3]. As shown in Fig. 3, the inherent structure of the graphs are not clear in the image matrices found by the unsupervised method. However, the latent structure is clearly captured in the image matrices found by our semi-supervised framework including the ring structure $(1 - 2 - 4 - 3 - 1)$, star structure $(1 - 3, 2 - 3, 4 - 3)$ and chain structure $(1 - 2 - 3 - 4)$.

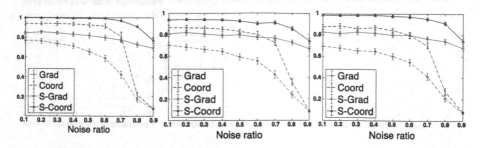

Fig. 4. Sensitivity to noise on ring (left), hierarchy (middle), and chain structure (right)

5.4   Sensitivity to Noise

In this experiment, we evaluate the performance as background noise increases. We generated data sets of ring, hierarchy and chain structures with 100 vertices and 5 blocks. We increased the background noise ratio for each data set from 0 to 0.9 and recorded the performance of different algorithms. The results shown in Fig. 4 are the sample mean and 95% confidence interval for the mean of 10 different data sets, 3 different constraint sets (100 must-link and 100 cannot-link constraints) for each data set and 20 different initializations (600 executions). The results shown in Fig. 4 demonstrate that the addition of background information to blockmodelling improves the noise resistance of the baseline methods. We can see that when the noise ratio increases, the performance of (*Coord*) and (*Grad*) drops significantly in comparison to the corresponding proposed semi-supervised (*S-Coord* and *S-Grad*) versions of the algorithms, respectively.

5.5   Sensitivity to the Number of Constraints

In this experiment we evaluate to what degree the supervision constraints affect the quality of blockmodel solutions and runtime of the algorithms. We generated data sets with ring, hierarchy and chain structures containing 100 vertices and 5 blocks. We increased the number of supervision constraints, derived from the ground truth labels, from zero to ten percent of the total number of possible pairwise constraints[4]. The results are shown in Fig. 5. Clearly, our semi-supervised algorithms (*S-Coord* and *S-Grad*) improve the solution quality significantly when

---

[3] We observed similar images from other samples in our experiments as well.

[4] The total amount of pairwise information for a graph of size n is $n \times (n - 1)/2$.

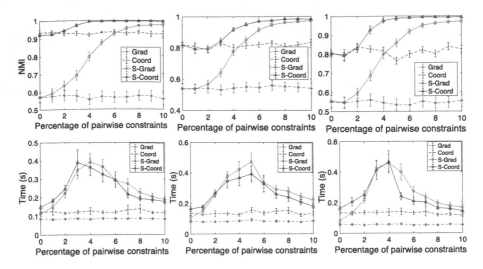

Fig. 5. Sensitivity to the amount of constraints: NMI (top row, higher is better) and runtime (bottom row) for ring (left), hierarchy (middle), and chain (right) structures

more and more supervision information is available, whereas the unsupervised algorithms ignore the information.

Incorporating the pairwise information, however, increases the runtime of the algorithms so that according to Fig. 5, in the worst case, *S-Coord* and *S-Grad* are around 2 to 4.5 times slower than their corresponding unsupervised algorithms. This increased runtime is partly due to the cost of updating the Lagrange multipliers in each iteration but mainly, depending on the constraints, the possibility of requiring more iterations to converge. However, more pairwise constraints do not always increase the runtime. After some amount (around 4%), adding more constraints decreases the runtime because the stronger information requires fewer iterations to converge.

6 Conclusion

In this paper we proposed a semi-supervised blockmodelling framework that is able to incorporate background knowledge to better find the latent structure and position assignments in complex networks. Our framework is based on the method of Lagrange multipliers and can be coupled with existing iterative optimization approaches for blockmodeling. It has been shown in our experiments on real and synthetic data sets that our framework improves the quality of the solution and noise resistance of the blockmodelling algorithms. An interesting direction for future research is to exploit other types of domain knowledge in semi-supervised blockmodelling and also further scaling it.

References

1. Berry, M.W., Browne, M., Langville, A.N., Pauca, V.P., Plemmons, R.J.: Algorithms and applications for approximate nonnegative matrix factorization. Comput. Stat. Data Anal. **52**(1), 155–173 (2007)
2. Blondel, V.D., Guillaume, J.-L., Lambiotte, R., Lefebvre, E.: Fast unfolding of communities in large networks. J. Stat. Mech. Theory Exp. **2008**(10), P10008 (2008)
3. Chan, J., Lam, S., Hayes, C.: Generalised blockmodelling of social and relational networks using evolutionary computing. Soc. Netw. Anal. Mining **4**(1), 155 (2014)
4. Chan, J., Leckie, C., Bailey, J., Ramamohanarao, K.: TRIBAC: discovering interpretable clusters and latent structures in graphs. In: Proceedings of the 15th IEEE International Conference on Data Mining, pp. 737–742 (2015)
5. Chan, J., Liu, W., Kan, A., Leckie, C., Bailey, J., Ramamohanarao, K.: Discovering latent blockmodels in sparse and noisy graphs using non-negative matrix factorisation. In: Proceedings of the 22nd ACM International Conference on Information and Knowledge Management, pp. 811–816. ACM (2013)
6. Danon, L., Diaz-Guilera, A., Duch, J., Arenas, A.: Comparing community structure identification. J. Stat. Mech.: Theory Exp. **2005**(09), P09008 (2005)
7. Davidson, I., Ravi, S.: Intractability and clustering with constraints. In: Proceedings of the 24th International Conference on Machine Learning, pp. 201–208. ACM (2007)
8. Davidson, I., Ravi, S., Shamis, L.: A SAT-based framework for efficient constrained clustering. In: Proceedings of SIAM International Conference on Data Mining, pp. 94–105 (2010)
9. Duong, K.-C., Vrain, C., et al.: Constrained clustering by constraint programming. Artif. Intell. **244**, 70–94 (2017)
10. Eaton, E., Mansbach, R.: A spin-glass model for semi-supervised community detection. In: AAAI, pp. 900–906 (2012)
11. Fiala, J., Paulusma, D.: The computational complexity of the role assignment problem. In: Baeten, J.C.M., Lenstra, J.K., Parrow, J., Woeginger, G.J. (eds.) ICALP 2003. LNCS, vol. 2719, pp. 817–828. Springer, Heidelberg (2003). https://doi.org/10.1007/3-540-45061-0_64
12. Fortunato, S.: Community detection in graphs. Phys. Rep. **486**(3), 75–174 (2010)
13. Ganji, M., Bailey, J., Stuckey, P.J.: Lagrangian constrained clustering. In: Proceedings of SIAM International Conference on Data Mining, pp. 288–296. SIAM (2016)
14. Ganji, M., Bailey, J., Stuckey, P.J.: A declarative approach to constrained community detection. In: Beck, J.C. (ed.) CP 2017. LNCS, vol. 10416, pp. 477–494. Springer, Cham (2017). https://doi.org/10.1007/978-3-319-66158-2_31
15. Ganji, M., Bailey, J., Stuckey, P.J.: Lagrangian constrained community detection. In: Proceedings of AAAI (2018, to appear)
16. Ganji, M., et al.: Image constrained blockmodelling: a constraint programming approach. In: Proceedings of SIAM International Conference on Data Mining (2018, to appear)
17. Karrer, B., Newman, M.E.: Stochastic blockmodels and community structure in networks. Phys. Rev. E **83**(1), 016107 (2011)
18. Klein, D., Kamvar, S.D., Manning, C.D.: From instance-level constraints to space-level constraints: making the most of prior knowledge in data clustering. Technical report, Stanford (2002)

19. Long, B., Zhang, Z., Philip, S.Y.: A general framework for relation graph clustering. Knowl. Inf. Syst. **24**(3), 393–413 (2010)
20. Reichardt, J., White, D.R.: Role models for complex networks. Eur. Phys. J. B-Condens. Matter Complex Syst. **60**(2), 217–224 (2007)
21. Wang, F., Li, T., Wang, X., Zhu, S., Ding, C.: Community discovery using non-negative matrix factorization. Data Mining Knowl. Discov. **22**(3), 493–521 (2011)
22. Wasserman, S., Faust, K.: Social Network Analysis: Methods and Applications, vol. 8. Cambridge University Press, Cambridge (1994)
23. Zhang, Y., Yeung, D.-Y.: Overlapping community detection via bounded nonnegative matrix tri-factorization. In: Proceedings of the 18th ACM SIGKDD International Conference on Knowledge Discovery and Data Mining, pp. 606–614. ACM (2012)

Kernel Methods

Large-Scale Nonlinear Variable Selection via Kernel Random Features

Magda Gregorová[1,2]($\boxtimes$), Jason Ramapuram[1,2], Alexandros Kalousis[1,2], and Stéphane Marchand-Maillet[2]

[1] Geneva School of Business Administration, HES-SO, Geneva, Switzerland
magda.gregorova@hesge.ch
[2] University of Geneva, Geneva, Switzerland

Abstract. We propose a new method for input variable selection in nonlinear regression. The method is embedded into a kernel regression machine that can model general nonlinear functions, not being a priori limited to additive models. This is the first kernel-based variable selection method applicable to large datasets. It sidesteps the typical poor scaling properties of kernel methods by mapping the inputs into a relatively low-dimensional space of random features. The algorithm discovers the variables relevant for the regression task together with learning the prediction model through learning the appropriate nonlinear random feature maps. We demonstrate the outstanding performance of our method on a set of large-scale synthetic and real datasets. Code related to this paper is available at: https://bitbucket.org/dmmlgeneva/srff_pytorch.

1 Introduction

It has been long appreciated in the machine learning community that learning sparse models can bring multiple benefits such as better interpretability, improved accuracy by reducing the curse of dimensionality, computational efficiency at prediction times, reduced costs for gathering and storing measurements, etc. A plethora of sparse learning methods has been proposed for linear models [16]. However, developing similar methods in the nonlinear setting proves to be a challenging task.

Generalized additive models [15] can use similar sparse techniques as their linear counterparts. However, the function class of linear combinations of non-linear transformations is too limited to represent general nonlinear functions. Kernel methods [29] have for long been the workhorse of nonlinear modelling. Recently, a substantial effort has been invested into developing kernel methods with feature selection capabilities [5]. The most successful approaches within the filter methods are based on mapping distributions into the reproducing kernel Hilbert spaces (RKHS) [23]. Amongst the embedded methods, multiple algorithms use the feature-scaling weights proposed in [32]. The authors in [28] follow an alternative strategy based on the function and kernel partial derivatives.

All the kernel-based approaches above suffer from a common problem: they do not scale well for large data sets. The kernel methods allow for nonlinear

© Springer Nature Switzerland AG 2019
M. Berlingerio et al. (Eds.): ECML PKDD 2018, LNAI 11052, pp. 177–192, 2019.
https://doi.org/10.1007/978-3-030-10928-8_11

modelling by applying high dimensional (possibly infinite-dimensional) nonlinear transformations $\phi : \mathcal{X} \to \mathcal{H}$ to the input data. Due to what is known as the kernel trick, these transformations do not need to be explicitly evaluated. Instead, the kernel methods operate only over the inner products between pairs of data points that can be calculated quickly by the use of positive definite kernel functions $k : \mathcal{X} \times \mathcal{X} \to \mathbb{R}$, $k(\mathbf{x}, \tilde{\mathbf{x}}) = \langle \phi(\mathbf{x}), \phi(\tilde{\mathbf{x}}) \rangle$. Given that these inner products need to be calculated for all data-point pairs, the kernel methods are generally costly for datasets with a large number n of training points both in terms of computation and memory. This is further exacerbated for the kernel variable selection methods, which typically need to perform the $\mathcal{O}(n^2)$ kernel evaluations multiple times (per each input dimension, or with each iterative update).

In this work we propose a novel kernel-based method for input variable selection in nonlinear regression that can scale to datasets with large numbers of training points. The method builds on the idea of approximating the kernel evaluations by Fourier random features [24]. Instead of fixing the distributions generating the random features a priori, it learns them together with the predictive model such that they degenerate for the irrelevant input dimensions. The method falls into the category of embedded approaches that seek to improve predictive accuracy of the learned models through sparsity [18]. This is the first kernel-based variable selection method for general nonlinear functions that can scale to large datasets of tens of thousands of training data.

2   Background

We formulate the problem of nonlinear regression as follows: given a training set of n input-output pairs $\mathcal{S}_n = \{(\mathbf{x}_i, y_i) \in (\mathcal{X} \times \mathcal{Y}) : \mathcal{X} \subseteq \mathbb{R}^d, \mathcal{Y} \subseteq \mathbb{R}, i \in \mathbb{N}_n\}$ sampled i.i.d. according to some unknown probability measure ρ, our task is to estimate the regression function $f : \mathcal{X} \to \mathcal{Y}$, $f(\mathbf{x}) = \mathbb{E}(y|\mathbf{x})$ that minimizes the expected squared error loss $\mathcal{L}(f) = \mathbb{E}(y - f(\mathbf{x}))^2 = \int (y - f(\mathbf{x}))^2 \, d\rho(\mathbf{x}, y)$.

In the variable selection setting, we assume that the regression function does not depend on all the d input variables. Instead, it depends only on a subset $\mathcal{I}$ of these of size $l < d$, so that $f(\mathbf{x}) = f(\tilde{\mathbf{x}})$ if $x^s = \tilde{x}^s$ for all dimensions $s \in \mathcal{I}$.

We follow the standard theory of regularised kernel learning and estimate the regression function as the solution to the following problem

$$\hat{f} = \arg\min_{f \in \mathcal{H}} \hat{\mathcal{L}}(f) + \lambda \|f\|_{\mathcal{H}}^2. \tag{1}$$

Here the function hypothesis space $\mathcal{H}$ is a reproducing kernel Hilbert space (RKHS), $\|f\|_{\mathcal{H}}$ is the norm induced by the inner product in that space, and $\hat{\mathcal{L}}(f) = \frac{1}{n} \sum_i^n (y_i - f(\mathbf{x}_i))^2$ is the empirical loss replacing the intractable expected loss above.

From the standard properties of the RKHS, the classical result (e.g. [29]) states that the evaluation of the minimizing function $\hat{f}$ at any point $\tilde{\mathbf{x}} \in \mathcal{X}$

can be represented as a linear combination of the kernel functions k over the n training points

$$\hat{f}(\tilde{\mathbf{x}}) = \sum_{i}^{n} c_i \, k(\mathbf{x}_i, \tilde{\mathbf{x}}). \tag{2}$$

The parameters $\mathbf{c}$ are obtained by solving the linear problem

$$(\mathbf{K} + \lambda \mathbf{I}_n) \, \mathbf{c} = \mathbf{y}, \tag{3}$$

where $\mathbf{K}$ is the $n \times n$ kernel matrix with the elements $K_{ij} = k(\mathbf{x}_i, \mathbf{x}_j)$ for all $\mathbf{x}_i, \mathbf{x}_j \in \mathcal{S}_n$.

2.1 Random Fourier Features

Equations (2) and (3) point clearly to the scaling bottlenecks of the kernel regression. In principal, at training it needs to construct and keep in memory the $(n \times n)$ kernel matrix and solve an n dimensional linear system ($\propto \mathcal{O}(n^3)$). Furthermore, the whole training set $\mathcal{S}_n$ needs to be stored and accessed at test time so that the predictions are of the order $\mathcal{O}(n)$.

To address these scaling issues, the authors in [24] proposed to map the data into a low-dimensional Euclidean space $\mathbf{z} : \mathcal{X} \to \mathbb{R}^D$ so that the inner products in this space are close approximations of the corresponding kernel evaluation $\langle \mathbf{z}(\mathbf{x}), \mathbf{z}(\tilde{\mathbf{x}}) \rangle_{\mathbb{R}^D} \approx \langle \phi(\mathbf{x}), \phi(\tilde{\mathbf{x}}) \rangle_{\mathcal{H}} = k(\mathbf{x}, \tilde{\mathbf{x}})$. Using the nonlinear features $\mathbf{z}(\mathbf{x}) \in \mathbb{R}^D$ the evaluations of the minimising function can be approximated by

$$\hat{f}(\tilde{\mathbf{x}}) \approx \langle \mathbf{z}(\tilde{\mathbf{x}}), \mathbf{a} \rangle_{\mathbb{R}^D}, \tag{4}$$

where the coefficients $\mathbf{a}$ are obtained from solving the linear system

$$(\mathbf{Z}^T \mathbf{Z} + \lambda \mathbf{I}_D) \, \mathbf{a} = \mathbf{Z}^T \mathbf{y}, \tag{5}$$

where $\mathbf{Z}$ is the $(n \times D)$ matrix of the random features for all the data points. The above approximation requires the construction of the $\mathbf{Z}$ matrix and solving the D-dimensional linear problem, hence significantly reducing the training costs if $D \ll n$. Moreover, access to training points is no longer needed at test time and the predictions are of the order $\mathcal{O}(D) \ll \mathcal{O}(n)$.

To construct well-approximating features, the authors in [24] called upon Bochner's theorem which states that a continuous function $g : \mathbb{R}^d \to \mathbb{R}$ with $g(\mathbf{0}) = 1$ is positive definite if and only if it is a Fourier transform of some probability measure on $\mathbb{R}^d$. For translation-invariant positive definite kernels we thus have

$$k(\mathbf{x}, \tilde{\mathbf{x}}) = g(\mathbf{x} - \tilde{\mathbf{x}}) = g(\boldsymbol{\lambda}) = \int_{\mathbb{R}^d} e^{i \boldsymbol{\omega}^T \boldsymbol{\lambda}} \, d\mu(\boldsymbol{\omega}), \tag{6}$$

where $\mu(\boldsymbol{\omega})$ is the probability measure on $\mathbb{R}^d$. In the above, g is the characteristic function of the multivariate random variable $\boldsymbol{\omega}$ defined by the expectation $g(\boldsymbol{\lambda}) = \mathbb{E}_{\boldsymbol{\omega}}(e^{i \boldsymbol{\omega}^T \boldsymbol{\lambda}}) = \mathbb{E}_{\boldsymbol{\omega}}(e^{i \boldsymbol{\omega}^T (\mathbf{x} - \tilde{\mathbf{x}})}) = \mathbb{E}_{\boldsymbol{\omega}}(e^{i \boldsymbol{\omega}^T \mathbf{x}} e^{-i \boldsymbol{\omega}^T \tilde{\mathbf{x}}}) = k(\mathbf{x}, \tilde{\mathbf{x}})$.

It is straightforward to show that the expectation over the complex exponential can be decomposed into an expectation over an inner product $\mathbb{E}_{\boldsymbol{\omega}}(e^{i\boldsymbol{\omega}^T(\mathbf{x}-\tilde{\mathbf{x}})}) = \mathbb{E}_{\boldsymbol{\omega}}\langle\psi_{\boldsymbol{\omega}}(\mathbf{x}),\psi_{\boldsymbol{\omega}}(\tilde{\mathbf{x}})\rangle$ where the nonlinear mappings are defined as $\psi_{\boldsymbol{\omega}} : \mathcal{X} \to \mathbb{R}^2$, $\psi_{\boldsymbol{\omega}}(\mathbf{x}) = [\cos(\boldsymbol{\omega}^T\mathbf{x})\ \sin(\boldsymbol{\omega}^T\mathbf{x})]^T$. In [24] the authors proposed an even lower-dimensional transformation $\varphi_{\boldsymbol{\omega},b}(\mathbf{x}) : \mathcal{X} \to \mathbb{R}$

$$\varphi_{\boldsymbol{\omega},b}(\mathbf{x}) = \sqrt{2}\cos(\boldsymbol{\omega}^T\mathbf{x} + b), \tag{7}$$

where b is sampled uniformly from $[0, 2\pi]$ and that satisfies the expectation equality $\mathbb{E}_{\boldsymbol{\omega}}(e^{i\boldsymbol{\omega}^T(\mathbf{x}-\tilde{\mathbf{x}})}) = \mathbb{E}_{\boldsymbol{\omega},b}\langle\varphi_{\boldsymbol{\omega},b}(\mathbf{x}),\varphi_{\boldsymbol{\omega},b}(\tilde{\mathbf{x}})\rangle$. We chose to work with the mapping φ (dropping the subscripts $\boldsymbol{\omega}, b$ when there is no risk of confusion) in the remainder of the text. The approximating nonlinear feature $\mathbf{z}(\mathbf{x})$ for each data-point $\mathbf{x}$ is obtained by concatenating D instances of the random mappings $\mathbf{z}(\mathbf{x}) = [\varphi^1(\mathbf{x}),\ldots,\varphi^D(\mathbf{x})]^T$ with $\boldsymbol{\omega}$ and b sampled according to their probability distribution so that the expectation is approximated by the sample sum.

2.2   Variable Selection Methods

In this section we position our research with respect to other nonlinear methods for variable selection with an emphasis on kernel methods.

In the class of generalized additive models, lessons learned from the linear models can be reused to construct sparse linear combinations of the nonlinear functions of each variable or, taking into account also possible interactions, of all possible pairs, triplets, etc., e.g. [20,26,31,34]. Closely related to these are the multiple kernel learning methods that seek to learn a sparse linear combination of kernel bases, e.g. [2,3,19]. While these methods have shown some encouraging results, their simplifying additive assumption and the fast increasing complexity when higher-order interactions shall be considered (potentially 2^d additive terms for d input variables) clearly present a serious limitation.

Recognising these shortcomings, multiple alternative approaches for general nonlinear functions were explored in the literature. They can broadly be grouped into three categories [18]: filters, wrappers and embedded methods.

The filter methods consider the variable selection as a preprocessing step that is then followed by an independent algorithm for learning the predictive model. Many traditional methods based on information-theoretic or statistical measures of dependency (e.g. information gain, Fisher-score, etc.) fall into this category [6]. More recently, significant advancement has been achieved in formulating criteria more appropriate for non-trivial nonlinear dependencies [8,9,12,27,30, 33]. These are based on the use of (conditional) cross-covariance operators arising from embedding probability measures into the reproducing kernel Hilbert spaces (RKHS) [23]. However, they are still largely disconnected from the predictive model learning procedure and oblivious of the effects the variable selection has on the final predictive performance.

The wrapper methods perform variable selection on top of the learning algorithm treating it as a black box. These are practical heuristics (such as greedy forward or backward elimination) for the search in the 2^d space of all possible subsets of the d input variables [18]. Classical example in this category is the SVM with Recursive Feature Elimination [14]. The wrappers are universal methods that can be used on top of any learning algorithm but they can become expensive for large dimensionalities d, especially if the complexity of the underlying algorithm is high.

Finally, the embedded methods link the variable selection to the training of the predictive model with the view to achieve higher predictive accuracy stemming from the learned sparsity. Our method falls into this category. There are essentially just two branches of kernel-based methods here: methods based on feature rescaling [1,10,21,25,32], and derivative-based methods [11,28]. We discuss the feature rescaling methods in more detail in Sect. 3.2. The derivative based methods use regularizers over the partial derivatives of the function and exploit the derivative reproducing property [36] to arrive at an alternative finite-dimensional representation of the function. Though theoretically intriguing, these methods scale rather badly as in addition to the $(n \times n)$ kernel matrix they construct also the $(nd \times n)$ and $(nd \times nd)$ matrices of first and second order partial kernel derivatives and use their concatenations to formulate the sparsity constrained optimisation problem.

There exist two other large groups of *sparse* nonlinear methods. These address the sparsity in either the latent feature representation, e.g. [13], or in the data instances, e.g. [7]. While their motivation partly overlaps with ours (control of overfitting, lower computational costs at prediction), their focus is on a different notion of sparsity that is out of the scope of our discussion.

3   Towards Sparsity in Input Dimensions

As stated above, our objective in this paper is learning a regression function that is sparse with respect to the input variables. Stated differently, the function shall be insensitive to the values of the inputs in the $d - l$ large complement set $\mathcal{I}^c$ of the irrelevant dimensions so that $f(\mathbf{x}) = f(\tilde{\mathbf{x}})$ if $x^s = \tilde{x}^s$ for all $s \in \mathcal{I}$.

From Eq. (4) we observe that the function evaluation is a linear combination of the D random features φ. The random features (7) are in turn constructed from the input $\mathbf{x}$ through the inner product $\boldsymbol{\omega}^T \mathbf{x}$. Intuitively, if the function $\widehat{f}$ is insensitive to an input dimension s, the value of the corresponding input x^s shall not enter the inner product $\boldsymbol{\omega}^T \mathbf{x}$ generating the D random features. Formally, we require $\omega^s x^s = 0$ for all $s \in \mathcal{I}^c$ which is obviously achieved by $\omega^s = 0$. We therefore identify the problem of learning sparse predictive models with sparsity in vectors $\boldsymbol{\omega}$.

3.1   Learning Through Random Sampling

Though in Eq. (7) $\boldsymbol{\omega}$ appears as a parameter of the nonlinear transformation φ, it cannot be learned directly as it is the result of random sampling from the probability distribution $\mu(\boldsymbol{\omega})$. In order to ensure the same sparse pattern in the

D random samples of $\boldsymbol{\omega}$, we use a procedure similar to what is known as the reparametrization trick in the context of variational auto-encoders [17].

We begin by expanding Eq. (6) of the Bochner's theorem into the marginals across the d dimensions[1]

$$g(\boldsymbol{\lambda}) = \int_{\mathbb{R}^d} e^{i\boldsymbol{\omega}^T\boldsymbol{\lambda}} \, d\mu(\boldsymbol{\omega}) = \int_{\mathbb{R}} e^{i\omega^1\lambda^1} d\mu(\omega^1) \dots \int_{\mathbb{R}} e^{i\omega^d\lambda^d} d\mu(\omega^d). \qquad (8)$$

To ensure that $\omega^s = 0$ when $s \in \mathcal{I}^c$ in all the D random samples, the corresponding probability measure (distribution) $\mu(\omega^s)$ needs to degenerate to $\delta(\omega^s)$. The distribution $\delta(\omega^s)$ has all its mass concentrated at the point $\omega^s = 0$, and has the property $\int_\mathcal{X} h(\omega^s) \, d\delta(\omega^s) = h(0)$. In particular for h the complex exponential we have $\int_\mathcal{X} e^{i\omega^s\lambda^s} \, d\delta(\omega^s) = 1$ so that the value of λ^s has no impact on the product in Eq. (8), and therefore no impact on $g(\boldsymbol{\lambda})$.[2]

Reparametrization Trick. To ensure that all the D random samples of $\boldsymbol{\omega}$ have the same sparse pattern we need to be able to optimise through its random sampling. For each element ω of the vector $\boldsymbol{\omega}$, we parametrize the sampling distributions $\mu_\gamma(\omega)$ by its scale γ so that $\lim_{\gamma \to 0} \mu_\gamma(\omega) = \delta(\omega)$. We next express each of the univariate random variables ω as a deterministic transformation of the form $\omega = q_\gamma(\epsilon) = \gamma\epsilon$ (scaling) of an auxiliary random variable ϵ with a fixed probability distribution $\mu_1(c)$ with the scale parameter $\gamma = 1$. For example, for the Gaussian and Laplace kernels the auxiliary distribution $\mu_1(\epsilon)$ are the standard Gaussian and Cauchy respectively.

By the above reparametrization of the random variable $\boldsymbol{\omega}$ we disconnect the sampling operation over $\boldsymbol{\epsilon}$ from the rescaling operation $\boldsymbol{\omega} = \mathbf{q_\gamma}(\boldsymbol{\epsilon}) = \boldsymbol{\epsilon} \odot \boldsymbol{\gamma}$ with a deterministic parameter vector $\boldsymbol{\gamma}$. Sparsity in $\boldsymbol{\omega}$ (and therefore the learned model) can now be achieved by learning sparse parameter vector $\boldsymbol{\gamma}$.

Though in principle it would be possible to learn the sparsity in the sampled $\boldsymbol{\omega}$'s directly, this would mean sparsifying instead of one vector $\boldsymbol{\gamma}$ the D sampled vectors $\boldsymbol{\omega}$. Moreover, the procedure would need to cater for the additional constraint that all the samples have the same sparse pattern. While theoretically possible, we find our reparametrization approach more elegant and practical.

3.2   Link to Feature Scaling

In the previous section we have built our strategy for sparse learning using the inverse Fourier transform of the kernels and the degeneracy of the associated probability measures. When we plug the rescaling operation into the random feature mapping (7)

$$\varphi(\mathbf{x}) = \sqrt{2}\cos(\boldsymbol{\omega}^T\mathbf{x} + b) = \sqrt{2}\cos((\boldsymbol{\epsilon} \odot \boldsymbol{\gamma})^T\mathbf{x} + b) = \sqrt{2}\cos(\boldsymbol{\epsilon}^T(\boldsymbol{\gamma} \odot \mathbf{x}) + b), \quad (9)$$

we see that the parameters $\boldsymbol{\gamma}$ can be interpreted as weights scaling the input variables $\mathbf{x}$. This makes a link to the variable selection methods based on feature

---

[1] This is possible due to the independence of the d dimensions of the r.v. $\boldsymbol{\omega}$.

[2] And from (6) and (4) it neither impacts the kernel and regression function evaluation.

scaling. These are rather straightforward when the kernel is simply linear, or when the nonlinear transformations $\phi(\mathbf{x})$ can be evaluated explicitly (e.g. polynomial) [10, 32]. In essence, instead of applying the weights to the input features, they are applied to the associated model parameters and suitably constrained to approximate the zero-norm problem.

More complex kernels, for which the nonlinear features $\phi(\mathbf{x})$ cannot be directly evaluated (may be infinite dimensional), are considered in [1, 21, 25]. Here the scaling is applied within the kernel function $k(\boldsymbol{\gamma} \odot \mathbf{x}, \boldsymbol{\gamma} \odot \tilde{\mathbf{x}})$. The methods typically apply a two-step iterative procedure: they fix the rescaling parameters $\boldsymbol{\gamma}$ and learn the corresponding n-long model parameters vector $\mathbf{c}$ (Eq. (2)); fix $\mathbf{c}$ and learn the d-long rescaling vector $\boldsymbol{\gamma}$ under some suitable zero-norm approximating constraint. The naive formulation for $\boldsymbol{\gamma}$ is a nonconvex problem that requires calculating derivatives of the kernel functions with respect to $\boldsymbol{\gamma}$ (which depending on the kernel function may become rather expensive). In [1], the author proposed a convex relaxation based on linearization of the kernel function. Nevertheless, all the existing methods applying the feature scaling within the kernel functions scale badly with the number of instances as they need to recalculate the $(n \times n)$ kernel matrix and solve the corresponding optimisation (typically $\mathcal{O}(n^3)$) with every update of the weights $\boldsymbol{\gamma}$.

4   Sparse Random Fourier Features Algorithm

In this section we present our algorithm for learning with Sparse Random Fourier Features (SRFF).

Input   : training data $(\mathbf{X}, \mathbf{y})$; hyper-parameters λ, D, size of Δ simplex
Output   : model parameters $\mathbf{a}$, scale vector $\boldsymbol{\gamma}$
Initialise : $\boldsymbol{\gamma}$ evenly on simplex Δ, $\boldsymbol{\epsilon}_j \sim \mu_I(\boldsymbol{\epsilon})$ and $b_j \sim U[0, 2\pi]$, $\forall j \in \mathbb{N}_D$
Objective: $J(\mathbf{a}, \boldsymbol{\gamma}) = \|\mathbf{y} - \mathbf{Z}\mathbf{a}\|_2^2 + \lambda\|\mathbf{a}\|_2^2$

repeat   // Alternating descent
   begin Step 1: Solve for $\mathbf{a}$
     rescalings $\boldsymbol{\omega}_j = \boldsymbol{\gamma} \odot \boldsymbol{\epsilon}_j$, $\forall j \in \mathbb{N}_D$
     random features $\mathbf{z}(\mathbf{x}) = [\varphi^1(\mathbf{x}), \dots, \varphi^D(\mathbf{x})]$, $\forall \mathbf{x} \in \mathcal{S}_n$ // equation (9)
     $\mathbf{a} \leftarrow \arg\min_{\mathbf{a}} \|\mathbf{y} - \mathbf{Z}\mathbf{a}\|_2^2 + \lambda\|\mathbf{a}\|_2^2$ // equation (5)
   end
   begin Step 2: Solve for $\boldsymbol{\gamma}$
     $\boldsymbol{\gamma} \leftarrow \arg\min_{\boldsymbol{\gamma} \in \Delta} \|\mathbf{y} - \mathbf{Z}\mathbf{a}\|_2^2$ // projected gradient descent
   end
until *objective convergence*;
Algorithm 1. Sparse Random Fourier Features (SRFF) algorithm

Similarly to the feature scaling methods we propose a two-step alternative procedure to learn the model parameters $\mathbf{a}$ and the distribution scalings $\boldsymbol{\gamma}$. For a fixed $\boldsymbol{\gamma}$ we generate the random features for all the input training points $\mathcal{O}(nD)$, and solve the linear problem (5) $\mathcal{O}(D^3)$ to get the D-long model parameters $\mathbf{a}$.

Given that in our large-sample settings we assume $D \ll n$, this step is significantly cheaper than the corresponding step for learning the $\mathbf{c}$ parameters in the existing kernel feature scaling methods described in Sect. 3.2.

In the second step, we fix the model parameters $\mathbf{a}$ and learn the d-long vector of the distribution scalings $\boldsymbol{\gamma}$. We formulate the optimisation problem as the minimisation of the empirical squared error loss with $\boldsymbol{\gamma}$ constrained on the probability simplex Δ to encourage the sparsity.

$$\arg\min_{\boldsymbol{\gamma} \in \Delta} J(\boldsymbol{\gamma}), \qquad J(\boldsymbol{\gamma}) := ||\mathbf{y} - \mathbf{Za}||_2^2 \qquad (10)$$

Here the $(n \times D)$ matrix $\mathbf{Z}$ is constructed by concatenating the D random features φ with the $\boldsymbol{\gamma}$ rescaling (9).

We solve problem (10) by the projected gradient method with accelerated FISTA line search [4]. The gradient can be constructed from the partial derivatives as follows

$$\frac{\partial J(\boldsymbol{\gamma})}{\gamma^s} = -(\mathbf{y} - \mathbf{Z}\mathbf{a})^T \frac{\partial \mathbf{Z}}{\partial \gamma^s} \mathbf{a} \qquad \forall s \in \mathbb{N}_d$$

$$\frac{\partial Z_{ij}}{\partial \gamma^s} = -\sqrt{2}\sin(\boldsymbol{\epsilon}^T(\boldsymbol{\gamma} \odot \mathbf{x}) + b)\,\epsilon^s x^s, \qquad \epsilon^s = \omega_s/\gamma_s. \qquad (11)$$

Unlike in the other kernel feature scaling methods, the form of the gradient (11) is always the same irrespective of the choice of the kernel. The particular kernel choice is reflected only in the probability distribution from which the auxiliary variable $\boldsymbol{\epsilon}$ is sampled and has no impact on the gradient computations. In our implementation (https://bitbucket.org/dmmlgeneva/srff_pytorch), we leverage the automatic differentiation functionality of pytorch in order to obtain the gradient values directly from the objective formulation.

5   Empirical Evaluation

We implemented our algorithm in pytorch and made it executable optionally on CPUs or GPUs. All of our experiments were conducted on GPUs (single p100). The code including the settings of our experiments amenable for replication is publicly available at https://bitbucket.org/dmmlgeneva/srff_pytorch.

In our empirical evaluation we compare to multiple baseline methods. We included the nonsparse random Fourier features method (RFF) of [24] in our main SRFF code as a call option. For the naive mean and ridge regression we use our own matlab implementation. For the linear lasso we use the matlab PASPAL package [22]. For the nonlinear Sparse Additive Model (SPAM) [26] we use the R implementation of [35]. For the Hilbert-Schmidt Independence Criterion lasso method (HSIC) [33], and the derivative-based embedded method of [28] (Denovas) we use the authors' matlab implementation.

Except SPAM, all of the baseline sparse learning methods use a two step procedure for arriving at the final model. They first learn the sparsity using either predictive-model-dependent criteria (lasso, Denovas) or in a completely disconnected fashion (HSIC). In the second step (sometimes referred to as de-biasing [28]), they use a base non-sparse learning method (ridge, or kernel ridge) to learn a model over the selected variables (including hyper-parameter search and cross-validation). For HSIC, which is a filter method that does not natively predict the regression outputs, we use as the second step our implementation of the RFF. It searches through the candidate sparsity patterns HSIC produces and uses the validation mean square error as a criteria for the final model selection. In contrast to these, our SRFF method is a *single step procedure* that does not necessitate this extra re-learning phase.

Experimental Protocol. In all our experiments we use the same protocol. We randomly split the data into three independent subsets: train, validation and test. We use the train subset for training the models, we use the validation subset to perform the hyper-parameter search, and we use the test set to evaluate the predictive performance. We repeat all the experiments 30 times, each with a different random train/validation/test split.

We measure the predictive performance in terms of the root mean squared error (RMSE) over the test samples, averaged over the 30 random replications of the experiments. The regularization hyper-parameter λ (exists in ridge, lasso, Denovas, HSIC, RFF and SRFF) is searched within a 50-long data-dependent grid (automatically established by the methods). The smoothing parameter in Denovas is fixed to 10 following the authors' default [28]. We use the Gaussian kernel for all the experiments with the width σ set as the median of the Euclidean distances amongst the 20 nearest neighbour instances. We use the same kernel in all the kernel methods and the corresponding scale parameter $\gamma = 1/\sigma$ in the random feature methods for comparability of results. We fix the number of random features to $D = 300$ for all the experiments in both RFF and SRFF.

We provide the results of the baseline nonlinear sparse methods (SPAM, HSIC, Denovas) only for the smallest experiments. As explained in the previous sections, the motivation for our paper is to address the poor scaling properties of the existing methods. Indeed, none of the baseline kernel sparse methods scales to the problems we consider here. HSIC [33] creates a $(n \times n)$ kernel matrix per each dimension d and solves a linear lasso problem over the concatenated vectorization of these with memory requirements $(n^2 \times d)$ and complexity $\mathcal{O}(n^4)$. In our tests, it did not finish within 24 h running on 20 CPUs (Dual Core Intel Xeon E5-2680 v2/2.8 GHz) for the smallest training size of 1000 instances in our SE3 experiment. Within the same time span it did not arrive at a solution for any of the experiments with $n > 1000$. Denovas constructs, stores in memory, and operates over the $(n \times n)$, $(nd \times n)$ and $(nd \times nd)$ kernel matrix and the matrices of the first and second order derivatives. In our tests the method finished with an out-of-memory error (with 32 GB RAM) for the SE1 with 5k training samples and for SE2 problem already with 1k training instances. SPAM finished with errors for most of the real-data experiments.

5.1 Synthetic Experiments

We begin our empirical evaluation by exploring the performance over a set of synthetic experiments. The purpose of these is to validate our method under controlled conditions when we understand the true sparsity of the generating model. We also experiment with various nonlinear functions and increasing data sizes in terms of both the sample numbers n and the dimensionality d.

Table 1. Summary of synthetic experiments

Exp code	Train size	Test size	Total dims	Relevant dims	Generative function
SE1	1k–50k	1k	18	5	$y = \sin\left((x_1 + x_3)^2\right)\sin(x_7 x_8 x_9) + N(0, 0.1)$
SE2	1k–50k	1k	100	5	$y = \log\left((\sum_{s=11}^{15} x_s)^2\right) + N(0, 0.1)$
SE3	1k–50k	10k	1000	10	$y = 10(z_1^2 + z_3^2)e^{-2(z_1^2 + z_3^2)} + N(0, 0.01)$

We use the same size for the test and validation samples. In all the experiments, the data instances are generated from a standard normal distribution. In the functions, subscripts are dimensions, superscripts are exponents. For more detailed description of the generative function of SE3 see the appropriate section in the text.

SE1: The very first of our experiments is a rather small problem with only $d = 18$ input dimensions of which only 5 are relevant for the regression function. In Table 2 we compare our SRFF method to the baselines for the smallest sample setting with $n = 1000$. Most of the methods (linear, additive or non-sparse) do not succeed in learning a model for the complex nonlinear relationships between the inputs and outputs and fall back to predicting simple mean.

The general nonlinear models with sparsity (HSIC, Denovas and SRFF) divert from the simple mean prediction. They all discover and use in the predictive model the same sparse pattern (see Fig. 1 for SRFF). Denovas and SRFF achieve almost identical results which confirms that our method is competitive with the state of the art methods in terms of predictive accuracy and variable selection.[3]

In Table 3 we document how the increasing training size contributes to improving the predictive performance even in the case of several thousands instances. The performance of the SRFF method for the largest 50k sample is by about 6% better than for the 1k training size. For the other methods the problem remains out of their reach[4] and they stick to the mean prediction even

---

[3] The low predictive performance of HSIC is the result of the 2nd model fitting step. It could potentially be improved with an additional kernel learning step. However, as we keep the kernel fixed for all the other methods, we do not perform the kernel search for HSIC either.

[4] The class of linear functions is too limited and the nonlinear function with all the variables considered by RFF is too complex.

Table 2. SE1 - Test RMSE for $n = 1000$

	Mean	Ridge	Lasso	RFF	SPAM	HSIC	Denovas	SRFF
RMSE	0.287	0.287	0.287	0.287	0.0287	0.341	**0.272**	**0.272**
Std	0.009	0.009	0.009	0.009	0.009	0.060	0.009	0.009

Predictive performance in terms of root mean squared error (RMSE) over independent test sets for the SE1 dataset with training size $n = 1000$. The std line is the standard deviation of the RMSE across the 30 resamples.

for higher training sizes.[5] We do not provide any comparisons with the nonlinear sparse methods because, as explained above, they do not scale to the sample sizes we consider here.

Table 3. SE1 - Test RMSE for increasing train size n

n	Mean	Ridge	Lasso	RFF	SRFF
1k	0.287 (0.009)	0.287 (0.009)	0.287 (0.009)	0.287 (0.009)	**0.272** (0.009)
5k	0.284 (0.011)	0.284 (0.011)	0.284 (0.011)	0.284 (0.011)	**0.263** (0.010)
10k	0.285 (0.010)	0.285 (0.010)	0.285 (0.010)	0.286 (0.010)	**0.261** (0.011)
50k	0.283 (0.010)	0.283 (0.010)	0.283 (0.010)	0.283 (0.010)	**0.255** (0.009)

Predictive performance in terms of root mean squared error (RMSE) over independent test sets for the SE1 dataset with increasing training size n. The standard deviation of the RMSE across the 30 resamples is in the brackets.

The improved predictive performance for the larger training sizes goes hand in hand with the variable selection, Fig. 1. For the smallest 1k training sample, SRFF identifies only the 7th, 8th and 9th relevant dimensions. They enter the sine in the generative function in a product and therefore have a larger combined effect on the function outcome than the squared sum of dimensions 1 and 3. These two dimensions are picked up by the method from the larger training sets and this contributes to the increase in the predictive performance.

SE2: In the second experiment we increase the dimensionality to $d = 100$ and change the nonlinear function (see Table 1). The overall outcomes are rather similar to the SE1 experiment. Again, it's only the nonlinear sparse models that predict something else than mean, SPAM marginally better, HSIC marginally worse. Our SRFF method clearly outperforms all the other methods in the predictive accuracy. It also correctly discovers the 5 relevant variables with the median value of γ for these dimensions between 0.92–1.04 while the maximum

---

[5] The small variations in the error stem from using different training sets to estimate the mean.

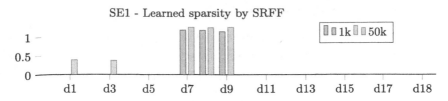

Fig. 1. Learned sparsity pattern γ by the SRFF method for the 1k and 50k training size in the SE1 experiment (the median of the 30 replications). The other nonlinear sparse methods learn the same pattern for the 1k problem but cannot solve the 50k problem.

for all the irrelevant variables is 0.06.[6] The advantage of SRFF over the baselines for large sample sizes (Tables 4 and 5) is even more striking than in the SE1 experiment (Tables 3 and 4).

Table 4. SE2 - Test RMSE for $n = 1000$

	Mean	Ridge	Lasso	RFF	SPAM	HSIC	SRFF
RMSE	2.216	2.216	2.216	2.216	2.162	2.357	**1.603**
Std	0.105	0.105	0.105	0.104	0.110	0.141	0.104

Predictive performance in terms of root mean squared error (RMSE) over independent test sets for the SE2 dataset with training size $n = 1000$. The std line is the standard deviation of the RMSE across the 30 resamples.

Table 5. SE2 - Test RMSE for increasing train size n

n	Mean	Ridge	Lasso	RFF	SRFF
1k	2.216 (0.105)	2.216 (0.105)	2.216 (0.105)	2.216 (0.105)	**1.603** (0.104)
5k	2.211 (0.079)	2.211 (0.079)	2.211 (0.079)	2.211 (0.079)	**1.278** (0.076)
10k	2.224 (0.115)	2.224 (0.115)	2.224 (0.115)	2.224 (0.115)	**1.272** (0.138)
50k	2.224 (0.082)	2.224 (0.082)	2.224 (0.082)	2.224 (0.082)	**1.273** (0.080)

Predictive performance in terms of root mean squared error (RMSE) over independent test sets for the SE2 dataset with increasing training size n. The standard deviation of the RMSE across the 30 resamples is in the brackets.

SE3: In this final synthetic experiment we increase the dimensionality to $d = 1000$ to further stretch our SRFF method. There are only 10 relevant input variables in this problem. The first 5 were generated as random perturbations of the random variable z_1, e.g. $x_1 = z_1 + N(0, 0.1)$, the second 5 by the same procedure from z_2, e.g. $x_5 = z_2 + N(0, 0.1)$. The remaining 990 input variables were generated by the same process from the other 198 standard normal z's.

---

[6] SPAM and HSIC discover the correct patterns as well but it does not help their predictive accuracy.

We summarise the results for the 1k and 50k training instances in Table 6. As in the other synthetic experiments, the baseline methods are not able to capture the nonlinear relationships of this extremely sparse problem and instead predict a simple mean. Our SRFF method achieves significantly better accuracy for the 1k training set, and it further considerably improves with 50k samples to train on. These predictive gains are possible due to SRFF correctly discovering the set of relevant variables. In the 1k case, the medians across the 30 data resamples of the learned γ parameters are between 0.37–0.71 for the 10 relevant variables and maximally 0.05 for the remaining 990 irrelevant variables. In the 50k case, the differences are even more clearly demarcated: 1.19–1.64 for the relevant, and 0.03 maximum for the irrelevant (bearing in mind that the total sum over the vector γ is the same in both cases).

Table 6. SE3 - Test RMSE for increasing train size n

n	Mean	Ridge	Lasso	RFF	SRFF
1k	0.676 (0.002)	0.676 (0.002)	0.676 (0.002)	0.676 (0.002)	**0.478** (0.031)
50k	0.677 (0.002)	0.677 (0.002)	0.677 (0.002)	0.677 (0.002)	**0.206** (0.004)

Predictive performance in terms of root mean squared error (RMSE) over independent test sets for the SE3 dataset with increasing training size n. The standard deviation of the RMSE across the 30 resamples is in the brackets.

5.2 Real Data Experiments

We experiment on four real datasets (Table 7): three from the LIACC[7] regression repository, and one Kaggle dataset[8]. The summary of these is presented in Table 8. The RFF results illustrate the advantage nonlinear modelling has over simple linear models. Our sparse nonlinear SRFF method clearly outperforms all the linear as well as the non-sparse nonlinear RFF method. Moreover, it is the only nonlinear sparse learning method that can handle problems of these large-scale datasets.

Table 7. Summary of real-data experiments

Data source	Dataset name	Exp code	Train size	Test size	Total dims
LIAC	Computer Activity	RCP	6k	1k	21
LIAC	F16 elevators	REL	6k	1k	17
LIAC	F16 ailernos	RAI	11k	1k	39
Kaggle	Ore mining impurity	RMN	50k	10k	21

We use the same size for the test and validation samples.

---

[7] http://www.dcc.fc.up.pt/~ltorgo/Regression/DataSets.html.
[8] https://www.kaggle.com/edumagalhaes/quality-prediction-in-a-mining-process.

Table 8. Real-data experiments - Test RMSE for increasing train size n

n	Mean	Ridge	Lasso	RFF	SRFF
RCP	18.518 (0.988)	9.686 (0.705)	9.689 (0.711)	8.194 (0.635)	**2.516** (0.184)
REL	1.044 (0.050)	0.514 (0.210)	0.468 (0.178)	0.446 (0.036)	**0.314** (0.032)
RAI	1.013 (0.034)	0.430 (0.018)	0.430 (0.017)	0.498 (0.038)	**0.407** (0.022)
RMN	1.014 (0.006)	0.987 (0.006)	0.987 (0.006)	0.856 (0.009)	**0.716** (0.008)

Predictive performance in terms of root mean squared error (RMSE) over independent test sets for the real datasets. The standard deviation of the RMSE across the 30 resamples is in the brackets.

6   Summary and Conclusions

We present here a new kernel-based method for learning nonlinear regression function with relevant variable subset selection. The method is unique amongst the state of the art as it can scale to tens of thousands training instances, way beyond what any of the existing kernel-based methods can handle. For example, while none of the tested sparse method worked over datasets with more than 1k instances, the CPU version of our SRFF finished the *full* validation search over 50 hyper-parameters λ in the 50k SE1 experiment within two hours on a laptop with a Dual Intel Core i3 (2nd Gen) 2350M/2.3 GHz and 16 GB RAM.

We focus here on nonlinear regression but the extension to classification problems is straightforward by replacing appropriately the objective loss function. We used the Gaussian kernel for our experiments as one of the most popular kernels in practice. But the principals hold for other shift-invariant kernels as well, and the method and the algorithm can be applied to them directly as soon as the corresponding probability measure $\mu(\boldsymbol{\omega})$ is recovered and the reparametrization of $\boldsymbol{\omega}$ explained in Sect. 3.1 can be applied.

Acknowledgements. This work was partially supported by the research projects HSTS (ISNET) and RAWFIE #645220 (H2020). The computations were performed at University of Geneva on the Baobab and Whales clusters. We specifically wish to thank Yann Sagon, the Baobab administrator, for his excellent work and continuous support.

References

1. Allen, G.I.: Automatic feature selection via weighted kernels and regularization. J. Comput. Graph. Stat. **22**(2), 284–299 (2013)
2. Bach, F.: Consistency of the group lasso and multiple kernel learning. J. Mach. Learn. Res. **9**, 1179–1225 (2008)
3. Bach, F.: High-dimensional non-linear variable selection through hierarchical kernel learning. ArXiv arXiv:0909.0844 (2009)
4. Beck, A., Teboulle, M.: A fast iterative shrinkage-thresholding algorithm for linear inverse problems. SIAM J. Imaging Sci. **2**(1), 183–202 (2009)

5. Bolón-Canedo, V., Sánchez-Maroño, N., Alonso-Betanzos, A.: A review of feature selection methods on synthetic data. Knowl. Inf. Syst. **34**(3), 483–519 (2013)
6. Bolón-Canedo, V., Sánchez-Maroño, N., Alonso-Betanzos, A.: Recent advances and emerging challenges of feature selection in the context of big data. Knowl. Based Syst. **86**, 33–45 (2015)
7. Chan, A.B., Vasconcelos, N., Lanckriet, G.R.G.: Direct convex relaxations of sparse SVM. In: International Conference on Machine Learning (2007)
8. Chen, J., Stern, M., Wainwright, M.J., Jordan, M.I.: Kernel feature selection via conditional covariance minimization. In: Advances in Neural Information Processing Systems (NIPS) (2017)
9. Fukumizu, K., Leng, C.: Gradient-based kernel method for feature extraction and variable selection. In: Advances in Neural Information Processing Systems (NIPS) (2012)
10. Grandvalet, Y., Canu, S.: Adaptive scaling for feature selection in SVMs. In: Advances in Neural Information Processing Systems (NIPS) (2002)
11. Gregorová, M., Kalousis, A., Marchand-Maillet, S.: Structured nonlinear variable selection. In: Conference on Uncertainty in Artificial Intelligence (UAI) (2018)
12. Gretton, A., Fukumizu, K., Teo, C.H., Song, L., Schölkopf, B., Smola, A.J.: A kernel statistical test of independence. In: Advances in Neural Information Processing Systems (NIPS) (2008)
13. Gurram, P., Kwon, H.: Optimal sparse kernel learning in the empirical kernel feature space for hyperspectral classification. IEEE J. Sel. Top. Appl. Earth Observ. Remote Sens. **7**(4), 1217–1226 (2014)
14. Guyon, I., Weston, J., Barnhill, S., Vapnik, V.: Gene selection for cancer classification using support vector machines. Mach. Learn. **46**(1–3), 389–422 (2002)
15. Hastie, T., Tibshirani, R.: Generalized Additive Models. Chapman and Hall, London (1990)
16. Hastie, T., Tibshirani, R., Wainwright, M.: Statistical Learning with Sparsity: The Lasso and Generalizations. CRC Press, Boca Raton (2015)
17. Kingma, D.P., Welling, M.: Auto-encoding variational Bayes. In: International Conference on Learning Representations (ICLR) (2014)
18. Kohavi, R., John, G.H.: Wrappers for feature subset selection. Artif. Intell. **97**(1–2), 273–324 (1997)
19. Koltchinskii, V., Yuan, M.: Sparsity in multiple kernel learning. Ann. Stat. **38**(6), 3660–3695 (2010)
20. Lin, Y., Zhang, H.H.: Component selection and smoothing in multivariate nonparametric regression. Ann. Stat. **34**(5), 2272–2297 (2006)
21. Maldonado, S., Weber, R., Basak, J.: Simultaneous feature selection and classification using kernel-penalized support vector machines. Inf. Sci. **181**(1), 115–128 (2011)
22. Mosci, S., Rosasco, L., Santoro, M., Verri, A., Villa, S.: Solving structured sparsity regularization with proximal methods. In: Balcázar, J.L., Bonchi, F., Gionis, A., Sebag, M. (eds.) ECML PKDD 2010. LNCS (LNAI), vol. 6322, pp. 418–433. Springer, Heidelberg (2010). https://doi.org/10.1007/978-3-642-15883-4_27
23. Muandet, K., Fukumizu, K., Sriperumbudur, B., Schölkopf, B.: Kernel mean embedding of distributions: a review and beyond. Found. Trends Mach. Learn. **10**(1–2), 1–141 (2017)
24. Rahimi, A., Recht, B.: Random features for large-scale kernel machines. In: Advances in Neural Information Processing Systems (NIPS) (2007)
25. Rakotomamonjy, A.: Variable selection using SVM-based criteria. J. Mach. Learn. Res. **3**, 1357–1370 (2003)

26. Ravikumar, P., Liu, H., Lafferty, J., Wasserman, L.: Spam: sparse additive models. In: Advances in Neural Information Processing Systems (NIPS) (2007)
27. Ren, S., Huang, S., Onofrey, J.A., Papademetris, X., Qian, X.: A scalable algorithm for structured kernel feature selection. In: Aistats (2015)
28. Rosasco, L., Villa, S., Mosci, S.: Nonparametric sparsity and regularization. J. Mach. Learn. Res. **14**(1), 1665–1714 (2013)
29. Schölkopf, B., Smola, A.J.: Learning with Kernels. The MIT Press, Cambridge (2002)
30. Song, L., Smola, A., Gretton, A., Borgwardt, K.M., Bedo, J.: Supervised feature selection via dependence estimation. In: Proceedings of the 24th International Conference on Machine Learning - ICML 2007 (2007)
31. Tyagi, H., Krause, A., Eth, Z.: Efficient sampling for learning sparse additive models in high dimensions. In: International Conference on Artificial Intelligence and Statistics (AISTATS) (2016)
32. Weston, J., Elisseeff, A., Scholkopf, B., Tipping, M.: Use of the zero-norm with linear models and kernel methods. J. Mach. Learn. Res. **3**, 1439–1461 (2003)
33. Yamada, M., Jitkrittum, W., Sigal, L., Xing, E.P., Sugiyama, M.: High-dimensional feature selection by feature-wise kernelized lasso. Neural Comput. **26**(1), 185–207 (2014)
34. Yin, J., Chen, X., Xing, E.P.: Group sparse additive models. In: International Conference on Machine Learning (ICML) (2012)
35. Zhao, T., Li, X., Liu, H., Roeder, K.: CRAN - Package SAM (2014)
36. Zhou, D.X.: Derivative reproducing properties for kernel methods in learning theory. J. Comput. Appl. Math. **220**(1–2), 456–463 (2008)

Fast and Provably Effective Multi-view Classification with Landmark-Based SVM

Valentina Zantedeschi[✉], Rémi Emonet, and Marc Sebban

Univ Lyon, UJM-Saint-Etienne, CNRS, Institut d Optique Graduate School,
Laboratoire Hubert Curien UMR 5516, 42023 Saint-Etienne, France
{valentina.zantedeschi,remi.emonet,marc.sebban}@univ-st-etienne.fr

Abstract. We introduce a fast and theoretically founded method for learning landmark-based SVMs in a multi-view classification setting which leverages the complementary information of the different views and linearly scales with the size of the dataset. The proposed method – called MVL-SVM – applies a non-linear projection to the dataset through multi-view similarity estimates w.r.t. a small set of randomly selected landmarks, before learning a linear SVM in this latent space joining all the views. Using the uniform stability framework, we prove that our algorithm is robust to slight changes in the training set leading to a generalization bound depending on the number of views and landmarks. We also show that our method can be easily adapted to a missing-view scenario by only reconstructing the similarities to the landmarks. Empirical results, both in complete and missing view settings, highlight the superior performances of our method, in terms of accuracy and execution time, w.r.t. state of the art techniques. Code related to this paper is available at: https://github.com/vzantedeschi/multiviewLSVM.

Keywords: Multi-view learning · Linear SVM
Landmark induced latent space · Uniform stability · Missing views

1 Introduction

Machine learning has mainly focused, during the past decades, on settings where training data is embedded in a single feature set. However, data collected nowadays is rarely of a single nature. They are rather observed in multiple, possibly heterogeneous views, where each view can take the form of a different source of information. Examples of multi-view datasets are documents translated in different languages, corpora of pictures with descriptive captions, clips with both audio and video streams and so on. Dealing with such scenarios led to the development of the multi-view learning setting [22,26,29] facing new challenges and requiring scientific breakthroughs. Basically, the need for designing multi-view algorithms relies on the observation that standard learning methods with good performance on single-view problems are, in most cases, inefficient in a multi-view setting [11,14,23]. Indeed, the views of an instance don't necessarily stand-alone because they might individually carry insufficient information about the

© Springer Nature Switzerland AG 2019
M. Berlingerio et al. (Eds.): ECML PKDD 2018, LNAI 11052, pp. 193–208, 2019.
https://doi.org/10.1007/978-3-030-10928-8_12

task at hand. Even worse, they can be noisy or missing for a part of the training set. Thus, learning a model jointly on the ensemble of views has been proved to be more expressive than view-specific models, because it exploits the possible complementarity between views [29].

The simplest solution to tackle multi-view problems consists in working on the concatenated space of views, i.e. treating each view as a subset of features. However, as the nature of the views can be heterogeneous, i.e. their corresponding features might lie in different input spaces, such a solution is often unfeasible. Moreover, it does not take into account the statistical specificities of each view and can suffer from the curse of dimensionality. A rich literature of methods has been proposed over the years to provide solutions for extracting information from multiple sources. Common multi-view state of the art approaches learn a set of single-view models either by *co-training* [5], in the attempt to capture both the commonalities and idiosyncrasies of the views, or by *co-regularization* [11,21] over the predictions, aiming at maximizing their agreement (see [22,26,29] for surveys). However, because of the computational overhead originated by training and testing with multiple models, these methods are generally slower than standard single-view algorithms.

A few techniques [14,15,18] have also been proposed suggesting to address the problem in a unified space common to all views, allowing us to learn a single model while exploiting the different sources of information. However, this interesting idea faces a major issue: the cost required to extract the complementary information usually results in algorithms nonetheless barely competitive in terms of execution time. Following this promising line of work, we propose in this paper a new latent space-based approach, called MVL-SVM, which leverages the complementary information and which is fast, scalable and provably effective. As shown in Fig. 1, we base our work on Support Vector Machines (SVMs) [9] which are well known for their robustness, simplicity, efficiency as well as their theoretical foundations via generalization guarantees. In order to keep the time complexity and memory usage low, we formulate our problem as a Linear SVM in a joint space created by comparing the instances, a view at a time, to a small set of randomly selected landmarks, also observed in multiple views. The instance/landmark comparison is carried out by mean of similarity functions, such as the RBF kernel, each defined on a view. Doing so, we solve a linearized joint problem over all views, in which the statistical characteristics of the views are recoded in similarity estimates with points spread over their spaces. Additionally, by applying non-linear mappings, we efficiently capture the non-linearities and multi-modalities of the view spaces while avoiding the drawbacks of Kernel SVMs (see [2,6,16,20]). Such benefits would not be possible without projecting the points on landmarks: the mapping ensures that the algorithm works on homogeneous features and it also controls the dimensionality of the projected space. To theoretically validate our method, we derive a tight generalization bound by proving its stability w.r.t. changes in the training set, utilizing the framework of the Uniform Stability [7]. Finally, we propose an imputation technique for adapting MVL-SVM to the missing-view context, which exploits the information coming from the landmarks.

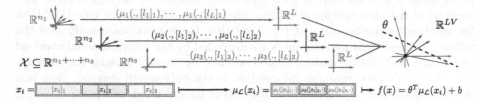

Fig. 1. Overview of the proposed MVL-SVM method. From V views (3 here) of possibly different nature, points are projected on randomly selected landmarks $l_1, \cdots, l_L$ using view specific non-linear mappings $\mu_1, \cdots, \mu_V$. Then, a linear separator is learned in $\mathbb{R}^{LV}$, the joint space of projections.

To sum up, our contribution is three-fold:

1. We introduce a simple, fast and scalable multi-view learning algorithm which benefits from a latent space constructed from similarities to a small set of landmarks. We also show that our approach can be adapted to a missing-view scenario.
2. Using the uniform stability framework, we show that our algorithm is robust to slight changes in the training set leading to a generalization bound that converges uniformly with the number of training examples and that directly depends on the number of views and landmarks.
3. Our experimental results highlight that MVL-SVM allows us to reach very competitive performance in much less time than state of the art methods, overcoming the main issue related to classic latent space-based approaches.

The remainder of the document is organized as follows: Sect. 2 is devoted to the related work; In Sect. 3, we present MVL-SVM's algorithm before deriving in Sect. 4 generalization guarantees in the form of an upper bound on the true risk; An extension of our approach to the missing-view scenario is presented in Sect. 5; Our experimental results are reported in Sect. 6.

2   Related Work

The key to effectively tackling multi-view problems is arguably exploiting the diversity between views. As mentioned earlier, the different views rarely contain, alone, sufficient information for the task at hand and leveraging their complementarity is imperative. We can distinguish two principal families of approaches which address multi-view problems: those which optimize a set of single-view learners and combine their predictions, and those which learn a single model in a common space shared by all views.

Co-training and co-regularization methods [5,11] belong to the first category. Basically, they train multiple view-specific models either by alternatively optimizing them, "teaching" one another, or by fostering their smoothness in predictions. The final step of such techniques consists in aggregating the predictions of the view-specific classifiers, for instance by majority vote [11,21] or by weighted majority vote [13,19]. Note that these methods usually face the following issues: their performances are degraded by the computational overload of training and testing multiple learners; also, by usually making the assumption that the views' common information is the only worth keeping, they boil down to denoising the single views from their uncorrelated information. Yet, it is worth noticing that the information relevant to the task is not necessarily the one the views share, but the one that can be extracted by aggregating the views' incomplete information.

The second category of multi-view learning algorithms contains methods working on Vector-valued Kernel Hilbert Spaces (vvRKHS) [17], whose reproducing kernel outputs, for a pair of multi-view points, a matrix of similarities, each component weighting the similarity of the points observed in a pair of views. These methods are extremely powerful, because they are able to keep the statistical specificities of each view and to extract the complementary information from the diversity of the sources. Of particular interest is Multi-view Metric Learning (MVML [14]) which combines vvRKHS with Metric Learning [3,27] and has proved to outperform Kernel-based state of the art methods, such as Multiple Kernel Learning [12]. MVML jointly learns a classifier and a kernel matrix encoding the within-view and between-view relationships. Although the computations are sped up by working on an approximated Gram matrix, obtained through the Nyström technique [25], this powerful approach is not sufficiently competitive in terms of execution time. To overcome this complexity issue of kernel-based methods, L^3-SVM has been recently proposed in [28] for single-view classification as a different way to take advantage of the discriminatory capabilities of kernels while being fast and scalable. Through clustering and projections on *landmarks*, this algorithm speeds up the learning process while training expressive classifiers, competitive with Kernel-SVMs. This algorithm also comes with a generalization bound on the true risk, even though it is derived independently from the number of clusters. In this paper, we aim at (i) benefiting from this promising landmarks-based SVM paradigm, (ii) adapting it to the multi-view scenario and (iii) deriving theoretical guarantees which take into account both the number of landmarks and views.

Another open problem in multi-view learning is how to deal with realizations of the points that are partially incomplete, i.e. some views of multiple instances are missing. In order to apply a multi-view algorithm, one might have to discard the points with missing views, which may result in a loss of performance, or to complete them using different techniques while trying not to introduce bias. Common practices consist in replacing the missing values with zeros or with the mean or median values of the considered feature. On the other hand, multi-view kernel specific techniques have been proposed to complete the Gram matrices of incomplete views. By making the assumption that similarities between points should be consistent from one view to another, the missing values of a view's Gram matrix are inferred by aligning its eigen-space to the ones of the other views. This can be done by Graph Laplacian regularization [24] (finding the matrix that minimizes its product with the Graph Laplacian matrix of a complete reference view) or by learning convex combinations of normalized kernel matrices [4]. A first limitation of such approaches comes from the fact that they cannot be applied on non square matrices. This prevents us from using them on matrices containing the similarities to a subset of points, like in landmarks-based SVM approaches. Beyond this constraint, the assumption that views are strongly similar and the constraint of having the points altogether observed in a view seem too strong. Another multi-view imputation technique relies on the existence of view generating functions for approximating the missing values. For example, in [1], the authors resort to translation functions for documents in multiple-languages. Unfortunately, depending on the application at hand, such functions are not always available.

In this paper, we make use of the information coming from a small set of randomly selected landmarks to impute the missing values. As for Laplacian imputation [24], we do not need to reconstruct the actual missing features of a point, but only its similarities w.r.t. the landmarks, which drastically simplifies the problem. Through Least Square minimization, we impute the missing similarities by learning the linear combinations of the landmarks projected in the latent space.

3   Multi-view Landmarks-Based SVM (MVL-SVM)

3.1   Notations and Problem Statement

We consider the problem of learning from a dataset $S = \{z_i = (x_i, y_i)\}_{i=1}^m$ of m instances i.i.d. according to a joint distribution D and observed in a multi-view space of V views, so that $x_i \in X \subseteq \mathbb{R}^{n_1 + \cdots + n_V}$, in which views are potentially of different dimensionality, and $y_i \in Y = \{-1, 1\}$. In the following, we will use the notation $[x_i]_v$ to refer to the realization of point x_i in the view v. Moreover, we denote $L = \{l_p\}_{p=1}^L \in X^L$, a set of L landmarks of the input space selected randomly from the training sample.

We aim at learning a classifier $f : X \to \mathbb{R}$ in the joint space defined by the different views as follows (see also Fig. 1):

$$f(x) = \theta^T \mu_L(x) + b \tag{1}$$

where $\theta \in \mathbb{R}^{LV}$ is a vector of weights, each associated to a view v of a landmark p and $\mu_L(x_i) = [\mu_1([x_i]_1, [l_1]_1), \ldots, \mu_1([x_i]_1, [l_L]_1), \ldots, \mu_V([x_i]_V, [l_L]_V)]$ can be interpreted as the mapping function from the input space X to a new landmark space $H \subseteq \mathbb{R}^{LV}$. The sign of the function is retained for prediction ($\hat{y} = sign(f(x))$), i.e. test examples need to be projected as well on the latent space. Notice that each point is compared to the set of landmarks one view at a time and that the problem is now linear in the space $\mathcal{H}$. To capture the non-linearities of the space, we rely on the choice of view-specific score functions $\mu_v : \mathbb{R}^{n_v} \times \mathbb{R}^{n_v} \to \mathbb{R}$ between representations of points in a given view.

The choice of projecting the dataset on selected landmarks is crucial for the discriminatory power of the resulting classifier. As a matter of fact, it enables to express the statistical peculiarities of a view space through similarity estimates and additionally it allows us to work on a latent space common to all views, which has multiple benefits: firstly, it allows to control the dimensionality of the space by choosing the number of landmarks; secondly, it enables to learn a unique classifier, avoiding the problem of combining the outputs of view-specific models; lastly, and most importantly, it loosens the assumptions on the relationship between view information, especially the one on their correlation.

3.2   Optimization Problem and Algorithm

As for standard SVM, our objective function consists in maximizing the margin between the class hyperplanes while minimizing a surrogate function of the classification error:

$$F(f) = \frac{1}{2} \|f\|^2 + \frac{c}{m} \sum_{i=1}^{m} \ell(f, z_i) \tag{2}$$

where $\ell(f, z) = \max(0, 1 - yf(x))$ is the hinge loss. We formulate the multi-view classification problem as a soft-margin SVM learning that we solve in its primal form:

$$\underset{\theta, b, \xi}{\arg\min} \frac{1}{2} \|\theta\|^2 + \frac{c}{m} \sum_{i=1}^{m} \xi_i$$
$$s.t. \quad y_i \left(\theta^T \mu_{\mathcal{L}}(x_i) + b \right) \geq 1 - \xi_i; \quad \xi_i \geq 0 \ \forall i = 1..m. \tag{3}$$

The main difference with standard-SVM is the working input space and its interpretation. Basically, we learn how to linearly combine the point-landmark similarities, describing how they should change over the views for a class. The pseudo-code of MVL-SVM is reported in Algorithm 1.

To recapitulate, our landmark-induced latent space allows us to efficiently extract the complementarity between views while capturing their statistical peculiarities. Moreover, MVL-SVM's flexibility makes it suited to deal with multiple not necessarily correlated views, potentially heterogeneous and of different dimensionality. This flexibility, combined with its scalability, makes MVL-SVM applicable to a wide set of problems.

4   Theoretical Results

Since the parameters θ and b are optimized from a finite set of training examples, a key question is how the learned model behaves at test time. Using the

Input: a sample $\mathcal{S} = \{z_i = (x_i, y_i)\}_{i=1}^m \subseteq \mathbb{R}^{n_1+\cdots+n_V} \times \{-1, 1\}$
and a set of view-specific score functions $\{\mu_v : \mathbb{R}^{n_v} \times \mathbb{R}^{n_v} \to \mathbb{R}\}_{v=1}^V$
1. Select $\mathcal{L} = \{l_p\}_{p=1}^L$ uniformly from $\{x_i\}_{i=1}^m$;
2. Project $\mathcal{S}$ on the latent space:
for $i = 1$ **to** m **do**
$\qquad \mu_{\mathcal{L}}(x_i) = [\mu_1([x_i]_1, [l_1]_1), \ldots, \mu_1([x_i]_1, [l_L]_1), \ldots, \mu_V([x_i]_V, [l_L]_V)]$
end for
3. Learn $\theta \in \mathbb{R}^{LV}$ as the minimizer of Problem (3);
4. Use $sign(\theta^T \mu_{\mathcal{L}}(x) + b)$ for prediction.

Algorithm 1. MVL-SVM algorithm.

theoretical framework of the Uniform Stability [7], we analyze in this section the generalization properties of our algorithm by deriving an upper bound on its true risk. We will see that the stability of our method and, consequently, its generalization capabilities, depend on the choice of the projection functions, the number of selected landmarks and the characteristic of the dataset, such as the number of views and the size of the training set.

4.1 MVL-SVM's Uniform Stability

An algorithm is said to enjoy uniform stability if it outputs similar solutions from slightly different datasets. Let S be the original dataset and S^i the set obtained after replacing the ith sample z_i of S by a new sample z_i' drawn according to the unknown underlying distribution $\mathcal{D}$. We say that an algorithm is uniformly stable if, on a new instance, the difference between the loss suffered by the solution f learned from S and the loss suffered by the solution f^i learned from S^i converges in $O(\frac{1}{m})$. More formally,

Definition 1 (Uniform Stability). *A learning algorithm A has uniform stability* $2\frac{\beta}{m}$ *w.r.t. the loss function ℓ with $\beta \in \mathbb{R}^+$ if*

$$\sup_{z \sim \mathcal{D}} |\ell(f, z) - \ell(f^i, z)| \leq 2\frac{\beta}{m}.$$

The uniform stability is directly implied by the triangle inequality if

$$\sup_{z \sim \mathcal{D}} |\ell(f, z) - \ell(f^{\setminus i}, z)| \leq \frac{\beta}{m}$$

where $f^{\setminus i}$ is learned on $S^{\setminus i}$, the set S without the ith instance z_i.

The notion of σ-admissibility is helpful for studying the uniform stability of an algorithm. In order for the algorithm to be stable, it is necessary to prove that, for a given point, the difference between its loss function evaluated for any two possible hypotheses is bounded by the difference of hypotheses' predictions, scaled by a constant.

Definition 2 (σ-admissibility). *A loss function $\ell(f, z)$ is σ-admissible w.r.t. f if it is convex w.r.t. its first argument and $\forall f_1, f_2$ and $\forall z = (x, y) \in \mathcal{Z}$:*

$$|\ell(f_1, z) - l(f_2, z)| \leq \sigma |f_1(x) - f_2(x)|.$$

In our case, and according to [7], we know that the hinge loss is 1-admissible. We can now present our main theoretical result.

Theorem 1. Uniform Stability *Given the inverse regularizer weight c (from Eq. (3)), MVL-SVM has uniform stability $\frac{cLVM^2}{m}$, where $M = 1$ if μ_v uses the RBF kernel.*

Proof. As $\ell(f, z)$ is 1-admissible, $\forall z = (x, y) \in \mathcal{Z}$,

$$|\ell(f^{\backslash i}, z) - \ell(f, z)| \leq |f^{\backslash i}(x) - f(x)| = |\Delta f(x)| \tag{4}$$

with $\Delta f = f^{\backslash i} - f$. By denoting $\Delta \theta = \theta^{\backslash i} - \theta$, we can derive, $\forall z = (x, y) \in \mathcal{Z}$,

$$\begin{aligned}
|\Delta f(x)| &= |\theta^{\backslash i} \mu_{\mathcal{L}}(x)^T - \theta \mu_{\mathcal{L}}(x)^T| \\
&= |(\theta^{\backslash i} - \theta) \mu_{\mathcal{L}}(x)^T| \\
&\leq \left\| \theta^{\backslash i} - \theta \right\| \|\mu_{\mathcal{L}}(x)\| \tag{5} \\
&\leq \|\Delta \theta\| \|\mu_{\mathcal{L}}(x)\| \\
&\leq \|\Delta \theta\| \sqrt{LV} \|\mu_{\mathcal{L}}(x)\|_\infty \tag{6} \\
&\leq \|\Delta \theta\| \sqrt{LV} \max_{l,v}(\mu_v([x]_v, [l]_v)) \\
&\leq \|\Delta \theta\| \sqrt{LV} M \tag{7}
\end{aligned}$$

with $M = \max_{l,v}(\mu_v([x]_v, [l]_v))$.

Equation (5) is due to the Cauchy-Swartz inequality and Eq. (6) is because $\|\mu_{\mathcal{L}}(x)\| \leq \sqrt{LV} \|\mu_{\mathcal{L}}(x)\|_\infty$ recalling that $\mu_{\mathcal{L}}(x) \in \mathbb{R}^{LV}$.

The value of M depends on the chosen scores functions $\{\mu_v\}_{v=1}^V$. For instance, if all μ_v are the RBF kernel $M = 1$.

From Lemma 21 of [7] we get:

$$2 \|\Delta \theta\|^2 \leq \frac{c}{m} |\Delta f(x_i)|.$$

Then, by instantiating Eq. (7) for $x = x_i$, we get

$$\|\Delta \theta\|^2 \leq \frac{c}{2m} |\Delta f(x_i)| \leq \frac{c}{2m} \|\Delta \theta\| \sqrt{LV} M$$

and as $\|\Delta \theta\| > 0$, we obtain

$$\|\Delta \theta\| \leq \frac{c}{2m} \sqrt{LV} M. \tag{8}$$

So, plugging Eq. (8) in Eq. (7), we get

$$\forall z = (x, y), \quad |\Delta f(x)| \leq \|\Delta \theta\| \sqrt{LV} M \leq \frac{cLVM^2}{2m}$$

which, with Eq. (4), gives the $\frac{cLVM^2}{m}$ uniform stability. □

Note that the stability of MVL-SVM depends on the number of landmarks L. Our method is stable only if $L \ll \frac{m}{V}$, which is not a strong condition considering that usually $m \gg V$. Moreover, this bound expresses that, the smaller L, the more stable the algorithm. This is consistent with the fact that L controls the dimensionality of the projected space in which the multi-view model is learned.

4.2   Generalization Bound

From [7], we know that:

Theorem 2. *Let A be an algorithm with uniform stability $\frac{2\beta}{m}$ w.r.t. a loss ℓ such that $0 \le \ell(f, z) \le E$, for f the minimizer of F and $\forall z \in \mathcal{Z}$. Then, for any i.i.d. sample S of size m and for any $\delta \in (0,1)$, with probability $1 - \delta$:*

$$R_D(f) \le \hat{R}_S(f) + \frac{2\beta}{m} + (4\beta + E)\sqrt{\frac{\ln \frac{1}{\delta}}{2m}}$$

where $R_D(f)$ is the true risk on distribution D and $\hat{R}_S(f)$ is the empirical risk on sample S.

Corollary 1. *The generalization bound of MVL-SVM derived using the Uniform Stability framework is as follows:*

$$R_D(f) \le \hat{R}_S(f) + \frac{cLVM^2}{m} + \left(2cLVM^2 + 1 + 2c\sqrt{LV}M\right)\sqrt{\frac{\ln \frac{1}{\delta}}{2m}}.$$

Proof. The constant E can be estimated by considering the following:

$$F(f) \le F(0)$$

$$\frac{1}{2}\|\theta\|^2 + \frac{c}{m}\sum_{i=1}^{m}\max(0, 1 - y_i(\theta\mu_{\mathcal{L}}(x_i)^T)) \le \frac{1}{2}\|0\|^2 + \frac{c}{m}\sum_{i=1}^{m}\max(0, 1 - y_i(0\mu_{\mathcal{L}}(x_i)^T)) \tag{9}$$

$$\frac{1}{2}\|\theta\|^2 \le c$$

$$\|\theta\|^2 \le 2c$$

Equation (9) is because $\forall a, b, c \in \mathbb{R}^+$, $a + b \le c$ implies that $b \le c$. Thus,

$$\ell(f, z) = \max(0, 1 - y\theta\mu_{\mathcal{L}}(x)^T)$$
$$\le 1 + |\theta\mu_{\mathcal{L}}(x)^T|$$
$$\le 1 + \|\theta\| \, \|\mu_{\mathcal{L}}(x)\| \tag{10}$$
$$\le 1 + 2c\sqrt{LV}M = E$$

Equation (10) comes again from the Cauchy-Swartz inequality.   □

5   Learning with Missing Views

Up to this section, we have made the implicit assumption that all the instances were observed in all the views. Because it is common in real-case scenarios that some points are observed only in a subset of views, we now illustrate how to adapt our formulation to the so-called missing-view setting.

The formulation from Eq. 3 is applicable only when all the points of the training and test sets are observed in all the views. To extend our method to the context of missing views, we apply a reconstruction step before learning. As we want to preserve the scalability of our approach, we do not impute missing values in the original input space: we rather design a dedicated method that imputes missing values by directly leveraging the information coming from the set of landmarks. We simply formulate our imputation as a Least Square over the known values as follows:

$$\arg \min_{R} \|M - RP\|_{\mathcal{F}'}^2 \tag{11}$$

with M the $m \times LV$ matrix of projection values, P the $L \times LV$ matrix of projected landmarks, R the unknown $m \times L$ reconstruction matrix and $\|.\|_{\mathcal{F}'}$ the Frobenius norm considering only the non-missing values (in our case, the missing values are those of M). The problem from Eq. (11) boils down to learning linear combinations of landmark similarities over all the views and, for this reason, all the views of the landmarks need to be known. Doing so, we avoid estimating the actual missing features and we directly impute the view-dependent similarities between points and landmarks.

It is worth noting that each point projection is reconstructed independently and that the system is always (over-)determined for each point, as at least one block of size L of the point projection is known (at least one view's features are given) and the number of unknowns is L.

6   Experimental Results

In this section, we report and analyze the performances of our method w.r.t. the state of the art algorithms, in terms of both classification accuracy and training and testing execution times. We perform two sets of experiments: (i) learning with complete views and (ii) learning with missing views. We will specifically study the behavior of MVL-SVM w.r.t. the number of landmarks keeping in mind that the larger the number of landmarks, the better the discriminatory power of the classifier, but the slower the learning process.

An implementation of our method, based on the Liblinear library [10], together with the other existing algorithms (when the codes are open-source) is available at https://github.com/vzantedeschi/multiviewLSVM.

6.1   Datasets, Methods and Experimental Setup

For these experiments, we employ two multi-class datasets that provide multi-view representations of the instances:

- Flower17[1] contains 1360 pictures of 17 categories of flowers, which come with 7 different distance matrices between pictures (*i.e.* the 7 views);
- uWaveGesture [8] is formed by 4478 vectors describing 8 different gestures as captured by 3 accelerometers (the 3 views).

In order to prove the significance of embedding the datasets in a single space, we compare methods that learn a single classifier on a latent space and methods that learn a set of single-views classifiers. Moreover, we principally compare MVL-SVM to SVM-based approaches, to highlight the interest of using landmark-mappings. Multi-class classification is carried-out through the one-vs-all procedure.

We report the results of the following baselines:

- **MVML** [14] that optimizes over both the classifier and the metric matrix, and which is designed to make the most of the between-view and within-view relationships;
- the co-regularization technique **SVM-2k** [11], which regularizes over the predictions enforcing their smoothness. Originally designed for 2 view learning, we adapted this algorithm to work with $V \geq 2$ views by learning a SVM-2k for every pair of views and combining their predictions using a majority vote;
- **SVMs** which consists in learning a Kernel-SVM per view and aggregating their predictions by majority vote.

All the previous methods, and ours, utilize the Radial Basis Function (RBF, squared exponential) kernel for comparing the points, with a radius that we fixed equal to the square root of the number of features. We make use of the 3 train-val-test splits provided for Flower17, and of the train-test split for uWaveGesture, tuning by cross-validation over the training set. We repeat each experiment 5 times, reporting the average test value and its standard deviation when it is not null. For MVL-SVM, at each iteration we randomly select a new set of landmarks to underline how the chosen landmarks affect the expressiveness of the latent space. We tune the hyper-parameters of the methods by grid-search over the following set-values: for MVML, we evaluate $\lambda \in \{10^{-8}, \ldots, 10\}$ and $\eta \in \{10^{-3}, \ldots, 10^2\}$, as indicated in the original paper; for SVM-2K, we consider c_1, c_2 and $d \in \{10^{-4}, \ldots, 1\}$ and fix $\varepsilon = 10^{-3}$; for both SVMs and MVL-SVM, we consider $c \in \{10^{-3}, \ldots, 10^4\}$.

6.2   Learning with Complete Views

In this first experiment, we compare the methods on complete datasets, where all the points are observed on all the views. In particular, we study the impact of the

---

[1] http://www.robots.ox.ac.uk/~vgg/data/flowers/17/.

dimensionality of our latent space, controlled by the number of landmarks, on the performances of MVL-SVM. As the rank of the Nyström-approximated Gram matrix of MVML and the number of landmarks of MVL-SVM are comparable, because they both measure the number of computed similarities, we draw them on the same axis and compare these two methods also on this criterion. We explore values from 10 to the size of the training set (validation set not included). Because of MVML's huge computational complexity (see Fig. 3), its results in Fig. 2 are truncated at a smaller approximation level.

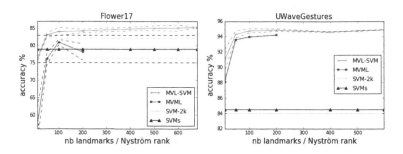

Fig. 2. Average test accuracies (with standard deviations) w.r.t. the number of landmarks/Nyström rank.

Figure 2 shows the test accuracies on both datasets. It is manifest how working on a latent space is of great benefit: both methods that exploit this idea show significant better test accuracies than those that learn view-specific classifiers, especially for the uWaveGestures dataset where views are very complementary. It is worth noting that MVL-SVM is able to reach the best performance even with a small number of landmarks (10 for uWaveGestures and 50 for Flower17).

Moreover, majority-vote techniques seem more sensitive to the choice of points selected for training (see Flower17) than MVL-SVM, which is consistently robust to the variations in the set of landmarks.

Figures 3 and 4 highlight the other important advantage of MVL-SVM: its fastness. At training time, MVL-SVM's execution time is linear in the number of landmarks and several magnitudes smaller than baselines' times. At test time, MVL-SVM is only slightly beaten by MVML, but it could be accounted to optimizations in the code. Notice how learning multiple learners (SVM-2k and SVMs) considerably slows down both training and test steps. Handling multiple models is, indeed, a heavy overhead.

Overall, MVL-SVM achieves significantly better test accuracy that the considered baselines, even with a limited number of landmarks, while training several order of magnitude faster and being comparably fast at test time.

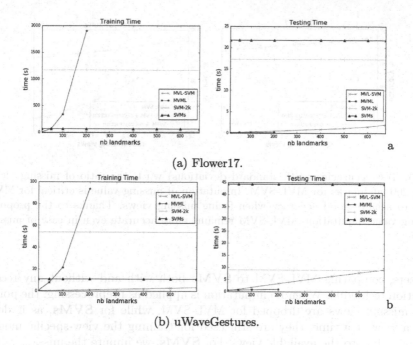

(a) Flower17.

(b) uWaveGestures.

Fig. 3. Training and test times w.r.t. the number of landmarks. MVL-SVM is very fast and scales linearly with the number of landmarks, unlike MVML.

6.3   Learning with Missing Views

With this second series of experiments, we aim at evaluating the validity of the imputation technique proposed in Sect. 5. We make use of the two previously described datasets that we modify for the current task: we drop random views of their points with a ratio of missing views over total number of views (mV) varying in the interval $[0, 0.5]$. For MVL-SVM, the number of landmarks L is fixed to 200. In Fig. 5, we draw the test accuracies in this new setting for both

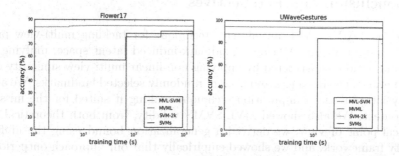

Fig. 4. Average test accuracies w.r.t. the training time. Compared to the other methods, MVL-SVM reaches high accuracy even with very low computational budget. The x axis is in logarithmic scale.

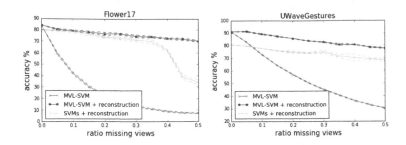

Fig. 5. Test accuracies (with standard deviations) w.r.t. the ratio of missing views, using 200 landmarks for MVL-SVM. Imputation of missing value is critical for MVL-SVM to achieve good accuracy when facing missing views. Thanks to the proposed missing value imputation, MVL-SVM remains more accurate even in case of missing views.

datasets, comparing MVL-SVM to **SVMs** both with and without any reconstruction technique. When no imputation is applied as preprocessing, the points with missing views are dropped for MVL-SVM, while for **SVMs**, as it deals with a view at a time, they are still used for training the view-specific models corresponding to the available views. For **SVMs**, we impute the missing values using Graph Laplacian imputation [24] by fixing the Gram matrix of the view with the most points as the reference view for reconstructing all the other views. Remark that the points missing from the reference view will not have their views reconstructed, which might explain the drop in accuracy of **SVMs** for a ratio bigger than 0.3 for Flower17.

Notice how preprocessing the dataset is fundamental for applying MVL-SVM to the missing-view scenario. This is not surprising as, using a latent space, we can train the model only on points observed in all the views. Even if the accuracy of both methods (with reconstruction) still slightly decays with the ratio of missing views, the gain in performances is dramatic.

7   Conclusion and Perspectives

We proposed MVL-SVM, an effective technique for tackling multi-view problems, training a linear-SVM on a landmark-induced latent space, unifying the view information, constructed by applying non-linear multi-view similarity estimates between the instances and a set of randomly selected landmarks. We additionally introduced an imputation technique making it suited for the missing-view context. We also showed MVL-SVM's validity, from both theoretical and empirical point of view: we derived a generalization bound using the uniform stability framework, and we showed empirically that our approach outperforms the considered baselines in terms of accuracy while being several order of magnitude faster. MVL-SVM rely on a set of landmarks that is shared for all views. According to the application at hand, it might be interesting to consider more landmarks in some of the views, and future work includes considering different landmarks in the views. Additionally, by using block-sparsity in the final linear-separator, automated landmark selection could be achieved, giving MVL-SVM an even better test-time execution speed. The missing view imputation technique can also be improved by considering a joint optimization of the reconstruction matrix R and the linear classifier (θ, b).

Acknowledgments. This work has been funded by the ANR projects LIVES (ANR-15-CE23-0026-03) and SOLSTICE (ANR-13-BS02-01).

References

1. Amini, M., Usunier, N., Goutte, C.: Learning from multiple partially observed views-an application to multilingual text categorization. In: Advances in Neural Information Processing Systems, pp. 28–36 (2009)
2. Bakır, G., Bottou, L., Weston, J.: Breaking SVM complexity with cross training. In: Advances in Neural Information Processing Systems, vol. 17, pp. 81–88 (2005)
3. Bellet, A., Habrard, A., Sebban, M.: A survey on metric learning for feature vectors and structured data (2013). http://arxiv.org/abs/1306.6709
4. Bhadra, S., Kaski, S., Rousu, J.: Multi-view kernel completion. Mach. Learn. **106**(5), 713–739 (2017)
5. Blum, A., Mitchell, T.: Combining labeled and unlabeled data with co-training. In: Proceedings of the Eleventh Annual Conference on Computational Learning Theory, pp. 92–100. ACM (1998)
6. Bordes, A., Bottou, L., Gallinari, P.: SGD-QN: careful quasi-newton stochastic gradient descent. J. Mach. Learn. Res. **10**, 1737–1754 (2009)
7. Bousquet, O., Elisseeff, A.: Stability and generalization. J. Mach. Learn. Res. **2**, 499–526 (2002)
8. Chen, Y., et al.: The UCR time series classification archive, July 2015. www.cs.ucr.edu/~eamonn/time_series_data/
9. Cortes, C., Vapnik, V.: Support-vector networks. Mach. Learn. **20**(3), 273–297 (1995)
10. Fan, R.E., Chang, K.W., Hsieh, C.J., Wang, X.R., Lin, C.J.: LIBLINEAR: a library for large linear classification. J. Mach. Learn. Res. **9**, 1871–1874 (2008)
11. Farquhar, J., Hardoon, D., Meng, H., Shawe-taylor, J.S., Szedmak, S.: Two view learning: SVM-2K, theory and practice. In: Advances in Neural Information Processing Systems, pp. 355–362 (2006)
12. Gönen, M., Alpaydın, E.: Multiple kernel learning algorithms. J. Mach. Learn. Res. **12**, 2211–2268 (2011)
13. Goyal, A., Morvant, E., Germain, P., Amini, M.-R.: PAC-Bayesian analysis for a two-step hierarchical multiview learning approach. In: Ceci, M., Hollmén, J., Todorovski, L., Vens, C., Džeroski, S. (eds.) ECML PKDD 2017. LNCS (LNAI), vol. 10535, pp. 205–221. Springer, Cham (2017). https://doi.org/10.1007/978-3-319-71246-8_13
14. Huusari, R., Kadri, H., Capponi, C.: Multi-view metric learning in vector-valued kernel spaces. In: Proceedings of the Twenty-First International Conference on Artificial Intelligence and Statistics, pp. 415–424 (2018)

15. Kadri, H., Ayache, S., Capponi, C., Koço, S., Dupé, F.X., Morvant, E.: The multi-task learning view of multimodal data. In: Asian Conference on Machine Learning, pp. 261–276 (2013)

16. Ladicky, L., Torr, P.: Locally linear support vector machines. In: Proceedings of the 28th International Conference on Machine Learning (ICML-11), pp. 985–992 (2011)

17. Micchelli, C.A., Pontil, M.: On learning vector-valued functions. Neural Comput. **17**(1), 177–204 (2005)

18. Minh, H.Q., Bazzani, L., Murino, V.: A unifying framework for vector-valued manifold regularization and multi-view learning. In: ICML (2), pp. 100–108 (2013)

19. Minh, H.Q., Bazzani, L., Murino, V.: A unifying framework in vector-valued reproducing kernel Hilbert spaces for manifold regularization and co-regularized multiview learning. J. Mach. Learn. Res. **17**(25), 1–72 (2016)

20. Steinwart, I.: Sparseness of support vector machines. J. Mach. Learn. Res. **4**, 1071–1105 (2003)

21. Sun, S.: Multi-view Laplacian support vector machines. In: Tang, J., King, I., Chen, L., Wang, J. (eds.) ADMA 2011. LNCS (LNAI), vol. 7121, pp. 209–222. Springer, Heidelberg (2011). https://doi.org/10.1007/978-3-642-25856-5_16

22. Sun, S.: A survey of multi-view machine learning. Neural Comput. Appl. **23**(7–8), 2031–2038 (2013)

23. Tang, J., Tian, Y., Zhang, P., Liu, X.: Multiview privileged support vector machines. IEEE Trans. Neural Netw. Learn. Syst. (2017)

24. Trivedi, A., Rai, P., Daumé III, H., DuVall, S.L.: Multiview clustering with incomplete views. In: NIPS Workshop (2010)

25. Williams, C.K., Seeger, M.: Using the Nyström method to speed up kernel machines. In: Advances in Neural Information Processing Systems, pp. 682–688 (2001)

26. Xu, C., Tao, D., Xu, C.: A survey on multi-view learning. CoRR abs/1304.5634 (2013), http://arxiv.org/abs/1304.5634

27. Yang, L., Jin, R.: Distance metric learning: a comprehensive survey. Michigan State Univ. **2**(2), 4 (2006)

28. Zantedeschi, V., Emonet, R., Sebban, M.: L^3-SVMs: landmarks-based linear local support vectors machines. arXiv preprint arXiv:1703.00284 (2017)

29. Zhao, J., Xie, X., Xu, X., Sun, S.: Multi-view learning overview: recent progress and new challenges. Inf. Fusion **38**, 43–54 (2017)

Nyström-SGD: Fast Learning of Kernel-Classifiers with Conditioned Stochastic Gradient Descent

Lukas Pfahler[✉] and Katharina Morik

Artificial Intelligence Group, TU Dortmund University, Dortmund, Germany
{lukas.pfahler,katharina.morik}@tu-dortmund.de

Abstract. Kernel methods are a popular choice for classification problems, but when solving large-scale learning tasks computing the quadratic kernel matrix quickly becomes infeasible. To circumvent this problem, the Nyström method that approximates the kernel matrix using only a smaller sample of the kernel matrix has been proposed. Other techniques to speed up kernel learning include stochastic first order optimization and conditioning. We introduce Nyström-SGD, a learning algorithm that trains kernel classifiers by minimizing a convex loss function with conditioned stochastic gradient descent while exploiting the low-rank structure of a Nyström kernel approximation. Our experiments suggest that the Nyström-SGD enables us to rapidly train high-accuracy classifiers for large-scale classification tasks. Code related to this paper is available at: https://bitbucket.org/Whadup/kernelmachine/.

1 Introduction

Kernel methods are a very powerful family of learning algorithms for classification problems. In addition to their empirical success, we can give strong statistical guarantees for the generalization error [3,18]. However, learning kernel classifiers traditionally involves computing a $N \times N$ kernel matrix and solving linear algebra problems like matrix inversion or eigen-decomposition requiring $\mathcal{O}(N^3)$ operations, which quickly becomes infeasible for large-scale learning.

Recently, Ma and Belkin [13] have proposed an algorithm for large scale kernel learning based on conditioned stochastic gradient descent. Conditioning is an established technique from first-order optimization where we change the coordinate system of the optimization problem to achieve faster convergence to a low-loss classifier. Another recent work by Rudi et al. [17] applies a combination of conditioning and Nyström sampling to learn kernel classifiers. Nyström sampling is a technique to approximate a kernel matrix by evaluating only a smaller sample of m rows of the full kernel matrix. This reduces both the time and space requirement to $\mathcal{O}(mN)$. Both these approaches are limited to minimizing mean-squared error loss (RMSE), where the empirical risk minimization problem reduces to solving a linear system. RMSE is not an ideal loss function for classification problems, as it punishes model outputs that lead to correct

© Springer Nature Switzerland AG 2019
M. Berlingerio et al. (Eds.): ECML PKDD 2018, LNAI 11052, pp. 209–224, 2019.
https://doi.org/10.1007/978-3-030-10928-8_13

classifications. Since different loss function exhibit different convergence behavior of the classification error, by choosing a good loss function it is possible to obtain even faster convergence to high-accuracy solutions [9].

We extend the work of Ma and Belkin to allow general convex loss functions that are better suited for learning classifiers and present *Nyström-SGD*, an algorithm that combines stochastic gradient descent, conditioning and Nyström sampling. We summarize our contributions as follows:

– We show that the conditioned SGD algorithm by Ma and Belkin can be extended to convex loss functions other than RMSE. While their derivation is based on rewriting the optimization problem as solving a linear system, we view it as convex optimization problem and extend results for optimization in Euclidean spaces [8] to reproducing kernel Hilbert spaces.
– We present Nyström-SGD, a learning algorithm that computes a Nyström approximation of the kernel matrix and runs conditioned SGD with this approximated kernel. We show that we can derive a useful conditioner from the approximation and that the resulting update rule of the iterative optimization algorithm can be implemented extremely efficiently.
– We show our method achieves competitive results on benchmark datasets. While computing the Nyström-approximation and the conditioner are computationally expensive steps, our experiments suggest that our approach yields lower loss solutions in less time than previous algorithms.

The remainder of the paper is structured as follows: We begin by revisiting reproducing kernel Hilbert spaces and introducing necessary notation. We continue by discussing related research, particularly covering techniques used to speed up kernel learning. In Sect. 4 we introduce conditioned stochastic gradient descent in reproducing kernel Hilbert spaces. In Sect. 5 we show how to speed up this approach by replacing the kernel matrix with the Nyström approximation, which results in very efficient update rules. In Sect. 6 we demonstrate the effectiveness of our approach in various experiments. This paper is concluded in Sect. 7.

2   Preliminaries on Reproducing Kernel Hilbert Spaces

We begin this section by reviewing *reproducing kernel Hilbert spaces* (RKHS) in which we seek to find linear classification models. The necessary notations will be introduced in this section.

We are interested in learning classification models given labeled data points $(x_i, y_i) \in \mathcal{X} \times \mathcal{Y}$ with $i = 1, ..., N$. We denote by $X = [x_1, ..., x_N]$ the data matrix.

Let $k : \mathcal{X} \times \mathcal{X} \to \mathbb{R}$ be a positive-definite kernel. A reproducing kernel Hilbert space associated to a kernel k is a Hilbert space with associated scalar product $\langle \cdot, \cdot \rangle_{\mathcal{H}}$. Data points $x \in \mathcal{X}$ are embedded in $\mathcal{H}$ via $x \mapsto k(x, \cdot)$. The reproducing property $\langle f, k(x, \cdot) \rangle_{\mathcal{H}} = f(x)$ holds for all $f \in \mathcal{H}$. Particularly it holds that $\langle k(x, \cdot), k(x', \cdot) \rangle_{\mathcal{H}} = k(x, x')$.

We learn functions from the model family

$$x \mapsto \langle f, k(x, \cdot) \rangle_{\mathcal{H}} \tag{1}$$

where $f, k(x, \cdot)$ are elements of the RKHS $\mathcal{H}$. We can think of f as a hyperplane in $\mathcal{H}$. By the representer theorem for kernel Hilbert spaces, we can express these functions as

$$\langle f, k(x, \cdot) \rangle_{\mathcal{H}} = \sum_{i=1}^{n} \alpha_i k(x_i, x). \tag{2}$$

The kernel analogue of the Gram matrix $X^T X$ is called kernel matrix and we write $K = k(x_i, x_j)|_{i,j=1,\ldots,N}$. For notational convenience we define $K_{Ni} = K_i$ and $K_{NB} = [K_{Ni}]_{i \in B}$. Similarly $K_{BB} = k(x_i, x_j)|_{i,j \in B}$. Furthermore we define $K_N. = [k(x_i, \cdot)]_{i=1,\ldots,N}$.

The kernel analogue of the covariance matrix $C = \frac{1}{N} \sum_{i=1}^{N} x_i x_i^T$ is the covariance operator

$$C = \frac{1}{N} \sum_{i=1}^{N} k(x_i, \cdot) \otimes k(x_i, \cdot). \tag{3}$$

Interestingly, kernel matrix and covariance operator share the same spectrum of eigenvalues and have closely connected eigenvectors and eigenfunctions [16,19]. If λ is an eigenvalue of K with normalized eigenvector $u \in \mathbb{R}^N$, i.e. $\lambda u = K u$, the corresponding normalized eigenfunction of C is

$$\ell = \frac{1}{\sqrt{\lambda}} \sum_{i=1}^{N} u_i k(x_i, \cdot). \tag{4}$$

It then holds that $\frac{\lambda}{N} \ell = C\ell$.

3 Related Work

Stochastic gradient descent is probably the most used optimization algorithm used in machine learning recently, particularly for deep network training. SGD training of kernel classifiers has been studied in the past. We separate two approaches: First, as the representer theorem allows us to represent kernel classifiers as a linear combination of kernel evaluations, it is possible to take the derivative of the loss with respect to those coefficients α and optimize using SGD. This was pioneered by Chapelle [4]. The second approach is to take the derivative with respect to the parameter vector in the reproducing kernel Hilbert space, as proposed by Shalev-Shwartz et al. [20]. The resulting SGD update rule can be expressed in terms of α as well and we will rely on this formulation in this work.

Conditioning is a technique to speed up first order optimization methods by transforming the geometry of the optimization problem using a conditioning matrix [8]. Recently it has also been applied to optimization for kernel methods, mostly for kernel ridge regression. In kernel ridge regression, learning reduces to solving a linear system. This linear system can be solved with iterative optimization methods. Avron et al. [2] speed up this optimization using a conditioner based on random feature maps. Cutajar et al. [6] evaluate a variety of different conditioners and achieve fast convergence. Recently Ma and Belkin [13] proposed to use a conditioner to speed up SGD training of kernel classifiers. Their conditioning operator changes the geometry of the RKHS and is very similar to the conditioner for Euclidean spaces proposed by Gonen and Shalev-Shwartz [8]. They prove that their conditioning operator accelerates convergence and that the speedup depends on the decay of eigenvalues of the kernel matrix. Furthermore they show that without conditioning, only a fraction of classifiers is reachable in a polynomial number of epochs, which limits the expressive power of classifiers learned with vanilla SGD.

With Nyström sampling [7,10,21] we avoid computing the full kernel matrix and instead compute a low-rank approximation based on a sample of rows of the kernel matrix. It is subjected to rigorous analysis for the mean squared error loss [12]. Recently Rudi et al. [17] show that it suffices to sample $\mathcal{O}(\sqrt{N})$ rows of the kernel matrix to achieve optimal statistical accuracy if we use Nyström sampling for kernel ridge regression. Their approach iteratively solves a linear system using conditioned conjugate gradient descent where the conditioner is obtain via Nyström sampling as well.

An alternative approach for making large-scale kernel learning tractable is sampling the kernel feature space instead of sampling the kernel matrix. By approximating the RKHS in finite dimensional Euclidean spaces, we reduce the kernel learning task to the linear case where efficient learning is easy. The most popular choice of features are random Fourier features [11,15].

4   Kernel Learning with Conditioned SGD

We continue by reviewing the conditioned stochastic gradient descent for Euclidean vector spaces, on which we will base our conditioned stochastic gradient descent algorithm in RKHS.

4.1   Conditioned Stochastic Gradient Descent

Let us review a preconditioning solution in Euclidean vector spaces proposed by Gonen and Shalev-Shwartz [8]. We want to minimize the empirical loss of a linear classifier with a convex loss function l via

$$\min_{w \in \mathbb{R}^d} \frac{1}{N} \sum_{i=1}^{N} l(\langle w, x_i \rangle, y_i) \tag{5}$$

using a stochastic (sub)gradient descent (SGD) approach. The iterative algorithm updates an intermediate solution by randomly choosing a minibatch of examples $B \subseteq \{1, ..., N\}$ and applying the iteration

$$w^{t+1} = \arg\min_w \frac{|B|}{2\eta} ||w - w^t||_*^2 + \sum_{i \in B} l\left(\langle w, x_i \rangle, y_i\right)$$
$$+ \sum_{i \in B} \langle w - w^t, l'(\langle w, x_i \rangle, y_i) \cdot x_i \rangle \tag{6}$$

where $|| \cdot ||_*$ is a vector norm and $l'(\langle w^t, x_i \rangle, y_i)$ is the (sub)derivative of $l(\hat{y}, y)$ with respect to the first argument $\hat{y}$ evaluated at $\langle w^t, x_i \rangle$. If we use the Euclidean norm, we get the classic stochastic gradient descent update rule

$$w^{t+1} = w^t - \frac{\eta}{|B|} \sum_{i \in B} l'(\langle w^t, x_i \rangle, y_i) \cdot x_i \tag{7}$$

If instead we use the norm $||w||_A = \sqrt{w^T A w}$ for a positive-definite matrix A, the update rule becomes

$$w^{t+1} = w^t - \frac{\eta}{|B|} \sum_{i \in B} l'(\langle w^t, x_i \rangle, y_i) \cdot A^{-1} x. \tag{8}$$

We call A the conditioning matrix. We want to select A such that the optimization algorithm converges to a minimum as fast as possible, while still allowing efficient updates. To this end Gonen and Shalev-Shwartz [8] propose to use a conditioning matrix with a low-rank structure. This allows efficient matrix-vector multiplication, hence the update rule has little overhead over the traditional SGD update. We denote by $C = U \Lambda U^T$ the eigen-decomposition of C where U is the orthonormal matrix of eigenvectors and $\Lambda = \text{diag}(\lambda_1, ..., \lambda_N)$ contains the eigenvalues in non-increasing order. Golen and Shalev-Shwartz propose the following conditioner

$$A^{-1} = I - \sum_{i=1}^{k} \left(1 - \frac{a}{\sqrt{\lambda_i}}\right) u_i u_i^T \tag{9}$$

$$a = \sqrt{\frac{1}{N-k} \sum_{i=k+1}^{N} \lambda_i} \tag{10}$$

that only uses the first k eigenvalues and -vectors of C. However computing a requires knowing all the eigenvalues. We settle for an upper bound: By bounding $\lambda_i \leq \lambda_{k+1}$ for all $i \geq k+1$ we have $a \leq \sqrt{\lambda_{k+1}} =: \tilde{a}$. This leads to the conditioner

$$A^{-1} = I - \sum_{i=1}^{k} \left(1 - \sqrt{\frac{\lambda_{k+1}}{\lambda_i}}\right) u_i u_i^T. \tag{11}$$

which is the foundation of our approach.

4.2 Kernel Learning with Conditioned Stochastic Gradient Descent

In the kernel setting, the empirical risk minimization objective becomes

$$\min_{f \in \mathcal{H}} \frac{1}{N} \sum_{i=1}^{N} l(\langle f, k(x_i, \cdot) \rangle_{\mathcal{H}}, y_i) \tag{12}$$

Shalev-Shwartz et al. show that the SGD update (7) carries over to the kernel setting [20] where $f^t \in \mathcal{H}$. We obtain

$$f^{t+1} = f^t - \frac{\eta}{|B|} \sum_{i \in B} l'(\langle f^t, k(x_i, \cdot) \rangle_{\mathcal{H}}, y_i) k(x_i, \cdot) \tag{13}$$

By setting $f^0 = \mathbf{0} = \sum_i 0 \cdot k(x_i, \cdot)$ we obtain an update rule that can be expressed in terms of α as in (2). By induction, every f^t can be written as a linear combination of kernel basis functions $k(x_i, \cdot)$. By substituting $k(x_i, \cdot) \mapsto e_i$ with the ith standard basis vector e_i, we obtain the update rule for α.

We propose to use a similar conditioner matrix as (11) for kernel reproducing Hilbert spaces. Let λ_i, ℓ_i be the eigenvalues and eigenfunctions of $\mathcal{C}$. We define the conditioning operator

$$\mathcal{A}^{-1} = \mathcal{I} - \sum_{i=1}^{k} \left(1 - \sqrt{\frac{\lambda_{k+1}}{\lambda_i}}\right) \ell_i(\cdot) \otimes \ell_i(\cdot) \tag{14}$$

which is very similar to the conditioner proposed by Ma and Belkin [13], but has an additional square-root[1]. We obtain the update rule for conditioned SGD in kernel reproducing Hilbert spaces

$$f^{t+1} = f^t - \frac{\eta}{|B|} \sum_{i \in B} l'(\langle f^t, k(x_i, \cdot) \rangle, y_i) \mathcal{A}^{-1} k(x_i, \cdot) \tag{15}$$

In this scenario, we can still derive efficient updates for α. We decompose $\mathcal{A}^{-1} = \mathcal{I} - \mathcal{D}$ and focus on $\mathcal{D}$ as $\mathcal{I}$ results in standard SGD updates. For notational convenience we define $\tilde{\lambda}_i = 1 - \sqrt{\lambda_{k+1}\lambda_i^{-1}}$. We see that

$$\mathcal{D}k(x_i, \cdot) = \sum_{j=1}^{k} \tilde{\lambda}_j \ell_j(\cdot) \langle \ell_j(\cdot), k(x_i, \cdot) \rangle_{\mathcal{H}} \tag{16}$$

$$= \sum_{j=1}^{k} \tilde{\lambda}_j \ell_j(\cdot) \sum_{l=1}^{N} \frac{1}{\sqrt{N\lambda_j}} u_{lj} \cdot k(x_l, x_i) \tag{17}$$

$$= \sum_{j=1}^{k} \frac{\tilde{\lambda}_j}{N\lambda_j} K_{Ni}^T u_j \sum_{l=1}^{N} u_{lj} k(x_l, \cdot) \tag{18}$$

$$= K_{Ni}^T \underbrace{\sum_{j=1}^{k} \frac{\tilde{\lambda}_j}{N\lambda_j} u_j u_j^T}_{=:D} K_{N\cdot} =: K_{Ni}^T D K_{N\cdot} \tag{19}$$

For notational convenience we define $D := U\tilde{\Lambda}U$ with $\tilde{\Lambda}$ is a diagonal matrix with coefficients as in (19). Thus we can rewrite the conditioned SGD iteration in terms of α as

---

[1] Inspection of the source-code released with [13] shows that the experiments were actually conducted with a conditioner that uses the same square-root.

$$\alpha^{t+1} = \alpha^t - \frac{\eta}{|B|} \sum_{i \in B} l'(\langle \alpha^t, K_{Ni} \rangle, y_i)(e_i - DK_{Ni}) \tag{20}$$

The update can still be computed efficiently, as it does not need additional kernel evaluations and D is a low-rank matrix which allows efficient matrix-vector multiplication.

5   Faster Training via Nyström Sampling

So far we have derived a conditioned stochastic gradient descent algorithm that operates in reproducing kernel Hilbert spaces. Unfortunately we either have to store $\mathcal{O}(N^2)$ elements of the kernel matrix or compute $\mathcal{O}(N^2)$ kernel evaluations with each pass over the data. We address this by applying Nyström sampling to obtain an approximation of the kernel matrix that only requires $\mathcal{O}(Nm)$ kernel evaluations and storage for a constant sample-size $m < N$. While Ma and Belkin [13] use a Nyström sampling approach to estimate the conditioning matrix, we additionally use it to speed up the learning.

The Nyström approximation is computed as follows [7,10,21]: We draw a sample of m landmark points L uniformly at random from X and compute K_{NL} and extract K_{LL}. Now the approximation is defined as

$$\tilde{K} = K_{NL} K_{LL}^{-1} K_{NL}^T \tag{21}$$

To obtain the conditioning matrix D, we compute the exact eigenvalues and eigenvectors of $\tilde{K}$. Let $V \Sigma V^T = K_{LL}$ be the eigen-decomposition of K_{LL}. We decompose

$$K_{NL} V \Sigma^{-1} =: QR \tag{22}$$

where Q is unitary and R is upper-triangular and then decompose

$$R \Sigma R^T =: \tilde{U} \Lambda \tilde{U}^T \tag{23}$$

where $\tilde{U}$ is unitary and Λ is diagonal. Now we can write

$$\tilde{K} = Q \tilde{U} \Lambda \tilde{U}^T Q^T =: U \Lambda U^T \tag{24}$$

where $U := Q\tilde{U}$ is unitary. Thus U contains the orthonormal eigenvectors of $\tilde{K}$ and Λ contains the corresponding eigenvalues. We can store $\tilde{K}$ in $\mathcal{O}(Nm)$ storage because of its low-rank structure. The computational cost of computing the eigenvalues and -vectors is dominated by computing the QR-decomposition that needs $\mathcal{O}(Nm^2)$ operations and the two eigen-decomposition operations that need $\mathcal{O}(m^3)$ operations with standard linear algebra routines. Since we know all the eigenvalues and vectors, we can set $k = m$. We now show that this has no negative effects on the runtime of SGD updates.

We note that predictions are computed as $\tilde{K}\alpha = U \Lambda U^T \alpha$, hence we never need to know $\alpha \in \mathbb{R}^N$ explicitly, but it suffices to provide $\Lambda U^T \alpha \in \mathbb{R}^m$. Thus

---

Algorithm 1. Nyström-SGD(Data-Set $(x_i, y_i)_{i=1,...,N}$, kernel k, and hyperparameters as described above)

Sample landmark points L and materialize K_{NL}
Decompose $V\Sigma V^T := K_{LL}$.
Decompose $QR := K_{NL}V\Sigma^{-1}$
Decompose $\tilde{U}\Lambda\tilde{U}^T := R\Sigma R^T$
Multiply $U := Q\tilde{U}$.
Set $\tilde{\Lambda}$ according to (19).
Initialize $w = 0 \in \mathbb{R}^m$.
for $i = 1, ..., t$ **do**
    $\hat{y} = U_B.w$ for random minibatch B.
    $w = w - \frac{\eta}{|B|} \sum\limits_{i \in B} l'(\hat{y}_i, y_i) \cdot (\Lambda - \tilde{\Lambda}\Lambda^2)U_i^T.$
return w

---

we can express the stochastic gradient updates not in α, but in $\Lambda U^T \alpha$. This dramatically accelerates the stochastic gradient update, which simplifies to

$$\Lambda U^T \alpha^{t+1} = \Lambda U^T \alpha t - \frac{\eta}{|B|} \sum_{i \in B} l'_i \Lambda U^T (e_i - D\tilde{K}_{Ni})$$

$$= \Lambda U^T \alpha^t - \frac{\eta}{|B|} \sum_{i \in B} l'_i \Lambda U^T (e_i - U\tilde{\Lambda}U^T U\Lambda U_i^T)$$

$$= \Lambda U^T \alpha^t - \frac{\eta}{|B|} \sum_{i \in B} l'_i (\Lambda - \tilde{\Lambda}\Lambda^2)U_i^T. \tag{25}$$

The update reduces to computing the product of a diagonal matrix with selected rows of U which costs $\mathcal{O}(|B|m)$ operations. The update rule without conditioning can be derived analogously. The full learning algorithm is depicted in Algorithm 1.

To obtain predictions for previously unseen data x, we extend the Nyström approximation kernel matrix by one row and compute $\hat{y}(x) = K_L(x)K_{LL}^{-1}K_{NL}^T\alpha$ where $K_L(x) = k(x_i, x)|_{i \in L}$. We can express this in terms of $\Lambda U^T \alpha$ as

$$\hat{y}(x) = K_L(x)(VR^T\tilde{U}\Lambda^{-1})(\Lambda U^T\alpha) \tag{26}$$

where $VR^T\tilde{U}\Lambda^{-1} \in \mathbb{R}^{m \times m}$ is a constant matrix that we compute only once and $\Lambda U^T \alpha$ is the output of the learner.

6 Experiments

In this section we empirically investigate the advantages of using conditioning and a Nyström kernel approximation for training kernel classifiers. We first present the setup of our experiments. Then we investigate convergence behavior of conditioned SGD with general loss functions. We present results on classification performance of our method and finally evaluate the runtime benefits of Nyström-SGD.

6.1 Design Choices

We first describe the basic setup of our experiment, including the datasets and kernels used.

Datasets. We conduct our experiments mostly on standard datasets used in many machine learning publications. We use standard datasets MNIST, CIFAR10, IMDB and SUSY. The FACT dataset [1] contains 16 high-level features derived from simulated sensor measurements of a telescope recording cosmic rays in two classes, gamma and proton. Important properties of the datasets are summarized in Table 1.

We apply standard pre-processing to all datasets: We reduce color-channels to a single greyscale value and normalize these to a $[0, 1]$ range. Text data is tokenized and transformed into a bag-of-words representation with relative word frequencies. The most-frequent 30 words are omitted, the following 10k most-frequent words are used as features.

Table 1. Datasets used in our evaluations

Dataset	TYPE	N	DIM	$|\mathcal{Y}|$	VAL.-N
MNIST	IMAGE	60K	784	10	10K
CIFAR10	IMAGE	50K	1024	10	10K
IMDB	BoW	25K	10,000	2	25K
SUSY	PHYSICS	4M	18	2	1M
FACT	PHYSICS	1.10M	16	2	369K

Kernels. The most-important user-choice for running the proposed learning algorithm is the choice of a suitable kernel function. For our evaluations, we rely on the following three kernel functions:

- **RBF-Kernel:** Probably the most-frequently used kernel for classification tasks is the radial basis function kernel defined as $k(x, x') = \exp(-\gamma\frac{1}{2}||x - x'||_2^2)$ where $\gamma > 0$ is a hyperparameter. We use $\gamma = \frac{1}{d}$ where d is the dimension of x in all our experiments.
- **Inverted-Polynomial-Kernel:** $k(x, x') = (2 - \langle \bar{x}, \bar{x}' \rangle)^{-1}$ where $\bar{x} = ||x||_2^{-1}\cdot x$ denotes the normalized input vector [22].
- **Arc-Cosine-Kernel:** The arc-cosine kernel assesses what fraction of half-spaces both x and x' lie in. $k(x, x') = \frac{1}{\pi}||x|| \cdot ||x'|| \cdot (\sin\theta + (\pi - \theta)\cos\theta)$ where $\theta = \arccos\frac{\langle x, x' \rangle}{||x|| \cdot ||x'||}$ [5].

The latter two kernels have connections to deep networks. These have been used for theoretical analyses of neural network learning: The arc-cosine kernel induces feature maps that correspond to neural networks with infinitely-many ReLu-nodes [5]; the inverted-polynomial kernel induces a class of predictor functions that is a superset of fully-connected networks with bounded weights [22].

Loss Functions. We experiment with different classification losses:

- **Mean-Squared Error.** Plugging in the mean-squared error loss $l(\hat{y}, y) = \frac{1}{2}(\hat{y} - y)^2$, our proposed algorithm is highly related to the approach proposed by Ma and Belkin [13], as discussed above. We believe that the Mean-Squared-Error is not ideal for classification problems as it unnecessarily punishes outputs that lead to correct classifications when $\hat{y}y > 1$.
- **Hinge-Loss.** The loss function used in support vector machines is defined as $l(\hat{y}, y) = \max(0, 1 - \hat{y}y)$ with $y \in \{-1, 1\}$. The hinge-loss does not have a Lipschitz-continuous subgradient, but the smoothed-hinge loss or Huber-loss can be used if a Lipschitz-smooth loss function is desired.
- **Squared Hinge-Loss** $l(\hat{y}, y) = \max(0, 1 - \hat{y}y)^2$. It consistently outperforms other loss functions in the evaluations by Janocha and Czarnecki [9]. It has a Lipschitz-smooth gradient and large gradient when classification is far off.

Hyperparameters. Arguably the most important hyperparameter is the step-size. Following the arguments of Ma and Belkin [13], we set $\eta = \frac{1}{\lambda_1}$ when no conditioner is used. For RMSE and Squared-Hinge, λ_1 is an upper bound of the Lipschitz constant of the gradient, using the inverse of the Lipschitz constant is a popular selection for the stepsize for convex optimization. When conditioning is used, we set $\eta = \frac{1}{\lambda_1}\sqrt{\frac{\lambda_1}{\lambda_k}}$ to account for the conditioning operator. We briefly experimented with using larger stepsizes, however for some combinations of kernel and dataset this leads to divergence. For instance, a multiplicative increase of factor 2 leads to divergence on MNIST with squared hinge loss and RMSE. We defer the analysis of other stepsize policies than constant steps to future work.

In most of our experiments, we use $m = 10,000$ as the sample size to estimate the eigenvalues or approximate the kernel matrix. For the SUSY dataset we settle for $m = 5000$ due to main memory constraints. This does not negatively impact classification performance. We use a batch size of 64 in all our experiments. When we use Nyström sampling, we set $k = m$, otherwise we set $k = 160$ following the results of Ma and Belkin.

6.2   Convergence Results

In this section we compare the optimization progress with conditioning to the standard SGD updates. We show that the conditioned SGD updates substantially accelerate convergence to low-loss solutions. Ma and Belkin demonstrated the effectiveness of conditioning for RMSE loss, we hypothesize that these results carry over to other convex losses.

To this end we run SGD and the conditioned SGD algorithm from Sect. 4.2 with all combinations of kernel, loss and the smaller three datasets. We run the optimization for 10 training epochs. As we can see in Table 2, conditioning consistently yields better loss values than unconditioned SGD.

Next we want to test our hypothesis, that there are loss functions better suited for training classifiers than RMSE. Following the results of Janocha and

Table 2. Loss after 10 epochs of training for different datasets and kernel functions. Runs using conditioning (columns marked 'yes') consistently reach lower losses than vanilla SGD ('no').

		RMSE		HINGE		HINGE^2	
	COND.	NO	YES	NO	YES	NO	YES
MNIST	RBF	2.97	1.46	2.06	0.64	2.97	0.79
	ARC	0.72	0.25	0.40	0.06	0.44	0.05
	INV	0.67	0.11	0.39	0.01	0.41	0.01
CIFAR	RBF	3.60	3.36	2.28	2.13	3.60	3.35
	ARC	3.28	2.69	2.11	1.89	3.24	2.45
	INV	3.21	2.23	2.08	1.67	3.15	1.95
IMDB	RBF	2.02	1.78	2.00	1.85	2.02	1.79
	ARC	1.71	0.72	1.63	0.47	1.55	0.50
	INV	1.55	0.63	1.63	0.46	1.55	0.49

Czarnecki [9], we expect to obtain faster convergence to classifiers with small empirical risk with squared hinge loss than with RMSE. For each dataset and kernel, we compare the training accuracies achieved with each loss function after 10 epochs and use the lowest accuracy as reference. Now we check after how many epochs the training accuracy exceeded this worst accuracy. Figure 1 shows the results for the arccosine kernel; we see that the squared hinge loss reaches the reference accuracy using only 6–7 of the 10 epochs. This suggests that using general convex loss functions is beneficial. Squared hinge loss causes large gradients for far-off predictions and zero gradient for correct predictions, thereby combining benefits of RMSE loss and hinge loss.

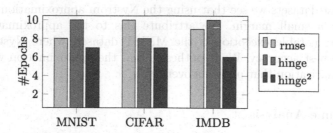

Fig. 1. Number of epochs needed to reach accuracy of slowest learner (lower is better)

6.3   Classification Performances

In this section we evaluate the classification performance achieved by running Nyström-SGD on the datasets. All our experiments do not use explicit regularization. However when using SGD-like methods where the initial solution is zero, we can regularize by limiting the number of training iterations. This early-stopping regularization is effective, as the norm of the solution grows with each gradient update, thus early solutions have lower norms. We report the best validation error, of the best choice of kernel and loss function. Nyström-SGD is compared to conditioned SGD with the full kernel matrix as well as Eigen-Pro, the approach by Ma and Belkin [13], that uses RMSE loss and the full kernel matrix. For the larger datasets, due to memory constraints we only run Nyström-SGD ourselves and rely on published results for comparison.

Table 3. Classification performance of different configurations. We tried different kernels (ARC: arccosine, INV: inverted polynomial) and loss functions (RMSE: root mean-squared error, HINGE: Hinge loss, HINGE^2: squared hinge loss) and report the number of epochs used for training (EP).

	NYSTRÖM SGD		CONDITIONED SGD		EIGEN PRO	
	ACC.	CONFIG	ACC.	CONFIG	ACC.	CONFIG
FACT	86.77	ARC, HINGE 10EP	-	-	-	-
SUSY	79.84	ARC, HINGE^2 5EP	-	-	**80.20**	[13]
MNIST	**98.77**	INV, HINGE^2 7EP	98.68	INV, HINGE^2 7EP	98.51	INV, RMSE 7EP
CIFAR	46.08	INV, HINGE^2 8EP	**48.19**	INV, HINGE^2 70EP	48.17	INV, RMSE 72EP
IMDB	88,08	ARC, HINGE^2 3EP	**88.52**	INV, HINGE^2 16EP	88.29	INV, RMSE 8EP

We depict our results in Table 3. On the Fact dataset we achieve 86.7% accuracy, which compares to the published 87.1% ROC AUC achieved with random forests [14]. The performance on CIFAR is far worse than state-of-the-art results achieved with convolution networks. We believe that choosing a kernel designed for image tasks will improve this performance.

Over most datasets we see that using the Nyström approximation decreases accuracy by a small margin. We attribute this to the approximation error induced. The notable exception is the MNIST dataset, where Nyström-SGD achieves the best accuracy. We hypothesize that the approximation works as a form of regularization that prevents overfitting.

6.4   Runtime Analysis

In this section, we analyze the trade-off between the initialization phase of the algorithm and the actual training, i.e. the SGD iterations.

Obviously computing the Nyström approximation of the kernel matrix and its eigenvalues and vectors are costly operations. We need to verify that it is worthwhile to do so. In Table 4 we report the runtime of the two phases in seconds for the different datasets. We use arccosine kernel and squared hinge loss, but these choices have very little impact on runtime. We use a machine with two Intel Xeon E5-2697v2 CPUs with 12 cores each and ~300 GB of memory.

Table 4. Runtime for different combinations of components: Using Conditioning (C) and using Nyström sampling (N). Using Nyström sampling dramatically reduces the cost of training epochs. Computing the exact conditioner for a Nyström approximation is the most expensive initialization.

C	N	PHASE	FACT	SUSY	MNIST	CIFAR	IMDB
Y	N	SETUP	-	-	130 s	113 s	255 s
		EPOCH	-	-	448 s	420 s	835 s
N	Y	SETUP	823 s	984 s	172 s	198 s	445 s
		EPOCH	114 s	390 s	12 s	11 s	3 s
Y	Y	SETUP	3400 s	4941 s	442 s	485 s	648 s
		EPOCH	114 s	390 s	12 s	11 s	3 s

First of all we note that the SGD epochs are significantly accelerated by using the Nyström approximation. The factor depends on the runtime of computing the kernel function and is higher when the dimension of d is higher. This speedup comes at the cost of computing the approximation once. This takes only unsubstantially longer than computing the approximate conditioner for the exact kernel matrix. We see that computing the exact conditioner for the Nyström operation is the most expensive initialization step. Depending on the dataset, it is as expensive as only computing the Nyström approximation and running between 20 and >60 epochs of SGD.

Table 5. Progress of the optimization compared between Nyström-SGD with and without conditioning. After 200 epochs of SGD training without conditioning (Vanilla), the accuracy of Nyström-SGD still is not matched.

	CONDITIONING 20 EPOCHS		VANILLA 20 EPOCHS		VANILLA 200 EPOCHS	
	TRAIN	TEST	TRAIN	TEST	TRAIN	TEST
FACT	**86.77**	**86.84**	84.28	84.3	84.99	85.10
SUSY	**80.04**	**80.27**	80.05	80.12	80.18	80.18
MNIST	**100.00**	**98.47**	96.26	95.91	98.78	97.98
CIFAR	**73.80**	**46.05**	35.52	33.16	44.53	38.06
IMDB	**97.88**	85.67	79.47	80.24	87.58	**86.26**

Thus for learning with Nyström approximation, we check if models trained only few epochs with conditioning outperform models trained more iterations without conditioning. As we can see in Table 5 this is generally the case. To achieve the same accuracy on the training data as the learner with conditioning after 20 epochs, the learner without conditioning requires more than 200 epochs of training. Thus in total, the cost of learning a high-accuracy prediction model with Nyström and conditioning is lower than without conditioning or without

Nyström. All but one classifiers trained with conditioning have better validation accuracy after 20 epochs than vanilla SGD after 200 epoches. The IMDB dataset is an exception: As we can see in Fig. 2, with conditioning we reach the solution with best validation accuracy in the second epoch and from there on start to overfit. The vanilla learner does not reach the overfitting stage in 200 epochs.

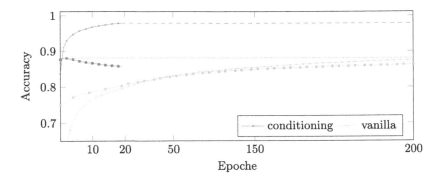

Fig. 2. Convergence behavior for training (x) and testing (squares) on IMDB data. In 200 epochs, vanilla SGD does not reach training accuracy of Nyström-SGD after 20 epochs. Nyström-SGD reaches solution with best validation accuracy after 1 epoch and subsequently begins to overfit. Vanilla SGD is still learning after 200 epochs and does not overfit yet.

Overall, these results are in line with the result of Ma and Belkin [13], that vanilla SGD can only reach a fraction of the function space in polynomial time.

7   Conclusion and Outlook

Building on results of Ma and Belkin [13], we have derived a conditioned stochastic gradient descent algorithm for learning kernel classifiers that minimize arbitrary convex loss functions. Then we have presented Nyström-SGD, a learning algorithm that combines conditioned SGD with Nyström approximation. This way we reduce the number of kernel evaluations needed to train a model; overall we need only $m \cdot N$ kernel evaluations. We compute a useful conditioner for optimizing models using the approximated kernel matrix by efficiently computing its exact eigen-decomposition. Exploiting the structure of conditioner and kernel approximation has allowed us to speed-up the stochastic gradient updates significantly.

In our experiments we have shown the benefits of conditioning and the advantages of choosing a suitable loss function for fast convergence to a high-accuracy classifier. Nyström-SGD has a computationally expensive setup phase that computes the approximation and conditioner in $\mathcal{O}(m^2N + m^3)$ and fast iterations with cost per epoch of $\mathcal{O}(Nm)$. Our experiments suggest that this expensive setup is worthwhile, as we rapidly converge to a high-accuracy solutions on a number of benchmark datasets.

In the future, we want to derive generalization bounds that quantify the influence of the sample size m similar to the ones presented by Rudi et al. [17]. Furthermore we want to investigate using approximate conditioning operators for the Nyström kernel approximation that are cheaper to compute by avoiding the computation of the full QR decomposition and the second eigen-decomposition.

Currently the sample for the Nyström approximation is drawn uniformly at random. More advanced sampling schemes have been proposed. These allow to draw smaller samples while maintaining the same level of accuracy. Incorporating these into our algorithm is certainly beneficial.

Acknowledgment. Part of the work on this paper has been supported by Deutsche Forschungsgemeinschaft (DFG) within the Collaborative Research Center SFB 876 "Providing Information by Resource-Constrained Analysis", project C3.

References

1. Anderhub, H., et al.: Design and operation of FACT - the first G-APD Cherenkov telescope. J. Instrum. **8**(06), P06008 (2013)
2. Avron, H., Clarkson, K.L., Woodruff, D.P.: Faster kernel ridge regression using sketching and preconditioning. SIAM J. Matrix Anal. Appl. **38**(4), 1116–1138 (2017)
3. Bartlett, P.L., Bousquet, O., Mendelson, S.: Local rademacher complexities. Ann. Stat. **33**(4), 1497–1537 (2005)
4. Chapelle, O.: Training a support vector machine in the primal. Neural Comput. **19**(5), 1155–1178 (2007)
5. Cho, Y., Saul, L.K.: Kernel methods for deep learning. In: Advances in Neural Information Processing Systems, pp. 342–350 (2009)
6. Cutajar, K., Osborne, M.A., Cunningham, J.P., Filippone, M.: Preconditioning Kernel Matrices. In: Balcan, M.F., Weinberger, K.Q. (eds.) Proceedings of The 33rd International Conference on Machine Learning, vol. 48, pp. 2529–2538. PMLR, New York (2016)
7. Drineas, P., Mahoney, M.W.: On the Nyström method for approximating a gram matrix for improved kernel-based learning. J. Mach. Learn. Res. **6**(12), 2153–2175 (2005)
8. Gonen, A., Shalev-Shwartz, S.: Faster SGD using sketched conditioning. Technical report (2015)
9. Janocha, K., Czarnecki, W.M.: On loss functions for deep neural networks in classification. Schedae Informaticae **25**, 1–10 (2017)
10. Kwok, J.T., Lu, B.l.: Making large-scale nystrom approximation possible. In: Proceedings of the 27th International Conference on Machine Learning, p. 12 (2010)
11. Le, Q.V., Sarlos, T., Smola, A.J.: Fastfood: approximate kernel expansions in log-linear time. In: Dasgupta, S., McAllester, D. (eds.) Proceedings of the 30th International Conference on Machine Learning, pp. 244–252. PMLR, Atlanta (2013)
12. Lin, J., Rosasco, L.: Optimal rates for learning with Nyström stochastic gradient methods. Technical report (2017)

13. Ma, S., Belkin, M.: Diving into the shallows: a computational perspective on large-scale shallow learning. In: Guyon, I., et al. (eds.) Advances in Neural Information Processing Systems 30. Curran Associates, Inc. (2017)

14. Noethe, M., Al., E.: FACT - performance of the first Cherenkov telescope observing with SiPMs. In: Proceedings of Science: 35th International Cosmic Ray Conference, vol. 301, pp. 0–7 (2017)

15. Rahimi, A., Recht, B.: Weighted sums of random kitchen sinks: replacing minimization with randomization in learning. In: Advances in Neural Information Processing Systems (NIPS) (2009)

16. Rosasco, L.: On learning with integral operators. J. Mach. Learn. Res. **11**, 905–934 (2010)

17. Rudi, A., Carratino, L., Rosasco, L.: FALKON: an optimal large scale kernel method. In: Guyon, I., et al. (eds.) Advances in Neural Information Processing Systems 30, pp. 3891–3901. Curran Associates, Inc. (2017)

18. Schölkopf, B., Smola, A.J.: Learning with Kernels. MIT Press, London (2002)

19. Schölkopf, B., Smola, A., Müller, K.-R.: Kernel principal component analysis. In: Gerstner, W., Germond, A., Hasler, M., Nicoud, J.-D. (eds.) ICANN 1997. LNCS, vol. 1327, pp. 583–588. Springer, Heidelberg (1997). https://doi.org/10.1007/BFb0020217

20. Shalev-Shwartz, S., Singer, Y., Srebro, N., Cotter, A.: Pegasos: primal estimated sub-gradient solver for SVM. Math. Program. **127**, 3–30 (2011)

21. Williams, C.K., Seeger, M.: Using the Nyström method to speed up kernel machines. In: NIPS, pp. 3–9 (2000)

22. Zhang, Y., Lee, J.D., Jordan, M.I.: l1-regularized neural networks are improperly learnable in polynomial time. In: Proceedings of the 33rd International Conference on Machine Learning, vol. 48 (2016)

Learning Paradigms

Hyperparameter Learning
for Conditional Kernel Mean Embeddings
with Rademacher Complexity Bounds

Kelvin Hsu[1,2](✉), Richard Nock[1,2,3](✉), and Fabio Ramos[1,2](✉)

[1] University of Sydney, Sydney, Australia
{Kelvin.Hsu,Fabio.Ramos}@sydney.edu.au, Richard.Nock@anu.edu.au
[2] Data61, CSIRO, Sydney, Australia
[3] Australian National University, Canberra, Australia

Abstract. Conditional kernel mean embeddings are nonparametric models that encode conditional expectations in a reproducing kernel Hilbert space. While they provide a flexible and powerful framework for probabilistic inference, their performance is highly dependent on the choice of kernel and regularization hyperparameters. Nevertheless, current hyperparameter tuning methods predominantly rely on expensive cross validation or heuristics that is not optimized for the inference task. For conditional kernel mean embeddings with categorical targets and arbitrary inputs, we propose a hyperparameter learning framework based on Rademacher complexity bounds to prevent overfitting by balancing data fit against model complexity. Our approach only requires batch updates, allowing scalable kernel hyperparameter tuning without invoking kernel approximations. Experiments demonstrate that our learning framework outperforms competing methods, and can be further extended to incorporate and learn deep neural network weights to improve generalization. (Source code available at: https://github.com/Kelvin-Hsu/cake).

1 Introduction

Conditional mean embeddings (CMEs) are attractive because they encode conditional expectations in a reproducing kernel Hilbert space (RKHS), bypassing the need for a parametrized distribution (Song et al. 2013). They are part of a broader class of techniques known as kernel mean embeddings, where nonparametric probabilistic inference can be carried out entirely within the RKHS because difficult marginalization integrals become simple linear algebra (Muandet et al. 2016). This very general framework is core to modern kernel probabilistic methods, including kernel two-sample testing (Gretton et al. 2007), kernel Bayesian inference (Fukumizu et al. 2013), density estimation (Kanagawa and Fukumizu 2014; Song et al. 2008), component analysis (Muandet et al. 2013),

Electronic supplementary material The online version of this chapter (https://doi.org/10.1007/978-3-030-10928-8_14) contains supplementary material, which is available to authorized users.

© Springer Nature Switzerland AG 2019
M. Berlingerio et al. (Eds.): ECML PKDD 2018, LNAI 11052, pp. 227–242, 2019.
https://doi.org/10.1007/978-3-030-10928-8_14

dimensionality reduction (Fukumizu et al. 2004), feature discovery (Jitkrittum et al. 2016), and state space filtering (Kanagawa et al. 2016).

Nevertheless, like most kernel based models, their performance is highly dependent on the hyperparameters chosen. For these models, the model selection process usually begins by selecting a kernel, whose parameters become part of the model *hyperparameters*, which may further include noise or regularization hyperparameters. Given a set of hyperparameters, training is performed by solving either a convex optimization problem, such as the case in support vector machines (SVMs) (Schölkopf and Smola 2002), or a set of linear equations, such as the case in Gaussian processes (GPs) (Rasmussen and Williams 2006), regularized least squares classifiers (RLSCs) (Rifkin et al. 2003), and CMEs. Unfortunately, hyperparameter tuning is not straight forward, and often cross validation (Song et al. 2013) or median length heuristics (Muandet et al. 2016) remain as the primary approaches for this task. The former can be computationally expensive and sensitive to the selection and number of validation sets, while the latter heuristic only applies to hyperparameters with a length scale interpretation and makes no reference to the conditional inference problem involved as it does not make use of the targets.

One notable success story in this domain are GPs, which employ their marginal likelihood as an objective for hyperparameter learning. The marginal likelihood arises from its Bayesian formulation, and exhibits certain desirable properties – in particular, the ability to automatically balance between data fit and model complexity. On the other hand, CMEs are not necessarily Bayesian, and hence they do not benefit from a natural marginal likelihood formulation, yet such a balance is critical when generalizing the model beyond known examples.

Can we formulate a learning objective for CMEs to balance data fit and model complexity, similar to the marginal likelihood of GPs? For CMEs with categorical targets and arbitrary input, we present such a learning objective as our main contribution. In particular, we: (1) derive a data-dependent model complexity measure $r(\theta, \lambda)$ for a CME with hyperparameters (θ, λ) based on the Rademacher complexity of a relevant class of CMEs, (2) propose a novel learning objective based on this complexity measure to control generalization risk by balancing data fit against model complexity, and (3) design a scalable hyperparameter learning algorithm under this objective using stochastic batch gradient updates. We show that this learning objective produces CMEs that generalize better than that learned from cross validation, empirical risk minimization (ERM), and median length heuristics on standard benchmarks, and apply such an algorithm to incorporate and learn neural network weights to improve generalization accuracy.

2   Background and Related Work

2.1   Conditional Mean Embeddings

To construct a conditional mean embedding operator $\mathcal{U}_{Y|X}$ corresponding to the distribution $\mathbb{P}_{Y|X}$, where $X : \Omega \to \mathcal{X}$ and $Y : \Omega \to \mathcal{Y}$ are measurable random variables, we first choose a kernel $k : \mathcal{X} \times \mathcal{X} \to \mathbb{R}$ for the input space $\mathcal{X}$ and another kernel $l : \mathcal{Y} \times \mathcal{Y} \to \mathbb{R}$ for the output space $\mathcal{Y}$. These kernels k and l each describe how similarity is measured within their respective domains $\mathcal{X}$ and $\mathcal{Y}$, and are symmetric positive definite such that they uniquely define the RKHS $\mathcal{H}_k$ and $\mathcal{H}_l$. The conditional mean embedding operator $\mathcal{U}_{Y|X}$ is then the operator $\mathcal{U} : \mathcal{H}_k \to \mathcal{H}_l$ for which $\mu_{Y|X=x} = \mathcal{U}k(x, \cdot)$, where $\mu_{Y|X=x} := \mathbb{E}[l(Y, \cdot)|X = x]$ is the CME (Song et al. 2009). In this sense, it sweeps out a family of conditional mean embeddings $\mu_{Y|X=x}$ in $\mathcal{H}_l$, each indexed by the input variable $x \in \mathcal{X}$. We then define cross covariance operators $C_{YX} := \mathbb{E}[l(Y, \cdot) \otimes k(X, \cdot)] : \mathcal{H}_k \to \mathcal{H}_l$ and $C_{XX} := \mathbb{E}[k(X, \cdot) \otimes k(X, \cdot)] : \mathcal{H}_k \to \mathcal{H}_k$. Alternatively, they can be seen as elements within the tensor product space $C_{YX} \in \mathcal{H}_l \otimes \mathcal{H}_k$ and $C_{XX} \in \mathcal{H}_k \otimes \mathcal{H}_k$.

Under the assumption that $k(x, \cdot) \in \text{image}(C_{XX})$, it can be shown that $\mathcal{U}_{Y|X} = C_{YX}C_{XX}^{-1}$. While this assumption is satisfied for finite domains $\mathcal{X}$ with a characteristic kernel k, it does not necessarily hold when $\mathcal{X}$ is a continuous domain (Fukumizu et al. 2004), which is the case for many classification problems. In this case, $C_{YX}C_{XX}^{-1}$ becomes only an approximation to $\mathcal{U}_{Y|X}$, and we instead regularize the inversion and use $\mathcal{U}_{Y|X} = C_{YX}(C_{XX} + \lambda I)^{-1}$, which also serves to avoid overfitting (Song et al. 2013). CMEs are useful for probabilistic inference since conditional expectations of a function $g \in \mathcal{H}_l$ can be expressed as inner products with the CME, $\mathbb{E}[g(Y)|X = x] = \langle \mu_{Y|X=x}, g \rangle$, provided that $\mathbb{E}[g(Y)|X = \cdot] \in \mathcal{H}_k$ (Song et al. 2009, Theorem 4).

Furthermore, as both C_{YX} and C_{XX} are defined via expectations, we can estimate them with their respective empirical means to derive a nonparametric estimate for $\mathcal{U}_{Y|X}$ based on finite collection of observations $\{x_i, y_i\} \in \mathcal{X} \times \mathcal{Y}$, $i \in \mathbb{N}_n := \{1, \ldots, n\}$,

$$\hat{\mathcal{U}}_{Y|X} = \Psi(K + n\lambda I)^{-1}\Phi^T, \tag{1}$$

where $K_{ij} := k(x_i, x_j)$, $\Phi := [\phi(x_1) \ldots \phi(x_n)]$, $\Psi := [\psi(y_1) \ldots \psi(y_n)]$, $\phi(x) := k(x, \cdot)$, and $\psi(y) := l(y, \cdot)$ (Song et al. 2013). The empirical CME defined by $\hat{\mu}_{Y|X=x} := \hat{\mathcal{U}}_{Y|X}k(x, \cdot)$ then stochastically converges to the CME $\mu_{Y|X=x}$ in the RKHS norm at a rate of $O_p((n\lambda)^{-\frac{1}{2}} + \lambda^{\frac{1}{2}})$, under the assumption that $k(x, \cdot) \in \text{image}(C_{XX})$ (Song et al. 2009, Theorem 6). This allows us to approximate the conditional expectation with $\langle \hat{\mu}_{Y|X=x}, g \rangle$ instead,

$$\mathbb{E}[g(Y)|X = x] \approx \langle \hat{\mu}_{Y|X=x}, g \rangle = \mathbf{g}^T(K + n\lambda I)^{-1}\mathbf{k}(x), \tag{2}$$

where $\mathbf{g} := \{g(y_i)\}_{i=1}^n$ and $\mathbf{k}(x) := \{k(x_i, x)\}_{i=1}^n$.

2.2 Hyperparameter Learning

Hyperparameter learning for CMEs is particularly difficult compared to marginal or joint embeddings, since the kernel $k = k_\theta$ with hyperparameters $\theta \in \Theta$ is to be learned jointly with a regularization hyperparameter $\lambda \in \Lambda = \mathbb{R}_+$. Grünewälder et al. (2012) proposed to hold out a validation set $\{k(x_{t_j}, \cdot), l(y_{t_j}, \cdot)\}_{j=1}^J$ and minimize $\frac{1}{J} \sum_{j=1}^J \left\| l(y_{t_j}, \cdot) - \hat{\mathcal{U}}_{Y|X}k(x_{t_j}, \cdot) \right\|_{\mathcal{H}_l}^2$ where $\hat{\mathcal{U}}_{Y|X}$ is estimated from the remaining training set using (1). This could also be repeated over multiple folds for cross validation. Song et al. (2013, p. 15) also uses this cross validation approach, but adds regularization $\lambda \|\mathcal{U}\|^2_{HS}$ to the validation objective. Validation sets are necessary for improving generalization to unseen examples. This is because the CME is already the solution that minimizes the objective from Grünewälder et al. (2012) over the operator space, so further optimization over the hyperparmeters using the same training set would lead to overfitting. Moreover, the cross validation objective changes depending on the particular split and number of folds. Additionally, by fitting a separate model for each fold during learning, they incur a large computational cost of $O(Jn^3)$ for J folds, and become prohibitive with large datasets. This spells a need for an alternative hyperparameter learning framework using a different objective.

When cross validation is too expensive, length scales can be set by the median heuristic (Muandet et al. 2016) via $\ell = \text{median}_{i,j}(\|x_i - x_j\|_2)$ for many stationary kernels. However, they cannot be used to set hyperparameters other than length scales, such as λ. In the setting of two sample testing, Gretton et al. (2012) note that they can possibly lead to poor performance. In the context of CMEs, they are also unable to leverage supervision from labels. Flaxman et al. (2016) proposed a Bayesian learning framework for marginal mean embeddings via inducing points, although it is unclear how this can be extended to CMEs. Fukumizu et al. (2009) also investigated the choice of kernel bandwidth for stationary kernels in the setting of binary classification and two sample testing using maximum mean discrepancy (MMD), but has yet to generalize to CMEs or multiclass settings.

2.3   Rademacher Complexity

Rademacher complexity (Bartlett and Mendelson 2002) measures the expressiveness of a function class F by its ability to shatter, or fit, noise. They are data-dependent measures, and are thus particularly well suited to learning tasks where generalization is vital, since complexity penalties that are not data dependent cannot be universally effective (Kearns et al. 1997). The Rademacher complexity (Bartlett and Mendelson 2002, Definition 2) of a function class F is defined by $\mathcal{R}_n(F) := \mathbb{E}[\sup_{f \in F} \|\frac{2}{n} \sum_{i=1}^{n} \sigma_i f(X_i)\|]$, where $\{\sigma_i\}_{i=1}^{n}$ are iid Rademacher random variables, taking values in $\{-1, 1\}$ with equal probability, and $\{X_i\}_{i=1}^{n}$ are iid random variables from the same distribution $\mathbb{P}_X$. Since $\{\sigma_i\}_{i=1}^{n}$ are distributed independently without knowledge of f, the intuition is to interpret $\{\sigma_i\}_{i=1}^{n}$ as labels that are simply noise. For a given set of inputs $\{X_i\}_{i=1}^{n}$, the term inside the norm is high when the sign of $f(X_i)$ matches the signs of σ_i averaged across $i \in \mathbb{N}_n$, meaning that f has managed to fit the noise well. We take this as the defining feature of what it means for a model f to be complex. The supremum then finds the f within F that fits the noise the best, intuitively representing the most complex f within F. The final expectation then averages this quantity across realizations of $\{X_i\}_{i=1}^{n}$ from $\mathbb{P}_X$.

Rademacher complexities are usually applied in the context where classifiers are trained by minimizing some empirical loss within a class of classifiers whose Rademacher complexity is bounded. In the context of multi-label learning, Yu et al. (2014) used trace norm regularization to bound the Rademacher complexity, achieving tight generalization bounds. Xu et al. (2016) extends the trace norm regularization approach by considering the local Rademacher complexity on a subset of the predictor class, where they instead minimize the tail sum of the predictor singular values. Local Rademacher complexity has also been employed for multiple kernel learning (Cortes et al. 2013; Kloft and Blanchard 2011) to learn convex combinations of fixed kernels for SVMs. Similarly, Pontil and Maurer (2013) also used trace norm regularization to bound the Rademacher complexity and minimize the truncated hinge loss. Nevertheless, while Rademacher complexities have been employed to restrict the function class considered for training weight parameters, they have not been applied to learn kernel hyperparameters itself.

3 Multiclass Conditional Embeddings

In this section we present a particular type of CMEs that are suitable for prediction tasks with categorical targets. We show that for CMEs with categorical targets and arbitrary inputs, we can further infer conditional probabilities directly, and not just conditional expectations. As there can be more than two target categories, we refer to these CMEs as multiclass conditional embeddings (MCEs).

For categorical targets, the output label space is finite and discrete, taking values only in $\mathcal{Y} = \mathbb{N}_m := \{1, \ldots, m\}$. Naturally, we choose the Kronecker delta kernel $\delta : \mathbb{N}_m \times \mathbb{N}_m \to \{0, 1\}$ as the output kernel l, where labels that are the same have unit similarity and labels that are different have no similarity. That is, for all pairs of labels $y_i, y_j \in \mathcal{Y}$, $\delta(y_i, y_j) = 1$ only if $y_i = y_j$ and is 0 otherwise. As δ is an integrally strictly positive definite kernel on $\mathbb{N}_m$, it is therefore characteristic (Sriperumbudur et al. 2010, Theorem 7). Therefore, by definition (Fukumizu et al. 2004), δ uniquely defines a RKHS $\mathcal{H}_\delta = \text{span}\{\delta(y, \cdot) : y \in \mathcal{Y}\}$, which is the closure of the span of its kernel induced features (Xu and Zhang 2009). For $\mathcal{Y} = \mathbb{N}_m$, this means that any $g : \mathbb{N}_m \to \mathbb{R}$ that is bounded on its discrete domain $\mathbb{N}_m$ is in the RKHS of δ, because we can always write $g = \sum_{y=1}^m g(y)\delta(y, \cdot) \in \text{span}\{\delta(y, \cdot) : y \in \mathcal{Y}\} \subseteq \mathcal{H}_\delta$. In particular, indicator functions on $\mathbb{N}_m$ are in $\mathcal{H}_\delta$, since $\mathbb{1}_c(y) := \mathbb{1}_{\{c\}}(y) = \delta(c, y)$, so that $\mathbb{1}_c = \delta(c, \cdot)$ are simply the canonical features of $\mathcal{H}_\delta$. Such properties do not necessarily hold for continuous target domains in general. For discrete target domains, this convenient property enables consistent estimations of decision probabilities.

Let $p_c(x) := \mathbb{P}[Y = c | X = x]$ be the *decision probability function* for class $c \in \mathbb{N}_m$, which is the probability of the class label Y being c when the example X is x. Importantly, note that there are no restrictions on the input domain $\mathcal{X}$ as long as a kernel k can be defined on it. For example, $\mathcal{X}$ could be the continuous Euclidean space $\mathbb{R}^d$, the space of images, or the space of strings. We begin by writing this probability as an expectation of indicator functions,

$$p_c(x) := \mathbb{P}[Y = c | X = x] = \mathbb{E}[\mathbb{1}_c(Y) | X = x]. \tag{3}$$

With $\mathbb{1}_c \in \mathcal{H}_\delta$, we let $g = \mathbb{1}_c$ in (2) and $\mathbf{1}_c := \{\mathbb{1}_c(y_i)\}_{i=1}^n$ to estimate the right hand side of (3) by

$$\hat{p}_c(x) = f_c(x) := \mathbf{1}_c^T (K + n\lambda I)^{-1} \mathbf{k}(x). \tag{4}$$

Let $\mathbf{Y} := \begin{bmatrix} \mathbf{1}_1 & \mathbf{1}_2 & \cdots & \mathbf{1}_m \end{bmatrix} \in \{0,1\}^{n \times m}$ be the one hot encoded labels of $\{y_i\}_{i=1}^n$. The vector of empirical decision probabilities over the classes $c \in \mathbb{N}_m$ is then

$$\hat{\mathbf{p}}(x) = \mathbf{f}(x) := \mathbf{Y}^T (K + n\lambda I)^{-1} \mathbf{k}(x) \in \mathbb{R}^m. \tag{5}$$

Since $\mathcal{U} = \hat{\mathcal{U}}_{Y|X}$ (1) is the solution to a regularized least squares problem in the RKHS from $k(x, \cdot) \in \mathcal{H}_k$ to $l(y, \cdot) \in \mathcal{H}_l$ (Grünewälder et al. 2012), CMEs are essentially kernel ridged regressors (KRRs) with targets in the RKHS. In this case, because $\mathcal{Y} = \mathbb{N}_m$ is discrete, $\mathcal{H}_\delta$ can be identified with $\mathbb{R}^m$. As a result, the rows of the MCE can also be seen as m KRRs (Friedman et al. 2001) on binary $\{0,1\}$-targets, where they all share the same input kernel k. Because they all share the same kernel to form the MCE, we can show that the empirical decision probabilities (4) do converge to the population decision probability.

Theorem 1 (Convergence of Empirical Decision Probability Function).
Assuming that $k(x, \cdot)$ is in the image of C_{XX}, the empirical decision probability function $\hat{p}_c : \mathcal{X} \to \mathbb{R}$ (4) converges uniformly to the true decision probability $p_c : \mathcal{X} \to [0,1]$ (3) at a stochastic rate of at least $O_p((n\lambda)^{-\frac{1}{2}} + \lambda^{\frac{1}{2}})$ for all $c \in \mathcal{Y} = \mathbb{N}_m$. See appendix (see supplementary material) for proof, including for all subsequent theorems.

In particular, the assumption $k(x, \cdot) \in \text{image}(C_{XX})$ is a statement on the input kernel k, not the output kernel l, which is a Kronecker delta $l = \delta$ for MCEs. It is worthwhile to note that this assumption is common for CMEs, and is not as restrictive as it may first appear, as it can be relaxed through introducing the regularization hyperparameter λ (1) in practice (Muandet et al. 2016; Song et al. 2013, 2009, pp. 74–75, Sects. 3 and 3.1 *resp.*).

Note that for finite n the probability estimates (4) may not necessarily lie in the range $[0,1]$ nor form a normalized distribution for finite n. Nonetheless, Theorem 1 guarantees that they approach one with increasing sample size. When normalized distributions are required, clip-normalized estimates can be used,

$$\tilde{p}_c(x) := \frac{\max\{\hat{p}_c(x), 0\}}{\sum_{j=1}^m \max\{\hat{p}_j(x), 0\}}. \tag{6}$$

This does not change the resulting prediction, since $\hat{y}(x) = \text{argmax}_{c \in \mathbb{N}_m} \hat{p}_c(x)$ $= \text{argmax}_{c \in \mathbb{N}_m} \tilde{p}_c(x)$. Theorem 1 also implies that eventually the effect of clip-normalization vanishes, where $\tilde{p}_c(x)$ approaches to both $\hat{p}_c(x)$ and thus $p_c(x)$ with increasing sample sizes.

Importantly, this enables MCEs to be naturally applied to perform probabilistic classification in multiclass settings with categorical targets. In contrast, in terms of probabilistic classification, support vector classifiers (SVCs) do not output probabilities and probabilistic extensions require difficult calibration, while Gaussian process classifiers (GPCs) require posterior approximations. Furthermore, in terms of the multiclass setting, multiclass extensions to SVCs and GPCs often employ the one versus all (OVA) or one versus one (OVO) scheme (Aly 2005), resulting in multiple separately trained binary classifiers with no guarantees of coherence between their outputs. Instead, training a single MCE is sufficient for producing consistent multiclass probabilistic estimates.

Similar to RLSC, MCEs are solutions to a regularized least squares problem in a RKHS (Grünewälder et al. 2012), resulting in a similar system of linear equations. Nevertheless, RLSCs primarily differ in the way they handle the labels, in which binary labels $\{-1, 1\}$ appear directly in the squared loss instead of its kernel feature $\delta(y_i, \cdot)$ or, equivalently, its one hot encoded form $\mathbf{y}_i$. Consequently, multiclass extensions for RLSC either require using the OVA scheme (Rifkin et al. 2003) which suffers from computational and coherence issues, or alternatively minimize the total loss across all binarized tasks for the overall least squares problem (Pahikkala et al. 2012). Although the latter attempts to link the classifiers together through its loss, both approaches still produce separate classifiers for each class. As a result, multiclass RLSC does not produce consistent estimates of class probabilities as in Theorem 1 for MCEs.

4   Hyperparameter Learning with Rademacher Complexity Bounds

In this section we derive learning theoretic bounds that motivate our proposed hyperparameter learning algorithm, and discuss how it can be extended in various ways to enhance scalability and performance. From here onwards, we denote θ as the kernel hyperparameters of the kernel $k = k_\theta$.

We begin by defining a loss function as a measure for performance. For decision functions of the form $\mathbf{f} : \mathcal{X} \to \mathcal{A} := \mathbb{R}^m$ whose entries are probability estimates, we employ a modified cross entropy loss,

$$\mathcal{L}_\epsilon(y, \mathbf{f}(x)) := -\log\left[\mathbf{y}^T \mathbf{f}(x)\right]_\epsilon^1 = -\log\left[f_y(x)\right]_\epsilon^1, \qquad (7)$$

to express risk, where we use the notation $[\,\cdot\,]_\epsilon^1 := \min\{\max\{\,\cdot\,, \epsilon\}, 1\}$ for $\epsilon \in (0, 1)$. It is worthwhile to point out that this choice only makes sense due to Theorem 1, as it allows us to interpret the outputs of the CME as asymptotic probability estimates. Note that we employ the loss on the original probability estimates (5), not the clip-normalized version (6). We employ this loss in virtue of Theorem 1, where we expect $\mathbf{f}(x)$ (5) to be approximations to the population decision probabilities. In contrast, direct outputs from SVCs, GPCs, or RLSCs are not consistent probability estimates and cannot take advantage of (7) easily.

However, simply minimizing the empirical loss $\frac{1}{n}\sum_{i=1}^n \mathcal{L}_\epsilon(y_i, \mathbf{f}_{\theta,\lambda}(x_i))$ over the hyperparameters (θ, λ) could lead to an overfitted model. We therefore employ Rademacher complexity bounds to control the model complexity of MCEs.

Let Θ and Λ be a space of kernel and regularization hyperparameters respectively. We define the class of MCEs over these hyperparameter spaces by

$$F_n(\Theta, \Lambda) := \{\mathbf{f}_{\theta,\lambda}(x) : \theta \in \Theta, \lambda \in \Lambda\}. \qquad (8)$$

We denote $W_{\theta,\lambda}^T \equiv \hat{\mathcal{U}}_{Y|X}^{(\theta,\lambda)}$ so that $\|W_{\theta,\lambda}\|_{\mathrm{tr}} = \|\hat{\mathcal{U}}_{Y|X}^{(\theta,\lambda)}\|_{HS}$ to reflect the dependence on (θ, λ) and also to emphasize the role it plays as the weights of the decision function. We first restrict the space of hyperparameters by the norms of $W_{\theta,\lambda}$ and $k_\theta(x, x)$ to obtain an upper bound to the Rademacher complexity of $F_n(\Theta, \Lambda)$.

Theorem 2 (MCE Rademacher Complexity Bound). *Suppose that the trace norm* $\|W_{\theta,\lambda}\|_{\mathrm{tr}} \le \rho$ *is bounded for all* $\theta \in \Theta, \lambda \in \Lambda$. *Further suppose that the canonical feature map is bounded in RKHS norm* $\|\phi_\theta(x)\|^2_{\mathcal{H}_{k_\theta}} = k_\theta(x,x) \le \alpha^2$, $\alpha > 0$, *for all* $x \in \mathcal{X}, \theta \in \Theta$. *For any set of training observations* $\{x_i, y_i\}^n_{i=1}$, *the Rademacher complexity of the class of MCEs* $F_n(\Theta, \Lambda)$ *(8) is bounded by*

$$\mathcal{R}_n(F_n(\Theta, \Lambda)) \le 2\alpha\rho. \tag{9}$$

Bartlett and Mendelson (2002) showed that the expected risk can be bounded with high probability using the empirical risk and the Rademacher complexity of the loss composed with the function class. For a Lipchitz loss, Ledoux and Talagrand (2013) further showed that the latter quantity can be bounded using the Rademacher complexity of the function class itself. We use these two results to arrive at the following probabilistic upper bound to our expected loss.

Theorem 3 (MCE ϵ-Specific Expected Risk Bound). *Assume the same assumptions as Theorem 2. For any integer* $n \in \mathbb{N}_+$, *any* $\epsilon \in (0, e^{-1})$, *and any set of training observations* $\{x_i, y_i\}^n_{i=1}$, *with probability of at least* $1 - \beta$ *over iid samples* $\{X_i, Y_i\}^n_{i=1}$ *of length* n *from* $\mathbb{P}_{XY}$, *every* $f \in F_n(\Theta, \Lambda)$ *satisfies*

$$\mathbb{E}[\mathcal{L}_{e^{-1}}(Y, f(X))] \le \frac{1}{n} \sum_{i=1}^n \mathcal{L}_\epsilon(Y_i, f(X_i)) + 4e\,\alpha\rho + \sqrt{\frac{8}{n} \log \frac{2}{\beta}}. \tag{10}$$

However, for hyperparameter learning, we would require a risk bound for specific choice of hyperparameters, not just for a set of hyperparameters. For some $\tilde{\theta} \in \Theta$ and $\tilde{\lambda} \in \Lambda$, we construct a subset of hyperparameters $\Xi(\tilde{\theta}, \tilde{\lambda}) \subseteq \Theta \times \Lambda$ defined by $\Xi(\tilde{\theta}, \tilde{\lambda}) := \{(\theta, \lambda) \in \Theta \times \Lambda : \|W_{\theta,\lambda}\|_{\mathrm{tr}} \le \|W_{\tilde{\theta},\tilde{\lambda}}\|_{\mathrm{tr}}, \sup_{x\in\mathcal{X}} k_\theta(x,x) \le \alpha^2(\tilde{\theta}) := \sup_{x\in\mathcal{X}} k_{\tilde{\theta}}(x,x)\}$. Clearly, this subset is non-empty, since $(\tilde{\theta}, \tilde{\lambda}) \in \Xi(\tilde{\theta}, \tilde{\lambda})$ is itself an element of this subset. Thus, we can assert that $\|W_{\theta,\lambda}\|_{\mathrm{tr}} \le \rho = \|W_{\tilde{\theta},\tilde{\lambda}}\|_{\mathrm{tr}}$ is bounded for all $(\theta, \lambda) \in \Xi(\tilde{\theta}, \tilde{\lambda})$, and that $\|\phi_\theta(x)\|^2_{\mathcal{H}_{k_\theta}} = k_\theta(x,x) \le \alpha^2 = \sup_{x\in\mathcal{X}} k_{\tilde{\theta}}(x,x)$ is bounded for all $x \in \mathcal{X}, (\theta, \lambda) \in \Xi(\tilde{\theta}, \tilde{\lambda})$.

We can now choose some arbitrary $\tilde{\theta} \in \Theta, \tilde{\lambda} \in \Lambda$ and apply Theorem 3 with $\rho = \|W_{\tilde{\theta},\tilde{\lambda}}\|_{\mathrm{tr}}$ and $\alpha^2 = \sup_{x\in\mathcal{X}} k_{\tilde{\theta}}(x,x)$ and by considering only the hyperparameters $(\theta, \lambda) \in \Xi(\tilde{\theta}, \tilde{\lambda})$. The probabilistic statement (10) then only holds for $(\theta, \lambda) \in \Xi(\tilde{\theta}, \tilde{\lambda})$. In particular, since $(\tilde{\theta}, \tilde{\lambda}) \in \Xi(\tilde{\theta}, \tilde{\lambda})$, it holds for $(\theta, \lambda) = (\tilde{\theta}, \tilde{\lambda})$. Applying this choice, the only hyperparameters that remain in the statement are $(\tilde{\theta}, \tilde{\lambda})$. We then replace these symbols with (θ, λ) again to avoid cluttered notation. Since they were chosen arbitrarily from $\Theta \times \Lambda$, we arrive at our final result.

Theorem 4 (MCE Expected Risk Bound for Hyperparameters). *For any integer* $n \in \mathbb{N}_+$ *and any set of training observations* $\{x_i, y_i\}^n_{i=1}$ *used to define* $\mathbf{f}_{\theta,\lambda}$ *(5), with probability* $1 - \beta$ *over iid samples* $\{X_i, Y_i\}^n_{i=1}$ *of length* n *from* $\mathbb{P}_{XY}$, *every* $\theta \in \Theta$, $\lambda \in \Lambda$, *and* $\epsilon \in (0, e^{-1})$ *satisfies*

$$\mathbb{E}[\mathcal{L}_{e^{-1}}(Y, \mathbf{f}_{\theta,\lambda}(X))] \le \frac{1}{n} \sum_{i=1}^n \mathcal{L}_\epsilon(Y_i, \mathbf{f}_{\theta,\lambda}(X_i)) + 4e\,r(\theta, \lambda) + \sqrt{\frac{8}{n} \log \frac{2}{\beta}}, \tag{11}$$

Algorithm 1. MCE Hyperparameter Learning with Stochastic Gradient Updates

1: **Input:** kernel family $k_\theta : \mathcal{X} \times \mathcal{X} \to \mathbb{R}$, dataset $\{x_i, y_i\}_{i=1}^n$, initial kernel hyperparameters θ_0, initial regularization hyperparameters λ_0, learning rate η, cross entropy loss threshold ϵ, batch size n_b
2: $\theta \leftarrow \theta_0, \lambda \leftarrow \lambda_0$
3: **repeat**
4:     Sample the next batch $\mathcal{I}_b \subseteq \mathbb{N}_n$ s.t. $|\mathcal{I}_b| = n_b$
5:     $Y \leftarrow \{\delta(y_i, c) : i \in \mathcal{I}_b, c \in \mathbb{N}_m\}$ $\in \{0, 1\}^{n_b \times m}$
6:     $K_\theta \leftarrow \{k_\theta(x_i, x_j) : i \in \mathcal{I}_b, j \in \mathcal{I}_b\}$ $\in \mathbb{R}^{n_b \times n_b}$
7:     $L_{\theta,\lambda} \leftarrow \text{cholesky}(K_\theta + n_b \lambda I_{n_b})$ $\in \mathbb{R}^{n_b \times n_b}$
8:     $V_{\theta,\lambda} \leftarrow L_{\theta,\lambda}^T \backslash (L_{\theta,\lambda} \backslash Y)$ $\in \mathbb{R}^{n_b \times m}$
9:     $P_{\theta,\lambda} \leftarrow K_\theta V_{\theta,\lambda}$ $\in \mathbb{R}^{n_b \times m}$
10:    $r(\theta, \lambda) \leftarrow \alpha(\theta)\sqrt{\text{trace}(V_{\theta,\lambda}^T K_\theta V_{\theta,\lambda})}$
11:    $q(\theta, \lambda) \leftarrow \frac{1}{n_b} \sum_{i=1}^{n_b} \mathcal{L}_\epsilon((Y)_i, (P_{\theta,\lambda})_i) + 4e\, r(\theta, \lambda)$
12:    $(\theta, \lambda) \leftarrow \text{GradientBasedUpdate}(q, \theta, \lambda; \eta)$
13: **until** maximum iterations reached
14: **Output:** kernel hyperparameters θ, regularization λ

---

where $r(\theta, \lambda) := \sqrt{\text{trace}(V_{\theta,\lambda}^T K_\theta V_{\theta,\lambda}) \sup_{x \in \mathcal{X}} k_\theta(x, x)}$ and $V_{\theta,\lambda} := (K_\theta + n\lambda I)^{-1}\mathbf{Y}$.

In particular, $r(\theta, \lambda)$ is an upper bound to the Rademacher complexity of a relevant class of MCEs based on the hyperparameters $\Xi(\theta, \lambda)$. We call $r(\theta, \lambda)$ the Rademacher complexity bound (RCB) and use it to measure the model complexity of a MCE with hyperparameters (θ, λ). Since the training set itself is a sample of length n drawn from $\mathbb{P}_{XY}$, the inequality (11) holds with probability $1 - \beta$ when the random variables (X_i, Y_i) are realized as the training observations (x_i, y_i). Motivated by this, we employ this upper bound as the learning objective for hyperparameter learning,

$$q(\theta, \lambda) := \frac{1}{n} \sum_{i=1}^n \mathcal{L}_\epsilon(y_i, \mathbf{f}_{\theta,\lambda}(x_i)) + 4e\, r(\theta, \lambda). \tag{12}$$

Importantly, the first term is an empirical risk that measures data fit, and the second term is the RCB that measures model complexity. Together, this learning objective achieves a balance between data fit and model complexity, similar to the corresponding property of a negative log marginal likelihood learning objective.

4.1 Extensions

Batch Stochastic Gradient Update. Since Theorem 4 holds for any $n \in \mathbb{N}_+$ and any set of data $\{x_i, y_i\}_{i=1}^n$ from $\mathbb{P}_{XY}$, the bound (11) also holds with high probability for a batch subset of the training data. However, the batch size cannot be too small, in order to keep the constant $\sqrt{8 \log(2/\beta)/n}$ relatively small. We therefore propose to use only a random batch subset of the data to perform each gradient update. This enables scalable hyperparameter learning through batch stochastic gradient updates, where each gradient update stochastically improves a different probabilistic upper bound of the generalization risk. Note that without Theorem 4, it is not straightforward to simply apply stochastic gradient updates to optimize q, since r depends on the dataset but is not written in terms of a summation over the data. We present this scalable hyperparameter learning approach via batch stochastic gradient updates in Algorithm 1, reducing the time complexity from $O(n^3)$ to $O(n_b^3)$, where n_b is the batch size. The Cholesky decomposition for the full training set requires $O(n^3)$ time and is necessary only for inference, instead of once every learning iteration. It can be further avoided by using random Fourier features (Rahimi and Recht 2008) or kernel herding (Chen et al. 2010) to approximate the already learned MCE. All further inference takes $O(n^2)$ time, or potentially less with approximation, using back substitution.

Batch Validation. While we simply instantiated (X_i, Y_i) to be the training observations in Theorem 4 to obtain (12), this does not have to be the case for batch updates. Instead, in each learning iteration, we could further split the batch into two sub-batches – one for training and one for validation. The training batch is used to form the MCE $\mathbf{f}_{\theta,\lambda}$ and RCB $r(\theta, \lambda)$, while we evaluate the empirical risk on the validation batch,

$$q^{(V)}(\theta, \lambda) := \frac{1}{n_{(V)}} \sum_{i=1}^{n_{(V)}} \mathcal{L}_\epsilon(y_i^{(V)}, \mathbf{f}_{\theta,\lambda}^{(T)}(x_i^{(V)})) + \tau\, r^{(T)}(\theta, \lambda), \tag{13}$$

where (T) and (V) denotes training and validation. Importantly, in contrast to standard cross validation, not all data is required for each update due to the presence of the RCB. Furthermore, although the multiplier on the RCB is $4e$, experiments show that generalization performance can improve if we use a smaller multiplier $\tau < 4e$, suggesting an upper bound tighter than (11) may exist. In practice, these two extensions work well together. Intuitively, by introducing a validation batch to measure empirical data fit, a smaller weight on the complexity penalty is required.

Conditional Embedding Network. Our learning algorithm does not restrict the way the kernel k_θ is constructed from its hyperparameters $\theta \in \Theta$. One particularly useful type of MCEs are those where the input kernel $k_\theta(x, x') = \langle \varphi_\theta(x), \varphi_\theta(x') \rangle$ is constructed from neural networks $\varphi_\theta : \mathcal{X} \to \mathbb{R}^p$ explicitly. We refer to MCEs constructed this way as conditional embedding networks (CENs). In these cases, the weights and biases of the neural network become the kernel hyperparameters θ of the CENs. We can therefore learn network weights and biases jointly under (12). CENs can also scale easily, since the $n \times n$ Cholesky decomposition required for full gradient updates in Algorithm 1 can be transformed into a $p \times p$ decomposition by the Woodbury matrix inversion identity (Higham 2002), reducing the time complexity to $O(p^3 + np^2)$. This allows scalable learning for $n \gg p$ even without using batch gradient updates. For inference, standard map reduce methods can be used. We direct the reader to Appendix C (see supplementary material) for detailed discussion on the various MCE architectures and their implementation as compared to Algorithm 1.

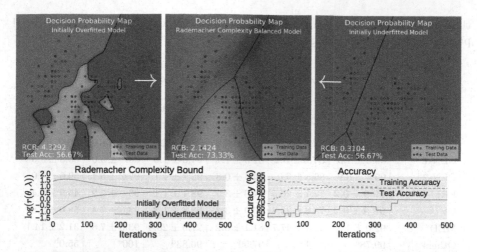

Fig. 1. Rademacher complexity balanced learning of hyperparameters for an isotropic Gaussian MCE using the first two attributes of the iris dataset. (Color figure online)

5   Experiments

Toy Example. The first two of four total attributes of the iris dataset (Fisher 1936) are known to have class labels that are non-separable by any means, in that the same example $x \in \mathbb{R}^2$ may be assigned different output labels $y \in \mathbb{N}_3 := \{1, 2, 3\}$. In these difficult scenarios, the notion of model complexity is extremely important, and the success of a learning algorithm greatly depends on how it balances training performance and model complexity to avoid both underfitting and overfitting.

Figure 1 demonstrates Algorithm 1 with full gradient updates ($n_b = n$) to learn hyperparameters of the MCE on the two attribute iris dataset. The kernel used is isotropic Gaussian with diagonal length scales $\Sigma = \ell^2 I_2$ and sensitivity $\alpha = \sigma_f$, so that the hyperparameters are $\theta = (\alpha, \ell)$ and λ. We evaluate the performance of the learning algorithm on a withheld test set using 20% of the available 150 data samples. Attributes are scaled into the unit range $[0, 1]$ and decision probability maps are plotted for the region $[-0.5, 1.05]^2$, where the red, green, and blue color channels represent the clip-normalized decision probability (6) for classes $c = 1, 2, 3$. We begin from two initial sets of hyperparameters, one originally overfitting and another underfitting the training data. Initially, both models perform sub-optimally with a test accuracy of 56.67%. We see that the RCB $r(\theta, \lambda)$ appropriately measures the amount of overfitting with high (resp. low) values for the overfitted (resp. underfitted) model. We then learn hyperparameters with Algorithm 1 for 500 iterations from both initializations at rate $\eta = 0.01$, where both models converges to a balanced model with a moderate RCB and an improved test accuracy of 73.33%. In particular, the initially overfitted model learns a simpler model at the expense of lower training performance, emphasizing the benefits of complexity based regularization, without which the learning would only maximize training performance at the cost of further overfitting. Meanwhile, the initially underfitted model learns to increase complexity to improve the sub-optimal performance on the training set.

Table 1. Test accuracy (%) on UCI datasets

Method	Banknote	Ecoli	Robot	Segment	Wine	Yeast
GMCE	99.9 ± 0.2	87.5 ± 4.4	96.7 ± 0.9	98.4 ± 0.8	97.2 ± 3.7	52.5 ± 2.1
GMCE-SGD	98.8 ± 0.9	84.5 ± 5.0	95.5 ± 0.9	96.1 ± 1.5	93.3 ± 6.0	60.3 ± 4.4
CEN-1	99.5 ± 1.0	87.5 ± 3.2	82.3 ± 7.1	94.6 ± 1.6	96.1 ± 5.0	55.8 ± 5.0
CEN-2	99.4 ± 0.9	86.3 ± 6.0	94.5 ± 0.8	96.7 ± 1.1	97.2 ± 5.1	59.6 ± 4.0
ERM	99.9 ± 0.2	72.1 ± 20.5	91.0 ± 3.7	98.1 ± 1.1	93.9 ± 5.2	45.9 ± 6.4
CV	99.9 ± 0.2	73.8 ± 23.8	90.9 ± 3.4	98.3 ± 1.3	93.3 ± 7.4	58.0 ± 5.8
MED	92.0 ± 4.3	42.1 ± 47.7	81.1 ± 6.2	27.3 ± 26.4	93.3 ± 7.8	31.2 ± 14.1
Others	99.78[a]	81.1[b]	97.59[c]	96.83[d]	100[e]	55.0[b]

UCI Datasets. We demonstrate the average performance of learning anisotropic Gaussian kernels and kernels constructed from neural networks on standard UCI datasets (Bache and Lichman 2013), summarized in Table 1. The former has a shallow but wide model architecture, while the latter has a deeper but narrower model architecture. The Gaussian kernel is learned with both full (GMCE) and batch stochastic gradient updates (GMCE-SGD) using a tenth ($n_b \approx \frac{n}{10}$) of the training set each training iteration, with sensitivity and length scales initialized to 1. For CENs, we randomly select two simple fully connected architectures with 16-32-8 (CEN-1) and 96-32 (CEN-2) hidden units respectively, and learn the conditional mean embedding without dropout under ReLU activation. Biases and standard deviations of zero mean truncated normal distributed weights are initialized to 0.1, and are to be learned with full gradient updates. For all experiments, λ is initialized to 1 and is learned jointly with the kernel. Optimization is performed with the Adam optimizer (Kingma and Ba 2016) in TensorFlow (Abadi et al. 2016) with a rate of $\eta = 0.1$ and $\epsilon = 10^{-15}$ under the learning objective $q(\theta, \lambda)$ (12). Learning is run for 1000 epochs to allow direct comparison. All attributes are scaled to the unit range. Each model is trained on 9 out of 10 folds and tested on the remaining fold, which are shuffled over all 10 combinations to obtain the test accuracy average and deviation. We compare our results to MCEs whose hyperparameters are tuned by ERM (without the RCB term in (12)), cross validation (CV), and the median heuristic (MED), as well as to other approaches using neural networks (Freire et al. 2009; Kaya et al. 2016 a; c), probabilistic binary trees (Horton and Nakai 1996, b), decision trees (Zhou et al. 2004, d), and regularized discriminant analysis (Aeberhard et al. 1992, e).

Table 1 shows that our learning algorithm outperforms other hyperparameter tuning algorithms, and performs similarly to competing methods. Our method achieves this without any case specific tuning or heuristics, but by simply placing a conditional mean embedding on training data and applying a complexity bound based learning algorithm. The stochastic gradient approach for Gaussian kernels performs similarly to the full gradient approach, supporting the claim of Theorem 4 for $n = n_b$. For CENs, we did not attempt to choose an optimal architecture for each dataset. The learning algorithm is tasked to train the same simple network for different datasets using 1000 epochs to achieve comparable performance.

Learning Pixel Relevance. We apply Algorithm 1 to learn length scales of anisotropic Gaussian, or automatic relevance determination (ARD), kernels on pixels of the MNIST digits dataset (LeCun et al. 1998). In the top left plot of Fig. 2, we train on datasets of varying sizes, from 50 to 5000 images, and show the accuracy on the standard test set of 10000 images. All hyperparameters are initialized to 1 before learning. We train both SVCs and GPCs under the OVA scheme, and use a Laplace approximation for the GPC posterior. In all cases MCEs outperform SVCs as it cannot learn hyperparameters without expensive cross validation. MCEs also outperform GPCs as more data becomes available. Under the OVA scheme, the GPC approach learns a set of kernel hyperparameters for each class, while our approach learns a consistent set of hyperparameters for all classes. Consequently, for 5000 data points, the computational time required for hyperparameter learning of GPCs is on the order of days even for isotropic Gaussian kernels, while Algorithm 1 is on the order of hours for anistropic Gaussian kernels even without batch updates. We also compare hyperparameter learning with and without the RCB. For small n below 750 samples, the latter outperforms the former (e.g. 86.69% and 86.96% for $n = 500$), while for large n the former outperforms the latter (e.g. 96.05% and 95.3% for $n = 5000$). This verifies that complexity based regularization becomes especially important as data size grows, when overfitting starts to decrease generalization performance. The images at the bottom of Fig. 2 show the pixel length scales learned through batch stochastic gradient updates ($n_b = 1200$) over all available training images the groups of digits shown, demonstrating the most discriminative regions.

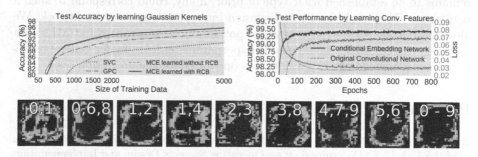

Fig. 2. Top: Test accuracy by learning Gaussian kernels (left) and deep convolutional features (right); Bottom: Learned pixel length scales under ARD kernels

Learning Convolutional Layers. We now apply Algorithm 1 to train a CEN with convolutional layers on MNIST. We employ an example architecture from the TensorFlow tutorial on deep MNIST classification (Abadi et al. 2016). This ReLU activated convolutional neural network (CNN) uses two convolutional layers, each with max pooling, followed by a fully connected layer with a drop out probability of 0.5. The original CNN then employs a final softmax regressor on the last hidden layer for classification. The CEN instead employs a linear kernel on the last hidden layer to construct the conditional mean embedding. We then train both networks from the same initialization using batch updates of $n_b = 6000$ images for 800 epochs, with learning rate $\eta = 0.01$. All biases and weight standard deviations are initialized to 0.1. The network weights and biases of the CEN are learned jointly with the regularization hyperparameter, initialized to $\lambda = 10$, under our learning objective (12), while the original CNN is trained under its usual cross entropy loss. The fully connected layer is trained with a drop out probability of 0.5 for both cases to allow direct comparison. The top right plot in Fig. 2 shows that CENs learn at a much faster rate, maintaining a higher test accuracy at all epochs. After 800 epochs, CEN reaches a test accuracy of 99.48%, compared to 99.26% from the original CNN. This demonstrates that our learning algorithm can perform end-to-end learning with convolutional layers from scratch, by simply replacing the softmax layer with a MCE. The resulting CEN can outperform the original CNN in both convergence rate and accuracy.

6   Conclusion and Future Work

We developed a scalable hyperparameter learning framework for CMEs with categorical targets based on Rademacher complexity bounds. These bounds reveal a novel data-dependent quantity $r(\theta, \lambda)$ that reflect its model complexity. We use this measure as an regularization term in addition to the empirical loss for hyperparameter learning. In parallel light to the case with regularized least squares, it remains to be established what type of prior, if any, could correspond to such a regularizer. This would lead to a Bayesian interpretation of our framework. We also envision that such a quantity could potentially be generalized to CMEs with arbitrary targets, which would enable hyperparameter learning for general conditional mean embeddings in a way that is optimized for the prediction task.

References

Abadi, M., et al.: TensorFlow: a system for large-scale machine learning. In: Proceedings of the 12th USENIX Symposium on Operating Systems Design and Implementation (OSDI), Georgia (2016)

Aeberhard, S., Coomans, D., De Vel, O.: Comparison of classifiers in high dimensional settings. Department of Mathematics and Statistics, James Cook University, North Queensland, Australia, Technical report (92-02) (1992)

Aly, M.: Survey on multiclass classification methods. Neural Netw. **19**, 1–9 (2005)

Bache, K., Lichman, M.: UCI machine learning repository (2013)

Bartlett, P.L., Mendelson, S.: Rademacher and Gaussian complexities: risk bounds and structural results. J. Mach. Learn. Res. **3**, 463–482 (2002)

Chen, Y., Welling, M., Smola, A.: Super-samples from kernel herding. In: The Twenty-Sixth Conference Annual Conference on Uncertainty in Artificial Intelligence (UAI-2010), pp. 109–116. AUAI Press (2010)

Cortes, C., Kloft, M., Mohri, M.: Learning kernels using local Rademacher complexity. In: Advances in Neural Information Processing Systems (2013)

Fisher, R.A.: The use of multiple measurements in taxonomic problems. Ann. Eugenics **7**(2), 179–188 (1936)

Flaxman, S., Sejdinovic, D., Cunningham, J.P., Filippi, S.: Bayesian learning of kernel embeddings. In: Proceedings of the Thirty-Second Conference on Uncertainty in Artificial Intelligence, pp. 182–191. AUAI Press (2016)

Freire, A.L., Barreto, G.A., Veloso, M., Varela, A.T.: Short-term memory mechanisms in neural network learning of robot navigation tasks: a case study. In: 6th Latin American Robotics Symposium (LARS) (2009)

Friedman, J., Hastie, T., Tibshirani, R.: The Elements of Statistical Learning. Springer Series in Statistics, vol. 1. Springer, New York (2001). https://doi.org/10.1007/978-0-387-84858-7

Fukumizu, K., Bach, F.R., Jordan, M.I.: Dimensionality reduction for supervised learning with reproducing kernel Hilbert spaces. J. Mach. Learn. Res. **5**, 73–99 (2004)

Fukumizu, K., Gretton, A., Lanckriet, G.R., Schölkopf, B., Sriperumbudur, B.K.: Kernel choice and classifiability for RKHS embeddings of probability distributions. In: Advances in Neural Information Processing Systems (2009)

Fukumizu, K., Song, L., Gretton, A.: Kernel Bayes' rule: Bayesian inference with positive definite kernels. J. Mach. Learn. Res. **14**(1), 3753–3783 (2013)

Gretton, A., Borgwardt, K.M., Rasch, M., Schölkopf, B., Smola, A.J.: A kernel method for the two-sample-problem. In: Advances in Neural Information Processing Systems, pp. 513–520 (2007)

Gretton, A., et al.: Optimal kernel choice for large-scale two-sample tests. In Advances in Neural Information Processing Systems (2012)

Grünewälder, S., Lever, G., Baldassarre, L., Patterson, S., Gretton, A., Pontil, M.: Conditional mean embeddings as regressors. In: Proceedings of the 29th International Conference on Machine Learning, ICML 2012, vol. 2 (2012)

Higham, N.J.: Accuracy and Stability of Numerical Algorithms. SIAM, Philadelphia (2002)

Horton, P., Nakai, K.: A probabilistic classification system for predicting the cellular localization sites of proteins. In: ISMB, vol. 4 (1996)

Jitkrittum, W., Szabó, Z., Chwialkowski, K.P., Gretton, A.: Interpretable distribution features with maximum testing power. In: Advances in Neural Information Processing Systems, pp. 181–189 (2016)

Kanagawa, M., Fukumizu, K.: Recovering distributions from Gaussian RKHS embeddings. In: AISTATS, pp. 457–465 (2014)

Kanagawa, M., Nishiyama, Y., Gretton, A., Fukumizu, K.: Filtering with state-observation examples via kernel Monte Carlo filter. Neural Comput. **28**(2), 382–444 (2016)

Kaya, E., Yasar, A., Saritas, I.: Banknote classification using artificial neural network approach. Int. J. Intell. Syst. Appl. Eng. **4**(1), 16–19 (2016)

Kearns, M., Mansour, Y., Ng, A.Y., Ron, D.: An experimental and theoretical comparison of model selection methods. Mach. Learn. **27**(1), 7–50 (1997)

Kingma, D., Ba, J.: Adam: a method for stochastic optimization. In: The International Conference on Learning Representations (ICLR) (2016)

Kloft, M., Blanchard, G.: The local Rademacher complexity of lp-norm multiple kernel learning. In: Advances in Neural Information Processing Systems, pp. 2438–2446 (2011)

LeCun, Y., Bottou, L., Bengio, Y., Haffner, P.: Gradient-based learning applied to document recognition. Proc. IEEE **86**(11), 2278–2324 (1998)

Ledoux, M., Talagrand, M.: Probability in Banach Spaces (2013)

Muandet, K., Balduzzi, D., Schölkopf, B.: Domain generalization via invariant feature representation. In: ICML (1), pp. 10–18 (2013)

Muandet, K., Fukumizu, K., Sriperumbudur, B., Schölkopf, B.: Kernel mean embedding of distributions: a review and beyonds. arXiv:1605.09522 [stat.ML] (2016)

Pahikkala, T., Airola, A., Gieseke, F., Kramer, O.: Unsupervised multi-class regularized least-squares classification. In: IEEE 12th International Conference on Data Mining (ICDM), pp. 585–594. IEEE (2012)

Pontil, M., Maurer, A.: Excess risk bounds for multitask learning with trace norm regularization. In: Conference on Learning Theory, pp. 55–76 (2013)

Rahimi, A., Recht, B.: Random features for large-scale kernel machines. In: Advances in Neural Information Processing Systems, pp. 1177–1184 (2008)

Rasmussen, C.E., Williams, C.K.I.: Gaussian Processes for Machine Learning. The MIT Press, Cambridge (2006)

Rifkin, R., Yeo, G., Poggio, T., et al.: Regularized least-squares classification. In: Nato Science Series Sub Series III Computer and Systems Sciences (2003)

Schölkopf, B., Smola, A.J.: Learning with Kernels: Support Vector Machines, Regularization, Optimization, and Beyond. MIT Press, Cambridge (2002)

Song, L., Fukumizu, K., Gretton, A.: Kernel embeddings of conditional distributions: a unified kernel framework for nonparametric inference in graphical models. IEEE Signal Process. Mag. **30**(4), 98–111 (2013)

Song, L., Huang, J., Smola, A., Fukumizu, K.: Hilbert space embeddings of conditional distributions with applications to dynamical systems. In: Proceedings of the 26th Annual International Conference on Machine Learning (2009)

Song, L., Zhang, X., Smola, A., Gretton, A., Schölkopf, B.: Tailoring density estimation via reproducing kernel moment matching. In: Proceedings of the 25th International Conference on Machine Learning, pp. 992–999. ACM (2008)

Sriperumbudur, B.K., Gretton, A., Fukumizu, K., Schölkopf, B., Lanckriet, G.R.: Hilbert space embeddings and metrics on probability measures. J. Mach. Learn. Res. **11**, 1517–1561 (2010)

Xu, C., Liu, T., Tao, D., Xu, C.: Local Rademacher complexity for multi-label learning. IEEE Trans. Image Process. **25**(3), 1495–1507 (2016)

Xu, Y., Zhang, H.: Refinement of reproducing kernels. J. Mach. Learn. Res. **10**, 107–140 (2009)

Yu, H.-f., Jain, P., Kar, P., Dhillon, I.: Large-scale multi-label learning with missing labels. In: Proceedings of the 31st International Conference on Machine Learning (ICML-2014), pp. 593–601 (2014)

Zhou, Z.-H., Wei, D., Li, G., Dai, H.: On the size of training set and the benefit from ensemble. In: Dai, H., Srikant, R., Zhang, C. (eds.) PAKDD 2004. LNCS (LNAI), vol. 3056, pp. 298–307. Springer, Heidelberg (2004). https://doi.org/10.1007/978-3-540-24775-3_38

Deep Learning Architecture Search by Neuro-Cell-Based Evolution with Function-Preserving Mutations

Martin Wistuba$^{(\boxtimes)}$

IBM Research, Dublin, Ireland
martin.wistuba@ibm.com

Abstract. The design of convolutional neural network architectures for a new image data set is a laborious and computational expensive task which requires expert knowledge. We propose a novel neuro-evolutionary technique to solve this problem without human interference. Our method assumes that a convolutional neural network architecture is a sequence of neuro-cells and keeps mutating them using function-preserving operations. This novel combination of approaches has several advantages. We define the network architecture by a sequence of repeating neuro-cells which reduces the search space complexity. Furthermore, these cells are possibly transferable and can be used in order to arbitrarily extend the complexity of the network. Mutations based on function-preserving operations guarantee better parameter initialization than random initialization such that less training time is required per network architecture. Our proposed method finds within 12 GPU hours neural network architectures that can achieve a classification error of about 4% and 24% with only 5.5 and 6.5 million parameters on CIFAR-10 and CIFAR-100, respectively. In comparison to competitor approaches, our method provides similar competitive results but requires orders of magnitudes less search time and in many cases less network parameters.

Keywords: Automated machine learning
Neural architecture search · Evolutionary algorithms

1 Introduction

Deep learning techniques have been the key to major improvements in machine learning in various domains such as image and speech recognition and machine translation. Besides more affordable computational power, the proposal of new kinds of architectures such as ResNet [8] and DenseNet [9] helped to increase the accuracy. However, the selection on which architecture to choose and how to wire different layers for a particular data set is not trivial and demands domain expertise and time from the human practitioner.

Within the last one or two years we observed an increase in research efforts by the machine learning community in order to automate the search for neural

© Springer Nature Switzerland AG 2019
M. Berlingerio et al. (Eds.): ECML PKDD 2018, LNAI 11052, pp. 243–258, 2019.
https://doi.org/10.1007/978-3-030-10928-8_15

network architectures. Researchers showed that both, reinforcement learning [31] and neuro-evolution [20], are capable of finding network architectures that are competitive to the state-of-the-art. Since these methods still require GPU years until this performance is reached, further work has been proposed to significantly decrease the run time [1,3,15,30,32].

In this paper, we want to present a simple evolutionary algorithm which reduces the search time to just hours. This is an important step since now, similar to hyperparameter optimization for other machine learning models, optimizing the network architecture becomes affordable for everyone. Our presented approach starts from a very simple network template which contains a sequence of neuro-cells. These neuro-cells are architecture patterns and the optimal pattern will be automatically detected by our proposed algorithm. This algorithm assumes that the cell initially contains only a single convolutional layer and then keeps changing it by function-preserving mutations. These mutations change the structure of the architecture without changing the network's predictions. This can be considered as a special initialization such that the network requires less computational effort for training.

Our contributions in this paper are three-fold:

1. We are the first to propose an evolutionary algorithm which optimizes neuro-cells with function-preserving mutations.
2. We expand the set of function-preserving operations proposed by Chen et al. [4] to depthwise separable convolutions, kernel widening, skip connections and layers with multiple in- and outputs.
3. We provide empirical evidence that our method is outperforming many competitors within only hours of search time. We analyze our proposed method and the transferability of neuro-cells in detail.

2   Related Work

Evolutionary algorithms and reinforcement learning are currently the two state-of-the-art techniques used by neural network architectures search algorithms. With Neural Architecture Search [31], Zoph et al. demonstrated in an experiment over 28 days and with 800 GPUs that neural network architectures with performances close to state-of-the-art architectures can be found. In parallel or inspired by this work, others proposed to use reinforcement learning to detect sequential architectures [1], reduce the search space to repeating cells [30,32] or apply function-preserving actions to accelerate the search [3].

Neuro-evolution dates back three decades. In the beginning it focused only on evolving weights [18] but it turned out to be effective to evolve the architecture as well [23]. Neuro-evolutionary algorithms gained new momentum due to the work by Real et al. [20]. In an extraordinary experiment that used 250 GPUs for almost 11 days, they showed that architectures can be found which provide similar good results as human-crafted image classification network architectures. Very recently, the idea of learning cells instead of the full network has also been adopted for evolutionary algorithms [15]. Miikkulainen et al. even propose to coevolve a set of cells and their wiring [17].

Other methods that try to optimize neural network architectures or their hyperparameters are based on model-based optimization [7,14,22,26], random search [2] and Monte-Carlo Tree Search [19,27].

3 Function-Preserving Knowledge Transfer

Chen et al. [4] proposed a family of function-preserving network manipulations in order to transfer knowledge from one network to another. Suppose a teacher network is represented by a function $f\left(\mathbf{x} \mid \boldsymbol{\theta}^{(f)}\right)$ where $\mathbf{x}$ is the input of the network and $\boldsymbol{\theta}^{(f)}$ are its parameters. Then an operation changing the network f to a student network g is called function-preserving if and only if the output for any given model remains unchanged:

$$\forall \mathbf{x}: \ f\left(\mathbf{x} \mid \boldsymbol{\theta}^{(f)}\right) = g\left(\mathbf{x} \mid \boldsymbol{\theta}^{(g)}\right). \tag{1}$$

Note that typically the number of parameters of f and g are different. We will use this approach in order to initialize our mutated network architectures. Then, the network is trained for some additional epochs with gradient-based optimization techniques. Using this initialization, the network requires only few epochs before it provides decent predictions. We briefly explain the proposed manipulations and our novel contributions to it. Please note that a fully connected layer is a special case of a convolutional layer.

3.1 Convolutions in Deep Learning

Convolutional layers are a common layer type used in neural networks for visual tasks. We denote the convolution operation between the layer input $X \in \mathbb{R}^{w \times h \times i}$ with a layer with parameters $W \in \mathbb{R}^{k_1 \times k_2 \times i \times o}$ by $X * W$. Here, i is the number of input channels, $w \times h$ the input dimension, $k_1 \times k_2$ the kernel size and o the number of output feature maps. Depthwise separable convolutions, or for short just separable convolutions, are a special kind of convolution factored into two operations. During the depthwise convolution a spatial convolution with parameters $W_d \in \mathbb{R}^{k_1 \times k_2 \times i}$ is applied for each channel separately. We denote this operation by using $\circledast$. This is in contrast to the typical convolution which is applied across all channels. In the next step the pointwise convolution, i.e. a convolution with a 1×1 kernel, traverses the feature maps which result from the first operation with parameters $W_p \in \mathbb{R}^{1 \times 1 \times i \times o}$. Comparing the normal convolution operation $X * W$ with the separable convolution $(X \circledast W_d) * W_p$, we immediately notice that in practice the former requires with $k_1 k_2 i o$ more parameters than the latter which only needs $k_1 k_2 i + i o$. Figure 1 provides a graphical representation of the network. If $X^{(l)}$ is the input for an operation in layer $l + 1$, e.g. a convolution, then we represent each channel $X^{(l)}_{:,:,i}$

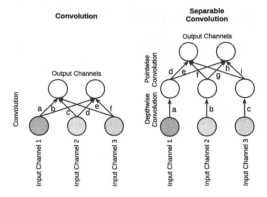

Fig. 1. Comparison of a standard convolution to a separable convolution. The separable convolution first applies a spatial convolution for each channel separately. Afterwards, a convolution with a 1×1 kernel is applied. Circles represent one channel of the feature map in the network, arrows a spatial convolution.

by a circle. Arrows represent a spatial convolution which is parameterized by some parameters indicated by a character (in our example characters a to i). We clearly see that the depthwise convolution within the depthwise separable convolution separately operates on channels and normal convolutions operate across channels.

3.2 Layer Widening

Assume the teacher network f contains a convolutional layer with a $k_1 \times k_2$ kernel which is represented by a matrix $W^{(l)} \in \mathbb{R}^{k_1 \times k_2 \times i \times o}$ where i is the number of input feature maps and o is the number of output feature maps or filters. Widening this layer means that we increase the number of filters to $o' > o$. Chen et al. [4] proposed to extend $W^{(l)}$ by replicating the parameters along the last axis at random. This means the widened layer of the student network uses the parameters

$$V^{(l)}_{\cdot,\cdot,\cdot,j} = \begin{cases} W^{(l)}_{\cdot,\cdot,\cdot,j} & j \leq o \\ W^{(l)}_{\cdot,\cdot,\cdot,r} & r \text{ uniformly sampled from } \{1,\dots,o\} \end{cases}. \tag{2}$$

In order to achieve the function-preserving property, the replication of some filters needs to be taken into account for the next layer $V^{(l+1)}$. This is achieved by dividing the parameters of $W^{(l+1)}_{\cdot,\cdot,j,\cdot}$ by the number of times the j-th filter has been replicated. If n_j is the number of times the j-th filter was replicated, the weights of the next layer for the student network are defined by

$$V^{(l+1)}_{\cdot,\cdot,j,\cdot} = \frac{1}{n_j} W^{(l+1)}_{\cdot,\cdot,j,\cdot}. \tag{3}$$

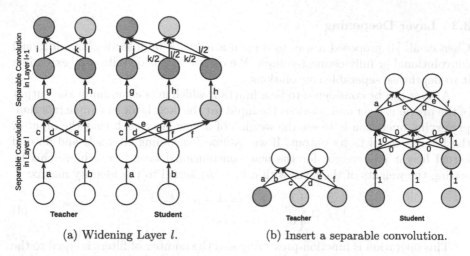

(a) Widening Layer l.   (b) Insert a separable convolution.

Fig. 2. Visualization of different function-preserving operations. Same colored circles represent identical feature maps. Circles without filling can have any value and are not important for the visualization. Activation functions are omitted to avoid clutter. (Color figure online)

We extended this mechanism to depthwise separable convolutional layers. A depthwise separable convolutional layer at depth l is widened as visualized in Fig. 2a. The pointwise convolution for the student is estimated according to Eq. 2. This results into replicated output feature maps indicated by two green colored circles in the figure. The depthwise convolution is identical to the one of the teacher network, i.e. the operations with parameters a and b. Independently of whether we used a depthwise separable or normal convolution in layer l, widening it requires adaptations in a following depthwise separable convolutional layer as visualized in Fig. 2a. The parameters of the depthwise convolution are replicated according to the replication of parameters in the previous layer similar to Eq. 2. In our example we replicated the operation with parameters f in the previous layer. Therefore, we have now replicated spatial convolutions with parameters h. Furthermore, the parameter of the pointwise convolution (in the example parameterized by i, j, k and l) depend on the replications in the previous layers analogously to Eq. 3. In our example we did not replicate the blue feature map, so the weights for this channel remain unchanged. However, we duplicated the green feature map which is transformed into the purple feature map depthwise convolution. Taking into account that this channel contributes now twice to the pointwise convolution, all corresponding weights (in the example k and l) are divided by two.

Widening the separable layer followed by another separable layer is the most complicated case. Other cases can be derived by dropping the depthwise convolutions from Fig. 2a.

3.3 Layer Deepening

Chen et al. [4] proposed a way to deepen a network by inserting an additional convolutional or fully connected layer. We complete this definition by extending it to depthwise separable convolutions.

A layer can be considered to be a function which gets as an input the output of the previous layer and provides the input for the next layer. A simple function-preserving operation is to set the weights of a new layer such that the input of the layer is equal to its output. If we assume i incoming channels and an odd kernel height and weight for the new convolutional layer, we achieve this by setting the weights of the layer with a $k_1 \times k_2$ kernel to the identity matrix:

$$V_{j,h}^{(l)} = \begin{cases} I_{i,i} & j = \frac{k_1+1}{2} \wedge h = \frac{k_2+1}{2} \\ 0 & \text{otherwise} \end{cases}. \tag{4}$$

This operation is function-preserving and the number of filters is equal to the number of input channels. More filters can be added by layer widening, however, it is not possible to use less than i filters for the new layer. Another restriction is that this operation is only possible for activation functions σ with

$$\sigma(\mathbf{x}) = \sigma(I\sigma(\mathbf{x})) \; \forall \mathbf{x}. \tag{5}$$

The ReLU activation function ReLU $(\mathbf{x}) = \max\{\mathbf{x}, \mathbf{0}\}$ fulfills this requirement.

We extend this operation to depthwise convolutions and visualize it in Fig. 2b. The parameters of the pointwise convolution V_p are initialized analogously to Eq. 4 and the depthwise convolution V_d is set to one:

$$V_p = I_{i,i} \tag{6}$$
$$V_d = 1. \tag{7}$$

As we see in Fig. 2b, this initialization ensures that both, the depthwise and pointwise convolution, just copy the input. New layers can be inserted at arbitrary positions with one exception. Under certain conditions an insertion right after the input layer is not function-preserving. For example if a ReLU activation is used, there exists no identity function for inputs with negative entries.

3.4 Kernel Widening

Increasing the kernel size in a convolutional layer is achieved by padding the tensor using zeros until it matches the desired size. The same idea can be applied to increase the kernel size of depthwise separable convolution by padding the depthwise convolution with zeros.

3.5 Insert Skip Connections

Many modern neural network architectures rely on skip connections [8]. The idea is to add the output of the current layer to the output of a previous. One simple example is

$$X^{(l+1)} = \sigma\left(X^{(l)} * V^{(l|1)} + X^{(l)}\right). \tag{8}$$

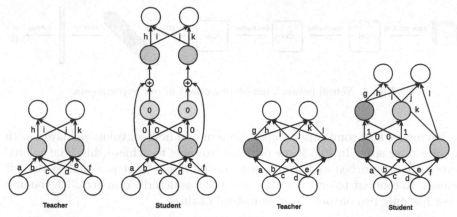

(a) Insert a skip with a convolution.   (b) Branch the colored layer and insert a
                                               convolution into the left branch.

Fig. 3. Visualization of different function-preserving operations. Same colored circles represent identical feature maps. Circles without filling can have any value and are not important for the visualization. Activation functions are omitted to avoid clutter. (Color figure online)

Therefore, we propose a function-preserving operation which allows inserting skip connection. We propose to add layer(s) and initialize them in a way such that the output is 0 independent on the input. This allows to add a skip because now adding the output of the previous layer to zero is an identity operation. We visualized a simple example in Fig. 3a based on Eq. 8. A new operation is added setting its parameters to zero, $V^{(l+1)} = \mathbf{0}$, achieving a zero output. Now, adding this output to the input is an identity operation.

3.6 Branch Layers

We also propose to branch layers. Given a convolutional layer $X^{(l)} * W^{(l+1)}$ it can be reformulated as

$$\text{merge}\left(X^{(l)} * V_1^{(l+1)},\ X^{(l)} * V_2^{(l+1)}\right), \tag{9}$$

where *merge* concatenates the resulting output. The student network's parameters are defined as

$$V_1^{(l+1)} = W_{\cdot,\cdot,\cdot,1:\lfloor o/2 \rfloor}^{(l+1)}$$
$$V_2^{(l+1)} = W_{\cdot,\cdot,\cdot,(\lfloor o/2 \rfloor+1):o}^{(l+1)}.$$

This operation is not only function-preserving, it also does not add any further parameters and in fact is the very same operation. However, combining this operation with other function-preserving operations allows to extend networks

Fig. 4. Neural network template as used in our experiments.

by having parallel convolutional operations or add new convolutional layers with smaller filter sizes. In Fig. 3b we demonstrate how to achieve this. The colored layer is first branched and then a new convolutional layer is added to the left branch. In contrast to only adding a new layer as described in Sect. 3.3, the new layer has only two output channels instead of three.

3.7 Multiple In- or Outputs

All the presented operations are still possible for networks where a layer might have inputs from different layers or provide output for multiple outputs. In that case only the affected weights need to be adapted according to the aforementioned equations.

4 Evolution of Neuro-Cells

The very basic idea of our proposed cell-based neuro-evolution is the following. Given is a very simple neural network architecture which contains multiple neuro-cells (see Fig. 4). The cells itself share their structure and the task is to find a structure that improves the overall neural network architecture for a given data set and machine learning task. In the beginning, a cell is identical to a convolutional layer and is changed during the evolutionary optimization process. Our evolutionary algorithm is using tournament selection to select an individual from the population: randomly, a fraction k of individuals is selected from the population. From this set the individual with highest fitness is selected for mutation. We define the fitness by the accuracy achieved by the individual on a hold-out data set. The mutation is selected at random which is applied to all neuro-cells such that they remain identical. The network is trained for some epochs on the training set and is then added to the population. Finally, the process starts all over again. After meeting some stopping criterion, the individual with highest fitness is returned.

4.1 Mutations

All mutations used are based on the function-preserving operations introduced in the last section. This means, a mutation does not change the fitness of an individual, however, it will increase its complexity. The advantage over creating the same network structure with randomly initialized weights is obviously that we start with a partially pretrained network. This enables us to train the network in less epochs. All mutations are applied only to the structure within a neuro-cell if not otherwise mentioned. Our neuro-evolutional algorithm considers the following mutations.

Insert Convolution. A convolution is added at a random position. Its kernel size is 3 × 3, the number of filters is equal to its input dimension. It is randomly decided whether it is a separable convolution instead.

Branch and Insert Convolution. A convolution is selected at random and branched according to Sect. 3.6. A new convolution is added according to the "Insert Convolution" mutation in one of the branches. For an example see Fig. 3b.

Insert Skip. A convolution is selected at random. Its output is added to the output of a newly added convolution (see "Insert Convolution") and is the input for the following layers. For an example see Fig. 3a.

Alter Number of Filters. A convolution is selected at random and widened by a factor uniformly at random sampled from [1.2, 2]. This mutation might also be applied to convolutions outside of a neuro-cell.

Alter Number of Units. Similar to the previous one but alters the number of units of fully connected layers. This mutation is only applied outside the neuro-cells.

Alter Kernel Size. Selects a convolution at random and increases its kernel size by two along each axis.

Branch Convolution. Selects a convolution at random and branches it according to Sect. 3.6.

The motivation of selecting this set of mutations is to enable the neuro-evolutionary algorithm to discover similar architectures as proposed by human experts. Adding convolutions allows to reach popular architectures such as VGG16 [21], combinations of adding skips and convolutions allow to discover residual networks [8]. Finally the combination of branching, change of kernel sizes and addition of (separable) convolutions allows to discover architectures similar to Inception [25], Xception [5] or FractalNet [13].

The optimization is started with only a single individual. We enrich the population by starting with an initialization step which creates 15 mutated versions of the first individual. Then, individuals are selected based on the previously described tournament selection process.

5 Experiments

In the experimental section we will run our proposed method for the task of image classification on the two data sets CIFAR-10 and CIFAR-100. We conduct the following experiments. First, we analyze the performance of our neuro-evolutional approach with respect to classification error and compare it to various competitor approaches. We show that we achieve a significant search time improvement at costs of slightly larger error. Furthermore, we give insights how the evolution and the neuro-cells progress and develop during the optimization process. Additionally, we discuss the possibility of transferring detected cells to novel data sets. Finally, we compare the performance of two different random approaches in order to prove our method's benefit.

5.1   Experimental Setup

The network template used in our experiments is sketched in Fig. 4. It starts with a small convolution, followed twice by a neuro-cell and a max pooling layer. Then, another neuro-cell is added, followed by a larger convolution, a fully connected layer and the final softmax layer. Each max pooling layer has a stride of two and is followed by a drop-out layer with drop-out rate 70%. The fully connected layer is followed by a drop-out layer with rate 50%. In this section, whenever we sketch or mention a convolutional layer, we actually mean a convolutional layer followed by batch normalization [11] and a ReLU activation. The neuro-cell is initialized with a single convolution with 128 filters and a kernel size of 3×3. A weight decay of 0.0001 is used.

We evaluate our method and compare it to competitor methods on CIFAR-10 and CIFAR-100 [12]. We use standard preprocessing and data augmentation. All images are preprocessed by subtracting from each channel its mean and dividing it by its standard deviation. The data augmentation involves padding the image to size 40×40 and then cropping it to dimension 32×32 as well as flipping images horizontally at random. We split the official training partitions into a partition which we use to train the networks and a hold-out partition to evaluate the fitness of the individuals.

For the neuro-evolutionary algorithm we select a tournament size equal to 15% of the population but at least two. The initial network is trained for 63 epochs, every other network is trained for 15 epochs with Nesterov momentum and a cosine learning rate schedule with initial learning rate 0.05, $T_0 = 1$ and $T_{mul} = 2$ [16]. We define the fitness of an individual by the accuracy of the corresponding network on the hold-out partition. After the search budget is exhausted, the individual with highest fitness is trained on the full training split until convergence using CutOut [6]. Finally, the error on test is reported.

5.2   Search for Networks

In Table 1 we report the mean and standard deviation of our approach across five runs and compare it to other approaches.

The first block contains several architectures proposed by human experts. DenseNet [9] is clearly the best among them, reaching an error of 4.51% with only 800 thousand parameters. Using about 25 million parameters, the error decreases to 3.42%.

The second block contains several architecture search methods based on reinforcement learning. Most of them are able to find very competitive networks but at the cost of very high search times. NASNet [32] finds the best-performing

Table 1. Classification error on CIFAR-10 and CIFAR-100 including spent search time in GPU days. The first block presents the performance of state-of-the-art human-designed architectures. The second block contains results of various automated architecture search methods based on reinforcement learning. The third block contains results for automated methods based on evolutionary algorithms. The final block presents our results. For our method, we report the mean of five repetitions for the classification error and the number of parameters, the best run and the run with least network parameters.

Method	Duration	CIFAR-10		CIFAR-100	
		Error	Params	Error	Params
ResNet [8] reported by [10]	N/A	6.41	1.7 M	27.22	1.7 M
FractalNet [13]	N/A	5.22	38.6M	23.30	38.6 M
Wide ResNet (depth = 16) [29]	N/A	4.81	11.0 M	22.07	11.0 M
Wide ResNet (depth = 28) [29]	N/A	4.17	36.5 M	20.50	36.5 M
DenseNet-BC ($k = 12$) [9]	N/A	4.51	**0.8 M**	22.27	0.8 M
DenseNet-BC ($k = 24$) [9]	N/A	3.62	15.3 M	17.60	15.3 M
DenseNet-BC ($k = 40$) [9]	N/A	**3.42**	25.6 M	17.18	25.6 M
NAS no stride/pooling [31]	22,400	5.50	4.2 M	-	-
NAS predicting strides [31]	22,400	6.01	**2.5 M**	-	-
NAS max pooling [31]	22,400	4.47	7.1 M	-	-
NAS max pooling + more filters [31]	22,400	3.65	37.4 M	-	-
NASNet [32]	2,000	**3.41**	3.3 M	-	-
MetaQNN [1]	100	6.92	11.2 M	27.14	11.2 M
BlockQNN [30]	96	3.6	?	18.64	?
Efficient Architecture Search [3]	10	4.23	23.4 M	-	-
Large-Scale Evolution [20]	2,600	5.4	5.4 M	23.0	40.4 M
Hierarchical Evolution [15]	300	**3.75**	15.7 M	-	-
CGP-CNN (ResSet) [24]	27.4	6.05	**2.6 M**	-	-
CoDeepNEAT [17]	?	7.30	?	-	-
Ours (mean)	0.5	4.02	5.6 M	23.92	6.5 M
Ours (mean)	1	3.89	7.0 M	22.32	6.7 M
Ours (best)	0.5	**3.57**	5.8 M	22.08	6.8 M
Ours (best)	1	3.58	7.2 M	21.74	5.3 M
Ours (least params)	0.5	4.19	**3.8 M**	28.15	5.0 M
Ours (least params)	1	3.77	5.8 M	21.74	5.3 M

network which is on par with DenseNet but requires less parameters. However, the authors report that they required about 5.5 GPU years in order to reach this performance. Efficient Architecture Search [3] still achieves an error of 4.23% but reduces the search time drastically to ten days.

The third block contains various automated approaches based on evolutionary methods. Hierarchical Evolution [15] finds the best performing architecture among them in 300 GPU days. Methodologically, our approach also belongs into this category. We want to highlight in particular the search time required by our proposed method. Within only 12 and 24 h, respectively, a network architecture is found which gives better predictions than most competitors and is very close to the best methods. After 12 h of search, we report a mean classification error over five repetitions of 4.02 ± 0.376 and 23.92 ± 2.493 on CIFAR-10 and CIFAR-100, respectively. Extending the search by another 12 h, the error reduces to 3.89 ± 0.231 and 22.32 ± 0.429.

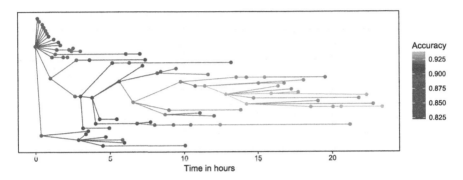

Fig. 5. Evolutionary algorithm over time. Each dot represents an individual, connections represent the ancestry. After the initialization, the algorithm quickly focuses on ancestors from only one initial individual. (Color figure online)

In order to give insights into the optimization process, we visualized one run on CIFAR-10 in Figs. 5 and 6. Figure 5 visualizes the fitness of each individual but also its ancestry by a phylogenetic tree [28]. The x-axis represents the time, the y-axis has no meaning. The color indicates the fitness, dots represent individuals and the ancestry is represented by edges. We notice that within the first 10 h the fitness is increasing quickly. Afterwards, progress is slow but steady. Figure 6 provides in parallel insight which stages the final neuro-cell underwent. Over time the cell develops multiple computation branches, finally adding some skip connections. Notice, that branching the 7×7 convolution as first shown at Hour 19 has no purpose. However, this might have changed for a longer run when e.g. another layer was added in one of these branches.

5.3   Neuro-Cell Transferability

An interesting aspect is whether a neuro-cell detected on one data set can be reused in a different architecture and for a different data set. For this reason, we expanded the template from Fig. 4 by duplicating the number of cells to the one shown in Fig. 7. We used the cells and other hyperparameters detected in our 12 h

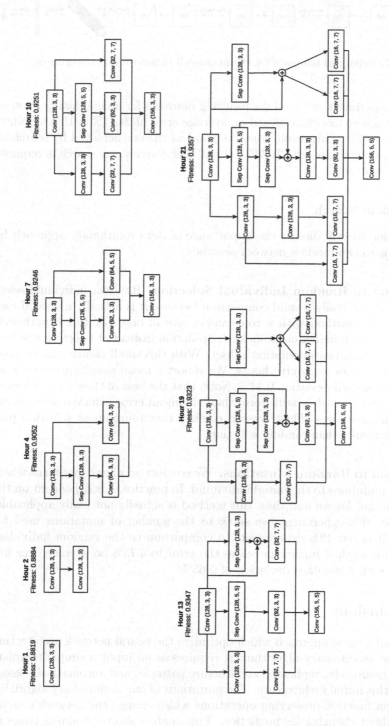

Fig. 6. Evolutionary process of the best neuro-cell found during one run on CIFAR-10. Some intermediate states are skipped.

Fig. 7. Expanded template for the neuro-cell transferability experiment.

CIFAR-10 experiment and used the resulting networks for image classification on CIFAR-100. These models achieved an average error of 24.77% with a standard deviation of 1.61%. This result is not as good as the one achieved by searching for the best architecture for CIFAR-100 but therefore no new search is required for the new data set.

5.4   Random Search

In this section we will discuss the importance of our evolutionary approach by comparing it to two random network searches.

Comparison to Random Individual Selection. Random individual selection is in fact not really a valid comparison because it is actually a special case of our proposed method with a tournament size of one. For this experiment, we select a random individual from the population instead of selecting the best individual of a random population subset. With this small change, we run our algorithm five times for twelve hours. We report a mean classification error of 4.55% with standard deviation 0.34%. Note, that the best of these runs achieved an error of 4.04% which is still worse than the mean error achieved when using larger tournament sizes. Thus, we can confirm that tournament selection provides better results than random selection.

Comparison to Random Mutations. We conduct another experiment where we apply k mutations to the initial individual. In practice, k is dependent on the data set and not known and thus, this method is actually not really applicable. However, for this experiment, we set k to the number of mutations used for the best cell in our 12 h experiment. In comparison to the random individual selection, this method further increases the error to 4.73% on average over five repetitions with a standard deviation of 0.63%.

6   Conclusions

We proposed a novel approach which optimizes the neural network architecture based on an evolutionary algorithm. It requires as an input a simple template containing neuro-cells, replicated architecture patterns, and automatically keeps improving this initial architecture. The mutations of our evolutionary algorithm are based on function-preserving operations which change the network's architecture without changing its prediction. This enables shorter training times in comparison to a random initialization. In comparison to the state-of-the-art, we report very competitive results and show outstanding results with respect to the search time. Our approach is up to 50,000 times faster than some of the competitor methods with an error rate at most 0.6% higher than the best competitor on CIFAR-10.

References

1. Baker, B., Gupta, O., Naik, N., Raskar, R.: Designing neural network architectures using reinforcement learning. In: Proceedings of the International Conference on Learning Representations, ICLR 2017, Toulon, France, 24–26 April (2017)
2. Bergstra, J., Bengio, Y.: Random search for hyper-parameter optimization. J. Mach. Learn. Res. **13**, 281–305 (2012)
3. Cai, H., Chen, T., Zhang, W., Yu, Y., Wang, J.: Reinforcement learning for architecture search by network transformation. CoRR abs/1707.04873 (2017)
4. Chen, T., Goodfellow, I.J., Shlens, J.: Net2Net: accelerating learning via knowledge transfer. In: Proceedings of the International Conference on Learning Representations, ICLR 2016, San Juan, Puerto Rico, 2–4 May (2016)
5. Chollet, F.: Xception: deep learning with depthwise separable convolutions. CoRR abs/1610.02357 (2016)
6. Devries, T., Taylor, G.W.: Improved regularization of convolutional neural networks with cutout. CoRR abs/1708.04552 (2017)
7. Diaz, G.I., Fokoue-Nkoutche, A., Nannicini, G., Samulowitz, H.: An effective algorithm for hyperparameter optimization of neural networks. IBM J. Res. Dev. **61**(4), 9 (2017)
8. He, K., Zhang, X., Ren, S., Sun, J.: Deep residual learning for image recognition. In: 2016 IEEE Conference on Computer Vision and Pattern Recognition, CVPR 2016, Las Vegas, NV, USA, 27–30 June 2016, pp. 770–778 (2016)
9. Huang, G., Liu, Z., van der Maaten, L., Weinberger, K.Q.: Densely connected convolutional networks. In: 2017 IEEE Conference on Computer Vision and Pattern Recognition, CVPR 2017, Honolulu, HI, USA, 21–26 July 2017, pp. 2261–2269 (2017)
10. Huang, G., Sun, Y., Liu, Z., Sedra, D., Weinberger, K.Q.: Deep Networks with Stochastic Depth. In: Leibe, B., Matas, J., Sebe, N., Welling, M. (eds.) ECCV 2016, Part IV. LNCS, vol. 9908, pp. 646–661. Springer, Cham (2016). https://doi.org/10.1007/978-3-319-46493-0_39
11. Ioffe, S., Szegedy, C.: Batch normalization: accelerating deep network training by reducing internal covariate shift. In: Proceedings of the 32nd International Conference on Machine Learning, ICML 2015, Lille, France, 6–11 July 2015, pp. 448–456 (2015)
12. Krizhevsky, A.: Learning multiple layers of features from tiny images. Technical report (2009)
13. Larsson, G., Maire, M., Shakhnarovich, G.: Fractalnet: Ultra-deep neural networks without residuals. In: Proceedings of the International Conference on Learning Representations, ICLR 2017, Toulon, France, 24–26 April (2017)
14. Liu, C., et al.: Progressive neural architecture search. CoRR abs/1712.00559 (2017)
15. Liu, H., Simonyan, K., Vinyals, O., Fernando, C., Kavukcuoglu, K.: Hierarchical representations for efficient architecture search. In: Proceedings of the International Conference on Learning Representations, ICLR 2018, Vancouver, Canada (2018)

16. Loshchilov, I., Hutter, F.: SGDR: Stochastic gradient descent with warm restarts. In: Proceedings of the International Conference on Learning Representations, ICLR 2017, Toulon, France, 24–26 April (2017)

17. Miikkulainen, R., et al.: Evolving deep neural networks. CoRR abs/1703.00548 (2017)

18. Miller, G.F., Todd, P.M., Hegde, S.U.: Designing neural networks using genetic algorithms. In: Proceedings of the 3rd International Conference on Genetic Algorithms, June 1989, pp. 379–384. George Mason University, Fairfax, Virginia, USA (1989)

19. Negrinho, R., Gordon, G.J.: Deeparchitect: Automatically designing and training deep architectures. CoRR abs/1704.08792 (2017)

20. Real, E., et al.: Large-scale evolution of image classifiers. In: Proceedings of the 34th International Conference on Machine Learning, ICML 2017, Sydney, NSW, Australia, 6–11 August 2017, pp. 2902–2911 (2017)

21. Simonyan, K., Zisserman, A.: Very deep convolutional networks for large-scale image recognition. CoRR abs/1409.1556 (2014)

22. Snoek, J., Larochelle, H., Adams, R.P.: Practical bayesian optimization of machine learning algorithms. In: Advances in Neural Information Processing Systems 25: 26th Annual Conference on Neural Information Processing Systems 2012, Proceedings of a meeting held 3–6 December 2012, Lake Tahoe, Nevada, United States, pp. 2960–2968 (2012)

23. Stanley, K.O., Miikkulainen, R.: Evolving neural networks through augmenting topologies. Evol. Comput. 10(2), 99–127 (2002)

24. Suganuma, M., Shirakawa, S., Nagao, T.: A genetic programming approach to designing convolutional neural network architectures. In: Proceedings of the Genetic and Evolutionary Computation Conference, GECCO 2017, Berlin, Germany, 15–19 July 2017, pp. 497–504 (2017)

25. Szegedy, C., et al.: Going deeper with convolutions. In: IEEE Conference on Computer Vision and Pattern Recognition, CVPR 2015, Boston, MA, USA, 7–12 June 2015, pp. 1–9 (2015)

26. Wistuba, M.: Bayesian optimization combined with successive halving for neural network architecture optimization. In: Proceedings of AutoML@PKDD/ECML 2017, Skopje, Macedonia, 22 September 2017, pp. 2–11 (2017)

27. Wistuba, M.: Finding competitive network architectures within a day using UCT. CoRR abs/1712.07420 (2017)

28. Yu, G., Smith, D.K., Zhu, H., Guan, Y., Lam, T.T.Y.: ggtree: an R package for visualization and annotation of phylogenetic trees with their covariates and other associated data. Methods Ecol. Evol. 8(1), 28–36 (2016)

29. Zagoruyko, S., Komodakis, N.: Wide residual networks. In: Proceedings of the British Machine Vision Conference 2016, BMVC 2016, York, UK, 19–22 September 2016 (2016)

30. Zhong, Z., Yan, J., Liu, C.: Practical network blocks design with q-learning. CoRR abs/1708.05552 (2017)

31. Zoph, B., Le, Q.V.: Neural architecture search with reinforcement learning. In: Proceedings of the International Conference on Learning Representations, ICLR 2017, Toulon, France, 24–26 April (2017)

32. Zoph, B., Vasudevan, V., Shlens, J., Le, Q.V.: Learning transferable architectures for scalable image recognition. CoRR abs/1707.07012 (2017)

VC-Dimension Based Generalization Bounds for Relational Learning

Ondřej Kuželka[1](✉), Yuyi Wang[2], and Steven Schockaert[3]

[1] Department of Computer Science, KU Leuven, Leuven, Belgium
ondrej.kuzelka@kuleuven.be
[2] DISCO Group, ETH Zurich, Zurich, Switzerland
yuwang@ethz.ch
[3] School of Computer Science and Informatics, Cardiff University, Cardiff, UK
SchockaertS1@cardiff.ac.uk

Abstract. In many applications of relational learning, the available data can be seen as a sample from a larger relational structure (e.g. we may be given a small fragment from some social network). In this paper we are particularly concerned with scenarios in which we can assume that (i) the domain elements appearing in the given sample have been uniformly sampled without replacement from the (unknown) full domain and (ii) the sample is complete for these domain elements (i.e. it is the full substructure induced by these elements). Within this setting, we study bounds on the error of sufficient statistics of relational models that are estimated on the available data. As our main result, we prove a bound based on a variant of the Vapnik-Chervonenkis dimension which is suitable for relational data.

1 Introduction

In one of the most common settings in statistical relational learning (SRL), we are given a fragment of a relational structure (i.e. a *training example*) from which we want to learn a model for making predictions about the unseen parts of the structure. For example, the relational structure could correspond to a large social network and the training example to a fragment of the social network specifying the relationships that hold among a small sample of the users, along with their attributes. Clearly, in order to provide any guarantees on the accuracy of these predictions, we need to make (simplifying) assumptions about how the training structures are obtained. In this paper, we follow the setting from [8,9], where it is assumed that these structures are all obtained as fragments induced by domain elements sampled uniformly without replacement.

Electronic supplementary material The online version of this chapter (https://doi.org/10.1007/978-3-030-10928-8_16) contains supplementary material, which is available to authorized users.

M. Berlingerio et al. (Eds.): ECML PKDD 2018, LNAI 11052, pp. 259–275, 2019.
https://doi.org/10.1007/978-3-030-10928-8_16

The specific problem that we consider in this paper is to bound the error that we make when estimating probabilities of first-order theories from the training example, or more specifically, the probability that a first-order theory Φ is satisfied in a small randomly sampled fragment of the relational structure. While this setting has already been studied in [7–9], one important remaining problem, which will be the focus of this paper, relates to how the theory Φ is obtained. Typically, Φ is chosen from some hypothesis class, based on the same training example that is used to estimate its probability. The bounds that were derived in [7] for such cases depend on the size of this hypothesis class. Unfortunately, this can quickly lead to vacuous bounds in many cases. In fact, in some applications, the most natural hypothesis classes are either infinite or so large that they are effectively infinite for all practical purposes. This is the case, for instance, whenever we want to use constructs involving numerical expressions. To address this issue, in this paper we derive bounds which depend on the VC-dimension of the hypothesis class, instead of its size. In this way, we can also obtain, in many cases, tighter bounds than the ones we derived in [7]. To the best of our knowledge, the bounds we introduce in this paper are the first VC-dimension based bounds for relational learning problems.

2   Preliminaries

In this paper we consider function-free language $\mathcal{L}$, which is built from a finite set of constants $Const$, a set of variables Var and a set of predicates $Rel = \bigcup_i Rel_i$, where Rel_i contains the predicates of arity i. Throughout this paper we assume that the sets $Const$, Var and Rel are fixed. For $a_1, ..., a_k \in Const \cup Var$ and $R \in Rel_k$, we call $R(a_1, ..., a_k)$ an *atom*. If $a_1, .., a_k \in Const$, this atom is called *ground*. A *literal* is an atom or its negation. A formula is called *closed* if all variables are bound by a quantifier. Note that although the set $Const$ is required to be finite, it can have arbitrary size, so that we could, for instance, represent all 64-bit floating point numbers. From an application point of view, this allows us to consider formulas involving numerical expressions. For example, we could have a predicate Sum, whose intended meaning is that $Sum(x, y, z)$ holds iff $z = x + y$ where $+$ represents floating-point addition.

2.1   Relational Learning Setting

Relational Examples. The learning setting considered in this paper follows the one that was introduced in [8,9]. The central notion is that of a *relational example* (or simply *example* if there is no cause for confusion), which is defined as a pair $(\mathcal{A}, \mathcal{C})$, with $\mathcal{C}$ a set of constants and $\mathcal{A}$ a set of ground atoms which only use constants from $\mathcal{C}$. A relational example is intended to provide a complete description of a possible world, hence any ground atom over $\mathcal{C}$ which is not contained in $\mathcal{A}$ is implicitly assumed to be false. Note that this is why we have to explicitly specify $\mathcal{C}$, as opposed to simply considering the set of constants appearing in $\mathcal{A}$. For instance, the relational example $(\{sm(alice)\}, \{alice\})$ is different from $(\{sm(alice)\}, \{alice, bob\})$, as in the latter case we know that *bob* does not smoke (i.e. the atom $sm(bob)$ is known to be false since it is not specified to be true) whereas in the former case we have no knowledge about *bob*. We denote by $\Omega(\mathcal{L}, k)$ the set of all possible relational examples $\Upsilon = (\mathcal{A}, \mathcal{C})$ where $\mathcal{A}$ only contains ground atoms from $\mathcal{L}$ and $|\mathcal{C}| = k$.

Example 1. Let us assume that the only predicate in $\mathcal{L}$ is $sm/1$ and the only constant in is *alice*. Then $\Omega(\mathcal{L}, 1) = \{(sm(alice), \{alice\}), (\emptyset, \{alice\})\}$.

Let $\Upsilon = (\mathcal{A}, \mathcal{C})$ be a relational example and $\mathcal{S} \subseteq \mathcal{C}$. The fragment $\Upsilon\langle \mathcal{S} \rangle = (\mathcal{B}, \mathcal{S})$ is defined as the restriction of Υ to the constants in $\mathcal{S}$, i.e. $\mathcal{B}$ is the set of all atoms from $\mathcal{A}$ which only contain constants from $\mathcal{S}$.

Example 2. Let

$$\Upsilon = (\{fr(alice, bob), fr(bob, alice), fr(bob, eve), fr(eve, bob), sm(alice)\},$$
$$\{alice, bob, eve\}),$$

i.e. the only smoker is *alice* and the friendship structure is:

Then $\Upsilon\langle\{alice, bob\}\rangle = (\{sm(alice), fr(alice, bob), fr(bob, alice)\}, \{alice, bob\})$.

In the considered setting, we are given a single relational example $\Upsilon = (\mathcal{A}, \mathcal{C})$, and this example is assumed to have been sampled from a larger relational example $\aleph = (\mathcal{A}_\aleph, \mathcal{C}_\aleph)$. The intended meaning is that $\aleph$ covers the entire domain which we would like to model and Υ is the fragment of the domain which is known at training time. Throughout this paper, we will assume that $\mathcal{C}_\aleph$ is finite. As in [7,8] we assume that Υ as sampled from $\aleph$ by the following process.

Definition 1 (Sampling from a global example). *Let $\aleph = (\mathcal{A}_\aleph, \mathcal{C}_\aleph)$ be a relational example called the* global example. *Let $n \in \mathbb{N} \setminus \{0\}$ and let $Unif(\mathcal{C}_\aleph, n)$ denote uniform distribution on size-n subsets of $\mathcal{C}_\aleph$. Training relational examples Υ are sampled from the global example $\aleph$ by first sampling $\mathcal{C}_\Upsilon \sim Unif(\mathcal{C}_\aleph, n)$ and defining $\Upsilon = \aleph\langle \mathcal{C}_\Upsilon \rangle$.*

Probabilities of Formulas. In a given relational example, any closed formula α is classically either true or false. To assign probabilities to formulas in a meaningful way, considering that we typically only have a single relational example available for training, we can consider how often the formula is satisfied in small fragments of the given relational example.

Definition 2 (Probability of a formula [8]). *Let $\Upsilon = (\mathcal{A}, \mathcal{C})$ be a relational example and $k \in \mathbb{N}$. The probability of a closed formula α is defined as follows[1]:*

$$Q_{\Upsilon, k}(\alpha) = P_{\mathcal{S} \sim Unif(\mathcal{C}, k)} [\Upsilon\langle \mathcal{S} \rangle \models \alpha]$$

where $Unif(\mathcal{C}, k)$ denotes uniform distribution on size-k subsets of $\mathcal{C}$.

---

[1] We will use Q for probabilities of formulas as defined in this section, to avoid confusion with other "probabilities" we deal with in the text.

Clearly $Q_{\Upsilon,k}(\alpha) = \frac{1}{|\mathcal{C}_k|} \cdot \sum_{\mathcal{S} \in \mathcal{C}_k} \mathbb{1}(\Upsilon\langle\mathcal{S}\rangle \models \alpha)$ where $\mathcal{C}_k$ is the set of all size-k subsets of $\mathcal{C}$. The above definition can straightforwardly be extended to probabilities of sets of formulas (which we will also call *theories* interchangeably): if Φ is a set of formulas, we then have $Q_{\Upsilon,k}(\Phi) = Q_{\Upsilon,k}(\bigwedge \Phi)$ where $\bigwedge \Phi$ denotes the conjunction of all formulas in Φ.

Example 3. Let $sm/1$ be a unary predicate denoting that someone is a smoker, e.g. $sm(alice)$ means that $alice$ is a smoker. Let us consider the following example:

$$\Upsilon = (\{fr(alice, bob), sm(alice), sm(eve)\}, \{alice, bob, eve\}),$$

and formulas $\alpha = \forall X : sm(X)$ and $\beta = \exists X, Y : fr(X, Y)$. Then, for instance, $Q_{\Upsilon,1}(\alpha) = 2/3$, $Q_{\Upsilon,2}(\alpha) = 1/3$ and $Q_{\Upsilon,2}(\beta) = 1/3$.

It is not difficult to check that under the sampling assumption from Definition 1, for any theory Φ it holds that $Q_{\aleph,k}(\Phi) = \mathbb{E}_\Upsilon [Q_{\Upsilon,k}(\Phi)]$ [8].

Representing Theories as Functions. By definition, to compute $Q_{\Upsilon,k}(\Phi)$, we only need to know for which of the elements of $\Omega(\mathcal{L}, k)$ it holds that Φ is satisfied. To make this view explicit, we will formulate the results in this paper in terms of functions from $\Omega(\mathcal{L}, k)$ to $\{0, 1\}$. For a given theory Φ, the associated function f_Φ is defined for $\Gamma \in \Omega(\mathcal{L}, k)$ as $f_\Phi(\Gamma) = 1$ if $\Gamma \vdash \Phi$ and $f_\Phi(\Gamma) = 0$ otherwise. The advantage of this formulation is that our results then directly apply to settings where other representation frameworks than classical logic are used for representing the theory. For example, a theory could be implicitly represented by a neural network with a hard-thresholding output unit. For notational convenience, we also write $\Gamma \models f$ if $f(\Gamma) = 1$. We then naturally extend the definition of $Q_{\Upsilon,k}$ to functions: $Q_{\Upsilon,k}(f) = P_{\mathcal{S} \sim Unif(\mathcal{C},k)} [\Upsilon\langle\mathcal{S}\rangle \models f] = P_{\mathcal{S} \sim Unif(\mathcal{C},k)} [f(\Upsilon\langle\mathcal{S}\rangle) = 1]$.

2.2   VC-Dimension

The next definition describes the classical notion of VC-dimension [15], specialized to our relational learning setting that is used throughout this paper to measure the complexity of hypothesis classes.

Definition 3 (VC-dimension). *Let k be a positive integer and let $\mathcal{H}$ be a hypothesis class of functions $f : \Omega(\mathcal{L}, k) \to \{0, 1\}$. Let $\mathcal{X} = \{\Upsilon_1, \Upsilon_2, \ldots, \Upsilon_d\} \subseteq \Omega(\mathcal{L}, k)$. We say that $\mathcal{H}$ shatters $\mathcal{X}$ if for every $\mathcal{Y} \subseteq \mathcal{X}$, there is $f \in \mathcal{H}$ such that $f(\Upsilon) = 1$ for all $\Upsilon \in \mathcal{Y}$ and $f(\Upsilon) = 0$ for all $\Upsilon \in \mathcal{X} \setminus \mathcal{Y}$. The VC dimension of $\mathcal{H}$ is the largest integer d such that there exists a subset of $\Omega(\mathcal{L}, k)$ with cardinality d that is shattered by $\mathcal{H}$.*

The next definition formalizes what we mean when we say that two functions are equivalent w.r.t. a given global example.

Definition 4. *We say two functions f and g are k-equivalent w.r.t. a global example $\aleph$ if for any size-k set $\mathcal{S}$ it holds that $f(\aleph\langle\mathcal{S}\rangle) = g(\aleph\langle\mathcal{S}\rangle)$.*

Naturally the above two definitions can also be applied to theories, e.g. two theories Φ and Θ are k-equivalent w.r.t. a global example $\aleph$ if their associated functions f_Φ and f_Θ are k-equivalent. The following observation will play an important role in the proofs.

Remark 1. *The maximum number of hypotheses that are mutually non-equivalent w.r.t. a given (finite) global example $\aleph$ is finite.*

A consequence of this observation is that even for infinite hypothesis classes, in principle, there are only finitely many different hypotheses that need to be considered. However, given that we typically do not know the size of the global example, in practice it is not possible to rely on the number of non-equivalent hypotheses to apply the bounds from [7] to infinite hypothesis classes. In contrast, the bounds that we introduce in this paper can still be applied in such cases, as long as the hypothesis class has a finite VC-dimension.

The ability to deal with infinite hypothesis classes makes it possible, for instance, to learn theories based on differentiable architectures [12,18] or based on graph kernels [17].

3   Motivation

The main aim of this paper is to derive bounds on how accurately we can estimate $Q_{\aleph,k}(f)$ from a given training relational example Υ, where f is viewed as a logical formula. The need for such probability estimates naturally arises, among others, in the setting of relational marginal problems, which were studied in [8]. In that setting, we are given a set of formulas $\Theta = \{\alpha_1, \ldots, \alpha_{|\Theta|}\}$, a set of constants $\mathcal{C}$ and a training relational example $\Upsilon = (\mathcal{A}_\Upsilon, \mathcal{C}_\Upsilon)$. The task is to use the probabilities of $\alpha_1, \ldots, \alpha_{|\Theta|}$ that are estimated from the training relational example Υ to perform inference on the domain $\mathcal{C}$. Specifically, the task is to find a maximum entropy distribution on the set of all relational examples of the form $\Psi = (\mathcal{A}_\Psi, \mathcal{C})$, such that $\mathbb{E}[Q_{\Psi,k}(\alpha_i)] = \widehat{Q}_{\Upsilon,k}(\alpha_i)$ for all $\alpha_i \in \Theta$. Here, $\widehat{Q}_{\Upsilon,k}(\alpha_i)$ is an estimate of $\mathbb{E}[Q_{\Psi,k}(\alpha_i)]$ which is based on $Q_{\Upsilon,k}(\alpha_i)$. If $|\mathcal{C}| \leq |\mathcal{C}_\Upsilon|$ then this estimate is simply given by $\widehat{Q}_{\Upsilon,k}(\alpha_i) = Q_{\Upsilon,k}(\alpha_i)$. In general, however, the value $Q_{\Upsilon,k}(\alpha_i)$ needs to be adjusted to account for the difference in the size of the training relational example domain $\mathcal{C}_\Upsilon$ and the domain $\mathcal{C}$ over which we want to perform inference. The resulting distribution is similar to a Markov logic network, and can be used in applications for similar purposes[2]; it is an exponential family distribution of the following form:

$$P(\Psi) = \frac{1}{Z} \exp\left(\sum_{\alpha_i \in \Theta} w_i \cdot Q_{\Psi,k}(\alpha_i) \right).$$

---

[2] The relational marginal problems that we consider in this paper are referred to as Model A in [8]. Another type of relational marginal problems, referred to as Model B in [8], leads to distributions that are exactly Markov logic networks.

In the case $|\mathcal{C}| = |\mathcal{C}_{\Upsilon}|$, the weights w_i can be obtained by solving a maximum likelihood problem which is the dual of the maximum entropy problem. Ideally, we would use $Q_{\aleph,k}(\alpha_i)$ as the estimates of $Q_{\Psi,k}(\alpha_i)$ in the maximum entropy problems. Since, in reality, we do not have access to $Q_{\aleph,k}(\alpha_i)$, we need to use the estimates based on $Q_{\Upsilon,k}(\alpha_i)$. The results we present in this paper shed light on the impact of this simplification. We refer the reader to [8] for more details.

Estimates of $Q_{\aleph,k}(f)$ also play a central role in the analysis of PAC-reasoning [6,14] for relational domains as studied in [7]. This analysis also relies on the sampling assumptions from Definition 1. Specifically, in that setting, a training relational example Υ and a test relational example Ψ are sampled from $\aleph$ and the learner's task is to find a set of first-order logic formulas that will not produce too many errors on Ψ when using a restricted form of classical reasoning. To obtain guarantees on the number of literals that are incorrectly inferred using this form or reasoning, we essentially need to bound the difference of $Q_{\Upsilon,k}(\Phi)$ and $Q_{\aleph,k}(\Phi)$ (which allows us to bound the difference with $Q_{\Psi,k}(\Phi)$), which is exactly the problem we also study in this paper. In contrast to [7], however, we are interested in bounds that are based on the VC-dimension of the hypothesis space.

4   Summary of the Results

Intuitively, what we need to find is a suitable bound on the quantity $|Q_{\aleph,k}(f) - Q_{\Upsilon,k}(f)|$, i.e. we want to bound the error we make when estimating the overall probability of f (i.e. the value $Q_{\aleph,k}(f)$) from a training fragment of the global example. In most application settings, however, f itself is also chosen using the training relational example Υ, e.g. by choosing the hypothesis f that maximizes $Q_{\Upsilon,k}(f)$ among the functions from some hypothesis class $\mathcal{H}$. This means that we cannot find a suitable bound for $|Q_{\aleph,k}(f) - Q_{\Upsilon,k}(f)|$ without taking the hypothesis class $\mathcal{H}$ into account. The classical solution, which we will also follow, is to instead bound the quantity $\sup_{f \in \mathcal{H}} |Q_{\aleph,k}(f) - Q_{\Upsilon,k}(f)|$. The main result of this paper takes the form of two theorems that provide probabilistic bounds on this latter quantity. The proof of these theorems is presented in Sect. 6.

The first theorem bounds the expected value of $\sup_{f \in \mathcal{H}} |Q_{\aleph,k}(f) - Q_{\Upsilon,k}(f)|$ when $\mathcal{C}_{\Upsilon}$ is viewed as a random variable. Interestingly, this bound is essentially the same as the classical bound for the i.i.d. setting [13], except that the value of n from the classical bound is replaced by $\lfloor n/k \rfloor$, which is perhaps not surprising as it is the maximum number of non-overlapping size-k subsets of $\mathcal{C}_{\Upsilon}$.

Theorem 1. *Let $\aleph = (\mathcal{A}_{\aleph}, \mathcal{C}_{\aleph})$ be a global example and $\mathcal{C}_{\Upsilon}$ be sampled uniformly from all size-n subsets of $\mathcal{C}_{\aleph}$ and let us define $\Upsilon = \aleph \langle \mathcal{C}_{\Upsilon} \rangle$. Then for any hypothesis class $\mathcal{H}$ of functions $f : \Omega(\mathcal{L}, k) \to \{0, 1\}$ with finite VC-dimension d, the following holds:*

$$\mathbb{E}\left[\sup_{f \in \mathcal{H}} |Q_{\aleph,k}(f) - Q_{\Upsilon,k}(f)| \right] \leq 2 \cdot \sqrt{\frac{2d \log\left(2e \lfloor n/k \rfloor / d\right)}{\lfloor n/k \rfloor}}$$

The second theorem provides a tail bound for $P\left[\sup_{f\in\mathcal{H}}|Q_{\aleph,k}(f) - Q_{\Upsilon,k}(f)| \geq \varepsilon\right]$. We note that the bound on expected error from Theorem 1 cannot be derived from Theorem 2, although a different bound on expected error with looser constants could be derived from Theorem 2.

Theorem 2. *Let* $\aleph = (\mathcal{A}_{\aleph}, \mathcal{C}_{\aleph})$ *be a global example and* $\mathcal{C}_{\Upsilon}$ *be sampled uniformly from all size-n subsets of* $\mathcal{C}_{\aleph}$ *and let us define* $\Upsilon = \aleph\langle\mathcal{C}_{\Upsilon}\rangle$. *Then for any hypothesis class* $\mathcal{H}$ *of functions* $f : \Omega(\mathcal{L}, k) \rightarrow \{0, 1\}$ *with finite VC-dimension d, the following holds for any* $0 < \varepsilon \leq 1$:

$$P\left[\sup_{f\in\mathcal{H}}|Q_{\aleph,k}(f) - Q_{\Upsilon,k}(f)| \geq \varepsilon\right]$$
$$\leq \exp\left(-\frac{\lfloor n/k\rfloor \varepsilon^2}{4}\right) + \varepsilon\sqrt{8\pi\lfloor n/k\rfloor}\left(\frac{2e\lfloor n/k\rfloor}{d}\right)^d \cdot \exp\left(-\frac{\lfloor n/k\rfloor \varepsilon^2}{8}\right)$$

Up to somewhat looser constants, the tail bound from Theorem 2 can be shown to also have the same form as the existing VC tail bounds [16]. In particular, the bound implies the following simpler, albeit looser bound:

$$P\left[\sup_{f\in\mathcal{H}}|Q_{\aleph,k}(f) - Q_{\Upsilon,k}(f)| \geq \varepsilon\right]$$
$$\leq \left(1 + \sqrt{8\pi\lfloor n/k\rfloor}\left(\frac{2e\lfloor n/k\rfloor}{d}\right)^d\right) \cdot \exp\left(-\frac{\lfloor n/k\rfloor \varepsilon^2}{8}\right).$$

5 Related Work

There have been several works studying theoretical properties of various statistical relational learning settings. Dhurandhar and Dobra [3] derived Hoeffding-type inequalities for classifiers trained with relational data. However, there are several important differences with our work. First, their bounds are not VC-type bounds. Moreover, their results, based on restricting the independent interactions of data points, cannot be applied in our setting, which is more general than the one they consider. Certain other statistical properties of learning have also been studied for SRL models. For instance, Xiang and Neville [19] studied consistency of estimation in a certain relational learning setting.

From a different perspective, abstracting from the relational logic setting, our results can also be seen as bounds for uniform deviations of U-statistics [4] under sampling *without* replacement. Not many results are known for this particular setting in the literature. One exception is the work of Nandi and Sen [11] who only derived bounds on variance in this setting. It is not possible to derive our results from theirs. In particular, we need Chernoff-type bounds whereas the variance bounds from their work would only give us Chebyshev-type bounds. A more thoroughly studied setting is the estimation of U-statistics under sampling *with* replacement. Clémencon, Lugosi and Vayatis [1] derived among others[3] VC-inequalities in a setting similar to ours, but under sampling with replacement, which makes their analysis simpler. However, such an assumption would not make sense in the relational learning setting where it would mean, for instance, that we would end up with multiple copies of the same individual (e.g. ending up with social networks in which the same person can occur multiple times).

6   Derivation of the Bounds

In this section, we prove Theorems 1 and 2 using a series of lemmas. First, in Sect. 6.1, we define a sampling process for generating vectors containing $\lfloor n/k \rfloor$ size-k fragments of Υ. The sampling process has two important properties. First, the fragments in each of the vectors are distributed as size-k fragments sampled i.i.d. from $\aleph$ (assuming Υ is sampled as in Definition 1). Second, the average of the estimates of $Q_{\aleph,k}(f)$ computed from the vectors converges to $Q_{\Upsilon,k}(f)$. These two properties allow us to use the sampling process to derive a bound on expected value of the random variable $\sup_{f \in \mathcal{H}} |Q_{\aleph,k}(f) - Q_{\Upsilon,k}(f)|$ in Sect. 6.2, which finishes the proof of Theorem 1.

The proof of Theorem 2 is a bit more involved. First, in Sect. 6.3, we derive bounds on the moment-generating function of a random variable that can be obtained if we only know its tail bounds. Then, in Sect. 6.4, we combine the results from the preceding sections to prove Theorem 2. In particular, we use the bound moment-generating function to obtain a tail bound on the estimates of $Q_{\aleph,k}(f)$ by exploiting a trick that is sometimes called *average of sums-of-i.i.d blocks* [1].

6.1   Extracting Independent Samples

In this section we describe a sampling process that allows us to obtain $\lfloor n/k \rfloor$ samples from Υ that are distributed as i.i.d. samples from $\aleph$, assuming Υ is sampled as in Definition 1.

Lemma 1. *Let* $\aleph = (\mathcal{A}_{\aleph}, \mathcal{C}_{\aleph})$ *be a global example. Let* $0 \leq n \leq |\mathcal{C}_{\aleph}|$, $q \geq 1$ *and* $1 \leq k \leq n$ *be integers. Let* $\mathbf{X} = (\mathcal{S}_1, \mathcal{S}_2, \ldots, \mathcal{S}_{\lfloor \frac{n}{k} \rfloor})$ *be a vector of subsets of* $\mathcal{C}_{\aleph}$, *each sampled uniformly and independently of the others from all size-k subsets of* $\mathcal{C}_{\aleph}$. *Next let* $\mathcal{I}' = \{1, 2, \ldots, |\mathcal{C}_{\aleph}|\}$ *and let* $\mathbf{Y}_j = (\mathcal{S}'_{j,1}, \mathcal{S}'_{j,2}, \ldots, \mathcal{S}'_{j, \lfloor \frac{n}{k} \rfloor})$, *for* $1 \leq j \leq q$, *be vectors sampled by the following process:*

1. *Sample* $\mathcal{C}_{\Upsilon}$ *uniformly from all size-n subsets of* $\mathcal{C}_{\aleph}$.
2. *For j from 1 to q:*
   (a) Sample subsets $\mathcal{I}'_1, \ldots, \mathcal{I}'_{\lfloor \frac{n}{k} \rfloor}$ *of size k from* $\mathcal{I}'$.

---

[3] The main results of [1] are bounds that assume a certain 'low-noise' condition. Although they only derived bounds for the case $k = 2$ (in our notation), the results directly related to ours can be extended for larger k's as well.

(b) *Sample an injective function* $g : \bigcup_{i=1}^{\lfloor n/k \rfloor} \mathcal{I}'_i \to \mathcal{C}_\Upsilon$ *uniformly from all such functions.*

(c) *Define* $\mathcal{S}'_{j,i} = g(\mathcal{I}'_i)$ *for all* $0 \le i \le \lfloor \frac{n}{k} \rfloor$.

Then the following holds:

1. *The random vectors* $\mathbf{X}$ *and* $\mathbf{Y}_j$ *have the same distribution for any* $1 \le j \le q$.
2. *For any function* $f : \Omega(\mathcal{L}, k) \to [0, 1]$ *it holds:*

$$P\left[\left| Q_{\Upsilon, k}(f) - \frac{1}{q \lfloor n/k \rfloor} \sum_{j=1}^{q} \sum_{i=1}^{\lfloor n/k \rfloor} f\left(\Upsilon \langle \mathcal{S}'_{j,i} \rangle \right) \right| \ge \epsilon \right] \le 2 \exp\left(-2q\varepsilon^2 \right)$$

Proof. The first part of the proof follows immediatelly from Lemma 3 in [8] (which, for completeness, we reprove in the online[4] appendix as Lemma 5 (see supplementary material)). For the second part, we may first notice that, after $\mathcal{C}_\Upsilon$ is sampled and fixed, $Q_{\Upsilon, k}(f) = \mathbb{E}\left[f\left(\Upsilon \langle \mathcal{S}'_{j,i} \rangle \right) \right]$, as the probability of $\mathcal{S}'_{j,i}$ being a particular size-k subset of $\mathcal{C}_\Upsilon$ is the same for all such subsets. The second part can then be shown by applying Hoeffding inequality to q i.i.d. samples $\frac{1}{\lfloor n/k \rfloor} \sum_{i=1}^{\lfloor n/k \rfloor} f(\Upsilon \langle \mathcal{S}_{j,i} \rangle)$, $j = 1, 2, \ldots, q$, which have the same expected value $Q_{\Upsilon, k}(f)$. □

At this point, one might wonder if the above lemma already gives us a way to find VC-type bounds for relational data, based on the following strategy: sample $\lfloor n/k \rfloor$ size-k fragments from a given training relational example Υ using the procedure defined in Lemma 1 and use this set of fragments as our training data. Although this would allow us to use standard bounds that are known for learning from i.i.d. data [15], there are two problems with this approach. The first problem is that in reality we do not always know the size of the global example $\aleph$ and hence we do not know how to get a sample of $\lfloor n/k \rfloor$ size-k sets that behaves as an independent sample from $\aleph$ (noting that we need to know the size of $\aleph$ to define the set $\mathcal{I}'$ in Lemma 1). The second problem is that there are cases where only sampling the $\lfloor n/k \rfloor$ samples is sub-optimal from the point of view of statistical power, as we illustrate in the next example.

Example 4. Consider a global structure which takes the form of a large directed graph, and assume that we are interested in estimating the probability that the formula $\exists X, Y : edge(X, Y)$ holds for a fragment of the structure induced by two randomly sampled nodes. Assume furthermore that the given graph was generated by sampling (directed) edges independently with some probability p. The probability that $\exists X, Y : edge(X, Y)$ holds for any two nodes will thus correspond to some value p^* close to $1 - (1-p)^2$. As we will see, given a training fragment induced by n nodes from this graph, we can only generate $\lfloor \frac{n}{2} \rfloor$ samples that behave like i.i.d. samples. In this case, a more accurate estimate of p^* can be obtained by using all size-2 fragments of the training fragment.

---

[4] https://arxiv.org/abs/1804.06188.

Nonetheless, the strategy based on sampling $\lfloor n/k \rfloor$ size-k fragments may actually be optimal in the worst case as we illustrate in the next example.

Example 5. Let us again consider the setting from Example 4, which we can now describe more formally. In particular, assume that $\aleph = (\mathcal{A}_\aleph, \mathcal{C}_\aleph)$ represents a large directed graph. Let $k = 2$ and $\varPhi = \{\exists X, Y : edge(X, Y)\}$. Let $\varUpsilon$ be a relational example sampled uniformly from $\aleph$ (i.e. $\varUpsilon = \aleph\langle\mathcal{C}_\varUpsilon\rangle$ where $\mathcal{C}_\varUpsilon$ is sampled uniformly from all size-n subsets of $\mathcal{C}_\aleph$). Let us now, in contrast to the assumption underlying Example 4, assume that the directed graph was constructed using the following process. For all nodes v, we flip a biased coin with probability of heads being q. If it lands heads, we add a directed edge from v to all other nodes. In this case[5], $Q_{\aleph,k}(\varPhi) = p' \approx 1 - (1 - q)^2$. The main difference with the setting from Example 4 is that estimating p' now effectively corresponds to estimation of a property of nodes, as we are also able to recover p' by observing how many nodes have at least one outgoing edge. However, this also means that the effective sample size in this case only grows linearly with the number of vertices (as opposed to quadratically in Example 4). This, at least asymptotically (up to a multiplicative constant), is a worst-case scenario as the number of independent samples that we are able to obtain using Lemma 1 also grows linearly with the number of vertices in the sample $\varUpsilon$ (i.e. linearly with $|\mathcal{C}_\varUpsilon|$).

6.2   Bounding Expected Error

In this section we use the results from Sect. 6.1 to obtain a bound on the expected value of $\sup_{f\in H} |Q_{\aleph,k}(f) - Q_{\varUpsilon,k}(f)|$.

Lemma 2. *Let $\aleph = (\mathcal{A}_\aleph, \mathcal{C}_\aleph)$ be a global example and $\mathcal{C}_\varUpsilon$ be sampled uniformly from all size-n subsets of $\mathcal{C}_\aleph$ and let us define $\varUpsilon = \aleph\langle\mathcal{C}_\varUpsilon\rangle$. Let $\mathbf{Y}_j = (\mathcal{S}'_{j,1}, \ldots, \mathcal{S}'_{j,\lfloor\frac{n}{k}\rfloor})$, where $1 \leq j \leq q$, be random vectors sampled as in Lemma 1. Then for any hypothesis class $\mathcal{H}$ of functions $f : \Omega(\mathcal{L}, k) \to \{0, 1\}$ with finite VC-dimension d, the following holds:*

$$\mathbb{E}\left[\sup_{f\in\mathcal{H}} |Q_{\aleph,k}(f) - Q_{\varUpsilon,k}(f)|\right] \leq \lim_{q\to\infty} \mathbb{E}\left[\sup_{f\in\mathcal{H}} \left|Q_{\aleph,k}(f) - \frac{1}{q\cdot\lfloor\frac{n}{k}\rfloor}\sum_{i=1}^{q}\sum_{\mathcal{S}\in\mathbf{Y}_i} f(\varUpsilon\langle\mathcal{S}\rangle)\right|\right]$$

Proof. We have

$$\mathbb{E}\left[\sup_{f\in\mathcal{H}} |Q_{\aleph,k}(f) - Q_{\varUpsilon,k}(f)|\right]$$

$$= \lim_{q\to\infty} \mathbb{E}\left[\sup_{f\in\mathcal{H}} \left|Q_{\aleph,k}(f) - \left(\frac{1}{q\lfloor\frac{n}{k}\rfloor}\sum_{i=1}^{q}\sum_{\mathcal{S}\in\mathbf{Y}_i} f(\varUpsilon\langle\mathcal{S}\rangle)\right)\right.\right.$$

---

[5] More formally, the following holds, assuming $\aleph$ is generated by the respective random processes. In the setting from Example 4 we have $\mathbb{E}_\aleph[Q_{\aleph,k}(\varPhi)] = 1 - (1 - p)^2$ and in the setting from this example we have $\mathbb{E}_\aleph[Q_{\aleph,k}(\varPhi)] = 1 - (1 - q)^2$.

$$+ \left(\frac{1}{q \lfloor \frac{n}{k} \rfloor} \sum_{i=1}^{q} \sum_{\mathcal{S} \in \mathbf{Y}_i} f(\Upsilon\langle\mathcal{S}\rangle) \right) - Q_{\Upsilon,k}(f) \Bigg| \Bigg]$$

$$\leq \lim_{q \to \infty} \mathbb{E} \left[\sup_{f \in \mathcal{H}} \left| Q_{\aleph,k}(f) - \left(\frac{1}{q \lfloor \frac{n}{k} \rfloor} \sum_{i=1}^{q} \sum_{\mathcal{S} \in \mathbf{Y}_i} f(\Upsilon\langle\mathcal{S}\rangle) \right) \right| \right]$$

$$+ \lim_{q \to \infty} \mathbb{E} \left[\sup_{f \in \mathcal{H}} \left| \left(\frac{1}{q \lfloor \frac{n}{k} \rfloor} \sum_{i=1}^{q} \sum_{\mathcal{S} \in \mathbf{Y}_i} f(\Upsilon\langle\mathcal{S}\rangle) \right) - Q_{\Upsilon,k}(f) \right| \right] \quad (1)$$

To finish the proof, we show that the last summand in (1) is zero. To this end, first note that it follows from Remark 1 that the supremum only needs to be taken over a finite number t of hypotheses, one from each equivalence class of functions that are equal on all size-k subsets of $\mathcal{C}_{\aleph}$. Together with Lemma 1 and the union bound on the finitely many equivalence classes, we find

$$P \left[\sup_{f \in \mathcal{H}} \left| \left(\frac{1}{q \lfloor \frac{n}{k} \rfloor} \sum_{i=1}^{q} \sum_{\mathcal{S} \in \mathbf{Y}_i} f(\Upsilon\langle\mathcal{S}\rangle) \right) - Q_{\Upsilon,k}(f) \right| \geq \varepsilon \right] \leq 2 \cdot t \cdot \exp\left(-2q\varepsilon^2\right)$$

Then it follows using $\mathbb{E}[X] = \int_0^1 P[X \geq x] dx$ (assuming $P[X \in [0;1]] = 1$) that

$$\mathbb{E} \left[\sup_{f \in \mathcal{H}} \left| \left(\frac{1}{q \lfloor \frac{n}{k} \rfloor} \sum_{i=1}^{q} \sum_{\mathcal{S} \in \mathbf{Y}_i} f(\Upsilon\langle\mathcal{S}\rangle) \right) - Q_{\Upsilon,k}(f) \right| \right] \leq \int_0^1 2 \cdot t \cdot \exp\left(-2qx^2\right) dx.$$

Finally, noticing that $\lim_{q \to \infty} \int_0^1 2 \cdot t \cdot \exp\left(-2qx^2\right) dx = 0$ finishes the proof. $\square$

Lemma 3. *Suppose* $\mathbf{Y}_j = (\mathcal{S}'_{j,1}, \ldots, \mathcal{S}'_{j,\lfloor \frac{n}{k} \rfloor})$ *is a random vector sampled as in Lemma 1. Then for any hypothesis class of functions* $f : \Omega(\mathcal{L}, k) \to \{0,1\}$ *with VC-dimension* d *we have:*

$$P \left[\sup_{f \in \mathcal{H}} \left| Q_{\aleph,k}(f) - \frac{1}{\lfloor n/k \rfloor} \sum_{\mathcal{S} \in \mathbf{Y}_j} f(\Upsilon\langle\mathcal{S}\rangle) \right| \geq \varepsilon \right] \leq 4 \left(\frac{2e\lfloor n/k \rfloor}{d} \right)^d \exp\left(-\frac{\lfloor n/k \rfloor \varepsilon^2}{8}\right)$$

and

$$\mathbb{E} \left[\sup_{f \in \mathcal{H}} \left| Q_{\aleph,k}(f) - \frac{1}{\lfloor n/k \rfloor} \sum_{\mathcal{S} \in \mathbf{Y}_j} f(\Upsilon\langle\mathcal{S}\rangle) \right| \right] \leq 2 \sqrt{\frac{2d \log\left(2e\lfloor n/k \rfloor/d\right)}{\lfloor n/k \rfloor}}$$

Proof. Since $\mathcal{S}'_{j,1}, \ldots, \mathcal{S}'_{j,\lfloor \frac{n}{k} \rfloor}$ are sampled in an i.i.d. way, the classical VC inequality applies [16]. The expected value bound can be derived from the bound (6.4) in [13][6]. $\square$

---

[6] The specific form that we use here can be found in the lecture notes of Philippe Rigollet https://bit.ly/2H89wPn.

We are now ready to prove Theorem 1.

Proof (of Theorem 1). Let $\mathbf{Y}_j = (\mathcal{S}'_{j,1}, \ldots, \mathcal{S}'_{j,\lfloor \frac{n}{k} \rfloor})$, where $1 \leq j \leq q$ for a given integer q, be random vectors sampled as in Lemma 1. First, using Lemma 2 for the first step, we find

$$
\mathbb{E}\left[\sup_{f \in H} |Q_{\aleph,k}(f) - Q_{\Upsilon,k}(f)|\right]
$$

$$
\leq \lim_{q \to \infty} \mathbb{E}\left[\sup_{f \in H} \left|Q_{\aleph,k}(f) - \frac{1}{q}\sum_{j=1}^{q} \frac{1}{\lfloor n/k \rfloor} \sum_{S \in \mathbf{Y}_j} f(\Upsilon\langle S \rangle)\right|\right]
$$

$$
= \lim_{q \to \infty} \mathbb{E}\left[\sup_{f \in H} \left|\frac{1}{q}\sum_{j=1}^{q} \left(Q_{\aleph,k}(f) - \frac{1}{\lfloor n/k \rfloor} \sum_{S \in \mathbf{Y}_j} f(\Upsilon\langle S \rangle)\right)\right|\right]
$$

$$
\leq \lim_{q \to \infty} \mathbb{E}\left[\frac{1}{q}\sup_{f \in H} \sum_{j=1}^{q} \left|Q_{\aleph,k}(f) - \frac{1}{\lfloor n/k \rfloor} \sum_{S \in \mathbf{Y}_j} f(\Upsilon\langle S \rangle)\right|\right]
$$

$$
\leq \lim_{q \to \infty} \mathbb{E}\left[\frac{1}{q}\sum_{j=1}^{q} \sup_{f \in H} \left|Q_{\aleph,k}(f) - \frac{1}{\lfloor n/k \rfloor} \sum_{S \subset \mathbf{Y}_j} f(\Upsilon\langle S \rangle)\right|\right]
$$

$$
= \lim_{q \to \infty} \frac{1}{q}\sum_{j=1}^{q} \mathbb{E}\left[\sup_{f \in H} \left|Q_{\aleph,k}(f) - \frac{1}{\lfloor n/k \rfloor} \sum_{S \in \mathbf{Y}_j} f(\Upsilon\langle S \rangle)\right|\right]
$$

$$
= \mathbb{E}\left[\sup_{f \in H} \left|Q_{\aleph,k}(f) - \frac{1}{\lfloor n/k \rfloor} \sum_{S \in \mathbf{Y}_1} f(\Upsilon\langle S \rangle)\right|\right] \tag{2}
$$

Note that the last equality is a consequence of Lemma 1, from which it among others follows that all $\mathbf{Y}_j$'s have the same distribution. In other words, all the q expected values are equal. Finally, we can use Lemma 3 to bound (2) which finishes the proof.   □

It is also possible to get rid of the logarithmic factor in the bound on expected error. However, as mentioned in [2], such bounds are worse up to very large training set sizes due to the increased constant factors.

6.3   From Tail Bounds to Moment-Generating Functions

In this section, we derive bounds on the moment-generating function of a random variable from its tail bounds.

Lemma 4. *For a non-negative random variable X, if there exist constants $C \geq e$ and $B > 0$ such that*

$$
P[X \geq t] \leq C \exp(-t^2/B) \qquad \forall t \geq 0,
$$

then for any $\lambda > 0$

$$\mathbb{E}\left[\exp\left(\lambda X\right)\right] \leq 1 + \lambda C \sqrt{\pi B} \exp\left(\frac{\lambda^2 B}{4}\right)$$

Proof. We have:

$$\mathbb{E}\left[X^p\right] = \int_0^\infty \mathrm{P}\left(X^p \geq u\right) du = \int_0^\infty \mathrm{P}\left(X^p \geq t^p\right) \cdot p \cdot t^{p-1} dt$$

$$= \int_0^\infty \mathrm{P}\left(X \geq t\right) \cdot p \cdot t^{p-1} dt \leq \int_0^\infty C \cdot e^{-t^2/B} \cdot p \cdot t^{p-1} dt$$

Next, for the moment-generating function, we have

$$\mathbb{E}\left[\exp\left(\lambda X\right)\right] \leq 1 + \sum_{p=1}^\infty \frac{\lambda^p \mathbb{E}\left[X^p\right]}{p!} \leq 1 + \sum_{p=1}^\infty \frac{\lambda^p \int_0^\infty C \cdot e^{-t^2/B} \cdot p \cdot t^{p-1} dt}{p!}$$

$$\leq 1 + C \int_0^\infty e^{-t^2/B} \cdot \sum_{p=1}^\infty \frac{\lambda^p p \cdot t^{p-1}}{p!} dt$$

$$= 1 + C\lambda \int_0^\infty e^{-t^2/B} \cdot \sum_{p=0}^\infty \frac{\lambda^p \cdot t^p}{p!} dt$$

$$= 1 + C\lambda \int_0^\infty e^{-t^2/B} \cdot e^{t\lambda} dt = 1 + C\lambda \int_0^\infty e^{-\frac{\left(t - \frac{1}{2}\lambda B\right)^2}{B} + \frac{\lambda^2 B^2}{4}} dt$$

$$= 1 + C\lambda e^{\frac{\lambda^2 B^2}{4}} \int_0^\infty e^{-\frac{\left(t - \frac{1}{2}\lambda B\right)^2}{B}} dt$$

$$= 1 + \frac{1}{2} C\lambda \sqrt{\pi B} e^{\frac{\lambda^2 B^2}{4}} \left(\mathrm{erf}\left(\frac{\lambda\sqrt{B}}{2}\right) + 1\right)$$

$$\leq 1 + C\lambda \sqrt{\pi B} \exp\left(\frac{\lambda^2 B}{4}\right)$$

Note that it is easy to check that all the series in the above derivation converge absolutely. The Fubini-Tonelli theorem justifies the change of order of summation and integration. □

6.4   From Moment-Generating Functions to Tail Bounds

We can now finish the proof of our main result, Theorem 2.

Proof (of Theorem 2). Let $\mathbf{Y}_j = (\mathcal{S}'_{j,1}, \ldots, \mathcal{S}'_{j,\lfloor \frac{n}{k} \rfloor})$, for $1 \leq j \leq q$, be random vectors sampled as in Lemma 1. For convenience, let us also define

$$R_\Upsilon^{(q)}(f) = \frac{1}{q} \sum_{j=1}^q \frac{1}{\lfloor n/k \rfloor} \sum_{\mathcal{S} \in \mathbf{Y}_j} f(\Upsilon\langle \mathcal{S}\rangle).$$

First, we have

$$P\left[\sup_{f\in\mathcal{H}}|Q_{\aleph,k}(f)-Q_{\Upsilon,k}(f)|\geq\varepsilon\right]$$

$$=P\left[\sup_{f\in\mathcal{H}}\left\{\left|Q_{\aleph,k}(f)-R_\Upsilon^{(q)}(f)+R_\Upsilon^{(q)}(f)-Q_{\Upsilon,k}(f)\right|\right\}\geq\varepsilon\right]$$

$$\leq P\left[\sup_{f\in\mathcal{H}}\left\{\left|Q_{\aleph,k}(f)-R_\Upsilon^{(q)}(f)\right|+\left|R_\Upsilon^{(q)}(f)-Q_{\Upsilon,k}(f)\right|\right\}\geq\varepsilon\right]$$

$$\leq P\left[\sup_{f\in\mathcal{H}}\left\{\left|Q_{\aleph,k}(f)-R_\Upsilon^{(q)}(f)\right|\right\}+\sup_{f\in\mathcal{H}}\left\{\left|R_\Upsilon^{(q)}(f)-Q_{\Upsilon,k}(f)\right|\right\}\geq\varepsilon\right]$$

It follows from the fact that the supremum needs to be taken only over the finitely many equivalence classed of $\mathcal{H}$ on $\aleph$ and from Lemma 1 (see the discussion in the proof of Lemma 5) that for any $\varepsilon^*>0$ and $\delta^*>0$ there is an integer q_0 such that for all $q\geq q_0$:

$$P\left[\sup_{f\in\mathcal{H}}\left\{\left|R_\Upsilon^{(q)}(f)-Q_{\Upsilon,k}(f)\right|\right\}\geq\varepsilon^*\right]\leq\delta^*.$$

Hence, for any $\varepsilon^*>0$, $\delta^*>0$ and a suitably large $q\geq q_0$ we have

$$P\left[\sup_{f\in\mathcal{H}}\left\{\left|Q_{\aleph,k}(f)-R_\Upsilon^{(q)}(f)\right|\right\}+\sup_{f\in\mathcal{H}}\left\{\left|R_\Upsilon^{(q)}(f)-Q_{\Upsilon,k}(f)\right|\right\}\geq\varepsilon\right]$$

$$\leq P\left[\sup_{f\in\mathcal{H}}\left\{\left|Q_{\aleph,k}(f)-R_\Upsilon^{(q)}(f)\right|\right\}\geq\varepsilon-\varepsilon^*\right]+\delta^*.$$

Taking the limit $q_0\to\infty$ we obtain

$$P\left[\sup_{f\in\mathcal{H}}|Q_{\aleph,k}(f)-Q_{\Upsilon,k}(f)|\geq\varepsilon\right]\leq\lim_{q\to\infty}P\left[\sup_{f\in\mathcal{H}}\left|Q_{\aleph,k}(f)-R_\Upsilon^{(q)}(f)\right|\geq\varepsilon\right]$$

Next we need to bound the right-hand side of the above inequality. For any q we have

$$P\left[\sup_{f\in\mathcal{H}}\left|Q_{\aleph,k}(f)-R_\Upsilon^{(q)}(f)\right|\geq\varepsilon\right]$$

$$=P\left[\sup_{f\in\mathcal{H}}\left|Q_{\aleph,k}(f)-\frac{1}{q}\sum_{j=1}^q\frac{1}{\lfloor n/k\rfloor}\sum_{S\in\mathbf{Y}_j}f(\Upsilon\langle S\rangle)\right|\geq\varepsilon\right]$$

$$=P\left[\sup_{f\in\mathcal{H}}\left|\frac{1}{q}\sum_{j=1}^q\left(Q_{\aleph,k}(f)-\frac{1}{\lfloor n/k\rfloor}\sum_{S\in\mathbf{Y}_j}f(\Upsilon\langle S\rangle)\right)\right|\geq\varepsilon\right]$$

$$\leq P\left[\sup_{f\in\mathcal{H}}\frac{1}{q}\sum_{j=1}^{q}\left|Q_{\aleph,k}(f)-\frac{1}{\lfloor n/k\rfloor}\sum_{\mathcal{S}\in\mathbf{Y}_j}f(\Upsilon\langle\mathcal{S}\rangle)\right|\geq\varepsilon\right]$$

$$\leq P\left[\frac{1}{q}\sum_{j=1}^{q}\sup_{f\in\mathcal{H}}\left|Q_{\aleph,k}(f)-\frac{1}{\lfloor n/k\rfloor}\sum_{\mathcal{S}\in\mathbf{Y}_j}f(\Upsilon\langle\mathcal{S}\rangle)\right|\geq\varepsilon\right]$$

Let us denote

$$T_j=\sup_{f\in\mathcal{H}}\left|Q_{\aleph,k}(f)-\frac{1}{\lfloor n/k\rfloor}\sum_{\mathcal{S}\in\mathbf{Y}_j}f(\Upsilon\langle\mathcal{S}\rangle)\right|.$$

Combining Lemmas 3 and 4, we can bound $\mathbb{E}\left[\exp\left(\lambda T_j\right)\right]$ as

$$\mathbb{E}\left[\exp\left(\lambda T_j\right)\right]\leq 1+4\lambda\sqrt{\frac{8\pi}{\lfloor n/k\rfloor}}\left(\frac{2e\lfloor n/k\rfloor}{d}\right)^d\exp\left(\frac{2\lambda^2}{\lfloor n/k\rfloor}\right).$$

Let us denote $T=\frac{1}{q}\sum_{j=1}^{q}T_j$. We use the observation from [5] that due to Jensen's inequality and linearity of expectation

$$\mathbb{E}\left[\exp\left(\lambda T\right)\right]\leq\frac{1}{q}\sum_{j=1}^{q}\mathbb{E}\left[\exp\left(\lambda T_j\right)\right]=\mathbb{E}\left[\exp\left(\lambda T_1\right)\right].$$

Next we obtain a bound on $P[T\geq\varepsilon]$ from the bound on $\mathbb{E}\left[\exp\left(\lambda T\right)\right]$. In particular, for positive λ, we have

$$P[T\geq\varepsilon]=P[e^{\lambda\cdot X}\geq e^{\lambda\cdot\varepsilon}]\leq e^{-\lambda\cdot\varepsilon}\mathbb{E}\left[e^{\lambda\cdot T}\right]$$

$$\leq e^{-\lambda\cdot\varepsilon}\left(1+4\lambda\sqrt{\frac{8\pi}{\lfloor n/k\rfloor}}\left(\frac{2e\lfloor n/k\rfloor}{d}\right)^d\exp\left(\frac{2\lambda^2}{\lfloor n/k\rfloor}\right)\right).$$

where the Markov inequality was used for the third step. Since the above bound holds for any q, it also holds in the limit. Next, we can plug in $\lambda:=\frac{\varepsilon\cdot\lfloor n/k\rfloor}{4}$ and obtain:

$$P[T\geq\varepsilon]\leq\exp\left(-\frac{\lfloor n/k\rfloor\varepsilon^2}{4}\right)+\varepsilon\sqrt{8\pi\lfloor n/k\rfloor}\left(\frac{2e\lfloor n/k\rfloor}{d}\right)^d\cdot\exp\left(-\frac{\lfloor n/k\rfloor\varepsilon^2}{8}\right).$$

$\square$

7   Concluding Remarks

We have derived VC-dimension based bounds which can be applied in relational learning settings where one may assume that the training data (i.e. some given relational structure) was obtained from a larger relational structure by sampling without replacement. This includes many of the typical application settings in which, for instance, Markov logic networks are used. The considered bounds are useful, among others, for the analysis of relational marginal problems [8] and PAC-reasoning in relational domains [7].

There are several interesting avenues for future work. First, in this paper, we have not studied the realizable learning case for which, at least in the classical i.i.d. case, one can obtain faster convergence rates. It would be interesting to extend our results into the realizable case. Similarly, it would be of interest to study bounds under low-noise conditions [1], which sit somewhere between the realizable case and the case studied in this paper. Another natural direction for future work would be to extend the PAC-Bayesian setting into relational learning, as the bounds that are derived in this setting tend to be tighter in practice [10].

Acknowledgements. OK's work was partially supported by the Research Foundation - Flanders (project G.0428.15). SS is supported by ERC Starting Grant 637277.

References

1. Clémençon, S., Lugosi, G., Vayatis, N.: Ranking and empirical minimization of u-statistics. Annal. Stat. **36**, 844–874 (2008)
2. Devroye, L., Györfi, L., Lugosi, G.: A Probabilistic Theory of Pattern Recognition. Stochastic Modelling and Applied Probability, vol. 31. Springer, New York (1996). https://doi.org/10.1007/978-1-4612-0711-5
3. Dhurandhar, A., Dobra, A.: Distribution-free bounds for relational classification. Knowl. Inf. Syst. **31**(1), 55–78 (2012)
4. Hoeffding, W.: A class of statistics with asymptotically normal distribution. Annal. Math. Stat. **19**, 293–325 (1948)
5. Hoeffding, W.: Probability inequalities for sums of bounded random variables. J. Am. Stat. Assoc. **58**(301), 13–30 (1963)
6. Juba, B.: Implicit learning of common sense for reasoning. In: Proceedings of the 23rd International Joint Conference on Artificial Intelligence, pp. 939–946 (2013)
7. Kuželka, O., Wang, Y., Davis, J., Schockaert, S.: PAC-reasoning in relational domains. In: Proceedings of the 34th Conference on Uncertainty in Artificial Intelligence, UAI 2018 (2018)
8. Kuželka, O., Wang, Y., Davis, J., Schockaert, S.: Relational marginal problems: theory and estimation. In: Proceedings of the Thirty-Second AAAI Conference on Artificial Intelligence (AAAI-18) (2018)
9. Kuželka, O., Davis, J., Schockaert, S.: Induction of interpretable possibilistic logic theories from relational data. In: Proceedings of the 26th International Joint Conference on Artificial Intelligence, pp. 1153–1159 (2017)
10. Langford, J., Shawe-Taylor, J.: PAC-Bayes & margins. In: Proceedings of the Annual Conference on Neural Information Processing Systems, pp. 423–430 (2002)
11. Nandi, H., Sen, P.: On the properties of u-statistics when the observations are not independent: Part two unbiased estimation of the parameters of a finite population. Calcutta Stat. Assoc. Bull. **12**(4), 124–148 (1963)
12. Rocktäschel, T., Riedel, S.: End-to-end differentiable proving. In: Proceedings of the Annual Conference on Neural Information Processing Systems, pp. 3791–3803 (2017)

13. Shalev-Shwartz, S., Ben-David, S.: Understanding Machine Learning: From Theory to Algorithms. Cambridge University Press, New York (2014)
14. Valiant, L.G.: Knowledge infusion. In: Proceedings of the 21st National Conference on Artificial Intelligence, pp. 1546–1551 (2006)
15. Vapnik, V.: The Nature of Statistical Learning Theory. Springer, New York (2000). https://doi.org/10.1007/978-1-4757-3264-1
16. Vapnik, V., Chervonenkis, A.Y.: On the uniform convergence of relative frequencies of events to their probabilities. Theory Probab. Appl. **16**(2), 264 (1971)
17. Vishwanathan, S.V.N., Schraudolph, N.N., Kondor, R., Borgwardt, K.M.: Graph kernels. J. Mach. Learn. Res. **11**, 1201–1242 (2010)
18. Šourek, G., Aschenbrenner, V., Železný, F., Kuželka, O.: Lifted relational neural networks. In: Proceedings of the NIPS Workshop on Cognitive Computation: Integrating Neural and Symbolic Approaches (2015)
19. Xiang, R., Neville, J.: Relational learning with one network: an asymptotic analysis. In: Proceedings of the Fourteenth International Conference on Artificial Intelligence and Statistics, pp. 779–788 (2011)

Robust Super-Level Set Estimation Using Gaussian Processes

Andrea Zanette, Junzi Zhang, and Mykel J. Kochenderfer[⊠]

Stanford University, Stanford, CA, USA
{zanette,junziz,mykel}@stanford.edu

Abstract. This paper focuses on the problem of determining as large a region as possible where a function exceeds a given threshold with high probability. We assume that we only have access to a noise-corrupted version of the function and that function evaluations are costly. To select the next query point, we propose maximizing the expected volume of the domain identified as above the threshold as predicted by a Gaussian process, robustified by a variance term. We also give asymptotic guarantees on the exploration effect of the algorithm, regardless of the prior misspecification. We show by various numerical examples that our approach also outperforms existing techniques in the literature in practice.

Keywords: Active learning · Gaussian processes · Level set estimation

1 Introduction

Many scientific and engineering problems involve determining the maximum value a function f over a region Ω. However, some applications require determining a large subregion of Ω where the function under consideration exceeds a given threshold t. This problem of super-level set estimation arises naturally in the context of safety control, signal coverage, and environmental monitoring [1].

Formally, we consider the problem of finding the region where a function is above some threshold t with probability at least δ:

$$\{\mathbf{x} \in \Omega \mid P\left(f(\mathbf{x}) > t\right) > \delta\}. \tag{1}$$

We assume that we only have access to a noise-corrupted version of the function and that function evaluations are costly. In order to fully specify the set in Eq. (1), a probabilistic model P for the function must be assumed. In this work, we use Gaussian processes [2], a standard model that directly provides confidence intervals and can easily incorporate new information from the samples.

Electronic supplementary material The online version of this chapter (https://doi.org/10.1007/978-3-030-10928-8_17) contains supplementary material, which is available to authorized users.

© Springer Nature Switzerland AG 2019
M. Berlingerio et al. (Eds.): ECML PKDD 2018, LNAI 11052, pp. 276–291, 2019.
https://doi.org/10.1007/978-3-030-10928-8_17

The problem is closely related to Bayesian optimization, but the associated techniques are not directly transferable to the problem of level set estimation. A major issue is that it is unclear whether one should focus on identifying points around the threshold for better separation, or aim at points far from the threshold to accelerate the discovery of interested regions. Similar to other exploration-exploitation trade-offs in active learning, we give affirmative answers to both by proposing Robust Maximum Improvement for Level-set Estimation (RMILE), an algorithm to maximize the expected volume of the domain where a function exceeds the threshold with high probability, robustified by a exploration-driven variance term.

Furthermore, we discuss a criterion for establishing convergence of a generic acquisition function on finite grids that is robust to misspecification of the models. In particular, we show how this criterion applies to our algorithm.

Related Work. Some relevant techniques for addressing costly function evaluations are found in the field of Bayesian optimization, which have received growing attention in recent years [3]. Bayesian optimization aims at finding the maximum or minimum of a black box function by repeatedly updating prior beliefs over the function in a Bayesian fashion. This framework is particularly well suited for global optimization of costly functions because it makes effective use of all the information (*i.e.*, the samples) acquired during the process by carefully selecting query points according to an acquisition function. Examples of acquisition functions include the probability of improvement [4], expected improvement [5], the Gaussian process upper confidence bound [6], and information-based policies [7].

In the literature of level set estimation, when level sets are estimated from an existing dataset, several approaches are available [8–10]. In the context of active learning, a topic of growing interest [11–13], one technique is known as the Straddle heuristic [14], where the expected value and variance given by a Gaussian process are combined to characterize the uncertainty at each candidate point, based on which the next query point is chosen. A refinement of the Straddle heuristic was suggested as the LSE algorithm, which is an online classification method based on confidence intervals with some information-theoretic convergence guarantees [1]. This idea is further developed in [15], with theoretical guarantees that offer a unifying framework of Bayesian optimization and level set estimation. In a different direction, a Gaussian process-based algorithm addressing time-varying level set estimation was proposed [16], where a global expected error estimate is adopted as the acquisition function.

A problem similar to ours was considered by [17], where the authors partition the space using a predefined coarse grid and the goal is to find sub-regions with an average "score" above some threshold. The computation of the scores relies on Bayesian quadrature, and no theoretical guarantee on the algorithm is provided. It was later extended to find regions matching general patterns not restricted to the excess of some threshold [18]. Although some preliminary analysis about exploration-exploitation trade-off is given, there is no discussion about the limit behavior of the algorithm.

In some applications, one is directly concerned with the online estimation of the volume of the super-level set with some given threshold [19,20]. However, existing methods are still focused on uncertainty reduction mechanisms similar to the Straddle heuristic.

This paper is structured as follows. Section 2 provides the necessary preliminaries for our work. Section 3 discusses the relation between super-level set estimation and binary classification, and proposes the RMILE algorithm. The asymptotic behavior of RMILE is discussed in Sect. 4, followed by various numerical experiments in Sect. 5 and some conclusive remarks in Sect. 6.

2   Preliminaries

We assume that we only have noisy measurements of the function $f(\mathbf{x})$. In other words, we have access to $f(\mathbf{x}) + \epsilon$ where the noise ϵ is normally distributed $\epsilon \sim \mathcal{N}(0, \sigma_\epsilon^2)$ and is independent from the sampling location or the function value. We consider a discrete domain Ω with finitely many points where we would like to classify the function as either above the threshold or below it, with some degree of confidence. We assume that we are allowed to query the function at very few points with noisy evaluations, in which case the unseen regions of the domain can only be classified with some probability, for example, $f(\mathbf{x}) > t$ with probability at least δ. In the design of our algorithm, we assume that the function $f(\mathbf{x})$ is a sample from a Gaussian process (GP). However, our asymptotic analysis in Sect. 4 is model independent, i.e., it holds without any additional probabilistic assumptions on the function measurements.

2.1   Gaussian Processes

A GP $\{f(\mathbf{x}) \mid \mathbf{x} \in \Omega\}$ is a collection of random variables, any finite subset of which is distributed according to a multivariate Gaussian specified by the mean function $\mu(\mathbf{x})$ and the kernel $k(\mathbf{x}, \mathbf{x}')$. Suppose that we have a prior mean $\mu_0(\mathbf{x})$ and kernel $k_0(\mathbf{x}, \mathbf{x}')$ for the GP, and n (noisy) measurements $\{(\mathbf{x}_i, y_i)\}_{i=1}^n$, where $y_i = f(\mathbf{x}_i) + \epsilon_i$ and $\epsilon_i \sim N(0, \sigma_\epsilon^2)$ for $i = 1, \ldots, n$. The posterior of $\{f(\mathbf{x}) \mid \mathbf{x} \in \Omega\}$ is still a GP, and its mean and kernel functions can be computed analytically as follows:

$$
\begin{aligned}
\mu_n(\mathbf{x}) &= \mu_{\mathbf{x}_{1:n}, y_{1:n}}(\mathbf{x}) = \mu_0(\mathbf{x}) + k_n(\mathbf{x})^T (K_n + \sigma^2 I)^{-1} (y_{1:n} - \mu_0(\mathbf{x}_{1:n})), \\
k_n(\mathbf{x}, \mathbf{x}') &= k_{\mathbf{x}_{1:n}}(\mathbf{x}, \mathbf{x}') = k_0(\mathbf{x}, \mathbf{x}') - k_n(\mathbf{x})^T (K_n + \sigma^2 I)^{-1} k_n(\mathbf{x}'),
\end{aligned}
\tag{2}
$$

where $k_n(\mathbf{x}) = [k_0(\mathbf{x}, \mathbf{x}_1), \ldots, k_0(\mathbf{x}, \mathbf{x}_n)]^T$, $K_n = [k_0(\mathbf{x}_i, \mathbf{x}_j)]_{i,j=1}^n$. In particular, the posterior variance at $\mathbf{x}$ is $\sigma_n^2(\mathbf{x}) = k_n(\mathbf{x}, \mathbf{x})$. Intuitively, as the number of measurements increases, the actual f will be gradually revealed.

2.2   Framework for Super-Level Set Estimation

Algorithm 1 is the conceptual framework of the algorithms for level set estimation, which is adopted in most, if not all of the related literature. Here the last step in Algorithm 1 follows Eq. (1), in which P_{GP} is the probability measure defined according to the posterior Gaussian process (i.e., conditioned on the filtration $\{(\mathbf{x}_i, y_i)\}_{i=1,...,n}$). The estimated super-level set is denoted I_{GP}. We remark that one can also decide the membership of the estimated level set in an online fashion, as is done in [1,15].

Algorithm 1. Framework for Estimating the Super-level Set

Input: prior mean μ_0, kernel k_0, objective function f, threshold t, tolerance δ.
  for $i = 1, 2, \ldots$ **do**
    $\mathbf{x}^+ \leftarrow$ SELECTPOINT(GP)
    Query the objective function at $\mathbf{x}^+$ to obtain y^+
    Update $GP^+ \leftarrow GP$ using $(\mathbf{x}^+, y^+)$
  Output: the estimated super-level set $I_{GP} := \{\mathbf{x} \in \Omega \mid P_{GP}(f(\mathbf{x}) > t) > \delta\}$.

2.3   Notation and Assumptions

At timestep n, we have observed $\{(\mathbf{x}_i, y_i)\}_{i=1,...,n}$, where $\mathbf{x}_i$ is the ith sampling location and y_i is the resulting noisy observation. We denote with the subscript GP (e.g., μ_{GP}, σ_{GP} and k_{GP}) the quantities conditioned on the filtration $\{(\mathbf{x}_i, y_i)\}_{i=1,...,n}$. We use the subscript GP^+ to denote such quantities still conditioned on the filtration $\{(\mathbf{x}_i, y_i)\}_{i=1,...,n}$ and additionally on the sampling location denoted $\mathbf{x}^+$. Notice that, while $\mu_{GP}(\mathbf{x})$ is a deterministic quantity that can be computed with the predictive Eq. (2), $\mu_{GP^+}(\mathbf{x})$ depends on the random outcome y^+ at $\mathbf{x}^+$ and is therefore a random variable.

Unless otherwise specified, we always restrict ourselves to a finite fixed grid $\mathbf{z}_1, \ldots, \mathbf{z}_m \in \Omega$ as the set of all candidate sampling locations in Ω. Here $\mathbf{z}_{1:m}$ are all distinct. We will then slightly abuse notation to use Ω to denote the set of grid points $\mathbf{z}_{1:m}$, with $|\Omega| = m$. We also assume without loss of generality that the prior kernel k_0 is positive definite.

3   Super-Level Set Estimation

This section begins with some remarks on the relation between super-level set estimation and binary classification, and then describes our RMILE algorithm.

3.1   Relation to Binary Classification

A GP uniquely specifies a probability distribution for the unseen examples $\mathbf{x} \in \Omega$ and can be used to infer the region where the threshold is exceeded with probability at least δ. At any point $\mathbf{x}$, the posterior distribution of the random variable $f(\mathbf{x})$ is still normal [2]. Let $\mu_{GP}(\mathbf{x})$ and $\sigma_{GP}^2(\mathbf{x})$ denote its posterior mean and variance, respectively. Then the condition $P(f(\mathbf{x}) > t) > \delta$ in Eq. (1) can be reformulated as follows:

$$\mu_{GP}(\mathbf{x}) - \beta\sigma_{GP}(\mathbf{x}) > t, \tag{3}$$

where β is a fixed coefficient that depends on δ. For a normal distribution, if $\delta = 97.5\%$ then $\beta \approx 1.96$. The user is free to select δ according to the application. If human safety is at stake, δ should be sufficiently high to avoid misclassification, although this will also result in a smaller I_{GP}.

Define as FP the set of *false positives*, that is, the points $\mathbf{x}$ such that $f(\mathbf{x}) \leq t$ but are classified as $\in I_{GP}$, and FN the set of *false negatives*, which are all the points $\mathbf{x}$ such that $f(\mathbf{x}) > t$ but are classified as not in I_{GP}. The following is a straightforward observation:

Lemma 1. *The classification rule identified by Eq. (3) minimizes the expected weighted misclassification error:*

$$\mathbb{E}\left(\delta\mathbb{1}\{\mathbf{x} \in FP\} + (1 - \delta)\mathbb{1}\{\mathbf{x} \in FN\}\right), \tag{4}$$

among all deterministic classification rules under the posterior probability measure given by the Gaussian processes. Here, $\mathbf{x} \in \Omega$ is arbitrary and fixed.

In the above expression, the expectation is conditioned on the filtration, *i.e.*, on the observed samples $\{(\mathbf{x}_i, y_i)\}_{i=1,\ldots,n}$. We can see that rule (3) penalize false positives much more than false negatives when δ is close to 1. As a result, they are relatively conservative in the inclusion of points in the estimated supe-level set, and thus balance with our "radical" acquisition function to be introduced below.

3.2   Robust Maximum Improvement in Level-Set Estimation

Our idea is to develop a method that aims to find the largest possible area where the function exceeds a given threshold with high probability.

Let I_{GP} be the set of points currently classified as above the threshold using the posterior GP. If we could sample at an arbitrary $\mathbf{x}^+$ and incorporate the feedback $y(\mathbf{x}^+) = f(\mathbf{x}^+) + \epsilon$ into the GP, then we would obtain a new Gaussian process, GP^+, which is a function of the (random) outcome $y(\mathbf{x}^+)$ and the sampling location $\mathbf{x}^+$. Then GP^+ can be used to infer an updated classification I_{GP^+}. We thus consider maximizing the volume of I_{GP^+}, *i.e.*, $|I_{GP^+}| := \sum_{\mathbf{x} \in \Omega} \mathbb{1}\{P_{GP^+}(f(\mathbf{x}) > t) > \delta\}$ to find the "optimal" sampling location. Equivalently, we would like to find the point $\mathbf{x}^+$ that would yield the maximum improvement $|I_{GP^+}| - |I_{GP}|$ in expectation among all candidate sampling locations. Formally, the next sampling point is chosen as the solution to:

$$\arg\max_{\mathbf{x}^+ \in \Omega} \mathbb{E}_{y^+} |I_{GP^+}| - |I_{GP}|, \tag{5}$$

where the expectation is taken with respect to the random outcome y^+ (which is shorthand for $y(\mathbf{x}^+)$) resulting from sampling at $\mathbf{x}^+$, and is conditioned on the filtration $\{(\mathbf{x}_i, y_i)\}_{i=1,\ldots,n}$. This criterion is similar to the expected improvement developed in the context of Bayesian optimization [5], and is also closely related to the criteria of [17,18]. However, their framework differ from ours, and they all suffer from potential lack of exploration. In particular, [17,18] focus on region-level detection instead of point-wise detection, as we focus on here. Although Eq. (5) does fall as a special case of the acquisition function proposed in [18] when a single point is chosen as the "linear functional" there, their explanation for exploration inside each region becomes meaningless as each region then contains only a single point. As a result, no convergence guarantees have been established for these algorithms, despite their empirical success in certain problems. In particular, it remains an open issue how an intrinsic exploration strategy can be included. To this end, we modify criterion (5) by introducing a trade-off between the posterior variances, which we prove to ensure certain asymptotic convergence, as we discuss below.

While Eq. (5) defines a reasonable acquisition function that seeks improvement in the discovery of points lying in the super-level set with high probability, it may suffer from potential model misspecification and lack of exploration. Consider an extreme case when all the points are classified as above the threshold by the chosen prior at the beginning. Then Eq. (5) may lead the procedure to stall and repeatedly sample at locations with the largest function values to maintain the largest super-level set specified by the prior. To remedy this issue, we modify the criterion in Eq. (5) so that the algorithm cannot "get stuck" indefinitely. To achieve this goal, we guarantee a minimum positive exploration bonus everywhere by introducing a marginal variance term. Let $|I_{GP}^\epsilon| = \sum_{\mathbf{x} \in \Omega} \mathbb{1}\{P_{GP}(f(\mathbf{x}) > t - \epsilon) > \delta\}$ for $\epsilon > 0$, which is essentially a shift of the threshold from t to $t - \epsilon$ (the shift is mostly for technical reasons to simplify the analysis). Our final *acquisition function* is then defined as:

$$E_{GP}(\mathbf{x}^+) = \max\left\{\mathbb{E}_{y^+} |I_{GP^+}| - |I_{GP}^\epsilon|, \gamma \sigma_{GP}(\mathbf{x}^+)\right\}, \qquad (6)$$

for some small constants $\epsilon > 0, \gamma > 0$. Intuitively, the additional variance term ensures that the algorithm moves to a region with a higher variance when the expected improvement is sufficiently reduced at the current point.

3.3 Efficient Implementation

At each candidate sample point $\mathbf{x}^+ \in \Omega$, to evaluate Eq. (6), we would need to sample $f(\mathbf{x}^+)$ from the current GP, from which we can compute the posterior classification and repeat this procedure in a Monte Carlo fashion to estimate the expected improvement.

Nevertheless, it is possible to avoid sampling from the GP here. To see this, notice that by Eq. (2), the variance $\sigma_{GP^+}^2(\mathbf{x})$ is unaffected by a new observation y^+ as it only depends on the sampling location $\mathbf{x}^+$. On the other hand, the posterior mean $\mu_{GP^+}(\mathbf{x})$ is linearly correlated with the sample y^+ (to indicate this dependency we would rewrite it as $\mu_{GP+}(\mathbf{x}; y^+))$. Therefore, it is possible to compute the outcome y^+ that would change the classification for point $\mathbf{x}$ under consideration, that is, we only need to determine the "limit" value for the new sample y^+ that turns the indicator $\mathbb{1}\{P_{GP+}(f(\mathbf{x}) > t) > \delta\}$ on or off in the computation of $\mathbb{E}_{y^+}|I_{GP+}|$. As a result, we obtain the following expression:

Lemma 2. $\mathbb{E}_{y^+}|I_{GP+}|$ *obtained by sampling at* $\mathbf{x}^+$ *can be computed analytically as follows:*

$$\sum_{\mathbf{x} \in \Omega} \Phi\left(\frac{\sqrt{\sigma_{GP}^2(\mathbf{x}^+) + \sigma_\epsilon^2}}{|\mathrm{Cov}_{GP}(f(\mathbf{x}), f(\mathbf{x}^+))|} \times (\mu_{GP}(\mathbf{x}) - \beta\sigma_{GP+}(\mathbf{x}) - t)\right) \qquad (7)$$

where $\Phi(\cdot)$ *is the cumulative distribution function (CDF) of the standard normal random variables, and*

$$\sigma_{GP+}^2(\mathbf{x}) = \mathrm{Cov}_{GP+}(f(\mathbf{x}), f(\mathbf{x})) = \sigma_{GP}^2(\mathbf{x}) - \frac{\mathrm{Cov}_{GP}^2(f(\mathbf{x}), f(\mathbf{x}^+))}{\sigma_{GP}^2(\mathbf{x}^+) + \sigma_\epsilon^2}, \qquad (8)$$

and $\mathrm{Cov}_{GP}(f(\mathbf{x}), f(\mathbf{x}^+)) = k_{GP}(\mathbf{x}, \mathbf{x}^+)$ *is the (current) posterior covariance between* $f(\mathbf{x})$ *and* $f(\mathbf{x}^+)$.

In the above derivation, we are implicitly assuming that $f(\mathbf{x})$ can be modeled as a sample from the GP. The posterior covariance $\mathrm{Cov}_{GP}(f(\mathbf{x}), f(\mathbf{x}^+))$ can be calculated, for example, by using Eq. (2). For a fixed grid, such computation can be rearranged so that the full posterior covariance matrix is stored and updated at each iteration through rank-one updates. Different trade-offs between computational and memory complexity are also possible. Lemma 2 is also shown in [17,18], but to reduce notational confusion in cross-referencing, we provide a different and more direct proof in the appendix (see supplementary material).

The RMILE algorithm follows the super-level set estimation framework described in Algorithm 1. At each iteration, it calls SELECTPOINT (Algorithm 2). For a fixed sampling location $\mathbf{x}^+$, the algorithm computes the acquisition function (6) using Eq. (7).

Although it is possible to identify the level set at any time during the execution, we do not enforce an online classification scheme as in [1,15]. Instead, the classification is done offline using all available information, which makes the algorithm work better in practice, as can be seen in the numerical experiments.

Algorithm 2. RMILE (*SelectPoint(GP)*)

Input: Current GP, constants $\gamma > 0$, $\epsilon > 0$.
for $\mathbf{x}^+ \in \Omega$ **do**
  Compute $E_{GP}(\mathbf{x}^+) = \max\{\mathbb{E}_{y^+}|I_{GP+}| - |I_{GP}^\epsilon|, \gamma\sigma_{GP}(\mathbf{x}^+)\}$
  Output: $\mathbf{x}^* = \arg\max_{\mathbf{x}^+ \in \Omega} E_{GP}(\mathbf{x}^+)$.

3.4 Connection to Uncertainty/Variance Reduction

So far we have assumed that the threshold is known a priori. In fact, the design process in engineering typically involves several "iterations" where the conceptual idea is revised, leading to changes in the requirements or the threshold. In such cases, one would like to obtain a model of the function that can later be used to identify different regions, say $I_{GP}^{(t_1)}, I_{GP}^{(t_2)}, I_{GP}^{(t_3)}$ corresponding to different thresholds t_1, t_2, t_3 (this notation should not be confused with I_{GP}^ϵ previously used to indicate a shift in the threshold). For three thresholds, this can be easily done by redefining the objective function to maximize:

$$\mathbb{E}_{y^+}\left(|I_{GP+}^{(t_1)}| + |I_{GP+}^{(t_2)}| + |I_{GP+}^{(t_3)}|\right) - \left(|I_{GP}^{(t_1-\epsilon)}| + |I_{GP}^{(t_2-\epsilon)}| + |I_{GP}^{(t_3-\epsilon)}|\right), \qquad (9)$$

so that the algorithm is biased towards identifying all three of them. If, for example, $f(\mathbf{x}) > \max(t_1, t_2, t_3)$, then that point will contribute three times as much to the objective function.

We can naturally extend the idea of Eq. (9) and look for all thresholds in a given range $[a, b]$. That is, the finite summation in Eq. (9) can be replaced by an integral over all possible thresholds: $\mathbb{E}_{y^+} \int_{-\infty}^{\infty} |I_{GP+}^{(t)}| - |I_{GP}^{(t-\epsilon)}| dt$. Interestingly, if one considers the extreme case when $\epsilon = 0$, then our algorithm reduces to a type of variance minimization:

Lemma 3. *If the acquisition function is redefined as:*

$$E_{GP}^{var}(\mathbf{x}^+) := \mathbb{E}_{y^+} \int_{-\infty}^{\infty} |I_{GP+}^{(t)}| - |I_{GP}^{(t)}| dt, \qquad (10)$$

then Algorithm 2 minimizes the l_1-norm of the posterior standard deviation, i.e., the next query point $\mathbf{x}^+$ is selected as $\mathbf{x}^+ = \arg\min_{\mathbf{x}^+ \in \Omega} \sum_{\mathbf{x} \in \Omega} \sigma_{GP+}(\mathbf{x})$.

Since we can recast the objective function as $\sum_{\mathbf{x} \in \Omega} \sigma_{GP+}(\mathbf{x}) - \sigma_{GP}(\mathbf{x})$, the acquisition function (10) chooses the point that maximizes the reduction in the standard deviation across the domain. This is similar to the acquisition function used in [15] for an appropriate choice of the parameters.

4 Asymptotic Behavior on Finite Grids

In the absence of noise, a well-designed algorithm should avoid re-sampling at the same location since additional information is not acquired. In other words, on finite grids every point should be sampled at most once before an algorithm terminates.

In the case of noisy measurements, however, an algorithm may need to re-sample at the same location multiple times in order to get a more accurate estimate of the function value. Intuitively, so long as an algorithm samples each point of the grid infinitely often, the underlying function should be gradually revealed [2]. Below, we first validate some reasonable assumptions that guarantee the asymptotic convergence of a generic acquisition function in Algorithm 1, and then show that RMILE satisfies these conditions.

Lemma 4. *Let $E_{GP}(\mathbf{x}^+)$ be an acquisition function that depends on the posterior GP at a potential query point $\mathbf{x}^+$, such that for sufficiently small $\sigma_{GP}^2(\mathbf{x}^+)$, there exists a function $u(\cdot)$ which only depends on the posterior variance $\sigma_{GP}^2(\mathbf{x}^+)$, with*

$$E_{GP}(\mathbf{x}^+) \leq u(\sigma_{GP}(\mathbf{x}^+)), \qquad \lim_{\sigma(\mathbf{x}^+)\to 0^+} u(\sigma_{GP}(\mathbf{x}^+)) = 0. \qquad (11)$$

In addition, assume that there exists a global lower bound $l(\cdot)$, such that

$$E_{GP}(\mathbf{x}^+) \geq l(\sigma_{GP}(\mathbf{x}^+)), \qquad \lim_{\sigma(\mathbf{x}^+)\to 0^+} l(\sigma_{GP}(\mathbf{x}^+)) = 0, \qquad (12)$$

and assume that $l(\sigma_{GP}(\mathbf{x}^+))$ is strictly increasing in $\sigma_{GP}(\mathbf{x}^+)$.

If Algorithm 1 selects the next query point as $\arg\max_{\mathbf{x}^+} E_{GP}(\mathbf{x}^+)$ and is run without termination, then there cannot be a point in the grid that is sampled only finitely many times.

The lemma does not assume that the true function can be represented as a sample from a Gaussian process; only the upper and lower bounds on the acquisition function are needed as a function of the posterior marginal variance. The intuition is that $E_{GP}(\mathbf{x}^+)$ can fluctuate as the sampling process progresses; however, as the variance $\sigma_{GP}(\mathbf{x}^+)$ of a point is progressively reduced, one more sample at $\mathbf{x}^+$ should bring less and less improvement as measured by $E_{GP}(\mathbf{x}^+)$. This implies that the algorithm will move to a location where $E_{GP}(\cdot)$ is higher. The proof can be found in the appendix (see supplementary material).

We are now ready to verify the robustness of our algorithm: as we show next, it satisfies the assumption of Lemma 4. Let $\bar{\sigma}^2 := \max_{i=1,\dots,m} \sigma_0^2(\mathbf{z}_i)$, where $\mathbf{z}_i$ are the grid points in Ω.

Lemma 5. *For the acquisition function (6) with $\gamma > 0$, $\epsilon > 0$, we have:*

- $u(\sigma_{GP}(\mathbf{x}^+)) = \max\left(|\Omega|\Phi\left(\frac{\sigma_\epsilon}{\bar{\sigma}\sigma_{GP}(\mathbf{x}^+)}(-\epsilon/2)\right), \gamma\sigma_{GP}(\mathbf{x}^+)\right)$
- $l(\sigma_{GP}(\mathbf{x}^+)) = \gamma\sigma_{GP}(\mathbf{x}^+)$

Also the lower bound $l(\sigma_{GP}(\mathbf{x}^+))$ is monotonically increasing and

$$\lim_{\sigma(\mathbf{x}^+)\to 0^+} u(\sigma(\mathbf{x}^+)) = \lim_{\sigma(\mathbf{x}^+)\to 0^+} l(\sigma(\mathbf{x}^+)) = 0.$$

The roles of ϵ and γ are important in terms of the asymptotic behavior. More precisely, the modification that leads to Eq. (6) ensures a minimum exploration bonus given by $\gamma\sigma_{GP}(\mathbf{x}^+)$ and is a crucial difference compared to [17, 18].

5 Numerical Experiments

This section empirically assesses the proposed procedure on numerical experiments. We use a standard squared exponential kernel

$$k(\mathbf{x}, \mathbf{x}') = \sigma_{ker}^2 \exp(-\|\mathbf{x} - \mathbf{x}'\|_2^2/(2l^2)) \qquad (13)$$

We start by examining the effectiveness of the robust modifications, and then proceed to comparing our proposed approach with state-of-the art algorithms. Although in principle the model noise level σ_ϵ can be different from the algorithm noise level (which we also denote as σ_ϵ with slight abuse of notation), we typically take them to be the same as is done conventionally in the literature, unless otherwise stated (*e.g.*, in the next subsection). We also emphasize that to make the comparisons fair, the performance of all the algorithms is evaluated with classification criterion (3), instead of the criteria proposed in the original papers (*e.g.*, posterior mean).

5.1   Robustification Effects

This section shows how the robust adjustment parameters ϵ and γ in Eq. (6) help improve the performance of the algorithm. We compare two sets of parameters: (a) $\epsilon = \gamma = 10^{-8}$; (b) $\epsilon = 0$, $\gamma = -\infty$. Notice that with parameter set (b), Eq. (6) reduces to Eq. (5), *i.e.*, the one without guaranteed convergence.

To stabilize the performance, we first sample at 3 points chosen uniformly at random as seeds, and compute the resulting posterior distribution as the prior. To showcase the robustness of our algorithm, we keep the prior mean to be 0, and $\ln(\sigma_{ker}) = 4$, $\ln(l) = 1$ throughout the experiments in this subsection. For each problem, we run 25 simulations of our algorithm.

We consider the negative Himmelblau's function (Fig. 2) defined in $[-5, 5] \times [-5, 5]$, a commonly used multi-modal function. We take a uniform grid of 30×30 points, and the threshold is set to $t = -50$. Here we consider two sets of noise levels: (1) a small noise setting with both model and algorithm noise levels $\sigma_\epsilon = 0.1$; (2) a misspecified large noise setting, with model noise level 30 and algorithm noise level 3. The results are shown in Fig. 1. Here we label parameter set (a) as "RMILE" and parameter set (b) as "MILE". We can see that in both cases, the robust version outperforms the vanilla one, and the difference is more dramatic in the second (harder) case.

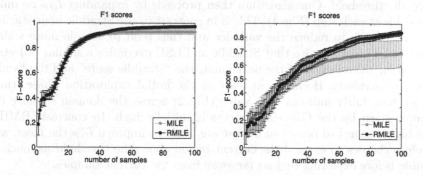

Fig. 1. Himmelblau's function. Left: small noise. Right: large misspecified noise.

Our algorithm is quite robust to the parameter choices of ϵ and γ, so long as they are positive. In particular, we obtained almost the same performance as above when setting $\epsilon = \gamma = 10^{-2}$.

For simplicity, hereafter we set $\epsilon = 10^{-12}$ and $\gamma = 10^{-10}$. We compare our approach against the Straddle heuristic [14] and the LSE algorithm [1], which are the most relevant algorithms to our work. Another relevant approach is the TRUVAR algorithm, which has been found to perform similarly to LSE in numerical experiments for level-set estimation [15].

5.2   2D Synthetic Examples

We consider a sinusoidal function $\sin(10x_1) + \cos(4x_2) - \cos(3x_1x_2)$, whose contours are plotted in Fig. 4 defined in the box $[0,1] \times [0,2]$. We superimpose a grid of 30×60 points uniformly separated and run 25 simulations for our algorithm RMILE along with LSE and the Straddle heuristic. The normally distributed noise has standard deviation $\ln(\sigma_\epsilon) = -1$, and the prior has uniform mean $= 0$ with $\ln(\sigma_{ker}) = 1.0$ and $\ln(l) = -1.5$. The threshold is set to $t = 1$. These are supposed to be representative of the prior knowledge that the user may have about the function at hand, but are not necessarily the hyper-parameters that maximize the likelihood of some held-out data under the Gaussian process.

Similarly, we also consider again the Himmelblau's function. We run 25 simulations on a 50×50 grid for our algorithm, the LSE algorithm and the Straddle heuristic. We assume that the true Himmelblau's function can be evaluated with some normally distributed noise with standard deviation $\ln(\sigma_\epsilon) = 2.0$ and mean zero. The threshold chosen for the experiment is $t = -100$, with a prior mean of -100, prior standard deviation $\ln(\sigma_{ker}) = 4$, and $\ln(l) = 0$.

The advantage of the procedure with respect to the state of the art is demonstrated by the F_1-score on the sinusoidal and Himmelblau's function (Fig. 3). We also show precision and recall separately for the numerical experiments on Himmelblau's function in Fig. 3 bottom left and bottom right, respectively.

In both numerical experiments it is relatively easy to find an initial point above the threshold. Our algorithm then proceeds by expanding I_{GP} as much as possible at each step (Fig. 4). This is in contrast to the Straddle heuristic and LSE which seek to reduce the variance and thus tend to sample more widely in the initial phase. Notice that Straddle and LSE maximize a similar objective function for the selection of the next point, the "Straddle score" and the "ambiguity", respectively. However, at least in the initial exploration phase, these metrics have fairly uniform high value (Fig. 2) across the domain because the variance given by the Gaussian process is initially high. In contrast, RMILE gives higher scores to points in Ω that are likely to improve I_{GP} the most, and therefore chooses to expand the current region above the threshold as much as possible before exploring regions far away from the current samples.

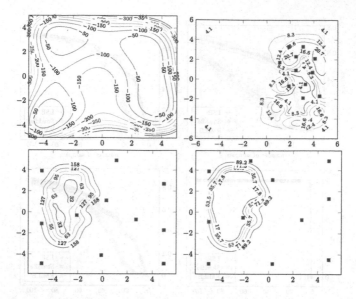

Fig. 2. Top left: true contours of the Himmelblau's function. Top right: value of the expected improvement for RMILE. Bottom left: Straddle score. Bottom right: ambiguity for the LSE algorithm. The locations of the first 11 samples for the first run are superimposed.

Thus, our algorithm is more suitable especially when a very limited exploration budget is available and one cannot afford to reconstruct a good model of the function. Although we use the well-established F_1-score for comparison, it may not always be the most appropriate and fair metric for our proposed problem. In particular, as we noted in Lemma 1, our classification rule penalizes false positives far more than false negatives.

5.3   Simulation Experiments: Aircraft Collision Avoidance

We evaluate our method in the task of estimating the sensor requirements for an aircraft collision avoidance system. We consider pairwise encounters between aircrafts, the behavior of which is dictated by a joint policy produced by modeling the problem as a Markov decision process and solving for optimal actions using value iteration. Observation noise is applied over two state variables, the relative angle and heading between the two aircrafts. The noise for each variable is sampled independently from a normal distribution with mean zero and standard deviation varying depending on assumed sensor precision. For each sample, 500 pairwise encounters are simulated, and the estimated probability of a near mid-air collision (NMAC) is returned. We apply the negative logit transformation to the output to map it to the real line. We look for a threshold $t = 1$, and the origin is given as a seed. Again, RMILE samples in a more structured way, progressively expanding I_{GP} while balancing the reduction of the variance in the promising region with some exploration (Fig. 5).

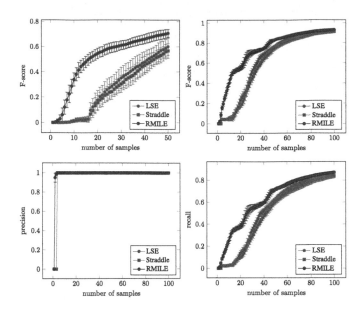

Fig. 3. F_1-score on the sinusoidal function (top left), and Himmelblau's function (top right). The means and confidence intervals of precision (bottom left) and recall (bottom right) refer to the Himmelblau's function experiment.

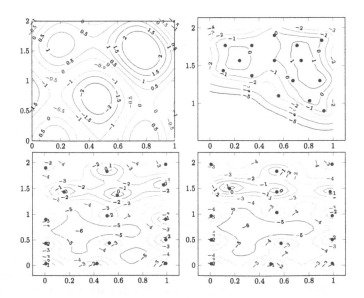

Fig. 4. Top left: true contours of the sinusoidal function. Location of the first 15 samples along with the contours given by the GP for $\mu_{GP}(\mathbf{x}) - 1.96\sigma_{GP}(\mathbf{x})$ for RMILE (top right), Straddle (bottom left) and LSE (bottom right).

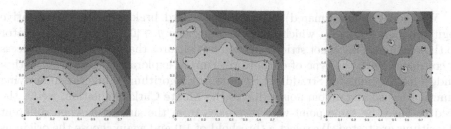

Fig. 5. Aircraft collision avoidance. Contours for $\mu_{GP}(\mathbf{x}) - 1.96\sigma_{GP}(\mathbf{x})$ given by the posterior Gaussian process. Left: RMILE algorithm. Middle: LSE algorithm. Right: Straddle heuristic. The yellow region is the area of interest. (Color figure online)

(a) $\mu_{GP}(\mathbf{x}) - 1.96\sigma_{GP}(\mathbf{x})$ contours. RMILE on the left, LSE in the middle, and Straddle on the right. The yellow region is the area of interest.

(b) $\mu_{GP}(\mathbf{x})$ contours. RMILE on the left, LSE in the middle, Straddle on the right.

(c) $1.96\sigma_{GP}(\mathbf{x})$ contours. RMILE on the left, LSE in the middle, Straddle on the right.

Fig. 6. Auto example: required sensor precision. (Color figure online)

5.4 Simulation Experiments: Required Sensor Precision

We assess our method on estimating actuator performance requirements in an automotive setting. We seek to determine the necessary precision for longitudinal and lateral acceleration maneuvers of simulated vehicles such that the likelihood of hard braking events is below a threshold. In these experiments, we simulate a single, five-second scenario involving twenty vehicles for 100 steps. The vehicles are propagated according to a bicycle model, with longitudinal behavior generated by the Intelligent Driver Model [21] and lane changing behavior dictated by the MOBIL [22] model. The two input parameters are sampled from a normal distribution, the standard deviation of which models the actuator precision.

We model the (estimated) probability of hard braking using the negative logit function $-\log \frac{y}{1-y}$, which maps the outcome $y \in [0,1]$ from the simulator to the real line. This is not strictly needed but ensures that the Gaussian process is consistent with the type of output. We run an exploratory simulation with a budget of 20 points for Straddle, LSE and our algorithm. While the underlying function has some random noises due to the Monte Carlo simulation, we fix the seed to have the same point-wise responses from the simulator when different algorithms are tested. We select a threshold of 1.0 and again choose the origin as the initial seed. In Fig. 6a we plot the contours for $\mu_{GP}(\mathbf{x}) - 1.96\sigma_{GP}(\mathbf{x})$. It can be seen that LSE and the Straddle heuristic both try to reduce the uncertainty by spreading out the sample points. Crucially, our algorithm places more samples together to compensate for the noise and reduce the variance (Fig. 6c) in the promising region (Fig. 6b) above $t = 1.0$. This allows the classifier to use the posterior GP to make a more confident prediction and identify the area of interest with high confidence (yellow region in Fig. 6a).

6   Conclusions

We have considered the problem of level set estimation where only a noise-corrupted version of the function is available with a very limited exploration budget. The aim is to discover as rapidly as possible a region where the threshold is exceeded with high probability. We propose to select the next query point that maximizes the expected volume of the domain of points above the threshold in a one-step lookahead procedure, and derive analytical formulae to compute this quantity in closed forms. We give a simple criterion to verify convergence of generic acquisition functions and verify that our algorithm satisfies such requirements. Our algorithm also compares favorably with the state of the art on numerical experiments. In particular, it uses information gained from a few samples more effectively, making it suitable when a very limited exploration budget is available. At the same time, it retains asymptotic convergence guarantees, making it especially compelling in the case of misspecified models.

Acknowledgments. Blake Wulfe provided the simulator for the simulations experiments. The authors are grateful to the reviewers for their comments.

References

1. Gotovos, A., Casati, N., Hitz, G., Krause, A.: Active learning for level set estimation. In: International Joint Conference on Artificial Intelligence (IJCAI), pp. 1344–1350 (2013)
2. Rasmussen, C.E., Williams, C.K.I.: Gaussian Processes for Machine Learning. MIT Press, Cambridge (2006)
3. Shahriari, B., Swersky, K., Wang, Z., Adams, R.P., de Freitas, N.: Taking the human out of the loop: a review of Bayesian optimization. Proc. IEEE **104**(1), 148–175 (2016)

4. Kushner, H.J.: A new method of locating the maximum point of an arbitrary multipeak curve in the presence of noise. J. Basic Eng. **86**(1), 97–106 (1964)
5. Mockus, J., Tiesis, V., Zilinskas, A.: The application of Bayesian methods for seeking the extremum. In: Dixon, L.C.W., Szego, G.P. (eds.) Towards Global Optimization, vol. 2. North-Holland Publishing Company, Amsterdam (1978)
6. Srinivas, N., Krause, A., Kakade, S.M., Seeger, M.: Gaussian process optimization in the bandit setting: no regret and experimental design. In: International Conference on Machine Learning (ICML), pp. 1015–1022 (2010)
7. Shahriari, B., Wang, Z., Hoffman, M.W., Bouchard-Cote, A., de Freitas, N.: An entropy search portfolio for Bayesian optimization. In: Advances on Neural Information Processing Systems (NIPS) (2014)
8. Willet, R.M., Nowak, R.D.: Minimax optimal level-set estimation. IEEE Trans. Image Process. **16**(12), 2965–2979 (2007)
9. Soni, A., Haupt, J.: Level set estimation from compressive measurements using box constrained total variation regularization. In: IEEE International Conference on Image Processing (ICIP), pp. 2573–2576 (2012)
10. Krishnamurthy, K., Bajwa, W.U., Willett, R.: Level set estimation from projection measurements: performance guarantees and fast computation. SIAM J. Imaging Sci. **6**(4), 2047–2074 (2013)
11. Krause, A., Singh, A., Guestrin, C.: Near-optimal sensor placements in Gaussian processes: theory, efficient algorithms and empirical studies. J. Mach. Learn. Res. **9**, 235–284 (2008)
12. Martino, L., Vicent, J., Camps-Valls, G.: Automatic emulator and optimized lookup table generation for radiative transfer models. In: IEEE International Geoscience and Remote Sensing Symposium, pp. 1457–1460 (2017)
13. Busby, D.: Hierarchical adaptive experimental design for Gaussian process emulators. Reliab. Eng. Syst. Saf. **94**(7), 1183–1193 (2009)
14. Bryan, B., Nichol, R.C., Genovese, C.R., Schneider, J., Miller, C.J., Wasserman, L.: Active learning for identifying function threshold boundaries. In: Advances in Neural Information Processing Systems (NIPS), pp. 163–170 (2005)
15. Bogunovic, I., Scarlett, J., Krause, A., Cevher, V.: Truncated variance reduction: a unified approach to Bayesian optimization and level-set estimation. In: Advances in Neural Information Processing Systems (NIPS), pp. 1507–1515 (2016)
16. Yang, J., Wang, Z., Wu, Z.: Level set estimation with dynamic sparse sensing. In: IEEE Global Conference on Signal and Information Processing (GlobalSIP), pp. 487–491 (2014)
17. Ma, Y., Garnett, R., Schneider, J.: Active area search via Bayesian quadrature. In: Artificial Intelligence and Statistics (AISTATS), pp. 595–603 (2014)
18. Ma, Y., Sutherland, D., Garnett, R., Schneider, J.: Active pointillistic pattern search. In: Artificial Intelligence and Statistics (AISTATS), pp. 672–680 (2015)
19. Bect, J., Ginsbourger, D., Li, L., Picheny, V., Vazquez, E.: Sequential design of computer experiments for the estimation of a probability of failure. Stat. Comput. **22**(3), 773–793 (2012)
20. Chevalier, C., Bect, J., Ginsbourger, D., Vazquez, E., Picheny, V., Richet, Y.: Fast parallel kriging-based stepwise uncertainty reduction with application to the identification of an excursion set. Technometrics **56**(4), 455–465 (2014)
21. Treiber, M., Hennecke, A., Helbing, D.: Congested traffic states in empirical observations and microscopic simulations. Phys. Rev. E **62**(2), 1805 (2000)
22. Kesting, A., Treiber, M., Helbing, D.: General lane-changing model MOBIL for car-following models. Transp. Res. Rec. **1999**(1), 86–94 (2007)

Scalable Nonlinear AUC Maximization Methods

Majdi Khalid$^{(\boxtimes)}$, Indrakshi Ray, and Hamidreza Chitsaz

Computer Science Department, Colorado State University, Fort Collins, USA
majdi.khaled@gmail.com, indrakshi.ray@colostate.edu, chitsaz@chitsazlab.org

Abstract. The area under the ROC curve (AUC) is a widely used measure for evaluating classification performance on heavily imbalanced data. The kernelized AUC maximization machines have established a superior generalization ability compared to linear AUC machines because of their capability in modeling the complex nonlinear structures underlying most real-world data. However, the high training complexity renders the kernelized AUC machines infeasible for large-scale data. In this paper, we present two nonlinear AUC maximization algorithms that optimize linear classifiers over a finite-dimensional feature space constructed via the k-means Nyström approximation. Our first algorithm maximizes the AUC metric by optimizing a pairwise squared hinge loss function using the truncated Newton method. However, the second-order batch AUC maximization method becomes expensive to optimize for extremely massive datasets. This motivates us to develop a first-order stochastic AUC maximization algorithm that incorporates a scheduled regularization update and scheduled averaging to accelerate the convergence of the classifier. Experiments on several benchmark datasets demonstrate that the proposed AUC classifiers are more efficient than kernelized AUC machines while they are able to surpass or at least match the AUC performance of the kernelized AUC machines. We also show experimentally that the proposed stochastic AUC classifier is able to reach the optimal solution, while the other state-of-the-art online and stochastic AUC maximization methods are prone to suboptimal convergence. Code related to this paper is available at: https://sites.google.com/view/majdikhalid/.

1 Introduction

The area under the ROC Curve (AUC) [11] has a wide range of applications in machine learning and data mining such as recommender systems, information retrieval, bioinformatics, and anomaly detection [1,5,22,25,26]. Unlike error rate, the AUC metric does not consider the class distribution when assessing the performance of classifiers. This property renders the AUC a reliable measure to evaluate classification performance on heavily imbalanced datasets [7], which are not uncommon in real-world applications.

The optimization of the AUC metric aims to learn a score function that scores a random positive instance higher than any negative instance. Therefore, the AUC metric is a threshold-independent measure. In fact, it evaluates a

© Springer Nature Switzerland AG 2019
M. Berlingerio et al. (Eds.): ECML PKDD 2018, LNAI 11052, pp. 292–307, 2019.
https://doi.org/10.1007/978-3-030-10928-8_18

classifier over all possible thresholds, hence eliminating the effect of imbalanced class distribution. The objective function maximizing the AUC metric optimizes a sum of pairwise losses. This objective function can be solved by learning a binary classifier on pairs of positive and negative instances that constitute the difference space. Intuitively, the complexity of such algorithms increases linearly with respect to the number of pairs. However, linear ranking algorithms like RankSVM [4,21], which can optimize the AUC directly, have shown a learning complexity independent from the number of pairs.

However, the kernelized versions of RankSVM [4,13,20] are superior to linear ranking machines in terms of producing higher AUC classification accuracy. This is due to its ability to model the complex nonlinear structures that underlie most real-world data. Analogous to kernel SVM, the kernelized RankSVM machines entail computing and storing a kernel matrix, which grows quadratically with the number of instances. This hinders the efficiency of kernelized RankSVM machines for learning on large datasets.

The recent approaches attempt to scale up the learning for AUC maximization from different perspectives. The first approach adopts online learning techniques to optimize the AUC on large datasets [9,10,17,18,32]. However, online methods result in inferior classification accuracy compared to batch learning algorithms. The authors of [15] develop a sparse batch nonlinear AUC maximization algorithm, which can scale to large datasets, to overcome the low generalization capability of online AUC maximization methods. However, sparse algorithms are prone to the under-fitting problem due to the sparsity of the model, especially for large datasets. The work in [28] imputes the low generalization capability of online AUC maximization methods to the optimization of the surrogate loss function on a limited hypothesis space. Therefore, it devises a nonparametric algorithm to maximize the real AUC loss function. However, learning such nonparametric algorithm on high dimensional space is not reliable.

In this paper, we address the inefficiency of learning nonlinear kernel machines for AUC maximization. We propose two learning algorithms that learn linear classifiers on a feature space constructed via the k-means Nyström approximation [31]. The first algorithm employs a linear batch classifier [4] that optimizes the AUC metric. The batch classifier is a Newton-based algorithm that requires the computation of all gradients and the Hessian-vector product in each iteration. While this learning algorithm is applicable for large datasets, it becomes expensive for training enormous datasets embedded in a large dimensional feature space. This motivates us to develop a first-order stochastic learning algorithm that incorporates the scheduled regularization update [3] and scheduled averaging [23] to accelerate the convergence of the classifier. The integration of these acceleration techniques allows the proposed stochastic method to enjoy the low complexity of classical first-order stochastic gradient algorithms and the fast convergence rate of second-order batch methods.

The remainder of this paper is organized as follows. We begin by reviewing closely related work in Sect. 2. In Sect. 3, we define the AUC problem and present related background. The proposed methods are presented in Sect. 4.

The experimental results are shown in Sect. 5. Finally, we conclude the paper and point out the future work in Sect. 6.

2   Related Work

The maximization of the AUC metric is a bipartite ranking problem, a special type of ranking algorithm. Hence, most ranking algorithms can be used to solve the AUC maximization problem. The large-scale kernel RankSVM is proposed in [20] to address the high complexity of learning kernel ranking machines. However, this method still depends quadratically on the number of instances, which hampers its efficiency. Linear RankSVM [2,4,14,21,27] is more applicable to scaling up in comparison to the kernelized variations. However, linear methods are limited to linearly separable problems. Recent study [6] explores the Nyström approximation to speed up the training of the nonlinear kernel ranking function. This work does not address the AUC maximization problem. It also does not consider the k-means Nyström method and only uses a batch ranking algorithm. Another method [15] attempts to speed up the training of nonlinear AUC classifiers by learning a sparse model constructed incrementally based on chosen criteria [16]. However, the sparsity can deteriorate the generalization ability of the classifier.

Another class of research proposes using online learning methods to reduce the training time required to optimize the AUC objective function [9,10,17,18, 32]. The work in [32] addresses the complexity of pairwise learning by deploying a first-order online algorithm that maintains a buffer of fixed size for positive and negative instances. The work in [17] proposes a second-order online AUC maximization algorithm with a fixed-sized buffer. The work [10] maintains the first-order and second-order statistics for each instance instead of the buffering mechanism. Recently the work in [30] formulates the AUC maximization problem as a convex-concave saddle point problem. The proposed algorithm in [30] solves a pairwise squared hinge loss function without the need to access the buffered instances or the second-order information. Therefore, it shows linear space and time complexities per iteration with respect to the number of features.

The work in [12] proposes a budget online kernel method for nonlinear AUC maximization. For massive datasets, however, the size of the budget needs to be large to reduce the variance of the model and to achieve an acceptable accuracy, which in turns increases the training time complexity. The work [8] attempts to address the scalability problem of kernelized online AUC maximization by learning a mini-batch linear classifier on an embedded feature space. The authors explore both Nyström approximation and random Fourier features to construct an embedding in an online setting. Despite their superior efficiency, online linear and nonlinear AUC maximization algorithms are susceptible to suboptimal convergence, which leads to inferior AUC classification accuracy.

Instead of maximizing a surrogate loss function, the authors of [28] attempt to optimize the real AUC loss function using a nonparametric learning algorithm. However, learning the nonparametric algorithm on high dimensional datasets is not reliable.

3   Preliminaries and Background

3.1   Problem Setting

Given a training dataset $\mathcal{S} = \{x_i, y_i\} \in \mathbb{R}^{n \times d}$, where n denotes the number of instances and d refers to the dimension of the data, generated from unknown distribution $\mathcal{D}$. The label of the data is a binary class label $y = \{-1, 1\}$. We use n^+ and n^- to denote the number of positive and negative instances, respectively. The maximization of the AUC metric is equivalent to the minimization of the following loss function:

$$\mathcal{L}(f; \mathcal{S}) = \frac{1}{n} \sum_{i=1}^{n^+} \sum_{j=1}^{n^-} I(f(x_i^+) \leq f(x_j^-)), \tag{1}$$

for a linear classifier $f(x) = w^T x$, where $I(\cdot)$ is an indicator function that outputs 1 if its argument is true, and 0 otherwise. The discontinuous nature of the indicator function makes the pairwise minimization problem (1) hard to optimize. It is common to replace the indicator function with its convex surrogate function as follows,

$$\mathcal{L}(f; \mathcal{S}) = \frac{1}{n} \sum_{i=1}^{n^+} \sum_{j=1}^{n^-} \ell(f(x_i^+) - f(x_j^-))^p. \tag{2}$$

This pairwise loss function $\ell(f(x_i^+) - f(x_j^-))$ is convex in w, and it upper bounds the indicator function. The pairwise loss function is defined as hinge loss when $p = 1$, and is defined as squared hinge loss when $p = 2$. The optimal linear classifier w for maximizing the AUC metric can be obtained by minimizing the following objective function:

$$\min_{w} \frac{1}{2} ||w||^2 + C \sum_{i=1}^{n^+} \sum_{j=1}^{n^-} max(0, 1 - w^T (x_i^+ - x_j^-))^p, \tag{3}$$

where $||w||$ is the Euclidean norm and C is the regularization hyper-parameter. Notice that the weight vector w is trained on the pairs of instances $(x^+ - x^-)$ that form the difference space. This linear classifier is efficient in dealing with large-scale applications, but its modeling capability is limited to the linear decision boundary.

The kernelized AUC maximization can also be formulated as an unconstrained objective function [4, 20]:

$$\min_{\beta \in \mathbb{R}^n} \frac{1}{2} \beta^T K \beta + C \sum_{(i,j) \in A} max(0, 1 - ((K\beta)_i - (K\beta)_j)^p, \tag{4}$$

where K is the kernel matrix, and A is a sparse matrix that contains all possible pairs $A \equiv \{(i, j) | y_i > y_j\}$. In the batch setting, the computation of the kernel costs $\mathcal{O}(n^2 d)$ operations, while storing the kernel matrix requires $\mathcal{O}(n^2)$ memory. Moreover, the summation over pairs costs $\mathcal{O}(n \log n)$ [20]. These complexities make kernel machines costly to train compared to the linear model that has linear complexity with respect to the number of instances.

3.2 Nyström Approximation

The Nyström approximation [19,31] is a popular approach to approximate the feature maps of linear and nonlinear kernels. Given a kernel function $K(\cdot, \cdot)$ and landmark points $\{u_l\}_{l=1}^v$ generated or randomly chosen from the input space S, the Nyström method approximates a kernel matrix G as follows,

$$G \approx \bar{G} = EW^{-1}E^T,$$

where $W_{ij} = \kappa(u_i, u_j)$ is a kernel matrix computed on landmark points and W^{-1} is its pseudo-inverse. The matrix $E_{ij} = \kappa(x_i, u_j)$ is a kernel matrix representing the intersection between the input space and the landmark points. The matrix W is factorized using singular value decomposition or eigenvalue decomposition as follows: $W = U \Sigma^{-1} U^T$, where the columns of the matrix U hold the orthonormal eigenvectors while the diagonal matrix Σ holds the eigenvalues of W in descending order. The Nyström approximation can be utilized to transform the kernel machines into linear machines by nonlinearly embedding the input space in a finite-dimensional feature space. The nonlinear embedding for an instance x is defined as follows,

$$\varphi(x) = U_r \, \Sigma_r^{-\frac{1}{2}} \phi^T(x),$$

where $\phi(x) = [\kappa(x, u_1), \ldots, \kappa(x, u_v)]$, the diagonal matrix Σ_r holds the top r eigenvalues, and U_r is the corresponding eigenvectors. The rank-r, $r \leq v$, is the best rank-r approximation of W. We use the k-means algorithm to generate the landmark points [31]. This method has shown a low approximation error compared to the standard method, which selects the landmark points based on uniform sampling without replacement from the input space. The complexity of the k-means algorithm is linear $\mathcal{O}(nvd)$, while the complexity of singular value decomposition or eigenvalue decomposition is $\mathcal{O}(v^3)$. Therefore, the complexity of the k-means Nyström approximation is linear in the input space.

4 Nonlinear AUC Maximization

In this section, we present the two nonlinear algorithms that maximize the AUC metric over a finite-dimensional feature space constructed using the k-means Nyström approximation [31]. First, we solve the pairwise squared hinge loss function in a batch learning mode using the truncated Newton solver [4]. For the second method, we present a stochastic learning algorithm that minimizes the pairwise hinge loss function.

The main steps of the proposed nonlinear AUC maximization methods are shown in Algorithm 1. In the embedding steps, we construct the nonlinear

---

Algorithm 1. Nonlinear AUC Maximization

Embedding Steps:

Compute the centroid points $\{u_l\}_{l=1}^{v}$

Form the matrix W: $W_{ij} = \kappa(u_i, u_j)$

Compute the eigenvalue decomposition: $W = U\Sigma U^T$

Form the matrix E: $E_i = \phi(x_i) = [\kappa(x_i, u_1), \ldots, \kappa(x_i, u_v)]$

Construct the feature space: $\varphi(X) = U_r \Sigma_r^{-\frac{1}{2}} E^T$

Training:

Learn the batch model described in Algorithm 2 or the stochastic model detailed in Algorithm 3

Prediction:

Map a test point x: $\varphi(x) = U_r \Sigma_r^{-\frac{1}{2}} \phi^T(x)$

Score value: $w^T \varphi(x)$

---

mapping (embedding) based on a given kernel function and landmark points. The landmark points are computed by the k-means clustering algorithm applied to the input space. Once the landmark points are obtained, the matrix W and its decomposition are computed. The original input space is then mapped non-linearly to a finite-dimensional feature space in which the nonlinear problem can be solved using linear machines.

The AUC optimization (3) can be solved for w in the embedded space as follows,

$$\min_{w} \frac{1}{2}\|w\|^2 + C \sum_{i=1}^{n^+} \sum_{j=1}^{n^-} max(0, 1 - w^T(\varphi(x_i^+) - \varphi(x_j^-)))^p, \qquad (5)$$

where $\varphi(x)$ is a nonlinear feature mapping for x. The minimization of (5) can be solved using truncated Newton methods [4] as shown in Algorithm 2. The matrix A in Algorithm 2 is a sparse matrix of size $r \times n$, where r is the number of pairs. The matrix A holds all possible pairs in which each row of A has only two nonzero values. That is, if $(i, j) \mid y_i > y_j$, the matrix A has a k-th row such that $A_{ki} = 1, A_{kj} = -1$. However, the complexity of this Newton batch learning is dependent on the number of pairs. The authors of [4] also proposed the PSVM+ algorithm, which avoids the direct computation of pairs by reformulating the pairwise loss function in such a way that the calculations of the gradient and the Hessian-vector product are accelerated.

Algorithm 2. Batch Nonlinear AUC Maximization

Input: embedded data $\tilde{X}$
Output: the ranking model w
initial vector $w \leftarrow 0$
while stopping criterion is not satisfied **do**
    $D = max(0, 1 - A(w^T \tilde{X}))$
    Compute gradient $g = w - (CD^T A\tilde{X})^T$
    Compute a search direction s_t by applying conjugate gradient to solve
    $\nabla^2 F(w_k)s = -\nabla F(w_k)$
    Update $w_{k+1} = w_k + s_k$
end while

Nevertheless, the optimization of PRSVM+ to maximize the AUC metric still requires $O(n\hat{d} + 2n + \hat{d})$ operations to compute each of the gradient and the Hessian-vector product in each iteration, where $\hat{d}$ is the dimension of the embedded space. This makes the training of PRSVM+ expensive for massive datasets embedded using a large number of landmark points. A large set of landmark points is desirable to improve the approximation of the feature maps; hence boosting the generalization ability of the involved classifier.

To address this complexity, we present a first-order stochastic method to maximize the AUC metric on the embedded space. Specifically, we optimize a pairwise hinge loss function using stochastic gradient descent accelerated by scheduling both the regularization update and averaging techniques. The proposed stochastic algorithm can be seen as an averaging variant of the SVMSGD2 method proposed in [3]. Algorithm 3 describes the proposed stochastic AUC maximization method. The algorithm randomly selects a positive and negative instance and updates the model in each iteration as follows,

$$w_{t+1} = w_t + \frac{1}{\lambda(t + t_0)}\ell'(w_t^T x_t)x_t,$$

where $\ell'(z) = max(0, 1 - z)$ is a hinge loss function, the vector x_t holds the difference $\varphi(x_i^+) - \varphi(x_j^-)$, w_t is the solution after t iterations, and $\lambda(t + t_0)$ is the learning rate, which decreases in each iteration. The hyper-parameter λ can be tuned on a validation set. The positive constant t_0 is set experimentally, and it is utilized to prevent large steps in the first few iterations [3]. The model is regularized each $rskip$ iterations to accelerate its convergence. We also foster the acceleration of the model by implementing an averaging technique [23,29]. The intuitive idea behind the averaging step is to reduce the variance of the model that stems from its stochastic nature. We regulate the regularization update and averaging steps to be performed each $askip$ and $rskip$ iterations as follows,

$$w_{t+1} = w_{t+1} - rskip(t + t_0)^{-1}w_{t+1}$$

$$\tilde{w}_{q+1} = \frac{q\tilde{w}_q + w_{t+1}}{q},$$

where $\tilde{w}$ is the averaged solution after q iterations with respect to the *askip*. The advantage of regulating the averaging step is to reduce the per iteration complexity, while effectively accelerating the convergence.

The presented first-order stochastic AUC maximization requires $\mathcal{O}(\hat{d}a)$ operations per iteration in addition to the $\mathcal{O}(\hat{d})$ operations needed for each of the regularization update and averaging steps that occur per *rskip* and *askip* iterations respectively, where a denotes the average number of nonzero coordinates in the embedded difference vector x_t.

Algorithm 3. Stochastic Nonlinear AUC Maximization

Input: embedded data $\tilde{X}$, λ, t_0, T, rskip, askip
Output: the ranking model w
$w_1 \leftarrow 0$ and $\tilde{w}_1 \leftarrow 0$, $rcount = rskip$, $acount = askip$, $q = 1$
for $t = 1, \ldots, T$ **do**
    Randomly pick a pair $i_t \in 1, \ldots, n^+$, $j_t \in 1, \ldots, n^-$
    $x_t = \tilde{x}_{i_t} - \tilde{x}_{j_t}$
    $w_{t+1} = w_t + \frac{1}{\lambda(t+t_0)}\ell'(w_t^T x_t)x_t$
    $rcount = rcount - 1$
    if rcount ≤ 0 **then**
        $w_{t+1} = w_{t+1} - rskip(t + t_0)^{-1}w_{t+1}$
        $rcount = rskip$
    end if
    $acount = acount - 1$
    if acount ≤ 0 **then**
        $\tilde{w}_{q+1} = \frac{q\tilde{w}_q + w_{t+1}}{q}$
        $q = q + 1$
        $acount = askip$
    end if
end for
set $w = \tilde{w}_q$
return w

---

5 Experiments

In this section, we evaluate the proposed methods on several benchmark datasets and compare them with kernelized AUC algorithm and other state-of-the-art online AUC maximization algorithms. The experiments are implemented in MATLAB, while the learning algorithms are written in C language via MEX files. The experiments were performed on a computer equipped with an Intel 4 GHz processor with 32 G RAM.

5.1 Benchmark Datasets

The datasets we use in our experiments can be downloaded from LibSVM website[1] or UCI[2]. The datasets that are not split (i.e., spambase, magic04, connect-4, skin, and covtype) into training and test sets; we randomly divide them into 80%–20% for training and testing. The features of each dataset are standardized to have zero mean and unit variance. The multi-class datasets (e.g., covtype and usps) are converted into class-imbalanced binary data by grouping the instances into two sets, where each set has the same number of class labels. To speed up the experiments that include the kernelized AUC algorithm, we train all the compared methods on 80k instances, randomly selected from the training set. The other experiments are performed on the entire training data. The characteristics of the datasets along with their imbalance ratios are shown in Table 1.

Table 1. Benchmark datasets

Data	#training	#test	#feat	Ratio
spambase	3,680	921	57	1.53
usps	7,291	2,007	256	1.40
magic04	15,216	3,804	10	1.84
protein	17,766	6,621	357	2.11
ijcnn1	49,990	91,701	22	9.44
connect-4	54,045	13,512	126	3.06
acoustic	78,823	19,705	50	3.31
skin	196,045	49,012	3	3.83
cod-rna	331,152	157,413	8	2.0
covtype	464,809	116,203	54	10.65

5.2 Compared Methods and Model Selection

We compare the proposed methods with kernel RankSVM and linear RankSVM, which can be used to solve the AUC maximization problem. We also include two state-of-the-art online AUC maximization algorithms. The random Fourier method that approximates the kernel function is also involved in the experiments where the resulting classifier is solved by linear RankSVM.

1. **RBF-RankSVM:** This is the nonlinear kernel RankSVM [20]. We use Gaussian kernel $K(x, y) = exp(-\gamma||x - y||^2)$ to model the nonlinearity of the data. The best width of the kernel γ is chosen by 3-fold cross validation

---

[1] https://www.csie.ntu.edu.tw/~cjlin/libsvmtools/datasets/.
[2] http://archive.ics.uci.edu/ml/index.php.

on the training set via searching in $\{2^{-6}, \ldots, 2^{-1}\}$. The regularization hyper-parameter C is also tuned by 3-fold cross validation by searching in the grid $\{2^{-5}, \ldots, 2^5\}$. The searching grids are selected based on [20]. We also train the RBF-RankSVM on $1/5$ subsamples, selected randomly.

2. **Linear RankSVM (PRSVM+):** This is the linear RankSVM that optimizes the squared hinge loss function using truncated Newton [4]. The best regularization hyper-parameter C is chosen from the grid $\{2^{-15}, \ldots, 2^{10}\}$ via 3-fold cross validation.

3. **RFAUC:** This uses the random Fourier features [24] to approximate the kernel function. We use PRSVM+ to solve the AUC maximization problem on the projected space. The hyper-parameters C and γ are selected via 3-fold cross validation by searching on the grids $\{2^{-15}, \ldots, 2^{10}\}$ and $\{1, 10, 100\}$, respectively.

4. **NOAM:** This is the sequential variant of online AUC maximization [32] trained on a feature space constructed via the k-means Nyström approximation. The hyper-parameters are chosen as suggested by [32] via 3-fold cross validation. The number of positive and negative buffers is set to 100.

5. **NSOLAM:** This is the stochastic online AUC maximization [30] trained on a feature space constructed via the k-means Nyström approximation. The hyper-parameters of the algorithm (i.e., the learning rate and the bound on the weight vector) are selected via 3-fold cross validation by searching in the grids $\{1 : 9 : 100\}$ and $\{10^{-1}, \ldots, 10^5\}$, respectively. The number of epochs is set to 15.

6. **NBAUC:** This is the proposed batch AUC maximization algorithm trained on the embedded space. We solve it using the PRSVM+ algorithm [4]. The hyper-parameter C is tuned similarly to the Primal RankSVM.

7. **NSAUC:** This is the proposed stochastic AUC maximization algorithm trained on the embedded space. The hyper-parameter λ is chosen from the grid $\{10^{-10}, \ldots, 10^{-7}\}$ via 3-fold cross validation.

For those algorithms that involve the k-means Nyström approximation (i.e., our proposed methods, NOAM, and NSOLAM), we compute 1600 landmark points using the k-means clustering algorithm, which is implemented in C language. We select a Gaussian kernel function to be used with the k-means Nyström approximation. The bandwidth of the Gaussian function is set to be the average squared distance between the first 80k instances and the mean computed over these 80k instances. For a fair comparison, we also set the number of random Fourier features to 1600.

5.3   Results for Batch Methods

The comparison of batch AUC maximization methods in terms of AUC classification accuracy on the test set is shown in Table 2, while Table 3 compares these batch methods in terms of training time. For connect-4 dataset, the results of RBF-RankSVM are not reported because the training runs over five days.

We observe that the proposed NBAUC outperforms the competing batch methods in terms of AUC classification accuracy. The AUC performance of RBF-RankSVM might be improved for some datasets if the best hyper-parameters are selected on a more restricted grid of values. Nevertheless, the training of NBAUC is several orders of magnitude faster than RBF-RankSVM. The fast training of NBAUC is clearly demonstrated on the large datasets.

The proposed NBAUC shows a robust AUC performance compared to RFAUC on most datasets. This can be attributed to the robust capability of the k-means Nyström method in approximating complex nonlinear structures. It also indicates that a better generalization can be attained by capitalizing on the data to construct the feature maps, which is the main characteristic of the Nyström approximation, while the random Fourier features are oblivious to the data.

We also observe that the AUC performance of both RBF-RankSVM and its variant applied to random subsamples outperform the linear RankSVM, except for the protein dataset. However, RBF-RankSVM methods require longer training, especially for large datasets. We see that the linear RankSVM performs better than the kernel AUC machines on the protein dataset. This implies that the protein dataset is linearly separable. However, the AUC performance of the proposed method NBAUC is even better than linear RankSVM on this dataset.

Table 2. Comparison of AUC performance for batch classifiers on the benchmark datasets.

Data	RBF-RankSVM	RBF-RankSVM(subsample)	Linear RankSVM	RFAUC	**NBAUC**
spambase	98.00	96.02	97.47	97.75	98.04
usps	99.08	98.54	90.27	97.42	99.24
magic04	92.18	91.34	84.47	92.83	93.06
protein	80.97	77.60	83.30	58.43	84.33
ijcnn1	99.68	99.35	91.56	98.86	99.57
connect-4	-	91.32	88.20	91.10	94.09
acoustic	93.60	93.02	87.38	91.82	94.14
skin	99.92	99.92	94.81	100	99.98
cod-rna	99.07	99.07	98.85	99.12	99.12
covtype	93.94	94.05	87.75	95.99	96.03

5.4 Results for Stochastic Methods

We now compare our stochastic algorithm NSAUC with the state-of-the-art online AUC maximization methods, NOAM and NSOLAM. We also include the results of the proposed batch algorithm NBAUC for reference. The k-means Nyström approximation is implemented separately for each algorithm as

Table 3. Comparison of training time (in seconds) for batch classifiers on the benchmark datasets.

Data	RBF-RankSVM	RBF-RankSVM(subsample)	Linear RankSVM	RFAUC	**NBAUC**
spambase	3.08	0.10	0.13	3.59	7.71
usps	492.30	0.83	1.42	6.77	27.68
magic04	518.04	3.71	0.08	21.51	25.46
protein	2614.7	4.81	4.47	14.20	73.81
ijcnn1	15,434	282	0.57	80.17	88.87
connect-4	-	12,701	3.42	62.60	164.48
acoustic	134,030	5,610	1.88	92.74	151.78
skin	2037.30	78.20	0.20	73.18	23.71
cod-rna	5,715	255.4	0.44	83.01	113.66
covtype	133,270	11,670	2.54	273.67	220.90

introduced in Sect. 4. We experiment on the following large datasets: ijcnn1, connect-4, acoustic, skin, cod-rna, and covtype. Table 4 shows the comparison of the proposed methods with the online AUC maximization algorithms. Notice that the reported training time in Table 4 indicates only the time cost of the learning steps with excluding the embedding steps.

We can see that the proposed NSAUC achieves a competitive AUC performance compared to the proposed NBAUC, but with less training time. On the

Table 4. Comparison of AUC classification accuracy and training time (in seconds) for the proposed algorithms with other online AUC maximization algorithms. The training time does not include the embedding steps.

Data	Metric	NOAM	NSOLAM	**NSAUC**	**NBAUC**
ijcnn1	AUC	98.16	98.86	99.69	99.57
	Training time	6.24	6.88	4.80	40.70
connect-4	AUC	85.96	90.60	94.04	94.08
	Training time	6.97	7.39	10.74	36.96
acoustic	AUC	89.90	91.00	94.04	94.14
	Training time	10.80	10.82	23.80	59.34
skin	AUC	99.98	99.01	99.98	99.98
	Training time	6.26	5.66	6.60	10.32
cod-rna	AUC	98.29	99.10	99.19	99.18
	Training time	42.09	47.06	34.23	148.46
covtype	AUC	91.29	92.25	96.00	96.60
	Training time	61.75	63.59	49.17	1110.44

largest dataset covtype, the AUC performance of NSAUC is on par with NBAUC, while it only requires 49.17 s for training compared to more than 18 min required by NBAUC. In contrast to the online methods, the proposed NSAUC is able to converge to the optimal solution obtained by the batch method NBAUC. We attribute the robust performance of NSAUC to the effectiveness of scheduling both the regularization update and averaging.

We observe that the proposed NSAUC requires longer training time on some datasets (e.g., connect-4 and acoustic) compared to the online methods; however, the difference in the training time is not significant. In addition, we see that NSOLAM performs better than NOAM in terms of AUC classification accuracy. This implies the advantage of optimizing the pairwise squared hinge loss function, performed by NSOLAM, over the pairwise hinge lose function, carried out by NOAM, for one-pass AUC maximization.

5.5   Study on the Convergence Rate

We investigate the convergence of NSAUC and its counterpart NSOLAM with respect to the number of epochs. We also include NSVMSGD2 algorithm [3] that minimizes the pairwise hinge loss function on a feature space constructed via the k-means Nyström approximation, described in Sect. 4. The algorithm NSVMSGD2 is analogous to the proposed algorithm NSAUC, but with no averaging step. The AUC performances of these stochastic methods upon varying the number of epochs are depicted in Fig. 1. We vary the number of epochs according to the grid $\{1, 2, 3, 4, 5, 10, 20, 50, 100, 200, 300, 400\}$, and run the stochastic algorithms using the same setup described in the previous subsection. In all subfigures, the x-axis represents the number of epochs, while the y-axis is the AUC classification accuracy on the test data.

The results show that the proposed NSAUC converges to the optimal solution on all datasets. We can also see that the AUC performance of NSAUC outperforms its non-averaging variant NSVMSGD2 on four datasets (i.e., ijcnn1, cod-rna, acoustic, and connect-4), while its training time is on par with that of NSVMSGD2. This indicates the effectiveness of incorporating the scheduled averaging technique. Furthermore, the AUC performance of NSAUC does not fluctuate with varying the number of epochs on all datasets. This implies that choosing the best number of epochs would be easy.

In addition, we can observe that the AUC performance of NSOLAM does not show significant improvement after the first epoch. The reason is that NSOLAM reaches a local minimum (i.e., a saddle point) in a single pass and gets stuck there.

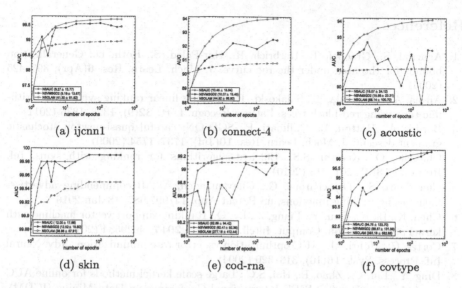

(a) ijcnn1   (b) connect-4   (c) acoustic

(d) skin   (e) cod-rna   (f) covtype

Fig. 1. AUC classification accuracy of stochastic AUC algorithms with respect to the number of epochs. We randomly pick a positive and negative instance for each iteration in NSAUC and NSVMSGD2, where n iterations correspond to one epoch. The values in parentheses denote the averaged training time (in seconds) along with the standard deviation over all epochs. The training time excludes the computational time of the embedding steps. The x-axis is displayed in log-scale.

6   Conclusion and Future Work

In this paper, we have proposed scalable batch and stochastic nonlinear AUC maximization algorithms. The proposed algorithms optimize linear classifiers on a finite-dimensional feature space constructed via the k-means Nyström approximation. We solve the proposed batch AUC maximization algorithm using truncated Newton optimization, which minimizes the pairwise squared hinge loss function. The proposed stochastic AUC maximization algorithm is solved using a first-order gradient descent that implements scheduled regularization update and scheduled averaging to accelerate the convergence of the classifier. We show via experiments on several benchmark datasets that the proposed AUC maximization algorithms are more efficient than the nonlinear kernel AUC machines, while their AUC performances are comparable or even better than the nonlinear kernel AUC machines. Moreover, we show experimentally that the proposed stochastic AUC maximization algorithm outperforms the state-of-the-art online AUC maximization methods in terms of AUC classification accuracy with a marginal increase in the training time for some datasets. We demonstrate empirically that the proposed stochastic AUC algorithm converges to the optimal solution in a few epochs, while other online AUC maximization algorithms are susceptible to suboptimal convergence. In the future, we plan to use the proposed algorithms in solving large-scale multiple-instance learning.

References

1. Agarwal, S., Graepel, T., Herbrich, R., Har-Peled, S., Roth, D.: Generalization bounds for the area under the roc curve. J. Mach. Learn. Res. **6**(Apr), 393–425 (2005)
2. Airola, A., Pahikkala, T., Salakoski, T.: Training linear ranking svms in linearithmic time using red-black trees. Pattern Recogn. Lett. **32**(9), 1328–1336 (2011)
3. Bordes, A., Bottou, L., Gallinari, P.: SGD-QN: careful quasi-newton stochastic gradient descent. J. Mach. Learn. Res. **10**(Jul), 1737–1754 (2009)
4. Chapelle, O., Keerthi, S.S.: Efficient algorithms for ranking with svms. Inf. Retrieval **13**(3), 201–215 (2010)
5. Chaudhuri, S., Theocharous, G., Ghavamzadeh, M.: Recommending advertisements using ranking functions, uS Patent App. 14/997,987, 18 Jan 2016
6. Chen, K., Li, R., Dou, Y., Liang, Z., Lv, Q.: Ranking support vector machine with kernel approximation. Comput. Intell. Neurosci. **2017**, 4629534 (2017)
7. Cortes, C., Mohri, M.: AUC optimization vs. error rate minimization. Adv. Neural Inf. Process. Syst. **16**(16), 313–320 (2004)
8. Ding, Y., Liu, C., Zhao, P., Hoi, S.C.: Large scale kernel methods for online AUC maximization. In: 2017 IEEE International Conference on Data Mining (ICDM), pp. 91–100. IEEE (2017)
9. Ding, Y., Zhao, P., Hoi, S.C., Ong, Y.S.: An adaptive gradient method for online AUC maximization. In: AAAI, pp. 2568–2574 (2015)
10. Gao, W., Jin, R., Zhu, S., Zhou, Z.H.: One-pass AUC optimization. In: ICML, vol. 3, pp. 906–914 (2013)
11. Hanley, J.A., McNeil, B.J.: The meaning and use of the area under a receiver operating characteristic (ROC) curve. Radiology **143**(1), 29–36 (1982)
12. Hu, J., Yang, H., King, I., Lyu, M.R., So, A.M.C.: Kernelized online imbalanced learning with fixed budgets. In: AAAI, pp. 2666–2672 (2015)
13. Joachims, T.: A support vector method for multivariate performance measures. In: Proceedings of the 22nd International Conference on Machine Learning, pp. 377–384. ACM (2005)
14. Joachims, T.: Training linear SVMs in linear time. In: Proceedings of the 12th ACM SIGKDD International Conference on Knowledge Discovery and Data Mining, pp. 217–226. ACM (2006)
15. Kakkar, V., Shevade, S., Sundararajan, S., Garg, D.: A sparse nonlinear classifier design using AUC optimization. In: Proceedings of the 2017 SIAM International Conference on Data Mining, pp. 291–299. SIAM (2017)
16. Keerthi, S.S., Chapelle, O., DeCoste, D.: Building support vector machines with reduced classifier complexity. J. Mach. Learn. Res. **7**(Jul), 1493–1515 (2006)
17. Khalid, M., Ray, I., Chitsaz, H.: Confidence-weighted bipartite ranking. In: Li, J., Li, X., Wang, S., Li, J., Sheng, Q.Z. (eds.) ADMA 2016. LNCS (LNAI), vol. 10086, pp. 35–49. Springer, Cham (2016). https://doi.org/10.1007/978-3-319-49586-6_3
18. Kotlowski, W., Dembczynski, K.J., Huellermeier, E.: Bipartite ranking through minimization of univariate loss. In: Proceedings of the 28th International Conference on Machine Learning (ICML-11), pp. 1113–1120 (2011)
19. Kumar, S., Mohri, M., Talwalkar, A.: Ensemble nystrom method. In: Advances in Neural Information Processing Systems, pp. 1060–1068 (2009)
20. Kuo, T.M., Lee, C.P., Lin, C.J.: Large-scale kernel RankSVM. In: Proceedings of the 2014 SIAM International Conference on Data Mining, pp. 812–820. SIAM (2014)

21. Lee, C.P., Lin, C.J.: Large-scale linear rankSVM. Neural Comput. **26**(4), 781–817 (2014)
22. Liu, T.Y.: Learning to rank for information retrieval. Found. Trends Inf. Retrieval **3**(3), 225–331 (2009)
23. Polyak, B.T., Juditsky, A.B.: Acceleration of stochastic approximation by averaging. SIAM J. Control Optim. **30**(4), 838–855 (1992)
24. Rahimi, A., Recht, B.: Random features for large-scale kernel machines. In: Advances in Neural Information Processing Systems, pp. 1177–1184 (2008)
25. Rendle, S., Balby Marinho, L., Nanopoulos, A., Schmidt-Thieme, L.: Learning optimal ranking with tensor factorization for tag recommendation. In: Proceedings of the 15th ACM SIGKDD International Conference on Knowledge Discovery and Data Mining, pp. 727–736. ACM (2009)
26. Root, J., Qian, J., Saligrama, V.: Learning efficient anomaly detectors from K-NN graphs. In: Artificial Intelligence and Statistics, pp. 790–799 (2015)
27. Sculley, D.: Large scale learning to rank. In: NIPS Workshop on Advances in Ranking, pp. 58–63 (2009)
28. Szörényi, B., Cohen, S., Mannor, S.: Non-parametric Online AUC Maximization. In: Ceci, M., Hollmén, J., Todorovski, L., Vens, C., Džeroski, S. (eds.) ECML PKDD 2017. LNCS (LNAI), vol. 10535, pp. 575–590. Springer, Cham (2017). https://doi.org/10.1007/978-3-319-71246-8_35
29. Xu, W.: Towards optimal one pass large scale learning with averaged stochastic gradient descent. arXiv preprint arXiv:1107.2490 (2011)
30. Ying, Y., Wen, L., Lyu, S.: Stochastic online AUC maximization. In: Advances in Neural Information Processing Systems, pp. 451–459 (2016)
31. Zhang, K., Tsang, I.W., Kwok, J.T.: Improved nyström low-rank approximation and error analysis. In: Proceedings of the 25th International Conference on Machine Learning, pp. 1232–1239. ACM (2008)
32. Zhao, P., Jin, R., Yang, T., Hoi, S.C.: Online AUC maximization. In: Proceedings of the 28th International Conference on Machine Learning (ICML-11), pp. 233–240 (2011)

Matrix and Tensor Analysis

Lambert Matrix Factorization

Arto Klami, Jarkko Lagus, and Joseph Sakaya$^{(\boxtimes)}$

Department of Computer Science, University of Helsinki, Helsinki, Finland
{arto.klami,jarkko.lagus,joseph.sakaya}@cs.helsinki.fi

Abstract. Many data generating processes result in skewed data, which should be modeled by distributions that can capture the skewness. In this work we adopt the flexible family of Lambert W distributions that combine arbitrary standard distribution with specific nonlinear transformation to incorporate skewness. We describe how Lambert W distributions can be used in probabilistic programs by providing stable gradient-based inference, and demonstrate their use in matrix factorization. In particular, we focus in modeling logarithmically transformed count data. We analyze the weighted squared loss used by state-of-the-art word embedding models to learn interpretable representations from word co-occurrences and show that a generative model capturing the essential properties of those models can be built using Lambert W distributions.

Keywords: Skewed data · Matrix factorization · Lambert distribution

1 Introduction

Real-valued data is often modeled with probabilistic models relying on normal likelihood, which captures simple additive noise well. Many realistic data generating processes, however, correspond to noise with wider tails. If the noise is still symmetric there are well-understood likelihoods to choose from. For example, student-t, Cauchy and stable distributions can be used for modeling heavy-tailed noise, and the machine learning community has used them for providing *robust* variants of various analysis methods [2,26].

When only one of the tails is heavy, the distribution is called *skewed* – a distribution with heavy left tail has negative skew and a distribution with heavy right tail has positive skew. Such data occurs in many applications: income and wealth are highly skewed, stock market returns are negatively skewed, and many monotonic transformations of symmetric noise result in skewness. There are, however, way fewer tools for modeling skewed data. The skew-normal distribution and its extensions to student-t [4] and Cauchy [3] are the main alternatives, but they are computationally cumbersome and limited in expressing skewness.

We build on the work of Goerg [11], who proposed a flexible family of skewed distributions by combining arbitrary continuous base distribution $F_Z(z)$ with forward transformation $g(z) = ze^{\gamma z}$. He coined it the Lambert $W \times F_Z$ distribution because the corresponding backward transformation $g^{-1}(x)$ uses the

© Springer Nature Switzerland AG 2019
M. Berlingerio et al. (Eds.): ECML PKDD 2018, LNAI 11052, pp. 311–326, 2019.
https://doi.org/10.1007/978-3-030-10928-8_19

Lambert W function [8]. Here γ controls the skewness independently of the other properties of the distribution, which makes the family a promising solution for general-purpose modeling of skewed data.

Maximum likelihood estimation of Lambert $W \times F_Z$ distribution is possible with an iterated generalized method of moments [11]. We are, however, interested in using the distribution as a building block in more complex systems, such as mixtures, matrix factorizations or arbitrary probabilistic programs, which requires more general inference algorithms. The construction using differentiable transformation of standard density is in principle directly applicable for gradient-based learning, including also posterior inference strategies such as variational approximations [28] and Hamiltonian Monte Carlo [6]. Unfortunately, the mapping $g(z)$ in [11] is not bijective because the range of the Lambert W function is limited, which causes computational issues. We show that an equivalent probability density can be obtained by using a bijective transformation that deviates from W only in regions with small probability, and we provide a computationally efficient distribution that can be used for modeling arbitrary skewed data.

We are particularly interested in the family because it provides a natural basis for fully generative interpretation of representation learning techniques used for estimating word embeddings from co-occurrences of words, such as GloVe [23] and Swivel [25]. Embeddings capturing semantic and syntactic relationships between words can be extracted by modeling logarithmic co-occurrence counts with a linear model. The state-of-the-art models do this by minimizing a weighted squared error between the model and the data, weighting the individual elements by the observed data itself. This allows the model to focus more on modeling co-occurrences that are considered more reliable.

Generative alternatives have been proposed for skip-gram models that learn embeddings based on individual sentences [5,27], but generative variants for the models operating directly on the co-occurrence counts of the whole corpus are missing. Some attempts have been made, but none of these manage to replicate the properties obtained by weighting the least squares loss-based model on the data itself. For example, Vilnis and McCallum [29] model the variance by summing individual terms for each row and column and hence cannot give higher uncertainty for uncommon co-occurrences, Li et al. [19] use empirical Bayes priors centered at the observed data failing to produce a proper generative model, and Jameel and Schockaert [16] model the variances by explicitly analyzing the residuals and do not provide a closed-form model.

We show that the weighted least squares error used by GloVe [23] and Swivel [25] can be interpreted as a negative log likelihood of a distribution that has a number of interesting properties. It is locally Gaussian, it is left skewed, and its variance and mean are negatively correlated. We show that the Lambert $W \times F$ family of distributions allows replicating these properties when we use as the base distribution F a normal distribution controlled by a single parameter μ influencing both the mean and variance. Building on this, we propose a simple probabilistic model that can be used for extracting interpretable embeddings from count data. It combines the above likelihood with an additional latent variable indicating presence or absence of co-occurrence, with probability that is controlled by the same parameter μ that controls the location and scale of the data generating distribution.

We demonstrate the Lambert matrix factorization technique first in standard matrix factorization application with skewed noise to illustrate that the gradient-based learning scheme works as intended. We then proceed to showcase the algorithm in modeling log-transformed count data, demonstrating also that it achieves accuracy comparable to its closest non-probabilistic comparison method Swivel [25] on standard evaluation metrics for word embeddings.

2   Lambert $W \times F$ Distribution

The Lambert $W \times F$ distribution family [11] is defined through an input/output system driven by some standard distribution $F_Z(z)$ with zero mean and unit variance. Latent inputs z drawn from F_Z are transformed with skewing function

$$x = g(z) = ze^{\gamma z}, \tag{1}$$

where γ is a parameter controlling the skewness. While the specific choice of the transformation may seem arbitrary, is has several desirable properties that are difficult to achieve at the same time: it is easy to differentiate, reduces to identity for $\gamma = 0$, and allows easy change of skew direction ($\gamma > 0$ and $\gamma < 0$ correspond to right and left skewness, respectively). It hence provides an interesting opportunity for probabilistic programming since it allows skewing any base distribution by a learnable quantity.

The probability density of x can be expressed in terms of the base density $p_Z(z)$ using change of variables as

$$p_X(x|\gamma) = p_Z(g^{-1}(x)) \left| \frac{dg^{-1}(x)}{dx} \right|, \tag{2}$$

where the absolute value of the Jacobian of the inverse transformation $z = g^{-1}(x)$ accounts for the change of volume under the nonlinear transformation. The inverse of (1) is

$$z = g^{-1}(x) = W_0(\gamma x)/\gamma, \tag{3}$$

where $W_0(\cdot)$ is the principal branch of the Lambert W function that gives rise to the name of the distribution family. The W function is illustrated in Fig. 2 (top left), showing its two real branches $W_0(x)$ and $W_{-1}(x)$. The function has numerous applications in differential calculus, quantum mechanics, fluid dynamics etc. Consequently, its properties have been studied in detail [8].

The derivatives of W (for $x \neq -1/e$) can be computed using

$$\frac{dW(x)}{dx} = \frac{1}{x + e^{W(x)}} \tag{4}$$

and consequently the density in (2) becomes

$$p_Z(W(\gamma x)/\gamma) \frac{1}{\gamma x + e^{W(\gamma x)}}.$$

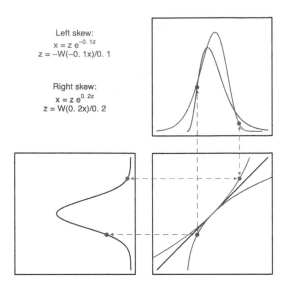

Fig. 1. Illustration of the Lambert $W \times F$ distribution. Draws from a standard distribution (bottom left), here $\mathcal{N}(0,1)$, are transformed to draws from Lambert $W \times F$ distribution (top right) by $g(z) = ze^{\gamma z}$, where γ controls skewness. For $\gamma < 0$ the transformation is convex and the resulting distribution is left skew (red), whereas for $\gamma > 0$ the transformation is concave and the distribution is right skew (blue). (Color figure online)

We illustrate the density and the construction based on a standard distribution (here normal) in Fig. 1. For practical modeling purposes the skewed distribution would then be transformed with further location-scale transformation

$$y = \sigma x + \mu, \tag{5}$$

which results in the final density

$$p_Y(\mu, \sigma, \gamma) = \frac{p_Z(\frac{1}{\gamma}W(\frac{\gamma(y-\mu)}{\sigma}))}{\gamma(y-\mu) + \sigma e^{W(\frac{\gamma(y-\mu)}{\sigma})}}.$$

We denote this density with $\mathcal{W}_F(\mu, \sigma, \gamma)$, where F tells the base distribution.

Goerg [11] provides extensive analysis of the family, showing, for example, closed-form expressions for various moments of the distribution when $F = \mathcal{N}$, and that for wide range of γ around zero the parameter is related to the actual skewness of the distribution as $\text{skew}(X) = 6\gamma$. This implies that relatively small values of γ are likely to be most useful.

2.1 Practical Implementation for Probabilistic Programming

Using Lambert $W \times F$ distributions in probabilistic programs is in principle straightforward. Since the transformation is differentiable, we can construct a transformed distribution by combining any base distribution with a chain of two transformations: First we skew the draws using (1) and then perform the location-scale transformation (5). The backwards transformations and their derivatives required for evaluating the likelihood are known, and sampling from the distribution is easy. However, for a practical implementation for probabilistic programs we need to pay attention to some details.

Computation of W. The Lambert W function does not have a closed-form expression and hence requires iterative computation, typically carried out by the Halley's method. An initial guess $w \approx W(x)$ is iteratively refined using

$$w_{j+1} = w_j - \frac{w_j e^{w_j} - x}{e^{w_j}(w_j + 1) - \frac{(w_j+2)(w_j e^{w_j} - x)}{2w_j+2}},$$

where roughly 10 iterations seem to suffice in most cases. While this reasonably efficient procedure is available in most scientific computing languages, using an iterative algorithm for every likelihood evaluation is not very desirable. However, $W(x)$ is a scalar function that in practical applications needs to be evaluated only for a relatively narrow range of inputs. Hence it is easy to pre-compute and use tabulated values during learning, converting a trivial pre-computation effort into fast evaluation with practically arbitrarily high precision. The derivatives of $W(x)$ can be computed using the same pre-computation table due to (4).

Finite Support. The support of the Lambert $W \times F$ distribution described in [11] is finite. For $\gamma < 0$ the support is $[-\infty, -\frac{1}{\gamma e})$ and for $\gamma > 0$ it is $(-\frac{1}{\gamma e}, \infty]$. This imposes strict limits for the set of possible μ, σ and γ values compatible with given data y. This limitation is a property of the distribution itself and will hold for reasonable parameter values that describe the data well. However, during learning it poses unnecessary computational difficulties since naive unconstrained optimization over the parameters would often result in zero probability. To avoid this we would need to conduct optimization over a constraint set determined by interplay between μ, σ and γ.

Instead of constrained optimization we modify the density itself to have small probability for data outside the support of the original formulation, by altering the forward and backward transformations in a way that allows computation of the log-likelihood for arbitrary values. This solution has the advantage that it not only removes need for constrained optimization but it also makes the distribution more robust for outliers, allowing individual data points to fall outside the support because of additional noise components not taken into account by the model.

Our modified formulation uses a piece-wise extension of $W(x)$ defined as

$$W_e(x) = \begin{cases} W(x) & \text{if } x \geq d \\ W(d) + \frac{x-d}{x+e^{W(d)}} & \text{if } x < d \end{cases}$$

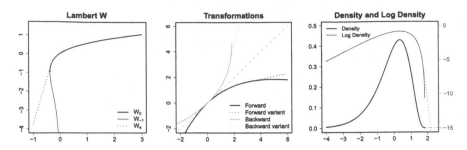

Fig. 2. Illustration of our modified Lambert W function (left) that replaces the W_{-1} branch with linear continuation of W_0, and its implication for the forward and backwards transformations (middle) for rather large skewness $\gamma = -0.2$. The forward transformations start deviating only at $z = 3$. When using $N(0,1)$ as the base distribution only 0.1% of samples fall into that region. The corresponding probability densities (right) are consequently nearly identical, the only difference being that the modified version has support over the whole real line but with extremely low probability for $x > 2$ excluded altogether by the original formulation.

with suitable choice of d that is between $-e^{-1}$ and 0 – in our experiments we used $d = -0.9e^{-1}$ but other values close to $-e^{-1}$ result in similar behavior. This modification replaces the secondary branch $W_{-1}(x)$ (and small part of the main branch) with linear extension that matches the main branch $W_0(x)$ and its derivative at d. While this extension clearly breaks down characteristic properties of the Lambert W function, making it inapplicable in most other uses of the function, the Lambert $W \times F$ family of distributions is mostly concerned about values for which $W_e(x) = W(x)$.

The resulting transformations are illustrated in Fig. 2 to demonstrate how they only influence the very low-probability tail of the distribution, giving finite but very small probabilities for samples that would be impossible under standard Lambert $W \times F$ distribution, enabling gradient-based learning.

On Skewness. With $\gamma = 0$ the backward transformation (3) involves division by zero. Any practical implementation should hence implement the non-skewed scenario as a special case. When learning skewness we also want to avoid extreme values for improved stability. We re-parameterize $\gamma = \gamma_m \tanh \hat{\gamma}$ so that $\hat{\gamma}$ can take any real value while limiting $|\gamma| < \gamma_m$. We use $\gamma_m = 0.2$ in our experiments, corresponding to roughly maximum skewness of 1.2.

2.2   Lambert Matrix Factorization

Probabilistic matrix factorization refers to the model

$$x|\theta, \beta, \phi \sim p(x|\theta^T \beta, \phi), \qquad \theta \sim p(\theta), \qquad \beta \sim p(\beta),$$

where a bilinear term $\theta^T \beta$ models the mean of the generating distribution and ϕ refers to all other parameters of the distribution, such as the variance in case of normal likelihood. This formulation covers practical models proposed for recommender engines [21], probabilistic interpretation of PCA [15] and CCA [17], group factor analysis [18], and non-negative matrix factorization [22]. All these models differ merely in the choice of priors and likelihoods.

Lambert $W \times F$ distributions as presented here are designed to be used with arbitrary probabilistic models, and hence are applicable for matrix factorization as well. We use $\mathcal{W}_F(\theta^T \beta, \sigma, \gamma)$ as the likelihood, modeling the mean still with a bilinear term. We can then either treat $\phi = \{\sigma, \gamma\}$ as global parameters to modulate the scale and skewness of the noise, or assume more complicated structure for them as well. In fact, we will later make also σ a function of $\theta^T \beta$.

3   Modeling Dyadic Data

As a practical application of the Lambert $W \times F$ distribution in matrix factorization, we consider the problem of learning interpretable distributed representations for dyadic data, observed as N triplets (i_n, j_n, C_n) interpreted as object i occurring together with item j with associated scalar count c for the number of such co-occurrences. The classical example concerns learning word embeddings based on observing that word j occurs c times in the context of word i, providing for each word a vectorial representation θ_i. Another example considers media consumption, where user i has, for example, listened to the song j for c times and high counts are interpreted as expressing interest for the song.

Such data can be thought of having been generated by drawing N independent observations from a joint density $p(i, j)$ over pairs of i and j, corresponding to each count C_{ij} following a Binomial distribution $\text{Binom}(N, p_{ij})$. Since N is here typically very large and each p_{ij} is small since they need to sum up to one over IJ entries, the distribution can accurately be approximated also by $\text{Poisson}(\lambda_{ij})$ where $\lambda_{ij} = N p_{ij}$. We will later use both interpretations when deriving specific elements of our proposed generative model.

To extract meaningful vectorial representations from such data, we should model ratios of co-occurrence probabilities with linear models. In practice this is typically achieved by modeling point-wise mutual information (or its empirical estimates) with

$$\theta_i^T \beta_j \approx \log \frac{p(j, i)}{p(i)p(j)} \approx \log \frac{C_{ij} N}{C_i C_j}.$$

This can equivalently be written as matrix factorization with row and column biases for the logarithmic counts themselves, letting the bias terms learn the negative logarithmic counts of the marginals [23]:

$$\theta_i^T \beta_j + a_i + b_j - \log N \approx \log C_{ij}.$$

The state of the art techniques solve the above problem by minimizing the squared error between the approximation and the logarithmic counts. Importantly, they weight the individual terms of the loss by a term that increases roughly linearly as a function of C_{ij}, the actual count, and hence exponentially as a function of $\log C_{ij}$ that is being modeled. Putting more effort in modeling large counts is intuitively reasonable, but it makes generative interpretations of these models difficult.

3.1   Generative Embedding Model for Dyadic Data

We propose to model such dyadic data with the following probabilistic program. It assumes bilinear form with bias terms for a location parameter μ_{ij} and uses the same parameter for controlling three aspects of the data: the probability of observing a co-occurrence, the expected logarithmic count, and the variance.

The first stage is achieved by drawing a binary variable h_{ij} from a Bernoulli distribution with parameter $f(\mu_{ij}) = 1 - e^{-e^{\mu_{ij}}}$ (See Sect. 3.2). For $h_{ij} = 1$ the model then generates logarithmically transformed observed data $y_{ij} = \log C_{ij}$ from a Lambert $W \times F$ distribution with mean μ_{ij} and $\sigma_{ij}(\mu_{ij})$. Finally, the observed count is transformed through an exponential linear unit to guarantee positive output (since $\log C \geq 0$ for $C > 0$).

The full model is

$$\mu_{ij} = \theta_i^T \beta_j + a_i + b_j - \log N$$
$$h_{ij} \sim \text{Bernoulli}(f(\mu_{ij}))$$
$$y_{ij}|h_{ij} = 0 = -\infty$$
$$y_{ij}|h_{ij} = 1 = \begin{cases} \bar{y}_{ij} & \text{if } \bar{y}_{ij} \geq 0 \\ \alpha(e^{\bar{y}_{ij}} - 1) & \text{if } \bar{y}_{ij} < 0 \end{cases}, \text{ where} \tag{6}$$
$$\bar{y}_{ij} \sim \mathcal{W}_N(\mu_{ij}, \sigma(\mu_{ij}), \gamma),$$

with suitable priors assigned for the parameters. In practice the constant $\log N$ can be incorporated as a part of either bias term. The model includes several non-obvious elements that are critical in modeling log-transformed counts, listed below and motivated in detail in the following subsections.

1. A single location parameter μ controls (a) the probability $f(\mu)$ of observing a co-occurrence, (b) the expected logarithmic count μ of co-occurrences, and (c) the variance $\sigma(\mu_{ij})$ of the logarithmic counts
2. The likelihood function is left-skewed, belonging to the Lambert $W \times F$ family and using normal distribution as the base distribution
3. The variance $\sigma(\mu_{ij})^2$ is maximal around $\mu = 1$ and decreases exponentially towards both infinities.

3.2   Missing Co-occurrences

Many dyadic data sets are very sparse, so that most co-occurrences are missing. Various solutions to this challenge have been proposed: GloVe [23] ignored the missing co-occurrences completely by setting the corresponding weight in loss function to zero, likelihood evaluations in hierarchical Poisson factorization scale linearly in the observed co-occurrences [12] but only for models that are linear in the actual count space, and Swivel [25] models missing co-occurrences based on an alternative loss that penalizes for over-estimating the point-wise mutual information and uses parallel computation for fast evaluation.

We replace these choices with a Bernoulli model that determines whether a particular co-occurrence is seen, so that the probability of generating a zero comes directly from the assumed process of Binomial draws for each pair. By approximating $\text{Binom}(N, p_{ij})$ with $\text{Poisson}(\lambda_{ij})$ we immediately see that $p(C_{ij} = 0) = e^{-\lambda_{ij}}$. Since we model the expected logarithmic count by μ_{ij} we note that $\lambda_{ij} = e^{\mu_{ij}}$ and hence $p(C_{ij} = 0) = e^{-e^{\mu_{ij}}}$ and $f(\mu_{ij}) = p(C_{ij} = 1) = 1 - e^{-e^{\mu_{ij}}}$. In other words, the underlying generative assumption directly implies that the same parameter controlling the expected value should control also the probability of success in the Bernoulli model.

The log-likelihood of this model can be contrasted with the loss in Swivel [25] assumed for measuring the error for the pairs that do not co-occur. They use a logistic loss between the model and data that corresponds to having observed a single co-occurrence, in an attempt to allow some leeway for the model. The loss Swivel uses, here written for logarithmic counts instead of point-wise mutual information for easier comparison, is $\log(1 + e^{\mu_{ij}})$. Our model, in turn, results in negative logarithmic likelihood $-\log p(C_{ij} = 0) = e^{\mu_{ij}}$, and hence it penalizes more heavily for observing no co-occurrences for large μ.

3.3   Likelihood

Let us next take a look at the likelihood function, which we assume to be Lambert $W \times N$ with left-skewness indicated by $\gamma < 0$. Furthermore, we assume the scale $\sigma(\mu)$ of the distribution is controlled by the same parameter μ that controls the mean μ. This has close connection with the weighted least squares error [23,25],

$$-\frac{1}{2} \sum_{i,j} h(e^{y_{ij}}) \|y_{ij} - \mu_{ij}\|^2,$$

where y_{ij} is the logarithmic count and $h(\cdot)$ is a function that grows according to the observed count itself. The loss is superficially similar to the negative log-likelihood of normal distribution, merely replacing the constant precision τ with one parameterized by the observed data y.

Since the weight depends on the observed data the loss does not match an actual normal distribution. However, if we assume it corresponds to negative log-likelihood of some distribution then we can interpret the loss function directly as a log-likelihood by normalizing it suitably. Alternatively, we can search for equiprobability contours of the loss function around any given μ by solving for

$$\frac{1}{2} \tau(x)(x - \mu)^2 = c$$

for some constant $c > 0$. Plugging in $\tau(x) = e^{\alpha x}$ results in two solutions, one at each side of μ:

$$x = \mu + \frac{2}{\alpha} W \left(\frac{a\sqrt{c}}{\sqrt{2}} \sqrt{e^{-\alpha\mu}} \right) \quad \text{and} \quad x = \mu + \frac{2}{\alpha} W \left(-\frac{a\sqrt{c}}{\sqrt{2}} \sqrt{e^{-\alpha\mu}} \right).$$

For small c, via linearization of the Lambert W, these correspond to

$$x = \mu \pm \frac{\sqrt{2}}{\sqrt{e^{\alpha\mu}}} \sqrt{c}$$

which matches the equiprobability contours of normal distribution with precision $\tau = \sqrt{e^{\alpha\mu}}$. For larger values the solution on the left side is further away from μ than the one on the right, corresponding to negative skewness. Furthermore, the separation of the distances is controlled by the Lambert W function that is used as the backward transformation in the Lambert $W \times F$ family of distributions.

The derivation shows that the likelihoods induced by GloVe (which uses $h(y_{ij}) = e^{3/4 y_{ij}}$) and Swivel ($h(y_{ij}) = 0.1 + 0.25 e^{1/2 y_{ij}}$) correspond to negatively skewed likelihoods where the variance and mean are controlled by the same parameter. The derivation does not imply direct correspondence with the Lambert $W \times F$ family since the $W(x)$ function is here used in slightly different fashion. Nevertheless, empirical comparison with the GloVe and Swivel losses normalized into likelihoods and the Lambert W family with suitable skewness reveal that the family is able to replicate the behavior accurately (Fig. 3). Note that GloVe and Swivel losses only determine relative precisions, not absolute ones; for the illustration we re-scaled them to match the overall scaling of our model.

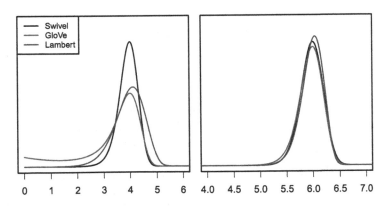

Fig. 3. Comparison of the densities induced by GloVe and Swivel and the explicit density modeled by Lambert W distribution in our model, using $\gamma = -0.1$. The x-axis corresponds to logarithmic count. For small expected counts (left) GloVe induces likelihood that is even more left-skewed (due to pushing precision to zero for very small values), but for expected log-counts around 6 (right) the distributions are almost identical for all three methods.

3.4   Variance Scaling

The final missing piece concerns the function $\sigma(\mu)$ used for scaling the variance. As mentioned above, GloVe and Swivel assume here simple exponential formulas for the precision, $e^{3/4y}$ and $0.1 + 0.25e^{1/2y}$ respectively. To derive a justified variance scaling we start with the series expansion

$$\text{Var}[\log C] \approx \frac{1}{\mathbb{E}[C]^2}\text{Var}[C],$$

which holds for large values. Since $C \propto e^{\mu}$, we see the variance of $y = \log C$ should indeed decrease exponentially with the mean, matching the general intuition of GloVe and Swivel.

Many of the observations, however, are for C in single digits, for which the approximation does not hold. Whenever $C = 0$ the variance becomes infinite, but our generative model takes care of these instances with the h variable. Hence, what we need is a formula for the variance of $\log C$ when $C > 0$, applicable for small C. We are not aware of simple analytic expressions for this, but we can easily analyze the empirical behavior of large collection of samples drawn from the Binomial distribution assumed to generate the data. Figure 4 shows that for large μ the variance follows the analytic expression $e^{-\mu}$ and for large negative μ it follows e^{μ}. The two regions smoothly transition to each other in the middle,

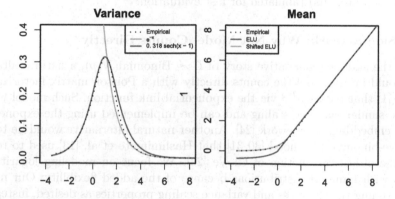

Fig. 4. Logarithm of data drawn from Binomial model gives raise to specific non-linear mapping from the mean parameter to variance and mean of the distribution. The left plot depicts the variance of $\log C$ for $C > 0$ and we see the empirical variance (dotted black line) is well captured by the hyperbolic secant function (red line). For positive parameters also the analytic expression (green line) is a good match. The right plot depicts the mean of C for $C > 0$. The empirical curve (dotted black line) can be matched by re-scaled and shifted exponential linear unit (ELU), but in practice standard ELU (red line) that gives negative values for negative inputs works better since it does not map the zero observations to minus infinity in the parameter space. (Color figure online)

with maximum value of 0.318 reached at $\mu = 1$. This can be approximated by an average of e^{μ} and $e^{-\mu}$, resulting in hyperbolic cosine (for precision) or hyperbolic secant (for variance). If we further match the maximal variance at $\mu = 1$, we get

$$\sigma(\mu) = \sqrt{\frac{0.318}{\cosh(\mu - 1)}}$$

that is almost exact replicate of the empirical variance except that for large μ the inputs are off by one. We allow for this slight overestimation of variance to avoid excessively narrow distributions for very large counts.

3.5   Computation

Computing the likelihood requires evaluating the loss for all co-occurrences, including ones that are not observed in the data. For efficient maximum likelihood estimation we adopt the parallel computation strategy presented in Swivel [25]. The data is split into shards, $k \times k$ submatrices, that roughly maintain the frequency characteristics of the original data, giving rise to efficient mini-batch training. Further speedups can be obtained by performing asynchronous updates in parallel so that each worker fetches and stores parameters on a central parameter server in a lock-free fashion [9]. Because the Lambert function $W(x)$ and its derivative are required for only a narrow range of inputs, they can be pre-computed and tabulated for fast evaluation.

3.6   Side Remark: Why Not Model Counts Directly?

Given the assumed generative story of $C \sim \text{Binomial}(N, p)$, a natural alternative would be to model the counts directly with a Poisson matrix factorization model [1] that passes $\theta^T \beta$ via the exponential link function. Such model would achieve similar variance scaling and can be implemented using the exponential family embeddings framework [24]. Another natural alternative would be to use negative-binomial likelihood [30,31] that Hashmimoto et al. [13] used to reproduce the objective function of GloVe [23]. We focus on modeling logarithmic counts with Lambert distributions because of the added flexibility: Our model allows tuning the skewness and variance scaling properties as desired, instead of being forced to adopt the exact choices Poisson and negative binomial distributions induce that may be suboptimal for the data.

4   Experiments

4.1   Lambert Matrix Factorization for Skewed Data

As a sanity check we run the model on artificial data sampled from the model itself. Given randomly generated θ and β, we sample data from the process

$$z \sim \mathcal{N}(0, 1), \qquad x = \theta^T \beta + z e^{\gamma z},$$

using $\gamma = -0.05$ to produce slightly left skewed data. We then perform gradient-based learning for θ and β using fixed γ to evaluate the log-likelihood of the observed data under various choices of skewness. The first plot in Fig. 5 illustrates how the likelihood is maximized with the right value, and this is verified by performing inference over γ; we find $\hat{\gamma} = -0.058$ as the learnt parameter.

4.2 Representation Learning for Co-occurrence Data

Next we evaluate the importance of individual elements of our model in generative description of dyadic data. We sample a total of $N = 400,000$ tokens for $I = 1000$ and $J = 200$ with low-rank (10 factors) linear structure for p_{ij}. We then model the resulting dyadic data $\{C_{ij}, i, j\}$ using variants of the model (6). In particular, we compare the full model against alternatives that omit one or more of the elements considered crucial parts of the generative model. We try both replacing the Lambert distribution with symmetric normal distribution and replacing the $\sigma(\mu)$ function with constant σ to remove the variance scaling.

We train the models on data where only 70% of the co-occurrences – missing or not – are observed and the rest are left as test data, to demonstrate the added benefit of working with generative models. Figure 5 plots the training and test log-likelihoods for the variants mentioned above, showing how the algorithm converges rapidly. Both scaling the variances and using left-skewed distributions are found to be important – each element alone improves the likelihoods and the full model combining both is the best.

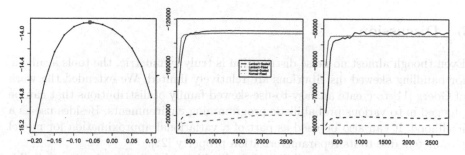

Fig. 5. *Left:* Demonstration of Lambert W for artificial data generated from the model with $\gamma = -0.05$. *Center and right:* Progress of training (center) and test (right log-likelihoods as function of the iteration. The full model with skewed distribution and variance scaling (black solid line) outperforms alternatives that are either forced to use symmetric normal likelihood (red lines) or constant variance for all entries (dashed lines). (Color figure online)

Table 1. Performance of the proposed Lambert matrix factorization and the comparison method Swivel in standard word embedding tasks. Higher is better for all metrics.

METRIC	SWIVEL	LMF
Word Similarity, Finkelstein et al. [10]	0.704	**0.714**
Word Relatedness, Finkelstein et al. [10]	0.578	**0.608**
MEN dataset, Bruni et al. [7]	0.674	**0.680**
Stanford Rare Word, Luong et al. [20]	**0.403**	0.397
SimLex-999, Hill et al. [14]	0.291	**0.332**

4.3 Word Representations

Finally, we compare the proposed model against its closest non-generative comparison Swivel [25] on a word embedding task. Swivel uses two losses, one for positive co-occurrences and one for the missing ones, whereas our model performs maximum likelihood inference for the model (6) and hence combines both elements in a single loss. For efficient computation for large data both models use the sharding technique [25]. We use a one gigabyte snapshot of the Wikipedia dump data with vocabulary size of 50 K and shard size of 1024 to infer embeddings of 300 dimensions. We measure performance using standard evaluation metrics, presenting the results in Table 1. The proposed method matches the accuracy of Swivel in all comparisons, showing that our probabilistic program matches the key properties of word embedding models.

5   Discussion

Even though almost no noise distribution is truly symmetric, the tools available for handling skewed distributions are relatively limited. We extended the work of Goerg [11] to create an easy-to-use skewed family of distributions that can be plugged in to various probabilistic programming environments. Besides use as a likelihood, it can also be used as part of a variational approximation for model parameters due to its reparameterization property [28].

We demonstrated the flexibility of the likelihood family by using it as a building block in purely generative model for learning embeddings from dyadic data. We analyzed the loss functions proposed for this task by GloVe [23] and Swivel [25] and showed that their squared errors weighted by the observed data itself can be interpreted as left-skewed distributions where the skewing is performed with the Lambert W function.

Our main goal was to introduce the probability density and its computational facilities, and to provide a proof-of-concept application. The proposed model roughly matches the accuracy of Swivel in various word embedding evaluation tasks, but we did not seek to provide maximally accurate embeddings. Instead, we see the most likely use cases for our solution in more complex models that build on the same principles but include more hierarchy instead of just learning simple vectorial representation, for example in form of time-evolving embeddings [5] and mixtures for polysemous embeddings [27]. We also note that even though we here performed maximum a posteriori analysis for easier comparison, the machinery presented here would directly allow full posterior analysis as well.

Acknowledgements. The project was supported by Academy of Finland (grants 266969 and 313125) and Tekes (Scalable Probabilistic Analytics).

References

1. Ailem, M., Role, F., Nadif, M.: Sparse poisson latent block model for document clustering. IEEE Trans. Knowl. Data Eng. **29**(7), 1563–1576 (2017)
2. Archambeau, C., Delannay, N., Verleysen, M.: Robust probabilistic projections. In: Proceedings of the 23rd International Conference on Machine Learning, pp. 33–40 (2006)
3. Arnold, B., Beaver, R.J.: The skew-Cauchy distribution. Stat. Probab. Lett. **49**, 285–290 (2000)
4. Azzalini, A., Capitanio, A.: Distributions generated by perturbation of symmetry with emphasis on a multivariate skew t distribution. J. Roy. Stat. Soc. Ser. B **65**, 367–389 (2003)
5. Bamler, R., Mandt, S.: Dynamic word embeddings. In: Proceedings of the 34th International Conference on Machine Learning (2017)
6. Betancourt, M.: A conceptual introduction to Hamiltonian Monte Carlo. Technical report. arXiv:1701.02434 (2017)
7. Bruni, E., Boleda, G., Baroni, M., Tran, N.K.: Distributional semantics in technicolor. In: Proceedings of the 50th Annual Meeting of the Association for Computational Linguistics: Long Papers - Volume 1, ACL 2012, pp. 136–145. Association for Computational Linguistics (2012)
8. Corless, R., Gonnet, G., Hare, D., Jeffrey, D., Knuth, D.: On the Lambert W function. Adv. Comput. Mathe. **5**(1), 329–359 (1993)
9. Dean, J., et al.: Large scale distributed deep networks. Adv. Neural Inf. Process. Syst. **25**, 1223–1231 (2012)
10. Finkelstein, L., et al.: Placing search in context: the concept revisited. In: Proceedings of the 10th International Conference on World Wide Web, pp. 406–414. ACM (2001)
11. Goerg, G.M.: Lambert W random variables - a new family of generalized skewed distributions with applications to risk estimation. Ann. Appl. Stat. **5**(3), 2197–2230 (2011)
12. Gopalan, P., Hofman, J.M., Blei, D.M.: Scalable recommendation with hierarchical poisson factorization. In: Proceedings of the 31st Conference on Uncertainty in Artificial Intelligence, pp. 326–335 (2015)
13. Hashimoto, T.B., Alvarez-Melis, D., Jaakkola, T.S.: Word, graph and manifold embedding from markov processes. CoRR abs/1509.05808 (2015)
14. Hill, F., Reichart, R., Korhonen, A.: Simlex-999: evaluating semantic models with genuine similarity estimation. Comput. Linguist. **41**, 665–695 (2015)
15. Ilin, A., Raiko, T.: Practical approaches to principal component analysis in the presence of missing data. J. Mach. Learn. Res. **11**, 1957–2000 (2010)
16. Jameel, S., Schockaert, S.: D-GloVe: a feasible least squares model for estimating word embedding densities. In: Proceedings of the 26th International Conference on Computational Linguistics, pp. 1849–1860 (2016)

17. Klami, A., Virtanen, S., Kaski, S.: Bayesian canonical correlation analysis. J. Mach. Learn. Res. **14**, 965–1003 (2013)
18. Klami, A., Virtanen, S., Leppäaho, E., Kaski, S.: Group factor analysis. IEEE Trans. Neural Netw. Learn. Syst. **26**(9), 2136–2147 (2015)
19. Li, S., Zhu, J., Miao, C.: A generative word embedding model and its low rank positive semidefinite solution. In: Proceedings of the 2015 Conference on Empirical Methods in Natural Language Processing, pp. 1599–1609 (2015)
20. Luong, T., Socher, R., Manning, C.D.: Better word representations with recursive neural networks for morphology. In: Proceedings of the 17th Conference on Computational Natural Language Learning, CoNLL 2013, pp. 104–113 (2013)
21. Mnih, A., Salakhutdinov, R.R.: Probabilistic matrix factorization. In: Advances in Neural Information Processing Systems, pp. 1257–1264 (2008)
22. Paisley, J., Blei, D., Jordan, M.: Bayesian nonnegative matrix factorization with stochastic variational inference. In: Handbook of Mixed Membership Models and Their Applications. Chapman and Hall (2014)
23. Pennington, J., Socher, R., Manning, C.D.: Glove: Global vectors for word representation. In: Empirical Methods in Natural Language Processing (EMNLP), pp. 1532–1543 (2014)
24. Rudolph, M., Ruiz, F., Mandt, S., Blei, D.: Exponential family embeddings. In: Advances in Neural Information Processing Systems, pp. 478–486 (2016)
25. Shazeer, N., Doherty, R., Evans, C., Waterson, C.: Swivel: Improving embeddings by noticing what's missing. arXiv:1602.02215 (2016)
26. Teimouri, M., Rezakhah, S., Mohammdpour, A.: Robust mixture modelling using sub-Gaussian alpha-stable distribution. Technical report. arXiv:1701.06749 (2017)
27. Tian, F., Dai, H., Bian, J., Gao, B.: A probabilistic model for learning multi-prototype word embeddings. In: Proceedings of the 25th International Conference on Computational Linguistics, pp. 151–160 (2014)
28. Titsias, M., Lázaro-Gredilla, M.: Doubly stochastic variational bayes for non-conjugate inference. In: Proceedings of International Conference on Machine Learning (ICML) (2014)
29. Vilnis, L., McCallum, A.: Word representations via Gaussian embeddings. In: Proceedings of International Conference on Learning Representations (2015)
30. Zhou, M., Carin, L.: Negative binomial process count and mixture modeling. IEEE Trans. Pattern Anal. Mach. Intell. **37**(2), 307–320 (2015)
31. Zhou, M., Hannah, L., Dunson, D., Carin, L.: Beta-negative binomial process and Poisson factor analysis. In: Proceedings of Artificial Intelligence and Statistics, pp. 1462–1471 (2012)

Identifying and Alleviating Concept Drift in Streaming Tensor Decomposition

Ravdeep Pasricha$^{(\boxtimes)}$, Ekta Gujral, and Evangelos E. Papalexakis

Department of Computer Science and Engineering, University of California Riverside,
900 University Avenue, Riverside, CA, USA
{rpasr001,egujr001}@ucr.edu, epapalex@cs.ucr.edu

Abstract. Tensor decompositions are used in various data mining applications from social network to medical applications and are extremely useful in discovering latent structures or *concepts* in the data. Many real-world applications are dynamic in nature and so are their data. To deal with this dynamic nature of data, there exist a variety of online tensor decomposition algorithms. A central assumption in all those algorithms is that the number of latent concepts remains fixed throughout the entire stream. However, this need not be the case. Every incoming batch in the stream may have a different number of latent concepts, and the difference in latent concepts from one tensor batch to another can provide insights into how our findings in a particular application behave and deviate over time. In this paper, we define "concept" and "concept drift" in the context of streaming tensor decomposition, as the manifestation of the variability of latent concepts throughout the stream. Furthermore, we introduce *SeekAndDestroy* (The method name is after (and a tribute to) Metallica's song from their first album (who also owns the copyright for the name)), an algorithm that detects concept drift in streaming tensor decomposition and is able to produce results robust to that drift. To the best of our knowledge, this is the first work that investigates concept drift in streaming tensor decomposition. We extensively evaluate *SeekAndDestroy* on synthetic datasets, which exhibit a wide variety of realistic drift. Our experiments demonstrate the effectiveness of *SeekAndDestroy*, both in the detection of concept drift and in the alleviation of its effects, producing results with similar quality to decomposing the entire tensor in one shot. Additionally, in real datasets, *SeekAndDestroy* outperforms other streaming baselines, while discovering novel useful components. Code related to this paper is available at: https://github.com/ravdeep003/conceptDrift.

Keywords: Tensor analysis · Streaming · Concept drift
Unsupervised learning

1 Introduction

Data comes in many shapes and sizes. Many real world applications deal with data that is multi-aspect (or multi-dimensional) in nature. An example of multi-aspect data would be interactions between different users in a social network over

© Springer Nature Switzerland AG 2019
M. Berlingerio et al. (Eds.): ECML PKDD 2018, LNAI 11052, pp. 327–343, 2019.
https://doi.org/10.1007/978-3-030-10928-8_20

period of time. Interactions like who messages whom, who liked whose posts or who shared (re-tweet) whose post. This can be modeled as a three-mode tensor, user-user being two modes of the tensor and time being the third mode, where each data point can be considered as an interaction between two users.

Tensor decomposition has been used in many data mining applications and is an extremely useful tool for finding latent structures in tensor in an unsupervised fashion. There exist a wide variety of tensor decomposition models and algorithms available, interested readers can refer to [9,13] for details. In this paper, our main focus is on CP/PARAFAC decomposition [7] (henceforth refered to as CP for brevity), which decomposes a tensor into a sum of rank-one tensors, each one being a latent factor (or *concept*) in the data. CP has been widely used in many applications, due to its ability to uniquely uncover latent components in a variety of unsupervised multi-aspect data mining applications [13].

In today's world data is not static, data keeps on evolving over time. In real world applications like stock market and e-commerce websites hundred of transaction (if not thousands) takes place every second, or in applications like social media where every second, thousands of new interactions take place forming new communities of users who interact with each other. In this example, we consider each *community* of people within the graph as a *concept*.

There has been a considerable amount of work in dealing with online or streaming CP decomposition [6,11,16], where the goal is to absorb the updates to the tensor in the already computed decomposition, as they arrive, and avoid recomputing the decomposition every time new data arrives. However, despite the already existing work in the literature, a central issue has been left, to the best of our knowledge, entirely unexplored. All of the existing online/streaming tensor decomposition literature assumes that the concepts in the data (whose number is equal to the rank of the decomposition) remains *fixed* throughout the lifetime of the application. What happens if the number of components changes? What if a new component is introduced, or an existing component splits into two or more new components? This is an instance of *concept drift* in unsupervised tensor analysis, and this paper is a look at this problem from first principles.

Our contributions in this paper are the following:

- **Characterizing concept drift in streaming tensors:** We define concept and concept drift in time evolving tensors and provide a quantitative method to measure the concept drift.
- **Algorithm for detecting and alleviating concept drift in streaming tensor decomposition:** We provide an algorithm which detects drift in the streaming data and also updates the previous decomposition without any assumption on the rank of the tensor.
- **Experimental evaluation on real and synthetic data:** We extensively evaluate our method on both synthetic and real datasets and out-perform state of the art methods in cases where the rank is not known a priori and perform on par in other cases.
- **Reproducibility:** Our implementation is made publicly available[1] for reproducibility of experiments.

---

[1] https://github.com/ravdeep003/conceptDrift.

2   Problem Formulation

2.1   Tensor Definition and Notations

Tensor $\underline{\mathbf{X}}$ is collection of stacked matrices $(\mathbf{X}_1, \mathbf{X}_2, \ldots \mathbf{X}_K)$ with dimension $\mathbb{R}^{I \times J \times K}$, where I and J represents rows and columns of matrix and K represents number of views. In other words, a tensor is a higher order abstraction of a matrix. For simplicity, we call the term "dimension" as "mode" of tensor, where "modes" are the numbers of views used to index the tensor. The rank($\underline{\mathbf{X}}$) is the minimum number of rank-1 tensors computed from its latent components which are required to re-produce $\underline{\mathbf{X}}$ as their sum. Table 1 represents the notations used throughout the paper.

Table 1. Table of symbols and their description

Symbols	Definition
$\underline{\mathbf{X}}, \mathbf{X}, \mathbf{x}, x$	Tensor, Matrix, Column vector, Scalar
$\mathbb{R}$	Set of Real Numbers
$\circ$	Outer product
$\|\mathbf{A}\|_F, \|\mathbf{a}\|_2$	Frobenius norm, ℓ_2 norm
$\mathbf{X}(:,r)$	r^{th} column of $\mathbf{X}$
$\odot$	Khatri-Rao product (column-wise Kronecker product [13])

Tensor Batch: A batch is a (N-1)-mode partition of tensor $\underline{\mathbf{X}} \in \mathbb{R}^{I \times J \times K}$ where size is varied only in one mode and other modes remain unchanged. Here, tensor $\underline{\mathbf{X}}_{new}$ is of dimension $\mathbb{R}^{I \times J \times t_{new}}$ and existing tensor $\underline{\mathbf{X}}_{old}$ is of dimension $\mathbb{R}^{I \times J \times t_{old}}$. The full tensor $\underline{\mathbf{X}} = [\underline{\mathbf{X}}_{old}; \underline{\mathbf{X}}_{new}]$ where its temporal mode $K = t_{old} + t_{new}$. The tensor $\underline{\mathbf{X}}$ can be partitioned into horizontal $\underline{\mathbf{X}}(\mathrm{I},:,:)$, lateral $\underline{\mathbf{X}}(:,\mathrm{J},:)$, and frontal $\underline{\mathbf{X}}(:,:,\mathrm{K})$ mode.

CP Decomposition: The most popular and extensively used tensor decompositions is the Canonical Polyadic or CANDECOMP/PARAFAC decomposition, referred to as CP decomposition henceforth. Given a 3-mode tensor $\underline{\mathbf{X}}$ of dimension $\mathbb{R}^{I \times J \times K}$, and rank at most R can be written

$$\underline{\mathbf{X}} = \sum_{r=1}^{R} (a_r \odot b_r \odot c_r) \iff \underline{\mathbf{X}}(i,j,k) = \sum_{r=1}^{R} A(i,r)B(j,r)C(k,r)$$

$\forall\, i \in \{1, 2, \ldots, I\},\, j \in \{1, 2, \ldots, J\},\, k \in \{1, 2, \ldots, K\}$ and $\mathbf{A} \in \mathbb{R}^{I \times R}, \mathbf{B} \in \mathbb{R}^{J \times R}$ and $\mathbf{C} \in \mathbb{R}^{K \times R}$. For tensor approximation, we adopted minimizing least square criteria as $\mathcal{L} \approx \min \frac{1}{2}\|\underline{\mathbf{X}} - \mathbf{A}(\mathbf{C} \odot \mathbf{B})^T\|_F^2$ where $\|\underline{\mathbf{X}}\|_F^2$ is the sum of squares of its all elements and $\|.\|_F$ is *Frobenius* (norm). The CP model is nonconvex in $\mathbf{A}, \mathbf{B}$ and $\mathbf{C}$. We refer interested readers to popular surveys [9,13] on tensor decompositions and its applications for more details.

2.2   Problem Definition

Let us consider a social media network like Facebook, where a large number of users ($\approx 684K$) update information every single minute, and Twitter, where about $\approx 100K$ users tweet every minute[2]. Here, we have interactions arriving continuously at high velocity, where each interaction consists of User Id, Tag Ids, Device, and Location information etc. How can we capture such dynamic user interactions? How to identify concepts which can signify a potential newly emerging community, complete disappearance of interactions, or a merging of one or more communities to a single one? When using tensors to represent such dynamically evolving data, our problem falls under "streaming" or "online" tensor analysis. Decomposing streaming or online tensors is challenging task, and concept drift in incoming data makes the problem significantly more difficult, especially in applications where we care about characterizing the concepts in the data, in addition to merely approximating the streaming tensor adequately.

Before we conceptualize the problem that our paper deals with, we define certain terms which are necessary to set up the problem. Consider $\underline{\mathbf{X}}$ and $\underline{\mathbf{Y}}$ be two incremental batches of a streaming tensors of rank R and F respectively. Let $\underline{\mathbf{X}}$ be the initial tensor at time t_0 and $\underline{\mathbf{Y}}$ be the batch of the streaming tensor which arrives at time t_1 such as $t_1 > t_0$. The CP decomposition for these two tensors is given as follows:

$$\underline{\mathbf{X}} \approx \sum_{r=1}^{R} \mathbf{A}(:,r) \circ \mathbf{B}(:,r) \circ \mathbf{C}(:,r) \tag{1}$$

$$\underline{\mathbf{Y}} \approx \sum_{r=1}^{F} \mathbf{A}(:,r) \circ \mathbf{B}(:,r) \circ \mathbf{C}(:,r) \tag{2}$$

Concept: In case of tensors, we define *concept* as one latent component; a sum of R such components make up the tensor. In above equations tensor $\underline{\mathbf{X}}$ and $\underline{\mathbf{Y}}$ has R and F concepts respectively.

Concept Overlap: We define *concept overlap* as the set of latent concepts that are common or shared between two streaming CP decompositions. Consider Fig. 1 where R and F both are equal to three, which means both tensors $\underline{\mathbf{X}}$ and $\underline{\mathbf{Y}}$ have three concepts. Each concept of $\underline{\mathbf{X}}$ corresponds to each concept of $\underline{\mathbf{Y}}$. This means that there are three concepts that overlap between $\underline{\mathbf{X}}$ and $\underline{\mathbf{Y}}$. The minimum and maximum number of concept overlaps between two tensors can be zero and $\min(R, F)$ respectively. Thus, the value of concept overlap lies between 0 and $\min(R, F)$. In Sect. 3 we propose an algorithm for detecting such overlap.

$$0 \leq \text{Concept Overlap} \leq \min(R, F) \tag{3}$$

New Concept: If there exists a set of concepts which are not similar to any of the concepts already present in the most recent tensor batch, we call all such concepts in that set as *new concepts*. Consider Fig. 2(a), where $\underline{\mathbf{X}}$ has two

---

[2] https://mashable.com/2012/06/22/data-created-every-minute/.

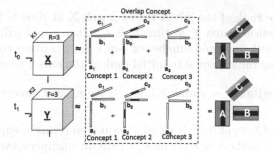

Fig. 1. Complete overlap of concepts

concepts ($R = 2$) and $\underline{Y}$ has three concepts ($F = 3$). We see that at time t_1 tensor $\underline{Y}$ batch has three concepts, out of which, two match with tensor $\underline{X}$ concepts and one concept (namely concept 3) does not match with any concept of $\underline{X}$. In this scenario we say that concept 1 and 2 are *overlapping* concepts and concept 3 is a *new concept*.

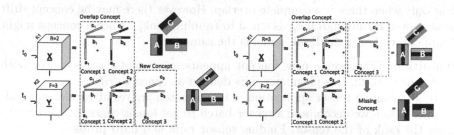

Fig. 2. (a) Concept appears (b) Concept disappears

Missing Concept: If there exists a set of concepts which was present at time t_0, but was missing at future time t_1, we call the concepts in the set *missing concepts*. For example, consider Fig. 2(b), at time t_0, the CP decomposition of $\underline{X}$ has three concepts, and at time t_1 CP decomposition of $\underline{Y}$ has two concepts. Two concepts of $\underline{X}$ and $\underline{Y}$ match with each other and one concept, present at t_0, is missing at t_1; we label that concept, as *missing concept*.

Running Rank: Running Rank (runningRank) at time t is defined as the total number of unique concepts (or latent components) seen until time t. Running Rank is different from tensor rank of a tensor batch. It may or may not be equal to rank of the current tensor batch. Consider Fig. 1, runningRank at time t_1 is three, since the total unique number of concepts seen until t_1 is three. Similarly runningRank of Fig. 2(b) at time t_1 is three, even though rank of $\underline{Y}$ is two, since the number unique concepts seen until t_1 is three.

Let us assume rank of the initial tensor batch $\underline{\mathbf{X}}$ at time t_0 is R and rank of the subsequent tensor batch $\underline{\mathbf{Y}}$ at time t_1 is F. Then runningRank at time t_1 is sum of running rank at t_0 and number of new concepts discovered from t_0 to t_1. At time t_0 running rank is equal to initial rank of the tensor batch in this case R.

$$\text{runningRank}_{t_1} = \text{runningRank}_{t_0} + num(\text{newConcept})_{t_1-t_0} \tag{4}$$

Concept Drift: Concept drift is usually defined in terms of supervised learning [3,14,15]. In [14], authors define concept drift in unsupervised learning as the change in probability distribution of a random variable over time. We define concept drift in the context of latent concepts, which is based on rank of the tensor batch. We first give an intuitive description of concept in terms of running rank, and then define concept drift.

Intuition: Consider running rank at time t_1 be runningRank_{t_1} and running at time t_2 be runningRank_{t_2}. If runningRank_{t_1} is not equal to runningRank_{t_2}, then there is a concept drift i.e. either a new concept has appeared, or a concept has disappeared. However, this definition does not capture every single case. Assume if runningRank_{t_1} is equal to runningRank_{t_2}. In this case, there is no drift only when there is a complete overlap. However there may be concept drift present even if runningRank_{t_1} is equal to runningRank_{t_2}, since a concept might disappear while runningRank remains the same.

Definition: Whenever a new concept appears, a concept disappears, or both from time $t1$ to $t2$, this phenomenon is defined as *concept drift*.

In a streaming tensor application, a tensor batch arrives at regular intervals of time. Before we decompose a tensor batch to get latent concepts, we need to know the rank of the tensor. Finding tensor rank is a hard problem [8] and it is beyond the scope of this paper. There has been considerable amount of work which approximates rank of a tensor [10,12]. In this paper we employ AutoTen [12] to compute a low rank of a tensor. As new advances in tensor rank estimation happen, our proposed method will also benefit.

Problem 1. **Given** (a) tensor $\underline{\mathbf{X}}$ of dimensions $I \times J \times K_1$ and rank R, (b) $\underline{\mathbf{Y}}$ of dimensions $I \times J \times K_2$ of rank F at time t_0 and t_1 respectively as shown in figure 3. Compute $\underline{\mathbf{X}}_{new}$ of dimension $I \times J \times (K_1 + K_2)$ of rank equal to runningRank at time t_1 as shown in equation (5) using factor matrices of $\underline{\mathbf{X}}$ and $\underline{\mathbf{Y}}$.

$$\underline{\mathbf{X}}_{new_{t_1}} \approx \sum_{r=1}^{\text{runningRank}} \mathbf{A}(:,r) \circ \mathbf{B}(:,r) \circ \mathbf{C}(:,r) \tag{5}$$

---

3   Proposed Method

Consider a social media application where thousands of connections are formed every second, for example, who follows whom or who interacts with whom. These connections formed can be viewed as forming communities. Over a period of time communities disappear, new communities appear or some communities reappear after sometime. Number of communities at any given point of time is dynamic. There is no way of knowing what communities will appear or disappear in future. When this data stream is captured as a tensor, communities refer to latent concepts and appearing and disappearing of communities over a period of a time is referred to as concept drift. Here we need a dynamic way of figuring out number of communities in a tensor batch rather than assuming constant number of communities in all tensor batches.

To the best of our knowledge, there is no algorithmic approach that detects concept drift in streaming tensor decomposition. As we mentioned in Sect. 1, there has been considerable amount of work [6,11,16] which deals with streaming tensor data and applies batch decomposition on incoming slices and combine the results. But these methods don't take change of rank in consideration, which could reveal new latent concept in the data sets. Even if we know the rank (latent concept) of the complete tensor, the tensor batches of that tensor might not have same rank as the complete tensor.

In this paper we propose *SeekAndDestroy*, a streaming CP decomposition algorithm that does not assume rank is fixed. *SeekAndDestroy* detects the rank of every incoming batch in order to decompose it, and finally, updates the existing decomposition after detecting and alleviating concept drift, as defined in Sect. 2.

An integral part of *SeekAndDestroy* is detecting different concepts and identifying concept drift in streaming tensor. In order to do this successfully, we need to solve following problems:

P1: Finding the rank of a tensor batch.
P2: Finding New Concept, Concept Overlap and Missing Concept between two consecutive tensor batch decomposition.
P3: Updating the factor matrices to incorporate the new and missing concepts along with concept overlaps.

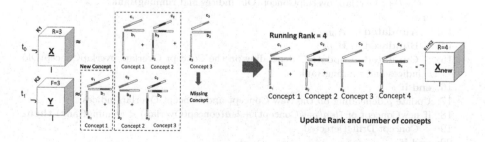

Fig. 3. Problem formulation

Finding Number of Latent Concepts: Finding the rank of the tensor is beyond the scope of this paper, thus we employ AutoTen [12]. Furthermore, in Sect. 4, we perform our experiments on synthetic data where we know the rank (and use that information as given to us by an "oracle") and repeat those experiments using AutoTen, comparing the error between them; the gap in quality signifies room for improvement that *SeekAndDestroy* will reap, if rank estimation is solved more accurately in the future.

Finding Concept Overlap: Given a rank of tensor batch, we compute its latent components using CP decomposition. Consider Fig. 3 as an example. At time t_1, the number of latent concepts we computed is represented by F, and we already had R components before new batch **Y** arrived. In this scenario, there could be three possible cases: (1) $R = F$ (2) $R > F$ (3) $R < F$.

For each one of the cases mentioned above, there may be new concepts appear at t_1, or concepts disappear from t_0 to t_1, or there could be shared concepts between two decompositions. In Fig. 3. we see that, even though R is equal to F, we have one new concept, one missing concept and two shared/overlapping concepts. Now, at time t_1, we have four unique concepts, which means our runningRank at t_1 is four.

Algorithm 1. *SeekAndDestroy* for Detecting & Alleviating Concept Drift

---

Input: Tensor $\underline{\mathbf{X}}_{new}$ of size $I \times J \times K_{new}$, Factor matrices $\mathbf{A}_{old}, \mathbf{B}_{old}, \mathbf{C}_{old}$ of size $I \times R$, $J \times R$ and $K_{old} \times R$ respectively, runningRank, mode.

Output: Factor matrices $\mathbf{A}_{updated}, \mathbf{B}_{updated}, \mathbf{C}_{updated}$ of size $I \times$ runningRank, $J \times$ runningRank and $(K_{new} + K_{old}) \times$ runningRank, ρ, runningRank.

1: $batchRank \leftarrow getRankAutoten(\underline{\mathbf{X}}_{new}, runningRank)$
2: $[\mathbf{A}, \mathbf{B}, \mathbf{C}, \boldsymbol{\lambda}] = CP(\underline{\mathbf{X}}_{new}, batchRank)$.
3: $\mathbf{colA}, \mathbf{colB}, \mathbf{colC} \leftarrow$ Compute Column Normalization of $\mathbf{A}, \mathbf{B}, \mathbf{C}$.
4: $\mathbf{normMatA}, \mathbf{normMatB}, \mathbf{normMatC} \leftarrow$ Absorb $\boldsymbol{\lambda}$ and Normalize $\mathbf{A}, \mathbf{B}, \mathbf{C}$.
5: $rhoVal \leftarrow colA .* colB .* colC$
6: [newConcept, conceptOverlap, $overlapConceptOld$] $\leftarrow$
    $findConceptOverlap(\mathbf{A}_{old}, \mathbf{normMatA})$
7: **if** newConcept **then**
8:     runningRank $\leftarrow$ runningRank $+ len($newConcept$)$
9:     $\mathbf{Aupdated} \leftarrow \left[\mathbf{A}_{old} \ \mathbf{normMatA}(:, \text{newConcept})\right]$
10:    $\mathbf{Bupdated} \leftarrow \left[\mathbf{B}_{old} \ \mathbf{normMatB}(:, \text{newConcept})\right]$
11:    $\mathbf{Cupdated} \leftarrow$ update $\mathbf{C}$ depending on the New Concept,
       Concept Overlap, overlapConceptOld indices and runningRank
12: **else**
13:    $\mathbf{Aupdated} \leftarrow \mathbf{A}_{old}$
14:    $\mathbf{Bupdated} \leftarrow \mathbf{B}_{old}$
15:    $\mathbf{Cupdated} \leftarrow$ update $\mathbf{C}$ depending on the Concept Overlap, overlapConceptOld
       indices and runningRank
16: **end if**
17: Update ρ depending on the New Concept and Concept Overlap indices
18: **if** newConcept or $(len($newConcept$) + len($conceptOverlap$) <$ runningRank$)$ **then**
19:    Concept Drift Detected
20: **end if**

In order to discover which concepts are shared, new, or missing we use the *Cauchy-Schwarz inequality* which states for two vectors $\mathbf{a}$ and $\mathbf{b}$ we have $\mathbf{a}^T\mathbf{b} \leq ||\mathbf{a}||_2||\mathbf{b}||_2$. Algorithm 2 provides the general outline of technique used in finding concepts. It takes a column-normalized matrices $\mathbf{A}_{\mathbf{old}}$ and $\mathbf{A}_{\mathbf{batch}}$ of size $I \times R$ and $I \times batchRank$ respectively as input. We compute the dot product for all permutations of columns between two matrices, as shown below

$$\mathbf{A}_{\mathbf{old}}^T(:, col_i) \cdot \mathbf{A}_{\mathbf{batch}}(:, col_j)$$

col_i and col_j are the respective columns. If the computed dot product is higher than the threshold value, the two concepts match, and we consider them as shared/overlapping between $\mathbf{A}_{\mathbf{old}}$ and $\mathbf{A}_{\mathbf{batch}}$. If the dot product between a column in $\mathbf{A}_{\mathbf{batch}}$ and with all the columns in $\mathbf{A}_{\mathbf{old}}$ has a value less than the threshold, we consider it as a new concept. This solves problem **P2**. In the experimental evaluation, we demonstrate the behavior of *SeekAndDestroy* with respect to that threshold.

SeekAndDestroy: This is our overall proposed algorithm, which detects concept drift between the two consecutive tensor batch decompositions, as illustrated in Algorithm 1 and updates the decomposition in a fashion robust to the drift. *SeekAndDestroy* takes factor matrices $(\mathbf{A}_{\mathbf{old}}, \mathbf{B}_{\mathbf{old}}, \mathbf{C}_{\mathbf{old}})$ of previous tensor batch (say at time t_0), running rank at $t_0(\mathbf{runningRank}_{t_0})$ and new tensor batch $(\underline{\mathbf{X}}_{new})$ (say at time t_1) as inputs. Subsequently, *SeekAndDestroy* computes the tensor rank for the batch (**batchRank**) for $\underline{\mathbf{X}}_{new}$ using AutoTen.

Using the estimated rank **batchRank**, *SeekAndDestroy* computes the CP decomposition of $\underline{\mathbf{X}}_{new}$, which returns factor matrices $\mathbf{A}, \mathbf{B}, \mathbf{C}$. We normalize the columns of A, B, C to unit ℓ_2 norm and we store the normalized matrices into **normMatA**, **normMatB**, and **normMatC**, as shown by lines 3–4 of Algorithm 1. Both $\mathbf{A}_{\mathbf{old}}$ and normalized matrix $\mathbf{A}$ are passed to $findConceptOverlap$ function as described above. This returns the indexes of new concept and indexes of overlapping concepts from both matrices. Those indexes inform *SeekAndDestroy*, while updating the factor matrices, where to append the overlapped concepts. If there are new concepts, we update A and B factor matrices simply by adding new columns from normalized factor matrices of $\underline{\mathbf{X}}_{new}$ as shown in lines 9–10 of Algorithm 1. Furthermore, we update the running rank by adding number of new concept discovered to the previous running rank. If there is only overlapping concepts and no new concepts, then $\mathbf{A}$ and $\mathbf{B}$ factor matrices does not change.

Updating Factor Matrix C: In this paper, for simplicity of exposition, we are focusing on streaming data that are increasing only on one mode. However, our proposed method readily generalizes to cases where more than one modes grow over time.

In order to update the "evolving" factor matrix ($\mathbf{C}$ in our case), we use a different technique from the one used to update $\mathbf{A}$ and $\mathbf{B}$. If there is a new concept discovered in **normMatC** then

$$\mathbf{C}_{updated} = \begin{bmatrix} \mathbf{C}_{old} & zeroCol \\ zerosM & \mathbf{normMatC}(:, newConcept) \end{bmatrix} \tag{6}$$

where $zeroCol$ is of size $K_{old} \times len(newConcept)$, $zerosM$ is of size $K_{new} \times R$ and $\mathbf{C}_{updated}$ is of size $(K_{old} + K_{new}) \times$ runningRank.

If there are overlapping concepts, then we update $\mathbf{C}$ accordingly as shown below; in this case $\mathbf{C}_{updated}$ is again of size $(K_{old} + K_{new}) \times$ runningRank.

$$\mathbf{C}_{updated} = \begin{bmatrix} \mathbf{C}_{old}(:, overlapConceptOld) \\ \mathbf{normMatC}(:, conceptOverlap) \end{bmatrix} \tag{7}$$

If there are missing concepts we append an all-zeros matrix (column vector) to those indexes.

The Scaling Factor ρ: When we reconstruct the tensor from updated factor (normalized) matrices, we need a way to re-scale the columns of those factor matrices. In our approach we compute element wise product on normalized columns of factor matrices ($\mathbf{A}$, $\mathbf{B}$, $\mathbf{C}$) of $\underline{\mathbf{X}}_{new}$ as shown in line 5 of Algorithm 1. We use the same technique as the one used in updating C matrix, in order to match the values between two consecutive intervals, and we add this value to previously computed values. If it is a missing concept, we simply add zero to it. While reconstructing the tensor we take the average of vector over the number of batches received and we re-scale the components as follows

$$\underline{\mathbf{X}}_r = \sum_{r=1}^{\text{runningRank}} \rho_r \mathbf{A}_{\text{upd.}}(:, r) \circ \mathbf{B}_{\text{upd.}}(:, r) \circ \mathbf{C}_{\text{upd.}}(:, r).$$

4   Experimental Evaluation

We evaluate our algorithm on the following criteria:

Q1: Approximation Quality: We compare *SeekAndDestroy*'s reconstruction accuracy against state-of-the-art streaming baselines, in data that we generate synthetically so that we observe different instances of concept drift. In cases where *SeekAndDestroy* outperforms the baselines, we argue that this is due to the detection and alleviation of concept drift.

Q2: Concept Drift Detection Accuracy: We evaluate how effectively *SeekAndDestroy* is able to detect concept drift in synthetic cases, where we control the drift patterns.

Q3: Sensitivity Analysis: As shown in Sect. 3, *SeekAndDestroy* expects the matching threshold as a user input. Furthermore, its performance may depend on the selection of the batch size. Here, we experimentally evaluate *SeekAndDestroy*'s sensitivity along those axes.

Algorithm 2. Find Concept Overlap

Input: Factor matrices $\mathbf{A}_{old}$, normMatA of size $I \times R$, $I \times batchRank$ respectively.
Output: newConcept, conceptOverlap, overlapConceptOld
1: $THRESHOLD \leftarrow 0.6$
2: **if** $R == batchRank$ **then**
3:    Generate all the permutations for [1:R]
4:    **foreach** *permutation* **do**
| Compute dot product of $\mathbf{A}_{old}$ *and* normMatA(:, **permutation**)
     end
5: **else if** $R > batchRank$ **then**
6:    Generate all the permutations(1:R, batchRank)
7:    **foreach** *permutation* **do**
| Compute dot product of $\mathbf{A}_{old}(:, permutation)$ *and* normMatA
     end
8: **else if** $R < batchRank$ **then**
9:    Generate all the permutations (1:batchRank, R)
10:    **foreach** *permutation* **do**
| Compute dot product of $\mathbf{A}_{old}$ *and* normMatA(:, **permutation**)
     end
11: **end if**
12: Select the best permutation based on the maximum sum.
13: If dot product value of a column is less than threshold its a New Concept
14: If dot product value of a column is more than threshold then its a Concept Overlap.
15: Return column index's of New Concept and Concept Overlap for both matrices

---

Q4: Effectiveness on Real Data: In addition to measuring *SeekAndDestroy*'s performance in real data, we also evaluate its ability to identify useful and interpretable latent concepts in real data, which elude other streaming baselines.

4.1 Experimental Setup

We implemented our algorithm in Matlab using tensor toolbox library [2] and we evaluate our algorithm on both synthetic and real data. We use [12] method available in literature to find rank of incoming batch.

In order to have full control of the drift phenomena, we generate synthetic tensors with different ranks for every tensor batch, we control the batch rank of the tensor with factor matrix **C**. Table 2 shows the specification of the datasets created. For instance dataset **SDS2** has an initial tensor batch whose tensor rank is 2 and last tensor batch whose tensor rank is 10 (full rank). The batches in between the initial and final tensor batch can have any rank between initial and final rank (in this case 2–10). The reason we assign the final batch rank as the full rank is to make sure the tensor created is not rank deficient. We make the synthetic tensor generator available as part of our code release.

In order for us to obtain robust estimates of performance, we require all experiments to either (1) run for 1000 iterations, or (2) the standard deviation converges to a second significant digit (whichever occurs first). For all reported results, we use the median and the standard deviation.

Table 2. Table of Datasets analyzed

DataSet	Dimension	Initial rank	Full rank	Batch size	Matching threshold
SDS1	$100 \times 100 \times 100$	2	5	10	0.6
SDS2			10		
SDS3	$300 \times 300 \times 300$	2	5	50	0.6
SDS4			10		
SDS5	$500 \times 500 \times 500$	2	5	100	0.6
SDS6			10		

4.2 Evaluation Metrics

We evaluate *SeekAndDestroy* and the baselines methods using *relative error*. Relative Error provides the measure of effectiveness of the computed tensor with respect to the original tensor and is defined as follows (lower is better):

$$RelativeError = \left(\frac{||\underline{\mathbf{X}}_{original} - \underline{\mathbf{X}}_{computed}||_F}{||\underline{\mathbf{X}}_{original}||_F} \right) \tag{8}$$

4.3 Baselines for Comparison

To evaluate our method, we compare *SeekAndDestroy* with two state-of-the-art streaming baselines: OnlineCP [16] and SamBaTen [6]. Both baselines assume that the rank remains fixed throughout the entire stream. When we evaluate the approximation accuracy of the baselines, we run two different versions of each method, with different input ranks: (1) *Initial Rank*, which is the rank of the initial batch, same as the one that *SeekAndDestroy* uses, and (2) *Full Rank*, which is the "oracle" rank of the full tensor, if we assume we could compute that in the beginning of the stream. Clearly, *Full Rank* offers a great advantage to the baselines since it provides information from the future.

4.4 Q1: Approximation Quality

The first dimension that we evaluate is the approximation quality. More specifically, we evaluate whether *SeekAndDestroy* is able to achieve good approximation of the original tensor (in the form of low error) in case where concept drift is occurring in the stream. Table 3 contains the general results of *SeekAndDestroy*'s accuracy, as compared to the baselines. We observe that *SeekAndDestroy* outperforms the two baselines, in the pragmatic scenario where they are given the same starting rank as *SeekAndDestroy* (Initial Rank). In the non-realistic, "oracle" case, OnlineCP performs better than SamBaTen and *SeekAndDestroy*, however this case is a very advantageous lower bound on the error for OnlineCP.

Through extensive experimentation we made the following interesting observation: in the cases where most of the concepts in the stream appear in the beginning of the stream (e.g., in batches 2 and 3), *SeekAndDestroy* was able to further outperform the baselines. This is due to the fact that, if *SeekAndDestroy* has already "seen" most of the possible concepts early-on in the stream, it is more likely to correctly match concepts in later batches of the stream, since there already exists an almost-complete set of concepts to compare against. Indicatively, in this case *SeekAndDestroy* achieved 0.1176 ± 0.0305 where as OnlineCP achieved 0.1617 ± 0.0702.

4.5   Q2: Concept Drift Detection Accuracy

The second dimension along which we evaluate *SeekAndDestroy* is its ability to successfully detect concept drift. Figure 4 shows the rank discovered by *SeekAndDestroy* at every point of the stream, plotted against the actual rank. We observe that *SeekAndDestroy* is able to successfully identify changes in rank, which, as we have already argued, signify concept drift. Furthermore, Table 4(b) shows three example runs that demonstrate the concept drift detection accuracy.

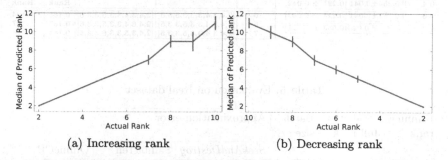

(a) Increasing rank   (b) Decreasing rank

Fig. 4. *SeekAndDestroy* is able to successfully detect concept drift, which is manifested as changes in the rank throughout the stream

Table 3. Approximation error for *SeekAndDestroy* and the baselines. *SeekAndDestroy* outperforms the baselines in the realistic case where all methods start with the same rank

DataSet	OnlineCP (Initial Rank)	OnlineCP (Full Rank)	SamBaTen (Initial Rank)	SamBaTen (Full Rank)	*SeekAndDestroy*
SDS1	0.2782 ± 0.0221	0.197 ± 0.086	$\mathbf{0.261 \pm 0.048}$	0.317 ± 0.058	0.283 ± 0.075
SDS2	0.2537 ± 0.0125	0.168 ± 0.507	$\mathbf{0.244 \pm 0.028}$	0.480 ± 0.051	0.253 ± 0.0412
SDS3	0.2731 ± 0.0207	0.205 ± 0.164	0.385 ± 0.021	0.445 ± 0.164	$\mathbf{0.266 \pm 0.081}$
SDS4	0.245 ± 0.013	0.171 ± 0.537	0.299 ± 0.045	0.402 ± 0.049	$\mathbf{0.221 \pm 0.0423}$
SDS5	0.2719 ± 0.0198	0.206 ± 0.022	0.559 ± 0.046	0.519 ± 0.0219	$\mathbf{0.256 \pm 0.105}$
SDS6	0.238 ± 0.013	0.171 ± 0.374	0.510 ± 0.036	0.547 ± 0.0276	$\mathbf{0.208 \pm 0.0433}$

4.6   Q3: Sensitivity Analysis

The results we have presented so far for *SeekAndDestroy* have used a matching
threshold of 0.6. The threshold was chosen because it is intuitively larger than
a 50% match, which is a reasonable matching threshold. In this experiment, we
investigate the sensitivity of *SeekAndDestroy* to the matching threshold parameter. Table 4(a) shows exemplary approximation errors for thresholds of 0.4, 0.6,
and 0.8. We observe that (1) the choice of threshold is fairly robust for values
around 50%, and (2) the higher the threshold, the better the approximation,
with threshold of 0.8 achieving the best performance.

Table 4. (a) Experimental results for error of approximation of incoming batch
with different matching threshold values. Dataset SDS2 and SDS4 are of dimension
$\mathbb{R}^{100 \times 100 \times 100}$ and $\mathbb{R}^{300 \times 300 \times 300}$, respectively. We see that the threshold is fairly robust
around 0.5, and a threshold of 0.8 achieves the highest accuracy (b) Experimental
results on SDS1 for error of approximation of incoming slices with known and predicted rank

Threshold	SDS2	SDS4
0.4	0.253±0.041	0.221 ± 0.042
0.6	0.253±0.041	0.221 ± 0.042
0.8	0.101 ±0.040	0.033 ± 0.011

Running Rank	Actual Rank	Predicted Rank	Approx. Error Actual Rank	Predicted Rank
6	[2,4,3,4,3,3,5,3,3,5]	[2,4,3,4,3,3,5,3,3,6]	0.185	0.194
6	[2,4,3,4,3,3,5,3,3,5]	[2,4,3,4,3,3,5,3,3,6]	0.185	0.197
7	[2,4,3,4,3,3,5,3,3,5]	[2,4,3,5,3,3,6,3,3,6]	0.185	0.278

Table 5. Evaluation on Real dataset

Running rank	Predicted full rank	Batch size	Approximation error		
			SeekAndDestroy	SambaTen	OnlineCP
7 ± 0.88	4 ± 0.57	22	**0.68 ± 0.002**	0.759 ± 0.059	0.941 ± 0.001

4.7   Q4: Effectiveness on Real Data

To evaluate effectiveness of our method on real data, we use the Enron time-
evolving communication graph dataset [1]. Our hypothesis is that in such complex real data, there should exists concept drift in streaming tensor decomposition. In order to validate that hypothesis, we compare the approximation error
incurred by *SeekAndDestroy* against the one incurred by the baselines, shown
in Table 5. We observe that the approximation error of *SeekAndDestroy* is lower
than the two baselines. Since the main difference between *SeekAndDestroy* and
the baselines is that *SeekAndDestroy* takes concept drift into consideration, and
strives to alleviate its effects, this result (1) provides further evidence that there
exists concept drift in the Enron data, and (2) demonstrates *SeekAndDestroy*'s
effectiveness on real data.

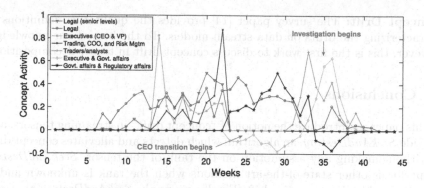

Fig. 5. Timeline of concepts discovered in Enron

The final rank for Enron as computed by *SeekAndDestroy* was 7, indicating the existence of 7 time-evolving communities in the dataset, as shown in Fig. 5. This number of communities is higher than what previous tensor-based analysis has uncovered [1,5]. However, analyzing the (static) graph using a highly-cited non-tensor based method [4], we were able to detect 7 communities, therefore *SeekAndDestroy* may be discovering subtle communities that have eluded previous tensor analysis. In order to verify that, we delved deeper into the communities and we plot their temporal evolution (taken from matrix **C**) along with their annotations (when inspecting the top-5 senders and receivers within each community). Indeed, a subset of the communities discovered matches with the ones already known in the literature [1,5]. Additionally, *SeekAndDestroy* was able to discover community #3, which refers to a group of executives, including the CEO. This community appears to be active up until the point that the CEO transition begins, after which point it dies out. This behavior is indicative of concept drift, and *SeekAndDestroy* was able to successfully discover and extract it.

5   Related Work

Tensor Decomposition: Tensor decomposition techniques are widely used for static data. With the explosion of big data, data grows at a rapid speed and an extensive study required on the online tensor decomposition problem. Sidiropoulos [11] introduced two well-known PARAFAC based methods namely RLST (recursive least square) and SDT (simultaneous diagonalization tracking) to address the online 3-mode tensor decomposition. Zhou et al. [16] proposed OnlineCP for accelerating online factorization that can track the decompositions when new updates arrived for N-mode tensors. Gujral et al. [6] proposed Sampling-based Batch Incremental Tensor Decomposition algorithm which updates online computation of CP/PARAFAC and performs all computations in the reduced summary space. However, no prior work addresses concept drift.

Concept Drift: The survey paper [14] provides the qualitative definitions of characterizing the drifts on data stream models. To the best of our knowledge, however, this is the first work to discuss concept drift in tensor decomposition.

6   Conclusions

In this paper we introduce the notion of "concept drift" in streaming tensors. and provide *SeekAndDestroy*, an algorithm which detects and alleviates concept drift it without making any assumption on the rank of the tensor. *SeekAndDestroy* outperforms other state-of-the-art methods when the rank is unknown and is effective in detecting concept drift. Finally, we apply *SeekAndDestroy* on a real time-evolving dataset, discovering novel drifting concepts.

Acknowledgements. Research was supported by the Department of the Navy, Naval Engineering Education Consortium under award no. N00174-17-1-0005, the National Science Foundation EAGER Grant no. 1746031, and by an Adobe Data Science Research Faculty Award. Any opinions, findings, and conclusions or recommendations expressed in this material are those of the author(s) and do not necessarily reflect the views of the funding parties.

References

1. Bader, B., Harshman, R., Kolda, T.: Analysis of latent relationships in semantic graphs using DEDICOM. In: Workshop for Algorithms on Modern Massive Data Sets (2006)
2. Bader, B.W., Kolda, T.G., et al.: MATLAB Tensor Toolbox Version 2.6, February 2015. http://www.sandia.gov/~tgkolda/TensorToolbox/
3. Bifet, A., Gama, J., Pechenizkiy, M., Zliobaite, I.: Handling concept drift: importance, challenges and solutions. PAKDD-2011 Tutorial, Shenzhen, China (2011)
4. Blondel, V.D., Guillaume, J.L., Lambiotte, R., Lefebvre, E.: Fast unfolding of communities in large networks. J. Stat. Mech. Theory Exp. **2008**(10), P10008 (2008)
5. Papalexakis, E.E., Faloutsos, C., Sidiropoulos, N.D.: ParCube: sparse parallelizable tensor decompositions. In: Flach, P.A., De Bie, T., Cristianini, N. (eds.) ECML PKDD 2012. LNCS, vol. 7523, pp. 521–536. Springer, Heidelberg (2012). https://doi.org/10.1007/978-3-642-33460-3_39
6. Gujral, E., Pasricha, R., Papalexakis, E.E.: SamBaTen: sampling-based batch incremental tensor decomposition. arXiv preprint arXiv:1709.00668 (2017)
7. Harshman, R.: Foundations of the PARAFAC procedure: models and conditions for an "explanatory" multimodal factor analysis (1970)
8. Håstad, J.: Tensor rank is NP-complete. J. Algorithms **11**(4), 644–654 (1990)
9. Kolda, T.G., Bader, B.W.: Tensor decompositions and applications. SIAM Rev. **51**(3), 455–500 (2009)
10. Mørup, M., Hansen, L.K.: Automatic relevance determination for multi-way models. J. Chemom. **23**(7–8), 352–363 (2009)
11. Nion, D., Sidiropoulos, N.: Adaptive algorithms to track the PARAFAC decomposition of a third-order tensor. Signal Process. **57**(6), 2299–2310 (2009)

12. Papalexakis, E.E.: Automatic unsupervised tensor mining with quality assessment. In: Proceedings of the 2016 SIAM International Conference on Data Mining, pp. 711–719. SIAM (2016)
13. Papalexakis, E.E., Faloutsos, C., Sidiropoulos, N.D.: Tensors for data mining and data fusion: models, applications, and scalable algorithms. ACM Trans. Intell. Syst. Technol. (TIST) **8**(2), 16 (2017)
14. Webb, G.I., Hyde, R., Cao, H., Nguyen, H.L., Petitjean, F.: Characterizing concept drift. Data Min. Knowl. Disc. **30**(4), 964–994 (2016)
15. Webb, G.I., Lee, L.K., Petitjean, F., Goethals, B.: Understanding concept drift. CoRR abs/1704.00362 (2017). http://arxiv.org/abs/1704.00362
16. Zhou, S., Vinh, N.X., Bailey, J., Jia, Y., Davidson, I.: Accelerating online CP decompositions for higher order tensors. In: Proceedings of the 22nd ACM SIGKDD International Conference on Knowledge Discovery and Data Mining, pp. 1375–1384. ACM (2016)

MASAGA: A Linearly-Convergent Stochastic First-Order Method for Optimization on Manifolds

Reza Babaneẓhad[(✉)], Issam H. Laradji, Alireza Shafaei, and Mark Schmidt

Department of Computer Science, University of British Columbia,
Vancouver, BC, Canada
{rezababa,issamou,shafaei,schmidtm}@cs.ubc.ca

Abstract. We consider the stochastic optimization of finite sums over a Riemannian manifold where the functions are smooth and convex. We present MASAGA, an extension of the stochastic average gradient variant SAGA on Riemannian manifolds. SAGA is a variance-reduction technique that typically outperforms methods that rely on expensive full-gradient calculations, such as the stochastic variance-reduced gradient method. We show that MASAGA achieves a linear convergence rate with uniform sampling, and we further show that MASAGA achieves a faster convergence rate with non-uniform sampling. Our experiments show that MASAGA is faster than the recent Riemannian stochastic gradient descent algorithm for the classic problem of finding the leading eigenvector corresponding to the maximum eigenvalue. Code related to this paper is available at: https://github.com/IssamLaradji/MASAGA.

Keywords: Variance reduced stochastic optimization ·
Riemannian manifold

1 Introduction

The most common supervised learning methods in machine learning use empirical risk minimization during the training. The minimization problem can be expressed as minimizing a finite sum of loss functions that are evaluated at a single data sample. We consider the problem of minimizing a finite sum over a Riemannian manifold,

$$\min_{x \in \mathcal{X} \subseteq \mathcal{M}} f(x) = \frac{1}{n} \sum_{i=1}^{n} f_i(x),$$

Electronic supplementary material The online version of this chapter (https://doi.org/10.1007/978-3-030-10928-8_21) contains supplementary material, which is available to authorized users.

M. Berlingerio et al. (Eds.): ECML PKDD 2018, LNAI 11052, pp. 344–359, 2019.
https://doi.org/10.1007/978-3-030-10928-8_21

where $\mathcal{X}$ is a geodesically convex set in the Riemannian manifold $\mathcal{M}$. Each function f_i is geodesically Lipschitz-smooth and the sum is geodesically strongly-convex over the set $\mathcal{X}$. The learning phase of several machine learning models can be written as an optimization problem of this form. This includes principal component analysis (PCA) [39], dictionary learning [34], Gaussian mixture models (GMM) [10], covariance estimation [36], computing the Riemannian centroid [11], and PageRank algorithm [33].

When $\mathcal{M} \equiv \mathbb{R}^d$, the problem reduces to convex optimization over a standard Euclidean space. An extensive body of literature studies this problem in deterministic and stochastic settings [5, 22, 23, 28, 29]. It is possible to convert the optimization over a manifold into an optimization in a Euclidean space by adding $x \in \mathcal{X}$ as an optimization constraint. The problem can then be solved using projected-gradient methods. However, the problem with this approach is that we are not explicitly exploiting the geometrical structure of the manifold. Furthermore, the projection step for the most common non-trivial manifolds used in practice (such as the space of positive-definite matrices) can be quite expensive. Further, a function could be non-convex in the Euclidean space but geodesically convex over an appropriate manifold. These factors can lead to poor performance for algorithms that operate with the Euclidean geometry, but algorithms that use the Riemannian geometry may converge as fast as algorithms for convex optimization in Euclidean spaces.

Stochastic optimization over manifolds and their convergence properties have received significant interest in the recent literature [4, 14, 30, 37, 38]. Bonnabel [4] and Zhang et al. [38] analyze the application of stochastic gradient descent (SGD) for optimization over manifolds. Similar to optimization over Euclidean spaces with SGD, these methods suffer from the aggregating variance problem [40] which leads to sublinear convergence rates.

When optimizing finite sums over Euclidean spaces, variance-reduction techniques have been introduced to reduce the variance in SGD in order to achieve faster convergence rates. The variance-reduction techniques can be categorized into two groups. The first group is memory-based approaches [6, 16, 20, 32] such as the stochastic average gradient (SAG) method and its variant SAGA. Memory-based methods use the memory to store a stale gradient of each f_i, and in each iteration they update this "memory" of the gradient of a random f_i. The averaged stored value is used as an approximation of the gradient of f.

The second group of variance-reduction methods explored for Euclidean spaces require full gradient calculations and include the stochastic variance-reduced gradient (SVRG) method [12] and its variants [15, 19, 25]. These methods only store the gradient of f, and not the gradient of the individual f_i functions. But, these methods occasionally require evaluating the full gradient of f as part of their gradient approximation and require two gradient evaluations per iteration. Although SVRG often dramatically outperforms the classical gradient descent (GD) and SGD, the extra gradient evaluation typically lead to a slower convergence than memory-based methods. Furthermore, the extra gradient calculations of SVRG can lead to worse performance than the classical SGD during the early iterations where SGD has the most advantage [9]. Thus, when the bottleneck of the process is the gradient computation itself, using memory-based methods like SAGA can improve performance [3,7]. Furthermore, for several applications it has been shown that the memory requirements can be alleviated by exploiting special structures in the gradients of the f_i [16,31,32].

Several recent methods have extended SVRG to optimize the finite sum problem over a Riemannian manifold [14,30,37], which we refer to as RSVRG methods. Similar to the case of Euclidean spaces, RSVRG converges linearly for geodesically Lipschitz-smooth and strongly-convex functions. However, these methods also require the extra gradient evaluations associated with the original SVRG method. Thus, they may not perform as well as potential generalizations of memory-based methods like SAGA.

In this work we present *MASAGA*, a variant of SAGA to optimize finite sums over Riemannian manifolds. Similar to RSVRG, we show that it converges linearly for geodesically strongly-convex functions. We also show that both MASAGA and RSVRG with a non-uniform sampling strategy can converge faster than the uniform sampling scheme used in prior work. Finally, we consider the problem of finding the leading eigenvector, which minimizes a quadratic function over a sphere. We show that MASAGA converges linearly with uniform and non-uniform sampling schemes on this problem. For evaluation, we consider one synthetic and two real datasets. The real datasets are MNIST [17] and the Ocean data [18]. We find the leading eigenvector of each class and visualize the results. On MNIST, the leading eigenvectors resemble the images of each digit class, while for the Ocean dataset we observe that the leading eigenvector represents the background image in the dataset.

In Sect. 2 we present an overview of essential concepts in Riemannian geometry, defining the geodesically convex and smooth function classes following Zhang *et al.* [38]. We also briefly review the original SAGA algorithm. In Sect. 3, we introduce the MASAGA algorithm and analyze its convergence under both uniform and non-uniform sampling. Finally, in Sect. 4 we empirically verify the theoretical linear convergence results.

2   Preliminaries

In this section we first present a review of Riemannian manifold concepts, however, for a more detailed review we refer the interested reader to the literature [1,27,35]. Then, we introduce the class of functions that we optimize over such manifolds. Finally, we briefly review the original SAGA algorithm.

2.1   Riemannian Manifold

A Riemannian manifold is denoted by the pair $(\mathcal{M}, G)$, that consists of a smooth manifold $\mathcal{M}$ over $\mathbb{R}^d$ and a metric G. At any point x in the manifold $\mathcal{M}$, we define $\mathcal{T}_{\mathcal{M}}(x)$ to be the tangent plane of that point, and G defines an inner product in this plane. Formally, if p and q are two vectors in $\mathcal{T}_{\mathcal{M}}(x)$, then $\langle p, q \rangle_x = G(p, q)$. Similar to Euclidean space, we can define the norm of a vector and the angle between two vectors using G.

To measure the distance between two points on the manifold, we use the geodesic distance. Geodesics on the manifold generalize the concept of straight lines in Euclidean space. Let us denote a geodesic with $\gamma(t)$ which maps $[0,1] \rightarrow \mathcal{M}$ and is a function with constant gradient,

$$\frac{d^2}{dt^2}\gamma(t) = 0.$$

To map a point in $\mathcal{T}_{\mathcal{M}}(x)$ to $\mathcal{M}$, we use the exponential function $\mathrm{Exp}_x : \mathcal{T}_{\mathcal{M}}(x) \rightarrow \mathcal{M}$. Specifically, $\mathrm{Exp}_x(p) = z$ means that there is a geodesic curve $\gamma_x^z(t)$ on the manifold that starts from x (so $\gamma_x^z(0) = x$) and ends at z (so $\gamma_x^z(1) = z = \mathrm{Exp}_x(p)$) with a velocity of p ($\frac{d}{dt}\gamma_x^z(0) = p$). When the Exp function is defined for every point in the manifold, we call the manifold geodesically-complete. For example, the unit sphere in $\mathcal{R}^n$ is geodesically complete. If there is a unique geodesic curve between any two points in $\mathcal{M}' \in \mathcal{M}$, then the Exp_x function has an inverse defined by the Log_x function. Formally the $\mathrm{Log}_x \equiv \mathrm{Exp}_x^{-1} : \mathcal{M}' \rightarrow \mathcal{T}_{\mathcal{M}}(x)$ function maps a point from $\mathcal{M}'$ back into the tangent plane at x. Moreover, the geodesic distance between x and z is the length of the unique shortest path between z and x, which is equal to $\|\mathrm{Log}_x(z)\| = \|\mathrm{Log}_z(x)\|$.

Let u and $v \in \mathcal{T}_{\mathcal{M}}(x)$ be linearly independent so they specify a two dimensional subspace $\mathcal{S}_x \in \mathcal{T}_{\mathcal{M}}(x)$. The exponential map of this subspace, $\mathrm{Exp}_x(\mathcal{S}_x) = \mathcal{S}_{\mathcal{M}}$, is a two dimensional submanifold in $\mathcal{M}$. The sectional curvature of $\mathcal{S}_M$ denoted by $\mathrm{K}(\mathcal{S}_{\mathcal{M}}, x)$ is defined as a Gauss curvature of $\mathcal{S}_M$ at x [41]. This sectional curvature helps us in the convergence analysis of the optimization method. We use the following lemma in our analysis to give a trigonometric distance bound.

Lemma 1 *(Lemma 5 in [38]). Let a, b, and c be the side lengths of a geodesic triangle in a manifold with sectional curvature lower-bounded by $K_{\min}$. Then*

$$a^2 \leq \frac{c\sqrt{|K_{\min}|}}{\tanh(\sqrt{|K_{\min}|}c)}b^2 + c^2 - 2bc\cos(\angle(b,c)).$$

Another important map used in our algorithm is the parallel transport. It transfers a vector from a tangent plane to another tangent plane along a geodesic. This map is denoted by $\Gamma_x^z : \mathcal{T}_{\mathcal{M}}(x) \rightarrow \mathcal{T}_{\mathcal{M}}(z)$, and maps a vector from the tangent plane $\mathcal{T}_{\mathcal{M}}(x)$ to a vector in the tangent plane $\mathcal{T}_{\mathcal{M}}(z)$ while preserving the norm and inner product values.

$$\langle p, q \rangle_x = \langle \Gamma_x^z(p), \Gamma_x^z(q) \rangle_z$$

Grassmann Manifold. Here we review the Grassmann manifold, denoted $\mathrm{Grass}(p, n)$, as a practical Riemannian manifold used in machine learning. Let p and n be positive integers with $p \leq n$. $\mathrm{Grass}(p, n)$ contains all matrices in $\mathbb{R}^{n \times p}$ with orthonormal columns (the class of orthogonal matrices). By the definition of an orthogonal matrix, if $M \in \mathrm{Grass}(p, n)$ then we have $M^\top M = I$, where $I \in \mathbb{R}^{p \times p}$ is the identity matrix. Let $q \in \mathcal{T}_{\mathrm{Grass}(p,n)}(x)$, and $q = U \Sigma V^\top$ be its p-rank singular value decomposition. Then we have:

$$\mathrm{Exp}_x(tq) = xV \cos(t\Sigma)V^\top + U \sin(t\Sigma)V^\top.$$

The parallel transport along a geodesic curve $\gamma(t)$ such that $\gamma(0) = x$ and $\gamma(1) = z$ is defined as:

$$\Gamma_x^z(tq) = (-xV \sin(t\Sigma)U^\top + U \cos(t\Sigma)U^\top + I - UU^\top)q.$$

2.2   Smoothness and Convexity on Manifold

In this section, we define convexity and smoothness of a function over a manifold following Zhang et al. [38]. We call $\mathcal{X} \in \mathcal{M}$ geodesically convex if for any two points y and z in $\mathcal{X}$, there is a geodesic $\gamma(t)$ starting from y and ending in z with a curve inside of $\mathcal{X}$. For simplicity, we drop the subscript in the inner product notation.

Algorithm 1. The Original SAGA Algorithm

1: **Input:** Learning rate η.
2: Initialize $x_0 = 0$ and memory $M^{(0)}$ with gradient of x_0.
3: **for** $t = 1, 2, 3, \ldots$ **do**
4:     $\hat{\mu} = \frac{1}{n} \sum_{j=1}^n M^t[j]$
5:     Pick i_t uniformly at random from $\{1 \ldots n\}$.
6:     $\nu_t = \nabla f_{i_t}(x_t) - M^t[i_t] + \frac{1}{n} \sum_{j=1}^n M^t[j]$
7:     $x_{t+1} = x_t - \eta(\nu_t)$
8:     Set $M^{t+1}[i_t] = \nabla f_{i_t}(x_t)$ and $M^{t+1}[j] = M^t[j]$ for all $j \neq i_t$.
9: **end for**

Formally, a function $f : \mathcal{X} \to \mathbb{R}$ is called geodesically convex if for any y and z in $\mathcal{X}$ and the corresponding geodesic γ, for any $t \in [0, 1]$ we have:

$$f(\gamma(t)) \leq (1 - t)f(y) + tf(z).$$

Similar to the Euclidean space, if the Log function is well defined we have the following for convex functions:

$$f(z) + \langle g_z, \mathrm{Log}_z(y) \rangle \leq f(y),$$

where g_z is a subgradient of f at x. If f is a differentiable function, the Riemannian gradient of f at z is a vector g_z which satisfies $\frac{d}{dt}|_{t=0} f(\mathrm{Exp}_z(tg_z)) = \langle g_z, \nabla f(z) \rangle_z$, with ∇f being the gradient of f in $\mathbb{R}^n$. Furthermore, we say that f is geodesically μ-strongly convex if there is a $\mu > 0$ such that:

$$f(z) + \langle g_z, \mathrm{Log}_z(y) \rangle + \frac{\mu}{2} \|\mathrm{Log}_z(y)\|^2 \leq f(y).$$

Let $x^* \in \mathcal{X}$ be the optimum of f. This implies that there exists a subgradient at x^* with $g_{x^*} = 0$ which implies that the following inequalities hold:

$$\|\mathrm{Log}_z(x^*)\|^2 \leq \frac{2}{\mu}(f(z) - f(x^*))$$

$$\langle g_z, \mathrm{Log}_z(x^*)\rangle + \frac{\mu}{2}\|\mathrm{Log}_z(x^*)\|^2 \leq 0$$

Finally, an f that is differentiable over $\mathcal{M}$ is said to be a Lipschitz-smooth function with the parameter $L > 0$ if its gradient satisfies the following inequality:

$$\|g_z - \Gamma_y^z[g_y]\| \leq L\|\mathrm{Log}_z(y)\| = L\ \mathrm{d}(z, y),$$

where $\mathrm{d}(z, y)$ is the distance between z and y. For a geodesically smooth f the following inequality also holds:

$$f(y) \leq f(z) + \langle g_z, \mathrm{Log}_z(y)\rangle + \frac{L}{2}\|\mathrm{Log}_z(y)\|^2.$$

2.3 SAGA Algorithm

In this section we briefly review the SAGA method [6] and the assumptions associated with it. SAGA assumes f is μ-strongly convex, each f_i is convex, and each gradient ∇f_i is Lipschitz-continuous with constant L. The method generates a sequence of iterates x_t using the SAGA Algorithm 1 (line 7). In the algorithm, M is the memory used to store stale gradients. During each iteration, SAGA picks one f_{i_t} randomly and evaluates its gradient at the current iterate value, $\nabla f_{i_t}(x_t)$. Next, it computes ν_t as the difference between the current $\nabla f_{i_t}(x_t)$ and the corresponding stale gradient of f_{i_t} stored in the memory plus the average of all stale gradients (line 6). Then it uses this vector ν_t as an approximation of the full gradient and updates the current iterate similar to the gradient descent update rule. Finally, SAGA updates the stored gradient of f_{i_t} in the memory with the new value of $\nabla f_{i_t}(x_t)$.

Let $\rho_{\mathrm{saga}} = \frac{\mu}{2(n\mu+L)}$. Defazio et al. [6] show that the iterate value x_t converges to the optimum x^* linearly with a contraction rate $1 - \rho_{\mathrm{saga}}$,

$$\mathbb{E}\left[\|x_t - x^*\|^2\right] \leq (1 - \rho_{\mathrm{saga}})^t C,$$

where C is a positive scalar.

3 Optimization on Manifold with SAGA

In this section we introduce the MASAGA algorithm (see Algorithm 2). We make the following assumptions:

1. Each f_i is geodesically L-Lipschitz continuous.
2. f is geodesically μ-strongly convex.
3. f has an optimum in $\mathcal{X}$, i.e., $x^* \in \mathcal{X}$.

4. The diameter of $\mathcal{X}$ is bounded above, *i.e.*, $\max_{u,v \in \mathcal{X}} \mathrm{d}(u,v) \leq D$.
5. Log_x is defined when $x \in \mathcal{X}$.
6. The sectional curvature of $\mathcal{X}$ is bounded, *i.e.*, $K_{\min} \leq K_\mathcal{X} \leq K_{\max}$.

These assumptions also commonly appear in the previous work [14,30,37,38]. Similar to the previous work [14,37,38], we also define the constant ζ which is essential in our analysis:

$$\zeta = \begin{cases} \frac{\sqrt{|K_{\min}|}D}{\tanh(\sqrt{|K_{\min}|}D)} & \text{if } K_{\min} < 0 \\ 1 & \text{if } K_{\min} \geq 0 \end{cases}$$

In MASAGA we modify two parts of the original SAGA: (i) since gradients are in different tangent planes, we use parallel transport to map them into the same tangent plane and then do the variance reduction step (line 6 of Algorithm 2), and (ii) we use the Exp function to map the update step back into the manifold (line 7 of Algorithm 2).

3.1 Convergence Analysis

We analyze the convergence of MASAGA considering the above assumptions and show that it converges linearly. In our analysis, we use the fact that MASAGA's estimation of the full gradient ν_t is unbiased (like SAGA), *i.e.*, $\mathbb{E}[\nu_t] = \nabla f(x_t)$. For simplicity, we use ∇f to denote the Riemannian gradient instead of g_x. We assume that there exists an incremental first-order oracle (IFO) [2] that gets an $i \in \{1, ..., n\}$, and an $x \in \mathcal{X}$, and returns $(f_i(x), \nabla f_i(x)) \in (\mathbb{R} \times \mathcal{T}_\mathcal{M}(x))$.

Theorem 1. *If each f_i is geodesically L-smooth and f is geodesically μ-strongly convex over the Riemannian manifold $\mathcal{M}$, the MASAGA algorithm with the constant step size $\eta = \frac{2\mu + \sqrt{\mu^2 - 8\rho(1+\alpha)\zeta L^2}}{4(1+\alpha)\zeta L^2}$ converges linearly while satisfying the following:*

$$\mathbb{E}\left[\mathrm{d}^2(x_t, x^*)\right] \leq (1-\rho)^t \Upsilon^0,$$

where $\rho = \min\{\frac{\mu^2}{8(1+\alpha)\zeta L^2}, \frac{1}{n} - \frac{1}{\alpha n}\}$, $\Upsilon^0 = 2\alpha\zeta\eta^2 \sum_{i=1}^n \|M^0[i] - \Gamma_{x^*}^{x_0}[\nabla f_i(x^*)]\|^2 + \mathrm{d}^2(x_0, x^*)$ is a positive scalar, and $\alpha > 1$ is a constant.

Algorithm 2. MASAGA Algorithm

1: **Input:** Learning rate η and $x_0 \in \mathcal{M}$.
2: Initialize memory $M^{(0)}$ with gradient of x_0.
3: **for** $t = 1, 2, 3, \ldots$ **do**
4:     $\hat{\mu} = \frac{1}{n} \sum_{j=1}^n M^t[j]$
5:     Pick i_t uniformly at random from $\{1 \ldots n\}$.
6:     $\nu_t = \nabla f_{i_t}(x_t) - \Gamma_{x_t}^{x_t}[M^t[i_t] - \hat{\mu}]$
7:     $x_{t+1} = \mathrm{Exp}_{x_t}(-\eta(\nu_t))$
8:     Set $M^{t+1}[i_t] = \Gamma_{x_t}^{x_0}[\nabla f_{i_t}(x_t)]$ and $M^{t+1}[j] = M^t[j]$ for all $j \neq i_t$.
9: **end for**

Proof. Let $\delta_t = \mathrm{d}^2(x_t, x^*)$. First we find an upper-bound for $\mathbb{E}\left[\|\nu_t\|^2\right]$.

$$
\begin{aligned}
\mathbb{E}\left[\|\nu_t\|^2\right] &= \mathbb{E}\left[\|\nabla f_{i_t}(x_t) - \Gamma_{x_0}^{x_t}\left[M^t[i_t] - \hat{\mu}\right]\|^2\right] \\
&= \mathbb{E}\left[\|\nabla f_{i_t}(x_t) - \Gamma_{x^*}^{x_t}\left[\nabla f_{i_t}(x^*)\right] - \Gamma_{x_0}^{x_t}\left[M^t[i_t] - \Gamma_{x^*}^{x_0}\left[\nabla f_{i_t}(x^*)\right] - \hat{\mu}\right]\|^2\right] \\
&\leq 2\mathbb{E}\left[\|\nabla f_{i_t}(x_t) - \Gamma_{x^*}^{x_t}\left[\nabla f_{i_t}(x^*)\right]\|^2\right] \\
&\quad + 2\mathbb{E}\left[\|\Gamma_{x_0}^{x_t}\left[M^t[i_t] - \Gamma_{x^*}^{x_0}\left[\nabla f_{i_t}(x^*)\right] - \hat{\mu}\right]\|^2\right] \\
&\leq 2\mathbb{E}\left[\|\nabla f_{i_t}(x_t) - \Gamma_{x^*}^{x_t}\left[\nabla f_{i_t}(x^*)\right]\|^2\right] \\
&\quad + 2\mathbb{E}\left[\|M^t[i_t] - \Gamma_{x^*}^{x_0}\left[\nabla f_{i_t}(x^*)\right]\|^2\right] \\
&\leq 2L^2\delta_t + 2\mathbb{E}\left[\|M^t[i_t] - \Gamma_{x^*}^{x_0}\left[\nabla f_{i_t}(x^*)\right]\|^2\right]
\end{aligned}
$$

The first inequality is due to $(a+b)^2 \leq 2a^2 + 2b^2$ and the second one is from the variance upper-bound inequality, *i.e.*, $\mathbb{E}\left[x^2 - \mathbb{E}\left[x\right]^2\right] \leq \mathbb{E}\left[x^2\right]$. The last inequality comes from the geodesic Lipschitz smoothness of each f_i. Note that the expectation is taken with respect to i_t.

$$
\begin{aligned}
\mathbb{E}\left[\delta_{t+1}\right] &\leq \mathbb{E}\left[\delta_t - 2\left\langle\nu_t, \mathrm{Exp}_{x_t}^{-1}(-x^*)\right\rangle + \zeta\eta^2\|\nu_t\|^2\right] \\
&= \delta_t - 2\eta\left\langle\nabla f(x_t), \mathrm{Exp}_{x_t}^{-1}(-x^*)\right\rangle + \zeta\eta^2\mathbb{E}\left[\|\nu_t\|^2\right] \\
&\leq \delta_t - \eta\mu\delta_t + \zeta\eta^2\mathbb{E}\left[\|\nu_t\|^2\right] \\
&\leq (1-\mu\eta)\delta_t + \zeta\eta^2\left[2L^2\delta_t + 2\mathbb{E}\left[\|M^t[i_t] - \Gamma_{x^*}^{x_0}\left[\nabla f_{i_t}(x^*)\right]\|^2\right]\right] \\
&= (1-\mu\eta + 2\zeta L^2\eta^2)\delta_t + 2\zeta\eta^2\Psi_t
\end{aligned}
$$

The first inequality is due to the trigonometric distance bound, the second one is due to the strong convexity of f, and the last one is due to the upper-bound of ν_t. Ψ_t is defined as follows:

$$
\Psi_t = \frac{1}{n}\sum_{i=1}^{n}\|M^t[i] - \Gamma_{x^*}^{x_0}\left[\nabla f_i(x^*)\right]\|^2.
$$

We define the Lyaponov function

$$
\Upsilon^t = \delta_t + c\Psi_t
$$

for some $c > 0$. Note that $\Upsilon^t \geq 0$, since both δ_t and Ψ_t are positive or zero. Next we find an upper-bound for $\mathbb{E}\left[\Psi_{t+1}\right]$.

$$
\begin{aligned}
\mathbb{E}\left[\Psi_{t+1}\right] &= \frac{1}{n}\left(\frac{1}{n}\sum_{i=1}^{n}\|\nabla f_i(x_t) - \Gamma_{x^*}^{x_t}\left[\nabla f_i(x^*)\right]\|^2\right) \\
&\quad + (1 - \frac{1}{n})\left(\frac{1}{n}\sum_{i=1}^{n}\|M^t[i] - \Gamma_{x^*}^{x_0}\left[\nabla f_i(x^*)\right]\|^2\right) \\
&= \frac{1}{n}\left(\frac{1}{n}\sum_{i=1}^{n}\|\nabla f_i(x_t) - \Gamma_{x^*}^{x_t}\left[\nabla f_i(x^*)\right]\|^2\right) + (1 - \frac{1}{n})\Psi_t \\
&\leq \frac{L^2}{n}\delta_t + (1 - \frac{1}{n})\Psi_t
\end{aligned}
$$

The inequality is due to the geodesic Lipschitz smoothness of f_i. Then, for some positive $\rho \leq 1$ we have the following inequality:

$$\mathbb{E}\left[\Upsilon^{t+1}\right] - (1-\rho)\Upsilon^t \leq (1 - \mu\eta + 2\zeta L^2\eta^2 - (1-\rho) + \frac{cL^2}{n})\delta_t$$

$$+ (2\zeta\eta^2 - c(1-\rho) + c(1 - \frac{1}{n}))\Psi_t. \tag{1}$$

In the right hand side of Inequality 1, δ_t and Ψ_t are positive by construction. If the coefficients of δ_t and Ψ_t in the right hand side of the Inequality 1 are negative, we would have $\mathbb{E}\left[\Upsilon^{t+1}\right] \leq (1-\rho)\Upsilon^t$. More precisely, we require

$$2\zeta\eta^2 - c(1-\rho) + c(1 - \frac{1}{n}) \leq 0 \tag{2}$$

$$1 - \mu\eta + 2\zeta L^2\eta^2 - (1-\rho) + \frac{cL^2}{n} \leq 0 \tag{3}$$

To satisfy Inequality 2 we require $\rho \leq \frac{1}{n} - \frac{2\zeta\eta^2}{c}$. If we set $c = 2\alpha n\zeta\eta^2$ for some $\alpha > 1$, then $\rho \leq \frac{1}{n} - \frac{1}{\alpha n}$, which satisfies our requirement. If we replace the value of c in Inequality 3, we will get:

$$\rho - \mu\eta + 2\zeta L^2\eta^2 + 2\alpha\zeta L^2\eta^2 \leq 0$$

$$\eta \in (\eta^- = \frac{2\mu - \sqrt{\mu^2 - 8\rho(1+\alpha)\zeta L^2}}{4(1+\alpha)\zeta L^2}, \eta^+ = \frac{2\mu + \sqrt{\mu^2 - 8\rho(1+\alpha)\zeta L^2}}{4(1+\alpha)\zeta L^2})$$

To ensure the term under the square root is positive, we also need $\rho < \frac{\mu^2}{8(1+\alpha)\zeta L^2}$. Finally, if we set $\rho = \min\{\frac{\mu^2}{8(1+\alpha)\zeta L^2}, \frac{1}{n} - \frac{1}{\alpha n}\}$ and $\eta = \eta^+$, then we have:

$$\mathbb{E}\left[\Upsilon^{t+1}\right] \leq (1-\rho)^{t+1}\Upsilon^0,$$

where Υ^0 is a scalar. Since $\Psi_t > 0$ and $\mathbb{E}\left[\delta_{t+1}\right] \leq \mathbb{E}\left[\Upsilon^{t+1}\right]$, we get the required bound:

$$\mathbb{E}\left[\delta_{t+1}\right] \leq (1-\rho)^{t+1}\Upsilon^0.$$

Corollary 1. *Let* $\beta = \frac{n\mu^2}{8\zeta L^2}$, *and* $\bar{\alpha} = \beta + \sqrt{\frac{\beta^2}{4} + 1} > 1$. *If we set* $\alpha = \bar{\alpha}$ *then we will have* $\rho = \frac{\mu^2}{8(1+\bar{\alpha})\zeta L^2} = \frac{1}{n} - \frac{1}{\bar{\alpha}n}$. *Furthermore, to reach an* ϵ *accuracy, i.e.,* $\mathbb{E}\left[d^2(x_T, x^*)\right] < \epsilon$, *we require that the total number of MASAGA (Algorithm 2) iteration* T *satisfy the following inequality:*

$$T \geq (\frac{8(1+\bar{\alpha})\zeta L^2}{\mu^2}) \log(\frac{1}{\epsilon}). \tag{4}$$

Note that this bound is similar to the bound of Zhang *et al.* [37]. To make it clear, notice that $\bar{\alpha} \leq 2\beta + 1$. Therefore, if we plug this upper-bound into Inequality 4 we get

$$T = \mathcal{O}(\frac{(2\beta + 2)\zeta L^2}{\mu^2}) \log(\frac{1}{\epsilon}) = \mathcal{O}(\frac{n\mu^2}{8\zeta L^2} \frac{\zeta L^2}{\mu^2} + \frac{\zeta L^2}{\mu^2}) \log(\frac{1}{\epsilon}) - \mathcal{O}(n + \frac{\zeta L^2}{\mu^2}) \log(\frac{1}{\epsilon}).$$

The $\frac{L^2}{\mu^2}$ term in the above bound is the squared condition number that could be prohibitively large in machine learning applications. In contrast, the original SAGA and SVRG algorithms only depend on $\frac{L}{\mu}$ on convex function within linear spaces. In the next section, we improve upon this bound through non-uniform sampling techniques.

3.2   MASAGA with Non-uniform Sampling

Using non-uniform sampling for stochastic optimization in Euclidean spaces can help stochastic optimization methods achieve a faster convergence rate [9,21,31]. In this section, we assume that each f_i has its own geodesically L_i-Lipschitz smoothness as opposed to a single geodesic Lipschitz smoothness $L = \max\{L_i\}$. Now, instead of uniformly sampling f_i, we sample f_i with probability $\frac{L_i}{n\bar{L}}$, where $\bar{L} = \frac{1}{n}\sum_{i=1}^{n} L_i$. In machine learning applications, we often have $\bar{L} \ll L$. Using this non-uniform sampling scheme, the iteration update is set to

$$x_{t+1} = \mathrm{Exp}_{x_t}(-\eta(\frac{\bar{L}}{L_{i_t}}\nu_t)),$$

which keeps the search direction unbiased, i.e., $\mathbb{E}\left[\frac{\bar{L}}{L_{i_t}}\nu_t\right] = \nabla f(x_t)$. The following theorem shows the convergence of the new method.

Theorem 2. *If f_i is geodesically L_i-smooth and f is geodesically μ-strongly convex over the manifold $\mathcal{M}$, the MASAGA algorithm with the defined non-uniform sampling scheme and the constant step size $\eta = \frac{2\mu+\sqrt{\mu^2-8\rho(\bar{L}+\alpha L)\frac{\varsigma}{\gamma}\bar{L}}}{4(\bar{L}+\alpha L)\frac{\varsigma}{\gamma}\bar{L}}$ converges linearly as follows:*

$$\mathbb{E}\left[\mathrm{d}^2(x_t, x^*)\right] \le (1-\rho)^t \Upsilon^0,$$

where $\rho = \min\{\frac{\gamma\mu^2}{8(1+\alpha)\varsigma L \bar{L}}, \frac{\gamma}{n} - \frac{\gamma}{\alpha n}\}$, $\gamma = \frac{\min\{L_i\}}{L}$, $L = \max\{L_i\}$, $\bar{L} = \frac{1}{n}\sum_{i=1}^{n} L_i$, and $\alpha > 1$ is a constant, and $\Upsilon^0 = \frac{2\alpha\varsigma\eta^2}{\gamma}\sum_{i=1}^{n}\frac{\bar{L}}{L_i}\|M^0[i] - \Gamma_{x^*}^{x_0}[\nabla f_i(x^*)]\|^2 + \mathrm{d}^2(x_0, x^*)$ are positive scalars.

Proof of the above theorem could be found in the supplementary material.

Corollary 2. *Let $\beta = \frac{n\mu^2}{8\varsigma L L}$, and $\bar{\alpha} = \beta + \sqrt{\frac{\beta^2}{4}+1} > 1$. If we set $\alpha = \bar{\alpha}$ then we have $\rho = \frac{\gamma\mu^2}{8(1+\bar{\alpha})\varsigma L L} = \frac{\gamma}{n} - \frac{\gamma}{\bar{\alpha}n}$. Now, to reach an ϵ accuracy, i.e., $\mathbb{E}\left[\mathrm{d}^2(x_T, x^*)\right] < \epsilon$, we require:*

$$T = \mathcal{O}(n + \frac{\varsigma L \bar{L}}{\gamma\mu^2})\log(\frac{1}{\epsilon}), \tag{5}$$

where T is the number of the necessary iterations.

Observe that the number of iterations T in Equality 5 depends on $\bar{L}L$ instead of L^2. When $\bar{L} \ll L$, the difference could be significant. Thus, MASAGA with non-uniform sampling could achieve an ϵ accuracy faster than MASAGA with uniform sampling.

Similarly we can use the same sampling scheme for the RSVRG algorithm [37] and improve its convergence. Specifically, if we change the update rule of Algorithm 1 of Zhang *et al.* [37] to

$$x_{t+1}^{s+1} = \mathrm{Exp}_{x_t^{s+1}} -\eta(\frac{\bar{L}}{L_{i_t}}\nu_t^{s+1}),$$

then Theorem 1 and Corollary 1 of Zhang *et al.* [37] will change to the following ones.

Theorem 3 *(Theorem 1 of [37] with non-uniform sampling). If we use non-uniform sampling in Algorithm 1 of RSVRG [37] and run it with the option I as described in the work, and let*

$$\alpha = \frac{3\zeta\eta\bar{L}^2}{\mu - 2\zeta\eta\bar{L}^2} + \frac{(1 + 4\zeta\eta^2 - 2\eta\mu)^m(\mu - 5\zeta\eta\bar{L}^2)}{\mu - 2\zeta\eta\bar{L}^2} < 1,$$

where m is the number of the inner loop iterations, then through S iterations of the outer loop, we have

$$\mathbb{E}\left[\mathrm{d}^2(\tilde{x}^S, x^*)\right] \le (\alpha)^S \mathrm{d}^2(\tilde{x}^0, x^*)$$

The above theorem can be proved through a simple modification to the proof of Theorem 1 in RSVRG [37].

Corollary 3 *(Corollary 1 of [37] with non-uniform sampling). With non-uniform sampling in Algorithm 1 of RSVRG, after $\mathcal{O}(n + \frac{\zeta\bar{L}^2}{\gamma\mu^2})\log(\frac{1}{\epsilon})$ IFO calls, the output x_a satisfies*

$$\mathbb{E}\left[f(x_a) - f(x^*)\right] \le \epsilon.$$

Note that through the non-uniform sampling scheme we improved the RSVRG [37] convergence by replacing the L^2 term with a smaller $\bar{L}^2$ term.

4   Experiments: Computing the Leading Eigenvector

Computing the leading eigenvector is important in many real-world applications. It is widely used in social networks, computer networks, and metabolic networks for community detection and characterization [24]. It can be used to extract a feature that "best" represents the dataset [8] to aid in tasks such as classification, regression, and background subtraction. Furthermore, it is used in the PageRank algorithms which requires computing the principal eigenvector of the matrix describing the hyperlinks in the web [13]. These datasets can be huge (the web has more than three billion pages [13]). Therefore, speeding up the leading eigenvector computation will have a significant impact on many applications.

We evaluate the convergence of MASAGA on the problem of computing the leading eigenvalue on several datasets. The problem is written as follows:

$$\min_{\{x \mid x^\top x = 1\}} f(x) = -\frac{1}{n} x^\top \left(\sum_{i=1}^{n} z_i z_i^\top \right) x, \tag{6}$$

which is a non-convex objective in the Euclidean space $\mathbb{R}^d$, but a (strongly-) convex objective over the Riemannian manifold. Therefore, MASAGA can achieve a linear convergence rate on this problem. We apply our algorithm on the following datasets:

- **Synthetic.** We generate Z as a 1000×100 matrix where each entry is sampled uniformly from $(0, 1)$. To diversify the Lipschitz constants of the individual z_i's, we multiply each z_i with an integer obtained uniformly between 1 and 100.
- **MNIST** [17]. We randomly pick $10,000$ examples corresponding to digits 0–9 resulting in a matrix $Z \in \mathbb{R}^{10,000 \times 784}$.
- **Ocean.** We use the ocean video sequence data found in the UCSD background subtraction dataset [18]. It consists of 176 frames, each resized to a 94×58 image.

We compare MASAGA against RSGD [37] and RSVRG [4]. For solving geodesically smooth convex functions on the Riemannian manifold, RSGD and RSVRG achieve sublinear and linear convergence rates respectively. Since the manifold for Eq. 6 is that of a sphere, we have the following functions:

$$P_X(H) = H - \text{trace}(X^\top H)X, \qquad \nabla_r f(X) = P_X(\nabla f(X)),$$

$$\text{Exp}_X(U) = \cos(\|U\|)X + \frac{\sin(\|U\|)}{\|U\|} U, \qquad \Gamma_y^x(U) = P_y(U), \tag{7}$$

where P corresponds to the tangent space projection function, $\nabla_r f$ the Riemannian gradient function, Exp the exponential map function, and Γ the transport function. We evaluate the progress of our algorithms at each epoch t by computing the relative error between the objective value and the optimum as $\frac{f(x^t) - f^*}{|f^*|}$. We have made the code available at https://github.com/IssamLaradji/MASAGA.

For each algorithm, a grid-search over the learning rates $\{10^{-1}, 10^{-2}, ..., 10^{-9}\}$ is performed and plot the results of the algorithm with the best performance in Fig. 1. This plot shows that MASAGA is consistently faster than RSGD and RSVRG in the first few epochs. While it is expected that MASAGA beats RSGD since it has a better convergence rate, the reason MASAGA can outperform RSVRG is that RSVRG needs to occasionally re-compute the full gradient. Further, at each iteration MASAGA requires a single gradient evaluation instead of the two evaluations required by RSVRG. We see in Fig. 1 that non-uniform (NU) sampling often leads to faster progress than uniform (U) sampling, which is consistent with the theoretical analysis. In the NU sampling case,

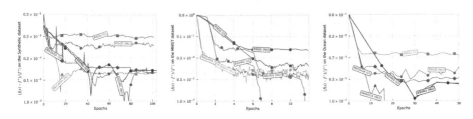

Fig. 1. Comparison of MASAGA (ours), RSVRG, and RSGD for computing the leading eigenvector. The suffix (U) represents uniform sampling and (NU) the non-uniform sampling variant.

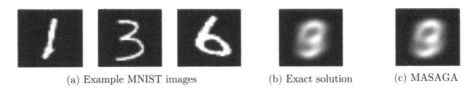

(a) Example MNIST images   (b) Exact solution   (c) MASAGA

Fig. 2. The obtained leading eigenvectors of all MNIST digits.

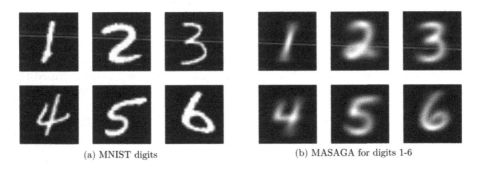

(a) MNIST digits   (b) MASAGA for digits 1-6

Fig. 3. The obtained leading eigenvectors of the MNIST digits 1–6.

we sample a vector z_i based on its Lipschitz constant $L_i = ||z_i||^2$. Note that for problems where L_i is not known or costly to compute, we can estimate it by using Algorithm 2 of Schmidt *et al.* [31].

Figures 2 and 3 show the leading eigenvectors obtained for the MNIST dataset. We run MASAGA on 10,000 images of the MNIST dataset and plot its solution in Fig. 2. We see that the exact solution is similar to the solution obtained by MASAGA, which represent the most common strokes among the MNIST digits. Furthermore, we ran MASAGA on 500 images for digits 1–6 independently and plot its solution for each class in Fig. 3. Since most digits of the same class have similar shapes and are fairly centered, it is expected that the leading eigenvector would be similar to one of the digits in the dataset.

Figure 4 shows qualitative results comparing MASAGA, RSVRG, and RSGD. We run each algorithm for 20 iterations and plot the results. MASAGA's and

(a) Example Ocean frames

(b) Exact solution   (c) RSGD   (d) MASAGA   (e) RSVRG

Fig. 4. The obtained leading eigenvectors of the ocean dataset after 20 iterations.

RSVRG's results are visually similar to the exact solution. However, the RSGD result is visually different than the exact solution (the difference is in the center-left of the two images).

5   Conclusion

We introduced MASAGA which is a stochastic variance-reduced optimization algorithm for Riemannian manifolds. We analyzed the algorithm and showed that it converges linearly when the objective function is geodesically Lipschitz-smooth and strongly convex. We also showed that using non-uniform sampling improves the convergence speed of both MASAGA and RSVRG algorithms. Finally, we evaluated our method on a synthetic dataset and two real datasets where we empirically observed linear convergence. The empirical results show that MASAGA outperforms RSGD and is faster than RSVRG in the early iterations. For future work, we plan to extend MASAGA by deriving convergence rates for the non-convex case of geodesic objective functions. We also plan to explore accelerated variance-reduction methods and block coordinate descent based methods [26] for Riemannian optimization. Another potential future work of interest is a study of relationships between the condition number of a function within the Euclidean space and its corresponding condition number within a Riemannian manifold, and the effects of sectional curvature on it.

References

1. Absil, P.A., Mahony, R., Sepulchre, R.: Optimization Algorithms on Matrix Manifolds. Princeton University Press, Princeton (2009)
2. Agarwal, A., Bottou, L.: A lower bound for the optimization of finite sums. arXiv preprint (2014)
3. Bietti, A., Mairal, J.: Stochastic optimization with variance reduction for infinite datasets with finite sum structure. In: Advances in Neural Information Processing Systems, pp. 1622–1632 (2017)

4. Bonnabel, S.: Stochastic gradient descent on Riemannian manifolds. IEEE Trans. Autom. Control **58**(9), 2217–2229 (2013)
5. Cauchy, M.A.: Méthode générale pour la résolution des systèmes d'équations simultanées. Comptes rendus des séances de l'Académie des sciences de Paris **25**, 536–538 (1847)
6. Defazio, A., Bach, F., Lacoste-Julien, S.: SAGA: a fast incremental gradient method with support for non-strongly convex composite objectives. In: Advances in Neural Information Processing Systems (2014)
7. Dubey, K.A., Reddi, S.J., Williamson, S.A., Poczos, B., Smola, A.J., Xing, E.P.: Variance reduction in stochastic gradient Langevin dynamics. In: Advances in Neural Information Processing Systems, pp. 1154–1162 (2016)
8. Guyon, C., Bouwmans, T., Zahzah, E.h.: Robust principal component analysis for background subtraction: systematic evaluation and comparative analysis. In: Principal Component Analysis. InTech (2012)
9. Harikandeh, R., Ahmed, M.O., Virani, A., Schmidt, M., Konečný, J., Sallinen, S.: StopWasting my gradients: practical SVRG. In: Advances in Neural Information Processing Systems, pp. 2251–2259 (2015)
10. Hosseini, R., Sra, S.: Matrix manifold optimization for Gaussian mixtures. In: Advances in Neural Information Processing Systems, pp. 910–918 (2015)
11. Jeuris, B., Vandebril, R., Vandereycken, B.: A survey and comparison of contemporary algorithms for computing the matrix geometric mean. Electron. Trans. Numer. Anal. **39**(EPFL-ARTICLE-197637), 379–402 (2012)
12. Johnson, R., Zhang, T.: Accelerating stochastic gradient descent using predictive variance reduction. In: Advances in Neural Information Processing Systems (2013)
13. Kamvar, S., Haveliwala, T., Golub, G.: Adaptive methods for the computation of pagerank. Linear Algebra Appl. **386**, 51–65 (2004)
14. Kasai, H., Sato, H., Mishra, B.: Riemannian stochastic variance reduced gradient on Grassmann manifold. arXiv preprint arXiv:1605.07367 (2016)
15. Konečný, J., Richtárik, P.: Semi-stochastic gradient descent methods. arXiv preprint (2013)
16. Le Roux, N., Schmidt, M., Bach, F.: A stochastic gradient method with an exponential convergence rate for strongly-convex optimization with finite training sets. In: Advances in Neural Information Processing Systems (2012)
17. LeCun, Y., Bottou, L., Bengio, Y., Haffner, P.: Gradient-based learning applied to document recognition. Proc. IEEE **86**(11), 2278–2324 (1998)
18. Mahadevan, V., Vasconcelos, N.: Spatiotemporal saliency in dynamic scenes. IEEE Trans. Pattern Anal. Mach. Intell. **32**(1), 171–177 (2010). https://doi.org/10.1109/TPAMI.2009.112
19. Mahdavi, M., Jin, R.: MixedGrad: an $o(1/t)$ convergence rate algorithm for stochastic smooth optimization. In: Advances in Neural Information Processing Systems (2013)
20. Mairal, J.: Optimization with first-order surrogate functions. arXiv preprint arXiv:1305.3120 (2013)
21. Needell, D., Ward, R., Srebro, N.: Stochastic gradient descent, weighted sampling, and the randomized Kaczmarz algorithm. In: Advances in Neural Information Processing Systems, pp. 1017–1025 (2014)
22. Nemirovski, A., Juditsky, A., Lan, G., Shapiro, A.: Robust stochastic approximation approach to stochastic programming. SIAM J. Optim. **19**(4), 1574–1609 (2009)
23. Nesterov, Y.: A method for unconstrained convex minimization problem with the rate of convergence $O(1/k^2)$. Doklady AN SSSR **269**(3), 543–547 (1983)

24. Newman, M.E.: Modularity and community structure in networks. Proc. Natl. Acad. Sci. **103**(23), 8577–8582 (2006)
25. Nguyen, L., Liu, J., Scheinberg, K., Takáč, M.: SARAH: a novel method for machine learning problems using stochastic recursive gradient. arXiv preprint arXiv:1703.00102 (2017)
26. Nutini, J., Laradji, I., Schmidt, M.: Let's Make Block Coordinate Descent Go Fast: Faster Greedy Rules, Message-Passing, Active-Set Complexity, and Superlinear Convergence. ArXiv e-prints, December 2017
27. Petersen, P., Axler, S., Ribet, K.: Riemannian Geometry, vol. 171. Springer, Cham (2016). https://doi.org/10.1007/978-3-319-26654-1
28. Polyak, B.T., Juditsky, A.B.: Acceleration of stochastic approximation by averaging. SIAM J. Contr. Optim. **30**(4), 838–855 (1992)
29. Robbins, H., Monro, S.: A stochastic approximation method. Ann. Math. Statist. **22**(3), 400–407 (1951). https://doi.org/10.1214/aoms/1177729586
30. Sato, H., Kasai, H., Mishra, B.: Riemannian stochastic variance reduced gradient. arXiv preprint arXiv:1702.05594 (2017)
31. Schmidt, M., Babanezhad, R., Ahmed, M., Defazio, A., Clifton, A., Sarkar, A.: Non-uniform stochastic average gradient method for training conditional random fields. In: Artificial Intelligence and Statistics, pp. 819–828 (2015)
32. Shalev-Shwartz, S., Zhang, T.: Stochastic dual coordinate ascent methods for regularized loss minimization. J. Mach. Learn. Res. **14**, 567–599 (2013)
33. Sra, S., Hosseini, R.: Geometric optimization in machine learning. In: Minh, H.Q., Murino, V. (eds.) Algorithmic Advances in Riemannian Geometry and Applications. ACVPR, pp. 73–91. Springer, Cham (2016). https://doi.org/10.1007/978-3-319-45026-1_3
34. Sun, J., Qu, Q., Wright, J.: Complete dictionary recovery over the sphere. In: 2015 International Conference on Sampling Theory and Applications (SampTA), pp. 407–410. IEEE (2015)
35. Udriste, C.: Convex Functions and Optimization Methods on Riemannian Manifolds, vol. 297. Springer, Dordrecht (1994). https://doi.org/10.1007/978-94-015-8390-9
36. Wiesel, A.: Geodesic convexity and covariance estimation. IEEE Trans. Signal Process. **60**(12), 6182–6189 (2012)
37. Zhang, H., Reddi, S.J., Sra, S.: Riemannian SVRG: fast stochastic optimization on Riemannian manifolds. In: Advances in Neural Information Processing Systems, pp. 4592–4600 (2016)
38. Zhang, H., Sra, S.: First-order methods for geodesically convex optimization. In: Conference on Learning Theory, pp. 1617–1638 (2016)
39. Zhang, T., Yang, Y.: Robust principal component analysis by manifold optimization. arXiv preprint arXiv:1708.00257 (2017)
40. Zhao, P., Zhang, T.: Stochastic optimization with importance sampling for regularized loss minimization. In: International Conference on Machine Learning, pp. 1–9 (2015)
41. Ziller, W.: Riemannian manifolds with positive sectional curvature. In: Dearricott, O., Galaz-Garcia, F., Kennard, L., Searle, C., Weingart, G., Ziller, W. (eds.) Geometry of Manifolds with Non-negative Sectional Curvature. LNM, vol. 2110, pp. 1–19. Springer, Cham (2014). https://doi.org/10.1007/978-3-319-06373-7_1

Block CUR: Decomposing Matrices
Using Groups of Columns

Urvashi Oswal[1][(✉)], Swayambhoo Jain[2], Kevin S. Xu[3], and Brian Eriksson[4]

[1] University of Wisconsin-Madison, Madison, WI 53706, USA
uoswal@wisc.edu
[2] Technicolor Research, Palo Alto, CA 94306, USA
[3] The University of Toledo, Toledo, OH 43606, USA
[4] Adobe, San Jose, CA 95110, USA

Abstract. A common problem in large-scale data analysis is to approximate a matrix using a combination of specifically sampled rows and columns, known as CUR decomposition. In many real-world environments, the ability to sample specific individual rows or columns of the matrix is limited by either system constraints or cost. In this paper, we consider matrix approximation by sampling predefined *blocks* of columns (or rows) from the matrix. We present an algorithm for sampling useful column blocks and provide novel guarantees for the quality of the approximation. We demonstrate the effectiveness of the proposed algorithms for computing the Block CUR decomposition of large matrices in a distributed setting with multiple nodes in a compute cluster and in a biometric data analysis setting using real-world user data from content testing.

Keywords: Matrix decomposition · Leverage score
Random sampling

1 Introduction

The ability to perform large-scale data analysis is often limited by two opposing forces. The first force is the need to store data in a matrix format for the purpose of analysis techniques such as regression or classification. The second force is the inability to store the data matrix completely in memory due to the size of the matrix in many application settings. This conflict gives rise to storing factorized matrix forms, such as SVD or CUR decompositions [5].

We consider a matrix A with m rows and n columns, *i.e.,* $A \in \mathbb{R}^{m \times n}$. Using a truncated k number of singular vectors (where $k < \min\{m,n\}$), the singular value decomposition (SVD) provides the best rank-k approximation to the original matrix. The singular vectors often do not preserve the structure in original data. Preserving the original structure in the data may be desirable

---

This research was performed while U.O., K.S.X., and B.E. were at Technicolor.

M. Berlingerio et al. (Eds.): ECML PKDD 2018, LNAI 11052, pp. 360–376, 2019.
https://doi.org/10.1007/978-3-030-10928-8_22

due to many reasons including interpretability in case of biometric data or for storage efficiency in case of sparse matrices. This has led to the introduction of the CUR decomposition, where the factorization is performed with respect to a subset of rows and columns of the matrix itself. This specific decomposition describes the matrix A as the product of a subset of matrix rows R and a subset of matrix columns C (along with a matrix U that fits $A \approx CUR$).

Significant prior work has examined how to efficiently choose the rows and columns in the CUR decomposition and has derived worst-case error bounds (*e.g.,* [11]). These methods have been applied successfully to many real-world problems including genetics [13], astronomy [19], and mass spectrometry imaging [18]. Unfortunately, a primary assumption of current CUR techniques, that individual rows and columns of the matrix can be queried, is either impossible or quite costly in many real-world problems and instead require a block approach.

In this paper, we consider the following two applications which represent the two main motivating factors for considering block decompositions.

Biometric Data Analysis. In applications where the ordering of rows or columns is meaningful, such as images, video, or speech data matrices, sampling contiguous blocks of columns adds contextual information that is necessary for interpretability of the factorized representation. One emerging application is audience reaction analysis of video content using biometrics. We focus on the scenario where users watch video content while wearing sensors, and changes in biometric sensors indicate changes in reaction to the content. For example, increases in heart rate or a spike in electrodermal activity (EDA) may be associated with an increase in content engagement [8,15]. Unfortunately, there is significant cost in acquiring each user's reaction to lengthy content so instead we collect full responses (corresponding to some rows of the matrix) from only a limited number of users. For remaining users, we propose to collect responses for only a few important scenes of the video (corresponding to column blocks of the matrix) as shown in Fig. 1a and then *approximate* their full response. An individual time sample in this use case cannot be queried in isolation due to the lack of context that caused that biometric reaction. Instead, collections of time segments (i.e., blocks) must be presented to the user. In this setting, block sampling can be viewed as a *restriction* which leads to more interpretable solutions.

Distributed Storage Systems. Large-scale datasets often require distributed storage, a regime where there can be substantial overhead involved in querying individual rows or columns of a matrix. In these regimes, it is more efficient to retrieve predefined *blocks* of rows or columns at one time corresponding to the rows or columns stored on the same node, as shown in Fig. 1b, in order to minimize the overhead in terms of latency while keeping the throughput constant. In doing so, one forms a Block CUR decomposition, with more details provided in Sect. 4.2. Current CUR decomposition techniques do not take advantage of this predefined block structure.

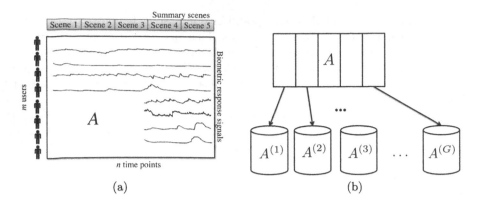

Fig. 1. Applications: (a) Biometric data analysis. Blocks of columns or time instances correspond to scenes in a video and provide context for biometric reaction. (b) Distributed storage of a large matrix across multiple nodes in a compute cluster. Blocks are allocated to each of the G nodes.

Main Contributions. Using these insights into real-world applications of CUR decomposition, this paper makes a series of contributions. We propose a randomized Block CUR algorithm for subset selection of rows and blocks of columns and derive novel worst-case error bounds for this randomized algorithm. On the theory side, we present new theoretical results related to approximating matrix multiplication and generalized ℓ_2 regression in the block setting. These results are the fundamental building blocks used to derive the error bounds for the randomized algorithm. The sample complexity bounds feature a non-trivial dependence on the matrix partition, *i.e.*, the distribution of information in the blocks of the matrix. This dependence cannot be obtained by simply extending the analysis of the original individual column CUR setting to the Block CUR setting. As a result, our analysis finds a sample complexity improvement on the order of the *block stable rank* of a matrix (See Table 1 in Sect. 3).

On the practical side, this algorithm performs fast block sampling taking advantage of the natural storage of matrices in distributed environments (See Table 2 in Sect. 4.2). We demonstrate empirically that the proposed Block CUR algorithms can achieve a significant speed-up when used to decompose large matrices in a distributed data setting. We conduct a series of CUR decomposition experiments using Apache Spark on Amazon Elastic Map-Reduce (Amazon EMR) using both synthetic and real-world data. In this distributed environment, we find that our Block CUR approach achieves a speed-up of 2x to 6x for matrices larger than 12000×12000. This is compared with previous CUR approaches that sample individual rows and columns and while achieving the same matrix approximation error rate. We also perform experiments with real-world user biometric data from a content testing environment and present interesting use cases where our algorithms can be applied to user analytics tasks.

2 Setup and Background

2.1 Notation

Let I_k denote the $k \times k$ identity matrix and 0 denote a zero matrix of appropriate size. We denote vectors (matrices) with lowercase (uppercase) bold symbols like a (A). The i-th row (column) of a matrix is denoted by A_i (A^i). We represent the i-th block of rows of a matrix by $A_{(i)}$ and the i-th block of columns of a matrix by $A^{(i)}$. Let $[n]$ denote the set $\{1, 2, \ldots, n\}$. Let $\rho = \mathrm{rank}(A) \leq \min\{m, n\}$ and $k \leq \rho$. The singular value decomposition (SVD) of A can be written as $A = U_{A,\rho} \Sigma_{A,\rho} V_{A,\rho}^T$ where $U_{A,\rho} \in \mathbb{R}^{m \times \rho}$ contains the ρ left singular vectors; $\Sigma_{A,\rho} \in \mathbb{R}^{\rho \times \rho}$ is the diagonal matrix of singular values, $\sigma_i(A)$ for $i = 1, \ldots, \rho$; and $V_{A,\rho}^T \in \mathbb{R}^{\rho \times n}$ is an orthonormal matrix containing the ρ right singular vectors of A. We denote $A_k = U_{A,k} \Sigma_{A,k} V_{A,k}^T$ as the best rank-k approximation to A in terms of Frobenius norm. The pseudoinverse of A is defined as $A^\dagger = V_{A,\rho} \Sigma_{A,\rho}^{-1} U_{A,\rho}^T$. Also, note that $CC^\dagger A = U_C U_C^T A$ is the projection of A onto the column space of C, and $AR^\dagger R = AV_{R,k} V_{R,k}^T$ is the projection of A onto the row space of R. The Frobenius norm and spectral norm of a matrix are denoted by $\|A\|_F$ and $\|A\|_2$ respectively. The square of the Frobenius norm is given by $\|A\|_F^2 = \sum_{i=1}^m \sum_{j=1}^n A_{i,j}^2 = \sum_{i=1}^k \sigma_i^2(A)$. The spectral norm is given by $\|A\|_2 = \max_i \sigma_i(A)$.

2.2 The CUR Problem and Other Related Work

The need to factorize a matrix using a collection of rows and columns of that matrix has motivated the CUR decomposition literature. CUR decomposition is focused on sampling rows and columns of the matrix to provide a factorization that is close to the best rank-k approximation of the matrix. One of the most fundamental results for a CUR decomposition of a given matrix $A \in \mathbb{R}^{m \times n}$ was obtained in [5]. We re-state it here for the sake of completion and setting the appropriate context for our results to be stated in the next section. This relative error bound result is summarized in the following theorem.

Theorem 1 (*Theorem 2 from [5] applied to A^T*). *Given $A \in \mathbb{R}^{m \times n}$ and an integer $k \leq \min\{m, n\}$, let $r = O(\frac{k^2}{\varepsilon^2} \ln(\frac{1}{\delta}))$ and $c = O(\frac{r^2}{\varepsilon^2} \ln(\frac{1}{\delta}))$. There exist randomized algorithms such that, if c columns are chosen to construct C and r rows are chosen to construct R, then with probability $\geq 1 - \delta$, the following holds:*

$$\|A - CUR\|_F \leq (1 + \varepsilon)\|A - A_k\|_F$$

where $\varepsilon, \delta \in (0, 1)$, $U = W^\dagger$ and W is the scaled intersection of C and R.

This theorem states that as long as enough rows and columns of the matrix are acquired (r and c, respectively), then the CUR decomposition will be within a constant factor of the error associated with the best rank-k approximation of that matrix. Central to the proposed randomized algorithm was the concept of sampling columns of the matrix based on a *leverage score*. The leverage score measures the contribution of each column to the approximation of A.

Definition 2. *The **leverage score** of a column is defined as the squared row norm of the top-k right singular vectors of $\boldsymbol{A}$ corresponding to the column:*

$$\ell_j = \|\boldsymbol{V}_{A,k}^T \boldsymbol{e}_j\|_2^2, \quad j \in [n],$$

where $\boldsymbol{V}_{A,k}$ consists of the top-k right singular vectors of $\boldsymbol{A}$ as its rows, and $\boldsymbol{e}_j$ is the j-th column of identity matrix which picks the j-th column of $\boldsymbol{V}_{A,k}^T$.

The CUR algorithm involves randomly sampling r rows using probabilities generated by the calculated leverage scores to obtain the matrix $\boldsymbol{R}$, and thereafter sampling c columns of $\boldsymbol{A}$ based on leverage scores of the $\boldsymbol{R}$ matrix to obtain $\boldsymbol{C}$. The key technical insight in [5] is that the leverage score of a column measures "how much" of the column lies in the subspace spanned by the top-k left singular vectors of $\boldsymbol{A}$; therefore, this method of sampling is also known as *subspace* sampling. By sampling columns that lie in this subspace more often, we get a relative-error low rank approximation of the matrix. The concept of sampling the important columns of a matrix based on the notion of subspace sampling first appeared in context of fast ℓ_2 regression in [4] and was refined in [5] to obtain performance error guarantees for CUR matrix decomposition.

These guarantees were subsequently improved in follow-up work [11]. Modified versions of this problem have been studied for adaptive sampling [16], divide-and-conquer algorithms for parallel computations [10], and input-sparsity algorithms [2]. The authors of [16] propose an adaptive sampling algorithm that requires only $c = O(k/\varepsilon)$ columns to be sampled when the entire matrix is known and its SVD can be computed. The authors of [2] proposed an optimal, deterministic CUR algorithm. In [1], the authors prove the lower bound of the column selection problem; at least $c = k/\varepsilon$ columns are selected to achieve the $(1 + \varepsilon)$ ratio.

These prior results require sampling of arbitrary rows and columns of the matrix $\boldsymbol{A}$, which may be either unrealistic or inefficient in many practical applications. In this paper, we focus on the problem of efficiently sampling pre-defined blocks of columns (or rows) of the matrix to provide a factorization that is close to the best rank-k approximation of the matrix in the block sampling environment for biometric and distributed computation, explore the performance advantages of block sampling over individual column sampling, and provide the first non-trivial theoretical error guarantees for Block CUR decomposition. In the following section, we propose and analyze a randomized algorithm for sampling blocks of the matrix based on *block leverage scores*.

3   The Block CUR Algorithm

A block may be defined as a collection of s columns or rows. For clarity of exposition, without loss of generality, we consider column blocks, but the techniques and derivations also hold for row blocks by applying them to the transpose of the matrix. For ease of exposition, we also assume equal-sized blocks but one could

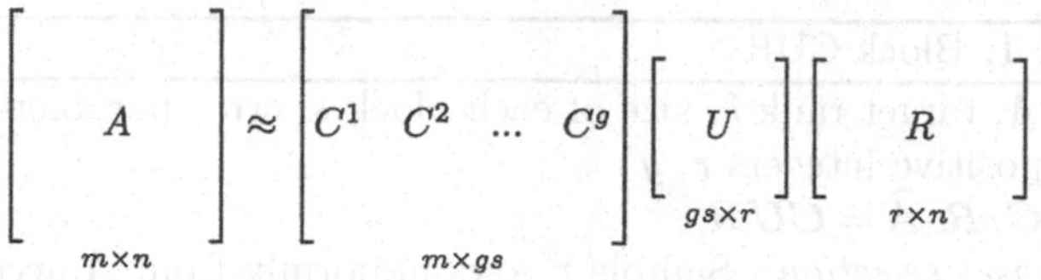

$$\left[\begin{array}{c} \\ A \\ \\ \end{array}\right] \approx \left[\begin{array}{cccc} C^1 & C^2 & \cdots & C^g \end{array}\right]\left[\begin{array}{c} U \\ \\ {\scriptstyle gs \times r} \end{array}\right]\left[\begin{array}{c} R \\ \\ {\scriptstyle r \times n} \end{array}\right]$$

$$\scriptstyle m \times n \qquad\qquad\qquad m \times gs$$

Fig. 2. Example Block CUR decomposition, where $C^t \in \mathbb{R}^{m \times s}$ for $t \in [g]$ is sampled from $\{A^{(j_t)} : j_t \in [G]\}$.

extend the methods to blocks of varying sizes. Let $G = \lceil n/s \rceil$ be the number of possible blocks in A. We consider the blocks to be predefined due to natural constraints or cost, such as data partitioning in a distributed compute cluster.

The Block CUR algorithm approximates the underlying matrix A using g blocks of columns and r rows, as represented in Fig. 2. Given the new regime of submatrix blocks, we begin by defining a *block leverage score* for each block of columns.

Definition 3. *The **block leverage score** of a group of columns is defined as the sum of the squared row norms of the top-k right singular vectors of A corresponding to the columns in the block:*

$$\ell_g(A, k) = \|V_{A,k}^T E_g\|_F^2, \quad g \in [G],$$

where $V_{A,k}$ consists of the top-k right singular vectors of A, and E_g consists of the corresponding block of columns in the identity matrix which picks the columns of $V_{A,k}^T$ corresponding to the elements in block g.

Much like the individual column leverage scores defined in [5], the block leverage scores measure how much a particular column block contributes to the approximation of the matrix A.

3.1 Algorithm Details

The Block CUR Algorithm, detailed in Algorithm 1, takes as input the matrix A and returns as output an $r \times n$ matrix R consisting of a small number of rows of A and an $m \times c$ matrix C consisting of a small number of column blocks from A. In Algorithm 1, for $t \in [g]$, block $j_t \in [G]$ is sampled with some probability p_{j_t} and scaled using matrix $S \in \mathbb{R}^{n \times gs}$. The (j_t, t)-th non-zero $s \times s$ block of S is defined as $S_{j_t,t} = I_s/\sqrt{gp_{j_t}}$ where $g = c/s$ is the number of blocks picked by the algorithm. This sampling matrix picks the blocks of columns and scales each block to compute $C = AS$. A similar sampling and scaling matrix S_R is defined to pick the blocks of rows and scale each block to compute $R = S_R^T A$.

In addition to block sampling of columns, another advantage of this algorithm is not requiring the computation of a full SVD of A. In many large-scale applications, it may not be feasible to compute the SVD of the entire matrix. In these cases, algorithms requiring knowledge of the leverage scores cannot be

---

Algorithm 1. Block CUR

Input : A, target rank k, size of each block s, error parameter ε,
        positive integers r, g

Output: $C, R, \widehat{A} = CUR$

1. *Row subset selection:* Sample r rows uniformly from A according to
   $p_i = 1/m$ for $i \in [m]$ and compute $R = S_R^T A$.
2. *Column block subset selection:* For $t \in [g]$, select a block of columns
   $j_t \in [G]$ independently with probability $p_{j_t} = \dfrac{\ell_i(R,r)}{r} = \dfrac{\|V_{R,r}^T E_i\|_F^2}{r}$
   for $i \in [G]$ and update S, where $V_{R,r}$ consists of the top-r right singular
   vectors of R, and E_i picks the columns $V_{R,r}^T$ corresponding to the
   elements in block i. Compute $C = AS$.
3. *CUR approximation:* $\widehat{A} = CUR$ where $U = W^\dagger$, and $W = RS$ is the
   scaled intersection of R and C.

---

used. Instead, we use an estimate of the block leverage scores called the *approximate block leverage scores*. A subset of the rows (corresponding to users) are chosen uniformly at random, and the block scores are calculated using the top-k right singular vectors of this row matrix instead of the entire matrix. This step is not the focus of this paper so it can also be replaced with other fast approximate calculations of leverage scores involving sketching or additional sampling [3,17]. The advantage of using our approximate leverage scores is that the same set of rows is used to approximate the scores and also to compute the CUR approximation. Hence no additional sampling or sketching steps are required.

The running time of Algorithm 1 is driven by the time required to compute the SVD of R, i.e., $\mathcal{O}(SVD(R))$, and the time to construct R, C and U. Construction of R requires $\mathcal{O}(rn)$ time, construction of C takes $\mathcal{O}(mc)$ time, construction of W requires $\mathcal{O}(rc)$ time and construction of U takes $\mathcal{O}(r^2c)$ time.

3.2   Theoretical Results and Discussion

The main technical contribution of the paper is a novel relative-error bound on the quality of approximation using blocks of columns or rows to approximate a matrix $A \in \mathbb{R}^{m \times n}$. Before stating the main result, we define two important quantities that measure important properties of the matrix A that are fundamental to the quality of approximation. We first define a property of matrix rank relative to the collection of matrix blocks. Specifically, we focus on the concept of *matrix stable rank* from [14] and define the *block stable rank* as the minimum stable rank across all matrix blocks.

Definition 4. *Let $V_{A,k}$ consist of the top-k right singular vectors of A. Then the block stable rank is defined as*

$$\alpha_A = \min_{g \in [G]} \frac{\|V_{A,k}^T E_g\|_F^2}{\|V_{A,k}^T E_g\|_2^2},$$

where E_g consists of the corresponding block of columns in the identity matrix that picks the columns of $V_{A,k}^T$ corresponding to the elements in block g.

Intuitively, the above definition gives a measure of how informative the worst matrix column block is. The second property is a notion of *column space incoherence*. When we sample rows uniformly at random, we can give relative error approximation guarantees when the matrix A satisfies an incoherence condition. This avoids pathological constructions of rows of A that cannot be sampled at random.

Definition 5. *The top-k* **column space incoherence** *is defined as*

$$\mu := \mu(U_{A,k}^T) = \frac{m}{k} \max_i \|U_{A,k}^T e_i\|_2^2,$$

where e_i picks the i-th column of $U_{A,k}^T$.

The column space incoherence is used to provide a guarantee for fast approximation without computing the SVD of the entire matrix A. Equipped with these definitions, we state the main result that provides a relative-error guarantee for the Block CUR approximation in Theorem 6.

Theorem 6. *Given $A \in \mathbb{R}^{m \times n}$ with incoherent top-k column space, i.e., $\mu \leq \mu_0$, let $r = O\left(\mu_0 \frac{k^2}{\varepsilon^2} \ln(\frac{1}{\delta})\right)$ and $g = O\left(\frac{r^2}{\alpha_R \varepsilon^2} \ln(\frac{1}{\delta})\right)$. There exist randomized algorithms such that, if r rows and g column blocks are chosen to construct R and C, respectively, then with probability $\geq 1 - \delta$, the following holds:*

$$\|A - CUR\|_F \leq (1 + \varepsilon)\|A - A_k\|_F,$$

where $\varepsilon, \delta \in (0, 1)$ and $U = W^\dagger$ is the pseudoinverse of scaled intersection of C and R.

We provide a sketch of the proof in Sect. 3.3 and highlight the main technical challenges in proving the claim. We defer proof details to an extended version of the paper [12]. Next, we discuss the differences between our technique and prior CUR algorithms. This includes additional assumptions required, algorithmic trade-offs, and discussion of sampling and computational complexity.

Block Stable Rank. Theorem 6 tells us that the number of blocks required to achieve an ε relative error depends on the structure of the blocks (through α_R). Intuitively, this is saying the groups that provide more information improve the approximation faster than less informative groups. The α_R term depends on the stable or numerical rank (a stable relaxation of exact rank) of the blocks. The stable rank $\alpha = \|A\|_F^2 / \|A\|_2^2$ is a relaxation of the rank of the matrix; in fact, it is stable under small perturbations of the matrix A [14]. For instance, the stable rank of an approximately low rank matrix tends to be low. The α_R term defined in Theorem 6 is the minimum stable rank of the column blocks. Thus, the α_R term gives a dependence of the block sampling complexity on the stable ranks of the blocks. It is easy to check that $1 \leq \alpha_R \leq s$. In the best case, when

Table 1. Table comparing the sample complexity needed for given ε using our Block CUR result and a bound obtained by trivial extension of traditional CUR. For ease of comparison, we show the results with full SVD computation ignoring incoherence assumption stated as Corollary 1 in [12] (recall $1 \leq \alpha_R \leq s$).

Results	r	g
Traditional CUR extended to block setting	$\mathcal{O}\left(\frac{k^2}{\varepsilon^2}\log\left(\frac{1}{\delta}\right)\right)$	$\mathcal{O}\left(\frac{k^4}{\varepsilon^6}\log^3\left(\frac{1}{\delta}\right)\right)$
Our Block CUR	$\mathcal{O}\left(\frac{k^2}{\varepsilon^2}\log\left(\frac{1}{\delta}\right)\right)$	$\mathcal{O}\left(\frac{k^4}{\alpha_R\varepsilon^6}\log^3\left(\frac{1}{\delta}\right)\right)$

all the groups have full stable rank with equal singular values, α_R achieves its maximum. The worst case $\alpha_R = 1$ is achieved when a group or block is rank-1. That is, sampling groups of rank s gives us a lot more information than groups of rank 1, which leads to a reduction in the total sampling complexity.

Incoherence. The column space incoherence (Definition 5) is used to provide a guarantee for approximation without computing the SVD of the entire matrix A. However, if it is possible to compute the SVD of the entire matrix, then the rows can be sampled using row leverage scores, and the incoherence assumption can be dropped. The relative error guarantee, independent of incoherence, for the full SVD Block CUR approximation is stated as Corollary 1 in the extended version of the paper [12] and follows by a similar analysis as for Theorem 6. Other than block sampling, the setup of this result is equivalent to the traditional column sampling result stated in Lemma 8. Next, we compare the block sampling result with extensions of traditional column sampling.

Sample Complexity: Comparison with Extensions of Traditional CUR Results. In order to compare the sample complexity of our block sampling results with trivial block extensions of traditional column sampling results, we focus our attention on the similar leverage score based CUR result in Theorem 1. A simple extension to block setting could be obtained by considering a larger row space in which blocks are expanded to vectors. This would lead to a sample complexity bound obtained by Theorem 1. The sampling complexity of the Block CUR derived in Theorem 6 tells us the number of sampling operations or queries that need to be made to memory in order to construct the R and C matrices. As shown in Table 1 the column block sample complexity obtained by traditional CUR extensions results is always greater than or equal to those required by our Block CUR result because $1 \leq \alpha_R \leq s$. This happens since traditional CUR-based results are obtained by completely ignoring the block structure of the matrix.

More recent adaptive column sampling-based algorithms such as [2,16] require only $c = O(k/\varepsilon)$ columns to be sampled. These results assume full computation of the SVD is possible and are use ideas like deterministic, Batson/Srivastava/Spielman (BSS) sampling and adaptive sampling on top of leverage scores. By extending these advanced techniques to block sampling, it may be possible to obtain tighter bounds, but it does not bring new insight into the problem of sampling blocks. Therefore we defer this extension to future work.

3.3   Proof Sketch of Main Result

We provide a sketch of the proof of Theorem 6 and defer details to the extended version [12]. The proof of the main result rests on two important lemmas. These results are important in their own right and could be useful wherever the block sampling issue arises. The first result concerns approximate block multiplication.

Block Multiplication Lemma. The following lemma shows that the product of two matrices A and B can be approximated by the product of the smaller sampled and scaled block matrices. This is the key in proving the main result.

Lemma 7. *Let* $A \in \mathbb{R}^{m \times n}$, $B \in \mathbb{R}^{n \times p}$, $\varepsilon, \delta \in (0, 1)$, *and* α_A *be defined as* $\alpha_A := \min_{i \in [G]} \frac{\|A^{(i)}\|_F^2}{\|A^{(i)}\|_2^2}$. *Construct* $C_{m \times gs}$ *and* $R_{gs \times n}$ *using sampling probabilities* p_i *that satisfy*

$$p_i \geq \beta \frac{\|A^{(i)}\|_F^2}{\sum_{j=1}^{G} \|A^{(j)}\|_F^2},$$

for all $i \in [G]$ *and where* $\beta \in (0, 1]$. *Then, with probability at least* $1 - \delta$,

$$\|AB - CR\|_F \leq \frac{1}{\delta \sqrt{\beta g \alpha_A}} \|A\|_F \|B\|_F.$$

The proof details are provided in the extended version of the paper [12]. The main difficulty in proving this claim is to account for the block structure. Even though one could trivially extend individual column sampling analysis to this setting by serializing the blocks, this would lead to trivial bounds as they do not leverage the block structure. Our results exploit this knowledge and hence introduce a dependence of the sample complexity on the block stable rank of the matrix. Using the block multiplication lemma we prove Lemma 8, which states a non-boosting approximation error result for Algorithm 1.

Lemma 8. *Given* $A \in \mathbb{R}^{m \times n}$ *with incoherent top-k column space, i.e.,* $\mu \leq \mu_0$, *let* $r = O(\mu_0 \frac{k^2}{\varepsilon^2})$ *and* $g = O(\frac{r^2}{\alpha_R \varepsilon^2})$. *If rows and column blocks are chosen according to Algorithm 1, then with probability at least 0.7, the following holds:*

$$\|A - CUR\|_F \leq (1 + \varepsilon)\|A - A_k\|_F,$$

where $\varepsilon \in (0, 1)$, $U = W^\dagger$ *is the pseudoinverse of the scaled intersection of* C *and* R.

The proof of Lemma 8 follows standard techniques in [5] with modifications necessary for block sampling (see [12] for the proof details). Finally, the result in Theorem 6 follows by applying standard boosting methods to Lemma 8 and running Algorithm 1 $t = \ln(\frac{1}{\delta})$ times. By choosing the solution with minimum error and observing that $0.3 < 1/e$, we have that the relative error bound holds with probability greater than $1 - e^{-t} = 1 - \delta$.

Remark. As a consequence of Lemma 8, we show that if enough blocks are sampled with high probability, then $\|A - AS(RS)^\dagger R\|_F \le (1+\varepsilon)\|A - AR^\dagger R\|_F$. This gives a guarantee on the approximate solution obtained by solving a block-sampled regression problem $\min_{X \in \mathbb{R}^{m \times r}} \|(AS) - X(RS)\|_F$ instead of the entire least squares problem. As a special case of the above result, when $R = A$ we get a bound for the block column subset selection problem. If $g = \mathcal{O}(\frac{k^2}{\alpha_A \varepsilon^2} \log(\frac{1}{\delta}))$ blocks are chosen, then with probability at least $1 - \delta$ we have $\|A - CC^\dagger A\|_F \le (1 + \varepsilon)\|A - A_k\|_F$.

4   Experiments

4.1   Experiments with Biometric Data

One emerging application is audience reaction analysis of video content using biometrics. Specifically, users watch video content while wearing sensors, with changes in biometric sensors indicating changes in reaction to the content. For example, increases in heart rate or a spike in electrodermal activity may be associated with an increase in content engagement. In prior work, biometric signal analysis techniques have been developed to determine valence [15] (e.g., positive vs. negative reactions to films) and content segmentation [9]. Unfortunately these experiments require a large number of users to sit through the entire video content, which can be both costly and time-consuming.

We consider the observed biometric signals as a matrix with m users (as rows) and n biometric time samples (as columns). Matrix approximation techniques, such as CUR decomposition, point to the ability to infer the complete matrix by showing the entire content to only a subset of users (i.e., rows), while the remaining users see only selected scenes of the content (i.e., column blocks). To replicate a user's true reaction to content, individual columns cannot be sampled (e.g., showing the user 0.25 s of video content) given the lack of scene context. Instead, longer scenes must be shown to the user to gather a representative response. Therefore, the Block CUR decomposition proposed in this paper is directly applicable.

The biometric experiment setup is as follows. We attached 24 subjects with the Empatica E3 wearable sensor [6] that measures electro-dermal activity (EDA) at 4 Hz. The subjects were shown a 41-min episode of the television series "NCIS", in the genres of action and crime. The resulting biometric data matrix was 24 × 9929. Our goal is to use Block CUR decomposition to show only a subset of users the entire content, and to then impute the biometric data for users that have viewed only a small number of selected scenes from the content.

Results. We refer to the biometric data matrix as A and plot the EDA traces (rows) corresponding to four users in Fig. 3a. To demonstrate the low rank nature of the data, we plot the Frobenius norm of A covered by A_k as a function of k in Fig. 3b. We find that for this data, only 5 singular vectors are needed to capture 80% of the total Frobenius norm of the complete matrix.

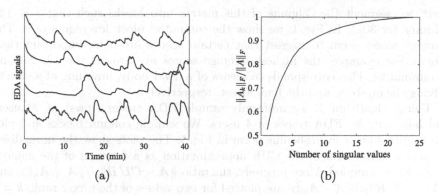

(a)

(b)

Fig. 3. Panel (a) shows EDA data for four users watching the NCIS video and (b) demonstrates the low rank nature of A.

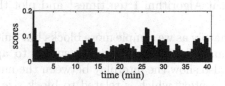

Fig. 4. Block leverage scores for EDA with $k = 5$ and $s = 120$ columns (30 s).

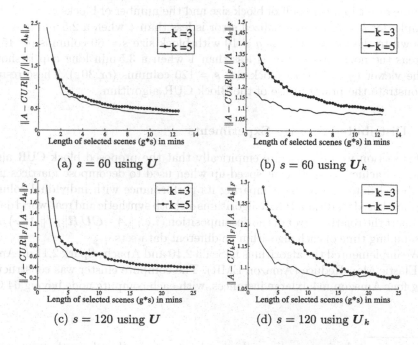

(a) $s = 60$ using U

(b) $s = 60$ using U_k

(c) $s = 120$ using U

(d) $s = 120$ using U_k

Fig. 5. Error plots for two values of target rank, $k = 3, 5$.

Next, we segment the columns of this matrix into blocks such that $s = 120$ columns (or 30 s). In Fig. 4, we show the computed block leverage scores. The leverage scores seem to suggest that certain scenes are more important than others. For example, the highest leverage scores are around the 12, 26, and 38 min marks. This corresponds to scenes of a dead body, unveiling of a clue to solving the mystery, and the final arrest, respectively.

Using Algorithm 1, we uniformly sample EDA traces (rows) of 20 users and hold out the EDA traces of 4 users. We sample column blocks and plot the resulting error in Frobenius norm in Fig. 5. The plots show the normalized Frobenius norm error of the CUR approximation as a function of the number of blocks, g, sampled. More precisely, the ratio $\|A - CUR\|_F / \|A - A_k\|_F$ and $\|A - CU_kR\|_F / \|A - A_k\|_F$ are plotted for two values of the target rank, $k = 3$ and 5 and two values of block size, $s = 60$ and 120 columns per block (15 and 30 s), respectively. U_k is the rank-k approximation of U, and using CU_kR leads to an exactly rank-k matrix approximation since this may be a restriction in some applications. We repeat Algorithm 1 ten times[1] and plot the mean normalized error over 10 trials.

The error drops sharply as we sample more blocks but quickly flattens demonstrating that a summary of the movie could suffice to approximate the full responses. The plots also show the interplay between the number of blocks sampled and the issue of context which is related to block size. To give the viewer some context we would want to make the scene as long as possible but we want to show them only a summary of the content to reduce the cost. These conflicting aims result in a trade-off of block size and the number of blocks sampled. For example, for $k = 5$, the normalized error is less than 1 when a 2.5 min long clip is shown to the viewer, that is $g = 10$ with block size $s = 60$ columns (or 15 s), whereas the normalized error is less than 1 when a 3.5 min long clip is shown to the viewer ($g = 7$) with block size $s = 120$ columns (or 30 s). These results demonstrate the practical use of the Block CUR algorithm.

4.2   Distributed Storage Experiments

In this section we demonstrate empirically that the proposed block CUR algorithm can achieve a significant speed-up when used to decompose matrices in a distributed data setting by comparing its performance with individual column sampling-based traditional CUR algorithms on both synthetic and real-world data. We report the relative-error of the decomposition (*i.e.*, $\|A - CUR\|_F / \|A\|_F$) and the sampling time of each algorithm on different data-sets.

We implemented the algorithms in Scala 2.10 and Apache Spark 2.11 on Amazon Elastic Map-Reduce (Amazon EMR). The compute cluster was constructed using four Amazon m4.4xlarge instances, with each compute node having 64 GB

---

[1] Plots generated using *sampling without replacement* even though our theory supports *sampling with replacement* since sampling the same blocks is inefficient in practice.

Table 2. Table comparing the number of sampling operations needed for given ε using our Block CUR result based on block sampling and traditional CUR-based on individual column sampling (note this is not the same as the vectorized block columns in Table 1). This leads to speed-up since it is more efficient to retrieve predefined blocks than querying individual rows or columns in these regimes. The α_R term we introduce satisfies the bound $1 \leq \alpha_R \leq s$.

Method	Number of sampling operations
Traditional CUR	$\mathcal{O}\left(\frac{k^2}{\varepsilon^2}\log(\frac{1}{\delta}) + \frac{k^4}{\varepsilon^6}\log^3(\frac{1}{\delta})\right)$
Block CUR	$\mathcal{O}\left(\frac{k^2}{\varepsilon^2}\log(\frac{1}{\delta}) + \frac{k^4}{\alpha_R \varepsilon^6}\log^3(\frac{1}{\delta})\right)$

of RAM. Using Spark, we store the data sets as a resilient distributed dataset (RDD), a collection of elements partitioned across the nodes of the cluster (see Fig. 2). In other words, Spark partitions the data into many blocks and distributes these blocks across multiple nodes in the cluster. Using block sampling, we can approximate the matrix by sampling only a subset of the important blocks. Meanwhile, individual column sampling would require looking up all the partitions containing specific columns of interest as shown in Table 2. Our experiments examine the runtime speed-up from our block sampling CUR that exploits the partitioning of data.

Synthetic Experiments. The synthetic data is generated by $A = UV$ where $U \in \mathbb{R}^{m \times k}$ and $V \in \mathbb{R}^{k \times n}$ are random matrices with i.i.d. Gaussian random entries, resulting in a low rank matrix A. We perform CUR decomposition on matrices of size $m \times n$ with $m = n$, target rank k, and number of blocks G (set here across all experiments to be 100). The leverage scores are calculated by computing the SVD of the rows sampled uniformly with $R \in \mathbb{R}^{r \times n}$. We sample one-sixth of the rows. Figure 6 shows the plots for relative error achieved with respect to the runtime required to sample C and R matrices for both Block CUR and traditional CUR algorithms. To focus on the speed-up achieved by taking into account the block storage of data we compare running times of only the sampling operations of the algorithms (which excludes the time required to compute the SVD). We note that other steps in both algorithms can be updated to include faster variants such as the approximation of leverage scores by sketching or sampling [3]. We vary g, the number of blocks chosen, from 1 to 6. The number of columns chosen is thus $c = gs$, where s denotes the number of columns in a block and varies from 50 to 200. We repeat each algorithm twice for the specified number of columns, with each realization as a point in the plot. The Block CUR algorithm samples the c columns in g blocks, while traditional CUR algorithm samples the c columns one at a time.

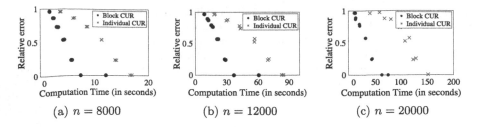

Fig. 6. Performance on synthetic $n \times n$ matrices with rank $n/10$.

Consistently, these results show that block sampling achieves the relative error much faster than the individual column sampling, with performance gains increasing as the size of the matrix grows, as shown in Fig. 6. While the same amount of data is being transmitted regardless of whether block or individual column sampling is used, block sampling is much faster because it needs to contact fewer executors to retrieve blocks of columns rather than the same number of columns individually. In the worst case, sampling individual columns may need to communicate with all of the executors, while block sampling only needs to communicate with g executors. Thus, by exploiting the partitioning of the data, the Block CUR approach is able to achieve roughly the same quality of approximation as traditional column-based CUR, as measured by relative error, with significantly less computation time.

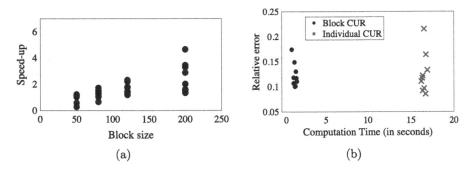

Fig. 7. Performance on $900 \times 10,000$ Arcene dataset with block size 12. (a) Runtime speed-up from block sampling compared to individual column sampling for varying block sizes. (b) Block CUR achieves similar relative errors as individual CUR with much lower computation time.

Real-World Experiments. We also conduct experiments on the Arcene dataset [7] which has 900 rows and 10,000 columns. We compare the runtime for both block and traditional CUR decomposition. We again find consistent improvements for the block approach compared with individual column sampling. With block size $s = 12$, sampling up to 10 groups led to an average speed up of 11.2 over individual column sampling, as shown in Fig. 7. The matrix is very low rank, and sampling a few groups gave small relative errors.

5   Conclusion

In this paper we extended the problem of CUR matrix decomposition to the block setting which is naturally relevant to distributed storage systems and biometric data analysis. We proposed a novel algorithm and derived its performance bounds. We demonstrated its practical utility on real-world distributed storage systems and audience analytics. Some future directions for this work include calculating the leverage scores quickly or adaptively, and considering the algorithms and error bounds when the matrix has a pre-specified structure like sparsity.

References

1. Boutsidis, C., Drineas, P., Magdon-Ismail, M.: Near-optimal column-based matrix reconstruction. SIAM J. Comput. **43**(2), 687–717 (2014)
2. Boutsidis, C., Woodruff, D.P.: Optimal CUR matrix decompositions. In: Proceedings of the 46th Annual ACM Symposium on Theory of Computing, pp. 353–362. ACM (2014)
3. Drineas, P., Magdon-Ismail, M., Mahoney, M.W., Woodruff, D.P.: Fast approximation of matrix coherence and statistical leverage. J. Mach. Learn. Res. **13**(12), 3475–3506 (2012)
4. Drineas, P., Mahoney, M.W., Muthukrishnan, S.: Sampling algorithms for l_2 regression and applications. In: Proceedings of the 17th Annual ACM-SIAM Symposium on Discrete Algorithms, pp. 1127–1136. Society for Industrial and Applied Mathematics (2006)
5. Drineas, P., Mahoney, M.W., Muthukrishnan, S.: Relative-error CUR matrix decompositions. SIAM J. Matrix Anal. Appl. **30**(2), 844–881 (2008)
6. Garbarino, M., Lai, M., Bender, D., Picard, R., Tognetti, S.: Empatica E3–a wearable wireless multi-sensor device for real-time computerized biofeedback and data acquisition. In: Proceedings of the 4th International Conference on Wireless Mobile Communication and Healthcare, pp. 39–42, November 2014
7. Guyon, I., Gunn, S.R., Ben-Hur, A., Dror, G.: Result analysis of the NIPS 2003 feature selection challenge. In: Advances in Neural Information Processing Systems, pp. 545–552 (2004)
8. Jain, S., Oswal, U., Xu, K.S., Eriksson, B., Haupt, J.: A compressed sensing based decomposition of electrodermal activity signals. IEEE Trans. Biomed. Eng. **64**(9), 2142–2151 (2017)
9. Lian, W., Rao, V., Eriksson, B., Carin, L.: Modeling correlated arrival events with latent Semi-Markov processes. In: Proceedings of the International Conference on Machine Learning (2014)
10. Mackey, L.W., Jordan, M.I., Talwalkar, A.: Divide-and-conquer matrix factorization. In: Advances in Neural Information Processing Systems, pp. 1134–1142 (2011)
11. Mahoney, M.W., Drineas, P.: CUR matrix decompositions for improved data analysis. Proc. Nat. Acad. Sci. **106**(3), 697–702 (2009)
12. Oswal, U., Jain, S., Xu, K.S., Eriksson, B.: Block CUR: decomposing large distributed matrices. arXiv preprint arXiv:1703.06065 (2017)
13. Paschou, P., et al.: PCA-correlated SNPs for structure identification in worldwide human populations. PLoS Genet. **3**(9), e160 (2007)
14. Rudelson, M., Vershynin, R.: Sampling from large matrices: an approach through geometric functional analysis. J. ACM **54**(4), 21 (2007)

15. Silveira, F., Eriksson, B., Sheth, A., Sheppard, A.: Predicting audience responses to movie content from electro-dermal activity signals. In: Proceedings of the ACM International Joint Conference on Pervasive and Ubiquitous Computing, pp. 707–716. ACM (2013)

16. Wang, S., Zhang, Z.: A scalable CUR matrix decomposition algorithm: lower time complexity and tighter bound. In: Advances in Neural Information Processing Systems, pp. 647–655 (2012)

17. Xu, M., Jin, R., Zhou, Z.H.: CUR algorithm for partially observed matrices. In: Proceedings of the International Conference on Machine Learning, pp. 1412–1421 (2015)

18. Yang, J., Rübel, O., Mahoney, M.W., Bowen, B.P.: Identifying important ions and positions in mass spectrometry imaging data using CUR matrix decompositions. Anal. Chem. **87**(9), 4658–4666 (2015)

19. Yip, C.W., et al.: Objective identification of informative wavelength regions in galaxy spectra. Astron. J. **147**(5), 110 (2014)

Online and Active Learning

SpectralLeader: Online Spectral Learning for Single Topic Models

Tong Yu[1]([⊠]), Branislav Kveton[2], Zheng Wen[3], Hung Bui[4],
and Ole J. Mengshoel[1]

[1] Electrical and Computer Engineering, Carnegie Mellon University, Pittsburgh, USA
tongy1@andrew.cmu.edu, ole.mengshoel@sv.cmu.edu
[2] Google Research, Mountain View, USA
bkveton@google.com
[3] Adobe Research, San Jose, USA
zwen@adobe.com
[4] DeepMind, Mountain View, USA
bui.h.hung@gmail.com

Abstract. We study the problem of learning a latent variable model online from a stream of data. Latent variable models are popular because they can explain observed data through unobserved concepts. These models have traditionally been studied in the offline setting. In the online setting, online expectation maximization (EM) is arguably the most popular approach for learning latent variable models. Although online EM is computationally efficient, it typically converges to a local optimum. In this work, we develop a new online learning algorithm for latent variable models, which we call SpectralLeader. SpectralLeader converges to the global optimum, and we derive a sublinear upper bound on its n-step regret in a single topic model. In both synthetic and real-world experiments, we show that SpectralLeader performs similarly to or better than online EM with tuned hyper-parameters.

Keywords: Online learning · Spectral method · Topic models

1 Introduction

Latent variable models explain observed data through unobserved concepts. They have been successfully applied in a wide variety of fields, such as speech recognition, natural language processing, and computer vision [5,15,16,20]. Despite their successes, latent variable models are typically studied in the offline setting. However, in many practical problems, a learning agent needs to learn a latent variable model online. With online algorithms, we can update the model efficiently and do not need to store all the past data. For instance, a recommender system may want to learn to cluster its users online based on their real-time behavior. This paper aims to develop algorithms for such online learning problems.

B. Kveton—This work was done while the author was at Adobe Research.

Several existing algorithms learn latent variable models online by extending expectation maximization (EM) algorithm. Those algorithms are known as online EM, and include stepwise EM [6,13] and incremental EM [14]. Similar to offline EM, each iteration of online EM includes an E-step to fill in the values of latent variables based on their estimated distribution, and an M-step to update the model parameters. The main difference is that each step of online EM only uses data received recently, rather than the whole dataset. This ensures that online EM is computationally efficient and can be used to learn latent variable models online. However, similar to offline EM, online EM algorithms have one major drawback: they may converge to a local optimum and hence suffer from a non-diminishing performance loss.

To overcome these limitations, we develop an online learning algorithm that performs almost as well as the globally optimal latent variable model, which we call `SpectralLeader`. Specifically, we propose an online learning variant of the spectral method [3], which can learn the parameters of latent variable models offline with guarantees of convergence to a global optimum. Our online learning setting is defined as follows. We have a sequence of n topic models, one at each time $t \in [n]$. The prior distribution of topics can change arbitrarily over time, while the conditional distribution of words is stationary. At time t, the learning agent observes a document of words, which is sampled i.i.d. from the model at time t. The goal of the agent is to predict a sequence of model parameters with low cumulative regret with respect to the best solution in hindsight, which is constructed based on the sampling distribution of the words over n steps.

This paper makes several contributions. First, it is the first paper to formulate online learning with the spectral method as a regret minimization problem. Second, we propose `SpectralLeader`, an online learning variant of the spectral method for single topic models [3]. To reduce computational and space complexities of `SpectralLeader`, we introduce reservoir sampling. Third, we prove a sublinear upper bound on the n-step regret of `SpectralLeader`. Finally, we compare `SpectralLeader` to stepwise EM in extensive synthetic and real-world experiments. We observe that stepwise EM is sensitive to the setting of its hyper-parameters. In all experiments, `SpectralLeader` performs similarly to or better than stepwise EM with optimized hyper-parameters.

2   Related Work

The spectral method by tensor decomposition has been widely applied in different latent variable models, such as mixtures of tree graphical models [3], mixtures of linear regressions [7], hidden Markov models (HMM) [4], latent Dirichlet allocation (LDA) [2], Indian buffet process [18], and hierarchical Dirichlet process [19]. The spectral method first empirically estimates low-order moments of observations and then applies decomposition methods with a unique solution to recover the model parameters. One major advantage of the spectral method is that it learns globally optimal solutions [3].

Traditional online learning methods for latent variable models usually extend traditional iterative methods for learning latent variable models in the offline setting [6,11,13,14]. Offline EM calculates sufficient statistics based on all the data, while in online EM the sufficient statistics are updated with recent data in each iteration [6,13,14]. Online algorithms are used to learn LDA on streaming data [1,11]. These online algorithms either have no convergence analysis or converge to local minima, while we aim to develop an algorithm with a theoretical guarantee of convergence to a global optimum.

An online spectral learning method has also been developed [12], with a focus on improving computational efficiency, by conducting optimization of multilinear operations in SGD and avoiding directly forming tensors. Online stochastic gradient for tensor decomposition has been analyzed [9] in a different online setting: they do not look at the online problem as regret minimization and the analysis focuses on convergence to a local minimum. In contrast, we develop an online spectral method with a theoretical guarantee of convergence to a global optimum. Further, our method is robust in the non-stochastic setting where the topics of documents are correlated over time. This non-stochastic setting has not been previously studied in the context of online spectral learning [12].

3 Spectral Method for Single Topic Models

This section introduces the offline spectral method in latent variable models. Specifically, we describe how the method works in the single topic model [3].

In the single topic model, the goal is to learn the latent topics of documents from the observed words in each document. Without loss of generality, we describe the spectral method and `SpectralLeader` (Sect. 5) in the setting where each document contains three words. The extension to more than three words is straightforward (Sect. 7). Let the number of distinct topics be K and the size of the vocabulary be d. Then our model can be viewed as a mixture model, where the three observed words $\mathbf{x}^{(1)}$, $\mathbf{x}^{(2)}$, and $\mathbf{x}^{(3)}$ are conditionally i.i.d. given topic C, which is drawn from some distribution over topics. Later in Sect. 4, we study a more general setting where the distribution of topic can change over time. Each word is one-hot encoded, that is $\mathbf{x}^{(l)} = e_i$ if and only if $\mathbf{x}^{(l)}$ represents word i, where $e_1, \ldots, e_d$ is the standard coordinate basis in $\mathbb{R}^d$. Define $[n] = \{1, \ldots, n\}$. The model is parameterized by the probability of each topic j, $\omega_j = P(C = j)$ for $j \in [K]$, and the conditional probability of all words $u_j \in [0,1]^d$ given topic j. The ith entry of u_j is $u_j(i) = P(\mathbf{x}^{(l)} = e_i | C = j)$ for $i \in [d]$. To recover the model parameters, it suffices to construct a third order tensor $\bar{M}_3$ as

$$\mathbb{E}[\mathbf{x}^{(1)} \otimes \mathbf{x}^{(2)} \otimes \mathbf{x}^{(3)}] = \sum_{1 \leq i,j,k \leq d} P(\mathbf{x}^{(1)} = e_i, \mathbf{x}^{(2)} = e_j, \mathbf{x}^{(3)} = e_k)\, e_i \otimes e_j \otimes e_k.$$

We recover the parameters of the topic model by decomposing $\bar{M}_3$ as

$$\bar{M}_3 = \sum_{i=1}^{K} \omega_i u_i \otimes u_i \otimes u_i. \tag{1}$$

Unfortunately, such a decomposition is generally NP-hard [3]. Instead, we can decompose an orthogonal decomposable tensor. One way to make $\bar{M}_3$ orthogonal decomposable is by whitening. We can define a whitening matrix as $\bar{W} = UA^{-1/2}$, where $A \in \mathbb{R}^{K \times K}$ is the diagonal matrix of the positive eigenvalues of $\bar{M}_2 = \mathbb{E}[\mathbf{x}^{(1)} \otimes \mathbf{x}^{(2)}] = \sum_{i=1}^{K} \omega_i u_i \otimes u_i$ and $U \in \mathbb{R}^{d \times K}$ is the matrix of K eigenvectors associated with those eigenvalues. After whitening, instead of decomposing $\bar{M}_3$, we can decompose $\bar{T} = \mathbb{E}[\bar{W}^\top \mathbf{x}^{(1)} \otimes \bar{W}^\top \mathbf{x}^{(2)} \otimes \bar{W}^\top \mathbf{x}^{(3)}]$ as $\bar{T} = \sum_{i=1}^{K} \lambda_i v_i \otimes v_i \otimes v_i$ by the *power iteration method* [3]. Finally, the model parameters are recovered as $\omega_i = \frac{1}{\lambda_i^2}$ and $u_i = \lambda_i (\bar{W}^\top)^+ v_i$, where $(\bar{W}^\top)^+$ is the pseudoinverse of $\bar{W}^\top$. In practice, only a noisy realization of $\bar{T}$ is typically available, which is constructed from empirical counts. Such tensors can be decomposed approximately and the errors of such decompositions are analyzed in Theorem 5.1 of Anandkumar *et al.* [3].

4   Online Learning for Single Topic Models

We study the following online learning problem in the single topic model discussed in Sect. 3. We have a sequence of n topic models, one at each time $t \in [n]$. The prior distribution of topics can change arbitrarily over time, while the conditional distribution of words is stationary. We denote by $\mathbf{x}_t = (\mathbf{x}_t^{(l)})_{l=1}^3$ a tuple of one-hot encoded words in the document at time t, which is sampled i.i.d. from the model at time t. Non-stationary distributions of topics are common in practice. For instance, in the recommender system example in Sect. 1, user clusters tend to be correlated over time. The clusters can be viewed as topics.

We represent the distribution of words at time t by a cube $P_t = \mathbb{E}[\mathbf{x}_t^{(1)} \otimes \mathbf{x}_t^{(2)} \otimes \mathbf{x}_t^{(3)}] \in [0, 1]^{d \times d \times d}$. In particular, the probability of observing the triplet of words (i, j, k) at time t is

$$P_t(i, j, k) = \sum_{c=1}^{K} P_t(c) P(\mathbf{x}_t^{(1)} = e_i | c) P(\mathbf{x}_t^{(2)} = e_j | c) P(\mathbf{x}_t^{(3)} = e_k | c), \qquad (2)$$

where $P_t(c)$ is the prior distribution of topics at time t. This prior distribution can change arbitrarily with t.

The learning agent predicts the distribution of words $\hat{M}_{3,t-1} \in [0, 1]^{d \times d \times d}$ at time t and is evaluated by its per-step loss $\ell_t(\hat{M}_{3,t-1})$. The agent aims to minimize its cumulative loss, which measures the difference between the predicted distribution $\hat{M}_{3,t-1}$ and the observations $\mathbf{x}_t^{(1)} \otimes \mathbf{x}_t^{(2)} \otimes \mathbf{x}_t^{(3)}$ over time.

But what should the loss be? In this work, we define the *loss* at time t as

$$\ell_t(M) = \|\mathbf{x}_t^{(1)} \otimes \mathbf{x}_t^{(2)} \otimes \mathbf{x}_t^{(3)} - M\|_F^2, \qquad (3)$$

where $\|.\|_F$ is the *Frobenius norm*. For any tensor $M \in \mathbb{R}^{d \times d \times d}$, we define its Frobenius norm as $\|M\|_F = \sqrt{\sum_{i,j,k=1}^{d} M(i,j,k)^2}$. This choice can be justified as follows. Let

$$\bar{M}_{3,n} = \frac{1}{n} \sum_{t=1}^{n} P_t = \frac{1}{n} \sum_{t=1}^{n} \mathbb{E}[\mathbf{x}_t^{(1)} \otimes \mathbf{x}_t^{(2)} \otimes \mathbf{x}_t^{(3)}] \tag{4}$$

be the average of distributions from which $\mathbf{x}_t^{(1)} \otimes \mathbf{x}_t^{(2)} \otimes \mathbf{x}_t^{(3)}$ are generated in n steps. Then

$$\bar{M}_{3,n} = \operatorname*{argmin}_{M \in [0,1]^{d \times d \times d}} \sum_{t=1}^{n} \mathbb{E}[\ell_t(M)], \tag{5}$$

as shown in Lemma 3 in Sect. 6.4. In other words, the loss function is chosen such that a natural *best solution in hindsight*, $\bar{M}_{3,n}$ in (5), is the minimizer of the cumulative loss.

With the definition of the loss function and the best solution in hindsight, the goal of the learning agent is to minimize the regret

$$R(n) = \sum_{t=1}^{n} \mathbb{E}[\ell_t(\hat{M}_{3,t-1}) - \ell_t(\bar{M}_{3,n})], \tag{6}$$

where $\ell_t(\hat{M}_{3,t-1})$ is the loss of our estimated model at time t and $\ell_t(\bar{M}_{3,n})$ is the loss of the best solution in hindsight, respectively. Minimizing the regret in the online setting guarantees that the learnt model can provide more and more accurate predictions over time.

Unlike traditional online algorithms that minimize the negative log-likelihood [13], we minimize the parameter recovery loss. In the offline setting, the spectral method minimizes the recovery loss in a wide range of models [3,7,17].

5   Algorithm SpectralLeader

We propose SpectralLeader, an online learning algorithm for minimizing the regret in (6). Its pseudocode is in Algorithm 1. At each time t, the input is observation $(\mathbf{x}_t^{(l)})_{l=1}^{3}$. We maintain reservoir samples $((\mathbf{x}_z^{(l)})_{l=1}^{3})_{z \in S_{t-1}}$ from the previous $t-1$ time steps, where S_{t-1} is the time indices of these samples.

The algorithm operates as follows. First, in line 1 we construct the second-order moment from the reservoir samples, where $\Pi_2(3)$ is the set of all 2-permutations of $\{1,2,3\}$. Then we estimate A_{t-1} and U_{t-1} by eigendecomposition, and construct the whitening matrix W_{t-1} in line 2. After whitening, we build the third-order tensor T_{t-1} from whitened words $((W_{t-1}^{\top}\mathbf{x}_z^{(l)})_{l=1}^{3})_{z \in S_{t-1}}$ in line 3, where $\Pi_3(3)$ is the set of all 3-permutations of $\{1,2,3\}$. Then in line 4 with the power iteration method [3], we decompose T_{t-1} and get its eigenvalues

---

Algorithm 1. SpectralLeader at time t.

Input: Observations $(\mathbf{x}_t^{(l)})_{l=1}^3$

1 $M_{2,t-1} \leftarrow \frac{1}{|\mathcal{S}_{t-1}||\Pi_2(3)|} \sum_{z \in \mathcal{S}_{t-1}} \sum_{\pi \in \Pi_2(3)} \mathbf{x}_z^{(\pi(1))} \otimes \mathbf{x}_z^{(\pi(2))}$

2 $W_{t-1} \leftarrow U_{t-1} A_{t-1}^{-1/2}$, where $A_{t-1} \in \mathbb{R}^{K \times K}$ is the diagonal matrix of K positive eigenvalues of $M_{2,t-1}$ and $U_{t-1} \in \mathbb{R}^{d \times K}$ is the matrix of eigenvectors associated with these positive eigenvalues

3 $T_{t-1} \leftarrow$
   $\frac{1}{|\mathcal{S}_{t-1}||\Pi_3(3)|} \sum_{z \in \mathcal{S}_{t-1}} \sum_{\pi \in \Pi_3(3)} W_{t-1}^\top \mathbf{x}_z^{(\pi(1))} \otimes W_{t-1}^\top \mathbf{x}_z^{(\pi(2))} \otimes W_{t-1}^\top \mathbf{x}_z^{(\pi(3))}$

4 Obtain $(\lambda_{t-1,i})_{i=1}^K$ and $(v_{t-1,i})_{i=1}^K$ from T_{t-1} by power iteration method

5 $\omega_{t-1,i} \leftarrow \frac{1}{\lambda_{t-1,i}^2}$, $u_{t-1,i} \leftarrow \lambda_{t-1,i}(W_{t-1}^\top)^+ v_{t-1,i}$ for all $i \in [K]$

6 Generate a random number $a \in [0,1]$

7 **if** $t \leq m_r$ **then**

8 $\quad\lfloor \; \mathcal{S}_t \leftarrow \mathcal{S}_{t-1} \cup \{t\}$

9 **else if** $a \leq m_r/(t-1)$ **then**

10 $\quad$ Remove a random element of $\mathcal{S}_{t-1}$

11 $\quad\lfloor \; \mathcal{S}_t \leftarrow \mathcal{S}_{t-1} \cup \{t\}$

12 **else**

13 $\quad\lfloor \; \mathcal{S}_t \leftarrow \mathcal{S}_{t-1}$

Output: Model parameters $\omega_{t-1,i}$ and $u_{t-1,i}$

$(\lambda_{t-1,i})_{i=1}^K$ and eigenvectors $(v_{t-1,i})_{i=1}^K$. Finally, in line 5 we recover the parameters of the model, the probability of topics $(\omega_{t-1,i})_{i=1}^K$ and the conditional probability of words $(u_{t-1,i})_{i=1}^K$. After recovering the parameters, we update the set of reservoir samples in lines 6 to 13. We keep m_r reservoir samples $\mathbf{x}_z$, $z \in [t-1]$. When $t \leq m_r$, the new observation $(\mathbf{x}_t^{(l)})_{l=1}^3$ is added to the reservoir. When $t > m_r$, $(\mathbf{x}_t^{(l)})_{l=1}^3$ replaces a random observation in the reservoir with probability $m_r/(t-1)$.

In SpectralLeader, we use reservoir sampling for computational efficiency reasons. Without reservoir sampling, the operations in lines 1 and 3 of Algorithm 1 would depend on t because all past observations are used to construct $M_{2,t-1}$ and T_{t-1}. Besides, the whitening operation in line 3 would depend on t because all past observations are whitened by a matrix W_{t-1} that changes with t. With reservoir sampling, we approximate $M_{2,t-1}$, T_{t-1}, and W_{t-1} with m_r reservoir samples. We discuss how to set m_r in Sect. 6.2.

6   Analysis

In this section, we bound the regret of SpectralLeader. In Sect. 6.1, we analyze the regret of SpectralLeader without reservoir sampling in the noise-free setting. In this setting, at time t the agent knows the distribution of words $(P_z)_{z=1}^{t-1}$.

The regret is due to not knowing P_t at time t. In Sect. 6.2, we analyze the regret of `SpectralLeader` with reservoir sampling in the noise-free setting. In this setting, the agent knows $(P_z)_{z \in S_{t-1}}$ at time t, which is a random sample of $(P_z)_{z=1}^{t-1}$. In comparison to Sect. 6.1, the additional regret is due to reservoir sampling. In Sect. 6.3, we discuss the regret of `SpectralLeader` with reservoir sampling in the noisy setting. In this setting, the agent approximates each distribution P_z with its single empirical observation $(\mathbf{x}_z^{(l)})_{l=1}^3$, for any $z \in S_{t-1}$. In comparison to Sect. 6.2, the additional regret is due to noisy observations. All supplementary lemmas are stated and proved in Sect. 6.4.

6.1   Noise-Free Setting

We first analyze an idealized variant of `SpectralLeader`, where the agent knows $(P_z)_{z=1}^{t-1}$ at time t. In this setting, the algorithm is similar to Algorithm 1, except that lines 1 and 3 are replaced, respectively, by $\bar{M}_{2,t-1} = \frac{1}{t-1} \sum_{z=1}^{t-1} \mathbb{E}[\mathbf{x}_z^{(1)} \otimes \mathbf{x}_z^{(2)}]$ and $\bar{T}_{t-1} = \frac{1}{t-1} \sum_{z=1}^{t-1} \mathbb{E}[\bar{W}_{t-1}^\top \mathbf{x}_z^{(1)} \otimes \bar{W}_{t-1}^\top \mathbf{x}_z^{(2)} \otimes \bar{W}_{t-1}^\top \mathbf{x}_z^{(3)}]$.

We denote by $\bar{W}_{t-1}$ the corresponding whitening matrix in line 2, and by $\bar{\omega}_{t-1,i}$ and $\bar{u}_{t-1,i}$ the estimated model parameters. In this noise-free setting, the power iteration method in line 4 is exact. Therefore, the prediction of the learning agent at time t satisfies $\hat{M}_{3,t-1} = \sum_{i=1}^K \bar{\omega}_{t-1,i} \bar{u}_{t-1,i} \otimes \bar{u}_{t-1,i} \otimes \bar{u}_{t-1,i} = \bar{M}_{3,t-1}$ for any t, according to (1).

Theorem 1. *Let* $\hat{M}_{3,t-1} = \bar{M}_{3,t-1}$ *at all times* $t \in [n]$. *Then*

$$R(n) \leq 4\sqrt{d^3} \log n.$$

Proof. From Lemma 4, $\sum_{t=1}^n \mathbb{E}[\ell_t(\bar{M}_{3,n})] \geq \sum_{t=1}^n \mathbb{E}[\ell_t(\bar{M}_{3,t})]$. Now note that $\hat{M}_{3,t-1} = \bar{M}_{3,t-1}$ at any time t, and therefore

$$R(n) = \sum_{t=1}^n \mathbb{E}[\ell_t(\bar{M}_{3,t-1}) - \ell_t(\bar{M}_{3,n})] \leq \sum_{t=1}^n \mathbb{E}[\ell_t(\bar{M}_{3,t-1}) - \ell_t(\bar{M}_{3,t})].$$

At any time t and for any $\mathbf{x}_t$,

$$\ell_t(\bar{M}_{3,t-1}) - \ell_t(\bar{M}_{3,t}) \leq 4\|\bar{M}_{3,t-1} - \bar{M}_{3,t}\|_F$$

$$= 4 \left\| \frac{1}{t-1} \sum_{t'=1}^{t-1} P_{t'} - \frac{1}{t} \sum_{t'=1}^{t} P_{t'} \right\|_F$$

$$= \frac{4}{t} \left\| \frac{1}{t-1} \sum_{t'=1}^{t-1} P_{t'} - P_t \right\|_F \leq \frac{4\sqrt{d^3}}{t},$$

where the first inequality is by Lemma 5 and the second inequality is from the fact that all entries of P_t are in $[0, 1]$ at any time $t \in [n]$. Therefore, $R(n) \leq \sum_{t=1}^n \frac{4\sqrt{d^3}}{t} \leq 4\sqrt{d^3} \log n$. This concludes our proof. $\square$

6.2   Reservoir Sampling in Noise-Free Setting

We further analyze `SpectralLeader` with reservoir sampling in the noise-free setting. As discussed in Sect. 5, without reservoir sampling, the construction time of the decomposed tensor at time t would grow linearly with t, which is undesirable. In this setting, the algorithm is similar to Algorithm 1, except that lines 1 and 3 are replaced, respectively, by $\hat{M}_{2,t-1} = \frac{1}{|S_{t-1}|}\mathbb{E}[\sum_{z \in S_{t-1}} \mathbf{x}_z^{(1)} \otimes \mathbf{x}_z^{(2)}]$ and $\tilde{T}_{t-1} = \frac{1}{|S_{t-1}|}\mathbb{E}[\sum_{z \in S_{t-1}} \tilde{W}_{t-1}^{\top}\mathbf{x}_z^{(1)} \otimes \tilde{W}_{t-1}^{\top}\mathbf{x}_z^{(2)} \otimes \tilde{W}_{t-1}^{\top}\mathbf{x}_z^{(3)}]$, where S_{t-1} are indices of the reservoir samples at time t. We denote by $\tilde{W}_{t-1}$ the corresponding whitening matrix in line 2, and by $\tilde{w}_{t-1,i}$ and $\tilde{u}_{t-1,i}$ the estimated model parameters. As in Sect. 6.1, the power iteration method in line 4 is exact, and therefore the prediction of the learning agent at time t satisfies $\hat{M}_{3,t-1} = \sum_{i=1}^{K} \tilde{w}_{t-1,i}\,\tilde{u}_{t-1,i} \otimes \tilde{u}_{t-1,i} \otimes \tilde{u}_{t-1,i} = \tilde{M}_{3,t-1}$ for any t. The main result of this section is stated below.

Theorem 2. *Let all corresponding entries of $\tilde{M}_{3,t-1}$ and $\bar{M}_{3,t-1}$ be close with a high probability,*

$$P(\exists t, i, j, k : |\tilde{M}_{3,t-1}(i,j,k) - \bar{M}_{3,t-1}(i,j,k)| \geq \epsilon) = \delta \tag{7}$$

for some small $\epsilon \in [0,1]$ and $\delta \in [0,1]$. Let $\hat{M}_{3,t-1} = \tilde{M}_{3,t-1}$ at all times $t \in [n]$. Then

$$R(n) \leq 4\sqrt{d^3}\epsilon n + 4\sqrt{d^3}\delta n + 4\sqrt{d^3}\log n.$$

Proof. From the definition of $R(n)$ in (6) and the bound in Theorem 1,

$$R(n) = \sum_{t=1}^{n}\mathbb{E}[\ell_t(\tilde{M}_{3,t-1}) - \ell_t(\bar{M}_{3,t-1})] + \sum_{t=1}^{n}\mathbb{E}[\ell_t(\bar{M}_{3,t-1}) - \ell_t(\bar{M}_{3,n})]$$

$$\leq \sum_{t=1}^{n}\mathbb{E}[\ell_t(\tilde{M}_{3,t-1}) - \ell_t(\bar{M}_{3,t-1})] + 4\sqrt{d^3}\log n.$$

We bound the first term above as follows. Suppose that the event in (7) does not happen. Then $\ell_t(\tilde{M}_{3,t-1}) - \ell_t(\bar{M}_{3,t-1}) \leq 4\sqrt{d^3}\epsilon$, from Lemma 5 and the fact that all corresponding entries of $\tilde{M}_{3,t-1}$ and $\bar{M}_{3,t-1}$ are ϵ-close. Now suppose that the event in (7) happens. Then $\ell_t(\tilde{M}_{3,t-1}) - \ell_t(\bar{M}_{3,t-1}) \leq 4\sqrt{d^3}$, from Lemma 5 and the fact all entries of $\tilde{M}_{3,t-1}$ and $\bar{M}_{3,t-1}$ are in $[0,1]$. Finally, note that the event in (7) happens with probability δ. Now we chain all inequalities and obtain $R(n) \leq 4\sqrt{d^3}\epsilon n + 4\sqrt{d^3}\delta n + 4\sqrt{d^3}\log n$. □

Note that the reservoir at time t, $S_{t-1} \in [t-1]$, is a random sample of size m_r for any $t > m_r + 1$. Therefore, from Hoeffding's inequality [10] and the union bound, we get that

$$\delta = P(\exists t, i, j, k : |\tilde{M}_{3,t-1}(i,j,k) - \bar{M}_{3,t-1}(i,j,k)| \geq \epsilon)$$

$$\leq 2\sum_{t=m_r+2}^{n} d^3 \exp[-2\epsilon^2 m_r] \leq 2d^3 n \exp[-2\epsilon^2 m_r].$$

In addition, let the size of the reservoir be $m_r = \epsilon^{-2} \log(d^3 n)$. Then the regret bound in Theorem 2 simplifies to $R(n) < 4\sqrt{d^3}\epsilon n + 4\sqrt{d^3}\log n + 8$. This bound can be sublinear in n only if $\epsilon = o(1)$. Moreover, the definition of m_r and $m_r \le n$ imply that $\epsilon \ge \sqrt{\log(d^3 n)/n}$. As a result of these constraints, the range of reasonable values for ϵ is $[\sqrt{\log(d^3 n)/n}, o(1))$.

For any $\epsilon \in [\sqrt{\log(d^3 n)/n}, o(1))$, the regret $R(n)$ is sublinear in n, where ϵ is a tunable parameter. At lower values of ϵ, $R(n) = O(\sqrt{n})$ but the reservoir size approaches n. At higher values of ϵ, the reservoir size is $O(\log n)$ but $R(n)$ approaches n.

6.3 Reservoir Sampling in Noisy Setting

Finally, we discuss the regret of `SpectralLeader` with reservoir sampling in the noisy setting. In this setting, the analyzed algorithm is Algorithm 1. The predicted distribution at time t is $\hat{M}_{3,t-1} = \sum_{i=1}^{K} \omega_{t-1,i}\, u_{t-1,i} \otimes u_{t-1,i} \otimes u_{t-1,i}$.

From the definition of $R(n)$ and our earlier analysis, $R(n)$ can be decomposed and bounded from above as

$$R(n) \le \sum_{t=1}^{n} \mathbb{E}[\ell_t(\hat{M}_{3,t-1}) - \ell_t(\tilde{M}_{3,t-1})] + 4\sqrt{d^3}\epsilon n + 4\sqrt{d^3}\log n + 8 \qquad (8)$$

when the size of the reservoir is $m_r = \epsilon^{-2}\log(d^3 n)$.

Suppose that $m_r \to \infty$ as $n \to \infty$, for instance by setting $\epsilon = n^{-\frac{1}{4}}$. Under this assumption, $M_{2,t-1}$ in `SpectralLeader` approaches $\tilde{M}_{2,t-1}$ (Sect. 6.2) because $M_{2,t-1}$ is an empirical estimator of $\tilde{M}_{2,t-1}$ on m_r observations. By Weyl's and Davis-Kahan theorems [8,21], the eigenvalues and eigenvectors of $M_{2,t-1}$ approach those of $\tilde{M}_{2,t-1}$ as $m_r \to \infty$, and thus the whitening matrix W_{t-1} in `SpectralLeader` approaches $\tilde{W}_{t-1}$ (Sect. 6.2). Since T_{t-1} in `SpectralLeader` is an empirical estimator of $\tilde{T}_{t-1}$ (Sect. 6.2) on m_r whitened observations and $W_{t-1} \to \tilde{W}_{t-1}$, we have $T_{t-1} \to \tilde{T}_{t-1}$ as $m_r \to \infty$. In our online setting (Sect. 4), over time the data samples are generated by topic models with the same conditional distribution of words. Thus, the reservoir samples are actually generated by a topic model with this conditional distribution and an arbitrary distribution of topics. Therefore, Theorem 5.1 of Anandkumar et al. [3] applies: the eigenvalues and eigenvectors of T_{t-1} approach those of $\tilde{T}_{t-1}$ as $T_{t-1} \to \tilde{T}_{t-1}$. This implies that $\hat{M}_{3,t-1} \to \tilde{M}_{3,t-1}$, as all quantities that $\hat{M}_{3,t-1}$ and $\tilde{M}_{3,t-1}$ depend on approach each other as $m_r \to \infty$. Therefore, $\lim_{n\to\infty}\lim_{t\to n}(\ell_t(\hat{M}_{3,t-1}) - \ell_t(\tilde{M}_{3,t-1})) = 0$ and the regret bound in (8) is $o(n)$, sublinear in n, as $n \to \infty$.

6.4 Technical Lemmas

Lemma 3. *Let* $\ell_t(M) = \|\mathbf{x}_t^{(1)} \otimes \mathbf{x}_t^{(2)} \otimes \mathbf{x}_t^{(3)} - M\|_F^2$. *Then*

$$\bar{M}_{3,n} = \underset{M \in [0,1]^{d \times d \times d}}{\operatorname{argmin}} \sum_{t=1}^{n} \mathbb{E}[\ell_t(M)], \qquad (9)$$

where $\bar{M}_{3,n}$ is defined in (4).

Proof. It is sufficient to show that

$$\bar{M}_{3,n}(i,j,k) = \underset{y \in [0,1]}{\operatorname{argmin}} \frac{1}{n} \sum_{t=1}^{n} \mathbb{E}[(\mathbf{x}_t^{(1)} \otimes \mathbf{x}_t^{(2)} \otimes \mathbf{x}_t^{(3)}(i,j,k) - y)^2] \tag{10}$$

for any (i,j,k), where $\bar{M}_{3,n}(i,j,k)$ and $\mathbf{x}_t^{(1)} \otimes \mathbf{x}_t^{(2)} \otimes \mathbf{x}_t^{(3)}(i,j,k)$ are the (i,j,k)-th entries of tensors $\bar{M}_{3,n}$ and $\mathbf{x}_t^{(1)} \otimes \mathbf{x}_t^{(2)} \otimes \mathbf{x}_t^{(3)}$, respectively. To prove the claim, let $f(y) = \frac{1}{n} \sum_{t=1}^{n} \mathbb{E}[(\mathbf{x}_t^{(1)} \otimes \mathbf{x}_t^{(2)} \otimes \mathbf{x}_t^{(3)}(i,j,k) - y)^2]$. Then

$$\frac{\partial}{\partial y} f(y) = 2y - \frac{2}{n} \sum_{t=1}^{n} \mathbb{E}[\mathbf{x}_t^{(1)} \otimes \mathbf{x}_t^{(2)} \otimes \mathbf{x}_t^{(3)}(i,j,k)].$$

Now we put the derivative equal to zero and get $y = \bar{M}_{3,n}(i,j,k)$. $\square$

Lemma 4. *For any n, $\sum_{t=1}^{n} \mathbb{E}[\ell_t(\bar{M}_{3,n})] \geq \sum_{t=1}^{n} \mathbb{E}[\ell_t(\bar{M}_{3,t})]$.*

Proof. We prove this claim by induction. First, suppose that $n = 0$. Then trivially $\mathbb{E}[\ell_t(\bar{M}_{3,0})] \geq \mathbb{E}[\ell_t(\bar{M}_{3,0})]$. Second, by induction hypothesis, we have that

$$\sum_{t-1}^{n-1} \mathbb{E}[\ell_t(\bar{M}_{3,n-1})] \geq \sum_{t=1}^{n-1} \mathbb{E}[\ell_t(\bar{M}_{3,t})]. \tag{11}$$

Then

$$\sum_{t=1}^{n} \mathbb{E}[\ell_t(\bar{M}_{3,n})] = \sum_{t=1}^{n-1} \mathbb{E}[\ell_t(\bar{M}_{3,n})] + \mathbb{E}[\ell_n(\bar{M}_{3,n})]$$

$$\geq \sum_{t=1}^{n-1} \mathbb{E}[\ell_t(\bar{M}_{3,n-1})] + \mathbb{E}[\ell_n(\bar{M}_{3,n})] \geq \sum_{t=1}^{n} \mathbb{E}[\ell_t(\bar{M}_{3,t})],$$

where the first inequality is from (9) and the second inequality is from (11). $\square$

Lemma 5. *For any tensors $M \in [0,1]^{d \times d \times d}$ satisfying $\sum_{i,j,k=1}^{d} M(i,j,k) = 1$, and $M' \in [0,1]^{d \times d \times d}$ satisfying $\sum_{i,j,k=1}^{d} M'(i,j,k) = 1$, we have*

$$\ell_t(M) - \ell_t(M') \leq 4\|M - M'\|_F.$$

Proof. The proof follows from elementary algebra

$\ell_t(M) - \ell_t(M')$

$\quad = (\ell_t^{\frac{1}{2}}(M) + \ell_t^{\frac{1}{2}}(M'))(\ell_t^{\frac{1}{2}}(M) - \ell_t^{\frac{1}{2}}(M'))$

$\quad \leq (\ell_t^{\frac{1}{2}}(M) + \ell_t^{\frac{1}{2}}(M'))\|M - M'\|_F$

$\quad = \left(\|\mathbf{x}_t^{(1)} \otimes \mathbf{x}_t^{(2)} \otimes \mathbf{x}_t^{(3)} - M\|_F + \|\mathbf{x}_t^{(1)} \otimes \mathbf{x}_t^{(2)} \otimes \mathbf{x}_t^{(3)} - M'\|_F \right) \|M - M'\|_F$

$\quad \leq \left(2\|\mathbf{x}_t^{(1)} \otimes \mathbf{x}_t^{(2)} \otimes \mathbf{x}_t^{(3)}\|_F + \|M\|_F + \|M'\|_F \right) \|M - M'\|_F$

$\quad = (2 + \|M\|_F + \|M'\|_F) \|M - M'\|_F \leq 4\|M - M'\|_F.$

The first equality is from $\alpha^2 - \beta^2 = (\alpha+\beta)(\alpha-\beta)$. The first inequality is from the reverse triangle inequality. The second inequality is from the triangle inequality. The third equality is from the fact that only one entry of $\mathbf{x}_t^{(1)} \otimes \mathbf{x}_t^{(2)} \otimes \mathbf{x}_t^{(3)}$ is 1 and all the rest are 0, by the definition of $(\mathbf{x}_t^{(l)})_{l=1}^3$ in Sect. 4. The third inequality is from $\|M\|_F = \sqrt{\sum_{i,j,k=1}^d M(i,j,k)^2} \leq \sqrt{\sum_{i,j,k=1}^d |M(i,j,k)|} = 1$, and similarly $\|M'\|_F \leq 1$, which follows from the fact that tensors M and M' represent distributions with all entries in $[0,1]$ and summing up to 1.   □

7 Extensions of `SpectralLeader`

`SpectralLeader` can be extended as follows. First, it can easily be extended to documents with $L \geq 3$ words. Then, at time t, T_{t-1} in Algorithm 1 is calculated by averaging over all $\binom{L}{3}3!$ ordered triplets of words [3,22]. The analysis essentially remains the same and the regret bound still holds. Second, `SpectralLeader` can be extended to more complicated latent variable models. For example, we can use `SpectralLeader` in Algorithm 1 to learn Gaussian mixture models (GMM) online, by redefining $M_{2,t-1}$ and T_{t-1} according to Theorem 3.2 of Anandkumar *et al.* [3]. The current analysis does not apply to GMM, since P_t in (2) is not bounded in GMM. We leave the analysis of `SpectralLeader` in such more complicated models for future work.

8 Experiments

In this section, we evaluate `SpectralLeader` and compare it empirically with stepwise EM [6]. We experiment with both stochastic and non-stochastic synthetic problems, as well as with two real-world problems.

Our chosen baseline is stepwise EM [6], an online EM algorithm. We choose this baseline as it outperforms other online EM algorithms [13], such as incremental EM [14]. Stepwise EM has two key tuning parameters: the step-size reduction power α and the mini-batch size m [6,13]. The smaller the α, the faster the old sufficient statistics are forgotten. The mini-batch size m is the number of documents to calculate the sufficient statistics for each update of stepwise EM. With larger m, we can usually add stability to stepwise EM. We compared `SpectralLeader` to stepwise EM with varying α and m.

All compared algorithms are evaluated by their models at time t, $\theta_{t-1} = ((\omega_{t-1,i})_{i=1}^K, (u_{t-1,i})_{i=1}^K)$, which are learned from the first $t-1$ steps. We report two metrics: *average negative predictive log-likelihood up to step n*, $\mathcal{L}_n^{(1)} = \frac{1}{n}\sum_{t=2}^n \left(-\log\sum_{i=1}^K P_{\theta_{t-1}}(C=i)\prod_{l=1}^L P_{\theta_{t-1}}(\mathbf{x}=\mathbf{x}_t^{(l)} \mid C=i)\right)$, where L is the number of observed words in each document; and *average recovery error up to step n*, $\mathcal{L}_n^{(2)} = \frac{1}{n}\sum_{t=2}^n \|M_{3,*} - \hat{M}_{3,t-1}\|_F^2$. The latter metric is the average difference between the distribution in hindsight $M_{3,*}$ and the predicted distribution $\hat{M}_{3,t-1}$ at time t, and measures the parameter reconstruction error. Specifically, $M_{3,*} = \sum_{i=1}^K \omega_{*,i} u_{*,i} \otimes u_{*,i} \otimes u_{*,i}$ and $\hat{M}_{3,t-1} = \sum_{i=1}^K \omega_{t-1,i} u_{t-1,i} \otimes u_{t-1,i} \otimes u_{t-1,i}$,

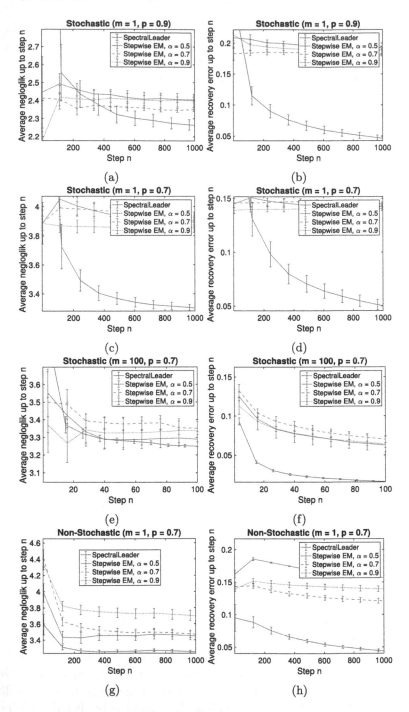

Fig. 1. Evaluation on stochastic synthetic problems and non-stochastic synthetic problems. We compare `SpectralLeader` to stepwise EM with varying step-size reduction power α and mini-batch size m. The first column shows results under metric $\mathcal{L}_n^{(1)}$ and the second column shows results under metric $\mathcal{L}_n^{(2)}$.

where $\theta^* = ((\omega_{*,i})_{i=1}^K, (u_{*,i})_{i=1}^K)$ are the parameters of the unknown model. In synthetic problems, we know θ^*. In real-world problems, we learn θ^* by the spectral method since we have all data in advance. The recovery error is related to the regret, through the relation of our loss function and the Frobenius norm in Lemma 5. Note that EM in the offline setting minimizes the negative log-likelihood, while the spectral method in the offline setting minimizes the recovery error of tensors. All reported results are averaged over 10 runs.

8.1 Synthetic Problems

Stochastic Synthetic Problems. In this stochastic setting, the topic of the document at all times t is sampled i.i.d. from a fixed distribution. This setting represents a scenario where the sequence of topics is not correlated. The number of distinct topics is $K = 3$, the vocabulary size is $d = 3$, and each document has 3 observed words. In practice, some topics are typically more popular than others. Therefore, we sample topics as follows. At each time, the topic C is randomly sampled from the distribution where $P(C = 1) = 0.15$, $P(C = 2) = 0.35$, and $P(C = 3) = 0.5$. Given the topic, the conditional probability of words is $P(\mathbf{x} = e_i | C = j) = p$ when $i = j$, and $P(\mathbf{x} = e_i | C = j) = \frac{1-p}{2}$ when $i \neq j$. With smaller p, the conditional distribution of words given different topic becomes similar, and the difficulty of distinguishing different topics increases. For $m = 1$, we evaluate on two problems where $p = 0.7$ and $p = 0.9$. For $m = 100$, we further focus on the more difficult problem where $p = 0.7$. We show the results before the different methods converge: for $m = 1$, we report results before $n = 1000$, and for $m = 100$ we report both results before $n = 100$.

Results for the stochastic setting are reported in Fig. 1. We observe three trends. First, under metric $\mathcal{L}_n^{(1)}$, stepwise EM is very sensitive to its parameters α and m, while SpectralLeader is competitive or even better, compared to stepwise EM with its best α and m. For example, the best α is 0.7 in Fig. 1a, and the best α is 0.9 in Fig. 1c. Even for the same problem with different m, the best α is different: the best α is 0.9 in Fig. 1c, while the best α is 0.5 in Fig. 1e. In all cases, SpectralLeader performs the best. Second, similar to [13], stepwise EM improves when the mini-batch size increases to $m = 100$. But SpectralLeader still performs better compared to stepwise EM with its best α. Third, SpectralLeader performs much better than stepwise EM under metric $\mathcal{L}_n^{(2)}$. These results indicate that a careful grid search of α and m is usually needed to optimize stepwise EM. Such grid search in the online setting is nearly impossible, since future data are unknown in advance. In contrast, SpectralLeader is very competitive without any parameter tuning.

Non-Stochastic Synthetic Problems. The non-stochastic setting is the same as the stochastic setting, except that topics of the documents are strongly correlated over time. We look at an extreme case of correlated topics in the steaming data. In each batch of 100 steps, sequentially we have 15 documents from topic 1, 35 documents from topic 2, and 50 documents from topic 3. We focus on the more difficult problem where $p = 0.7$.

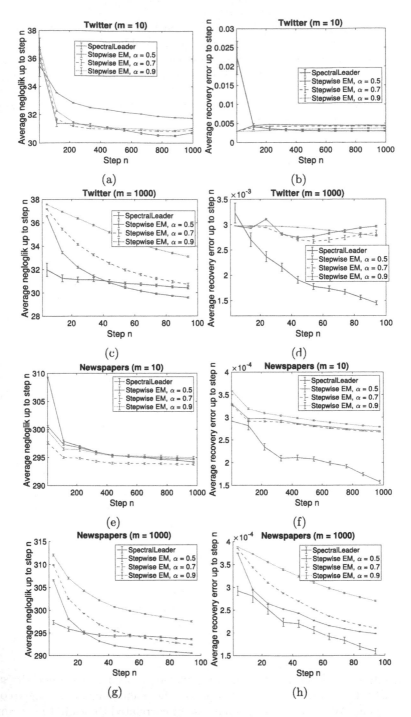

Fig. 2. Evaluation on real-world datasets. We compare `SpectralLeader` to stepwise EM with varying step-size reduction power α and mini-batch size m. The first column shows results under metric $\mathcal{L}_n^{(1)}$ and the second column shows results under metric $\mathcal{L}_n^{(2)}$.

Our results in this non-stochastic setting are reported in Figs. 1g and h. For stepwise EM, the α leading to lowest negative log-likelihood is 0.5. This result matches well the fact that the smaller the α, the faster the old sufficient statistics are forgotten, and the faster stepwise EM adapts to the non-stochastic setting. `SpectralLeader` is even better than stepwise EM with $\alpha = 0.5$. Note that $\alpha = 0.5$ is the smallest valid value of α for stepwise EM [13].

8.2 Real World Problems

In this section, we evaluate on Newspapers data[1] over multiple years and Twitter data[2] during the 2016 United States elections. They provide streaming data with timestamps and the distributions of topics change over time. After preprocessing, we retain the 500 most frequent words in the vocabulary. We set $K = 5$. We evaluate all algorithms on 100 K documents.[3] We compare `SpectralLeader` to stepwise EM with multiple α, and mini-batch sizes $m = 10$ and $m = 1000$. We show the results before the different methods converge: for $m = 10$, we report results before $n = 1000$, and for $m = 1000$ we report results before $n = 100$. To handle large-scale streaming data, such as 5 M words in Newspapers data, we introduce reservoir sampling, and set the window size of reservoir to 10,000.

Our results are reported in Fig. 2. We observe four major trends. First, under metric $\mathcal{L}_n^{(2)}$, `SpectralLeader` performs better than stepwise EM. Second, under metric $\mathcal{L}_n^{(1)}$, for $m = 10$ versus $m = 1000$, the optimal α for stepwise EM are different on both datasets. Third, when $m = 10$, under $\mathcal{L}_n^{(1)}$, `SpectralLeader` performs competitive with or better than stepwise EM with its best α. Fourth, when $m = 1000$, under $\mathcal{L}_n^{(1)}$, `SpectralLeader` is not as good as stepwise EM with its best α. However, directly using `SpectralLeader` without tuning any parameters can still provide good performance. These results suggest that, even when the mini-batch size is large, `SpectralLeader` is still very useful under the log-likelihood metric. In practice, we can quickly achieve reasonable results with `SpectralLeader` without any parameter tuning.

9   Conclusions

We develop `SpectralLeader`, a novel online learning algorithm for latent variable models. In an instance of a single topic model, we define a novel per-step loss function, prove that `SpectralLeader` converges to a global optimum, and derive a sublinear regret bound for `SpectralLeader`. Our experimental results suggest that `SpectralLeader` performs similarly to or better than a fine-tuned online EM. In future work, we want to extend our method to more complicated latent-variable models, such as HMMs and LDA [3].

---

[1] Please see https://www.kaggle.com/snapcrack/all-the-news.

[2] Please see https://www.kaggle.com/kinguistics/election-day-tweets.

[3] The per-step computational cost of `SpectralLeader` is larger than that of stepwise EM, when the number of topics, number of observed words and vocabulary size increase. We leave improving the efficiency of `SpectralLeader` as future work.

Acknowledgments. This work is supported, in part, by funding from Adobe and Intel to CMU.

References

1. Amoualian, H., Clausel, M., Gaussier, E., Amini, M.R.: Streaming-lda: a copula-based approach to modeling topic dependencies in document streams. In: KDD, pp. 695–704. ACM (2016)
2. Anandkumar, A., Foster, D.P., Hsu, D.J., Kakade, S.M., Liu, Y.K.: A spectral algorithm for latent Dirichlet allocation. In: NIPS, pp. 917–925 (2012)
3. Anandkumar, A., Ge, R., Hsu, D.J., Kakade, S.M., Telgarsky, M.: Tensor decompositions for learning latent variable models. JMLR **15**(1), 2773–2832 (2014)
4. Anandkumar, A., Hsu, D., Kakade, S.M.: A method of moments for mixture models and hidden Markov models. In: COLT, p. 33-1 (2012)
5. Bishop, C.M.: Pattern Recognition and Machine Learning. Springer, New York (2006)
6. Cappé, O., Moulines, E.: On-line expectation-maximization algorithm for latent data models. J. R. Stat. Soc. Ser. B (Stat. Methodol.) **71**(3), 593–613 (2009)
7. Chaganty, A.T., Liang, P.: Spectral experts for estimating mixtures of linear regressions. In: ICML, pp. 1040–1048 (2013)
8. Davis, C., Kahan, W.M.: The rotation of eigenvectors by a perturbation. III. SIAM J. Numer. Anal. **7**(1), 1–46 (1970)
9. Ge, R., Huang, F., Jin, C., Yuan, Y.: Escaping from saddle points - online stochastic gradient for tensor decomposition. In: COLT, pp. 797–842 (2015)
10. Hoeffding, W.: Probability inequalities for sums of bounded random variables. J. Am. Stat. Assoc. **58**(301), 13–30 (1963)
11. Hoffman, M., Bach, F.R., Blei, D.M.: Online learning for latent Dirichlet allocation. In: NIPS, pp. 856–864 (2010)
12. Huang, F., Niranjan, U., Hakeem, M.U., Anandkumar, A.: Online tensor methods for learning latent variable models. JMLR **16**, 2797–2835 (2015)
13. Liang, P., Klein, D.: Online EM for unsupervised models. In: NAACL HLT, pp. 611–619. Association for Computational Linguistics (2009)
14. Neal, R.M., Hinton, G.E.: A view of the EM algorithm that justifies incremental, sparse, and other variants. In: Learning in Graphical Models, pp. 355–368. Springer, Dordrecht (1998). https://doi.org/10.1007/978-94-011-5014-9_12
15. Nowozin, S., Lampert, C.H.: Structured learning and prediction in computer vision. Found. Trends® Comput. Graph. Vis. **6**(3–4), 185–365 (2011)
16. Rabiner, L.R.: A tutorial on hidden Markov models and selected applications in speech recognition. Proc. IEEE **77**(2), 257–286 (1989)
17. Shaban, A., Farajtabar, M., Xie, B., Song, L., Boots, B.: Learning latent variable models by improving spectral solutions with exterior point method. In: UAI, pp. 792–801. AUAI Press, Arlington (2015)
18. Tung, H.Y., Smola, A.J.: Spectral methods for Indian buffet process inference. In: NIPS, pp. 1484–1492 (2014)
19. Tung, H.Y.F., Wu, C.Y., Zaheer, M., Smola, A.J.: Spectral methods for nonparametric models (2017). arXiv preprint arXiv:1704.00003
20. Wallach, H.M.: Topic modeling: beyond bag-of-words. In: ICML, pp. 977–984. ACM (2006)

21. Weyl, H.: Das asymptotische verteilungsgesetz der eigenwerte linearer partieller differentialgleichungen (mit einer anwendung auf die theorie der hohlraumstrahlung). Mathematische Annalen **71**(4), 441–479 (1912)
22. Zou, J.Y., Hsu, D.J., Parkes, D.C., Adams, R.P.: Contrastive learning using spectral methods. In: NIPS, pp. 2238–2246 (2013)

Online Learning of Weighted Relational Rules for Complex Event Recognition

Nikos Katzouris[1(✉)], Evangelos Michelioudakis[1,2], Alexander Artikis[1,3], and Georgios Paliouras[1]

[1] National Center for Scientific Research (NCSR) "Demokritos", Athens, Greece
[2] National and Kapodistrian University of Athens, Athens, Greece
[3] University of Pireaus, Pireaus, Greece
{nkatz,a.artikis,paliourg}@iit.demokritos.gr

Abstract. Systems for symbolic complex event recognition detect occurrences of events in time using a set of event definitions in the form of logical rules. The Event Calculus is a temporal logic that has been used as a basis in event recognition applications, providing among others, connections to techniques for learning such rules from data. We advance the state-of-the-art by combining an existing online algorithm for learning crisp relational structure with an online method for weight learning in Markov Logic Networks (MLN). The result is an algorithm that learns complex event patterns in the form of Event Calculus theories in the MLN semantics. We evaluate our approach on a challenging real-world application for activity recognition and show that it outperforms both its crisp predecessor and competing online MLN learners in terms of predictive performance, at the price of a small increase in training time. Code related to this paper is available at: https://github.com/nkatzz/OLED.

Keywords: Online structure and weight learning
Markov logic networks · Event calculus

1 Introduction

Complex event recognition systems [7] process sequences of *simple events*, such as sensor data, and recognize *complex events* of interest, i.e. events that satisfy some temporal pattern. Systems for symbolic event recognition [3] typically use a knowledge base of first-order rules to represent complex event patterns. Learning such patterns from data is highly desirable, since their manual development is usually a difficult and error-prone task. The Event Calculus (EC) [23] is a temporal logical formalism that has been used as a basis in event recognition applications [2], providing among others, direct connections to machine learning, via Inductive Logic Programming (ILP) [8] and Statistical Relational Learning (SRL) [9].

Event recognition applications typically deal with noisy data streams [1]. Algorithms that learn from such streams are required to work in an online fashion, building a model with a single pass over the input [14]. Although a number

© Springer Nature Switzerland AG 2019
M. Berlingerio et al. (Eds.): ECML PKDD 2018, LNAI 11052, pp. 396–413, 2019.
https://doi.org/10.1007/978-3-030-10928-8_24

of online relational learners have been proposed [12,18,31], learning theories in the EC is a challenging task that most relational learners, including the aforementioned ones, cannot fully undertake [19,28]. As a result, two online algorithms have been proposed, both capable of learning complex event patterns in the form of EC theories, from relational data streams. The first, OLED (Online Learning of Event Definitions) [21], adapts the Hoeffding bound-based [16] framework of [10] for online decision tree learning to an ILP setting. The second algorithm, OSLα (Online Structure Learning with Background Knowledge Axiomatization) [27], builds on the method of [18] for learning structure and weights in Markov Logic Networks (MLN) [29], towards online learning of EC theories in the MLN semantics.

Both these algorithms have shortcomings. OLED is a crisp learner, therefore its performance could be improved via SRL techniques that combine logic with probability. On the other hand, OSLα uses an efficient online weight learning technique, based on the AdaGrad algorithm [13], but its structure learning component is sub-optimal: It tends to generate large sets of rules, many of which are of low heuristic value with a marginal contribution to the quality of the learned model. The maintenance cost of such large rule sets during learning results in poor efficiency, with no clear gain in the predictive accuracy, while it also negatively affects model interpretability.

In this work we present a new algorithm that attempts to combine the best of these two learners: OLED's structure learning strategy, which is more conservative than OSLα's and typically explores much smaller rule sets, with OSLα's weight learning technique. We show that the resulting algorithm outperforms both its predecessors in terms of predictive performance, at the price of a tolerable increase in training time. We empirically validate our approach on a benchmark dataset of activity recognition.

2   Related Work

Machine learning techniques for event recognition are attracting attention in the Complex Event Processing community [26]. However, existing approaches are relatively ad-hoc and they have several limitations [19], including limited support for background knowledge utilization and uncertainty handling. In contrast, we adopt an event recognition framework that allows access to well-established (statistical) relational learning techniques [8,9], which can overcome such limitations. Moreover, using the EC allows for efficient event recognition via dedicated reasoners, such as RTEC [2].

However, learning with the EC is challenging for most ILP and SRL algorithms. A main reason for that is the non-monotonicity of the Negation as Failure operator that the EC uses for commonsense reasoning, which makes the divide-and-conquer-based search of most ILP algorithms inappropriate [19,20,28]. Non-monotonic ILP algorithms can handle the task [5,28], but they scale poorly, as they learn whole theories from the entirety of the training data, while improvements to such algorithms that allow some form of incremental processing to enhance efficiency [4,24] cannot handle data arriving over time. Contrary to such approaches, OLED, the ILP algorithm we build upon, scales adequately and learns online [21].

A line of related work tries to "upgrade" non-monotonic ILP learners to an SRL setting, via weight learning techniques with probabilistic semantics [6,11]. However, the resulting algorithms suffer from the same limitations, related to scalability, as their crisp predecessors. In the field of Markov Logic Networks (MLN), which is the SRL framework we adopt, OSLα is the sole algorithm capable of learning structure and weights for EC theories.

Online learning settings, as the one we assume in this work, are under-explored both in ILP and in SRL. A few ILP approaches have been proposed. In [12] the authors propose an online algorithm, which generates a rule from each misclassified example in a stream, aiming to construct a theory that accounts for all the examples in the stream. In [31] the authors propose an online learner based on Aleph[1] and Winnow [25]. The algorithm maintains a set of rules, corresponding to Winnow's features. Rules are weighted and are used for the classification of incoming examples, via the weighted majority of individual rule verdicts for each example, while their weights are updated via Winnow's mistake-driven weight update scheme. New rules are greedily generated by Aleph from misclassified examples. A similar approach is put forth by OSL [18], an online learner for MLN, which OSLα builds upon to allow for learning with the EC. OSL greedily generates new rules to account for misclassified examples that stream in, and then relies on weight learning to identify which of these rules are relevant. Common to these online learners is that they tend to generate unnecessarily large rule sets, which are hard to maintain. This is precisely the issue that we address in this work.

3   Background

We assume a first-order language, where predicates, terms, atoms, literals (possibly negated atoms), rules (clauses) and theories (collections of rules) are defined as in [9], while not denotes Negation as Failure. A rule is represented by $\alpha \leftarrow \delta_1, \ldots, \delta_n$, where α is an atom, (the head of the rule), and $\delta_1, \ldots, \delta_n$ is a conjunction of literals (the body of the rule). A term is ground if it contains no variables. We follow [9] and adopt a Prolog-style syntax. Therefore, predicates and ground terms in logical expressions start with a lower-case letter, while variable terms start with a capital letter.

The Event Calculus (EC) [23] is a temporal logic for reasoning about events and their effects. Its ontology consists of *time points* (integer numbers); *fluents*, i.e. properties that have different values in time; and events, i.e. occurrences in time that may alter fluents' values. The axioms of the EC incorporate the commonsense *law of inertia*, according to which fluents persist over time, unless they are affected by an event. We use a simplified version of the EC that has been

---

[1]   http://www.cs.ox.ac.uk/activities/machinelearning/Aleph/aleph.

shown to suffice for event recognition [2]. The basic predicates and its domain-independent axioms are presented in Table 1(a) and (b) respectively. Axiom (1) in Table 1(b) states that a fluent F holds at time T if it has been initiated at the previous time point, while Axiom (2) states that F continues to hold unless it is terminated. Definitions for initiatedAt/2 and terminatedAt/2 predicates are given in an application-specific manner by a set of *domain-specific* axioms.

As a running example we use the task of activity recognition, as defined in the CAVIAR project[2]. The CAVIAR dataset consists of 28 videos of actors performing a set of activities. Manual annotation (performed by the CAVIAR team) provides ground truth for two activity types. The first type corresponds to simple events and consists of the activities of a person at a certain video frame/time point, such as *walking*, or *standing still*. The second activity type corresponds to complex events and consists of activities that involve more than

Table 1. (a), (b) The basic predicates and the domain-independent axioms of EC. (c) Example data from activity recognition. For example, at time point 1 person with id_1 is *walking*, her (X, Y) coordinates are $(201, 454)$ and her direction is $270°$. The annotation for the same time point states that persons with id_1 and id_2 are not moving together, in contrast to the annotation for time point 2. (d) An example of two domain-specific axioms in the EC. E.g. the first clause dictates that *moving together* between two persons X and Y is initiated at time T if both X and Y are walking at time T, their euclidean distance is less than 25 pixel positions and their difference in direction is less than $45°$. The second clause dictates that *moving together* between X and Y is terminated at time T if one of them is standing still at time T (exhibits an inactive behavior) and their euclidean distance at T is greater that 30.

(a)

Predicate	Meaning
happensAt(E, T)	Event E occurs at time T.
initiatedAt(F, T)	At time T, a period of time for which fluent F holds is initiated.
terminatedAt(F, T)	At time T, a period of time for which fluent F holds is terminated.
holdsAt(F, T)	Fluent F holds at time T.

(b)

Domain-Independent Axioms

holdsAt$(F, T + 1) \leftarrow$ (1)
    initiatedAt(F, T)

holdsAt$(F, T + 1) \leftarrow$ (2)
    holdsAt(F, T),
    not terminatedAt(F, T)

(c)

Narrative for time 1:	Narrative for time 2:
happensAt$(walk(id_1), 1)$.	happensAt$(walk(id_1), 2)$.
happensAt$(walk(id_2), 1)$.	happensAt$(walk(id_2), 2)$.
holdsAt$(coords(id_1, 201, 454), 1)$.	holdsAt$(coords(id_1, 201, 454), 2)$.
holdsAt$(coords(id_2, 230, 440), 1)$	holdsAt$(coords(id_2, 227, 440), 2)$
holdsAt$(direction(id_1, 270), 1)$	holdsAt$(direction(id_1, 275), 2)$
holdsAt$(direction(id_2, 270), 1)$	holdsAt$(direction(id_2, 278), 2)$

Annotation for time 1:
not holdsAt$(move(id_1, id_2), 1)$

Annotation for time 2:
holdsAt$(move(id_1, id_2), 2)$

(d)

Two Domain-specific axioms:

initiatedAt$(move(X, Y), T) \leftarrow$
    happensAt$(walk(X), T)$,
    happensAt$(walk(Y), T)$,
    $distLessThan(X, Y, 25, T)$,
    $dirLessThan(X, Y, 45, T)$

terminatedAt$(move(X, Y), T) \leftarrow$
    happensAt$(inactive(X), T)$,
    $distMoreThan(X, Y, 30, T)$

---

[2] http://homepages.inf.ed.ac.uk/rbf/CAVIARDATA1/.

one person, e.g. two people *meeting each other*, or *moving together*. The goal is to recognize complex events as combinations of simple events and additional contextual knowledge, such as a person's direction and position.

Table 1(c) presents some example CAVIAR data, consisting of a narrative of simple events in terms of happensAt/2, expressing people's short-term activities, and context properties in terms of holdsAt/2, denoting people' coordinates and direction. Table 1(c) also shows the annotation of complex events (long-term activities) for each time-point in the narrative. Negated complex events' annotation is obtained via the closed-world assumption (although both positive and negated annotation atoms are presented in Table 1(c), to avoid confusion). Table 1(d) presents two domain-specific axioms in the EC.

The learning task we address in this work is to learn definitions of complex events, in the form of domain-specific axioms in the EC, i.e. initiation and termination conditions of complex events, as in Table 1(d). The training data consists of a set of *Herbrand interpretations*, i.e. sets of narrative atoms, annotated by complex event instances, as in Table 1(c). Given such a training set $\mathcal{I}$ and some background knowledge B, the goal is to find a theory H that accounts for as many positive and as few negative examples as possible, throughout $\mathcal{I}$. Given an interpretation $I \in \mathcal{I}$, a positive (resp. negative) example is a complex event atom $\alpha \in I$ (resp. $\alpha \notin I$). We assume an online learning setting, in the sense that a learning algorithm is allowed only a single-pass over $\mathcal{I}$.

3.1 Online Learning of Markov Logic Networks with OSLα

OSLα [27] builds on the OSL [18] algorithm for online learning of Markov Logic Networks (MLN). An MLN is a set of weighted first-order logic rules. Along with a set of domain constants, it defines a ground Markov network containing one feature for each grounding of a rule in the MLN, with the corresponding weight. Learning an MLN consists of learning its structure (the rules in the MLN) and their weights.

OSLα works by constantly updating the rules in an MLN in the face of new interpretations that stream-in, by adding new rules and updating the weights of existing ones. At time t, OSL receives an interpretation I_t and uses its current hypothesis, H_t, to "make a prediction", i.e. to infer the truth values of *query atoms*, given the *evidence atoms* in I_t. In the learning problem that we address in this work, query atoms are instances of initiatedAt/2 and terminatedAt/2 predicates, the "target" predicates defining complex events, while evidence atoms are instances of predicates that declare simple events' occurrences, or other contextual knowledge, like happensAt/2 and *distLessThan*/4 respectively (see also Table 1(c)).

To infer the query atoms' truth values given the interpretation I_t, OSLα uses Maximum Aposteriori (MAP) inference [18], which amounts to finding the truth assignment to the query atoms that maximizes the sum of the weights of H_t's rules satisfied by I_t. This is a weighted MAX-SAT problem, whose solution OSLα efficiently approximates using LP-relaxed Integer Linear Programming, as in [17]. The inferred truth values of the query atoms, resulting from MAP inference may differ from the true ones, dictated by the annotation in I_t. The mistakes may be either False Negatives (*FNs*) or False Positives (*FPs*). Since *FNs* are query atoms which are not entailed by an existing rule in the inferred interpretation, in the next step OSLα constructs a set of new rules as a remedy. To this end, OSLα represents the current interpretation I_t as a hypergraph with the constants of I_t as the nodes and each true ground atom δ in I_t as a hyperedge connecting the nodes (constants) that appear as δ's arguments. Each inferred *FN* query atom α is used as a "seed" to generate new rules, by finding all paths in the hypergraph, up to a user-defined maximum length, that connect α's constants. Each such path consists of hyperedges, corresponding to a conjunction of true ground atoms connected by their arguments. The path is turned into a rule with this conjunction in the body and the seed *FN* atom in the head, and a lifted version of this rule is obtained by variabilizing the ground arguments. By construction, each such rule logically entails the seed *FN* atom it was generated from.

This technique of *relational pathfinding* may generate a large number of rules from each inferred *FN* query atom, and many of these rules are not useful. Thus, OSLα relies on L_1-regularized weight learning, which in the long run, pushes the weights of non-useful rules to zero. Weight learning is also OSLα's way to handle *FP* query atoms in the inferred state, which are due to erroneously satisfied rules in the current theory H_t. These rules are penalized by reducing their weights. OSLα uses an AdaGrad-based [13] weight learning technique, which supports L_1-regularization.

AdaGrad is a subgradient-based method for online convex optimization, i.e. at each step it updates a feature vector, based on the subgradient of a convex loss function of the features. Seeing the rules in an MLN theory $H = \{r_1, \ldots, r_n\}$ as the feature vector that needs to be optimized, the authors in [18] use a simple variant of the hinge-loss as a loss function, whose subgradient is the vector $-\langle \Delta g_1, \ldots \Delta g_n \rangle$, where Δg_i denotes the difference between the true groundings of the i-th rule in the actual true state and the MAP-inferred state respectively. Based on this difference, AdaGrad updates the weight w_i^t of the i-th rule in the theory at time t by:

$$w_i^{t+1} = sign(w_i^t - \frac{\eta}{C_i^t} \Delta g_i^t) \ max\{0, |w_i^t - \frac{\eta}{C_i^t} \Delta g_i^t| - \lambda \frac{\eta}{C_i^t}\} \tag{1}$$

where t-superscripts in terms denote the respective values at time t, η is a learning rate parameter, λ is a regularization parameter and $C_i^t = \delta + \sqrt{\sum_{j=1}^t (\Delta g_i^j)^2}$ is a term that expresses the rule's quality so far, as reflected by the accumulated sum of Δg_i's, amounting to the i-th rule's past mistakes (plus a $\delta \geq 0$ to avoid division by zero in η/C_i^t). The C_i^t term gives an adaptive flavour to the algorithm, since the magnitude of a weight update via the term $|w_i^t - \frac{\eta}{C_i^t} \Delta g_i^t|$ in Eq. (1), is affected by the rule's previous history, in addition to its current mistakes, expressed by Δg_i^t. The regularization term in Eq. (1), $\lambda \frac{\eta}{C_i^t}$, is the amount by which the i-th rule's weight is discounted when $\Delta g_i^t = 0$. This is to eventually push to zero the weights of irrelevant rules, which have very few, or even no groundings in the training interpretations.

In contrast to its predecessor, OSL, OSLα uses specialized techniques for pruning large parts of the hypergraph structure to enhance efficiency when learning with the EC. However, its rule generation technique remains an important bottleneck. Repeatedly searching for paths in the hypergraph structure is expensive in its own right, but it also tends to blindly generate large sets of rules, which, in turn, increases the cost of MAP inference during learning.

4   Learning Weighted Rules with WoLED

Aiming to improve the efficiency of online learning with the EC in MLN, we propose WoLED (Weighted Online Learning of Event Definitions), an extension of the OLED crisp online ILP learner [21], to an MLN setting. OLED draws inspiration from the VFDT (Very Fast Decision Trees) algorithm [10], whose online strategy is based on the Hoeffding bound [16]. Given a random variable X with range in $[0, 1]$ and an observed mean $\overline{X}$ of its values after n independent observations, the Hoeffding Bound states that, with probability $1 - \delta$, the true mean $\hat{X}$ of the variable lies in an interval $(\overline{X} - \epsilon, \overline{X} + \epsilon)$, where $\epsilon = \sqrt{\frac{ln(1/\delta)}{2n}}$.

OLED learns a rule r with a hill-climbing process, where literals are gradually added to r's body yielding *specializations* of r of progressively higher quality, which is assessed by some scoring function G, based on the positive and negative examples that a rule entails. This strategy is common in ILP, where at each specialization step (addition of a literal), a number of candidate specializations of the parent rule are evaluated on the entire training set and the best one is selected. OLED adapts this strategy to an online setting, using the Hoeffding bound, as follows: Assume that after having evaluated r and a number of its candidate specializations on n examples, r_1 is r's specialization with the highest mean G-score $\overline{G}$ and r_2 is the second-best one, i.e. $\Delta \overline{G} = \overline{G}(r_1) - \overline{G}(r_2) > 0$. Then by the Hoeffding bound we have that for the true mean of the scores' difference $\Delta \hat{G}$ it holds that $\Delta \hat{G} > \Delta \overline{G} - \epsilon$, with probability $1 - \delta$, where $\epsilon = \sqrt{\frac{ln(1/\delta)}{2n}}$. Hence, if $\Delta \overline{G} > \epsilon$, then $\Delta \hat{G} > 0$, implying that r_1 is indeed the best specialization, with probability $1 - \delta$. In order to decide which specialization to select, it thus suffices to evaluate the specializations on examples from the stream until $\Delta \overline{G} > \epsilon$. Each of these examples is processed once, thus giving rise to a single-pass rule learning strategy.

Positive (resp. negative) examples in this setting are true (resp. false) instances of target predicates, which are present in (resp. generated via the closed-world assumption from) the incoming training interpretations. Such instances correspond to query atoms in an MLN setting (see Sect. 3.1), therefore we henceforth refer to positive and negative examples as true and false query atoms respectively.

The literals used for specialization are drawn from a *bottom rule* [8], denoted by $\perp$, a most-constrained rule that entails a single true query atom. A bottom rule is usually too restrictive to be used for classification and its purpose is to define a space of potentially good rules, consisting of those that θ-subsume $\perp$. OLED moves through this search space in a top-down fashion, starting from an empty-bodied rule $head(\perp) \leftarrow true$ and considering at each time specializations that result by the addition of a fixed number of literals from $\perp$ to the body of a parent rule (the default is one literal).

4.1   WoLED

Our approach is based on combining OLED's rule learning technique with OSLα's weight learning strategy. Contrary to OSLα's approach, where once a rule is generated, its structure remains fixed and the only way to improve its quality is by tuning its weight, WoLED progressively learns both the structure of a rule and its weight, by jointly optimizing them together. To this end, a rule is gradually specialized via the online hill-climbing strategy described earlier, while constantly updating the weight of each such specialization. Given a rule r and a set of r's specializations S_r, WoLED learns weights for r and each $r' \in S_r$, and uses Hoeffding tests to identify over time, a rule $r_1 \in S_r$ of higher quality, as assessed by a combination of r_1's structure and weight. Once that happens, r is replaced by r_1 and the process continues for as long as new specializations of r_1 improve its quality. To learn weights, WoLED replaces OLED's crisp logical inference with MAP inference and uses OSLα's mistake-driven weight learning technique described in Sect. 3.1, which updates a rule's weight based on the query atoms that the rule misclassifies in the MAP-inferred state.

WoLED's high-level strategy is illustrated in Fig. 1. At each point in time WoLED maintains a theory $H_t = \{r_1, \ldots, r_n\}$, and for each rule r_i, a set of current specializations $\{r_i^1, \ldots, r_i^n\}$ and an associated bottom rule $\perp_{r_i}$. At time t, WoLED receives the t-th training interpretation I_t and it subsequently goes through a five-step process. In the first step it performs MAP inference on I_t, using the current theory H_t and the background knowledge. The MAP-inferred interpretation is checked for *FN* query atoms. If such an atom α exists, meaning that no existing rule in H_t entails it, WoLED proceeds to a "theory expansion" step, where it starts growing a new rule r_{i+1} to account for that. This amounts to using the *FN* atom α as a "seed" to generate a bottom rule $\perp_{r_{i+1}}$ and adding to H the empty-bodied rule $head(\perp_{r_{i+1}}) \leftarrow true$. From that point on this rule is gradually specialized using literals from $\perp_{r_{i+1}}$. The generation of $\perp_{r_{i+1}}$ is guided by a set of *mode declarations* [8] (see Fig. 1), a form of language bias specifying the signatures of literals that are allowed to be placed in the head or the body of a rule, in addition to the types of their arguments (e.g. time, or id in the declarations of Fig. 1) and whether they correspond to variables or constants (indicated respectively by "+" and "#" in the declarations of Fig. 1).

After a potential theory expansion a weight update step follows, where the weights of each rule $r \in H_t$ and the weights of each one of its current specializations are updated, based on their mistakes on the MAP-inferred state generated

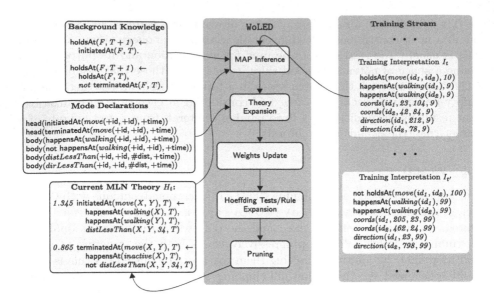

Fig. 1. Illustration of WoLED's high-level strategy.

previously. A Hoeffding test for each rule $r \in H_t$ follows, and if a test succeeds for some rule r, it is "expanded", i.e. r's structure and weight are replaced by the ones of its best-scoring specialization, r_1. The final step is responsible for pruning the current theory, i.e. removing low-quality rules. The current version of the weighted theory is output and the training loop continues (see Fig. 1).

We next go into some details of WoLED's functionality, using the pseudocode in Algorithm 1. The input to Algorithm 1 consists of the background knowledge, a rule-scoring function G, based on the true positive (TP), false positive (FP) and false negative (FN) examples entailed by a rule, AdaGrad's parameters η, δ_a, λ, discussed in Sect. 3.1, the confidence parameter for the Hoeffding test δ_h and a set of mode declarations. Additionally, a minimum G-score value for acceptable rules, a "warm-up" parameter N_{min} and a "specialization depth" parameter d, to be explained shortly.

The MAP-inferred state is generated in lines 2–3 of Algorithm 1, using the *active fragment* of the theory. The latter consists of those rules in the theory that have been evaluated on at least N_{min} examples (line 2), where N_{min} is the input "warm-up" parameter. This is to avoid using rules which are too premature and useless for inference, such as an empty-bodied rule that has just been created.

In lines 4–8 of Algorithm 1, new rules are generated for each FN atom in the MAP-inferred state, as described earlier. In addition to its corresponding bottom rule to draw literals for specialization, each rule is also equipped with a number of counters, which accumulate the TP, FP and FN instances entailed by the rule over time, to use for calculating its G-score; and an accumulator, denoted by $Subgradients(r)$, which is meant to store the history (sum) of the

Algorithm 1. WoLED
Input: $\mathcal{I}$: A stream of training interpretations; M: Mode declarations; B: Background knowledge; G: Rule evaluation function; δ_h: Confidence for the Hoeffding test; η, λ, δ_a: AdaGrad's parameters; d: Specialization depth; N_{min}: Warm-up period; $Score_{min}$: G-score quality threshold; H: A (possibly empty) set of first order rules.

```
 1: for each I ∈ I do
 2:     H_active = {r ∈ H  : r has been evaluated on at least N_min examples}
 3:     I_MAP := the MAP-inferred interpetation of H_active ∪ B on I.
 4:     for each true query atom α ∈ I ∖ I_MAP do
 5:         Generate a bottom rule ⊥ from I ∪ B, using the mode declarations and α as a seed atom.
 6:         r := head(⊥) ← true
 7:         TPs(r), FPs(r), FNs(r), Subgradients(r) := 0; ⊥_r := ⊥
 8:         H ← H ∪ {r}
 9:     for each r ∈ H do
10:         ρ_d(r) := {head(r) ← body(r) ∧ D | D ⊂ body(⊥_r) and |D| ≤ d}
11:         UpdateWeights(r, I_MAP)
12:         for each r′ ∈ ρ_d(r) do
13:             UpdateWeights(r′, I_MAP)
14:         ΔḠ := Ḡ(r₁) − Ḡ(r₂), where r₁, r₂ ∈ ρ_d(r) are r's best and second-best specializations.
```
15: $\quad\quad \epsilon := \sqrt{\frac{ln(1/\delta_h)}{2N_r}}$, where N_r is the sum of r's groundings so far.

16: $\quad\quad \bar{\epsilon} :=$ the mean value of $\epsilon = \sqrt{\frac{ln(1/\delta_h)}{2N_{r_k}}}$ observed so far any rule r_k.
```
17:         if [ΔḠ > ε or ΔḠ < ε < ε̄] and Ḡ(r₁) > Ḡ(r) then
18:             Replace r by r₁.
```
19: $\quad\quad \bar{N}_s :=$ the average number of $\mathcal{O}(\frac{1}{\epsilon^2}ln\frac{1}{\delta_h})$ examples for which the Hoeffding test has succeeded so far during the learning process.
```
20:         if r is unchanged for a period longer than N̄_s and Score_min − Ḡ(r) > ε then
21:             Remove r from H.
22: return H
```

rule's mistakes throughout the learning process (see the term C_i^t in Eq. (1)). As explained in Sect. 3.1, a rule's mistakes w.r.t. each incoming interpretation is a coordinate in the the subgradient vector of AdaGrad's loss function (hence the name "$Subgradients(r)$" for their accumulator) and they affect the magnitude of a rule's weight update. Therefore, $Subgradients(r)$ is used for updating r's weight with AdaGrad.

Updating the weights of each rule $r \in H$, as well as the weights of r's candidate specializations follows, in lines 9–13 of Algorithm 1 (r's specializations are denoted by $\rho_d(r)$ (line 10), where d is the specialization depth parameter mentioned earlier, controlling the "depth" of allowed specializations, which are considered at each time). The weight-update process is presented in Algorithm 2. It uses AdaGrad's strategy discussed in Sect. 3.1. The difference between r's true groundings in the true state and the MAP-inferred one is first calculated (line 2), it's square is added to r's $Subgradients$ accumulator (line 3 – see the C_i^t term in Eq. (1), Sect. 3.1) and then r's weight is updated (line 5). Algorithm 2 is also responsible for updating the TP, FP, FN counters for a rule r, which are used to calculate its G-score.

The Hoeffding test follows to decide if a rule should be specialized (lines 14–18, Algorithm 1). A rule is specialized either if the test succeeds ($\Delta\bar{G} > \epsilon$), or if a tie-breaking condition is met ($\Delta\bar{G} < \epsilon < \tau$), where τ is a threshold set to

---

Algorithm 2. UpdateWeights($r, \eta, \delta_a, \lambda, I_{MAP}, I_{TRUE}$):

Input: r: a rule; η, λ, δ_a: AdaGrad's learning rate, regularization parameter and smoothness parameter respectively; I_{MAP}, I_{TRUE}: the MAP-inferred state and the true state respectively for a training interpretation I.

1:    $w_r :=$ the weight of rule r.
2:    $\Delta g_r :=$ the difference in true groundings of rule r in the inferred state I_{MAP} and the true state I_{TRUE}.
3:    $Subgradients(r) \leftarrow Subgradients(r) + (\Delta g_r)^2$.
4:    $C_r := \delta_\alpha + \sqrt{Subgradients(r)}$
5:    $w_r \leftarrow sign(w_r - \frac{\eta}{C_r}\Delta g_r) \, max\{0, |w_r - \frac{\eta}{C_r}\Delta g_r| - \lambda\frac{\eta}{C_r}\}$
6:    $TPs(r) \leftarrow TPs(r) + |\{\alpha \in I_{MAP} \cap I_{TRUE} : \alpha$ is a grounding of $head(r)\}|$.
7:    $FPs(r) \leftarrow FPs(r) + |\{\alpha \in I_{MAP} \setminus I_{TRUE} : \alpha$ is a grounding of $head(r)\}|$.
8:    $FNs(r) \leftarrow FNs(r) + |\{\alpha \in I_{TRUE} \setminus I_{MAP} : \alpha$ is a grounding of $head(r)\}|$.

---

the mean value of ϵ observed so far. Also, to ensure that no rule r is replaced by a specialization of lower quality, we demand that $\bar{G}(r_1) > \bar{G}(r)$, where r_1 is r's best-scoring specialization indicated by the Hoeffding test.

The final step is responsible for removing rules of low quality (lines 19–21, Algorithm 1). A rule is removed if it remains unchanged (is not specialized) for a significantly large period of time, set to the average number of $\mathcal{O}(\frac{1}{\epsilon^2}ln\frac{1}{\delta_h})$ examples for which the Hoeffding test has succeeded so far, and there is enough confidence, via an additional Hoeffding test, that its mean G-score is lower than a minimum acceptable G-score $Score_{min}$.

5   Experimental Evaluation

We present an experimental evaluation of WoLED on CAVIAR, a benchmark dataset for activity recognition (see Sect. 3 for CAVIAR's description). All experiments were conducted on Debian Linux machine with a 3.6 GHz processor and 16 GB of RAM. The code and the data to reproduce the experiments are available online[3]. WoLED is implemented in Scala, using the Clingo answer set solver[4] for grounding and the lpsolve[5] Linear Programming solver for probabilistic MAP-inference, on top of the LoMRF[6] platform, an implementation of MLN. The OSLα version to which we compare in these experiments also relies on lpsolve, but uses LoMRF's custom grounder.

5.1   Comparison with Related Online and Batch Learners

In our first experiment we compare WoLED with (i) OSLα and OSL [18], discussed in Sect. 3.1; (ii) The crisp version of OLED; (iii) EC$_{crisp}$, a hand-crafted set of crisp rules for CAVIAR; (iv) MaxMargin, an MLN consisting of EC$_{crisp}$'s rules, with weights

---

[3] https://github.com/nkatzz/OLED.
[4] https://potassco.org/clingo/.
[5] https://sourceforge.net/projects/lpsolve/.
[6] https://github.com/anskarl/LoMRF.

optimized by the the Max-Margin weight learning method of [17]; (v) XHAIL, a batch, crisp ILP learner using a combination of inductive and adbuctive logic programming. The rules used by EC_{crisp} and MaxMargin may be found in [30].

MaxMargin was selected because it was shown to achieve good results on CAVIAR [30], while XHAIL was selected because it is one of the few existing ILP algorithms capable of learning theories in the EC.

A fragment of the CAVIAR dataset has been used in previous work to evaluate $OSL\alpha$ and MaxMargin's performance [27, 30]. To compare to these approaches we therefore used this fragment in our first experiment. The target complex events in this dataset are related to two persons *meeting each other* or *moving together* and the training data consist of the parts of CAVIAR where these complex events occur. The fragment dataset contains a total of 25,738 training interpretations. The results with OLED were achieved using a significance parameter $\delta_h = 10^{-2}$ for the Hoeffding test, a rule pruning threshold $Score_{min}$ (see also Algorithm 1) of 0.8 for *meeting* and 0.7 for *moving* and a warm-up parameter of $N_{min} = 1000$ examples. WoLED also used this parameter configuration, in addition to $\eta = 1.0, \lambda = 0.01, \delta_\alpha = 1.0$ for weight learning with AdaGrad. These parameters were reported in [27] and were also used with $OSL\alpha/OSL$.

The results were obtained using 10-fold cross validation and are presented in Table 2(a) in the form of *precision, recall* and f_1-*score*. These statistics were micro-averaged over the instances of recognized complex events from each fold. Table 2(a) also presents average training times per fold for all approaches except EC_{crisp}, where there is no training involved, average theory sizes for OLED, $OSL\alpha$, and XHAIL, as well as the fixed theory size of EC_{crisp} and MaxMargin. The reported theory sizes are in the form of total number of literals in a theory. The online methods were allowed only a single-pass over the training data.

WoLED achieves the best F_1-score for *meeting* and the second-best F_1-score for *moving*, right after the batch weight optimizer MaxMargin. This is a notable result. Moreover, this gain in predictive accuracy comes with a tolerable decrease in efficiency of approximately half a minute, as compared to OLED's training times, which are the best among all learners. This extra overhead in training time for WoLED is due to the cost of the probabilistic MAP-inference, which replaces OLED's crisp logical inference to allow for weight optimization. Regarding theory sizes, WoLED outputs hypotheses comparable in size with the hand-crafted knowledge base, and much more compressed as opposed to $OSL\alpha$. This is another notable result. XHAIL learns the most compressed hypotheses, since it is a batch learner, which also explains its increased training times. MaxMargin, also has high training times, paying the price of batch (weight) optimization.

OSL was unable to process the dataset within 25 h, at which time training was terminated. The reasons for that are related to it being unable to take advantage of the background knowledge, thus it is practically unable to learn with the Event Calculus [27]. $OSL\alpha$ overcomes OSL's difficulties, but learns unnecessarily large theories, which differ in size by several orders of magnitude from all others learners'. In turn, this affects $OSL\alpha$'s training times, which are also increased. In contrast, WoLED achieves improved predictive accuracy and compressed theories with minimal training overhead.

Table 2. Experimental results on (a) the CAVIAR fragment of [30] (top) and (b) the complete CAVIAR dataset (bottom).

		Method	Precision	Recall	F_1-score	Theory size	Time (sec)
(a)	*Moving*	EC_{crisp}	**0.909**	0.634	0.751	28	–
		OLED	0.867	0.724	0.789	34	**28**
		WoLED	0.882	0.835	0.857	30	59
		OSLα	0.837	0.590	0.692	3316	1300
		OSL	-	-	-	-	>25 h
		MaxMargin	0.844	**0.941**	**0.890**	28	1692
		XHAIL	0.779	0.914	0.841	**14**	7836
	Meeting	EC_{crisp}	0.687	0.855	0.762	23	–
		OLED	**0.947**	0.760	0.843	31	**22**
		WoLED	0.892	0.888	**0.889**	29	52
		OSLα	0.902	0.863	0.882	1231	180
		OSL	-	-	-	-	>25 h
		MaxMargin	0.919	0.813	0.863	23	1133
		XHAIL	0.804	**0.927**	0.861	**15**	7248
(b)	*Moving*	OLED	0.682	0.787	0.730	**38**	**63**
		WoLED	**0.783**	**0.821**	**0.801**	51	108
		EC_{crisp}	0.721	0.639	0.677	28	–
	Meeting	OLED	0.701	0.886	0.782	**41**	**43**
		WoLED	**0.808**	**0.877**	**0.841**	56	98
		EC_{crisp}	0.644	0.855	0.735	23	–

Figure 2 presents the holdout evaluation [15] for the online learners compared in this experiment. Holdout evaluation consists of assessing the quality of an online learner on a holdout test set, at regular time intervals during learning, thus obtaining a learning curve of its performance over time. Figure 2 presents average F_1-scores, obtained by performing holdout evaluation on each fold of the tenfold cross-validation process: at each fold, each learner's theory is evaluated on the fold's test set every 1000 time points and the F_1-scores from each evaluation point are averaged over all ten folds.

WoLED and OLED have an adequate performance, with relatively smooth learning curves, while they eventually converge to stable theories of acceptable performance. Moreover, WoLED outperforms both OLED and OSLα in most of the evaluation process.

In contrast to the online behaviour of WoLED and OLED, OSLα's performance exhibits abrupt fluctuations. For *moving* in particular, OSLα's average F_1-score reaches its peak (0.87) after processing data from approximately 10,000 time points, and then it drops significantly until the final average F_1-score value of 0.69 reported in Table 2. This behavior may be attributed to OSLα's rule generation strategy. Contrary to WoLED, which uses Hoeffding tests to select rules with significant heuristic value, OSLα greedily adds new rules to the current

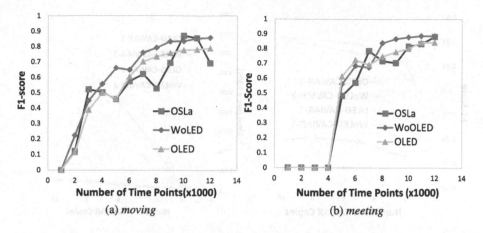

Fig. 2. Online holdout evaluation on CAVIAR.

theory, so as to locally improve its performance, without taking into account the new rules' quality on larger portions of the data. Overall, this results in poor online performance, since rules with no quality guarantees on the training set may be responsible for a large number of mistakes on unseen data, by e.g. fitting the noise in the training data. OSLα relies solely on weight learning to minimize the weights of low-quality rules in the long run. However, OSLα's holdout evaluation indicates that in principle this requires larger training sets, since, at least in the case of *moving*, OSLα's theories exhibit no sign of convergence. On the other hand, OSLα's increased training times reported in Table 2, due to the ever-increasing cost of maintaining unnecessarily large theories, indicate that training on larger datasets is impractical.

5.2   Evaluation on Larger Data Volumes

In this section we evaluate WoLED on larger data volumes, starting with the entire CAVIAR dataset, which consists of 282,067 interpretations, in contrast to 25,738 interpretations in the CAVIAR fragment. Due to the increased training times of OSLα, XHAIL and MaxMargin, we did not experiment with these algorithms. The target complex events were *meeting* and *moving* as previously. The additional training data (i.e. those not contained in the CAVIAR fragment) were negative instances for both complex events (recall that the parts of CAVIAR where *meeting* and *moving* occur were already contained in the CAVIAR fragment). This way, the dataset used in this experiment is much more imbalanced than the one used in the previous experiment. The parameter configuration for the two learners was as reported in Sect. 5.1. The results were obtained via tenfold cross-validation and are presented in Table 2(b).

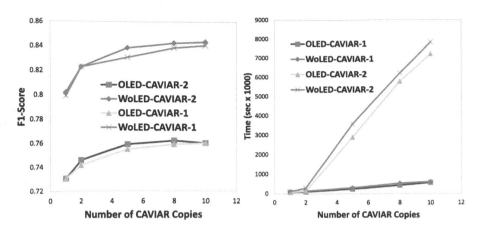

Fig. 3. Evaluation on larger data volumes for the *moving* complex event.

The average F_1-score for both algorithms is decreased, as compared to the previous experiment, due to the increased number of false positives, caused by the large number of additional negative instances. WoLED significantly outperforms OLED for both complex events, at the price of a tolerable increase in training times.

To test our approach further we used larger datasets generated from CAVIAR in two different settings. In the first setting, to which we henceforth refer as CAVIAR-1, we generated datasets by sequentially appending copies of the original CAVIAR dataset, incrementing the time-stamps in each copy. Therefore, training with datasets in the CAVIAR-1 setting amounts to re-iterating over the original dataset a number of times. In the second setting, to which we refer as CAVIAR-2, datasets were also obtained from copies of CAVIAR, but this time the time-stamps in the data were left intact and each copy differed from the others in the constants referring to the tracked entities (persons, objects) that appear in simple and complex events. In each copy of the dataset, the coordinates of each entity p differ by a fixed offset from the coordinates of the entity of the original dataset that p mirrors. The copies were "merged", grouping together by time-stamp the data from each copy. Therefore, the number of constants in each training interpretation in datasets of the CAVIAR-2 setting is multiplied by the number of copies used to generate the dataset. We performed experiments on datasets obtained from 2, 5, 8 and 10 CAVIAR copies for both settings. The target concept was *moving* and on each dataset we used tenfold cross-validation to measure F_1-scores and training times.

The results are presented in Fig. 3. F_1-scores improve with larger data volumes for both learners, slightly more so with the datasets in the CAVIAR-2 setting, while WoLED achieves better F_1-scores than OLED thanks to weight learning. Training times increase slowly with the data size in the "easier" CAVIAR-1 setting, where both learners require 8–10 minutes on average to learn from the largest dataset in this setting. In contrast, training times increase abruptly in the harder CAVIAR-2 setting, where learning from the largest dataset requires more than 2.5 h on average for both learners. This is due to the additional domain constants in the datasets of the CAVIAR-2 setting, which result in exponentially larger ground theories.

6   Conclusions and Future Work

We presented an algorithm for online learning of event definitions in the form of Event Calculus theories in the MLN semantics. We extended an online ILP algorithm to a statistical relational learning setting via online weight optimization. We evaluated our approach on an activity recognition application, showing that it outperforms both its crisp predecessor and competing algorithms for online learning in MLN. There are several directions for further work. We plan to improve scalability using parallel/distributed learning, along the lines of [22]. We also plan to evaluate different algorithms for online weight optimization and develop methodologies for online hyper-parameter adaptation.

Acknowledgements. This project has received funding from the European Union's Horizon 2020 research and innovation programme under grant agreement No 780754.

References

1. Alevizos, E., Skarlatidis, A., Artikis, A., Paliourasm, G.: Probabilistic complex event recognition: a survey. ACM Computing Surveys (2018) (to appear)
2. Artikis, A., Sergot, M., Paliouras, G.: An event calculus for event recognition. IEEE Trans. Knowl. Data Eng. **27**(4), 895–908 (2015)
3. Artikis, A., Skarlatidis, A., Portet, F., Paliouras, G.: Logic-based event recognition. Knowl. Eng. Rev. **27**(4), 469–506 (2012)
4. Athakravi, D., Corapi, D., Broda, K., Russo, A.: Learning through hypothesis refinement using answer set programming. In: Zaverucha, G., Santos Costa, V., Paes, A. (eds.) ILP 2013. LNCS (LNAI), vol. 8812, pp. 31–46. Springer, Heidelberg (2014). https://doi.org/10.1007/978-3-662-44923-3_3
5. Corapi, D., Russo, A., Lupu, E.: Inductive logic programming as abductive search. In ICLP-2010, pp. 54–63 (2010)
6. Corapi, D., Sykes, D., Inoue, K., Russo, A.: Probabilistic rule learning in nonmonotonic domains. In: Leite, J., Torroni, P., Ågotnes, T., Boella, G., van der Torre, L. (eds.) CLIMA 2011. LNCS (LNAI), vol. 6814, pp. 243–258. Springer, Heidelberg (2011). https://doi.org/10.1007/978-3-642-22359-4_17
7. Cugola, G., Margara, A.: Processing flows of information: from data stream to complex event processing. ACM Comput. Surv. (CSUR) **44**(3), 15 (2012)
8. De Raedt, L.: Logical and Relational Learning. Springer, Heidelberg (2008). https://doi.org/10.1007/978-3-540-68856-3
9. De Raedt, L., Kersting, K., Natarajan, S., Poole, D.: Statistical relational artificial intelligence: logic, probability, and computation. Synth. Lect. Artif. Intell. Mach. Learn. **10**(2), 1–189 (2016)
10. Domingos, P., Hulten, G.: Mining high-speed data streams. In: ACM SIGKDD, pp. 71–80. ACM (2000)

11. Dragiev, S., Russo, A., Broda, K., Law, M., Turliuc, C.: An abductive-inductive algorithm for probabilistic inductive logic programming. In: Proceedings of the 26th International Conference on Inductive Logic Programming (Short papers), London, UK, 2016, pp. 20–26 (2016)
12. Dries, A., De Raedt, L.: Towards clausal discovery for stream mining. In: De Raedt, L. (ed.) ILP 2009. LNCS (LNAI), vol. 5989, pp. 9–16. Springer, Heidelberg (2010). https://doi.org/10.1007/978-3-642-13840-9_2
13. Duchi, J., Hazan, E., Singer, Y.: Adaptive subgradient methods for online learning and stochastic optimization. J. Mach. Learn. Res. **12**, 2121–2159 (2011)
14. Gama, J.: Knowledge Discovery from Data Streams. CRC Press, Boca Raton (2010)
15. Gama, J., Sebastião, R., Rodrigues, P.P.: On evaluating stream learning algorithms. Mach. Learn. **90**(3), 317–346 (2013)
16. Hoeffding, W.: Probability inequalities for sums of bounded random variables. J. Am. Stat. Assoc. **58**(301), 13–30 (1963)
17. Huynh, T.N., Mooney, R.J.: Max-Margin weight learning for markov logic networks. In: Buntine, W., Grobelnik, M., Mladenić, D., Shawe-Taylor, J. (eds.) ECML PKDD 2009, Part I. LNCS (LNAI), vol. 5781, pp. 564–579. Springer, Heidelberg (2009). https://doi.org/10.1007/978-3-642-04180-8_54
18. Huynh, T.N., Mooney, R.J.: Online structure learning for markov logic networks. In: Gunopulos, D., Hofmann, T., Malerba, D., Vazirgiannis, M. (eds.) ECML PKDD 2011, Part II. LNCS (LNAI), vol. 6912, pp. 81–96. Springer, Heidelberg (2011). https://doi.org/10.1007/978-3-642-23783-6_6
19. Katzouris, N.: Scalable relational learning for event recognition. PhD Thesis, University of Athens (2017). http://users.iit.demokritos.gr/nkatz/papers/nkatz-phd.pdf
20. Katzouris, N., Artikis, A., Paliouras, G.: Incremental learning of event definitions with inductive logic programming. Mach. Learn. **100**(2–3), 555–585 (2015)
21. Katzouris, N., Artikis, A., Paliouras, G.: Online learning of event definitions. Theory Pract. Log. Program. **16**(5–6), 817–833 (2016)
22. Katzouris, N., Artikis, A., Paliouras, G.: Parallel Online Learning of Event Definitions. In: Lachiche, N., Vrain, C. (eds.) ILP 2017. LNCS (LNAI), vol. 10759, pp. 78–93. Springer, Cham (2018). https://doi.org/10.1007/978-3-319-78090-0_6
23. Kowalski, R., Sergot, M.: A logic-based calculus of events. New Gener. Comput. **4**(1), 67–95 (1986)
24. Law, M., Russo, A., Broda, K.: Iterative learning of answer set programs from context dependent examples. Theory Pract. Log. Program. **16**(5–6), 834–848 (2016)
25. Littlestone, N.: Learning quickly when irrelevant attributes abound: a new linear-threshold algorithm. Mach. Learn. **2**(4), 285–318 (1988)
26. Margara, A., Cugola, G., Tamburrelli, G.: Learning from the past: automated rule generation for complex event processing. In: Proceedings of the 8th ACM International Conference on Distributed Event-Based Systems, pp. 47–58. ACM (2014)
27. Michelioudakis, E., Skarlatidis, A., Paliouras, G., Artikis, A.: OSLα: online structure learning using background knowledge axiomatization. In: Frasconi, P., Landwehr, N., Manco, G., Vreeken, J. (eds.) ECML PKDD 2016, Part I. LNCS (LNAI), vol. 9851, pp. 232–247. Springer, Cham (2016). https://doi.org/10.1007/978-3-319-46128-1_15
28. Ray, O.: Nonmonotonic abductive inductive learning. J. Appl. Log. **7**(3), 329–340 (2009)

29. Richardson, M., Domingos, P.: Markov logic networks. Mach. Learn. **62**(1–2), 107–136 (2006)
30. Skarlatidis, A., Paliouras, G., Artikis, A., Vouros, G.: Probabilistic event calculus for event recognition. ACM Trans. Comput. Log. (TOCL) **16**(2), 11 (2015)
31. Srinivasan, A., Bain, M.: An empirical study of on-line models for relational data streams. Mach. Learn. **106**(2), 243–276 (2017)

Toward Interpretable Deep Reinforcement Learning with Linear Model U-Trees

Guiliang Liu[✉], Oliver Schulte, Wang Zhu, and Qingcan Li

School of Computing Science, Simon Fraser University, Burnaby, Canada
{gla68,zhuwangz,qingcanl}@sfu.ca, oschulte@cs.sfu.ca

Abstract. Deep Reinforcement Learning (DRL) has achieved impressive success in many applications. A key component of many DRL models is a neural network representing a Q function, to estimate the expected cumulative reward following a state-action pair. The Q function neural network contains a lot of implicit knowledge about the RL problems, but often remains unexamined and uninterpreted. To our knowledge, this work develops the first mimic learning framework for Q functions in DRL. We introduce Linear Model U-trees (LMUTs) to approximate neural network predictions. An LMUT is learned using a novel on-line algorithm that is well-suited for an active play setting, where the mimic learner observes an ongoing interaction between the neural net and the environment. Empirical evaluation shows that an LMUT mimics a Q function substantially better than five baseline methods. The transparent tree structure of an LMUT facilitates understanding the network's learned strategic knowledge by analyzing feature influence, extracting rules, and highlighting the super-pixels in image inputs. Code related to this paper is available at: https://github.com/Guiliang/uTree_mimic_mountain_car.

1 Introduction: Mimic a Deep Reinforcement Learner

Deep Reinforcement Learning has mastered human-level control policies in a wide variety of tasks [14]. Despite excellent performance, the learned knowledge remains implicit in neural networks and hard to explain: there is a trade-off between model performance and interpretability [11]. One of the frameworks for addressing this trade-off is mimic learning [1], which seeks a sweet spot between accuracy and efficiency, by training an interpretable mimic model to match the predictions of a highly accurate model. Many works [2,5,7] have developed types of mimic learning to distill knowledge from deep models to a mimic model with tree representation, but for *supervised* learning only. However, DRL is an unsupervised process, where agents continuously interact with an environment, instead of learning from a static training/testing dataset.

O. Schulte—Supported by a Discovery Grant from the Natural Sciences and Engineering Council of Canada. This research utilized Titan X GPUs donated by the NVIDIA Corporation.

© Springer Nature Switzerland AG 2019
M. Berlingerio et al. (Eds.): ECML PKDD 2018, LNAI 11052, pp. 414–429, 2019.
https://doi.org/10.1007/978-3-030-10928-8_25

This work develops a novel mimic learning framework for Reinforcement Learning. We examine two different approaches to *generating data for RL mimic learning*. Within the first *Experience Training* setting, which allows applying traditional batch learning methods to train a mimic model, we record all state action pairs during the training process of DRL and complement them with Q values as soft supervision labels. Storing and reading the training experience of a DRL model consumes much time and space, and the training experience may not even be available to a mimic learner. Therefore our second *Active Play* setting generates streaming data through interacting with the environment using the mature DRL model. The active play setting requires an on-line algorithm to dynamically update the model as more learning data is generated.

U-tree [13,20] is a classic online reinforcement learning method which represents a Q function using a tree structure. To strengthen its generalization ability, we add a linear model to each leaf node, which defines a novel Linear Model U-Tree (LMUT). To support the active play setting, we introduce a novel on-line learning algorithm for LMUT, which applies Stochastic Gradient Descent to update the linear models, given some memory of recent input data stored on each leaf node. We conducted an empirical evaluation in three benchmark environments with five baseline methods. Two natural evaluation metrics for an RL mimic learner are: (1) fidelity [7]: how well the mimic model matches the predictions of the neural net, as in supervised learning, and (2) *play performance*: how well the average return achieved by a controller based on the mimic model matches the return achieved by the neural net. Play performance is the most relevant metric for reinforcement learning. Perfect fidelity implies a perfect match in play performance. However, our experiments show that approximate fidelity does not imply a good match in play performance. This is because RL mimic learning must strike a balance between coverage: matching the neural net across a large section of the state space, and optimality: matching the neural net on the states that are most important for performance. In our experiments, LMUT learning achieves a good balance: the best match to play performance among the mimic methods, and competitive fidelity to the neural net predictions. The transparent tree structure of LMUT makes the DRL neural net interpretable. To analyze the mimicked knowledge, we calculate the importance of input features and extract rules for typical examples of agent behavior. For image inputs, the super-pixels in input images are highlighted to illustrate the key regions.

Contributions. The main contributions of this paper are as follow: (1) To our best knowledge, the first work that extends interpretable mimic learning to Reinforcement Learning. (2) A novel on-line learning algorithm for LMUT, a novel model tree to mimic a DRL model. (3) We show how to interpret a DRL model by analyzing the knowledge stored in the tree structure of LMUT.

The paper is organized as follow. Section 2 covers the background and related work of DRL, mimic learning and U-tree. Section 3 introduces the mimic learning framework and Sect. 4 shows how to learn a LMUT. Empirical evaluation is performed in Sects. 5 and 6 discusses the interpretability of LMUT.

2   Background and Related Work

Reinforcement Learning and the Q-function. Reinforcement Learning constructs a policy for an agent to interact with its environment and maximize cumulative reward [18]. Such an environment can be formalized as a Markov Decision Process (MDP) with 4-tuple (S, A, P, R), where at timestep t, an agent observes a state $s_t \in S$, chooses a action $a_t \in A$ and receives a reward $r_t \in R$ and the next observation $s_{t+1} \in S$ from environment. A Q function represents the value of executing action a_t under state s_t [18]. Given a policy π, the value is the expectation of the sum of discounted reward $Q_t(s_t, a_t) = \mathbb{E}_\pi(\sum_{k=0}^{\infty} \gamma^k r_{t+k+1})$. Temporal difference learning updates the current Q-value estimates towards the observed reward and estimated utility of the resulting state s_{t+1}. A Deep Q-Network (DQN) [14] uses a neural network architecture to approximate the Q function for complex feature and action spaces. Parameter (θ) updates minimize the differentiable loss function:

$$\mathcal{L}(\theta_i) \approx (r_t + \gamma \max_{a_t} Q(s_{t+1}, a_{t+1}|\theta_i) - Q(s_t, a_t|\theta_i))^2 \tag{1}$$

$$\theta_{i+1} = \theta_i + \alpha \nabla_\theta \mathcal{L}(\theta_i) \tag{2}$$

Mimic Learning. Recent works on mimic learning [1,5,7] have demonstrated that models like shallow feed-forward neural networks or decision trees can mimic the function of a deep neural net. In the *oracle framework*, soft output labels are collected by passing inputs to a large, complex and accurate deep neural network [22] Then we train a mimic model with the soft output as supervisor. The results indicate that training a mimic model with soft output achieves substantial improvement in accuracy and efficiency, over training the same model type directly with hard targets from the dataset.

U-Tree Learning. A tree structure is transparent and interpretable, allowing rule extraction and measuring feature influence [5]. U-tree [13] learning was developed as an online reinforcement learning algorithm for a tree structure representation. A U-tree takes a set of observed feature/action values as input and maps it to a state value (or Q-value). [20] introduces the continuous U-tree (CUT) for continuous state features. CUT learning generates a tree-based discretization of the input signal and estimates state transition probabilities in every leaf node [20]. Dynamic programming is applied to solve the resulting Markov Decision Process (MDP). Although CUTs have been successfully applied in test environments like Corridor and Hexagonal Soccer, constructing a CUT from raw data is rather slow and consumes significant computing time and space.

3   Mimic Learning for Deep Reinforcement Learning

Unlike supervised learning, a DRL model is not trained with static input/output data pairs; instead it interacts with the environment by selecting actions to perform and adjusting its policy to maximize the expectation of cumulative reward. We now present two settings to mimic the Q functions in DRL models.

Experience Training generates data for batch training, following [1,5]. To construct a mimic dataset, we record all the observation signals I and actions a during the DRL process. A signal I is a vector of continuous features that represents a state (one-hot representation for discrete features). Then, by inputting them to a mature DRL model, we obtain their corresponding soft output Q and use the entire input/output pairs $\{(\langle I_1, a_1 \rangle, \hat{Q}_1(I_1, a_1)), (\langle I_2, a_2 \rangle, \hat{Q}_2(I_2, a_2)), ..., (\langle I_T, a_T \rangle, \hat{Q}_T(I_T, a_T))\}$ as the **experience training dataset**.

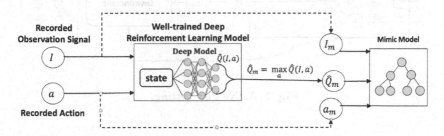

Fig. 1. Experience Training setting

Active Play generates mimic data by applying a mature DRL model to interact with the environment. Similar to [19], our active learner ℓ has three components: (q, f, I). The first component q is a querying function $q(I)$ that gives the current observed signal I, selects an action a. The querying function controls ℓ's interaction with the environment so it must consider the balance between exploration and exploitation. Our querying function is the ϵ-greedy scheme [14] (ϵ decaying from 1 to 0). The second component f is the deep model that produces Q values: $f : (I, a) \rightarrow range(\hat{Q})$.

As shown in Fig. 2, the mimic training data is generated in the following steps: *Step 1:* Given a starting observation signal I_t on time step t, we select an action $a_t = q(I_t)$, and obtain a soft output Q value $\hat{Q}_t = f(I_t, a_t)$. *Step 2:* After performing a_t, the environment provides a reward r_t and the next state observation I_{t+1}. We record a labelled **transition** $T_t = \{I_t, a_t, r_t, I_{t+1}, \hat{Q}_t(I_t, a_t)\}$ where the soft label $\hat{Q}_t(I_t, a_t)$ comes from the well trained DRL model. A transition is the basic observation unit for U-tree learning. *Step 3:* We set I_{t+1} as the next starting observation signal, repeat above steps until we have training data for the active learner ℓ to finish sufficient updates over mimic model m. This process produces an infinite data stream (transitions T) in sequential order. We use minibatch online learning, where the learner returns a mimic model M after some fixed batchsize B of queries.

Compared to Experience Training, Active Play does not require recording data during the training process of DRL models. This is important for the following reasons. (1) Many mimic learners have access only to the trained deep models. (2) Training a DRL model often generates a large amount of data, which requires much memory and is computationally challenging to process. (3) The

Experience Training data includes frequent visits to suboptimal states, which makes it difficult for the mimic learner to obtain an optimal return, as our evaluation illustrates.

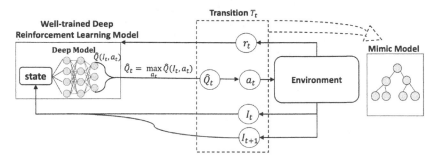

Fig. 2. Active Play setting.

4 Learning Linear Model U-Trees

A neural network with continuous activation functions computes a continuous function. A regression tree can approximate a continuous function arbitrarily closely, given enough leaves. Continuous U-Trees (CUTs) are essentially regression trees for value functions, and therefore a natural choice for a tree structure representation of a DRL Q function. However, their ability to generalize is limited, and CUT learning converges slowly. We introduce a *novel extension of CUT*, Linear Model U-Tree (LMUT), that allows CUT leaf nodes to contain a linear model, rather than simple constants. Being strictly more expressive than a regression tree, a linear model tree can also approximate a continuous function arbitrarily closely, with typically many fewer leaves [4]. Smaller trees are more interpretable, and therefore more suitable for mimic learning.

As shown in Fig. 3 and Table 1, each leaf node of a LMUT defines a partition cell of the input space, which can be interpreted as a discrete state s for the decision process. Within each partition cell, LMUT also records the reward r and the transition probabilities p of performing action a on the current state s, as shown in the Leaf Node 5 of Fig. 3. So LMUT learning builds a Markov Decision Process (MDP) from the interaction data between environment and deep model. Compared to a linear Q-function approximator [18], a LMUT defines an ensemble of linear Q-functions, one for each partition cell. Since each Q-value prediction Q_N^{UT} comes from a single linear model, the prediction can be explained by the feature weights of the model.

We now discuss how to train an LMUT. Similar to [20], we separate the training into two phases: (1) Data Gathering Phase and (2) Node Splitting Phase.

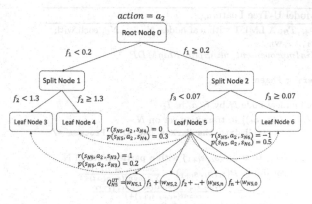

Fig. 3. An example of Linear Model U-Tree (LMUT).

Table 1. Partition cell

Node name	Partition cell
Leaf Node 3	$f_1 < 0.2,$
	$f_2 < 1.3$
Leaf Node 4	$f_1 < 0.2,$
	$f_2 \geq 1.3$
Leaf Node 5	$f_1 \geq 0.2,$
	$f_2 < 0.07$
Leaf Node 6	$f_1 \geq 0.2,$
	$f_2 \geq 0.07$

4.1 Data Gathering Phase

Data Gathering Phase assigns transitions to leaf nodes and prepares them for fitting linear models and splitting nodes. Given an input transition T, we pass it through feature splits down to a leaf node. As an option, an LMUT can dynamically build an MDP, in which case it updates transition probabilities, rewards and average Q values on the leaf nodes. The complete Data Gathering Phase process is detailed in part I (the first for loop) of Algorithm 1.

4.2 Node Splitting Phase

After node updating, LMUT scans the leaf nodes and updates their linear model with Stochastic Gradient Descent (SGD). If SGD achieves insufficient improvement on node N, LMUT determines a new split and adds the resulting leaves to the current partition cell. For computational efficiency, our node splitting phase considers only a single split for each leaf given a single minibatch of new transitions. Part II of Algorithm 1 shows the detail of the node splitting phase. LMUT applies a minibatch stagewise fitting approach to learn linear models in the leaves of the U-tree. Like other stagewise approaches [10], this approach provides smoothed weight estimates where nearby leaves tend to have similar weights. We use Stochastic Gradient Descent to implement the weight updates.

Algorithm 1. Linear Model U-Tree Learning

Input: Transitions $T_1, \ldots, T_B$; A LMUT with leaf nodes $N_1, \ldots, N_L$, each with
weight vector $\mathbf{w_1}, \ldots, \mathbf{w_L}$
Hyperparameters: *MinImprovement, minSplit, FlagMDP*

```
/* Part I: Data Gathering Phase */
for t = 1 to B do
```
Find the partition cell on leaf node N by I_t, a_t in T_t
Add $T_t = \langle I_t, a_t, r_t, I_{t+1}, \hat{Q}_t(I_t, a_t) \rangle$ to transition set on N
if *FlagMDP*   `/* Update the Markov Decision Process */`
then

Map observation (I_t, I_{t+1}) to state (s_t, s_{t+1}) within partition cell of N
Update Transitions Probability $P(s_t, a_t, s_{t+1}) = \frac{count(s_t, a_t, s_{t+1}) + 1}{\sum_i count(s_t, a_t, s_i) + 1}$
Update Reward $R(s_t, a_t, s_{t+1}) = \frac{R(s_t, a_t, s_{t+1}) * count(s_t, a_t, s_{t+1}) + r_t}{count(s_t, a_t, s_{t+1}) + 1}$
Compute $Q_{avg}^{UT}(s_t, a_t) = \frac{\hat{Q}_t(s_t, a_t) * count(s_t, a_t) + \hat{Q}_t(I_t, a_t)}{count(s_t, a_t) + 1}$
Increment $count(s_t, a_t)$ and $count(s_t, a_t, s_{t+1})$ by 1

end
end

```
/* Part II: Node Splitting Phase */
for i = 1 to L do
```
$w_i, err_i := $ WeightUpdate$(\mathbf{T_{N_i}}, \mathbf{w_i})$ `/* Update the weights by SGD */`
if $err \le MinImprovement$ **then**

for *distinction D in GetDistinction(N_i)* **do**

Split Node N_i to FringeNodes by distinction D
Compute distribution of Q function $\sigma_{N_i}(Q)$ on Node N_i
for *each node F in FringeNodes* **do**

Compute distribution $\sigma_F(Q)$
`/* This function is discussed in Splitting Criterion */`
$p = $ SplittingCriterion$(\sigma_{N_i}(Q), \sigma_F(Q))$
if $p \ge minSplit$ **then**
$\quad BestD = D$
$\quad minSplit = p$
end
Remove all the fringe nodes

end

end
if *BestD* **then**

Split Node N_i by *BestD* to define ChildNodes $N_{i,1}, \ldots, N_{i,C}$
Assign Transitions set $\mathbf{T_{N_i}}$ to ChildNodes
for $c = 1$ to C **do**

`/* Initial Child Node weights = Parent Node weights */`
$w_{i,c} := w_i$
$w_{i,c}, err_{i,c} := $ WeightUpdate$(\mathbf{T_{N_i,c}}, \mathbf{w_{i,c}})$

end

end

end
end

Stochastic Gradient Descent (SGD) Weight Updates is a straight-forward well-established online weight learning method for a single linear regression model. The weights and bias of linear regression on leaf node N are updated by applying SGD over all Transitions assigned to N. For a transition $T_t = \langle I_t, a_t, r_t, I_{t+1}, \hat{Q}(I_t, a_t) \rangle$, we take I_t as input and $\hat{Q}_t \equiv \hat{Q}(I_t, a_t)$ as label. We build a separate LMUT for each action, so the linear model on N is function of the J state features: $Q^{UT}(I_t|w_N, a_t) = \sum_{j=1}^{J} I_{tj} w_{Nj} + w_{N0}$. We update the weights w_N on leaf node N by applying SGD with loss function $\mathcal{L}(w_N) = \sum_t 1/2(\hat{Q}_t - Q^{UT}(I_t|w_N, a_t))^2$. The updates are computed with a single pass over each minibatch.

Algorithm 2. SGD Weight Update at a leaf node

Input: Transitions $T_1, \ldots, T_m$, node N = leaf node with weight vector w_0
Output: updated weight vector w, training error err
Hyperparameters: number of iterations E; step size α
$w := w_0$;
for $e = 1$ to E **do**
    for $t=1$ to m **do**
        $w := w + \alpha \nabla_w \mathcal{L}(w)$;
    end
end
Compute training error $err = 1/m \sum_{t=1}^{m} (\hat{Q}_t - Q^{UT}(I_t|w, a_t))^2$

Splitting Criterion is used to find the best split on the leaf node, if SGD achieves limited improvement. We have tried three splitting criteria including working response of SGD, Kolmogorov–Smirnov (KS) test and Variance Test. The first method aims to find the best split to improve working response of the parent linear model on the data for its children. But as reported in [10], the final result becomes less intelligible. The second method Kolmogorov–Smirnov (KS) test is a non-parametric statistical test that measures the differences in empirical cumulative distribution functions between the child datasets. The final Variance criterion selects a split that generates child nodes whose Q values contain the least variance. The idea is similar to the variance reduction method applied in CART tree. Like Uther and Veloso [20], we found that the Variance test works well with less time complexity than KS test ($O(n)$ v.s. $O(n^2)$), so we select the Variance test as the splitting criterion. Exploring the different possible splits efficiently is the main scalability challenge in LMUT learning (cf. [20]).

5 Empirical Evaluation

We evaluate the mimic performance of LMUT by comparing it with five other baseline methods under three evaluation environments. Empirical evaluation measures the mimic match in regression and game playing, under experience training and active play learning.

5.1 Evaluation Environment

The evaluation environments include **Mountain Car, Cart Pole** and **Flappy Bird**. Our environments are simulated by OpenAI Gym toolkit [3]. Mountain Car and Cart Pole are two benchmark tasks for reinforcement learning [17]. Mountain Car is about accelerating a car to the top of the hill and Cart Pole is about balancing a pole in the upright position. Mountain Car and Cart Pole have a discrete action space and a continuous feature space. Flappy Bird is a mobile game that controls a bird to fly between pipes. Flappy Bird has two discrete actions, and its observation consists of four consecutive images [14]. We follow the Deep Q-Learning (DQN) method to play this game. During the image preprocessing, the input images are first rescaled to 80*80, transferred to gray image and then binary images. With 6,400 features, the state space of Flappy Bird is substantially more complex than that for Cart Pole and Mountain Car.

5.2 Baseline Methods

Batch Methods. We fit the input/output training pairs $(\langle I, a \rangle, \hat{Q}(I, a))$ using batch tree learners. A **CART** regression tree [12] predicts for each leaf node, the mean Q-value over the samples assigned to the leaf. M5 [15] is a tree training algorithm with more generalization ability. It first constructs a piecewise constant tree and then prunes to build a linear regression model for the instances in each leaf node. The WEKA toolkit [8] provides an implementation of M5. We include M5 with Regression-Tree option (**M5-RT**) and M5 tree with Model-Tree option (**M5-MT**) in our baselines. M5-MT builds a linear function on each leaf node, while M5-RT has only a constant value.

On-line Learning Methods. The recent **Fast Incremental Model Tree (FIMT)** [9] method is applied. Similar to M5-MT, it builds a linear model tree, but can perform explicit change detection and informed adaption for evolving data stream. We experiment with a basic version of FIMT and an advanced version with **Adaptive Filters** on leaf nodes (named **FIMT-AF**).

5.3 Fidelity: Regression Performance

We evaluate how well our LMUT approximates the soft output ($\hat{Q}$ values) from Q function in a Deep Q-Network (DQN). We report the standard regression metrics **Mean Absolute Error (MAE)**, and **Root Mean Square Error (RMSE)**.

Under the *Experience Training* setting, we compare the performance of CART, M5-RT, M5-MT, FIMT and FIMT-AF with our LMUT. The dataset sizes are 150 K transitions for Mountain Car, 70 K transitions for Car Pole, and 20 K transitions for Flappy Bird. Because of the high dimensionality of the Flappy Bird state space, storing 20 K transitions requires 32 GB main memory. Given an experience training dataset, we apply 10 fold cross evaluation to train and test our model.

Table 2. Result of Mountain Car

Method		Evaluation metrics		
		MAE	RMSE	Leaves
Experience Training	CART	0.284	0.548	1772.4
	M5-RT	0.265	0.366	779.5
	M5-MT	**0.183**	**0.236**	240.3
	FIMT	3.766	5.182	4012.2
	FIMT-AF	2.760	3.978	3916.9
	LMUT	0.467	0.944	620.7
Active Play	FIMT	3.735	5.002	1020.8
	FIMT-AF	2.312	3.704	712.4
	LMUT	0.475	1.015	453.0

Table 3. Result of cart pole

Method		Evaluation metrics		
		MAE	RMSE	Leaves
Experience Training	CART	15.973	34.441	55531.4
	M5-RT	25.744	48.763	614.9
	M5-MT	19.062	37.231	155.1
	FIMT	43.454	65.990	6626.1
	FIMT-AF	31.777	50.645	4537.6
	LMUT	**13.825**	**27.404**	658.2
Active Play	FIMT	32.744	62.862	2195.0
	FIMT-AF	28.981	51.592	1488.9
	LMUT	14.230	43.841	416.2

For the *Active Play* setting, batch training algorithms like CART and M5 are not applicable, so we experiment only with online methods, including FIMT, FIMT-AF and LMUT. We first train the mimic models with 30k consecutive transitions from evaluation environments, and evaluate them with another 10k transitions. The result for the three evaluation environments are shown in Tables 2, 3 and 4. The differences between the results of LMUT and the results of other models are statistically significant (t-test; $p < 5\%$).

Compared to the other two online learning methods (FIMT and FIMT-AF), LMUT achieves a better fit to the neural net predictions with a much smaller model tree, especially in the active play online setting. This is because both FIMT and FIMT-AF update their model tree continuously after each datum, whereas LMUT fits minibatches of data at each leaf. Neither FIMT nor FIMT-AF terminate on high-dimensional data.[1] So we omit the result of applying FIMT and FIMT-AF in the Flappy Bird environment. Comparing to batch methods, the CART tree model has significantly more leaves than our LMUT, but not better fit to the DQN than M5-RT, M5-MT and LMUT, which suggests overfitting. In the Mountain Car and Flappy Bird environments, model tree batch learning (M5-RT and M5-MT) performs better than LMUT, while LMUT achieves comparable fidelity, and leads in the Cart Pole environment. In conclusion, (1) our LMUT learning algorithm outperforms the state-of-the-art online model tree learner FIMT. (2) Although LMUT is an online learning method, it showed competitive performance to batch methods even in the batch setting.

Learning Curves. We apply consecutive testing [9] to analyze the performance of LMUT learning in more detail. We compute the correlation and testing error of LMUT as more transitions for learning are provided (From 0 to 30k) under the active play setting. To adjust the error scale across different game environments, we use **Relative Absolute Error (RAE)** and **Relative Square Error (RSE)**. We repeat the experiment 10 times and plot the shallow graph in Fig. 5. In the Mountain Car environment, LMUT converges quickly with its performance increasing smoothly in 5 k transitions. But for complex environments like Cart Pole and Flappy Bird, the evaluation metrics fluctuate during the learning process but approach the optimum within 30 k transitions.

---

[1] For example, in the Flappy Bird environment, FIMT takes 29 min and 10.8GB main memory to process 10 transitions on a machine using i7-6700HQ CPU.

Table 4. Result of Flappy Bird

Method		Evaluation metrics		
		MAE	RMSE	Leaves
Experience Training	CART	0.018	0.036	700.3
	M5-RT	0.027	0.041	226.1
	M5-MT	**0.016**	**0.030**	412.6
	LMUT	0.019	0.043	578.5
Active Play	LMUT	0.024	0.050	229.0

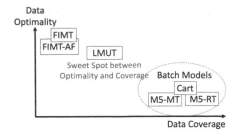

Fig. 4. Coverage v.s. Optimality

5.4 Matching Game Playing Performance

We now evaluate how well a model mimics Q functions in DQN by directly playing the games with them and computing the average reward per episode. (The games in OpenAI Gym toolkit are divided into episodes that start when a game begins and terminate when: (1) the player reaches the goal, (2) fails for a fixed number of times or (3) the game time passes a preset threshold). Specifically, given an input signal I_t, we obtain Q values from the mimic model and select an action $a_t = \max_a Q(I_t, a)$. By executing a_t in the current game environment, we receive a reward r_t and next observation signal I_{t+1}. This process is repeated until a game episode terminates. This experiment uses **Average Reward Per Episodes (ARPE)**, a common evaluation metric that has been applied by both DRL models [14] and OpenAI Gym tookit [3], to evaluate mimic models. In the *Experience Training* setting, the play performance of CART, M5-RT, M5-MT, FIMT, FIMT-AF and our LMUT are evaluated and compared by partial 10-fold cross evaluation, where we select 9 sections of data to train the mimic models and test them by directly playing another 100 games. For the *Active play*, only the online methods FIMT and FIMT-AF are compared, without the Flappy Bird environment (as discussed in Sect. 5.3). Here we train the mimic models with 30k transitions, and test them in another 100 games.

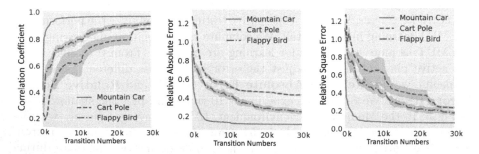

Fig. 5. Consecutive testing of LMUT

Table 5 shows the results for game playing performance. We first experiment with learning a Continuous U-Tree (CUT) *directly using reinforcement learning* [20] instead of mimic learning. CUT converges slowly with limited performance, especially in the high-dimensional Flappy Bird environment. This shows the difficulty of directly constructing a tree model from the environment.

We find that *among all mimic methods, LMUT achieves the Game Play Performance APER closest to the DQN.* Although the batch learning models have strong fidelity in regression, they do not perform as well in game playing as the DQN. Game playing observation shows that the batch learning models (CART, M5-RT, M5-MT) are likely to choose sub-optimal actions in some key scenarios (e.g., when a pole tilts to one side with high velocity in Cart Pole.). This is because the neural net controller selects many sub-optimal actions at the beginning of training, so the early training experience contains many sub-optimal state-action pairs. The batch models fit the entire training experience equally, while our LMUT fits more closely the most recently generated transitions from a mature controller. More recent transitions tend to correspond to optimal actions. The FIMT algorithms keep adapting to the most recent input only, and fail to build adequate linear models on their leaf nodes. Compared to them, LMUT achieves a sweet spot between optimality and coverage (Fig. 4).

Table 5. Game playing performance

Model		Game Environment		
		Mountain Car	Cart Pole	Flappy Bird
Deep Model	*DQN*	*−126.43*	*175.52*	*123.42*
Basic Model	CUT	−200.00	20.93	78.51
Experience Training	CART	−157.19	100.52	79.13
	M5-RT	−200.00	65.59	42.14
	M5-MT	−178.72	49.99	78.26
	FIMT	−190.41	42.88	N/A
	FIMT-AF	−197.22	37.25	N/A
	LMUT	−154.57	145.80	97.62
Active Play	FIMT	−189.29	40.54	N/A
	FIMT-AF	−196.86	29.05	N/A
	LMUT	**−149.91**	**147.91**	**103.32**

6   Interpretability

We discuss how to interpret a DRL model through analyzing the knowledge stored in the transparent tree structure of LMUT: computing feature influence, analyzing the extracted rules and highlighting the super-pixels.

6.1   Feature Influence

Feature importance is one of the most common interpretation tools for tree-based models [5,21]. In an LMUT model, splitting thresholds define partition cells for input signals. We evaluate the influence of a splitting feature by the total variance reduction of the Q values. Recall that our LMUT learning algorithm estimates a weight w_{Nf} for every node and feature f in the model tree (Algorithm 1 Part II). The magnitude of these node weights provides extra knowledge about feature importance. To measure the influence of f on N, we multiply f's standardized square weight by the expected Variance Reduction from splitting N on f:

$$Inf_f^N = (1 + \frac{w_{Nf}^2}{\sum_{j=1}^{J} w_{Nj}^2})(var_N - \sum_{c=1}^{C} \frac{Num_c}{\sum_{i=1}^{C} Num_i} var_c), \tag{3}$$

where Num_c is the number of training instances assigned to child node c and var_N is the variance of Q values at node N. The total influence of a feature Inf_f is the sum of Inf_f^N for all nodes N split by f in our LMUT. For Mountain Car and Cart Pole, we report the feature influences in Table 6. The most important feature for Mountain Car and Cart Pole are Velocity and Pole Angle respectively, which matches the common understanding of the domains. For Flappy Bird the observations are 80*80 pixel images, so LMUT uses pixels as splitting features. Figure 6 illustrates the pixels with feature influences $Inf_f > 0.008$ (the mean of all feature influences). Because locating the bird is very important. the most influential pixels are located on the top left where the bird is likely to be.

Table 6. Feature influence

	Feature	Influence
Mountain Car	Velocity	376.86
	Position	171.28
Cart Pole	Pole angle	30541.54
	Cart Velocity	8087.68
	Cart Position	7171.71
	Pole Velocity At Tip	2953.73

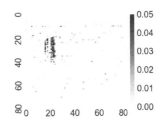

Fig. 6. Super pixels in Flappy Bird

6.2   Rule Extraction

Rule extraction is a common method to distill knowledge from tree models [2,7]. We extract and analyze rules for the Mountain Car and Cart Pole environment. Figure 7 (top) shows three typical examples of extracted rules in *Mountain Car* environment. The rules are presented in the form of partition cells (splitting intervals in LMUT). Each cell contains the range of velocity, position and a Q vector ($\mathbf{Q} = \langle Q_{move\_left}, Q_{no\_push}, Q_{move\_right} \rangle$) representing the average Q-value in the cell. The top left example is a state where the cart is moving toward the left hill with very small velocity. The extracted rule suggests pushing right

(Q_{move_right} has the largest value -29.4): the cart is almost stopped on the left, and by pushing right, it can increase its momentum. The top middle example illustrates a state where the car is approaching the top of the left hill with larger left side velocity (compared to the first example). In this case, however, the cart should be pushed left ($Q_{move\_left} = -25.2$ is the largest), in order to store more Gravitational Potential Energy and prepare for the final rush to the target. The rush will lead to the state shown in the top right image, where the cart should be pushed right to reach the target. Notice that the fewer steps are required to reach the target in a given state, the larger its Q-value.

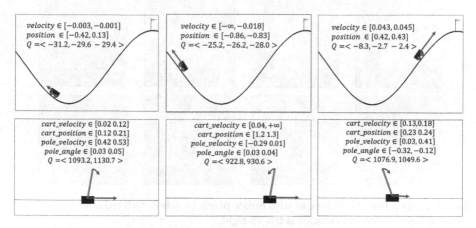

Fig. 7. Examples of rule extraction for Mountain Car and Cart Pole.

Figure 7 (bottom) shows three examples of extracted rules in the *Cart Pole* environment. Each cell contains the scope of cart position, cart velocity, pole angle, pole velocity and a Q vector ($\mathbf{Q} = \langle Q_{push\_left}, Q_{push\_right} \rangle$). The key for Cart Pole is using inertia and acceleration to balance the pole. The bottom left example illustrates the tree rule that the cart should be pushed right (i.e., ($Q_{push\_right} > Q_{push\_left}$), if the pole tilts to the right with a velocity less than 0.5. A similar scenario is the second example, where the pole is also tilting to the right but has velocity towards the left. The Q-values correctly indicate that we should push right to maintain this trend; even if the cart is close to the right-side border, which makes its Q values smaller than in the first example. The third example describes a case where a pole tilts to the left with velocity towards the right. The model correctly selects a left push to achieve a left acceleration.

6.3 Super-Pixel Explanation

In video games, DRL models take the raw pixels from four consecutive images as input. To mimic the deep models, our LMUT also learns on four continuous images and performs splits directly on raw pixels. Deep models for image input can be explained by super-pixels [16]. We highlight the pixels that have feature influence $Inf_f > 0.008$ (the mean of all feature influences) along the splitting path from root to the target partition cell. Figure 8 provides two examples of input images with their highlighted pixels at the beginning of game (top) and in the middle of game (bottom). Most splits are made on the first image which reflects the importance of the most recent input. The first image is often used to locate the pipes (obstacles) and the bird, while the remaining three images provide further information about the bird's location and velocity.

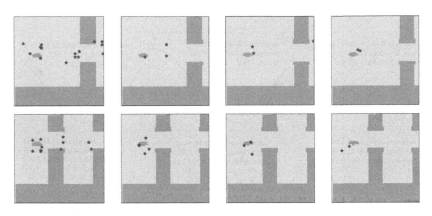

Fig. 8. Flappy Bird input images with Super-pixels (marked with red stars). The input order of four consecutive images is left to right.

7   Conclusion

This work introduced a mimic learning framework for a Reinforcement Learning Environment. A novel Linear Model U-tree represents an interpretable model with the expressive power to approximate a Q-value function learned by a deep neural net. We introduced a novel on-line LMUT mimic learning algorithm based on stochastic gradient descent. Empirical evaluation compared LMUT with five baseline methods on three different Reinforcement Learning environments. The LMUT model achieved the clearly best match to the neural network in terms of *its performance on the RL task*. We illustrated the ability of LMUT to extract the knowledge implicit in the neural network model, by (1) computing the influence of features, (2) analyzing the extracted rules and (3) highlighting the super-pixels. A direction for future work is to explore variants of our LMUT, for example by adding tree pruning, and by experimenting more extensively with hyperparameters. Another important topic is sampling strategies for the active play setting, which would illuminate the difference we observed between matching the neural net's play performance, vs. matching the function it represents.

References

1. Ba, J., Caruana, R.: Do deep nets really need to be deep? In: Advances in Neural Information Processing Systems, pp. 2654–2662 (2014)
2. Boz, O.: Extracting decision trees from trained neural networks. In: Proceedings SIGKDD, pp. 456–461. ACM (2002)
3. Brockman, G., et al.: OpenAI Gym (2016). arXiv preprint arXiv:1606.01540
4. Chaudhuri, P., Huang, M.C., Loh, W.Y., Yao, R.: Piecewise-polynomial regression trees. Statistica Sinica **4**(1), 143–167 (1994)
5. Che, Z., et al.: Interpretable deep models for ICU outcome prediction. In: AMIA Annual Symposium Proceedings, vol. 2016, p. 371. AMIA (2016)
6. Craven, M., Shavlik, J.W.: Extracting tree-structured representations of trained networks. In: Advances in Neural Information Processing Systems, pp. 24–30 (1996)
7. Dancey, D., Bandar, Z.A., McLean, D.: Logistic model tree extraction from artificial neural networks. IEEE Trans. Syst. Man Cybern. Part B (Cybern.) **37**(4), 794–802 (2007)
8. Hall, M., Frank, E., et al.: The weka data mining software: an update. ACM SIGKDD Explor. Newsl. **11**(1), 10–18 (2009)
9. Ikonomovska, E., Gama, J., Džeroski, S.: Learning model trees from evolving data streams. Data Min. Knowl. Discov. **23**(1), 128–168 (2011)
10. Landwehr, N., Hall, M., Frank, E.: Mach. Learn. **59**(1–2), 161–205 (2005)
11. Lipton, Z.C.: The mythos of model interpretability (2016). arXiv preprint arXiv:1606.03490
12. Loh, W.Y.: Classification and regression trees. Data Min. Knowl. Discov. **1**(1), 14–23 (2011). Wiley Interdisciplinary Reviews
13. McCallum, A.K., et al.: Learning to use selective attention and short-term memory in sequential tasks. Proc. SAB **4**, 315–325 (1996)
14. Mnih, V., Kavukcuoglu, K., Silver, D., et al.: Human-level control through deep reinforcement learning. Nature **518**(7540), 529–533 (2015)
15. Quinlan, R.J.: Learning with continuous classes. In: 5th Australian Joint Conference on Artificial Intelligence, pp. 343–348. World Scientific, Singapore (1992)
16. Ribeiro, M.T., Singh, S., Guestrin, C.: Why should i trust you?: explaining the predictions of any classifier. In: Proceedings SIGKDD, pp. 1135–1144. ACM (2016)
17. Riedmiller, M.: Neural fitted Q iteration – first experiences with a data efficient neural reinforcement learning method. In: Gama, J., Camacho, R., Brazdil, P.B., Jorge, A.M., Torgo, L. (eds.) ECML 2005. LNCS (LNAI), vol. 3720, pp. 317–328. Springer, Heidelberg (2005). https://doi.org/10.1007/11564096_32
18. Sutton, R.S., Barto, A.G.: Introduction to Reinforcement Learning, vol. 135. MIT Press, Cambridge (1998)
19. Tong, S., Koller, D.: Support vector machine active learning with applications to text classification. J. Mach. Learn. Res. **2**, 45–66 (2001)
20. Uther, W.T., Veloso, M.M.: Tree based discretization for continuous state space reinforcement learning. In: AAAI/IAAI, pp. 769–774 (1998)
21. Wu, M., Hughes, M., et al.: Beyond sparsity: tree regularization of deep models for interpretability. In: AAAI (2018)
22. Johansson, U., Sönströd, C., König, R.: Accurate and interpretable regression trees using oracle coaching. In: IEEE SSCI Symposium on Computational Intelligence and Data Mining (CIDM), pp. 194–201 (2014)

Online Feature Selection by Adaptive Sub-gradient Methods

Tingting Zhai[1] , Hao Wang[1] , Frédéric Koriche[2] , and Yang Gao[1(✉)]

[1] State Key Laboratory for Novel Software Technology,
Nanjing University, Nanjing 210023, China
zhtt.go@gmail.com,{wanghao,gaoy}@nju.edu.cn
[2] Center of Research in Information in Lens, Université d'Artois,
62307 Lens, France
koriche@cril.fr

Abstract. The overall goal of *online feature selection* is to iteratively select, from high-dimensional streaming data, a small, "budgeted" number of features for constructing accurate predictors. In this paper, we address the online feature selection problem using novel truncation techniques for two online sub-gradient methods: Adaptive Regularized Dual Averaging (ARDA) and Adaptive Mirror Descent (AMD). The corresponding truncation-based algorithms are called B-ARDA and B-AMD, respectively. The key aspect of our truncation techniques is to take into account the magnitude of feature values in the current predictor, together with their frequency in the history of predictions. A detailed regret analysis for both algorithms is provided. Experiments on six high-dimensional datasets indicate that both B-ARDA and B-AMD outperform two advanced online feature selection algorithms, OFS and SOFS, especially when the number of selected features is small. Compared to sparse online learning algorithms that use ℓ_1 regularization, B-ARDA is superior to ℓ_1-ARDA, and B-AMD is superior to Ada-Fobos. Code related to this paper is available at: https://github.com/LUCKY-ting/online-feature-selection.

Keywords: Online feature selection
Adaptive sub-gradient methods · High-dimensional streaming data

1 Introduction

Feature selection is an important topic of machine learning and data mining, for constructing sparse, accurate and interpretable models [7,9,13]. Given a batch of high-dimensional data instances, the overall goal is to find a small subset of relevant features, which are used to construct a low-dimensional predictive model. In modern applications involving streaming data, feature selection is not a "single-shot" offline operation, but an online process that iteratively updates the pool of relevant features, so as to track a sparse predictive model [16,20]. A prototypical example of online feature selection is the anti-spam filtering task,

M. Berlingerio et al. (Eds.): ECML PKDD 2018, LNAI 11052, pp. 430–446, 2019.
https://doi.org/10.1007/978-3-030-10928-8_26

in which the learner is required to classify each incoming message, using a small subset of features that is susceptible to evolve over time.

Conceptually, the online feature selection problem can be cast as a repeated prediction game between the learner and its environment. During each round t of the game, the learner starts by selecting a subset of at most B features over $\{1, \cdots, d\}$, where B is a predefined budget. Upon those selected features is built a predictive model $\boldsymbol{w}_t$ which, in the present paper, is assumed to be a linear function over $\mathbb{R}^d$. Then, a labelled example $(\boldsymbol{x}_t, y_t) \in \mathbb{R}^d \times \mathbb{R}$ is supplied by the environment, and the learner incurs a loss $f(\boldsymbol{w}_t; \boldsymbol{x}_t, y_t)$. The overall goal for the learner is to minimize its cumulative loss over T rounds of the game.

From a computational viewpoint, online feature selection is far from easy since, at each round t, the learner is required to solve a constrained optimization task, characterized by a budget (or ℓ_0 pseudo-norm) constraint on the model $\boldsymbol{w}_t$. Actually, this problem is known to be NP-hard for common loss functions advocated in classification and regression settings [10]. In order to alleviate this difficulty, two main approaches have been proposed in the literature. The first approach is to replace the nonconvex ℓ_0 constraint by a convex ℓ_1 constraint, or an ℓ_1 regularizer [3,4,6,8,11,15]. Though this approach is promoting the sparsity of solutions, it cannot guarantee that, at each iteration, the number of selected features is bounded by the predefined budget B. The second approach is divided in two main steps: first, solve a convex, unconstrained optimization problem, and next, seek a new solution that approximates the unconstrained solution while satisfying the ℓ_0 constraint. Based on this second approach, the OFS [16] and SOFS [20] strategies exploit truncation techniques for maintaining a budgeted number of features. However, OFS is oblivious to the history of predictions made so far, which might prove useful for assessing the frequencies of features. SOFS uses a suboptimal truncation rule that only considers the confidence of feature values in the current model, but ignores the magnitude of feature values which, again, could prove useful for estimating their relevance. Moreover, Wu et al. [20] did not provide any theoretical analysis for SOFS.

In this paper, we investigate the online feature selection problem using novel truncation techniques. Our contributions are threefold:

1. Two online feature selection algorithms, called Budgeted ARDA (B-ARDA) and Budgeted AMD (B-AMD), are proposed. B-ARDA and B-AMD perform truncation to eliminate irrelevant features. In our paper, the relevance of features is assessed by their frequency in the sequence of predictions, and their magnitude in the current predictor.
2. A detailed regret analysis for both algorithms is provided, which captures the intuition and rationale behind our truncation techniques.
3. Experiments on six high-dimensional datasets reveal the superiority of the proposed algorithms compared with both advanced feature selection algorithms and ℓ_1-based online learning algorithms.

The paper is organized as follows. Section 2 provides some related work in feature selection and online learning. Section 3 presents the notation used throughout the paper and elaborates on the problem setting. Our learning algorithms and their regret analysis are detailed in Sect. 4. Comparative experiments are given in Sect. 5. Finally, Sect. 6 concludes the paper.

2   Related Work

Feature selection is a well-studied topic in machine learning and data mining [1,7,23]. Existing feature selection approaches include batch (or offline) methods and online methods. Batch methods, examined for instance in [12–14,18], typically require an access to all available data, which makes them difficult to operate on sequential data. On the other hand, online methods are more suited to handle large-scale, and potentially streaming, information. Currently, there are two different "online modes" for selecting features. The first mode assumes that the number of examples is fixed but features arrive sequentially over time, such as in [17,19,22]. Contrastingly, the second mode assumes that the number of features is known in advance, but examples are supplied one by one, as studied for example in [16,20]. We focus here on the second online mode, which is more natural for real-world streaming data. According to this mode, online feature selection methods can be grouped into three categories, summarized in Table 1.

Table 1. A list of recent works in online feature selection

Sparsity strategy	References/methods
ℓ_1 constraint	[2,5]
ℓ_1 regularization	Fobos [3], TrunGrad [8], ℓ_1-RDA [21], CMD [6], ℓ_1-ARDA [4], Ada-Fobos [4], SOL [15]
ℓ_0 truncation	OFS [16], SOFS [20]

ℓ_1 *Constraint/Regularization.* Methods enforcing ℓ_1 constraints project the solution w after gradient descent update onto an ℓ_1 ball with radius r. Recent works, such as [2,5], focus on designing efficient projection algorithms. There are also many researches which aim at solving an ℓ_1-regularized convex optimization problem. Notably, in [3], Duchi et al. propose the Fobos algorithm, which first performs a sub-gradient descent in order to get an intermediate solution, and then seeks a new solution that stays close to the intermediate solution and has a low ℓ_1 norm complexity. The second stage can be solved efficiently by truncating coefficients below a threshold in the intermediate solution. In [8], Langford et al. claim that such truncation operation is too aggressive and propose an alternative truncated gradient technique (TrunGrad), which gradually shrinks the coefficients to zero by a small amount. In [6], Duchi et al. generalize the Online Mirror Descent (OMD) to regularized losses, and propose the Composite Mirror Descent (CMD) algorithm, which exploits the composite structure of the objective to get desirable effects. Their derived algorithms include Fobos as an special case. In [21], Xiao presents an ℓ_1-Regularized Dual Averaging algorithm (ℓ_1-RDA) which, at each iteration, minimizes the sum of three terms: a linear function obtained by averaging all previous sub-gradients, an ℓ_1 regularization term and an additional strongly convex regularization term. In [4], Duchi et al. propose ARDA and ACMD, which adaptively modify the proximal function in order to incorporate the information related to the geometry of data observed in earlier iterations. The derived algorithms, ℓ_1-ARDA and Ada-Fobos, achieve better performance than their non-adaptive versions, namely, ℓ_1-RDA and Fobos. In [15], Wang et al. present a framework for sparse online classification. Their methods perform feature selection by carefully tuning the ℓ_1 regularization parameter.

ℓ_0 *Truncation.* In contrast with the above approaches, Jin et al. [16] propose a truncation method that satisfies the budget (or ℓ_0) constraint at each iteration. Their OFS algorithm first projects the predictor w (obtained from gradient descent) onto an ℓ_2 ball, so that most of the numerical values of w are concentrated to their largest elements, and then keeps only the B largest weights in w. Wu et al. [20] further explore the truncation method for a confidence-weighted learning algorithm AROW, and proposed SOFS, which simply truncates the elements with least confidence after the update step in the diagonal version of AROW.

Our proposed online feature selection algorithms are also based on truncation techniques. Yet, our approaches differ from OFS and SOFS in the sense that truncation strategies are tailored to advanced adaptive sub-gradient methods, namely ARDA and AMD, which can perform more informative gradient descent, and which can find highly discriminative but rarely seen features. Moreover, we provide a detailed regret analysis for truncated versions of ARDA and AMD.

3   Notation and Problem Setting

In what follows, lowercase letters denote scalars or vectors, and uppercase letters represent matrices. An exception is the parameter B that captures our budget on the number of selected features. Let $[d]$ denote the set $\{1, \cdots, d\}$. We use I to denote the identity matrix, and $\text{diag}(v)$ to denote the diagonal matrix with vector v on the diagonal. For a linear predictor w_t chosen at iteration t, we use $w_{t,i}$ to denote its ith entry. As usual, we use $\langle v, w \rangle$ to denote the inner product between v and w, and for any $p \in [1, \infty]$, we use $||w||_p$ to denote the ℓ_p norm of w. We also use $||w||_0$ to denote the ℓ_0 pseudo-norm of w, that is, $||w||_0 = |\{i \in [d] : w_i \neq 0\}|$. For a convex loss function f_t, the sub-differential set of f_t at w is denoted by $\partial f_t(w)$, and g_t is used to denote a sub-gradient of f_t at w_t, i.e. $g_t \in \partial f_t(w_t)$. When f_t is differentiable at w, we use $\nabla f_t(w)$ to denote its unique sub-gradient (called gradient). Let $g_{1:t} = [g_1 \ g_2 \ \cdots \ g_t]$ be a $d \times t$ matrix obtained by concatenating the sub-gradients g_j from $j = 1$ to t. The ith row vector of $g_{1:t}$ is denoted by $g_{1:t,i}$. Let ψ_t be a strictly convex and continuously differentiable function defined, at each iteration t, on a closed convex set $C \subseteq \mathbb{R}^d$ and let $\mathcal{D}_{\psi_t}(x, y)$ denote the corresponding Bregman divergence, given by:

$$\mathcal{D}_{\psi_t}(x, y) = \psi_t(x) - \psi_t(y) - \langle \nabla \psi_t(y), x - y \rangle, \quad \forall x, y \in C.$$

By construction, we have $\mathcal{D}_{\psi_t}(x, y) \geq 0$ and $\mathcal{D}_{\psi_t}(x, x) = 0$ for all $x, y \in C$.

As mentioned above, the online feature selection problem can be formulated as a repeated prediction game between the learner and its environment. At iteration t, a new data point $\boldsymbol{x}_t \in \mathbb{R}^d$ is supplied to the learner, which is required to predict a label for $\boldsymbol{x}_t$ according to its current model $\boldsymbol{w}_t$. We assume that $\boldsymbol{w}_t$ is a sparse linear function in $\mathbb{R}^d$ such that $||\boldsymbol{w}_t||_0 \leq B$, where B is a predefined budget. Once the learner has committed to its prediction, the true label $y_t \in \mathbb{R}$ of $\boldsymbol{x}_t$ is revealed, and the learner suffers a loss $l(\boldsymbol{w}_t; (\boldsymbol{x}_t, y_t))$. We use here $l_t(\boldsymbol{w}_t) = l(\boldsymbol{w}_t; (\boldsymbol{x}_t, y_t))$, and we assume that $l_t(\boldsymbol{w}_t) = f_t(\boldsymbol{w}_t) + \varphi(\boldsymbol{w}_t)$, where $f_t(\boldsymbol{w}_t)$ is a convex loss function and $\varphi(\boldsymbol{w}_t)$ is a regularization function. The performance of the learner is measured according to its *regret*:

$$\mathcal{R}^T = \sum_{t=1}^{T} l_t(\boldsymbol{w}_t) - \min_{w \in \mathbb{R}^d : ||w||_0 \leq B} \sum_{t=1}^{T} l_t(\boldsymbol{w}),$$

where $||\boldsymbol{w}_t||_0 \leq B$ for all t. Our goal is to devise online feature selection strategies for which, regrets are sublinear in T. The nonconvex ℓ_0 constraint makes our problem more challenging than standard online convex optimization tasks.

4   B-ARDA and B-AMD

Advanced ARDA and AMD algorithms can take full advantage of the sub-gradient information observed in earlier iterations to perform more informative learning. Since ARDA and AMD are different methods, we need to develop specific truncation strategies for each of them.

4.1   B-ARDA and Its Regret Analysis

A straightforward approach for performing ℓ_0 truncations is to keep the B elements with largest magnitude (in absolute value) in the current predictor $\boldsymbol{w}_t$. Such a naive approach suffers from an important shortcoming: frequently occurring discriminative features tend to be removed. This flaw results from the updating rule of adaptive sub-gradient methods: frequent attributes are given *low* learning rates, while infrequent attributes are given *high* learning rates.

Thus, we need to consider a more sophisticated truncation approach which takes into account the frequencies of features, together with their magnitude. To this end, we present the pseudocode of B-ARDA described in Algorithm 1. Basically, B-ARDA starts with a standard ARDA iteration from Step 1 to Step 9, and provides an intermediate solution $\boldsymbol{z}_{t+1}$, for which $||\boldsymbol{z}_{t+1}||_0 \leq B$ may not hold; then at Step 10, the algorithm truncates $\boldsymbol{z}_{t+1}$ in order to find a new solution $\boldsymbol{w}_{t+1}$ so that $||\boldsymbol{w}_{t+1}||_0 \leq B$ is satisfied. In our truncation operation, we consider both the magnitude of elements in $\boldsymbol{z}_{t+1}$, and the frequency of features conveyed by the diagonal matrix $\boldsymbol{H}_t$.

Note that the update at Step 9 often takes a closed-form. For example, if we use the standard Euclidean regularizer $\varphi(\boldsymbol{w}) = \frac{\lambda}{2}||\boldsymbol{w}||_2^2$, we get that

$$\boldsymbol{z}_{t\,|\,1} = -\eta(\lambda\eta t\boldsymbol{I} + \boldsymbol{H}_t)^{-1}\sum_{i=1}^{t} \boldsymbol{g}_i.$$

Algorithm 1. B-ARDA

Input: Data stream $\{(\boldsymbol{x}_t, y_t)\}_{t=1}^{\infty}$, constant $\delta > 0$, step-size $\eta > 0$, budget B
Output: $\boldsymbol{w}_t$

1 $\boldsymbol{w}_1 = \boldsymbol{0}$, $\boldsymbol{g}_{1:0} = []$;
2 **for** $t = 1, 2, \cdots$ **do**
3   Receive $\boldsymbol{x}_t$;
4   Predict the label of $\boldsymbol{x}_t$ with $\boldsymbol{w}_t$;
5   Receive y_t and suffer loss $f_t(\boldsymbol{w}_t)$;
6   Receive sub-gradient $\boldsymbol{g}_t \in \partial f_t(\boldsymbol{w}_t)$;
7   Update $\boldsymbol{g}_{1:t} = [\boldsymbol{g}_{1:t-1}\ \boldsymbol{g}_t]$, $s_{t,i} = ||\boldsymbol{g}_{1:t,i}||_2$;
8   Set $\boldsymbol{H}_t = \delta\boldsymbol{I} + \mathrm{diag}(s_t)$, $\bar{\boldsymbol{g}}_t = \frac{1}{t}\sum_{i=1}^{t}\boldsymbol{g}_i$, $\psi_t(\boldsymbol{w}) = \frac{1}{2}\langle\boldsymbol{w}, \boldsymbol{H}_t\boldsymbol{w}\rangle$;
9   ARDA update:

$$z_{t+1} = \arg\min_{w \in \mathbb{R}^d}\left\{\eta\langle\bar{\boldsymbol{g}}_t, \boldsymbol{w}\rangle + \eta\varphi(\boldsymbol{w}) + \frac{1}{t}\psi_t(\boldsymbol{w})\right\} \tag{1}$$

10   Truncation operation:

$$\boldsymbol{w}_{t+1} = \arg\min_{w \in \mathbb{R}^d}\langle\boldsymbol{w} - \boldsymbol{z}_{t+1}, \boldsymbol{H}_t(\boldsymbol{w} - \boldsymbol{z}_{t+1})\rangle, \text{ subject to } ||\boldsymbol{w}||_0 \le B \tag{2}$$

The truncation operation at Step 10 can be efficiently solved by a simple greedy procedure. Let $\boldsymbol{v}_{t+1} \in \mathbb{R}^d$ be the vector with entries $v_{t+1,j} = H_{t,jj}z_{t+1,j}^2$. Based on this notation, if $||\boldsymbol{z}_{t+1}||_0 \le B$, $\boldsymbol{w}_{t+1} = \boldsymbol{z}_{t+1}$; otherwise, $\boldsymbol{w}_{t+1} = \boldsymbol{z}_{t+1}^B$, where

$$z_{t+1,i}^B = \begin{cases} z_{t+1,i} & \text{if } H_{t,ii}z_{t+1,i}^2 \text{ occurs in the } B \text{ largest values of } \boldsymbol{v}_{t+1}, \\ 0 & \text{otherwise.} \end{cases}$$

The following result demonstrates that our truncation strategy for ARDA can lead to a sublinear regret. The proof, built essentially on the work of [4], is included in Appendix 1 for completeness.

Theorem 1. *Let $\xi_t^2 = \langle\boldsymbol{w}_t - \boldsymbol{z}_t, \boldsymbol{H}_{t-1}(\boldsymbol{w}_t - \boldsymbol{z}_t)\rangle$, which is the factual truncation error at iteration $t-1$. Set $\max_t ||\boldsymbol{g}_t||_\infty \le \delta$ and $\max_t \xi_t \le \xi$. For any $\boldsymbol{w}^* \in \mathbb{R}^d$, B-ARDA achieves the following regret bound:*

$$\mathcal{R}_{B\text{-}ARDA}^T \le \frac{\delta}{2\eta}||\boldsymbol{w}^*||_2^2 + \left(\frac{1}{2\eta}||\boldsymbol{w}^*||_\infty + \eta\right)\sum_{i=1}^d ||\boldsymbol{g}_{1:T,i}||_2 + \xi\sqrt{2T\sum_{i=1}^d ||\boldsymbol{g}_{1:T,i}||_2}.$$

To see why the bound is sublinear, we notice from [4] that

$$\sum_{i=1}^d ||\boldsymbol{g}_{1:T,i}||_2 = \sqrt{d}\sqrt{\inf_{s:s\succeq0,\langle1,s\rangle\le d}\left\{\sum_{t=1}^T\langle\boldsymbol{g}_t, \mathrm{diag}(s)^{-1}\boldsymbol{g}_t\rangle\right\}} \le \sqrt{d}\sqrt{\sum_{t=1}^T ||\boldsymbol{g}_t||_2^2}.$$

For the maximum truncation error $\xi = 0$, we directly recover the regret bound of ARDA. If $\xi \ne 0$, we get bounds of the form:

1. if ξ is $O(||\boldsymbol{w}^*||_\infty\sqrt{\sum_{i=1}^d ||\boldsymbol{g}_{1:T,i}||_2/T})$, $\mathcal{R}_{\text{B-ARDA}}^T = O(||\boldsymbol{w}^*||_\infty \sum_{i=1}^d ||\boldsymbol{g}_{1:T,i}||_2)$.
2. if ξ is $\Omega(||\boldsymbol{w}^*||_\infty\sqrt{\sum_{i=1}^d ||\boldsymbol{g}_{1:T,i}||_2/T})$, $\mathcal{R}_{\text{B-ARDA}}^T = O(\xi\sqrt{2T\sum_{i=1}^d ||\boldsymbol{g}_{1:T,i}||_2})$.

In other words, the cumulative loss of B-ARDA using only B features converges to that of an optimal solution in hindsight as T approaches infinity. The value of ξ is determined by the budget parameter B; larger values of B produce a smaller ξ, while smaller values of B yield a larger ξ.

We mention in passing that the naive truncation method, described in the beginning of this section, may be implemented by replacing the Step 10 in Algorithm 1 with

$$\boldsymbol{w}_{t+1} = \arg\min_{\boldsymbol{w}\in\mathbb{R}^d} \langle \boldsymbol{w} - \boldsymbol{z}_{t+1}, \boldsymbol{w} - \boldsymbol{z}_{t+1}\rangle, \text{ subject to } ||\boldsymbol{w}||_0 \leq B.$$

The regret produced by such truncation is, however, *not* sublinear since:

$$\sum_{t=1}^T \langle \boldsymbol{g}_t, \boldsymbol{w}_t - \boldsymbol{z}_t\rangle \leq \sum_{t=1}^T ||\boldsymbol{g}_t||_2 ||\boldsymbol{w}_t - \boldsymbol{z}_t||_2 \leq \xi \sum_{t=1}^T ||\boldsymbol{g}_t||_2 \quad (\xi = ||\boldsymbol{w}_t - \boldsymbol{z}_t||_2).$$

4.2　B-AMD and Its Regret Analysis

We now focus on a truncation technique for the sub-gradient method AMD. Our approach is also considering both the magnitude of elements and the frequency of features. The pseudocode of B-AMD is presented in Algorithm 2, where $\mathcal{D}_{\psi_t}(\boldsymbol{w}, \boldsymbol{w}_t)$ is the Bregman divergence between $\boldsymbol{w}$ and $\boldsymbol{w}_t$. Note that we use AMD rather than ACMD since we do not use the composite structure of the objective function, but the truncation operation, to produce sparse solutions.

In essence, B-AMD performs an AMD iteration and then truncates the returned solution. Importantly, the AMD update at Step 9 admits a closed-form solution: $\boldsymbol{z}_{t+1} = \boldsymbol{w}_t - \eta\boldsymbol{H}_t^{-1}\boldsymbol{g}_t$. Similarly to B-ARDA, the truncation operation at Step 10 can be solved efficiently: if $||\boldsymbol{z}_{t+1}||_0 \leq B$, $\boldsymbol{w}_{t+1} = \boldsymbol{z}_{t+1}$; otherwise, $\boldsymbol{w}_{t+1} = \boldsymbol{z}_{t+1}^B$ where $\boldsymbol{z}_{t+1,i}^B = \boldsymbol{z}_{t+1,i}$ if $H_{t,ii}|\boldsymbol{z}_{t+1,i}|$ occurs in the B largest values of $\{H_{t,jj}|\boldsymbol{z}_{t+1,j}|, j \in [d]\}$, and $\boldsymbol{z}_{t+1,i}^B = 0$, otherwise.

The next theorem provides a regret bound for B-AMD, and conveys the rationale for the designed truncation. The proof is given in Appendix 2.

Theorem 2. *Set* $\xi_t = \sum_{i=1}^d H_{t,ii}|\boldsymbol{z}_{t+1,i} - \boldsymbol{w}_{t+1,i}|$. *For any* $\boldsymbol{w}^* \in \mathbb{R}^d$, *B-AMD achieves the following regret bound:*

$$\mathcal{R}_{\text{B-AMD}}^T \leq \frac{1}{\eta}||\boldsymbol{w}^*||_\infty \sum_{t=1}^T \xi_t + \left(\frac{1}{2\eta}\max_{t\leq T}||\boldsymbol{w}^* - \boldsymbol{w}_t||_\infty^2 + \eta\right)\sum_{i=1}^d ||\boldsymbol{g}_{1:T,i}||_2,$$

where the first term of right-hand side is obtained from truncation.

Informally, the regret bound in Theorem 2 indicates that the cumulative loss of B-AMD converges toward the cumulative loss of the optimal $\boldsymbol{w}^*$ as T tends toward infinity, and the gap between the two is mainly dominated by the sum of truncation errors, that is, $\sum_{t=1}^T \xi_t$. This observation implies that we should try to minimize ξ_t at each round in order to reduce the gap. If the truncation error is set to $\xi_t = 0$ for any t, the regret bound of AMD is immediately recovered.

Algorithm 2. B-AMD

Input: Data stream $\{(\boldsymbol{x}_t, y_t)\}_{t=1}^{\infty}$, constant δ, step-size η, budget B

Output: $\boldsymbol{w}_t$

1 $\boldsymbol{w}_1 = \boldsymbol{0}$, $\boldsymbol{g}_{1:0} = []$;

2 **for** $t = 1, 2, \cdots$ **do**

3   Receive $\boldsymbol{x}_t$;

4   Predict the label of $\boldsymbol{x}_t$ with $\boldsymbol{w}_t$;

5   Receive y_t and suffer loss $f_t(\boldsymbol{w}_t)$;

6   Receive sub-gradient $\boldsymbol{g}_t \in \partial f_t(\boldsymbol{w}_t) + \partial \varphi(\boldsymbol{w}_t)$;

7   Update $\boldsymbol{g}_{1:t} = [\boldsymbol{g}_{1:t-1} \; \boldsymbol{g}_t]$, $s_{t,i} = \|\boldsymbol{g}_{1:t,i}\|_2$;

8   Set $\boldsymbol{H}_t = \delta \boldsymbol{I} + \text{diag}(\boldsymbol{s}_t)$, $\psi_t(\boldsymbol{w}) = \frac{1}{2}\langle \boldsymbol{w}, \boldsymbol{H}_t \boldsymbol{w}\rangle$;

9   AMD update:

$$z_{t+1} = \arg\min_{\boldsymbol{w} \in \mathbb{R}^d} \{\eta \langle \boldsymbol{g}_t, \boldsymbol{w}\rangle + \mathcal{D}_{\psi_t}(\boldsymbol{w}, \boldsymbol{w}_t)\} \tag{3}$$

10   Truncation operation:

$$\boldsymbol{w}_{t+1} = \arg\min_{\boldsymbol{w} \in \mathbb{R}^d} \sum_{i=1}^{d} H_{t,ii}|w_i - z_{t+1,i}|, \text{ subject to } \|\boldsymbol{w}\|_0 \leq B \tag{4}$$

---

5 Experiments

This section reports two experimental studies[1]. In the first experiment, we compare B-ARDA and B-AMD with OFS and SOFS; in the second one, we compare our algorithms with ℓ_1-ARDA and Ada-Fobos, which achieve feature selection by carefully tuning the ℓ_1 regularization parameter. Although the theoretical analysis of our algorithms holds for many convex losses and regularization functions, we use here the squared hinge loss and ℓ_2 regularizer, that is, $f_t(\boldsymbol{w}_t) = (\max\{0, 1 - y_t\langle \boldsymbol{w}_t, \boldsymbol{x}_t\rangle\})^2$ and $\varphi(\boldsymbol{w}_t) = \frac{\lambda}{2}\|\boldsymbol{w}_t\|_2^2$.

5.1 Datasets

Our experiments were performed on six high-dimensional binary classification datasets, selected from different domains. Their statistics are presented in Table 2, where "data density" is the maximal number of non-zero features per instance divided by the total number of features. Arcene's task is to distinguish cancer versus normal patterns from mass-spectrometric data. Dexter and farm_ads are text classification problems in a bag-of-words representation. Gisette aims to separate the highly confusable digits '4' and '9'. The above four datasets are available in UCI repository. Pcmac and basehock are a subset extracted from 20newsGroup[2]. Pcmac is to separate documents from

---

[1] Our codes are available at https://github.com/LUCKY-ting/online-feature-selection.

[2] http://www.cad.zju.edu.cn/home/dengcai/Data/TextData.html.

Table 2. A summary of datasets

Dataset	# features (d)	# train (n)	# test	density
arcene	10000	100	100	71.25%
dexter	20000	300	300	1.65%
gisette	5000	6000	1000	29.6%
basehock	26214	1197	796	6.48%
pcmac	26214	1168	777	4.5%
farm_ads	54877	3313	830	4.19%

"ibm.pc.hardware" and "mac.hardware", and basehock is to distinguish "baseball" versus "hockey".

5.2   Comparison with Online Feature Selection Algorithms

We first compared B-ARDA and B-AMD with OFS [16] and SOFS [20] on datasets in Table 2. For OFS, B-ARDA and B-AMD algorithms, the regularization parameter λ and the step-size η were obtained by choosing values in $\{10^{-1}, 10^{-1.5}, \cdots, 10^{-8}\}$, and taking the best performance in the training set. A similar interval was used for selecting the best parameter $1/\gamma$ for SOFS. We set $\delta = 10^{-2}$ for B-ARDA and B-AMD on all datasets. Based on these empirically optimal parameter values, we vary the budget B in order to plot the test accuracy versus the number of selected features.

In order to make our results reliable under the optimal parameter setting, each algorithm was run 10 times, each time with τ passes on the training examples. Namely, each pass is done with a random permutation of the training set, and the classifier output at the end of τ passes is evaluated on a separated test set. The number of passes τ was set as $\lceil \frac{2d}{n} \rceil$ for each dataset. Figures 1 and 2 display the average test accuracy of all algorithms for varying feature budgets.

Based on Fig. 1, we can observe that B-ARDA achieves the highest test accuracy for every budget parameter B. By contrast, B-AMD is outperformed by B-ARDA, but remains better than SOFS. By coupling Figs. 1 and 2, we observe that the performance gap between B-ARDA and the other algorithms decreases as the budget B increases. The results for B-AMD are mixed: for small values of B, this strategy is outperformed by OFS, due to a large truncation error; but when the budget is gradually increasing, B-AMD outperforms OFS at some value of B. For example, on the gisette and farm_ads datasets, B-AMD outperforms OFS at $B \geq 1000$ and $B \geq 2000$, respectively. SOFS achieves poor accuracy for small budgets, but its performance is approaching B-ARDA and B-AMD by increasing B. This steams from the fact SOFS tends to keep more features to achieve an accuracy that is competitive with that of B-ARDA and B-AMD. We can clearly see that B-ARDA, B-AMD and SOFS are all outperforming OFS for large values of B. To sum up, when a small number of features is desired,

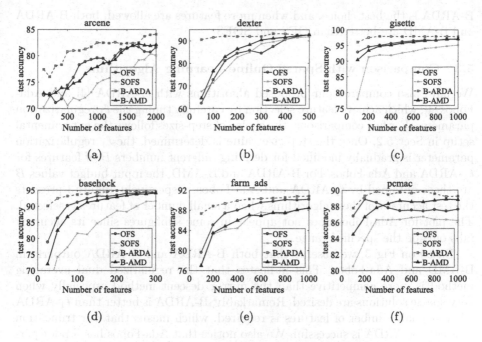

Fig. 1. Test performance w.r.t. OFS and SOFS (small feature budgets)

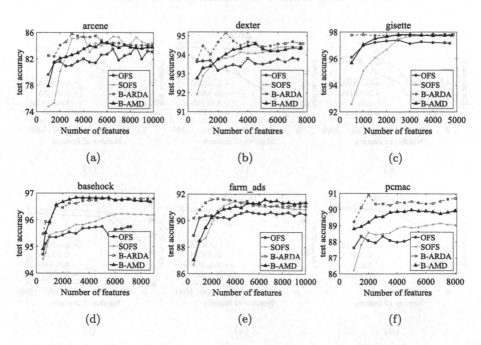

Fig. 2. Test performance w.r.t. OFS and SOFS (large feature budgets)

B-ARDA is the best choice, and when more features are allowed, both B-ARDA and B-AMD are better than OFS and SOFS.

5.3 Comparison with Sparse Online Learning Algorithms

We have also compared our proposed algorithms with ℓ_1-ARDA [4] and Ada-Fobos [4], which achieve feature selection by carefully tuning the ℓ_1 regularization parameter. For fair comparisons, the choice of step-sizes follows the experimental setup in Sect. 5.2. Once the step-size value is determined, the ℓ_1 regularization parameter is gradually modified for deriving different numbers B of features for ℓ_1-ARDA and Ada-Fobos. For B-ARDA and B-AMD, the input budget values B are those obtained by ℓ_1-ARDA and Ada-Fobos, respectively. Figure 3 presents the test accuracy of these algorithms when a small number of features is selected. The plot for Ada-Fobos does not appear in some subfigures since its accuracy falls outside the specified range.

Based on Fig. 3, we observe that both B-ARDA and ℓ_1-ARDA outperform B-AMD and Ada-Fobos. This indicates that the regularized dual averaging method is more competitive than the mirror descent method especially when very sparse solutions are desired. Remarkably, B-ARDA is better than ℓ_1-ARDA when a small number of features is required, which means that our truncation strategy for ARDA is successful. We also notice that Ada-Fobos has a poor performance for small budgets; by contrast, B-AMD is much better. We do not present the plots for large number of features due to space constraints, but

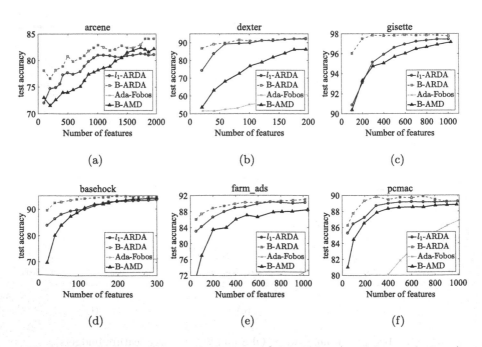

Fig. 3. Test performance w.r.t. ℓ_1-ARDA and Ada-Fobos (small feature budgets)

we report the observed results: as the number of features increases, the performance gaps among these algorithms are gradually shrinking, and finally, these algorithms empirically attain a similar test accuracy. Yet, from a practical viewpoint, it is much simpler to select a desired number of features for B-ARDA and B-AMD. For ℓ_1-ARDA and Ada-Fobos, the number of features cannot be determined in advance: it is empirically conditioned by the choice of the regularization parameter.

6   Conclusion

In this paper, two novel online feature selection algorithms, called B-ARDA and B-AMD, have been proposed and analyzed. Both algorithms perform feature selection via truncation techniques, which take into account the magnitude of feature values in the current predictor, together with the frequency of features in the observed data stream. By taking as input a desired budget, both algorithms are easy to control, especially in comparison with ℓ_1-based feature selection techniques. We have shown on six high-dimensional datasets that B-ARDA outperforms advanced OFS and SOFS especially when a small number of features is required; when more features are allowed, both B-ARDA and B-AMD are better than OFS and SOFS. Compared with ℓ_1-ARDA and Ada-Fobos that achieve feature selection by carefully tuning the ℓ_1 regularization parameter, B-ARDA is shown to superior to ℓ_1-ARDA and B-AMD superior to Ada-Fobos, which corroborates the interest of our truncation strategies. A natural perspective of research is to investigate whether our approach may be extended to "structured" feature selection tasks, such as group structures.

Acknowledgments. The authors would like to acknowledge support for this project from the National Key R&D Program of China (2017YFB0702600, 2017YFB0702601), the National Natural Science Foundation of China (Nos. 61432008, 61503178) and the Natural Science Foundation of Jiangsu Province of China (BK20150587).

Appendix 1: Proof of Theorem 1

Proof. Denote the Mahalanobis norm with respect to a symmetric matrix A by $||\cdot||_A = \sqrt{\langle\cdot, A\cdot\rangle}$. If A is positive definite, $||\cdot||_A$ is a norm.

Recall that a function ψ is μ-*strongly convex* with respect to a norm $||\cdot||$ if the following inequality holds for any w_1 and w_2:

$$\psi(w_1) - \psi(w_2) - \langle\nabla\psi(w_2), w_1 - w_2\rangle \geq \frac{\mu}{2}||w_1 - w_2||^2.$$

Note that since $\psi_t(w) = \frac{1}{2}\langle w, H_t w\rangle$, and H_t is positive definite, $t\varphi(w) + \frac{\psi_t(w)}{\eta}$ is $\frac{1}{\eta}$-strongly convex with respect to the norm $||\cdot||_{H_t}$.

Let ψ_t^* be the conjugate dual of $t\varphi(w) + \frac{\psi_t(w)}{\eta}$, that is,

$$\psi_t^*(v) = \sup_{w\in\mathbb{R}^d}\left\{\langle v, w\rangle - t\varphi(w) - \frac{\psi_t(w)}{\eta}\right\}.$$

Since $\psi_t^*(v)$ is a linear function of v, we have

$$\nabla \psi_t^*(v) = \arg\min_{w \in \mathbb{R}^d} \left\{ -\langle v, w \rangle + t\varphi(w) + \frac{\psi_t(w)}{\eta} \right\}.$$

Owing to the duality of strong convexity and strong smoothness, ψ_t^* is η-smooth with respect to $||\cdot||_{H_t^{-1}}$. From the definition of smoothness, we have for any v_1 and v_2,

$$\psi_t^*(v_1) - \psi_t^*(v_2) - \langle \nabla \psi_t^*(v_2), v_1 - v_2 \rangle \le \frac{\eta}{2} ||v_1 - v_2||_{H_t^{-1}}^2. \tag{5}$$

Let $v_t = \sum_{i=1}^t g_i$. For any $w^* \in \mathbb{R}^d$, we have

$$\sum_{t=1}^T (l_t(w_t) - l_t(w^*)) = \sum_{t=1}^T (f_t(w_t) + \varphi(w_t) - f_t(w^*) - \varphi(w^*))$$

$$= \varphi(w_1) - \varphi(w_{T+1}) + \sum_{t=1}^T (f_t(w_t)$$
$$+ \varphi(w_{t+1}) - f_t(w^*) - \varphi(w^*))$$

$$\le \sum_{t=1}^T (\langle g_t, w_t - w^* \rangle + \varphi(w_{t+1}) - \varphi(w^*))$$

$$\le \sum_{t=1}^T (\langle g_t, w_t \rangle + \varphi(w_{t+1})) + \frac{1}{\eta} \psi_T(w^*)$$

$$+ \sup_{w \in \mathbb{R}^d} \left\{ -\sum_{t=1}^T \langle g_t, w \rangle - T\varphi(w) - \frac{1}{\eta} \psi_T(w) \right\}$$

$$\le \sum_{t=1}^T (\langle g_t, w_t \rangle + \varphi(w_{t+1})) + \frac{1}{\eta} \psi_T(w^*) + \psi_T^*(-v_T).$$

According to the update Eq. (6) of B-ARDA, we have

$$\psi_T^*(-v_T) = -\langle v_T, z_{T+1} \rangle - T\varphi(z_{T+1}) - \frac{1}{\eta} \psi_T(z_{T+1})$$

$$\le_1 -\langle v_T, z_{T+1} \rangle - (T-1)\varphi(z_{T+1}) - \frac{1}{\eta} \psi_{T-1}(z_{T+1}) - \varphi(z_{T+1})$$

$$\le \sup_{w \in \mathbb{R}^d} \left\{ -\langle v_T, w \rangle - (T-1)\varphi(w) - \frac{1}{\eta} \psi_{T-1}(w) \right\} - \varphi(z_{T+1})$$

$$= \psi_{T-1}^*(-v_T) - \varphi(z_{T+1})$$

$$\le_2 \psi_{T-1}^*(-v_{T-1}) - \langle \nabla \psi_{T-1}^*(-v_{T-1}), g_T \rangle + \frac{\eta}{2} ||g_T||_{H_{T-1}^{-1}}^2 - \varphi(z_{T+1})$$

$$= \psi_{T-1}^*(-v_{T-1}) - \langle z_T, g_T \rangle + \frac{\eta}{2} ||g_T||_{H_{T-1}^{-1}}^2 - \varphi(z_{T+1}),$$

where $\leq_1$ follows from the monotonicity of $\psi_t(\boldsymbol{w})$, that is, $\psi_{t+1}(\boldsymbol{w}) \geq \psi_t(\boldsymbol{w})$ and $\leq_2$ follows from (5).

Combining the above inequality and using the fact that $\varphi(\boldsymbol{w}_t) \leq \varphi(\boldsymbol{z}_t)$ for any t, we obtain that

$$\sum_{t=1}^{T}(l_t(\boldsymbol{w}_t) - l_t(\boldsymbol{w}^*)) \leq \frac{1}{\eta}\psi_T(\boldsymbol{w}^*) + \psi_{T-1}^*(-\boldsymbol{v}_{T-1}) + \frac{\eta}{2}\|\boldsymbol{g}_T\|_{H_{T-1}^{-1}}^2$$

$$+ \langle \boldsymbol{g}_T, \boldsymbol{w}_T - \boldsymbol{z}_T \rangle + \sum_{t=1}^{T-1}(\langle \boldsymbol{g}_t, \boldsymbol{w}_t \rangle + \varphi(\boldsymbol{w}_{t+1})).$$

Since $\{\boldsymbol{w} \in \mathbb{R}^d : \|\boldsymbol{w}\|_0 \leq B\}$ is a subset of $\mathbb{R}^d$, the above upper bound holds for $\mathcal{R}_{\text{B-ARDA}}^T$.

By repeating the above process, we get that

$$\mathcal{R}_{\text{B-ARDA}}^T \leq \frac{1}{\eta}\psi_T(\boldsymbol{w}^*) + \psi_0^*(-\boldsymbol{v}_0) + \frac{\eta}{2}\sum_{t=1}^{T}\|\boldsymbol{g}_t\|_{H_{t-1}^{-1}}^2 + \sum_{t=1}^{T}\langle \boldsymbol{g}_t, \boldsymbol{w}_t - \boldsymbol{z}_t \rangle$$

$$\leq_1 \frac{1}{\eta}\psi_T(\boldsymbol{w}^*) + \frac{\eta}{2}\sum_{t=1}^{T}\|\boldsymbol{g}_t\|_{H_{t-1}^{-1}}^2 + \sum_{t=1}^{T}\|\boldsymbol{w}_t - \boldsymbol{z}_t\|_{H_{t-1}}\|\boldsymbol{g}_t\|_{H_{t-1}^{-1}}$$

$$\leq_2 \frac{1}{\eta}\psi_T(\boldsymbol{w}^*) + \frac{\eta}{2}\sum_{t=1}^{T}\|\boldsymbol{g}_t\|_{H_{t-1}^{-1}}^2 + \xi\sum_{t=1}^{T}\|\boldsymbol{g}_t\|_{H_{t-1}^{-1}}$$

$$\leq \frac{1}{\eta}\psi_T(\boldsymbol{w}^*) + \frac{\eta}{2}\sum_{t=1}^{T}\|\boldsymbol{g}_t\|_{H_{t-1}^{-1}}^2 + \xi\sqrt{T\sum_{t=1}^{T}\|\boldsymbol{g}_t\|_{H_{t-1}^{-1}}^2}, \qquad (6)$$

where we used $\psi_0^*(-\boldsymbol{v}_0) = 0$ and the Hölder's inequality (for dual norms) for $\leq_1$. For $\leq_2$, we used $\xi_t = \|\boldsymbol{w}_t - \boldsymbol{z}_t\|_{H_{t-1}}$ and the assumption that $\xi_t \leq \xi$ for $t = 1, 2, \cdots T$. We now give a bound for each term.

$$\psi_T(\boldsymbol{w}^*) = \frac{\delta}{2}\|\boldsymbol{w}^*\|_2^2 + \frac{1}{2}\langle \boldsymbol{w}^*, \text{diag}(\boldsymbol{s}_T)\boldsymbol{w}^* \rangle \leq \frac{\delta}{2}\|\boldsymbol{w}^*\|_2^2 + \frac{1}{2}\|\boldsymbol{w}^*\|_\infty \sum_{i=1}^{d}\|\boldsymbol{g}_{1:T,i}\|_2$$

With the assumption $\max_t \|\boldsymbol{g}_t\|_\infty \leq \delta$, we can use Lemma 4 in [4] and get

$$\sum_{t=1}^{T}\|\boldsymbol{g}_t\|_{H_{t-1}^{-1}}^2 \leq \sum_{t=1}^{T}\langle \boldsymbol{g}_t, \text{diag}(\boldsymbol{s}_t)^{-1}\boldsymbol{g}_t \rangle \leq 2\sum_{i=1}^{d}\|\boldsymbol{g}_{1:T,i}\|_2$$

The main result follows by plugging these local bounds into (6).

Appendix 2: Proof of Theorem 2

Proof. For any $w^* \in \mathbb{R}^d$, we have

$$
\begin{aligned}
\eta(f_t(w_t) &+ \varphi(w_t) - f_t(w^*) - \varphi(w^*)) \\
&\leq \langle \eta g_t, w_t - w^* \rangle = \langle \eta g_t, w_t - z_{t+1} + z_{t+1} - w^* \rangle \\
&= \langle \eta g_t + H_t(z_{t+1} - w_t), z_{t+1} - w^* \rangle + \langle \eta g_t, w_t - z_{t+1} \rangle \\
&\quad + \langle H_t(w_t - z_{t+1}), z_{t+1} - w^* \rangle \\
&\leq_1 \langle \eta g_t, w_t - z_{t+1} \rangle + \langle H_t(w_t - z_{t+1}), z_{t+1} - w^* \rangle \\
&= \eta \langle \sqrt{\eta} g_t, \frac{1}{\sqrt{\eta}}(w_t - z_{t+1}) \rangle + \mathcal{D}_{\psi_t}(w^*, w_t) - \mathcal{D}_{\psi_t}(w^*, z_{t+1}) - \mathcal{D}_{\psi_t}(z_{t+1}, w_t) \\
&\leq_2 \frac{\eta^2}{2} \|g_t\|_{H_t^{-1}}^2 + \frac{1}{2}\|w_t - z_{t+1}\|_{H_t}^2 - \mathcal{D}_{\psi_t}(z_{t+1}, w_t) \\
&\quad + \mathcal{D}_{\psi_t}(w^*, w_t) - \mathcal{D}_{\psi_t}(w^*, z_{t+1}) \\
&= \frac{\eta^2}{2} \|g_t\|_{H_t^{-1}}^2 + \mathcal{D}_{\psi_t}(w^*, w_t) - \mathcal{D}_{\psi_t}(w^*, z_{t+1}) \\
&= \frac{\eta^2}{2} \|g_t\|_{H_t^{-1}}^2 + \mathcal{D}_{\psi_t}(w^*, w_t) - \mathcal{D}_{\psi_t}(w^*, w_{t+1}) + (\mathcal{D}_{\psi_t}(w^*, w_{t+1}) - \mathcal{D}_{\psi_t}(w^*, z_{t+1})) \\
&\leq_3 \frac{\eta^2}{2} \|g_t\|_{H_t^{-1}}^2 + \mathcal{D}_{\psi_t}(w^*, w_t) - \mathcal{D}_{\psi_t}(w^*, w_{t+1}) + \xi_t \|w^*\|_\infty,
\end{aligned}
$$

where $\leq_1$ follows from the KKT optimality condition for (3), i.e. for any $w \in \mathbb{R}^d$,

$$
\langle \eta g_t + H_t(z_{t+1} - w_t), w - z_{t+1} \rangle \geq 0.
$$

In $\leq_2$, Fenchel-Yong inequality is used, and $\leq_3$ follows from

$$
\begin{aligned}
\mathcal{D}_{\psi_t}(w^*, w_{t+1}) &- \mathcal{D}_{\psi_t}(w^*, z_{t+1}) \\
&= \frac{1}{2}\left(\|w_{t+1}\|_{H_t}^2 - \|z_{t+1}\|_{H_t}^2 \right) + \langle w^*, H_t(z_{t+1} - w_{t+1}) \rangle \\
&\leq \langle w^*, H_t(z_{t+1} - w_{t+1}) \rangle \leq \|w^*\|_\infty \sum_{i=1}^{d} H_{t,ii} |z_{t+1,i} - w_{t+1,i}| = \xi_t \|w^*\|_\infty.
\end{aligned}
$$

Summing over $t = 1, 2, \cdots T$, we have that

$$
\begin{aligned}
\mathcal{R}_{\text{B-AMD}}^T &\leq \frac{\eta}{2} \sum_{t=1}^{T} \|g_t\|_{H_t^{-1}}^2 + \frac{1}{\eta}\|w^*\|_\infty \sum_{t=1}^{T} \xi_t + \frac{1}{\eta} \mathcal{D}_{\psi_1}(w^*, w_1) \\
&\quad + \frac{1}{\eta} \sum_{t=1}^{T-1} (\mathcal{D}_{\psi_{t+1}}(w^*, w_{t+1}) - \mathcal{D}_{\psi_t}(w^*, w_{t+1})) \\
&\leq \frac{\eta}{2} \sum_{t=1}^{T} \|g_t\|_{H_t^{-1}}^2 + \frac{1}{\eta}\|w^*\|_\infty \sum_{t=1}^{T} \xi_t + \frac{1}{2\eta} \max_{t \leq T} \|w^* - w_t\|_\infty^2 \sum_{i=1}^{d} \|g_{1:T,i}\|_2 \\
&\leq_1 \eta \sum_{i=1}^{d} \|g_{1:T,i}\|_2 + \frac{1}{\eta}\|w^*\|_\infty \sum_{t=1}^{T} \xi_t + \frac{1}{2\eta} \max_{t \leq T} \|w^* - w_t\|_\infty^2 \sum_{i=1}^{d} \|g_{1:T,i}\|_2,
\end{aligned}
$$

where the last inequality follows from Lemma 4 in [4].

References

1. Brown, G., Pocock, A.C., Zhao, M., Luján, M.: Conditional likelihood maximisation: a unifying framework for information theoretic feature selection. J. Mach. Learn. Res. **13**, 27–66 (2012)
2. Condat, L.: Fast projection onto the simplex and the ℓ_1 ball. Math. Program. **158**(1–2), 575–585 (2016)
3. Duchi, J.C., Singer, Y.: Efficient online and batch learning using forward backward splitting. J. Mach. Learn. Res. **10**, 2899–2934 (2009)
4. Duchi, J.C., Hazan, E., Singer, Y.: Adaptive subgradient methods for online learning and stochastic optimization. J. Mach. Learn. Res. **12**, 2121–2159 (2011)
5. Duchi, J.C., Shalev-Shwartz, S., Singer, Y., Chandra, T.: Efficient projections onto the ℓ_1-ball for learning in high dimensions. In: Proceedings of ICML, pp. 272–279 (2008)
6. Duchi, J.C., Shalev-Shwartz, S., Singer, Y., Tewari, A.: Composite objective mirror descent. In: Proceedings of COLT, pp. 14–26 (2010)
7. Guyon, I., Elisseeff, A.: An introduction to variable and feature selection. J. Mach. Learn. Res. **3**, 1157–1182 (2003)
8. Langford, J., Li, L., Zhang, T.: Sparse online learning via truncated gradient. J. Mach. Learn. Res. **10**, 777–801 (2009)
9. Rao, N.S., Nowak, R.D., Cox, C.R., Rogers, T.T.: Classification with the sparse group lasso. IEEE Trans. Signal Process. **64**(2), 448–463 (2016)
10. Shalev-Shwartz, S., Srebro, N., Zhang, T.: Trading accuracy for sparsity in optimization problems with sparsity constraints. SIAM J. Optim. **20**(6), 2807–2832 (2010)
11. Shalev-Shwartz, S., Tewari, A.: Stochastic methods for ℓ_1-regularized loss minimization. J. Mach. Learn. Res. **12**, 1865–1892 (2011)
12. Song, L., Smola, A.J., Gretton, A., Bedo, J., Borgwardt, K.M.: Feature selection via dependence maximization. J. Mach. Learn. Res. **13**, 1393–1434 (2012)
13. Tan, M., Tsang, I.W., Wang, L.: Towards ultrahigh dimensional feature selection for big data. J. Mach. Learn. Res. **15**(1), 1371–1429 (2014)
14. Tan, M., Wang, L., Tsang, I.W.: Learning sparse SVM for feature selection on very high dimensional datasets. In: Proceedings of ICML, pp. 1047–1054 (2010)
15. Wang, D., Wu, P., Zhao, P., Wu, Y., Miao, C., Hoi, S.C.H.: High-dimensional data stream classification via sparse online learning. In: Proceedings of ICDM, pp. 1007–1012 (2014)
16. Wang, J., Zhao, P., Hoi, S.C., Jin, R.: Online feature selection and its applications. IEEE Trans. Knowl. Data Eng. **26**(3), 698–710 (2014)
17. Wang, J., et al.: Online feature selection with group structure analysis. IEEE Trans. Knowl. Data Eng. **27**(11), 3029–3041 (2015)
18. Woznica, A., Nguyen, P., Kalousis, A.: Model mining for robust feature selection. In: Proceedings of SIGKDD, pp. 913–921 (2012)
19. Wu, X., Yu, K., Ding, W., Wang, H., Zhu, X.: Online feature selection with streaming features. IEEE Trans. Pattern Anal. Mach. Intell. **35**(5), 1178–1192 (2013)
20. Wu, Y., Hoi, S.C.H., Mei, T., Yu, N.: Large-scale online feature selection for ultrahigh dimensional sparse data. ACM Trans. Knowl. Discov. Data **11**(4), 48:1–48:22 (2017)

21. Xiao, L.: Dual averaging methods for regularized stochastic learning and online optimization. J. Mach. Learn. Res. **11**, 2543–2596 (2010)
22. Yu, K., Wu, X., Ding, W., Pei, J.: Scalable and accurate online feature selection for big data. ACM Trans. Knowl. Discov. Data **11**(2), 16:1–16:39 (2016)
23. Yu, L., Liu, H.: Efficient feature selection via analysis of relevance and redundancy. J. Mach. Learn. Res. **5**, 1205–1224 (2004)

Frame-Based Optimal Design

Sebastian Mair[✉], Yannick Rudolph, Vanessa Closius, and Ulf Brefeld

Leuphana University, Lüneburg, Germany
{mair,brefeld}@leuphana.de

Abstract. Optimal experimental design (OED) addresses the problem of selecting an optimal subset of the training data for learning tasks. In this paper, we propose to efficiently compute OED by leveraging the geometry of data: We restrict computations to the set of instances lying on the border of the convex hull of all data points. This set is called the frame. We (i) provide the theoretical basis for our approach and (ii) show how to compute the frame in kernel-induced feature spaces. The latter allows us to sample optimal designs for non-linear hypothesis functions without knowing the explicit feature mapping. We present empirical results showing that the performance of frame-based OED is often on par or better than traditional OED approaches, but its solution can be computed up to twenty times faster.

Keywords: Active learning · Fast approximation · Frame
Optimal experimental design · Regression

1 Introduction

Consider a supervised learning task with n unlabeled data points $\mathcal{X}$. Obtaining labels for all instances is prohibitive, but there is a budget k that allows to label $k \ll n$ points. The goal is to select the best subset of $\mathcal{X}$ of size k such that the learned model is optimal with respect to some optimality measure. This problem is known as Optimal Experimental Design (OED) [13].

In the classical setting of OED, the learning task is a linear regression $\mathbf{y} = \mathbf{Xw} + \varepsilon$ with target vector $\mathbf{y}$, design matrix $\mathbf{X} \in \mathbb{R}^{n \times d}$, parameters $\mathbf{w}$, and i.i.d. Gaussian noise $\varepsilon \sim \mathcal{N}(\mathbf{0}, \sigma^2 \mathbf{I})$. There are many optimality criteria that can be employed for optimal designs. A common choice is to minimize the covariance of the parameter estimation given by

$$
\text{Cov}[\mathbf{w}] = \sigma^2 \left(\sum_{\substack{\mathbf{x} \in S \\ |S|=k}} \mathbf{x}\mathbf{x}^\top \right)^{-1}.
$$

Minimizing the above quantity is equivalent to maximizing the confidence of the learned parameters. However, finding the subset S of size k that actually

© Springer Nature Switzerland AG 2019
M. Berlingerio et al. (Eds.): ECML PKDD 2018, LNAI 11052, pp. 447–463, 2019.
https://doi.org/10.1007/978-3-030-10928-8_27

minimizes the covariance turns into a combinatorial problem that, depending on n and k, is often infeasible. There are two scenarios to cope with the situation. The first one builds upon the assumption that experiments can be repeated many times such that the design matrix contains duplicate rows; hence, this approach requires obtaining multiple outcomes for the same experiment. The second and more relevant scenario does not allow for repeating the experiment.

Surrogates have been suggested to quantify the optimality of a subset. Popular choices exploit the determinant (D-optimality), the spectral-norm (E-optimality), or the trace (A-optimality) of the covariance matrix of the k points [23]. Nevertheless, intrinsically the problem remains combinatorial and very demanding and the only remedy being a pre-selection of promising candidate points to reduce the complexity of the task.

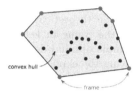

Fig. 1. Illustration of the frame.

In this paper, we exploit the geometry of the data and propose to use the frame as such a candidate set of points. The frame is the smallest subset of the data that realizes the same convex hull as all data points. Thus, the frame consists of the extreme points of the data set. Figure 1 shows an example. We show that restricting the optimization problem to the frame yields competitive results in terms of optimality and predictive performance but comes with a much smaller computational cost. To leverage OED for non-linear problems, we devise a novel approach to compute the frame in kernel-induced feature spaces; this allows us to sample random designs for non-linear regression models without knowing the explicit feature mapping. Our approach of computing the frame can be seen as a transposed LASSO that selects data points instead of features. We discuss the relation to LASSO [27] in greater detail and also address the connection to active learning.

The remainder is structured as follows: Sect. 2 contains the main contribution of frame-based optimal experimental design and Sect. 3 reports on empirical results. Section 4 reviews related work and Sect. 5 concludes.

2   Frame-Based Optimal Design

2.1   Preliminaries

We consider a discrete input set $\mathcal{X} = \{\mathbf{x}_1, \ldots, \mathbf{x}_n\}$ consisting of n data points in d dimensions. The convex hull $\mathrm{conv}(\mathcal{X})$ is the intersection of all convex sets containing $\mathcal{X}$. Furthermore, $\text{conv}(\mathcal{X})$ is the set of all convex combinations of points in $\mathcal{X}$. A central concept of this paper is the frame that is introduced in the following definition.

Definition 1. *Let $\mathcal{X}$ be a discrete input set. The minimal cardinality subset of $\mathcal{X}$, which produces the same convex hull as $\mathcal{X}$, is called the frame $\mathcal{F}$, i.e., $\text{conv}(\mathcal{F}) = \text{conv}(\mathcal{X})$.*

Hence, the frame consists of the extreme points of $\mathcal{X}$. Those points cannot be represented as convex combinations of other points rather than themselves. By $q = |\mathcal{F}|$, we refer to the size of the frame and we call the portion of points in $\mathcal{X}$ belonging to the frame $\mathcal{F}$ the *frame density* q/n.

2.2   Optimal Experimental Design

In the classical setting of OED, the task is a linear regression

$$y = Xw + \varepsilon, \tag{1}$$

where $\mathbf{y}$ is the vector of targets $y_i \in \mathbb{R}$, $\mathbf{X} \in \mathbb{R}^{n \times d}$ is the pool of n experiments $\mathbf{x}_i \in \mathbb{R}^d$, $\mathbf{w} \in \mathbb{R}^d$ are the model parameters and $\varepsilon \sim \mathcal{N}(0, \sigma^2 I)$ is a vector of i.i.d. Gaussian noise. The maximum likelihood estimate of the parameters $\mathbf{w}$ has a closed-form and is given by

$$\mathbf{w}^* = \underset{\mathbf{w}}{\text{argmin}} \sum_{i=1}^{n} (\mathbf{w}^\top \mathbf{x}_i - y_i)^2 = (\mathbf{X}^\top \mathbf{X})^{-1} \mathbf{X}^\top \mathbf{y}.$$

The goal of OED is to choose a subset S of size k out of the n points for which the estimation of $\mathbf{w}$ is optimal in some sense. As common, we require $d \le k \ll n$. Optimality can be measured in several ways. One idea is to increase the confidence of learning the parameters by minimizing the covariance of the parameter estimation For the regression problem stated above, the covariance matrix is given by

$$\text{Cov}_S[\mathbf{w}] = \sigma^2 \left(\sum_{\mathbf{x} \in S} \mathbf{x}\mathbf{x}^\top \right)^{-1},$$

where $S \subset \mathcal{X}$ is the selected subset with $|S| = k$. This leads to a combinatorial optimization problem as follows

$$\min_{\lambda} f\left(\sum_{i=1}^{n} \lambda_i \mathbf{x}_i \mathbf{x}_i^\top \right) \text{ s.t. } \sum_{i=1}^{n} \lambda_i \le k \text{ and } \lambda_i \in \{0,1\} \ \forall i. \tag{2}$$

Here, $f : \mathbb{S}_d^+ \to \mathbb{R}$ is an optimality criterion that assigns a real number to every feasible experiment (positive semi-definite matrices). The setting can be seen as maximizing the information we obtain from executing the experiment with fixed effort. The most popular choices for f are D-, E-, and A-optimality [23] given by

$$f_D(\Sigma) = (\det(\Sigma))^{-1/d} \qquad \text{(D-optimality)}$$
$$f_E(\Sigma) = \|\Sigma^{-1}\|_2 \qquad \text{(E-optimality)}$$
$$f_A(\Sigma) = d^{-1} \operatorname{tr}(\Sigma^{-1}) \qquad \text{(A-optimality)}$$

Unfortunately, the combinatorial optimization problem in Eq. (2) cannot be solved efficiently. A remedy is to use a continuous relaxation, which is efficiently solvable:

$$\min_{\lambda} f\left(\sum_{i=1}^{n} \lambda_i \mathbf{x}_i \mathbf{x}_i^{\top}\right) \text{ s.t. } \sum_{i=1}^{n} \lambda_i \le k \text{ and } \lambda_i \in [0, 1] \; \forall i. \qquad (3)$$

The following lemma characterizes the solution of the optimization problem above.

Lemma 2. *Let $\boldsymbol{\lambda}^{\star}$ be the optimal solution of Problem (3). Then $\|\boldsymbol{\lambda}^{\star}\|_1 = k$.*

However, the support of $\boldsymbol{\lambda}^{\star}$ is usually much larger than k and the solution needs to be sparsified in order to end up with k experiments. Approaches therefor include pipage rounding schemes [1], sampling [28], regret minimization [2], and greedy removal strategies [19,28].

2.3   Restricting Optimal Experimental Design to the Frame

D-optimal design minimizes the determinant of the error covariance matrix. Its dual problem is known as Minimum Volume Confidence Ellipsoid [10]. Geometrically, the optimal solution is an ellipsoid that encloses the data with minimum volume. For E-optimality, the dual problem can be interpreted as minimizing the diameter of the confidence ellipsoid [5]. In A-optimal design the goal is to find the subset of points that optimizes the total variance of parameter estimation. Figure 2 depicts the support of the optimal solution $\boldsymbol{\lambda}^{\star}$ for D-, E-, and A-optimal designs as well as their confidence ellipsoids derived from their dual problems. The right hand figure shows the frame of the same data. The confidence ellipsoids clearly touch the points at the border of the data while the interior points are enclosed. Hence, we propose to discard all interior points entirely in the optimization and restrict the optimization to the frame, that is, to the points lying on the border of the convex hull.

Non-linear regression can be done by applying a feature mapping $\phi : \mathcal{X} \to \mathcal{X}'$ to the data. The model then becomes $y_i = \mathbf{w}^{\top}\phi(\mathbf{x}_i)$, which is still linear in parameters. Considering the dual of the regression problem we can employ kernels that implicitly do a feature mapping. However, the regression is still a linear model, but in feature space $\mathcal{X}'$. Knowing the frame in $\mathcal{X}'$ would allow us to sample random designs rendering a naive version of non-linear or kernelized OED possible. In Subsect. 2.5, we show how to compute the frame in kernel-induced feature spaces.

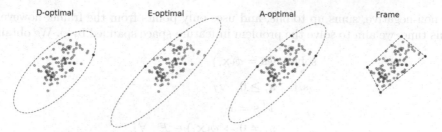

Fig. 2. Example of D-, E-, A-optimal designs and the frame on toy data.

2.4 Computing the Frame

As outlined in Definition 1, the frame $\mathcal{F}$ is the minimum cardinality subset of the data which yields the same convex hull as the data $\mathcal{X}$. Hence, it is given as a solution of the following problem:

$$
\begin{aligned}
\mathcal{F} = \operatorname*{argmin}_{\{\mathbf{z}_1,\ldots,\mathbf{z}_m\} \subseteq \mathcal{X}} \quad & |\{\mathbf{z}_1,\ldots,\mathbf{z}_m\}| \\
\text{s.t.} \quad & \forall \mathbf{x} \in \mathcal{X} : \mathbf{x} = \sum_j \mathbf{s}_j \mathbf{z}_j \text{ with } \mathbf{s}^\top \mathbf{1} = 1 \text{ and } \mathbf{s}_j \geq 0.
\end{aligned}
\tag{4}
$$

We briefly review prior work [18] employing quadratic programming to compute a representation of every data point using only points from the frame. The representation of $\mathbf{x}_i$ is given by a convex combination of frame points as

$$
\begin{aligned}
\operatorname*{solve}_{\mathbf{s}} \quad & \mathbf{X}^\top \mathbf{s} = \mathbf{x}_i \\
\text{s.t.} \quad & \mathbf{s}_j \geq 0 \quad \forall j \\
& \mathbf{s}^\top \mathbf{1} = 1 \\
& \mathbf{s}_j \neq 0 \Rightarrow \mathbf{x}_j \in \mathcal{F} \quad \forall j.
\end{aligned}
\tag{5}
$$

Equation (5) can be rewritten as a non-negative least-squares problem with an additional condition that only points on the frame are allowed to contribute to the solution $\mathbf{s}$. Mair et al. [18] also show that the NNLS algorithm of Lawson and Hanson [15] solves the resulting optimization problem. After computing the representation $\mathbf{s}$ for every point, the frame is recovered by unifying the support of every $\mathbf{s}$. This yields the full frame since every frame point can only be represented by itself.

2.5 Computing the Frame in Kernel-Induced Feature Spaces

Let $\phi : \mathbb{R}^d \to \mathbb{R}^D$ be a feature mapping, $\Phi \in \mathbb{R}^{D \times n}$ be the mapped design matrix, and $\mathbf{K}$ be the kernel matrix induced by a kernel $k(\mathbf{x}, \mathbf{z}) = \phi(\mathbf{x})^\top \phi(\mathbf{z})$. As before, the idea is to solve a linear system subject to the constraints that the solution $\mathbf{s}$

is non-negative, sums up to one, and uses only points from the frame; however, this time, we aim to solve the problem in feature space spanned by ϕ. We obtain

$$\underset{\mathbf{s}}{\text{solve}} \quad \varPhi\mathbf{s} = \phi(\mathbf{x}_i)$$

$$\text{s.t.} \quad \mathbf{s}_j \geq 0 \quad \forall j$$

$$\mathbf{1}^\top\mathbf{s} = 1$$

$$\mathbf{s}_j \neq 0 \Rightarrow \phi(\mathbf{x}_j) \in \mathcal{F} \quad \forall j.$$

The constraint $\mathbf{1}^\top\mathbf{s} = 1$ can be incorporated into the system of linear equations by augmenting $\varPhi$ with a row of ones and $\phi(\mathbf{x})$ with a static 1. Let $\psi(\mathbf{x}) = (\phi(\mathbf{x})^\top, 1)^\top$ and $\varPsi = (\psi(\mathbf{x}_1), \ldots, \psi(\mathbf{x}_n)) \in \mathbb{R}^{(D+1)\times n}$, we obtain

$$\text{solve } \varPsi\mathbf{s} = \psi(\mathbf{x}_i) \text{ s.t. } \mathbf{s}_j \geq 0 \wedge \mathbf{s}_j \neq 0 \Rightarrow \phi(\mathbf{x}_j) \in \mathcal{F}.$$

The approach can be kernelized by multiplying from the left with $\varPsi^\top$:

$$\varPsi^\top\varPsi\mathbf{s} = \varPsi^\top\psi(\mathbf{x}_i) \text{ s.t. } \mathbf{s}_j \geq 0 \wedge \mathbf{s}_j \neq 0 \Rightarrow \phi(\mathbf{x}_j) \in \mathcal{F}.$$

Since there is always a solution [18], we can equivalently solve the non-negative least squares problem

$$\underset{\mathbf{s}\geq 0}{\text{argmin}} \quad \frac{1}{2}\|\varPsi^\top\varPsi\mathbf{s} - \varPsi^\top\psi(\mathbf{x}_i)\|_2^2 \tag{6}$$

$$\text{s.t.} \quad \mathbf{s}_j \neq 0 \Rightarrow \phi(\mathbf{x}_j) \in \mathcal{F}.$$

A kernel can now be applied by exploiting the relationship between $\varPsi$, $\varPhi$, and $\mathbf{K}$ as follows

$$\varPsi^\top\varPsi = \varPhi^\top\varPhi + \mathbb{1}_{nn} = \mathbf{K} + \mathbb{1}_{nn} =: \mathbf{L} \tag{7}$$

$$\varPsi^\top\psi(\mathbf{x}_i) = \varPhi^\top\phi(\mathbf{x}_i) + \mathbb{1}_{n1} = \mathbf{K}_{\cdot i} + \mathbb{1}_{n1} = \mathbf{L}_{\cdot i}, \tag{8}$$

where $\mathbb{1}_{nm} \in \mathbb{R}^{n\times m}$ denotes the matrix of ones. The resulting problem becomes

$$\underset{\mathbf{s}\geq 0}{\text{argmin}} \quad \frac{1}{2}\|\mathbf{L}\mathbf{s} - \mathbf{L}_{\cdot i}\|_2^2 \tag{9}$$

$$\text{s.t.} \quad \mathbf{s}_i \neq 0 \Rightarrow \phi(\mathbf{x}_i) \in \mathcal{F}.$$

A standard non-negative least squares problem can be solved, for example, by the algorithm of Lawson and Hanson [15]. Bro and De Jong [6] increase the efficiency by caching the quantities in Eq. (7). This renders the problem in Eq. (9) feasible. To demonstrate this, we first show that whenever an inner product between two points is maximized, one of the points is an extreme point and thus belongs to the frame of the data.

Lemma 3. *Let $\mathcal{X}$ be a finite set of discrete points, then*

$$\forall \mathbf{x} \in \mathcal{X}: \quad \underset{\mathbf{x}'\in\mathcal{X}}{\text{argmax}} \langle \mathbf{x}, \mathbf{x}'\rangle \in \mathcal{F}.$$

Algorithm 1. Kernel-Frame

Data: kernel matrix $\mathbf{K}$
Result: indices of ext. points $\mathcal{E}$
$\mathbf{L} = \mathbf{K} + \mathbb{1}_{nn}$
$\mathcal{E} = \emptyset$
for $i = 1, 2, \ldots, n$ **do**
$\quad$ $\mathbf{s}_i = \text{bro-dejong}(\mathbf{L}, \mathbf{L}[:, i])$
$\quad$ $\mathcal{P}_i = \{\, j \in \{1, 2, \ldots, n\} \mid (\mathbf{s}_i)_j > 0 \,\}$
$\quad$ $\mathcal{E} = \mathcal{E} \cup \mathcal{P}_i$

Proof. Linearity and convexity of the inner product imply that its maximum is realized by an extreme point of the domain. Since the domain is $\mathcal{X}$, the maximum belongs to its frame $\mathcal{F}$. $\qquad\square$

Theorem 4. *The active-set method from Bro and De Jong [6] solves the problem in Eq. (9).*

Proof. The algorithm selects points that contribute to the solution $\mathbf{s}$ by maximizing the negative gradient of the objective. The selection is implemented by the criterion $j = \text{argmax}_j\ [\mathbf{L}_{\cdot i} - \mathbf{L}\mathbf{s}]_j$, where j is the index of the selected point. Thus, the selection process is maximizing a linear function and Lemma 3 assures that this point belongs to the frame. $\qquad\square$

Solving problem (6) for the i-th data point $\mathbf{x}_i$ yields either its index i in case $\mathbf{x}_i$ is a point on the frame or, if $\mathbf{x}_i$ is an interior point, the solution is the index set $\mathcal{P}_i$ of points on the frame that recover $\mathbf{x}_i$ as a convex combination. The entire frame is recovered by solving Eq. (6) for all points in $\mathcal{X}$ as stated in Corollary 5 and depicted in Algorithm 1.

Corollary 5. *Let $k(\mathbf{x}, \mathbf{z}) = \phi(\mathbf{x})^\top \phi(\mathbf{z})$ be a kernel and $\mathcal{X} = \{\mathbf{x}_1, \ldots, \mathbf{x}_n\}$ be a data set. Then Algorithm 1 yields the frame of $\{\phi(\mathbf{x}_1), \ldots, \phi(\mathbf{x}_n)\}$.*

Proof. Algorithm 1 computes the solution $\mathbf{s}_i$ for every mapped data point $\phi(\mathbf{x}_i)$. Theorem 4 ensures that the positive positions of every $\mathbf{s}_i$ $(i = 1, 2, \ldots, n)$ refers to points on the frame. Hence, taking the union of those positions recovers the frame indices $\mathcal{E}$. $\qquad\square$

Note first that the frame in kernel-induced feature space can be found without knowing the explicit feature map ϕ and second that the *for*-loop in Algorithm 1 can be trivially parallelized.

2.6   Frame Densities for Common Kernels

To analyze the frame sizes in kernel-induced feature spaces, we focus on rbf and polynomial kernels. The former is given by $k(\mathbf{x}, \mathbf{z}) = \exp(-\gamma \|\mathbf{x} - \mathbf{z}\|_2^2)$, where $\gamma > 0$ is a scaling parameter. The induced feature mapping ϕ of the rbf kernel has an infinite dimensionality. Corollary 6 shows, that this kernel always yields a full frame, that is: every point belongs to the frame and the frame density is consequently equal to one.

Corollary 6. *Let X be the data set of distinct points and k be the rbf kernel with parameter $\gamma \neq 0$. Then every point belongs to the frame $\mathcal{F}$ in feature space.*

Proof. Gaussian gram matrices have full rank [24]. Hence, the images $\phi(\mathbf{x}_1), \ldots,$ $\phi(\mathbf{x}_n)$ in feature space are linearly independent. Thus, every image can only be represented by itself and, every point belongs to the frame $\mathcal{F}$. $\square$

The polynomial kernel is given by $k(\mathbf{x}, \mathbf{z}) = (\mathbf{x}^\top \mathbf{z} + c)^p$ with degree $p \in \mathbb{N}$ and constant $c \in \mathbb{R}_0^+$. A feature of the polynomial kernel is an explicit representation of the implicit feature mapping ϕ. E.g., for the homogeneous polynomial kernel with $c = 0$, we have

$$\phi_{\mathbf{m}}(\mathbf{x}) = \sqrt{\frac{p!}{\prod_{i=1}^n m_i!}} \prod_{i=1}^n x_i^{m_i}$$

for all multi-indices $\mathbf{m} = (m_1, \ldots, m_n) \in \mathbb{N}^n$ satisfying $\sum_{i=1}^n m_i = p$. That is, new features consist of monomials of the input features x_i, while the multi-indices $\mathbf{m}$ denote their respective degrees. The condition $\sum_{i=1}^n m_i = p$ assures that all possible combinations are uniquely accounted for and leads to a feature space dimension of size

$$D = \binom{p+d-1}{p} = \frac{(p+d-1)!}{p!(d-1)!}.$$

For the explicit mapping corresponding to the heterogeneous kernel (where $c \neq 0$) that realizes a feature space with dimensionality

$$D = \binom{p+d}{p},$$

as well as for more details, we refer to [24,25]. For the polynomial kernel we obtain a full frame if the dimension D of the feature space exceeds the number of data points n.

Corollary 7. *Let X be the normalized and distinct data set of size n in d dimensions and k be the polynomial kernel with degree p and offset $c = 0$. If $n \leq \frac{(p+d-1)!}{p!(d-1)!}$, then every point belongs to the frame $\mathcal{F}$ in feature space.*

Proof. The polynomial feature map yields linearly independent feature vectors of size $\frac{(p+d-1)!}{p!(d-1)!}$ for a data set with unique observations. Hence, if the number of data points is lower than the dimensionality of the mapping, all points belong to the frame $\mathcal{F}$. $\square$

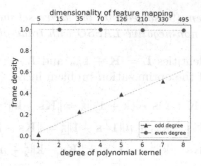

Fig. 3. Frame density for various polynomial degrees on synthetic data of size $n = 2500$ in $d = 5$ dimensions. The initial frame density is 1%. The data set is introduced in Sect. 3.

Although a formal proof regarding the influence of the degree p of the homogeneous polynomial kernel is missing, we would like to provide some intuition: We empirically apply a homogeneous polynomial kernel to a synthetic data set with $n = 2500$ points in $d = 5$ dimensions with an initial frame density of 1%. Figure 3 shows the resulting frame densities. For odd degrees, the frame density is growing with increasing values of p. This is due to the increasing dimensionality in feature space. However, for even degrees the frame is always full. We conclude with the following conjecture:

Conjecture 8. Let $\mathcal{X}$ be the normalized and distinct data set of size n in d dimensions and k be the polynomial kernel with degree p and offset $c = 0$. If p is even, then every point belongs to the frame $\mathcal{F}$ in feature space.

2.7 Computing the Frame and LASSO

LASSO [27] solves regression tasks by combining a squared loss with an ℓ_1-regularizer on the parameters. Thus, LASSO simultaneously performs a regression and variable selection such that the influence of redundant variables is set to zero and a sparse parameter vector is obtained. The corresponding optimization problem for a regression scenario as in Eq. (1) is given by

$$\min_{\mathbf{w}} \|\mathbf{Xw} - \mathbf{y}\|_2^2 + \lambda \|\mathbf{w}\|_1,$$

where $\lambda \geq 0$ is a trade-off parameter. A special case is obtained by restricting the parameters to be positive, yielding a *non-negative LASSO*:

$$\min_{\mathbf{w} \geq 0} \|\mathbf{Xw} - \mathbf{y}\|_2^2 + \lambda \|\mathbf{w}\|_1 \quad \Longleftrightarrow \quad \min_{\mathbf{w} \geq 0} \|\mathbf{Xw} - \mathbf{y}\|_2^2 + \lambda \mathbf{1}^{\top} \mathbf{w}.$$

Computing the frame can be seen as a transposed version of the LASSO problem in which not variables but data points are selected. The following proposition shows that the problem in Eq. (9) is equivalent to a non-negative LASSO, if one ignores the constraint that elicits only frame points to contribute to the solution.

Proposition 9. *Problem (9) solved with the active-set method from Bro and De Jong is equivalent to a non-negative LASSO with trade-off parameter* $\lambda = n$.

Proof. By using the identities $\mathbf{L} = \mathbf{K} + \mathbb{1}_{nn}$ and $\mathbf{L}_{\cdot i} = \mathbf{K}_{\cdot i} + 1 = \mathbf{k} + 1$, we rewrite the objective of the optimization problem in Eq. (9) as follows:

$$\|\mathbf{L}\mathbf{s} - \mathbf{L}_{\cdot i}\|_2^2 = \|(\mathbf{K} + \mathbb{1})\mathbf{s} - (\mathbf{k} + 1)\|_2^2 = \|\mathbf{K}\mathbf{s} - \mathbf{k}\|_2^2 + \|\mathbb{1}\mathbf{s} - 1\|_2^2$$
$$= \|\mathbf{K}\mathbf{s} - \mathbf{k}\|_2^2 + n\|1^\top \mathbf{s} - 1\|_2^2 = \|\mathbf{K}\mathbf{s} - \mathbf{k}\|_2^2 + n1^\top \mathbf{s} - n$$
$$\equiv \|\mathbf{K}\mathbf{s} - \mathbf{k}\|_2^2 + n1^\top \mathbf{s} = \|\mathbf{K}\mathbf{s} - \mathbf{k}\|_2^2 + n\|\mathbf{s}\|_1.$$

Hence, the objective is an ℓ_1-regularized least-squares problem. In combination with the non-negativity constraint, we obtain a non-negative LASSO.   □

3   Experiments

In this section, we empirically investigate frame-based optimal experimental design. Throughout this section, we compare the performance of the following different approaches. *Uniform-data* samples the subset S uniformly at random without replacement from all data points $\mathcal{X}$. A second approach *uniform-frame* uses the same strategy but samples points from the frame $\mathcal{F}$ instead of $\mathcal{X}$. If the size of $|S|$ exceeds the size of the frame, *uniform-frame* always draws the full frame and randomly selects the remaining points from $\mathcal{X} \setminus \mathcal{F}$. The *greedy* baseline chooses the points in S one after another according to their contribution to the objective of D-optimal design. The baselines $\{D,E,A\}$-*optimal* use the continuous relaxations of the $\{D,E,A\}$-optimal design criteria, respectively. After solving the optimization problem, we sample the subset S according to a strategy outlined by Wang et al. [28]. Analogously, $\{D,E,A\}$-*optimal-frame* restricts the computation of the previous three baselines to the frame. Finally, the *Fedorov* baseline selects S according to the Fedorov Exchange algorithm [13] and optimizes D-optimality. We initialize this baseline using random samples from $\mathcal{X}$, random samples from $\mathcal{F}$, and with the output of *greedy*.

The continuous relaxations are optimized using sequential quadratic programming [20]; the number of iterations is limited to 250. We report on average performances over 100 repetitions, error bars indicate one standard deviation and a vertical line denotes the frame size when included. The greedy algorithm is executed once and we conduct only 10 repetitions for every Fedorov Exchange initialization due to its extensive runtime. We want to shed light on the following list of questions.

Is the restriction to the frame competitive in terms of performance? The first experiment thus studies the performance of optimal designs of the proposed approaches on the real-world data set Concrete [29]. Concrete consists of a design pool of $n = 1030$ instances with $d = 8$ dimensions and has a frame density of 48%. The task is to predict the compressive strength of different types of concrete.

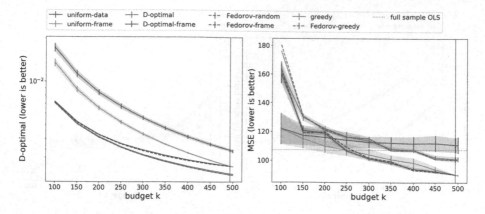

Fig. 4. Results for D-optimal designs on concrete.

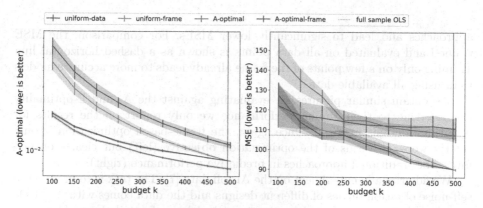

Fig. 5. Results for A-optimal designs on concrete.

We measure the performance in terms of the D-optimality criterion as well as the mean squared error (MSE), given by $\text{MSE} = \frac{1}{n}\|\mathbf{y} - \mathbf{X}\mathbf{w}\|_2^2$. For the latter, we train an ordinary least squares regression on the selected points and evaluate on the remaining $n - k$ points.

Figure 4 (left) shows the results with respect to the D-optimality criterion. Sampling uniformly from the frame (*uniform-frame*) performs consistently better than sampling from all data (*uniform-data*). Thus, exploiting the frame allows to sample better designs without solving any optimization problem other than computing the frame. The situation changes once the designs are optimized in addition. Frame-based approaches (**-optimal-frame*) are close to their competitors computed on all data (**-optimal*) but no longer better. Interior points thus do contribute, if only marginally, to the optimization.

However, Fig. 4 (right) shows that the slight improvement in the objective function does not carry over for the predictive performance. By contrast, the frame-based approaches (**-optimal-frame*) consistently outperform the other

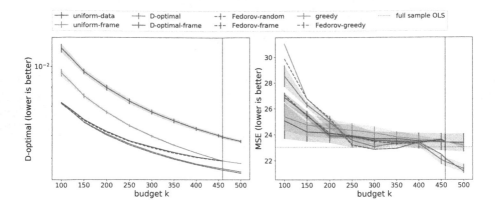

Fig. 6. Results for D-optimal designs on airfoil.

approaches and lead to significantly lower MSEs. For comparison, the MSE trained and evaluated on all data points is shown as a dashed horizontal line. Training only on a few points of the frame already leads to more accurate models than using all available data.

We obtain similar pictures for evaluating against the A- and E optimality criteria. Due to their similar performance we only report on the results for A-optimal designs in Fig. 5. Once again, the frame-based optimization is only sightly worse in terms of the optimization objective (left) but clearly outperforms the traditional approaches in predictive performance (right).

We additionally experiment on the Airfoil data [7]. The task is to predict the self-noise of airfoil blades of different designs and the data comes with $n = 1503$ experiments describing tests in a wind tunnel with $d = 5$ attributes and the data has a frame density of 31%.

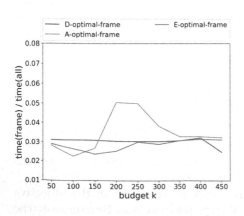

Fig. 7. Timing results on airfoil.

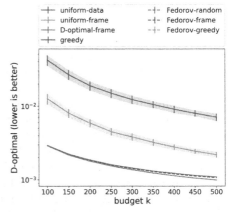

Fig. 8. Results on California Housing.

The results for D-optimal designs are shown in Fig. 6. Once again, the frame-based approaches perform slightly worse or on par in terms of the optimality criterion. However, the predictive performance measured in MSE is no longer superior. The errors are similar to those using uniform samples of the data. Thus, the dataset shows that even though the optimality criterion is well approximated, an error reduction is not guaranteed. However, this does not pose as a limitation to our approach as D-optimal design does not guarantee a reduction either.

Is the restriction to the frame efficient? We now report on the efficiency of our approach on Airfoil. Figure 7 illustrates the relative time of the frame-based approaches in comparison to their traditional analogues that are computed on all data. The y-axis thus shows *time(frame)/time(all)*. We can report a drastically faster computation taking only 2–5% of the time of the traditional variants. We credit this finding to Airfoil's frame density of 31%. That is, restricting the data to the frame already discards 69% of the data and the resulting optimization problems are much smaller.

Naturally, the smaller the frame size the faster the computation as we leave out more and more interior points. We thus experiment on the California Housing data [22] where the task is to estimate the median housing prices for different census blocks. This data comes with $n = 20,640$ instances in $d = 8$ dimensions but possesses a frame density of only 8%.

Figure 8 depicts the result with respect to the D-optimal criterion. The figure again shows that naively sampling from the frame (*uniform-frame*) is significantly better than a drawing random samples from all data (*uniform-data*). All other tested algorithms perform even better and realize almost identical curves. The D-optimal baseline could not be computed in a reasonable amount of time due to the size of the data. Only restricting the computations to the frame rendered the computation feasible.

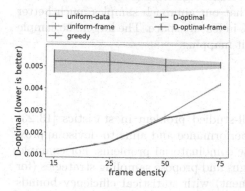

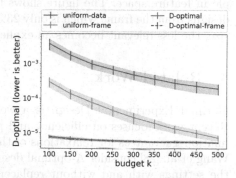

Fig. 9. Effect of the frame size on synthetic data.

Fig. 10. Results on synthetic data using a polynomial kernel of degree $p = 3$.

What is the impact of the frame density? We already mentioned that the frame density q/n influences the efficiency of frame-based approaches. A frame density of z implies that the $(1 - z)$-th part of the data are interior points and can thus be ignored in subsequent computations.

To show this influence empirically, we control the frame density on synthetic data from López [17]. The data we use consists of $n = 2,500$ instances in $d = 5$ dimensions and comes in five different sets realizing frame densities of 1%, 15%, 25%, 50% and 75% respectively. Figure 9 shows the resulting D-optimality criteria for the different frame densities. Up to a frame density of 50%, randomly sampling from the frame (*uniform-frame*) performs on par with all other approaches, thus showing the efficiency of our proposal. For higher frame densities the performance of *uniform-frame* diverges towards *uniform-data*. Nevertheless, restricting D-OED to the frame stays on par with its peers. This experiment suggests, that the smaller the frame density the better the competitiveness of frame-based OED.

Does sampling in kernel-induced feature spaces work? In our last set of experiments, we consider sampling random designs on synthetic data for non-linear regression problems. We use synthetic data as described above with a frame density of 1%. We employ a homogeneous ($c = 0$) polynomial kernel with a degree of $p = 3$ that allows for obtaining the explicit feature mapping ϕ which is needed for all approaches except *uniform-**.

Figure 10 illustrates the results. Approaches optimizing the D-optimal design criterion (D-*) perform equally well, irrespectively of whether they sample from the frame or not. This result confirms the competitiveness of restricting OED to the frame. However, both approaches rely on the explicit feature map.

Strategies that are purely based on sampling (*uniform-**) do not need an explicit mapping. Sampling at random from all data (*uniform-data*) trivially does not rely on anything but a list of indexes. Finally, sampling from the frame (*uniform-frame*) uses the proposed kernel frame algorithm (Algorithm 1) to sample in feature space. The figure shows that our approach samples much better designs from the frame which is only 23% in feature space. The larger the sample size, the less relevant becomes an explicit mapping.

4   Related Work

Optimal Experimental Design is a well-studied problem in statistics [13,23]. Recent work focuses on efficiency and performance and aims to devise approximation guarantees for relaxations of the combinatorial problems. For example, Wang et al. [28] consider A-optimal designs and propose sampling strategies (for the settings with and without replacement) with statistical efficiency bounds as well as a greedy removal approach. Allen-Zhu et al. [2] propose a regret-minimization strategy for the setting without replacement which works for most optimality criteria. Mariet and Sra [19] use elementary symmetric polynomials (ESP) for OED and introduce ESP-design, a interpolation between A- and D-optimal design that includes both as special cases. They provide approximation guarantees for sampling and greedy removal strategies.

OED has close ties to many other problems. D-optimality, for example, is related to volume sampling [3,9,16] and determinantal point processes [14]; both are used in many applications to sample informative and diverse subsets.

The problem setting we consider is moreover related to active learning [8,26]. Common active learning strategies sequentially select data points based on some uncertainty criterion or heuristic. Data points are for instance selected based on the confidence of the model to an assigned label or according to the maximal model update in the worst case. Usually active learning iteratively selects instances and then re-trains to include the newly gained label into the model generation. In contrast to such iterative active learning scenarios with feedback, OED corresponds to selecting a single optimal batch prior to labeling and learning.

The frame can be straight forwardly obtained by convex hull algorithms. However, many of them are motivated and limited to two- or three-dimensional settings. Quickhull [4] works in higher dimensionalities but quickly becomes infeasible. If the enumeration of vertices is dropped, convex hull algorithms can be turned into methods that directly (and only) compute the frame. Common approaches for examples include linear programming to test whether a point is part of the frame or not [11,12,21]. Recent methods use quadratic programming to efficiently compute the frame [18].

5   Conclusion

We proposed to leverage the geometry of the data to efficiently compute optimal designs. Our contribution was motivated by the observation that traditional OED variants optimize enclosing ellipsoids that are supported by extreme data points. Hence, we proposed to restrict the computations to the frame which is the smallest subset of the data that yields the same convex hull as all data. We devised an optimization problem to compute the frame to sample random designs in kernel-induced feature spaces and provided a theoretical foundation for the eligibility of different kernel functions. Our contribution can be viewed as a transposed version of LASSO that selects data points instead of features.

Empirically, we showed that restricting optimal design to the frame yields competitive designs with respect to D-, E-, and A-optimality criteria on several real-world data sets. Interior data points are ignored by our frame-based approaches and we observed computational speed-ups of up to a factor of twenty. Our contribution rendered OED problems feasible on data at large scales for moderate frame densities.

References

1. Ageev, A.A., Sviridenko, M.I.: Pipage rounding: a new method of constructing algorithms with proven performance guarantee. J. Comb. Optim. **8**(3), 307–328 (2004)
2. Allen-Zhu, Z., Li, Y., Singh, A., Wang, Y.: Near-optimal design of experiments via regret minimization. In: International Conference on Machine Learning, pp. 126–135 (2017)
3. Avron, H., Boutsidis, C.: Faster subset selection for matrices and applications. SIAM J. Matrix Anal. Appl. **34**(4), 1464–1499 (2013)
4. Barber, C.B., Dobkin, D.P., Huhdanpaa, H.: The quickhull algorithm for convex hulls. ACM Trans. Math. Softw. (TOMS) **22**(4), 469–483 (1996)
5. Boyd, S., Vandenberghe, L.: Convex Optimization. Cambridge University Press, New York (2004)
6. Bro, R., De Jong, S.: A fast non-negativity-constrained least squares algorithm. J. Chemom. **11**(5), 393–401 (1997)
7. Brooks, T.F., Pope, D.S., Marcolini, M.A.: Airfoil self-noise and prediction (1989)
8. Chaudhuri, K., Kakade, S.M., Netrapalli, P., Sanghavi, S.: Convergence rates of active learning for maximum likelihood estimation. In: Advances in Neural Information Processing Systems, pp. 1090–1098 (2015)
9. Dereziński, M., Warmuth, M.K.: Subsampling for ridge regression via regularized volume sampling. arXiv preprint arXiv:1710.05110 (2017)
10. Dolia, A.N., De Bie, T., Harris, C.J., Shawe-Taylor, J., Titterington, D.M.: The minimum volume covering ellipsoid estimation in kernel-defined feature spaces. In: Fürnkranz, J., Scheffer, T., Spiliopoulou, M. (eds.) ECML 2006. LNCS (LNAI), vol. 4212, pp. 630–637. Springer, Heidelberg (2006). https://doi.org/10.1007/11871842_61
11. Dulá, J.H., Helgason, R.V.: A new procedure for identifying the frame of the convex hull of a finite collection of points in multidimensional space. Eur. J. Oper. Res. **92**(2), 352–367 (1996)
12. Dulá, J.H., López, F.J.: Competing output-sensitive frame algorithms. Comput. Geom. **45**(4), 186–197 (2012)
13. Fedorov, V.V.: Theory of Optimal Experiments. Elsevier, New York (1972)
14. Kulesza, A., et al.: Determinantal point processes for machine learning. Found. Trends® Mach. Learn. **5**(2–3), 123–286 (2012)
15. Lawson, C.L., Hanson, R.J.: Solving least squares problems, vol. 15. SIAM (1995)
16. Li, C., Jegelka, S., Sra, S.: Polynomial time algorithms for dual volume sampling. In: Advances in Neural Information Processing Systems, pp. 5045–5054 (2017)
17. Lopez, F.J.: Generating random points (or vectors) controlling the percentage of them that are extreme in their convex (or positive) hull. J. Math. Model. Algorithms **4**(2), 219–234 (2005)
18. Mair, S., Boubekki, A., Brefeld, U.: Frame-based data factorizations. In: International Conference on Machine Learning, pp. 2305–2313 (2017)
19. Mariet, Z.E., Sra, S.: Elementary symmetric polynomials for optimal experimental design. In: Advances in Neural Information Processing Systems, pp. 2136–2145 (2017)
20. Nocedal, J., Wright, S.J.: Sequential quadratic programming. In: Nocedal, J., Wright, S.J. (eds.) Numerical Optimization. Springer Series in Operations Research and Financial Engineering. Springer, New York (2006). https://doi.org/10.1007/978-0-387-40065-5_18

21. Ottmann, T., Schuierer, S., Soundaralakshmi, S.: Enumerating extreme points in higher dimensions. Nord. J. Comput. **8**(2), 179–192 (2001)
22. Pace, R.K., Barry, R.: Sparse spatial autoregressions. Stat. Probab. Lett. **33**(3), 291–297 (1997)
23. Pukelsheim, F.: Optimal Design of Experiments. SIAM (2006)
24. Schölkopf, B., Smola, A.J.: Learning with Kernels: Support Vector Machines, Regularization, Optimization, and Beyond. MIT Press, Cambridge (2002)
25. Smola, A.J., Schölkopf, B., Müller, K.R.: The connection between regularization operators and support vector kernels. Neural Netw. **11**(4), 637–649 (1998)
26. Sugiyama, M., Nakajima, S.: Pool-based active learning in approximate linear regression. Mach. Learn. **75**(3), 249–274 (2009)
27. Tibshirani, R.: Regression shrinkage and selection via the lasso. J. R. Stat. Soc. Ser. B (Methodol.) **58**, 267–288 (1996)
28. Wang, Y., Yu, A.W., Singh, A.: On computationally tractable selection of experiments in measurement-constrained regression models. J. Mach. Learn. Res. **18**(143), 1–41 (2017)
29. Yeh, I.C.: Modeling of strength of high-performance concrete using artificial neural networks. Cem. Concr. Res. **28**(12), 1797–1808 (1998)

Hierarchical Active Learning
with Proportion Feedback on Regions

Zhipeng Luo$^{(\boxtimes)}$ and Milos Hauskrecht

Department of Computer Science, University of Pittsburgh,
Pittsburgh, PA 15260, USA
{ZHL78,milos}@pitt.edu

Abstract. Learning of classification models in practice often relies on
human annotation effort in which humans assign class labels to data
instances. As this process can be very time-consuming and costly, finding
effective ways to reduce the annotation cost becomes critical for building
such models. To solve this problem, instead of soliciting instance-based
annotation we explore *region*-based annotation as the feedback. A region
is defined as a hyper-cubic subspace of the input feature space and it
covers a subpopulation of data instances that fall into this region. Each
region is labeled with a number in [0, 1] (in binary classification setting),
representing a human estimate of the positive (or negative) class proportion in the subpopulation. To learn a classifier from region-based feedback we develop an active learning framework that hierarchically divides
the input space into smaller and smaller regions. In each iteration we
split the region with the highest potential to improve the classification
models. This iterative process allows us to gradually learn more refined
classification models from more specific regions with more accurate proportions. Through experiments on numerous datasets we demonstrate
that our approach offers a new and promising active learning direction
that can outperform existing active learning approaches especially in
situations when labeling budget is limited and small. Code related to
this paper is available at: https://github.com/patrick-luo/hierarchical-
active-learning.git.

Keywords: Active learning · Proportion label · Classification

1 Introduction

Learning of classification models from real-world data often requires non-trivial
human annotation effort on labeling data instances. As this annotation process
is often time-consuming and costly, the key challenge then is to find effective
ways to reduce the annotation effort while guaranteeing that models built from
the limited feedback are accurate enough to be applied in practice. One popular
machine learning solution to address the annotation problem is active learning. It
aims to sequentially select examples to be labeled next by evaluating the possible

© Springer Nature Switzerland AG 2019
M. Berlingerio et al. (Eds.): ECML PKDD 2018, LNAI 11052, pp. 464–480, 2019.
https://doi.org/10.1007/978-3-030-10928-8_28

impact of the examples on the solution. Active learning has been successfully applied in domains as diverse as computer vision, natural language processing and bio-medical data mining [8,14,15].

Despite enormous progress in active learning research in recent years, the majority of current active learning solutions focus on *instance-based* methods that query and label individual data instances. Unfortunately, this may limit its applicability when targeting complex real-world classification tasks. There are two reasons for this. First, when the labeling budget is severely restricted, a small number of labeled data may not properly cover or represent the entire input space. In other words, the data selected by active learning are likely to suffer from *sampling bias* problem. To mitigate this issue Dasgupta [2] has developed a hierarchical active learning approach to sample instances in a more robust way which is driven by not only the current sampled data, but also the underlying structure in the data.

Second, instance-based learning framework often assumes instances are easy to label for humans. But it is not always true. Consider two realms of applications: (1) in political elections where the privacy is a concern, collecting one's feedback is hard or infeasible [9,10,16]; (2) in medical domain patient records can be very complex as each record has numerous entries which require careful reviewing [3,11]. For example, when a physician diagnoses a patient (e.g. for possible heart condition) he/she must review the patient record that consists of complex collections of results, symptoms and findings (such as *age, BMI, glucose levels, HbA1c blood test, blood pressure, etc.*). The review and the assessment of these records w.r.t. a specific condition may become extremely time-consuming as it often requires physicians to peruse through a large quantity of data [4,5].

In light of this, novel active learning methods based on *group* queries have been proposed: AGQ+ [3], RIQY [11] and HALG [7]. The basic idea here is to (1) embody similar instances together as a *group*, (2) induce the most compact *region* which are conjunctive patterns of the input feature space to represent the group and (3) solicit a *generic label* on the region instead of on any specific instance. The region label is a number in [0, 1] (known as *proportion label* [6,9,10,16]) which represents a human estimate of the proportion of instances in *positive* or *negative* class in the subpopulation of instances in that region. This line of work has shown empirically that active learning with proportion feedback on generic regions works more efficiently than instance-based active learning.

Our Contribution. In this work, we develop and explore a new region-based active learning framework called HALR (**H**ierarchical **A**ctive **L**earning with proportion feedback on **R**egions) that learns instance-level classifiers from region queries and region-proportion feedback. In particular, our framework *actively* builds a hierarchical tree of regions with the aim to refine the leaf regions to be as *pure* as possible after very *few* splits and queries made. Briefly, our method starts from an unbounded region that covers the entire input feature space and this region initializes as the root of the tree. Then we grow this tree incrementally by splitting the most *uncertain* leaf region into two sub-regions. Whenever the new regions are generated, their proportion labels are either directly assigned

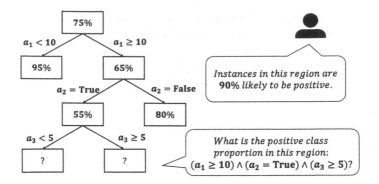

Fig. 1. An example of building a hierarchical tree of regions which is conceptually equivalent to a decision tree. The left shows a snapshot of the tree structure after $t = 3$ splits, generated from the root region on the top level. Each rectangle represents a certain region and the percentage number means its proportion label. Each link is a value constraint on some dimension a_i and it is inherited to all the descendant regions. To query the proportion label of a new region (say the right one on the lowest level), we describe it by using conjunctive patterns shown on the bottom right and a human annotator will assign a label to it according to its description. The label of the complementary region (the one on the left) will be inferred according to the constraint between its parent's label and sibling's label.

by a human annotator, or inferred by the proportion constraint. The general picture is illustrated in Fig. 1. At the end our algorithm outputs a hierarchical tree of labeled regions that can be either (1) directly used as a decision tree classifier, or alternatively, (2) be used to learn many different parametric binary classification models from proportion labels as proposed by [6,9,10,16], or by simply sampling instance labels [7] according to the known class proportion in each region and feeding them to standard instance-level learning algorithms.

The crucial part of our algorithm is to develop a strategy to split the leaf regions *without* knowing any labeled instances. To meet this challenge we design a competition procedure which dynamically tests and chooses one of two heuristic strategies to split the regions. The first one is *unsupervised* which is based on clustering. The second one is *supervised* and it relies on classification model that assigns class probabilities to every data instance. We will show that these heuristics can actively compete and also assist each other to drive our splits.

The remainder of the paper is organized as follows. First, we will review past work closely related to our framework. Second, we will explain the details of our proposed framework from Sects. 3 to 7. After that we will test our approach on a number datasets and compare its performance to multiple other active learning approaches. Finally, we will discuss the experiment results.

2   Related Work

2.1   Hierarchical Active Learning

Our hierarchical learning framework is motivated by Dasgupta *et al.*'s work [2, 13] that leverages a pre-compiled hierarchical clustering to drive the instance selection procedure. They start learning from a few coarse-grained clusters and gradually split clusters that are impure (in terms of class labels) to smaller ones such that the label entropy is reduced. In terms of training models, not only the labeled instances but also the ones with predicted labels in the *sufficiently pure* clusters are used for learning. While their approach is able to reduce the sampling bias, learning with predicted labeled data can be risky especially when the class distribution is severely unbalanced, as the instances from the minor class are hardly sampled. In our work we overcome this limitation by directly querying and learning from regions of which the proportion labels are friendlier to the minor class. Another difference worth noting is that we do not pre-compile a hierarchy of regions which can be done totally unsupervisedly (e.g. build a K-Dimension tree beforehand). Instead, we build the tree dynamically where each of the splits is determined by not only the unsupervised heuristic but also a supervised heuristic which reflects the current belief of the base model.

2.2   Learning from Group Proportion Feedback

Multiple works [6,9,10,16] study the problem of learning instance-level classifiers from apriori given groups/regions and their class proportion labels. The motivation scenarios can be political election, online purchasing or spam filtering. For example, we can easily obtain the percentage of voting results on election in each county and use these group proportions to predict individual's voting preference. These real life examples have greatly encouraged the development of learning algorithms that can eat proportion feedback. There are two main categories of the algorithms. The first one uses the proportion label as a proxy that approximates to the sufficient statistics required by the final likelihood function [9,10]. The second category develops models that generate consistent instance labels with the group proportions [6,16]. What beyond the scope of the above works is that they assume the groups are formed and labeled *apriori*, and thus they do not study the problem of how to form the groups and how to obtain the proportion labels for these groups.

2.3   Active Learning from Group Proportion Feedback

AGQ+ [3] and RIQY [11] are the early works that explore active learning strategies with group/region proportion feedback instead of instance-based feedback. The motivation for the group queries is that in many practical domains, annotators may prefer to work with region-based queries which are shorter (in terms of feature space), less confusing and more intuitive. As an example consider the heart disease classification task presented in [11]:

An Instance Query Example. An instance query for the heart disease problem covers all features of the patient case: *"Consider a patient with (sex = female) ∧ (age = 39) ∧ (chest pain type = 3) ∧ (fasting blood sugar = 150 mg/dL) ... (20 more features omitted). Does the patient has a heart disease?"* The label is a binary (true, false) response.

A Group Query Example. In contrast, a group query using conjunctive patterns which represent a region of the input feature space may be only associated with a subset of the features: *"Consider a population of patients with (sex = female) ∧ (40 < age < 50) ∧ (chest pain type = 3) ∧ (fasting blood sugar within [130,150] mg/dL) ... (not necessarily using all the features). What is the chance that a patient from this population has a heart disease?"*. The label is an empirical estimate of the proportion of cases in the population who suffer from the heart disease, say *"75% patients within this region suffer from the disease"*.

In terms of group formation, both AGQ+ and RIQY build groups by (1) choosing the most uncertain instance x_u from the unlabeled data pool according to the current classification model, and (2) aggregating a number of instances as a group G_u in a close neighborhood of x_u. The region description of the group G_u is then automatically learned using decision tree algorithm. After the proportion label of the group G_u is annotated, all the instances inside the group G_u are either assigned hard labels (RIQY) or weighted labels (AGQ+). Finally the classification model is re-trained using all the labeled data. The major limitation of the methods is that their group selection approach is ad-hoc, driven by instance-based selection and enriched by nearby data instances. As a consequence, this approach may fail to discover meaningful regions.

A more recent approach that addresses some limitations of the early group active-learning methods is HALG [7]. HALG uses a hierarchical clustering, similarly to Dasgupta *et al*'s work, to generate clusters of instances which are then approximated by regions. As this hierarchy of regions is pre-clustered, their active learning algorithm, which selects groups/regions to be split and labeled next, can only make decisions within this fixed hierarchy. While this novel group formation approach is able to capture the structure of the unlabeled data (unsupervised heuristic), the fixed hierarchy can significantly limit the behavior of seeking the class information which is important to the model (supervised heuristic). That is, the unsupervised heuristic used in HALG overly dominates its supervised heuristic. To overcome this issue, our proposed HALR method *dynamically* refines regions by *directly* dividing the input feature space into sub-spaces (still in a hierarchical fashion) and further, our active region refinement is explicitly controlled and balanced between the supervised and unsupervised heuristics.

3   Our Framework

Our HALR framework is summarized in Algorithm 1. It aims to actively build a hierarchical tree of regions with proportion labels and then uses this tree to learn an instance-level binary classification model. We assume the classification model is a probabilistic one (e.g. Logistic Regression or an Support Vector Machine

---

Algorithm 1. Hierarchical Active Learning Framework (HALR)

Input: An unlabeled data pool $\mathcal{U}$; A labeling budget
Output: A binary classification model $P(y|\boldsymbol{x}; \hat{\boldsymbol{\theta}})$
1: $T \leftarrow$ Build a 1-node tree whose root region is the entire feature space of $\mathcal{U}$;
2: Query the proportion label of T's root;
3: Leaf nodes $L^{(1)} \leftarrow \{T\text{'s root}\}$;
4: Active learning time $t \leftarrow 1$;
5: **repeat**
6:    Train the base model $P(y|\boldsymbol{x}; \hat{\boldsymbol{\theta}}^{(t)})$ with current leaf nodes $L^{(t)}$;
7:    Choose a most *uncertain* region R_* in $L^{(t)}$ to be split;
8:    Divide R_* into two sub-regions (it is co-decided by probabilistic clustering
      and probabilistic classification (based on $P(y|\boldsymbol{x}; \hat{\boldsymbol{\theta}}^{(t)})$));
9:    Query or infer the proportion labels of the sub-regions derived from R_*;
10:   $L^{(t+1)} \leftarrow \{L^{(t)} - R_*\} \cup \{R_*'s \text{ sub-regions}\}$;
11:   $t \leftarrow t + 1$
12: **until** the labeling budget runs out
13: **return** $P(y|\boldsymbol{x}; \hat{\boldsymbol{\theta}}^{(t)})$

---

with Platt's transformation). Such a model is treated as our base model which will be used to provide supervised heuristic and decisions to guide the tree-building process. Our algorithm works as follows. The tree is initialized with a root region covering the entire input space and as well as all the unlabeled data $\mathcal{U}$ (line 1). The root region is assigned a proportion label which can be interpreted as the *prior* probability of classes (line 2). The tree is gradually refined through active learning cycles (Line 5–12) which iteratively replace leaf regions with more refined sub-regions. In each cycle, we (1) select the most *uncertain* leaf region R_* to split; (2) divide it two sub-regions using a condition that placed on one the input dimension; (3) query or infer the proportion labels of the new sub-regions and (4) replace R_* with the new sub-regions in the tree. Every time the new regions are generated and labeled, the base classification model will be re-learned with all the labeled leaf regions. The whole process resembles decision tree learning algorithm, but in our case we do not have any labeled instances to drive the splits. In the following we will define region concept (Sect. 4) and uncertainty of regions (Sect. 5) and then explain how we split the most uncertain region (Sect. 6).

4   The Concept of Regions

Our base learning task is to learn a binary classification model and our active learning scenario is a pool-based one [12] which assumes the unlabeled data are abundant. That is, a pool of n unlabeled training instances $\mathcal{U}$ are randomly drawn from a fixed marginal distribution $p(\boldsymbol{x})$ of an unknown joint distribution of $p(\boldsymbol{x}, y)$. Each instance $\boldsymbol{x}$ is a vector of d features, each of which can be symbolic or numeric. So the input feature space is a d-dimensional one where each dimension is either discrete or continuous and the domain depends on the natural definition of that feature. x also has a binary class label $y \in \{0, 1\}$ which is never queried individually. In our framework, however, the class information is given only on aggregated instances which are described as regions. Initially, there is only region that is defined as the entire feature space of $\mathcal{U}$. Because there is no value constraint on any of the dimensions, this first region is unbounded and it conceptually contains all the instances from $\mathcal{U}$. When a binary split is made on some value v from some dimension a, there will be two sub-regions generated with one value constraint on the dimension a either $<v$ or $\geq v$. This type of binary splits will recursively divide the sub-regions and in the end a hierarchical tree of regions will be generated where the leaf regions do not overlap with each other but co-partition the whole feature space and data in $\mathcal{U}$. Each region is thus a hyper-cubic subspace defined by conjunctive patterns. For example a region of patients may be described as: *(gender = male) ∧ (heart rate 80–100) ∧ (temperature 100–110 F)... (other dimensions unbounded)*.

In terms of the region feedback, the human assessment is made via a proportion label which is an estimate of the proportion of the *positive* or *negative* class in the population of instances that fall into the definition of that region. For example, given the region of patients described above, physicians could say *"70% of patients in the population defined by a region suffer from a heart disease"*. Or alternatively, we can interpret the proportion label as an instance-level likelihood: *"Each patient in the population is 70% likely to have a heart disease"*. Initially, the root region is assigned a proportion label which corresponds to the *prior* probability of classes. So in this sense, the proportion label of each sub-region can be understood as a *conditional* probability of classes given the value constraints on some of the input dimensions.

5   The Uncertainty of Regions

Given the definition of regions we now want to define a score that would help us to decide which region should be split next in each active learning cycle. One sensible way is to use the uncertainty (or impurity) of regions. This idea has been successfully used in decision tree learning process. Here, the impurity is measured in terms of the entropy (C4.5) or the Gini-Index (CART) scores. With the help of the impurity measure one can build a decision tree recursively where in each step one leaf region is split along one of the input dimensions. By comparing all possible splits for all eligible leaf regions, the best region and the best split that leads to the maximum reduction in uncertainty, or the maximum information gain, can be identified. Unfortunately, this process applied in the decision tree learning to assess uncertainty and gain requires instance labels and hence, it cannot be replicated in our framework where instance labels are unknown.

Another issue to consider in the development of the region splitting criteria is that the information gain ignores the region size. Here the region size is defined as the empirical number of instances contained in a region. Intuitively, the largest benefit from the split should be realized when not only the impure regions but also large regions are split. In light of this, we propose a new *uncertainty* score that takes into account both the size and the proportion label in deciding which region should be split next.

Suppose that at time t there are $N^{(t)}$ leaf regions $L^{(t)} = \{(R_i, \mu_i)\}_{i=1}^{N^{(t)}}$ where each region $R_i = \{x_{ij}\}_{j=1}^{n_i}$ has n_i instances and has been assigned a label $\mu_i \in [0,1]$ representing the *positive* class proportion, our goal is to choose the most uncertain region R_* to split. The uncertainty of each region R_i is defined as the expected number of wrong labels (denoted by w_i) if we randomly guess the class labels of all instances in R_i based on its proportion label μ_i. In particular, the procedure to calculate uncertainty is explained as follows:

i. For each instance in R_i, sample its label as an independent Bernoulli process with the parameter $= \mu_i$. This creates n_i sampled labels;

ii. Calculate the distribution of w_i, i.e. the number of mismatches between the sampled labels and the true labels. Although the true labels are unknown, each true label can be assumed to follow an independent Bernoulli distribution with the parameter $= \mu_i$. Therefore, the probability of mismatch for each instance also follows in independent Bernoulli distribution with parameter $= P(mismatch) = P[false\ positive] + P[false\ negative] = 2\mu_i(1-\mu_i)$. Then apparently w_i follows a Binomial distribution $Bin(n_i, 2\mu_i(1-\mu_i))$;

iii. And use the expectation $\mathbb{E}(w_i) = 2\mu_i(1-\mu_i)n_i$ as the uncertainty of R_i.

This uncertainty defined above clearly shows that larger n_i or more uncertain μ_i (closer to 0.5) leads to more uncertainty of region R_i. Please note here $2\mu_i(1-\mu_i)$ matches exactly the definition of Gini-Index, so throughout this paper we will choose Gini-Index as the gain measurement for later use. Finally we select $R_* = \arg\max_{R_i \in L^{(t)}} \mathbb{E}(w_i)$ to be the most uncertain region to split at current active learning cycle t.

6   The Split of Regions

Now given the region R_*, we need to determine what input dimension to split and what value should be used to define the split. Since there are no labeled instances in our framework, we resort to two heuristics to drive the split.

6.1   Unsupervised Heuristic

The first heuristic is *unsupervised*. It is based on probabilistic clustering. Clustering is a simple yet often effective guidance. The assumption behind it is that similar data instances tend to carry similar class labels and it has been used frequently in semi-supervised learning [17]. In other words, dissimilar data are likely to fall into different classes and so the region splits should be driven by the underlying structure of data. To implement this idea, we perform a 2-means probabilistic clustering on the instances $\{x_{*j}\}_{j=1}^{n_*}$ in R_*, assuming there is mix of two cluster centers in $\{x_{*j}\}$ and the probabilities of cluster membership are given by Expectation and Maximization (EM) algorithm. Thus each instance

$\boldsymbol{x}_{*j}$ will have an **Unsupervised** probabilistic label p_j^U indicating the chance of belonging to one of the two clusters. Given these instance-level labels, standard decision tree splitting procedure based on information gain can be now directly applied to split R_*. Here we use Gini-Index and say this procedure gives us the empirically optimal split of R_* from value v^U on dimension a^U based on the set of probabilistic unsupervised labels $\{p_j^U\}$.

6.2   Supervised Heuristic

Our second heuristic is *supervised* and it relies on the base classification model. In various active learning algorithms the base model plays an important role in determining which data should be queried next. An example is the classic Uncertainty Sampling approach [12]. The base model reflects the current belief of the class distribution on instances and thus its guidance on the region splitting cannot be ignored. Formally at learning time t, the base model is learned as $P(y|\boldsymbol{x}; \hat{\boldsymbol{\theta}}^{(t)})$, so each instance $\boldsymbol{x}_{*j}$ will also have a **Supervised** probabilistic label p_j^S reflecting the likelihood of belonging to one of the two classes. Here $p_j^S = P(y = 1|\boldsymbol{x}_{*j}; \hat{\boldsymbol{\theta}}^{(t)})$. Similarly, given these instance-level labels Gini-Index-based gain can again be applied to split R_* and say it gives the best split from value v^S on dimension a^S.

6.3   Combination of the Two Heuristics

Table 1 summarizes the pros and cons of the two heuristics. Initially when the supervision is scarce, the base model trained can be very likely to make biased decisions. This problem was formally stated as *sampling bias* by Dasgupta *et al.* [2] and they leverage hierarchical clustering to assist the base model. In our framework we use clustering too as an unsupervised heuristic to alleviate the bias issue. However, the unsupervised heuristic may not always work well in the long run. Hence the best option appears to be the combination of the two heuristics.

Table 1. Comparison of the two heuristics

	Unsupervised heuristic	Supervised heuristic
Pros	Relies on the semi-supervised assumption which is often effective	Gives instance-level estimates which directly reflect the class distribution
Cons	But this assumption may not hold all the time	But initially these estimates are poor simply because the supervision is little

To combine and also to evaluate the two heuristics, we introduce a competition procedure described in Algorithm 2. The general idea is to perform a test split on each of the proposed splits separately and compare their actual gains. Larger gain is better and so the final split will take whatever the corresponding heuristic suggests. We also maintain a list H that records the winning history of the heuristics in the past splits and this H will be used to test whether the supervised heuristic is doing significant better than the unsupervised one in the long run. If the test result is significant, it marks that our base model is good enough to make splitting decisions alone and from then on, every region split will only be determined by the supervised heuristic. That is, Algorithm 2 will *not* be called any more once we believe the supervised heuristic is performing significantly better and the final split will directly take the supervised proposal.

Algorithm 2. The competition procedure of choosing heuristic

> **Input:** Unsupervised split (a_U, v_U); Supervised split (a_S, v_S); Winning history of heuristics H
> **Output:** The final split (a_F, v_F); updated history H; Binomial test result of supervised heuristic
> 1: Binomial test result $r \leftarrow$ *Not significant*
> 2: **if** $a_U = a_S$ and $v_U = v_S$ **then**
> 3: $a_F \leftarrow a_S$; $v_F \leftarrow v_S$;
> 4: **else**
> 5:   Do a test split on (a_U, v_U) and get its gain G_U;
> 6:   Do a test split on (a_S, v_S) and get its gain G_S;
> 7:   **if** $G_U > G_S$ **then**
> 8:   Append *"Unsupervised heuristic wins"* to H;
> 9: $a_F \leftarrow a_U$; $v_F \leftarrow v_U$;
> 10:   **else**
> 11:   Append *"Supervised heuristic wins"* to H;
> 12: $a_F \leftarrow a_S$; $v_F \leftarrow v_S$;
> 13:   Test result $r \leftarrow$ Binomial test (Algorithm 3) on H;
> 14:   **end if**
> 15: **end if**
> 16: **return** (a_F, v_F), H and r

Test Split. The test split and the calculation of the gain procedure called in Line 5 or 6 in Algorithm 2 is identical to the evaluation of a standard decision tree splitting. Here we show how to calculate the gain G_S of the test split on R_* proposed by the supervised heuristic. The gain of G_U can be calculated similarly.

   i. Split R_* from value v_S on dimension a_S into two sub-regions R^L and R^R;
   ii. Route each instance in R_* to R^L or R^R by testing the feature value of the instance on dimension a_S either $< v_S$ or $\geq v_S$;
   iii. Query the proportion label of one sub-region. Say R^L is given a label μ^L;
   iv. Infer the label μ^R of R^R. This does not require a human assessment because of the proportion label constraint: $n^L \mu^L + n^R \mu^R = n_* \mu_*$ with $n^L + n^R = n_*$,

where n^L, n^R and n_* are the number of instances contained in R^L, R^R and R_*, and μ_* is the label of R_*. Simply $\mu_R = (n_*\mu_* - n^L\mu^L)/n^R$;

v. Apply Gini-Index to calculate the gain (or uncertainty reduction):

$$G_S = GI(\mu_*) - \frac{n^L}{n_*}GI(\mu^L) - \frac{n^R}{n_*}GI(\mu^R)$$

where $GI(\mu) = 2\mu(1 - \mu)$.

Algorithm 3. Binomial test of the supervised heuristic

Input: Winning history of heuristics H; Window size W; Significance level α
Output: *Significant* or *Not significant*
1: **if** $length(H) < W$ **then**
2:     **return** *Not significant*
3: **end if**
4: H_0: winning chance of supervised heuristic $p_S \le 0.5$ in the last W trials in H;
5: H_A: $p_S > 0.5$
6: Test statistic $B^* \leftarrow$ number of supervised wins in the last W outcomes;
7: $p\_value \leftarrow$ do binomial test on B^*;
8: **return** *Significant* if $p\_value < \alpha$ else *Not significant*

Binomial Test. Algorithm 3 provides the details of the Binomial test that decides whether the supervised heuristic is doing significantly better than the unsupervised one. The null hypothesis H_0 means the supervised heuristic is doing equally well or worse than the unsupervised heuristic in the latest W trials. In other words, the winning chance of the supervised heuristic p_S is ≤ 0.5. Under H_0 the number of supervised wins B^* follows a Binomial distribution $Bin(W, 0.5)$ and we do a right-tailed test of B^* to carry out the p-value. We reject H_0 if the p-value is less than a given confidence level α and choose the alternative.

To make the test stronger, or to be more conservative, multiple such tests with different window sizes can be done simultaneously. To ensure the same family wise error rate α, Bonferroni correction can be applied. In our implementation, we combine a short term window $W_S = 5$ and a long term window $W_L = 10$ with the same family wise $\alpha = 0.05$. The purpose of performing two tests together is to ensure the supervised heuristic is indeed doing stably well both in the most recent time and in the long run.

7   Learning a Model from Labeled Regions

Now the last remaining question is how to learn a general instance-level model from labeled regions. As introduced in Related Work section, various algorithms can be applied to learn instance-level classification models from proportion labels [6,9,10,16]. Hence, at any time t the base classification model $P(y|\boldsymbol{x};\boldsymbol{\theta})$ can be learned from the set of leaf regions $L^{(t)} = \{(R_i, \mu_i)\}$ where each region R_i has been labeled as μ_i and contains a certain number of training instances.

Apart from the complex learning methods, we adopt another simple but effective method based on *instance sampling* such that instance-based learning algorithms can be used (introduced by HALR [7]). The idea is to create a sample of labeled instances $S = \{(\boldsymbol{x}_k, y_k)\}_{k=1}^K$ from $L^{(t)}$. The $\{\boldsymbol{x}_k\}_{k=1}^K$ part in S is sort of fixed while each of the label y_k is sampled from Bernoulli distribution with the parameter equal to μ_i, which is the proportion label of region R_i that contains $\boldsymbol{x}_k$. Now given S, the parameter vector of the base model can be learned through maximum likelihood estimation (MLE), denoted by $\hat{\boldsymbol{\theta}}$. $\hat{\boldsymbol{\theta}}$ may vary because of the randomness in S, however under some moderate MLE assumptions required by Central Limit Theorem, $\hat{\boldsymbol{\theta}}$ asymptotically follows a normal distribution $\mathcal{N}(\boldsymbol{\theta}, \boldsymbol{\Sigma})$ conditioned on $\{\boldsymbol{x}_k\}$, where $\boldsymbol{\theta}$ is the converged parameter when $K \to \infty$ and the variance $\boldsymbol{\Sigma}$ is the inverse of Fisher information matrix $\mathcal{I}_K(\boldsymbol{\theta})$ depending on the actual finite sample size K. In practice, the asymptotic property can be satisfied by sampling multiple times the label of each $\boldsymbol{x}_k$ and aggregating them up into S. In our experiments each instance label is sampled from 5 to 10 times depending on datasets and then S is large enough to give a small $\boldsymbol{\Sigma}$ (estimated as $\hat{\boldsymbol{\Sigma}}$ by $\hat{\boldsymbol{\theta}}$).

8   Experiments

We conduct an empirical study to evaluate our proposed approach on 8 general binary classification data sets collected from UCI machine learning repository [1]. The purpose of this study is to research how efficiently (in terms of number of queries) our framework can learn classification models in cost-sensitive tasks.

8.1   Data Sets

The 8 data sets come from a variety of real life applications:

  i. **Seismic:** Predict if seismic bumps are in hazardous state.
 ii. **Ozone:** Detect ozone level for some days.
iii. **Messidor:** Predict if Messidor images contain signs of diabetic retinopathy.
 iv. **Spam:** Detect spam emails in commercial emails.
  v. **Music:** Classify the geographical origin of music.
 vi. **Wine:** Predict wine quality based on its properties.
vii. **SUSY:** Distinguish a physical signal from background process.

Table 2 suggests various properties of the datasets. Some have been used in previous work (*Wine*) [11,15]; some are high-dimensional (*Ozone, Spam, Music*); and some are unbalanced in class distribution (*Seismic, Ozone, Wine unbalance*).

Table 2. 8 UCI data sets

Dataset	# of data	# of features	Major class %	Feature type
Seismic	2584	18	93%	Numeric, Symbolic
Ozone	1847	72	93%	Numeric
Messidor	1151	19	53%	Numeric, Symbolic
Spam	4601	57	60%	Numeric, Symbolic
Music	1059	68	53%	Numeric
Wine	4898	11	67%	Numeric
Wine$_{ub}$	1895	11	95%	Numeric
SUSY	5000	18	55%	Numeric

8.2   Methods Tested

We compare our method (HALR) to 3 different methods:

i. **DWUS:** Density-Weighted Uncertainty Sampling is an instance-based method that combines both the uncertainty score and the structure of data [12].
ii. **RIQY:** The state-of-the-art method with proportion feedback on regions [11].
iii. **HS:** Hierarchical Sampling by Dasgupta [2].

8.3   Experimental Settings

Data Split. We split each data set into three disjoint parts: the initial labeled dataset (about 1%–2% of all available data), a test dataset (about 25% of data) and an unlabeled dataset $\mathcal{U}$ (the rest) used as training data. DWUS and RIQY require the initial labeled data to start training, but not our method nor HS.

Region Proportion Label Feedback. To simulate the effect of a human oracle in determining the label of a region, RIQY has originally introduced the way of region queries, which is to simply count the class proportion from labels of the empirical instances that fall into the region.

Evaluation Metrics. We adopt Area Under the Receiver Operating Characteristic curve (AUC) to evaluate the generalized classification quality of Logistic Regression on the test data. Our graphs will plot the AUC scores iteratively after each $t \leq 200$ queries are posed, which is large enough for all methods to converge. Also we assume all kinds of queries consume the same unit cost, although in practice sometimes a instance query is cheaper or oppositely in our cases a region query is more feasible and efficient. To reduce the experiment variations all results are averaged over 20 runs in different random splits.

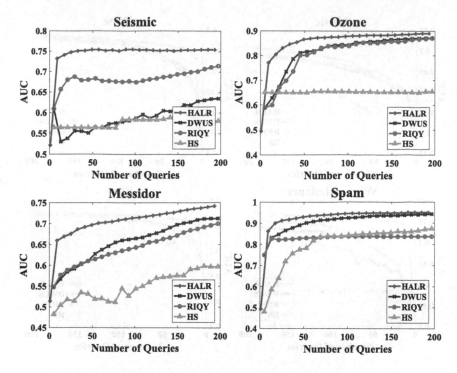

Fig. 2. Performances of different methods on the first 4 datasets (Color figure online)

8.4   Experiment Results

The main results are shown in Figs. 2 and 3. Overall, our HALR (in red line) is able to outperform other methods on majority of the datasets and is close to the best performing method on the remaining sets. There are two primary strengths: first, initially when the labeling budget is severely limited, learning with region-based feedback is superior to learning with the same number of labeled instances, simply because generic region-based queries can carry richer class information than specific instance queries. Second, the initial steep slopes and early convergence in our learning curves lend great credence to our active learning strategy that it is capable of splitting the most uncertain region in the right way and consequently it can accelerate the base model convergence rate.

Unbalanced Class. For data sets *Seismic, Ozone* and *Wine unbalance* (simulated from *Wine*) with unbalanced class distribution, our method performs even better as it could capture the minor class information via proportion labels. In contrast, instance-based methods (e.g. DWUS) may find them slowly; hierarchical sampling (HS) completely failed due to the reason that it always determines the labels of unlabeled instances by majority vote in those *pure enough* (but not entirely pure) clusters, which may totally lose the minor class information.

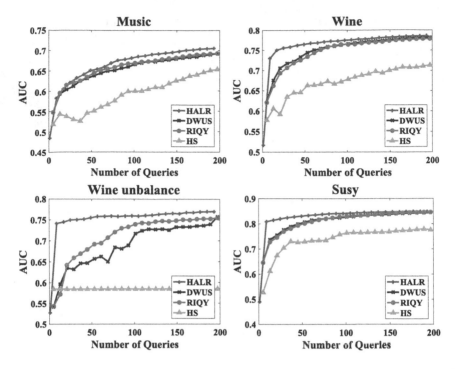

Fig. 3. Performances of different methods on the last 4 datasets (Color figure online)

Complexity of Region Description. Here we show how complex on average our region description could be, in terms of number of features used in the conjunctive patterns. In particular, we calculate *feature reduction rate* for each region R, which is defined as $1 - \frac{\#features\ to\ describe\ R}{\#(All\ features)}$. The results in Table 3 show the average reduction rate among 20 repetitions. This table suggests that region-based queries only use less than half or even 10% of the full dimensional information for human to annotate. This property considerably simplifies the interaction with human annotators when objects are high-dimensional, as region-based queries will present only the relevant features for querying.

Table 3. The averaged feature reduction rate (FRR) of region queries

Dataset	FRR	Dataset	FRR
Wine	59%	Spam	76%
Ozone	90%	Music	90%
Messidor	66%	SUSY	74%
Seismic	77%	Wine$_{ub}$	58%

9   Conclusions

We develop a new learning framework HALR that can actively learn instance-based classification models from proportion feedback on regions. The regions used in our framework are formed by hierarchical division of the input feature space. In each of the splits, we choose the most uncertain region to divide which considers both the size and the label purity of the region. Then the actual splits are co-decided by both unsupervised and supervised heuristics. Our empirical experiment results show that the regions can be refined to be pure in very few splits and thus they are able to improve the base model quality rapidly. In terms of application, our framework is best suited when providing region-based feedback is more feasible or easier than instance-based queries, as we only present the relevant and partial feature information for querying.

Acknowledgements. The work presented in this paper was supported by NIH grants R01GM088224 and R01LM010019. The content of the paper is solely the responsibility of the authors and does not necessarily represent the official views of NIH.

References

1. Asuncion, A., Newman, D.: UCI machine learning repository (2007)
2. Dasgupta, S., Hsu, D.: Hierarchical sampling for active learning. In: Proceedings of the 25th International Conference on Machine Learning, pp. 208–215. ACM (2008)
3. Du, J., Ling, C.X.: Asking generalized queries to domain experts to improve learning. IEEE Trans. Knowl. Data Eng. **22**(6), 812–825 (2010)
4. Hauskrecht, M., et al.: Outlier-based detection of unusual patient-management actions: an ICU study. J. Biomed. Inform. **64**, 211–221 (2016)
5. Hauskrecht, M., Batal, I., Valko, M., Visweswaran, S., Cooper, G.F., Clermont, G.: Outlier detection for patient monitoring and alerting. J. Biomed. Inform. **46**(1), 47–55 (2013)
6. Kück, H., de Freitas, N.: Learning about individuals from group statistics. CoRR abs/1207.1393 (2012). http://arxiv.org/abs/1207.1393
7. Luo, Z., Hauskrecht, M.: Hierarchical active learning with group proportion feedback. In: Proceedings of the Twenty-Seventh International Joint Conference on Artificial Intelligence, IJCAI 2018, pp. 2532–2538 (2018)
8. Nguyen, Q., Valizadegan, H., Hauskrecht, M.: Learning classification models with soft-label information. J. Am. Med. Inform. Assoc. **21**(3), 501–508 (2014)
9. Patrini, G., Nock, R., Rivera, P., Caetano, T.: (Almost) no label no cry. In: Advances in Neural Information Processing Systems, pp. 190–198 (2014)
10. Quadrianto, N., Smola, A.J., Caetano, T.S., Le, Q.V.: Estimating labels from label proportions. J. Mach. Learn. Res. **10**, 2349–2374 (2009)
11. Rashidi, P., Cook, D.J.: Ask me better questions: active learning queries based on rule induction. In: Proceedings of the 17th ACM SIGKDD International Conference on Knowledge Discovery and Data Mining, pp. 904–912. ACM (2011)
12. Settles, B.: Active learning. Synth. Lect. Artif. Intell. Mach. Learn. **6**(1), 1–114 (2012)
13. Urner, R., Wulff, S., Ben-David, S.: PLAL: cluster-based active learning. In: Conference on Learning Theory, pp. 376–397 (2013)

14. Valizadegan, H., Nguyen, Q., Hauskrecht, M.: Learning classification models from multiple experts. J. Biomed. Inform. **46**(6), 1125–1135 (2013)
15. Xue, Y., Hauskrecht, M.: Active learning of classification models with likert-scale feedback. In: SIAM Data Mining Conference. SIAM (2017)
16. Yu, F., Liu, D., Kumar, S., Tony, J., Chang, S.F.: \proptoSVM for learning with label proportions. In: ICML, pp. 504–512 (2013)
17. Zhu, X., Lafferty, J., Ghahramani, Z.: Combining active learning and semi-supervised learning using Gaussian fields and harmonic functions. In: ICML 2003 Workshop on the Continuum from Labeled to Unlabeled Data in Machine Learning and Data Mining, vol. 3 (2003)

Pattern and Sequence Mining

An Efficient Algorithm for Computing Entropic Measures of Feature Subsets

Frédéric Pennerath[1,2]([✉])

[1] Université de Lorraine, CentraleSupélec, CNRS, LORIA, 57000 Metz, France
frederic.pennerath@centralesupelec.fr
[2] Université Paris Saclay, CentraleSupélec, CNRS, LORIA, 57000 Metz, France

Abstract. Entropic measures such as conditional entropy or mutual information have been used numerous times in pattern mining, for instance to characterize valuable itemsets or approximate functional dependencies. Strangely enough the fundamental problem of designing efficient algorithms to compute entropy of subsets of features (or mutual information of feature subsets relatively to some target feature) has received little attention compared to the analog problem of computing frequency of itemsets. The present article proposes to fill this gap: it introduces a fast and scalable method that computes entropy and mutual information for a large number of feature subsets by adopting the divide and conquer strategy used by *FP-growth* – one of the most efficient frequent itemset mining algorithm. In order to illustrate its practical interest, the algorithm is then used to solve the recently introduced problem of mining *reliable approximate functional dependencies*. It finally provides empirical evidences that in the context of non-redundant pattern extraction, the proposed method outperforms existing algorithms for both speed and scalability. Code related to this chapter is available at: https://github.com/P-Fred/HFP-Growth.

Keywords: Pattern mining · Entropic measures
Algorithm efficiency · Approximate functional dependency
Pattern redundancy

1 Introduction

Entropic measures such as conditional entropy or mutual information have been used numerous times in pattern mining, for instance to characterize valuable itemsets [5–7] or approximate functional dependencies [3,8,9]. In such setting, one considers datasets where data are described by nominal features, i.e. features with a finite number of possible values. These data are interpreted as IID samples of some distribution for which the set $\mathcal{F}$ of features are seen as categorical random variables. For every considered subset $\mathcal{X} \subseteq \mathcal{F}$ of features, the entropy $H(\mathcal{X})$ can be approximated by an empirical estimation $\hat{H}(\mathcal{X}) = -\sum_t \sigma_{\mathcal{D}}(t) \log_2 (\sigma_{\mathcal{D}}(t))$ where frequencies $\sigma_{\mathcal{D}}(t)$ are computed for all value combinations t (latter called tuples)

© Springer Nature Switzerland AG 2019
M. Berlingerio et al. (Eds.): ECML PKDD 2018, LNAI 11052, pp. 483–499, 2019.
https://doi.org/10.1007/978-3-030-10928-8_29

of $\mathcal{X}$ observed in the dataset. A similar expression allows to empirically estimate mutual information $I(\mathcal{X}; \mathcal{Y}) = H(\mathcal{X}) + H(\mathcal{Y}) - H(X \cup \mathcal{Y})$ between $\mathcal{X}$ and a target feature subset $\mathcal{Y}$, usually restricted to a single target feature.

Entropy $H(\mathcal{X})$ measures the amount of uncertainty when guessing samples of $\mathcal{X}$, or equivalently, the quantity of information conveyed by it. Similarly mutual information $I(\mathcal{X}; \mathcal{Y})$ is the amount of information shared between feature subsets $\mathcal{X}$ and $\mathcal{Y}$. Both quantities have interesting properties. In particular $\mathcal{X} \mapsto H(\mathcal{X})$ and $\mathcal{X} \mapsto I(\mathcal{X}; \mathcal{Y})$ are non negative monotonic functions in lattice $(2^{\mathcal{F}}, \subseteq)$ of feature subsets. This property builds up a formal analogy with the anti-monotonic property of itemset frequency so that some frequent itemset mining techniques such as the levelwise search used by Apriori [1] have been transposed for the computation of entropic measures (see for instance [8]). Despite this analogy the problem of designing fast and scalable algorithms to compute entropy of feature subsets has received little attention compared to the problem of computing frequency of itemsets, for which many algorithms have been proposed [2]. The present article addresses this problem as it introduces a new algorithm to compute entropy and mutual information for a large number of feature subsets, adopting the same divide and conquer strategy used by *FP-growth* [4] – one of the most efficient frequent itemset mining algorithm [2].

In order to illustrate its practical interest, the algorithm is then used to solve specifically the recently introduced problem of mining *reliable approximate functional dependencies* [9]. Given a target feature Y, the problem consists in finding the top-k feature sets $\mathcal{X}_{1 \leq i \leq k}$ which have the k highest *reliable fractions of information* relatively to Y. This score denoted $\hat{F}_0(\mathcal{X}; Y)$ is a robust estimation of the normalized mutual information between $\mathcal{X}$ and Y that is unbiased and equal to 0 in case of independence between $\mathcal{X}$ and Y. This prevents from misinterpreting strong observed dependencies between $\mathcal{X}$ and Y that are not statistically representative because they are based on a too small number of data. In the same article, an algorithm is proposed to mine exactly or approximatively the top-k reliable approximate functional dependencies (RAFD) using a parallelized beam search strategy coupled to a branch and bound pruning optimization.

While authors of [9] focus on small values for k (mainly $k = 1$), we are interested by much larger values, typically $k = 10^4$. This interest seems counterintuitive as the only presumable effect of increasing k is to produce more uninteresting patterns with lower scores. In reality, top-k patterns provide highly redundant pieces of information as similar patterns are likely to have similar scores. Increasing k provides a substantial list of top-k patterns from which can be extracted a reduced set of still highly scored but non redundant patterns called *Locally Optimal Patterns* (LOP) [10,11]: Given a pattern scoring function and given a neighbourhood function that maps every pattern to a set of neighbouring patterns, a pattern P is *locally optimal* if its score is maximal within P's neighbourhood. The neighbourhood generally used is a δ-metric neighbourhood: two sets of features $\mathcal{X}_1$ and $\mathcal{X}_2$ are neighbours if their distance $d(\mathcal{X}, \mathcal{X}')$ defined as the cardinalily $|\mathcal{X} \Delta \mathcal{X}'|$ of their symmetric difference is not greater than δ. Once top-k patterns have been mined, LOPs can easily be extracted by checking for every top-k pattern if some better ranked pattern is one of its neighbour (however this naive algorithm with a complexity in $\Theta(k^2)$ only works for relatively small values of k. For more elaborate algorithm, see [10]).

For sake of illustration, the amount of redundancy of a top-k pattern $\mathcal{X}$ can be assessed by the minimal distance $\delta_{min}(\mathcal{X})$ between $\mathcal{X}$ and any better scored pattern: the higher the distance, the more original the pattern. The first columns of Table 1 provide the histogram of δ_{min} of top-2 to top-10000 patterns, computed on some datasets used in the evaluation section. For almost every dataset, between 97 % to 100 % of top-k patterns differ only with one single feature from a better scored pattern. On the other side, the last four columns of Table 1 provide the rank distribution of LOPs (for $\delta = 1$) in the sorted list of top-k patterns.

Table 1. Histograms of δ_{min} and LOPs' rank among the top-10000 patterns.

Dataset	Distance δ_{min}					Rank of LOP			
	1	2	3	4	>1	1–10	11–100	101–1000	1001–10000
german	9968	29	2		32	4	8	13	7
lymphography	9964	33	1	1	36	4	5	15	12
vehicle	9943	54	1	1	57	6	20	17	14
sonar	9753	233	5	7	247	7	27	77	136
penbased	9719	279	0	1	281	10	76	154	41
segment	9961	38	0	0	39	5	5	22	7
specftheart	9902	95	0	1	98	3	8	36	51
twonorm	5157	4840	2	0	4843	10	90	900	3843
wdbc	9865	130	2	1	135	9	19	41	66

One notices that a significant part of LOPs have large ranks. It is thus essential to be able to mine top-k patterns with large values for k. Main contributions of this paper are:

1. An algorithm to compute entropy and mutual information of feature subsets, resulting from a non straightforward adaptation of the frequent itemset mining algorithm *FP-growth* [4].
2. An adaptation of the previous algorithm to address the problem of discovering the top-k *Reliable Approximate Functional Dependencies* [9]. The algorithm mines large numbers of patterns with low memory footprint so that it becomes possible to extract "hard-to-reach" locally optimal patterns.
3. Empirical evidences that for both problems, proposed algorithms outperform existing methods for both speed and scalability.

The rest of the paper is structured as follows: Sect. 2 considers the general problem of fast and scalable computation of entropic measures on sets of features and introduces the algorithm *HFP-growth*. Section 3 introduces algorithm *IFP-growth* to compute mutual information relatively to some target feature and applies it to the discovery of Reliable Approximate Functional Dependencies. Section 4 presents comparative tests performed to evaluate speed and scalability of HFP-growth and IFP-growth, before Sect. 5 concludes.

2   An Algorithm to Compute Entropy of Feature Subsets

2.1   Definitions and Problem Statement

Let's first define properly the required notions of entropy and data partitions. Given a dataset $\mathcal{D}$ of n data described by a set $\mathcal{F}$ of nominal features, data are interpreted as IID samples of some distribution where every feature $X \in \mathcal{F}$ is seen as a categorical random variable defined over some domain denoted $\mathcal{D}_X$. Given a subset $\mathcal{X} = \{X_1, \ldots, X_k\}$ of features listed in some arbitrary order, its *joint distribution* $P_{\mathcal{X}}$ is defined over the cartesian product $T(\mathcal{X}) = \mathcal{D}_{X_1} \times \cdots \times \mathcal{D}_{X_k}$ containing all k-tuples $(x_1, \ldots, x_k)$ for all $x_1 \in \mathcal{D}_{X_1}, \ldots, x_k \in \mathcal{D}_{X_k}$. Assuming the undefined form $0 \times \log_2(0)$ is equal to zero, the *entropy* $\mathrm{H}(\mathcal{X})$ of this joint distribution is defined as:

$$\mathrm{H}(\mathcal{X}) \stackrel{\text{def}}{=} \mathbb{E}\left(\log_2\left(\frac{1}{P_{\mathcal{X}}}\right)\right) = - \sum_{t \in T(\mathcal{X})} P_{\mathcal{X}}(t) \log_2(P_{\mathcal{X}}(t)) \tag{1}$$

The *empirical entropy* $\hat{\mathrm{H}}(\mathcal{X})$ estimates this entropy from the available samples, replacing probability $P_{\mathcal{X}}(t)$ of tuple t with its *relative frequency* $\sigma_{\mathcal{D}}(t)$ in $\mathcal{D}$. Entropy is a monotonic function: given two feature subsets $\mathcal{X}_1$ and $\mathcal{X}_2$, $\mathcal{X}_1 \subseteq \mathcal{X}_2$ implies $\hat{\mathrm{H}}(\mathcal{X}_1) \le \hat{\mathrm{H}}(\mathcal{X}_2)$. The entropy is minimal and equal to zero for the empty set; it is maximal for the whole set $\mathcal{F}$ of features. In order to formalize the problem of computing efficiently the entropy of a large number of feature subsets, one considers the analog problem of computing frequency of frequent patterns. To this end, one defines the *relative entropy* $\hat{h}(\mathcal{X})$ as the ratio of $\hat{\mathrm{H}}(\mathcal{X})$ over the maximal possible entropy $\hat{\mathrm{H}}(\mathcal{F})$ of all features so that its values are always between 0 and 1. One then says a subset $\mathcal{X}$ of features is *definite* relatively to some threshold $h_{max} \in [0, 1]$ if $\hat{h}(\mathcal{X}) \le h_{max}$ (Definite subsets of binary features are also called low-entropy sets in [6]). The considered problem is then the following:

Problem 1. Given a dataset $\mathcal{D}$ of nominal features and a threshold $h_{max} \in [0, 1]$, the *problem of mining definite feature subsets* consists in computing the empirical entropy of every definite subset of features relatively to h_{max} and $\mathcal{D}$.

While this problem can naively be solved by implementing an APriori like algorithm [1] based on formula 1, this method is highly unefficient, not only because the APriori approach is not the best strategy but also because expression 1 requires for every feature subset $\mathcal{X}$ to compute frequencies of all possible tuples whose number increases exponentially with the size of $\mathcal{X}$. In order to provide a more efficient algorithm, empirical entropy should be defined as a function of data partitions. A data partition is any partition of the dataset $\mathcal{D}$. Any subset $\mathcal{X}$ of features can be mapped to a data partition. For this purpose, let's say two data d_1 and d_2 are *equivalent* relatively to $\mathcal{X}$ if for every feature X of $\mathcal{X}$, their respective values $X(d_1)$ and $X(d_2)$ are equal. The set of equivalence classes defines the said *data partition* denoted $\mathcal{P}(\mathcal{X})$ of $\mathcal{X}$. The empirical entropy $\hat{\mathrm{H}}(\mathcal{X})$ can thus be rewritten as the entropy $\mathrm{H}(\mathcal{P}(\mathcal{X}))$ of its data partition defined as:

$$\mathrm{H}(\mathcal{P}(\mathcal{X})) \stackrel{\mathrm{def}}{=} \sum_{P \in \mathcal{P}(\mathcal{X})} \frac{|P|}{|\mathcal{D}|} \log_2 \left(\frac{|\mathcal{D}|}{|P|} \right) = \log_2(n) - \frac{\sum_{P \in \mathcal{P}(\mathcal{X})} |P| \log_2 (|P|)}{n} \quad (2)$$

The set of data partitions builds a lattice with intersection and union operators that necessarily induces an ordering relation called refinement relation: a partition $\mathcal{P}_1$ is a *refinement of* (or is *included in*) a partition $\mathcal{P}_2$ if every part of $\mathcal{P}_1$ is included in any part of $\mathcal{P}_2$. The intersection $\mathcal{P}_1 \cap \mathcal{P}_2$ is the most general refinement of $\mathcal{P}_1$ and $\mathcal{P}_2$:

$$\mathcal{P}_1 \cap \mathcal{P}_2 = \{ P_1 \cap P_2 \,/\, P_1 \in \mathcal{P}_1, P_2 \in \mathcal{P}_2, P_1 \cap P_2 \neq \emptyset \} \quad (3)$$

It is easy to prove by double inclusion that the data partition of a set $\mathcal{X}$ of features is the intersection of data partitions of all features of $\mathcal{X}$:

$$\mathcal{P}(\mathcal{X}) = \bigcap_{X \in \mathcal{X}} \mathcal{P}(\{X\}) \quad (4)$$

This latter property is essential to design an efficient mining algorithm. Indeed let's assume there exists some encoding of data partitions that enables an efficient procedure to compute the intersection of two data partitions and its entropy. Under this hypothesis, it gets possible to enumerate efficiently in a depth first search manner the definite feature subsets, by intersecting the data partition $\mathcal{P}(\mathcal{X})$ of the current pattern $\mathcal{X}$ with the data partition $\mathcal{P}(\{Y\})$ of the next feature $Y \notin \mathcal{X}$ to add and then by pruning the current branch as soon as the entropy of resulting partition $\mathcal{P}(\mathcal{X} \cup \{Y\})$ is larger than $H_{limit} = h_{max} \times \hat{\mathrm{H}}(\mathcal{F})$. The next section explains the algorithm in details.

2.2   Algorithm HFP-growth

The presented algorithm is called *HFP-growth* since it adopts FP-growth's data structure called *FP-tree* along with its divide and conquer strategy [4]. The symbol of entropy "H" emphasizes the fact HFP-growth computes entropy of definite features sets instead of frequency of frequent itemsets. The next paragraphs develop the three main components of HFP-growth: first, the *HFP-tree* data structure that is an adaptation of FP-trees, then the algorithmic primitives to process the HFP-tree which are the most different part compared to FP-growth, finally the global algorithm with its divide and conquer approach.

Table 2. Dataset

F_1	F_2	F_3
a	c	f
a	c	g
a	d	f
a	e	f
b	c	f
b	c	g
b	d	f
b	e	f

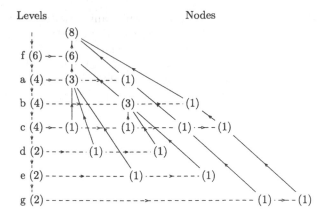

Fig. 1. Equivalent FP-tree

HFP-Tree. FP-growth is a frequent itemset mining algorithm that stores the whole dataset in memory thanks to a compact data structure called *FP-tree*. Many variants of FP-tree exist. Only the simplest most essential form is presented here: An FP-tree is mainly a lexicographic tree, also called trie, that encodes a dataset viewed as a collection of itemsets. In order for the trie to have the smallest memory footprint and thus the smallest number of nodes, the unfrequent items are removed and the remaining frequent items are sorted in decreasing order of frequency. The FP-tree represented on Fig. 1 is built from dataset shown on Table 2. An FP-tree also provides for every item i a single linked list enumerating nodes representing item i in the lexicographic tree, traversing the trie from left to right. These lists are called *levels* hereafter. Levels are represented with dashed lines on Fig. 1. A node n represents the itemset containing items of levels intersecting the branch from the root node up to n. For instance rightmost node of g's level represents itemset bcg. Every node essentially stores pointers to its parent node and to its right sibling in its level, along with a counter set to the number of data containing node's itemset.

HFP-tree is an FP-tree with some differences as shown on Fig. 2:

– An HFP-tree has additional components called *groups* of levels. Every group corresponds to some feature X. X's group is the entry point for levels, one for each possible value of X. Levels attached to X's group thus represent the parts of $\mathcal{P}(\{X\})$.

– A group also stores entropy $\hat{H}(\{X\})$ of X computed at startup when reading the dataset. Features whose entropy $\hat{H}(\{X\})$ is greater than H_{limit} are not considered as they cannot be part of a definite feature set. This allows to sort groups in increasing order of entropy so that nodes representing the most definite features are close to the root node whereas leaf nodes at the bottom of the tree represent the most fluctuating features: this trick reduces the number of nodes, while remaining compatible with the intersection procedure

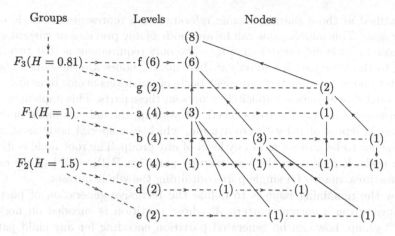

Fig. 2. An HFP-tree

explained later. In particular it does not interlace levels of different groups as standard FP-tree would do (like f and g levels of group F_3 on Fig. 1).

- A node stores in addition to fields already mentioned for FP-tree, two additional node pointers called *master* and *heir* that are essential for processing the tree as explained in the next subsection.
- Another difference compared to FP-growth is that only one HFP-tree is built to represent the input dataset. This characteristic is very convenient as HFP-growth does not dynamically allocate memory but at startup. If HFP-growth succeeds in building its tree (and in practice it always does on current datasets less than few gigabytes), it will eventually complete without running out of memory after potentially several hours of processing. In comparison FP-growth clones many FP-trees during its run. However this advantage has to be mitigated as some later FP-growth implementations have managed to avoid FP-tree cloning.

Processing HFP-Tree. The interest of HFP-tree is to enable a fast computation of data partitions when feature subsets are enumerated in a depth search order consistent with the way features are indexed. To explain why, let's describe in a first stage how a data partition $\mathcal{P}(\mathcal{X})$ is encoded in the HFP-tree for some given subset $\mathcal{X}$ of features without explaining how this encoding can be built. In the followings, index i of feature X_i refers to the feature of the ith group: feature X_1 matches the first group, i.e the closest of the tree root, feature X_2 matches the 2nd closest group and so on. Let's assume $\mathcal{X} = \{X_{i_1}, \ldots, X_{i_k}\}$ with $i_1 < \cdots < i_k$. Individual nodes in X_{i_k}'s levels represent parts of $\mathcal{P}\left(\cup_{i=1}^{i_k}\{X_i\}\right) = \cap_{i=1}^{i_k}\mathcal{P}(\{X_i\})$. Since $\mathcal{P}\left(\cup_{i=1}^{i_k}\{X_i\}\right)$ is a refinement of $\mathcal{P}(\mathcal{X})$, for any given part P of $\mathcal{P}(\mathcal{X})$, there are several nodes of X_{i_k}'s group that are part of P, or put another way, data members of P are covered by different nodes of X_{i_k}'s group. These nodes representing P might spread among several levels of the i_k^{th} group but they can be identified as those sharing a same reference to a representative node called *master node*. This master node can be any node of any previous or current group (i.e whose index is not greater than i_k). The only requirement is that two nodes belong to the same part if and only if their master nodes are the same. The use of master nodes to implicitly represent parts of partitions avoids dynamic memory allocation of objects to explicitly represent these parts. This implementation technique substantially improves speed. At startup, the only available partition encoding is represented by the root node, which is a special node as it is the only one not to be a member of any level of any group. The root node is its own master node. It represents the zero entropy partition $\mathcal{P}(\emptyset) = \{\mathcal{D}\}$ of the empty set of features, made of a single part containing the whole dataset.

Now the remaining issue is to define the recursive generation of partition encodings: given a current pattern $\mathcal{X}$ whose partition is encoded on nodes of the $i_k{}^{\text{th}}$ group, how can be generated partition encoding for any child pattern $\mathcal{X} \cup \{X_{i_{k+1}}\}$ of $\mathcal{X}$? Put another way, let's consider some new feature X_j with $j > i_k$ and let's assume nodes of the previous $(j-1)^{\text{th}}$ group encode partition of $\mathcal{X}$, then the jth group must generate two different exploration branches:

- Either one adds feature X_j to current pattern $\mathcal{X}$. In this case, one has to encode on nodes of the jth group, partition $\mathcal{P}(\mathcal{X} \cup \{X_j\}) = \mathcal{P}(\mathcal{X}) \cap \mathcal{P}(\{X_j\})$ using (1) the available encoding of partition $\mathcal{P}(\mathcal{X})$ by nodes of the $(j-1)^{\text{th}}$ group and (2) the partition $\mathcal{P}(\{X_j\})$ whose parts are levels of the jth group. This is called the `intersect` operation applied to the jth group.
- Or feature X_j is not added to $\mathcal{X}$. In this case, one simply has to forward the available encoding of $\mathcal{P}(\mathcal{X})$ from nodes of the $(j-1)^{\text{th}}$ group to nodes of the current j^{th} group. This is called the `skip` operation applied to the jth group.

The `skip` operation can easily be parallelized as it simply consists for every node n of every level of the jth group to declare its master node to be the master node of its parent node. The `intersect` operation is more subtle as every part of the $j-1^{\text{th}}$ group might be intersected by different levels of the jth group. For every level L of the jth group, one has to gather nodes of L whose parents are member of the same part of $\mathcal{P}(\mathcal{X})$, say otherwise, whose parents have the same master node. These subsets define new parts of $\mathcal{P}(\mathcal{X} \cup \{X_j\})$ as illustrated on Fig. 3.

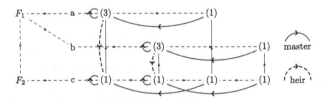

Fig. 3. Running `intersect` on c's level after skipping F_3 and intersecting F_1

More precisely, every time the parent of a node n of L has a master m not processed yet, a new part P of $\mathcal{P}(\mathcal{X} \cup \{X_j\})$ is discovered. n is then designated as the master of P and its reference is saved in master m as its *heir* node. When other nodes are found such that the master m of their parents is the same as the one for n, they receive the heir of m as their master. After processing nodes of a level, all discovered parts are complete so that the sum appearing in expression 2 can be updated by their cardinalities. After processing all levels of the group j, its nodes completely encode partition $\mathcal{P}(\mathcal{X} \cup \{X_j\})$ and since the sum of expression 2 is completed, intersect can return entropy $\hat{H}(\mathcal{X} \cup \{X_j\})$.

Algorithm. Once the HFP-tree has been built, HFP-growth uses a divide and conquer strategy based on the two previous operations skip and intersect. This strategy is similar with the one of FP-growth but while skip and intersect require a top down recursion (i.e. from the top of the tree to the bottom), FP-growth's algorithm requires to process levels in a bottom up order, starting from the deepest levels in the FP-tree. Let m be the number of groups in the HFP-tree, i.e. the number of features of $\mathcal{F}$ after removing feature X with entropy less than H_{limit}. The algorithm is based on a recursive mine function as shown by pseudocode 1. For every feature index i from 1 to $m-1$, it recursively calls mine on the next $i+1^{\text{th}}$ feature twice: first after applying the procedure intersect on the ith group in the case the returned entropy is not greater than H_{limit}, second after appling skip to the ith group systematically.

Inputs : A dataset $\mathcal{D}$ and a threshold $h_{max} \in [0,1]$
Output: List of definite feature sets $\mathcal{X}$ with their entropy $\hat{H}(\mathcal{X})$

HFP-tree $\mathcal{T}$, $H_{limit} \leftarrow$ build-HFP-tree$(\mathcal{D}, h_{max})$;
output$(\emptyset, 0)$;
mine$(\emptyset, 1)$

function mine$(\mathcal{X}, j)$ is
  if $j \leq$ number m of groups of $\mathcal{T}$ then
    $H \leftarrow$ intersect$(\mathcal{T}, j)$;
    if $H \leq H_{limit}$ then
      output$(\mathcal{X} \cup \{X_j\}, H)$;
      mine$(\mathcal{X} \cup \{X_j\}, j+1)$
    end
    skip$(\mathcal{T}, j)$;
    mine$(\mathcal{X}, j+1)$
  end
end

Algorithm 1. The HFP-growth algorithm

3   An Algorithm to Compute Mutual Information of Feature Subsets

In this section one shows how HFP-growth can be adapted to compute efficiently mutual information of feature subsets with some target feature, along with an unbiased variant of it introduced in [9]. Because the main subject of this paper is the efficient computation of entropic measures, one could limit the study to mutual information since computing the unbiased variant does not fundamentally change the algorithm. However this provides at the same time a sounder statistical problem to solve and an existing algorithm to compare with. The resulting algorithm is called *IFP-growth*, "I" standing for information.

3.1   Problem Statement

Mutual information $I(\mathcal{X}; \mathcal{Y}) = H(\mathcal{X}) + H(\mathcal{Y}) - H(X \cup Y)$ estimates the amount of information shared by two feature subsets $\mathcal{X}$ and $\mathcal{Y}$. It is equal to zero in case $\mathcal{X}$ and $\mathcal{Y}$ are independent. Mutual information is particularly interesting in supervised problems where $\mathcal{Y}$ is restricted to a single target feature Y. In such problems, one searches to predict Y from highly dependent feature subsets $\mathcal{X}$, i.e. with a large mutual information $I(\mathcal{X}; \{Y\})$. Mutual information can be estimated empirically from data partitions according to:

$$\hat{I}(\mathcal{X}; \mathcal{Y}) = \hat{H}(\mathcal{P}(\mathcal{Y})) + \hat{H}(\mathcal{P}(\mathcal{X})) - \hat{H}(\mathcal{P}(\mathcal{X}) \cap \mathcal{P}(\mathcal{Y})) \qquad (5)$$

Because $H(\mathcal{Y})$ is an upper bound for $I(\mathcal{X}; \mathcal{Y})$, one often uses a normalized mutual information $F(\mathcal{X}; \mathcal{Y}) = \frac{I(\mathcal{X};\mathcal{Y})}{H(\mathcal{Y})}$ within the range $[0, 1]$ called *fraction of information ratio* in [3,9]. As stated in the introduction, mutual information $\mathcal{X} \mapsto \hat{I}(\mathcal{X}, \mathcal{Y})$ is a non decreasing function of $\mathcal{X}$. Obviously the more features in $\mathcal{X}$, the more predictable $\mathcal{Y}$ from $\mathcal{X}$. However one should not forget that a dataset $\mathcal{D}$ is a limited sampling of the real joint distribution of $\mathcal{X}$ and $\mathcal{Y}$. For large feature sets $\mathcal{X}$, data partition of $\mathcal{P}(\mathcal{X} \cup \mathcal{Y})$ is the intersection $\cap_{X \in \mathcal{X} \cup \mathcal{Y}} \mathcal{P}(\{X\})$ of many data partitions. Therefore parts of $\mathcal{P}(\mathcal{X} \cup \mathcal{Y})$ get statistically very small when the size of $\mathcal{X}$ increases. Within these small parts, strong but spurious dependencies appear between $\mathcal{X}$ and $\mathcal{Y}$ even when $\mathcal{X}$ and $\mathcal{Y}$ are drawn from independent distributions. New entropic measures have since been proposed in [9,12–14]. These measures are similar in spirit with mutual information but robust to the previous "just by chance" artefact. One of these mesures considered in [9] is the *reliable fraction of information* $\hat{F}_0(\mathcal{X}; \mathcal{Y})$ that is unbiased, i.e whose expected value is 0 when $\mathcal{X}$ and $\mathcal{Y}$ are independent. More precisely $\hat{F}_0(\mathcal{X}; \mathcal{Y}) = \hat{F}(\mathcal{X}; \mathcal{Y}) - \frac{\hat{m}_0(\mathcal{X},\mathcal{Y})}{\hat{H}(\mathcal{Y})}$ where $\hat{m}_0(\mathcal{X}, \mathcal{Y})$ is the expected value of $\hat{I}(\mathcal{X}; \mathcal{Y})$ under hypothesis of independence between $\mathcal{X}$ and $\mathcal{Y}$. This bias is computed using a permutation model defined over contingency tables of $\mathcal{X}$ and $\mathcal{Y}$ respectively. Since the elements in these tables are nothing else than the cardinalities of parts in $\mathcal{P}(\mathcal{X})$ and $\mathcal{P}(\mathcal{Y})$, $\hat{m}_0(\mathcal{X}, \mathcal{Y})$ can be rewritten as $\hat{m}_0(\mathcal{P}(\mathcal{X}), \mathcal{P}(\mathcal{Y}))$, i.e as a function depending only on $\mathcal{P}(\mathcal{X})$ and $\mathcal{P}(\mathcal{Y})$. The exact expression of $\hat{m}_0(\mathcal{P}(\mathcal{X}), \mathcal{P}(\mathcal{Y}))$ is given by a sum of expected values for hypergeometric distributions. The detailed equation and derivation details are provided in [9] with reference to [12,14]. The expression of $\hat{F}_0(\mathcal{X};\mathcal{Y})$ to compute is finally:

$$\hat{F}_0(\mathcal{X};\mathcal{Y}) = 1 + \frac{\hat{H}(\mathcal{P}(\mathcal{X})) - \hat{H}(\mathcal{P}(\mathcal{X}) \cap \mathcal{P}(\mathcal{Y})) - \hat{m}_0(\mathcal{P}(\mathcal{X}), \mathcal{P}(\mathcal{Y}))}{\hat{H}(\mathcal{P}(\mathcal{Y}))} \quad (6)$$

It is worth noting $\mathcal{X} \mapsto \hat{F}_0(\mathcal{X};\mathcal{Y})$ is not a monotonic function as mutual information is. It has the expected nice property of penalizing with low scores not only short non informative sets of features $\mathcal{X}$ but also long informative but not statistically representative patterns. Finding the top-k feature sets $\mathcal{X}$ with highest score $\hat{F}_0(\mathcal{X};\{Y\})$ relatively to a target feature Y is thus a sound and non trivial optimization problem addressed in [9]: the resulting associations $\mathcal{X} \rightarrow Y$ are called the top-k *reliable approximate functional dependencies* (RAFD). A mining algorithm is also proposed in [9] whose implementation is called dora. This algorithm finds these dependencies using a beam search strategy coupled to a branch and bound pruning: it backtracks the current branch as soon as the upper bound $1 - \hat{m}_0(\mathcal{X};\{Y\})/\hat{H}(\{Y\})$ of scores accessible from the current pattern $\mathcal{X}$ is not greater than score $\hat{F}_0(\mathcal{X}_k;\{Y\})$ of the worst top-k pattern $\mathcal{X}_k$ found so far. In order to process difficult datasets, the algorithm can also solve a relaxed version of this problem: it consists in replacing the previous pruning condition by predicate $\alpha \times (1 - \hat{m}_0(\mathcal{X};\{Y\})/\hat{H}(\{Y\})) \leq \hat{F}_0(\mathcal{X}_k;\{Y\})$ for some parameter $\alpha \in]0,1]$, a value $\alpha = 1$ corresponding to the exact resolution. This approximation amounts to find k feature subsets $(\hat{\mathcal{X}}_i)_{1 \leq i \leq k}$ so that the lowest score $\hat{F}_0(\hat{\mathcal{X}}_k;\{Y\})$ of these patterns is not lower than $\alpha \times \tilde{F}_0(\mathcal{X}_k;\{Y\})$ where $\mathcal{X}_k$ is the pattern with the lowest score among the real top-k patterns $(\mathcal{X}_i)_{1 \leq i \leq k}$. In summary, mining *reliable approximate functional dependencies* is defined by a dataset $\mathcal{D}$, a target feature Y, a number k and an α coefficient. A new algorithm to address this problem is proposed in the next section.

3.2   Algorithm IFP-Growth

Two preliminary remarks can be done about Eq. 6: first $\mathcal{P}(\mathcal{Y})$ and a fortiori $\hat{H}(\mathcal{P}(\mathcal{Y}))$ are constants independent of $\mathcal{X}$ so that they can be computed once forever at startup. Second $\hat{F}_0(\mathcal{X};\mathcal{Y})$ could be computed directly with two parallel HFP-trees: one whose groups encode $\mathcal{P}(\{X\})$ for every feature $X \in \mathcal{F} \setminus \{Y\}$, the second whose groups encode $\mathcal{P}(\{X\} \cap \mathcal{P}(\{Y\}))$. However this approach is memory-costly while it is possible to get a solution with a unique HFP-tree. The resulting algorithm *IFP-growth* is summarized by pseudocode 2.

A first trick of IFP-growth is to put the group of target feature Y at the bottom of HFP-tree, independently of its entropy. This last group is processed differently than the others. Another trick is to switch the order HFP-growth

Inputs : A dataset $\mathcal{D}$, target feature Y, number k, coef. $\alpha \in [0, 1]$
Output: List of top-k feature sets $(\mathcal{X}_i)_{1 \leq i \leq k}$ with their score $\hat{F}_0(\mathcal{X}_i)$

$F_0^{kth} \leftarrow 0$; $Q \leftarrow$ build-min-priority-queue() ;
HFP-tree $\mathcal{T}$, $H_Y \leftarrow$ build-IFP-tree$(\mathcal{D}, Y)$;
mine$(\emptyset, 1)$; Output content of Q

function mine $(\mathcal{X}, j)$ **is**
   if $j <$ *number m of groups of* $\mathcal{T}$ **then**
      skip$(\mathcal{T}, j)$; mine$(\mathcal{X}, j+1)$;
      $H_X \leftarrow$ intersect$(\mathcal{T}, j)$ $m_0 \leftarrow$ compute-bias$(\mathcal{T}, j)$;
      if $1 - m_0/H_Y > F_0^{kth}/\alpha$ **then**
        | mine$(\mathcal{X} \cup \{X_j\}, j+1)$
      end
   else
      $H_{XY} \leftarrow$ intersect$(\mathcal{T}, j)$;
      $F_0 \leftarrow 1 + (H_X - H_{XY} - m_0)/H_Y$;
      Q.insert$((\mathcal{X}, F_0))$;
      if Q.size $> k$ **then**
        | Q.pop-min() ; $F_0^{kth} \leftarrow \min(Q).F_0$
      end
   end
end

Algorithm 2. The IFP-growth algorithm

calls intersect and skip: IFP-growth first develops the skip 's branch before intersect 's one. When intersect is applied to $i_k{}^{th}$ group in order to generate encoding of $\mathcal{P}(\{X_{i_1}, \ldots, X_{i_k}\})$, combination of these two changes allow:

- First to save in global variables the entropy $\hat{H}(\mathcal{X})$ (that intersect just returned) and the bias $\hat{m}_0(\mathcal{X}; \{Y\})$ that can be computed from constant $\mathcal{P}(\mathcal{Y})$ and from $\mathcal{P}(\mathcal{X})$ (whose encoding has also been computed by intersect).
- Then to call recursively the skip operation on the successive groups up to reaching the last group of Y without modifying values of the two global variables. Calling intersect then returns $\hat{H}(\mathcal{X} \cup \{Y\})$. At this points all terms of Eq. 6 are available to compute $\hat{F}_0(\mathcal{X}_k; \{Y\})$ and see if this score is sufficient for it to be inserted in the priority queue Q storing the top-k patterns.

As shown by pseudocode 2, the branch and bound pruning strategy with coefficient α can seamlessly be integrated in IFP-growth. However IFP-growth pruning is assumed to be less efficent than dora's one since dora uses a beam search strategy converging quickly to good top-k pattern candidates while HFP-tree dictates more rigidly the order in which subsets must be enumerated.

4   Empirical Evaluation

For sake of comparison, datasets considered in [9] are reused (see the KEEL data repository at http://www.keel.es). However in order to limit their number, only the most challenging datasets given on Table 3 are considered, defined as those whose processing by either IFP-growth or dora requires more than 10 s for $k = 1$. In order to be fair with dora whose beam search strategy is compatible with an intensive use of multithreading, tests are run on an Intel Xeon Silver 4414 biprocessor with a total of 20 hyper-threaded cores. The memory footprint of every running algorithm is monitored (internal Java Virtual Machine's heap size for dora and processus heap size for other algorithms) and limited to 45 GB. Source codes of HFP-growth, IFP-growth and HApriori can be downloaded from https://github.com/P-Fred/HFP-Growth whereas dora is available from http://eda.mmci.uni-saarland.de/prj/dora.

Evaluation of HFP-Growth. Since HFP-growth cannot be straightforwardly compared with an existing algorithm, one studies to what extent the well known performance gap between FP-growth and the baseline algorithm APriori [2] is reproduced between HFP-growth and the APriori counterpart, specially implemented for this purpose. This latter algorithm, hereafter called H-APriori, uses the same levelwise pruning method as APriori to remove candidate subsets having at least one predecessor that is not definite. It then computes in one pass over the dataset the entropies of all remaining candidates using formula 1. HFP-growth's processing times (resp. memory footprints) as functions of h_{max} are given on Fig. 4 (resp. Fig. 5). The time (resp. memory) gain factor defined as the ratio of H-APriori's processing time (resp. memory footprint) over HFP-growth's one is provided on Fig. 6 (resp. Fig. 7).

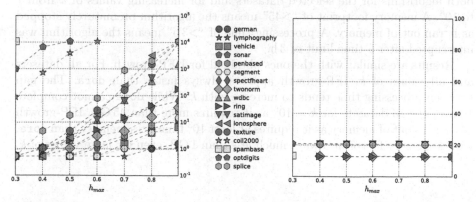

Fig. 4. HFP-growth's processing times (in seconds)

Fig. 5. HFP-growth's memory footprints (in megabytes)

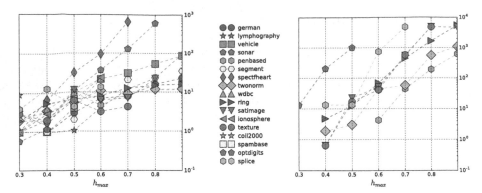

Fig. 6. Time gain factors   **Fig. 7.** Memory gain factors

HFP-growth appears 1 to 3 orders of magnitude faster than H-APriori, with a speedup factor increasing with threshold h_{max}. Memory gain is even bigger: HFP-growth has a low constant memory footprint of few megabytes (values on Fig. 5 appear to be rounded by the OS's heap allocation mechanism) whereas H-APriori shortly requires many gigabytes of memory to store candidate patterns. While HFP-growth can run as long as necessary, H-APriori often runs out of memory before completing.

Evaluation of IFP-Growth. In order to limit the number of tests to compare IFP-growth and dora, one only considers the exact problem of discovering RAFD, i.e for $\alpha = 1$. However similar conclusions can be drawn for values of α less than one. Table 3 summarizes processing times and memory footprints of both algorithms for the selected datasets and for increasing values of k from 1 to 10^6. A memory footprint of ">45" means the algorithm prematurely stopped as it ran out of memory. A processing time of ">3 h" means the algorithm was interrupted after a time limit of 3 h.

Results are similar with the ones obtained for HFP-growth: For all datasets and all values of k, IFP-growth appears always faster than dora. The gap between processing time tends to increase with k. While dora cannot complete 15 over the 16 datasets for $k = 10^6$ as it requires more than 45 GB, IFP-growth never runs out of memory as it requires at least 10^3 times less memory than dora, with a memory footprint always much less than 1 GB, even for large values of k.

Table 3. Comparison of processing times and memory footprints of IFP-growth (IFPG) and dora for various datasets and values for k. Datasets are sorted from the easiest to the most difficult (according to IFPG's processing time for $k = 10^6$). Tuples under dataset names provide the main dataset characteristics: for a tuple (n, m, c), n is the number of data, m is the number of features and c is the number of classes for the target feature.

Dataset (n, m, c)	Algo.	Processing time (s.)				Memory footprint (GB)			
		$k = 1$	$k = 10^2$	$k = 10^4$	$k = 10^6$	$k = 1$	$k = 10^2$	$k = 10^4$	$k = 10^6$
german	IFPG	21	22	23	22	0.005	0.006	0.007	0.010
(1000,20,2)	dora	110	110	130	150	14	15	16	18
lymphography	IFPG	46	50	73	110	0.005	0.005	0.010	0.036
(148,18,4)	dora	69	80	120	410	14	16	17	27
vehicle	IFPG	130	140	160	160	0.006	0.007	0.008	0.36
(846,18,4)	dora	820	840	880	???	26	26	26	> 45
sonar	IFPG	220	260	320	470	0.005	0.005	0.007	0.12
(208,60,2)	dora	1300	1600	2000	???	38	40	42	> 45
penbased	IFPG	30	38	225	500	0.023	0.023	0.024	0.030
(10992,16,10)	dora	160	260	5000	???	15	16	43	> 45
segment	IFPG	12	20	110	600	0.008	0.009	0.015	0.044
(2310,19,7)	dora	43	84	390	???	15	15	22	> 45
spectfheart	IFPG	1800	2000	2200	2300	0.005	0.006	0.007	0.13
(267,44,2)	dora	???	???	???	???	> 45	> 45	> 45	> 45
twonorm	IFPG	160	170	370	4600	0.019	0.019	0.020	0.10
(7400,21,3)	dora	2800	3600	???	???	43	43	> 45	> 45
wdbc	IFPG	120	300	710	6330	0.007	0.007	0.008	0.12
(569,30,2)	dora	510	770	1800	???	18	21	32	> 45
ring	IFPG	970	1100	1900	6700	0.019	0.019	0.020	0.140
(7400,20,2)	dora	> 3h.	???	???	???	???	> 45	> 45	> 45
satimage	IFPG	780	1300	3700	> 3h.	0.026	0.026	0.028	???
(6435,36,7)	dora	4500	> 3h.	???	???	31	???	> 45	> 45
ionosphere	IFPG	640	2300	> 3h.	> 3h.	0.010	0.010	???	???
(351,33,2)	dora	1900	> 3h.	???	???	30	???	> 45	> 45
texture	IFPG	3000	4500	> 3h.	> 3h.	0.025	0.025	???	???
(5500,40,11)	dora	> 3h.	> 3h.	???	???	???	???	> 45	> 45
coil2000	IFPG	> 3h.	> 3h.	> 3h.	> 3h.	???	???	???	???
(9822,85,2)	dora	> 3h.	> 3h.	> 3h.	???	???	???	???	> 45
spambase	IFPG	> 3h.	> 3h.	> 3h.	> 3h.	???	???	???	???
(4597,57,2)	dora	> 3h.	> 3h.	> 3h.	???	???	???	???	> 45
optdigits	IFPG	> 3h.	> 3h.	> 3h.	> 3h.	???	???	???	???
(5620,64,10)	dora	???	???	???	???	> 45	> 45	> 45	> 45
splice	IFPG	> 3h.	> 3h.	> 3h.	> 3h.	???	???	???	???
(3190,60,3)	dora	???	???	???	???	> 45	> 45	> 45	> 45

5   Conclusion

In this paper, an efficient algorithm computes entropy of definite feature subsets likewise efficient algorithms exist to compute frequency of frequent itemsets. This algorithm proves to be much faster and scalable than its counterpart based on a levelwise approach. This algorithm is extended to compute mutual information and reliable fraction of information relatively to some target feature. Again, when applied to the discovery of the top-k reliable approximate functional dependencies, this algorithm shows an important gain of time and space efficiency compared to the existing algorithm. While these algorithms have been instanciated on two specific problems, they should be considered as generic algorithmic building blocks enabling to solve various data mining problems relying on entropic measures such as entropy or mutual information. These algorithms can also be easily adapted for computing any other non entropic measures whose expression depends mainly on data partitions. This is the case of the first scores proposed in the context of approximate functional dependencies. The level of performance of these algorithms also enable a more systematic search of sets of non redundant informative subsets of features. Another promising application is the extraction of Bayesian networks from data as these models can be seen as solutions of entropy-based optimization problems.

Ackowledgements. This work has been partially supported by the European Interreg Grande Région project "GRONE".

References

1. Agrawal, R., Srikant, R.: Fast algorithms for mining association rules in large databases. In: Proceedings of 20th International Conference on Very Large Data Bases, pp. 487–499. Morgan Kaufmann, Santiago de Chile (1994)
2. Bayardo, R., Goethals, B., Zaki, M.J. (eds.) Proceedings of the IEEE ICDM Workshop on Frequent Itemset Mining Implementations. Brighton (2004). http://fimi.ua.ac.be/
3. Dalkilic, M.M., Roberston, E.L.: Information dependencies. In: Proceedings of the 19th ACM SIGMOD-SIGACT-SIGART Symposium on Principles of Database Systems, pp. 245–253. ACM, Dallas (2000)
4. Han, J., Pei, J., Yin, Y., Mao, R.: Mining frequent patterns without candidate generation: a frequent-pattern tree approach. Data Min. Knowl. Discov. 8(1), 53–87 (2004)
5. Heikinheimo, H., Hinkkanen, E., Mannila, H., Mielikinen, T., Seppnen, J.K.: Finding low-entropy sets and trees from binary data. In: Proceedings of the 13th ACM SIGKDD International Conference on Knowledge Discovery and Data Mining, p. 350. ACM, San Jose (2007)
6. Heikinheimo, H., Vreeken, J., Siebes, A., Mannila, H.: Low-entropy set selection. In: Proceedings of the SIAM International Conference on Data Mining, pp. 569–580. SIAM, Sparks (2009)
7. Knobbe, A.J., Ho, E.K.Y.: Maximally informative k-itemsets and their efficient discovery. In: Proceedings of the 12th ACM SIGKDD International Conference on Knowledge Discovery and Data Mining, pp. 237–244. ACM, Philadelphia (2006)

8. Mampaey, M.: Mining non-redundant information-theoretic dependencies between itemsets. In: Bach Pedersen, T., Mohania, M.K., Tjoa, A.M. (eds.) DaWaK 2010. LNCS, vol. 6263, pp. 130–141. Springer, Heidelberg (2010). https://doi.org/10. 1007/978-3-642-15105-7_11

9. Mandros, P., Boley, M., Vreeken, J.: Discovering reliable approximate functional dependencies. In: Proceedings of the 23rd ACM SIGKDD International Conference on Knowledge Discovery and Data Mining, pp. 355–363. ACM, Halifax (2017)

10. Pennerath, F.: Fast extraction of locally optimal patterns based on consistent pattern function variations. In: Balcázar, J.L., Bonchi, F., Gionis, A., Sebag, M. (eds.) ECML PKDD 2010. LNCS (LNAI), vol. 6323, pp. 34–49. Springer, Heidelberg (2010). https://doi.org/10.1007/978-3-642-15939-8_3

11. Pennerath, F., Napoli, A.: The model of most informative patterns and its application to knowledge extraction from graph databases. In: Buntine, W., Grobelnik, M., Mladenić, D., Shawe-Taylor, J. (eds.) ECML PKDD 2009. LNCS (LNAI), vol. 5782, pp. 205–220. Springer, Heidelberg (2009). https://doi.org/10.1007/978-3-642-04174-7_14

12. Romano, S., Bailey, J., Nguyen, V., Verspoor, K.: Standardized mutual information for clustering comparisons: one step further in adjustment for chance. In: Proceedings of the 31st International Conference on Machine Learning, vol. 32, pp. 1143–1151. PMLR, Bejing (2014)

13. Romano, S., Vinh, N.X., Bailey, J., Verspoor, K.: A framework to adjust dependency measure estimates for chance. In: Proceedings of the 2016 SIAM International Conference on Data Mining, pp. 423–431. SIAM, Miami (2016)

14. Vinh, N.X., Epps, J., Bailey, J.: Information theoretic measures for clusterings comparison: variants, properties, normalization and correction for chance. J. Mach. Learn. Res. 11, 2837–2854 (2010)

Anytime Subgroup Discovery in Numerical Domains with Guarantees

Aimene Belfodil[1,2(✉)], Adnene Belfodil[1(✉)], and Mehdi Kaytoue[1,3]

[1] Univ Lyon, INSA Lyon, CNRS, LIRIS UMR 5205, 69621 Lyon, France
{aimene.belfodil,adnene.belfodil,mehdi.kaytoue}@insa-lyon.fr
[2] Mobile Devices Ingénierie, 100 Avenue Stalingrad, 94800 Villejuif, France
[3] Infologic, 99 avenue de Lyon, 26500 Bourg-Lès-Valence, France

Abstract. Subgroup discovery is the task of discovering patterns that accurately discriminate a class label from the others. Existing approaches can uncover such patterns either through an exhaustive or an approximate exploration of the pattern search space. However, an exhaustive exploration is generally unfeasible whereas approximate approaches do not provide guarantees bounding the *error* of the best pattern quality nor the exploration progression (*"How far are we of an exhaustive search"*). We design here an algorithm for mining numerical data with three key properties w.r.t. the state of the art: (i) It yields progressively interval patterns whose quality improves over time; (ii) It can be interrupted anytime and always gives a guarantee bounding the *error* on the top pattern quality and (iii) It always bounds a distance to the exhaustive exploration. After reporting experimentations showing the effectiveness of our method, we discuss its generalization to other kinds of patterns. Code related to this paper is available at: https://github.com/Adnene93/RefineAndMine.

Keywords: Subgroup discovery · Anytime algorithms · Discretization

1 Introduction

We address the problem of discovering patterns that accurately discriminate one class label from the others in a numerical dataset. Subgroup discovery (SD) [27] is a well established pattern mining framework which strives to find out data regions uncovering such interesting patterns. When it comes to numerical attributes, a pattern is generally a conjunction of restrictions over the attributes, e.g., pattern $50 \leq age < 70 \land smoke\_per\_day \geq 3$ fosters lung cancer incidence. To look for such patterns (namely interval patterns), various approaches are usually implemented. Common techniques perform *a discretization* transforming the

---

A. Belfodil and A. Belfodil—Both authors contributed equally to this work.

Electronic supplementary material The online version of this chapter (https://doi.org/10.1007/978-3-030-10928-8_30) contains supplementary material, which is available to authorized users.

M. Berlingerio et al. (Eds.): ECML PKDD 2018, LNAI 11052, pp. 500–516, 2019.
https://doi.org/10.1007/978-3-030-10928-8_30

numerical attributes to categorical ones in a pre-processing phase before using the wide spectrum of existing mining techniques [2,3,20,22]. This leads, however, to a loss of information even if an exhaustive enumeration is performed on the transformed data [2]. Other approaches explore the whole search space of all restrictions either exhaustively [6,14,18] or heuristically [5,23]. While an exhaustive enumeration is generally unfeasible in large data, the various state-of-the-art algorithms that heuristically explore the search space provide no provable guarantee on how they *approximate* the top quality patterns and on how far they are from an exhaustive search. Recent techniques set up a third and elegant paradigm, that is direct sampling approaches [3,4,13]. Algorithms falling under this category are non-enumerative methods which directly sample solutions from the pattern space. They simulate a distribution which rewards high quality patterns with respect to some interestingness measure. While [3,4] propose a direct two-step sampling procedure dedicated for categorical/boolean datasets, authors in [13] devise an interesting framework which add a third step to handle the specificity of numerical data. The proposed algorithm addresses the discovery of dense neighborhood patterns by defining a new density metric. Nevertheless, it does not consider the discovery of discriminant numerical patterns in labeled numerical datasets. Direct sampling approaches abandon the completeness property and generate only approximate results. In contrast, anytime pattern mining algorithms [5,16] are enumerative methods which exhibits the anytime feature [29], a solution is always available whose quality improves gradually over time and which converges to an exhaustive search if given enough time, hence ensuring completeness. However, to the best of our knowledge, no existing anytime algorithm in SD framework, makes it possible to ensure guarantees on the patterns discriminative power and the remaining distance to an exhaustive search while taking into account the nature of numerical data.

To achieve this goal, we propose a novel anytime algorithm, `RefineAndMine`, tailored for discriminant interval patterns discovery in numerical data. It starts by mining interval patterns in a coarse discretization, followed by successive refinements yielding increasingly finer discretizations highlighting potentially new interesting patterns. Eventually, it performs an exhaustive search, if given enough time. Additionally, our method gives two provable guarantees at each refinement. The first evaluates how close is the best found pattern so far to the optimal one in the whole search space. The second measures how already found patterns are diverse and cover well all the interesting regions in the dataset.

The outline is as follows. We recall in Sect. 2 basic definitions. Next, we define formally the problem in Sect. 3. Subsequently We introduce in Sect. 4 our mining algorithm before formulating the guarantees it provides in Sect. 5. We empirically evaluate the efficiency of `RefineAndMine` in Sect. 6 and discuss its potential improvements in Sect. 7. Additional materials are available in our companion page[1]. For more details and proofs, please refer to the technical report[2].

---

[1] https://github.com/Adnene93/RefineAndMine.
[2] https://goo.gl/NWtXfp.

2 Preliminaries

Input. A *labeled numerical dataset* $(\mathcal{G}, \mathcal{M})$ is given by a finite set (of objects) $\mathcal{G}$ partitioned into two subsets $\mathcal{G}^+$ and $\mathcal{G}^-$ enclosing respectively positive (target) and negative instances; and a sequence of numerical attributes $\mathcal{M} = (m_i)_{1 \leq i \leq p}$ of size $p = |\mathcal{M}|$. Each *attribute* m_i is an application $m_i : \mathcal{G} \to \mathbb{R}$ that associates to each object $g \in \mathcal{G}$ a value $m_i(g) \in \mathbb{R}$. We can also see $\mathcal{M}$ as a mapping $\mathcal{M} : \mathcal{G} \to \mathbb{R}^p, g \mapsto (m_i(g))_{1 \leq i \leq p}$. We denote $m_i[\mathcal{G}] = \{m_i(g) \mid g \in \mathcal{G}\}$ (More generally, for a function $f : E \to F$ and a subset $A \subseteq E$, $f[A] = \{f(e) \mid e \in A\}$). Figure 1 (left table) presents a 2-dimensional labeled numerical dataset and its representation in the Cartesian plane (filled dots represent positive instances).

Interval Patterns and Their Extents. When dealing with numerical domains in SD, we generally consider for intelligibility *interval patterns* [18]. An *Interval pattern* is a conjunction of restrictions over the numerical attributes; i.e. a set of conditions *attribute* $\gtrless v$ with $\gtrless \in \{=, \leq, <, \geq, >\}$. Geometrically, interval patterns are *axis-parallel hyper-rectangles*. Figure 1 (center-left) depicts pattern (non-hatched rectangle) $c_2 = (1 \leq m_1 \leq 4) \wedge (0 \leq m_2 \leq 3) \triangleq [1, 4] \times [0, 3]$.

Interval patterns are naturally partially ordered thanks to "hyper-rectangle inclusion". We denote the *infinite partially ordered set* (*poset*) of all interval patterns by $(\mathcal{D}, \sqsubseteq)$ where $\sqsubseteq$ (same order used in [18]) denotes the dual order $\supseteq$ of hyper-rectangle inclusion. That is pattern $d_1 \sqsubseteq d_2$ iff d_1 encloses d_2 ($d_1 \supseteq d_2$). It is worth mentioning that $(\mathcal{D}, \sqsubseteq)$ forms a *complete lattice* [26]. For a subset $S \subseteq \mathcal{D}$, the join $\bigsqcup S$ (i.e. smallest upper bound) is given by the rectangle intersection. Dually, the meet $\bigsqcap S$ (i.e the largest lower bound) is given by the smallest hyper-rectangle enclosing all patterns in S. Note that the top (resp. bottom) pattern in $(\mathcal{D}, \sqsubseteq)$ is given by $\top = \emptyset$ (resp. $\bot = \mathbb{R}^p$). Figure 1 (right) depicts two patterns (hatched) $e_1 = [1, 5] \times (1, 4]$ and $e_2 = [0, 4) \times [2, 6]$, their meet (non hatched) $e_1 \sqcap e_2 = [0, 5] \times (1, 6]$ and their join (black) $e_1 \sqcup e_2 = [1, 4) \times [2, 4]$.

A pattern $d \in \mathcal{D}$ is said to cover an object $g \in \mathcal{G}$ iff $\mathcal{M}(g) \in d$. To use the same order $\sqsubseteq$ to define such a relationship, we associate to each $g \in \mathcal{G}$ its corresponding pattern $\delta(g) \in \mathcal{D}$ which is the degenerated hyper-rectangle $\delta(g) = \{\mathcal{M}(g)\} = \times_{i=1}^{p}[m_i(g), m_i(g)]$. The cover relationship becomes $d \sqsubseteq \delta(g)$. The *extent* of a pattern is the set of objects supporting it. Formally, there is a function $ext : \mathcal{D} \to \wp(\mathcal{G}), d \mapsto \{g \in \mathcal{G} \mid d \sqsubseteq \delta(g)\} = \{g \in \mathcal{G} \mid \mathcal{M}(g) \in d\}$ (where $\wp(\mathcal{G})$ denotes the set of all subsets of $\mathcal{G}$). Note that if $d_1 \sqsubseteq d_2$ then $ext(d_2) \subseteq ext(d_1)$. We define also the positive (resp. negative) extent as follows: $ext^+(d) = ext(d) \cap \mathcal{G}^+$ (resp. $ext^-(d) = ext(d) \cap \mathcal{G}^-$). With the mapping $\delta : \mathcal{G} \to \mathcal{D}$ and the complete lattice $(\mathcal{D}, \sqsubseteq)$, we call the triple $\mathbb{P} = (\mathcal{G}, (\mathcal{D}, \sqsubseteq), \delta)$ the *interval pattern structure* [10, 18].

Measuring the Discriminative Power of a Pattern. In SD, a quality measure $\phi : \mathcal{D} \to \mathbb{R}$ is usually defined to evaluate at what extent a pattern *well-discriminates* the positive instances in $\mathcal{G}^+$ from those in $\mathcal{G}^-$. Two atomic measures are generally employed to quantify the quality of a pattern d: the *true positive rate* $tpr : d \to |ext^+(d)|/|\mathcal{G}^+|$ and the *false positive rate* $fpr : d \to |ext^-(d)|/|\mathcal{G}^-|$. Several measures exist in the literature [12, 21]. A

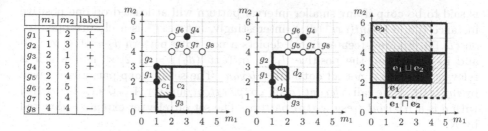

Fig. 1. (left to right) **(1)** a labeled numerical dataset. **(2)** closed c_1 vs non-closed c_2 interval patterns. **(3)** cotp d_1 vs non cotp d_2. **(4)** meet and join of two patterns.

measure is said to be *objective* or *probability based* [12] if it depends solely on the number of co-occurrences and non co-occurrences of the pattern and the target label. In other words, those measures can be defined using only tpr, fpr and potentially other constants (e.g. $|\mathcal{G}|$). Formally, $\exists \phi^* : [0,1]^2 \to \mathbb{R}$ s.t. $\phi(d) = \phi^*(tpr(d), fpr(d))$. Objective measures depends only on the pattern *extent*. Hence, we use interchangeably $\phi(ext(d))$ and $\phi(d)$. An objective quality measure ϕ is said to be *discriminant* if its associated measure ϕ^* is *increasing* with tpr (fpr being fixed) and *decreasing* with fpr (tpr being fixed). For instance, with $\alpha^+ = |\mathcal{G}^+|/|\mathcal{G}|$ and $\alpha^- = |\mathcal{G}^-|/|\mathcal{G}|$ denoting labels prevalence, $wracc^*(tpr, fpr) = \alpha^+ \cdot \alpha^- \cdot (tpr - fpr)$ and $informedness^*(tpr, fpr) = tpr - fpr$ are discriminant measures.

Compressing the Set of Interesting Patterns Using Closure. Since discriminant quality measures depend only on the extent, *closed patterns* can be leveraged to reduce the number of resulting patterns [10]. A pattern $d \in \mathcal{D}$ is said to be closed (w.r.t. pattern structure $\mathbb{P}$) if and only if it is the most restrictive pattern (i.e. the smallest hyper-rectangle) enclosing its extent. Formally, $d = int(ext(d))$ where int mapping (called *intent*) is given by: $int : \wp(\mathcal{G}) \to \mathcal{D}, A \mapsto \prod_{g \in A} \delta(g) = \bigtimes_{i=1}^{p} [\min_{g \in A} m_i(g), \max_{g \in A} m_i(g)]$. Figure 1 (center-left) depicts the closed interval pattern (hatched rectangle) $c_1 = [1,2] \times [1,3]$ which is the closure of $c_2 = [1,4] \times [0,3]$ (non hatched rectangle). Note that since $\mathcal{G}$ is finite, the set of all closed patterns is finite and is given by $int[\wp(\mathcal{G})]$.

A More Concise Set of Patterns Using Relevance Theory. Figure 1 (center-right) depicts two interval patterns, the hatched pattern $d_1 = [1,2] \times [1,3]$ and the non-hatched one $d_2 = [1,4] \times [1,4]$. While both patterns are *closed*, d_1 has better discriminative power than d_2 since they both cover exactly the same positive instances $\{g_1, g_2, g_3\}$; yet, d_2 covers more negative instances than d_1. *Relevance theory* [11] formalizes this observation and helps us to remove some clearly uninteresting closed patterns. In a nutshell, a closed pattern $d_1 \in \mathcal{D}$ is said to be *more relevant than* a closed pattern $d_2 \in \mathcal{D}$ iff $ext^+(d_2) \subseteq ext^+(d_1)$ and $ext^-(d_1) \subseteq ext^-(d_2)$. For ϕ discriminant, if d_1 is more relevant than d_2 then $\phi(d_1) \geq \phi(d_2)$. A closed pattern d is said to be *relevant* iff there is no other closed pattern c that is more relevant than d. It follows that if a closed pattern is relevant then it is *closed on the positive* (cotp for short). An interval pattern is said to be `cotp` if any smaller interval pattern will at least drop one positive instance (i.e. $d = int(ext^+(d))$). interestingly, $int \circ ext^+$ is a closure operator on $(\mathcal{D}, \sqsubseteq)$. Figure 1 (center-right) depicts a non `cotp` pattern $d_2 = [1,4] \times [1,4]$ and its closure on the positive $d_1 = int(ext^+(d_2)) = [1,2] \times [1,3]$ which is relevant. Note that not all `cotp` are *relevant*. The set of `cotp` patterns is given by $int[\wp(\mathcal{G}^+)]$. We call *relevant (resp. `cotp`) extent*, any set $A \subseteq \mathcal{G}$ s.t. $A = ext(d)$ with d is a *relevant (resp. `cotp`) pattern*. The set of relevant extents is denoted by $\mathcal{R}$.

3   Problem Statement

Correct Enumeration of Relevant Extents. First, consider the (simpler) problem of enumerating all relevant extents in $\mathcal{R}$. For a (relevant extents) enumeration algorithm, three properties need generally to hold. An algorithm which output is the set of solutions $\mathcal{S}$ is said to be (1) *complete* if $\mathcal{S} \supseteq \mathcal{R}$, (2) *sound* if $\mathcal{S} \subseteq \mathcal{R}$ and (3) *non redundant* if each solution in $\mathcal{S}$ is outputted *only once*. It is said to be *correct* if the three properties hold. Guyet et al. [15] proposed a *correct algorithm* that enumerate relevant extents induced by the *interval pattern structure* in two steps: (1) Start by a *DFS complete* and *non redundant* enumeration of all `cotp` patterns (extents) using `MinIntChange` algorithm [18]; (2) Post-process the found `cotp` patterns by removing non relevant ones using [11] characterization (this step adds the *soundness* property to the algorithm).

Problem Statement. Given a discriminant objective quality measure ϕ, we want to design an *anytime enumeration algorithm* such that: (1) given enough time, outputs all relevant extents in $\mathcal{R}$, (2) when interrupted, provides a guarantee bounding the difference of quality between the top-quality found extent and the top possible quality w.r.t. ϕ; and (3) outputs a second guarantee ensuring that the resulting patterns are diverse.

Formally, let $\mathcal{S}_i$ be the set of outputted solutions by the anytime algorithm at some step (or instant) i (at $i+1$ we have $\mathcal{S}_i \subseteq \mathcal{S}_{i+1}$). We want that (1) when i is big enough, $\mathcal{S}_i \supseteq \mathcal{R}$ (only *completeness* is required). For (2) and (3), we define two metrics[3] to compare the results in $\mathcal{S}_i$ with the ones in $\mathcal{R}$. The first metric, called *accuracy* (Eq. 1), evaluates the difference between top pattern quality ϕ in $\mathcal{S}_i$ and $\mathcal{R}$ while the second metric, called *specificity* (Eq. 2), evaluates how diverse and complete are patterns in $\mathcal{S}_i$.

$$accuracy_\phi(\mathcal{S}_i, \mathcal{R}) = \sup_{A \in \mathcal{R}} \phi(A) - \sup_{B \in \mathcal{S}_i} \phi(B) \tag{1}$$

$$specificity(\mathcal{S}_i, \mathcal{R}) = \sup_{A \in \mathcal{R}} \inf_{B \in \mathcal{S}_i} (|A \triangle B|/|\mathcal{G}|) \tag{2}$$

The idea behind *specificity* is that each extent A in $\mathcal{R}$ is "approximated" by the most similar extent in $\mathcal{S}_i$; that is the set $B \in \mathcal{S}_i$ minimizing the *metric distance* $A, B \mapsto |A \triangle B|/|\mathcal{G}|$ in $\wp(\mathcal{G})$. The *specificity*[4] is then the highest

---

[3] The metrics names fall under the taxonomy of [29] for anytime algorithms.

[4] The *specificity* is actually a directed Hausdorff distance [17] from $\mathcal{R}$ to $\mathcal{S}_i$.

possible distance (pessimistic). Note that $specificity(S_i, R) = 0$ *is equivalent to* $S_i \supseteq R$. Clearly, the lower these two metrics are, the closer we get to the desired output R. While $accuracy_\phi$ and $specificity$ can be *evaluated* when a complete exploration of R is possible, our aim is to *bound* the two aforementioned measures independently from R providing a *guarantee*. In other words, the anytime algorithm need to output additionally to S_i, the two following measures: (2) $\overline{accuracy_\phi}(S_i)$ and (3) $\overline{specificity}(S_i)$ s.t. $accuracy_\phi(S_i, R) \leq \overline{accuracy_\phi}(S_i)$ and $specificity(S_i, R) \leq \overline{specificity}(S_i)$. These two bounds need to decrease overtime providing better information on R through S_i.

4   Anytime Interval Pattern Mining

Discretizations and Pattern Space. Our algorithm relies on the enumeration of a chain of discretization from the coarsest to the finest. A *discretization* of $\mathbb{R}$ is any *partition* of $\mathbb{R}$ using intervals. In particular, let $C = \{c_i\}_{1 \leq i \leq |C|} \subseteq \mathbb{R}$ be a finite set with $c_i < c_{i+1}$ for $i \in \{1, ..., |C| - 1\}$. Element of C are called *cut points* or *cuts*. We associate to C a *finite discretization* denoted by $dr(C)$ and given by $dr(C) = \{(-\infty, c_1)\} \cup \{[c_i, c_{i+1}) \mid i \in \{1, ..., |C| - 1\}\} \cup \{[c_{|C|}, +\infty)\}$.

Generally speaking, let $p \in \mathbb{N}^*$ and let $C = (C_k)_{1 \leq k \leq p} \in \wp(\mathbb{R})^p$ representing *sets of cut points* associated to each dimension k (i.e. $C_k \subseteq \mathbb{R}$ finite $\forall k \in \{1, ..., p\}$). The partition $dr(C)$ of $\mathbb{R}^p$ is given by: $dr(C) = \prod_{k=1}^{p} dr(C_k)$. Figure 2 depicts two discretizations. Discretizations are ordered using the natural order between partitions[5]. Moreover, cut-points sets are ordered by $\leq$ as follows: $C^1 \leq C^2 \equiv (\forall k \in \{1, ..., p\}) \ C_k^1 \subseteq C_k^2$ with $C^i = (C_k^i)_{1 \leq k \leq p}$. Clearly, if $C^1 \leq C^2$ then discretization $dr(C^1)$ *is coarser than* $dr(C^2)$.

Let $C = (C_k)_{1 \leq k \leq p}$ be the cut-points. Using the elementary hyper-rectangles (i.e. cells) in the discretization $dr(C)$, one can build a (finite) subset of descriptions $D_C \subseteq D$ which is the set of all possible descriptions (hyper-rectangles) that can be built using these cells. Formally: $D_C = \{\sqcap S \mid S \subseteq dr(C)\}$. Note that $\top = \emptyset \in D_C$ since $\sqcap \emptyset = \sqcup D = \top$ by definition. Proposition 1 states that $(D_C, \sqsubseteq)$ is a complete sub-lattice of $(D, \sqsubseteq)$.

Proposition 1. $(D_C, \sqsubseteq)$ *is a finite (complete) sub-lattice of* $(D, \sqsubseteq)$ *that is:* $\forall d_1, d_2 \in D_C : d_1 \sqcup d_2 \in D_C$ *and* $d_1 \sqcap d_2 \in D_C$. *Moreover, if* $C^1 \leq C^2$ *are two cut-points sets, then* $(D_{C^1}, \sqsubseteq)$ *is a (complete) sub-lattice of* $(D_{C^2}, \sqsubseteq)$.

Finest Discretization for a Complete Enumeration of Relevant Extents. There exist cut points $C \subseteq \wp(\mathbb{R})^p$ such that the space $(D_C, \sqsubseteq)$ holds all relevant extents (i.e. $ext[D_C] \supseteq R$). For instance, if we consider $C = (m_k[G])_{1 \leq k \leq p}$, the description space $(D_C, \sqsubseteq)$ holds all relevant extents. However, is there coarser discretization that holds all the relevant extents? The answer is affirmative. One can show that the only interesting cuts are those separating between positive and negative instances (called boundary cutpoints by [9]). We call such cuts, *relevant cuts*. They are denoted by $C^{rel} =$

---

[5] Let E be a set, a partition P_2 of E *is finer than* a partition P_1 (or P_1 *is coarser than* P_2) and we denote $P_1 \leq P_2$ if any subset in P_1 is a subset of a subset in P_2.

$(C_k^{rel})_{1 \leq k \leq p}$ and we have $ext[\mathcal{D}_{C^{rel}}] \supseteq \mathcal{R}$. Formally, for each dimension k, a value $c \in m_k[\mathcal{G}]$ is a *relevant cut in* C_k^{rel} for attribute m_k iff: $(c \in m_k[\mathcal{G}^+]$ and $prev(c, m_k[\mathcal{G}]) \in m_k[\mathcal{G}^-])$ or $(c \in m_k[\mathcal{G}^-]$ and $prev(c, m_k[\mathcal{G}]) \in m_k[\mathcal{G}^+])$ where $next(c, A) = \inf\{a \in A \mid c < a\}$ (resp. $prev(c, A) = \sup\{a \in A \mid a < c\}$) is the following (resp. preceding) element of c in A. Finding *relevant cuts* C_k^{rel} is of the same complexity of sorting $m_k[\mathcal{G}]$ [9]. In the dataset depicted in Fig. 1, relevant cuts are given by $C^{rel} = (\{2,3,4,5\}, \{4,5\})$. Discretization $dr(C_2^{rel})$ is depicted in Fig. 2 (center).

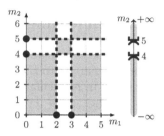

Fig. 2. (left) Discretization $dr((C_1, C_2))$ in $\mathbb{R}^2$ with $C_1 = \{2,3\}$ and $C_2 = \{4,5\}$ and **(right)** discretization $dr((C_2))$ in $\mathbb{R}$. Adding a cut point in any C_k will create finer discretization.

Anytime Enumeration of Relevant Extents. We design an *anytime* and *interruptible* algorithm dubbed RefineAndMine. This method, presented in Algorithm 1, relies on the enumeration of a chain of discretizations on the data space, from the coarsest to the finest. It begins by searching relevant cuts in pre-processing phase (line 2). Then, it builds a *coarse discretization* (line 3) containing a small set of relevant cut-points. Once the initial discretization built, cotp patterns are mined thanks to MinIntChange Algorithm (line 4) [18]. Then as long as the algorithm is not interrupted (or within the computational budget), we add new cut-points (line 6) building finer discretizations. For each added cut-point (line 8), only new interval patterns are searched for (mined descriptions d are new but their extents $ext(d)$ are not necessarily new) . That is cotp patterns which left or right bound is *cut* on the considered attribute *attr* (i.e. $d.I_{attr} \in \{[cut, a), [cut, +\infty), [a, cut), (-\infty, cut) \mid a \in C_{attr}^{cur}\}$ with $d.I_{attr}$ is the $attr^{th}$ interval of d). This can be done by a slight modification of MinIntChange method. RefineAndMine terminates when the set of relevant cuts is exhausted (i.e. $C^{cur} = C^{rel}$) ensuring a *complete* enumeration of relevant extents $\mathcal{R}$.

The initial discretization (Line 3) can be done by various strategies (see [28]). A simple, yet efficient, choice is the equal frequency discretization with a fixed number of cuts. Other strategies can be used, e.g. [9]. Adding new cut-points (Line 6) can also be done in various ways. One strategy is to add a random relevant cut on a random attribute to build the next discretization. Section 5.3 proposes another more elaborated strategy that heuristically guide RefineAndMine to rapidly find good quality patterns (observed experimentally).

5   Anytime Interval Pattern Mining with Guarantees

Algorithm RefineAndMine starts by mining patterns in a coarse discretization. It continues by mining more patterns in increasingly finer discretizations until the search space is totally explored (final complete lattice being $(\mathcal{D}_{C^{rel}}, \sqsubseteq)$). According to Proposition 1, the description spaces built on discretizations are complete sub-lattices of the total description space. A similar idea involves performing successive enumeration of growing pattern languages (projections) [6]. In our case, it is a successive enumeration of growing complete sub-lattices. For the sake of generality, in the following of this section $(\mathcal{D}, \sqsubseteq)$ denotes a *complete lattice*, and for all $i \in \mathbb{N}^*$, $(\mathcal{D}_i, \sqsubseteq)$ denotes *complete sub-lattices* of $(\mathcal{D}, \sqsubseteq)$ such that $\mathcal{D}_i \subseteq \mathcal{D}_{i+1} \subseteq \mathcal{D}$. For instance, in RefineAndMine, the total complete lattice is $(\mathcal{D}_{C^{rel}}, \sqsubseteq)$ while the $(\mathcal{D}_i, \sqsubseteq)$ are $(\mathcal{D}_{C^{cur}}, \sqsubseteq)$ at each step. Following Sect. 3 notation, the outputted set $\mathcal{S}_i$ at a step i contains the set of all *cotp extents* associated to $\mathcal{D}_i$. Before giving the formulas of $\overline{accuracy_\phi}(\mathcal{S}_i)$ and $\overline{specificity}(\mathcal{S}_i)$, we give some necessary definitions and underlying properties. At the end of this section, we show how RefineAndMine can be adapted to efficiently compute these two bounds for the case of interval patterns.

Similarly to the interval pattern structure [18], we define in the general case a *pattern structure* $\mathbb{P} = (\mathcal{G}, (\mathcal{D}, \sqsubseteq), \delta)$ on the *complete lattice* $(\mathcal{D}, \sqsubseteq)$ where $\mathcal{G}$ is a non empty finite set (partitioned into $\{\mathcal{G}^+, \mathcal{G}^-\}$) and $\delta : \mathcal{G} \to \mathcal{D}$ is a mapping associating to each object its description (recall that in interval pattern structure, δ is the degenerated hyper-rectangle representing a single point). The extent *ext* and intent *int* operators are then respectively given by $ext : \mathcal{D} \to \wp(\mathcal{G}), d \mapsto \{g \in \mathcal{G} \mid d \sqsubseteq \delta(g)\}$ and $int : \wp(\mathcal{G}) \to \wp(\mathcal{G}), A \mapsto \bigsqcap_{g \in A} \delta(g)$ with $\bigsqcap$ represents the meet operator in $(\mathcal{D}, \sqsubseteq)$ [10].

5.1   Approximating Descriptions in a Complete Sub-lattice

Upper and Lower Approximations of a Pattern. We start by approximating each pattern in $\mathcal{D}$ using two patterns in $\mathcal{D}_i$. Consider for instance Fig. 3 where $\mathcal{D}$ is the space of interval patterns in $\mathbb{R}^2$ while $\mathcal{D}_C$ is the space

---

Algorithm 1. RefineAndMine

Input: $(\mathcal{G}, \mathcal{M})$ a numerical datasets with $\{\mathcal{G}^+, \mathcal{G}^-\}$ partition of $\mathcal{G}$
1 **procedure** RefineAndMine()
2   Compute relevant cuts C^{rel}
3   Build an initial set of cut-points $C^{cur} \leq C^{rel}$
4   Mine cotp patterns in $\mathcal{D}_{C^{cur}}$ (and their extents) using MinIntChange
5   **while** $C^{cur} \neq C^{rel}$ *and within* **computational budget do**
6   Choose the next relevant cut $(attr, cut)$ with $cut \in C^{rel}_{attr} \setminus C^{cur}_{attr}$
7   Add the relevant cut cut to C^{cur}
8   Mine **new cotp patterns** (and their extents) in $\mathcal{D}_{C^{cur}}$

containing only rectangles that can be built over discretization $dr(C)$ with $C = (\{1, 4, 6, 8\}, \{1, 3, 5, 6\})$. Since the hatched rectangle $d = [3, 7] \times [2, 5.5] \in \mathcal{D}$ does not belong to $\mathcal{D}_C$, two descriptions in $\mathcal{D}_C$ can be used to encapsulate it. The first one, depicted by a gray rectangle, is called the *upper approximation* of d. It is given by the smallest rectangle in $\mathcal{D}_C$ enclosing d. Dually, the second approximation represented as a black rectangle and coined *lower approximation* of d, is given by the greatest rectangle in $\mathcal{D}_C$ enclosed by d. This two denominations comes from Rough Set Theory [25] where lower and upper approximations form together a *rough set* and try to capture the undefined rectangle $d \in \mathcal{D}\backslash\mathcal{D}_C$. Definition 1 formalizes these two approximations in the general case.

Definition 1. *The upper approximation mapping $\overline{\psi_i}$ and lower approximation mapping $\underline{\psi_i}$ are the mappings defined as follows:*

$$\overline{\psi_i} : \mathcal{D} \to \mathcal{D}_i, d \mapsto \bigsqcup \{c \in \mathcal{D}_i \mid c \sqsubseteq d\} \qquad \underline{\psi_i} : \mathcal{D} \to \mathcal{D}_i, d \mapsto \bigsqcap \{c \in \mathcal{D}_i \mid d \sqsubseteq c\}$$

The existence of these two mappings is ensured by the fact that $(\mathcal{D}_i, \sqsubseteq)$ is a complete sublattice of $(\mathcal{D}, \sqsubseteq)$. *Theorem 4.1* in [8] provides more properties for the two aforementioned mappings. Proposition 2 restates an important property.

Proposition 2. $\forall d \in \mathcal{D} : \overline{\psi_i}(d) \sqsubseteq d \sqsubseteq \underline{\psi_i}(d)$. *The term lower and upper-approximation here are reversed to fit the fact that in term of* extent *we have* $\forall d \in \mathcal{D}: ext(\underline{\psi_i}(d)) \subseteq ext(d) \subseteq ext(\overline{\psi_i}(d))$.

A Projected Pattern Structure. Now that we have the upper-approximation mapping $\overline{\psi_i}$, one can associate a new pattern structure $\mathbb{P}_i = (\mathcal{G}, (\mathcal{D}_i, \sqsubseteq), \overline{\psi_i} \circ \delta)^6$ to the pattern space $(\mathcal{D}_i, \sqsubseteq)$. It is worth mentioning, that while extent ext_i mapping associated to $\mathbb{P}_i$ is equal to ext, the intent int_i of $\mathbb{P}_i$ is given by $int_i :$ $\wp(\mathcal{G}) \to \mathcal{D}_i, A \mapsto \overline{\psi_i}(int(A))$. Note that, the set of cotp patterns associated to $\mathbb{P}_i$ are given by $int_i[\wp(\mathcal{G}^+)] = \overline{\psi_i}[int[\wp(\mathcal{G}^+)]]$. That is, the upper approximation of a cotp pattern in $\mathbb{P}$ is a cotp pattern in $\mathbb{P}_i$.

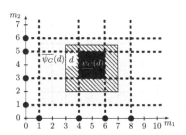

Fig. 3. Description $d = [3, 7] \times [2, 5.5]$ in $\mathcal{D}$ (hatched) and $C = (\{1, 4, 6, 8\}, \{1, 3, 5, 6\})$. Upper approximation of d in $\mathcal{D}_C$ is $\overline{\psi_C}(d) = [1, 8) \times [1, 6)$ (gray rectangle) while lower approximation of d is $\underline{\psi_C}(d) = [4, 6) \times [3, 5)$ (black rectangle).

---

[6] $\mathbb{P}_i$ is said to be a projected pattern structure of $\mathbb{P}$ by the projection $\overline{\psi_i}$ [7].

Encapsulating Patterns Using Their Upper-Approximations. We want to encapsulate any description by knowing only its upper-approximation. Formally, we want some function $f : \mathcal{D}_i \to \mathcal{D}_i$ such that $(\forall d \in \mathcal{D}) \overline{\psi_i}(d) \sqsubseteq d \sqsubseteq f(\overline{\psi_i}(d))$. Proposition 3 define such a function f (called *core*) and states that the *core* is the tightest (w.r.t. $\sqsubseteq$) possible function f.

Proposition 3. *The function* $core_i$ *defined by:*

$$core_i : \mathcal{D}_i \to \mathcal{D}_i, c \mapsto core(c) = \underline{\psi_i} \left(\bigsqcup \left\{ d \in \mathcal{D} \mid \overline{\psi_i}(d) = c \right\} \right)$$

verifies the following property: $\forall d \in \mathcal{D} : \overline{\psi_i}(d) \sqsubseteq d \sqsubseteq \underline{\psi_i}(d) \sqsubseteq core_i(\overline{\psi_i}(d))$. *Moreover, for* $f : \mathcal{D}_i \to \mathcal{D}_i$, $(\forall d \in \mathcal{D}) d \sqsubseteq f(\overline{\psi_i}(d)) \Leftrightarrow (\forall c \in \mathcal{D}_i) core_i(c) \sqsubseteq f(c)$.

Note that, while the *core* operator definition depends clearly on the complete lattice $(\mathcal{D}, \sqsubseteq)$, its computation should be done independently from $(\mathcal{D}, \sqsubseteq)$.

We show here how to compute the *core* in `RefineAndMine`. In each step and for cut-points $C = (C_k) \subseteq \wp(\mathbb{R})^p$, the finite lattice $(\mathcal{D}_C, \sqsubseteq)$ is a sub-lattice of the finest finite lattice $(\mathcal{D}_{C^{rel}}, \sqsubseteq)$ (since $C \leq C^{rel}$). Thereby, the *core* is computed according to this latter as follows: Let $d \in \mathcal{D}_C$ with $d.I_k = [a_k, b_k)$ for all $k \in \{1, ..., p\}$. The left (resp. right) bound of $core_C(d).I_k$ for any k is equal to $next(a_k, C_k)$ (resp. $prev(b_k, C_k)$) if $next(a_k, C_k^{rel}) \notin C_k$ (resp. $prev(b_k, C_k^{rel}) \notin C_k$). Otherwise, it is equal to a_k (resp. b_k). Consider the step $C = (\{2, 3\}, \{4, 5\})$ in `RefineAndMine` (its associated discretization is depicted in Fig. 2 (left)) and recall that the relevant cuts set is $C^{rel} = (\{2, 3, 4, 5\}, \{4, 5\})$. The core of the bottom pattern $\bot = \mathbb{R}^2$ at this step is $core_{C^{cur}}(\bot) = (-\infty, 3) \times \mathbb{R}$. Indeed, there is three descriptions in $\mathcal{D}_{C^{rel}}$ which upper approximation is $\bot$, namely $\bot$, $c_1 = (-\infty, 4) \times \mathbb{R}$ and $c_2 = (-\infty, 5) \times \mathbb{R}$. Their lower approximations are respectively $\bot$, $(-\infty, 3) \times \mathbb{R}$ and $(-\infty, 3) \times \mathbb{R}$. The join (intersection) of these three descriptions is then $core_{C^{cur}}(\bot) = (-\infty, 3) \times (-\infty, +\infty)$. Note that particularly for interval patterns, the *core* has *monotonicity*, that is $(\forall c, d \in \mathcal{D}_C) c \sqsubseteq d \Rightarrow core_C(c) \sqsubseteq core_C(d)$.

5.2 Bounding Accuracy and Specificity Metrics

At the ith step, the outputted extents $\mathcal{S}_i$ contains the set of `cotp` extents in $\mathbb{P}_i$. Formally, $int_i[\mathcal{S}_i] \supseteq int_i[\wp(\mathcal{G}^+)]$. Theorems 1 and 2 gives respectively the *bounds* $\overline{accuracy_\phi}$ and *specificity*.

Theorem 1. *Let* $\phi : \mathcal{D} \to \mathbb{R}$ *be a discriminant objective quality measure. The accuracy metric is bounded by:*

$$\overline{accuracy_\phi}(\mathcal{S}_i) = \sup_{c \in int_i[\mathcal{S}_i]} \left[\phi^* \left(tpr(c), fpr(core_i(c)) \right) - \phi^* \left(tpr(c), fpr(c) \right) \right]$$

Moreover $\overline{accuracy_\phi}(\mathcal{S}_{i+1}) \leq \overline{accuracy_\phi}(\mathcal{S}_i)$.

Theorem 2. *The specificity metric is bounded by:*

$$\overline{specificity}(\mathcal{S}_i) = \sup_{c \in int_i[\mathcal{S}_i]} \left((|ext(c)| - |ext(core_i^+(c))|)/(2 \cdot |\mathcal{G}|)) \right)$$

where $core_i^+(c) = int_i(ext^+(core_i(c)))$, *that is* $core_i^+(c)$ *is the closure on the positive of* $core_i(c)$ *in* $\mathbb{P}_i$. *Moreover* $\overline{specificity}(\mathcal{S}_{i+1}) \leq \overline{specificity}(\mathcal{S}_i)$.

5.3 Computing and Updating Bounds in RefineAndMine

We show below how the different steps of the method RefineAndMine (see Algorithm 1) should be updated in order to compute the two bounds $\overline{accuracy}$ and $\overline{specificity}$. For the sake of brevity, we explain here a naive approach to provide an overview of the algorithm. Note that here, *core* (resp. $core^+$) refers to $core_{C^{cur}}$ (resp. $core_{C^{cur}}^+$).

Compute the Initial Bounds (Line 4). As MinIntChange enumerates all cotp patterns $d \in \mathcal{D}_{C^{cur}}$, RefineAndMine stores in a key-value structure (i.e. map) called BoundPerPosExt the following entries:

$$ext^+(d) : \left(\phi(d), \phi^* \left(tpr(d), fpr(core(d)) \right), (|ext(d)| - |ext(core^+(d))|)/(2 \cdot |\mathcal{G}|) \right)$$

The *error-bounds* $\overline{accuracy}_\phi$ and $\overline{specificity}$ are then computed at the end by a single pass on the entries of BoundPerPosExt using Theorems 1 and 2.

Update the Bounds After Adding a New Cut-Point (Line 8). In order to compute the new *error-bounds* $\overline{accuracy}_\phi$ and $\overline{specificity}$ which decrease according to Theorems 1 and 2, one need to add/update some entries in the structure BoundPerPosExt. For that, only two types of patterns should be looked for:

1. The new cotp patterns mined by RefineAndMine, that is those which left or right bound on attribute *attr* is the added value *cut*. Visiting these patterns will add potentially new entries in BoundPerPosExt or update ancient ones.
2. The old cotp which core changes (i.e. becomes less restrictive) in the new discretization. One can show that these patterns are those which left bound is $prev(cut, C_{attr}^{cur})$ or right bound is $next(cut, C_{attr}^{cur})$ on attribute *attr*. Visiting these patterns will only update ancient entries of BoundPerPosExt by potentially decreasing both second and third value.

Adding a New Cut-Point (Line 7). We have implemented for now a strategy which aims to decrease the $\overline{accuracy}_\phi$. For that, we search in BoundPerPosExt for the description d having the maximal value $\phi^* \left(tpr(d), fpr(core(d)) \right)$. In order to decrease $\overline{accuracy}_\phi$, we increase the size of $core(d)$ (to potentially increase $fpr(core(d))$). This is equivalent to choose a cut-point in the border region $C_{attr}^{rel} \backslash C_{attr}^{cur}$ for some attribute *attr* such that $cut \in d.I_{attr} \backslash core(d).I_{attr}$. Consider that we are in the step where the current discretization C^{cur} is the one depicted in Fig. 2. Imagine that the bottom pattern $\bot = \mathbb{R}^2$ is the one associated to the maximal value $\phi^* \left(tpr(\bot), fpr(core(\bot)) \right)$. The new cut-point should be chosen in $\{4, 5\}$ for $attr = 1$ (recall that $core(\bot) = (-\infty, 3) \times (-\infty, +\infty)$). Note that if for such description there is no remaining relevant cut in its *border regions* for all $attr \in \{1, ..., p\}$ then $core(d) = d$ ensuring that d is the *top pattern*.

Table 1. Benchmark datasets and their characteristics: number of numerical attributes, number of rows, number of all possible intervals, the considered class and its prevalence

Dataset	Num	Rows	Intervals	Class	α	Dataset	Num	Rows	Intervals	Class	α
ABALONE_02_M	2	4177	56×10^6	M	0.37	GLASS_02_1	2	214	161×10^6	1	0.33
ABALONE_03_M	3	4177	74×10^9	M	0.37	GLASS_04_1	4	214	5×10^{15}	1	0.33
CREDITA_02_+	2	666	1×10^9	+	0.45	HABERMAN_03_2	3	306	47×10^6	2	0.26
CREDITA_04_+	4	666	3×10^{15}	+							

6   Empirical Study

In this section we report quantitative experiments over the implemented algorithms. For reproducibility purpose, the source code is made available in our companion page[7] which also provide a wider set of experiments. Experiments were carried out on a variety of datasets (Table 1) involving ordinal or continuous numerical attributes from the UCI repository.

First, we study the effectiveness of `RefineAndMine` in terms of the speed of convergence to the optimal solution, as well as regarding the evolution over time of the accuracy of the provided bounding quality's guarantee. To this end, we report in Fig. 4, the behavior of `RefineAndMine` (i.e. quality and bounding guarantee) according to the execution time to evaluate the time/quality trade-off of the devised approach. $\overline{accuracy}$ as presented in Theorem 1 is the difference between the quality and its bounding measure. The experiments were conducted by running both `RefineAndMine` and the exhaustive enumeration algorithm (`MinIntChange` performed considering $\mathcal{D}_{C^{rel}}$) on the benchmark datasets using *informedness* measure. The exhaustive algorithm execution time enables the estimation of the computational overhead incurred by `RefineAndMine`. We interrupt a method if its execution time exceeds two hours. Note that, in the experiments, we choose to disable the computation of specificity since the latter is only optional and does not affect the effectiveness of the algorithm. This in contrast to the quality bound computation which is essential as it guides `RefineAndMine` in the cut-points selection strategy. The experiments give evidence of the effectiveness of `RefineAndMine` both in terms of finding the optimal solution as well as in providing stringent bound on the top quality pattern in a prompt manner. Two important milestones achieved by `RefineAndMine` during its execution are highlighted in Fig. 4. The first one, illustrated by the *green dotted line*, points out the required time to find the best pattern. The second milestone (*purple line*) is reached when the quality's and the bound's curves meet, this ensures that the best quality was already found by `RefineAndMine`. Interestingly, we observe that for most configurations the second milestone is attained by `RefineAndMine` promptly and well before the exhaustive method termination time. This is explained by the fact that the adopted cut points selection strategy aims to decrease as early as possible the $\overline{accuracy}$ metric. Finally,

---

[7] Companion page: https://github.com/Adnene93/RefineAndMine.

`RefineAndMine` requires in average 2 times of the requested execution time (*red dotted line*) by the exhaustive algorithm. This overhead is mostly incurred by the quality guarantee computation.

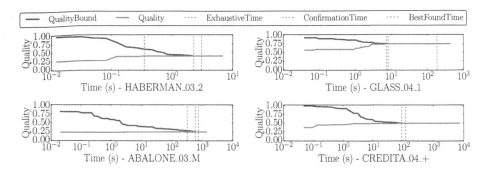

Fig. 4. Evolution over time of top pattern quality and its bounding guarantee provided by `RefineAndMine`. Execution time is reported in log scale. The last figure reports that the exhaustive enumeration algorithm was not able to finish within 2 h

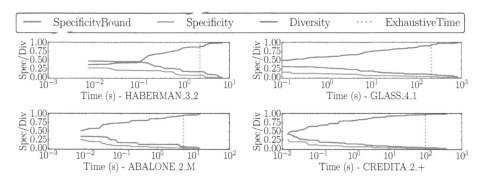

Fig. 5. Efficiency of `RefineAndMine` in terms of retrieving a diverse patterns set. Execution time is reported in log scale. The ground-truth for each benchmark dataset corresponds to the obtained Top10 diversified patterns set with a similarity threshold of 0.25 and a minimum *tpr* of 15%.

We illustrate in Fig. 5 the behavior of `RefineAndMine` in terms of finding diverse set of high quality patterns covering different parts of the dataset. To evaluate how quickly the devised approach finds a diverse patterns set, we run the exhaustive approach over the benchmark datasets to constitute a top-k diverse patterns set heuristically as following: the patterns extracted by the exhaustive search algorithm are sorted according to the quality measure and the best pattern is kept in the returned top-k list. Next, the complete patterns list are iterated over, and the top-k list is augmented by a pattern if and only if its similarity with all the patterns of the current content of the top-k list is lower than a given threshold (a Jaccard index between extents). This process is interrupted if the desired number of patterns of the top-k list is reached or no remaining dissimilar pattern is available. Similar post-processing techniques were used by [5,20]. Once this ground truth top-k list is constituted over some benchmark dataset, we run RefineAndMine and measure the *specificity quantity* of the obtained results set *Sol* with the top-k list. *specificity* metric is rewritten in Eq. 3 to accommodate the desired evaluation objective of these experiments. Still, it remains upper-bounded by the general formula of $\overline{specificity}$ given in Theorem 2. This in order to evaluate at what extent the visited patterns by RefineAndMine well-cover the ground-truth patterns which are scattered over different parts of some input dataset. We report in Fig. 5 both *specificity* and its bounding guarantee $\overline{specificity}$, as well as, a diversity metric defined in Eq. 4. Such a metric was defined in [5] to evaluate the ability of an approximate algorithm to retrieve a given ground-truth (i.e. diversified top-k discriminant patterns set). This diversity metric relies on a similarity rather than a distance (as in specificity), and is equal to 1 when all patterns of the top-k list are fully discovered.

$$specificity(\text{top-k}, Sol) = \sup_{d \in \text{top-k}} \inf_{c \in Sol} (|ext(d) \Delta ext(c)|/|\mathcal{G}|) \tag{3}$$

$$diversity(\text{top-k}, Sol) = \underset{d \in \text{top-k}}{avg} \sup_{c \in Sol} (Jaccard(ext(d), ext(c))) \tag{4}$$

In most configurations, we notice that RefineAndMine is able to uncover approximately 80% (given by *diversity*) of the ground truth's patterns in less than 20% of the time required by the exhaustive search algorithm. For instance, in ABALONE_02_M, we observe that after 2 s (12% of the required time for the exhaustive algorithm), the patterns outputted by RefineAndMine approximate 92% of the ground truth. Moreover, we observe that the specificity and $\overline{specificity}$ decrease quickly with time, guaranteeing a high level of diversity.

For a comparative study, we choose to compare RefineAndMine with the closest approach following the same paradigm (anytime) in the literature, that is the recent MCTS4DM technique [5]. MCTS4DM is depicted by the authors as an algorithm which enables the anytime discovery of a diverse patterns set of high quality. While MCTS4DM ensures interruptibility and an exhaustive exploration if given enough time and memory budget, it does not ensures any theoretical guarantees on the distance from optimality and on the diversity. We report in Fig. 6 a comparative evaluation between the two techniques. To realize this study, we investigate the ability of the two methods in retrieving the ground truth patterns, this by evaluating the quality of their respective diversified top-k lists against the ground truth using the diversity metric (Eq. 4). We observe that RefineAndMine outperforms MCTS4DM both in terms of finding the best pattern, and of uncovering diverse patterns set of high qualities. This is partially due to the fact that our method is specifically tailored for mining discriminant patterns in numerical data, in contrast to MCTS4DM which is agnostic of the interestingness measure and the description language. Note that, to enable a fair comparison of the two approaches, we report the full time spent by the methods including the overhead induced by the post-computation of the diversified top-k patterns set.

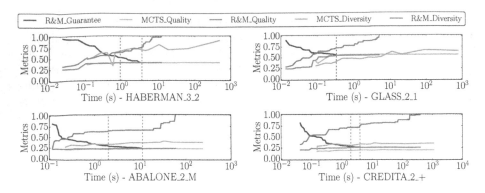

Fig. 6. Comparative experiments between RefineAndMine ($R\&M$) and MCTS4DM. Execution time is reported in log scale. The ground-truth for each benchmark dataset corresponds to the obtained Top10 diversified patterns set with a similarity threshold of 0.25 and no minimum support size threshold.

7   Discussions and Conclusion

We introduced a novel anytime pattern mining technique for uncovering discriminant patterns in numerical data. We took a close look to discriminant interestingness measures to focus on hyper-rectangles in the dataset fostering the presence of some class. By leveraging the properties of the quality measures, we defined a guarantee on the accuracy of RefineAndMine in approximating the optimal solution which improves over time. We also presented a guarantee on the specificity of RefineAndMine –which is agnostic of the quality measure– ensuring its diversity and completeness. Empirical evaluation gives evidence of the effectiveness both in terms of finding the optimal solution (w.r.t. the quality measure ϕ) and revealing local optimas located in different parts of the data.

This work paves the way for many improvements. RefineAndMine can be initialized with more sophisticated discretization techniques [9,19]. We have to investigate additional cut-points selection strategies. While we considered here discriminant pattern mining, the enumeration process (i.e. *successive refinement of discretizations*) can be tailored to various other quality measures in subgroup discovery. For example, the accuracy bound guarantee definition can be extended to handle several other traditional measures such as Mutual Information, χ^2 and *Gini split* by exploiting their (quasi)-convexity properties w.r.t. *tpr* and *fpr* variables [1,24]. Other improvements include the adaptation of RefineAndMine for high-dimensional datasets and its generalization for handling additional types of attributes (categorical, itemsets, etc.). The latter is facilitated by the generic notions from Sect. 5 and the recent works of Buzmakov et al. [6].

Aknowledgement. This work has been partially supported by the project *ContentCheck* **ANR-15-CE23-0025** funded by the French National Research Agency, the Association Nationale Recherche Technologie (**ANRt**) French program and the **APRC Conf Pap - CNRS** project. The authors would like to thank the reviewers for their valuable remarks. They also warmly thank Loïc Cerf, Marc Plantevit and Anes Bendimerad for interesting discussions.

References

1. Abudawood, T., Flach, P.: Evaluation measures for multi-class subgroup discovery. In: Buntine, W., Grobelnik, M., Mladenić, D., Shawe-Taylor, J. (eds.) ECML PKDD 2009. LNCS (LNAI), vol. 5781, pp. 35–50. Springer, Heidelberg (2009). https://doi.org/10.1007/978-3-642-04180-8_20
2. Atzmueller, M., Puppe, F.: SD-Map – a fast algorithm for exhaustive subgroup discovery. In: Fürnkranz, J., Scheffer, T., Spiliopoulou, M. (eds.) PKDD 2006. LNCS (LNAI), vol. 4213, pp. 6–17. Springer, Heidelberg (2006). https://doi.org/10.1007/11871637_6
3. Boley, M., Lucchese, C., Paurat, D., Gärtner, T.: Direct local pattern sampling by efficient two-step random procedures. In: KDD, pp. 582–590 (2011)
4. Boley, M., Moens, S., Gärtner, T.: Linear space direct pattern sampling using coupling from the past. In: KDD, pp. 69–77 (2012)
5. Bosc, G., Boulicaut, J., Raïssi, C., Kaytoue, M.: Anytime discovery of a diverse set of patterns with monte carlo tree search. DMKD 32(3), 604–650 (2018)
6. Buzmakov, A., Kuznetsov, S.O., Napoli, A.: Fast generation of best interval patterns for nonmonotonic constraints. In: Appice, A., Rodrigues, P.P., Santos Costa, V., Gama, J., Jorge, A., Soares, C. (eds.) ECML PKDD 2015. LNCS (LNAI), vol. 9285, pp. 157–172. Springer, Cham (2015). https://doi.org/10.1007/978-3-319-23525-7_10
7. Buzmakov, A., Kuznetsov, S.O., Napoli, A.: Revisiting pattern structure projections. In: Baixeries, J., Sacarea, C., Ojeda-Aciego, M. (eds.) ICFCA 2015. LNCS (LNAI), vol. 9113, pp. 200–215. Springer, Cham (2015). https://doi.org/10.1007/978-3-319-19545-2_13
8. Denecke, K., Wismath, S.L.: Galois connections and complete sublattices. In: Denecke, K., Erné, M., Wismath, S.L. (eds.) Galois Connections and Applications, vol. 565, pp. 211–229. Springer, Dordrecht (2004). https://doi.org/10.1007/978-1-4020-1898-5_4
9. Fayyad, U.M., Irani, K.B.: Multi-interval discretization of continuous-valued attributes for classification learning. In: IJCAI, pp. 1022–1029 (1993)
10. Ganter, B., Kuznetsov, S.O.: Pattern structures and their projections. In: Delugach, H.S., Stumme, G. (eds.) ICCS-ConceptStruct 2001. LNCS (LNAI), vol. 2120, pp. 129–142. Springer, Heidelberg (2001). https://doi.org/10.1007/3-540-44583-8_10
11. Garriga, G.C., Kralj, P., Lavrac, N.: Closed sets for labeled data. J. Mach. Learn. Res. 9, 559–580 (2008)
12. Geng, L., Hamilton, H.J.: Interestingness measures for data mining: a survey. ACM Comput. Surv. 38(3), 9 (2006)
13. Giacometti, A., Soulet, A.: Dense neighborhood pattern sampling in numerical data. In: SIAM, pp. 756–764 (2018)
14. Grosskreutz, H., Rüping, S.: On subgroup discovery in numerical domains. Data Min. Knowl. Discov. 19(2), 210–226 (2009)
15. Guyet, T., Quiniou, R., Masson, V.: Mining relevant interval rules. CoRR abs/1709.03267 (2017), http://arxiv.org/abs/1709.03267
16. Hu, Q., Imielinski, T.: ALPINE: progressive itemset mining with definite guarantees. In: SIAM, pp. 63–71 (2017)

17. Huttenlocher, D.P., Klanderman, G.A., Rucklidge, W.: Comparing images using the hausdorff distance. IEEE Trans. Pattern Anal. Mach. Intell. **15**(9), 850–863 (1993)
18. Kaytoue, M., Kuznetsov, S.O., Napoli, A.: Revisiting numerical pattern mining with formal concept analysis. In: IJCAI, pp. 1342–1347 (2011)
19. Kurgan, L., Cios, K.J.: Discretization algorithm that uses class-attribute interdependence maximization. In: IC-AI, pp. 980–987 (2001)
20. van Leeuwen, M., Knobbe, A.J.: Diverse subgroup set discovery. Data Min. Knowl. Discov. **25**(2), 208–242 (2012)
21. Lenca, P., Meyer, P., Vaillant, B., Lallich, S.: On selecting interestingness measures for association rules: user oriented description and multiple criteria decision aid. Eur. J. Oper. Res. **184**(2), 610–626 (2008)
22. Lucas, T., Silva, T.C.P.B., Vimieiro, R., Ludermir, T.B.: A new evolutionary algorithm for mining top-k discriminative patterns in high dimensional data. Appl. Soft Comput. **59**, 487–499 (2017)
23. Mampaey, M., Nijssen, S., Feelders, A., Knobbe, A.J.: Efficient algorithms for finding richer subgroup descriptions in numeric and nominal data. In: ICDM, pp. 499–508 (2012)
24. Morishita, S., Sese, J.: Traversing itemset lattice with statistical metric pruning. In: ACM SIGMOD-SIGACT-SIGART, pp. 226–236 (2000)
25. Pawlak, Z.: Rough sets. Int. J. Parallel Program. **11**(5), 341–356 (1982)
26. Roman, S.: Lattices and Ordered Sets. Springer, New York (2008). https://doi.org/10.1007/978-0-387-78901-9
27. Wrobel, S.: An algorithm for multi-relational discovery of subgroups. In: Komorowski, J., Zytkow, J. (eds.) PKDD 1997. LNCS, vol. 1263, pp. 78–87. Springer, Heidelberg (1997). https://doi.org/10.1007/3-540-63223-9_108
28. Yang, Y., Webb, G.I., Wu, X.: Discretization methods. In: Maimon, O., Rokach, L. (eds.) Data Mining and Knowledge Discovery Handbook, 2nd edn, pp. 101–116. Springer, Boston (2010). https://doi.org/10.1007/978-0-387-09823-4_6
29. Zilberstein, S.: Using anytime algorithms in intelligent systems. AI Mag. **17**(3), 73–83 (1996)

Discovering Spatio-Temporal Latent Influence in Geographical Attention Dynamics

Minoru Higuchi[1], Kanji Matsutani[2], Masahito Kumano[1],
and Masahiro Kimura[1(✉)]

[1] Department of Electronics and Informatics, Ryukoku University, Otsu, Japan
kimura@rins.ryukoku.ac.jp
[2] Tokai Regional Headquarters, NTT West Corporation, Nagoya, Japan

Abstract. We address the problem of modeling the occurrence process of events for visiting attractive places, called points-of-interest (POIs), in a sightseeing city in the setting of a continuous time-axis and a continuous spatial domain, which is referred to as modeling geographical attention dynamics. By combining a Hawkes process with a time-varying Gaussian mixture model in a novel way and incorporating the influence structure depending on time slots as well, we propose a probabilistic model for discovering the spatio-temporal influence structure among major sightseeing areas from the viewpoint of geographical attention dynamics, and aim to accurately predict POI visit events in the near future. We develop an efficient method of inferring the parameters in the proposed model from the observed sequence of POI visit events, and present an analysis method for the geographical attention dynamics. Using real data of POI visit events in a Japanese sightseeing city, we demonstrate that the proposed model outperforms conventional models in terms of predictive accuracy, and uncover the spatio-temporal influence structure among major sightseeing areas in the city from the perspective of geographical attention dynamics.

Keywords: Geographical attention dynamics · Point process model
Spatio-temporal influence structure

1 Introduction

With the development of smart mobile devices, location acquisition technologies and social media, a large amount of event data with spatio-temporal information has become available and offers an opportunity to better understand people's location preferences and mobility patterns in a sightseeing city [3]. In location-based social networking services (LBSNs) such as Foursquare and Facebook Places, check-in sequences of users to *points-of-interest (POIs)* are observed, where a finite number of venues are listed as POIs in advance. Clearly, there

© Springer Nature Switzerland AG 2019
M. Berlingerio et al. (Eds.): ECML PKDD 2018, LNAI 11052, pp. 517–534, 2019.
https://doi.org/10.1007/978-3-030-10928-8_31

exist infinitely many attractive places in the city, including various geographical points giving beautiful views and street spots with artistic atmosphere. In photo-sharing services such as Flickr, observations of where and when people took photos are obtained. In order to take into account infinitely many attractive places on a continuous spatial domain, we consider extending the definition of POI. In particular, we also refer to the geographical locations of such photos that were taken on sightseeing tours and uploaded to a photo-sharing site as POIs, and aim at precisely investigating people's experiences in visiting attractive places in the city. Namely, in our definition, it is supposed that any POI offers an attractive place in a sense. Note that a complete list of all POIs cannot be obtained in advance.

Recently, researchers [2,8,21] have examined the next POI recommendation problem, that is, the problem of predicting which POI a user is most likely to visit at the next discrete time-step given the current check-in POI, where it is assumed that a finite set of POIs is specified in advance and historical check-in sequences of users in an LBSN are provided. However, these studies were unable to fully capture the continuous structure of space-time, and thus Liu et al. [14] extended them to the case of a continuous time-axis by integrating temporal interval assessment. On the other hand, since online items posted on social media sites such as Facebook and Twitter gain their popularity by the amount of attention received (e.g., the number of Facebook shares and the number of retweets), several studies have been made on modeling the attention dynamics of online items in a continuous time-axis [11,16,19,22]. Zhou et al. [24] presented a point process model in a discretized time-axis and a continuous spatial domain by fusing a time-varying Gaussian mixture model with a non-homogeneous Poisson process, and successfully estimated the spatial distribution of Toronto's ambulance demand at a specified discrete time-step (i.e., each two-hour interval), where each Gaussian component shows a representative geographical area. In the case of dealing with events of visiting POIs in a sightseeing city, such a component may correspond to a major sightseeing area. However, this study is unable to extract the influence structure among components from the viewpoint of visiting POIs, while such knowledge can become important for tourism marketing.

For a given sightseeing city, we consider the problem of modeling the occurrence process of events for visiting POIs in a continuous time-axis and a continuous spatial domain, which is referred to as that of modeling the *geographical attention dynamics*, and aim to provide deep insights into the properties of people's location preferences and mobility patterns on sightseeing tours in the city. What we observe is both a time-sequence of events (see Fig. 1a) and their locations (see Fig. 1b). Given a season, the sightseeing city should have a finite number of major sightseeing areas $C_1, \ldots, C_K$ (see Fig. 1c), where these represent major tourism topics, and are allowed to geographically intersect each other. In the same way as the attention dynamics of online items in social media, we first assume that the occurrences of previous events increase the possibility of future events. In particular, POI-visit events should exhibit a geographically self-exciting nature, where an event that happened in an area C_k may cause its

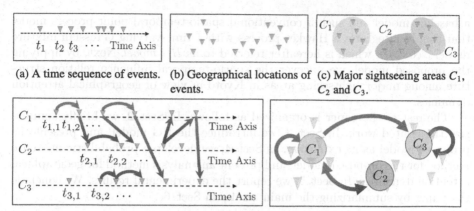

(a) A time sequence of events.   (b) Geographical locations of   (c) Major sightseeing areas C_1,
events. C_2 and C_3.

(d)   Geographical   self-excitations   and   mutual-   (e) Main spatio-temporal influence rela-
excitations.   tions among C_1, C_2 and C_3.

Fig. 1. An illustration of geographical attention dynamics. Down-pointing triangles indicate the time points or geographical locations for POI visit events. Arrows in (d) indicates triggering relations between events. For example, (d) illustrates that the event in C_1 at time $t_{1,2}$ triggered the event in C_2 at time $t_{2,1}$. Arrows in (e) represent the main influence relations among latent components (major sightseeing areas).

subsequent events in the same area C_k. Also, it is natural to suppose that the geographical attention dynamics has a geographically mutually-exciting nature, where an event that happened in an area C_k can trigger the subsequent events in any other area C_ℓ (see Fig. 1d). Moreover, the temporal decay rate of such effect should vary according to area C_k. Thus, based on the data of many people's POI visit events in the season, it is desirable to identify major sightseeing areas $C_1, \ldots, C_K$ and find the spatio-temporal influence relations among $C_1, \ldots, C_K$ in terms of geographical mutual-excitation (see Fig. 1e).

In this paper, we propose a probabilistic model for discovering the spatio-temporal influence structure among major sightseeing areas from the perspective of geographical attention dynamics in a continuous space-time, and aim at accurately predicting the future POI visit events. To this end, we combine a Hawkes process [13], which is a counting process [1] frequently utilized to capture mutual excitations between events, with a time-varying Gaussian mixture model in a novel way. Also, we incorporate the influence structure depending on time slots into our model since it is known that users' activities in LBSNs are often influenced by time [10,20] and such temporal properties may rely on sightseeing cities as well as seasons. We develop an efficient method of inferring the parameters in the proposed model from the observed sequence of POI visit events, and provide an analysis method for the geographical attention dynamics in terms of spatio-temporal influence relations among major sightseeing areas. Using real data of POI visit events in Japanese sightseeing city "Kyoto" obtained from a photo-sharing site, we evaluate the proposed method. First, for predicting the future POI visit events, we show the effectiveness of the proposed model compared to a baseline model and such a conventional spatio-temporal point process model that simply integrates a Hawkes process with a time-varying Gaussian mixture model (see [24]), which is hereafter referred to as *HG model*. Next, by applying the proposed model, we uncover the spatio-temporal influence relation structure among major sightseeing areas in Kyoto in view of geographical attention dynamics.

The rest of the paper is organized as follows: In Sect. 2, we briefly summarize the related work. In Sect. 3, we introduce the HG model, and present the proposed model as its extension. In Sect. 4, we develop a probabilistic inference method for the proposed model, and present an analysis method for geographical attention dynamics. In Sect. 5, we report the experimental results. We conclude the paper by summarizing the main results in Sect. 6.

2   Related Work

Several studies have been made on predicting POI visit events in the near future. As described in Sect. 1, Chen et al. [2], Feng et al. [8] and Zhang et al. [21] investigated the next POI recommendation problem for a finite number of given POIs in a discretized time-axis. To address the problem in a continuous time-axis, Liu et al. [14] presented a method of exploiting temporal interval assessment. To construct an accurate predictive model of event data with mark information such as POI in a continuous time-axis for a finite number of given marks, Du et al. [5] extended a marked Hawkes process, and proposed such a marked temporal point process that incorporates a recurrent neural network. However, unlike our current approach, these works have a limitation in treating a continuous spatial domain, that is, it is difficult to handle the situation where there may be infinitely many POIs and a complete list of all POIs is unavailable. To model ambulance demand in a discretized time-axis and a continuous spatial domain, Zhou et al. [24] integrated a time-varying Gaussian mixture model with a non-homogeneous Poisson process. Note that this model can simply be extended to the HG model, which is a model for a continuous space-time. However, as already mentioned, the HG model also has a limitation in analyzing the spatio-temporal influence structure among latent components. In this paper, by properly extending the HG model, we propose a probabilistic model for the geographical attention dynamics in a continuous space-time, and aim at discovering the spatio-temporal influence structure among major sightseeing areas in a given season.

There have been many investigations related to modeling continuous-time events generated by users in social media. As described in Sect. 1, Wang et al. [19], Shen et al. [16], Gao et al. [11] and Zhao et al. [22] considered individually modeling the attention dynamics of online items posted on social media sites in order to predict their future popularity. Thus, unlike our current approach, these works are unable to properly analyze the relations among all the online items involved. On the other hand, Gomez-Rodriguez et al. [12] and Danesh-mand et al. [4] examined the problem of extracting the social influence network structure among users from the observed information cascades (i.e., the observed sequences of events for sharing the same online items). Multivariate Hawkes processes are often leveraged to model event sequences forming information cascades in social networks [6,23]. They are also extended to model coevolution dynamics of information diffusion and social network growth [7]. However, these studies exploited who shared which online item, and assume that a finite set of online items and users is given in advance. In this paper, we also focus on a Hawkes process, but unlike the studies mentioned above, we try to infer the spatio-temporal influence relation structure among POIs and predict the future POI visit events without knowing who visited which POI from the viewpoint of privacy protection. We also note that multivariate Hawkes processes cannot simply be applied to our problem since a complete list of all POIs and users is unavailable in advance.

3   Model

For predictive modeling of geographical attention dynamics, we consider modeling the occurrence process of events for visiting POIs in a sightseeing city during a time period $[0, T')$ corresponding to one of its tourist seasons in the setting of a continuous space-time, where T' (> 0) is assumed to be a few months, and the corresponding continuous spatial domain is denoted by $\Omega \subset \mathbb{R}^2$.

3.1   Preliminaries

For any $t \in (0, T')$, let N_t be the total number of events during time period $[0, t)$, and for each $n = 1, \ldots, N_t$, we represent the nth event as tuple $(t_n, \boldsymbol{x}_n)$, meaning that location $\boldsymbol{x}_n = (x_{n,1}, x_{n,2}) \in \Omega$ was visited and registered as a POI at time $t_n \in [0, T')$ on a sightseeing tour. We also denote the sequence of events (i.e., the history) up to but not including time t as

$$\mathcal{H}_t = \{(t_n, \boldsymbol{x}_n); n = 1, \ldots, N_t\}.$$

Based on the previous work [24], we focus on modeling the event occurrence process as a spatio-temporal point process with intensity function $\lambda(t) f(\boldsymbol{x} \mid t)$ for $\forall t \in (0, T')$ and $\forall \boldsymbol{x} = (x_1, x_2) \in \Omega$, where $\lambda(t)$ is the intensity function of a temporal point process and $f(\boldsymbol{x} \mid t)$ is a time-varying Gaussian mixture for the spatial distribution. Namely, $\lambda(t) f(\boldsymbol{x} \mid t) \, dt \, d\boldsymbol{x}$ is the conditional probability of observing an event within a small domain $[t, t+dt) \times \{[x_1, x_1 + dx_1) \times [x_2, x_2 + dx_2)\}$ given the history $\mathcal{H}_t$ (see [1,5,7]). Note that for $0 < \forall T \leq T'$, the probability density of $\mathcal{H}_T$ is given by

$$p(\mathcal{H}_T) = \exp\left\{-\int_0^T \lambda(t) \, dt\right\} \prod_{n=1}^{N_T} \{\lambda(t_n) f(\boldsymbol{x}_n \mid t_n)\}. \tag{1}$$

3.2   Spatial Distribution

We begin with defining $f(\boldsymbol{x} \mid t)$ for any $(t, \boldsymbol{x}) \in (0, T') \times \Omega$. Since people's travel behaviors should vary by time of day, several studies [10, 20] separated each day into different time slots to improve POI recommendations in LBSNs. We also adopt this idea, and decompose each day into such M time slots $TS_1, \ldots, TS_M$ that are appropriate for the city and the season to be considered[1]. Let $h : [0, T') \rightarrow \{1, \ldots, M\}$ be the *time slot function*, meaning that each time $t \in [0, T')$ belongs to time slot $TS_{h(t)}$.

Previous work [24] fixed the mixture component distributions across all time slots to overcome data sparsity issues, and tried to capture an accurate spatial structure. In the same way as [24], we define the spatial distribution $f(\boldsymbol{x} \mid t)$ by

$$f(\boldsymbol{x} \mid t, \Theta) = \sum_{k=1}^{K} \phi_{h(t),k} \, g(\boldsymbol{x} \mid \boldsymbol{\mu}_k, \Sigma_k), \quad \forall \boldsymbol{x} \in \Omega, \tag{2}$$

where K is the number of components, and $g(\boldsymbol{x} \mid \boldsymbol{\mu}_k, \Sigma_k)$ is the 2-dimensional Gaussian density with mean vector $\boldsymbol{\mu}_k$ and covariance matrix Σ_k for $k = 1, \ldots, K$. The mixing coefficients $\{\phi_{m,k}\}$ for time slot TS_m satisfy $0 < \phi_{m,k} < 1$ together with $\sum_{k=1}^{K} \phi_{m,k} = 1$ for $m = 1, \ldots, M$. Also, the parameters for $f(\boldsymbol{x} \mid t)$ are aggregated into the parameter set $\Theta = \{\phi_{m,k}, \boldsymbol{\mu}_k, \Sigma_k; m = 1, \ldots, M, k = 1, \ldots, K\}$. We can consider that each Gaussian component C_k essentially represents a geographical area corresponding to a major tourism topic, and is identified with a major sightseeing area. Thus, we leverage this identification and try to analyze the influence relations among those major sightseeing areas.

3.3   Spatio-Temporal Point Process

Next, we consider modeling $\lambda(t)$ for any $t \in (0, T')$.

Baseline Model. One of the simplest models for a temporal point process is a Poisson process, where $\lambda(t)$ is assumed to be independent of history $\mathcal{H}_t$ and given by

$$\lambda(t \mid \alpha) = \alpha, \quad \forall t \in (0, T'). \tag{3}$$

Here, α is a positive constant. Thus, the spatio-temporal point process model defined by intensity function $\lambda(t \mid \alpha) \, f(\boldsymbol{x} \mid t, \Theta)$ (see Eqs. (2) and (3)) is regarded as a baseline.

---

[1] Although it is desirable to automatically detect such a decomposition from data, we here assume that $TS_1, \ldots, TS_M$ are specified in advance. Our future work will involve developing this kind of method.

Conventional Model. As described in the previous sections, a Hawkes process is frequently used to model continuous-time events and capture mutually-exciting interactions between events, and has been investigated for various applications (see [13,22]). Thus, $\lambda(t)$ can be modeled as a Hawkes process,

$$\lambda(t \mid \alpha, \beta, \gamma) = \alpha + \beta \sum_{(t_n, \boldsymbol{x}_n) \in \mathcal{H}_t} \exp\{-\gamma(t - t_n)\}, \quad \forall t \in (0, T'). \qquad (4)$$

where α, β and γ are positive constants. Here, the spatio-temporal point process defined by intensity function $\lambda(t \mid \alpha, \beta, \gamma) f(\boldsymbol{x} \mid t, \Theta)$ (see Eqs. (2) and (4)) is referred to as *HG model*. Note that the HG model can be regarded as a conventional model presented in the previous work [24].

Proposed Model. By incorporating both component dependent temporal influence decay (see Sect. 1) and time-slot varying influence degree (see Sect. 3.2), we extend the HG model, and aim to discover the spatio-temporal influence structure among components from the viewpoint of geographical attention dynamics and to more accurately predict POI visit events in the near future. The proposed model is defined as the spatio-temporal point process with intensity function $\lambda(t \mid Z_t, \alpha, \beta, \gamma) f(\boldsymbol{x} \mid t, \Theta)$ (see Eqs. (2) and (5)). Here, $\lambda(t)$ is modeled as

$$\lambda(t \mid Z_t, \alpha, \beta, \gamma) = \alpha + \sum_{(t_n, \boldsymbol{x}_n) \in \mathcal{H}_t} \beta_{h(t_n)} \exp\{-\gamma_{z(\boldsymbol{x}_n \mid t_n)}(t - t_n)\}, \quad \forall t \in (0, T'), \quad (5)$$

where $z(\boldsymbol{x}_n \mid t_n)$ denotes the component ID of location $\boldsymbol{x}_n$ drawn from Gaussian mixture $f(\boldsymbol{x} \mid t_n, \Theta)$ at time t_n, i.e., $z(\boldsymbol{x}_n \mid t_n) = k$ if and only if $\boldsymbol{x}_n \in C_k$ at time t_n, for $n = 1, \ldots, N_t$. Z_t is defined as

$$Z_t = \{z(\boldsymbol{x}_n \mid t_n); \, n = 1, \ldots, N_t\}.$$

Also, for the city during the current season, $\alpha > 0$ expresses its underlying attractiveness, $\beta_m > 0$ represents the influence degree of time slot TS_m for $m = 1, \ldots, M$, and $\gamma_k > 0$ indicates the temporal influence decay rate of component C_k for $k = 1, \ldots, K$. Parameters β and γ are defined as $\beta = (\beta_1, \ldots, \beta_M)$ and $\gamma = (\gamma_1, \ldots, \gamma_K)$, respectively. Here, based on the additivity for independent Poisson processes (see [6,13,15]), for any $t \in (0, T')$, we introduce a set of latent variables,

$$Y_t = \{y_n; \, n = 1, \ldots, N_t\},$$

such that the nth event $(t_n, \boldsymbol{x}_n)$ was triggered by the y_nth event $(t_{y_n}, \boldsymbol{x}_{y_n})$, where $y_n = 0, 1, \ldots, n-1$, and $y_n = 0$ means that the nth event was triggered by the underlying attractiveness, i.e., the background intensity α. Namely, it is known that the point process with intensity function $\lambda(t_n \mid Z_{t_n}, \alpha, \beta, \gamma)$ at time t_n is the superposition of the Poisson processes with intensity functions $\lambda(t_n; y_n \mid Z_{t_n}, \alpha, \beta, \gamma)$, $(y_n = 0, 1, \ldots, n-1)$, where

$$\lambda(t_n; y_n \mid Z_{t_n}, \alpha, \beta, \gamma) = \begin{cases} \alpha & \text{if } y_n = 0 \\ \beta_{h(t_{y_n})} \exp\left\{-\gamma_{z(\boldsymbol{x}_{y_n} \mid t_{y_n})}(t_n - t_{y_n})\right\} & \text{if } 1 \le y_n < n \end{cases}$$

$$(6)$$

for $n = 1, \ldots, N_t$. We consider extracting the influence relation $R_{k,\ell}$ from component C_ℓ to component C_k for $k, \ell = 1, \ldots, K$ by leveraging Z_T and Y_T.

4   Learning Method

For the observed data $\mathcal{H}_T$ with $0 < T < T'$, we develop a method of inferring the parameters Θ, Z_T, α, β, γ and Y_T in the proposed model, and provide a method for prediction and analysis of the geographical attention dynamics.

4.1   Inference

We present an inference method of the proposed model from $\mathcal{H}_T$.

First, we estimate Θ by maximizing the likelihood function $p(\mathcal{H}_T \mid \Theta, Z_T, \alpha, \beta, \gamma)$. By Eq. (1), it is sufficient to maximize function $\mathcal{L}(\Theta) = \prod_{(t_n, \boldsymbol{x}_n) \in \mathcal{H}_T} f(\boldsymbol{x}_n \mid t_n, \Theta)$. We employ an EM algorithm. Note that the number K of components is assumed to be fixed in this paper although it can also be estimated from the observed data by exploiting some techniques such as affinity propagation [9] and birth-and-death Markov chain Monte Carlo [18,24]. Let $\bar{\Theta}$ be the current estimate of Θ. Then, the update rule "$\hat{\Theta} = \left\{ \{\hat{\phi}_{m,k}\}, \{\hat{\boldsymbol{\mu}}_k\}, \{\hat{\Sigma}_k\} \right\} \leftarrow \bar{\Theta} = \left\{ \{\bar{\phi}_{m,k}\}, \{\bar{\boldsymbol{\mu}}_k\}, \{\bar{\Sigma}_k\} \right\}$" is obtained as follows[2]:

$$\hat{\phi}_{m,k} = \frac{1}{|\mathcal{H}_T^m|} \sum_{(t_n, \boldsymbol{x}_n) \in \mathcal{H}_T^m} \frac{\bar{\phi}_{m,k}\, g(\boldsymbol{x}_n \mid \bar{\boldsymbol{\mu}}_k, \bar{\Sigma}_k)}{f(\boldsymbol{x}_n \mid t_n, \bar{\Theta})},$$

$$\hat{\boldsymbol{\mu}}_k = \frac{1}{\sum_{n=1}^{N_T} \bar{a}_{n,k}} \sum_{n=1}^{N_T} \bar{a}_{n,k}\, \boldsymbol{x}_n, \quad \hat{\Sigma}_k = \frac{1}{\sum_{n=1}^{N_T} \bar{a}_{n,k}} \sum_{n=1}^{N_T} \bar{a}_{n,k}\, (\boldsymbol{x}_n - \hat{\boldsymbol{\mu}}_k)\,(\boldsymbol{x}_n - \hat{\boldsymbol{\mu}}_k)^{\mathrm{T}}$$

for $m = 1, \ldots, M$ and $k = 1, \ldots, K$, where the superscript T stands for a matrix transpose, each 2-vector is treated as a 2×1 matrix, and

$$\mathcal{H}_T^m = \{(t, \boldsymbol{x}) \in \mathcal{H}_T;\, h(t) = m\}, \quad \bar{a}_{n,k} = \frac{\bar{\phi}_{h(t_n),k}\, g(\boldsymbol{x}_n \mid \bar{\boldsymbol{\mu}}_k, \bar{\Sigma}_k)}{f(\boldsymbol{x}_n \mid t_n, \bar{\Theta})}.$$

Also, $|S|$ denotes the number of elements in a set S. With this method, we get the estimate Θ^* of Θ. Then, for each $(t_n, \boldsymbol{x}_n) \in \mathcal{H}_T$ and $k = 1, \ldots, K$, the posterior probability $\psi_k(\boldsymbol{x}_n \mid t_n)$ of location $\boldsymbol{x}_n$ at time t_n is given by

$$\psi_k(\boldsymbol{x}_n \mid t_n) = P(z(\boldsymbol{x}_n \mid t_n) = k \mid t_n, \boldsymbol{x}_n, \Theta^*) = \frac{\phi^*_{h(t_n),k}\, g(\boldsymbol{x}_n \mid \boldsymbol{\mu}^*_k, \Sigma^*_k)}{f(\boldsymbol{x}_n \mid t_n, \Theta^*)}, \tag{7}$$

and thus the estimate Z_T^* of Z_T can be obtained by

$$z^*(\boldsymbol{x}_n \mid t_n) = \operatorname*{argmax}_{1 \le k \le K} \psi_k(\boldsymbol{x}_n \mid t_n).$$

---

[2] For simplicity, no priors are here assumed for Θ. Note that it is clearly possible to give some natural priors.

Next, we develop a Bayesian method of estimating α, β and γ based on Eq. (1). To this end, we introduce the latent variables Y_T and try to infer Y_T as well (see Sect. 3.3). We consider leveraging the joint likelihood $p(\mathcal{H}_T, Y_T \mid \Theta^*, Z_T^*, \alpha, \beta, \gamma)$,

$$
p(\mathcal{H}_T, Y_T \mid \Theta^*, Z_T^*, \alpha, \beta, \gamma)
$$
$$
\propto \exp\left\{ -T\alpha - \sum_{m=1}^{M} \beta_m \, G_m(\gamma \mid Z_T^*) \right\} \prod_{(t_n, x_n) \in \mathcal{H}_T} \lambda(t_n; y_n \mid Z_T^*, \alpha, \beta, \gamma), \quad (8)
$$

where

$$
G_m(\gamma \mid Z_T^*) = \sum_{n=1}^{N_T} \frac{1}{\gamma_{z(x_n \mid t_n)}} \left(1 - \exp\left\{ -\gamma_{z(x_n \mid t_n)} (T - t_n) \right\} \right) I(h(t_n) = m).
$$

Here, $I(v)$ is an indicator function such that $I(v) = 1$ if v is true, $I(v) = 0$ otherwise. Suppose that α, β and γ are independently generated from the following priors (i.e., gamma distributions):

$$
\alpha \sim \text{Gamma}(\nu_\alpha, \eta_\alpha), \quad \beta_m \sim \text{Gamma}(\nu_\beta, \eta_\beta), \quad \gamma_k \sim \text{Gamma}(\nu_\gamma, \eta_\gamma), \quad (9)
$$

for $m = 1, \ldots, M$ and $k = 1, \ldots, K$, where $\nu_\alpha, \eta_\alpha, \nu_\beta, \eta_\beta, \nu_\gamma, \eta_\gamma > 0$ are hyperparameters. Then, $p(\mathcal{H}_T, Y_T \mid \Theta^*, Z_T^*, \alpha, \beta, \gamma)$ can be analytically marginalized over α and β for priors (see Eqs. (8) and (9)), and we have

$$
p(\mathcal{H}_T, Y_T \mid \Theta^*, Z_T^*, \gamma, \nu_\alpha, \eta_\alpha, \nu_\beta, \eta_\beta)
$$
$$
= \int_{\mathbb{R}_+ \times \mathbb{R}_+^M} p(\mathcal{H}_T, Y_T \mid \Theta^*, Z_T^*, \alpha, \beta, \gamma) \, p(\alpha \mid \nu_\alpha, \eta_\alpha) \, p(\beta \mid \nu_\beta, \eta_\beta) \, d\alpha \, d\beta
$$
$$
\propto \exp\left\{ -\sum_{n=1}^{N_T} \gamma_{z(x_{y_n} \mid t_n)} (t_n - t_{y_n}) \right\} \frac{\Gamma(L_0 + \nu_\alpha)}{(T + \eta_\alpha)^{L_0 + \nu_\alpha}} \frac{\eta_\alpha^{\nu_\alpha}}{\Gamma(\nu_\alpha)}
$$
$$
\times \prod_{m=1}^{M} \left\{ \frac{\Gamma(L_m + \nu_\beta)}{\{G_m(\gamma \mid Z_T^*) + \eta_\beta\}^{L_m + \nu_\beta}} \frac{\eta_\beta^{\nu_\beta}}{\Gamma(\nu_\beta)} \right\}, \quad (10)
$$

where $\mathbb{R}_+$ denotes the space of positive real numbers, $\Gamma(s)$ is the gamma function,

$$
L_0 = \sum_{n=1}^{N_T} I(y_n = 0)
$$

indicates the number of events triggered by the background intensity, and

$$
L_m = \sum_{n=2}^{N_T} I(h(t_{y_n}) = m) \, I(y_n \geq 1)
$$

indicates the number of events triggered by the preceding events within time slot TS_m. By iterating the following three steps, we obtain the estimates α^*, β^* and γ^* of α, β and γ, respectively: (1) Gibbs sampling for Y_T. (2) Metropolis-Hastings sampling for γ. (3) Sampling for α and β, and updating of hyper-parameters. Moreover, based on the superposition theorem of independent Poisson processes (see [6,15]), we estimate the posterior probability $\xi_{n,i} = P(y_n = i \mid \mathcal{H}_T, \Theta^*, Z_T^*, \alpha^*, \beta^*, \gamma^*)$ as

$$\xi_{n,i} = \frac{\lambda(t_n; y_n = i \mid Z_T^*, \alpha^*, \beta^*, \gamma^*)}{\sum_{j=0}^{n-1} \lambda(t_n; y_n = j \mid Z_T^*, \alpha^*, \beta^*, \gamma^*)} \tag{11}$$

for $n = 1, \ldots, N_T$, $i = 0, 1, \ldots, n-1$ (see Eq. (6)). Note that $\{\xi_{n,i}\}$ provide the posterior distribution of Y_T. Below, we will describe the above three steps (1), (2) and (3) in detail.

Gibbs Sampling for Y_T: Given the current samples of Y_T, a new value of y_n for $n = 1, \ldots, N_T$ is sampled from $\{0, \ldots, n-1\}$ using the Gibbs sampler of the conditional probability (see Eq. (10)),

$$P(y_n = i \mid \mathcal{H}_T, Y_T^{-n}, \Theta^*, Z_T^*, \gamma, \nu_\alpha, \eta_\alpha, \nu_\beta, \eta_\beta) \propto p(y_n = i, \mathcal{H}_T \mid Y_T^{-n}, \Theta^*, Z_T^*, \gamma, \nu_\alpha, \eta_\alpha, \nu_\beta, \eta_\beta)$$

$$\propto \begin{cases} \dfrac{L_0^{-n} + \nu_\alpha}{T + \eta_\alpha} & \text{if } i = 0 \\[2ex] \dfrac{L_{h(t_i)}^{-n} + \nu_\beta}{G_{h(t_i)}(\gamma \mid Z_T^*) + \eta_\beta} \exp\left\{ -\gamma_{z(x_i \mid t_i)} (t_n - t_i) \right\} & \text{if } i = 1, \ldots, n, \end{cases}$$

where the superscript $-n$ stands for the set or value excluding the nth event.

Metropolis-Hastings Sampling for γ: Due to the nonconjugacy of γ, we consider leveraging a Metropolis-Hastings algorithm to obtain the invariant distribution of γ for current samples of Y_T. Here, we exploit a normal distribution $q(\gamma' \mid \gamma)$ as a proposal distribution for candidate γ', Using the symmetric property, $q(\gamma' \mid \gamma) = q(\gamma \mid \gamma')$, the acceptance probability of γ' is obtained by

$$Q(\gamma' \mid \gamma) = \min\left\{ 1, \frac{p(\gamma' \mid \mathcal{H}_T, Y_T, \Theta^*, Z_T^*, \nu_\alpha, \eta_\alpha, \nu_\beta, \eta_\beta)}{p(\gamma \mid \mathcal{H}_T, Y_T, \Theta^*, Z_T^*, \nu_\alpha, \eta_\alpha, \nu_\beta, \eta_\beta)} \right\}$$

Note that $Q(\gamma' \mid \gamma)$ is easily computed by using the relation,

$$p(\gamma \mid \mathcal{H}_T, Y_T, \Theta^*, Z_T^*, \nu_\alpha, \eta_\alpha, \nu_\beta, \eta_\beta) \propto p(\mathcal{H}_T, Y_T \mid \gamma, \Theta^*, Z_T^*, \nu_\alpha, \eta_\alpha, \nu_\beta, \eta_\beta) p(\gamma \mid \nu_\gamma, \eta_\gamma)$$

(see Eqs. (9) and (10)). We accept γ' according to $Q(\gamma' \mid \gamma)$. By iterating these operations, we obtain a sample for γ.

Sampling for α and β, and Updating of Hyper-Parameters: Given the current samples for Y_T and γ, we sample α and β by the expected values of the posterior distributions $p(\alpha \mid \mathcal{H}_T, Y_T, \nu_\alpha, \eta_\alpha) = \text{Gamma}(L_0 + \nu_\alpha, T + \eta_\alpha)$ and $p(\beta_m \mid \mathcal{H}_T, Y_T, Z_T^*, \gamma, \nu_\beta, \eta_\beta) = \text{Gamma}(L_m + \nu_\beta, G_m(\gamma \mid Z_T^*) + \eta_\beta)$ as follows:

$$\alpha = \frac{L_0 + \nu_\alpha}{T + \eta_\alpha}, \quad \beta_m = \frac{L_m + \nu_\beta}{G_m(\gamma \mid Z_T^*) + \eta_\beta}$$

for $m = 1, \ldots, M$ (see Eqs. (8) and (9)). Next, we update the hyperparameters ν_α, η_α, ν_β, η_β, ν_γ and η_γ through the maximum likelihood estimations (see Eqs. (8), (9) and (10)). Here, the objective function for ν_α and η_α is given by

$$\mathcal{L}_\alpha(\nu_\alpha, \eta_\alpha) = \ln \frac{\Gamma(L_0 + \nu_\alpha)}{(T + \eta_\alpha)^{L_0 + \nu_\alpha}} + \ln \frac{\eta_\alpha^{\nu_\alpha}}{\Gamma(\nu_\alpha)}.$$

Also, the objective functions for ν_β and η_β is given by

$$\mathcal{L}_\beta(\nu_\beta, \eta_\beta) = \sum_{m=1}^{M} \ln \frac{\Gamma(L_m + \nu_\beta)}{\{G_m(\gamma \,|\, Z_T^*) + \eta_\beta\}^{L_m + \eta_\beta}} + M \ln \frac{\eta_\beta^{\nu_\beta}}{\Gamma(\nu_\beta)},$$

and the objective function for ν_γ and η_γ is given by

$$\mathcal{L}_\gamma(\nu_\gamma, \eta_\gamma) = \sum_{k=1}^{K} \left(\ln \gamma_k^{\nu_\gamma - 1} - \eta_\gamma \gamma_k \right) + K \ln \frac{\eta_\gamma^{\nu_\gamma}}{\Gamma(\nu_\gamma)}.$$

Based on Newton's method, we obtain the update rules for these hyperparameters.

4.2 Prediction and Analysis

Using the proposed model inferred from $\mathcal{H}_T$, we provide a framework for predicting the future events and analyzing the geographical attention dynamics.

We predict the events occurring in $[T, T')$ by simulating the proposed model under Ogata's thinning algorithm [15] based on the intensity function given by

$$\lambda^*(t) f^*(x \,|\, t)$$
$$= \left(\alpha^* + \sum_{(t_i, x_i) \in \mathcal{H}_t} \beta_{h(t_i)}^* \sum_{k=1}^{K} \psi_k(x_i \,|\, t_i) \exp\left\{-\gamma_k^*(t - t_i)\right\} \right) \sum_{k=1}^{K} \phi_{h(t),k}^* g(x \,|\, \mu_k^*, \Sigma_k^*) \quad (12)$$

for any $t \in [T, T')$ (see Eqs. (2), (5) and (7)).

We analyze the geographical attention dynamics in the following way. We first examine the estimated parameters α^*, β^*, γ^* and Θ^* in detail. Next, we extract the influence relation $R_{k,\ell}$ from latent component C_ℓ to latent component C_k by

$$R_{k,\ell} = \sum_{n=2}^{N_T} \sum_{i=1}^{n-1} \xi_{n,i} \psi_k(x_n \,|\, t_n) \psi_\ell(x_i \,|\, t_i)$$

for $k, \ell = 1, \ldots, K$ (see Eq. (11)), and analyze it. Here, note that each component C_k is identified with a major sightseeing area, and each $\xi_{n,i}$ measures the spatio-temporal influence from the ith event (t_i, x_i) to the nth event (t_n, x_n).

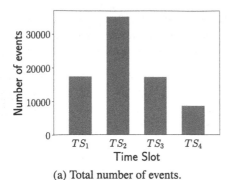

(a) Total number of events.

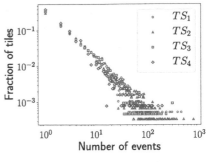

(b) Distribution for number of events.

Fig. 2. Statistical analysis for the number of events within each time slot.

5   Experiments

Using real data of POI visit events in "Kyoto", the ancient capital of Japan (a famous sightseeing city), we first evaluate the proposed model in terms of prediction performance. Next, by applying the proposed analysis method, we try to examine the properties of the geographical attention dynamics in Kyoto.

5.1   Datasets

We collected such photos that were taken within Kyoto city in 2014 and uploaded to photo-sharing site Flickr[3]. By regarding those photos as a set of photos taken on sightseeing tours, we constructed real data for POI visit events in Kyoto. The total number of those photos was 78, 239. By taking into account Kyoto's attractive seasons represented by cherry blossoms and autumn leaves, we focus on the spring data from March 1 to May 7 and the autumn data from October 1 to December 7. Also, from the perspective of Kyoto's sightseeing, we divide one day into $M = 4$ time slots, and set time slots TS_1, TS_2, TS_3 and TS_4 as 6 am to 11 am, 11 am to 4 pm, 4 pm to 9 pm and 9 pm to 6 am, respectively. Figure 2a indicates the number of events within each TS_m. Unsurprisingly, it is seen that many events occurred in daytime TS_2, and a relatively small number of events occurred in night-time and early-morning TS_4.

For each of the spring and autumn data, we constructed seven datasets $\mathcal{D}_1, \ldots, \mathcal{D}_7$ in the following way: We let the training period $[0, T)$ and the test period $[T, T')$ be two months and one day, respectively. In the case of the spring data, for example, for dataset $\mathcal{D}_1$, training period $[0, T)$ is March 1 to April 30 and test period $[T, T')$ is May 1, and for dataset $\mathcal{D}_2$, training period $[0, T)$ is March 2 to May 1 and test period $[T, T')$ is May 2. In the case of the autumn data, for example, for dataset $\mathcal{D}_1$, training period $[0, T)$ is October 1 to November 30 and test period $[T, T')$ is December 1, and for dataset $\mathcal{D}_2$, training period $[0, T)$ is October 2 to December 1 and test period $[T, T')$ is December 2.

---

[3] https://www.flickr.com/.

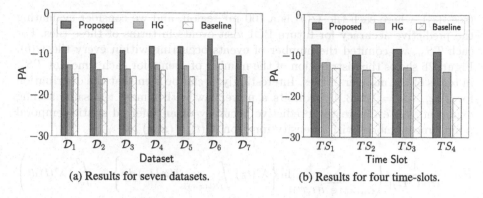

Fig. 3. Predictive accuracy for the spring data.

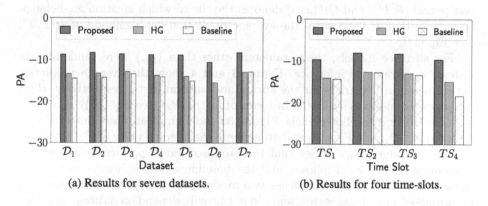

Fig. 4. Predictive accuracy for the autumn data.

5.2 Evaluation of Prediction Performance

For predicting future POI visit events, we compared the proposed model (see Eq. (5)) with the HG model (see Eq. (4)) and the baseline model (see Eq. (3)). Here, the parameters α, β and γ for the HG model were estimated by a commonly used method for learning a Hawkes process (i.e., a maximal likelihood method based on an EM algorithm (see [7])), and the parameter α for the baseline model was also estimated in the same way. For inferring α, β, γ and $\{\xi_{n,i}\}$ in the proposed model, we in particular implemented 1,000 iterations with 200 burn-in. In view of Kyoto's sightseeing, the number of components was set as $K = 8$, and eight representative tourist spots were always used as the initial positions of parameters $\{\mu_k; k = 1, \ldots, 8\}$ in parameter inference, for all three models.

By taking the issue of spatial resolution limitation into consideration, we decompose an appropriate rectangular region covering Kyoto's spatial domain Ω into a collection of 250×400 consecutive *tiles* $\{\Omega(b); b = 1, \ldots, 250 \times 400\}$

(see [17]), where each tile $\Omega(b)$ is a 100 m^2 region, and we consider evaluating the predictive accuracy for future POI visit events in terms of these tiles. For each TS_m, we counted the number of events occurring within every tile $\Omega(b)$. Figure 2b shows the distribution of the number of events for each time slot TS_m in terms of the number of tiles. Interestingly, it can be seen that the distribution for TS_m, ($m = 1, 2, 3, 4$), exhibits a power law with almost the same scaling exponent. We evaluate the predictive accuracy of an inferred spatio-temporal point process model with intensity function $\lambda^*(t) f^*(x \mid t)$ by

$$PA = \frac{1}{|\mathcal{H}(T,T')|} \left(\sum_{(t_n, x_n) \in \mathcal{H}(T,T')} \ln\left\{ \lambda^*(t_n) \int_{\Omega(b(x_n))} f(x \mid t_n) \, dx \right\} - \int_T^{T'} \lambda^*(t) \, dt \right)$$

(13)

(see [24])[4], where $\mathcal{H}(T, T') = \mathcal{H}_{T'} \setminus \mathcal{H}_T$ stands for the set of events occurring in test period $[T, T']$, and $\Omega(b(x_n))$ denotes the tile to which location x_n belongs. Here, note that PA measures the average prediction log-likelihood of $\mathcal{H}(T, T')$ (see Eq. (1)).

For all three models, the parameters other than $\{\mu_k\}$ were randomly initialized in parameter inference. Figures 3 and 4 show the average prediction performance on five trials for the spring and autumn data, respectively. Here, the proposed model is evaluated in terms of metric PA (see Eq. (13)), compared with the HG and baseline models. Figures 3a and 4a indicate the value of PA for each dataset $\mathcal{D}_j$, and Figs. 3b and 4b indicate the average value of PA restricted to each time slot TS_m. We see that the proposed model performs the best, the conventional HG model follows, and the baseline model is always worse than these two models. Unlike the other two models, the prediction performance of the proposed model was stable, and did not heavily depend on datasets and time slots. These results imply that it is significant to incorporate both component dependent temporal influence decay and time-slot varying influence degree, and demonstrate the effectiveness of the proposed model.

5.3   Analysis of Geographical Attention Dynamics

By applying the proposed method, we examine the properties of the geographical attention dynamics in Kyoto. Here, we only report the analysis results for the spring data (see Fig. 5).

Figure 5a displays the geographical locations of the latent Gaussian components $C_1, \ldots, C_8$ estimated, which represent the major sightseeing areas of Kyoto in the spring. Figure 5b gives a visualization result of the estimated parameters $\{\phi^*_{m,k}\}$, where each $\phi^*_m = (\phi^*_{m,1}, \ldots, \phi^*_{m,8})$ indicates the popularity distribution among components within time slot TS_m. We observe that C_1, C_2 and C_3 are always popular, and C_5 and C_6 are also popular in some time slots to a

---

[4] Other performance metrics based on industry practices (see [24]) can also be used to evaluate it. We confirmed that the results for such a metric were similar to those for PA.

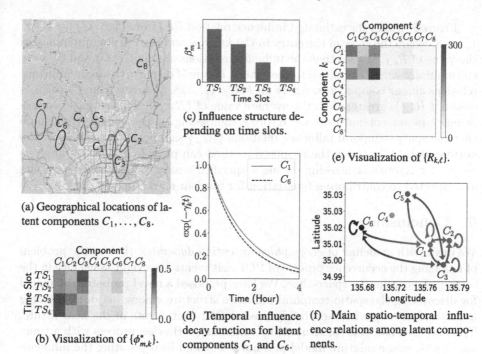

(a) Geographical locations of latent components $C_1, \ldots, C_8$.

(b) Visualization of $\{\phi^*_{m,k}\}$.

(c) Influence structure depending on time slots.

(d) Temporal influence decay functions for latent components C_1 and C_6.

(e) Visualization of $\{R_{k,\ell}\}$.

(f) Main spatio-temporal influence relations among latent components.

Fig. 5. Analysis results for the spring data.

certain degree. Here, C_1, C_2 and C_3 correspond to neighborhoods of Kyoto Imperial Palace (Kyoto Gosho)[5]. Heian-jingu Shrine[6] and Kiyomizu-dera Temple[7], respectively, and they are located near Kyoto's downtown. Also, C_5 corresponds to a neighborhood of Kinkaku-ji Temple[8] featuring a shining golden pavilion, which is located in a suburban area of Kyoto city. C_6 corresponds to Arashiyama area[9], which is a touristy district on the western outskirts of Kyoto, and famous as a place of scenic beauty.

Figure 5c shows the estimated parameters $\beta^* = (\beta^*_1, \ldots, \beta^*_4)$, which represent the influence structure depending on time slots. We can see that events occurred during morning TS_1 were the most influential, while events occurred during night-time and early-morning TS_4 were the least influential. For the estimated parameters $\{\gamma^*_k\}$, there was little difference among C_1, C_2, C_3 and C_5. Figure 5d displays the temporal influence decay functions estimated for components C_1 and C_6. This implies that the influence of events occurred in C_6 more rapidly decayed than that of events occurred in C_1.

---

[5] https://www.japan.travel/en/spot/1168/.

[6] https://www.japan.travel/en/spot/1195/.

[7] https://www.japan.travel/en/spot/2199/.

[8] https://kyoto.travel/en/shrine_temple/132.

[9] https://www.japan.travel/en/spot/1142/.

Figure 5e shows the estimated influence relation $R_{k,\ell}$ from C_ℓ to C_k for $k, \ell = 1, \ldots, 8$, where the color of the entry in the kth row and the ℓth column indicates the value of $R_{k,\ell}$. We see that the influence relations among C_1, C_2 and C_3 were substantially strong compared to the others. Figure 5f displays the main influence relations among components, where for $k, \ell = 1, \ldots, 8$, an arrow from C_ℓ to C_k is drawn if $R_{k,\ell}$ is greater than the average value of $\{R_{k',\ell'}\}$. This reveals people's primary movement patterns for Kyoto's sightseeing in the spring. Here, note that the spatio-temporal influence relations $\{R_{k,\ell}\}$ significantly changed for the autumn data. Like these, the proposed method can provide interesting analysis results for Kyoto's sightseeing during a specified season. These analysis results are expected to contribute a foundation for tourism marketing.

6   Conclusion

We dealt with modeling of *geographical attention dynamics*, that is, the problem of modeling the occurrence process of POI visit events for a sightseeing city in the setting of a continuous space-time. We have proposed a novel probabilistic model for discovering the spatio-temporal influence structure among major sightseeing areas, and attempted to accurately predict POI visit events in the near future. The proposed model is constructed by combining a Hawkes process with a time-varying Gaussian mixture model in a novel way and incorporating the influence structure depending on time slots as well. We developed an efficient method of inferring the parameters in the proposed model from the observed sequence of POI visit events, and provided an analysis method for the geographical attention dynamics. Using real data of Kyoto, a Japanese sightseeing city, we demonstrated that the proposed model significantly outperforms the conventional HG model and the baseline model in terms of predictive accuracy, and revealed the spatio-temporal influence relation structure among major sightseeing areas in Kyoto from the viewpoint of geographical attention dynamics.

In this paper, we focused on Kyoto's data obtained from Flickr, a photo-sharing site. Clearly, it is possible to apply the proposed method to other sightseeing cities and geographical regions including several sightseeing cities. Our immediate future work is to evaluate the proposed method for various sightseeing cities around the world and to explore POIs of variable geographical scales. We also supposed that the latent Gaussian components extracted by the proposed method represent major tourism topics and can be identified with major sightseeing areas. Our future work includes exploring spatial distributions other than Gaussian mixture. In several photo-sharing services, there are many photos that are annotated not only with GPS locations and time-stamps but also with text documents, and a method of detecting spatio-temporally exclusive topics from those data is investigated (see [17]). By applying such a method, we also plan to develop a framework of easily interpreting those latent components in terms of tourism topics.

Acknowledgments. This work was supported in part by JSPS KAKENIII Grant Number JP17K00433.

References

1. Aalen, O., Borgan, O., Gjessing, H.: Survival and Event History Analysis: A Process Point of View. Springer, New York (2008)
2. Chen, C., Yang, H., Lyu, M., King, I.: Where you like to go next: successive point-of-interest recommendation. In: Proceedings of IJCAI 2013, pp. 2605–2611 (2013)
3. Chen, S., et al.: Interactive visual discovering of movement patterns from sparsely sampled geo-tagged social media data. IEEE Trans. Vis. Comput. Graph. **22**(1), 270–279 (2016)
4. Daneshmand, H., Gomez-Rodriguez, M., Song, L., Schölkopf, B.: Estimating diffusion network structures: recovery conditions, sample complexity and soft-thresholding algorithm. In: Proceedings of ICML 2014, pp. 793–801 (2014)
5. Du, N., Dai, H., Upadhyay, U., Gomez-Rodriguez, M., Song, L.: Recurrent marked temporal point processes: embedding event history to vector. In: Proceedings of KDD 2016, pp. 1555–1564 (2016)
6. Farajtabar, M., Du, N., Gomez-Rodriguez, M., Valera, I., Zha, H., Song, L.: Shaping social activity by incentivizing users. In: Proceedings of NIPS 2014, pp. 2474–2482 (2014)
7. Farajtabar, M., Wang, Y., Gomez-Rodriguez, M., Li, S., Zha, H., Song, L.: COEVOLVE: a joint point process model for information diffusion and network evolution. J. Mach. Learn. Res. **18**(41), 1–49 (2017)
8. Feng, S., Li, X., Zeng, Y., Cong, G., Chee, Y., Yuan, Q.: Personalized ranking metric embedding for next new poi recommendation. In: Proceedings of IJCAI 2015, pp. 2069–2075 (2015)
9. Frey, B., Dueck, D.: Clustering by passing messages between data points. Science **315**(5814), 972–976 (2007)
10. Gao, H., Tang, J., Hu, X., Liu, H.: Exploring temporal effects for location recommendation on location-based social networks. In: Proceedings of RecSys 2013, pp. 93–100 (2013)
11. Gao, S., Ma, J., Chen, Z.: Modeling and predicting retweeting dynamics on microblogging platforms. In: Proceedings of WSDM 2015, pp. 107–116 (2015)
12. Gomez-Rodriguez, M., Leskovec, J., Krause, A.: Inferring networks of diffusion and influence. In: Proceedings of KDD 2010, pp. 1019–1028 (2010)
13. Hawkes, A.: Spectra of some self-exciting and mutually exiting point process. Biometrika **58**(1), 83–90 (1971)
14. Liu, Y., Liu, C., Liu, B., Qu, M., Xiong, H.: Unified point-of-interest recommendation with temporal interval assessment. In: Proceedings of KDD 2016, pp. 1015–1024 (2016)
15. Ogata, Y.: On lewis' simulation method for point processes. IEEE Trans. Inf. Theory **27**(1), 23–31 (1981)
16. Shen, H., Wang, D., Song, C., Barabási, A.L.: Modeling and predicting popularity dynamics via reinforced poisson processes. In: Proceedings of AAAI 2014, pp. 291–297 (2014)
17. Shin, S., et al.: STExNMF: spatio-temporally exclusive topic discovery for anomalous event detection. In: Proceedings of ICDM 2017, pp. 435–444 (2017)
18. Stephens, M.: Bayesian analysis of mixture models with an unknown number of components: an alternative to reversible jump methods. Ann. Stat. **28**(1), 40–74 (2000)
19. Wang, D., Song, C., Barabási, A.L.: Quantifying long-term scientific impact. Science **342**(6154), 127–132 (2013)

20. Yuan, Q., Cong, G., Ma, Z., Sun, A., Magnenat-Thalmann, N.: Time-aware point-of-interest recommendation. In: Proceedings of SIGIR 2013, pp. 363–372 (2013)
21. Zhang, J., Chow, C.: Spatiotemporal sequential influence modeling for location recommendations: a gravity-based approach. ACM Trans. Intell. Syst. Technol. **7**(1), 11:1–11:25 (2015)
22. Zhao, Q., Erdogdu, M., He, H., Rajaraman, A., Leskovec, J.: SEISMIC: a self-exciting point process model for predicting tweet popularity. In: Proceedings of KDD 2015, pp. 1513–1522 (2015)
23. Zhou, K., Zha, H., Song, L.: Learning social infectivity in sparse low-rank networks using multi-dimensional hawkes processes. In: Proceedings of AISTATS 2013, pp. 641–649 (2013)
24. Zhou, Z., Matteson, D., Woodard, D., Henderson, S., Micheas, A.: A spatio-temporal point process model for ambulance demand. J. Am. Stat. Assoc. **110**(509), 6–15 (2015)

Mining Periodic Patterns
with a MDL Criterion

Esther Galbrun[1]([⊠]), Peggy Cellier[2], Nikolaj Tatti[1,4], Alexandre Termier[2],
and Bruno Crémilleux[3]

[1] Department of Computer Science, Aalto University, Espoo, Finland
{esther.galbrun,nikolaj.tatti}@aalto.fi
[2] Univ. Rennes, {INSA, Inria}, CNRS, IRISA, Rennes, France
{peggy.cellier,alexandre.termier}@irisa.fr
[3] Normandie Univ., UNICAEN, ENSICAEN, CNRS – UMR GREYC, Caen, France
bruno.cremilleux@unicaen.fr
[4] F-Secure, Helsinki, Finland

Abstract. The quantity of event logs available is increasing rapidly, be
they produced by industrial processes, computing systems, or life tracking, for instance. It is thus important to design effective ways to uncover
the information they contain. Because event logs often record repetitive
phenomena, mining periodic patterns is especially relevant when considering such data. Indeed, capturing such regularities is instrumental in
providing condensed representations of the event sequences.

We present an approach for mining periodic patterns from event logs
while relying on a Minimum Description Length (MDL) criterion to evaluate candidate patterns. Our goal is to extract a set of patterns that
suitably characterises the periodic structure present in the data. We
evaluate the interest of our approach on several real-world event log
datasets. Code related to this paper is available at: https://github.com/
nurblageij/periodic-patterns-mdl.

Keywords: Periodic patterns · MDL · Sequence mining

1 Introduction

Event logs are among the most ubiquitous types of data nowadays. They can be
machine generated (server logs, database transactions, sensor data) or human
generated (ranging from hospital records to life tracking, a.k.a. quantified self),
and are bound to become ever more voluminous and diverse with the increasing
digitisation of our lives and the advent of the Internet of Things (IoT). Such
logs are often the most readily available sources of information on a system or
process of interest. It is thus critical to have effective and efficient means to
analyse them and extract the information they contain.

Many such logs monitor repetitive processes, and some of this repetitiveness
is recorded in the logs. A careful analysis of the logs can thus help understand the

© Springer Nature Switzerland AG 2019
M. Berlingerio et al. (Eds.): ECML PKDD 2018, LNAI 11052, pp. 535–551, 2019.
https://doi.org/10.1007/978-3-030-10928-8_32

characteristics of the underlying recurrent phenomena. However, this is not an easy task: a log usually captures many different types of events. Events related to occurrences of different repetitive phenomena are often mixed together as well as with noise, and the different signals need to be disentangled to allow analysis. This can be done by a human expert having a good understanding of the domain and of the logging system, but is tedious and time consuming.

Periodic pattern mining algorithms [17] have been proposed to tackle this problem. These algorithms can discover periodic repetitions of sets or sequences of events amidst unrelated events. They exhibit some resistance to noise, when it takes the form of slight variations in the inter-occurrence delay [2] or of the recurrence being limited to only a portion of the data [16]. However, such algorithms suffer from the traditional plague of pattern mining algorithms: they output too many patterns (up to several millions), even when relying on condensed representations [15].

Recent approaches have therefore focused on optimising the quality of the extracted *pattern set* as a whole [5], rather than finding individual high-quality patterns. In this context, the adaptation of the Minimal Description Length (MDL) principle [8,18] to pattern set mining has given rise to a fruitful line of work [3,4,20,21]. The MDL principle is a concept from information theory based on the insight that any structure in the data can be exploited to compress the data, and aiming to strike a balance between the complexity of the model and its ability to describe the data.

The most important structure of the data on which we focus here, i.e. of event logs, is the periodic recurrence of some events. For a given event sequence, we therefore want to identify a set of patterns that captures the periodic structure present in the data, and we devise a MDL criterion to evaluate candidate pattern sets for this purpose. First, we consider a simple type of model, representing event sequences with cycles over single events. Then, we extend this model so that cycles over distinct events can be combined together. By simply letting our patterns combine not only events but also patterns recursively, we obtain an expressive language of periodic patterns. For instance, it allows us to express the following daily routine:

Starting Monday at 7:30 AM, wake up, then, 10 min later, prepare coffee, repeat every 24 h for 5 days, repeat this every 7 days for 3 months

as a pattern consisting of two nested cycles, respectively with 24 h and 7 days periods, over the events "waking up" and "preparing coffee".

In short, we propose a novel approach for mining periodic patterns using a MDL criterion. The main component of this approach—and our main contribution— is the definition of an expressive pattern language and the associated encoding scheme which allows to compute a MDL-based score for a given pattern collection and sequence. We design an algorithm for putting this approach into practise and perform an empirical evaluation on several event log datasets. We show that we are able to extract sets of patterns that compress the input sequences and to identify meaningful patterns.

We start by reviewing the main related work, in Sect. 2. In Sect. 3, we introduce our problem setting and a simple model consisting of cycles over single events, which we extend in Sect. 4. We present an algorithm for mining periodic patterns that compress in Sect. 5 and evaluate our proposed approach over several event log datasets in Sect. 6. We reach conclusions in Sect. 7.

We focus here on the high-level ideas, and refer the interested reader to our report [6] that includes technical details, additional examples and experiments.

2   Related Work

The first approaches for mining periodic patterns used extremely constrained definitions of the periodicity. In [17], *all* occurrences must be regularly spaced; In [9,10], some missing occurrences are permitted but all occurrences must follow the same regular spacing. As a result, these approaches are extremely sensitive to even small amounts of noise in the data. Ma *et al.* [16] later proposed a more robust approach, which can extract periodic patterns in the presence of gaps of arbitrary size in the data. While the above approaches require time to be discretized as a preprocessing (time steps of hour or day length, for example), several solutions have been proposed to directly discover candidate periods from raw timestamp data, using the Fast Fourier Transform [2] or statistical models [14,22]. All of the above approaches are susceptible to producing a huge number of patterns, making the exploitation of their results difficult. The use of a *condensed representation* for periodic patterns [15] allows to significantly reduce the number of patterns output, without loss of information, but falls short of satisfactorily addressing the problem.

Considering pattern mining more in general, to tackle this pervasive issue of the overwhelming number of patterns extracted, research has focused on extracting *pattern sets* [5]: finding a (small) set of patterns that together optimise some interest criterion. One such criterion is based on the Minimum Description Length (MDL) principle [7]. Simply put, it states that *the best model is the one that compresses the data best.* Following this principle, the KRIMP algorithm [21] was proposed, to select a subset of frequent itemsets that yields the best lossless compression of a transactional database. This algorithm was later improved [19] and the approach extended to analyse event sequences [3,13,20]. Along a somewhat different approach, Kiernan and Terzi proposed to use MDL to summarize event sequences [12].

To the best of our knowledge, the only existing method that combines periodic pattern mining and a MDL criterion was proposed by Heierman *et al.* [11]. This approach considers a single regular episode at a time and aims to select the best occurrences for this pattern, independently of other patterns. Instead, we use a MDL criterion in order to select a good collection of periodic patterns.

3   Preliminary Notation and Problem Definition

Next, we formally define the necessary concepts and formulate our problem, focusing on simple cycles.

Event Sequences and Cycles. Our input data is a collection of timestamped occurrences of some events, which we call an *event sequence*. The events come from an alphabet Ω and will be represented with lower case letters. We assume that an event can occur only once per time step, so the data can be represented as a list of timestamp–event pairs, such as

$$S_1 = \langle (2,c), (3,c), (6,a), (7,a), (7,b), (19,a), (30,a), (31,c), (32,a), (37,b) \rangle .$$

Whether timestamps represent days, hours, seconds, or something else depends on the application, the only requirement is that they be expressed as positive integers. We denote as $S^{(\alpha)}$ the event sequence S restricted to event α, that is, the subset obtained by keeping only occurrences of event α.

We denote as $|S|$ the number of timestamp–event pairs contained in event sequence S, i.e. its *length*, and $\Delta(S)$ the time spanned by it, i.e. its *duration*. That is, $\Delta(S) = t_{\text{end}}(S) - t_{\text{start}}(S)$, where $t_{\text{end}}(S)$ and $t_{\text{start}}(S)$ represent the largest and smallest timestamps in S, respectively.

Given such an event sequence, our goal is to extract a representative collection of cycles. A *cycle* is a periodic pattern that takes the form of an ordered list of occurrences of an event, where successive occurrences appear at the same distance from one another. We will not only consider perfect cycles, where the inter-occurrence distance is constant, but will allow some variation.

A cycle is specified by indicating:

- the repeating event, called *cycle event* and denoted as α,
- the number of repetitions of the event, called *cycle length*, r,
- the inter-occurrence distance, called *cycle period*, p, and
- the timestamp of the first occurrence, called *cycle starting point*, τ.

Cycle lengths, cycle periods and cycle starting points take positive integer values (we choose to restrict periods to be integers for simplicity and interpretability). More specifically, we require $r > 1$, $p > 0$ and $\tau \geq 0$.

In addition, since we allow some variation in the actual inter-occurrence distances, we need to indicate an offset for each occurrence in order to be able to reconstruct the original subset of occurrences, that is, to recover the original timestamps. For a cycle of length r, this is represented as an ordered list of $r - 1$ signed integer offsets, called the *cycle shift corrections* and denoted as E. Hence, a cycle is a 5-tuple $C = (\alpha, r, p, \tau, E)$.

For a given cycle $C = (\alpha, r, p, \tau, E)$, with $E = \langle e_1, \ldots, e_{r-1} \rangle$ we can recover the corresponding occurrences timestamps by reconstructing them recursively, starting from τ: $t_1 = \tau$, $t_k = t_{k-1} + p + e_{k-1}$. Note that this is different from first reconstructing the occurrences while assuming perfect periodicity as $\tau, \tau + p$, $\tau + 2p, \ldots, \tau + (r-1)p$, then applying the corrections, because in the former case the corrections actually accumulate.

Then, we overload the notation and denote the time spanned by the cycle as $\Delta(C)$. Denoting as $\sigma(E)$ the sum of the shift corrections in E, $\sigma(E) = \sum_{e \in E} e$, we have $\Delta(C) = (r-1)p + \sigma(E)$. Note that this assumes that the correction maintains the order of the occurrences. This assumption is reasonable since an alternative cycle that maintains the order can be constructed for any cycle that does not.

We denote as $cover(C)$ the corresponding set of reconstructed timestamp–event pairs $cover(C) = \{(t_1, \alpha), (t_2, \alpha), \ldots, (t_r, \alpha)\}$. We say that a cycle covers an occurrence if the corresponding timestamp–event pair belongs to the reconstructed subset $cover(C)$.

Since we represent time in an absolute rather than relative manner and assume that an event can only occur once at any given timestamp, we do not need to worry about overlapping cycles nor about an order between cycles. Given a collection of cycles representing the data, the original list of occurrences can be reconstructed by reconstructing the subset of occurrences associated with each cycle, regardless of order, and taking the union. We overload the notation and denote as $cover(\mathcal{C})$ the set of reconstructed timestamp–event pairs for a collection $\mathcal{C}$ of cycles $\mathcal{C} = \{C_1, \ldots, C_m\}$, that is $cover(\mathcal{C}) = \bigcup_{C \in \mathcal{C}} cover(C)$.

For a sequence S and cycle collection $\mathcal{C}$ we call *residual* the timestamp–event pairs not covered by any cycle in the collection: $residual(\mathcal{C}, S) = S \setminus cover(\mathcal{C})$.

We associate a cost to each individual timestamp–event pair $o = (t, \alpha)$ and each cycle C, respectively denoted as $L(o)$ and $L(C)$, which we will define shortly. Then, we can reformulate our problem of extracting a representative collection of cycles as follows:

Problem 1. Given an event sequence S, find the collection of cycles $\mathcal{C}$ minimising the cost

$$L(\mathcal{C}, S) = \sum_{C \in \mathcal{C}} L(C) + \sum_{o \in residual(\mathcal{C}, S)} L(o) \, .$$

Code Lengths as Costs. This problem definition can be instantiated with different choices of costs. Here, we propose a choice of costs motivated by the MDL principle. Following this principle, we devise a scheme for encoding the input event sequence using cycles and individual timestamp–event pairs. The cost of an element is then the length of the code word assigned to it under this scheme, and the overall objective of our problem becomes finding the collection of cycles that results in the shortest encoding of the input sequence, i.e. finding the cycles that compress the data most. In the rest of this section, we present our custom encoding scheme. Note that all logarithms are to base 2.

For each cycle we need to specify its event, length, period, starting point and shift corrections, that is $L(C) = L(\alpha) + L(r) + L(p) + L(\tau) + L(E)$. It is important to look more closely at the range in which each of these pieces of information takes value, at what values—if any—should be favoured, and at how the values of the different pieces depend on one another.

To encode the cycles' events, we can use codes based on the events' frequency in the original sequence, so that events that occur more frequently in the event sequence will receive shorter code words: $L(\alpha) = -\log(fr(\alpha)) = -\log(|S^{(\alpha)}| / |S|)$. This requires that we transmit the number of occurrences of each event in the original event sequence. To optimise the overall code length, the length of the code word associated to each event should actually depend on the frequency of the event in the selected collection of cycles. However, this would require keeping track of these frequencies and updating the code lengths dynamically. Instead, we use the frequencies of the events in the input sequence as a simple proxy.

Clearly, a cycle with event α cannot have a length greater than $|S^{(\alpha)}|$. Once the cycle event α and its number of occurrences are known, we can encode the cycle length with a code word of length $L(r) = \log(|S^{(\alpha)}|)$, resulting in the same code length for large numbers of repetitions as for small ones.

Clearly, a cycle spans at most the time of the whole sequence, i.e. $\Delta(C) \leq \Delta(S)$, so that knowing the cycle length, the shift corrections, and the sequence time span, we can encode the cycle period with a code word of length

$$L(p) = \log\left(\left\lfloor \frac{\Delta(S) - \sigma(E)}{r - 1} \right\rfloor\right).$$

Next, knowing the cycle length and period as well as the sequence time span, we can specify the value of the starting point with a code word of length

$$L(\tau) = \log(\Delta(S) - \sigma(E) - (r - 1)p + 1).$$

Finally, we encode the shift corrections as follows: each correction e is represented by $|e|$ ones, prefixed by a single bit to indicate the direction of the shift, with each correction separated from the previous one by a zero. For instance, $E = \langle 3, -2, 0, 4 \rangle$ would be encoded as 0*111*0*11*1000*1111*0 with value digits, separating digits and sign digits, in italics, bold and normal font, respectively (the sign bit for zero is arbitrarily set to 0 in this case). As a result, the code length for a sequence of shift corrections E is $L(E) = 2|E| + \sum_{e \in E} |e|$.

Putting everything together, we can write the cost of a cycle C as

$$L(C) = \log(|S|) + \log\left(\left\lfloor \frac{\Delta(S) - \sigma(E)}{r - 1} \right\rfloor\right)$$
$$+ \log(\Delta(S) - \sigma(E) - (r - 1)p + 1) + 2|E| + \sum_{e \in E} |e|.$$

On the other hand, the cost of an individual occurrence $o = (t, \alpha)$ is simply the sum of the cost of the corresponding timestamp and event:

$$L(o) = L(t) + L(\alpha) = \log(\Delta(S) + 1) - \log(|S^{(\alpha)}| / |S|).$$

Note that if our goal was to actually encode the input sequence, we would need to transmit the smallest and largest timestamps ($t_{\text{start}}(S)$ and $t_{\text{end}}(S)$), the size of the event alphabet ($|\Omega|$), as well as the number of occurrences of each event ($|S^{(\alpha)}|$ for each event α) of the event sequence. We should also transmit the number of cycles in the collection ($|C|$), which can be done, for instance with a code word of length $\log(|S|)$. However, since our goal is to compare collections of cycles, we can simply ignore this, as it represents a fixed cost that remains constant for any chosen collection of cycles.

Finally, consider that we are given an ordered list of occurrences $\langle t_1, t_2, \ldots, t_l \rangle$ of event α, and we want to determine the best cycle with which to cover all these occurrences at once. Some of the parameters of the cycle are determined, namely the repeating event α, the length r, and the timestamp of the first occurrence τ. All we need to determine is the period p that yields the shortest code length for the cycle. In particular, we want to find p that minimises $L(E)$. The shift corrections are such that $E_k = (t_{k+1} - t_k) - p$. If we consider the list of inter-occurrence distances $d_1 = t_2 - t_1, d_2 = t_3 - t_2, \ldots, d_{l-1} = t_l - t_{l-1}$, the problem of finding p that minimises $L(E)$ boils down to minimising $\sum_{d_i} |d_i - p|$. This is achieved by letting p equal the geometric median of the inter-occurrence distances, which, in the one-dimensional case, is simply the median. Hence, for this choice of encoding for the shift corrections, the optimal cycle covering a list of occurrences can be determined by simply computing the inter-occurrences distances and taking their median as the cycle period.

4   Defining Tree Patterns

So far, our pattern language is restricted to cycles over single events. In practise, however, several events might recur regularly together and repetitions might be nested with several levels of periodicity. To handle such cases, we now introduce a more expressive pattern language, that consists of a hierarchy of cyclic blocks, organised as a tree.

Instead of considering simple cycles specified as 5-tuples $C = (\alpha, r, p, \tau, E)$ we consider more general patterns specified as triples $P = (T, \tau, E)$, where T denotes the tree representing the hierarchy of cyclic blocks, while τ and E respectively denote the starting point and shift corrections of the pattern, as with cycles.

Pattern Trees. Each *leaf node* in a pattern tree represents a simple block containing one event. Each *intermediate node* represents a cycle in which the children nodes repeat at a fixed time interval. In other words, each intermediate node represents cyclic repetitions of a sequence of blocks. The root of a pattern tree is denoted as B_0. Using list indices, we denote the children of a node B_X as B_{X1}, B_{X2}, etc. All children of an intermediate node except the first one are associated to their distance to the preceding child, called the *inter-block distance*, denoted as d_X for node B_X. Inter-block distances take non-negative integer values. Each intermediate node B_X is associated with the period p_X and length r_X of the corresponding cycle. Each leaf node B_Y is associated with the corresponding occurring event α_Y.

An example of an abstract pattern tree is shown in Fig. 1. We call *height* and *width* of the pattern tree—and by extension of the associated pattern— respectively the number of edges along the longest branch from the root to a leaf node and the number of leaf nodes in the tree.

For a given pattern, we can construct a tree of event occurrences by expanding the pattern tree recursively, that is, by appending to each intermediate node the corresponding number of copies of the associated subtree, recursively. We call this expanded tree the *expansion tree* of the pattern, as opposed to the contracted *pattern tree* that more concisely represents the pattern.

We can enumerate the event occurrences of a pattern by traversing its expansion tree and recording the encountered leaf nodes. We denote as $occs^*(P)$ this list of timestamp–event pairs reconstructed from the tree, prior to correction.

As for the simple cycles, we will not only consider perfect patterns but will allow some variations. For this purpose, a list of shift corrections E is provided with the pattern, which contains a correction for each occurrence except the first one, i.e. $|E| = |occs^*(P)| - 1$. However, as for simple cycles, corrections accumulate over successive occurrences, and we cannot recover the list of corrected occurrences $occs(P)$ by simply adding the individual corrections to the elements of $occs^*(P)$. Instead, we first have to compute the accumulated corrections for each occurrence.

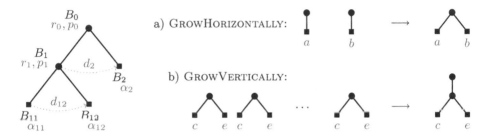

Fig. 1. Abstract pattern tree and examples of growing patterns through combinations.

Encoding the Patterns. To transmit a pattern, we need to encode its pattern tree, as well as its starting point and shift corrections. Furthermore, to encode the pattern tree, we consider separately its event sequence, its cycle lengths, its top-level period, and the other values, as explained below.

First we encode the event in the leaves of the pattern tree, traversing the tree from left to right, depth-first. We denote as A the string representing its event sequence. We encode each symbol s in the string A using a code of length $L(s)$, where $L(s)$ depends on the frequency of s, adjusted to take into account the additional symbols '(' and ')', used to delimit blocks. In particular, we set the code lengths for the extended alphabet such that $L('(') = L(')') = -\log(1/3)$ for the block delimiters, and $L(\alpha) = -\log(|S^{(\alpha)}|/(3|S|))$ for the original events.

Next, we encode the cycle lengths, i.e. the values r_X associated to each intermediate node B_X encountered while traversing the tree depth-first and from left to right, as a sequence of values, and denote this sequence R. For a block B_X the number of repetitions of the block cannot be larger than the number of occurrences of the least frequent event participating in the block, denoted as $\rho(B_X)$. We can thus encode the sequence of cycle lengths R with code of length

$$L(R) = \sum_{r_X \in R} L(r_X) = \sum_{r_X \in R} \log(\rho(B_X)).$$

Knowing the cycle lengths R and the structure of the pattern tree from its event sequence A, we can deduce the total number of events covered by the pattern. The shift corrections for the pattern consist of the correction to each event occurrence except the first one. This ordered list of values can be transmitted using the same encoding as for the simple cycles.

In simple cycles, we had a unique period characterising the distances between occurrences. Instead, with these more complex patterns, we have a period p_X for each intermediate node B_X, as well as an inter-block distance d_X for each node B_X that is not the left-most child of its parent.

First, we transmit the period of the root node of the pattern tree, B_0. In a similar way as with simple cycles, we can deduce the largest possible value for p_0 from r_0 and E. Since we do not know when the events within the main cycle occur, we assume what would lead to the largest possible value for p_0, that is, we assume that all the events within each repetition of the cycle happen at once, so that each repetition spans no time at all.

We denote as D the collection of all the periods (except p_0) and inter-block distances in the tree, that need to be transmitted to fully describe the pattern. To put everything together, the code used to represent a pattern $P = (T, \tau, E)$ has length

$$L(P) = L(A) + L(R) + L(p_0) + L(D) + L(\tau) + L(E) \, .$$

5   Algorithm for Mining Periodic Patterns that Compress

Recall that for a given input sequence S, our goal is to find a collection of patterns $\mathcal{C}$ that minimises the cost

$$L(\mathcal{C}, S) = \sum_{P \in \mathcal{C}} L(P) + \sum_{o \in residual(\mathcal{C}, S)} L(o) \, .$$

It is useful to compare the cost of different patterns, or sets of patterns, on a subset of the data, i.e. compare $L(\mathcal{C}', S')$ for different sets of patterns $\mathcal{C}'$ and some subsequence $S' \subseteq S$. In particular, we might compare the cost of a pattern P to the cost of representing the same occurrences separately. This means comparing

$$L(\{P\}, cover(P)) = L(P) \quad \text{and} \quad L(\emptyset, cover(P)) = \sum_{o \in cover(P)} L(o) \, .$$

If $L(\{P\}, cover(P)) < L(\emptyset, cover(P))$, we say that pattern P is *cost-effective*. In addition, we compare patterns in terms of their cost-per-occurrence ratio defined, for a pattern P, as $L(P)/|cover(P)|$, and say that a pattern is more *efficient* when this ratio is smaller.

A natural way to build patterns is to start with the simplest patterns, i.e. cycles over single events, and combine them together into more complex, possibly multi-level multi-event patterns.

Algorithm 1. Mining periodic patterns that compress.

Require: A multi-event sequence S, a number k of top candidates to keep
Ensure: A collection of patterns $\mathcal{P}$

```
 1: I ← ExtractCycles(S, k)
 2: C ← ∅; V ← I; H ← I
 3: while H ≠ ∅ or V ≠ ∅ do
 4:     V' ← CombineVertically(H, P, S, k)
 5:     H' ← CombineHorizontally(V, P, S, k)
 6:     C ← C ∪ H ∪ V; V ← V'; H ← H'
 7: P ← GreedyCover(C, S)
 8: return P
```

Assume that we have a pattern tree T_I which occurs multiple times in the event sequence. In particular, assume that it occurs at starting points τ_1, τ_2, ..., τ_{r_J} and that this sequence of starting points itself can be represented as a cycle of length r_J and period p_J. In such a case, the occurrences of T_I might be combined together and represented as a nested pattern tree. GROWVERTICALLY is the procedure which takes as input a collection $\mathcal{C}_I$ of patterns over a tree T_I and returns the nested pattern obtained by combining them together as depicted in Fig. 1(b).

On the other hand, given a collection of patterns that occur close to one another and share similar periods, we might want to combine them together into a concatenated pattern by merging the roots of their respective trees. GROWHORIZONTALLY is the procedure which takes as input a collection of patterns and returns the pattern obtained by concatenating them together in order of increasing starting points as depicted in Fig. 1(a).

As outlined in Algorithm 1, our proposed algorithm consists of three stages: *(i)* extracting cycles (line 1), *(ii)* building tree patterns from cycles (lines 2–6) and *(iii)* selecting the final pattern collection (line 7). We now present each stage in turn at a high-level.

Extracting Cycles. Considering each event in turn, we use two different routines to mine cycles from the sequence of timestamps obtained by restricting the input sequence to the event of interest, combine and filter their outputs to generate the set of initial candidate patterns. The first routine, EXTRACTCYCLESDP, uses dynamic programming. Indeed, if we allow neither gaps in the cycles nor overlaps between them, finding the best set of cycles for a given sequence corresponds to finding an optimal segmentation of the sequence, and since our cost is additive over individual cycles, we can use dynamic programming to solve it optimally [1]. The second routine, EXTRACTCYCLESTRI, extracts cycles using a heuristic which allows for gaps and overlappings. It collects triples (t_0, t_1, t_2) such that $||t_2 - t_1| - |t_1 - t_0|| \leq \ell$, where ℓ is set so that the triple can be beneficial when used to construct longer cycles. Triples are then chained into longer cycles. Finally, the set $\mathcal{C}$ of cost-effective cycles obtained by merging the output of the two routines is filtered to keep only the k most efficient patterns for each occurrence for a user-specified k, and returned.

Building Tree Patterns from Cycles. The second stage of the algorithm builds tree patterns, starting from the cycles produced in the previous stage. That is, while there are new candidate patterns, the algorithm performs combination rounds, trying to generate more complex patterns through vertical and horizontal combinations. If desired, this stage can be skipped, thereby restricting the pattern language to simple cycles.

In a round of vertical combinations performed by COMBINEVERTICALLY (line 4), each distinct pattern tree represented among the new candidates in $\mathcal{H}$ is considered in turn. Patterns over that tree are collected and EXTRACTCYCLESTRI is used to mine cycles from the corresponding sequence of starting points. For each obtained cycle, a nested pattern is produced by combining the corresponding candidates using GROWVERTICALLY (see Fig. 1(b)).

In a round of horizontal combinations performed by COMBINEHORIZONTALLY (line 5), a graph G is constructed, with vertices representing candidate patterns and with edges connecting pairs of candidates $\mathcal{K} = \{P_I, P_J\}$ for which the concatenated pattern $P_N = $ GROWHORIZONTALLY$(\mathcal{K})$ satisfies $L(\{P_N\}, cover(\mathcal{K})) < L(\mathcal{K}, cover(\mathcal{K}))$. A new pattern is then produced for each clique of G, by applying GROWHORIZONTALLY to the corresponding set of candidate patterns.

Selecting the Final Pattern Collection. Selecting the final set of patterns to output among the candidates in $\mathcal{C}$ is very similar to solving a weighted set cover problem. Therefore, the selection is done using a simple variant of the greedy algorithm for this problem, denoted as GREEDYCOVER (line 7).

6   Experiments

In this section, we evaluate the ability of our algorithm to find patterns that compress the input event sequences. We make the code and the prepared datasets publicly available.[1] To the best of our knowledge, no existing algorithm carries out an equivalent task and we are therefore unable to perform a comparative evaluation against competitors. To better understand the behaviour of our algorithm, we first performed experiments on synthetic sequences. We then applied our algorithm to real-world sequences including process execution traces, smartphone applications activity, and life-tracking. We evaluate our algorithm's ability to compress the input sequences and present some examples of extracted patterns.

For a given event sequence, the main objective of our algorithm is to mine and select a good collection of periodic patterns, in the sense that the collection should allow to compress the input sequence as much as possible. Therefore, the main measure that we consider in our experiments is the *compression ratio*, defined as the ratio between the length of the code representing the input sequence with the considered collection of patterns and the length of the code representing the input sequence with an empty collection of patterns, i.e. using

---

[1] https://github.com/nurblageij/periodic-patterns-mdl.

only individual event occurrences, given as a percentage. For a given sequence S and collection of patterns $\mathcal{C}$ the compression ratio is defined as

$$\%L = 100 \cdot L(\mathcal{C}, S)/L(\emptyset, S) ,$$

with smaller values associated to better pattern collections.

Table 1. Statistics of the event log sequences used in the experiments.

	$|S|$	$\Delta(S)$	$|\Omega|$	$\left|S^{(\alpha)}\right|$		$L(\emptyset, S)$	RT (s)	
				med	max		cycles	overall
3zap	181644	181643	443	22	36697	4154277	2094	35048
bugzilla	16775	16774	91	6	3332	303352	112	522
samba	28751	7461	119	44	2905	520443	214	2787
sacha	65977	221445	141	231	4389	1573140	2963	14377
ubiqLog (31 sequences)								
min	413	11391	10	23	194	6599	1	1
median	23859	87591	87	52	2131	486633	232	1020
max	167863	17900307	241	129	6101	3733349	2297	28973

Datasets. Our first two datasets come from a collaboration with STMicroelectronics and are execution traces of a set-top box based on the STiH418 SoC[2] running STLinux. Both traces are a log of system actions (interruptions, context switches and system calls) taken by the KPTrace instrumentation system developed at STMicroelectronics. The 3zap sequence corresponds to 3 successive changes of channel ("zap"), while the bugzilla sequence corresponds to logging a display blackout bug into the bug tracking system of ST. For our analysis of these traces, we do not consider timestamps, only the succession of events.

The ubiqLog dataset was obtained from the UCI Machine learning repository.[3] It contains traces collected from the smartphones of users over the course of two months. For each of 31 users we obtain a sequence recording what applications are run on that user's smartphone. We consider absolute timestamps with a granularity of one minute.

The samba dataset consists of a single sequence recording the emails identifying the authors of commits on the git repository of the samba network file system[4] from 1996 to 2016. We consider timestamps with a granularity of one day. We aggregated together users that appeared fewer than 10 times.

The sacha dataset consists of a single sequence containing records from the *quantified awesome* life log[5] recording the daily activities of its author between

---

[2] STiH418 description: http://www.st.com/resource/en/data_brief/stih314.pdf.
[3] https://archive.ics.uci.edu/ml/datasets/UbiqLog+(smartphone+lifelogging).
[4] https://git.samba.org/.
[5] http://quantifiedawesome.com/records.

November 2011 and January 2017. The daily activities are associated to start and end timestamps, and are divided between categories organised into a hierarchy. Categories with fewer than 200 occurrences were aggregated to their parent category. Each resulting category is represented by an event. Adjacent occurrences of the same event were merged together. We consider absolute timestamps with a granularity of 15 min.

Table 1 presents the statistics of the sequences used in our experiments.

We indicate the length ($|S|$) and duration ($\Delta(S)$) of each sequence, the size of its alphabet ($|\Omega|$), as well as the median and maximum length of the event subsequences ($|S^{(\alpha)}|$). We also indicate the code length of the sequence when encoded with an empty collection of patterns ($L(\emptyset, S)$), as well as the running time of the algorithm (RT, in seconds) for mining and selecting the patterns, as well as for the first stage of mining cycles for each separate event.

Measures. Beside the compression ratio ($\%L$) achieved with the selected pattern collections, we consider several other characteristics (see Table 2). For a given pattern collection $\mathcal{C}$, we denote the set of residuals $residual(\mathcal{C}, S)$ simply as $\mathcal{R}$ and look at what fraction of the code length is spent on them, denoted as $L:\mathcal{R} = \sum_{o \in \mathcal{R}} L(o)/L(\mathcal{C}, S)$. We also look at the number of patterns of differ-

Table 2. Summary of results for the separate event sequences.

	$\%L$	$L:\mathcal{R}$	s	/	v	/	h	/	m	c^+		$\%L$	$L:\mathcal{R}$	s	/	v	/	h	/	m	c^+
				3zap											bugzilla						
$\mathcal{C}_S$	56.32	0.41	11852/		– /		– /		–	2325	$\mathcal{C}_S$	48.58	0.12	262/		– /		– /		–	1652
$\mathcal{C}_V$	55.14	0.40	10581/581/		– /		–			2325	$\mathcal{C}_V$	48.56	0.12	259 /	1 /		– /		–		1652
$\mathcal{C}_H$	47.84	0.35	3459/		– /4912/				–	2325	$\mathcal{C}_H$	42.43	0.12	133 /		– /	70 /		–		1652
$\mathcal{C}_{V+H}$	47.40	0.34	3499 /419/4302/						–	2325	$\mathcal{C}_{V+H}$	42.39	0.12	130 /	1 /	72 /		–			1652
$\mathcal{C}_F$	46.99	0.34	3499 / 91 /4154/268							2325	$\mathcal{C}_F$	42.41	0.13	124 /	1 /	70 /	2				1652
				samba											sacha						
$\mathcal{C}_S$	28.42	0.14	429 /		– /		– /		–	2657	$\mathcal{C}_S$	74.34	0.37	9602/		– /		– /		–	304
$\mathcal{C}_F$	28.37	0.13	409 /	0 /	17 /	0				2657	$\mathcal{C}_F$	68.64	0.35	3957/	0 /2996/		0				582

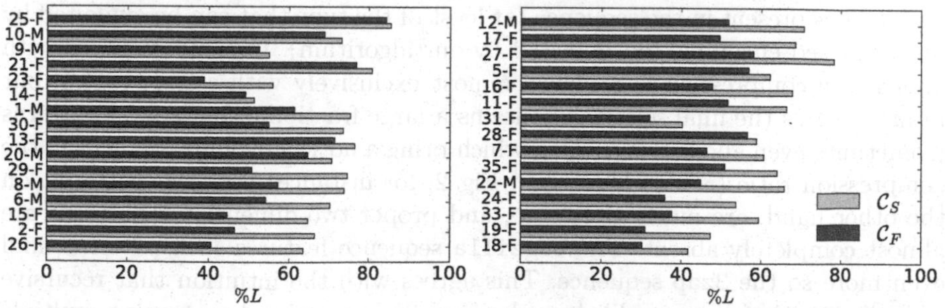

Fig. 2. Compression ratios for the sequences from the ubiqLog dataset.

ent types in $\mathcal{C}$: (s) simple cycles, i.e. patterns with width $= 1$ and height $= 1$, (v) vertical patterns, with width $= 1$ and height > 1, (h) horizontal patterns, with width > 1 and height $= 1$, and (m) proper two-dimensional patterns, with width > 1 and height > 1. Finally, we look at the maximum cover size of patterns in $\mathcal{C}$, denoted as c^+.

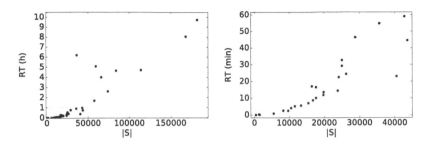

Fig. 3. Running times for mining the different sequences (in hours, left) and zooming in on shorter sequences (in minutes, right).

Results. In addition to the final collection of patterns returned by the algorithm after potentially a few rounds of combinations (denoted as $\mathcal{U}_F$), we also consider intermediate collections of patterns, namely a collection selected among cycles mined during the first stage of the algorithm (denoted as $\mathcal{C}_S$), including patterns from the first round of horizontal combinations ($\mathcal{C}_H$), of vertical combinations ($\mathcal{C}_V$), and both, i.e. at the end of the first round of combinations ($\mathcal{C}_{V+H}$).

A summary of the results for the separate event sequences 3zap, bugzilla, samba and sacha, is presented in Table 2. Figure 2 shows the compression ratios achieved on event sequences from the ubiqLog dataset.

We see that the algorithm is able to find sets of patterns that compress the input event sequences. The compression ratio varies widely depending on the considered sequence, from a modest 84% for some sequences from ubiqLog to a reduction of more than two thirds, for instance for samba. To an extent, the achieved compression can be interpreted as an indicator of how much periodic structure is present in the sequence (at least of the type that can be exploited by our proposed encoding and detected by our algorithm). In some cases, as with samba, the compression is achieved almost exclusively with simple cycles, but in many cases the final selection contains a large fraction of horizontal patterns (sometimes even about two thirds), which bring a noticeable improvement in the compression ratio (as can be seen in Fig. 2, for instance). Vertical patterns, on the other hand, are much more rare, and proper two-dimensional patterns are almost completely absent. The bugzilla sequence features such patterns, and even more so the 3zap sequence. This agrees with the intuition that recursive periodic structure is more likely to be found in execution logs tracing multiple recurrent automated processes.

Figure 3 shows the running times for sequences from the different datasets. The running times vary greatly, from only a few seconds to several hours. Naturally, mining longer sequences tends to require longer running times.

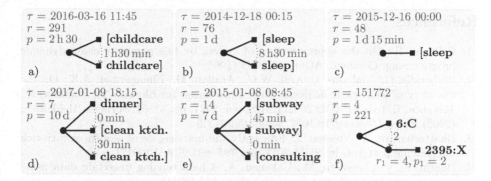

Fig. 4. Example patterns from sacha (a–e) and 3zap (f).

Example Patterns. Finally, we present some examples of patterns obtained from the sacha and 3zap sequences, in Fig. 4. The start and end of an activity A are denoted as "[A" and "A]" respectively. The patterns from the sacha sequence are simple and rather obvious, but they make sense when considering everyday activities. The fact that we are able to find them is a clear sign that the method is working. The 3zap pattern is a typical system case: the repetition of a context switch (6:C) followed by several activations of a process (2395:X).

Most of the discovered patterns are fairly simple. We suspect that this is due to the nature of the data: there are no significantly complex patterns in these event log sequences. In any case, the expressivity of our proposed pattern language comes at no detriment to the simpler, more common patterns, but brings the potential benefit of identifying sequences containing exceptionally regular structure.

7   Conclusion

In this paper, we propose a novel approach for mining periodic patterns with a MDL criterion, and an algorithm to put it into practise. Through our experimental evaluation, we show that we are able to extract sets of patterns that compress the input event sequences and to identify meaningful patterns.

How to take prior knowledge into account is an interesting question to explore. Making the algorithm more robust to noise and making it more scalable using for instance parallelisation, are some pragmatic directions for future work, as is adding a visualisation tool to support the analysis and interpretation of the extracted patterns in the context of the event log sequence.

Acknowledgements. The authors thank Hiroki Arimura and Jilles Vreeken for valuable discussions. This work has been supported by Grenoble Alpes Metropole through the Nano2017 Itrami project, by the QCM-BioChem project (CNRS Mastodons) and by the Academy of Finland projects "Nestor" (286211) and "Agra" (313927).

References

1. Bellman, R.: On the approximation of curves by line segments using dynamic programming. Commun. ACM **4**(6), 284 (1961)
2. Berberidis, C., Vlahavas, I., Aref, W.G., Atallah, M., Elmagarmid, A.K.: On the discovery of weak periodicities in large time series. In: Elomaa, T., Mannila, H., Toivonen, H. (eds.) PKDD 2002. LNCS, vol. 2431, pp. 51–61. Springer, Heidelberg (2002). https://doi.org/10.1007/3-540-45681-3_5
3. Bhattacharyya, A., Vreeken, J.: Efficiently summarising event sequences with rich interleaving patterns. In: SDM 2017, pp. 795–803. SIAM (2017)
4. Bonchi, F., van Leeuwen, M., Ukkonen, A.: Characterizing uncertain data using compression. In: SDM 2011, pp. 534–545. SIAM (2011)
5. De Raedt, L., Zimmermann, A.: Constraint-based pattern set mining. In: SDM 2007, pp. 237–248. SIAM (2007)
6. Galbrun, E., Cellier, P., Tatti, N., Termier, A., Crémilleux, B.: Mining periodic patterns with a MDL criterion. ArXiv e-prints (2018). arXiv:1807.01706 [cs.DB]
7. Grünwald, P.: Model selection based on minimum description length. J. Math. Psychol. **44**(1), 133–152 (2000)
8. Grünwald, P.: The Minimum Description Length Principle. MIT Press, Cambridge (2007)
9. Han, J., Dong, G., Yin, Y.: Efficient mining of partial periodic patterns in time series database. In: ICDE 1999, pp. 106–115 (1999)
10. Han, J., Gong, W., Yin, Y.: Mining segment-wise periodic patterns in time-related databases. In: KDD 1998, pp. 214–218 (1998)
11. Heierman III, E.O., Cook, D.J.: Improving home automation by discovering regularly occurring device usage patterns. In: ICDM 2003, pp. 537–540 (2003)
12. Kiernan, J., Terzi, E.: Constructing comprehensive summaries of large event sequences. ACM Trans. Knowl. Discov. Data **3**(4), 21:1–21:31 (2009)
13. Lam, H.T., Moerchen, F., Fradkin, D., Calders, T.: Mining compressing sequential patterns. In: SDM 2012, pp. 319–330. SIAM (2012)
14. Li, Z., Wang, J., Han, J.: Mining event periodicity from incomplete observations. In: KDD 2012, pp. 444–452. ACM (2012)
15. Lopez-Cueva, P., Bertaux, A., Termier, A., Méhaut, J.-F., Santana, M.: Debugging embedded multimedia application traces through periodic pattern mining. In: International Conference on Embedded Software, EMSOFT 2012 (2012)
16. Ma, S., Hellerstein, J.L.: Mining partially periodic event patterns with unknown periods. In: ICDE 2001, pp. 205–214. IEEE Computer Society (2001)
17. Özden, B., Ramaswamy, S., Silberschatz, A.: Cyclic association rules. In: ICDE 1998, pp. 412–421. IEEE Computer Society (1998)
18. Rissanen, J.: Modeling by shortest data description. Automatica **14**(5), 465–471 (1978)
19. Smets, K., Vreeken, J.: Slim: Directly mining descriptive patterns. In: SDM 2012, pp. 236–247. SIAM (2012)
20. Tatti, N., Vreeken, J.: The long and the short of it: summarising event sequences with serial episodes. In: KDD 2012, pp. 462–470. ACM (2012)

21. Vreeken, J., van Leeuwen, M., Siebes, A.: Krimp: mining itemsets that compress. Data Min. Knowl. Discov. **23**(1), 169–214 (2011)
22. Yuan, Q., Zhang, W., Zhang, C., Geng, X., Cong, G., Han, J.: PRED: periodic region detection for mobility modeling of social media users. In: WSDM 2017, pp. 263–272. ACM (2017)

Revisiting Conditional Functional Dependency Discovery: Splitting the "C" from the "FD"

Joeri Rammelaere[(✉)] and Floris Geerts

University of Antwerp, Antwerp, Belgium
{joeri.rammelaere,floris.geerts}@uantwerp.be

Abstract. Many techniques for cleaning dirty data are based on enforcing some set of integrity constraints. Conditional functional dependencies (CFDs) are a combination of traditional Functional dependencies (FDs) and association rules, and are widely used as a constraint formalism for data cleaning. However, the discovery of such CFDs has received limited attention. In this paper, we regard CFDs as an extension of association rules, and present three general methodologies for (approximate) CFD discovery, each using a different way of combining pattern mining for discovering the conditions (the "C" in CFD) with FD discovery. We discuss how existing algorithms fit into these three methodologies, and introduce new techniques to improve the discovery process. We show that the right choice of methodology improves performance over the traditional CFD discovery method CTane. Code related to this paper is available at: https://github.com/j-r77/cfddiscovery, https://codeocean.com/2018/06/20/discovering-conditional-functional-dependencies/code.

1 Introduction

Many organizations are faced with problems arising from poor data quality, such as inaccurate or inconsistent values. In order to clean such dirty data, many techniques make use of logical rules called integrity constraints, such that values are dirty if and only if they violate a rule. These constraints are typically supplied by human experts, or discovered from the data by algorithms. Dedicated repair algorithms then modify the data such that all constraints are satisfied. In this paper we focus on the *automatic discovery of constraints*.

Among the variety of proposed constraints, Conditional functional dependencies (CFDs) have been used extensively for data cleaning. Such CFDs are a generalization of traditional functional dependencies (FDs) and association rules (ARs). CFDs are more flexible than FDs, since they can capture dependencies that hold only on a subset of the data, and more expressive and succinct than ARs, since a CFD can also identify associations that hold on the attribute level.

Electronic supplementary material The online version of this chapter (https://doi.org/10.1007/978-3-030-10928-8_33) contains supplementary material, which is available to authorized users.

To discover CFDs for data cleaning, when typically only dirty data is available, it is necessary to discover *approximate* CFDs. That is, to discover CFDs that allow a certain amount of violations, in line with discovering confident association rules. To discover approximate CFDs, two algorithms have been proposed, based on the concept of *equivalence partitions*: CTane [9] and an unnamed method which we dub FindCFD [5]. These algorithms combine existing techniques for discovering FDs and ARs. While research on discovering traditional FDs has resurged in recent years, especially in the database community, the discovery of approximate CFDs has received less attention.

In this paper, we recast CFDs as an extension of association rules, and discuss CFD discovery from a more general perspective. We distinguish *three general methodologies* for discovering *confident* CFDs[1], as typically used for data cleaning, based on distinct ways of combining FD discovery with itemset mining. The first methodology is used by the CTane algorithm [9], and performs an integrated traversal of the lattice containing all possible CFDs. Additionally, we introduce two new methodologies, which explicitly consider CFD discovery as *a combination of FD discovery and pattern mining*. We introduce an *itemset-centric* approach, where patterns are mined at the top level, and FDs are subsequently discovered on the corresponding subsets of the data; and an *FD-centric* approach, which at the top level traverses the search space of FDs, and then mines those patterns for which the FD holds, generalizing the approach taken in FindCFD [5]. Moreover, in the FD-centric approach, we identify techniques for speeding up the pattern mining process, using information from the FD discovery process at the top level.

Both new methodologies are described in a flexible way, enabling the use of *any* FD discovery method based on *equivalence partitions*, and *any* itemset mining method based on *tidlists*, for each of the separate steps. As such, the methodologies we describe, represent in fact a *family* of algorithms. This has as a direct advantage that *CFD discovery can benefit directly from advances in FD and itemset discovery*. We also present a general pruning strategy for CFDs, such that each methodology can use an arbitrary strategy for traversing the search space of CFDs, e.g., breadth-first or depth-first. Both CTane and FindCFD were originally presented using a breadth-first strategy, because of pruning.

We show experimentally that both of our proposed methods typically outperform the integrated approach to CFD discovery, which is used by CTane. The FD-centric approach performs substantially better in most cases, especially on data with a higher number of attributes. We also identify situations in which the itemset-centric approach provides the best performance, namely when using a very low minimum support threshold. Moreover, the appropriate use of depth-first search strategies further improve runtime for the different methodologies.

---

[1] Other interestingness measures can be plugged in, if they can be computed from equivalence partitions. This is the case for most popular measures.

2   Related Work

Conditional functional dependencies (CFDs) are widely used in the context of constraint-based data quality (see [7,12] for recent surveys). CFDs were introduced in [8] as an extension of Functional dependencies (FDs), and three discovery algorithms have been proposed since: CTane and FastCFD [9][2], and Find-CFD [5]. Other work considers constant CFDs only [6]. Each of these discovery methods is rooted in FD discovery. Our three general approaches to CFD discovery can incorporate any FD discovery method making use of equivalence partitions, e.g., Tane [11], FUN [15], FD_Mine [20], and DFD [1]. Such methods support the discovery of *approximate* dependencies, and are well suited for integration with pattern mining. An overview and experimental evaluation of functional dependency discovery is presented in [16], where it is shown that Tane is the most performant algorithm on a considerable range of data sizes. CFD discovery can also be viewed as the discovery of special conjunctive queries [10], but at the cost of a more time-consuming discovery process.

Although interesting measures for FDs based on statistical tests have recently been proposed [13], we consider approximate CFDs defined in terms of support and confidence as these are most widely used in the data quality context.

Association rules (ARs) were first introduced in [2] for supermarket basket analysis. Discovery of ARs is based on mining frequent patterns, which has received much attention since. Of particular interest to our approaches for CFD discovery are so-called vertical itemset mining algorithms, which employ a vertical data layout for efficient frequency computation, such as Eclat [23]. Such algorithms are well-suited for integration with FD discovery, since the vertical data layout relates naturally to the equivalence partitions used in FD discovery, as shown in the following sections. For an overview of itemset and association rule mining, we refer to [22]. We view CFDs as a kind of ARs. An in-depth discussion relating FDs, CFDs, and ARs can be found in [14].

3   Preliminaries

We consider a relation schema R defined over a set $\mathcal{A}$ of attributes, where each attribute $A \in \mathcal{A}$ has a finite domain $dom(A)$. For an instance D of R, and tuple $t \in D$, we denote the projection of t onto a set of attributes X by $t[X]$. Each tuple $t \in D$ is assumed to have a unique identifier tid, e.g., a natural number.

A *conditional functional dependency* (CFD) [8] φ over R is a pair $(X \rightarrow A, t_p)$, where (i) X is a set of attributes in $\mathcal{A}$, and A is a single attribute in $\mathcal{A}$; (ii) $X \rightarrow A$ is a standard functional dependency (FD); and (iii) t_p is a *pattern tuple* with attributes in X and A, where for each B in $X \cup \{A\}$, $t_p[B]$ is either a constant 'b' in $dom(B)$, or an unnamed variable '$\_$'. A CFD $\varphi = (X \rightarrow A, t_p)$ in which $t_p[A] = $ '$\_$' is called *variable*, otherwise it is *constant*. For constant CFDs, $t_p[X]$ consists of constants only. Such a constant CFD is equivalent to a traditional association rule, and an FD is a CFD with t_p consisting solely of variables '$\_$'.

---

[2] FastCFD does not support the discovery of *approximate* CFDs.

The semantics of a CFD $\varphi = (X \rightarrow A, t_p)$ on an instance D is defined as follows. A tuple $t \in D$ is said to *match* a pattern tuple t_p in attributes X, denoted by $t[X] \asymp t_p[X]$, if for all $B \in X$, either $t_p[B] = \text{`\_'}$, or $t[B] = t_p[B]$. The tuple t *violates* a variable CFD $\varphi = (X \rightarrow A, t_p)$ if $t[X] \asymp t_p[X]$ and there exists another tuple t' in D such that $t[X] = t'[X]$ and $t[A] \neq t'[A]$. A tuple t *violates* a constant CFD $\varphi = (X \rightarrow A, t_p)$ if $t[X] = t_p[X]$ and $t[A] \neq t_p[A]$ hold. The set of all tids of tuples in D that violate a CFD φ is denoted by $\mathsf{VIO}(\varphi, D)$. If $\mathsf{VIO}(\varphi, D) = \emptyset$, then D *satisfies* φ, which is also denoted by $D \models \varphi$.

We present CFD discovery algorithms in this paper using concepts from itemset mining. We consider *itemsets* as sets of attribute-value pairs of the form (A, v), with $A \in \mathcal{A}$, and v a value in $\mathsf{dom}(A)$ or '_'. An instance D thus corresponds to a *transaction* database, with each tuple corresponding to a transaction of length $|\mathcal{A}|$. An item (A, v) with $v \in \mathsf{dom}(A)$ is *supported* in a tuple t if $t[A] = v$. Items $(A, \_)$ are supported by every transaction. A tuple supports an *itemset* I in D if it supports all items $i \in I$. The *cover* of an itemset I in D, denoted by $\mathsf{cov}(I, D)$ and also called I's *tidlist*, is the set of tids of tuples in D that support I. The *support* of I in D, denoted by $\mathsf{supp}(I, D)$, is equal to the number of tids in I's cover in D.

We can now write a CFD $\varphi = (X \rightarrow A, t_p)$ compactly as an *association rule* $I \rightarrow j$, between an itemset I and a single item j, where $I = \bigcup_{B \in X}\{(B, t_p[B])\}$ and $j = (A, t_p[A])$. In line with the notion of approximate FDs [11], we define the *confidence* of a CFD $\varphi = I \rightarrow j$ as $\mathsf{conf}(\varphi, D) = 1 - \frac{|D'|}{\mathsf{supp}(I, D)}$, where $D' \subset D$ is a minimal subset such that $D \setminus D' \models \varphi$. For a constant CFD, $|D'| = |\mathsf{VIO}(\varphi, D)|$, and hence $\mathsf{conf}(\varphi, D) = (\mathsf{supp}(I, D) - |\mathsf{VIO}(\varphi, D)|)/\mathsf{supp}(I, D) = \mathsf{supp}(I \cup \{j\}, D)/\mathsf{supp}(I, D)$ reduces to the standard confidence of an association rule. For variable CFDs, $|D'|$ is the minimum number of tuples that need to be altered or removed for φ to be satisfied. For example, if a violation set for a variable CFD contains two tuples with different A-values, the CFD can be made to hold by altering just one of the tuples. A CFD φ is called *exact* if $\mathsf{conf}(\varphi, D) = 1$, and *approximate* otherwise.

Finally, we consider CFD discovery algorithms based on the concept of *equivalence partitions*, as used in the Tane algorithm [11]. More specifically, given an itemset I consisting of attribute-value pairs, we say that two tuples s and t in D are *equivalent relative to* I if, for all $(B, v) \in I$, $s[B] = t[B] \asymp v$. For a tuple $s \in D$, $[s]_I$ denotes the *equivalence class* consisting of the tids of all tuples $t \in D$ that are equivalent with s relative to I. The *(equivalence) partition of* I, denoted by $\Pi(I)$, is the collection of $[s]_I$ for $s \in D^3$. For a single constant item, $\Pi((A, v)) = \{\mathsf{cov}((A, v), D)\}$, i.e., it consists of (A, v)'s tidlist. For a single variable item, $\Pi((A, \_)) = \{\mathsf{cov}((A, v)) \mid v \in \mathsf{dom}(A)\}$, i.e., it consists of all tidlists grouped together with regards to the A-values of the corresponding tuples. For an itemset I, $\Pi(I) = \bigcap_{i \in I} \Pi(i)$ in which equivalence classes are pairwise intersected. The *size* of $\Pi(I)$, denoted by $|\Pi(I)|$, is the number of equivalence classes in $\Pi(I)$. We use $\|\Pi(I)\|$ to denote the number of tids in $\Pi(I)$, equal to the support of I. Finally, we note that the CFD $I \rightarrow j$ holds iff $|\Pi(I)| = |\Pi(I \cup \{j\})|$.

---

[3] Strictly speaking this is only a partition of D when I contains variable items $(A, \_)$.

Problem Statement. *Given an instance D of a schema R, support threshold δ, and confidence threshold ε, the* approximate CFD discovery problem *is to find all CFDs φ over R with* $\mathsf{supp}(\varphi, D) \geq \delta$ *and* $\mathsf{conf}(\varphi, D) \geq 1 - \varepsilon$.

Example 1. We use the "play tennis" dataset from [18], shown in Table 1. One of the approximate CFDs φ on this dataset is $\{(\mathsf{Windy}, \mathsf{false}), (\mathsf{Outlook}, \_)\} \rightarrow (\mathsf{Play}, \_)$. Let $I = \{(\mathsf{Windy}, \mathsf{false}), (\mathsf{Outlook}, \_), (\mathsf{Play}, \_)\}$ and $j = (\mathsf{Play}, \_)$. The relevant equivalence partitions are $\Pi(I \setminus \{j\}) = \{\{1, 8, 9\}, \{3, 13\}, \{4, 5, 10\}\}$ and $\Pi(I) = \{\{1, 9\}, \{8\}, \{3, 13\}, \{4, 5, 10\}\}$. The sizes of the equivalence partitions are $|\Pi(I \setminus \{j\})| = 3$ and $|\Pi(I)| = 4$, and both partitions have support $\|\Pi(I \setminus \{j\})\| = \|\Pi(I)\| = 8$. The supported tuples t, i.e., where $t[\mathsf{Windy}] = \mathsf{false}$, are shaded grey in Table 1, with different shades corresponding to the different equivalence classes in $\Pi(I)$. The CFD can be made to hold exactly by removing the tuple with tid 8, such that $\Pi(I \setminus \{j\}) = \Pi(I)$, and hence its confidence is $1 - (|D'|/\|\Pi(I)\|) = 1 - (1/8) = 0.875$. Finally, $\mathsf{VIO}(\varphi, D) = \{t_1, t_8, t_9\}$.   ◇

Table 1. Running example based on the play tennis dataset [18]

tid	Outlook	Temperature	Humidity	Windy	Play
1	sunny	hot	high	false	dont
2	sunny	hot	high	true	dont
3	overcast	hot	high	false	play
4	rain	mild	high	false	play
5	rain	cool	normal	false	play
6	rain	cool	normal	true	dont
7	overcast	cool	normal	true	play
8	sunny	mild	high	false	dont
9	sunny	cool	normal	false	play
10	rain	mild	normal	false	play
11	sunny	mild	normal	true	play
12	overcast	mild	high	true	play
13	overcast	hot	normal	false	play
14	rain	mild	high	true	dont

4   Three Approaches for CFD Discovery

We present three general approaches for the discovery of approximate CFDs with high supports. These approaches differ in the way that the (itemset) search lattice is explored. First, we generalize the *integrated* approach [9], in which the combined search lattice of constant and variable ('_') patterns is traversed at once. For the other two, new approaches, we *decouple* the lattices for constant and variable patterns. We present the *Itemset-First* approach, followed by the *FD-First* approach. Both of these approaches consist of two separate algorithms, which either explore a lattice containing only constant patterns, or containing only variable patterns. After discussing the three methodologies, we derive the

---

Algorithm 1. Integrated CFD discovery algorithm

1: **procedure** MINE-INTEGRATED(D, δ, ε)
2:     $\mathcal{L} \leftarrow \{(A, v) \mid A \in \mathcal{A}, v \in \mathsf{dom}(A) \cup \{\_\}, \mathsf{supp}((A, v), D) \geq \delta\}$
3:     Compute $\Pi(\{i\}, D)$ for all $i \in \mathcal{L}$
4:     Initialize *fringe* with $\mathcal{L}$ depending on search strategy
5:     $\Sigma \leftarrow \emptyset$
6:     **while** *fringe* **not** empty **do**
7:         $I \leftarrow \text{POP}(fringe)$
8:         **for all** $j \in I$ **do**
9:             **if** $\mathsf{conf}(I \setminus \{j\} \rightarrow j, D) \geq 1 - \varepsilon$ **then**
10:                 $\Sigma \leftarrow \Sigma \cup \{I \setminus \{j\} \rightarrow j\}$
11:         insert children of I into *fringe* if $\mathsf{supp}(I, D) \geq \delta$
12:     **return** Σ

---

general time complexity of CFD discovery. As mentioned in the introduction, we describe our algorithms *independent* from the search strategy used. To achieve uniform pruning across all approaches and search strategies, we present pruning strategies based on a generalization of free itemsets [3] and a lookup table.

4.1 Integrated CFD Discovery

We start by describing the integrated approach MINE-INTEGRATED for discovering CFDs, as implemented by CTane [9]. Its pseudocode is shown in Algorithm 1. Algorithms based on this methodology traverse the entire search lattice for CFDs, consisting of both constant *and* variable patterns. The first level $\mathcal{L}$ of this lattice is initialized on line 2. For each singleton item, its equivalence partition is computed from the data; only sufficiently frequent constant items are retained.

The lattice is subsequently traversed, typically in either a breadth-first or depth-first manner[4]. Regardless of the choice of traversal, we refer to the set of current lattice elements considered as the *fringe*. Whenever an itemset I in the fringe is visited (line 7), all CFDs of the form $I \setminus \{j\} \rightarrow j$, for $j \in I$, are generated, and their confidence is computed from the equivalence partitions $\Pi(I \setminus \{j\})$ and $\Pi(I)$. If the confidence exceeds the threshold, then the CFD is added to the result Σ. An efficient algorithm for computing confidence is presented in Tane [11], and is based on the *error* of an equivalence class. More precisely, for all $\mathsf{eq} \in \Pi(I \setminus \{j\})$, let $\Pi(I)^{\mathsf{eq}}$ denote those $\mathsf{eq'} \in \Pi(I)$ with $\mathsf{eq'} \subset \mathsf{eq}$. In other words, $\Pi(I)^{\mathsf{eq}}$ contains all equivalence classes over I that match the same (constant) pattern as eq on the attributes $I \setminus \{j\}$. We define

$$\mathsf{error}(\mathsf{eq}, \Pi(I)) = ||\Pi(I)^{\mathsf{eq}}|| - \max_{\mathsf{eq'} \in \Pi(I)^{\mathsf{eq}}} |\mathsf{eq'}|.$$

---

[4] The CTane algorithm as presented in [9] employs a breadth-first traversal.

Generalizing the argument given in [11] for variable patterns to arbitrary (constant and variable) patterns, the confidence can then be computed as:

$$\mathsf{conf}(I \setminus \{j\} \rightarrow j) = 1 - \frac{\sum_{\mathsf{eq} \in \Pi(I \setminus \{j\})} \mathsf{error}(\mathsf{eq}, \Pi(I))}{\mathsf{supp}(I \setminus \{j\})}.$$

Example 2. We consider the CFD $\{(\mathsf{Windy}, \mathsf{false}), (\mathsf{Outlook}, \_)\} \rightarrow (\mathsf{Play}, \_)$ from our running example, and let $I = \{(\mathsf{Windy}, \mathsf{false}), (\mathsf{Outlook}, \_), (\mathsf{Play}, \_)\}$ and $j = (\mathsf{Play}, \_)$. We compute the error for each of the 3 equivalence classes in $\Pi(I \setminus \{j\}) = \{\{1, 8, 9\}, \{3, 13\}, \{4, 5, 10\}\}$. For $\mathsf{eq} = \{3, 13\}$ and $\mathsf{eq} = \{4, 5, 10\}$, we have $|\Pi(I)^{\mathsf{eq}}| = 1$, since the tuples within these equivalence classes have the same values for attribute Play. Hence, there is only one $\mathsf{eq}' \in \Pi(I)^{\mathsf{eq}}$, and $\|\Pi(I)^{\mathsf{eq}}\| = \max_{\mathsf{eq}' \in \Pi(I)^{\mathsf{eq}}} |\mathsf{eq}'|$, leading to an error of 0. This leaves us with $\mathsf{eq} = \{1, 8, 9\}$, for which $\Pi(I)^{\mathsf{eq}} = \{\{1, 9\}, \{8\}\}$. Indeed, this is the equivalence class containing the violations of the CFD. We compute the error as $\mathsf{error} = \|\{\{1, 9\}, \{8\}\}\| - \max(|\{1, 9\}|, |\{8\}|) = 1$, resulting in a confidence of $1 - (\mathsf{error}/\|\Pi(I)\|) = 1 - (1/8) = 0.875$, as mentioned in Example 1.   ◇

Finally, if I is sufficiently frequent, the children of I in the lattice are generated and inserted into the fringe (line 11). This is done by joining I with all itemsets J in the fringe that are (i) at the same level in the lattice, i.e., $|J| = |I|$; and (ii) such that J and I differ in only one item. A child M is then obtained as $I \cup J$, and $\Pi(M)$ is computed by intersecting $\Pi(I)$ with $\Pi(J)$. The Tane algorithm provides a linear algorithm for computing such an intersection, making use of a lookup table. Using a similar technique, confidence can be computed in linear time (see details in the online appendix [21]).

4.2   Itemset-First Discovery

The second, and new, approach to CFD discovery starts with an itemset mining step. The pseudocode of algorithm MINE-ITEMSET-FIRST is shown in Algorithm 2. The search lattice $\mathcal{L}$ is initialized (line 2) using only items with constant values. We therefore only require the cover of each item in $\mathcal{L}$ (the equivalence partition of a constant item corresponds to its cover). The lattice is traversed using an arbitrary search strategy and generated itemsets are inserted into the fringe.

When visiting itemset I in this approach, we initialize a separate FD searching algorithm (line 8). The item lattice for this FD search ($\mathcal{L}^{\mathrm{FD}}$) now consists only of those items in D with a variable pattern ('$\_$'), and whose attribute is not already present in $\mathsf{attrs}(I)$, the set of attributes in the items in I. In other words, we extend the constant pattern I with variable patterns to obtain CFDs. During the traversal of $\mathcal{L}^{\mathrm{FD}}$ the equivalence partition of each item is computed on D^I, the dataset D projected on I, i.e., using only tuples with a tid in $\mathsf{cov}(I, D)$. The algorithm FIND-FDs is then invoked (line 10), which can be any FD-discovery algorithm using equivalence partitions, to discover all FDs with confidence $\geq 1 - \varepsilon$ on D^I. The resulting FDs are augmented with the pattern I, and added to the

---

Algorithm 2. Itemset-First CFD discovery algorithm

1: **procedure** MINE-ITEMSET-FIRST(D, δ, ε)
2:     $\mathcal{L} \leftarrow \{(A, v) \mid A \in \mathcal{A}, v \in \mathsf{dom}(A), \mathsf{supp}((A, v), D) \geq \delta\}$
3:     Compute $\mathsf{cov}(\{i\}, D)$ for all $i \in \mathcal{L}$
4:     Initialize *fringe* with $\mathcal{L}$ depending on search strategy
5:     $\Sigma \leftarrow \emptyset$
6:     **while** *fringe* **not** empty **do**
7:         $I \leftarrow \mathsf{POP}(fringe)$
8:         $\mathcal{L}^{\mathrm{FD}} \leftarrow \{(A, \_) \mid A \in \mathcal{A} \setminus \mathsf{attrs}(I)\}$
9:         Compute $\Pi(\{k\}, D^I)$ for all $k \in \mathcal{L}^{\mathrm{FD}}$
10:         $\Sigma^{\mathrm{FD}} \leftarrow \mathsf{FIND\text{-}FDS}(\mathcal{L}^{\mathrm{FD}}, D^I, I, \varepsilon)$
11:         $\Sigma \leftarrow \Sigma \cup \{I \cup J \rightarrow j \mid J \rightarrow j \in \Sigma^{\mathrm{FD}}\}$
12:         insert children of I into *fringe* if their support $\geq \delta$
13:     **return** Σ

---

set Σ of CFDs (line 11). Since an FD is supported by all tuples in D^I, and $|D^I| \geq \delta$ is guaranteed by the support threshold on I, FIND-FDS is oblivious to the threshold δ. Pseudocode of FIND-FDS is available in the online appendix [21].

Example 3. In the running example, the itemset step will, for instance, visit the item (Windy, false), with $\mathsf{cov}((\mathsf{Windy}, \mathsf{false}), D) = \{1, 3, 4, 5, 8, 9, 10, 13\}$. Subsequently, an FD search is performed using only those tids in $\mathsf{cov}((\mathsf{Windy}, \mathsf{false}), D)$. Hence, within the FD search, the fringe is initialized with all variable items except for (Windy, _), and the equivalence partitions of these single items are computed only over the tids $\{1, 3, 4, 5, 8, 9, 10, 13\}$. The FD (Outlook, _) $\rightarrow$ (Play, _) is then found to hold, with sufficient confidence, and the CFD $\{(\mathsf{Windy}, \mathsf{false}), (\mathsf{Outlook}, \_)\} \rightarrow (\mathsf{Play}, \_)$ is added to the result. After exhausting the FD lattice for (Windy, false), the itemset mining step is resumed.   ◇

Similar to the integrated approach, the final step when visiting an itemset I is to insert its children into the fringe, if they are sufficiently frequent. The only difference, similar to the initialization of $\mathcal{L}$, is that we again only consider constant items, with equivalence partitions boiling down to the cover of the items. The cover of each child itemset M can then be computed using a straightforward intersection of $\mathsf{cov}(I, D)$ and $\mathsf{cov}(J, D)$, for the itemsets J in the fringe with $|J| = |I|$, and such that J and I differ in only one item.

4.3   FD-First Discovery

The third and final approach to CFD discovery, MINE-FD-FIRST, is shown in pseudocode in Algorithm 3. This approach is a generalization of the FindCFD algorithm [5], which starts with FD discovery. The search lattice $\mathcal{L}$ is thus initialized (line 2) using only variable items, i.e., one item (A, _) for each attribute $A \in \mathcal{A}$. As before, equivalence partitions are computed, after which a fringe is created and a breadth or depth-first traversal of the lattice follows.

Algorithm 3. FD-First CFD discovery algorithm

1: **procedure** MINE-FD-FIRST(D, δ, ε)
2:      $\mathcal{L} \leftarrow \{(\mathsf{A}, \_) \mid \mathsf{A} \in \mathcal{A}\}$
3:      Compute $\Pi(\{i\}, D)$ for all $i \in \mathcal{L}$
4:      Initialize *fringe* with $\mathcal{L}$ depending on search strategy
5:      $\Sigma \leftarrow \emptyset$
6:      **while** *fringe* **not** empty **do**
7:          $I \leftarrow \text{POP}(fringe)$
8:          **for all** $j \in I$ **do**
9:              **if** $\mathsf{conf}(I \setminus \{j\} \to j, D) \geq 1 - \varepsilon$ **then**
10:                 $\Sigma \leftarrow \Sigma \cup \{I \setminus \{j\} \to j\}$
11:             **if** $\mathsf{conf}(I \setminus \{j\} \to j, D) < 1$ **then**
12:                 $\mathcal{L}^{\text{Pat}} \leftarrow \{(\mathsf{A}, v) \mid \mathsf{A} \in \mathsf{attrs}(I), v \in \mathsf{dom}(\mathsf{A})\}$
13:                 Compute $\mathsf{cov}(\{i\}, \Pi(I))$ for all $i \in \mathcal{L}^{\text{Pat}}$
14:                 $\Sigma \leftarrow \Sigma \cup \text{MINE-PATTERNS}(\mathcal{L}^{\text{Pat}}, I \setminus \{j\} \to j, \Pi(I), \delta, \varepsilon)$
15:         insert children of I into *fringe*
16:     **return** Σ

For every item I in the lattice, we now consider all FDs of the form $I \setminus \{j\} \to j$ for $j \in I$ (line 8). If the FD is found to be sufficiently confident, it is added to the result Σ. However, if the FD does not fully hold on the data, we additionally run an itemset mining algorithm to find all constant patterns for which the FD is sufficiently confident. During this itemset mining, the lattice $\mathcal{L}^{\text{Pat}}$ of constant items is explored. This lattice is initialized on line 12.

The key to the MINE-FD-FIRST method's efficiency is that the support and confidence of a considered CFD $I \setminus \{j\} \to j$ can be computed based on the information contained in $\Pi(I)$. Indeed, each equivalence class $\mathsf{eq} \in \Pi(I)$ corresponds to a unique constant pattern over the attributes $\mathsf{attrs}(I)$. By assigning a unique identifier to each class, we define the cover of an item(set) J w.r.t. the equivalence partition of I, denoted as $\mathsf{cov}(J, \Pi(I))$, as the set of identifiers of equivalence classes in which the item occurs. We call such a cover a *pidlist* (for partition id). Since typically $|\mathsf{cov}(J, \Pi(I))| \ll |\mathsf{cov}(J, D)|$, efficiency is increased.

Example 4. Consider the FD $\{(\mathsf{Windy}, \_), (\mathsf{Outlook}, \_)\} \to (\mathsf{Play}, \_)$ corresponding to the itemset $I = \{(\mathsf{Windy}, \_), (\mathsf{Outlook}, \_), (\mathsf{Play}, \_)\}$, with equivalence class $\Pi(I) = \{\{1, 9\}, \{2\}, \{3, 13\}, \{4, 5, 10\}, \{6, 14\}, \{7, 12\}, \{8\}, \{11\}\}$. The constant pattern $(\mathsf{Windy}, \mathsf{false})$ can now be represented by its pidlist. That is, $\mathsf{cov}((\mathsf{Windy}, \mathsf{false}), \Pi(I)) = \{1, 3, 4, 7\}$. Since $\mathsf{supp}((\mathsf{Windy}, \mathsf{false}), D) = 8$, we have reduced the size of its cover by half.   ◇

The subprocedure MINE-PATTERNS now starts by initializing a fringe containing all frequent single (constant) items over the attributes in $I \setminus \{j\}$. For each item, its pidlist has been computed from $\Pi(I)$ (line 13). Procedure MINE-PATTERNS then traverses the constant itemset lattice, generating the pidlists of new itemsets by intersecting the pidlists of two of their parents in the lattice.

The support of an itemset M can be easily computed from its pidlist as follows,

$$\text{supp}(M, \Pi(I)) = \sum_{pid \,\in\, \text{cov}(M,\Pi(I))} |\Pi(I)[pid]|,$$

where $\Pi(I)[pid]$ denotes the equivalence class with identifier pid. Only itemsets M with $\text{supp}(M, \Pi(I)) \geq \delta$ are considered as possible patterns for a CFD. Whenever an itemset M is processed in MINE-PATTERNS, we validate the CFD $(I \setminus \{j\}) \oplus M \to j$, where $\oplus$ *replaces* those variable items in $(I \setminus \{j\})$ which have a constant counterpart in M, i.e., $(I \setminus \{j\}) \oplus M = M \cup \{(\mathsf{A}, \_) \in I \setminus \{j\} \mid \mathsf{A} \notin \text{attrs}(M)\}$. If the CFD is sufficiently confident, it is added to the result.

Pseudocode for algorithm MINE-PATTERNS is available in the online appendix [21]. As before, any itemset mining algorithm based on tidlists and any search strategy can be employed by MINE-PATTERNS. After the itemset mining step has finished, MINE-FD-FIRST continues by processing the remaining FDs in I, of the form $(I \setminus \{l\} \to l)$ with $l \neq j$, one by one. Finally, after all FDs in I have been processed, the children of I are added to the fringe. Since MINE-FD-FIRST only considers FDs at this level, a support check is not necessary.

We remark that the algorithm FindCFD [5] takes a similar approach, but, to our knowledge, does not perform an exhaustive search through the pattern lattice, i.e., the power set of $\mathcal{L}^{\text{Pat}}$. Indeed, if an FD does not hold, this algorithm examines the equivalence partitions to obtain a *constant* CFD, without any variable patterns. As such, FindCFD discovers only FDs and constant CFDs, whereas MINE-FD-FIRST discovers general CFDs containing variables *and* constants. The fact that FindCFD does not discover all CFDs is also noted in [4].

4.4 Time Complexity

We now discuss the time complexity of our three CFD discovery methodologies. Most of the computation concerns two operations: computing equivalence partitions (or tidlists), and validating CFDs. Both operations can be performed in $\mathcal{O}(|D|)$ time. For every element I in the lattice, the equivalence partition is computed once, and $|I|$ CFDs are validated. We simplify this as $|I|$ operations per lattice element. Given that there are $|\mathcal{A}|$ attributes in the dataset, a total of $2^{|\mathcal{A}|}$ combinations of attributes exist: at level i in the lattice, there are $\binom{|\mathcal{A}|}{i}$ attribute combinations of size i. Let d denote the average size of $\text{dom}(A)$, for $A \in \mathcal{A}$. Including variable patterns, there are at most $(d+1)^i$ itemsets containing an attribute combination of size i. The number of operations is then:

$$\sum_{i=1}^{|\mathcal{A}|} \binom{|\mathcal{A}|}{i}(d+1)^i i$$

Computing this expression gives a total of $|\mathcal{A}|(d+1)(d+2)^{|\mathcal{A}|-1}$ operations, each of which is $\mathcal{O}(|D|)$. Hence, the time complexity of the algorithms is:

$$\mathcal{O}(|\mathcal{A}| \times d^{|\mathcal{A}|} \times |D|).$$

While each of our three methods performs roughly the same number of operations, the difference between them is in the time required to perform these operations. Indeed, a tidlist intersection and an equivalence partition intersection are both $\mathcal{O}(|D|)$, but in practice the tidlist intersection is faster. The Itemset-First method most efficiently computes the projected databases on which it then performs an FD-search, while the FD-First method performs much of its intersections and validation on the pidlists, which are on average much smaller than $|D|$. These differences account for the improved performance of Itemset-First and FD-First over the Integrated approach, as experimentally shown in Sect. 5.

4.5   Pruning

We conclude by discussing pruning. Clearly, any CFD discovery algorithm can exploit the anti-monotonicity of support, to prune away all infrequent itemsets and their supersets. However, existing CFD discovery algorithms also provide pruning based on redundancy with respect to the antecedent of CFDs. Redundancy is defined using the concept of a preceding set:

Definition 5 (Preceding set). *Consider a database instance D and an itemset I containing attribute-value pairs. An itemset J is a preceding set of I, denoted $J \prec I$, if $J \neq I$ and for all $(\mathsf{A}, v) \in J$, either $(\mathsf{A}, v) \in I$, or $v =$ '_' and $(\mathsf{A}, a) \in I$, where a is a constant value in $\mathsf{dom}(\mathsf{A})$.*

Example 6. In our running example, the itemsets $\{(\mathsf{Windy}, \mathsf{false}), (\mathsf{Outlook}, \_)\}$ and $\{(\mathsf{Windy}, \_), (\mathsf{Outlook}, \_), (\mathsf{Play}, \_)\}$, among others, are preceding sets of the itemset $\{(\mathsf{Windy}, \mathsf{false}), (\mathsf{Outlook}, \_), (\mathsf{Play}, \_)\}$.   ◇

Definition 7 (CFD Redundancy). *Consider a database instance D and a CFD $\varphi : I \to j$ with $\mathsf{conf}(\varphi, D) \geq 1 - \varepsilon$. Then, φ is redundant if there exists a CFD $\varphi' : M \to n$ with $M \prec I$ and $\{n\} \preceq \{j\}$, and $\mathsf{conf}(\varphi', D) = \mathsf{conf}(\varphi, D)$.*

Example 8. In our example, the CFD $(\mathsf{Temperature}, \mathsf{Cool}) \to (\mathsf{Humidity}, \mathsf{Normal})$ holds exactly. This implies the redundancy of, for example, the CFDs

$$\{(\mathsf{Temperature}, \mathsf{Cool}), (\mathsf{Humidity}, \mathsf{Normal}), (\mathsf{Windy}, \_)\} \to (\mathsf{Play}, \_)$$
$$\{(\mathsf{Temperature}, \mathsf{Cool}), (\mathsf{Windy}, \_)\} \to (\mathsf{Humidity}, \_).   ◇$$

Such redundancy can be eliminated efficiently in CTane (and Tane), since it employs a breadth-first traversal of the integrated search lattice, and hence all immediately preceding sets of an itemset are directly available in the level above the current one in the lattice. Pruning is then performed by associating with every itemset I in the lattice a set $\mathcal{C}^+(I)$ of candidate consequents for I and its supersets. Initially, we set $\mathcal{C}^+(I) = \{(\mathsf{A}, v) \in \mathcal{I} \mid \text{if } (\mathsf{A}, v') \in I \text{ then } v = v'\}$, i.e., all items except those for which I already contains a different item with the same attribute. Whenever a CFD is found to hold, the relevant $\mathcal{C}^+$ sets are updated, removing candidate consequents which will lead to redundant CFDs. Clearly, if

$\mathcal{C}^+(I) = \emptyset$, then I and all its supersets can be removed from the search space. Updating the sets $\mathcal{C}^+$ is performed as follows in CTane:

1. If $D \models I \rightarrow j$, set $\mathcal{C}^+(M) = \mathcal{C}^+(M) \cap I$ for all M with $j \in M$ and $M \preceq I$;
2. When generating a new itemset X in the lattice, set $\mathcal{C}^+(X) = \mathcal{C}^+(X) \cap \mathcal{C}^+(I)$ for all $I \prec X$ with $|(X \setminus I)| = 1$.

To generalize this strategy across our different approaches and search strategies, where not all preceding sets may be readily available in the search lattice, we introduce two techniques. Firstly, we use a lookup table indexed by the consequent of a rule[5], and store a list of all CFDs with that consequent that hold exactly on D. When a confident CFD $I \rightarrow j$ is found, it then suffices to verify whether a preceding set of I is present in the table at index j. If a preceding set M is found, the CFD is redundant, and pruning is performed by setting $\mathcal{C}^+(I \cup \{j\}) = \mathcal{C}^+(I \cup \{j\}) \cap M$.

Table 2. Statistics of the UCI datasets used in the experiments. We report the number of tuples, distinct constant items, and attributes.

Dataset	$|\mathcal{D}|$	$|\mathcal{I}|$	$|\mathcal{A}|$
Adult	48842	202	11
Mushroom	8124	119	15
Nursery	12960	32	9

Our second pruning technique generalizes the concept of free itemsets [3] (also called generators [17]). An itemset M is called free if, for all $J \subset M$, it holds that $\mathsf{supp}(J, D) \neq \mathsf{supp}(M, D)$. Moreover, it is known that all subsets of a free set are also free. We extend this concept to equivalence classes:

Definition 9 (Eq-Free Itemset). *An itemset I is Eq-Free in an instance D if, for all $J \subset I$, $|\Pi(I, D)| \neq |\Pi(J, D)|$ or $\|\Pi(I, D)\| \neq \|\Pi(J, D)\|$.*

We now observe that, if a CFD $\varphi : I \rightarrow j$ holds on D, then the itemset $I \cup \{j\}$ is not *Eq-Free*. Indeed, it must necessarily hold that $|\Pi(I, D)| = |\Pi((I \cup \{j\}), D)|$ and $\|\Pi(I, D)\| = \|\Pi((I \cup \{j\}), D)\|$. Hence, in order to obtain non-redundant CFDs, we additionally need to verify the Eq-Freeness of the antecedent of every considered CFD. To implement this check efficiently, we use a lookup table as in the Talky-G algorithm for mining free itemsets [19].

5   Experiments

We experimentally validate the proposed techniques on real-life datasets from the UCI repository (http://archive.ics.uci.edu/ml/), described in Table 2. The mushroom dataset was restricted to its first 15 attributes, as runtimes became

---

[5] We store constant CFDs $I \rightarrow (A, v)$ both at indices (A, v) and $(A, \_)$.

too high when considering more attributes. The algorithms have been implemented in C++, the source code and used datasets are available for research purposes[6]. The program was tested on an Intel Xeon Processor (3.8 GHZ) with 32 GB of memory running Ubuntu. Our algorithms run entirely in main memory.

In Sect. 4, we have described the three approaches to CFD discovery in full generality, i.e., using any FD discovery algorithm based on equivalence partitions, any itemset mining algorithm using tidlists, and any search strategy. We begin the experimental section by describing specific instantiations of our approaches:

Integrated uses a depth-first implementation of the CTane algorithm

Itemset-First uses a breadth-first version of Eclat for the itemset mining step, and a depth-first Tane implementation for the FD discovery step

FD-First uses both a depth-first Tane step and depth-first itemset mining

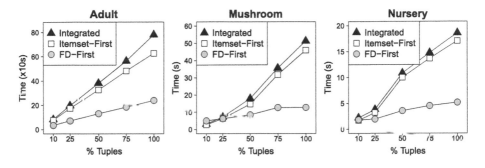

Fig. 1. Scalability of three CFD discovery algorithms in number of tuples.

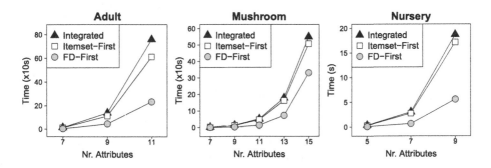

Fig. 2. Scalability of three CFD discovery algorithms in number of attributes.

All our depth-first implementations use a reverse pre-order traversal. We selected these three instantiations as the *best ones* – in terms of efficiency – out of a total of 18 different combinations. The runtime results of all instantiations are available in the online appendix [21].

---

[6] https://bit.ly/2yFNksO.

Since CFD (and FD) discovery is inherently exponential in the number of attributes of a dataset, we sometimes reduce the overall runtimes of the algorithms by enforcing a limit on the size of rules, called the maximum antecedent size. We compare the runtime of the three methodologies in function of the number of tuples and attributes of the data, the minimum support threshold, and the maximum antecedent size. The confidence threshold was found to have a negligible influence on runtime, and hence all experiments are run with $\varepsilon = 0$. Runtime plots in function of confidence can be found in the online appendix [21]. We emphasize that all methods return the exact same result in every experiment.

5.1   Number of Tuples

We first investigate the scalability of each approach in terms of the number of tuples. For this experiment, we consider only the first $X\%$ tuples of each dataset, with X ranging from 10% to 100%. The minimum support threshold was fixed at 10% of the number of tuples considered, and the maximum antecedent size was fixed at 6. The obtained runtimes are displayed in Fig. 1. We see that the FD-First approach scales better than the other approaches, and is faster overall.

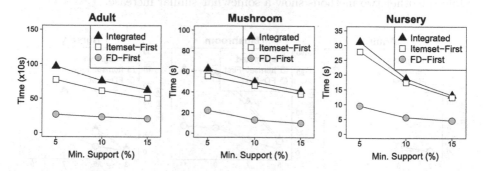

Fig. 3. Scalability of three CFD discovery algorithms in minimum support threshold.

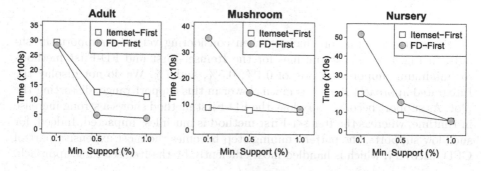

Fig. 4. Scalability of Itemset-First and FD-First discovery algorithms for very low minimum support thresholds.

5.2   Number of Attributes

Similar to the previous experiment, we now investigate the performance of the three algorithms in terms of the number of attributes, by considering only the first X attributes. In Fig. 2, the runtimes are shown on each dataset for increasing values of X. The minimum support threshold and maximum antecedent size were again fixed at 10% and 6, respectively. While each of the algorithms shows an exponential rise in runtime as the number of attributes increases, the FD-First method clearly outperforms the other approaches. The Integrated method is the slowest overall, and suffers most of all from the increasing number of attributes.

5.3   Minimum Support

We next fix the dimensionality of the data, using all tuples and attributes, and study the influence of the minimum support threshold on runtime. The results for the three datasets are shown in Fig. 3, for minimum support thresholds of 5%, 10%, and 15% of the total number of tuples. Overall, the support threshold has less impact than the number of attributes. The FD-First method shows the lowest increase in runtime as support decreases, and is clearly the fastest method, while the other two methods show a somewhat similar increase.

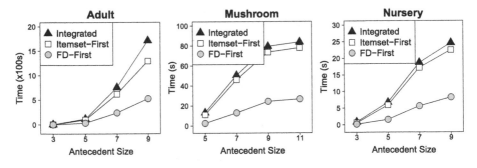

Fig. 5. Scalability of three CFD discovery algorithms in maximal size of antecedent.

However, the situation changes when considering very low support thresholds. In Fig. 4, we show runtimes for the Itemset-First and FD-First methods for minimum supports ranging of 0.1%, 0.5%, and 1%. We do not display the Integrated approach, since it is much slower in this support range, distorting the plot. As support becomes very low, the FD-First method shows a strong increase in runtime, whereas the Itemset-First method is much less impacted. Indeed, for such low supports, the pattern mining step becomes the most expensive part of CFD discovery, which is handled most efficiently by the Itemset-First approach.

5.4 Maximal Antecedent Size

We conclude the experimental section by investigating the impact of the maximal antecedent size threshold on the runtime of the algorithms. The results are shown in Fig. 5. The minimum support threshold was again fixed at 10%. We see an exponential increase in runtime, similar to that observed when the number of attributes was increased. The FD-First approach again performs best on every dataset, and shows the lowest increase in runtime as antecedent size increases.

6 Conclusion

We have presented the discovery of Conditional functional dependencies (CFDs) as a form of association rule mining, and classified the possible approaches into three categories, based on how these approaches combine pattern mining and functional dependency discovery. Two of these approaches have not been considered before. Moreover, we discuss how discovery and pruning can be performed independent of methodology and search strategy, either breadth-first or depth-first. We show experimentally that both our new approaches outperform the existing CTane algorithm, and identify situations in which either of these methods achieve the best performance. Most crucially, we have shown that the field of CFD discovery still offers opportunities for improvement. This is highly relevant in view of the popularity of CFDs in data cleaning. As future work, we plan to investigate parallelized or distributed discovery and develop incremental discovery methods to accommodate for dynamic, changing data.

References

1. Abedjan, Z., Schulze, P., Naumann, F.: DFD: efficient functional dependency discovery. In: CIKM, pp. 949–958. ACM (2014)
2. Agrawal, R., Imieliński, T., Swami, A.: Mining association rules between sets of items in large databases. In: ACM SIGMOD Record, vol. 22, pp. 207–216. ACM (1993)
3. Boulicaut, J.-F., Bykowski, A., Rigotti, C.: Approximation of frequency queries by means of free-sets. In: Zighed, D.A., Komorowski, J., Żytkow, J. (eds.) PKDD 2000. LNCS (LNAI), vol. 1910, pp. 75–85. Springer, Heidelberg (2000). https://doi.org/10.1007/3-540-45372-5_8
4. Chiang, F.: Data Quality Through Active Constraint Discovery and Maintenance. Ph.D. thesis, University of Toronto (Canada) (2012)
5. Chiang, F., Miller, R.J.: Discovering data quality rules. PVLDB **1**(1), 1166–1177 (2008)
6. Diallo, T., Novelli, N., Petit, J.M.: Discovering (frequent) constant conditional functional dependencies. IJDMMM **4**(3), 205–223 (2012)
7. Fan, W., Geerts, F.: Foundations of Data Quality Management. Synthesis Lectures on Data Management. Morgan & Claypool Publishers, San Rafael (2012)
8. Fan, W., Geerts, F., Jia, X., Kementsietsidis, A.: Conditional functional dependencies for capturing data inconsistencies. TODS **33**(2), 6 (2008)

9. Fan, W., Geerts, F., Li, J., Xiong, M.: Discovering conditional functional dependencies. TKDE **23**(5), 683–698 (2011)
10. Goethals, B., Page, W.L., Mannila, H.: Mining association rules of simple conjunctive queries. In: SDM, pp. 96–107. SIAM (2008)
11. Huhtala, Y., Kärkkäinen, J., Porkka, P., Toivonen, H.: TANE: an efficient algorithm for discovering functional and approximate dependencies. Comput. J. **42**(2), 100–111 (1999)
12. Ilyas, I.F., Chu, X.: Trends in cleaning relational data: consistency and deduplication. Found. Trends Databases **5**(4), 281–393 (2015)
13. Mandros, P., Boley, M., Vreeken, J.: Discovering reliable approximate functional dependencies. In: KDD, pp. 355–363. ACM (2017)
14. Medina, R., Nourine, L.: A unified hierarchy for functional dependencies, conditional functional dependencies and association rules. In: Ferré, S., Rudolph, S. (eds.) ICFCA 2009. LNCS (LNAI), vol. 5548, pp. 98–113. Springer, Heidelberg (2009). https://doi.org/10.1007/978-3-642-01815-2_9
15. Novelli, N., Cicchetti, R.: FUN: an efficient algorithm for mining functional and embedded dependencies. In: Van den Bussche, J., Vianu, V. (eds.) ICDT 2001. LNCS, vol. 1973, pp. 189–203. Springer, Heidelberg (2001). https://doi.org/10.1007/3-540-44503-X_13
16. Papenbrock, T.: Functional dependency discovery: an experimental evaluation of seven algorithms. PVLDB **8**(10), 1082–1093 (2015)
17. Pasquier, N., Bastide, Y., Taouil, R., Lakhal, L.: Discovering frequent closed itemsets for association rules. In: ICDT, pp. 398–416 (1999)
18. Quinlan, J.R.: Induction of decision trees. Mach. Learn. **1**, 81–106 (1986)
19. Szathmary, L., Valtchev, P., Napoli, A., Godin, R.: Efficient vertical mining of frequent closures and generators. In: Adams, N.M., Robardet, C., Siebes, A., Boulicaut, J.-F. (eds.) IDA 2009. LNCS, vol. 5772, pp. 393–404. Springer, Heidelberg (2009). https://doi.org/10.1007/978-3-642-03915-7_34
20. Yao, H., Hamilton, H.J., Butz, C.J.: Fd_mine: discovering functional dependencies in a database using equivalences. In: ICDM, pp. 729–732. IEEE (2002)
21. Full version. https://bit.ly/2II2oWq
22. Zaki, M.J., Meira Jr., W.: Data Mining and Analysis: Fundamental Concepts and Algorithms. Cambridge University Press, Cambridge (2014)
23. Zaki, M.J., Parthasarathy, S., Ogihara, M., Li, W., et al.: New algorithms for fast discovery of association rules. In: KDD, pp. 283–286 (1997)

Sqn2Vec: Learning Sequence Representation via Sequential Patterns with a Gap Constraint

Dang Nguyen[1(✉)], Wei Luo[1], Tu Dinh Nguyen[1], Svetha Venkatesh[1], and Dinh Phung[2]

[1] Center for Pattern Recognition and Data Analytics,
School of Information Technology, Deakin University, Geelong, Australia
{d.nguyen,wei.luo,tu.nguyen,svetha.venkatesh}@deakin.edu.au
[2] Faculty of Information Technology, Monash University,
Clayton Campus, Melbourne, VIC 3800, Australia
dinh.phung@monash.edu

Abstract. When learning sequence representations, traditional pattern-based methods often suffer from the data sparsity and high-dimensionality problems while recent neural embedding methods often fail on sequential datasets with a small vocabulary. To address these disadvantages, we propose an *unsupervised* method (named **Sqn2Vec**) which first leverages *sequential patterns* (SPs) to increase the vocabulary size and then learns *low-dimensional continuous* vectors for sequences via a neural embedding model. Moreover, our method enforces a *gap constraint* among symbols in sequences to obtain meaningful and discriminative SPs. Consequently, **Sqn2Vec** produces significantly better sequence representations than a comprehensive list of state-of-the-art baselines, particularly on sequential datasets with a relatively small vocabulary. We demonstrate the superior performance of **Sqn2Vec** in several machine learning tasks including sequence classification, clustering, and visualization.

1 Introduction

Many real-world applications such as web mining, text mining, bio-informatics, system diagnosis, and action recognition have to deal with sequential data. The core task of such applications is to apply machine learning methods, for example, K-means or Support Vector Machine (SVM) to sequential data to find insightful patterns or build effective predictive models. However, this task is challenging since machine learning methods typically require inputs as fixed-length vectors, which are not applicable to sequences.

Electronic supplementary material The online version of this chapter (https://doi.org/10.1007/978-3-030-10928-8_34) contains supplementary material, which is available to authorized users.

© Springer Nature Switzerland AG 2019
M. Berlingerio et al. (Eds.): ECML PKDD 2018, LNAI 11052, pp. 569–584, 2019.
https://doi.org/10.1007/978-3-030-10928-8_34

A well-known solution in data mining is to use *sequential patterns* (SPs) as features [6]. This approach first mines SPs from the dataset, and then represents each sequence in the dataset as a feature vector with binary components indicating whether this sequence contains a particular sequential pattern. We can see the dimension of the feature space is huge since the number of SPs is often large. Consequently, this leads to the high-dimensionality and data sparsity problems.

To reduce the dimension of the feature space, many researchers have tried to extract only *interesting SPs* under an *unsupervised* setting [5,9,17] or *discriminative SPs* under a *supervised* setting [3,6,19]. The methods discover interesting SPs, e.g., closed SPs [17], compressing SPs [9], and relevant SPs [5] without using the sequence labels. Although these methods can reduce the number of generated patterns, thus solving the high-dimensionality problem, they still suffer from the data sparsity problem. The methods discover discriminative SPs using different measures, e.g., information gain [6], support-cohesion [19], and behavioral constraint [3], which involve the sequence labels. Although these methods often show good performances in sequence classification, they usually require the labels for all training examples, which is often unrealistic in many real applications.

Recently, neural embedding approaches have been introduced to learn *low-dimensional continuous* embedding vectors for sequences using neural networks in a fully *unsupervised* manner. These methods primarily focus on text, where they learn embedding vectors for documents [2,10] and show significant improvements over non-embedding methods, e.g., Bag of-Words, in several applications such as document classification and sentiment analysis. However, they have two limitations. First, they mostly learn embedding vectors based on atoms in data (i.e., words), but do not consider sets of atoms (i.e., phrases). Second, they often perform poorly on datasets with a relatively small vocabulary [8]. In our experiments, the performances of document embedding methods dramatically reduce on sequential datasets whose the vocabulary size is less than 300.

Our Approach. To overcome the disadvantages of traditional pattern-based methods and recent embedding methods, we propose a novel *unsupervised* method (named **Sqn2Vec**) for learning sequence embeddings. In particular, we first extract a set of SPs which satisfy a gap constraint from the dataset. We then adapt a document embedding model to learn a vector for each sequence by predicting not only its belonging symbols but its SPs as well. By doing this, we can learn *low-dimensional continuous* vectors for sequences, which solves the weakness of pattern-based methods. We also take into account sets of atoms (i.e., SPs) during the learning process, which solves the weakness of embedding methods. More importantly, by considering both singleton symbols and SPs, we can increase the vocabulary size, which results in our better embeddings on sequential datasets with a small vocabulary. Moreover, since **Sqn2Vec** is fully unsupervised, it can be directly used for learning sequence embeddings in domains where labeled examples are difficult to obtain and the learned representations are well-generalized to different tasks such as sequence classification, clustering, and visualization.

To summarize, we make the following contributions:

1. We propose **Sqn2Vec**, an *unsupervised* embedding method, for learning *low-dimensional continuous* feature vectors for sequences.
2. We propose two models in **Sqn2Vec**, which *learn* sequence embeddings by predicting its belonging singleton symbols and SPs. The learned embeddings are meaningful and discriminative.
3. We demonstrate **Sqn2Vec** in both sequence classification and sequence clustering tasks, where it significantly outperforms the state-of-the-art baselines on 10 real-world sequential datasets.

2 Related Work

2.1 Sequential Pattern Based Methods for Sequence Representation

SPs have been widely used to construct feature vectors for sequences [6], which are essential inputs for different machine learning tasks. However, using SPs as features often suffers from the data sparsity and high-dimensionality problems. Recent SP-based methods have tried to extract only interesting or discriminative SPs. Several approaches have been proposed for mining interesting SPs. For example, Lam et al. [9] discovered compressing SPs which can optimally compress the dataset w.r.t an encoding scheme. In [5], a probabilistic approach was developed to mine relevant SPs which are able to reconstruct the dataset. Although interesting SPs can help to reduce the number of generated patterns (i.e., the feature space), they still suffer from the data sparsity problem.

To discover discriminative SPs, existing approaches have used the sequence labels during the mining process. For example, they use the label information to compute information gain [6], support-cohesion [19], or behavioral constraint [3]. Although discriminative SPs are useful for classification, they require sequence labels, making the mining process *supervised*. Related to sequence classification, SPs have been also used to build a set of predictive rules for classification, often called *sequential classification rules*. These rules represent the strong associations between SPs and labels, which can be used directly for prediction (i.e., they are used as rule-based classifiers) [19] or indirectly for prediction (i.e., they are used as features in other classifiers) [4].

2.2 Embedding Methods for Sequence Representation

Most existing approaches for sequence embedding learning mainly focus on text, where they learn embedding vectors for documents [2,10]. These methods have shown impressive successes in many natural language processing tasks such as document classification and sentiment analysis. They, however, are not suitable for sequential data in bio-informatics, navigation systems, and action recognition since different from text, these sequential datasets have a very small vocabulary size (i.e., the small number of distinct symbols). For example, the human DNA sequences only consist of four nucleotides A, C, G, and T. Related to sequence classification, several deep neural network models (also called *supervised embedding methods*) have been introduced such as long short term memory (LSTM) networks and bidirectional LSTM (Bi-LSTM) networks [16]. Since these methods require labels, their embeddings are not general enough to effectively apply to unsupervised tasks such as sequence clustering.

As far as we know, learning embedding vectors for sequences based on SPs has not been studied yet. In this paper, we propose the first approach which utilizes information of both singleton symbols and SPs to learn sequence embeddings. Different from discriminative pattern-based methods and supervised embedding methods, our method is fully *unsupervised*. Moreover, our method leverages SPs to capture the sequential relations among symbols as SP-based methods while it learns dense representations as embedding methods.

3   Framework

3.1   Problem Definition

Given a set of symbols $\mathcal{I} = \{e_1, e_2, ..., e_M\}$, a *sequential dataset* $\mathcal{D} = \{S_1, S_2, ..., S_N\}$ is a set of sequences where each sequence S_i is an ordered list of symbols [18]. The symbol at the position j in S_i is denoted as $S_i[j]$ and $S_i[j] \in \mathcal{I}$.

Our goal is to learn a mapping function $f : \mathcal{D} \to \mathbb{R}^d$ such that every sequence $S_i \in \mathcal{D}$ is mapped to a d-dimensional continuous vector. The mapping needs to capture the similarity among the sequences in $\mathcal{D}$, in the sense that S_i and S_j are similar if $f(S_i)$ and $f(S_j)$ are close to each other on the vector space, and vice versa. The matrix $\mathbf{X} = [f(S_1), f(S_2), ..., f(S_N)]$ then contains feature vectors of sequences, which can be direct inputs for many traditional machine learning and data mining tasks, particularly classification and clustering.

3.2   Learning Sequence Embeddings Based on Sequential Patterns

To learn sequence embeddings, one direct solution is to apply document embedding models [2,10] to the sequential dataset, where each sequence is treated as a document and symbols are treated as words. However, as we discussed in Sect. 1, existing document embedding methods are not suitable for sequential datasets in bio-informatics or system diagnosis since these datasets have a relatively small vocabulary (i.e., the very small number of symbols).

To improve the performances of document embedding models on such kind of sequential data, we propose to learn sequence embeddings based on SPs instead of singleton symbols. By doing this, we can increase the vocabulary size since the number of SPs is much larger than the number of symbols.

Sequential Pattern Discovery. Following the notations in [18], we define a sequential pattern as follows. Let $\mathcal{I} = \{e_1, e_2, ..., e_M\}$ be a set of symbols and $\mathcal{D} = \{S_1, S_2, ..., S_N\}$ be a sequential dataset.

Definition 1 (Subsequence). *Given two sequences* $S_1 = \{e_1, e_2, ..., e_n\}$ *and* $S_2 = \{e'_1, e'_2, ..., e'_m\}$, S_1 *is said to be a subsequence of* S_2 *or* S_1 *is contained in*

S_2 (denoted $S_1 \subseteq S_2$), if there exists a one-to-one mapping $\phi : [1,n] \rightarrow [1,m]$, such that $S_1[i] = S_2[\phi(i)]$ and for any positions i, j in S_1, $i < j \Rightarrow \phi(i) < \phi(j)$. In other words, each position in S_1 is mapped to a position in S_2, and the order of symbols is preserved.

Definition 2 *(Subsequence occurrence)*. *Given a sequence $S = \{e'_1, e'_2, ..., e'_m\}$ and a subsequence $X = \{e_1, e_2, ..., e_n\}$ of S, a sequence of positions $o = \{i_1, ..., i_m\}$ is an occurrence of X in S if $1 \leq i_k \leq m$ and $X[k] = S[i_k]$ for each $1 \leq k \leq n$, and $i_k < i_{k+1}$ for each $1 \leq k < n$.*

Example 1. $X = \{g,t\}$ (or $X = gt$ for short) is a subsequence of $S = gaagt$. There are two occurrences of X in S, namely $o_1 = \{1,5\}$ and $o_2 = \{4,5\}$.

Definition 3 *(Subsequence support)*. *Given a sequential dataset $\mathcal{D}$, the support of a subsequence X is defined as $sup(X) = \frac{|\{S_i \in \mathcal{D} | X \subseteq S_i\}|}{|\mathcal{D}|}$, i.e., the fraction of sequences in $\mathcal{D}$, which contain X.*

Definition 4 *(Sequential pattern)*. *Given a minimum support threshold $\delta \in [0,1]$, a subsequence X is said to be a sequential pattern if $sup(X) \geq \delta$.*

A sequential pattern can capture the sequential relation among symbols, but it does not pay attention on the gap among its elements. In bio-data and text data, this gap is very important because SPs whose symbols are far away from each other are often less meaningful than those whose symbols are close in the sequences. For example, consider a text dataset with two sentences S_1 = "machine learning is a field of computer science" and S_2 = "machine learning gives computer systems the ability to learn". Although two SPs X_1 = {machine, learning} and X_2 = {machine, computer} are found in both S_1 and S_2, X_2 is less meaningful than X_1 due to the large gap between "machine" and "computer". In other words, the two words "machine" and "computer" are in two different contexts. We believe that if we restrict the distance between two neighboring elements in a sequential pattern, then this pattern is more meaningful and discriminative. We define a sequential pattern satisfying a gap constraint as follows.

Definition 5 *(Gap constraint and satisfaction)*. *A gap is a positive integer, $\triangle > 0$. Given a sequence $S = \{e'_1, e'_2, ..., e'_m\}$ and an occurrence $o = \{i_1, ..., i_m\}$ of a subsequence X of S, if $i_{k+1} \leq i_k + \triangle$ ($\forall i_k \in [1, m-1]$), then we say that o satisfies the $\triangle$-gap constraint. If there is at least one occurrence of X satisfies the $\triangle$-gap constraint, we say that X satisfies the $\triangle$-gap constraint.*

Example 2. Among two occurrences of $X = gt$ in $S = gaagt$, namely $o_1 = \{1,5\}$ and $o_2 = \{4,5\}$, only o_2 satisfies the 1-gap constraint (i.e., $\triangle = 1$) since $5 \leq 4 + \triangle$. We say that X satisfies the 1-gap constraint because at least one of its occurrences does.

Definition 6 *(Sequential pattern satisfying a $\triangle$-gap constraint)*. *Given a sequential dataset $\mathcal{D}$, a gap constraint $\triangle > 0$, and a minimum support threshold*

$\delta \in [0, 1]$, *the support of a subsequence X in $\mathcal{D}$ with the $\triangle$-gap constraint, denoted* $sup(X, \triangle)$, *is the fraction of sequences in $\mathcal{D}$, where X appears as a subsequence satisfying the $\triangle$-gap constraint. X is called a sequential pattern which satisfies the $\triangle$-gap constraint if $sup(X, \triangle) \geq \delta$.*

Note that we consider the subsequences with length 1 (i.e., they contain only one symbol) satisfy any $\triangle$-gap constraint. Hereafter, we call a subsequence X a sequential pattern with the meaning that X is a sequential pattern satisfying a $\triangle$-gap constraint.

Example 3. Let consider an example sequential dataset as shown in Fig. 1(a). Assume that $\triangle = 1$ and $\delta = 0.7$. The subsequence $X = ag$ is contained in three sequences S_1, S_2, and S_4, and it also satisfies the 1-gap constraint in these three sequences. Thus, its support is $sup(X, \triangle) = 3/4 = 0.75$. We say that $X = ag$ is a sequential pattern since $sup(X, \triangle) \geq \delta$. With $\triangle = 1$ and $\delta = 0.7$, there are in total five SPs discovered from the dataset, as shown in Fig. 1(b), and each sequence now can be represented by a set of SPs, as shown in Fig. 1(c).

Seq	Symbols
S_1	{c, a, g, a, a, g, t}
S_2	{t, g, a, c, a, g}
S_3	{g, a, a, t}
S_4	{a, g}

(a)

SP	Symbols	*sup*
X_1	{a}	1.00
X_2	{g}	1.00
X_3	{t}	0.75
X_4	{a, g}	0.75
X_5	{g, a}	0.75

(b)

Seq	SPs
S_1	$\{X_1, X_2, X_3, X_4, X_5\}$
S_2	$\{X_1, X_2, X_3, X_4, X_5\}$
S_3	$\{X_1, X_2, X_3, X_5\}$
S_4	$\{X_1, X_2, X_4\}$

(c)

Fig. 1. Two forms of a sequence: a set of single symbols and a set of SPs. Table (a) shows a sequential dataset with four sequences where each of them is a set of symbols. Table (b) shows five SPs discovered from the dataset (here, $\triangle = 1$ and $\delta = 0.7$). Table (c) shows each sequence represented by a set of SPs.

Sequence Embedding Learning. After associating each sequence with a set of SPs, we follow the Paragraph Vector-Distributed Bag-of-Words (PV-DBOW) model introduced in [10] to learn embedding vectors for sequences. Given a target sequence S_t whose representation needs to be learned, and a set of SPs $\mathcal{F}(S_t) = \{X_1, X_2, ..., X_l\}$ contained in S_t, our goal is to maximize the log probability of predicting the SPs $X_1, X_2, ..., X_l$ which appear in S_t:

$$\max \sum_{i=1}^{l} \log \Pr(X_i \mid S_t) \tag{1}$$

Furthermore, $\Pr(X_i \mid S_t)$ is defined by a softmax function:

$$\Pr(X_i \mid S_t) = \frac{\exp(g(X_i) \cdot f(S_t))}{\sum_{X_j \in \mathcal{F}(\mathcal{D})} \exp(g(X_j) \cdot f(S_t))}, \tag{2}$$

where $g(X_i) \in \mathbb{R}^d$ and $f(S_t) \in \mathbb{R}^d$ are the embedding vectors of the sequential pattern $X_i \in \mathcal{F}(S_t)$ and the sequence S_t respectively, and $\mathcal{F}(\mathcal{D})$ is the set of all SPs discovered from the dataset $\mathcal{D}$.

Calculating the summation $\sum_{X_j \in \mathcal{F}(\mathcal{D})} \exp(g(X_j) \cdot f(S_t))$ in Eq. 2 is very expensive since the number of SPs in $\mathcal{F}(\mathcal{D})$ is often very large. To solve this problem, we approximate it using the negative sampling technique [13]. The idea is that instead of iterating over all SPs in $\mathcal{F}(\mathcal{D})$, we randomly select a relatively small number of SPs which are not contained in the target sequence S_t (these SPs are called *negative SPs*). We then attempt to distinguish the SPs contained in S_t from the negative SPs by minimizing the following binary objective function of logistic regression:

$$\mathcal{O}_1 = - \left[\log \sigma(g(X_i) \cdot f(S_t)) + \sum_{n=1}^{K} \mathbb{E}_{X^n \sim \mathcal{P}(X)} \log \sigma(-g(X^n) \cdot f(S_t)) \right], \quad (3)$$

where $\sigma(x) = \frac{1}{1+e^{-x}}$ is a sigmoid function, $\mathcal{P}(X)$ is the set of negative SPs, X^n is a negative sequential pattern draw from $\mathcal{P}(X)$ for K times, and $g(X^n) \in \mathbb{R}^d$ is the embedding vector of X^n.

We minimize $\mathcal{O}_1$ in Eq. 3 using stochastic gradient descent (SGD) where the gradients are derived as follows:

$$\frac{\partial \mathcal{O}_1}{\partial g(X^n)} = -\sigma(g(X^n) \cdot f(S_t) - \mathbb{I}_{X_i}[X^n]) \cdot f(S_t)$$

$$\frac{\partial \mathcal{O}_1}{\partial f(S_t)} = -\sum_{n=0}^{K} \sigma(g(X^n) \cdot f(S_t) - \mathbb{I}_{X_i}[X^n]) \cdot g(X^n), \quad (4)$$

where $\mathbb{I}_{X_i}[X^n]$ is an indicator function to indicate whether X^n is a sequential pattern $X_i \in \mathcal{F}(S_t)$ (i.e., the negative sequential pattern appears in the target sequence S_t) and when $n = 0$, then $X^n = X_i$.

3.3   Sqn2Vec Method for Learning Sequence Embeddings

When associating a sequence S_t with a set of SPs, S_t may not contain any SPs. In this case, we cannot learn a meaningful embedding vector for S_t. To avoid this problem, we propose two models which combine information of both single symbols and SPs to learn embedding vectors for sequences. These two models named **Sqn2Vec-SEP** and **Sqn2Vec-SIM** are presented next.

Sqn2Vec-SEP Model to Learn Sequence Embeddings. Given a sequence S_t, we separately learn an embedding vector $f_1(S_t)$ for S_t based on its symbols using the document embedding model PV-DBOW [10] and an embedding vector $f_2(S_t)$ for S_t based on its SPs (see Sect. 3.2). We then take the average of two embedding vectors to obtain the final embedding vector $f(S_t) = \frac{f_1(S_t)+f_2(S_t)}{2}$ for that sequence. The basic idea of **Sqn2Vec-SEP** is illustrated in Fig. 2.

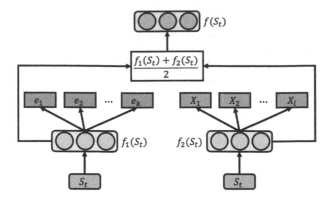

Fig. 2. Sqn2Vec-SEP model. Given a target sequence S_t, we learn the embedding vector $f_1(S_t)$ to predict its belonging symbols and learn the embedding vector $f_2(S_t)$ to predict its belonging SPs. We then take the average of $f_1(S_t)$ and $f_2(S_t)$ to obtain the final embedding vector $f(S_t)$ for S_t.

Sqn2Vec-SIM Model to Learn Sequence Embeddings. In the **Sqn2Vec-SEP** model, the sequence embeddings only capture the latent relationships between sequences and symbols and those between sequences and SPs separately. To overcome this weakness, we further propose the **Sqn2Vec-SIM** model which uses information of both single symbols and SPs of a sequence simultaneously. The overview of this model is shown in Fig. 3. More specifically, given a sequence S_t, our goal is to minimize the following objective function:

$$\mathcal{O}_2 = - \left[\sum_{e_i \in \mathcal{I}(S_t)} \log \Pr(e_i \mid S_t) + \sum_{X_i \in \mathcal{F}(S_t)} \log \Pr(X_i \mid S_t) \right], \qquad (5)$$

where $\mathcal{I}(S_t)$ is the set of singleton symbols contained in S_t and $\mathcal{F}(S_t)$ is the set of SPs contained in S_t.

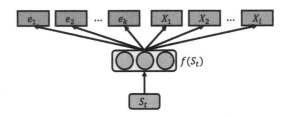

Fig. 3. Sqn2Vec-SIM model. Given a target sequence S_t, $\mathcal{I}(S_t) = \{e_1, e_2, ..., e_k\}$ is the set of symbols contained in S_t and $\mathcal{F}(S_t) = \{X_1, X_2, ..., X_l\}$ is the set of SPs contained in S_t. We learn the embedding vector $f(S_t)$ for S_t to predict both its belonging symbols and SPs.

Equation 5 can be simplified to:

$$\mathcal{O}_2 = - \sum_{p_i \in \mathcal{I}(S_t) \cup \mathcal{F}(S_t)} \log \Pr(p_i \mid S_t), \tag{6}$$

where $p_i \subseteq S_t$ is a symbol or a sequential pattern.

Following the same procedure in Sect. 3.2, we learn the embedding vector $f(S_t)$ for S_t, and the embedding vectors of two sequences S_i and S_j are close to each other if they contain similar symbols and SPs.

4 Experiments

4.1 Sequence Classification

The first experiment focuses on sequence classification, in which we compare our method with 11 baselines using sequential data from four application domains: text mining, action recognition, navigation analysis, and system diagnosis.

Datasets. We use eight benchmark datasets which are widely used for sequence classification. Their characteristics are summarized in Table 1. The *reuters* dataset is the four largest subsets of the Reuters-21578 dataset, consisting of news stories [19]. The three datasets *aslbu*, *aslgt*, and *auslan2* are derived from the videos of American and Australian Sign Language expressions [6]. The *context* dataset presents different locations of mobile devices carried by end-users [12]. The two datasets *pioneer* and *skating* are used in action recognition, which were introduced in [6]. The final dataset *unix* contains the command-line histories in a Unix system of nine end-users [19]. All datasets were used to evaluate the accuracy of sequence classification in [4–6,9,19].

Table 1. Statistics of eight sequential datasets.

Dataset	# sequences	# symbols	min. len	max. len	avg. len	# classes
reuters	1,010	6,380	4	533	93.84	4
aslbu	424	250	2	54	13.05	7
aslgt	3,464	94	2	176	43.67	40
auslan2	200	16	2	18	5.53	10
context	240	94	22	246	88.39	5
pioneer	160	178	4	100	40.14	3
skating	530	82	18	240	48.12	7
unix	5,472	1,697	1	1,400	32.34	4

Baselines. For a comprehensive comparison, we employ 11 state-of-the-art up-to-date baselines which can be categorized into four main groups:

- **Unsupervised SP-based methods:** We compare our method – **Sqn2Vec** with two state-of-the-art methods GoKrimp [9] and ISM [5]. We adopt their classification performances from their corresponding papers. We also employ another unsupervised baseline which constructs a binary feature vector for each sequence, with components indicating whether this sequence contains a sequential pattern with a $\triangle$-gap constraint (see Sect. 3.2). We name this baseline SP-BIN.
- **Supervised SP-based methods:** We select three representative and up-to-date baselines, namely BIDE-DC [6], SCIP [19], and MiSeRe [4]. We adopt the classification results of BIDE-DC reported in its supplemental appendix[1], those of SCIP from Table 10 in [19] and Fig. 12 in [4], and those of MiSeRe from Fig. 8 in [4].
- **Unsupervised embedding methods:** By considering a sequence as a document and symbols as words, we apply two recent state-of-the-art document embedding models for learning sequence embeddings, which are PV-DBOW [10] and Doc2Vec-C [2]. We also learn embedding vectors for sequences based on SPs (see Sect. 3.2), which we name Doc2Vec-SP.
- **Supervised embedding methods:** We implement two deep recurrent neural network models for sequence classification, LSTM and Bi-LSTM [16].

Our method **Sqn2Vec** has two different models which use different combinations of symbols and SPs. The **Sqn2Vec-SEP** model learns sequence embedding vectors from symbols and SPs separately (see Sect. 3.3) while the **Sqn2Vec-SIM** model learns sequence embedding vectors from symbols and SPs simultaneously (see Sect. 3.3).

Evaluation Metrics. After the feature vectors of sequences are constructed or learned, we feed them to an SVM with linear kernel [1] to classify the sequence labels. We use the linear-kernel SVM (a simple classifier) since we focus on the sequence embedding learning, not on a classifier, and this classifier was also used in [4,5,9,19]. The hyper-parameter C of SVM is set to 1, the same setting used in previous studies [5,9,19]. Each dataset is randomly split into 9 folds for training and 1 fold for testing. We repeat the classification process on each dataset 10 times and report the average classification accuracy. The standard deviation is not reported since all methods are very stable (their standard deviations are less than 10^{-1}).

Parameter Settings. Our method **Sqn2Vec** has three important parameters: the minimum support threshold δ, the gap constraint $\triangle$ for discovering SPs and the embedding dimension d for learning sequence embeddings. Since we develop **Sqn2Vec** in a fully unsupervised learning fashion, the values for δ, $\triangle$, and d

---

[1] https://sites.google.com/site/dfradkin/kais2014-separateAppendix.pdf.

are assigned without using sequence labels. We set $d = 128$ (a common value used in embedding methods [7,14]), set $\triangle = 4$ (a small gap which is able to capture the context of each symbol [15]), and set δ following the *elbow method* in [15]. Figure 4 illustrates the elbow method. From the figure, we can see that when the δ value decreases, the number of SPs slightly increases until a δ value where it significantly increases. This δ value, highlighted in red in the figure and chosen by the elbow method without considering the labels of sequences, is used in our experiments. In Sect. 4.1, we analyze the potential impact of selecting three parameters δ, $\triangle$, and d on the classification performance.

For a fair comparison, we use the same minimum support thresholds and gap constraints for our method and the baseline SP-BIN. We also set the embedding dimension required by three baselines PV-DBOW, Doc2Vec-C, and Doc2Vec-SP to 128, the same as one used in our method. For Doc2Vec-C, we use the source code[2] provided by the author with the same parameter settings except $d = 128$. We implement LSTM and Bi-LSTM with the following details[3]: the dimension of symbol embedding is 128, the number of LSTM hidden units is 100, the number of epochs is 50, the mini batch size is 64, the dropout rate for symbol embedding and LSTM is 0.2, and the optimizer is Adam.

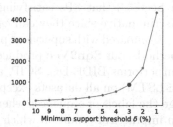

Fig. 4. The number of SPs discovered from the *reuters* dataset per δ (here, $\triangle = 4$). The red dot indicates the δ value selected via the elbow method. (Color figure online)

Results and Discussion. From Table 2, we can see two models in our method **Sqn2Vec** clearly results in better classification on all datasets compared with unsupervised embedding methods. **Sqn2Vec-SEP** achieves 2–97%, 5–232%, and 1–17% improvements over PV-DBOW, Doc2Vec-C, and Doc2Vec-SP respectively. As discussed in Sect. 1, two document embedding methods PV-DBOW and Doc2Vec-C perform poorly on sequential datasets with the small number of symbols, namely *aslbu*, *aslgt*, *auslan2*, *context*, *pioneer*, and *skating*. Especially, on the dataset *auslan2* whose the vocabulary size is only 16, their performances dramatically reduce, where they are 97–232% and 106–247% worse than **Sqn2Vec-SEP** and **Sqn2Vec-SIM** respectively. In contrast, on *unix* and the text dataset *reuters*, where the vocabulary size is large enough, their performances are quite good. Doc2Vec-C even achieves the second best result on *reuters*.

On all the datasets with the small vocabulary size, Doc2Vec-SP significantly outperforms PV-DBOW and Doc2Vec-C. This demonstrates that learning sequence embeddings from SPs is more effective than learning sequence embeddings from symbols, as discussed in Sect. 3.2. Two our models (**Sqn2Vec-SEP** and **Sqn2Vec-SIM**) are always superior than Doc2Vec-SP. This proves that our

---

[2] https://github.com/mchen24/iclr2017.
[3] We use the parameter settings suggested by Keras (https://keras.io) for LSTM.

proposal to incorporate the information of both singleton symbols and SPs into the sequence embedding learning is a better strategy than learning the sequence embeddings from SPs only (see Sect. 3.3).

For most cases, our method **Sqn2Vec** is better than unsupervised pattern-based methods. On three datasets *auslan2*, *skating*, and *unix*, **Sqn2Vec-SIM** outperforms these approaches by large margins (achieving 7–17%, 12–32%, and 33–34% gains over SP-BIN, GoKrimp, and ISM). Interestingly, our developed baseline SP-BIN, which uses SPs with a $\triangle$-gap constraint, is generally better than two state-of-the-art methods GoKrimp and ISM. This verifies our intuition in Sect. 3.2 that SPs satisfying a $\triangle$-gap constraint are more meaningful and discriminative since they can capture the context of each symbol.

Compared with supervised pattern-based methods and supervised embedding methods, our **Sqn2Vec** produces comparable performances on most datasets. It outperforms BIDE-DC, SCIP, and deep recurrent neural networks (LSTM and Bi-LSTM) on all datasets except *aslgt* and *unix*. Note that these methods leverage the labels of sequences when they construct/learn sequence representations, an impractical condition which actually benefits the supervised methods.

Table 2. Accuracy of our **Sqn2Vec** and 11 baselines on eight sequential datasets. Bold font marks the best performance in a column. The last row denotes the δ values used by our method for each dataset; they are determined using the elbow method (see Fig. 4). "–" means the accuracy is not available in the original paper.

Method	reuters	aslbu	aslgt	auslan2	context	pioneer	skating	unix
SP-BIN	94.85	**63.49**	78.76	28.00	90.83	**100.00**	31.70	82.06
GoKrimp	–	59.86	81.90	29.50	82.08	**100.00**	25.80[a]	74.33
ISM	–	60.20	75.70	24.60	78.30	**100.00**	25.60	–
BIDE-DC	–	59.03	**82.30**	30.67	51.53	97.92	28.11	57.74
SCIP	96.63	59.00	65.00	32.00	00.00	97.00	26.00	88.96
MiSeRe	–	63.00	74.00	32.00	81.00	**100.00**	31.00	–
PV-DBOW	95.84	46.51	70.98	16.00	86.25	84.38	26.79	90.75
Doc2Vec-C	97.62	37.21	49.25	9.50	37.50	62.50	16.98	86.35
Doc2Vec-SP	97.13	60.93	78.67	30.50	87.08	98.75	31.32	83.18
LSTM	90.40	50.70	69.74	25.00	71.25	93.75	27.74	90.86
Bi-LSTM	91.39	56.74	72.74	29.50	78.75	94.38	32.83	**91.79**
Sqn2Vec-SEP	**98.02**	62.56	81.18	31.50	**91.67**	99.38	**36.60**	90.78
Sqn2Vec-SIM	94.95	62.09	78.90	**33.00**	87.50	**100.00**	33.96	89.40
δ (%)	3%	2%	5%	3%	20%	4%	10%	7%

[a] As GoKrimp uses another version of *skating*, its result for *skating* is copied from [5].

Parameter Sensitivity. We examine how the different choices of three parameters δ, $\triangle$, and d affect the classification performance of **Sqn2Vec-SEP** on five datasets *reuters*, *aslbu*, *aslgt*, *auslan2*, and *pioneer*. Figure 5 shows the classification results as a function of one chosen parameter when the others are set to their default values. From Fig. 5(a), we can see our method is very stable on two datasets *reuters* and *aslgt*, where its classification performance just slightly changes with different δ values. On three datasets *aslbu*, *auslan2*, and *pioneer*, our prediction performance shows an increasing trend as δ is decreased. Another observation is that the values for δ selected by the elbow method often lead to the best or close to the best accuracy.

From Fig. 5(b), we also observe the performance of our **Sqn2Vec-SEP** is consistent on *reuters* and *aslgt*, where the gap constraint $\triangle$ is gain of relatively little relevant to the predictive task. On contrary, there is a first-increasing and then-decreasing accuracy line on two datasets *aslbu* and *pioneer*. One possible explanation is that if we set $\triangle$ large, the generated SPs are less meaningful as we discussed in Sect. 3.2. On *auslan2*, the increase of accuracy converges when $\triangle$ reaches 4.

Figure 5(c) suggests that the predictive performance increases on two datasets *aslgt* and *auslan2* when d is increased whereas there is a first-increasing and then-decreasing accuracy line on *aslbu* and *pioneer*. This finding differs from those in document embedding methods, where the embedding dimension generally shows a positive effect on document classification [2]. Again, our predictive performance is steady on *reuters*, which is shown by a straight accuracy line.

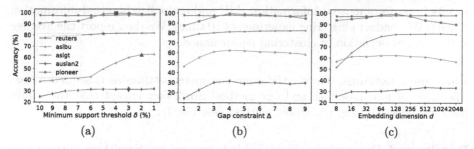

Fig. 5. Parameter sensitivity in sequence classification on five datasets *reuters*, *aslbu*, *aslgt*, *auslan2*, and *pioneer*. The minimum support thresholds δ selected via the elbow method and used in our experiments are indicated by red markers. (Color figure online)

4.2 Sequence Clustering

The second experiment illustrates how the latent representations learned by our proposed method can help the sequence clustering task, wherein we compare its performance with those of four baselines using text data.

Datasets. We use two text datasets for sequence clustering, namely *webkb* [15] and *news* [19]. *webkb* contains the content of webpages collected from computer departments of various universities. *news* is a subset of the dataset 20newsgroup, which is generated by selecting the five largest groups of documents. These two datasets are normalized (i.e., stop words are removed and the remaining words are stemmed), and can be downloaded from this website[4]. Their properties are summarized in Table 3.

---

[4] http://ana.cachopo.org/datasets-for-single-label-text-categorization.

Table 3. Statistics of two text datasets.

Dataset	# sequences	# symbols	min. len	max. len	avg. len	# classes
webkb	4,168	7,770	1	20,628	134.35	4
news	4,976	27,881	1	6,779	140.07	5

Baselines. We compare our **Sqn2Vec** with state-of-the-art embedding methods in text (PV-DBOW, Doc2Vec-C, and Doc2Vec-SP) and an unsupervised pattern-based method SP-BIN. We choose SP-BIN because it always outperforms other unsupervised pattern-based methods in sequence classification. These baselines are introduced in Sect. 4.1. We exclude supervised methods since they require sequence labels during the learning process, thus inappropriate for our unsupervised learning task – clustering.

Evaluation Metrics. To evaluate the clustering performance, the embedding vectors provided by each method are input to a clustering algorithm. Here, we use K-means (a simple clustering method) to group data and assess the clustering results in terms of mutual-information (MI) and normalized mutual-information (NMI). We conduct clustering experiments 10 times. We then report the average and standard deviation of clustering performance.

Parameter Settings. We use the same parameter settings as in sequence classification for four baselines and our method **Sqn2Vec**, except that the values for δ are selected using the elbow method (see Fig. 4).

Results and Discussion. From Table 4, we can see **Sqn2Vec** outperforms all the competitive baselines in terms of both MI and NMI. Compared with state-of-the-art document embedding methods, **Sqn2Vec-SEP** outperforms PV-DBOW, Doc2Vec-C, and Doc2Vec-SP by 24–32%, 38–40%, and 5–8% when clustering the *webkb* dataset. Similar improvements can be observed when clustering the *news* dataset, where the gains obtained by **Sqn2Vec-SEP** over PV-DBOW, Doc2Vec-C, and Doc2Vec-SP, around 8–14%, 72–77%, and 34–40%. Compared with the pattern-based method, the improvements are more significant, where **Sqn2Vec-SEP** outperforms SP-BIN by 163–184% on *webkb* and 767–1,090% on *news*.

4.3   Sequence Visualization

Figure 6 visualizes the document reprsentations learned by SP-BIN, PV-DBOW, Doc2Vec-C, Doc2Vec-SP, and our **Sqn2Vec-SEP** on the *news* dataset. We can see the documents from the same categories are clearly clustered using the embeddings generated by PV-DBOW and **Sqn2Vec-SEP**. On the other hand, SP-BIN, Doc2Vec-C, and Doc2Vec-SP do not distinguish different categories clearly.

Table 4. MI and NMI scores of our method **Sqn2Vec** and four baselines on two text datasets. The MI score is a non-negative value while the NMI score lies in the range [0, 1]. Bold font marks the best performance in a column.

Dataset	*webkb*		*news*	
Method	MI	NMI	MI	NMI
SP-BIN	0.19 (0.06)	0.16 (0.05)	0.10 (0.04)	0.09 (0.03)
PV-DBOW	0.41 (0.12)	0.34 (0.10)	1.04 (0.11)	0.72 (0.06)
Doc2Vec-C	0.39 (0.02)	0.30 (0.01)	0.69 (0.05)	0.44 (0.03)
Doc2Vec-SP	0.50 (0.14)	0.40 (0.10)	0.85 (0.12)	0.58 (0.07)
Sqn2Vec-SEP	**0.54 (0.10)**	**0.42 (0.07)**	**1.19 (0.15)**	**0.78 (0.06)**
Sqn2Vec-SIM	0.51 (0.11)	0.41 (0.08)	1.08 (0.22)	0.71 (0.12)
δ (%)	3%		3%	

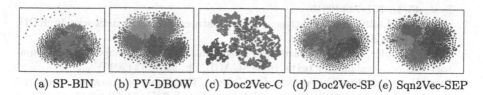

    (a) SP-BIN   (b) PV-DBOW   (c) Doc2Vec-C (d) Doc2Vec-SP (e) Sqn2Vec-SEP

Fig. 6. Visualization of document embeddings on *news* using t-SNE [11]. Different colors represent different categories. (Color figure online)

5 Conclusion

We have introduced **Sqn2Vec** – an unsupervised method for learning sequence embeddings from information of both singleton symbols and SPs. Our method is capable of capturing both the sequential relation among symbols and the semantic similarity among sequences. Our comprehensive experiments on 10 standard sequential datasets demonstrated the meaningful and discriminative representations learned by our approach in both sequence classification and sequence clustering tasks. In particularly, **Sqn2Vec** significantly outperforms several state-of-the-art baselines including pattern-based methods, embedding methods, and deep neural network models. Our approach can be applied to different real-world applications such as text mining, bio-informatics, action recognition, and system diagnosis. One of our future works is to integrate SPs into deep neural network models, e.g., LSTM and Bi-LSTM, to improve the classification performance.

Acknowledgment. Dinh Phung and Tu Dinh Nguyen gratefully acknowledge the partial support from the Australian Research Council (ARC).

References

1. Chang, C.-C., Lin, C.-J.: LIBSVM: a library for support vector machines. ACM Trans. Intell. Syst. Technol. **2**(3), 1–27 (2011)
2. Chen, M.: Efficient vector representation for documents through corruption. In: ICLR (2017)
3. De Smedt, J., Deeva, G., De Weerdt, J.: Behavioral constraint template-based sequence classification. In: Ceci, M., Hollmén, J., Todorovski, L., Vens, C., Džeroski, S. (eds.) ECML PKDD 2017. LNCS (LNAI), vol. 10535, pp. 20–36. Springer, Cham (2017). https://doi.org/10.1007/978-3-319-71246-8_2
4. Egho, E., Gay, D., Boullé, M., Voisine, N., Clérot, F.: A user parameter-free approach for mining robust sequential classification rules. Knowl. Inf. Syst. **52**(1), 53–81 (2017)
5. Fowkes J., Sutton, C.: A subsequence interleaving model for sequential pattern mining. In: KDD, pp. 835–844 (2016)
6. Fradkin, D., Mörchen, F.: Mining sequential patterns for classification. Knowl. Inf. Syst. **45**(3), 731–749 (2015)
7. Grover, A., Leskovec, J.: node2vec: scalable feature learning for networks. In: KDD, pp. 855–864 (2016)
8. Jin, L., Schuler, W.: A comparison of word similarity performance using explanatory and non-explanatory texts. In: NACACL, pp. 990–994 (2015)
9. Lam, H.T., Mörchen, F., Fradkin, D., Calders, T.: Mining compressing sequential patterns. Stat. Anal. Data Mining ASA Data Sci. J. **7**(1), 34–52 (2014)
10. Le, Q., Mikolov, T.: Distributed representations of sentences and documents. In: ICML, pp. 1188–1196 (2014)
11. Van Der Maaten, L., Hinton, G.: Visualizing data using t-SNE. J. Mach. Learn. Res. **9**, 2579–2605 (2008)
12. Mäntyjärvi, J., Himberg, J., Kangas, P., Tuomela, U., Huuskonen, P.: Sensor signal data set for exploring context recognition of mobile devices. In: PerCom, pp. 18–23 (2004)
13. Mikolov, T., Sutskever, I., Chen, K., Corrado, G., Dean, J.: Distributed representations of words and phrases and their compositionality. In: NIPS, pp. 3111–3119 (2013)
14. Nguyen, D., Luo, W., Nguyen, T.D., Venkatesh, S., Phung, D.: Learning graph representation via frequent subgraphs. In: SDM, pp. 306–314 (2018)
15. Rousseau, F., Kiagias, E., Vazirgiannis, M.: Text categorization as a graph classification problem. In: ACL, pp. 1702–1712 (2015)
16. Tai, K.S., Socher, R., Manning, C.: Improved semantic representations from tree-structured long short-term memory networks. In: ACL, pp. 1556–1566 (2015)
17. Wang, J., Han, J.: BIDE: efficient mining of frequent closed sequences. In: ICDE, pp. 79–90 (2004)
18. Zaki, M., Meira, W.: Data Mining and Analysis: Fundamental Concepts and Algorithms. Cambridge University Press, Cambridge (2014)
19. Zhou, C., Cule, B., Goethals, B.: Pattern based sequence classification. IEEE Trans. Knowl. Data Eng. **28**(5), 1285–1298 (2016)

Mining Tree Patterns with Partially Injective Homomorphisms

Till Hendrik Schulz[1(✉)], Tamás Horváth[1,2,3], Pascal Welke[1],
and Stefan Wrobel[1,2,3]

[1] Department of Computer Science, University of Bonn, Bonn, Germany
schulzth@cs.uni-bonn.de
[2] Fraunhofer IAIS, Schloss Birlinghoven, Sankt Augustin, Germany
[3] Fraunhofer Center for Machine Learning, Sankt Augustin, Germany

Abstract. One of the main differences between inductive logic programming (ILP) and graph mining lies in the pattern matching operator applied: While it is mainly defined by relational homomorphism (i.e., subsumption) in ILP, subgraph isomorphism is the most common pattern matching operator in graph mining. Using the fact that subgraph isomorphisms are injective homomorphisms, we bridge the gap between ILP and graph mining by considering a natural transition from homomorphisms to subgraph isomorphisms that is defined by *partially injective homomorphisms*, i.e., which require injectivity only for subsets of the vertex pairs in the pattern. Utilizing positive complexity results on deciding homomorphisms from bounded tree-width graphs, we present an algorithm mining frequent trees from *arbitrary* graphs w.r.t. partially injective homomorphisms. Our experimental results show that the predictive performance of the patterns obtained is comparable to that of ordinary frequent subgraphs. Thus, by preserving much from the advantageous properties of homomorphisms and subgraph isomorphisms, our approach provides a trade-off between efficiency and predictive power.

1 Introduction

Despite the facts that graphs can be considered as relational structures and graph patterns as function-free first-order goal clauses, *inductive logic programming* (ILP) [9] and *graph mining* are typically regarded as independent research fields. One of the reasons for this separation lies in the relative simplicity of the vocabularies for graphs as relational structures. Another important difference is that while the pattern matching operator in ILP is defined by *subsumption* (cf. [9]), a weakening of *first-order implication*, it is mainly the *subgraph isomorphism* in graph mining. For first-order function-free clauses, subsumptions are in fact homomorphisms between relational structures (see, e.g., [7]). Thus, ILP (mainly) applies *relational homomorphisms*, while graph mining deploys *subgraph isomorphisms*.

© Springer Nature Switzerland AG 2019
M. Berlingerio et al. (Eds.): ECML PKDD 2018, LNAI 11052, pp. 585–601, 2019.
https://doi.org/10.1007/978-3-030-10928-8_35

Our goal is to propose a binary feature space for *predictive graph mining* that is spanned by *frequent patterns*. On the one hand, frequent patterns w.r.t. *homomorphism* result, due to the lack of injectivity, in a loss in predictive performance compared to *subgraph isomorphism* when used for classification tasks. On the other hand, however, homomorphism is decidable in polynomial time for a broad class of patterns for which subgraph isomorphism remains persistently NP-complete (e.g., while the existence of a homomorphism from a path into a graph can be decided in polynomial time, this problem is NP-complete for subgraph isomorphism). As a trade-off between expressiveness and complexity, our goal is to preserve from the rigidity of subgraph isomorphism as much as possible, while utilizing the efficiency of homomorphisms for tractable graph classes. The difference between these two pattern matching operators is that any subgraph isomorphism is in fact an *injective* homomorphism. We therefore consider a natural transition from homomorphisms to subgraph isomorphisms that is defined by *partially injective homomorphisms*, i.e., which require injectivity only for a *subset* of the vertex pairs.

For loop-free graphs, any partially injective homomorphism can polynomially be reduced to an *ordinary* homomorphism by extending the graphs with additional edges corresponding to the injectivity constraints. To distinguish between original and constraint edges, we use edge colors. It holds that distinct sets of injectivity constraints define different partially injective homomorphisms problems between the same pattern and target graphs. In particular, the empty (resp. maximum) set of injectivity constraints corresponds to ordinary homomorphism (resp. subgraph isomorphism). By means of partially injective homomorphisms we can thus relax the rigid conception of having the binary choice between homomorphism and subgraph isomorphism, and are flexible to *dynamically* choose the degree of injectivity in the pattern matching operator. To the best of our knowledge, the application of *dynamic* pattern matching operators is an entirely novel characteristic in pattern mining, distinguishing it from all other traditional pattern matching operators used in ILP and graph mining.

Our approach can *efficiently* be applied to all pattern classes *from* which homomorphisms can be decided efficiently. For this work we consider the class of *bounded tree-width* graphs [11] (cf. [3] for the positive result on the complexity of homomorphisms from bounded tree-width graphs). More precisely, we restrict the patterns to *trees* and require the tree together with the additional constraint edges to form a graph of *tree-width* at most k, where $k > 0$ is some (small) constant. While this kind of partially injective homomorphisms from trees into arbitrary graphs is decidable in polynomial time, ordinary subgraph isomorphism from a tree remains NP-complete.

Using this idea, we propose an algorithm mining *frequent trees* w.r.t. partially injective homomorphisms. The rationale behind the choice of tree patterns is that the predictive performance achieved with frequent trees compares well to that of frequent connected subgraphs [13]. As the set of injectivity constraints depends on the particular pattern at hand, the output of the mining algorithm contains not only the tree patterns, but also the injectivity constraints.

A complete enumeration of *all* frequent patterns is, however, practically infeasible for the potentially huge number of injectivity constraint sets. We overcome this problem by considering only patterns which are *k-trees* [12], i.e., edge maximal graphs of tree-width at most k. Utilizing the algorithmic definition of k-trees (cf. [12]), we arrive at a natural refinement operator for the corresponding pattern mining problem, allowing for an *efficient* frequent pattern enumeration.

We have empirically evaluated the predictive performance and the runtime of our approach on real-world and artificial datasets. The predictive performance obtained by frequent subtrees w.r.t. partially injective homomorphism was very close to that of ordinary frequent subtrees and hence, to that of ordinary frequent subgraphs [13], and was achieved already for tree-width at most 3. Regarding the runtime, our algorithm is slower on molecular graphs than GASTON [10] and FSG [4] which seem to be specifically designed for this kind of graphs. However, already on slightly more complex structures beyond chemical graphs, our method always terminates and is faster by at least 1 (up to 3) orders of magnitude than GASTON and FSG (when they terminate at all). Thus, our approach offers a trade-off between runtime and predictive power via the choice of tree-width.

The rest of the paper is organized as follows. We collect the necessary notions in Sect. 2, introduce the concept of partially injective homomorphisms in Sect. 3, and present our mining algorithm in Sect. 4. We report our empirical results in Sect. 5 and conclude in Sect. 6. For page limitations, proofs are omitted in this short version.

2   Notions and Notation

In this section we collect the necessary notions from graph theory (see, e.g., [5]) and fix the notation. For a set S, let $[S]^2 = \{X \subseteq S : |X| = 2\}$. The set $\{u, v\} \in [S]^2$ is denoted by uv. An *undirected* (resp. *directed*) *labeled graph* over an alphabet Σ is a triple $G = (V, E, \ell)$ consisting of a set V of vertices, a set $E \subseteq [V]^2$ (resp. $E \subseteq V \times V$) of edges, and a labeling function $\ell : V \cup E \to \Sigma$ assigning a label from Σ to each vertex and edge. We often denote the set of vertices of G by $V(G)$ and the set of edges by $E(G)$. For simplicity we present our result for undirected graphs, by noting that any undirected graph G can be regarded as a directed graph such that all edges $uv \in E(G)$ are replaced by the directed edges (u, v) and (v, u). Accordingly, unless otherwise stated, by graphs we always mean undirected graphs.

A *homomorphism* from a graph $G = (V, E, \ell)$ into a graph $G' = (V', E', \ell')$ is a function $\varphi : V \to V'$ preserving all edges and labels, i.e., (i) $\varphi(u)\varphi(v) \in E'$ for all $uv \in E$, (ii) $\ell(v) = \ell'(\varphi(v))$ for all $v \in V$, and (iii) $\ell(uv) = \ell'(\varphi(u)\varphi(v))$ for all $uv \in E$. If, in addition, φ is *injective* then it is a *subgraph isomorphism* from G to G'. The graphs G and G' above are *isomorphic* if there exists a bijection φ between $V(G)$ and $V(G')$ such that φ and its inverse φ^{-1} are both homomorphisms. To enforce the morphisms above to satisfy certain injectivity constraints, the edges of the graphs, in addition to their labels, will have some color as well. In such cases homomorphisms (and hence, subgraph isomorphisms and isomorphisms) are required to preserve the edge colors as well. The definitions of homomorphism, subgraph isomorphism, and isomorphism between directed graphs are analogous with the additional constraint that φ must preserve not only the edges, but also their directions. As the generalization of our method from unlabeled graphs to labeled graphs is straightforward, for simplicity we present our algorithms for the unlabeled case.

We will pay a special attention to graphs of bounded *tree-width* [11]. For the definition and basic properties of bounded tree-width graphs, the reader is referred e.g. to [5]. Given a pattern graph H and a target graph G, it is NP-complete to decide whether there exists a homomorphism from H to G. If, however, the tree-width of H is bounded by some constant, then this problem can be decided in polynomial time for any graph G (see, e.g, [3]). For an integer $k > 0$, a k-*tree* [12] is defined recursively as follows: (i) A clique (i.e., fully connected graph) of $k + 1$ vertices is a k-tree. (ii) Given a k-tree G with n vertices, a k-tree with $n + 1$ vertices is obtained from G by adding a new vertex v to G and connecting v to all vertices of a k-clique (i.e., a clique of size k) of G. It is well-known that a k-tree has always tree-width k and that it is edge maximal w.r.t. this property, i.e., adding any further edge to a k-tree results in a graph of tree-width $k + 1$ (cf. [12]). Furthermore, all k-trees have $k|V(G)| - \binom{k+1}{2}$ edges. Finally, a *partial* k-*tree* is a subgraph of a k-tree and hence has tree-width at most k.

Let $\preceq$ be a preorder (i.e., reflexive and transitive relation) on a set S. For $a, b \in S$ we say that $a \prec b$ iff $a \preceq b$ and $a \not\succeq b$, and define the *equivalence* relation $\equiv$ on S by $a \equiv b$ iff $a \preceq b$ and $b \preceq a$. A function $\rho : S \to 2^S$ is called a *refinement operator* for $(S, \preceq)$ if for every $a \in S$ we have $\rho(a) \subseteq \{b \in S | a \preceq b\}$. Furthermore, ρ is (i) *locally finite* if $\rho(a)$ is finite for every $a \in S$, (ii) *complete* if for every $a, b \in S$ with $a \prec b$ there exist $a \equiv c_0, c_1, \ldots, c_n \equiv b$ with $c_i \in \rho(c_{i-1})$ for all $i = 1, \ldots, n$, (iii) *proper* if for all $a \in S$, $\rho(a) \subseteq \{b \in S | a \prec b\}$, and (iv) *ideal* if ρ is locally finite, complete and proper. For basic properties of refinement operators, the reader is referred e.g. to [9].

3   Partially Injective Homomorphisms

In this section we define *partially injective homomorphisms*, the central notion for this work, and discuss some of their properties. On the one hand, while the homomorphism and subgraph isomorphism problems are both NP-complete in general, their complexity behaves differently on special subproblems. In particular, homomorphism can be decided in polynomial time for a broad range of pattern classes for which subgraph isomorphism remains NP-complete. As an example, whereas the homomorphism problem from graphs of bounded tree-width can be decided in polynomial time [3], subgraph isomorphism remains NP-complete not only from trees (which have tree-width 1), but even from paths. On the other hand, however, as we empirically demonstrate in Sect. 5, frequent patterns generated w.r.t. subgraph isomorphism yield a much higher predictive performance when used for classification purposes than those w.r.t. homomorphism. As subgraph isomorphisms are injective homomorphisms, these empirical

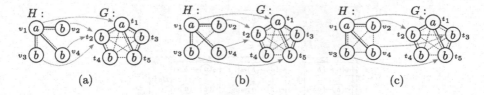

(a)   (b)   (c)

Fig. 1. Examples for different partially injective homomorphisms from H into G. Solid lines represent blue edges whereas dashed ones represent red edges.

results clearly highlight the importance of the *injectivity* of the pattern match-ing operator in practice. Motivated by our experiments, we propose a trade-off between complexity and predictive performance by bridging the gap between homomorphisms and subgraph isomorphisms. It follows from the remarks above that a natural transition from homomorphisms to subgraph isomorphisms can be obtained by *partial* injectivity, i.e., by requiring the injectivity constraint *not* for all vertex pairs in the pattern, but only for a subset of them. Formally, a pattern graph H can be embedded into a target graph G by a *partially injective homo-morphism* satisfying a set $\mathcal{C} \subseteq [V(H)]^2$ of injectivity constraints, denoted by $H \xrightarrow{\mathcal{C}} G$, if there exists a homomorphism φ from H to G such that $\varphi(u) \neq \varphi(v)$ for all $uv \in \mathcal{C}$. In case of ordinary homomorphisms, i.e., when $\mathcal{C} = \emptyset$, the above notation reduces to $H \to G$. We refer to the corresponding decision problem, denoted $\text{PIHom}(H, G, \mathcal{C})$, as PIHom *problem* and call the pair $(H, \mathcal{C})$ a PIHom *pattern*.

Our key idea is to consider such PIHom problems that can *polynomially* be reduced to *efficiently* decidable ordinary homomorphism problems. For the efficiency, we consider graphs of bounded tree-width, utilizing positive complex-ity results on deciding homomorphisms from this graph class [3]. Regarding the reduction, for H, G, and $\mathcal{C}$ above we transform H and G into two edge colored graphs by the following steps:

1. Color all (original) edges of H and G in *blue*,
2. for all $uv \in E(H) \cup \mathcal{C}$, connect u and v by a red edge, and
3. for *all* $u, v \in V(G)$ with $u \neq v$, connect u and v by a red edge.

Let the graphs obtained be denoted by $H\langle\mathcal{C}\rangle$ and $G\langle\top\rangle$. Since there is a one-to-one correspondence between the pair $(H, \mathcal{C})$ and the graph $H\langle\mathcal{C}\rangle$, we sometimes don't distinguish between the two notions. As homomorphisms between colored graphs preserve also the edge colors, the proof of the claim below is immediate from the definitions, by noting that all graphs considered in this work are loop free:

Proposition 1. *For $H, G, \mathcal{C}$ and $H\langle\mathcal{C}\rangle, G\langle\top\rangle$ above we have $H \xrightarrow{\mathcal{C}} G$ if and only if $H\langle\mathcal{C}\rangle \to G\langle\top\rangle$.*

Due to loop-freeness it suffices to consider the injectivity constraints only for unconnected vertex pairs in the original pattern graph H (i.e., we can assume w.l.o.g. that $\mathcal{C} \cap E(H) = \emptyset$). Note also that for H and G above, each subset

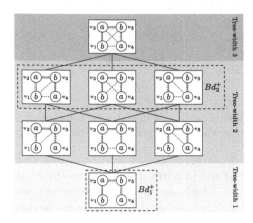

Fig. 2. Visualization of maximal borders in the lattice of PIHOM patterns for a path.

$C \subseteq [V]^2 \setminus E(H)$ defines a distinct problem $\text{PIHOM}(H, G, C)$ and that the set of all PIHOM patterns over H form a (complete) lattice $(\mathcal{L}_H, \preceq)$ with

$$\mathcal{L}_H = \{(H, C) : C \subseteq [V(H)]^2 \setminus E(H)\}$$

and with partial order $\preceq$ defined as follows: For all $C_1, C_2 \subseteq [V(H)]^2 \setminus E(H)$, $(H, C_1) \preceq (H, C_2)$ iff $C_1 \subseteq C_2$. The least element $(H, \emptyset)$ (resp. greatest element $(H, [V]^2 \setminus E(H))$) of $\mathcal{L}_H$ corresponds to ordinary homomorphism (resp. subgraph isomorphism) from H. For any target graph G, $(\mathcal{L}_H, \preceq)$ is closed downwards in the sense that $H \xrightarrow[C_1]{} G$ and $C_2 \subseteq C_1$ implies $H \xrightarrow[C_2]{} G$.

Figure 1 visualizes three partially injective homomorphisms from H into G. The graphs H and G are identical in all three examples, while the sets of injectivity constraints differ. We explicitly visualize the constraints in the depiction of patterns. For instance, the partially injective homomorphism depicted in Fig. 1(b) requires the homomorphism φ from H to G to satisfy the injectivity constraint $\varphi(v_2) \neq \varphi(v_3)$.

We will consider lattices of PIHOM *tree* patterns, i.e., when the first component in the PIHOM pattern (H, C) is a tree. In particular, when H is a tree with n vertices, the cardinality of the corresponding PIHOM pattern lattice $\mathcal{L}_H$ is $2^{O(n^2)}$, i.e., *exponential* in the size of H. Figure 2 illustrates such a lattice $(\mathcal{L}_H, \preceq)$ for a path H of length 3. Each depicted graph corresponds to a PIHOM tree pattern with a specific set of injectivity constraints.

4   Pattern Mining

In this section we define the problem of mining frequent maximally constrained PIHOM tree patterns. We first claim that there exists no locally finite and complete refinement operator for a natural representative set of this kind of patterns if we want to avoid redundancy. We therefore relax the problem definition and propose an efficient pattern mining algorithm tolerating redundancies in the output. The restriction of patterns to *trees* is motivated by complexity issues and by the remarkable predictive performance which is achieved by using frequent subtrees w.r.t. subgraph isomorphism [13]. In general, given a *tree* H and target graph G, $\text{PIHOM}(H, G, \mathcal{C})$ can be decided in polynomial time if $\mathcal{C} = \emptyset$ (i.e., for ordinary homomorphism), but is NP-complete whenever $\mathcal{C} = [V(H)]^2 \setminus E(H)$ (i.e., for subgraph isomorphism). We bridge this complexity gap by considering a distinguished subset of $\mathcal{L}_H$ for which the corresponding PIHOM problems are decidable in polynomial time for any target graph G. More precisely, for a tree H and some constant $k > 0$ we consider the pattern set $\mathcal{L}_H^k = \{(H, \mathcal{C}) \in \mathcal{L}_H : H\langle \mathcal{C}\rangle \text{ has tree-width at most } k\}$. Proposition 1 together with the positive complexity result on deciding homomorphisms from graphs of bounded tree-width [3] implies that for all $(H, \mathcal{C}) \in \mathcal{L}_H^k$ and for all target graphs G, $\text{PIHOM}(H, G, \mathcal{C})$ can be decided in polynomial time.

Our experiments (cf. Sect. 5) clearly demonstrate that, besides the structural gap between homomorphisms and subgraph isomorphisms discussed earlier, there is a large gap between their predictive performances as well. Furthermore, this gap vanishes as the number of injectivity constraints increases. Motivated by this empirical observation and the negative result formulated in Sect. 4.1 below, we will pay a special attention to the *maximal* elements of $\mathcal{L}_H^k$, i.e., to such PIHOM patterns $(H, \mathcal{C}) \in \mathcal{L}_H^k$ for which $H\langle \mathcal{C}\rangle$ is a complete graph whenever $|V(H)| \leq k + 1$; o/w it is a k-tree. In other words, we will consider the *positive border* Bd_k^+ of $(\mathcal{L}_H^k, \preceq)$ w.r.t. the following *interestingness predicate*: $(H, \mathcal{C}) \in \mathcal{L}_H$ is *interesting* if $H\langle \mathcal{C}\rangle$ has tree-width at most k.

As an example, all patterns $(H, \mathcal{C})$ with $|\mathcal{C}| = 2$ in Fig. 2 are maximally constrained w.r.t. tree-width 2, as they have tree-width 2 and adding any further constraint would increase their tree-width. (A 4-clique has tree-width 3.) The positive borders for $k = 1$ and $k = 2$ are denoted by Bd_1^+ and Bd_2^+, respectively. Clearly, $Bd_1^+ = \{(H, \emptyset)\}$ since adding any further injectivity constraint would result in a pattern of tree-width 2. In the example in Fig. 2 there are e.g. seven constraint sets for the path H connecting v_1 and v_4. For all of these constraint sets $\mathcal{C}$, $H\langle \mathcal{C}\rangle$ has tree-width at most 2. Yet, we are most interested in the three patterns with $|\mathcal{C}| = 2$ (i.e., Bd_2^+), as they have the highest degree of partial injectivity.

4.1   PIHom Core Patterns: A Negative Result

Using the concepts introduced above, in this section we consider the pattern language $\mathcal{L}^k$ defined by the union of the $\mathcal{L}_H^k$s over all tree patterns H. Given a database $\mathcal{D}$ of graphs, our goal is to generate a subset S of $\mathcal{L}^k$ on the basis of $\mathcal{D}$ that will span the binary feature space in which the elements of $\mathcal{D}$, as well as further unseen graphs will be embedded. For this application purpose, it is desirable to avoid "redundancies" among the patterns in S. There are various ways of defining redundancy. Perhaps the most natural one is the following definition: S contains *no* two *model equivalent* patterns, where the patterns $(H_1, \mathcal{C}_1), (H_2, \mathcal{C}_2) \in \mathcal{L}^k$ are model equivalent, denoted $(H_1, \mathcal{C}_1) \equiv_m (H_2, \mathcal{C}_2)$, if the equivalence $H_1 \xrightarrow[\mathcal{C}_1]{} G \iff H_2 \xrightarrow[\mathcal{C}_2]{} G$ holds for all graphs G.

For complexity reasons, we regard a relaxed notion of redundancy and show that even this weaker form raises severe algorithmic issues. More precisely, consider the order $\preceq_h$ on $\mathcal{L}^k$ defined as follows: For all $(H_1, \mathcal{C}_1), (H_2, \mathcal{C}_2) \in \mathcal{L}^k$, $(H_1, \mathcal{C}_1) \preceq_h (H_2, \mathcal{C}_2)$ iff $H_1 \langle \mathcal{C}_1 \rangle \to H_2 \langle \mathcal{C}_2 \rangle$. One can easily see that $\preceq_h$ is a preorder on $\mathcal{L}^k$. Using this definition, a set $S \subseteq \mathcal{L}^k$ is regarded as *non-redundant* if it contains no two *homomorphism equivalent* patterns, where $(H_1, \mathcal{C}_1), (H_2, \mathcal{C}_2) \in \mathcal{L}^k$ are homomorphism equivalent, denoted $(H_1, \mathcal{C}_1) \equiv_h (H_2, \mathcal{C}_2)$, iff $(H_1, \mathcal{C}_1) \preceq_h (H_2, \mathcal{C}_2)$ and $(H_2, \mathcal{C}_2) \preceq_h (H_1, \mathcal{C}_1)$. Clearly, $\equiv_m$ and $\equiv_h$ are both equivalence relations and $\equiv_h$ is finer than $\equiv_m$, i.e., the partition of $\mathcal{L}^k$ induced by $\equiv_h$ is a refinement of that induced by $\equiv_m$. Thus, model equivalent patterns are not necessarily homomorphism equivalent, implying that $\equiv_h$ may allow a certain amount of redundancies w.r.t. $\equiv_m$.

Let $\mathcal{L}^k / \equiv_h$ denote the set of equivalence classes induced by $\equiv_h$. For all $c \in \mathcal{L}^k / \equiv_h$, one can consider the *core* of c, a canonical representative element of c, defined as follows: Select an arbitrary pattern $(H, \mathcal{C}) \in c$ and take the smallest subgraph $H' \langle \mathcal{C}' \rangle$ of $H \langle \mathcal{C} \rangle$ such that $(H, \mathcal{C}) \equiv_h (H', \mathcal{C}')$. It holds that $H' \langle \mathcal{C}' \rangle$ (and hence, $(H', \mathcal{C}')$) can be computed by a greedy algorithm removing the redundant edges one by one. The properties of tree-width together with the results of [1, 3] imply that $(H', \mathcal{C}')$ is a core of c, it always exists and is unique modulo isomorphism independently of the choice of $(H, \mathcal{C})$, and can be calculated in time polynomial in the size of $(H, \mathcal{C})$. Let $\mathcal{L}_c^k$ be the set of such cores. The reason of using cores is that in such cases when the pattern matching operator is defined by (relational) homomorphism (e.g. in ILP [9]), the equivalence classes in $\mathcal{L}^k / \equiv_h$ may contain infinitely many patterns, due to the fact that $\preceq_h$ is not anti-symmetric.

While subgraph isomorphism as the pattern matching operator allows for a very natural refinement operator on the pattern language, this is typically not the case for homomorphism, caused also by the difference in the anti-symmetry. Indeed, in case of subgraph isomorphism, the pattern language along with the partial order defined by subgraph isomorphism can directly be translated into an *ideal* refinement operator (assuming that all patterns are connected); just extend the pattern at hand in every possible way either by a single edge or by a single vertex connected to one of the old vertices. In contrast, the preorder on $\mathcal{L}^k$ defined by homomorphism does not impose such an algebraic structure that could be turned into a (natural) algorithmic definition of a refinement operator on $\mathcal{L}_c^k$. In fact, $\mathcal{L}_c^k$ may contain cores having infinitely many "direct" refinements, implying the following negative result:

Theorem 1. *For all $k \geq 1$, there exists no finite and complete refinement operator for the preordered set $(\mathcal{L}_c^k, \preceq_h)$.*

The negative result formulated in Theorem 1 does not imply that PIHOM core patterns cannot be enumerated *efficiently*; this question is an open problem.[1] However, it indicates that traditional pattern generation paradigms based on refinement operators are not applicable to $(\mathcal{L}_c^k, \preceq_h)$. In the next section we therefore relax our problem setting and tolerate further redundancies in the output pattern set.

4.2   The Problem Definition

Theorem 1 implies that we have to consider a different pattern language in place of $\mathcal{L}_c^k$ if we want to generate the output patterns by using some algorithmically appropriate refinement operator. To achieve this goal, we consider PIHOM tree patterns that are *maximally* constrained w.r.t. tree-width k, i.e., the set $\mathcal{L}_{max}^k \subseteq \mathcal{L}^k$ defined as follows: For all $(H, \mathcal{C}) \in \mathcal{L}^k$, $(H, \mathcal{C}) \in \mathcal{L}_{max}^k$ iff $|V(H)| \leq k + 1$ and $H\langle\mathcal{C}\rangle$ is a complete graph or $|V(H)| > k + 1$ and $H\langle\mathcal{C}\rangle$ is a k-tree. The partially injective homomorphisms obtained in this way are as close as possible to subgraph isomorphism subject to bounded tree-width, resulting in a pattern set of higher predictive performance as shown empirically in Sect. 5. Using this definition, we consider the following pattern generation problem:

FREQUENT MAXIMALLY CONSTRAINED TREE MINING (FMCTM) PROBLEM:
   Given a finite set $\mathcal{D}$ of graphs and integers $t, h, k > 0$, *list* all $(H, \mathcal{C}) \in \mathcal{L}_{max}^k$ such that $|V(H)| \leq h$ and freq$((H, \mathcal{C}), \mathcal{D}) \geq t$, where freq$((H, \mathcal{C}), \mathcal{D})$ denotes the (absolute) *frequency* of the PIHOM tree pattern $(H, \mathcal{C})$ in $\mathcal{D}$, i.e., freq$((H, \mathcal{C}), \mathcal{D}) = |\{G \in \mathcal{D} : H \xrightarrow[\mathcal{C}]{} G\}|$.

PIHOM tree patterns satisfying the frequency constraint in the definition above will be referred to as *frequent* PIHOM tree patterns, or simply frequent patterns. As mentioned earlier, one of the most important distinguishing features of the problem setting above is that the pattern matching operator is *not* static (i.e., fixed in advance), but *dynamic*, in contrast to all traditional frequent graph mining algorithms. Clearly, whether a tree pattern is frequent or not directly depends on the underlying set of constraints applied in the embedding operator. It is therefore necessary to output a frequent tree along with the injectivity constraints defining the (dynamic) pattern matching operator, i.e., the output is always a pair $(H, \mathcal{C})$, instead of H only. Notice that in this work we are not explicitly interested in the semantical properties of the patterns but rather in the binary feature space spanned by them.

The parameter h in the problem definition provides an upper bound on the size of the output patterns. It ensures that the algorithm solving the FMCTM problem will always terminate. It is not difficult to see that without h, the output of the FMCTM problem may contain infinitely many frequent patterns. We also note that the output $\mathcal{O}$ of the FMCTM problem may contain frequent patterns

---

[1] In the long version of the paper we show that if the pattern language can further be restricted structurally then frequent cores w.r.t. a database cannot be generated in output polynomial time (unless P = NP).

Algorithm 1. LISTING FREQUENT MAXIMALLY CONSTRAINED TREES

---

input: graph dataset $\mathcal{D}$, integers $t, k, h > 0$
output: all t-frequent patterns of $\mathcal{L}_{\max}^k$ with size at least 1 and at most h

ENUMERATE$((H, \mathcal{C}))$:
  1: $R :=$ REFINEMENTS$((H, \mathcal{C}), k)$
  2: **for all** $(H', \mathcal{C}') \in R$ **do**
  3:     **if** $|V(H')| \leq h \wedge (H', \mathcal{C}') \notin \mathcal{O} \wedge \text{freq}((H', \mathcal{C}'), \mathcal{D}) \geq t$ **then**
  4:         **print** $(H', \mathcal{C}')$ and add it to $\mathcal{O}$
  5:         ENUMERATE$((H', \mathcal{C}'))$

MAIN:
  1: $\mathcal{O} := \emptyset$
  2: ENUMERATE$((\bot, \emptyset))$ // $\bot$ denotes the empty graph

---

that are homomorphism equivalent. Furthermore, the elements of $\mathcal{O}$ are not necessarily cores. In case the output is required to be a non-redundant subset of $\mathcal{L}_c^k$, after the computation of $\mathcal{O}$, one can first remove all patterns from it that are redundant w.r.t. homomorphism equivalence and then calculate the core for each pattern remaining in $\mathcal{O}$. Since all patterns in $\mathcal{O}$ have bounded tree-width, both steps can be performed in time polynomial in the size of $\mathcal{O}$.

4.3   The Mining Algorithm

In this section, we present our algorithm solving the FMCTM problem and prove that it is correct and enumerates the output patterns in incremental polynomial time. For the reader's convenience we ignore a number of (standard) optimization, implementation, and further simplification issues from this short paper and provide only a simplified version of the algorithm implemented.

We guarantee efficiency (i.e., incremental polynomial time) by considering the *partial* order $\preceq_{si}$ on $\mathcal{L}_{\max}^k$, instead of the *preorder* $\preceq_h$, where $\preceq_{si}$ is defined as follows: For all $(H, \mathcal{C}), (H', \mathcal{C}') \in \mathcal{L}_{\max}^k$, $(H, \mathcal{C}) \preceq_{si} (H', \mathcal{C}')$ iff there exists a subgraph isomorphism from $H\langle\mathcal{C}\rangle$ into $H'\langle\mathcal{C}'\rangle$. Clearly, $\text{freq}((H, \mathcal{C}), \mathcal{D}) \geq \text{freq}((H', \mathcal{C}'), \mathcal{D})$ whenever $H\langle\mathcal{C}\rangle \preceq_{si} H'\langle\mathcal{C}'\rangle$, i.e., frequency is *anti-monotonic* on the poset $(\mathcal{L}_{\max}^k, \preceq_{si})$. Thus, maximal PIHOM tree patterns are closed downwards w.r.t. frequency. While the poset $(\mathcal{L}_{\max}^k, \preceq_{si})$ allows for efficient pattern enumeration, the output may contain patterns that are homomorphism equivalent. That is, the price we have to pay for the positive complexity result is that the output may contain some redundant patterns.

Algorithm 1 is based on the recursive function ENUMERATE generating the output patterns in a DFS manner. Its input consists of the same parameters $\mathcal{D}$, t, h, and constant k as the FMCTM problem. The output patterns already generated are stored in the global variable $\mathcal{O}$. The algorithm calls ENUMERATE with the empty pattern $(\bot, \emptyset)$, where $\bot$ denotes the empty graph (line 2 of MAIN). As a first step (line 1 of ENUMERATE), function REFINEMENTS generates the set of refinements for the input pattern (H, C); the process governing how new candidate patterns are generated is determined by the refinement operator sketched below. If a newly generated candidate pattern (H', C') (i) fulfills the size constraint (i.e, $|V(H')| \leq h$), (ii) has not been generated before (i.e., $(H', C') \notin O$), and (iii) is t-frequent (i.e., freq$((H', C'), \mathcal{D}) \geq t$), we print it, store it in O, and call ENUMERATE recursively for this new frequent pattern (lines 3–5).

Refinement Operator. Function REFINEMENTS in Algorithm 1 returns the set R of *refinements* for a pattern $(H, C) \in \mathcal{L}_{\max}^k$ and $k > 0$. All patterns $(H', C') \in R$ are required to satisfy the following conditions: (i) H' is a supertree of H obtained by extending H with a new vertex and edge, (ii) $C \subseteq C'$, and (iii) $(H', C') \in \mathcal{L}_{\max}^k$, i.e, it is maximal w.r.t. tree-width k. That is, trees of size n are extended into trees of size $n + 1$ by condition (i). Furthermore, condition (iii) implies that $H'\langle C' \rangle$ is a k-tree if $|V(H')| > k + 1$; o/w it is a complete graph.

The algorithmic characterization of k-trees (cf. Sect. 2) gives rise to the following natural refinement operator on $\mathcal{L}_{\max}^k$: A pattern (H', C') of size $n + 1$ is among the refinements of a pattern $(H, C) \in \mathcal{L}_{\max}^k$ of size n iff (H', C') can be obtained from (H, C) in the following way: If $n = 0$ (i.e., $(H, C) = (\bot, \emptyset)$), we define the refinements of (H, C) by the set of graphs consisting of a single vertex (and no edges, as we consider loop-free graphs).[2] Otherwise, i.e., for $n > 0$, we proceed as follows: (i) Introduce a new vertex u. (ii) If $n \leq k$, then connect u to a vertex $v \in V(H)$ and add an injectivity constraint uv' to C for every $v' \in V(H) \setminus \{v\}$; o/w select a k-clique C in $H\langle C \rangle$, connect u to a vertex v of C in H, and add an injectivity constraint uv' to C for every $v' \in V(C) \setminus \{v\}$.

We are ready to state our main result:

Theorem 2. *For any $\mathcal{D}$, t, k, and h, Algorithm 1 is correct and generates the output patterns in incremental polynomial time.*

5 Experimental Evaluation

This section is concerned with the empirical evaluation of our approach. In particular, we evaluate the predictive performance of frequent maximally constrained PIHOM tree patterns and the runtime of the algorithm presented in Sect. 4. For our experiments we used a modified variant of Algorithm 1 making it practically feasible. We omit these technical details from this short version.

In the experiments below we first analyze how the degree of injectivity in PIHOM tree patterns impacts the resulting predictive performance. To answer this question, we fix a set of tree patterns and embed the data into feature spaces spanned by the tree patterns w.r.t. partially injective homomorphisms with increasing degree of injectivity, ranging from ordinary homomorphism to subgraph isomorphism. Our results clearly indicate a strong correlation between degree of injectivity and predictive performance. We then compare the predictive

---

[2] In case of labeled graphs, the number of such singleton graphs is equal to that of the different vertex labels in $\mathcal{D}$.

performance of the patterns generated by our practical algorithm with that of ordinary frequent subtrees on different benchmark datasets. We found that the predictive performance achieved by PIHOM tree patterns for already moderate amounts of injectivity constraints compares favorably to that of ordinary frequent subtrees, which, in turn, are very close to frequent subgraphs in predictive power. Finally, we provide runtime measures comparing our algorithm to state-of-the-art graph miners like GASTON [10] and FSG [4]. We show that while these algorithms are practical only for very restricted graph types, our approach performs well on *arbitrary* graph datasets. In particular, while our implementation was generally slower on the real-world benchmark chemical graph datasets considered, it clearly outperforms GASTON [10] and FSG [4] on artificial datasets containing only slightly more complex graph structures beyond molecular graphs.

Datasets. To evaluate the predictive performance, we use four benchmark molecular graph datasets. This choice is mainly of practical nature. In particular, the molecules considered are of a fairly simple structure [8], allowing for the application of state-of-the-art frequent subgraph mining systems that are generally very efficient on molecular graph data. This enables us to compare the predictive performance of our method to "gold-standard" results achieved by frequent subgraphs (w.r.t subgraph isomorphism).

The experiments are conducted on the well-established real-world datasets NCI1, NCI109, MUTAG, and PTC annotated for different binary target properties. NCI1 and NCI109 contain 4,110 (resp. 4,127) compounds. The graphs in these two datasets have 30 vertices and 32 edges on average. The MUTAG dataset contains merely 188 compounds; the graphs in this dataset have 18 vertices and 20 edges on average. Finally, the PTC dataset contains a total of 344 graphs, with an average number of 26 vertices and edges.

As most frequent subgraph mining systems like GASTON [10] are specifically designed to cope with graphs of simple structure (such as molecular graphs), runtime comparisons on only chemical graph datasets are not expressive enough. We therefore consider also artificial graph datasets generated according to the Erdős-Rényi random graph model. Each such dataset consists of 50 graphs with an average size of 25, corresponding to the chemical datasets above. The structural complexity of these graphs G corresponds to the edge/vertex ratio $q = |E(G)|/|V(G)|$. In our experiments, only connected graphs are considered.

Experimental Setup. Using 10-fold cross-validation, the predictive performance was measured in terms of AUC obtained by support vector machines (SVM) with the RBF kernel. In all experiments we used LIBSVM [2]. We applied SVM to the images of the graphs in the binary feature space spanned by the frequent patterns generated. The parameters of SVM (i.e. C and γ) were fixed throughout all classification tasks for a given database to avoid overfitting by chance. For all frequent pattern generation methods, we chose a frequency threshold value of 5%. Prior experiments have shown that patterns of fairly low sizes achieve the overall best predictive performances. We therefore limit the size of patterns to at most 9 vertices.

5.1   Degree of Injectivity vs. Predictive Performance

We first report our results investigating the influence of the degree of injectivity in PIHOM tree patterns on the predictive performance. To exclude possible side-effects caused by *different* pattern sets, for all datasets in our experiments we first fix a set S of tree patterns by selecting some random subset of the frequent trees generated w.r.t. subgraph isomorphism. Then, for all $H \in S$ we consider a PIHOM tree pattern $H_k = (H, C_k)$ that is maximally constrained w.r.t. tree-width k if $k = 1, \ldots, 4$; o/w it is fully constrained, i.e., $C_\top = [V(H)]^2 \setminus E(H)$. Since H is a tree, $C_1 = \emptyset$ and hence H_1 is the least element in the lattice $(\mathcal{L}_H, \preceq)$, corresponding to ordinary homomorphism from H. Analogously, $H_\top$ contains all possible injectivity constraints and is therefore the greatest element of $(\mathcal{L}_H, \preceq)$, corresponding to ordinary subgraph isomorphism from H. In this way we can simulate monotonically increasing degrees of injectivity in the pattern matching operator for a tree pattern H, leading from ordinary homomorphisms to subgraph isomorphisms. The degree of injectivity in the pattern matching operator is directly governed by k. For all $k \in \{1, \ldots, 4, \top\}$ we take the feature set $S_k = \{H_k : H \in S\}$ and embed all target graphs into the binary feature space spanned by S_k in the usual manner.

Figure 3 shows the predictive performances achieved for different degrees of injectivity defined by k. All four datasets indicate a significant difference between employing homomorphism ($k = 1$) and subgraph isomorphism ($k = \top$) as the

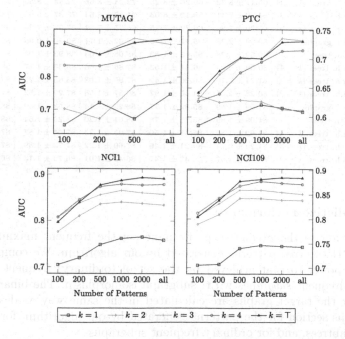

Fig. 3. Predictive performances in AUC for different degrees of injectivity governed by tree-width k. $k = \top$ corresponds to ordinary subgraph isomorphism.

pattern matching operators. While the results for MUTAG and PTC are fairly volatile due to their small size, with $k = 3$ and subgraph isomorphism yielding the best results, the datasets NCI1 and NCI109 clearly show a direct correlation between the degree of injectivity and predictive performance: Increasing values of k in almost every case yield an improvement in predictive performance. Notice that the gap between $k = 1$ (i.e., ordinary homomorphism) and $k = 2$ is already substantial for the datasets NCI1 and NCI109. For MUTAG and PTC it seems that $k = 3$ is necessary to significantly improve the predictive performance over homomorphism. In summary, a fairly low degree of injectivity already suffices to considerably outperform the predictive power of ordinary homomorphism.

Table 1. Prediction measures stated as AUC values in % for different tree-width choices k in contrast to freq. subgraphs and subtrees (s.g.i.: subgraph isomorphism, p.i.h.: partially injective homomorphism). Values are calculated by a SVM on feature vectors w.r.t. pattern graphs of size at most $|V|$.

Dataset	Frequent patterns	$|V| = 5$	$|V| = 6$	$|V| = 7$	$|V| = 8$	$|V| = 9$
MUTAG	s.g.i. graphs	73.18 ± 16.46	87.94 ± 10.70	88.29 ± 7.78	90.90 ± 6.83	91.99 ± 6.65
	s.g.i. trees	73.18 ± 16.46	84.07 ± 11.71	86.87 ± 6.23	89.84 ± 6.34	91.63 ± 5.89
	p.i.h. trees ($k = 4$)	73.18 ± 16.46	83.39 ± 11.20	87.10 ± 5.55	89.32 ± 7.66	90.49 ± 6.98
	p.i.h. trees ($k = 3$)	75.43 ± 13.69	83.51 ± 11.11	88.17 ± 6.38	89.15 ± 7.88	90.21 ± 8.36
	p.i.h. trees ($k = 2$)	61.44 ± 15.28	71.72 ± 10.08	76.03 ± 9.03	76.35 ± 12.28	76.92 ± 14.90
PTC	s.g.i. graphs	65.11 ± 8.32	66.82 ± 7.65	71.81 ± 8.27	73.69 ± 9.13	73.07 ± 9.34
	s.g.i. trees	65.11 ± 8.32	66.82 ± 7.65	71.91 ± 8.18	73.76 ± 9.09	73.08 ± 9.39
	p.i.h. trees ($k = 4$)	65.11 ± 8.32	69.26 ± 8.25	72.94 ± 8.55	72.54 ± 8.53	72.38 ± 8.02
	p.i.h. trees ($k = 3$)	64.30 ± 7.41	68.12 ± 8.03	71.87 ± 8.64	73.37 ± 9.01	73.24 ± 8.55
	p.i.h. trees ($k = 2$)	65.26 ± 7.80	66.79 ± 7.75	66.75 ± 7.31	67.53 ± 7.26	67.84 ± 6.53
NCI1	s.g.i. graphs	80.59 ± 2.28	85.97 ± 1.89	88.32 ± 1.11	89.20 ± 0.96	89.33 ± 1.13
	s.g.i. trees	80.59 ± 2.28	85.65 ± 1.83	88.08 ± 1.06	88.93 ± 1.03	89.14 ± 1.19
	p.i.h. trees ($k = 4$)	80.59 ± 2.28	86.41 ± 1.53	87.86 ± 1.08	88.35 ± 1.17	88.30 ± 1.29
	p.i.h. trees ($k = 3$)	81.74 ± 2.10	86.26 ± 1.42	87.85 ± 1.08	88.65 ± 1.23	88.77 ± 1.32
	p.i.h. trees ($k = 2$)	80.28 ± 2.51	84.45 ± 1.97	86.37 ± 1.58	87.27 ± 1.52	87.68 ± 1.35
NCI109	s.g.i. graphs	81.09 ± 1.89	85.82 ± 1.77	88.27 ± 1.14	88.53 ± 1.60	88.55 ± 1.77
	s.g.i. trees	81.09 ± 1.89	85.75 ± 1.78	88.01 ± 1.20	88.29 ± 1.54	88.37 ± 1.72
	p.i.h. trees ($k = 4$)	81.09 ± 1.89	86.38 ± 1.50	87.63 ± 1.73	87.64 ± 1.87	87.54 ± 2.03
	p.i.h. trees ($k = 3$)	81.24 ± 0.74	86.21 ± 1.46	87.60 ± 1.22	87.72 ± 1.89	87.77 ± 2.02
	p.i.h. trees ($k = 2$)	80.15 ± 2.07	84.43 ± 1.47	86.42 ± 1.01	86.79 ± 1.13	86.72 ± 1.66

5.2   Predictive Performance

We also evaluate the predictive performance of the frequent maximally constrained PIHOM tree patterns generated by our algorithm. We compare their predictive power to that achieved by the set of (ordinary) frequent subtrees, as well as frequent *subgraphs* w.r.t. subgraph isomorphism. The binary feature vectors for the target graphs are calculated in the same way as described in the previous section for the patterns generated by our algorithm, for ordinary frequent subtrees, and for ordinary frequent subgraphs.

Table 1 shows the predictive performance of our approach for different pattern sizes (cf. the last five columns). The full sets of frequent subgraphs perform slightly better than its subset formed by frequent subtrees. Interestingly, already for tree-width $k = 3$, the absolute difference (in AUC) to frequent subgraphs (cf. row "s.g.i. graphs") and to ordinary frequent subtrees (cf. row "s.g.i. trees") is marginal for all cases, but one (MUTAG with $|V| = 5$). In all cases, the differences are statistically insignificant (for $p = 0.05$). Hence, our approach offers an attractive trade-off between runtime (depending on k) and predictive power.

Table 2. Runtimes (in sec.) of our algorithm in comparison to GASTON and FSG on molecular and artificial datasets. Σ denotes the set of vertex labels. The cases in which GASTON ran out of the necessary memory are marked by "*mem err*". The maximum size of patterns to be found was set to 10 and the minimum support was set to 5% in all cases. Note that FSG does not restrict its search to trees but considers general graphs, in contrast to all other algorithms.

	MUTAG	PTC	NCI1	NCI109	Erdős-Rényi random graphs							
					$|\Sigma| = 1$				$|\Sigma| = 2$			
					$q = 1.0$	$q = 1.5$	$q = 2.0$	$q = 3.0$	$q = 1.0$	$q = 1.5$	$q = 2.0$	$q = 3.0$
GASTON (EL)	0.1	0.2	2.6	2.7	2.8	54.6	*mem err*	*mem err*	1.8	20.1	1168.9	56464.6
GASTON (RE)	0.3	1.0	8.5	8.6	5.0	39.5	1163.0	31061.4	5.7	44.9	1120.2	24778.0
FSG	0.7	4.1	30.2	29.9	194.2	10584.9	10888.4	10852.5	19.9	82.8	2375.7	58816.2
PIH Miner ($k = 3$)	0.8	3.4	27.3	24.1	0.3	1.3	2.8	8.4	5.3	16.3	149.9	622.8

5.3 Runtimes

We measured the runtimes of our algorithm and compared them to those achieved with the state-of-the-art graph miners GASTON [10] (using the graph counting methods *embedding lists* (EL) and *recomputed embeddings* (RE)) and FSG [4] on real-world and artificial datasets. While molecules, having roughly as many edges as vertices (i.e., having $q \approx 1.0$), q is up to 3.0 in the artificial datasets.

Table 2 shows the running times for each algorithm and dataset. As expected, GASTON and FSG perform very well on molecular graphs, compared to our algorithm. However, for slightly more complex structures (i.e. $q \geq 2.0$) the two traditional graph miners become quickly infeasible for the correlation between the number of embeddings and runtime. Our approach (referred to as *PIH Miner* in Table 2) clearly outperforms FSG and GASTON on both artificial datasets for $q \geq 2.0$.

6 Concluding Remarks

To bridge the gap between homomorphisms (ILP) and subgraph isomorphisms (graph mining), we proposed partially injective homomorphisms, a new kind of *dynamic* pattern matching operator. We considered the efficiently enumerable fragments of frequent maximally constrained PIHOM tree patterns w.r.t.

bounded tree-width and showed on benchmark molecular graph datasets that their predictive performance is close to that of ordinary frequent subtrees (and hence, to that of subgraphs as well). Since our algorithm does not assume any structural properties on the input graphs, it is effectively applicable also to such transaction graphs where most state-of-the-art pattern mining algorithms become infeasible.

Our approach raises several questions. Perhaps the most interesting one is the extension to more general relational vocabularies, bridging another gap between graph mining and ILP in the richness of the relational vocabularies. To formulate this generalization, we note that any $\text{PIHOM}(H, G, \mathcal{C})$ problem can polynomially be reduced to θ-subsumption between DATALOG goal clauses (or equivalently, Boolean conjunctive queries) as follows: Using a relational vocabulary Σ consisting of binary and unary predicate symbols only, represent $(H, \mathcal{C})$ as a DATALOG goal clause Q_H with a body composed of a set of literals for H and another set of literals of a distinguished binary predicate for the injectivity constraints in $\mathcal{C}$. Each vertex of H is represented by a unique variable in Q_H. For a target graph G, represent the PIHOM pattern $(G, [V(G)]^2)$ in a similar way by a clause Q_G over the same Σ. It holds that Q_H θ-subsumes Q_G iff $H \underset{c}{\rightarrow} G$. From this reduction and our empirical results it is immediate that the predictive performance of DATALOG goal clauses as patterns can also be improved by adding injectivity constraints (i.e., literals of a distinguished binary predicate) to their bodies. Considering our approach developed for graphs from the above logical viewpoint, it would be interesting to generalize it to relational vocabularies containing predicate symbols of *arbitrary* arities by considering such fragments of Boolean conjunctive queries for which θ-subsumption is in P (cf. [6]).

References

1. Chandra, A.K., Merlin, P.M.: Optimal implementation of conjunctive queries in relational data bases. In: Proceedings of the 9th Annual ACM Symposium on Theory of Computing, (STOC), pp. 77–90. ACM Press (1977)
2. Chang, C.-C., Lin, C.-J.: LIBSVM: a library for support vector machines. ACM Trans. Intell. Syst. Technol. **2**(3), 27 (2011)
3. Dalmau, V., Kolaitis, P.G., Vardi, M.Y.: Constraint satisfaction, bounded treewidth, and finite-variable logics. In: Van Hentenryck, P. (ed.) CP 2002. LNCS, vol. 2470, pp. 310–326. Springer, Heidelberg (2002). https://doi.org/10.1007/3-540-46135-3_21
4. Deshpande, M., Kuramochi, M., Wale, N., Karypis, G.: Frequent substructure-based approaches for classifying chemical compounds. Trans. Knowl. Data Eng. **17**(8), 1036–1050 (2005)
5. Diestel, R.: Graph Theory. Springer, Berlin (2000)
6. Gottlob, G., Leone, N., Scarcello, F.: Hypertree decompositions and tractable queries. J. Comput. Syst. Sci. **64**(3), 579–627 (2002)
7. Horváth, T., Turán, G.: Learning logic programs with structured background knowledge. Artif. Intell. **128**(1–2), 31–97 (2001)
8. Horváth, T., Ramon, J.: Efficient frequent connected subgraph mining in graphs of bounded tree-width. Theor. Comput. Sci. **411**(31), 2784–2797 (2010)

9. Nienhuys-Cheng, S.-H., de Wolf, R.: Foundations of Inductive Logic Programming. LNCS, vol. 1228. Springer, Heidelberg (1997). https://doi.org/10.1007/3-
   540-62927-0
10. Nijssen, S., Kok, J.N.: The Gaston tool for frequent subgraph mining. Electron.
    Notes Theor. Comput. Sci. **127**(1), 77–87 (2005)
11. Robertson, N., Seymour, P.D.: Graph minors. II. Algorithmic aspects of tree-width.
    J. Algorithms **7**(3), 309–322 (1986)
12. Rose, D.J.: On simple characterizations of k-trees. Discret. Math. **7**(3–4), 317–322
    (1974)
13. Welke, P., Horváth, T., Wrobel, S.: Probabilistic frequent subtrees for efficient
    graph classification and retrieval. Mach. Learn. **107**(11), 1847–1873 (2018)

Probabilistic Models and Statistical Methods

Variational Bayes for Mixture Models with Censored Data

Masahiro Kohjima$^{(\boxtimes)}$, Tatsushi Matsubayashi, and Hiroyuki Toda

NTT Service Evolution Laboratories, NTT Corporation, Yokosuka, Kanagawa, Japan
{kohjima.masahiro,matsubayashi.tatsushi,toda.hiroyuki}@lab.ntt.co.jp

Abstract. In this paper, we propose a variational Bayesian algorithm for mixture models that can deal with censored data, which is the data under the situation that the exact value is known only when the value is within a certain range and otherwise only partial information is available. The proposed algorithm can be applied to any mixture model whose component distribution belongs to exponential family; it is a natural generalization of the variational Bayes that deals with "standard" samples whose values are known. We confirm the effectiveness of the proposed algorithm by experiments on synthetic and real world data.

Keywords: Variational Bayes · Mixture models · Censoring Censored data

1 Introduction

Censoring is the situation that the exact value is known only when the value is within a certain range and otherwise only partial information is available [1]. It is quite common such as data about lifetimes, for example, human life span, mechanical failure [2], active period of smartphone apps [3] and users' contract periods of telecommunication and e-commerce service [4,5]. Most observation periods are limited and many samples may not expire within the period. Therefore, their true lifetimes are unknown and they become censored samples (Fig. 1a). Obviously, methods that can handle censored data are essential in fields such as insurance design, facility maintenance and marketing.

One promising approach for dealing with censored data is to adapt *mixture models* since the probability density underlying censored data often exhibits multimodality. For example, the distribution of machine failure is formed by initial failures and aging failures. For the estimation of mixture models from "standard" data (all samples of which have known values), variational Bayes (VB) [6] is a commonly used algorithm. However, VB for mixture models using censored data has received little attention in the literature.

In this study, we propose the VB algorithm that can deal with Censored data for Mixture models (VBCM). VBCM estimates the variational distribution of parameters and that of two types of latent variables that are indicative of the cluster assignments and the (unobserved) values of censored samples. It is shown

© Springer Nature Switzerland AG 2019
M. Berlingerio et al. (Eds.): ECML PKDD 2018, LNAI 11052, pp. 605–620, 2019.
https://doi.org/10.1007/978-3-030-10928-8_36

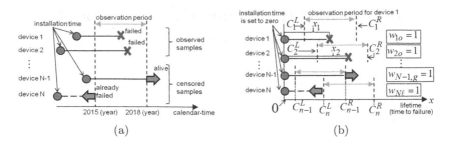

Fig. 1. Example of censored data which consist of observed samples and censored samples. (a) calendar-time and (b) lifetime representation.

that the update equations in VBCM use the expectation of sufficient statistics w.r.t. truncated distribution and cumulative density function, neither of which appear in the standard VB (VB with "standard" data). We also confirm that VBCM, inheriting a property from VB, the quantity used for model selection is analytically tractable and yields predictive distributions in analytic form. We derive the VBCM algorithm for (i) Gaussian mixture models (GMM) and more general (ii) mixture models whose components belong to the exponential family (EFM). VBCM can be seen as the generalization of VB since, if there are no censored samples, it reduces to VB.

We further extend the derived VBCM algorithm to develop a stochastic algorithm. As stated above, the update equation includes expected sufficient statistics; by replacing these with stochastically computed values clearly turns VBCM into a stochastic algorithm. By combining with the existing stochastic VB approach [7–9], we develop a stochastic VBCM, which can be seen, theoretically, as a natural gradient algorithm [10].

We conduct experiments on both synthetic and real data. The results show the superiority of the proposed algorithm in over existing VB and EM based algorithms where test log-likelihood is used as the performance metric.

The contributions of this paper are summarized below:

- We develop VBCM, a VB algorithm that can deal with censored data for mixture models. We also show that the objective of VBCM, which can be used for model selection, is analytically tractable and its predictive distribution have analytic form.
- We extend VBCM to the stochastic algorithm. It is theoretically proven that the algorithm is a (stochastic) natural gradient algorithm.
- We confirm the effectiveness of the proposed VBCM algorithm by numerical experiments using both synthetic data and real data.

The rest of this paper is organized as follows. In Sect. 2, we introduce related works. Mixture models and the generative process of censored data are illustrated in Sect. 3. Section 4 presents the proposed VBCM algorithm for GMM and its extension to EFM and to a stochastic algorithm is provided in Sect. 5. Section 6 is devoted to the experimental evaluation and Sect. 7 concludes the paper.

2   Related Works

Researchers continue to develop new algorithms for survival analysis based on promising machine learning algorithms. For example, Pölsterl et al. extend the support vector machine for censored data [11] and Fernandez et al. propose a method based on Gaussian processes [12]. Our study follows this context and we consider how to extend VB for analyzing censored data.

To model censored data, we introduce latent variables to handle the values of censored samples. The idea is not new as is shown by Dempster's paper [13]. Starting with mixture models that process censored data, one more latent variable indicating cluster assignment is used to develop the expectation-maximization algorithm called EMCM [14,15]. We use these latent variables to develop a new VB algorithm with Censored data for Mixture models (VBCM).

We also derive stochastic algorithm variant of VBCM; it stochastically computes expected sufficient statistics, an necessary step for parameter update in VBCM. Stochastic algorithms are also an active research topic in machine learning and stochastic VB algorithms have been published e.g., [7,9]. The above papers use the term "stochastic" to indicate mainly the use of parts of stochastically selected/generated training data. We follow this concept in developing a stochastic VBCM.

3   Mixture Models for Censored Data

3.1   Data Description

Here, we describe censored data. We explain using failure data shown in Fig. 1a since the data is the representative of censored data. Let N be the number of devices. For each device, its installation time is known. The failure time of a device is exactly known if it fails within the known observation period. If the device is alive at the end of the period, we know only that its failure time exceeds the end of period. Similarly, if the device is already broken at the beginning of the period, we know only its failure time precedes the beginning. We call the devices whose failure time is exactly observed as *observed samples* and the devices whose failure time is *not* exactly observed as *censored samples*.

We consider that the censored data is to be processed so that we can predict lifetime (time to failure), as shown in Fig. 1b. We denote the range of the observation period, where the lifetime of the i-th device is exactly observed, as $C_i = (C_i^L, C_i^R)$. This definition covers the setting that the period length depends on a device unlike Fig. 1a. We also denote the lifetime of the i-th observed sample (device) as x_i. Let $w_i = (w_{io}, w_{i\ell}, w_{ig})$ be the indicator variable which is set to $w_i = (1, 0, 0)$ if the lifetime of the i-th device is observed, $w_i = (0, 1, 0)$ if it precedes C_i^L, and $w_i = (0, 0, 1)$ if it is exceeds C_i^R. Then, the observed variables are $W = \{w_i\}_{i=1}^N, X = \{x_i\}_{i \in \mathcal{I}_o}$, where $\mathcal{I}_o$ is the set of observed samples. Similarly, we define the set of samples whose values precede (exceed) C_i^L (C_i^R) as $\mathcal{I}_\ell$ ($\mathcal{I}_g$). Note that target of this study is not limited to lifetime estimation and includes any type of problem wherein the data contains censored samples.

3.2   Models

We use mixture models to estimate the distribution underlying the censored data. The probability density function (PDF) of mixture models is defined as the linear summation of component distribution f,

$$P(x|\Theta = \{\boldsymbol{\pi}, \boldsymbol{\varphi}\}) = \sum_{k=1}^{K} \pi_k f(x|\varphi_k), \tag{1}$$

where K is the number of components and Θ represents the model parameters. $\boldsymbol{\pi} = \{\pi_1, \cdots, \pi_K\}$ and $\boldsymbol{\varphi} = \{\varphi_1, \cdots, \varphi_K\}$ represent the k-th component's mixing ratio and parameters, respectively. Examples of component f are (i) Gaussian distribution, f_G and, more generally, (ii) a distribution belonging to the exponential family, f_E, whose PDFs are defined as

$$f_G(x|\varphi_k := (\mu_k, \lambda_k)) = \mathcal{N}(x|\mu_k, \lambda_k^{-1}) = \sqrt{\frac{\lambda_k}{2\pi}} \exp\left(-\frac{\lambda_k}{2}(x - \mu_k)^2\right), \tag{2}$$

$$f_E(x|\varphi_k := (\eta_k)) = h(x) \exp\left(\eta_k \cdot T(x) - A(\eta_k)\right), \tag{3}$$

where μ_k and λ_k are the mean and precision, and η_k is the natural parameter. $T(x)$ is *sufficient statistics* and $A(\eta_k)$ is the log-normalizer. The exponential family includes various type of distributions such as exponential distribution and Poisson distribution:

$$\mathrm{Exp}(x|\lambda_k) = \lambda_k \exp(-\lambda_k x), \quad \mathrm{Pois}(x|\lambda_k) = \frac{\lambda_k^x \exp(-\lambda_k)}{x!}. \tag{4}$$

By assigning specific values to $T(x)$ and $A(\eta_k)$, the above densities are represented by Eq. (3) (See Table 1). We call the mixture models whose components are Gaussian and exponential family the Gaussian mixture models (GMM) and exponential family mixtures (EFM), respectively. Our notation f is the component distribution without distinction as to which distribution is adopted. We also denote the cumulative density function (CDF) of f as F:

$$F(C|\varphi_k) = \int_{-\infty}^{C} f(x|\varphi_k) dx. \tag{5}$$

The generative process of censored data using mixture models consists of 5 steps. At first, parameters $\Theta = \{\boldsymbol{\pi}, \boldsymbol{\varphi}\}$ follow prior distribution $P(\Theta) = P(\boldsymbol{\pi})P(\boldsymbol{\varphi})$. $P(\boldsymbol{\pi})$ is the Dirichlet distribution

$$P(\boldsymbol{\pi}) = \mathrm{Dirichlet}(\boldsymbol{\pi}|\alpha_0) = \frac{\Gamma(K\alpha_0)}{\Gamma(\alpha_0)^K} \prod_{k=1}^{K} \pi_k^{\alpha_0 - 1}, \tag{6}$$

and $P(\boldsymbol{\varphi})$ is a conjugate prior of component distribution. It is the Gaussian-gamma distribution when the component distribution is Gaussian:

$$P(\boldsymbol{\mu}, \boldsymbol{\lambda}) = \prod_{k=1}^{K} \mathrm{NormGam}(\mu_k, \lambda_k|\mu_0, \tau_0, a_0, b_0) \tag{7}$$

$$= \prod_{k=1}^{K} \mathcal{N}(\mu_k|\mu_0, (\tau_0 \lambda_k)^{-1}) \underbrace{\frac{1}{\Gamma(a_0)} b_0^{a_0} \lambda_k^{a_0 - 1} \exp(-b_0 \lambda_k)}_{\mathrm{Gam}(\lambda_k|a_0, b_0)}. \tag{8}$$

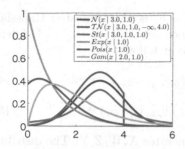

Fig. 2. Probability distributions

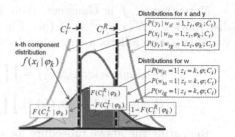

Fig. 3. Distributions for W, X, Y

The conjugate prior for exponential family distribution exists and is defined as

$$P(\boldsymbol{\eta}) = \prod_{k=1}^{K} g(\eta_k|\xi_0, \nu_0) = \prod_{k=1}^{K} \mathcal{Z}(\xi_0, \nu_0) \exp\left(\eta_k \cdot \xi_0 - \nu_0 A(\eta_k)\right), \qquad (9)$$

where the normalizer $\mathcal{Z}(\xi_0, \nu_0)$ is defined depending on the component distribution used (See Table 1). It follows that the prior for GMM is $P(\boldsymbol{\pi})P(\boldsymbol{\mu}, \boldsymbol{\lambda})$ and for EFM is $P(\boldsymbol{\pi})P(\boldsymbol{\eta})$. Note that $\alpha_0, \mu_0, \tau_0, a_0, b_0, \xi_0, \nu_0$ are the hyperparameters[1].

In the second step, for each i-th sample, latent variable $z_i = \{z_{i1}, \cdots, z_{iK}\}$ which indicates the belonging component[2], follows a multinomial distribution:

$$P(z_i|\boldsymbol{\pi}) = \text{Mult}(z_i|\boldsymbol{\pi}) = \prod_{k=1}^{K} \pi_k^{z_{ik}}. \qquad (10)$$

The third step is to generate observed variable w_i, which indicates an exact value is observed or is less/larger than C_i^L/C_i^R. Variable w_i follows a multinomial distribution whose parameters are defined from the CDF of component distribution since e.g., $F(C_i^L|\varphi_k)$ is the probability that random variable following $f(\cdot|\varphi_k)$ takes a value less than C_i^L (See Fig. 3):

$$P(w_i|z_i, \boldsymbol{\varphi}; C_i) = \prod_{k=1}^{K} \left\{ F(C_i^L|\varphi_k)^{w_{i\ell}} \left(F(C_i^R|\varphi_k) - F(C_i^L|\varphi_k)\right)^{w_{io}} \left(1 - F(C_i^R|\varphi_k)\right)^{w_{ig}} \right\}^{z_{ik}} \qquad (11)$$

In the final step, if $w_{io} = 1$, observed variable $x_i \in (C_i^L, C_i^R)$ follows *truncated* component distribution:

$$P(x_i|w_{io} = 1, z_i, \boldsymbol{\varphi}; C_i) = \prod_{k=1}^{K} f_{tr}(x_i|\varphi_k, C_i^L, C_i^R)^{z_{ik} w_{io}}, \qquad (12)$$

where *truncated* distribution $f_{tr}(x|\varphi, a, b)$ is defined as the distribution whose PDF is given by:

$$f_{tr}(x|\varphi, a, b) = \begin{cases} \left(F(b|\varphi) - F(a|\varphi)\right)^{-1} f(x|\varphi) & \text{(if } x \in (a, b]) \\ 0 & \text{(otherwise)} \end{cases} \qquad (13)$$

---

[1] We set $\alpha_0 = 1.0$, $\mu_0 = 0.0$, $\tau_0 = 0.001$, $a_0 = b_0 = 1.0$, $\xi_0 = \nu_0 = 1.0$ in the experiments.

[2] If the i-th sample belongs to the k-th component, $z_{ik} = 1$ and $z_{ik'} = 0$ for all $k' \neq k$.

Note that, if f is Gaussian, then $f_{tr}(x|\varphi, a, b)$ is the truncated Gaussian distribution $\mathcal{TN}(x|\mu, \sigma^2, a, b)$ (See Fig. 2). When $w_{i\ell} = 1$ or $w_{ig} = 1$, latent variable y_i also follows a truncated distribution as follows:

$$P(y_i|w_{i\ell} = 1, z_i, \varphi; C_i) = \prod_{k=1}^{K} f_{tr}(y_i|\varphi_k, -\infty, C_i^L)^{z_{ik}w_{i\ell}}, \qquad (14)$$

$$P(y_i|w_{ig} = 1, z_i, \varphi; C_i) = \prod_{k=1}^{K} f_{tr}(y_i|\varphi_k, C_i^R, \infty)^{z_{ik}w_{ig}}. \qquad (15)$$

Repeating the above procedure for all i generates X, W, Z, Y. The distributions used for generating X, W, Y are illustrated in Fig. 3. Using Eqs. (6)–(15), the joint probability of the variables and the parameters is given by

$$P(X, W, Y, Z, \Theta) = P(\Theta)P(Z|\pi)P(W|Z, \varphi)P(X, Y|W, Z, \varphi)$$

$$= P(\Theta)\left(\prod_{i=1}^{N} P(z_i|\pi)P(w_i|z_i, \varphi)\right)\left(\prod_{i\in\mathcal{I}_o} P(x_i|w_{io} = 1, z_i, \varphi)\right) \cdot \qquad (16)$$

$$\left(\prod_{i\in\mathcal{I}_\ell} P(y_i|w_{i\ell} = 1, z_i, \varphi)\right)\left(\prod_{i\in\mathcal{I}_g} P(y_i|w_{ig} = 1, z_i, \varphi)\right),$$

where dependency on C_i is omitted. Figure 4 shows a graphical model. In the experiments, we also use the following likelihood of observed variables:

$$P(X, W|\Theta) = \int P(Z|\pi)P(W|Z, \varphi)P(X, Y|W, Z, \varphi)dY\, dZ \qquad (17)$$

$$= \left(\prod_{i\in\mathcal{I}_o} P(x_i|\Theta)\right)\left(\prod_{i\in\mathcal{I}_\ell} \int_{-\infty}^{C_i^L} P(y|\Theta)dy\right)\left(\prod_{i\in\mathcal{I}_g} \int_{C_i^R}^{\infty} P(y|\Theta)dy\right).$$

Table 1. Examples of exponential family and their conjugate priors.

dist.	η	$h(x)$	$T(x)$	$A(\eta)$	prior $g(\eta|\xi, \nu)$	$Z(\xi, \nu)$	$E_{g(\eta|\xi,\nu)}[A(\eta)], \bar{\eta}$
$\mathrm{Exp}(x|\lambda)$	$-\lambda$	1	x	$-\log(-\eta)$	$\mathrm{Gam}(\lambda|\nu+1, \xi)$	$\frac{\xi^{(\nu+1)}}{\Gamma^{(\nu+1)}}$	$\log(\xi) - \psi(\nu+1), -\frac{\nu+1}{\xi}$
$\mathrm{Pois}(x|\lambda)$	$\log(\lambda)$	$1/x!$	x	$\exp(\eta)$	$\frac{\xi}{\nu}\mathrm{Gam}(\lambda|\xi+1, \nu)$	$\frac{\nu^{(\xi)}}{\Gamma^{(\xi)}}$	$\frac{\xi}{\nu}, \psi(\xi) - \log(\nu)$

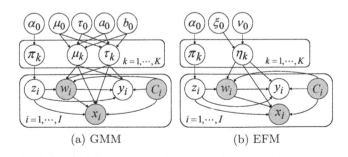

(a) GMM   (b) EFM

Fig. 4. Graphical models of (a) Gaussian mixtures and (b) mixture of exponential family. Shaded nodes indicate observed variables.

4 Proposed Variational Bayes

4.1 General Formulation and Derivation for GMM

Variational Bayes (VB) is a method that estimates variational distribution $q(\pi, \varphi, Z, Y)$, which approximates a posterior of parameters and latent variables [6,16]. VB needs to decide the factorization form of the variational distribution. Our VB algorithm considers the distribution in which each parameter is independent of the latent variables but Y and Z are dependent: $q(\pi, \varphi, Z, Y) = q(\pi)q(\varphi)q(Y, Z)$. This factorization form yields a tractable algorithm. Under the above factorization form, a variational distribution can be estimated by maximizing the functional $\mathcal{L}$, which is defined as

$$\mathcal{L}[q] = \mathbb{E}_{q(\pi)q(\varphi)q(Y,Z)}[\log P(X, W, Y, Z, \Theta) - \log q(\pi)q(\varphi)q(Y, Z)]. \tag{18}$$

We detail this functional in Sect. 4.2. From the optimality condition inherent in the variational method, the variational distribution that maximizes $\mathcal{L}$ must satisfy $q(\pi) \propto \exp\left(\mathbb{E}_{q(\varphi)q(Y,Z)}[\log P(X, W, Y, Z, \theta)]\right)$, $q(\varphi) \propto \exp\left(\mathbb{E}_{q(\pi)q(Y,Z)}[\log P(X, W, Y, Z, \theta)]\right)$ and $q(Y, Z) \propto \exp\left(\mathbb{E}_{q(\pi)q(\varphi)}[\log P(X, W, Y, Z, \theta)]\right)$.

For the Gaussian mixture models (GMM), $q(\pi), q(\mu, \lambda), q(Y, Z)$ follow Dirichlet, Gaussian-gamma and mixture of truncated Gaussian, respectively:

$$q(\pi) = \text{Dirichlet}(\pi|\alpha_k), \quad \alpha_k = \alpha_0 + \bar{N}_k, \quad \bar{N}_k = \sum_{i=1}^{N} \bar{z}_{ik}, \tag{19}$$

$$q(\mu, \lambda) = \prod_{k=1}^{K} \text{NormGam}\left(\mu_k, \lambda_k | \bar{\mu}_k, \tau_k, a_k, b_k\right), \tag{20}$$

$$\bar{\mu}_k = \tau_k^{-1}(\tau_0 \mu_0 + \bar{N}_k \bar{x}_k), \quad \tau_k = \tau_0 + \bar{N}_k, \tag{21}$$

$$a_k = a_0 + \frac{\bar{N}_k}{2}, \quad b_k = b_0 + \frac{1}{2}\left\{\tau_0 \mu_0^2 - \tau_k \bar{\mu}_k^2 + \bar{N}_k \bar{x}_k^{(2)}\right\}, \tag{22}$$

$$\bar{x}_k = \frac{1}{\bar{N}_k}\left\{\sum_{i \in \mathcal{I}_o} \bar{z}_{ik} x_i + \sum_{i \in \mathcal{I}_\ell} \bar{z}_{ik} \bar{y}_{ik}^\ell + \sum_{i \in \mathcal{I}_g} \bar{z}_{ik} \bar{y}_{ik}^g\right\}, \tag{23}$$

$$\bar{x}_k^{(2)} = \frac{1}{\bar{N}_k}\left\{\sum_{i \in \mathcal{I}_o} \bar{z}_{ik} x_i^2 + \sum_{i \in \mathcal{I}_\ell} \bar{z}_{ik} \bar{y}_{ik}^{\ell(2)} + \sum_{i \in \mathcal{I}_g} \bar{z}_{ik} \bar{y}_{ik}^{g(2)}\right\}, \tag{24}$$

$$q(Y, Z) = q(Z)q(Y|Z), \quad q(Z) = \prod_{i=1}^{M} \text{Mult}(z_i | s_{ik}), \quad s_{ik} \propto \exp(\gamma_{ik}), \tag{25}$$

$$\gamma_{ik} = \mathbb{E}_q[\log \pi_k] - \frac{1}{2}\tau_k^{-1} + \frac{1}{2}\mathbb{E}_q[\log \lambda_k] - \frac{1}{2}\log \bar{\lambda}_k + w_{io}\log f(x_i|\bar{\mu}_k, \bar{\lambda}_k^{-1})$$
$$+ w_{i\ell}\log F(C_i^L|\bar{\mu}_k, \bar{\lambda}_k^{-1}) + w_{ig}\log\left(1 - F(C_i^R|\bar{\mu}_k, \bar{\lambda}_k^{-1})\right), \tag{26}$$

$$q(Y|Z) = \prod_{i \in \mathcal{I}_\ell} \prod_{k=1}^{K} \mathcal{TN}(y_j|\bar{\mu}_k, \bar{\lambda}_k^{-1}, -\infty, C_i^L)^{z_{ik}} \prod_{i \in \mathcal{I}_g} \prod_{k=1}^{K} \mathcal{TN}(y_j|\bar{\mu}_k, \bar{\lambda}_k^{-1}, C_i^R, \infty)^{z_{ik}}, \tag{27}$$

where $\bar{z}_{ik} = s_{ik}$, $\bar{\lambda}_k = a_k/b_k$, $\mathbb{E}_q[\log \pi_k] = \psi(\alpha_k) - \psi(\sum_{k'} \alpha_{k'})$, $\mathbb{E}_q[\log \lambda_k] = \psi(a_k) - \log(b_k)$ and $\psi(\cdot)$ is the digamma function. Moreover, $\bar{y}_{ik}^\ell, \bar{y}_{ik}^g, \bar{y}_{ik}^{\ell(2)}, \bar{y}_{ik}^{g(2)}$ are the moments of truncated Gaussian distribution:

$$\bar{y}_{ik}^\ell = \mathbb{E}_{y_i \sim \mathcal{TN}(y_i|\bar{\mu}_k, \bar{\lambda}_k^{-1}, -\infty, C_i^L)}[y_i] = \bar{\mu}_k - \frac{1}{\tau_k}\frac{f(C_i^L|\bar{\mu}_k, \tau_k^{-1})}{F(C_i^L|\bar{\mu}_k, \tau_k^{-1})}, \tag{28}$$

$$\bar{y}_{ik}^g = \mathbb{E}_{y_i \sim \mathcal{TN}(y_i|\bar{\mu}_k, \bar{\lambda}_k^{-1}, C_i^R, \infty)}[y_i] = \bar{\mu}_k + \frac{1}{\tau_k} \frac{f(C_i^R|\bar{\mu}_k, \tau_k^{-1})}{1 - F(C_i^R|\bar{\mu}_k, \tau_k^{-1})}, \tag{29}$$

$$\bar{y}_{ik}^{\ell(2)} = \mathbb{E}_{y_i \sim \mathcal{TN}(y_i|\bar{\mu}_k, \bar{\lambda}_k^{-1}, -\infty, C_i^L)}[y_i^2] = \bar{\mu}_k^2 + \frac{1}{\tau_k} - \frac{1}{\tau_k} \frac{(C_i^L + \bar{\mu}_k)f(C_i^L|\bar{\mu}_k, \tau_k^{-1})}{F(C_i^L|\bar{\mu}_k, \tau_k^{-1})}, \tag{30}$$

$$\bar{y}_{ik}^{g(2)} = \mathbb{E}_{y_i \sim \mathcal{TN}(y_i|\bar{\mu}_k, \bar{\lambda}_k^{-1}, C_i^R, \infty)}[y_i^2] = \bar{\mu}_k^2 + \frac{1}{\tau_k} + \frac{1}{\tau_k} \frac{(C_i^R + \bar{\mu}_k)f(C_i^R|\bar{\mu}_k, \tau_k^{-1})}{1 - F(C_i^R|\bar{\mu}_k, \tau_k^{-1})}. \tag{31}$$

The above terms are derived using the mean and variance of truncated Gaussian distribution $\mathcal{TN}(x|\mu, \sigma^2, a, b)$ [17]:

$$\text{Mean} = \mu + \sigma \frac{\phi(\alpha) - \phi(\beta)}{\Phi(\beta) - \Phi(\alpha)}, \ \text{Variance} = \sigma^2 \left[1 + \frac{\alpha\phi(\alpha) - \beta\phi(\beta)}{\Phi(\beta) - \Phi(\alpha)} - \left(\frac{\phi(\alpha) - \phi(\beta)}{\Phi(\beta) - \Phi(\alpha)} \right)^2 \right],$$

where $\alpha = \frac{a-\mu}{\sigma}$, $\beta = \frac{b-\mu}{\sigma}$ and, ϕ and Φ is the PDF and CDF of standard normal.

VBCM for GMM repeats the above parameter update following Eqs. (19)–(27). Algorithm 1 shows a pseudo code. The objective $\mathcal{L}$ monotonically increases with the updates and the algorithm converges to (local) minima.

There are two main differences from standard VB and VBCM: (i) Use of CDF in updating $q(Y, Z)$ in Eq. (26). (ii) Use of the moments of truncated distribution in Eqs. (28)–(31). Both differences are needed to deal with censored samples. We emphasize that, if the data contains no censored samples, VBCM reduces to VB. Therefore, VBCM can be seen as the generalized form of VB.

Algorithm 1. VBCM: VB Algorithm for Mixture Model with Censored Data

---

Initialize $\{\alpha_k, \bar{\mu}_k, \tau_k, a_k, b_k\}$
repeat
    // VB-E Step
    for $i = 1$ to N **do**
        Set $s_{ik} \propto \exp(\gamma_{ik})$.
        Set $\bar{z}_{ik} = s_{ik}$.
        // For censored samples
        if $i \in \mathcal{I}_\ell$ **then** Compute $\bar{y}_{ik}^\ell, \bar{y}_{ik}^{\ell(2)}$ **end if**
        if $i \in \mathcal{I}_g$ **then** Compute $\bar{y}_{ik}^g, \bar{y}_{ik}^{g(2)}$ **end if**
    end for
    // VB-M Step
    Set $\alpha_k = \alpha_0 + \bar{N}_k$, $\bar{\mu}_k = \tau_k^{-1}(\tau_0\mu_0 + \bar{N}_k\bar{x}_k)$, $\tau_k = \tau_0 + \bar{N}_k$, $a_k = a_0 + (\bar{N}_k + 1)/2$,
    $b_k = b_0 + (\tau_0\mu_0^2 - \tau_k\bar{\mu}_k^2 + \bar{N}_k\bar{x}_k^{(2)})/2$.
until Converge

4.2   Evidence Lower Bound and Model Selection

The objective functional used in VBCM (and thus VB) is an important quantity since it can be used for *model selection*, i.e., determination of the number of components, K. Because the functional is a lower bound of (logarithm of) *Evidence*, It can be seen as quantifying the goodness of the model [18]. Although

Evidence is not analytically tractable, the functional for VB, which is referred to as Evidence Lower BOund (ELBO), is analytically tractable in many cases.

Also, we show that ELBO, which is defined as Eq. (18), is analytically tractable for VBCM. By expanding the term $\log P(X, W, Y, Z|\Theta)$ and $\log q(Y, Z)$, ELBO is written as

$$
\mathcal{L} = \sum_k \bar{N}_k \left\{ \mathbb{E}_q[\log \pi_k] - \frac{1}{2}\tau_k^{-1} + \frac{1}{2}\mathbb{E}_q[\log \lambda_k] - \frac{1}{2}\log \bar{\lambda}_k \right\}
$$

$$
+ \sum_{i,k} \bar{z}_{ik} \left\{ w_{io} \log f(x_i|\bar{\mu}_k, (\bar{\lambda}_k)^{-1}) + w_{i\ell} \log F(C_i^L|\bar{\mu}_k, (\bar{\lambda}_k)^{-1}) \right. \tag{32}
$$

$$
\left. + w_{ig} \log(1 - F(C_i^R|\bar{\mu}_k, (\bar{\lambda}_k)^{-1})) - \log s_{ik} \right\} + \mathbb{E}_q[\log P(\Theta) - \log q(\boldsymbol{\pi})q(\boldsymbol{\varphi})].
$$

Since the final term $\mathbb{E}_q[\log P(\Theta) - \log q(\boldsymbol{\pi})q(\boldsymbol{\varphi})]$ has analytic form, ELBO for VBCM is an analytically tractable quantity. In the experiments, we use ELBO in model selection.

4.3 Predictive Distribution

VB, and also VBCM, estimates the variational distribution of the parameter $q(\Theta) = q(\boldsymbol{\pi})q(\boldsymbol{\varphi})$, in other words, the parameter uncertainty is estimated. Such uncertainty is used for prediction following the Bayesian approach by using *predictive distribution*, which is defined as

$$
P_{\mathrm{VB}}^{pr}(x|X, W) = \int P(x|\Theta)q(\Theta)d\Theta. \tag{33}
$$

This distribution differs from the model distribution in general. When the model is GMM, the predictive distribution is the mixture of Student-t's distribution [19]:

$$
\mathrm{GMM} : P_{\mathrm{VB}}^{pr}(x|X, W) = \sum_k \frac{\alpha_k}{\sum_{k'} \alpha_{k'}} \mathrm{St}\left(x \middle| \bar{\mu}_k, \left(\frac{a_k \tau_k}{b_k(\tau_k + 1)}\right)^{-1}, 2a_k\right), \tag{34}
$$

where St is the PDF of Student-t's distribution:

$$
\mathrm{St}(x|\mu, \sigma^2, \nu) = \frac{\Gamma(\nu/2 + 1/2)}{\Gamma(\nu/2)} \frac{1}{\sigma\sqrt{\pi\nu}} \left[1 + \frac{1}{\nu}\left(\frac{x - \mu}{\sigma}\right)^2\right]^{-\nu/2 - 1/2} \tag{35}
$$

As shown in Fig. 2, Student-t's distribution has, compared to Gaussian distributions, a long tail, and converges to Gaussian in the limit $\nu \to \infty$. Therefore, when the number of samples belonging to a component, N_k, is small, which means $2a_k$ is small, the component distribution has a long tail reflecting parameter uncertainty; it converges to Gaussian as N_k goes to infinity. Note that this property does not hold in the non-Bayesian approach using plug-in distribution

$$
P_{\mathrm{EM}}^{pr}(x|X, W) = P(x|\Theta^{ML}), \quad \Theta^{ML} = \arg\max \log P(X, W|\Theta) \tag{36}
$$

In the experiments we use the above predictive distributions for evaluation.

5   Extensions

5.1   Generalization for Exponential Family Mixtures (EFM)

In this section, we generalize the algorithm presented in the previous section. Here, we first show the VB algorithm for any mixture model whose component belongs to the exponential family. Following the previous section, the variational distributions of EFM $q(\boldsymbol{\pi})$, $q(\boldsymbol{\eta})$, and $q(Y, Z)$ are given by Dirichlet, conjugate posterior and mixture of truncated exponential family distribution, respectively.

$$q(\boldsymbol{\pi}) = \mathrm{Dirichlet}(\boldsymbol{\pi}|\alpha_k), \quad \alpha_k = \alpha_0 + \bar{N}_k, \quad \bar{N}_k = \sum\nolimits_{i=1}^{N} \bar{z}_{ik}, \tag{37}$$

$$q(\boldsymbol{\eta}) = \prod_{k=1}^{K} g\left(\eta_k | \xi_k, \nu_k\right), \quad \xi_k = \xi_0 + \bar{T}_k, \quad \nu_k = \nu_0 + \bar{N}_k \tag{38}$$

$$\bar{T}_k = \sum\nolimits_{i=1}^{N} \bar{z}_{ik} \bar{T}_{ik}, \quad \bar{T}_{ik} = w_{io} T(x_i) + w_{i\ell} \bar{T}_{ik}^{\ell} + w_{ig} \bar{T}_{ik}^{g} \tag{39}$$

$$q(Y, Z) = q(Z) q(Y|Z), \quad q(Z) = \prod_{i=1}^{M} \mathrm{Mult}(z_i | s_{ik}), \quad s_{ik} \propto \exp(\gamma_{ik}), \tag{40}$$

$$\gamma_{ik} = \mathbb{E}_q[\log \pi_k] - \mathbb{E}_q[A(\eta_k)] + A(\bar{\eta}_k) + w_{io} \log f(x_i|\bar{\eta}_k)$$
$$+ w_{i\ell} \log F(C_i^L|\bar{\eta}_k) + w_{ig} \log\left(1 - F(C_i^R|\bar{\eta}_k)\right) \tag{41}$$

$$q(Y|Z) = \prod_{i \in \mathcal{I}_\ell} \prod_k f_{tr}(y_i|\bar{\eta}_k, -\infty, C_i^L)^{z_{ik}} \prod_{i \in \mathcal{I}_g} \prod_k f_{tr}(y_i|\bar{\eta}_k, C_i^R, \infty)^{z_{ik}}, \tag{42}$$

where $\bar{\eta}_k$ and $\mathbb{E}_q[A(\eta_k)]$ for exponential/Poisson distribution is shown in Table 1. $\bar{T}_{ik}^{\ell}, \bar{T}_{ik}^{g}$ is the expectation of sufficient statistics w.r.t. truncated exponential family:

$$\bar{T}_{ik}^{\ell} = \mathbb{E}_{y_i \sim f_{tr}(y_i|\bar{\eta}_k, -\infty, C_i^L)}[T(y_i)], \quad \bar{T}_{ik}^{g} = \mathbb{E}_{y_i \sim f_{tr}(y_i|\bar{\eta}_k, C_i^R, \infty)}[T(y_i)], \tag{43}$$

Similar to GMM, (i) CDF is used in Eq. (41) and (ii) expectation of sufficient statistics w.r.t. truncated distribution is used in Eq. (43). VBCM for EFM is the algorithm that performs iterative updating following Eqs. (37)–(42). It is guaranteed that objective $\mathcal{L}$ increases monotonically and converges.

We used the analytic form of the moments of truncated distribution in GMM (Eqs. (28)–(31)). However, truncated distributions are not well supported in popular statistical libraries[3] and it is desirable to develop a general algorithm that can be easily applied to any (exponential family) mixture model. Therefore, we investigate the approach that eliminates the need to know its analytic form in the next subsection.

5.2   Stochastic Algorithm

The key idea that permits VBCM to be used without demanding the analytic form of the expectation of sufficient statistics w.r.t. truncated distribution is

---

[3] For example, scipy.stats in python only supports truncated normal distribution and truncated exponential distribution.

Algorithm 2. SVBCM: Stochastic VB for mixture models with censored data

1: Initialize $\{\alpha_k, \bar{\mu}_k, \xi_k, \nu_k\}$.
2: **for** $t = 1$ to ∞ **do**
3: Set $s_{tk} \propto \exp(\gamma_{tk})$.
4: Sample $\tilde{z}_t \sim \text{Mult}(z_t|s_t)$.
5: **if** $i \in \mathcal{I}_o$ **then** Set $\tilde{T}_{tk} = T(x_t)$ **end if**
6: **if** $i \in \mathcal{I}_\ell$ **then** Compute $\tilde{T}_{tk} \approx \bar{T}_{tk}^\ell$ **end if**
7: **if** $i \in \mathcal{I}_g$ **then** Compute $\tilde{T}_{tk} \approx \bar{T}_{tk}^g$ **end if**
8: Compute $\tilde{\alpha}_k = \alpha_0 + N\tilde{z}_{tk}, \tilde{\xi}_k = \xi_0 + N\tilde{z}_{tk}\tilde{T}_{tk}, \tilde{\nu}_k = \nu_0 + N\tilde{z}_{tk}$.
9: Set $\alpha_k = (1 - \rho_t)\alpha_k + \rho_t\tilde{\alpha}_k, \xi_k = (1 - \rho_t)\xi_t + \rho_t\tilde{\xi}_k, \nu_k = (1 - \rho_t)\nu_k + \rho_t\tilde{\nu}_k$.
10: **end for**

---

to use stochastic sampling such as importance sampling [20]. This approach yields the VBCM stochastic algorithm, called here stochastic VBCM (SVBCM). Developing the stochastic VB algorithm itself is a important topic [7–9]. The term "stochastic" mainly means, in existing studies, that parameter update uses a part of stochastically selected/generated training data; its use makes the algorithm more flexible. Therefore, we develop SVBCM which combine the existing approach with stochastic computation of the expectation of sufficient statistics.

The proposed SVBCM algorithm is described in Algorithm 2. The algorithm considers that, each (algorithmic) time step t, sees the arrival of new data (x_t, w_t). The new data may be observed samples or censored samples. When the new data arrive, the parameters of $q(z_t)$, s_t, is computed and sample $\tilde{z}_t$ using $q(z_t)$. If the new data are observed samples, $\tilde{T}_{tk}$ is set to $T(x_t)$ and otherwise $\tilde{T}_{tk}$ is set to the value approximating $\bar{T}_{tk}^\ell$ or $\bar{T}_{tk}^g$ which is computed using any sampling scheme such as importance sampling. Next, we compute temporal parameters $\tilde{\alpha}, \tilde{\xi}, \tilde{\nu}$. These are interpreted as those computed using Eqs. (37), (38) if all training data are new data, i.e., $x_i = x_t, w_i = w_t$ ($\forall i$). Finally, the parameters of $q(\pi)$, $q(\eta)$ are updated by taking the weighted average of $\tilde{\alpha}, \tilde{\xi}, \tilde{\nu}$ and the current parameter. The weight ρ_t has the role of controlling learning speed. Following [8], we use $\rho_t = (\varsigma + t)^{-\kappa}$ in the experiments, where $\varsigma \geq 0$, and $\kappa \in (0.5, 1)$ are the control parameters.

Algorithm 2 is valid in the mini-batch setting where multiple data $\{(x_t, w_t)\}_{t=1}^S$ arrive at once by modifying the update of temporal parameter as follows: $\tilde{\alpha}_k = \alpha_0 + (N/S)\sum_t \tilde{z}_{tk}, \tilde{\xi}_k = \xi_0 + (N/S)\sum_t \tilde{z}_{tk}\tilde{T}_{tk}, \tilde{\nu}_k = \nu_0 + (N/S)\sum_t \tilde{z}_{tk}$, where S is the mini-batch size. If $S = N$ and $\rho_t = 1.0$ ($\forall t$) the algorithm is almost equivalent to the batch algorithm.

The update of variational distribution $q(\pi)$, $q(\eta)$ at final step of the algorithm can be seen as the natural gradient [10] of *per-sample* ELBO $\mathcal{L}^{per}(x_t, w_t)$:

$$\mathcal{L}^{per}(x_t, w_t) = \sum_k \left\{ \tilde{z}_{tk}\mathbb{E}_q[\log \pi_k] - \{\tilde{z}_{tk} + (\nu_0 - \nu_k)/N\}\mathbb{E}_q[A(\eta_k)] + \tilde{z}_{tk}A(\bar{\eta}_k) \right\} \quad (44)$$

$$+ \sum_k \tilde{z}_{tk}\left\{ w_{io}\log f(x_i|\bar{\mu}_k, (\bar{\lambda}_k)^{-1}) + w_{i\ell}\log F(C_i^L|\bar{\mu}_k, (\bar{\lambda}_k)^{-1}) + w_{ig}\log(1 - F(C_i^R|$$

$$\bar{\mu}_k, (\bar{\lambda}_k)^{-1})) - \log s_{ik} \right\} + \frac{1}{N}\sum_k \left\{ \bar{\eta}_k(\xi_0 - \xi_k) + \log\frac{Z(\xi_0, \nu_0)}{Z(\xi_k, \nu_k)} + \mathbb{E}_q\left[\log\frac{P(\pi)}{q(\pi)}\right] \right\}.$$

Note that summation over samples equals ELBO, $\sum_{i=1}^{N} \mathcal{L}^{per}(x_i, w_i) = \mathcal{L}$.

Theorem 1. The update of stochastic VBCM is the (stochastic) natural gradient of per-sample ELBO $\mathcal{L}^{per}(x_t, w_t)$.

Proof. We use the relation[4] $\bar{\eta} = -\frac{\partial}{\partial \xi} \log \mathcal{Z}(\xi, \nu)$, $\mathbb{E}_{q(\eta)}[A(\eta)] = \frac{\partial}{\partial \nu} \log \mathcal{Z}(\xi, \nu)$.
Since $\frac{\partial \log F(C|\eta)}{\partial \eta} = \frac{1}{F(C|\eta)} \int_{-\infty}^{C} \frac{\partial f(x|\eta)}{\partial \eta} dx = \mathbb{E}_{f_{tr}(x|\eta, -\infty, C)}[T(x)] - \mathbb{E}_{f(x|\eta)}[T(x)]$,
the partial derivative of $\mathcal{L}$ w.r.t. ξ_k and ν_k are written as

$$N\frac{\partial \mathcal{L}^{per}}{\partial \xi_k} = -(\nu_0 + N\bar{z}_{tk} - \nu_k)\frac{\partial^2 \log \mathcal{Z}(\xi_k, \nu_k)}{\partial \xi_k \partial \nu_k} - (\xi_0 + N\bar{z}_{tk}\bar{T}_{tk} - \xi_k)\frac{\partial^2 \log \mathcal{Z}(\xi_k, \nu_k)}{\partial \xi_k \partial \xi_k},$$

$$N\frac{\partial \mathcal{L}^{per}}{\partial \nu_k} = -(\nu_0 + N\bar{z}_{tk} - \nu_k)\frac{\partial^2 \log \mathcal{Z}(\xi_k, \nu_k)}{\partial \nu_k \partial \nu_k} - (\xi_0 + N\bar{z}_{tk}\bar{T}_{tk} - \xi_k)\frac{\partial^2 \log \mathcal{Z}(\xi_k, \nu_k)}{\partial \xi_k \partial \nu_k}.$$

It follows that the update for ξ_k, η_k is the stochastic natural gradient of $\mathcal{L}^{per}$:

$$\begin{pmatrix} \xi_k \\ \nu_k \end{pmatrix} - \rho_t N \begin{pmatrix} \frac{\partial^2 \log q(\eta_k)}{\partial \xi_k \partial \xi_k} & \frac{\partial^2 \log q(\eta_k)}{\partial \xi_k \partial \nu_k} \\ \frac{\partial^2 \log q(\eta_k)}{\partial \nu_k \partial \xi_k} & \frac{\partial^2 \log q(\eta_k)}{\partial \nu_k \partial \nu_k} \end{pmatrix}^{-1} \begin{pmatrix} \frac{\partial \mathcal{L}^{per}}{\partial \xi_k} \\ \frac{\partial \mathcal{L}^{per}}{\partial \nu_k} \end{pmatrix} = \begin{pmatrix} (1-\rho_t)\xi_k + \rho_t(\xi_0 + N\bar{z}_{tk}\bar{T}_{tk}) \\ (1-\rho_t)\nu_k + \rho_t(\nu_0 + N\bar{z}_{tk}) \end{pmatrix}$$

The proof of α_k is analogous.   □

Therefore, SVBCM can estimate variational distributions efficiently.

5.3   Predictive Distribution

The predictive distribution of EFM is shown at the end of this section. It is known the predictive distribution Eq. (33) for EFM is given by

$$\text{EFM} : P_{\text{VB}}^{pr}(x|X, W) = \sum_k \frac{\alpha_k}{\sum_{k'} \alpha_{k'}} \frac{\mathcal{Z}(\xi_k, \nu_k)}{\mathcal{Z}(\xi_k + T(x), \nu_k + 1)} h(x) \qquad (45)$$

where $\mathcal{Z}$ is the normalizer of conjugate prior (Table 1). When the models are EMM/PMM, whose components are exponential/Poisson distribution, the above predictive distribution is the mixture of gamma-gamma/negative binomial distribution (e.g., [21]),

$$\text{EMM} : P_{\text{VB}}^{pr}(x|X, W) = \sum_k \frac{\alpha_k}{\sum_{k'} \alpha_{k'}} \text{GG}(x|\nu_k + 1, \xi_k, 1), \qquad (46)$$

$$\text{PMM} : P_{\text{VB}}^{pr}(x|X, W) = \sum_k \frac{\alpha_k}{\sum_{k'} \alpha_{k'}} \text{NB}(x|\xi_k, (\nu_k + 1)^{-1}), \qquad (47)$$

where GG and NB is the PDF of the gamma-gamma distribution and negative binomial distribution, respectively[5]. The above predictive distributions are used in the experiments[6].

---

[4] Derived from $\mathcal{Z}(\xi, \nu)$ is a normalizer, $\mathcal{Z}(\xi, \nu) = \left(\int \exp\{\eta\xi - \nu A(\eta)\} d\eta \right)^{-1}$.

[5] $\text{GG}(x|a, r, b) = \frac{r^a}{\Gamma(a)} \frac{\Gamma(a+b)}{\Gamma(b)} \frac{x^{b-1}}{(r+x)^{(a+b)}}$, $\text{NB}(k|r, p) = \frac{\Gamma(k+r)}{k!\Gamma(r)} p^k (1-p)^r$.

[6] We found that gamma-gamma distributions are not implemented in standard statistical libraries and so the experiment uses the plug-in distribution $f(x|\bar{\eta}_k)$ as the component of predictive distribution of EMM.

6   Experiments

6.1   Setting

This section confirms the properties and predictive performance of VBCM.

Synthetic Data: We prepared three synthetic data set, gmm, emm and pmm, using manually made *true* distributions, GMM, EMM and PMM. We set the true number of components to $K^* = 2$ and set the true mixing ratio to $\pi^* = (1/2, 1/2)$. The true component parameters were set to $(\mu_1, \mu_2) = (-3.0, 3.0)$, $(\lambda_1, \lambda_2) = (1.0, 1.0)$ for gmm, $(\lambda_1, \lambda_2) = (0.3, 3.0)$ for emm, and $(\lambda_1, \lambda_2) = (1.0, 5.0)$ for pmm, respectively. We randomly generated 10 pairs of training and test data using the true distributions. The thresholds used in censoring gmm, emm and pmm were set to $(C_i^L, C_i^R) = (-4.0, 4.0)$, $(-\infty, 4.0)$ and $(-\infty, 6.0)$ for all i, respectively. The number of samples in each test data was 10000.

Real Survival Data: We also used two publicity available survival data sets, Rossi et al. [22]'s criminal recidivism data (rossi) and north central cancer treatment group (NCCTG)'s lung cancer data (ncctg)[7]. rossi is data pertaining to 432 convicts who were released from Maryland state prisons in the 1970s and who were followed for one year after release. The data contains the week of first arrest after release or censoring and we used the week information as the observed variable. ncctg is the survival data of patients with advanced lung cancer in NCCTG and we used the survival time information. rossi and ncctg contain, approximately, 74% and 28% censored samples, respectively. We standardized the data and prepared five data sets by dividing the data into five, using 80% of the data as training and the remaining 20% as test.

Baseline: We compare VBCM with three existing algorithms, EM, VB and EMCM. EM and VB use observed samples and do not use censored samples. EMCM is the EM algorithm that can use censored samples for learning mixture models [14,15]. We applied all algorithms with GMM to gmm, rossi and ncctg, with EMM to emm, and with PMM to pmm. We used the predictive distribution described by Eq. (33) for VB and VBCM and Eq. (36) for EM and EMCM.

Evaluation Metric: To evaluate predictive performance, we use the test log likelihood metric (larger values are better). Since test data contains censored data, based on Eq. (17), test log-likelihood is defined as

$$\text{test.} = \frac{1}{N_{te}} \log \Big\{ \prod_{i=1}^{N_{te}} P^{pr}(x_i^{te}|X,W)^{w_{io}^{te}} \Big(\int_{-\infty}^{C_i^L} P^{pr}(y|X,W)dy \Big)^{w_{il}^{te}} \Big(\int_{C_i^R}^{\infty} P^{pr}(y|X,W)dy \Big)^{w_{ig}^{te}} \Big\}$$

where $\{x_i^{te}\}_{i=1}^{N_{te}}, \{w_i^{te}\}_{i=1}^{N_{te}}$ is the test data and N_{te} is the number of test data.

6.2   Result

Algorithm Behavior: Figure 5(a)(b) shows the convergence behavior of ELBO. It is confirmed that the VBCM converges within 20 iterations and stochastic algorithm (SVBCM) with mini-batch size $S = 10, 20$ also converges to the

---

[7] Available at e.g., R survival package and python lifeline package.

nearly optimal result[8]. These indicate the effectiveness of VBCM and SVBCM as a estimation algorithm for mixture models. Figure 5(c) shows that ELBO is maximized when K corresponds to the true $K = K^* = 2$. This indicates the validity of ELBO as a model selection criterion.

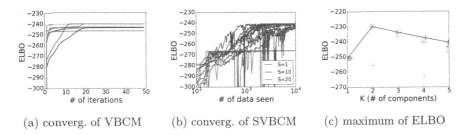

(a) converg. of VBCM   (b) converg. of SVBCM   (c) maximum of ELBO

Fig. 5. Behavior of VBCM for gmm ($N = 100, K = 5$). The convergence behaviors of (a) VBCM and (b) stochastic VBCM (SVBCM) with different runs are shown. (c) ELBO at converged parameter in each VBCM run for various K. Symbols "$+$" are plotted with small horizontal perturbations to avoid overlap and the blue line connects maximum values. ELBO is maximized when K is equals to the true value $K = K^* = 2$.

Synthetic Data Result: Figure 6 (a)(b)(c) shows the predictive performance for synthetic data. VBCM outperforms the existing algorithms. More precisely, the performance of VBCM is much better than VB and EM. This implies that the use of censored samples strengthens the superiority of VBCM. This is also confirmed by Fig. 7. Figure 7a shows that VB yielded components whose mean was nearer to the origin than the true value while VBCM predictions were close to the true distributions. This occurs because VB ignores the censored samples. Comparing VBCM with EMCM, VBCM is superior when the number of training data is small. This seems to be caused by using the Bayesian approach for making predictions; it integrates parameter uncertainty.

Real Survival Data Result: Figures 6(d)(e) show the predictive performance with real survival data. Similar to the synthetic data result, VBCM outperforms the existing algorithms. Since rossi contains more censored samples than ncctg, VBCM and EMCM attain better performance with rossi than VB or EM. The difference is small in ncctg and the performances of VBCM and VB are comparable when the number of training data (ratio) is small; VBCM and EMCM have comparable performance when the number of training data is large. These results imply that VBCM inherits the good points of VB and EMCM.

---

[8] The other control parameters were $\varsigma_0 = 10.0, \kappa = 0.6$. The stochastic algorithm uses randomly chosen training data repeatedly.

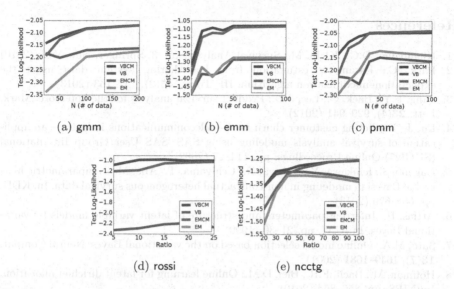

(a) gmm   (b) emm   (c) pmm

(d) rossi   (e) ncctg

Fig. 6. Predictive performance of (a)(b)(c) synthetic data experiment and (d)(e) real data experiment. x axis is the number of training data samples N or the ratio (%) of training data used and y axis is the test log-likelihood. Larger values are better. VBCM (proposed) outperforms existing algorithms.

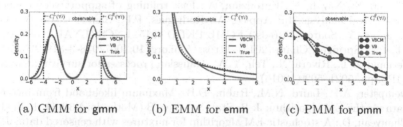

(a) GMM for gmm   (b) EMM for emm   (c) PMM for pmm

Fig. 7. Predictive distributions of proposed VBCM and VB using Gaussian mixture model (GMM), exponential mixture model (EMM) and poisson mixture model (PMM). The result of VBCM is close to the true distribution.

7 Conclusion

In this study, we proposed VBCM, a generalized VB algorithm that can well handle censored data using mixture models. The proposed algorithm can be applied to any mixture model whose component distribution belongs to exponential family and experiments on three synthetic data and two real data sets confirmed its effectiveness. Promising future work is to extend VBCM to support Dirichlet process mixture models [23].

References

1. Kleinbaum, D.G., Klein, M.: Survival Analysis, vol. 3. Springer, New York (2010)
2. Park, C.: Parameter estimation from load-sharing system data using the expectation–maximization algorithm. IIE Trans. **45**(2), 147–163 (2013)
3. Jung, E.Y., Baek, C., Lee, J.D.: Product survival analysis for the app store. Mark. Lett. **23**(4), 929–941 (2012)
4. Lu, J.: Predicting customer churn in the telecommunications industry—an application of survival analysis modeling using SAS. SAS User Group International (SUGI27) Online Proceedings, pp. 114–27 (2002)
5. Nagano, S., Ichikawa, Y., Takaya, N., Uchiyama, T., Abe, M.: Nonparametric hierarchal Bayesian modeling in non-contractual heterogeneous survival data. In: KDD, pp. 668–676 (2013)
6. Attias, H.: Inferring parameters and structure of latent variable models by variational Bayes. In: UAI, pp. 21–30 (1999)
7. Sato, M.A.: Online model selection based on the variational Bayes. Neural Comput. **13**(7), 1649–1681 (2001)
8. Hoffman, M., Bach, F.R., Blei, D.M.: Online learning for latent dirichlet allocation. In: NIPS, pp. 856–864 (2010)
9. Hoffman, M.D., Blei, D.M., Wang, C., Paisley, J.: Stochastic variational inference. J. Mach. Learn. Res. **14**(1), 1303–1347 (2013)
10. Amari, S.I.: Natural gradient works efficiently in learning. Neural Comput. **10**(2), 251–276 (1998)
11. Pölsterl, S., Navab, N., Katouzian, A.: Fast training of support vector machines for survival analysis. In: Appice, A., Rodrigues, P.P., Santos Costa, V., Gama, J., Jorge, A., Soares, C. (eds.) ECML PKDD 2015. LNCS (LNAI), vol. 9285, pp. 243–259. Springer, Cham (2015). https://doi.org/10.1007/978-3-319-23525-7_15
12. Fernández, T., Rivera, N., Teh, Y.W.: Gaussian processes for survival analysis. In: NIPS, pp. 5021–5029 (2016)
13. Dempster, A.P., Laird, N.M., Rubin, D.B.: Maximum likelihood from incomplete data via the EM algorithm. J. R. Stat. Soc. Ser. B (Methodol.) **39**(1), 1–38 (1977)
14. Chauveau, D.: A stochastic EM algorithm for mixtures with censored data. J. Stat. Plan. Inference **46**(1), 1–25 (1995)
15. Lee, G., Scott, C.: EM algorithms for multivariate gaussian mixture models with truncated and censored data. Comput. Stat. Data Anal. **56**(9), 2816–2829 (2012)
16. Jordan, M.I., Ghahramani, Z., Jaakkola, T.S., Saul, L.K.: An introduction to variational methods for graphical models. Mach. Learn. **37**(2), 183–233 (1999)
17. Johnson, N., Kotz, S., Balakrishnan, N.: Continuous Univariate Probability Distributions, vol. 1. Wiley, Hoboken (1994)
18. MacKay, D.J.: Information Theory, Inference and Learning Algorithms. Cambridge University Press, Cambridge (2003)
19. Murphy, K.P.: Conjugate Bayesian analysis of the univariate Gaussian: a tutorial (2007). http://www.cs.ubc.ca/~murphyk/Papers/bayesGauss.pdf
20. Bishop, C.M.: Pattern Recognition and Machine Learning. Springer, New York (2006)
21. Bernardo, J.M., Smith, A.F.: Bayesian Theory, vol. 405. Wiley, Hoboken (2009)
22. Rossi, P.H., Berk, R.A., Lenihan, K.J.: Money, Work, and Crime: Experimental Evidence. Academic Press, Cambridge (1980)
23. Blei, D.M., Jordan, M.I.: Variational inference for dirichlet process mixtures. Bayesian Anal. **1**(1), 121–143 (2006)

Exploration Enhanced Expected Improvement for Bayesian Optimization

Julian Berk[✉], Vu Nguyen, Sunil Gupta, Santu Rana, and Svetha Venkatesh

Centre for Pattern Recognition and Data Analytics, Deakin University,
Geelong, Australia
jmberk@deakin.edu.au

Abstract. Bayesian optimization (BO) is a sample-efficient method for global optimization of expensive, noisy, black-box functions using probabilistic methods. The performance of a BO method depends on its selection strategy through an acquisition function. This must balance improving our understanding of the function in unknown regions (exploration) with locally improving on known promising samples (exploitation). Expected improvement (EI) is one of the most widely used acquisition functions for BO. Unfortunately, it has a tendency to over-exploit, meaning that it can be slow in finding new peaks. We propose a modification to EI that will allow for increased early exploration while providing similar exploitation once the system has been suitably explored. We also prove that our method has a sub-linear convergence rate and test it on a range of functions to compare its performance against the standard EI and other competing methods. Code related to this paper is available at: https://github.com/jmaberk/BO_with_E3I.

1 Introduction

There are numerous situations, both in research and industry, where it is necessary to know the optimal input to a black box function but sampling it is either difficult or expensive. *Bayesian optimization* is one of the most evaluation efficient methods for finding the input, x^*, that will produce the optimal value of such systems [11]. It has been successfully applied to many problems in a wide range of fields including materials science, biomedical science, and even other computer science problems. An example of an application in materials science is the development of new polymer fibres [9]. In biomedical science, it has been used for many applications including studying how age effects time perception [19] and synthetic gene design [4]. Another application is the selection of hyperparameters for other machine learning algorithms [17].

Bayesian optimization methods work by fitting a probabilistic model to the available data (evaluation locations for the objective and corresponding function values). This model provides a distribution of all possible functions within a set range of possible inputs, $\mathcal{X} \in \mathbb{R}^d$. The most common model among these is the *Gaussian process* [15].

© Springer Nature Switzerland AG 2019
M. Berlingerio et al. (Eds.): ECML PKDD 2018, LNAI 11052, pp. 621–637, 2019.
https://doi.org/10.1007/978-3-030-10928-8_37

The predictions of the probabilistic model above are then used to make intelligent decisions about where to evaluate the objective function next, so that its optimum is found by using a reduced number of function evaluations. These intelligent decisions are made through an *acquisition function*. This maps $\forall x \in \mathcal{X}$ to some property that describes how useful sampling at that point will be in determining the black box function's true optima. Optimizing the acquisition function will therefore allow the best possible sample to be made from the black box function given the data and prior knowledge. Based on the decision of the acquisition function, the new sample is collected and evaluated. This sample can then be used to update the model, allowing the point after that to be determined from the updated acquisition function. This data-driven decision allows the optima of the function to be found in far fewer iterations than if samples had been taken at random [3].

The choice of acquisition function can greatly impact the number of iterations necessary to find the optimal input. As such, poor acquisition functions can lead to sup-optimal results or the need for a larger number of costly iterations. A good acquisition function needs to balance between trying to generalize from known good points (exploitation) and trying to search for new peaks in unexplored regions (exploration). There are currently many choices of acquisition functions that provide various degrees of exploration and exploitation. Two popular choices are *expected improvement* (EI) [7] and *Gaussian process upper confidence bound* (GP-UCB) [18], with EI being more exploitative without the need to choose hyperparameters and GP-UCB having more exploration but requiring the specification of several hyperparameters. These hyperparameters can reduce optimization performance if they are not suited to the problem and determining them is both computationally costly and potentially inaccurate. As such, EI is more popular.

We propose a modification to EI that will improve its exploration in the early stages of the experiment, but converge to its previous level of exploitation at later stages. This method is detailed in Sect. 3 along with a proof that it has a sublinear convergence rate. In Sect. 4 we discuss results from several experiments performed using our method. First, we verify that our method has increased exploration by testing it against competing methods on a synthetic function designed to favour high-exploration methods. We then test its performance in comparison to these methods on several benchmark functions and a machine learning hyperparameter tuning problem. Finally, we discuss results concerning the analytical properties of our method.

2   Bayesian Optimization and Expected Improvement

Below we first provide a background of Bayesian optimization and Gaussian processes. Then we discuss acquisition functions with a focus on expected improvement.

2.1   Bayesian Optimization

Bayesian optimization is an efficient method for optimizing noisy, expensive black box-functions [7]. More formally, the ultimate goal of the method is to find the input,

$$x^* = \underset{x \in \mathcal{X}}{\mathrm{argmax}} f(x), \tag{1}$$

that maximises the black-box function, $f(x)$, in the bounded input space, $\mathcal{X} \subset \mathbb{R}^d$. It is possible to directly draw potentially noisy samples from the function: $y_t = f(x_t) + \epsilon$ where ϵ a random noise term, $\epsilon \sim \mathcal{N}(0, \sigma_n)$ with some unknown σ_n. However, doing so is expensive so we wish to determine x^* in as few samples as possible. To do this, Bayesian optimization uses a statistical model for the black box function to construct a surrogate function that is cheaper to sample. The statistical model is generated from all current information about the system, including all prior knowledge and all t sampled input-output pairs, $D_t = \{x_i, y_i\}_{i=1}^t$.

The statistical model is often chosen to be a Gaussian process due to its flexibility and analytic properties [2]. While the statistical model is generally not accurate enough to directly locate the optima of the function, it gives a probabilistic estimate of the function with epistemic uncertainties. This means that it can be used to select the "best" new point to sample. This is done by finding the optima of a surrogate function called an acquisition function which emphasises characteristics that are desirable for the new point to have. Acquisition functions are discussed further in Sect. 2.3.

Once the new samples are found, they can then be used to improve the model, allowing us to find a new, potentially better point to sample. This process is iterated until a predetermined stopping condition has been met. As samples are costly, it is common to choose a maximum number of iterations as a stopping criteria, but other stopping criteria can be used as well, such as stopping when a satisfactory result has been found or when the possible improvement predicted by the model becomes too small. A more detailed review of Bayesian optimization can be found in [2].

2.2   Gaussian Process

A Gaussian process is a statistical model of the black-box function. It represents the function values, $f(x)$, at each point, $x \in \mathcal{X}$ as infinitely many correlated Gaussian random variables. As such, it is completely characterized by its mean and covariance functions, $m(x)$ and $k(x_i, x_j)$. More formally, we assume that $f(x) \sim \mathcal{GP}(m(x), k(x_i, x_j))$. The covariance function, also called a kernel, has a profound impact on the shape of the resulting Gaussian process. As such, the use of an appropriate kernel is vital.

One of the most popular kernels is the *square exponential kernel*. This is given by $k_{SE}(x_i, x_j) = \exp\left(-\frac{\|x_i - x_j\|^2}{2l^2}\right)$. Here, the length scale is completely determined by a single hyperparameter, l. This kernel was chosen because it is simple and translation invariant. This property is important for the generation of Thompson samples as discussed in Sect. 3. Other popular kernel functions are discussed in Rasmussen *et al.* [15].

Using our data with this kernel gives a posterior distribution over the function, $f(x)|D_t \sim \mathcal{GP}(m(x), \mathbf{K}_t)$. Here $\mathbf{K}_t = [k(x_i, x_j)_{\forall x_i, x_j \in D_t}]$ is the kernel matrix, which acts as the covariance matrix for the distribution. The posterior, in turn, can be used to calculate a predictive distribution, $p(f(x) \mid D_t, x) = \mathcal{N}(\mu_t(x), \sigma_t(x))$. The predictive distribution allows us to estimate the function value at any point x by calculating the predictive mean, $\mu_t(x)$, and variance, $\sigma_t^2(x)$. These are given by $\mu_t(x) = \mathbf{k}_*(\mathbf{K}_t + \sigma_n \mathbf{I})^{-1} \mathbf{y}$ and $\sigma_t^2(x) = k_t(x, x) - \mathbf{k}_*(\mathbf{K}_t + \sigma_n \mathbf{I})^{-1} \mathbf{k}_*^T$ with $\mathbf{k}_* = [k(x_1, x), k(x_2, x), \ldots, k(x_t, x)]$. Here $\mathbf{I}$ is the identity matrix with the same dimensions as $\mathbf{K}_t$ and σ_n is the function noise standard deviation.

2.3   Acquisition Functions

Once the Gaussian process has been built, it is used to select the optimal next point to sample from the black box function, x_t. However, exactly what qualifies a point to be x_t is non-trivial. As such, there are many potentially desirable properties that could be used to select x_t. Once a desired property is chosen, an *acquisition function*, $\alpha(x)$, is used to used to calculate it. This is generally far cheaper to evaluate than the black box function to the point where it is efficient to perform a global optimization on it $\forall x \in \mathcal{X}$ to determine a single sample of the black box function. More formally, the optimal next point is given by

$$x_t = \arg\max_{x \in X} \alpha(x) \tag{2}$$

Improvement Based Acquisition Functions. One of the most basic families of acquisition functions are the improvement based acquisition functions. These use the potential improvement over what is believed to be the current maxima, called the *incumbent*. The incumbent is often taken as the current best observed value, $y^+ = \max_{i \leq t}(y_i)$. The improvement is therefore given by $I(x) = \max(f(x) - y^+, 0)$.

Probability of Improvement. A simple acquisition function is the probability of improvement (PI) [8], which gives the probability that a given point will have an improvement over the incumbent. Despite being an intuitive and simple formulation, PI often favours points near the incumbent [2]. As a result, the algorithm tends to over exploit. This can lead to the algorithm failing to quickly find promising peaks away from the incumbent, reducing the optimization efficiency in cases where there is more that one peak in the black box function. This lack of exploration can be improved by maximizing the expected improvement instead of PI [6].

Expected Improvement. As $f(x)$ can be approximated by its Gaussian process predictive distribution, $I(x)$ can likewise be approximated as a function of this random variable. This allows us to take the expectation over this to find the expected amount of improvement at any point $x \in \mathcal{X}$, giving us the expected improvement acquisition function: $\alpha^{EI}(x) = \mathbb{E}[I(x)]$. This can be expressed with the Gaussian process predictive mean and variance in the following closed form [7]:

$$\alpha^{EI}(x) = \begin{cases} (\mu(x) - y^+)\Phi(z) + \sigma(x)\phi(z), & \text{if } \sigma(x) > 0 \\ 0 & \text{if } \sigma(x) = 0 \end{cases} \tag{3}$$

where $z = \frac{\mu(x) - y^+}{\sigma(x)}$, ϕ is the standard normal PDF, and Φ is the standard normal CDF. For a full analytical derivation of EI, we refer interested readers to [14].

EI has better exploration than PI but still tends to over-exploit in many situations, such as when it hits a local optimum. Despite this, EI is currently the most common acquisition functions due to its consistent performance without the need to choose additional hyperparameters.

A Heuristic Approach for Boosting the Exploration of EI: ζ-EI. It is a common belief that artificially increasing the incumbent by some positive ζ will reduce the value of the acquisition function near the currently sampled points, boosting exploration [10]. However, this method does not work well in practice as it is not easy to choose the right value of ζ. If this value is large, the algorithm will significantly over-explore. This often leads to inefficiency in optimization performance.

3   The Proposed E³I Method

In this section we will outline our modification to EI that will improve its exploration without causing it to significantly over-explore. We will then prove that, under some mild assumptions, it has a sub-linear regret bound.

3.1   Thompson Sampling

For our method, we wish to generate full random approximations of the black-box function. Hernández-Lobato *et al.* [5] have developed a method for doing this through Thompson sampling. For a shift invariant kernel such as the square exponential kernel we are using, Bochner's theorem [1] states that it has a Fourier dual, $s(w)$, which is equal to the spectral density of $k(x_i, x_j)$. Normalizing this as $\hat{s}(w) = s(w)/\beta$ allows us to represent the kernel as

$$k(x_i, x_j) = 2\beta \mathbb{E}_{\hat{s}(w)} \left[\cos(w^T x_i + b) \cos(w^T x_j + b) \right] \tag{4}$$

where $b \sim \mathcal{U}[0, 2\pi]$. If we draw V random samples of w and let $\phi(x) = \sqrt{\frac{2\beta}{V}} \{\cos(Wx + b), \sin(Wx + b)\}$, we can approximate the kernel with

$k(x_i, x_j) \approx \phi(x_i)^T \phi(x_j)$. By setting $\Phi = [\phi(x_1), \ldots, \phi(x_V)]$, we can also approximate the kernel matrix with $\mathbf{K} \approx \Phi\Phi^T + \sigma^2 \mathbf{I}$, where $\mathbf{I}$ is the $V \times V$ matrix identity. These estimates have been augmented by the random samples of w so they can be viewed as having random probable points added to them. This means that, if V is sufficiently large, the corresponding predictive mean can be viewed as a complete estimate of the black box function given the current data:

$$f(x) \approx g(x) = \phi(x)^T (\Phi\Phi^T + \sigma^2 \mathbf{I})^{-1} \Phi^T \mathbf{y} \tag{5}$$

The process for generating and finding the optima of these Thompson samples is outlined in Algorithm 1.

Algorithm 1. Thompson Sampling

Input: $D_{t-1} = \{x_i, y_i\}_{i=1}^{t-1}$, #random feature dimension, V, #Thompson samples, M

1: **for** $m = 1$ to M **do**
2:     Randomly generate $b \sim \mathcal{U}[0, 2\pi]$ and V weights, $w_i \sim \mathcal{N}(0, \mathbf{I}_{d \times d}) \, \forall i = 1 \ldots V$
3:     Let $W = [w_1, \ldots, w_V] \in \mathbb{R}^{V \times d}$
4:     Let $\phi(x) = \sqrt{\frac{2\beta}{V}} (\cos(Wx + b), \sin(Wx + b))$ and $\Phi = [\phi(x_1), \ldots, \phi(x_V)]$
5:     Thompson samples are given by $g_m(x) = \phi(x)^T (\Phi\Phi^T + \sigma^2 \mathbf{I})^{-1} \Phi^T \mathbf{y}$
6:     Use a global optimizer to find $g_m^* = \max_{x \in \mathcal{X}} g_m(x)$
7: **end for**
Output: $g_1^*, \ldots, g_M^*$

3.2   Exploration Enhanced Expected Improvement (E³I)

The Thompson sample functions have two useful properties. Firstly, without noise they will agree with the currently sampled points exactly (i.e. $g(x_i) = y_i, \forall(x_i, y_i) \in D_t$). This means that $g(x^+) = f(x^+) = y^+$. As such, either the maximum of $g(x)$ will occur at x^+, in which case $g^* = \max_x g(x) = y^+$, or it will occur elsewhere, in which case $g^* > y^+$. Secondly, the Thompson sample functions will also converge to the true function as the number of iterations increase. This means that g^* should converge towards y^+.

These two properties allow us to use g^* as the incumbent in EI instead of y^+. As $g^* \geq y^+$, the algorithm will have greater exploration. However, as the Thompson samples are sensible approximations of the underlying function, the method does not have the same risk of over-exploration as artificially increasing the incumbent does. As $g^* \to y^+$, it should also explore less at later stages in the algorithm when exploration is less important.

This approach assumes that any given Thompson sample is a good approximation of the black box function given the current data. Due to variations between Thompson samples, it is possible that any given Thompson sample may be an outlier, voiding this assumption. As such, we instead look at the distribution of possible Thompson samples. This makes the new acquisition function a function of this distribution. To obtain the best point, we take the expected value of this distribution, i.e. $\alpha^{E^3I}(x) = \mathbb{E}_g[\mathbb{E}_x[I(x,g^*)]]$. Unfortunately, determining this directly is difficult. As such, we instead generate M Thompson samples, $g_1^*, g_2^*, \ldots, g_M^*$, and find the sample mean instead. Setting $z = \frac{\mu(x)-g_m^*}{\sigma(x)}$ and $\tau(z) = z\Phi(z) + \phi(z)$ we get

$$\alpha^{E^3I}(x) = \frac{1}{M} \sum_{m=1}^{M} \mathbb{E}_x[I(x, g_m^*)] = \begin{cases} \frac{\sigma(x)}{M} \left[\sum_{m=1}^{M} \tau(z) \right], & \text{if } \sigma(x) > 0 \\ 0 & \text{if } \sigma(x) = 0 \end{cases} \quad (6)$$

We outline the E³I routine in Algorithm 2.

Algorithm 2. Bayesian optimization with E³I

Input: $D_{t-1} = \{x_i, y_i\}_{i=1}^{t-1}$, #Weights, V, #Thompson samples, M, #Iterations, T
1: **for** $t = 1$ to T **do**
2:     Generate the M Thompson sample optima, $g_1^*, \ldots, g_M^*$, using Algorithm 1
3:     Use a global optimizer to find $x_t = \arg\max_{x \in \mathcal{X}} \left(\alpha_t^{E^3I}(x) \right)$
4:     Query the black box function with x_t to get $y_t = f(x_t)$
5:     Augment the current data: $D_t = D_{t-1} \cup (x_t, y_t)$
6: **end for**
Output: $(x^*, y^*) = \arg\max_y D_T$

3.3 Convergence

The regret bound is one of the basic criteria for evaluating the performance of an optimization algorithm. EI has been shown to converge at a sub-linear rate under a variety of assumptions. Bull [3], and Ryzhov [16] both derive convergence rates in the absence of noise. Wang and de Freitas [20] were able to derive a convergence rate in the noisy setting, but they needed to use $\mu^+(x) = \max_x \mu(x)$ as the incumbent. As E³I utilizes a different incumbent, it is not compatible with this approach. Nguyen et al. [14] have shown that EI has a sub-linear convergence rate if a minimum improvement stopping condition is used. The proof is valid in the noisy setting and does not require a modified incumbent. As such we extend and employ it to show that this is also true for our method. Many of the lemmas used in their proof can be directly applied to our method.

We start our derivation for the regret bound of E³I as follows.

Lemma 1 (*Srinivas et al. [18]*). *Let $\delta \in (0,1)$ and assume that the noise variables, ϵ_t, are uniformly bounded by σ. Define $\beta_t = 2\|f\|_k^2 + 300\gamma_t \ln^3\left(\frac{t}{\delta}\right)$, then*

$$p\left(\forall t, \forall x \in \mathcal{X}, |\mu_t(x) - f(x)| \leq \sqrt{\beta_t}\sigma_t(x) \right) \geq 1 - \delta$$

Lemma 2. *The improvement function,* $I_{t,m}(x) = \max\left(0, f(x) - g^*_{t,m}\right)$, *and the acquisition function,* $\alpha_t^{E^3I}(x) = \frac{1}{M}\sum_{m=1}^{M}\mathbb{E}[I_{t,m}(x)]$ *satisfy the inequality* $\frac{1}{M}\sum_{m=1}^{M} I_{t,m}(x) - \sqrt{\beta_t}\sigma_{t-1}(x) \leq \alpha_t^{E^3I}(x)$.

Proof. In the case that $\sigma_{t-1}(x) = 0$, we have $\mathbb{E}[I_{t,m}(x)] = I_{t,m}(x)$, $\forall m \in [1, M]$. This means that $\alpha_t^{E^3I}(x) = \frac{1}{M}\sum_{m=1}^{M}\mathbb{E}[I_{t,m}(x)] = \frac{1}{M}\sum_{m=1}^{M} I_{t,m}(x)$ so the result is trivial. For $\sigma_{t-1}(x) > 0$, the proof is as follows. Using Lemma 1, $q_{t-1,m} = \frac{f(x) - g^*_{t-1,m}}{\sigma_{t-1}(x)}$, and $z_{t-1,m} = \frac{\mu_{t-1}(x) - g^*_{t-1,m}}{\sigma_{t-1}(x)}$, we can get the following result:

$$\alpha_t^{E^3I}(x) \geq \frac{1}{M}\sum_{m=1}^{M}\sigma_{t-1}(x)\tau\left(q_{t-1,m} - \sqrt{\beta_t}\right)$$

$$\geq \frac{1}{M}\sum_{m=1}^{M}\sigma_{t-1}(x)\left(q_{t-1,m} - \sqrt{\beta_t}\right) \qquad \text{by } \tau(z) \geq z$$

If $I_t(x) = 0$, the lemma becomes $\alpha_t^{E^3I}(x) \geq -\sqrt{\beta_t}\sigma_{t-1}(x)$. As $\alpha_t^{E^3I}(x) \geq 0$, $\sqrt{\beta_t} \geq 0$, and $\sigma_{t-1}(x) \geq 0$, this is always true. For $I_t(x) > 0$ we have $q_{t-1,m} = \frac{I_t(x)}{\sigma_{t-1}(x)}$. This gives us

$$\alpha_t^{E^3I}(x) \geq \frac{1}{M}\sum_{m=1}^{M} I_{t,m}(x) - \sqrt{\beta_t}\sigma_{t-1}(x)$$

which concludes our proof.   □

We now prove the main theorem:

Theorem 3. *Let $\kappa > 0$ be a predefined small constant as a stopping criteria, σ^2 be the measurement noise variance, $C \triangleq \log\left[\frac{1}{2\pi\kappa^2}\right]$, $\beta_t = 2\|f\|_k^2 + 300\gamma_t\ln^3\left(\frac{t}{\delta}\right)$ and $\delta \in (0,1)$. Then, with probability at least $1-\delta$, after T iterations the cumulative regret of E^3I using a collection of maxima samples g^*_m drawn from Thompson sampling as the incumbents obeys the following sublinear rate: $R_T \lesssim \sqrt{T\beta_T\gamma_T} \sim \mathcal{O}\left(\sqrt{T \times (\log T)^{d+4}}\right)$, where $\gamma_T \sim \mathcal{O}\left((\log T)^{d+1}\right)$ is the maximum information gain for the squared exponential kernel.*

Proof. Let $x_t = \operatorname*{argmax}_{x \in \mathcal{X}}\alpha_t^{E^3I}(x)$ be the choice at iteration t, the instantaneous regret is:

$$Mr_t = Mf(x^*) - Mf(x_t)$$

$$= Mf(x^*) - Mf(x_t) + \sum_{m=1}^{M} g^*_{t-1,m} - \sum_{m=1}^{M} g^*_{t-1,m}$$

$$= \underbrace{\sum_{m=1}^{M}\left[f(x^*) - g^*_{t-1,m}\right]}_{A_t} + \underbrace{\sum_{m=1}^{M}\left[f(x_t) + g^*_{t-1,m}\right]}_{B_t}$$

We need to connect this with the maximum information gain, γ_T. This can be done by bounding r_t with the GP posterior variance. We bound A_t with the using Lemma 2, Lemma 1, and the fact that $\alpha_t^{E^3I}(x^*) \leq \alpha_t^{E^3I}(x_t)$ to get

$$A_t = \sum_{m=1}^{M} \left[f(x^*) - g_{t-1,m}^* \right] = \sum_{m=1}^{M} I_{t,m}(x)$$

$$A_t \leq M \left[\alpha_t^{E^3I}(x^*) + \sqrt{\beta_t}\sigma_{t-1}(x^*) \right]$$

$$\leq M \left[\alpha_t^{E^3I}(x_t) + \sqrt{\beta_t}\sigma_{t-1}(x^*) \right] \qquad \text{by Lemma 2}$$

$$= M \left[\sigma_{t-1}(x_t)\tau(z_{t-1}(x_t)) + \sqrt{\beta_t}\sigma_{t-1}(x^*) \right] \qquad \text{by Lemma 1}$$

Likewise, we bound B_t with the following:

$$B_t = \sum_{m=1}^{M} \left[g_{t-1,m}^* - \mu_{t-1}(x_t) + \mu_{t-1}(x_t) - f(x_t) \right]$$

$$\leq \sum_{m=1}^{M} \left[\sigma_{t-1}(x_t)(-z_{t-1}(x_t)) + \sigma_{t-1}(x)\sqrt{\beta_t} \right] \qquad \text{by Lemma 1}$$

$$= M\sigma_{t-1}(x_t) \left[\tau(-z_{t-1}(x_t)) + \sqrt{\beta_t} - \tau(z_{t-1}(x_t)) \right] \qquad \text{by } z = \tau(z) - \tau(-z)$$

Combining these bounds and noting that the M term cancels out, we get

$$r_t \leq \left[\sigma_{t-1}(x_t) \left[\sqrt{\beta_t} + \tau(-z_{t-1}(x_t)) \right] + \sqrt{\beta_t}\sigma_{t-1}(x^*) \right]$$

Using the bound of $\tau(-z_{t-1}(x_t))$ in Lemma 9 from [14] and setting $C \triangleq \log \left[\frac{1}{2\pi\kappa^2} \right]$ we can simplify this to

$$r_t \leq \underbrace{\sigma_{t-1}(x_t) \left[\sqrt{\beta_t} + 1 + C \right]}_{L_t} + \underbrace{\sqrt{\beta_t}\sigma_{t-1}(x^*)}_{U_t}$$

We now look at the sum of the regret,

$$R_t = \sum_{t=1}^{T} r_t \leq \sum_{t=1}^{T} L_t + \sum_{t=1}^{T} U_t$$

Using the Cauchy-Schwartz inequality that $(a + b + c) \leq 3(a^2 + b^2 + c^2)$, that $\beta_T \geq \beta_t, \forall t \leq T$, and Lemma 7 from [14]) we can bound $\sum_{t=1}^{T} L_t$ with the following

$$\sum_{t=1}^{T} L_t \leq \sum_{t}^{T} \sigma_{t-1}^2(x_t)3(\beta_t + 1 + C)$$

$$\leq 3(\beta_T + 1 + C) \sum_{t}^{T} \sigma_{t-1}^2(x_t) \leq \frac{6(\beta_T + 1 + C)\gamma_T}{\log(1 + \sigma^{-2})}$$

Using the Cauchy-Schwartz inequality again we get

$$\sum_{t=1}^{T} L_t \le \sqrt{T} \sqrt{\sum_{t=1}^{T} L_t} \le \sqrt{\frac{6T(\beta_T + 1 + C)\gamma_T}{\log(1 + \sigma^{-2})}} \tag{7}$$

We can use Lemma 7 [14] and the Cauchy-Schwartz inequality on $\sum_{t=1}^{T} U_t$ as well to obtain a similar result:

$$\sum_{t=1}^{T} U_t \le \beta_T \sum_{t=1}^{T} \sigma_{t-1}(x^*) \le \sqrt{\frac{2T\beta_T\gamma_T}{\log(1 + \sigma^{-2})}} \tag{8}$$

Combining Eqs. (7) and (8) gives us our regret bound:

$$R_T \le \sqrt{\frac{2T\gamma_T}{\log(1 + \sigma^{-2})}} \left[\sqrt{3(\beta_T + 1 + C)} + \sqrt{\beta_T} \right]$$

The function of the maximum information gain, $\sqrt{T \times \gamma_T}$, will usually dominate this expression as $\beta_T \sim \mathcal{O}\left((\log T)^2\right)$. It is kernel dependent but for the squared exponential kernel used in this paper it is $\gamma_T \sim \mathcal{O}\left((\log T)^{d+1}\right)$. This means that our regret bound for this kernel is $R_T \sim \mathcal{O}\left(\sqrt{T \times (\log T)^{d+1}}\right)$, which vanishes in the limit of $\lim_{T \to \infty} \frac{R_T}{T} = 0$. We note that we achieve the similar form to the one in [14].   □

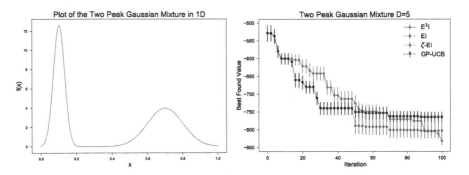

Fig. 1. A plot of the 1D Gaussian mixture function for illustration (left) and the performance of various methods on the 5D version of the same Gaussian mixture function (right). The higher dimensional function was used as little exploration is required in 1D. Lower is better. Note that GP-UCB and EI both get stuck on the initial lower value peak. ζ-EI manages to find the larger peak, but fails to exploit it. This suggests that it may be over-exploring due to an imperfect choice of ζ. On the contrary, it can be seen that E^{3}I is able to obtain a superior result through better late-stage exploitation.

4   Experiments

In this section we outline and discuss our experimental results. We apply our method to synthetic, benchmark and real world functions. We also test some other important properties of our method, such as the dependence on M and the convergence of the Thompson samples. **The code used for this paper can be found at** https://github.com/jmaberk/BO_with_E3I.

4.1   Experimental Setup

We performed several experiments comparing E^3I with the standard EI, ζ-EI with $\zeta = 0.01$ [2], and GP-UCB. In our algorithm, we scaled the inputs to be in the range $[0, 1]$ in all dimensions and standardized the sampled function values to have zero mean and a standard deviation of 1. This guarantees that our kernel magnitude will be scaled correctly for all functions. After this scaling, we use a square exponential kernel.

We ran our experiment 10 times per function with $d+1$ random initial points, where d is the number of input dimensions. The experiment was stopped after $T = 20d$ iterations. As we used a simple multi-start L-BFGS-B optimizer, we minimized the negative of all functions instead of maximizing them. As such, lower results are better.

4.2   Synthetic Multi-peak Function

As we expect our method to have higher exploration than EI, we will test it on functions which require exploration for better performance. In particular, we consider multi-peak functions. Methods with poor exploration can get stuck on sup-optimal peaks, significantly reducing performance. As such, we chose to use a two-peak Gaussian mixture function to verify the high exploration of our method. One peak was chosen to be wide ($\mathcal{N}(0.7, 0.01)$) while the other peak was chosen to be narrow but taller ($\mathcal{N}(0.1, 0.001)$) so that acquisition functions with poor exploration will tend to get stuck on the wider, smaller peak more often and hence not perform as well. We applied our suite of acquisition functions to the problem and have summarised our results in Fig. 1.

It is evident that our method is both better able to find the narrow peak faster, and that the performance gap increases as the number of dimensions increases.

4.3   Benchmark Functions

We also tested our method on several common multi-peak benchmark functions. These include the Levy (5D), Schwefel (4D), Shubert (2D), and Ackley (5D) functions[1]. The results for these are displayed in Fig. 2.

---

[1] All benchmark functions use the recommended parameters from https://www.sfu.ca/~ssurjano/optimization.html.

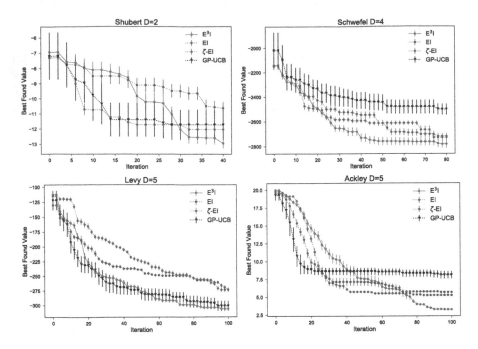

Fig. 2. Performance of various methods on a range of multi-peak benchmark functions. Lower is better. Note that our method generally seems to converge slowly at the early stages when other methods are exploiting, but it can beat the exploitative methods by finding better peaks. This is because E^3I tends to explore in the early stages and then tends to exploit later to hit the optimum.

These results show that our method is able to find a better optima more quickly than the other methods in these multi-peak test functions. GP-UCB also does very well with the Levy and Shubert functions, while the ζ-EI does fairly well on all experiments except the Levy function. This variance in performance is unsurprising, as both methods have parameters that control their level of exploration. If these are not suited to the problem, they can be detrimental to the algorithm's performance. Our methods increased exploration is automatically adjusted through the Thompson samples and therefore does not face this issue. As such, even in cases where these methods performed well, E^3I was able to show improvements over them.

4.4   Machine Learning Hyperparameter Tuning

Finally, we tested our method on a real-world application; the determination of optimal hyperparameters for a machine learning algorithm called *Bayesian Nonparametric Multi-label Classification* (BNMC) [12]. This algorithm is used to efficiently classify multi-labelled data by exploiting the correlation between the multiple labels and features. Its performance is dependant on six hyperparameters which we can tune with Bayesian optimization. These are the Dirichlet

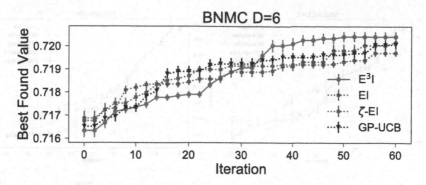

Fig. 3. The results for hyperparameter tuning a BNMC experiment. Note that, again, our algorithm performs poorly in the early stages while it explores for new optima but is able to find a better optima sooner than the other methods.

concentration parameters for both the feature and label, the learning rates for both SVI and SGD, the truncation threshold and the stick-breaking parameter. The data the algorithm was used on, called SceneData, consisted of 1196 test and 1211 training samples, each with 294 features. The results of this with our suite of acquisition functions is given in Fig. 3 with the F1 score used as the performance measure.

4.5 Sensitivity Analysis with Respect to the Number of Optima Samples, M

One of our key assumptions is that we can approximate the expectation over the distribution of Thompson samples as its sample mean. As $M \to \infty$, this will be true. However, using a very large M will increase computational costs. As such, we wish to find a value for M that will not compromise our results while also not being too expensive. To determine this, we ran the same experiment on both a 2D Shubert function and a 2D Schwefel function for a range of M values from 1 to 200. The results of which are summarised in Fig. 4. From these results, we can see that $M = 100$ seems to be an appropriate number of Thompson samples.

4.6 Computational Considerations

While our method has competitive performance with other methods, it has a considerable computational cost. Each Thompson sample requires both the inversion of a $V \times V$ matrix and a global optimization step which may increase exponentially with the dimension, d. The overall cost scales significantly with both the number of data points and the number of input dimensions, making it $\mathcal{O}(MNV^2)$ where N is the number of observations [13]. To give this some context, the average time per iteration for the 2D Shubert function earlier was 0.33 s with EI and 26 s for E^3I. Moving up to 4D with the Schwefel function, these become 1.4

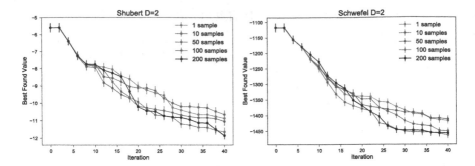

Fig. 4. Performance of E^{3}I on two functions with various number of Thompson samples, M. We can see that increasing M noticeably improves the results until $M = 50$, after which time little improvement is seen.

and 81 s respectively. These may seem high, but they are negligible when compared to the costs associated with sampling in many of the areas that Bayesian optimization is applied to.

One way to potentially reduce computational costs is to use a method by Wang et al. [21] to find the maxima of Thompson samples by sampling a Gumbel distribution. However, this method makes several assumptions that may lead to inaccurate results and as such is left for future work.

4.7 Empirical Convergence Analysis of Thompson Samples

One of the key assumptions of our method is that the Thompson sample functions will converge to the true function as T increases. This convergence was experimentally tested and the results are shown in Fig. 5. It is evident that

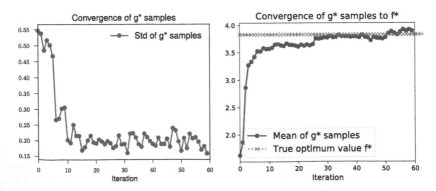

Fig. 5. The reduction of standard deviation between the Thompson sample function maxima (left) and the convergence of $\bar{g}^*$ to $f^* = \max_x(f(x))$ (right). We can see that as more samples are taken, the inter-sample variance is reducing and their mean is approaching f^*. This suggests that they are properly converging as the space is explored.

the sample mean, $\bar{g}^* = \frac{1}{M}\sum_{m=1}^{M} g_m^*$ is converging to f^* and that the sample standard deviation, calculated with $\sigma(g^*) = \sqrt{\frac{1}{M}\sum_{m=1}^{M}(g_m^* - \bar{g}^*)^2}$, is reducing with the number of iterations. These suggest that the Thompson samples are converging properly.

5 Conclusion

We have proposed a new approach for balancing exploration and exploitation in Bayesian optimization. Our approach makes use of Thompson sampling to guide the level of exploration. This results in the E^3I acquisition function.

Our method has been shown to perform better than competing methods on both several multi-peak test functions and on hyperparameter tuning for a BNMC experiment. We also show that it has a sub-linear regret bound.

The most important next step in improving E^3I is to resolve some of its computational issues. Beyond this, the effects of similar distribution-based approaches should be explored on other acquisition functions besides EI.

Acknowledgements. This research was supported by an Australian Government Research Training Program (RTP) Scholarship awarded to Mr Berk, and was partially funded by the Australian Government through the Australian Research Council (ARC). Prof Venkatesh is the recipient of an ARC Australian Laureate Fellowship (FL170100006).

Appendix A: E^3I Derivation

In this section, we provide the analytical derivation of E^3I, described in Eq. (6). In particular, we make use of the improvement function over the perceived optima sample generated from Thompson sampling, $I(x) = \max(f(x) - g^*, 0)$.

We wish to find the PDF of $I(x)$ so that later we can take its expectation. As we are modeling the system with a Gaussian process, we assume that $f(x) \sim \mathcal{N}(\mu(x), \sigma(x))$. This means that $f(x)$ has the PDF

$$p(f(x)) = \frac{1}{\sqrt{2\pi}\sigma(x)} \exp\left(\frac{-(f(x) - \mu(x))^2}{2\sigma^2(x)}\right) \tag{9}$$

Now that we have a PDF for $f(x)$, we can use it to find the PDF of $I(x)$ with the distribution function technique. Let us look at the CDF of $I(x)$ for $f(x) > g^*$ with the substitution $f(x) = I(x) + g^* \,\forall f(x) > g^*$:

$$CDF_{I(x)}(a) = \int_0^a \frac{1}{\sqrt{2\pi}\sigma(x)} \exp\left(\frac{-(I(x) + g^* - \mu(x))^2}{2\sigma^2(x)}\right) dI \tag{10}$$

Taking the partial derivative with respect to $I(x)$ this gives us its the PDF:

$$p(I(x)) = \frac{1}{\sqrt{2\pi}\sigma(x)} \exp\left(\frac{-(I(x) + g^* - \mu(x))^2}{2\sigma^2(x)}\right) \tag{11}$$

Now that we have the PDF, can take its expectation to derive our acquisition function:

$$\alpha^{E^3I}(x) = \mathbb{E}_{g^*}\left[\int_0^\infty \frac{I(x)}{\sqrt{2\pi}\sigma(x)} \exp\left(\frac{-(I(x) + g^* - \mu(x))^2}{2\sigma^2(x)}\right) dI(x)\right] \quad (12)$$

Unfortunately, g^* does not have a tractable algebraic expression. To circumvent this, we approximate the expectation over g^* with the sample mean. Assuming that we have M samples of g^* our acquisition function becomes

$$\alpha^{E^3I}(x) = \sum_{m=1}^M \int_0^\infty \frac{I_m(x)}{\sqrt{2\pi}\sigma(x)} \exp\left(\frac{-(I_m(x) + g_m^* - \mu(x))^2}{2\sigma^2(x)}\right) dI_m(x) \quad (13)$$

As each g_m^* is now a constant, the expression inside the summation is now functionally the same expression as found in this stage of the derivation of EI. This means that E^3I can be expressed as a sum of standard expected improvement acquisition functions with $z = \frac{\mu(x) - g_m^*}{\sigma(x)}$:

$$\alpha^{E^3I}(x) = \begin{cases} \frac{1}{M}\sum_{m=1}^M \left[(\mu(x) - g_m^*)\Phi(z) + \sigma(x)\phi(z)\right], & \text{if } \sigma(x) > 0 \\ 0 & \text{if } \sigma(x) = 0 \end{cases} \quad (14)$$

References

1. Bochner, S.: Lectures on Fourier Integrals: With an Author's Supplement on Monotonic Functions, Stieltjes Integrals and Harmonic Analysis. Princeton University Press, Princeton (1959). Translated from the Original German by Morris Tenenbaum and Harry Pollard
2. Brochu, E., Cora, V.M., de Freitas, N.: A tutorial on Bayesian optimisation of expensive cost functions, with application to active user modeling and hierarchical reinforcement learning. arxiv.org (2010)
3. Bull, A.D.: Convergence rates of efficient global optimization algorithms. J. Mach. Learn. Res. **12**(Oct), 2879–2904 (2011)
4. González, J., Longworth, J., James, D.C., Lawrence, N.D.: Bayesian optimization for synthetic gene design. arXiv preprint arXiv:1505.01627 (2015)
5. Hernández-Lobato, J.M., Hoffman, M.W., Ghahramani, Z.: Predictive entropy search for efficient global optimization of black-box functions. In: Advances in Neural Information Processing Systems, pp. 918–926 (2014)
6. Jalali, A., Azimi, J., Fern, X., Zhang, R.: A Lipschitz exploration-exploitation scheme for Bayesian optimization. In: Blockeel, H., Kersting, K., Nijssen, S., Železný, F. (eds.) ECML PKDD 2013. LNCS (LNAI), vol. 8188, pp. 210–224. Springer, Heidelberg (2013). https://doi.org/10.1007/978-3-642-40988-2_14
7. Jones, D.R., Schonlau, M., Welch, W.J.: Efficient global optimization of expensive black-box functions. J. Global Optim. **13**(4), 455–492 (1998)
8. Kushner, H.J.: A new method of locating the maximum point of an arbitrary multipeak curve in the presence of noise. J. Basic Eng. **86**(1), 97–106 (1964)

9. Li, C., et al.: Rapid Bayesian optimisation for synthesis of short polymer fiber materials. Sci. Rep. **7**(1), 5683 (2017). https://doi.org/10.1038/s41598-017-05723-0

10. Lizotte, D.J.: Practical Bayesian Optimization. Ph.D thesis, University of Alberta (2008)

11. Mockus, J.: Application of Bayesian approach to numerical methods of global and stochastic optimization. J. Global Optim. **4**(4), 347–365 (1994)

12. V. Nguyen, S. Gupta, S. Rana, C. Li, and S. Venkatesh. A Bayesian nonparametric approach for multi-label classification. In: Asian Conference on Machine Learning, pp. 254–269 (2016)

13. Nguyen, V., Gupta, S., Rana, S., Li, C., Venkatesh, S.: Predictive variance reduction search. In: NIPS Workshop on Bayesian Optimization, vol. 12 (2017)

14. Nguyen, V., Gupta, S., Rana, S., Li, C., Venkatesh, S.: Regret for expected improvement over the best-observed value and stopping condition. In: Asian Conference on Machine Learning, pp. 279–294 (2017)

15. Rasmussen, C.E., Williams, C.K.I.: Gaussian Processes for Machine Learning. MIT Press, Cambridge (2006)

16. Ryzhov, I.O.: On the convergence rates of expected improvement methods. Oper. Res. **64**(6), 1515–1528 (2016)

17. Snoek, J., Larochelle, H., Adams, R.P.: Practical Bayesian optimization of machine learning algorithms. In: NIPS, pp. 2951–2959 (2012)

18. Srinivas, N., Krause, A., Kakade, S., Seeger, M.: Gaussian process optimization in the bandit setting: no regret and experimental design. In: Proceedings of the 27th International Conference on Machine Learning, pp. 1015–1022 (2010)

19. Turgeon, M., Lustig, C., Meck, W.H.: Cognitive aging and time perception: roles of Bayesian optimization and degeneracy. Front. Aging Neurosci. **8**, 102 (2016). https://doi.org/10.3389/fnagi.2016.00102

20. Wang, Z., de Freitas, N.: Theoretical analysis of Bayesian optimisation with unknown Gaussian process hyper-parameters. In: NIPS Workshop on Bayesian Optimization (2014)

21. Wang, Z., Jegelka, S.: Max-value entropy search for efficient Bayesian optimization. In: International Conference on Machine Learning, pp. 3627–3635 (2017)

A Left-to-Right Algorithm for Likelihood Estimation in Gamma-Poisson Factor Analysis

Joan Capdevila[1,2(✉)], Jesús Cerquides[3], Jordi Torres[1,2], François Petitjean[4], and Wray Buntine[4]

[1] Universitat Politècnica de Catalunya (UPC), Barcelona, Spain
{jc,torres}@ac.upc.edu
[2] Barcelona Supercomputing Center (BSC), Barcelona, Spain
[3] Institut d'Investigació en Intel.ligència Artificial (IIIA-CSIC), Bellaterra, Spain
cerquide@iiia.csic.es
[4] Monash University, Victoria, Australia
{francois.petitjean,wray.buntine}@monash.edu

Abstract. Computing the probability of unseen documents is a natural evaluation task in topic modeling. Previous work has addressed this problem for the well-known Latent Dirichlet Allocation (LDA) model. However, the same problem for a more general class of topic models, referred here to as Gamma-Poisson Factor Analysis (GaP-FA), remains unexplored, which hampers a fair comparison between models. Recent findings on the exact marginal likelihood of GaP-FA enable the derivation of a closed-form expression. In this paper, we show that its exact computation grows exponentially with the number of topics and non-zero words in a document, thus being only solvable for relatively small models and short documents. Experimentation in various corpus also indicates that existing methods in the literature are unlikely to accurately estimate this probability. With that in mind, we propose L2R, a left-to-right sequential sampler that decomposes the document probability into a product of conditionals and estimates them separately. We then proceed by confirming that our estimator converges and is unbiased for both small and large collections. Code related to this paper is available at: https://github.com/jcapde/L2R, https://doi.org/10.7910/DVN/GDTAAC.

Keywords: Topic models · Gamma-Poisson · Factor Analysis
Left-to-right · Importance Sampling · Estimation methods

1 Introduction

Probabilistic topic models [1] have enabled the thematic exploration of document collections at a scale which would have been unfeasible for unassisted humans. Despite the growing interest in these models, there is some disagreement on

© Springer Nature Switzerland AG 2019
M. Berlingerio et al. (Eds.): ECML PKDD 2018, LNAI 11052, pp. 638–654, 2019.
https://doi.org/10.1007/978-3-030-10928-8_38

the methodologies to evaluate and compare them. Their unsupervised nature makes it difficult to propose a silver-bullet metric and this has led to a myriad of application-specific methods for evaluation ranging from document classification to word prediction. However, their probabilistic nature also suggests that computing the likelihood of a collection of held-out documents is a legitimate measure of the generalization capabilities of these models, independent of their final application.

Although previous work [4,15] has looked at this problem for the well-known Latent Dirichlet Allocation (LDA) [2], similar studies have not yet been conducted for related models like Gamma-Poisson (GaP) [5] and Poisson matrix factorisation (PMF) [7], and related forms of non-negative matrix factorisation [9]. GaP, PMF and their extensions, referred here to as GaP Factor Analysis (GaP-FA), represent a more general and expressive class of models [3] which explicitly take into account the document length. Because of this, GaP-FA have been successfully applied in many other domains beyond topic modeling [5,17]. For instance, they have been used to include implicit feedback in recommendation systems [7] or to perform statistical relational learning in sparse networks [16]. Therefore, the intrinsic evaluation of GaP-FA in terms of the likelihood of held-out data becomes relevant for much broader domains, even though in this paper we only consider applications of text analysis.

Computing the probability of a single document requires integrating out all document-level latent variables. This marginal distribution has no analytical solution in the original GaP model, but recent progress in the field has enabled the derivation of a closed-form expression by means of an augmented GaP model [6]. Nonetheless, we will show that the complexity of the exact solution grows exponentially with the number of topics and the number of non-zero words in the document. Moreover, the base of the exponential depends on the maximum count of any word in the document. This means that the exact marginal is only tractable in reasonably small scenarios. Thus, approximation methods to the marginal document likelihood are essential for evaluating GaP-FA under more realistic conditions.

Simple approximation methods, such as Direct Sampling (DS) or the Harmonic Mean (HM) method [12], are known to produce inaccurate estimations, particularly in high-dimensional setups. Despite this, their ease of implementation and low computational cost have promoted their use in LDA-like models [8,14]. As a result of this misuse, there is a need for more accurate and computationally efficient estimation methods. One approach which has been reported to output state-of-the-art results in LDA is the Left-to-right Sequential Sampler [4]. By leveraging the chain rule of probability, the algorithm decomposes the joint document probability into a product of conditionals, one conditional per word. Then, unbiased estimates can be built for each conditional given the posterior samples on the left-hand topics. However, three issues arise in GaP-FA due to the Gamma-Poisson construction: (1) The posterior distribution over the left-hand topic assignments is not tractable. (2) The computational cost of each exact conditional is exponential with the number of topics. (3) The time complexity grows quadratically with the number of non-zero words.

In this paper, we propose L2R, a left-to-right sequential sampler for GaP-FA that addresses (1) by means of Gibbs sampling the augmented model, (2) via Importance Sampling with proposal distributions conditioned on the left-hand samples (3) through a mathematical simplification that enables computing the conditional probability for all zero words at once. Moreover, we compare the accuracy of L2R to that of existing estimation methods in two different setups:

- in reasonably small scenarios, where the exact marginal can be assessed in moderate time and hence, conclusions about their accuracy can be drawn;
- in realistic scenarios, where the exact marginal and the vanilla left-to-right are computationally unfeasible and hence, only their convergence can be studied.

In the rest of this paper, we introduce some preliminary concepts about GaP-FA in Sect. 2. In Sect. 3, we formulate the problem in terms of computing the marginal document likelihood. We present an overview of existing estimation methods in Sect. 4. In Sect. 5, we describe the L2R algorithm. Finally, Sect. 6 contains the experimental work carried out in both scenarios.

2   Background

2.1   Gamma-Poisson Factor Analysis (GaP-FA)

Poisson Factor Analysis (PFA) is a type of discrete component or factor analysis [3] with Poisson likelihoods. This means that PFA assumes that the full count matrix $Y \in \mathbb{N}_0^{N \times W}$, where N refers to the number of documents and W to the vocabulary size, can be generated from a multivariate Poisson distribution parametrized through the product of two smaller matrices,

$$Y \sim \text{Pois}(\Theta\Phi) \tag{1}$$

where $\Theta \in \mathbb{R}_+^{N \times K}$ is the factor score matrix and $\Phi \in \mathbb{R}_+^{K \times W}$, the factor loading matrix. K refers to the dimension of the latent factors or topics in topic modeling. This method can be augmented with latent factor/topics counts x_{nwk} and express each count y_{nw} as the sum of the K independent counts,

$$y_{nw} = \sum_{k=1}^{K} x_{nwk}, \quad x_{nwk} \sim \text{Pois}(\theta_{nk}\phi_{kw}) \tag{2}$$

where y_{nw} are the observed word counts of the w-th word in the n-th document, and x_{nwk} corresponds to the hidden or latent counts in the k-th topic for the same document and word. θ_{nk} and ϕ_{kw} refer to the corresponding row/column entries in matrices Θ, Φ, respectively.

Several models have been developed from this by placing different types of priors over the factor score or loading matrices [17]. In this work, we restrict attention to methods that assume a Gamma distribution for each factor score. We refer to these models as Gamma-Poisson Factor Analysis (GaP-FA). This group includes

Non-negative Matrix Factorization (NMF) [9], gamma-Poisson (GaP) [5] and the three hierarchical models Γ-PFA, $\beta\Gamma$-PFA, $\beta\gamma\Gamma$-PFA presented in [17], among many others. For mathematical convenience, we consider the shape-scale parameterization of the Gamma distribution,

$$\theta_{i,k} \sim \mathrm{Ga}\left(r_k, \frac{p_k}{1-p_k}\right) \quad k = 1...K \tag{3}$$

where r_k corresponds to the shape and $\frac{p_k}{1-p_k}$ to the scale of the k-th factor. To satisfy the constraints of the Gamma, we must ensure that $r > 0$ and $0 < p < 1$.

Next, we review two compound probability distributions that are the result of assuming that the rate of a univariate and multivariate Poisson distribution is controlled by a Gamma random variable. These distributions will be useful in deriving marginal and conditional likelihoods for GaP-FA.

2.2   Negative Binomial (NB)

The Negative Binomial (NB) distribution is a discrete distribution for the number of successes in a sequence of i.i.d Bernoulli trials with probability p after observing a given number of r failures. The NB can be constructed by marginalizing a Poisson distribution whose rate θ is controlled by a gamma random variable parameterized as in Eq. (3) above. In other words,

$$\mathrm{NB}(x;r,p) = \int \mathrm{Pois}(x|\theta)\,\mathrm{Ga}\left(\theta;r,\frac{p}{1-p}\right)d\theta. \tag{4}$$

2.3   Negative Multinomial (NM)

The Negative Multinomial (NM) distribution [13] is the multivariate generalization of the NB distribution to W outcomes ($W > 1$), each occurring with probability q_w and for a given number of failures r.

As shown in [6], the NM can be built by marginalizing W independent Poisson distributions whose rate is controlled by a gamma random variable θ that is scaled by a vector $\phi_:$ of length W. This can be expressed mathematically as,

$$\mathrm{NM}\left(x_:;r,q_:=\frac{p\phi_:}{1-p+p\sum_w \phi_w}\right) = \int \prod_w \mathrm{Pois}(x_w|\theta\phi_w)\,\mathrm{Ga}\left(\theta;r,\frac{p}{1-p}\right)d\theta \tag{5}$$

where r are the number of failures and $q_: = \frac{p\phi_:}{1-p+p\sum_w \phi_w}$ is the vector of W success probabilities. When $\phi_:$ is a probability vector, which sums up to 1, the success probabilities of the NM become $q_: = p\phi_:$.

3   Problem Statement

A common and reasonable strategy to compute the probability of unseen documents in topic models is to use point estimates for the set of global parameters, instead of a fully Bayesian approach which would marginalize across all parameters [4,15]. This enables factorizing the held-out probability across documents in GaP-FA since documents are conditionally independent given the global parameters. As a result of this, the problem then reduces to calculating the *marginal document likelihood* for each held-out document independently by integrating out the document-level latent variables. We can express this for the GaP-FA model in Eq. (1) as,

$$p(\boldsymbol{Y}; \Phi, p, r) = \prod_n p(y_{n:}; \Phi, p, r) = \prod_n \int p(y_{n:}, \theta_{n:}; \Phi, p, r) \, d\theta_{n:} \qquad (6)$$

where the first equality expresses that documents $y_{n:}$ are independent given the set of global parameters $\Omega = \{\Phi, p, r\}$, and the second equality says that this probability is equal to the product across all marginal document likelihoods.

Next, we focus on deriving a closed-from expression for the *marginal document likelihood* in GaP-FA, $p(y_{n:}; \Phi, p, r)$, and show the computational issues that arise in its evaluation. It is important to note that the derivation and approximation of this marginal in the rest of this paper is equivalent for testing and training documents, so we will use $y_{n:}$ indistinctly to refer to both.

3.1 Exact Marginal Document Likelihood in GaP-FA

Following [6], the marginal likelihood of the n-th document in GaP-FA can be written from the augmented model in Eq. (2). Note that we can write the marginal as the sum of the marginal on $x_{n::}$ over all possible topic counts, which must add up to the observed counts $y_{n:}$ in the n-th document. This can be expressed formally as,

$$p(y_{n:}; \Phi, p, r) = \sum_{x_{n::} \in \mathbb{X}_{y_{n:}}} \prod_k p(x_{n:k}; r_k, p_k, \phi_{k:}) \qquad (7)$$

where the summation set $\mathbb{X}_{y_{n:}} = \{x_{n::} \in (\mathbb{N}_0)^{W \times K} \mid y_{n:} = \sum_{k=1}^{K} x_{n:k}\}$ corresponds to all the possible partitions of the topic counts in the n-th document into K parts. Factorization across the marginals on the topic counts is due to independence across these counts, as in Eq. (2).

Then, deriving a closed-form expression for the marginal document likelihood boils down to finding an analytical expression for $p(x_{n:k}; r_k, p_k, \phi_{k:})$. Following [6], this probability can be calculated by marginalizing out θ_{nk} in the augmented model. Moreover, a parametric distribution can be derived by noting that this marginal matches the NM distribution definition introduced in Eq. (5). In other words, the marginal distribution on the counts of the k-th topic can be written as,

$$p(x_{n:k}; \phi_{k:}, p_k, r_k) = \int \prod_w \text{Pois}(x_{nwk}|\theta_{nk}\phi_{kw})\text{Ga}\left(\theta_{nk}; r_k, \frac{p_k}{1-p_k}\right) d\theta_{nk}$$

$$= \text{NM}\left(x_{n:k}; r_k, q_{k:} = \frac{p_k\phi_{k:}}{1 - p_k + p_k\sum_w \phi_{kw}}\right) \quad (8)$$

where $\phi_{k:}, p_k, r_k$ are topic-dependent and $x_{n:k}, \theta_{nk}$ document-dependent too.

3.2 On the Time Complexity of the Exact Marginal

Evaluating Eq. (7) means summing the independent marginals on $x_{n:k}$ over all elements in the set $\mathbb{X}_{y_{n:}}$. As shown in Eq. (8), each marginal consists of a NM distribution which has a cost linear with the number words W in an unoptimized implementation of NM, or linear with the number of non-zero word counts W_c when all zero words are evaluated together. Therefore, the cost of each summand is linear with both the number of topics K and the number of non-zeros W_c, since K marginals need to be computed for each summand.

The number of sums in Eq. (7) equals the cardinality of the set $|\mathbb{X}_{y_{n:}}|$. As shown in [6], the cardinality is given by the product of the partitions in each word w. The latter consist of the number of partitions of a natural number, i.e. y_{nw}, into K parts, which is the combinatorial term of selecting $K-1$ objects from a collection of $y_{nw} + K - 1$. Therefore, the overall number of partitions for document n is $\prod_{\{w|y_{nw}\neq 0\}} \binom{y_{nw}+K-1}{K-1}$, where $\{w|y_{nw} \neq 0\}$ corresponds to the W_c non-zeros.

In the limit, one can show that this set grows exponentially with both the number of topics and the number of non-zeros $\mathcal{O}((y_{nmax})^{KW_c})$. We note that the base of the exponent is the maximum word count in the n-th document y_{nmax}. Therefore, the cost of summing over the set $|\mathbb{X}_{y_{n:}}|$ dominates the complexity of evaluating the exact marginal document likelihood.

As a result, the exact evaluation of the marginal document likelihood for GaP-FA is only tractable for reasonably small problems, such as in models with 5 topics, documents with 10 non-zero words and all words having 1 or 2 counts. However, the existence of this closed-from expression motivates the development of tailored estimation methods and to calibrate their outputs with the exact.

4 Related Work

Wallach et al. [15] presented several estimation methods for evaluating LDA in terms of held-out likelihood. Buntine [4] also compared the performance of these methods against the exact calculation for the same LDA model. The conclusion of both studies was that simple and commonly-used estimation methods fail to accurately estimate the document likelihood, specially in high-dimensional scenarios. But Wallach's Left-to-right algorithm was modified to a Sequential Sampler scheme and proven to be unbiased by Buntine. Given the quick convergences and unbiasedness properties of the Left-to-right Sequential Sampler, it can now be used as a gold standard for estimation in LDA with large number of samples.

To the best of our knowledge, no prior work exists for document likelihood estimation in GaP-FA. However, it is natural to wonder whether LDA methods can be directly applied in GaP-FA. As we have seen previously, the Gamma-Poisson construction differs from that of LDA and the time complexity of its marginal document likelihood is far more complex. The number of sums in LDA grows exponentially with the document length. Therefore, existing estimation methods [4,15] for LDA have to be amended accordingly. Next, we discuss the amendments and limitations imposed by GaP-FA.

In contrast with LDA, *Direct Sampling (DS)* or Importance Sampling with the prior as proposal cannot be formulated over the discrete variables of the augmented model $x_{n::}$, because the observed counts $y_{n:}$ follow a deterministic relationship with the topic counts $x_{n::}$. Therefore, DS has to be formulated over the continuous variables $\theta_{n:}$ as the Monte Carlo sampling of Eq. (6),

$$p(y_{n:}; \Phi, p, r) \approx \frac{1}{S} \sum_{s=1}^{S} p(y_{n:} | \theta_{n:}^{(s)}; \Phi, p, r) \qquad \text{where} \quad \theta_{n:}^{(s)} \sim p(\theta_{n:}; p, r), \quad (9)$$

where the likelihood $p(y_{n:} | \theta_{n,:}; \Phi, p, r)$ is W-variate Poisson with rates given by the vector $\theta_{n:} \Phi$ and $p(\theta_{n:}; p, r)$ is given by Eq. (3) for each topic $k < K$. Although this estimator is unbiased, the main caveat is that the proposal distribution ignores the observed counts and too many samples might be needed when the prior is far from the joint distribution.

An alternative to this formulation is to use samples from the posterior distribution and build an unbiased estimator through the *Harmonic Mean (HM)* method [12]. To sample the posterior, one needs to consider the augmented GaP-FA and perform Gibbs Sampling on the locally conjugated complete conditionals as in [17]. Although this method has been used in LDA-like topic models [8,14], the same authors expressed some reservations when introducing it due to the non-stable convergence and high variance. Note that this estimator cannot either be built on the discrete variables of the augmented model $x_{n::}$.

In fact, the deterministic relationship between the observed counts and the latent topic counts is what causes difficulties to tune other methods such Annealed Importance Sampling (AIS) [11], which transitions between the prior over the topic assignments and its posterior through a series of tempered distributions, or Chib-style estimators [10].

Related work in Poisson Factorization for topic modeling computes perplexity scores by holding out some random words in the document-term matrix instead of the full document [17]. A similar approach in LDA-like models consists of holding out the second half of a document, while the first half is added to the training data. The evaluation task, known as document completion [15], consists of computing the probability of the second half given the first. Although this task is known to be well correlated but biased for LDA, rigorous studies have not yet been conducted for GaP-FA. This work also paves the way for calibrating a document completion style method against the exact calculation and to develop specialized and unbiased sampling methods that approximate word prediction or document completion.

5   L2R: A Left-to-Right Algorithm for GaP-FA

In this section we present L2R, a tailored left-to-right sequential sampler [4] for GaP-FA. L2R builds on the general product rule of probability, in which any joint distribution can be decomposed into the product of several conditionals. By considering a left-to-right order of words, the joint probability of a document is decomposed by the product of W conditional probabilities where each is conditioned on the preceding left words. We can express this decomposition for GaP-FA as,

$$p(y_{n:}; \Phi, p, r) = \prod_{w=1}^{W} p(y_{nw}|y_{n<w}; \Phi, p, r) \tag{10}$$

where "$< w$" refers to words on the left side of w. Nonetheless, the exact calculation of these conditionals is still as intractable as the previous marginal likelihood. We now introduce the left topic counts $x_{n<w:}$ and marginalize them out as follows,

$$p(y_{n:}; \Phi, p, r) = \prod_{w=1}^{W} \sum_{x_{n<w:}} p(y_{nw}, x_{n<w:}|y_{n<w}; \Phi, p, r). \tag{11}$$

Given that the w-th word counts, y_{nw}, are conditionally independent from the left-hand side counts $y_{n<w}$ given their topic counts $x_{n<w:}$, the joint expression above can be split into two factors as,

$$p(y_{n:}; \Phi, p, r) = \prod_{w=1}^{W} \sum_{x_{n<w:}} p(y_{nw}|x_{n<w:}; \Phi, p, r) p(x_{n<w:}|y_{n<w}; \Phi, p, r). \tag{12}$$

This expression uncovers a sampling structure which suggests to draw samples from the posterior over the topic counts on the left-hand side of w and to evaluate the conditional probability of the current word count given these left samples. In other words, the two step process can be summarized as follows

$$x_{n<w:}^{(s)} \sim p(x_{n<w:}|y_{n<w}; \Phi, p, r) \tag{13}$$

$$p(y_{n:}; \Phi, p, r) \approx \prod_{w=1}^{W} \frac{1}{S} \sum_{s=1}^{S} p(y_{nw}|x_{n<w:}^{(s)}; \Phi, p, r) \tag{14}$$

Next, we present a method for drawing samples from the posterior over the topic counts in Eq. (13) and a strategy to approximate the inner conditionals in Eq. (14). This will enable us to address the first two issues mentioned in the Introduction. Then, we show that if we re-order documents in a particular way, we can avoid computing the product in Eq. (14) across all words in the vocabulary W, which addresses the third issue. Finally, we summarize all these contributions in the pseudo-code for the L2R algorithm and discuss its computational complexity.

5.1 Sampling the Left-Hand Topics

The posterior distribution in Eq. (13) does not have a closed-form expression due to the intractable normalizing constant. Therefore, a common thing to do is to build a Gibbs sampler to draw samples from it. However, the complete conditionals $p(x_{nw':}|x_{n<w:}^{\neg w'}, y_{n<w}; \Phi, p, r)\ \forall w' < w$ do not admit a computationally feasible sampler due to the conditioning on the observed counts $y_{nw'}$.

One way to sample from this posterior is to consider the augmented model in Eq. (2), but only over the left-hand side of w. This makes the model locally conjugate and it enables the derivation of the complete conditionals as,

$$p(\theta_{nk}|-) = \text{Ga}\left(\theta_{nk}; r_k + \sum_{w'<w} x_{n<w'k}, \frac{p_k}{1 - p_k + p_k \sum_{w'<w} \phi_{kw'}}\right)\ \forall k \le K \tag{15}$$

$$p(x_{nw':}|-) = \text{Mult}\left(x_{nw':}; y_{nw'}, \frac{\phi_{:w'}\theta_{n:}}{\sum_k \phi_{kw'}\theta_{nk}}\right)\ \forall w' < w \tag{16}$$

where "$|-)$" refers to all variables except the conditioned. These expressions can be integrated in a Gibbs sampling scheme in which we first sample Eq. (15) and then each of the left word counts as in Eq. (16), or vice-versa. However, only samples from the left-hand topics need to be recorded for the L2R algorithm.

5.2 Approximating the Conditional Probability

The inner conditional probability in Eq. (14) can be expressed as the sum of the marginal on $x_{nw:}$ over all possible topic counts, which must add up to the w-th word count y_{nw}. Given that topic counts are independent among them, the marginal also factorizes. We can write this as,

$$p(y_{nw}|x_{n<w:}^{(s)}; \Phi, p, r) = \sum_{x_{nwk} \in \mathbb{X}_{y_{nw}}} \prod_{k=1}^{K} p(x_{nwk}|x_{n<wk}^{(s)}; \phi_{k:}, p_k, r_k). \tag{17}$$

where the summation set $\mathbb{X}_{y_{nw}} = \{x_{nw:} \in (\mathbb{N} \cup 0)^K \mid y_{n:} = \sum_{k=1}^{K} x_{n:k}\}$ has cardinality $|\mathbb{X}_{y_{nw}}| = \binom{y_{nw}+K-1}{K-1}$.

The marginal above, which is conditioned to the left samples, can be derived by leveraging on the augmented model. By introducing θ_{nk}, the probability of the actual count x_{nwk} becomes conditionally independent of the left samples $x_{n<wk}^{(s)}$ given the introduced θ_{nk}. Therefore, the left samples influence the probability over θ_{nk}, but not that over x_{nwk} as shown,

$$p(x_{nwk}|-) = \int p(x_{nwk}|\theta_{nk}; \phi_{kw})p(\theta_{nk}|x_{n<wk}^{(s)}; \phi_{k:}, p_k, r_k)\,d\theta_{nk} \tag{18}$$

where "$-)$" refers to the set $\{x_{n<wk}^{(s)}, \phi_{k:}, p_k, r_k\}$.

In the integral above, we substitute the probability over θ_{nk} for the Poisson distribution in Eq. (2) and that over θ_{nk} for the complete conditional in Eq. (15). The resulting integral corresponds to the compound probability distribution in Eq. (4), which is a Negative Binomial (NB) parameterized as follows,

$$p(x_{nwk}|-) = \text{NB}\left(x_{nwk}; r_k + \sum_{w'<w} x_{nw'k}^{(s)}, \frac{\phi_{wk}p_k}{1 - p_k + p_k \sum_{w'\leq w} \phi_{w'k}}\right). \quad (19)$$

Although it is possible to compute the exact conditional probability through the closed-form expression given by Eq. (17), its computational cost still grows exponentially with the number of topics (note that the exponential growth is now independent of the number of non-zeros) and hence it is only tractable for a small number of topics or word counts y_{nw}.

Therefore, our alternative to the exact calculation consists in replacing the complicated sum in Eq. (17) with a Monte Carlo estimate. To do that, we propose to perform Importance Sampling with a proposal distribution which is conditioned on the left samples as follows,

$$q(x_{nw:}|x_{n<w:}^{(s)}; \phi_{:w}, p, r) = \text{Mult}(x_{nw:}; y_{nw}, \propto \phi_{:w}\mathbb{E}_{p(\theta_{n:}|x_{n<w:}^{(s)}, \Phi, r, p)}[\theta_{n:}]) \quad (20)$$

where expectation over $\theta_{n:}$ is computed w.r.t the complete conditional in Eq. (15). Given that this proposal is built taking into account the left-hand samples, the proposal will be close to the marginal $x_{nw:}$ as long as the left counts are good predictors of the target.

Finally, we estimate the conditional probability as,

$$
\begin{aligned}
x_{nw:}^{(s')} &\sim q(x_{nw:}|x_{n<w:}^{(s)}; \phi_{:w}, p, r) \\
p(y_{nw}|x_{n<w:}^{(s)}; \Phi, p, r) &\approx \frac{1}{S'} \sum_{s'} \frac{p(x_{nw:}^{(s')}|x_{n<w:}^{(s)}; \Phi, p, r)}{q(x_{nw:}^{(s')}|x_{n<w:}^{(s)}; \phi_{:w}, p, r)}
\end{aligned}
\quad (21)
$$

where S' corresponds to another set of samples which replace the intractable sum in Eq. (17). However, we will show in the experiments that with one single sample $S' = 1$, we can accurately approximate the exact in situations where the topics for the w-th word are likely to be predicted from the preceding topics, which is often the case if some thematic structure exists in the corpus.

5.3   Dealing with Zero Words

The left-to-right decomposition rule in Eq. (10) does not impose any specific word order to be valid. Besides, the inspection of the exact conditional formula from Eqs. (17) (19) reveals that words without counts contribute with a tractable term which only depends on the left-hand counts.

This suggests that if we re-order documents in such a way that all non-zero words precede zeros, we can re-use the posterior samples drawn for non-zero words to calculate the probability of zeros. Note that zeros do not contribute to the posterior sampling over the left-hand topics. This allows one to build a conditional probability for all words without counts $n \geq w_z$ that occur after the non-zeros $< w_z$. A closed-form expression can be derived for this probability which can be computed in linear time with the number of topics as,

$$p(y_{n \geq w_z} | x^{(s)}_{n < w_z}; \Phi, p, r) = \prod_k \left(\frac{1 - p_k + p_k \sum_{w' < w_z} \phi_{w'k}}{1 - p_k + p_k \sum_{w' \leq W} \phi_{w'k}} \right)^{r_k + \sum_{w' < w_z} x^{(s)}_{nw'k}} .$$

(22)

By re-ordering the document, reusing the posterior samples and the mathematical simplification shown above, we can speed up the algorithm from computing the conditional probability across all words in the vocabulary W to only those with non-zero counts W_c. Given that for most corpora, the vocabulary size is larger than the non-zero words per document ($W \geq W_c$), this makes a critical enhancement to the time-complexity of this algorithm as we show later.

5.4   Algorithm Pseudocode

In Algorithm 1, we present the pseudocode of L2R, summarizing the developments from the previous sections. The input data consists of the number of samples S used to approximate each of the factors in the left-to-right decomposition, the number of samples S' to draw from the proposal distribution in the case of sampled conditionals, the n-th document $y_{n:}$ sorted as in Sect. 5.3 and the point estimates for the global parameters $\Omega = \{\Phi, p, r\}$. The algorithm outputs the approximate marginal document likelihood $p(y_{n:}; \Phi, p, r)$.

From line 1 to 5, the algorithm approximates the conditionals distributions for non-zero words by computing the averaged probability across S samples for each word. To approximate this conditional probability, the algorithm uses the Importance Sampling scheme defined in Eq. (21).

From line 6 to 10, the algorithm approximates the conditionals for all words without counts following the same procedure as for non-zeros, except that the conditional for all non-zeros is computed at once in line 9 through its exact form given by Eq. (22).

The final estimate for marginal document likelihood is build from the product of the $W_c + 1$ probabilities in line 11.

5.5   On the Time Complexity of the L2R Algorithm

The time complexity of the L2R algorithm can be derived from the cost of the subprocesses in Algorithm 1. We first note that the cost of computing the conditionals for all non-zero words dominates over that of zeros because line 4 is linear with both the number of samples S' and the number of topics K, whereas

---

Algorithm 1. L2R algorithm

> **input** : $S, S', y_{n:}, \Omega = \{\Phi, p, r\}$
> **output**: $p(y_{n:}; \Omega)$

1 **for** $w \leftarrow 1$ **to** W_c **do**
2   **for** $s \leftarrow 1$ **to** S **do**
3          $x^{(s)}_{n<w:} \leftarrow \texttt{PostSamp}(x^{(s)}_{n<w:}, \Omega);$ *Eqs.* (15) (16)
4          $p(y_{nw}|x^{(s)}_{n<w:}; \Omega) \leftarrow \texttt{CondProb}(x^{(s)}_{n<w:}, \Omega, S');$ *Eq.* (21)
5      $p(y_{nw}|y_{n<w}; \Omega) = \frac{1}{S}\sum_s p(y_{nw}|x^{(s)}_{n<w:}; \Omega)$
6 $w_z \leftarrow W_c + 1$
7 **for** $s \leftarrow 1$ **to** S **do**
8      $x^{(s)}_{n<w_z:} \leftarrow \texttt{PostSamp}(x^{(s)}_{n<w_z:}, \Omega);$ *Eqs.* (15) (16)
9      $p(y_{n\geq w_z}|x^{(s)}_{n<w_z:}; \Omega) \leftarrow \texttt{CondProbZeros}(x^{(s)}_{n<w_z:}, \Omega);$ *Eq.* (22)
10 $p(y_{nw_z}|y_{n<w_z}; \Omega) = \frac{1}{S}\sum_s p(y_{n\geq w_z}|x^{(s)}_{n<w_z:}; \Omega)$
11 $p(y_{n:}; \Omega) \approx \prod_{w \leq w_z} p(y_{nw}|y_{n<w}; \Omega)$

---

line 9 is only linear with the latter. The cost of the posterior sampling process in line 3 and 8 is also linear with the number of topics K and non-zeros W_c. Therefore, the overall cost is given by $\mathcal{O}(W_c S(W_c + K + S'))$ which is quadratic in the number of non-zero words. Note also that without the optimization of zeros it would have been quadratic with the vocabulary size and without the approximate conditionals, exponential with the number of topics.

6 Empirical Results

In this section, we present the comparison results of the L2R algorithm against the exact marginal likelihood, Direct Sampling (DS) and the Harmonic Mean (HM) method. The code for L2R, DS and HM methods, as well as the processed corpora and trained GaP-FA have been made public[1].

6.1 Experiment Setup

We follow the setup in [4] which first compares methods against the exact in tractable scenarios and then looks at convergence in more realistic cases. In addition, we introduce a proper comparison metric, new document collections and a model which also infers the number of topics.

Experiments. We define two sets of experiments. The first consists of comparing the output probabilities of each estimator against the exact marginal likelihood. Given that the computational complexity of the marginal likelihood

---

[1] https://github.com/jcapde/L2R, https://doi.org/10.7910/DVN/GDTAAC.

Table 1. Document collections

Dataset	Vocabulary	Num. docs	Doc. length
NIPS	11,463	5,811	$1,899 \pm 513$
AP	10,473	2,246	194 ± 111
20NGs	11,928	18,846	123 ± 247
Reuters	8,843	19,043	79 ± 75
Twitter	6,344	10,523	25 ± 4
WS	4,679	12,309	9 ± 3

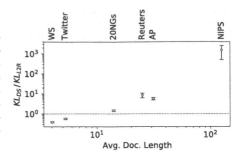

Fig. 1. KL Comparison of L2R vs. DS

is only tractable in *reasonably small scenarios* and it is dominated by the number of sums in Eq. (7), we restrict to scenarios in which the cardinality of the summation set is less than 10^9. To do this, we choose a maximum of 1000 documents from a downsized corpus whose word counts do not exceed this limit in a GaP-FA with a maximum of 5 topics. The second set of experiments consists of assessing the estimator's convergence in *more realistic conditions* for which the marginal likelihood is not tractable. In this setup, we use 1,000 evaluation documents for each collection in a GaP-FA with a maximum of 100 topics.

Comparison Measures. To compare several document probabilities to their true marginal, we propose to use the *Kullback-Leibler (KL)* or relative entropy, which is a proper divergence measure for probability distributions. We can interpret it as the number of extra bits required from using the estimated probabilities instead of the exact in decoding a codebook of length the number of evaluated documents. To study the convergence in realistic scenarios, we plot the *log-likelihood* for all evaluated documents as function of the number of samples.

Document Collections. Table 1 contains the 6 collections used in the experimentation. All datasets, except NIPS which was used as it is published, were pre-processed by removing stopwords, non-letters and words with two or less characters. We have also applied Porter Stemming and filtered out words that appeared less than 5 times or in more than 50% of documents. Then, vocabularies were cropped to the 100 most frequent words for experiments with the exact marginal and they were used as in Table 1 for experiments in realistic conditions. Note that collections in Table 1 are ordered decreasingly on the average document length, being datasets at the top commonly used as long-text corpus, while those at the bottom used in short-text studies.

Model Hyperparameters, Training and Samplers Parameters. Among all possible GaP-PFA models, we have chosen to train the $\beta\Gamma$-PFA model [17]. This model also corresponds to the GaP model [5] with inference on the number of topics by placing a Beta Process over the p hyperparameter in the Gamma-distributed factors from Eq. (3). This allows us to avoid model selection on a critical parameter such the number of topics. But, several other model hyper-parameters need to be specified, such as the maximum number of topics which was set as described above for the different experiments, the Dirichlet prior over Φ which was set $\alpha = 0.1$, the scale for gamma $r = 1$ and the Beta hyperparameters $c = 1$ and $\epsilon = 1/K_{max}$, as in [17]. The training of the global parameters Φ, p, r is performed with the complete collection following the Gibbs Sampling scheme in [17] which runs for 1000 iterations and discards a burn-in period of 500. Regarding the samplers, we varied the number of samples up to $S = 10,000$ for all methods, and we used $S' = 1$ for L2R to keep the same overall number of samples for all estimators.

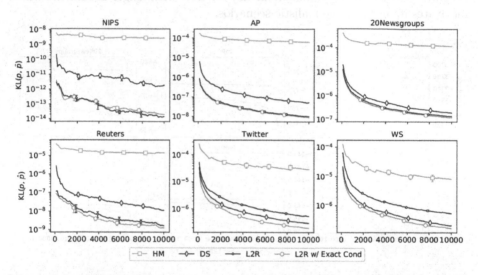

Fig. 2. Relative Entropy or KL between the estimated document probabilities and the exacts as a function of samples used. (Lower KL is better)

6.2   Experiments in Dimensioned Document Collections

Figure 2 shows the KL divergence between the exact and estimated probabilities as a function of the number of samples used by each estimation method. In this experiment, we have included the L2R with exact conditionals given by Eqs. (17)–(19) to compare against the proposed sampling. We have calculated the KL for all 4 estimation methods in the 6 collections with 1000 documents, except in NIPS and AP which only contained 1 and 460 documents with a tractable marginal, respectively. Each experiment was repeated 10 times and we plotted their mean and standard error.

Results show that L2R with exact conditionals achieves the lowest KL across all 6 datasets, followed very closely by the proposed L2R algorithm with $S' = 1$ which obtains the second lowest KL in 4 datasets. We note that L2R performs worse than DS in Twitter and WS datasets, which both are the shortest text datasets. This poor performance in short-text can also be explained by the fact that vanilla topic models struggle to learn predictive topic structure due to few word co-occurrence in a document, and hence the proposal in Eq. (20) is not close enough to the target to accurately estimate the conditionals with a single sample. Unreported experiments confirm that a larger S' makes the L2R estimates closer to those of the L2R with exact conditionals. In Fig. 1, we have compared the quality of L2R vs DS, as per the results obtained in the last sample of Fig. 2, as a function of the average document length of the downsized corpora. We observe a favorable tendency for L2R with longer documents which motivates its use in more realistic scenarios.

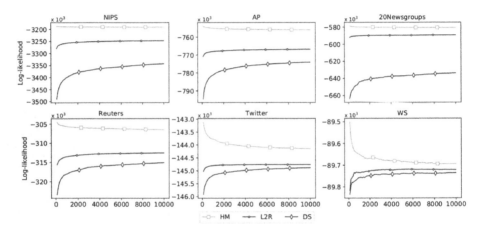

Fig. 3. Log-likelihood of the evaluated documents as a function of the number of samples.

6.3 Experiments in Realistic Document Collections

In Fig. 3, we plot the log-likelihood of 1,000 documents as a function of samples for the three methods that scale to the realistic scenario described above. Results show that L2R converges faster than DS in all 6 collections. The HM method also has a good convergence rate in the 4 datasets with longest documents, although the inaccuracy reported previously suggests that the method might be over-estimating the document likelihood like in LDA [4,15]. Therefore, the fast convergence and the fact that its estimates are sandwiched by estimators that tend to under- and over- estimate, validates L2R's use for document likelihood estimation in GaP-FA with just a few hundred samples.

7   Conclusions

In this paper, we have proposed L2R, a left-to-right algorithm for estimating the marginal document likelihood in GaP-FA. The accurate estimation in reasonably small scenarios and the quick convergence in realistic scenarios encourages its use for evaluating and comparing GaP-FA topic models in terms of unseen document likelihood.

Future work should explore new estimation methods capable of reducing the time complexity of L2R, which is quadratic in the number of non-zero words. Exploring the use of these methods and the exact calculation for other evaluation tasks like document completion or word prediction is another interesting avenue for future work.

Acknowledgements. This work was supported in part by Obra Social "LaCaixa", by the Australian Research Council under award DE170100037, by the SGR programs of the Catalan Government (2014-SGR-1051, 2014-SGR-118), by the Severo Ochoa Program SEV2015-0493 and by the the Spanish Ministry of Economy and Competitivity (MINECO) and the European Regional Development Fund (ERDF) under contracts TIN2015-65316 and Collectiveware TIN2015-66863-C2-1-R (MINECO/FEDER).

References

1. Blei, D.M.: Probabilistic topic models. Commun. ACM **55**(4), 77–84 (2012)
2. Blei, D.M., Ng, A.Y., Jordan, M.I.: Latent Dirichlet allocation. J. Mach. Learn. Res. **3**, 993–1022 (2003)
3. Buntine, W., Jakulin, A.: Discrete component analysis. In: Saunders, C., Grobelnik, M., Gunn, S., Shawe-Taylor, J. (eds.) SLSFS 2005. LNCS, vol. 3940, pp. 1–33. Springer, Heidelberg (2006). https://doi.org/10.1007/11752790_1
4. Buntine, W.L.: Estimating likelihoods for topic models. ACML **9**, 51–64 (2009)
5. Canny, J.: GaP: a factor model for discrete data. In: Proceedings of the 27th Annual International ACM SIGIR Conference on Research and Development in Information Retrieval, pp. 122–129. ACM (2004)
6. Filstroff, L., Lumbreras, A., Févotte, C.: Closed-form marginal likelihood in gamma-Poisson factorization. arXiv preprint arXiv:1801.01799 (2018)
7. Gopalan, P., Ruiz, F.J., Ranganath, R., Blei, D.: Bayesian nonparametric Poisson factorization for recommendation systems. In: Artificial Intelligence and Statistics, pp. 275–283 (2014)
8. Griffiths, T.L., Steyvers, M.: Finding scientific topics. Proc. Natl. Acad. Sci. **101**(suppl 1), 5228–5235 (2004)
9. Lee, D.D., Seung, H.S.: Algorithms for non-negative matrix factorization. In: Advances in Neural Information Processing Systems, pp. 556–562 (2001)
10. Murray, I., Salakhutdinov, R.R.: Evaluating probabilities under high-dimensional latent variable models. In: Advances in Neural Information Processing Systems, pp. 1137–1144 (2009)
11. Neal, R.M.: Annealed importance sampling. Stat. Comput. **11**(2), 125–139 (2001)
12. Newton, M.A., Raftery, A.E.: Approximate Bayesian inference with the weighted likelihood bootstrap. J. Roy. Stat. Soc. Seri. B (Methodological) **56**, 3–48 (1994)

13. Sibuya, M., Yoshimura, I., Shimizu, R.: Negative multinomial distribution. Ann. Inst. Stat. Math. **16**(1), 409–426 (1964)
14. Wallach, H.M.: Topic modeling: beyond bag-of-words. In: Proceedings of the 23rd International Conference on Machine Learning, pp. 977–984. ACM (2006)
15. Wallach, H.M., Murray, I., Salakhutdinov, R., Mimno, D.: Evaluation methods for topic models. In: Proceedings of the 26th Annual International Conference on Machine Learning, pp. 1105–1112. ACM (2009)
16. Zhao, H., Du, L., Buntine, W.: Leveraging node attributes for incomplete relational data. In: International Conference on Machine Learning, pp. 4072–4081 (2017)
17. Zhou, M., Hannah, L., Dunson, D.B., Carin, L.: Beta-negative binomial process and Poisson factor analysis. In: International Conference on Artificial Intelligence and Statistics, pp. 1462–1471 (2012)

Causal Inference on Multivariate and Mixed-Type Data

Alexander Marx[✉] and Jilles Vreeken

Max Planck Institute for Informatics and Saarland University,
Saarland Informatics Campus, 66123 Saarbrücken, Germany
{amarx,jilles}@mpi-inf.mpg.de

Abstract. How can we discover whether X causes Y, or vice versa, that Y causes X, when we are only given a sample over their joint distribution? How can we do this such that X and Y can be univariate, multivariate, or of different cardinalities? And, how can we do so regardless of whether X and Y are of the same, or of different data type, be it discrete, numeric, or mixed? These are exactly the questions we answer. We take an information theoretic approach, based on the Minimum Description Length principle, from which it follows that first describing the data over *cause* and then that of *effect* given *cause* is shorter than the reverse direction. Simply put, if Y can be explained more succinctly by a set of classification or regression trees conditioned on X, than in the opposite direction, we conclude that X causes Y. Empirical evaluation on a wide range of data shows that our method, CRACK, infers the correct causal direction reliably and with high accuracy on a wide range of settings, outperforming the state of the art by a wide margin. Code related to this paper is available at: http://eda.mmci.uni-saarland.de/crack.

1 Introduction

Telling cause from effect is one of the core problems in science. It is often difficult, expensive, or impossible to obtain data through randomized trials, and hence we often have to infer causality from, what is called, observational data [19]. We consider the setting where, given data over the joint distribution of two random variables X and Y, we have to infer the causal direction between X and Y. In other words, our task is to identify whether it is more likely that X causes Y, or vice versa, that Y causes X, or that the two are merely correlated.

In practice, X and Y do not have to be of the same type. The altitude of a location (real-valued), for example, determines whether it is a good habitat (binary) for a mountain hare. In fact, neither X nor Y have to be univariate. Whether or not a location is a good habitat for an animal is not just caused by a single aspect, but by a *combination* of conditions which are not necessarily of the

---

Electronic supplementary material The online version of this chapter (https://doi.org/10.1007/978-3-030-10928-8_39) contains supplementary material, which is available to authorized users.

M. Berlingerio et al. (Eds.): ECML PKDD 2018, LNAI 11052, pp. 655–671, 2019.
https://doi.org/10.1007/978-3-030-10928-8_39

same type. We are therefore interested in the general case where X and Y may be of any cardinality, and may be single or mixed-type. To the best of our knowledge there exists no method for this general setting. Causal inference based on conditional independence tests, for example, requires three variables, and cannot decide between $X \to Y$ and $Y \to X$ [19]. All existing methods that consider two variables are only defined for single-type pairs. Additive Noise Models (ANMs), for example, have only been proposed for univariate pairs of real-valued [20] or discrete variables [21], and similarly so for methods based on the independence of $P(X)$ and $P(Y \mid X)$ [25]. Trace-based methods require both X and Y to be strictly multivariate real-valued [4,10], and whereas ERGO [31] also works for univariate pairs, these again have to be real-valued. We refer to Sect. 3 for a more detailed overview of related work.

Our approach is based on algorithmic information theory. That is, we follow the postulate that if $X \to Y$, it will be easier—in terms of Kolmogorov complexity—to first describe X, and then describe Y given X, than the inverse direction [11,16,31]. Kolmogorov complexity is not computable, but can be approximated through the Minimum Description Length (MDL) principle [7,23], which we use to instantiate this framework. In addition, we develop a new causal indicator that is able to handle multivariate and mixed-type data.

In particular, we define an MDL score for coding forests, a model class where a model consists of classification and regression trees as this allows us to consider both discrete and continuous-valued data with one unified model. By allowing dependencies from X to Y, or vice versa, we can measure the difference in complexity between $X \to Y$ and $Y \to X$. Discovering a single optimal decision tree is already NP-hard, and hence we cannot efficiently discover the coding forest that describes the data most succinctly. We therefore propose CRACK, an efficient greedy algorithm for discovering good models directly from data. The inferences we make hence are all with respect to the class of coding trees, and the specific encoding we define. We discuss the implications with regard to identifiability, and through extensive empirical evaluation on synthetic, benchmark, and real-world data, we show that CRACK performs very well in practice—even under adversarial settings.

Our main contributions are as follows. We introduce the first framework for inferring the causal direction from univariate and multivariate single and mixed-type data—as opposed to existing methods that are only able to deal with either nominal or numeric data. We propose a new causal indicator based on the algorithmic Markov condition, instantiate it through MDL, and propose a fast algorithm to compute it. We provide extensive empirical evaluation of our method, in which we additionally introduce new multivariate cause-effect pairs with known ground truth. The paper is organized as usual.

2 Preliminaries

In this section, we introduce the notation and give brief primers to Kolmogorov complexity and the Minimum Description Length principle.

2.1   Notation

In this work we consider two sets of random variables X and Y. Further, we are given a data set D containing n i.i.d. samples drawn from the joint distribution of X and Y. For convenience, we call A the set of all random variables, where $A = X \cup Y$, with $|A| = m$ being the number of random variables in A. In the following, we will often refer to a random variable $A_i \in A$ as an attribute, regardless of whether it belongs to X or Y. An attribute A_i has a type, where $\text{type}(A_i) \in \{binary, categorical, numeric\}$. We will refer to binary and categorical attributes as *nominal* attributes. We write $\mathcal{X}_i$ to denote the domain of an attribute A_i. Respectively, the size of the domain of an attribute $|\mathcal{X}_i|$ is for discrete data simply the number of distinct values and for numeric data equal to $\frac{\max(A_i) - \min(A_i)}{\text{res}(A_i)} + 1$, where $\text{res}(A_i)$ is the resolution at which the data over attribute A_i was recorded. For example, a resolution of 1 means that we consider integers, of 0.01 means that A was recorded with a precision of up to a hundredth.

We will consider decision and regression trees. A tree T consist of $|T|$ nodes. We identify internal nodes as $v \in \text{int}(T)$, and leaf nodes as $l \in \text{lvs}(T)$. A leaf node l contains $|l|$ data points. All logarithms are to base 2, and we use $0 \log 0 = 0$.

2.2   Kolmogorov Complexity, a Brief Primer

The Kolmogorov complexity of a finite binary string x is the length of the shortest binary program p^* for a universal Turing machine $\mathcal{U}$ that generates x, and then halts [15]. Formally, we have $K(x) = \min\{|p| \mid p \in \{0,1\}^*, \mathcal{U}(p) = x\}$. Simply put, p^* is the most succinct *algorithmic* description of x, and the Kolmogorov complexity of x is the length of its ultimate lossless compression. Conditional Kolmogorov complexity, $K(x \mid y) \leq K(x)$, is then the length of the shortest binary program p^* that generates x, and halts, given y as input.

By definition, Kolmogorov complexity will make maximal use of any structure in x that can be expressed more succinctly algorithmically than by printing it verbatim. As such it is the theoretical optimal measure for complexity. However, due to the halting problem it is not computable [15]. Instead, we can approximate it from above through MDL [15].

2.3   MDL, a Brief Primer

The Minimum Description Length (MDL) principle [7,23] is a practical variant of Kolmogorov Complexity. Intuitively, instead of all programs, it considers only those programs that we know output x and halt. Formally, given a model class $\mathcal{M}$, MDL identifies the best model $M \in \mathcal{M}$ for data D as the one minimizing $L(D, M) = L(M) + L(D \mid M)$, where $L(M)$ is the length in bits of the description of M, and $L(D \mid M)$ is the length in bits of the description of data D given M. This is known as two-part MDL. There also exists one-part, or *refined* MDL, where we encode data and model together. Refined MDL is superior as it avoids arbitrary choices in the description language L, but in practice it is only computable for certain model classes. Given infinite data the model costs degenerate to an additive constant term, which is independent of the data. Hence given infinite data, two-part MDL converges to refined MDL. Note that in either case we are only concerned with code *lengths*—our goal is to measure the *complexity* of a dataset under a model class, not to actually compress it [7].

3 Related Work

Causal inference on observational data is a challenging problem, and has recently attracted a lot of attention [3,11,19,26]. Most existing proposals are highly specific in the type of causal dependencies and type of variables they can consider.

Classical constrained-based approaches, such as conditional independence tests, require three observed random variables [19,27], cannot distinguish Markov equivalent causal DAGs [30] and hence cannot decide between $X \rightarrow Y$ and $Y \rightarrow X$. Recently, there has been increased attention for methods that can infer the causal direction from only two random variables. Generally, they exploit certain properties of the joint distribution.

Additive Noise Models (ANMs) [26], for example, assume that the effect is a function of the cause and cause-independent additive noise. ANMs exist for univariate real-valued [9,20,26,32] and discrete data [21]. It is unclear how to extend this model for multivariate or mixed-type data. A related approach considers the asymmetry in the joint distribution of *cause* and *effect* for causal inference. The linear trace method (LTR) [10] and the kernelized trace method (KTR) [4] aim to find a structure matrix A and the covariance matrix Σ_X to express Y as AX. Both methods are restricted to multivariate continuous data. In addition, KTR assumes a deterministic, functional and invertible causal relation. Sgouritsa et al. [25] show that the marginal distribution of the cause is independent of the conditional distribution of the effect given the cause. To exploit this asymmetry, they propose the CURE algorithm, which is based on unsupervised reverse regression.

The algorithmic information-theoretic approach views causality in terms of Kolmogorov complexity. The key idea is that if X causes Y, the shortest description of the joint distribution $P(X,Y)$ is given by the separate descriptions of the distributions $P(X)$ and $P(Y \mid X)$ [11], and justifies additive noise model based causal inference [12]. However, as Kolmogorov complexity is not computable [15], causal inference using algorithmic information theory requires practical implementations, or notions of independence. For instance, the information-geometric approach [13] defines independence via orthogonality in information space for univariate continuous pairs. Vreeken [31] instantiates it with the cumulative entropy to infer the causal direction in continuous univariate and multivariate data. Mooij instantiates the first practical compression-based approach [18] using the Minimum Message Length. Budhathoki and Vreeken approximate $K(X)$ and $K(Y \mid X)$ through MDL, and propose ORIGO for causal inference on binary data [3]. Marx and Vreeken [16] propose SLOPE, an MDL based method employing local and global regression for univariate numeric data.

In contrast to all methods above, CRACK can consider pairs of any cardinality, univariate or multivariate, and of same, different, or even mixed-type.

4   Causal Inference by Compression

We pursue the goal of causal inference by compression. Below we give a short introduction to the key concepts.

4.1   Causal Inference by Complexity

The problem we consider is to infer, given data over two correlated variables X and Y, whether X caused Y, whether Y caused X, or whether X and Y are only correlated. As is common in this setting, we assume causal sufficiency [17]. That is, we assume there exists no hidden confounding variable Z that causes both X and Y. This scenario is relevant not only when we are given only two variables and hence no conditional independence tests can be applied, but also when we are given a partially directed causal skeleton and orientation rules based on conditional independence tests can not resolve the remaining undirected edges.

The algorithmic Markov condition, as recently postulated by Janzing and Schölkopf [11], states that factorizing the joint distribution over *cause* and *effect* into $P(cause)$ and $P(effect \mid cause)$, will lead to simpler—in terms of Kolmogorov complexity—models than factorizing it into $P(effect)$ and $P(cause \mid effect)$. Formally, they postulate that if X causes Y,

$$K(P(X)) + K(P(Y \mid X)) \leq K(P(Y)) + K(P(X \mid Y)) .$$

While in general the symmetry of information, $K(x) + K(y \mid x) = K(y) + K(x \mid y)$, holds up to an additive constant [15], Janzing and Schölkopf [11] showed it does *not* hold when X causes Y, or vice versa. Hence, we can trivially define

$$\Delta_{X \to Y} = K(P(X)) + K(P(Y \mid X)) , \tag{1}$$

as a causal indicator that uses this asymmetry to infer that $X \to Y$ as the most likely causal direction if $\Delta_{X \to Y} < \Delta_{Y \to X}$, and vice versa.

This indicator assumes access to the true distribution $P(\cdot)$. In practice, we only have access to empirical data. Moreover, following from the halting problem, Kolmogorov complexity is not computable. We can approximate it, however, via MDL [7,15], which also allows us to directly work with empirical distributions.

4.2   Causal Inference by MDL

For causal inference by MDL, we will need to approximate both the marginals $K(P(X))$ and $K(P(Y))$ as well as the conditionals $K(P(Y \mid X))$ and $K(P(X \mid Y))$. For the former, we need to consider the model classes $\mathcal{M}_X$ and $\mathcal{M}_Y$, while for the latter we need to consider class $\mathcal{M}_{Y|X}$ of models $M_{Y|X}$ that describe the data of Y dependent the data of X, and accordingly for the inverse direction.

That is, we are after the *causal* model $M_{X \to Y} = (M_X, M_{Y|X})$ from the class $\mathcal{M}_{X \to Y} = \mathcal{M}_X \times \mathcal{M}_{Y|X}$ that best describes the data Y by exploiting as much structure of X as possible to save bits. By MDL, we identify the optimal model $M_{X \to Y} \in \mathcal{M}_{X \to Y}$ for data over X and Y as the one minimizing

$$L(X, Y, M_{X \to Y}) = L(X, M_X) + L(Y, M_{Y|X} \mid X),$$

where the encoded length of the data of X under a given model is encoded using two-part MDL, similarly so for Y, if we consider the inverse direction.

To identify the most likely causal direction between X and Y by MDL we can now simply rewrite Eq. (1) to define the *Absolute Causal Indicator* (ACI) [3]

$$ACI_{X \to Y} = L(X, M_X) + L(Y, M_{Y|X} \mid X).$$

Akin to the Kolmogorov complexity based score, we infer that X is a likely cause of Y if $ACI_{X \to Y} < ACI_{Y \to X}$, Y is a likely cause of X if $ACI_{Y \to X} < ACI_{X \to Y}$.

4.3   Normalized Causal Indicator

The absolute causal indicator has nice theoretical properties that follow directly from the algorithmic Markov condition. However, by considering the absolute difference in encoded lengths between $X \to Y$ and $Y \to X$, it has an intrinsic bias towards data of higher marginal complexity. For example, when we gain 5 bits between encoding the data over Y conditioned on X, rather than independently, this is more impressive if $L(Y, M_Y)$ was 100 than when it was $1\,000\,000$ bits. This is particularly important in the mixed-data case, as the marginal complexity of a binary attribute will typically be much smaller than that of a attribute recorded at a higher resolution.

To address this shortcoming in ACI, we propose a novel, normalized indicator for causal inference on mixed-type data. We start with the ERGO indicator [31], which rather than the absolute difference considers the compression *ratios* of the target variables, i.e. iff $X \to Y$ then

$$\frac{L(X, M_{X|Y} \mid Y)}{L(X, M_X)} > \frac{L(Y, M_{Y|X} \mid X)}{L(Y, M_Y)}.$$

This score accounts for different marginal complexities of X and Y, and hence suffices for the univariate mixed-type data case. For the multivariate and mixed-type data case, we still face the same problem: if the variates of $Y_i \in Y$ are of different marginal complexities $L(Y_i, M_{Y_i})$, the gain in compression of one single Y_i may dominate the overall score simply because it has a larger marginal complexity than the others (e.g. because it has a larger domain).

We can compensate this by explicitly considering the compression ratios *per variate* $Y_i \in Y$, rather than the compression ratio over Y as a whole. Formally, we define our new *Normalized Causal Indicator* (*NCI*) as

$$NCI_{X \to Y} = \frac{1}{|Y|} \sum_{Y_i \in Y} \frac{L(Y_i, M_{Y_i|X} \mid X)}{L(Y_i, M_{Y_i})}.$$

As above, we infer $X \to Y$ if $NCI_{X \to Y} < NCI_{Y \to X}$ and vice versa.

Although free of bias from the marginal complexities of individual variates, we have to be careful to screen for redundancy within Y resp. X. By definition, the *NCI* counts the causal effect on each variate, and redundancies within Y (resp. X) hence exacerbate the measured effect. It is easy to detect redundancies within X resp. Y using standard independence tests, however.

In practice, we expect that *ACI* performs well on data where X and Y are of the same type, especially when $|X| = |Y|$ and the domain sizes of their attributes are balanced. Whenever the variates of X and Y are of different marginal complexities, e.g. because of unbalanced domains, dimensionality, and especially for mixed-type data, the experiments confirm that the *NCI* performs much better than the *ACI*.

5   MDL for Tree Models

To use MDL in practice, we need to specify an appropriate model class $\mathcal{M}$, and define how to encode both data and models in bits. Here, we need to be able to consider numeric, discrete and mixed-type data, be able to exploit dependencies between attributes of different types, and be able to encode the data of Y conditioned on the data of X. Classification and regression trees lend themselves very naturally to do all of this.

That is, we consider models M that contain a classification or regression tree T_i per attribute $A_i \in A$, where tree T_i encodes the data over A_i by exploiting dependencies on other attributes by means of splitting or regression. Together, the leaves of a tree encode the data of the attribute. Loosely speaking, the better we can fit the data in a leaf, the more succinctly we will be able to encode it.

A *valid* tree model M contains no cyclic dependencies between the trees $T_i \in M$, and hence a valid model can be represented by a DAG. Formally, we define $\mathcal{M}_A$ as the set of all valid tree models for data over a set of attributes A. We additionally define *conditional* tree models for Y given X, as the model class $\mathcal{M}_{Y|X}$ that consists of all valid tree models where we allow dependencies within Y, as well as from X to Y, but not from Y to X.

Cost of Data and Model. Now that we know the relevant model classes, we can define our MDL score. At the highest level, the number of bits to describe data over attributes A together with a valid model M for A as

$$L(A, M) = \sum_{T_i \in M} L(A_i, T_i) \,,$$

where we make use of the fact that M is a DAG, and we can hence serialize its dependencies.

In turn, the encoded cost of a tree T consists of two parts. First, we transmit its topology, and second the data in its leaves. For the topology, we indicate per node whether it is a leaf or an internal node, and if the latter, whether it is a split or regression node. Formally we hence have

$$L(A_i, T) = |T| + \sum_{v \in \text{int}(T)} (1 + L(v)) + \sum_{l \in \text{lvs}(T)} L(l) .$$

This leaves us to define the encoded cost of an internal node, $L(v)$, and the encoded cost of the data in the leaves, $L(l)$. We do this in turn.

Cost of a Node. A node $v \in T$ can be of two main types; it either defines a split, or a regression step. We consider these in turn. We consider both multiway and single splits. To encode a split node v, we need

$$L_{\text{split}}(v) = 1 + \log |A| + L(\varPhi_{\text{split}})$$

bits. We first encode whether it is a single or multiway split, then the attribute X_j, and last the conditions on which we split. For single way splits, $L(\varPhi_{\text{split}})$ corresponds to the cost of describing the value in the domain of X_j on which we split, which is $\log |\mathcal{X}_j|$ when X_j is categorical, and $\log |\mathcal{X}_j| - 1|$ when it is binary or numeric. For multiway splits on categorical attributes X_j we split on all values, which costs no further bits, while for numeric X_j we split on every value that occurs at least k times—with one residual split for all remaining data points. To encode k we use $L_{\mathbb{N}}$, the MDL optimal code for integers [24].

To encode a regression node n, we first encode the attribute we regress on, and then the parameters $\varPhi(v)$ of the regression, i.e.

$$L_{\text{reg}}(v) = \log |A| + \sum_{\phi \in \varPhi(v)} (1 + L_{\mathbb{N}}(s) + L_{\mathbb{N}}(\lfloor \phi \cdot 10^s \rfloor)) .$$

We encode each parameter $\phi \in \varPhi$ up to user defined precision, e.g. 0.001, by first encoding the corresponding number of significant digits s, e.g. 3, and then the shifted parameter value. In practice, for computational reasons, we use linear and quadratic regression, but note that this score is general for any regression technique with real-valued parameters.

Cost of a Leaf. In classification and regression trees, the actual data is stored in the leaves. To encode the data in a leaf of a nominal attribute, we can use refined MDL [14]. That is, we are guaranteed to be as close as possible to the number of bits we would need knowing the true model, even if the true generating model is not in our model class [7]. In particular, we encode the data of a nominal leaf using the stochastic complexity for multinomials as

$$L_{\text{nom}}(l) = |l| \cdot H(X_i \mid l) + \log \sum_{h_1 + \cdots + h_k = |l|} \frac{|l|!}{h_1! h_2! \cdots h_k!} ,$$

where H denotes the Shannon entropy. Kontkanen and Myllymäki [14] derived a recursive formula to calculate this in linear time.

For numeric data, refined MDL encodings have very high computational complexity [14]. In the interest of efficiency, we hence encode the data in numeric leaves with two-part MDL, using point models with a Gaussian, resp. uniform distribution. The former is especially fitting after regression, since such a step aims to minimizes the variance of Gaussian distributed error. A split or a regression node can reduce the variance and, or the domain size of data in the leaf, and each can therewith reduce the cost. The costs for a numeric leaf are

$$L_{\text{num}}(l \mid \sigma, \mu) = \frac{|l|}{2} \left(\frac{1}{\ln 2} + \log 2\pi\sigma^2 \right) - |l| \log \text{res}(X_i),$$

given empirical mean μ and variance σ or as uniform given min and max as

$$L_{\text{num}}(l \mid \min(l), \max(l)) = |l| \cdot \log \left(\frac{\max(l) - \min(l)}{\text{res}(X_i)} + 1 \right).$$

We encode the data as Gaussian if this costs fewer bits than encoding it as uniform. To indicate this decision, we use one bit and encode the minimum of both plus the corresponding parameters. As we consider empirical data, we can safely assume that all parameters lie in the domain of the given attribute. The encoded costs of a numeric leaf l hence are

$$L_{\text{num}}(l) = 1 + 2\log|\mathcal{X}_i| + \min\{L_{\text{num}}(l \mid \sigma, \mu), L_{\text{num}}(l \mid \min(l), \max(l))\}.$$

We now have a complete score. In the next section we discuss how to optimize it, but first we discuss some important causal aspects.

Identifiability and Limitations. Tree models are closely related to the algorithmic model of causality as postulated by Janzing and Schölkopf [11]. That is, every node X_i in a DAG can be computed by a program q_i with length $O(1)$ from its parents pa_i and additional input n_i—formally, $X_i = q_i(pa_i, n_i)$. Following the algorithmic Markov condition, the shortest description of X_i is through its parents.

In general, the MDL optimal tree model identifies the shortest description of a node A_i conditioned on a subset of attributes $S_i \subseteq A \backslash \{A_i\}$. In particular, by splitting or regressing on an attribute $A_j \in S_i$ it models program q_i given the parents as input. The remaining unexplained data that corresponds to the additional input or noise n_i is encoded in the leaves of the tree. In other words, tree T_i with the minimal costs relates to the tree where S_i contains only parents of A_i, and encodes exactly the relevant dependencies towards A_i.

Although tree models are very general, we can identify specific settings in which the model is identifiable. First, consider the case where X and Y are univariate and of a single type. If both are numeric, our model reduces to a simple regression model, for which we know the correct causal direction can be identified based on regression error for non-linear function with Gaussian noise and linear functions with either data or noise not being Gaussian distributed [1,16]. Similarly, for discrete data we can identify additive noise models using stochastic complexity [2]. Since we model dependencies according to the algorithmic model of causality, we can generalize these concepts for multivariate data.

Further, it is easy to see that our score is monotone under subset restriction. That is, if $X \subseteq Y$, we can define $Z = Y \setminus X$, and have $L(Y) = L(X \cup Z) = L(X) + L(Z \mid X) \geq L(X)$. Additionally, it is submodular, $L(X \cup Z) - L(X) \geq L(Y \cup Z) - L(Y)$. As it also is trivially 0 for empty input, it is an information measure, and hence we know by the results of Steudel [28] that under our score tree models themselves are identifiable.

In practice, we are limited by the optimality of our approximation of the Kolmogorov complexity. That is, any inferences we make are with respect to the encoding we defined above, rather than the much more generally defined Kolmogorov complexity. If the generating process does not use tree-models, or measures complexity differently, the inferences we draw based on our score may be wrong. The experiments show, however, that our scores are very reliable even in adversarial settings.

6   The Crack Algorithm

Finding the optimal decision tree for a single nominal attribute is NP-hard, and hence so is the optimization problem at hand. We introduce the CRACK algorithm, which stands for classification and regression based packing of data. CRACK is an efficient greedy heuristic for discovering a coding forest M from model class $\mathcal{M}$ for data over attributes A with low $L(A, M)$. It builds upon the well-known ID3 algorithm [22].

Greedy Algorithm. We give the pseudocode of CRACK as Algorithm 1. Before running the algorithm, we set the resolution per attribute. To be robust to noise, we set $res(A_i)$ for continuous attributes to the kth smallest distance between two adjacent values, with $k = 0.1 \cdot n$.

CRACK starts with an empty model consisting of only trivial trees, i.e. leaf nodes containing all records, per attribute (line 1). We iteratively discover that refinement of the current model that maximizes compression. To find the best refinement, we consider every attribute (4), and every legal additional split or regression of its corresponding tree (8). That is, a refinement is only legal when the dependency is allowed by the model family $\mathcal{M}$ (6–7) and the dependency graph remains acyclic.

The key subroutine of CRACK is REFINELEAF, in which we discover the optimal refinement of a leaf l in tree T_i. That is, it finds the optimal split of l over all candidate attributes A_j such that we minimize the encoded length. In case both A_i and A_j are numeric, REFINELEAF also considers the best linear and quadratic regression and decides for the variant with the best compression—choosing to split in case of a tie. In the interest of efficiency, we do not allow splitting or regressing multiple times on the same candidate.

Since we use a greedy heuristic to construct the coding trees, we have a worst case runtime of $O(2^m n)$, where m is the number of attributes and n is the number of rows. In practice, CRACK takes only a few seconds for all tested cause-effect pairs.

Causal Inference with Crack. To compute our causal indicators we run CRACK twice on D. First with model class $\mathcal{M}_{X|Y}$ to obtain $M_{X|Y}$ and second with $\mathcal{M}_{Y|X}$, to obtain $M_{Y|X}$. For $L(X \mid M_X)$ we assume a uniform prior and define $L(X \mid M_X) = -n\sum_{A_i \in X} \log \operatorname{res}(A_i)$ and do so analogue for Y. We refer to CRACK using NCI as CRACK$_N$, and as CRACK$_A$ using ACI.

Algorithm 1. CRACK$(A, \mathcal{M})$

---

   input : data over attributes A, model class $\mathcal{M}$
   output : tree model $M \in \mathcal{M}$ with low $L(A, M)$
1 $T_i \leftarrow$ TRIVIALTREE(A_i) for all $A_i \in A$;
2 $\mathcal{G} \leftarrow (V = \{v_i \mid i \in A\}, E = \emptyset)$;
3 **while** $L(A, M)$ decreases **do**
4   **for** $A_i \in A$ **do**
5      $O_i \leftarrow T_i$;
6   **for** $l \in \operatorname{lvs}(T_i), (i, j) \in \mathcal{G}$ **do**
7   **if** $E \cup (v_i, v_j)$ is acyclic **and** $j \notin \operatorname{path}(l)$ **then**
8          $T_i' \leftarrow$ REFINELEAF(T_i, l, j);
9   **if** $L(T_i') < L(O_i)$ **then**
10            $O_i \leftarrow T_i', e_i \leftarrow j$;
11    $k \leftarrow \arg\min_i\{L(O_i) - L(T_i)\}$;
12   **if** $L(O_k) < L(T_k)$ **then**
13      $T_k \leftarrow O_k, E \leftarrow E \cup (v_k, v_{e_k})$;
14 **return** $M \leftarrow \bigcup_i T_i$

---

7 Experiments

In this section, we evaluate CRACK empirically. We implemented CRACK in C++, and provide the source code including the synthetic data generator along with the tested datasets for research purposes.[1] The experiments concerning CRACK were executed single-threaded on a MacBook Pro with 2.6 GHz Intel Core i7 processor and 16 GB memory running Mac OS X. All tested data sets could be processed within seconds; with a maximum runtime of 3.8 s.

7.1 Synthetic Data

On synthetic data, we want to show the advantages of either score. In particular, we expect CRACK$_A$ to perform well on nominal data and numeric data with balanced domain sizes and dimensions, whereas we expect CRACK$_N$ to perform better on numeric data with varying domain sizes and mixed-type data.

We generate synthetic data with assumed ground truth $X \to Y$ with $|X| = k$ and $|Y| = l$, each having $n = 5\,000$ rows, in the following way. First, we randomly assign the type for each attribute in X. For nominal data, we randomly draw the number of classes between two (binary) and five and distribute the classes

---

[1] http://eda.mmci.uni-saarland.de/crack.

uniformly. Numeric data is generated following a normal distribution taken to the power of q by keeping the sign, leading to a sub-Gaussian ($q < 1.0$) or super-Gaussian ($q > 1.0$) distribution.[2]

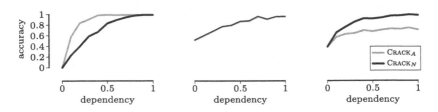

Fig. 1. Accuracy for *ACI* and *NCI* on nominal (left), numeric (middle) and mixed-type (right) data based on the dependency.

To create data with the true causal direction $X \to Y$, we introduce dependencies from X to Y, where we distinguish between splits and refinements. We call the probability threshold to create a dependency $\varphi \in [0, 1]$. For each $j \in \{1, \ldots, l\}$, we throw a biased coin based on φ for each $X_i \in X$ that determines if we model a dependency from X_i to Y_j. A split means that we find a category (nominal) or a split-point (numeric) on X_i to split Y_j into two groups, for which we model its distribution independently. As refinement, we either do a multiway split or model Y_j as a linear or quadratic function of X_i plus independent Gaussian noise.

Accuracy. First, we compare the accuracies of CRACK$_N$ and CRACK$_A$ with regard to single-type and mixed-type data. To do so, we generate 200 synthetic data sets with $|X| = |Y| = 3$ for each dependency level where $\varphi \in \{0.0, 0.1, \ldots 1.0\}$. Figure 1 shows the results for numeric, nominal and mixed-type data. For single-type data, the accuracy of both methods increases with the dependency, and reaches nearly 100% for $\varphi = 1.0$. At $\varphi = 0$, both approaches correctly do not decide instead of taking wrong decisions. As expected CRACK$_N$ strongly outperforms CRACK$_A$ on mixed-type data, reaching near 100% accuracy, whereas CRACK$_A$ reaches only 72%. On nominal data, CRACK$_A$ picks up the correct signal faster than CRACK$_N$.

Dimensionality. Next, we evaluate how sensitive both scores are w.r.t. the dimensionality of both X and Y, where we separately consider the cases of symmetric $k = l$ and asymmetric $k \neq l$ dimensionalities. Per setting, we consider the average accuracy over 200 independently generated data sets.

For the symmetric case, both methods are near to 100% on single-type data, whereas only CRACK$_N$ also reaches this target on mixed-type data, as can be seen in the appendix (see footnote 1). We now discuss the more interesting case for asymmetric pairs in detail. To test asymmetric pairs, we set the dimensionality of

---

[2] To ensure identifiability, we use super- and sub-Gaussians [9].

X to three, $|X| = 3$, and vary the dimensionality of Y from 1 to 11. To avoid bias, we choose the ground truth causal direction, i.e. $X \to Y$ and $Y \to X$, uniformly at random. We plot the results in Fig. 2. We observe that CRACK$_N$ has much less difficulty with the asymmetric data sets than CRACK$_A$. CRACK$_N$ performs near perfect and has a clear advantage over CRACK$_A$ on mixed-type and numeric data. CRACK$_A$ performs better on nominal only data for $l = 1$.

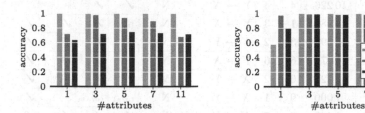

Fig. 2. Accuracy of *ACI* (left) and *NCI* (right) on for synthetically generated causal pairs of asymmetric cardinality, $|X| = 3$ and $|Y| \in \{1, 3, 5, 7, 11\}$ with ground truth $X \to Y$ or $Y \to X$ randomly chosen, for resp. nominal, numeric and mixed-type data.

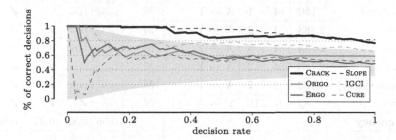

Fig. 3. [Higher is better] Decision rates of the multivariate methods CRACK, ORIGO and ERGO, and the univariate methods IGCI, CURE and SLOPE (dashed lines) on the univariate Tübingen causal benchmark pairs (100), weighted as defined.

7.2 Univariate Benchmark Data

To evaluate CRACK on univariate data, we apply it to the well-known Tübingen benchmark (v1.0) consisting of 100 univariate cause-effect pairs with known ground truth.[3] As these are mainly numeric pairs, with only a few categoric instances, we apply CRACK$_A$. We compare to the state of the art methods for multivariate pairs, ORIGO [3] and ERGO [31], and those specialized for univariate pairs, CURE [25], IGCI [13] and SLOPE [16] using their publicly available implementations and recommended parameter settings.

---

[3] https://webdav.tuebingen.mpg.de/cause-effect/.

Table 1. Comparison of LTR, ERGO, ORIGO and CRACK on 17 multivariate cause-effect pairs with known ground truth. The type is either "N" for numeric or "M" for mixed. A "✓" indicates a correct decision, a "–" an incorrect one and (n/a) that a method is not applicable.

						Decisions			
Causal pair	n	$\|X\|$	$\|Y\|$	Ground truth	Type	LTR	ERGO	ORIGO	CRACK
Climate	10 226	4	4	$Y \rightarrow X$	N	✓	✓	–	–
Ozone	989	1	3	$Y \rightarrow X$	N	(n/a)	✓	✓	✓
Car	392	3	2	$X \rightarrow Y$	N	–	✓	✓	✓
Radiation	72	16	16	$Y \rightarrow X$	N	–	–	–	✓
Symptoms	120	6	2	$X \rightarrow Y$	M	✓	✓	–	✓
Brightness	1 000	9	1	$X \rightarrow Y$	N	(n/a)	(n/a)	–	✓
Chemnitz	1 440	3	7	$X \rightarrow Y$	N	✓	✓	✓	✓
Precipitation	4 748	3	12	$X \rightarrow Y$	N	✓	–	–	✓
Stock 7	2 394	4	3	$X \rightarrow Y$	N	–	✓	–	✓
Stock 9	2 394	4	5	$X \rightarrow Y$	N	–	✓	–	✓
Haberman	306	3	1	$X \rightarrow Y$	M	✓	✓	–	–
Iris flower	150	4	1	$X \rightarrow Y$	M	(n/a)	(n/a)	–	✓
Canis	2 183	4	2	$X \rightarrow Y$	M	(n/a)	(n/a)	✓	✓
Lepus	2 183	4	3	$X \rightarrow Y$	M	(n/a)	(n/a)	✓	✓
Martes	2 183	4	2	$X \rightarrow Y$	M	(n/a)	(n/a)	✓	✓
Mammals	2 183	4	7	$X \rightarrow Y$	M	(n/a)	(n/a)	✓	✓
Octet	82	1	10	$Y \rightarrow X$	N	(n/a)	✓	✓	✓
Accuracy						0.56	0.82	0.47	0.88

For each approach, we sort the results by confidence. Accordingly, we calculate the decision rate, the percentage of correct inferences up to each k inferences, weighting the decisions as specified by the benchmark. We plot the results in Fig. 3 and show the 95% confidence interval of a fair coin flip as a grey area. Except to CRACK none of the multivariate methods is significant w.r.t. the fair coin flip. In particular, CRACK has an accuracy of over 90% for the first 41% of its decisions and reaches 77.2% overall—the final result of CRACK$_N$ is only 3% worse. CRACK also beats both CURE (52.5%) and IGCI (66.2%), which are methods specialized for univariate pairs. Perhaps most impressively, CRACK performs within the 95% confidence interval of the current state of the art on causal inference on univariate numeric pairs, SLOPE, which has an overall accuracy of 81.7%. SLOPE is at the advantage for univariate pairs as it can exploit non-deterministic structure in the data. While interesting, this idea sadly does not seem to be efficiently applicable to multivariate data.

7.3   Real World Data

Next, we apply $CRACK_N$ on multivariate mixed-type and single-type data, where we collected 17 cause effect pairs with known ground truth. We provide basic statistics for each pair in Table 1. The first six are part of the Tübingen benchmark [17], and the next four were provided by Janzing et al. [10]. Further, we extracted cause-effect pairs from the *Haberman* and *Iris* [5], *Mammals* [8] and *Octet* [6,29] data sets. More details are given in the appendix (see footnote 1).

We compare $CRACK_N$ with LTR [10], ERGO [31] and ORIGO [3]. ERGO and LTR do not consider categoric data, and are hence not applicable on all data sets. In addition, LTR is only applicable to strictly multivariate data sets. $CRACK_N$ is applicable to all data sets, infers 15/17 causal directions correctly, by which it has an overall accuracy of 88.2%. Importantly, the two wrong decisions have low confidences compared to the correct inferences.

In addition, we conduct an experiment to check whether or not our result is influenced by redundant variables within X or Y. Hence, we first apply a standard redundancy test (R, Hmisc, redun) to omit redundant attributes within X or Y ($R^2 \geq 0.95$). After the reduction step, we apply $CRACK_N$ to the non-redundant pairs. As result, we found that the *Climate* cause effect pair indeed contained redundant information and was inferred correctly after removing the redundant variables. For all other pairs, the prediction did not change. Hence, applying $CRACK_N$ after redundancy correction leads to an accuracy of 94.4%.

8   Conclusion

We considered the problem of inferring the causal direction from the joint distribution of two univariate or multivariate random variables X and Y consisting of single-, or mixed-type data. We point out weaknesses of known causal indicators and propose the normalized causal indicator for mixed-type data and data with highly unbalanced domains. Further, we propose a practical two-part MDL encoding based on classification and regression trees to instantiate the absolute and normalized causal indicators and provide CRACK, a fast greedy heuristic to efficiently approximate the optimal MDL score.

In the experiments, we evaluate the advantages of our proposed causal indicators and give advice on when to use them. On real world benchmark data, we are on par with the state of the art for univariate continuous data and beat the state of the art on multivariate data with a wide margin. For future work, we aim to investigate the application of CRACK for the discovery of causal networks as well as its application to biological networks.

Acknowledgements. The authors wish to thank Kailash Budhathoki for insightful discussions. Alexander Marx is supported by the International Max Planck Research School for Computer Science (IMPRS-CS). Both authors are supported by the Cluster of Excellence "Multimodal Computing and Interaction" within the Excellence Initiative of the German Federal Government.

References

1. Blöbaum, P., Janzing, D., Washio, T., Shimizu, S., Schölkopf, B.: Cause-effect inference by comparing regression errors. In: AISTATS (2018)
2. Budhathoki, K., Vreeken, J.: MDL for causal inference on discrete data. In: ICDM, pp. 751–756 (2017)
3. Budhathoki, K., Vreeken, J.: Origo: causal inference by compression. Knowl. Inf. Sys. **56**(2), 285–307 (2018)
4. Chen, Z., Zhang, K., Chan, L.: Nonlinear causal discovery for high dimensional data: a kernelized trace method. In: ICDM, pp. 1003–1008. IEEE (2013)
5. Dheeru, D., Karra Taniskidou, E.: UCI machine learning repository (2017)
6. Ghiringhelli, L.M., Vybiral, J., Levchenko, S.V., Draxl, C., Scheffler, M.: Big data of materials science: critical role of the descriptor. PRL **114**, 105503 (2015)
7. Grünwald, P.: The Minimum Description Length Principle. MIT Press, Cambridge (2007)
8. Heikinheimo, H., Fortelius, M., Eronen, J., Mannila, H.: Biogeography of European land mammals shows environmentally distinct and spatially coherent clusters. J. Biogeogr. **34**, 1053–1064 (2007)
9. Hoyer, P., Janzing, D., Mooij, J., Peters, J., Schölkopf, B.: Nonlinear causal discovery with additive noise models. In: NIPS, pp. 689–696 (2009)
10. Janzing, D., Hoyer, P., Schölkopf, B.: Telling cause from effect based on high-dimensional observations. In: ICML, pp. 479–486. JMLR (2010)
11. Janzing, D., Schölkopf, B.: Causal inference using the algorithmic markov condition. IEEE TIT **56**(10), 5168–5194 (2010)
12. Janzing, D., Steudel, B.: Justifying additive noise model-based causal discovery via algorithmic information theory. OSID **17**(2), 189–212 (2010)
13. Janzing, D., et al.: Information-geometric approach to inferring causal directions. AIJ **182–183**, 1–31 (2012)
14. Kontkanen, P., Myllymäki, P.: MDL histogram density estimation. In: AISTATS, pp. 219–226 (2007)
15. Li, M., Vitányi, P.: An Introduction to Kolmogorov Complexity and Its Applications. TCS. Springer, New York (2008). https://doi.org/10.1007/978-0-387-49820-1
16. Marx, A., Vreeken, J.: Telling Cause from Effect using MDL-based Local and Global Regression. In: ICDM, pp. 307–316. IEEE (2017)
17. Mooij, J., Peters, J., Janzing, D., Zscheischler, J., Schölkopf, B.: Distinguishing cause from effect using observational data: methods and benchmarks. JMLR **17**(32), 1–102 (2016)
18. Mooij, J., Stegle, O., Janzing, D., Zhang, K., Schölkopf, B.: Probabilistic latent variable models for distinguishing between cause and effect. In: NIPS (2010)
19. Pearl, J.: Causality: Models, Reasoning and Inference. Cambridge University Press, New York (2009)
20. Peters, J., Mooij, J., Janzing, D., Schölkopf, B.: Causal discovery with continuous additive noise models. JMLR **15**, 2009–2053 (2014)
21. Peters, J., Janzing, D., Schölkopf, B.: Causal inference on discrete data using additive noise models. IEEE TPAMI **33**(12), 2436–2450 (2011)
22. Quinlan, J.R.: Induction of decision trees. Mach. Learn. **1**(1), 81–106 (1986)
23. Rissanen, J.: Modeling by shortest data description. Automatica **14**(1), 465–471 (1978)

24. Rissanen, J.: A universal prior for integers and estimation by minimum description length. Ann. Stat. **11**(2), 416–431 (1983)
25. Sgouritsa, E., Janzing, D., Hennig, P., Schölkopf, B.: Inference of cause and effect with unsupervised inverse regression. AISTATS **38**, 847–855 (2015)
26. Shimizu, S., Hoyer, P.O., Hyvärinen, A., Kerminen, A.: A linear non-gaussian acyclic model for causal discovery. JMLR **7**, 2003–2030 (2006)
27. Spirtes, P., Glymour, C., Scheines, R.: Causation, Prediction, and Search. MIT press, Cambridge (2000)
28. Steudel, B., Janzing, D., Schölkopf, B.: Causal markov condition for submodular information measures. In: COLT, pp. 464–476. OmniPress (2010)
29. Van Vechten, J.A.: Quantum dielectric theory of electronegativity in covalent systems. I. Electronic dielectric constant. PhysRev **182**(3), 891 (1969)
30. Verma, T., Pearl, J.: Equivalence and synthesis of causal models. In: UAI, pp. 255–270 (1991)
31. Vreeken, J.: Causal inference by direction of information. In: SDM, pp. 909–917. SIAM (2015)
32. Zhang, K., Hyvärinen, A.: On the identifiability of the post-nonlinear causal model. In: UAI, pp. 647–655 (2009)

Recommender Systems

POLAR: Attention-Based CNN
for One-Shot Personalized Article
Recommendation

Zhengxiao Du$^{(\boxtimes)}$, Jie Tang, and Yuhui Ding

Department of Computer Science and Technology,
Tsinghua University, Beijing, China
{duzx16,dingyh15}@mails.tsinghua.edu.cn, jietang@tsinghua.edu.cn

Abstract. In this paper, we propose POLAR, an attention-based CNN combined with one-shot learning for personalized article recommendation. Given a query, POLAR uses an attention-based CNN to estimate the relevance score between the query and related articles. The attention mechanism can help significantly improve the relevance estimation. For example, on AMiner, this can help achieve a +5.0% improvement in terms of NDCG@3. One more challenge in personalized article recommendation is how to collect statistically sufficient training data for a recommendation model. POLAR combines a one-shot learning function into the recommendation model, which further gains significant improvements. For example, on AMiner, with only 1.6 feedbacks on average, POLAR achieves 2.7% improvement by NDCG@3. We evaluate the proposed POLAR on three different datasets: AMiner, Patent, and RARD. Experimental results demonstrate the effectiveness of the proposed model. Recently, we have successfully deployed POLAR into AMiner as the recommendation engine for article recommendation, which further confirms the effectiveness of the proposed model. Data related to this paper is available at: https://doi.org/10.6084/m9.figshare.7297319.

Keywords: Personalized recommendation · Term weighting · CNN
One-shot learning

1 Introduction

Nowadays the amount of academic articles has been quite large and increases dramatically every year. According to the statistics from NCSES[1], global publication output per year in science and engineering grew at an average annual rate of 6% from 2004 to 2014. How to recommend to users the articles they are most interested in has become a key problem for digital library service providers. Many academic search sites provide article recommendation on the information

---

[1] https://www.nsf.gov/statistics/.

© Springer Nature Switzerland AG 2019
M. Berlingerio et al. (Eds.): ECML PKDD 2018, LNAI 11052, pp. 675–690, 2019.
https://doi.org/10.1007/978-3-030-10928-8_40

page of a specific article to help users find related articles. However, these recommendations are often based on keyword similarity between the current article and candidates, which doesn't contain any form of personalization.

Typically, an article covers several different topics. For example, this paper, as the keywords show, covers *Personalized Recommendation, Term Weighting, CNN* and *One-shot Learning*. Users with different backgrounds and interests may prefer articles related to different topics. Recommendation results which ignore personalization don't take user diversity into account, and cannot satisfy most users.

Good personalization can be challenging. For academic search sites, many users are cold-start users, whose profiles are incomplete or missing and cannot provide much helpful information. Methods based on user feedback are typically preferred. However, the user feedback can be quite sparse and implicit. For new users only implicit feedback from the same session is available. Therefore, it is difficult to apply the traditional recommendation methods such as Content-based Recommendation [21] or Collaborative Filtering [2].

We define the personalized article recommendation problem as follows.

Definition 1. *Let* $D = \{d_1, d_2, \cdots, d_N\}$ *denote the set of candidate articles, where N is the candidate size. The input of our problem is a query article d_q, and a support set $S = \{(\hat{d}_i, \hat{y}_i)\}_{i=1}^{T}$ related to user u, where $\hat{d}_i$ is a support article and $\hat{y}_i$ represents the user feedback for $\hat{d}_i$. The output is a totally ordered set $R(d_q, S) \subset D$ with $|R| = k$, which is the top-k recommendation for u with respect to d_q.*

Text similarity, which plays a key role in recommender systems and information retrieval, poses another challenge. The bag-of-words model, on which most traditional methods are based, discards the information about word order and cooccurrence. Therefore these methods cannot capture the matching signals in phrase or higher levels. Recently, due to the development of word embeddings and neural networks, many neural similarity models that can directly deal with word sequences are proposed [20,29], but they often treat all the words in an article indiscriminately. Therefore, they cannot distinguish important parts of an article from stereotyped expressions such as *the paper describes* and *we find that*.

Our Contributions. To address these challenges, in this paper, we propose POLAR (PersOnaLized Article Recommendation framework), to combine the attention-based CNN with one-shot learning. Our main contributions can be summarized as follows:

1. We define the personalized article recommendation problem and show that it can be tackled in the framework of one-shot learning [14]. By transferring the method for classification to the ranking problem, we can overcome the sparsity of user feedback and improve the performance.
2. Based on the matching matrix in [20], we propose the attention matrix for text similarity, in which the importance of a term is calculated as the combination of the local and global weights.

3. Inspired by the success of convolutional neural network (CNN) [13] in image recognition, we build a CNN on the matching matrix and the attention matrix to capture the text similarity from word level to article level.

4. We conduct experiments on datasets of different sources and scales. Empirical results show that our framework can perform stably and significantly better than other comparative methods.

Organization. The rest of the paper is organized as follows: Sect. 2 reviews related work. Section 3 is devoted to our POLAR framework. Section 4 presents experimental results and Sect. 5 concludes the paper.

2   Related Work

Personalized Recommendation. Personalized recommender systems aim to recommend the most relevant items to a particular user in a given context. Content-based methods [21] compare item descriptions to the user profile to determine what to be recommend. Collaborative filtering methods [2] make rating prediction utilizing the past ratings of current user or similar users [16,22], or the combination of these two [5]. To combine the advantages of the former two groups of methods, hybrid methods [15] are further proposed to improve the user profile modeling.

One-Shot Learning. One-shot learning is important for classification in cases where few examples are available. The method in [14] models the knowledge learned in other classes as a prior probability function w.r.t. the model parameters. Given an exemplar of a novel class, they update the knowledge and generate a posterior density to recognize novel instances. In [10], a Siamese network is learned with several convolutional layers used before the fully-connected layers and the top-level energy function. Matching Nets [28] take as input not only the new sample, but a small support set which contains labeled examples. Embedding functions are implemented by an LSTM with read-attention over the support set. While all these models perform on image tasks, we take a step further and propose a one-shot learning framework to recommend articles.

Text Similarity. Traditional methods for measuring the similarity between two articles, such as BM25 [23] and TF-IDF [25], are based on the bag-of-words model. These methods often take as the similarity score the sum of weights of matched words in two articles. They don't perform well on identifying the matching of phrases and sentences.

Models based on neural networks can be categorized into two groups. The first group, called *representation based models*, get the distributed semantic representation of an article with neural networks and then take as the similarity score the similarity (often cosine similarity) between distributed representations of two articles. This group include DSSM [8], LSTM-RNN [19] and MV-LSTM [29]. However, these models often lack the ability to identify the specific matching signals. The second group of models, called *interaction based models*, use neural networks to learn the patterns in the word-level interaction of two articles, usually based on word embeddings, such as MatchPyramid [20] and K-NRM [31]. The DRMM [7] uses a multilayer perceptron over a histogram of word similarities to get the similarity score of two articles. These models lack the explicit expressions of word weights but rather depend on the characteristics of word embeddings. Representation based models and interaction based models have been combined in Duet [18] to improve the performance.

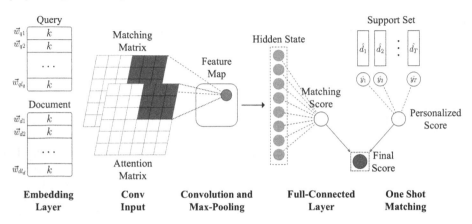

Fig. 1. The architecture of the overall framework. The articles are transformed into sequences of word embeddings through the embedding layer. The attention matrix and matching matrix are computed and sent to the CNN. The matching scores are combined with the support set to get the final scores.

3   Approach

3.1   Framework with One-Shot Learning

To get the ordered set R, for each article d_i in D our model computes a score $s(d_i|d_q, S)$, and k articles in D with the largest scores are selected as the top-k recommendation.

The recommendation problem for a specific user u can be considered as identifying whether u will accept an article or not and converted into binary classification. For each $(\hat{d}, \hat{y}) \in S$, $\hat{y}$ is binary(1 for relevant and 0 for irrelevant). S can be seen as the training set for classification, where $\hat{d}$ is a training instance and $\hat{y}$ is the corresponding label. It is probable to make an analogy between one-shot learning and our problem because S is of very limited size or even empty. Inspired by [28], our model computes $s(d_i|d_q, S)$ as follows:

$$s(d_i|d_q, S) = \begin{cases} c(d_q, d_i) & S = \varnothing \\ c(d_q, d_i) + \frac{1}{|S|} \sum_{(\hat{d},\hat{y})\in S} c(\hat{d}, d_i)\hat{y} & S \neq \varnothing \end{cases} \quad (1)$$

where $c(\cdot, \cdot)$ is our attention-based CNN for text similarity, which will be discussed in the following part. The first part of s is the *matching score* with the query article. The second part, the *personalized score*, is the normalized linear combination of the feedback in S with text similarity as coefficients, and equals zero when S is empty. The whole framework is illustrated in Fig. 1.

3.2 Matching Matrix and Attention Matrix

Each article d_i is a sequence of l_i terms $[t_{i1}, t_{i2}, \cdots, t_{il_i}]$ (We use *term* instead of *word* to show that the article has gone through preprocessing including tokenization and removal of stopwords). The matching matrix of article d_m and d_n, $M^{(m,n)} \in \mathbb{R}^{l_m \times l_n}$, is defined as follows:

$$M_{i,j}^{(m,n)} = \frac{w_{mi}^T \cdot w_{nj}}{\|w_{mi}\| \cdot \|w_{nj}\|} \tag{2}$$

where w_{mi} and w_{nj} are the word embeddings of term t_{mi} and t_{nj}. Since the cosine similarity of word embeddings can capture the semantic similarity [17], $M_{i,j}^{(m,n)}$ is the similarity between t_{mi} and t_{nj}.

Since all terms are treated equally in the matching matrix without any weighting, the matching matrix cannot reflect the term importance. Therefore the matching matrix cannot distinguish the matching signals of important terms from those of structural, unimportant terms.

Table 1. Attention mechanisms in CNN

Method	Description
Object parts selection	In *fine-grained classification* [30], image patches which contain parts of certain objects are selected through a supervised process to extract discriminative features
Attention matrix	In [32], an attention matrix is employed to give different attention weights to units in a feature map
Configurable convolution	For *visual question answering* task [4], configurable convolutional kernels are generated by transforming the question embeddings from the semantic space into the visual space, which implements the question-guided attention

To add the attention mechanism, we go over several applications of the attention mechanism in CNN in Table 1. We think the attention matrix, which can represent the importance of units in the feature map, quite suitable for our problem. The attention matrix, $A^{(m,n)} \in \mathbb{R}^{l_m \times l_n}$ is defined as follows.

$$A_{i,j}^{(m,n)} = r_{mi} \cdot r_{nj} \tag{3}$$

where r_{mi} and r_{nj} are the weights of term t_{mi} and t_{nj}. $\boldsymbol{M}^{(m,n)}$ and $\boldsymbol{A}^{(m,n)}$ are combined as the input of CNN.

3.3   Local Weight and Global Weight

Traditional methods for texts similarity often combine two types of term weights: the local weight, which depends on the specific document where the term occurs, and the global weight, which relies on the property of the whole corpus. Take the TF-IDF [25] method as an example. The TF (term frequency, how many times the term occurs in the given document) is the local weight and the Inverse DF (document frequency, how many documents the term occurs in) is the global weight.

We also combine the two weights in our model. The final weight of a term is the product of its local and global weights:

$$r_{ij} = \mu_{ij} \cdot \upsilon_{ij} \tag{4}$$

where μ_{ij} and υ_{ij} are respectively the local and global weights of the term t_{ij}.

Local Weight: How Relevant is the Term to the Subject of the Document? The local weight measures the relevance of a term to the subject of the document. For example, in the following text [6]:

Example 1. We propose a low-complexity audio-visual person authentication framework based on multiple features and multiple nearest-neighbor classifiers. The proposed MCCN method delivers a significant separation between the scores of client and impostors as observed on trials run on a unique database.

nearest-neighbor, *classifier* and *features* are obviously more important than *complexity* and *database*, and should have higher local weights, because they are more related to the topic of the text: *audio-visual authentication.*

Traditionally, the local weight is a math function of the frequency that the term occurs in a document, such as term frequency (TF) in TF-IDF [25] or the latter part in BM25 [23] ranking function:

$$\text{BM25}(d,q) = \sum_{i=1}^{n} \text{IDF}(q_i) \cdot \frac{\text{TF}(q_i,d)(k_1+1)}{\text{TF}(q_i,d) + k_1(1 - b + b\frac{|d|}{avgdl})} \tag{5}$$

where q_i is the i-th term of the query, $TF(q_i,d)$ is the term frequency of q_i in d and $avgdl$ is the average length of documents. k_1 and b are free parameters.

The basic idea of these methods is that the more important for a document a term is, the more frequently it occurs in the document. This is not always true. In Example 1, the important terms such as *authentication* and *classifier* occur only once, while the terms that occur more than once are stopwords like *of* and *on*. Therefore, a better mechanism for local weights is needed.

Local Weight Network. Inspired by [34], we propose a local weight network based on distributed word representations. The basic idea is that, because of the linearity of word embeddings, the subject of a document can be expressed as the mean of vectors of its terms. The difference between the mean vector and term vector can be seen as the *semantic difference* between the document and the term.

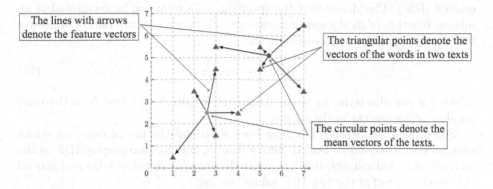

Fig. 2. A two-dimensional example of the feature vectors for local weights

To compute the local weight μ_{ij}, the feature vector $\boldsymbol{x}_{ij}$ is the difference between the word vector $\boldsymbol{w}_{ij}$ and the mean vector of d_i:

$$\boldsymbol{x}_{ij} = \boldsymbol{w}_{ij} - \overline{\boldsymbol{w}}_i \tag{6}$$

where

$$\overline{\boldsymbol{w}}_i = \frac{1}{n_i} \sum_{k=1}^{n_i} \boldsymbol{w}_{ik} \tag{7}$$

Figure 2 gives an illustration of the feature vector.

We employ a feed forward network to learn the patterns in the feature vector $\boldsymbol{x}_{ij}$ and produce the local weight. The network is a multilayer perceptron (MLP) with multiple hidden layers and gives outputs within an interval.

$$\boldsymbol{u}_{ij}^{(0)} = \boldsymbol{x}_{ij} \tag{8}$$

$$\boldsymbol{u}_{ij}^{(l)} = \text{ReLU}(\text{BN}(\boldsymbol{W}^{(l-1)} \cdot \boldsymbol{u}_{ij}^{(l-1)} + \boldsymbol{b}^{(l-1)})), l = 1, 2, \cdots, L \tag{9}$$

$$\mu_{ij} = \sigma(\boldsymbol{W}^{(L)} \cdot \boldsymbol{u}_{ij}^{(L)} + \boldsymbol{b}^{(L)}) + \alpha \tag{10}$$

where L is the number of hidden layers in the feed forward network, BN is Batch Normalization [9] between the affine transformation and ReLU non-linearity and σ is the Sigmoid function. α is a nonnegative hyperparameter to set a lower bound and avoid giving a term a local weight close to 0. The ratio of maximum value to minimum value of local weights is $1 + \frac{1}{\alpha}$. It indicates that the smaller α is, the wider the range of local weights is.

Global Weight: How Distinctive is the Term in the Whole Corpus? The global weight measures how distinctive and specific a term is. It is independent of the specific document, but depends on the whole corpus. For example, in a set of papers on computer science, *computer* and *software* are less specific than *medicine* and *neural* and should be given lower global weights. But in a medical document corpus, it may just be the reverse.

The most widespread form of global weights is the inverse document frequency (IDF). The idea is that the specificity of a term can be quantified as an inverse function of its document frequency. There are a whole family of inverse functions, and the most common one is:

$$\text{IDF}(t) = \log(\frac{N}{n_t}) \tag{11}$$

where t is the aim term, n_t is the document frequency of t and N is the total number of documents in the corpus.

Since the IDF measure has long been used and the use of other measures such as PageRank didn't lead to better results, here we also employ IDF as the measure of global weights. But to narrow the range of global weights and control the effect, instead of the raw IDF values, we use:

$$v_{ij} = [\text{IDF}(t_{ij})]^\beta \tag{12}$$

where β is a hyperparameter within the interval (0,1). The smaller β is, the narrower the range of global weights is.

3.4   Convolutional Neural Network

The matching matrix and attention matrix are combined by element-wise multiplication and sent to a CNN, which consists of several convolutional layers and max-pooling layers:

$$\boldsymbol{Z}^{(0,0)} = \boldsymbol{M} \otimes \boldsymbol{A} \tag{13}$$

$$\boldsymbol{Z}_{x,y}^{(l+1,k')} = \text{ReLU}(\text{BN}(\sum_{k=0}^{c_l-1} \sum_{i=0}^{r_k-1} \sum_{j=0}^{r_k-1} w_{i,j}^{(l+1,k)} \cdot z_{x+i,y+j}^{(l,k)} + b^{(l+1,k)})) \tag{14}$$

$$k' = 0, 1, \cdots, c_l, l = 0, 2, 4, \cdots,$$

$$\boldsymbol{Z}_{x,y}^{(l+1,k)} = \max_{0 \leq i < d_k} \max_{0 \leq j < d_k} z_{x \cdot d_k + i, y \cdot d_k + j}^{(l,k)}, l = 1, 3, 5, \cdots, \tag{15}$$

Similar to CNNs in image recognition [33], the filters in low-level convolutional layers can capture different matching signals between phrases, while the filters in high-level convolutional layers can capture the matching signals between sentences and paragraphs. The max-pooling layers can downsample the signals and reduce the spatial size of feature maps.

The output of the last max-pooling layer is then turned into a vector and passed through an MLP with several hidden layers, as described in Eq. 9. In this paper, we use only one hidden layer. For the final output, a single unit is connected to all the units of the last hidden layer.

3.5   Optimization and Training

The entire model, including the CNN and the local weight network, is trained end-to-end on the target task.

The hinge loss is used as the objective function for training. Given the triples $\{(d_q^{(i)}, d_+^{(i)}, d_-^{(i)})\}_{i=1}^N$, where article $d_+^{(i)}$ is ranked higher than article $d_-^{(i)}$ with respect to query article $d_q^{(i)}$, the loss is:

$$Loss = \sum_{i=1}^{N} \max(0, 1 - c(d_q^{(i)}, d_+^{(i)}) + c(d_q^{(i)}, d_-^{(i)})) \tag{16}$$

where $c(d_q, d)$ denotes the predicted matching score between d_q and d.

Since the size of some datasets we use is relatively small, for experiments on these datasets we train the model on a classifying task called citation prediction. Given the abstracts of two papers, the model needs to classify them as having citation relationship or not. Obviously, to complete the task, the model also needs to compute the relevance of two articles. In this case the loss function is the cross entropy. Given the triples $\{(d_1^{(i)}, d_2^{(i)}, y^{(i)}\}_{i=1}^N$, the loss is:

$$Loss = -\sum_{i=1}^{N} y^{(i)} \log(p^{(i)}) + (1 - y^{(i)}) \log(1 - p^{(i)})$$
$$p^{(i)} = \frac{1}{1 + e^{-c(d_1^{(i)}, d_2^{(i)})}} \tag{17}$$

where $p^{(i)}$ is the predicted probability that the i-th instance is positive and $y^{(i)}$ is the label of the i-th instance.

The optimization is done through standard backpropagation [24] and stochastic gradient descent method with mini-batches. For regularization, we use dropout [26] in the output of every hidden layer and early stopping strategy [3] to avoid over-fitting.

4   Experiments

In this section, to evaluate the proposed model, we conduct experiments on the article recommendation problem based on three datasets, in comparison with traditional methods and neural models.

4.1   Experiment Setup

Comparison Methods. The following are several traditional methods.

- **TF-IDF** [25]: The similarity score between a query and a document is computed by summing the weights of the query's terms which also occur in the document. The weight of a term is the product of its TF and IDF weights.

– **Doc2Vec** [12]: We get the distributed representation of each article via Paragraph Vector model. The similarity score between two articles is produced by the cosine similarity of their representations.
– **WMD** [11]: The Word Mover's Distance (WMD) is the minimum distance required to transport words from one document to another based on the word embeddings.

The following are several neural matching models.

– **MV-LSTM** [29]: The interactions between different positional sentence representations generated by a Bi-LSTM form a similarity matrix to generate the matching score.
– **MatchPyramid** [20]: A CNN is built on the standard matching matrix to get the matching score.
– **DRMM** [7]: The matching between the terms in the query and the document is expressed as a histogram, where only the counts of the matching score in different intervals are reserved. The histogram is sent to an MLP to get the matching score.
– **Duet** [18]: An interaction-based model and a representation-based model are combined to get the matching score of two articles.

Parameter Setting. In the Local Weight Network there are two hidden layers, with 64 and 32 hidden units respectively. The CNN has three convolutional layers and three max-pooling layers. The first and second convolutional layers both have 32 filters and the third convolutional layer has 16 filters. All convolutional filters are set to 3×3 and all max-pooling kernels are set to 2×2. The number of hidden units in the full-connected layer is set to 256. For the hyperparameters α and β, we set $\alpha = 1$ and $\beta = \frac{1}{4}$, which is discussed in Sect. 4.3.

The word embeddings in all the models above are 256 dimensions trained on Wikipedia via the skip-gram model, using hierarchical softmax and negative sampling [17].

Dataset. We evaluate the performance of the proposed model with two small, manually labeled datasets and a large-scale dataset based on user click.

The first dataset is based on papers from AMiner [27] and consists of 188 query papers with 10 candidate papers for each query. The second dataset is based on documents of patents coming from the Patent Full-Text Databases of the United States Patent and Trademark Office[2] and consists of 67 queries with 20 candidates for each query. In each dataset, we gather relevance judgments from college students or experts on patent analysis as the ground truth. The relevance is simply expressed as binary: relevant or irrelevant. Abstracts of the papers or the patent documents are used as texts and texts longer than 96 terms are truncated.

---

[2] http://patft.uspto.gov/.

Table 2. Results of relevance ranking (%). NG stands for NDCG

Method	AMiner			Patent			RARD		
	NG@3	NG@5	NG@10	NG@3	NG@5	NG@10	NG@1	NG@3	NG@5
TF-IDF	74.3	81.8	87.5	51.8	56.4	63.4	37.6	39.8	46.3
Doc2Vec	60.0	65.8	79.1	44.6	45.6	53.5	28.4	34.0	40.0
WMD	73.0	76.3	86.2	57.4	58.5	61.9	23.4	38.2	46.8
MV-LSTM	56.2	61.2	76.2	60.2	59.0	65.0	22.2	30.7	39.3
Duet	66.6	74.4	82.6	54.5	57.5	64.6	22.3	31.1	39.8
DRMM	75.0	79.9	87.1	55.0	56.2	64.7	33.1	36.3	40.6
MatchPyramid	73.5	80.0	86.8	56.4	61.4	64.4	29.1	36.2	42.8
POLAR	**80.3**	**85.2**	**90.1**	**67.8**	**69.5**	**73.6**	**42.8**	**46.3**	**51.5**

Since the sizes of two datasets are relatively small, we train the models on the citation prediction task, which is described in Sect. 3.5, for all comparison methods. The dataset for training is the Citation Network Dataset in AMiner [27].

The third dataset is Related-Article Recommendation Dataset (RARD) [1] from Sowiport, a digital library of social science articles that displays related articles to its users. The dataset contains 63923 distinct queries with user click log. Each query article has an average of 9.1 articles displayed. The displayed documents are generated by a recommender-as-a-service provider Mr. DLib, so they are of high relevance to the query. We choose 800 queries that have the most clicks for test and other queries are used for training. Since the abstracts of some articles are missing, the titles and the abstracts of articles are combined as texts. Texts longer than 64 terms are truncated.

4.2 Performance Comparison

Table 2 shows the ranking accuracy of different methods in terms of NDCG. For the fairness of comparison, all models don't involve user feedback, which will be discussed in Sect. 4.3.

From the evaluation results, we can observe that our proposed model POLAR can perform better than all the baselines. POLAR can outperform the best baselines 6.9%–13.2% on NDCG@3 and 3.3%–20.3% on NDCG@5. The average improvements of NDCG on each dataset are respectively 3.8%, 8.1% and 6.4%.

Among the traditional ranking models, TF-IDF is the most competitive one, in some cases even outperforming the best neural baselines by 5.5%. But we can also find that TF-IDF performs not very well on the patent dataset. The reason might be that documents of patents are often written by non-academic researchers and terms on the same topic might vary from person to person. Only taking the exact matching signals into account, TF-IDF might be unsuitable for such situation, while the methods based on word embeddings can perform better.

As for the neural ranking models, we can see that interaction based models, including DRMM and MatchPyramid, perform slightly better than representation based models. Although the Duet combines the interaction-based model and the representation-based model, it doesn't perform better than individual interaction-based models.

4.3   Analysis and Discussion

How One-Shot Learning Can Help. We utilize the datasets in the previous part to simulate the personalization problem. We select those queries that have more than one positive-labeled candidate. For every query, we randomly divide the labeled documents into two parts. The first part is used as the support set and the second part is used as the candidate set to recommend. Then we compare the proposed one shot framework (called POLAR-OS) with the best model that ignores support sets in the previous part (called POLAR-ALL). The support set is quite sparse compared with the size of candidates. For example, in the RARD dataset, the average size of support set for each query is only 1.5. In the AMiner dataset, the size of support set is only 1 for 45% queries and 2 for 47%. In the patent dataset, the sizes of support sets of 75% queries are no greater than 3.

The result is shown in Table 3. We can see that the performance can be improved with a small amount of feedback data. On average, POLAR-OS can outperform POLAR-ALL by 7.0% on NDCG@1 and 5.7% on NDCG@3.

Table 3. Performance for the model with one shot learning and without.

Method	AMiner		Patent		RARD	
	NDCG@1	NDCG@3	NDCG@1	NDCG@3	NDCG@1	NDCG@3
POLAR-ALL	76.1	79.2	52.3	66.2	36.5	36.5
POLAR-OS	**79.1**	**81.9**	**57.1**	**69.7**	**39.4**	**39.2**

How the Attention Matrix Can Help. To illustrate the improvements different parts of the attention matrix bring, we compare three versions of the proposed model with different attention matrices. To compute the attention matrix, POLAR-LOC uses only the local weights and POLAR-GLO uses only the global weights. POLAR-ALL uses both local weights and global weights. The performance in terms of NDCG@3 is shown in Fig. 3.

In most cases, POLAR-LOC, the model with the local weight network, perform better than POLAR-GLO. The reason might be that the local weight network is trainable, with greater ability to learn the importance of terms. IDF is only a statistical way to get approximate values. The complete model, POLAR-ALL, which combines the two weights, performs significantly better than either of them. This confirms that the local and global weights are complementary to each other.

To have a better understanding of how local and global weights work, we show the pixel images of four matrices in Fig. 4. From the images we can find that the local weights of most terms are low while the global weights of most terms

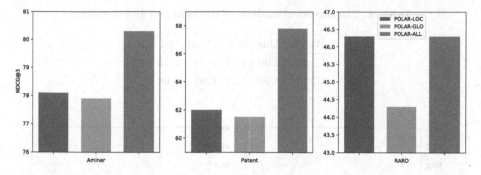

Fig. 3. The performance of different attention matrices

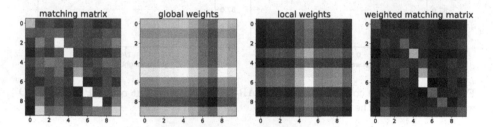

Fig. 4. The visualization result of four matrices used in the matching of a pair of texts. The brighter the pixel is, the larger value it has. The text pair is as follows (the words in brackets are removed stopwords): T1: novel robust stability criteria (for) stochastic hopfield neural networks (with) time delays. T2: new delay dependent stability criteria (for) neural networks (with) time varying delay.

are high. The statistical analysis of the local and global weights in Table 4 also supports this idea. Therefore, we can conclude that the global weights function by deemphasizing unimportant terms in the corpus with low weights, while the local weights function by highlighting key terms in specific articles.

Sensitivity Analysis of Hyperparameters. Since there are two hyperparameters α and β to control the effect of the local and global weights in our proposed model, we further study the effect of different choices of α and β. The result is shown in Fig. 5. In general, the variance in β has greater effect than that in α. In our model the global weights are predefined values which couldn't be changed once β is chosen, while the local weights are calculated by the local weight network, which can automatically adapt to different choices of α. Therefore it is important to choose the value of β. When β is close to 1, the global weights are equal to IDF values, which vary so greatly that the model will ignore the effect of cosine similarity. When β is close to 0, the global weights are almost uniform and have little effect. But the model with the value of α equal to 0 cannot perform well either,

Table 4. The statistical analysis of the local and global weights

Weight	Max	Min	Mean	Std
Local	2.00	1.00	1.20	0.15
Global	1.96	1.08	1.86	0.08

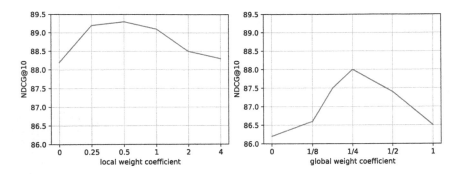

Fig. 5. Performance comparison for POLAR-LOC with different α and POLAR-GLO with different β on the AMiner dataset

because the local weight network can have too strong effect and be troubled by over-fitting.

5   Conclusion

In this paper, we study the problem of personalized article recommendation. We define the problem and propose a novel model POLAR to solve it. We utilize the framework of one shot learning to deal with the sparse user feedback and propose an attention based CNN model for text similarity. Experimental results show that the proposed model significantly outperforms both the traditional and the state-of-art neural baselines. The model has been used in AMiner to provide recommendation of similar papers.

For further work, we would like to combine our model with reinforcement learning (RL), to train a deeper and more powerful model in the online environment. We may also compare the performance of different attention mechanisms in CNN.

References

1. Beel, J., Carevic, Z., Schaible, J., Neusch, G.: RARD: the related-article recommendation dataset [data] (2017). https://doi.org/10.7910/DVN/HA8EAH
2. Breese, J.S., Heckerman, D., Kadie, C.: Empirical analysis of predictive algorithms for collaborative filtering. In: UAI, pp. 43–52 (1998)

3. Caruana, R., Lawrence, S., Giles, L.: Overfitting in neural nets: backpropagation, conjugate gradient, and early stopping. In: NIPS, pp. 381–387 (2000)
4. Chen, K., Wang, J., Chen, L., Gao, H., Xu, W., Nevatia, R.: ABC-CNN: an attention based convolutional neural network for visual question answering. CoRR abs/1511.05960 (2015)
5. Das, A.S., Datar, M., Garg, A., Rajaram, S.: Google news personalization: scalable online collaborative filtering. In: WWW, pp. 271–280 (2007). https://doi.org/10.1145/1242572.1242610
6. Das, A.: Audio visual person authentication by multiple nearest neighbor classifiers. In: Lee, S.-W., Li, S.Z. (eds.) ICB 2007. LNCS, vol. 4642, pp. 1114–1123. Springer, Heidelberg (2007). https://doi.org/10.1007/978-3-540-74549-5_116
7. Guo, J., Fan, Y., Ai, Q., Croft, W.B.: A deep relevance matching model for ad-hoc retrieval. In: CIKM, pp. 55–64 (2016). https://doi.org/10.1145/2983323.2983769
8. Huang, P.S., He, X., Gao, J., Deng, L., Acero, A., Heck, L.: Learning deep structured semantic models for web search using clickthrough data. In: CIKM, pp. 2333–2338 (2013). https://doi.org/10.1145/2505515.2505665
9. Ioffe, S., Szegedy, C.: Batch normalization: accelerating deep network training by reducing internal covariate shift. In: ICML, pp. 448–456 (2015)
10. Koch, G., Zemel, R., Salakhutdinov, R.: Siamese neural networks for one-shot image recognition. In: ICML Deep Learning Workshop, vol. 2 (2015)
11. Kusner, M.J., Sun, Y., Kolkin, N.I., Weinberger, K.Q.: From word embeddings to document distances. In: ICML, pp. 957–966 (2015)
12. Le, Q.V., Mikolov, T.: Distributed representations of sentences and documents. In: ICML, pp. 1188–1196 (2014)
13. Lecun, Y., Bottou, L., Bengio, Y., Haffner, P.: Gradient-based learning applied to document recognition. Proc. IEEE 86(11), 2278–2324 (1998). https://doi.org/10.1109/5.726791
14. Li, F., Fergus, R., Perona, P.: One-shot learning of object categories. IEEE Trans. Pattern Anal. Mach. Intell. 28(4), 594–611 (2006). https://doi.org/10.1109/TPAMI.2006.79
15. Li, L., Wang, D., Li, T., Knox, D., Padmanabhan, B.: SCENE: a scalable two-stage personalized news recommendation system. In: SIGIR, pp. 125–134 (2011). https://doi.org/10.1145/2009916.2009937
16. Marlin, B., Zemel, R.S.: The multiple multiplicative factor model for collaborative filtering. In: ICML, p. 73 (2004). https://doi.org/10.1145/1015330.1015437
17. Mikolov, T., Sutskever, I., Chen, K., Corrado, G., Dean, J.: Distributed representations of words and phrases and their compositionality. In: NIPS, pp. 3111–3119 (2013)
18. Mitra, B., Diaz, F., Craswell, N.: Learning to match using local and distributed representations of text for web search. In: WWW, pp. 1291–1299 (2017). https://doi.org/10.1145/3038912.3052579
19. Palangi, H., et al.: Deep sentence embedding using long short-term memory networks: analysis and application to information retrieval. IEEE/ACM Trans. Audio Speech Lang. Proc. 24(4), 694–707 (2016). https://doi.org/10.1109/TASLP.2016.2520371
20. Pang, L., Lan, Y., Guo, J., Xu, J., Wan, S., Cheng, X.: Text matching as image recognition. In: AAAI, pp. 2793–2799 (2016)
21. Pazzani, M.J., Billsus, D.: Content-based recommendation systems. In: Brusilovsky, P., Kobsa, A., Nejdl, W. (eds.) The Adaptive Web. LNCS, vol. 4321, pp. 325–341. Springer, Heidelberg (2007). https://doi.org/10.1007/978-3-540-72079-9_10

22. Rendle, S.: Factorization machines. In: ICDM, pp. 995–1000 (2010). https://doi.org/10.1109/ICDM.2010.127
23. Robertson, S.E., Walker, S., Jones, S., Hancock-Beaulieu, M., Gatford, M.: Okapi at TREC-3. In: TREC, pp. 109–126 (1994)
24. Rumelhart, D.E., Hinton, G.E., Williams, R.J.: Learning representations by back-propagating errors. Nature **323**(6088), 533–536 (1986)
25. Salton, G., Fox, E.A., Wu, H.: Extended boolean information retrieval. Commun. ACM **26**(11), 1022–1036 (1983). https://doi.org/10.1145/182.358466
26. Srivastava, N., Hinton, G.E., Krizhevsky, A., Sutskever, I., Salakhutdinov, R.: Dropout: a simple way to prevent neural networks from overfitting. JMLR **15**(1), 1929–1958 (2014)
27. Tang, J., Zhang, J., Yao, L., Li, J., Zhang, L., Su, Z.: ArnetMiner: extraction and mining of academic social networks. In: SIGKDD, pp. 990–998 (2008). https://doi.org/10.1145/1401890.1402008
28. Vinyals, O., Blundell, C., Lillicrap, T., Kavukcuoglu, K., Wierstra, D.: Matching networks for one shot learning. In: NIPS, pp. 3630–3638 (2016)
29. Wan, S., Lan, Y., Guo, J., Xu, J., Pang, L., Cheng, X.: A deep architecture for semantic matching with multiple positional sentence representations. In: AAAI, pp. 2835–2841 (2016)
30. Xiao, T., Xu, Y., Yang, K., Zhang, J., Peng, Y., Zhang, Z.: The application of two-level attention models in deep convolutional neural network for fine-grained image classification. In: CVPR, pp. 842–850 (2015). https://doi.org/10.1109/CVPR.2015.7298685
31. Xiong, C., Dai, Z., Callan, J., Liu, Z., Power, R.: End-to-end neural ad-hoc ranking with kernel pooling. In: SIGIR, pp. 55–64 (2017). https://doi.org/10.1145/3077136.3080809
32. Yin, W., Schütze, H., Xiang, B., Zhou, B.: ABCNN: attention-based convolutional neural network for modeling sentence pairs. TACL **4**, 259–272 (2016)
33. Zeiler, M.D., Fergus, R.: Visualizing and understanding convolutional networks. In: Fleet, D., Pajdla, T., Schiele, B., Tuytelaars, T. (eds.) ECCV 2014. LNCS, vol. 8689, pp. 818–833. Springer, Cham (2014). https://doi.org/10.1007/978-3-319-10590-1_53
34. Zheng, G., Callan, J.: Learning to reweight terms with distributed representations. In: SIGIR, pp. 575–584 (2015). https://doi.org/10.1145/2766462.2767700

Learning Multi-granularity Dynamic Network Representations for Social Recommendation

Peng Liu[✉], Lemei Zhang, and Jon Atle Gulla

Department of Computer Science, NTNU, Trondheim, Norway
{peng.liu,lemei.zhang,jon.atle.gulla}@ntnu.no

Abstract. With the rapid proliferation of online social networks, personalized social recommendation has become an important means to help people discover useful information over time. However, the cold-start issue and the special properties of social networks, such as rich temporal dynamics, heterogeneous and complex structures with millions of nodes, render the most commonly used recommendation approaches (e.g. Collaborative Filtering) inefficient. In this paper, we propose a novel multi-granularity dynamic network embedding (m-DNE) model for the social recommendation which is capable of recommending relevant users and interested items. In order to support online recommendation, we construct a heterogeneous user-item (HUI) network and incrementally maintain it as the social network evolves. m-DNE jointly captures the temporal semantic effects, social relationships and user behavior sequential patterns in a unified way by embedding the HUI network into a shared low dimensional space. Meanwhile, multi-granularity proximities which include the second-order proximity and the community-aware high-order proximity of nodes, are introduced to learn more informative and robust network representations. Then, with an efficient search method, we use the encoded representation of temporal contexts to generate recommendations. Experimental results on several real large-scale datasets show its advantages over other state-of-the-art methods.

Keywords: Social recommendation · Heterogeneous social network
Network embedding · Temporal context · Community detection

1 Introduction

In the last few decades, the rapid development of Web 2.0 and smart mobile devices have resulted in the dramatic proliferation of online social networks. According to Twitter statistics, the number of users is estimated to have surpassed 300 million generating more than 500 million tweets per day[1]. Faced with the abundance of user generated content, a key issue of social networking services is how to help users find their potential friends or interested items that

---

[1] https://www.omnicoreagency.com/twitter-statistics/.

© Springer Nature Switzerland AG 2019
M. Berlingerio et al. (Eds.): ECML PKDD 2018, LNAI 11052, pp. 691–708, 2019.
https://doi.org/10.1007/978-3-030-10928-8_41

match the users' preference as much as possible, by making use of both semantic information and social relationships. This is the problem of personalized social recommendation.

Collaborative filtering (CF) has been shown to be an effective approach to recommender systems. It makes predictions about user's interests based on preferences of other users. However, CF is generally designed for bipartite graphs which model interactions between users and items and thus cannot be easily applied over complex heterogeneous social networks. Besides, cold start issue becomes even more severe in online settings as the new users and new items will join in constantly over time. Many approaches [1,2] have been proposed to alleviate this problem, but they are not designed specifically for online environment.

Recently, network representation learning (NRL) has attracted a considerable amount of interest from various domains, with recommender systems being no exception [3,4]. The popularization of NRL in recommendation can be mainly attributed to the network embedding techniques which learn low-dimensional vertex representation by modelling vertex co-occurrence in individual user's interaction records, thus capturing the semantic relationships among vertices and boosting recommendation accuracy [4]. Cold start issues can be alleviated through mining the structure and relations among existing and newly arrived nodes. Despite these positive results, we argue that NRL for social recommendations still suffers from the following four challenges: (1) Different from widely used homogeneous networks, heterogeneous network which includes different-typed objects and links, is seldom studied but more commonly seen in real world. Besides, online networks often incorporate millions even billions of nodes and edges in real world, which brings more obstacles in dealing with them. (2) Most real-world networks are intrinsically dynamic with addition/deletion of edges and nodes. Meanwhile, similar as network structure, node attributes also change as new content patterns may emerge and outdated content patterns will fade. (3) So far, most previous network representation methods primarily preserve the local structure and content, such as the first- and second-order proximities of nodes, the global community structure, which is one of the most prominent features, is largely ignored. (4) Considering the online environment and frequently changing velocity of social networks, the scalability and updating complexity of learning algorithms should also play a pivotal role and be seriously reckoned. Recent researches only pay attention to several of the abovementioned challenges while still neglect one or more of them [5–9].

To address the problems raised above, we propose a novel multi-granularity dynamic network embedding (m-DNE) model for online social recommendation. Specifically, we firstly construct a heterogeneous user-item (HUI) network which is incrementally maintained as the social network evolves. Then, a low complexity incremental learning algorithm is applied to embed HUI into low-dimensional representation space with the use of multi-granularity proximity information (second-order and community-aware high-order proximities) of each vertex. Afterwards, an efficient search method and a time-decay mechanism are adopted to conduct recommendation tasks. To the best of our knowledge, we are the first to jointly model the temporal semantic effects, social relationships and user behavior sequential patterns in a unified way to address the issue of temporal dynamics, cold start and context awareness in an online social recommendation. Our experiments show that the proposed approach is superior to all baselines and state-of-the-art methods in social recommendation tasks.

In this paper, Sect. 2 introduces the related work. In Sect. 3, we define the key concepts and our problem. Sections 4, 5 and 6 present our model. We describe the experimental setup and results in Sect. 7 and Sect. 8 concludes the study.

2   Related Work

Social Recommender System. In recent years, many studies have demonstrated the success of utilizing rich social network information to improve the recommendation performance [2,4]. However, these efforts have not considered online updating or incremental processes. In order to capture the evolution of the recommender systems, Agarwal et al. [10] proposed a fast online bilinear factor model to learn item-specific factors through online regression by using a large amount of historical data to initialize the online models and thus reducing the dimensionality of the input features. Diaz-Aviles et al. [11] presented Stream Ranking Matrix Factorization, which utilizes a pairwise approach to matrix factorization for optimizing the personalized ranking of topics and follows a selective sampling strategy to perform incremental model updates based on active learning principles. Huang et al. [12] presented a practical scalable item-based collaborative filtering algorithm, with the characteristics such as robustness to implicit feedback problem. Subbian et al. [13] proposed a probabilistic neighbourhood-based algorithm for performing recommendations with streaming data. The recommendation strategies proposed by [12] and [13] focus on scalability and dynamic pruning in recommender systems. Our proposed framework considers the combination of the heterogeneous characteristics of social networks and graph-based updating schemes in online settings, and thus is substantially different from the above-mentioned systems.

Network Representation Learning. Recently, network representation learning which aims to learn low-dimensional node embedding is attracting increasing attentions. DeepWalk [6] models the second-order proximity for node embedding with path sampling, whose complexity is $O(|V|log|V|)$. Node2Vec [14] extends DeepWalk with a controlled path sampling process, which requires $O(|V|log|V| + |V|a^2)$ where a is the average degree of the graph. LINE [5] preserves both first- and second-order proximity with complexity of $O(a|E|)$. Compared with our methods, the above works have a lower or comparable complexity, but they are neither worked for heterogeneous networks nor aware of community structure. Metapath2vec [15] extends the network embedding methods to heterogeneous network by introducing metapath based random walk with the complexity of $O(a|E||V|)$. PTE [16] utilizes labels of words and constructs a large-scale heterogeneous text network to learn predictive embedding vectors for words with complexity of $O(a|E|)$. The above-mentioned approaches can model heterogeneous network but are still not community preserving. There is little work that tries to take into account community structure and dynamic environment. For example, Cavallari et al. [7] proposes a community embedding framework, ComE, which adopt global community structure to optimize node embedding results with relatively lower complexity of $O(|V| + |E|)$ but on homogeneous and static network. In [8], DANE performs network embedding in a dynamic environment also for homogeneous network with barely local structure of nodes, and thus ignores the importance of the high-order proximity. The online complexity of DANE is $O(|V|)$. M-NMF [9] constructs the modularity matrix, then applies non-negative matrix factorization to learn node embedding and community detection together with a higher complexity proportional to $O(|V|^2)$ based on static network. Our work is highly built upon LINE and Deep-Walk. The novelty lies in the idea of adopting the network embedding methods into the dynamic environment for online recommendation with comparatively low complexity $O(|V|log(|V|))$ in worst cases. As far as we know, it is the first attempt to improve the representation learning with incorporating the temporal community structure into the dynamic network embedding method.

3   Problem Formulation

In this section, we define the key concepts and present the problem statement of this study before the detailed description of our m-DNE model.

Definition 1 Heterogeneous User-Item (HUI) Network. *A heterogeneous user-item network can be represented by* $G_{mix} = G_{uu} \cup G_{pp} \cup G_{up}$, *which consists of the user-user relationship network* $G_{uu} = (\mathcal{U}, \varepsilon_{uu})$, *the item-item relationship network* $G_{pp} = (\mathcal{P}, \varepsilon_{pp})$ *and the user-item interaction network* $G_{up} = (\mathcal{U} \cup \mathcal{P}, \varepsilon_{up})$. *Among this,* $\mathcal{U} = \{u_1, u_2, ...u_n\}$ *is the set of users, where* u_i *is the user profile represented with a three tuple* $(uId, \mathcal{L}, \mathcal{D})$, *which indicates userID, user social links and a set of items associated with* u_i. $\mathcal{P} = \{p_1, p_2, ...p_n\}$ *is the set of items, where* p_i *is the item profile with a five tuple* $(iId, \mathcal{M}, \mathcal{H}, \mathcal{W}, \rho)$, *representing itemID, named entity, hashtag/category, content, create time respectively.* ε_{uu}, ε_{pp} *and* ε_{up} *are the sets of edges, which indicate different relation types.*

Definition 2 Community. *A community c is a group of vertices, including both users and items, in* G_{mix}, *and all vertices can be grouped into* $\mathcal{K}$ *communities* $\mathcal{C} = \{c_1, c_2, ..., c_{\mathcal{K}}\}$. *The communities can be overlapping, which is to say each vertex* $v \in \mathcal{U} \cup \mathcal{P}$, *can belong to different c to different degree.*

Finally, we formally define the problem investigated in our work. Given a time-stamped heterogeneous user-item network, we aim to provide online social recommendations stated as follows.

Problem 1 (Online Social Recommendation). *Given a heterogeneous user-item network* G_{mix} *at timestamp t and a querying user* $u \in \mathcal{U}$, *the task is to generate a ranked list of user or item recommendations that u would be interested in.*

4   Heterogeneous User-Item Network

4.1   HUI Network Construction

For notational simplicity, we ignore the time-subscript in this subsection. Given a set of users $\mathcal{U} = \{u_1, u_2, ...u_m\}$ and a set of items $\mathcal{P} = \{p_1, p_2, ...p_n\}$, to integrate the semantic effects, social relationships and the user behavior sequential patterns simultaneously, we construct a heterogeneous user-item network comprising two types of nodes and three types of edges, as shown in Fig. 1. The two types of nodes which consist of user and item nodes are formed by projecting the user set and item set respectively. The three types of edges are defined as follows: (1) Each user node u_i and each item node p_j are connected if user u_i shows an interest on item p_j. In the HUI network, such an edge is indicated by yellow solid lines. The associated item nodes of the user node u_i are denoted as $\mathcal{I}_p(u_i)$, the associated user nodes of the item node p_j are denoted as $\mathcal{I}_u(p_j)$. (2) Two user nodes u_i and u_j are connected with the property of user similarity $sim_u(u_i, u_j)$ if they have a social link, such as follower or followee. In the HUI network, such edge is indicated by grey dash lines. The adjacent user nodes of the user node u_i are denoted as $\mathcal{A}_u(u_i)$. (3) Two item nodes p_i and p_j are connected with the item similarity $sim_p(p_i, p_j)$ if they have a semantic link such as Named Entity or Hashtag. In the HUI network, such edge is indicated by orange dash lines. The adjacent item nodes of the item node p_i are denoted as $\mathcal{A}_p(p_i)$.

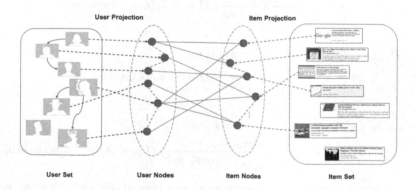

Fig. 1. The Heterogeneous User-Item (HUI) network.

We assume that $\mathcal{R}_i$ is a r-dimensional vector representing the social links of user u_i, where r is the total number of users, and the k-th dimension of vector $\mathcal{R}_i$ equals 1 only if there is an edge between u_i and u_k, otherwise 0. The user similarity $sim_u(u_i, u_j)$ between user u_i and user u_j can be defined as the cosine similarity between the two vectors. Likewise, we use the cosine similarity $sim_p(p_i, p_j)$ to measure the similarity between two item nodes p_i and p_j.

Directly applying random walk to the HUI network does not work due to different edge types, leading to a challenging problem. To this end, we propose a novel way to capture the different edge type characteristic into the transition probability matrix P, where three parameters α, β, γ with $\alpha + \beta + \gamma = 1$ are used to respectively control the relative importance of user behavior sequential patterns, social relationships and semantic effects. The values of α, β and γ will be varied depending on different datasets[2].

Definition 3. *A transition probability matrix* $P \in \mathbb{R}^{(m+n) \times (m+n)}$ *is constructed for the HUI network,*

$$P = \begin{pmatrix} P_u & P_{up} \\ P_{pu} & P_p \end{pmatrix} \tag{1}$$

which comprises four matrix blocks $P_u \in \mathbb{R}^{m \times m}$, $P_{up} \in \mathbb{R}^{m \times n}$, $P_{pu} \in \mathbb{R}^{n \times m}$ *and* $P_p \in \mathbb{R}^{n \times n}$ *respectively representing the transition probabilities of random walks between user nodes, from user nodes to item nodes, from item nodes to user nodes and between item nodes. That is*

$$
\begin{aligned}
P_{i,j} &= Prob(u_j | u_i), \qquad i < m, j < m \\
&= \frac{\beta}{\alpha + \beta} \times \frac{sim_u(u_i, u_j)}{\sum_{u_k \in \mathcal{A}_u(u_i)} sim_u(u_i, u_k)}, \; if \; u_j \in \mathcal{A}_u(u_i), \; otherwise \; 0.
\end{aligned}
\tag{2}
$$

$$
\begin{aligned}
P_{i,m+j} &= Prob(p_j | u_i), \qquad i < m, j < n \\
&= \frac{\alpha}{\alpha + \beta} \times \frac{w_{ij}}{\sum_{p_k \in \mathcal{I}_p(u_i)} w_{ik}}, \; if \; p_j \in \mathcal{I}_p(u_i), \; otherwise \; 0.
\end{aligned}
\tag{3}
$$

$$
\begin{aligned}
P_{m+i,j} &= Prob(u_j | p_i), \qquad i < n, j < m \\
&= \frac{\alpha}{\alpha + \gamma} \times \frac{w_{ji}}{\sum_{u_k \in \mathcal{I}_u(p_i)} w_{ki}}, \; if \; u_j \in \mathcal{I}_u(p_i), \; otherwise \; 0.
\end{aligned}
\tag{4}
$$

$$
\begin{aligned}
P_{m+i,m+j} &= Prob(p_j | p_i), \qquad i < n, j < n \\
&= \frac{\gamma}{\alpha + \gamma} \times \frac{sim_p(p_i, p_j)}{\sum_{p_k \in \mathcal{A}_p(p_i)} sim_p(p_i, p_k)}, \; if \; p_j \in \mathcal{A}_p(p_i), \; otherwise \; 0.
\end{aligned}
\tag{5}
$$

In the above definition, to incorporate user bias in our algorithm, we introduce w_{ij} which denotes the rating score that the user u_i assigns to item p_j, and it has different rating scales. For example, in the movie recommendation case, w_{ij} might correspond to an explicit rating given by user u_i to movie p_j or, in the case of twitter/music recommendation, w_{ij} is implicitly derived from user's interaction patterns, e.g., how many times user u_i has clicked/listened item p_j.

---

[2] In our experiments, we set $\alpha = \beta = \gamma = 1/3$ by grid-search over $\{1/9, 3/9, 4/9, 7/9\}$ which achieves the best recommendation performance for all datasets.

4.2   HUI Network Update

Assume at timestamp t, the current HUI network $G_{mix,t} = (\mathcal{V}_t, \varepsilon_t) = (\mathcal{U}_t, \varepsilon_{uu,t}, \mathcal{P}_t, \varepsilon_{pp,t}, \varepsilon_{up,t})$ contains the user node set $\mathcal{U}_t$, item node set $\mathcal{P}_t$ and their related edge sets $\varepsilon_{uu,t}$, $\varepsilon_{pp,t}$ and $\varepsilon_{up,t}$. Due to the evolution of the network, $\mathcal{U}_t$ and $\mathcal{P}_t$ will contain the sets of the newly attached nodes, denoted as $\Delta\mathcal{U}_t$ and $\Delta\mathcal{P}_t$ respectively, while there exists another subsets of $\mathcal{U}_t$ and $\mathcal{P}_t$ containing the nodes that have user or item profile changed at the current timestamp, which are denoted as $\Theta\mathcal{U}_t$ and $\Theta\mathcal{P}_t$. Similarly, subsets of $\varepsilon_{uu,t}$, $\varepsilon_{pp,t}$ and $\varepsilon_{up,t}$ contain the newly attached edges, separately denoted as $\Delta\varepsilon_{uu,t}$, $\Delta\varepsilon_{pp,t}$ and $\Delta\varepsilon_{up,t}$, while the edges with changed similarities or rating scores within $\varepsilon_{uu,t}$, $\varepsilon_{pp,t}$ and $\varepsilon_{up,t}$ at timestamp t are denoted as $\Theta\varepsilon_{uu,t}$, $\Theta\varepsilon_{pp,t}$ and $\Theta\varepsilon_{up,t}$.

It is necessary to update the HUI network from timestamp $t-1$ to t according to the evolving nodes ($\Delta\mathcal{U}_t \cup \Theta\mathcal{U}_t$, $\Delta\mathcal{P}_t \cup \Theta\mathcal{P}_t$) and edges ($\Delta\varepsilon_{uu,t} \cup \Theta\varepsilon_{uu,t}$, $\Delta\varepsilon_{pp,t} \cup \Theta\varepsilon_{pp,t}$, $\Delta\varepsilon_{up,t} \cup \Theta\varepsilon_{pp,t}$). This can be achieved by updating the two types of nodes and three types of edges in HUI network. Therefore, the active nodes at timestamp t (denoted as $\tilde{\mathcal{V}}_t$ and nodes in $\tilde{\mathcal{V}}_t$ are unique) are defined as follows:

$$\begin{aligned}
\tilde{\mathcal{V}}_t = \Delta\mathcal{U}_t \cup \Theta\mathcal{U}_t \cup \Delta\mathcal{P}_t \cup \Theta\mathcal{P}_t \cup \{u_i | \exists e_u \in \Delta\varepsilon_{uu,t} \cup \Theta\varepsilon_{uu,t}, e_u = (u_i, u_j)\} \\
\cup \{p_i | \exists e_p \in \Delta\varepsilon_{pp,t} \cup \Theta\varepsilon_{pp,t}, e_p = (p_i, p_j)\} \\
\cup \{u_k, p_f | \exists e_{up} \in \Delta\varepsilon_{up,t} \cup \Theta\varepsilon_{up,t}, e_{up} = (u_k, p_f)\}\}
\end{aligned} \tag{6}$$

The underlying principle of constructing the network and updating process can be analogous to the case of adopting sliding window schema to manage continuous data streams. The construction process of HUI network are based on the historical records, and the updating course of the network can be conducted only within several timestamps like a certain length sliding window. The worst case happens only when all nodes $\{v_i | v_i \in \mathcal{V}_t\}$ have changed within timestamp t. In such case, the retraining process of the whole HUI network is inevitable.

5   Multi-granularity Dynamic Network Embedding

Inspired by DeepWalk [6] and the idea of modelling document [17] in natural language processing, our model contains three main stages as shown in Fig. 2: heterogeneous random walk, community integration and model learning process, based on which, vertex representations will evolve after incremental learning. Given the length of random walk as h and the total number of random walks as l, the starting step will be performed at each of the active node $\tilde{\mathcal{V}}_t$ at timestamp t. Based on the updated transition probability matrix P, the random walk with restart on heterogeneous network proposed by [18] is employed to generate possible route sequences for active nodes, denoted as $S = \{s_1, s_2, ..., s_{|\tilde{\mathcal{V}}_t|}\}$. In the rest part of this section, we will illustrate the last two stages.

5.1 Community Integration

As the analogy between words in the text and vertices in walk sequences, we introduce the idea of processing streaming data in topic models to detect overlapping communities in heterogeneous dynamic networks. Before the introduction of community integration procedure, we make two assumptions on heterogeneous random walk sequences, graph vertices and communities as follows: (1) Each vertex in the HUI network can belong to multiple communities with different preferences of $Pr(c|v)$, and each vertex sequence also owns its community distribution. (2) A vertex in a specific sequence belongs to a distinct community, and the community is determined by the community's distribution over sequences $Pr(c|s)$ and the vertex's distribution over communities $Pr(v|c)$.

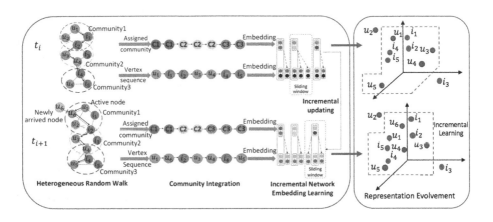

Fig. 2. The m-DNE model.

With the above assumptions and heterogeneous random walk sequences, we can assign community labels to vertices in particular sequence. More specifically, for a vertex v in a sequence s, we compute the conditional probability of a community c with the following equation:

$$Pr(c|v,s) = \frac{Pr(c,v,s)}{Pr(v,s)} \propto Pr(v|c)Pr(c|s) \qquad (7)$$

where $Pr(v|c)$ represents the role of v in community c, and $Pr(c|s)$ represents the community distribution in sequence s.

An ordinary way to estimate $Pr(v|c)$ and $Pr(c|s)$ is to use Gibbs Sampling. But it is not suitable for our updating progress. Thus, instead, we extend the Streaming Gibbs Sampling method proposed in [19] to achieve the conditional probability in our environment. According to the Bayesian Streaming Learning [19], if we fix the community distribution $C^{1:t-1}$ of the previous arrived sequences, then $C^{1:t}$ of the current timestamp can be achieved with $C^{1:t-1}$, and normal Gibbs Sampling on C^t. Therefore, the conditional distributions of $Pr(v|c)$ and $Pr(c|s)$ can be estimated as follows:

$$Pr(v|c) = \frac{N^t(v,c) + \beta_l}{\sum_{v' \in \mathcal{V}} N^t(v',c) + |\mathcal{V}|\beta_l}, \quad Pr(c|s) = \frac{N^t(c,s) + \alpha_l}{\sum_{c' \in \mathcal{C}} N^t(c',s) + \mathcal{K}\alpha_l} \quad (8)$$

where $N^t(v,c)$ is the number of times the vertex v assigned to community c at timestamp t and $N^t(c,s)$ is the number of vertices in sequence s are assigned to community c at t. Both $N^t(v,c)$ and $N^t(c,s)$ will be updated dynamically as community assignments change, and for different timestamps. β_l and α_l are smoothing factors in Latent Dirichlet Allocation [20]. With estimated $Pr(v|c)$ and $Pr(c|s)$, we assign a discrete community label c for each vertex v in sequence s.

5.2 Incremental Network Embedding Learning

To initialize the learning process on HUI network $G_{mix} = (\mathcal{V}, \varepsilon)$, given a certain vertex sequence $s = \{v_1, v_2, ..., v_{|s|}\}$, for each vertex v_i and its assigned community c_i, we will learn the representations of both vertices and communities by maximizing the average log probability of predicting context vertices using both v_i and c_i as formalized below:

$$\mathcal{L}(s) = \frac{1}{|s|} \sum_{i=1}^{|s|} \sum_{i-|W| \leq j \leq i+|W|} log Pr(v_j|v_i, c_i) \quad (9)$$

where v_j is the context node of the node v_i, and the probability $Pr(v_j|v_i, c_i)$ is defined using the softmax function:

$$Pr(v_j|v_i, c_i) = \frac{exp(v'_j \cdot \overline{v}_i)}{\sum_{v' \in \mathcal{V}} exp(v' \cdot \overline{v}_i)} \quad (10)$$

where v'_j is the context representation of its context node v_j. $\overline{v}_i$ is the average vector representation of the center node v_i and community label c_i defined as $\overline{v}_i = 1/2(v_i + c_i)$. In such case, the local context and the global community structure can be incorporated to enhance vertex representation learning. Then subsequently, during incremental learning process at each timestamp $t > 1$, the heterogeneous random walk procedure will start with active node set $\mathcal{V}_t$ to obtain possible route sequence set S.

To improve the computational efficiency of Eq. (10), in practical environment, we adopt hierarchical softmax[3], a computational efficient approximation of the full softmax in [21]. More precisely, given the average vector representation $\overline{v}_i$

---

[3] The hierarchical softmax needs to evaluate only about $log(|\mathcal{V}|)$ nodes instead of all the $|\mathcal{V}|$ nodes to obtain the probability distribution.

of v_i and c_i for target context v_j, let $L(v_j)$ be the length of its corresponding path, and let $b_n^{v_j} = 0$ when the path to v_j takes the left branch at the n-th layer and $b_n^{v_j} = 1$ otherwise. Then, the hierarchical softmax defines $Pr(v_j|v_i, c_i)$ as follows:

$$Pr(v_j|v_i, c_i) = \prod_{n=2}^{L(v_j)} ([\sigma(\overline{v}_i^T \theta_{n-1}^{v_j})]^{1-b_n^{v_j}} \cdot [1 - \sigma(\overline{v}_i^T \theta_{n-1}^{v_j})]^{b_n^{v_j}}) \qquad (11)$$

where $\sigma(z) = \frac{1}{1+exp(-z)}$. All parameters are trained by using the Stochastic Gradient Descent method. To derive how θ is update at each time step, the gradient for $\theta_{n-1}^{v_j}$ is computed as follows:

$$\frac{\partial \mathcal{L}(v_j, n)}{\partial \theta_{n-1}^{v_j}} = [1 - b_n^{v_j} - \sigma(\overline{v}_i^T \theta_{n-1}^{v_j})]\overline{v}_i \qquad (12)$$

To derive how the context embedding vectors are updated, the gradient for $\overline{v}_i$ is computed as follows:

$$\frac{\partial \mathcal{L}(v_j, n)}{\partial \overline{v}_i} = [1 - b_n^{v_j} - \sigma(\overline{v}_i^T \theta_{n-1}^{v_j})]\theta_{n-1}^{v_j} \qquad (13)$$

With this derivative, an embedding vector v_i and c_i in the context of node v_j can be updated as follows:

$$v_i \leftarrow v_i + \eta \sum_{n=2}^{L(v_j)} \frac{\partial \mathcal{L}(v_j, n)}{\partial \overline{v}_i}, \quad c_i \leftarrow c_i + \eta \sum_{n=2}^{L(v_j)} \frac{\partial \mathcal{L}(v_j, n)}{\partial \overline{v}_i} \qquad (14)$$

In Algorithm 1, we summarize the learning process using hierarchical softmax for proposed m-DNE model. The algorithm iterates through all possible route sequences and updates the embedding vectors until the procedure converges. In each iteration, given a current node, the algorithm first obtains its embedding vectors and computes its context embedding vector. Based on the derivative above, the binary tree in hierarchical sampling is updated followed by the embedding vector (line 9–14). Given the vector size of d, the leaf nodes number $|V|$, the sequence length $|s|$ within one iteration and window length $|W|$, then the time complexity for an iteration is $\mathcal{O}(d \cdot |W| \cdot |s| \cdot log(|V|))$.

Parallelizability. For real-world social networks, the frequency distribution of vertices in random walks follows a power law which results in a long tail of infrequent vertices [6]. Therefore, the updates of vertices' representation will be sparse in nature. Based on this, we adopt the lock-free solutions in the work [22] to parallelize asynchronous stochastic gradient descent (ASGD). Given that our updates are sparse and we do not acquire a lock to access the model shared parameters, ASGD will achieve an optimal rate of convergence.

Algorithm 1. Heterogeneous Softmax Algorithm for m-DNE

Input: Possible route sequence set S, window length $|W|$, embedding vector dimension d,
      sequence length $|s|$.
Output: The embedding representation v_i of v_i and representation c_i for c_i
1   Initialize the parameters randomly;
2   Shuffle the dataset;
3   **repeat**
4   Sample a route sequence $s = \{v_1, v_2, ..., v_{|s|}\}$ from S;
5   **for** $i = 1$ *to* $|s|$ **do**
6   Set $e \leftarrow 0$;
7   Compute the average representation $\overline{v_i} = 1/2(v_i + c_i)$;
8   **for** *each* $v_j \in s[i - |W|, i + |W|]$ **do**
9   **for** $n = 2$ *to* $L(v_j)$ **do**
10                  $q \leftarrow \sigma(\overline{v}_i \cdot \theta^{v_j}_{n-1})$;
11                  $g \leftarrow \eta \cdot (b_n^{v_j} - 1 - q)$;
12                  $e \leftarrow e + g \cdot \theta^{v_j}_{n-1}$;
13   Update $\theta^{v_j}_{n-1} \leftarrow \theta^{v_j}_{n-1} + g \cdot \overline{v}_i$;
14   **end**
15   Update $v_i \leftarrow v_i + \eta \cdot e$;
16   Update $c_i \leftarrow c_i + \eta \cdot e$;
17   **end**
18   **end**
19 **until** *convergence*;

---

6   Recommendation Using m-DNE

Recommendation procedure can be performed after obtaining the embeddings for each vertex. To recommend top-K friends to a user $u_i \in \mathcal{U}$ with D dimensional representation vector of $\overrightarrow{u_i} = (x_{i1}, x_{i2}, ..., x_{iD})$ and query time t, we compute the ranking score for user node u_j which does not have a direct link with u_i through the inner product of $\overrightarrow{u_i}$ and $\overrightarrow{u_j}$. Similar procedure can be found when recommending top-K items. Except that, to consider the freshness of the items such as tweets, we bring in the time decay function defined as $f(t_j, \lambda) = e^{-\lambda(t-t_j)}$, where t_j is the publication timestamp of item p_j and λ is employed to adjust the decay rate. Thus, the ranking score of item p_j can be obtained as follows: $S(u_i, p_j, t) = \overrightarrow{u_i} \cdot \overrightarrow{p_j} = f(t_j, \lambda) \sum_{n=1}^{D} x_{in} \cdot z_{jn}$. For computational efficiency, we adopt the Threshold-based Algorithm (TA) [23], which is capable of finding the top-k results by examining the minimum number of users/items.

7   Experiments

7.1   Experimental Setup

Dataset Description. For experimental study, we evaluate the proposed m-DNE model on three real-world datasets: Twitter, Last.fm and Flickr. We collected Twitter dataset from January to March 2017 with Twitter API[4], which includes users and their posts, and the Last.fm dataset for 1 month through

---

[4] https://dev.twitter.com/docs.

Last.fm API[5], which contains users and artists. We also adopted Flickr dataset[6] released online with friend relationships, images, and the activities of user comment image. In order to enrich the information about user and image, we extracted the timestamp of user comments, image uploading timestamp and image description with Flickr API[7]. For all datasets, the user-user links are constructed from bi-directional friendships between social network users, user-item links are constructed from the different activities of users (e.g., posting, listening or commenting items), and item-item links are constructed if the two artists/images share the same tag or the two posts have the same hashtag. The statistics of each dataset are summarized in Table 1.

Table 1. Some statistics of the datasets.

Datasets	#Users	#Items	#User-user links	#User-item links	#Item-item links
Twitter	69,830	6,284,665	429,836	69,131,820	30,795,807
Last.fm	41,258	10,361	235,417	11,486,510	1,820,649
Flickr	2,037,538	1,262,978	219,098,660	14,913,164	97,549,330

Baselines. We compared our model with five state-of-the art methods:

– **Weighted Regularized Matrix Factorization (WRMF).** A state-of-the-art offline matrix factorization model introduced by [24] is computed in batch mode, assuming the whole stream is stored and available for training.
– **Stream Ranking Matrix Factorization (RMFX).** It achieves partly online and much quicker updates of matrix factorization introduced in [11].
– **Metapath2vec (M2V)** [15]. It uses metapath-based random walks on heterogeneous graphs to obtain node representations. Following [15], we employ 5 meaningful meta-paths whose lengths are not longer than 4, "UIU", "UUIU", "UIIU", "UIUIU" and "UUIIU", since long meta-paths are likely to introduce noisy semantics. Here, $'U' = User$ and $'I' = Item$.
– **PTE** [16]. We build three bipartite heterogeneous networks: user-user, user-item and item-item, and retrain it as an unsupervised embedding methods.
– **M-NMF** [9]. It jointly models node and community embedding using non-negative matrix factorization.

Parameter Settings. WRMF setup is as follows: $\lambda_{WRMF} = 0.015$, $C = 1$, $epochs = 15$ for all datasets, which corresponds to a regularization parameter, a confidence weight that is put on positive observations, and the number of passes over observed data, respectively [24]. For RMFX, we set regularization constants λ_{RMFX}, learning rate η_0, and a learning rate schedule α equal to 0.1, 0.1, 1 for

---

[5] http://www.last.fm/api/.
[6] http://arnetminer.org/lab-datasets/flickr/flickr.rar.
[7] https://www.flickr.com/services/api/.

Twitter, 0.15, 0.05, 1.5 for Last.fm and 0.1, 0.15, 1 for Flickr using grid-search on stream data with cross validation [11]. Moreover, the number of iterations is set to the size of the reservoir. For all the embedding algorithms (metapath2vec, PTE, M-NMF and our model), the embedding dimensionality is set to 128, context window length is set to 8, walk length is set to 40, walks per vertex is set to 30, the neighborhood size is equal to 7 and the size of negative samples is equal to 5 for all datasets. For M-NMF, we followed the same tuning procedure in [9], and we found out that $\alpha = 0.1$ and $\beta = 5$ works at best for Twitter and Last.fm, while $\alpha = 10$ and $\beta = 5$ for Flickr. As for our m-DNE model for three datasets, we also set the dimension of community representation as 128. Following [20], the smoothing factors α_l and β_l are set to 2 and 0.5 respectively. We set decay rate $\lambda = 0.2$ for Twitter and 0.1 for Last.fm and Flickr. The number of communities $\mathcal{K}$ is set to 20 for m-DNE and M-NMF model [9]. We run experiments on Linux machines with eight 3.50 GHz Intel Xeon(R) CPUs and 16 GB memory.

Evaluation Criteria. Given a dataset $\mathcal{D}$ ordered according to time, including user and item profiles, we use the first 50% of $\mathcal{D}$ as historical data pool to train the models, while the rest half data mimics the streaming input called "candidate set". For evaluation, we first randomly select a reference time as "current time" in candidate set. Then, we test our recommendations for the following week starting from reference time, while the data before reference time in candidate set are used to tune the hyper-parameters. However, WRMF and RMFX cannot explicitly handle new user/item introduction during the testing phase. For a fair comparison, all testing sets only cover users/items existing in training set. During evaluation phase, all experimental results are averaged over 10 different runs for reliability, and there is no temporal overlapping between any testing set.

Since we are interested in measuring top-k recommendation instead of rating prediction, we measure the quality by looking at the *Recall@K* [25] and Average Reciprocal Hit-Rank (ARHR) [26], which are widely used for evaluating top-k recommender systems. We show the performance when $k = \{1, 5, 10\}$, as a larger value of k is usually ignored for a typical top-k recommendation [25].

7.2 Results

Recommendation Effectiveness. Table 2 summarizes the item and friend recommendation performance between our model and baselines. Besides, we also test our model without community attribute integration represented as DNE. From the results, we can observe that the *Recall@K* value grows gradually along with the increasing number of K, and the performance of item recommendation is better than friend recommendation. Besides, we can also observe on all datasets that: (1) Embedding-based algorithms (PTE, M2V, M-NMF, DNE and m-DNE) consistently perform better than non-embedding based benchmarks (WRMF, RMFX). It is because embedding-based algorithms can fully explore the network structure of the given information, which alleviates the issues of sparse and noisy signals. (2) The significant improvements show the promising benefit of the community integration and our incremental learning approach, which lead to the better performance of m-DNE than the other listed embedding methods.

Table 2. Top-k items and friends recommendation w.r.t. Recall@K (K = 1, 5, 10).

Method	Twitter			Last.fm			Flickr		
	Recall@1	Recall@5	Recall@10	Recall@1	Recall@5	Recall@10	Recall@1	Recall@5	Recall@10
Top-k items recommendation									
WRMF	0.152	0.229	0.301	0.226	0.293	0.387	0.204	0.261	0.356
RMFX	0.115	0.194	0.273	0.197	0.276	0.358	0.171	0.252	0.334
PTE	0.219	0.292	0.379	0.276	0.352	0.433	0.246	0.327	0.394
M2V	0.236	0.307	0.392	0.291	0.374	0.467	0.263	0.341	0.435
M-NMF	0.264	0.328	0.426	0.342	0.407	0.506	0.311	0.386	0.479
DNE	0.251	0.324	0.417	0.331	0.403	0.498	0.306	0.374	0.470
m-DNE	**0.309**	**0.385**	**0.472**	**0.395**	**0.471**	**0.557**	**0.368**	**0.449**	**0.531**
Top-k friends recommendation									
WRMF	0.113	0.175	0.266	0.172	0.247	0.314	0.148	0.220	0.281
RMFX	0.097	0.146	0.204	0.136	0.225	0.290	0.118	0.196	0.267
PTE	0.152	0.226	0.313	0.224	0.276	0.347	0.198	0.255	0.329
M2V	0.176	0.235	0.327	0.234	0.292	0.348	0.203	0.267	0.334
M-NMF	0.226	0.267	0.339	0.262	0.321	0.378	0.237	0.304	0.351
DNE	0.213	0.256	0.332	0.254	0.317	0.363	0.228	0.296	0.344
m-DNE	**0.243**	**0.294**	**0.371**	**0.298**	**0.352**	**0.406**	**0.275**	**0.329**	**0.390**

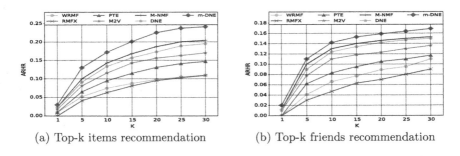

(a) Top-k items recommendation   (b) Top-k friends recommendation

Fig. 3. Recommendation performance w.r.t. ARHR.

Figure 3 compares the performance of alternative approaches taking ARHR as metric. During experiments, we vary the number of recommendations K from 1 to 30. As expected, our m-DNE model performs better with ARHR as well, and M-NMF ranks the second place followed by DNE, which shows the same orders in Table 2. In Fig. 3(a), as we recommend more items, since we have more chance to answer the true interested items correctly, ARHR grows gradually with increasing number K. The same trends appear in the friend recommendation task. To evaluate the efficiency of our model, we compare our m-DNE with other baselines on Twitter. As all baselines are not designed to handle dynamics except RMFX, we compare their cumulative running time over all time steps and plot it in a log scale. Each time step represents one day period. As can be seen in Fig. 4, m-DNE is much faster than the baselines which need to retrain and still show advantages compared with RMFX.

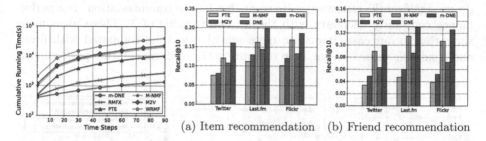

(a) Item recommendation   (b) Friend recommendation

Fig. 4. Cumulative running **Fig. 5.** Recommendations for cold-start cases.
time comparison.

Test for Cold Start Problem. We also conduct experiments to study the effectiveness of different algorithms in addressing cold-start issues. As pre-processing, the target users who have less than 20 available items and social links in total are selected. As there are not many interaction records between users and items available for cold-start cases, WRMF and RMFX which are based on collaborative filtering, are not suitable for cold-start experiments. Thus, we compare m-DNE with the baselines which can leverage social information to recommend cold-start cases. The experimental results are shown in Fig. 5, from which we have the following observations: (1) m-DNE model still performs best consistently in recommending cold-start cases; (2) by comparing with Table 2, the *Recall* value of all algorithms decreases. For instance, the *Recall* value of M-NMF rapidly drops from 42.6% to 12% for twitter item recommendation but still better than DNE model, while m-DNE deteriorate slightly, which validates that community-aware high order proximity and the ability to capture the dynamic properties of the network are key factors affecting the recommendation performance.

Sensitivity to Parameters. In this experiment, we study the influence of the embedding dimension d, the number of samples l and time decay rate λ by fixing the window size $|W| = 8$ and the random walk length $h = 40$. We vary one parameter each time to test the impact on recommendation performance with other parameters fixed. Because of the page limit, we only show the results on Twitter. But similar observations can be made on other datasets. Recommendation *Recall* value of m-DNE model is not highly sensitive to the dimension d, but still presents a tendency that its recommendation accuracy increases with the increasing number of d holistically, and it reaches peak when d is around 128. However, m-DNE is sensitive to l with the *Recall* score varying a lot. First, the performance of m-DNE increases quickly with the increasing number of l, this is because the model has not achieved convergence. Then, it does not change significantly when the number of samples becomes large enough, since m-DNE has converged. Thus, to achieve a satisfying trade off between effectiveness and efficiency of model training, we set $l = 30$ and $d = 128$ on all datasets. In Fig. 6(c), λ shows different influence on item/user recommendation tasks. For item recommendation, the performance reaches the peak when $\lambda = 0.2$ but drops significantly afterwards. However, for user recommendation, the performance constantly decreases with the increasing value of λ. These phenomena show that in our case, items are more sensitive to time compared with users, and a suitable value of λ can help to improve the recommendation performance.

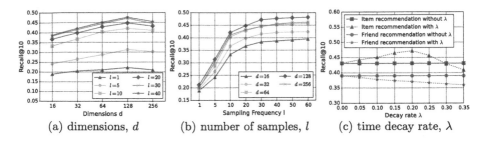

|(a) dimensions, d|(b) number of samples, l|(c) time decay rate, λ|

Fig. 6. Effect of different parameters on performance.

8　Conclusion

In this paper, we propose m-DNE, an efficient model which learns the embedding of heterogeneous social network by jointly modelling the temporal semantic effects, social relationships and user behavior sequential patterns in a unified way. Community-aware high-order proximity is applied to optimize the node representations. Besides, a parallel incremental learning algorithm and an efficient query processing technique are employed for recommendation efficiency. The experimental results show the effectiveness of our m-DNE on social recommendations. In the future, we will consider to integrate attributes from multiple social sites. Additionally, short-term user interest changes also need to be considered with the use of advanced deep learning models such as Recurrent Neural Network.

Acknowledgements. This work was supported by the Research Council of Norway (grant number 245469).

References

1. Sedhain, S., Sanner, S., Braziunas, D., Xie, L., Christensen, J.: Social collaborative filtering for cold-start recommendations. In: RecSys, pp. 345–348 (2014)
2. Kouki, P., Fakhraei, S., Foulds, J., Eirinaki, M., Getoor, L.: Hyper: a flexible and extensible probabilistic framework for hybrid recommender systems. In: RecSys, pp. 99–106. ACM (2015)
3. Grbovic, M., Radosavljevic, V., Djuric, N., Bhamidipati, N., Savla, J., Bhagwan, V., Sharp, D.: E-commerce in your inbox: product recommendations at scale. In: SIGKDD, pp. 1809–1818. ACM (2015)
4. Covington, P., Adams, J., Sargin, E.: Deep neural networks for youtube recommendations. In: RecSys, pp. 191–198. ACM (2016)
5. Tang, J., Qu, M., Wang, M., Zhang, M., Yan, J., Mei, Q.: Line: large-scale information network embedding. In: WWW, pp. 1067–1077 (2015)
6. Perozzi, B., Al-Rfou, R., Skiena, S.: Deepwalk: online learning of social representations. In: SIGKDD, pp. 701–710. ACM (2014)
7. Cavallari, S., Zheng, V.W., Cai, H., Chang, K.C.C., Cambria, E.: Learning community embedding with community detection and node embedding on graphs. In: CIKM, pp. 377–386. ACM (2017)
8. Li, J., Dani, H., Hu, X., Tang, J., Chang, Y., Liu, H.: Attributed network embedding for learning in a dynamic environment. In: CIKM, pp. 387–396 (2017)
9. Wang, X., Cui, P., Wang, J., Pei, J., Zhu, W., Yang, S.: Community preserving network embedding. In: AAAI, pp. 203–209 (2017)
10. Agarwal, D., Chen, B.C., Elango, P.: Fast online learning through offline initialization for time-sensitive recommendation. In: SIGKDD, pp. 703–712 (2010)
11. Diaz-Aviles, E., Drumond, L., Schmidt-Thieme, L., Nejdl, W.: Real-time top-n recommendation in social streams. In: RecSys, pp. 59–66. ACM (2012)
12. Huang, Y., Cui, B., Zhang, W., Jiang, J., Xu, Y.: Tencentrec: real-time stream recommendation in practice. In: SIGMOD, pp. 227–238. ACM (2015)
13. Subbian, K., Aggarwal, C., Hegde, K.: Recommendations for streaming data. In: CIKM, pp. 2185–2190. ACM (2016)
14. Grover, A., Leskovec, J.: node2vec: scalable feature learning for networks. In: SIGKDD, pp. 855–864. ACM (2016)
15. Dong, Y., Chawla, N.V., Swami, A.: metapath2vec: scalable representation learning for heterogeneous networks. In: SIGKDD, pp. 135–144. ACM (2017)
16. Tang, J., Qu, M., Mei, Q.: Pte: predictive text embedding through large-scale heterogeneous text networks. In: SIGKDD, pp. 1165–1174. ACM (2015)
17. Le, Q., Mikolov, T.: Distributed representations of sentences and documents. In: ICML, pp. 1188–1196 (2014)
18. Li, Y., Patra, J.C.: Genome-wide inferring gene-phenotype relationship by walking on the heterogeneous network. Bioinformatics 26(9), 1219–1224 (2010)
19. Gao, Y., Chen, J., Zhu, J.: Streaming Gibbs sampling for LDA model. arXiv preprint arXiv:1601.01142 (2016)
20. Blei, D.M., Ng, A.Y., Jordan, M.I.: Latent Dirichlet allocation. J. Mach. Learn. Res. 3(1), 993–1022 (2003)
21. Mikolov, T., Sutskever, I., Chen, K., Corrado, G.S., Dean, J.: Distributed representations of words and phrases and their compositionality. In: NIPS (2013)
22. Recht, B., Re, C., Wright, S., Niu, F.: Hogwild: a lock-free approach to parallelizing stochastic gradient descent. In: NIPS (2011)

23. Fagin, R., Lotem, A., Naor, M.: Optimal aggregation algorithms for middleware. J. Comput. Syst. Sci. **66**(4), 614–656 (2003)
24. Hu, Y., Koren, Y., Volinsky, C.: Collaborative filtering for implicit feedback datasets. In: ICDM, pp. 263–272. IEEE (2008)
25. Cremonesi, P., Koren, Y., Turrin, R.: Performance of recommender algorithms on top-n recommendation tasks. In: RecSys, pp. 39–46. ACM (2010)
26. Deshpande, M., Karypis, G.: Item-based top-n recommendation algorithms. TOIS **22**(1), 143–177 (2004)

GeoDCF: Deep Collaborative Filtering with Multifaceted Contextual Information in Location-Based Social Networks

Dimitrios Rafailidis[1,2]($\boxtimes$) and Fabio Crestani[3]

[1] Department of Computer Science, University of Mons, Mons, Belgium
dimitrios.rafailidis@umons.ac.be
[2] Department of Data Science and Knowledge Engineering,
Maastricht University, Maastricht, The Netherlands
[3] Faculty of Informatics, Università della Svizzera italiana, Lugano, Switzerland
fabio.crestani@usi.ch

Abstract. In this study we investigate the recommendation problem with multifaceted contextual information to overcome the scarcity of users' check-in data in Location-based Social Networks. To generate accurate personalized Point-of-Interest (POI) recommendations in the presence of data scarcity, we account for both users' and POIs' contextual information such as the social influence of friends, as well as the geographical and sequential transition influence of POIs on user's check-in behavior. We first propose a multi-view learning strategy to capture the multifaceted contextual information of users and POIs along with users' check-in data. Then, we feed the learned user and POI latent vectors to a deep neural framework, to capture their non-linear correlations. Finally, we formulate the objective function of our geo-based deep collaborative filtering model (GeoDCF) as a Bayesian personalized ranking problem to focus on the top-k recommendation task and we learn the parameters of our model via backpropagation. Our experiments on real-world datasets confirm that GeoDCF achieves high recommendation accuracy, significantly outperforming other state-of-the-art methods. Furthermore, we confirm the influence of both users' and POIs' contextual information on our GeoDCF model. The evaluation datasets are publicly available at: http://snap.stanford.edu/data/loc-gowalla.html, https://sites.google.com/site/yangdingqi/home/foursquare-dataset.

Keywords: Point-of-interest recommendation
Deep collaborative filtering · Multifaceted contextual information
Location-Based Social Networks

1 Introduction

With the emergence of Location-based Social Networks (LBSNs) such as Yelp and Foursquare, users can search for a Point-of-Interest (POI) e.g., a restaurant or a museum to visit, and share their location with their friends by making a

© Springer Nature Switzerland AG 2019
M. Berlingerio et al. (Eds.): ECML PKDD 2018, LNAI 11052, pp. 709–724, 2019.
https://doi.org/10.1007/978-3-030-10928-8_42

check-in at the POI they have visited. Such implicit source of feedback provides rich information about both users and POIs that can be leveraged to study the user's movement in urban cities, as well as enhance the quality of personalised POI recommendations. Most existing POI recommendation systems apply collaborative filtering techniques to suggest relevant POIs to users based on the assumption that similar-minded users are likely to visit similar POIs [7,26]. In practice, rather than explicit feedbacks of ratings for traditional recommendation systems, binary implicit feedbacks are usually available at LBSNs in the form of check-in data [20]. Several methods have been proposed to handle the case of users' implicit feedback, such as weighted matrix factorization [6], with square or cross-entropy loss functions to either minimize the rating error or predict if an unobserved item would be preferred or not by a user. However, provided that end-users are usually interested in the top-k recommendations, such loss functions do not focus on the top-k recommendation problem. To overcome this limitation, Bayesian Personalized Ranking (BPR) strategies use a pairwise ranking loss function, considering the relative ordering of items in a ranked list [17]. The pairwise ranking criterion of the BPR model is based on the assumption that a user prefers the observed items over the unobserved ones. This idea results in a pairwise ranking loss function that tries to discriminate between a small set of observed items and a very large set of unobserved ones. Due to the imbalance between the user's observed items and unobserved ones, the BPR model uniformly samples negative examples from the set of unobserved items to reduce the training time. However, both studies [6,17] ignore the multifaceted contextual information at LBSNs for POI recommendations [11,26].

POI recommendation strategies suffer from the data scarcity problem, as the number of POIs visited by a user is usually only a small portion of all the available POIs at a LBSN [2,9]. As a consequence, the data scarcity limits the performance of the collaborative filtering strategies when generating recommendations. To handle the data scarcity problem, POI recommendation strategies exploit the multifaceted contextual information of both users and POIs, such as the social influence of friends, as well as the geographical and sequential transition influence of POIs on user's check-in behavior. In particular, although user preferences are influenced by users' social relationships, the selections of social friends do not necessarily match [13,14]. As a consequence we have to learn the impact of friends' selections on users' check-in behavior [7]. Regarding POIs' geographical influence, user preferences are based on the user mobility and the geographical distances among POIs, as most users only visit POIs within small regions [10,11,21]. In addition, two users may behave differently with respect to time. For example, one often checks in at restaurants during lunch time, while the other likes bars and often checks in at midnight. The POI recommendation task becomes even more challenging, as there is a sequential transition influence of locations on users' check-in behaviors, where a user might like visiting POIs in a specific order e.g., office→, lunch, gym→home or home→bar [1,26].

Although POI recommendation strategies exploit different contextual factors, they do capture well the non-linear correlations of users' and POIs multifaceted information [20]. Also, they not necessarily focus on the ranking performance of the POI recommendation task, such as in the studies reported in [11,21]. To overcome the shortcomings of the existing methods we propose the GeoDCF model, making the following contributions: *(C1)* To account for the fact that the multifaceted information of users and POIs can significantly boost the quality of recommendations, we first introduce a multi-view joint factorization strategy. We compute the user and POI latent vectors by co-factorizing users' check-in behavior on POIs with users' and POIs' contextual information. *(C2)* To better capture the non-linear correlations of the user and POI latent vectors, we adopt a deep learning strategy by learning the model parameters via a backpropagation algorithm. *(C3)* To focus on the ranking performance, we formulate our model as a pairwise ranking task, by placing a BPR layer at the top of our deep learning architecture. Our experiments on benchmark datasets from the real-world LBSNs of Gowalla and Foursquare show that our GeoDCF model beats other baseline strategies. In addition, we experimentally show the impact of users' and POIs' contextual information on our model.

The remainder of the paper is organized as follows, Sect. 2 reviews the related work, and in Sect. 3 we formally define our pairwise ranking problem. Section 4 details the proposed GeoDCF model, Sect. 5 presents the experimental results and Sect. 6 concludes the study.

2   Related Work

In collaborative filtering with implicit feedback, such as weighted matrix factorization [6], some missing entries are treated as negative instances (negative sampling) trying to minimize a pointwise loss function such as the square or cross-entropy loss. Liu et al. [11] model geographical influence by incorporating neighboring characteristics into weighted matrix factorization to handle the implicit feedback of users' check-in data. Lian et al. [10] present a geographical weighted matrix factorization model that integrates geographical influence by modeling users' activity regions and the influence propagation on geographical space. Instead of using a pointwise loss function, Yuan et al. [23] focus on the top-k recommendation performance, presenting a model that incorporates geographical influence, assuming that neighborhood POIs of POIs previously visited by users should be ranked higher than distant ones. In a similar spirit, RankGeoFM is a ranking-based model that first learns users' preference rankings for POIs, and then includes the geographical influence of neighboring POIs to alleviate the recommendation accuracy [9].

Apart from the geographical influence on users' check-in behaviour, Ye et al. [21] also consider users' social correlation for POI recommendation, following a friend-based collaborative filtering strategy. In particular, they produce POI recommendations based on similar friends, where the similarity between friends is calculated based on their common check-in POIs and common friends. In [24], a friend-based collaborative filtering strategy is also used to leverage friends' check-ins, where the similarity between friends is computed based on the distance of their residences. In [12], a personalised ranking framework with multiple sampling criteria is proposed, leveraging both social correlation and geographical influence on users' check-in behavior. In particular, Manotumruksa et al. [12] apply a multi-center Gaussian model and a power-law distribution method, to capture the geographical influence and social correlation respectively when performing negative sampling for the non-visited POIs. In [7] a two-step POI recommendation framework is proposed, which first learns potential locations from users' friends and then, incorporates potential locations into weighted matrix factorization. Zhang et al. [26] employ an additive Markov chain to exploit the sequential transition influence between POIs, where the sequential probability of a user visiting a POI is based on the transition probability between all the user's visited POIs and a target non-visited POI.

Accordingly, in recommendation systems deep learning strategies use either a pointwise or a pairwise ranking loss function to handle user implicit feedback and capture user data non-linear correlations. In [8,18,22], various deep learning strategies are introduced to exploit user feedback with users' and items' side information. For example, Ying et al. [22] model implicit feedback in stacked denoising autoencoders with the side information of articles, such as the title and abstract of the articles. Ding et al. [3] design a ranking model for friend recommendations. However, the studies at [3,8,9,18] do not consider any contextual information when training their deep learning models, a key factor to generate accurate POI recommendations [7,21,25]. Recently, a deep recurrent neural network is proposed to capture the sequential transition influence on users' check-in behavior [4]. Nonetheless, the users' social relations are ignored at [4]. Yang et al. [20] introduce PACE, a deep neural architecture that jointly learns the embeddings of users and POIs to predict user preferences over POIs and various context associated with users and POIs. PACE first transforms the users' and POIs' contextual relations into graphs and then, employs neural embedding for POI recommendation as a bridge between collaborative filtering and semi-supervised learning. Instead of using a pairwise ranking function, PACE defines a pointwise function to handle the case of implicit feedback during the deep neural network learning. Consequently, PACE does not focus on the ranking performance when generating top-k POI recommendations.

3   Problem Formulation

Let $\mathcal{N}$ and $\mathcal{M}$ be the sets of users and POIs, where $n = |\mathcal{N}|$ and $m = |\mathcal{M}|$ are the numbers of users and POIs, respectively. Users' check-in data are tuples in the form of $(user, POI, time)$. In addition, each user u has a set of friends $\mathcal{A}_u$. Each POI is also associated with a pair of geographical latitude and longitude coordinates in the form of $(lat, long)$. In our problem we consider the following input matrices:

Definition 1 (Check-in matrix X). *"Based on the users' data we construct a binary check-in matrix $X \in \{0,1\}^{n \times m}$."*

Definition 2 (Social link matrix A). *"According to each user u's social relationships in $\mathcal{A}_u$, we compute a binary adjacency matrix $A \in \{0, 1\}^{n \times n}$."*

Definition 3 (Geographical similarity matrix G). *"Given the geographical coordinates $(lat, long)$, we first compute the angular distance $\delta(a, b)$ between each pair of POIs a and b based on the Haversine formula[1], and then we calculate the geographical similarity matrix $G \in \mathbb{R}_+^{m \times m}$. Each element of G is computed as $G(a, b) = \frac{1}{1+(\delta(a,b) \times r)}$, with $r = 6,371$ km being the earth radius."*

Definition 4 (Sequential transition matrix T). *"Provided that users' check-in data are timestamped, we calculate a transition matrix $T \in \mathbb{R}_+^{m \times m}$, where each element $T(a, b)$ corresponds to the frequency of successive POI visits, $a \rightarrow b$. In the sequential transition matrix T, we filter out successive POI visits in a long interval e.g., more than a day, as these successive POI visits are weakly or not correlated at all [26]."*

Given user preferences in the check-in matrix X, users' contextual information in A and POIs' contextual information in G and T, the goal of our model is to generate top-k POI recommendations for a user $u \in \mathcal{N}$. In our GeoDCF model we formulate the POI recommendation problem as a pairwise ranking task [17]. We define a check-in probability[2] x_{ui}, where $x_{ui} = X(u, i)$ denotes that user u has already visited POI i. Thus, we can define two disjoint sets, a set $\mathcal{X}_u^+$ of visited POIs that user u has already checked-in, and a set $\mathcal{X}_u^-$ of non-visited POIs. For the task of POI recommendation, we build a pairwise ranking model that is able to rank the visited POIs before the non-visited ones. For any pair of POIs i and j, with $i \in \mathcal{X}_u^+$ and $j \in \mathcal{X}_u^-$, the check-in probability x_{ui} should be greater than x_{uj}. To describe this relation we define a *partial relation* $i >_u j$. For each user $u \in \mathcal{N}$ the set of all partial relationships is computed as follows:

$$\mathcal{R}_u = \{i >_u j | i \in \mathcal{X}_u^+, j \in \mathcal{X}_u^-\} \tag{1}$$

We define our POI recommendation task as the following ranking problem:

Definition 5 (Problem). *"Given the set of all partial relationships $\mathcal{R}_u$ for each user $u \in \mathcal{N}$, the goal of GeoDCF is to maximize the ranking likelihood probability as follows:"*

$$\max \prod_{u \in \mathcal{N}} \prod_{(i,j) \in \mathcal{R}_u} P(i >_u j) \tag{2}$$

4 The GeoDCF Model

4.1 Model Overview

An overview of the proposed GeoDCF model is presented in Fig. 1. The inputs are the check-in matrix X and contextual matrices A, G and T (Sect. 3). In the

---

[1] https://en.wikipedia.org/wiki/Haversine_formula.

[2] Initially, the check-in probability is binary to construct matrix X, and after the model learning the entries of X are in the range of $[0, 1]$.

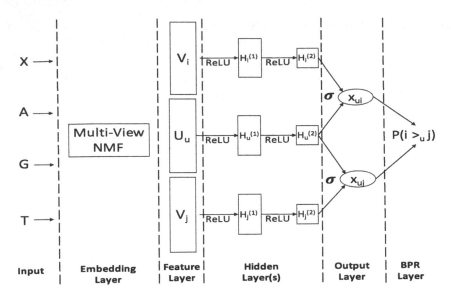

Fig. 1. Overview of GeoDCF. In this example, we use $h = 2$ hidden layers in our deep collaborative filtering strategy. For each user u POIs i and j denote a visited and a non-visited POI, respectively.

embedding layer the goal is to jointly learn the influence of the contextual information on user preferences and compute the latent matrices $U \in \mathbb{R}^{n \times d}$ and $V \in \mathbb{R}^{m \times d}$ of the preference matrix X, with d being the low dimensional embeddings. In the remaining layers of our architecture in Fig. 1, we perform *BPR learning* for the pairwise ranking task to generate POI recommendations. As defined in our pairwise ranking task in Eq. (2), for each user u we have pairs of partial relations $(i, j) \in \mathcal{R}_u$. In the **feature layer** we consider the POI latent vectors $V_i \in \mathbb{R}^d$ and $V_j \in \mathbb{R}^d$, that is the i-th and j-th rows of V, as well as the user latent vector $U_u \in \mathbb{R}^d$, the u-th row of U. Then, we design three neural networks[3], where each latent vector V_i, U_u and V_j is provided to the respective neural network. Given h **hidden layers**, we first try to capture the non-linear representations $H_i^{(q)}$, $H_u^{(q)}$ and $H_j^{(q)}$ of V_i, U_u and V_j in each neural network separately, with $q = 1, \ldots, h$. In the example of Fig. 1 we use $h = 2$ hidden layers. The **output layer** calculates the check-in probabilities x_{ui} and x_{uj} by combining the last hidden layers $H_i^{(h)}$, $H_u^{(h)}$ and $H_j^{(h)}$ with a sigmoid function $\sigma(x) = 1/(1 + e^{-x})$. Finally, the **BPR layer** predicts the probability of the partial relation $P(i >_u j)$.

---

[3] Alternatively, we could concatenate the three latent vectors and use a single neural network, increasing the computational cost of the learning process.

4.2   Embedding Layer

Given that we have to learn the influence of the contextual matrices A, G and T on user preferences in the check-in matrix X, at the embedding layer we formulate a multi-view joint factorization problem. In particular, we define the following joint loss function:

$$\min_{\Theta_e} \mathcal{L} = \mathcal{L}_X + \lambda_A \mathcal{L}_A + \lambda_G \mathcal{L}_G + \lambda_T \mathcal{L}_T \tag{3}$$

where the four loss functions $\mathcal{L}_X$, $\mathcal{L}_A$, $\mathcal{L}_G$ and $\mathcal{L}_T$ correspond to the joint factorizations of the input matrices X, A, G and T. Θ_e is the parameter set of the joint loss function $\mathcal{L}$, and parameters λ_A, λ_G and λ_T regularize the respective loss functions in Eq. (3). Note that in Eq. (3) a regularization parameter for $\mathcal{L}_X$ is omitted, as matrix X is the main check-in matrix with user preferences. The problem of the joint loss function in Eq. (3) is similar with the Multi-View Non-negative Matrix Factorization (MV-NMF) problem of [5]. MV-NMF tries to bring the latent matrices of different views as close as possible to a common consensus matrix. For example, if we assume that the four input matrices are only coupled at the POI dimension, we have a consensus matrix $V^* \in \mathbb{R}^{m \times d}$, with d being the low-dimensional latent embeddings. While jointly factorizing the input matrices, the goal of MV-NMF is to minimize the four reconstruction errors $||V^{(v)} - V^*||_F^2$ of the consensus matrix V^* and the respective POI latent matrices $V^{(v)} \in \mathbb{R}^{m \times d}$, with $v = 1, \ldots, 4$. Instead of having couplings at one dimension as in [5], in our setting the input matrices might be coupled at different dimensions, that is either at the user or POI dimensions. Thus, we extend [5], by introducing the user and POI consensus matrices $U^* \in \mathbb{R}^{n \times d}$ and $V^* \in \mathbb{R}^{m \times d}$ for the couplings at the user and POI dimensions, accordingly. We calculate the loss functions $\mathcal{L}_X$, $\mathcal{L}_A$, $\mathcal{L}_G$ and $\mathcal{L}_T$ of Eq. (3) as follows:

- $\mathcal{L}_X = ||X - UV^\top||_F^2 + \gamma_X ||U - U^*||_F^2 + \delta_X ||V - V^*||_F^2$, with the check-in matrix X being coupled with all the contextual matrices at the user or POI dimensions. $U \in \mathbb{R}^{n \times d}$ and $V \in \mathbb{R}^{m \times d}$ are the user and item latent matrices, when factorizing X.
- $\mathcal{L}_A = ||A - U_A V_A^\top||_F^2 + \gamma_A ||U_A - U^*||_F^2$. The social link matrix A is only coupled with the check-in matrix X at the user dimension, thus the reconstruction error of the POI consensus matrix V^* is omitted. Provided that A is symmetric we preserve the latent matrix $U_A \in \mathbb{R}^{n \times d}$, with $U_A = V_A$.
- $\mathcal{L}_G = ||G - U_G V_G^\top||_F^2 + \delta_G ||V_G - V^*||_F^2$. The geographical similarity matrix G is coupled with the check-in matrix X at the POI dimension. Given that G is symmetric we keep only the latent matrix $V_G \in \mathbb{R}^{n \times d}$, with $U_G = V_G$.
- $\mathcal{L}_T = ||T - U_T V_T^\top||_F^2 + \delta_T ||V_T - V^*||_F^2$. The sequential transition matrix T is coupled with the check-in matrix X at the POI dimension. In this case, we also preserve the latent matrix $V_T \in \mathbb{R}^{m \times d}$, with $U_T = V_T$.

The regularization parameters γ_X and γ_A control the reconstruction errors of the respective user latent matrices of each loss function and the user consensus matrix U^*. Accordingly, parameters δ_X, δ_G and δ_T are used to regularize the reconstruction errors of the respective POI latent matrices of each loss function and the POI consensus matrix V^*. To reduce the complexity of our model, in our implementation we set the regularization parameters for the consensus matrices to 0.01.

Summarizing, the parameter set Θ_e of the joint loss function $\mathcal{L}$ in Eq. (3) is set to $\Theta_e = \{U, V, U_A, V_G, V_T, U^*, V^*\}$, as A, G and T are symmetric matrices, with $U_A = V_A$, $U_G = V_G$ and $U_T = V_T$. However, the minimization problem of Eq. (3) is not convex with respect to all the variables of the parameter set Θ_e. To solve this problem, we follow an alternating optimization strategy, that is update one variable while fixing the remaining variables of Θ_e. According to the learning strategy of multiplicative rules [5], we compute the update rules of each variable for the alternating optimization algorithm. Due to lack of space we omit the presentation of the update rules, as they can be computed in a similar way as in the study at [5]. By solving the minimization problem of Eq. (3), the embedding layer computes the user and POI latent matrices U and V of the check-in matrix X with user preferences, by also accounting for the contextual information.

4.3   BPR Learning

Feature Layer. At the remaining layers of our architecture in Fig. 1 we adopt the BPR technique to produce top-k recommendations. Having computed the user and POI latent matrices U and V at the embedding layer, for each user $u \in \mathcal{N}$ we consider the partial relations $(i, j) \in \mathcal{R}_u$ based on Eq. (1). Then, in the feature layer we consider the low d-dimensional embeddings, that is the latent vectors V_i, U_u and V_j, which are then provided to the respective three neural networks, as shown in Fig. 1.

Hidden Layers. When training the GeoDCF model we aim to maximize the likelihood in Eq. (2), hence the loss function of GeoDCF becomes:

$$\min_{\Theta_b} \mathcal{L} = -\sum_{u \in \mathcal{N}} \sum_{(i,j) \in \mathcal{R}_u} P(i >_u j) + \lambda ||\Theta_b||^2 \qquad (4)$$

Θ_b is the parameter set, with $\Theta_b = \{W_i^{(q)}, W_u^{(q)}, W_j^{(q)}, b_i^{(q)}, b_u^{(q)}, b_j^{(q)}\}$, $\forall q = 1, \ldots, h$, where h is the number of hidden layers used in the three neural networks of Fig. 1. Matrices $W_i^{(q)}$, $W_u^{(q)}$ and $W_j^{(q)}$ are the weighting matrices of the q-th hidden layers to produce the deep learning representations of the latent vectors V_i, U_u and V_j. Variables $b_i^{(q)} b_u^{(q)}, b_j^{(q)}$ denote the respective biases of the q-th hidden layers of each neural network. As the size of hidden layers is important, in our architecture the bottom layer is the widest and each successive layer has a smaller number of hidden units. This way it learns more abstractive features of the d-dimensional embeddings and consequently better captures the non-linear correlations of the multifaceted contextual information with user preferences. For each neural network we implement the tower structure, halving the layer size for each successive layer. Hence, to implement the tower architecture we add the constraint of $2^h \leq d$ for the number of hidden layers h and the low d-dimensional embeddings of MV-NMF. For the hidden layers there are several choices of activation functions, like sigmoid, hyperbolic tangent $tanh(x)$ and rectifier linear unit function $ReLU(x)$. In our implementation, we used ReLU activation functions, with $ReLU(x) = \max(0, x)$, as they are non-saturated[4], well-suited for sparse data and making the model less likely to be overfitting [19]. Using ReLU activation functions, $\forall q = 1, \ldots, h$, the q-th hidden layers of the three neural networks produce the respective representations:

$$H_i^{(q)} = ReLU(W_i^{(q)} H_i^{(q-1)} + b_i^{(q-1)})$$
$$H_u^{(q)} = ReLU(W_u^{(q)} H_u^{(q-1)} + b_u^{(q-1)}) \tag{5}$$
$$H_j^{(q)} = ReLU(W_j^{(q)} H_j^{(q-1)} + b_j^{(q-1)})$$

with $H_i^{(0)} = V_i$, $H_u^{(0)} = U_u$ and $H_j^{(0)} = V_j$.

Output and BPR Layers. At the output layer, we use the hidden representations and the biases of the last hidden layers, that is the h-th layers of the three neural networks, which are then combined to compute the check-in probabilities x_{ui} and x_{uj} (Sect. 3). At the output layer we use the sigmoid function σ to ensure that the check-in probabilities x_{ui} and x_{uj} are in the range of $[0, 1]$. The check-in probabilities x_{ui} and x_{uj} are calculated as follows:

$$x_{ui} = \sigma(H_i^{(h)^\top} H_u^{(h)} + b_i^{(h)} + b_u^{(h)})$$
$$x_{uj} = \sigma(H_j^{(h)^\top} H_u^{(h)} + b_j^{(h)} + b_u^{(h)}) \tag{6}$$

Provided that x_{ui} and $x_{uj} \in [0, 1]$, at the BPR layer the partial relation between x_{ui} and x_{uj} is computed as $P(i >_u j) = (x_{ui} - x_{uj})/2 + 0.5$. Then, based on the computed probability $P(i >_u j)$, the prediction of a non-visited POI i is calculated by forwarding its low d-dimensional embedding V_i on the respective neural network as shown in Fig. 1 and then computing the check-in probability x_{ui}. The final top-k POI recommendations are generated by ranking the non-visited POIs based on the probability $P(i >_u j)$.

Model Training. In our implementation we used Tensorflow[5]. We computed the model parameters Θ_b via backpropagation with stochastic gradient descent. In particular, we employed mini-batch Adam, which adapts the learning rate for each parameter by performing smaller updates for frequent and larger updates for infrequent parameters. In each backpropagation iteration we performed negative sampling, as defined in BPR, to randomly select a subset of non-visited POIs as negative instances $j \in \mathcal{X}_u^-$. In our implementation we sampled five negative samples for each positive/observed sample, and set the batch size of mini-batch Adam to 512 with a learning rate 1e−4. Finally, to account for the fact that

---

[4] The saturation problems occurs when neurons stop learning and their output is near either 0 or 1, a problem that can be suffered by the sigmoid and tanh functions [19].
[5] www.tensorflow.org.

the initialization of the model parameters Θ_b plays an important role for the convergence and performance of our model, we followed a pretraining strategy. By applying single-view factorization of X and producing the respective latent matrices U and V, we first trained our model only using check-in data in X with random initializations until convergence - ignoring the contextual information in matrices A, G and T. Then, we used the trained parameters as the initialization of our model with the contextual information.

5   Experimental Evaluation

5.1   Datasets

In our experiments we used two publicly available datasets from Gowalla and Foursquare. The Gowalla check-in dataset[6] was generated from February 2009 to October 2010. Following [10,20] we filter out those users with fewer than 15 check-in POIs and those POIs with fewer than 10 visitors. The filtered dataset comprises 18,737 users, 32,510 POIs, 1,278,274 check-ins. The Gowalla check-in dataset includes all the contextual information, that is social correlation, as well as geographical and sequential transition information. The Foursquare dataset[7] includes check-in data from April 2012 to September 2013. We used the records generated within United States and eliminated those users with fewer than 10 check-in POIs, as well as those POIs with fewer than 10 visitors. The filtered dataset contains 24,941 users, 28,593 POIs and 1,196,248 check-ins. In the Foursquare dataset, geographical and sequential transition information is available, whereas users' social relations are missing.

5.2   Evaluation Protocol

To evaluate the top-k recommendation performance of the examined models we used the ranking-based metrics recall ($R@k$) and Normalized Discounted Cumulative Gain ($NDCG@k$). Recall $R@k$ is defined as the ratio of the relevant (checked-in) POIs in the top-k ranked list over all the relevant POIs for each user. The Normalized Discounted Cumulative Gain $NDCG@k$ metric considers the ranking of the relevant POIs in the top-k list. For each user the Discounted Cumulative Gain is defined as:

$$DCG@k = \sum_{l=1}^{k} \frac{2^{rel_l} - 1}{\log_2 l + 1} \tag{7}$$

where rel_l represents the relevance score of POI l, that is binary relevance in our case. We consider a POI as relevant if a user has checked-in, and irrelevant otherwise. $NDCG@k$ is the ratio of $DCG@k$ over the ideal $iDCG@k$ value for each user, that is the $DCG@k$ value given the check-in data in the test set.

---

[6] http://snap.stanford.edu/data/loc-gowalla.html.
[7] https://sites.google.com/site/yangdingqi/home/foursquare-dataset.

Following the evaluation protocol of [9, 20] we randomly select a percentage of 20% of the check-in data as a test set, while the remaining check-in data are used to train our model. We repeated our experiments five times, and in our results we report average recall and $NDCG$ over the five runs.

5.3 Compared Methods

In our experiments we compare the following methods:

- **RankGeoFM** [9]: a ranking-based model that first learns users' preference rankings for POIs, and then includes the geographical influence of neighboring POIs to generate top-k POI recommendations.
- **USG** [21]: a POI recommendation algorithm that considers both geographical influence and users' social correlation, following a friend-based collaborative filtering strategy with a pointwise loss function.
- **PACE** [20]: a deep learning strategy for jointly learning the embeddings of users and POIs to predict user preferences over POIs and all the available contextual information with a pointwise loss function.
- **MV-NMF:** a variant of our model, which ignores the deep learning strategy of GeoDCF by only performing multi-view NMF of the check-in matrix with the contextual information, as presented in Sect. 4.2. To generate recommendations we compute the factorized matrix as the product of the user and latent matrices $UV^\top$, and sort each row/user of the factorized matrix in a descending order. MV-NMF exploits all the available contextual information, and is a pointwise method.
- **GeoDCF:** the proposed model that first performs MV-NMF to calculate the user and POIs latent vectors and then performs BPR learning with our deep learning strategy.

The parameters of the examined methods have been determined via cross-validation and in our experiments we report the best results. The parameter analysis of the proposed method is further studied in Sect. 5.6.

5.4 Comparison with State-of-the-Art

In Fig. 2 we evaluate the performance of the examined models in terms of recall $R@k$ and $NDCG@k$, when varying the top-k POI recommendations. RankGeoFM and USG perform differently in the Gowalla and Foursquare datasets. Although both RankGeoFM and USG exploit users' check-in data and geographical information, in the Gowalla dataset USG achieves a better recommendation accuracy than RankGeoFM as USG also uses the available contextual information of users' social relations. Instead, users' social relations are missing from the Foursquare dataset (Sect. 5.1). As we can observe from Fig. 2 in the Foursquare dataset RankGeoFM beats USG. This occurs because RankGeoFM is a ranking-based method focusing on the ranking performance, while USG is a pointwise method. Regarding the most competitive method of PACE,

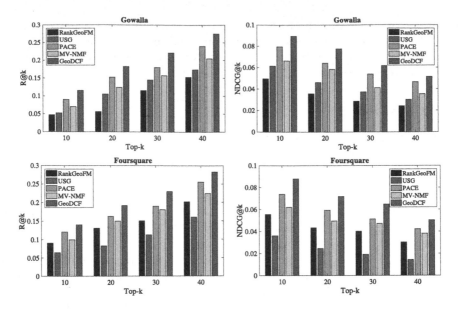

Fig. 2. Performance evaluation in terms of recall ($R@k$) and Normalized Discounted Cumulative Gain ($NDCG@k$) for the Gowalla and Foursquare datasets. Using the paired t-test, the proposed GeoDCF model outperforms all the baselines for $p < 0.05$.

Fig. 2 shows that PACE outperforms both RankGeoFM and USG by capturing the non-linear correlations of the available contextual information with its deep learning strategy. Compared to the proposed GeoDCF model, our MV-NMF variant performs poorly, as MV-NMF neither captures well the non-linear correlations of the users' and POIs' contextual information nor focuses on the ranking performance. Using the paired t-test we found out that compared to the second best method of PACE, our GeoDCF model achieves an average improvement of 18.96% and 17.81% in terms of recall and NDCG in all runs, at a significance level of $p < 0.05$. This occurs because PACE is a pointwise method and GeoDCF is a ranking-based model aiming to improve the top-k recommendation accuracy. Furthermore, GeoDCF also captures the non-linear correlations of the multifaceted contextual information with user preferences in our deep learning architecture.

5.5 Influence of Users' and POIs' Context

In Fig. 3 we evaluate separately the influence of users' and POIs' contextual information on our GeoDCF model. We denote *"check-in data"*, when GeoDCF only uses the check-in data to produce recommendations, ignoring any contextual information. Accordingly, *"check-in data+user context"* is a variant of the GeoDCF model which exploits check-in data and user context, that is users' social relations. Model *"check-in data+POI context"* is our variant when check-in

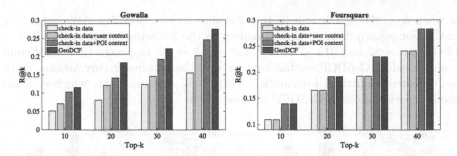

Fig. 3. Influence of users' and POIs' contextual information. Provided that in the Foursquare dataset users' social relations are missing, the variants *"check-in data"* and *"check-in data+user context"* have equal performance, and the variant *"check-in data+POI context"* has the same performance with GeoDCF.

data are only combined with POIs' contextual information, that is geographical and sequential transition information. As in the Foursquare dataset users' social correlations are missing, the variants *"check-in data"* and *"check-in data+user context"* have equal performance, and the variant *"check-in data+POI context"* has the same performance with GeoDCF. Clearly, as we can observe from Fig. 3 in both datasets the *"check-in data"* variant has the lowest performance, as it does not combine any contextual information with user preferences. This means that the contextual information of users or POIs can boost the recommendation accuracy. An interesting observation is that in the Gowalla dataset the *"check-in data+POI context"* variant outperforms the *"check-in data+user context"* variant, which indicates that POIs' context is more important than users' context in the POI recommendation task. This observation also complies with the observations of relevant studies such as [7, 20].

5.6   Parameter Analysis

The two most important parameters in our GeoDCF model are: (i) the number of low dimensional embeddings d at the embedding layer; and (ii) the number of hidden layers h of the neural networks. Given the constraint of $2^h \leq d$ of Sect. 4.3, we vary the number of low dimensional embeddings d to the power of 2. For $d = [1024, 512, 256]$ we vary the number of hidden layers h from 1 to 5 by a step of 1. As described in Sect. 4.3 the bottom layer with the low dimensional embeddings d is the widest and each successive layer has a smaller number of hidden units, to learn more abstractive features of the d-dimensional embeddings. To better capture the non-linear correlations of the multifaceted contextual information, we implement the tower structure for each neural network, that is halving the layer size for each successive layer. For example, for $d = 1024$ and $h = 3$ we have the following tower architecture $1024 \rightarrow 512 \rightarrow 256 \rightarrow 128$ or for $d = 512$ and $h = 2$ we have the architecture of $512 \rightarrow 256 \rightarrow 128$. Figure 4 shows the impact of the different deep learning architectures. We observe that the best architecture is when $d = 256$ and $h = 3$ in the Gowalla dataset, and $d = 512$ and $h = 4$ in the Foursquare dataset, corresponding to the following architectures: $256 \rightarrow 128 \rightarrow 64 \rightarrow 32$ and $512 \rightarrow 256 \rightarrow 128 \rightarrow 64 \rightarrow 32$, respectively. For different d and h values GeoDCF cannot capture well the non-linear correlations of the multifaceted contextual information with users' check-in data, which explains the low performance of GeoDCF in these cases.

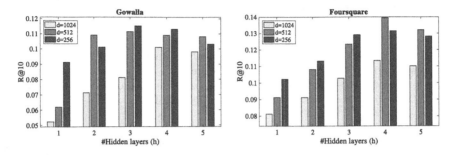

Fig. 4. Impact of different deep learning architectures when varying the number of low dimensional embeddings d at the embedding layer and the number of hidden layers h of the neural networks, subject to the constraint of $2^h \leq d$.

6   Conclusions

In this paper we presented GeoDCF, an efficient POI recommendation strategy to exploit the multifaceted information of users and POIs. The three key factors of the proposed model are the (i) exploit of the contextual information of users and POIs with a multi-view strategy at the embedding layer; (ii) capture of the non-linear correlations of the multifaceted contextual information with users' check-in data in our deep learning architecture; (iii) adding of a BPR layer at the top of our architecture to focus on the ranking performance. Our experimental evaluation on two benchmark datasets showed the superiority of GeoDCF to recently proposed baselines. Compared to the second best method, the proposed GeoDCF model achieved an average improvement of 18.96% and 17.81% in terms of recall and NDCG in all runs. We also evaluated GeoDCF with a variant of our model, which ignores the proposed deep learning architecture. Our experimental results demonstrated that GeoDCF outpeformed its variant. Clearly, the deep learning strategy can significantly boost the recommendation accuracy, by capturing the non-linear correlations of the contextual information and focusing on the ranking performance in the POI recommendation task. Finally, we evaluated the impact of users' and POIs' contextual information on our model separately. We showed that POIs' context contains more valuable information than users' social relations when generating POI recommendations, also confirmed by relevant studies [7,20]. Nowadays, users open multiple accounts on different social media platforms. An interesting future direction is to exploit user data from various social media platforms, following cross-domain strategies to produce POI recommendations. This is a challenging task as users behave differently in distinct social media platforms. For example, we plan to extend our GeoDCF model to generate POI recommendations for Foursquare users based on user data from Twitter and Instagram [15,16].

References

1. Cheng, C., Yang, H., Lyu, M.R., King, I.: Where you like to go next: successive point-of-interest recommendation. In: Proceedings of the International Joint Conference on Artificial Intelligence, pp. 2605–2611 (2013)
2. Cho, E., Myers, S.A., Leskovec, J.: Friendship and mobility: user movement in location-based social networks. In: Proceedings of the ACM International Conference on Knowledge Discovery and Data Mining, pp. 1082–1090 (2011)
3. Ding, D., Zhang, M., Li, S., Tang, J., Chen, X., Zhou, Z.: BayDNN: friend recommendation with Bayesian personalized ranking deep neural network. In: Proceedings of the ACM International Conference on Information and Knowledge Management, pp. 1479–1488 (2017)
4. Farseev, A., Samborskii, I., Filchenkov, A., Chua, T.: Cross-domain recommendation via clustering on multi-layer graphs. In: Proceedings of the ACM International Conference on Research and Development in Information Retrieval, pp. 195–204 (2017)
5. Gao, J., Han, J., Liu, J., Wang, C.: Multi-view clustering via joint nonnegative matrix factorization. In: Proceedings of the SIAM International Conference on Data Mining, pp. 252–260 (2013)
6. Hu, Y., Koren, Y., Volinsky, C.: Collaborative filtering for implicit feedback datasets. In: Proceedings of the IEEE International Conference on Data Mining, pp. 263–272 (2008)
7. Li, H., Ge, Y., Hong, R., Zhu, H.: Point-of-interest recommendations: learning potential check-ins from friends. In: Proceedings of the ACM International Conference on Knowledge Discovery and Data Mining, pp. 975–984 (2016)
8. Li, S., Kawale, J., Fu, Y.: Deep collaborative filtering via marginalized denoising auto-encoder. In: Proceedings of the ACM International Conference on Information and Knowledge Management, pp. 811–820 (2015)
9. Li, X., Cong, G., Li, X., Pham, T.N., Krishnaswamy, S.: Rank-GeoFM: a ranking based geographical factorization method for point of interest recommendation. In: Proceedings of the ACM International Conference on Research and Development in Information Retrieval, pp. 433–442 (2015)
10. Lian, D., Zhao, C., Xie, X., Sun, G., Chen, E., Rui, Y.: GeoMF: joint geographical modeling and matrix factorization for point-of-interest recommendation. In: Proceedings of the ACM International Conference on Knowledge Discovery and Data Mining, pp. 831–840 (2014)
11. Liu, Y., Wei, W., Sun, A., Miao, C.: Exploiting geographical neighborhood characteristics for location recommendation. In: Proceedings of the ACM International Conference on Information and Knowledge Management, pp. 739–748 (2014)
12. Manotumruksa, J., Macdonald, C., Ounis, I.: A personalised ranking framework with multiple sampling criteria for venue recommendation. In: Proceedings of the ACM International Conference on Information and Knowledge Management, pp. 1469–1478 (2017)

13. Rafailidis, D., Crestani, F.: Collaborative ranking with social relationships for top-n recommendations. In: Proceedings of the ACM International Conference on Research and Development in Information Retrieval, pp. 785–788 (2016)
14. Rafailidis, D., Crestani, F.: Joint collaborative ranking with social relationships in top-n recommendation. In: Proceedings of the ACM International Conference on Information and Knowledge Management, pp. 1393–1402 (2016)
15. Rafailidis, D., Crestani, F.: Top-n recommendation via joint cross-domain user clustering and similarity learning. In: Frasconi, P., Landwehr, N., Manco, G., Vreeken, J. (eds.) ECML PKDD 2016. LNCS (LNAI), vol. 9852, pp. 426–441. Springer, Cham (2016). https://doi.org/10.1007/978-3-319-46227-1_27
16. Rafailidis, D., Crestani, F.: A collaborative ranking model for cross-domain recommendations. In: Proceedings of the ACM International Conference on Information and Knowledge Management, pp. 2263–2266 (2017)
17. Rendle, S., Freudenthaler, C., Gantner, Z., Schmidt-Thieme, L.: BPR: Bayesian personalized ranking from implicit feedback. In: Proceedings of the International Conference on Uncertainty in Artificial Intelligence, pp. 452–461 (2009)
18. Wang, H., Wang, N., Yeung, D.: Collaborative deep learning for recommender systems. In: Proceedings of the ACM International Conference on Knowledge Discovery and Data Mining, pp. 1235–1244 (2015)
19. Xu, L., Choy, C., Li, Y.: Deep sparse rectifier neural networks for speech denoising. In: Proceedings of the IEEE International Workshop on Acoustic Signal Enhancement, pp. 1–5 (2016)
20. Yang, C., Bai, L., Zhang, C., Yuan, Q., Han, J.: Bridging collaborative filtering and semi-supervised learning: a neural approach for POI recommendation. In: Proceedings of the ACM International Conference on Knowledge Discovery and Data Mining, pp. 1245–1254 (2017)
21. Ye, M., Yin, P., Lee, W., Lee, D.L.: Exploiting geographical influence for collaborative point-of-interest recommendation. In: Proceeding of the ACM International Conference on Research and Development in Information Retrieval, pp. 325–334 (2011)
22. Ying, H., Chen, L., Xiong, Y., Wu, J.: Collaborative deep ranking: a hybrid pairwise recommendation algorithm with implicit feedback. In: Bailey, J., Khan, L., Washio, T., Dobbie, G., Huang, J.Z., Wang, R. (eds.) PAKDD 2016. LNCS (LNAI), vol. 9652, pp. 555–567. Springer, Cham (2016). https://doi.org/10.1007/978-3-319-31750-2_44
23. Yuan, F., Jose, J.M., Guo, G., Chen, L., Yu, H., Alkhawaldeh, R.S.: Joint geospatial preference and pairwise ranking for point-of-interest recommendation. In: Proceedings of the IEEE International Conference on Tools with Artificial Intelligence, pp. 46–53 (2016)
24. Zhang, J., Chow, C.: iGSLR: personalized geo-social location recommendation: a kernel density estimation approach. In: Proceeding of the ACM International Conference on Advances in Geographic Information Systems, pp. 324–333 (2013)
25. Zhang, J., Chow, C.: GeoSoCa: exploiting geographical, social and categorical correlations for point-of-interest recommendations. In: Proceedings of the ACM International Conference on Research and Development in Information Retrieval, pp. 443–452 (2015)
26. Zhang, J., Chow, C., Li, Y.: LORE: exploiting sequential influence for location recommendations. In: Proceedings of the ACM International Conference on Advances in Geographic Information Systems, pp. 103–112 (2014)

Personalized Thread Recommendation
for MOOC Discussion Forums

Andrew S. Lan[1]([⊠]), Jonathan C. Spencer[1], Ziqi Chen[2],
Christopher G. Brinton[1,3], and Mung Chiang[4]

[1] Princeton University, Princeton, USA
andrew.lan@princeton.edu
[2] HKUST, Clear Water Bay, Hong Kong
[3] Zoomi Inc., Wayne, USA
[4] Purdue University, West Lafayette, USA

Abstract. Social learning, i.e., students learning from each other through social interactions, has the potential to significantly scale up instruction in online education. In many cases, such as in massive open online courses (MOOCs), social learning is facilitated through discussion forums hosted by course providers. In this paper, we propose a probabilistic model for the process of learners posting on such forums, using point processes. Different from existing works, our method integrates topic modeling of the post text, timescale modeling of the decay in post excitation over time, and learner topic interest modeling into a single model, and infers this information from user data. Our method also varies the excitation levels induced by posts according to the thread structure, to reflect typical notification settings in discussion forums. We experimentally validate the proposed model on three real-world MOOC datasets, with the largest one containing up to 6,000 learners making 40,000 posts in 5,000 threads. Results show that our model excels at thread recommendation, achieving significant improvement over a number of baselines, thus showing promise of being able to direct learners to threads that they are interested in more efficiently. Moreover, we demonstrate analytics that our model parameters can provide, such as the timescales of different topic categories in a course.

Keywords: Discussion forums · Hawkes process
Personalized thread recommendation

1 Introduction

Online discussion forums have gained substantial traction over the past decade, and are now a significant avenue of knowledge sharing on the Internet. Attracting learners with diverse interests and backgrounds, some platforms (e.g., Stack

Electronic supplementary material The online version of this chapter (https://doi.org/10.1007/978-3-030-10928-8_43) contains supplementary material, which is available to authorized users.

M. Berlingerio et al. (Eds.): ECML PKDD 2018, LNAI 11052, pp. 725–740, 2019.
https://doi.org/10.1007/978-3-030-10928-8_43

Overflow, MathOverflow) target specific technical subjects, while others (e.g., Quora, Reddit) cover a wide range of topics from politics to entertainment.

More recently, discussion forums have become a significant component of online education, enabling students in online courses to learn socially as a supplement to their studying of the course content individually [2]; social interactions between learners have been seen to improve learning outcomes [4]. In particular, massive open online courses (MOOCs) often have tens of thousands of learners within single sessions, making the social interactions via these forums critical to scaling up instruction [3]. In addition to serving as a versatile complement to self-regulated learning [24], research has shown that learner participation on forums can be predictive of learning outcomes [26].

In this paper, we ask: *How can we model the activity of individual learners in MOOC discussion forums?* Such a model, designed correctly, presents several opportunities to optimize the learning process, including personalized news feeds to help learners sort through forum content efficiently, and analytics on factors driving participation.

1.1 Prior Work on Discussion Forums

Generic Online Discussion Sites. There is vast literature on analyzing user interactions in online social networks (e.g., on Facebook, Google+, and Twitter). Researchers have developed methods for tasks including link prediction [10,17], tweet cascade analysis [7,23], post topic analysis [21], and latent network structure estimation [14,15]. These methods are not directly applicable to modeling MOOC discussion forums since MOOCs do not support an inherent social structure; learners cannot become "friends" or "follow" one another.

Generic online discussion forums (e.g., Stack Overflow, Quora) have also generated substantial research. Researchers have developed methods for tasks including question-answer pair extraction [5], topic dynamics analysis [27], post structure analysis [25], and user grouping [22]. While these types of forums also lack explicit social structure, MOOC discussion forums exhibit several unique characteristics that need to be accounted for. First, topics in MOOC discussion forums are mostly centered around course content, assignments, and course logistics [3], making them far more structured than generic forums; thus, topic modeling can be used to organize threads and predict future activity. Second, there are no sub-forums in MOOCs: learners all post in the same venue even though their interests in the course vary. Modeling individual interest levels on each topic can thus assist learners in navigating through posts.

MOOC Forums. A few studies on MOOC discussion forums have emerged recently. The works in [19,20] extracted forum structure and post sentiment information by combining unsupervised topic models with sets of expert-specified course keywords. In this work, our objective is to model learners' forum behavior, which requires analyzing not only the content of posts but also individual learner interests and temporal dynamics of the posts.

In terms of learner modeling, the work in [8] employed Bayesian nonnegative matrix factorization to group learners into communities according to their posting behavior. This work relies on topic labels of each discussion post, though, which are either not available or not reliable in most MOOC forums. The work in [2] inferred learners' topic-specific seeking and disseminating tendencies on forums to quantify the efficiency of social learning networks. However, this work relies on separate models for learners and topics, whereas we propose a unified model. The work in [9] couples social network analysis and association rule mining for thread recommendation; while their approach considers social interactions among learners, they ignore the content and timing of posts.

As for modeling temporal dynamics, the work in [3] proposed a method that classifies threads into different categories (e.g., small-talk, course-specific) and ranks thread relevance for learners over time. This model falls short of making recommendations, though, since it does not consider learners individually. The work in [28] employed matrix factorization for thread recommendation and studied the effect of window size, i.e., recommending only threads with posts in a recent time window. However, this model uses temporal information only in post-processing, which limits the insights it offers. The work in [16] focuses on learner thread viewing rather than posting behavior, which is different from our study of social interactions since learners view threads independently.

The model proposed in [18] is perhaps most similar to ours, as it uses point processes to analyze discussion forum posts and associates different timescales with different types of posts to reflect recurring user behavior. With the task of predicting which Reddit sub-forum a user will post in next, the authors base their point processes model on self-excitations, as such behavior is mostly driven by a user's own posting history. Our task, on the contrary, is to recommend threads to learners taking a particular online course: here, excitations induced by other learners (e.g., explicit replies) can significantly affect a learner's posting behavior. As a result, the model we develop incorporates mutual excitation. Moreover, [18] labels each post based on the Reddit sub-forum it belongs to; no such sub-forums exist in MOOCs.

1.2   Our Model and Contributions

In this paper, we propose and experimentally validate a probabilistic model for learners posting on MOOC discussion forums. Our main contributions are as follows.

First, through point processes, our model captures several important factors that influence a learner's decision to post. In particular, it models the probability that a learner makes a post in a thread at a particular point in time based on four key factors: (i) the interest level of the learner on the topic of the thread, (ii) the timescale of the thread topic (which corresponds to how fast the excitation induced by new posts on the topic decay over time), (iii) the timing of the previous posts in the thread, and (iv) the nature of the previous posts regarding this learner (e.g., whether they explicitly reply to the learner). Through evaluation on three real-world datasets—the largest having more than 6,000 learners making more than 40,000 posts in more than 5,000 threads—we show that our model significantly outperforms several baselines in terms of thread recommendation, thus showing promise of being able to direct learners to threads they are interested in.

Second, we derive a Gibbs sampling parameter inference algorithm for our model. While existing work has relied on thread labels to identify forum topics, such metadata is usually not available for MOOC forum threads. As a result, we jointly analyze the post timestamp information and the text of the thread by coupling the point process model with a topic model, enabling us to learn the topics and other latent variables through a single procedure.

Third, we demonstrate several types of analytics that our model parameters can provide, using our datasets as examples. These include: (i) identifying the timescales (measured as half-lives) of different topics, from which we find that course logistics-related topics have the longest-lasting excitations, (ii) showing that learners are much (20–30 times) more likely to post again in threads they have already posted in, and (iii) showing that learners receiving explicit replies in threads are much (300–500 times) more likely to post again in these threads to respond to these replies.

2   Point Processes Forum Model

An online course discussion forum is generally comprised of a series of threads, with each thread containing a sequence of posts and comments on posts. Each post/comment contains a body of text, written by a particular learner at a particular point in time. A thread can further be associated with a topic, based on analysis of the text written in the thread. See our online technical report [11] for an example of a thread in a MOOC consisting of eight posts and comments and more intuitive explanations of the model setup. Moving forward, the terminology "posting in a thread" will refer to a learner writing either a post or a comment.

We postulate that a learner's decision to post in a thread at a certain point in time is driven by four main factors: (i) the learner's interest in the thread's topic, (ii) the timescale of the thread's topic, (iii) the number and timing of previous posts in the thread, and (iv) the learner's prior activity in the thread (e.g., whether there are posts that explicitly reply to the learner). The first factor is consistent with the fact that MOOC forums generally have no sub-forums: in the presence of diverse threads, learners are most likely to post in those covering topics they are interested in. The second factor reflects the observation that different topics exhibit different patterns of temporal dynamics. The third factor captures the common options for thread-ranking that online forums provide to users, e.g., by popularity or recency; learners are more likely to visit those at the top of these rankings. The fourth factor captures the common setup of notifications in discussion forums: learners are typically subscribed to threads automatically once they post in them, and notified of any new posts (especially those that explicitly reply to them) in these threads. To capture these dynamics, we model learners' posts in threads as events in temporal point processes [6], which will be described next.

Point Processes. A point process, the discretization of a Poisson process, is characterized by a rate function $\lambda(t)$ that models the probability that an event will happen in an infinitesimal time window dt [6]. Formally, the rate function at time t is given by

$$\lambda(t) = \mathbb{P}\left(\text{event in } [t, t + dt)\right) = \lim_{dt \to 0} \frac{N(t+dt)-N(t)}{dt}, \tag{1}$$

where $N(t)$ denotes the number of events up to time t [6]. Assuming the time period of interest is $[0, T)$, the likelihood of a series of events at times $t_1, \ldots, t_N < T$ is given by:

$$\mathcal{L}(\{t_i\}_{i=1}^N) = \left(\prod_{i=1}^N \lambda(t_i)\right) e^{-\int_0^T \lambda(\tau)d\tau}. \tag{2}$$

In this paper, we are interested in rate functions that are affected by excitations of past events (e.g., forum posts in the same thread). Thus, we resort to Hawkes processes [18], which characterize the rate function at time t given a series of past events at $t_1, \ldots, t_{N'} < t$ as

$$\lambda(t) = \mu + a \sum_{i=1}^{N'} \kappa(t - t_i),$$

where $\mu \geq 0$ denotes the constant background rate, $a \geq 0$ denotes the amount of excitation each event induces, i.e., the increase in the rate function after an event, and $\kappa(\cdot) : \mathbb{R}_+ \to [0,1]$ denotes a non-increasing decay kernel that controls the decay in the excitation of past events over time. In this paper, we use the standard exponential decay kernel $\kappa(t) = e^{-\gamma t}$, where γ denotes the decay rate. Through our model, different decay rates can be associated with different topics [18]; as we will see, this model choice enables us to categorize posts into groups (e.g., course content-related, small talk, or course logistics) based on their timescales, which leads to better model analytics.

Rate Function for New Posts. Let U, K, and R denote the number of learners, topics, and threads in a discussion forum, indexed by u, k, and r, respectively. We assume that each thread r functions independently, and that each learner's activities in each thread and on each topic are independent. Further, let z_r denote the topic of thread r, and let P_r denote the total number of posts in the thread, indexed by p; for each post p, we use u_p^r and t_p^r to denote the learner index and time of the post, and we use $p_i^r(u)$ to denote the ith post of learner u in thread r. Note that posts in a thread are indexed in chronological order, i.e., $p < p'$ if and only if $t_p^r < t_{p'}^r$. Finally, let $\gamma_k \geq 0$ denote the decay rate of each topic and let $a_{u,k}$ denote the interest level of learner u on topic k. We model the rate function that characterizes learner u posting in thread r (on topic $z_r = k$) at time t given all previous posts in the thread (i.e., posts with $t_p^r < t$) as

$$\lambda_{u,k}^r(t) = \begin{cases} a_{u,k} \sum_p e^{-\gamma_k(t-t_p^r)} & \text{if } t < t_{p_1^r(u)}^r \\ a_{u,k} \sum_{p:p<p_1^r(u)} e^{-\gamma_k(t-t_p^r)} \\ \quad + \alpha\, a_{u,k} \sum_{p:p \geq p_1^r(u), u \notin d_p^r} e^{-\gamma_k(t-t_p^r)} \\ \quad + \beta\alpha\, a_{u,k} \sum_{p:u \in d_p^r} e^{-\gamma_k(t-t_p^r)} & \text{if } t \geq t_{p_1^r(u)}^r. \end{cases} \tag{3}$$

In our model, $a_{u,k}$ characterizes the base level of excitation that learner u receives from posts in threads on topic k, which captures the different interest levels of learners on different topics. The exponential decay kernel models a topic-specific decay in excitation of rate γ_k from the time of the post.

Before $t^r_{p^r_1(u)}$ (the timestamp of the first post learner u makes in thread r), learner u's rate is given solely by the number and recency of posts in r ($t^r_{p^r_1(u)} = \infty$ if the learner never posts in this thread), while all posts occurring after $t^r_{p^r_1(u)}$ induce additional excitation characterized by the scalar variable α. This model choice captures the common setup in MOOC forums that learners are automatically subscribed to threads after they post in them. Therefore, we postulate that $\alpha > 1$, since new post notifications that come with thread subscriptions tend to increase a learner's chance of viewing these new posts, in turn increasing their likelihood of posting again in these threads. The observation of users posting immediately after receiving notifications is sometimes referred to as the "bursty" nature of posts on social media [7].

We further separate posts made after $t^r_{p^r_1(u)}$ by whether or not they constitute *explicit replies* to learner u. A post p' is considered to be an explicit reply to a post p in the same thread r if $t^r_{p'} > t^r_p$ and one of the following conditions is met: (i) p' makes direct reference (e.g., through name or the @ symbol) to the learner who made post p, or (ii) p' is the first comment under p.[1] d^r_p in (3) denotes the set of explicit recipients of p, i.e., if p is an explicit reply to learner u, then $u \in d^r_p$, while if p is not an explicit reply to any learners then $d^r_p = \emptyset$. This setup captures the common case of learners being notified of posts that explicitly reply to them in a thread. The scalar β characterizes the additional excitation these replies induce; we postulate that $\beta > 1$, i.e., the personal nature of explicit replies to learners' posts tends to further increase the likelihood of them posting again in the thread (e.g., to address these explicit replies).

Rate Function for Initial Posts. We must also model the process of generating the initial posts in threads. We characterize the rate function of these posts as time-invariant:

$$\lambda^r_{u,k}(t) = \mu_{u,k}, \tag{4}$$

where $\mu_{u,k}$ denotes the background posting rate of learner u on topic k. Separating the initial posts in threads from future posts in this way enables us to model learners' knowledge seeking (i.e., starting threads) and knowledge disseminating (i.e., posting responses in threads) behavior [2], through the background ($\mu_{u,k}$) and excitation levels ($a_{u,k}$), respectively.

Post Text Modeling. Finally, we must also model the text of each thread. Given the topic $z_r = k$ of thread r, we model W_r—the bag-of-words representation of the text in r across all posts—as being generated from the standard latent

---

[1] In this work, we restrict ourselves to these two concrete types of explicit replies; analyzing other, more ambiguous types is left for future work.

Dirichlet allocation (LDA) model [1], with topic-word distributions parameterized by ϕ_k. Details on the LDA model and the posterior inference step for ϕ_k
via collapsed Gibbs sampling in our parameter inference algorithm are omitted
for simplicity of exposition.

3   Parameter Inference

We now derive the parameter inference algorithm for our model. We perform
inference using Gibbs sampling, i.e., iteratively sampling from the posterior distributions of each latent variable, conditioned on the other latent variables. The
detailed steps are as follows:

1. Sample z_r. To sample from the posterior distribution of the topic of each
   thread, z_r, we put a uniform prior over each topic and arrive at the posterior

$$P(z_r = k \,|\, \ldots) \propto P(W_r \,|\, z_r) \prod_{k'} P(\{t_1^{r'}\}_{r':z_{r'}=k',u_1^r=u_1^{r'}} \,|\, \mu_{u_1^r,k'})$$
$$\cdot \prod_u P(\{t_p^r\}_{p:u_p^r=u} \,|\, a_{u,k}, \alpha, \beta, \gamma_k),$$

   where $\ldots$ denotes all variables except z_r. $P(W_r \,|\, z_r)$ denotes the likelihood of
   observing the text of thread r given its topic. $P(\{t_1^{r'}\}_{r':z_{r'}=k',u_1^r=u_1^{r'}} \,|\, \mu_{u_1^r,k'})$
   denotes the likelihood of observing the sequence of initial thread posts on
   topic k' made by the learner who also made the initial post in thread r;[2] this
   is given by substituting (4) into (2) as

$$P(\{t_1^{r'}\}_{r':z_{r'}=k',u_1^r=u_1^{r'}} \,|\, \mu_{u_1^r,k'}) = \mu_{u_1^r,k'}^{\sum_{r'} \mathbf{1}_{u_1^r=u_1^{r'},z_{r'}=k'}} e^{-\mu_{u_1^r,k'}T} \propto \mu_{u_1^r,k'}, \quad (5)$$

   where $\mathbf{1}_x$ denotes the indicator function that takes the value 1 when condition
   x holds and 0 otherwise. $P(\{t_p^r\}_{p:u_p^r=u} \,|\, a_{u,k}, \alpha, \beta, \gamma_k)$ denotes the likelihood
   of observing the sequence of posts made by learner u in thread r,[3] given by

$$P(\{t_p^r\}_{p:u_p^r=u} \,|\, a_{u,k}, \alpha, \beta, \gamma_k) = \left(\prod_{p:u_p^r=u} \lambda_{u,z_r}^r(t_p^r)\right) \left(e^{-\int_0^T \lambda_{u,z_r}^r(t)dt}\right), \quad (6)$$

   where the rate function $\lambda_{u,k}^r(t)$ for learner u in thread r (with topic k) is given
   by (3).
2. Sample γ_k. There is no conjugate prior distribution for the excitation decay
   rate variable γ_k. Therefore, we resort to a pre-defined set of decay rates $\gamma_k \in
   \{\gamma_s\}_{s=1}^S$. We put a uniform prior on γ_k over values in this set, and arrive at
   the posterior given by

$$P(\gamma_k = \gamma_s \,|\, \ldots) \propto \prod_{r:z_r=k} \prod_u P(\{t_p^r\}_{p:u_p^r=u} \,|\, a_{u,k}, \alpha, \beta, \gamma_s).$$

---

[2] If μ_1^r is not the initial poster in any thread r' with $z_{r'} = k'$, then $\{t_1^{r'}\} = \emptyset$.
[3] If u has not posted in r, then $\{t_p^r\} = \emptyset$.

3. Sample $\mu_{u,k}$. The conjugate prior of the learner background topic interest level variable $\mu_{u,k}$ is the Gamma distribution. Therefore, we put a prior on $\mu_{u,k}$ as $\mu_{u,k} \sim \mathrm{Gam}(\alpha_\mu, \beta_\mu)$ and arrive at the posterior distribution

$$P(\mu_{u,k} | \ldots) \propto \mathrm{Gam}(\alpha'_\mu, \beta'_\mu)$$

where

$$\alpha'_\mu = \alpha_\mu + \sum_r \mathbf{1}_{u_1^r = u, z_r = k}, \qquad \beta'_\mu = \beta_\mu + T.$$

4. Sample $a_{u,k}$, α, and β. The latent variables α and β have no conjugate priors. As a result, we introduce an auxiliary latent variable [14,23] e_p^r for each post p, where $e_{p'}^r = p$ means that post p is the "parent" of post p' in thread r, i.e., post p' was caused by the excitation that the previous post p induced. We first sample the parent variable for each post p according to

$$P(e_{p'}^r = p) \propto a^r(p, p') e^{-\gamma_{z_r}(t_{p'}^r - t_p^r)},$$

where $a^r(p, p') \in \{a_{u_{p'}^r, z_r}, \alpha a_{u_{p'}^r, z_r}, \beta a a_{u_{p'}^r, z_r}\}$ depending on the relationship between posts p and p' from our model, i.e., whether p' is the first post of $u_{p'}$ in the thread, and if not, whether p is an explicit reply to $u_{p'}$. In general, the set of possible parents of p is all prior posts $1, \ldots, p-1$ in r, but in practice, we make use of the structure of each thread to narrow down the set of possible parents for some posts.

With these parent variables, we can write $\mathcal{L}(\{t_p^r\}_{p:u_p^r = u})$, the likelihood of the series of posts learner u makes in thread r as

$$\mathcal{L} = \prod_r \mathcal{L}(\{t_p^r\}_{p=1}^{P_r}) = \prod_r \prod_u \mathcal{L}(\{t_p^r\}_{p:u_p^r = u}),$$

where $\mathcal{L}(\{t_p^r\}_{p:u_p^r = u})$ denotes the likelihood of the series of posts learner u makes in thread r. We can then expand the likelihood using the parent variables as

$$\mathcal{L}(\{t_p^r\}_{u_p^r = u}) = \prod_{p:p < p_1^r(u)} e^{-\frac{a_{u,z_r}}{\gamma_{z_r}}(1 - e^{-\gamma_{z_r}(T - t_p^r)})}$$

$$\left(\prod_{p':u_{p'}^r = u, e_{p'}^r = p} a_{u,z_r} e^{-\gamma_{z_r}(t_{p'}^r - t_p^r)}\right) \prod_{p:p \geq p_1^r(u), u \notin d_p^r} e^{-\frac{\alpha a_{u,z_r}}{\gamma_{z_r}}(1 - e^{-\gamma_{z_r}(T - t_p^r)})}$$

$$\left(\prod_{p':u_{p'}^r = u, e_{p'}^r = p} \alpha a_{u,z_r} e^{-\gamma_{z_r}(t_{p'}^r - t_p^r)}\right) \prod_{p:u \in d_p^r} e^{-\frac{\beta a a_{u,z_r}}{\gamma_{z_r}}(1 - e^{-\gamma_{z_r}(T - t_p^r)})}$$

$$\cdot \left(\prod_{p':u_{p'}^r = u, e_{p'}^r = p} \beta \alpha a_{u,z_r} e^{-\gamma_{z_r}(t_{p'}^r - t_p^r)}\right).$$

We now see that Gamma distributions are conjugate priors for $a_{u,k}$, α, and β. Specifically, if $a_{u,k} \sim \mathrm{Gam}(\alpha_a, \beta_a)$, its posterior is given by $P(a_{u,k} | \ldots) \sim \mathrm{Gam}(\alpha'_a, \beta'_a)$ where

$$\alpha'_a = \alpha_a + \sum_{r:z_r = k} \sum_p \mathbf{1}_{u_p^r = u},$$

$$\beta'_a = \beta_a + \sum_{r:z_r = k} \left(\sum_{p:p < p_1^r(u)} \frac{1}{\gamma_k}(1 - e^{-\gamma_k(T - t_p^r)})\right.$$

$$\left. + \sum_{p:p \geq p_1^r(u), u \notin d_p^r} \frac{\alpha}{\gamma_k}(1 - e^{-\gamma_k(T - t_p^r)}) + \sum_{p:u \in d_p^r} \frac{\beta \alpha}{\gamma_k}(1 - e^{-\gamma_k(T - t_p^r)})\right).$$

Similarly, if $\alpha \sim \text{Gam}(\alpha_\alpha, \beta_\alpha)$, the posterior is $P(\alpha|\ldots) \sim \text{Gam}(\alpha'_\alpha, \beta'_\alpha)$ where

$$\alpha'_\alpha = \alpha_\alpha + \sum_r \sum_p \sum_{p'} \mathbf{1}_{e^r_{p'}=p, p \geq p^r_1(u^r_{p'})},$$

$$\beta'_\alpha = \beta_\alpha + \sum_r \sum_u \left(\sum_{p:p \geq p^r_1(u), u \notin d^r_p} \frac{a_{u,z_r}}{\gamma_{z_r}} (1 - e^{-\gamma_{z_r}(T-t^r_p)}) \right.$$

$$\left. + \sum_{p:u \in d^r_p} \frac{\beta a_{u,z_r}}{\gamma_{z_r}} (1 - e^{-\gamma_{z_r}(T-t^r_p)}) \right).$$

Finally, if $\beta \sim \text{Gam}(\alpha_\beta, \beta_\beta)$, the posterior is $P(\beta|\ldots) \sim \text{Gam}(\alpha'_\beta, \beta'_\beta)$ where

$$\alpha'_\beta = \alpha_\beta + \sum_r \sum_p \sum_{p'} \mathbf{1}_{e^r_{p'}=p, u^r_{p'} \in d^r_p},$$

$$\beta'_\beta = \beta_\beta + \sum_r \sum_u \sum_{p:u \in d^r_p} \frac{\alpha a_{u,z_r}}{\gamma_{z_r}} (1 - e^{-\gamma_{z_r}(T-t^r_p)}).$$

We iterate the sampling steps 1–4 above after randomly initializing the latent variables according to their prior distributions. After a burn-in period, we take samples from the posterior distribution of each variable over multiple iterations, and use the average of these samples as its estimate.

4 Experiments

In this section, we experimentally validate our proposed model using three real-world MOOC discussion forum datasets. In particular, we first show that our model obtains substantial gains in thread recommendation performance over several baselines. Subsequently, we demonstrate the analytics on forum content and learner behavior that our model offers.

4.1 Datasets

We obtained three discussion forum datasets from 2012 offerings of MOOCs on Coursera: Machine Learning (ml), Algorithms, Part I (algo), and English Composition I (comp). The number of threads, posts and learners appearing in the forums, and the duration (the number of weeks with non-zero discussion forum activity) of the courses are given in Table 1.

Prior to experimentation, we perform a series of pre-processing steps. First, we prepare the text for topic modeling by (i) removing non-ascii characters, url links, punctuations and words that contain digits, (ii) converting nouns and verbs

Table 1. Basic statistics on the datasets.

Dataset	Threads	Posts	Learners	Weeks
ml	5,310	40,050	6,604	15
algo	1,323	9,274	1,833	9
comp	4,860	17,562	3,060	14

to base forms, (iii) removing stopwords,[4] and (iv) removing words that appear fewer than 10 times or in more than 10% of threads. Second, we extract the following information for each post: (i) the ID of the learner who made the post (u_p^r), (ii) the timestamp of the post (t_p^r), and (iii) the set of learners it explicitly replies to as defined in the model (d_p^r). For posts made anonymously, we do not include rates for them ($\lambda_{u,k}^r(t)$) when computing the likelihood of a thread, but we do include them as sources of excitation for non-anonymous learners in the thread.

4.2   Thread Recommendation

Experimental Setup. We now test the performance of our model on personalized thread recommendation. We run three different experiments, splitting the dataset based on the time of each post. The training set includes only threads initiated during the time interval $[0, T_1)$, i.e., $\{r : t_1^r \in [0, T_1)\}$, and only posts on those threads made before T_1, i.e., $\{p : t_p^r \leq T_1\}$. The test set contains posts made in time interval $[T_1, T_2)$, i.e., $\{p : t_p^r \in [T_1, T_2)\}$, but excludes new threads initiated during the test interval.

In the first experiment, we hold the length of the testing interval fixed to 1 day, i.e., $\Delta T = T_2 - T_1 = 1$ day, and vary the length of the training interval as $T_1 \in \{1 \text{ week}, \dots, W - 1 \text{ weeks}\}$, where W denotes the number of weeks that the discussion forum stays active. We set W to 10, 8, and 8 for ml, comp, and algo, respectively, to ensure the number of posts in the testing set is large enough. These numbers are less than those in Table 1 since learners drop out during the course, which leads to decreasing forum activity. In the second experiment, we hold the length of the training interval fixed at $W - 1$ weeks and vary the length of the testing interval as $\Delta T \in \{1 \text{ day}, \dots, 7 \text{ days}\}$. In the first two experiments, we fix $K = 5$, while in the third experiment, we fix the length of the training and testing intervals to 7 weeks and 1 week, respectively, and vary the number of latent topics as $K \in \{2, 3, \dots, 10, 12, 15, 20\}$.

For training, we set the values of the hyperparameters to $\alpha_a = \alpha_\mu = 10^{-4}$, and $\beta_a = \beta_\mu = \alpha_\alpha = \beta_\alpha = \alpha_\beta = \beta_\beta = 1$. We set the pre-defined decay rates $\{\gamma_s\}_{s=1}^S$ to correspond to half-lives (i.e., the time for the excitation of a post to decay to half of its original value) ranging from minutes to weeks. We run the inference algorithm for a total of 2,000 iterations, with 1,000 of these being burn-in iterations for good mixing.

Baselines. We compare the performance of our point process model (PPS) against four baselines: (i) Popularity (PPL), which ranks threads from most to least popular based on the total number of posts in each thread during the training time interval; (ii) Recency (REC), which ranks threads from newest to oldest based on the timestamp of their most recent post; (iii) Social influence (SOC), a variant of our PPS model that replaces learner topic interest levels with

---

[4] We use the stopword list in the Python natural language toolkit (http://www.nltk.org/) that covers 15 languages.

learner social influences (the "Hwk" baseline in [7]); and (iv) Adaptive matrix factorization (AMF), our implementation of the matrix factorization-based algorithm proposed in [28]. See our online technical report [11] for more explanations on the AMF baseline and a detailed, head-to-head comparison under the same experimental setting in [28].

To rank threads in our model for each learner, we calculate the probability that learner u will reply to thread r during the testing time interval as

$$P(u \text{ posts in } r) = \sum_k P(u \text{ posts in } r \mid z_r = k) \, P(z_r = k)$$
$$= \sum_k \left(1 - e^{-\int_{T_1}^{T_2} \lambda_{u,k}^r(t)dt} \right) P(z_r = k).$$

The rate function $\lambda_{u,k}^r(t)$ is given by (3). $P(z_r = k)$ is given by

$$P(z_r = k) \propto P(z_r = k \mid u_1^r) \, P(\mathcal{W}_r \mid z_r = k) \prod_u P(\{t_p^r\}_{p:u_p^r=u,t_p^r<T_1} \mid z_r = k),$$

where the likelihoods of the initial post and other posts are given by (2) and (5), and the thread text likelihood $P(\mathcal{W}_r | z_r = k)$ is given by the standard LDA model. The threads are then ranked from highest to lowest posting probability.

Evaluation Metric. We evaluate recommendation performance using the standard mean average precision for top-N recommendation (MAP@N) metric. This metric is defined by taking the mean (over all learners who posted during the testing time interval) of the average precision

$$AP_u@N = \sum_{n=1}^{N} \frac{P_u@n \cdot \mathbf{1}_{u \text{ posted in thread } r_u(n)}}{\min\{|\mathcal{R}_u|, N\}},$$

where $\mathcal{R}_u$ denotes the set of threads learner u posted in during the testing time interval $[T_1, T_2)$, $r_u(n)$ denotes the nth thread recommended to the learner, $P_u@n$ denotes the precision at n, i.e., the fraction of threads among the top n recommendations that the learner actually posted in, and $\mathbf{1}$ denotes the indicator function. We use $N = 5$ in the first two experiments, and vary $N \in \{3, 5, 10\}$ in the third experiment.

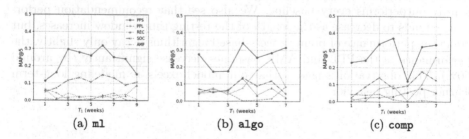

(a) ml   (b) algo   (c) comp

Fig. 1. Plot of recommendation performance over different lengths of the training time window T_1 on all datasets. Our model significantly outperforms every baseline.

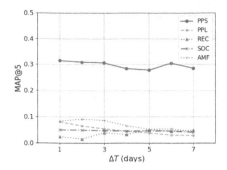

Fig. 2. Recommendation performance of the algorithms for varying testing window length ΔT on the `algo` dataset. The point process-based algorithms have highest performance and are more robust to ΔT.

Results and Discussion. Figure 1 plots the recommendation performance of our model and the baselines over different lengths of the training time window T_1 for each dataset. Overall, we see that our model significantly outperforms the baselines in each case, achieving 15%–400% improvement over the strongest baseline.[5] The fact that PPS outperforms the SOC baseline confirms our hypothesis that in MOOC forums, learner topic preference is a stronger driver of posting behavior than social influence, consistent with the fact that most forums do not have an explicit social network (e.g., of friends or followers). The fact that PPS outperforms the AMF baseline emphasizes the benefit of the temporal element of point processes in capturing the dynamics in thread activities over time, compared to the (mostly) static matrix factorization-based algorithms. Note also that as the amount of training data increases in the first several weeks, the recommendation performance tends to increase for the point processes-based algorithms while decreasing for PPL and REC. The observed fluctuations can be explained by the decreasing numbers of learners in the test sets as courses progress, since they tend to drop out before the end (see also Fig. 4).

Figure 2 plots the recommendation performance over different lengths of the testing time window ΔT for the `algo` dataset. As in Fig. 1, our model significantly outperforms every baseline. We also see that recommendation performance tends to decrease as the length of the testing time window increases, but while the performance of point process-based algorithms decay only slightly, the performance of the PPL and AMF baselines decrease significantly (by around 50%). This observation suggests that our model excels at modeling long-term learner posting behavior.

---

[5] Note that these findings are consistent across each dataset. Moving forward, we present one dataset in each experiment unless differences are noteworthy.

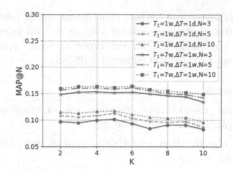

Fig. 3. Plot of recommendation performance of our model over the number of topics K on the ml dataset. The best performance is obtained at $K \approx 5$, though performance is stable for $K \leq 10$.

Finally, Fig. 3 plots the recommendation performance of the PPS model over different numbers of topics K for the ml dataset, for different choices of N, T_1 and ΔT. In each case, the performance rises slightly up to $K \approx 5$ and then drops for larger values (when overfitting occurs). Overall, the performance is relatively robust to K, for $K \leq 10$.

4.3   Model Analytics

Beyond thread recommendation, we also explore a few types of analytics that our trained model parameters can provide. For this experiment, we set $K = 10$ in order to achieve finer granularity in the topics; we found that this leads to more useful analytics.

Topic Timescales and Thread Categories. Table 2 shows the estimated half-lives γ_k and most representative words for five selected topics in the ml dataset that are associated with at least 100 threads. Figure 4 plots the total number of posts made on these topics each week during the course.

We observe topics with half-lives ranging from hours to weeks. We can use these timescales to categorize threads: course content-related topics (Topics 1 and 2) mostly have short half-lives of hours, small-talk topics (Topics 3 and 4) stay active for longer with half-lives of around one day, and course logistics topics (Topic 5) have much longer half-lives of around one week. Activities in threads on course content-related topics develop and decay rapidly, since they are most likely spurred by specific course materials or assignments. For example, posts on Topic 1 are about implementing gradient descent, which is covered in the second and third weeks of the course, and posts on Topic 2 are about neural networks, which is covered in the fourth and fifth weeks. Small-talk discussions are extremely common at the beginning and the end of the course, while course logistics discussions (e.g., concerning technical issues) are less frequent but steady in volume throughout the course.

Table 2. Estimated half-lives and highest constituent words (obtained by sorting the estimated topic-word distribution parameter vectors ϕ_k) for selected topics in the ml dataset with at least 100 threads. Different types of topics (course content-related, small-talk, or course logistics) exhibit different half-lives.

Topic	Half-life	Top words
1	4 h	gradient, row, element, iteration, return, transpose, logistic, multiply, initial, regularization
2	4 h	layer, classification, probability, neuron, unit, hidden, digit, nn, sigmoid, weight
3	1 day	interest, group, computer, Coursera, study, hello, everyone, student, learning, software
4	1 day	Coursera, deadline, professor, hard, score, certificate, review, experience, forum, material
5	1 week	screenshot, speed, player, subtitle, chrome, firefox, summary, reproduce, open, graph

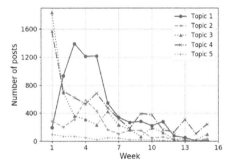

Fig. 4. Plot of the total number of posts on each topic week-by-week in the ml dataset. The week-to-week activity levels vary significantly across topics.

Table 3. Estimated levels of additional excitation brought by new activity notifications and explicit replies.

Dataset	ml	algo	comp
$\widehat{\alpha}$	29.0	23.3	33.6
$\widehat{\beta}$	19.2	12.2	10.6

Excitation from Notifications. Table 3 shows the estimated additional excitation induced by new activity notifications ($\widehat{\alpha}$) and explicit replies ($\widehat{\beta}$). In each course, we see that notifications increase the likelihood of participation significantly; for example, in ml, a learner's likelihood of posting after an explicit reply is 473 times higher than without any notification. Notice also that $\widehat{\beta}$ is lowest while $\widehat{\alpha}$ is highest in comp. This observation is consistent with the fact that in humanities courses like comp the discussions in each thread will tend to be longer [2], leading to more new activity notifications, while in engineering courses like `ml` and `algo` we would expect learners to more directly answer each other's questions, leading to more explicit replies.

5   Conclusions and Future Work

In this paper, we proposed a point processed-based probabilistic model for MOOC discussion forum posts, and demonstrated its performance in thread recommendation and analytics using real-world datasets. Possible avenues of future work include (i) jointly analyzing discussion forum data and time-varying learner grades [12,13] to better quantify the "flow of knowledge" between learners, (ii) incorporating up-votes and down-votes on the posts into the model, and (iii) leveraging the course syllabus to better model the emergence of new threads.

References

1. Blei, D., Ng, A., Jordan, M.: Latent Dirichlet allocation. J. Mach. Learn. Res. **3**, 993–1022 (2003)
2. Brinton, C.G., Buccapatnam, S., Wong, F., Chiang, M., Poor, H.V.: Social learning networks: efficiency optimization for MOOC forums. In: Proceedings of IEEE Conference on Computer Communications, pp. 1–9 (2016)
3. Brinton, C.G., Chiang, M., Jain, S., Lam, H., Liu, Z., Wong, F.: Learning about social learning in MOOCs: from statistical analysis to generative model. IEEE Trans. Learn. Technol. **7**(4), 346–359 (2014)
4. Brusilovsky, P., Somyürek, S., Guerra, J., Hosseini, R., Zadorozhny, V., Durlach, P.J.: Open social student modeling for personalized learning. IEEE Trans. Emerg. Topics Comput. **4**(3), 450–461 (2016)
5. Cong, G., Wang, L., Lin, C., Song, Y., Sun, Y.: Finding question-answer pairs from online forums. In: Proceedings of ACM SIGIR Conference on Research and Development in Information Retrieval, pp. 467–474, July 2008
6. Daley, D.J., Vere-Jones, D.: An Introduction to the Theory of Point Processes. Springer, New York (2003)
7. Farajtabar, M., Yousefi, S., Tran, L., Song, L., Zha, H.: A continuous-time mutually-exciting point process framework for prioritizing events in social media. arXiv preprint arXiv:1511.04145, November 2015
8. Gillani, N., Eynon, R., Osborne, M., Hjorth, I., Roberts, S.: Communication communities in MOOCs. arXiv preprint arXiv:1403.4640, March 2014
9. Kardan, A., Narimani, A., Ataiefard, F.: A hybrid approach for thread recommendation in MOOC forums. Int. J. Soc. Behav. Educ. Econ. Bus. Indus. Eng. **11**(10), 2195–2201 (2017)
10. Kim, M., Leskovec, J.: The network completion problem: Inferring missing nodes and edges in networks. In: Proceedings of ACM SIGKDD International Conference on Knowledge Discovery and Data Mining, pp. 47–58, August 2011
11. Lan, A.S., Spencer, J.C., Chen, Z., Brinton, C.G., Chiang, M.: Personalized thread recommendation for MOOC discussion forums. arXiv preprint arXiv:1806.08468, June 2018

12. Lan, A.S., Studer, C., Baraniuk, R.G.: Time-varying learning and content analytics via sparse factor analysis. In: Proceedings of ACM SIGKDD International Conference on Knowledge Discovery and Data Mining, pp. 452–461, August 2014

13. Lan, A.S., Studer, C., Waters, A.E., Baraniuk, R.G.: Joint topic modeling and factor analysis of textual information and graded response data. In: Proceedings of 6th International Conference on Educational Data Mining, pp. 324–325, July 2013

14. Linderman, S., Adams, R.: Discovering latent network structure in point process data. In: International Conference on Machine Learning, pp. 1413–1421, June 2014

15. Luo, D., et al.: Multi-task multi-dimensional Hawkes processes for modeling event sequences. In: Proceedings of International Joint Conference on Artificial Intelligence, pp. 3685–3691, August 2015

16. Mi, F., Faltings, B.: Adaptive sequential recommendation for discussion forums on MOOCs using context trees. In: Proceedings of International Conference on Educational Data Mining, pp. 24–31, June 2017

17. Miller, K., Jordan, M.I., Griffiths, T.L.: Nonparametric latent feature models for link prediction. In: Proceedings of Advances in Neural Information Processing Systems, pp. 1276–1284, December 2009

18. Mozer, M., Lindsey, R.: Neural Hawkes process memories. In: NIPS Symposium: Recurrent Neural Networks, December 2016

19. Ramesh, A., Goldwasser, D., Huang, B., Daumé III, H., Getoor, L.: Understanding MOOC discussion forums using seeded LDA. In: Proceedings of Conference Association for Computational Linguistics, pp. 28–33, June 2014

20. Ramesh, A., Kumar, S., Foulds, J., Getoor, L.: Weakly supervised models of aspect-sentiment for online course discussion forums. In: Proceedings of Conference Association for Computational Linguistics, pp. 74–83, July 2015

21. Ritter, A., Cherry, C., Dolan, B.: Unsupervised modeling of Twitter conversations. In: Proceedings of Language Technologies, pp. 172–180, July 2010

22. Shi, X., Zhu, J., Cai, R., Zhang, L.: User grouping behavior in online forums. In: Proceedings of ACM SIGKDD International Conference on Knowledge Discovery and Data Mining, pp. 777–786, June 2009

23. Simma, A., Jordan, M.I.: Modeling events with cascades of Poisson processes. In: Proceedings of Conference on Uncertainty in Artificial Intelligence, pp. 546–555, July 2010

24. Tomkins, S., Ramesh, A., Getoor, L.: Predicting post-test performance from online student behavior: a high school MOOC case study. In: Proceedings of International Conference on Educational Data Mining, pp. 239–246, June 2016

25. Wang, H., Wang, C., Zhai, C., Han, J.: Learning online discussion structures by conditional random fields. In: Proceedings of ACM SIGIR Conference on Research and Development in Information Retrieval, pp. 435–444, July 2011

26. Wang, X., Yang, D., Wen, M., Koedinger, K., Rosé, C.: Investigating how student's cognitive behavior in MOOC discussion forums affect learning gains. In: Proceedings of International Conference on Educational Data Mining, pp. 226–233, June 2015

27. Wu, H., et al.: Modeling dynamic multi-topic discussions in online forums. In: Proceedings of Conference on American Association for Artificial Intelligence, pp. 1455–1460, July 2010

28. Yang, D., Piergallini, M., Howley, I., Rose, C.: Forum thread recommendation for massive open online courses. In: Proceedings of International Conference on Educational Data Mining, pp. 257–260, July 2014

Inferring Continuous Latent Preference on Transition Intervals for Next Point-of-Interest Recommendation

Jing He[1], Xin Li[1(✉)], Lejian Liao[1], and Mingzhong Wang[2]

[1] BJ ER Center of HVLIP&CC, School of Computer Science,
Beijing Institute of Technology, Beijing, China
{skyhejing,xinli,liaolj}@bit.edu.cn
[2] School of Business, University of Sunshine Coast, Sippy Downs, Australia
mwang@usc.edu.au

Abstract. Temporal information plays an important role in Point-of-Interest (POI) recommendations. Most existing studies model the temporal influence by utilizing the observed check-in time stamps explicitly. With the conjecture that transition intervals between successive check-ins may carry more information for diversified behavior patterns, we propose a probabilistic factor analysis model to incorporate three components, namely, personal preference, distance preference, and transition interval preference. They are modeled by an observed third-rank transition tensor, a distance constraint, and a continuous latent variable, respectively. In our framework, the POI recommendation and the transition interval for user's very next move can be inferred simultaneously by maximizing the posterior probability of the overall transitions. Expectation Maximization (EM) algorithm is used to tune the model parameters. We demonstrate that the proposed methodology achieves substantial gains in terms of prediction on next move and its expected time over state-of-the-art baselines. Code related to this paper is available at: https://github.com/skyhejing/ECML-PKDD-2018.

Keywords: Point-of-Interest · Recommendation
Probabilistic factor analysis model

1 Introduction

Recently, many efforts have been devoted to next Point-of-Interest (POI) recommendation [8,18], which not only help users promptly identify their very next favorite POIs, but also benefit location-based social network (LBSNs) service providers to acquire more customers. However, to achieve accurate personalized POI recommendation is very challenging as it's well known that the check-in data

Electronic supplementary material The online version of this chapter (https://doi.org/10.1007/978-3-030-10928-8_44) contains supplementary material, which is available to authorized users.

© Springer Nature Switzerland AG 2019
M. Berlingerio et al. (Eds.): ECML PKDD 2018, LNAI 11052, pp. 741–756, 2019.
https://doi.org/10.1007/978-3-030-10928-8_44

of each user is in fact highly sparse. Existing work studied users' preferences on POIs by employing various context information, e.g., social relations, distance constraints and temporal information etc., to enhance the recommendation performance. To leverage on the temporal influence, some of existing works simply explore the temporal periodicity of human mobility, as the intuition behind is that human mobility often exhibits periodical patterns. For example, certain types of locations, such as office and Gym, are visited regularly at the same time slot [5,9,17].

Another research line of considering the temporal information is to explore the sequential behavior from successive check-ins. Cheng et al. utilized the factorized personalized Markov Chain (FPMC) [4] to capture the time-critical successive check-ins for next POI recommendation. Specifically, [12] investigated how to perform POI recommendations for a specific time period by learning users' evolving sequential preferences. Therefore, the expected time that a user will check in some places can be estimated by enumerating all possible time.

Furthermore, almost all existing work focused on utilizing the time stamps (absolute time) of check-ins, e.g., Monday morning, or 9:00 pm on Saturday. As reported in [9], user behavior shows clear periodic patterns. For example, people usually check in a workplace at 9:00 am. However, we argue that utilizing such patterns implicitly assumes that all check-ins follow a probability distribution, e.g., Gaussian with fixed mean and variance. In fact, people with different occupations follow various office hours. Consequently, the temporal patterns towards other activities should also vary. Thus, most existing work cannot take such diversity into consideration. Therefore, we argue that the time-intervals between check-ins do reflect the temporal behavior patterns of human activity with more flexibility.

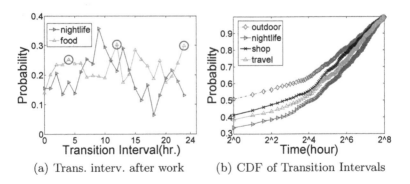

(a) Trans. interv. after work   (b) CDF of Transition Intervals

Fig. 1. Observations of transition intervals

Figure 1(a) plots the probability of next check-in locations in City of New York (NYC)[1] for "food" and "nightlife" given the previous check-in as "work" over different time intervals. There are three local maxima when the intervals are 4 h, 12 h, and 23 h for food check-in. The observation confirms a behavior pattern

---

[1] The check-in data is collected from Foursquare (See Sect. 4.1).

that people usually have breakfast 1 hour before, and have lunch and dinner at about 4 h and 12 h after they start to work, respectively. And the maximum of nightlife check-in indicates that people usually visit nightlife spots at about 10 h after the start of work. In summary, the working hour for each user may be different, but the transition intervals between check-in activities follow similar patterns. Figure 1(b) plots the cumulative distribution function (CDF) of the top-4 most popular location categories for "after work", in which the transition interval pattern is evident, e.g., the outdoor spots are often checked in after work within a short interval. We argue that such latent behavior patterns with time intervals should play a key role for next POI recommendation. The challenges come from two aspects, how to model such personalized latent behavior pattern and how to boost next POI recommendation with incorporation of the latent preferences.

In this paper, we propose a probabilistic approach to infer the continuous latent preference on transition intervals for next POI recommendation. Under our proposed framework, the next check-in and the time-intervals between successive check-ins can be obtained simultaneously. Thus, the checked-in time for a future activity can also be inferred from the time stamp of the previous check-in. Specifically, we model the overall transition preference with three components, namely, personal preference, spatial preference, and transition interval preference. They are modeled by a third-rank tensor, a distance constraint, and a continuous latent variable, respectively. Then, we develop a probabilistic factor analysis model to combine these three components. The probabilistic model performs statistical inference and Bayesian methods where the latent preference variables are assumed to follow Gaussian distributions with different means and variances. The transition probabilities, the estimated transition intervals, and the uncertainty of them are then obtained. During model learning phases, the Expectation Maximization (EM) is developed to optimize the proposed probability model. Experimental results demonstrate that the proposed model outperforms other state-of-the-art methods in terms of next POI recommendation and the expected transition time.

2   Related Work

Latest studies have integrated temporal influence into POI recommendation for further performance improvement.

[7] investigated the temporal cyclic patterns of user check-ins in terms of temporal non-uniformness and temporal consecutiveness and demonstrated its effectiveness to improve recommendation performance. [4] proposed a tensor-based FPMC-LR model for next POI recommendation by considering the sequential behaviors between check-ins. [6] proposed a personalized ranking metric embedding model (PRME) for next new POI recommendation by considering the order relationship between check-ins. [12] proposed a bi-weighted low-rank graph construction model to make recommendations for a specified future time period by considering users' evolving sequential preferences.

[11] explicitly modeled the check-in time as the modes of a fourth-rank tensor for next POI recommendation, in which the time interval is utilized to capture the intensity of relation between the two successive check-in locations. The approach is designed in a two-fold manner, which predicts the POI category first, and then determine the expected visitings. Their work assumed that the intensity of relation decays over the transition time, but the temporal periodicity exists. Its applicability is constrained by the availability and accuracy of the categorization.

However, none of the existing methods investigated the transition interval pattern explicitly. Alternatively, they utilized absolute time to capture the temporal preference. We argue that modelling the transition interval preference may help us to capture more diversified human mobility patterns. Moreover, existing work [9] modeled the latent behavior pattern in a discrete manner, and the number of patterns must be predefined. This motivates us to further investigate on the transition latent variables. In this paper, we extend the transition preference with the dimension of transition interval preference, and apply a continuous latent variable factor analysis approach to remove the requirement of predefinition.

3   Model Framework

Let $U = \{u_1, u_2, ..., u_M\}$ be a set of LBSN users, and $L = \{l_1, l_2, ..., l_N\}$ be a set of locations (POIs). Let l_u^t be the location visited by user u at time t, then the set of locations visited by u before time t is denoted by L_u^t, and $L_u^t = \{l_u^1, ..., l_u^{t-1}\}$. Let $\tau_{u,i,j}$ be the transition interval for u between location i and j, and the corresponding set of transition intervals for L_u^t is denoted by $T_u = \{\tau_{u,l_u^1,l_u^2}, ..., \tau_{u,l_u^{t-2},l_u^{t-1}}\}$. Our goal is to recommend user u next POI via the ranking of probabilities that he/she will move from the current location i to the next location j, with the transition interval $\tau_{u,i,j}$ inferred at the same time. Based on first-order Markov chain property, the transition probability is denoted as $x_{u,i,j} = p(j = l_u^t | i = l_u^{t-1})$. Thus, each user is associated with a specific transition matrix which in total generates a transition tensor $\chi \in \mathbb{N}^{|U| \times |L| \times |L|}$ with each $\chi_{u,i,j}$ representing the observed transition frequency of user u from location i to location j.

3.1   Preferences Modeling

Personal Preference. As the transitions among χ are partially observed, we follow most previous work [4,9] to apply the low-rank factorization model, a special case of Canonical Decomposition which models the pairwise interaction between all three modes of the tensor (user U, current location I, next location J), to fill up the missing values, given as:

$$\hat{\chi}_{u,i,j} = \boldsymbol{v}_u^{U,J} \cdot \boldsymbol{v}_j^{J,U} + \boldsymbol{v}_j^{J,I} \cdot \boldsymbol{v}_i^{I,J} + \boldsymbol{v}_u^{U,I} \cdot \boldsymbol{v}_i^{I,U} \tag{1}$$

where $\boldsymbol{v}_u^{U,J}$ and $\boldsymbol{v}_j^{J,U}$ denote the latent factor vectors for users and next locations, respectively. Other notions are defined in the same manner. The term $\boldsymbol{v}_u^{U,I} \cdot \boldsymbol{v}_i^{I,U}$ can be removed since it is independent of location j and does not affect the ranking result [14], thus, leading to a more compact expression for $\hat{\chi}_{u,i,j}$:

$$\hat{\chi}_{u,i,j} = \boldsymbol{v}_u^{U,J} \cdot \boldsymbol{v}_j^{J,U} + \boldsymbol{v}_j^{J,I} \cdot \boldsymbol{v}_i^{I,J} \qquad (2)$$

Transition Interval Preference. To extract the transition interval preference of user u for the next location, we first define the transition interval tensor Z. With the temporal intervals between successive POIs extracted for user u, the POI-POI transition interval matrix Z_u is constructed, which in turn generates a transition interval tensor $Z \in \mathbb{R}^{|U| \times |L| \times |L|}$ with each $\hat{z}_{u,i,j}$ representing the estimated transition interval of user u from location i to location j. Similar to yielding $\hat{\chi}_{u,i,j}$, $\hat{z}_{u,i,j}$ is generated by modeling the pairwise interaction between the modes of the transition interval tensor, given as:

$$\hat{z}_{u,i,j} = \boldsymbol{e}_u^{U,J} \cdot \boldsymbol{e}_j^{J,U} + \boldsymbol{e}_j^{J,I} \cdot \boldsymbol{e}_i^{I,J} \qquad (3)$$

where $\boldsymbol{e}_u^{U,J}$ and $\boldsymbol{e}_j^{J,U}$ denote the latent factor vectors for users and next locations in tensor Z, respectively. Other notions are defined in the same manner.

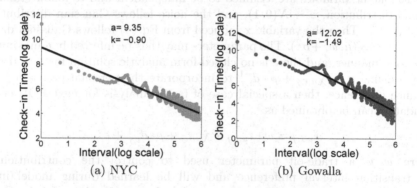

(a) NYC   (b) Gowalla

Fig. 2. Transition intervals v.s. the number of check-ins

To investigate the relation between the number of transitions and the transition intervals, we plot the check-in counts with respect to its transition intervals between two successive check-ins for Foursquare-NYC dataset and Gowalla dataset in log scale (Fig. 2). We can observe that the number of transitions decreases as the temporal interval increases, and the relation follows a power law distribution with exponent $k \approx -1$. Therefore, we define $z_{u,i,j}$ to specifically indicate the latent transition interval preference for user u to visit location j from the current location i by leveraging on the observations in Fig. 2 under conditions of uncertainty. We assume the latent variable $z_{u,i,j}$ is Gaussian with prior mean of $\hat{z}_{u,i,j}^{-1}$ and variance of σ_1^2, given as:

$$z_{u,i,j} \sim N(\hat{z}_{u,i,j}^{-1}, \sigma_1^2) \qquad (4)$$

Spatial Preference. Users' mobility is geographically constrained by the distance that one can travel within a limited time, and their preference to visit a location decreases as the geographical distance increases [5]. Most POIs which are likely to be explored are close to users' residence, workplace, and frequently visited POIs. Hence, the spatial behaviors of users can be utilized to enhance next POI recommendation. Here, we define the spatial influence $sp(d_{i,j})$ for any user to visit a location j which is $d_{i,j}$ km away from the current location i to leverage on the distance constraint: $sp(d_{i,j}) = \rho \cdot d_{i,j}^{-1}$. The optimal setting of ρ will be learned during the inference phase.

3.2   A Factor Analysis Latent Variable Model

In this paper, we adopt a so-called statistical *factor analysis* [2] latent variable model to incorporate the aforementioned three preferences, in which the combination is a linear function for observed data x:

$$x = w \cdot z + \mu + \epsilon \tag{5}$$

where z denotes the latent variable, ϵ is a z-independent noise process, w contains the *factor loading*, and μ permits the model to have non-zero mean. Conventionally, the latent variables are assumed to be independent and to follow standard normal distribution, $z \sim N(0, 1)$, and the noise follows Gaussian distribution, $\epsilon \sim N(0, \sigma^2)$. Then the variable x induced from Eq. (5) follows Gaussian distribution, $x \sim N(\mu, w^2 + \sigma^2)$. The parameters may thus be inferred in a maximum-likelihood manner, and there is no closed-form analytic solution for w and σ^2.

We define $\mu = \hat{\chi}_{u,i,j} + \rho \cdot d_{i,j}^{-1}$ to incorporate the personal preference and distance preference, then a special case of factor analysis for next POI recommendation can be obtained as:

$$\hat{x}_{u,i,j} = w \cdot z_{u,i,j} + \hat{\chi}_{u,i,j} + \rho \cdot d_{i,j}^{-1} + \epsilon \tag{6}$$

where w is a trade-off parameter used to control the contribution of the transition interval preference and will be learned during model inference phase. The set of all parameters for the proposed model is $\Theta :=$ $\{\rho, w, \sigma_1^2, \sigma_2^2, V_u^{U,J}, V_j^{J,U}, V_j^{J,I}, V_i^{I,J}, E_u^{U,J}, E_j^{J,U}, E_j^{J,I}, E_i^{I,J}\}$. In general, we believe that user behavior is personalized, and it is reasonable to assume that the check-ins from the same user share the same value of parameters $\{w, \sigma_1^2, \sigma_2^2\}$. Otherwise, they are not learnable for check-ins from the test set.

In our model, the noise represents random influences in the transitions that is not generated from the user preference, but arises from social network, weather, etc. Since the distribution over the noise is also Gaussian and defined as $\epsilon \sim N(0, \sigma_2^2)$, Eq. (6) implies a probability distribution over $x_{u,i,j}$ for a given $z_{u,i,j}$:

$$x_{u,i,j}|z_{u,i,j} \sim N(w \cdot z_{u,i,j} + \hat{\chi}_{u,i,j} + \rho \cdot d_{i,j}^{-1}, \sigma_2^2) \tag{7}$$

Given the above formulations, the marginal distribution for the observed $x_{u,i,j}$ is then obtained by integrating latent variables and is likewise Gaussian:

$$x_{u,i,j} \sim N(w \cdot \hat{z}_{u,i,j}^{-1} + \hat{\chi}_{u,i,j} + \rho \cdot d_{i,j}^{-1}, w^2 \cdot \sigma_1^2 + \sigma_2^2) \tag{8}$$

The conditional distribution of the latent variable $z_{u,i,j}$ given the observed $x_{u,i,j}$ can be derived by Bayes rule and is also Gaussian:

$$z_{u,i,j}|x_{u,i,j} \sim N(C, M) \tag{9}$$

where the posterior mean and posterior variance are specified by

$C = \frac{\sigma_1^2 w(x_{u,i,j} - \hat{x}_{u,i,j} - \rho d_{i,j}^{-1}) + \hat{z}_{u,i,j}^{-1}\sigma_2^2}{w^2\sigma_1^2 + \sigma_2^2}$ and $M = \frac{\sigma_1^2 \cdot \sigma_2^2}{w^2 \cdot \sigma_1^2 + \sigma_2^2}$ respectively.

3.3 Parameter Inference

Learning in probabilistic models can be simplified as maximizing the data log-likelihood with respect to all the model parameters. In our model, we consider the latent variables $\{z_{u,i,j}\}$ to be 'missing data' and complete data to comprise the observed transitions $x_{u,i,j}$ together with them. Assuming that users are independent as well as their check-in histories are independent, the corresponding complete-data log-likelihood is given as:

$$\mathcal{L}_C = \sum_{u \in U} \sum_{i \in L_u} \sum_{j \in L_u^t} \ln\{p(x_{u,i,j}, z_{u,i,j})\}$$

$$= \sum_{u \in U} \sum_{i \in L_u} \sum_{j \in L_u^t} \ln\{p(x_{u,i,j}|z_{u,i,j})p(z_{u,i,j})\} \tag{10}$$

where the components are derived from Eqs. (7) and (4):

$$p(x_{u,i,j}|z_{u,i,j}) = (2\pi\sigma_2^2)^{-\frac{1}{2}} \cdot$$
$$\exp\{-\frac{(x_{u,i,j} - w \cdot z_{u,i,j} - \hat{x}_{u,i,j} - \rho \cdot d_{i,j}^{-1})^2}{2\sigma_2^2}\} \tag{11}$$

$$p(z_{u,i,j}) = (2\pi\sigma_1^2)^{-\frac{1}{2}} \exp\{-\frac{(z_{u,i,j} - \hat{z}_{u,i,j}^{-1})^2}{2\sigma_1^2}\}. \tag{12}$$

Estimates for the model parameters Θ may be obtained by iterative maximization of $\mathcal{L}_C$, and a typical approach is to use Expectation-Maximization (EM) algorithm [15]. It is well known that EM algorithm iterates the two steps *expectation* (E-step) and *maximization* (M-step) until convergence, and it is guaranteed to increase the data likelihood to a local maximum. In the E-step, the expectation of $\mathcal{L}_C$, with respect to the posterior distribution of $z_{u,i,j}$ given the observed $x_{u,i,j}$, is computed by using the current estimate for the parameters Θ. In the M-step, new parameter values Θ' are determined by maximizing the expected complete-data log-likelihood. In our inference process, the corresponding steps are defined as follows:

E-step: we take the expectation of $\mathcal{L}_C$ with respect to the distributions $p(z_{u,i,j}|x_{u,i,j})$:

$$\langle\mathcal{L}_C\rangle = -\frac{1}{2}\sum_{u\in U}\sum_{i\in L_u}\sum_{j\in L_u^t}\{\ln(2\pi\sigma_2^2) \tag{13}$$

$$+ \frac{((x_{u,i,j} - \hat{\chi}_{u,i,j} - \rho d_{i,j}^{-1})^2 + w^2\langle z_{u,i,j}^2\rangle}{\sigma_2^2}$$

$$- \frac{2w\langle z_{u,i,j}\rangle(x_{u,i,j} - \hat{\chi}_{u,i,j} - \rho d_{i,j}^{-1})}{\sigma_2^2}$$

$$+ \ln(2\pi\sigma_1^2) + \frac{\langle z_{u,i,j}^2\rangle - 2\langle z_{u,i,j}\rangle\hat{z}_{u,i,j}^{-1} + \hat{z}_{u,i,j}^{-2}}{\sigma_1^2}\}$$

where we have omitted terms independent of the model parameters. The involved expectations are given as:

$$\langle z_{u,i,j}\rangle = \frac{\sigma_1^2 w(x_{u,i,j} - \hat{\chi}_{u,i,j} - \rho d_{i,j}^{-1}) + \hat{z}_{u,i,j}^{-1}\sigma_2^2}{w^2\sigma_1^2 + \sigma_2^2} \tag{14}$$

$$\langle z_{u,i,j}^2\rangle = \frac{\sigma_1^2 \cdot \sigma_2^2}{\sigma_2^2 + w^2 \cdot \sigma_1^2} + \langle z_{u,i,j}\rangle^2 \tag{15}$$

where $\langle z_{u,i,j}\rangle$ denotes the posterior mean of Eq. (9), and $\langle z_{u,i,j}^2\rangle$ is obtained in conjunction with the posterior variance of Eq. (9).

M-step: $\langle\mathcal{L}_C\rangle$ is maximized with respect to the model parameters Θ. This can be done by differentiating Eq. (13) and setting the partial derivatives to be zero, which gives a closed-form solution for parameters of $\{\rho, w, \sigma_1^2, \sigma_2^2\}$. For other parameters in factorization model, their values must be obtained via an iterative procedure as there are no closed-form solutions for them. Specifically, to obtain the revised parameters of $\{V_u^{U,J}, V_j^{J,U}, V_j^{J,I}, V_i^{I,J}, E_u^{U,J}, E_j^{J,U}, E_j^{J,I}, E_i^{I,J}\}$, we follow the widely used stochastic gradient decent (SGD) algorithm to optimize the partial derivations to be zero with respect to each parameter. That is to take the second derivative of $\mathcal{L}_C$. Then, the updating procedure is performed as:

$$\Theta' = \Theta + \alpha(\frac{\partial}{\partial\Theta}(\frac{\partial}{\partial\Theta}\langle\mathcal{L}_C\rangle)) \tag{16}$$

where $\alpha > 0$ is the learning rate.

To maximize the likelihood, the sufficient statistics of the posterior distribution are calculated from the E-step Eqs. (14) and (15), after which revised estimates of parameters are obtained from M-step. These steps are iterated in sequence until the algorithm is judged to be converged. The detailed algorithm and the parameter updating rules are shown in Algorithm 1.

Algorithm 1. Our Proposed Methodology

1: **Input:** check-in data D
2: **repeat**
3:     **E-Step:**
4:     $\langle z_{u,i,j} \rangle \leftarrow \frac{\sigma_1^2 w (\chi_{u,i,j} - \hat{\chi}_{u,i,j} - \rho d_{i,j}^{-1}) + \hat{z}_{u,i,j}^{-1} \sigma_2^2}{w^2 \sigma_1^2 + \sigma_2^2}$
5:     $\langle z_{u,i,j}^2 \rangle \leftarrow \frac{\sigma_1^2 \cdot \sigma_2^2}{\sigma_2^2 + w^2 \cdot \sigma_1^2} + \langle z_{u,i,j} \rangle^2$
6:     **M-Step:**
7:     $\sigma_1^2 \leftarrow \frac{1}{N_{u,d}} \sum_{u,d} (\langle z_{u,i,j}^2 \rangle - 2\langle z_{u,i,j} \rangle \cdot \hat{z}_{u,i,j}^{-1} + \hat{z}_{u,i,j}^{-2})$
8:     $\sigma_2^2 \leftarrow \frac{1}{N_{u,d}} \sum_{u,d} (w^2 \cdot \langle z_{u,i,j}^2 \rangle + (\hat{\chi}_{u,i,j} + \rho d_{i,j}^{-1})^2 + \chi_{u,i,j}^2 + 2w \cdot \langle z_{u,i,j} \rangle \cdot (\hat{\chi}_{u,i,j} +$
    $\rho d_{i,j}^{-1}) - 2w \cdot \chi_{u,i,j} \cdot \langle z_{u,i,j} \rangle) - 2\chi_{u,i,j} \cdot (\hat{\chi}_{u,i,j} + \rho d_{i,j}^{-1})$
9:     $\rho \leftarrow \frac{\sum_d d_{i,j}^{-1} \cdot (\chi_{u,i,j} - w\langle z_{u,i,j} \rangle) - \sum_d d_{i,j}^{-1} \cdot \hat{\chi}_{u,i,j}}{\sum_d d_{i,j}^{-2}}$
10:    $w \leftarrow \frac{\sum_{u,d} \langle z_{u,i,j} \rangle \cdot (\chi_{u,i,j} - \hat{\chi}_{u,i,j} - \rho d_{i,j}^{-1})}{\sum_{u,d} \langle z_{u,i,j}^2 \rangle}$
11:    $\gamma_1 \leftarrow \hat{\chi}_{u,i,j} + \rho \cdot d_{i,j}^{-1} + w\langle z_{u,i,j} \rangle - \chi_{u,i,j}$
12:    $v_u^{U,J} \leftarrow v_u^{U,J} + \alpha(2v_j^{J,U} \cdot \gamma_1)$
13:    $v_j^{J,U} \leftarrow v_j^{J,U} + \alpha(2v_u^{U,J} \cdot \gamma_1)$
14:    $v_j^{J,I} \leftarrow v_j^{J,I} + \alpha(2v_i^{I,J} \cdot \gamma_1)$
15:    $v_i^{I,J} \leftarrow v_i^{I,J} + \alpha(2v_j^{J,I} \cdot \gamma_1)$
16:    $e_u^{U,J} \leftarrow e_u^{U,J} + \alpha(2e_j^{J,U} \cdot (\hat{z}_{u,i,j} - \langle z_{u,i,j} \rangle^{-1}))$
17:    $e_j^{J,U} \leftarrow e_j^{J,U} + \alpha(2e_u^{U,J} \cdot (\hat{z}_{u,i,j} - \langle z_{u,i,j} \rangle^{-1}))$
18:    $e_j^{J,I} \leftarrow e_j^{J,I} + \alpha(2e_i^{I,J} \cdot (\hat{z}_{u,i,j} - \langle z_{u,i,j} \rangle^{-1}))$
19:    $e_i^{I,J} \leftarrow e_i^{I,J} + \alpha(2e_j^{J,I} \cdot (\hat{z}_{u,i,j} - \langle z_{u,i,j} \rangle^{-1}))$
20: **until** convergence
21: **return** Θ

---

4 Experiments

4.1 Datasets

We evaluate models on two real-world datasets which are acquired from Foursquare (NYC) and Gowalla and provided by [1,3] respectively. The statistics of the three datasets are listed in Table 1. Each dataset is split into two non-overlapping subsets to evaluate the model performance (for each user, the earliest 80% of check-ins as training set, and the remaining 20% check-ins as test set).

Table 1. Dataset statistics

	#User	#POI	#Check-in
Fours.-NYC	3401	106974	178143
Gowalla	1488	92679	226116

4.2   Evaluation Metrics

Different from the existing approaches, e.g., FPMC-LR, which took a set of POIs visited within a time interval as the previous/next POI to fit the training process to overcome the data sparsity, we take only two POIs to construct the <previous, next> POI pair during our training process, and to predict the "exact" next POI, instead of a set of next POIs as in FPMC-LR. Thus we only adopt recall to evaluate the performance for the "exact" next POI recommendation, as whatsoever the length of recommendation list is increasing, there exists only one correct solution for the "exact" next POI recommendation, and the precision cannot be higher than $1/|recommendation\ list|$. Therefore, we evaluate the performance of the next POI recommendation and the next new POI recommendation by defining recall as[2]:

$$Recall@N_{POI} = \frac{1}{|U|} \sum_{u \in U} \frac{|S_{N,u}^{POI} \cap S_{visited}^{POI}|}{|S_{visited}^{POI}|} \quad (17)$$

$$Recall@N_{POI}^{new} = \frac{1}{|U|} \sum_{u \in U} \frac{|S_{N,u}^{POI} \cap S_{visited}^{newPOI}|}{|S_{visited}^{newPOI}|} \quad (18)$$

where $S_{N,u}^{POI}$ is the list of top-N recommended POIs in descending order, $S_{visited}^{POI}$ denotes the visited POIs for user u, and $S_{visited}^{newPOI}$ denotes the locations that haven't been visited by a user yet in the training set but will be visited in the test set. $|U|$ denotes the number of the users, and N is the size of the next POI candidate list.

To make a fair comparison with the existing works, we further evaluate the performance of next POI recommendation by considering consecutive next check-ins within γ hours as the next location set (γ is set to 6 following [4,6]). Precision and recall are accordingly defined as:

$$Precision@N = \frac{|S_{N,u}^{POI} \cap S_{visited}^{\gamma}|}{N} \quad (19)$$

$$Recall@N = \frac{|S_{N,u}^{POI} \cap S_{visited}^{\gamma}|}{|S_{visited}^{\gamma}|} \quad (20)$$

where $S_{visited}^{\gamma}$ denotes the visited POIs for user u in the next γ hours. The precision and recall are computed by averaging all precision and recall for all samples in test set.

It is a relatively new research topic to predict the transition interval and evaluate the performance for such a model. We use the following two metrics to evaluate the performance of transition interval prediction.

---

[2] The precision is $Recall/|recomendation\ list|$ in our problem, where $|recomendation\ list|$ denotes the length of the recomm. list.

– Mean Absolute Percentage Error (MAPE) focuses on the difference between the estimated transition interval $\hat{z}_{u,i,j}$ and the actual time interval $T_{u,i,j}$ across all testing data:

$$MAPE = \frac{1}{|N_d|} \sum_{N_d} \frac{|T_{u,i,j} - \hat{z}_{u,i,j}|}{T_{u,i,j}} \tag{21}$$

where $|N_d|$ is the size of the test set. The model with smaller error is the better one.

– Precision for the POI recommendation is introduced to help the evaluation for the predicted transition interval of each movement. It is introduced because MAPE is susceptible to large errors and often take more weights from them, which makes MAPE less appropriate to evaluate the task of personalized POI recommendation. It is defined as:

$$Precision@T = \frac{1}{|U|} \sum_{u \in U} \frac{sum(\boldsymbol{S}_{T,u})}{|S_{visited}^{POI}|} \tag{22}$$

where $S_{T,u}$ equals to "1" if the difference between $\hat{z}_{u,i,j}$ and $T_{u,i,j}$ is less than a specified threshold T, i.e $|\hat{z}_{u,i,j} - T_{u,i,j}| < T$, or "0" otherwise.

4.3   Performance Comparison on Next POI Recommendation

Alongside with our model, the following state-of-the-art methods are evaluated and compared on the performance of next POI recommendation:

– **MF:** it factorizes the user-POI preference matrix in conventional recommender system [10].
– **PMF:** Probabilistic Matrix Factorization (PMF) [13] is a generalized matrix factorization model for traditional recommendation task.
– **FPMC-LR:** it extends factorized personalized Markov chain with the localized region constraint, which uses BPR as the optimization criterion [4].
– **PRME-G:** it considers the distance between current location and next location for metric embedding, which is the state-of-the-art personalized sequential POI recommendation [6].
– **LBP:** it jointly models next POI recommendation under the influence of users' latent behavior pattern, which is determined by the periodic property of human mobility along with time and the transition periodicity of location categories [9][3].

---

[3] LBP, FPMC-LR, PRME-G are not included in Table 4 for expected time prediction as we cannot find a way to predict transition interval valued in LBP. In fact, they all utilized BPR, a pairwise learning to rank algorithm, which takes the positive instance as pairs. However, there is no reasonable solution to define the positive instance or negative instance for the transition intervals. In this paper we use Canonical Decomposition, which is a special form of Tensor Factorization (TF), as an alternative approach for comparison.

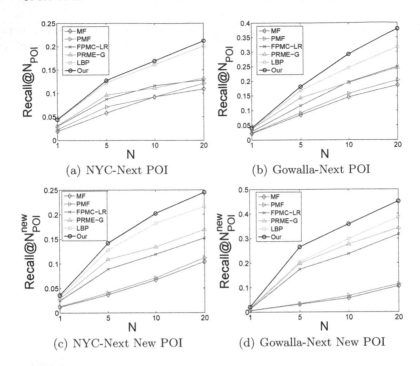

Fig. 3. Performance comparison on next POI recommendation

The parameters are tuned in the training set to find the optimal values, and they are subsequently used in the testing set. Figure 3 depicts the detailed results. The results show that:

- FPMC-LR, PRME-G, and our model all outperform MF and PMF significantly. It indicates that spatial influence plays an important role in next POI recommendation. Moreover, our model consistently outperforms FPMC-LR, PRME-G and LBP, which leads to the conclusion that the consideration of transition interval preference can better capture users' mobility in LBSNs.
- Our model has improved performance over baselines for next new POI recommendation. We argue that the gain also comes from the adoption of the transition interval preference. Since $\hat{\chi}_{u,i,j}$ models the observed transitions, the history of visiting "next new POI" is blank ($\hat{\chi}_{u,i,j} \approx 0$) in the training data. Therefore, the performance enhancement to $\hat{x}_{u,i,j}$ for next new POI recommendation is due to the part of $w \cdot z_{u,i,j}$ in Eq. (8)[4].

Tables 2 and 3 tabulate the comparison results when considering the next POI as a set of locations. It's obvious to see that our proposed model consistently outperforms other baselines in terms of precision and recall. And the improvement is even better than that in Fig. 3, which further verifies the effectiveness

---

[4] F-LR, P-G and LBP all apply the distance constraint similar to our distance preference model one way or another.

Table 2. Performance comparison of γ-hour next POI recommendation on NYC

Metrics	Precision\Recall					
	MF	PMF	F-LR	P-G	LBP	Our
top1	0.028\0.009	0.035\0.010	0.042\0.016	0.045\0.017	0.065\0.043	**.081\.053**
Imprv.	189%\489%	131%\430%	92.9%\231%	80.0%\212%	24.6%\23.3%	
top5	0.025\0.045	0.026\0.047	0.037\0.065	0.037\0.067	0.051\0.128	**.065\.157**
Imprv.	160%\249%	150%\234%	75.7%\142%	75.7%\134%	27.5%\22.7%	
top10	0.022\0.070	0.023\0.080	0.032\0.102	0.031\0.100	0.040\0.170	**.050\.211**
Imprv.	127%\201%	117%\164%	56.3%\107%	61.3%\111%	25.0%\24.1%	
top20	0.019\0.110	0.020\0.120	0.024\0.152	0.023\0.148	0.026\0.210	**.035\.264**
Imprv.	84%\140%	75.0%\120%	45.8%\73.7%	52.2%\78.4%	34.6%\25.7%	

Table 3. Performance comparison of γ-hour next POI recommendation on Gowalla

Metrics	Precision\Recall					
	MF	PMF	F-LR	P-G	LBP	Our
top1	0.008\0.003	0.009\0.004	0.013\0.006	0.014\0.007	0.018\0.009	**.023\.012**
Imprv.	188%\300%	156%\200%	76.9%\100%	64.3%\71.4%	27.8%\33.3%	
top5	0.035\0.101	0.038\0.105	0.059\0.123	0.063\0.131	0.095\0.195	**.121\.267**
Imprv.	246%\164%	218%\154%	105%\117%	92.1%\104%	27.4%\36.9%	
top10	0.033\0.132	0.036\0.143	0.052\0.199	0.058\0.223	0.081\0.302	**.109\.411**
Imprv.	230%\211%	203%\187%	110%\107%	87.9%\84.3%	34.6%\36.1%	
top20	0.030\0.232	0.031\0.251	0.042\0.293	0.045\0.321	0.061\0.394	**.082\.537**
Imprv.	173%\131%	165%\114%	95.2%\83.3%	82.2%\67.3%	34.4%\36.3%	

of incorporating transition interval preference for predicting POIs within a time slot, say for the next γ hours. We can also observe that precisions of γ-hour next POI recommendation fluctuate with increasing the length of recommendation list, which mainly accounts to the nature of the dataset. Our collected check-in data is a relatively sparse dataset. For example, the users may only contribute few check-ins in the next γ hour, the precision will inevitably decreases with the size of the candidate list increasing according to Eq. (19). Thus, we argue that recall is a more proper measurement to evaluate the performance for next POI recommendation.

4.4 Performance Comparison on Transition Interval

The performance comparison on transition intervals is performed between our model and the following baselines: MF, PMF, and FPMC. However, to our best knowledge, our model is the first work which is capable of providing the transition interval for next POI, and capable of obtaining the transition interval and next POI recommendation simultaneously. There is no way to get the transition interval values along with the POI recommendation directly in MF, PMF,

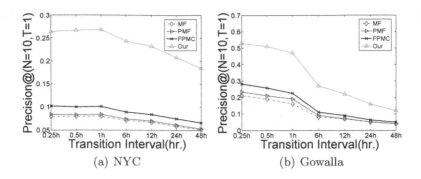

Fig. 4. Perf. comparison for transition interval prediction

and FPMC. In order to perform the comparison, we use the transition interval matrix/tensor (from MF, PMF, and FPMC) between POIs to generate the unobserved transition intervals. That is, we factorize the observed transition matrix and interval matrix from them separately, and then align the results to get their transition intervals for next POIs.

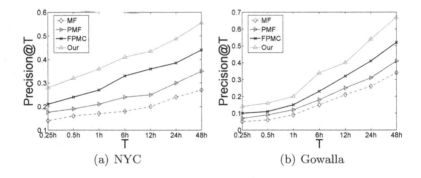

Fig. 5. Transition interval prediction v.s. T

Figure 4 shows the performance comparison on transition interval predictions. The results show that the proposed model always achieves the highest precision over baselines, which proves that our model is capable of providing effective POI recommendations to users as well as predicting how soon it will happen. We also compute MAPE between the predicted transition intervals and the ground truth of the test set (See Table 4). Lower values indicate more accurate predictions. It is evident that the proposed model outperforms the baselines by a significant margin. Figure 5 shows the performance comparison by relaxing the threshold T, and our method outperforms all the baselines again[5].

---

[5] Additional results reports in the Appendix of [16], which leads to similar evaluation conclusions.

Table 4. MAPE for our model and baselines on two test sets

	MF	PMF	FPMC	Our
Fours.-NYC	14.87	12.64	6.72	1.84
Gowalla	16.95	14.12	7.89	2.15

5 Conclusion

This paper proposes a probabilistic approach for next POI recommendation by exploring the transition interval patterns of each user, and the expected transition interval for next move can also be learned simultaneously. Specifically, the proposed model considers the transition interval preference as a key component of the overall transition behavior and utilizes a continuous latent variable to model such preference. The objective function is to maximize the posterior probability of the overall transitions. Expectation Maximization (EM) algorithm is used to estimate the model parameters. The experimental results on the real-world datasets show that the proposed model outperforms the state-of-the-art methods in terms of next POI recommendation, next new POI recommendation, and the expected transition interval prediction. Our proposed model can be easily extended to recommend next basket items and the purchase interval of them by redefining the transition tensor. In addition, we look forward to deploying the proposed model for commercial purposes, such as Weixin App, where the POI recommendation is able to benefit the users and location-based service providers. In the future, we would like to study how to integrate other contextual information into our model, e.g. social relationship and textual content of POIs, which may deeply exploit users' check-in behavior to enhance the recommendation performance.

Acknowledgments. This work has been partially supported by National Key R&D Program of China under Grant No. 2017FB0803300, NSFC under Grant No. 61772074.

References

1. Bao, J., Zheng, Y., Mokbel, M.F.: Location-based and preference-aware recommendation using sparse geo-social networking data. In: Proceedings of the 20th International Conference on Advances in Geographic Information Systems, pp. 199–208. ACM (2012)
2. Basilevsky, A.T.: Statistical Factor Analysis and Related Methods: Theory and Applications, vol. 418. Wiley, Hoboken (2009)
3. Cheng, C., Yang, H., King, I., Lyu, M.: Fused matrix factorization with geographical and social influence in location-based social networks. In: Twenty-Sixth AAAI Conference on Artificial Intelligence (2012)
4. Cheng, C., Yang, H., Lyu, M.R., King, I.: Where you like to go next: successive point-of-interest recommendation. In: Proceedings of the Twenty-Third International Joint Conference on Artificial Intelligence, pp. 2605–2611. AAAI Press (2013)

5. Cho, E., Myers, S.A., Leskovec, J.: Friendship and mobility: user movement in location-based social networks. In: Proceedings of the 17th ACM SIGKDD International Conference on Knowledge Discovery and Data Mining, KDD 2011, pp. 1082–1090. ACM, New York (2011)
6. Feng, S., Li, X., Zeng, Y., Cong, G., Chee, Y.M., Yuan, Q.: Personalized ranking metric embedding for next new POI recommendation. In: Proceedings of the 24th International Conference on Artificial Intelligence, pp. 2069–2075. AAAI Press (2015)
7. Gao, H., Tang, J., Hu, X., Liu, H.: Exploring temporal effects for location recommendation on location-based social networks. In: Proceedings of the 7th ACM Conference on Recommender Systems, pp. 93–100. ACM (2013)
8. Gao, H., Tang, J., Hu, X., Liu, H.: Content-aware point of interest recommendation on location-based social networks. In: Proceedings of the Twenty-Ninth AAAI Conference on Artificial Intelligence, Austin, Texas, USA, 25–30 January 2015, pp. 1721–1727 (2015)
9. He, J., Li, X., Liao, L., Song, D., Cheung, W.K.: Inferring a personalized next point-of-interest recommendation model with latent behavior patterns. In: Thirtieth AAAI Conference on Artificial Intelligence (2016)
10. Koren, Y., Bell, R.M., Volinsky, C., Jane, D.: Matrix factorization techniques for recommender systems, vol. 42, pp. 30–37, August 2009
11. Li, X., Jiang, M., Hong, H., Liao, L.: A time-aware personalized point-of-interest recommendation via high-order tensor factorization. ACM Trans. Inf. Syst. (TOIS) **35**(4), 31 (2017)
12. Liu, Y., Liu, C., Liu, B., Qu, M., Xiong, H.: Unified point-of-interest recommendation with temporal interval assessment. In: Proceedings of the 22nd ACM SIGKDD International Conference on Knowledge Discovery and Data Mining, pp. 1015–1024. ACM (2016)
13. Mnih, A., Salakhutdinov, R.: Probabilistic matrix factorization. In: Advances in Neural Information Processing Systems, pp. 1257–1264 (2007)
14. Rendle, S., Freudenthaler, C., Schmidt-Thieme, L.: Factorizing personalized Markov chains for next-basket recommendation. In: Proceedings of the 19th International Conference on World Wide Web, pp. 811–820. ACM (2010)
15. Rubin, D.B., Thayer, D.T.: EM algorithms for ML factor analysis. Psychometrika **47**(1), 69–76 (1982)
16. Supple. https://github.com/anonymityabcd/abcd1/blob/master/paper73sup.pdf (2018)
17. Yuan, Q., Cong, G., Ma, Z., Sun, A., Thalmann, N.M.: Time-aware point-of-interest recommendation. In: Proceedings of the 36th International ACM SIGIR Conference on Research and Development in Information Retrieval, pp. 363–372. ACM (2013)
18. Yuan, T., Cheng, J., Zhang, X., Qiu, S., Lu, H., et al.: Recommendation by mining multiple user behaviors with group sparsity. In: AAAI, pp. 222–228 (2014)

Transfer Learning

Feature Selection for Unsupervised Domain Adaptation Using Optimal Transport

Leo Gautheron[1,2(✉)], Ievgen Redko[1], and Carole Lartizien[1]

[1] Univ Lyon, INSA-Lyon, Université Claude Bernard Lyon 1,
UJM-Saint-Etienne CNRS, Inserm, CREATIS UMR 5220, U1206,
F-69621, Lyon, France
{leo.gautheron,ievgen.redko}@univ-st-etienne.fr,
carole.lartizien@creatis.insa-lyon.fr
[2] Univ Lyon, UJM-Saint-Etienne, CNRS, Institut d Optique Graduate School
Laboratoire Hubert Curien UMR 5516, F-42023, Saint-Etienne, France

Abstract. In this paper, we propose a new feature selection method for unsupervised domain adaptation based on the emerging *optimal transportation theory*. We build upon a recent theoretical analysis of optimal transport in domain adaptation and show that it can directly suggest a feature selection procedure leveraging the shift between the domains. Based on this, we propose a novel algorithm that aims to sort features by their similarity across the source and target domains, where the order is obtained by analyzing the coupling matrix representing the solution of the proposed optimal transportation problem. We evaluate our method on a well-known benchmark data set and illustrate its capability of selecting correlated features leading to better classification performances. Furthermore, we show that the proposed algorithm can be used as a preprocessing step for existing domain adaptation techniques ensuring an important speed-up in terms of the computational time while maintaining comparable results. Finally, we validate our algorithm on clinical imaging databases for computer-aided diagnosis task with promising results. Code related to this paper is available at: https://leogautheron.github.io/ and Data related to this paper is available at: https://github.com/LeoGautheron/ECML2018-FeatureSelectionOptimalTransport

1 Introduction

The majority of well-known machine learning algorithms used in real-world applications are built upon the common strategy often known as empirical risk minimization. This strategy suggests that a classifier that minimizes the loss over the observed samples is expected to generalize and thus to perform well on any other

Electronic supplementary material The online version of this chapter (https://doi.org/10.1007/978-3-030-10928-8_45) contains supplementary material, which is available to authorized users.

sample coming from the same probability distribution. However, this assumption is often violated in practice where a training sample may be different from new unseen data collected afterwards. For instance, one may consider the spam filtering problem. It is quite intuitive to suggest that a given user will be targeted with spam messages depending on its browsing history and that a classifier distinguishing between spam and non spam messages may not be equally efficient for two different users if it does not adapt correctly. In order to tackle this problem, a new learning paradigm called domain adaptation was proposed [4].

The main goal of domain adaptation is to provide methodological frameworks and algorithms that allow to reuse a classifier learned in one area, usually called *source domain*, in a different yet similar area usually called *target domain*. According to the domain adaptation theory presented in [3,4], the efficiency of a given adaptation algorithm depends on its capacity to reduce the discrepancy between the probability distributions of the considered source and target samples and on the existence of a good hypothesis (or classifier) that can minimize both source and target error functions. While finding this optimal hypothesis is a very difficult problem, most domain adaptation algorithms concentrate solely on reducing the discrepancy between two domains based on the observed samples. To this end, several papers [14,17,23,25] proposed to solve the domain adaptation problem by addressing it as a feature selection task. Indeed, for the general adaptation scenario, it is reasonable to assume that the shift between source and target domains may be caused by a changing behavior of a subset of features that characterize the data in both domains. In this case, identifying these features can help to reduce the discrepancy between the source and target domains samples and to allow efficient adaptation.

In this paper, we propose a new feature selection algorithm for unsupervised domain adaptation that allows to rank features based on their similarity across the source and target domains. Our key underlying idea is to solve the optimal transportation problem between the marginal distributions of features in the two domains in order to obtain a coupling matrix given by their joint probability distribution. The goal, then, is to use this coupling matrix to identify the most correlated features by analyzing the diagonal of the coupling matrix where higher coupling values indicate strong correlations between the source and target features. We note that contrary to the state-of-the-art methods that proceed by learning a new richer feature representation before identifying the invariant features, our method performs feature selection directly in the input space. This choice leads to more interpretable results and to a better understanding of the adaptation phenomenon as transformed features cannot directly point out to those descriptors that vary between the two domains. Furthermore, the shifted features identified by our method can be eliminated in order to speed-up domain adaptation algorithms whose running time often inherently depends on the dimensionality of the input data. This latter point is quite important as domain adaptation algorithms are often deployed for high-dimensional data arising from computer vision applications. Despite its advantages, our method does not aim to outperform the state-of-the-art classification results obtained by powerful feature transformation domain adaptation methods as most of them use a very rich class of mappings to find a new data representation. To this end, the foremost goal of this paper is to show that the proposed feature selection method is not a competitor of the state-of-the-art algorithms but is a complementary tool that provides important benefits both in terms of computational efficiency and better understanding of data. All the results presented in our paper are given in order to illustrate this rather than its superiority in terms of classification accuracy.

The rest of this paper is organized as follows: in Sect. 2 we present a short state-of-the-art on feature selection methods in domain adaptation. Section 3 is devoted to the introduction of basic elements related to the optimal transportation theory that are used later. In Sect. 4, we show how a theoretical analysis of domain adaptation with optimal transport can be used to derive a new adaptation algorithm based on feature selection. Based on this, we describe the proposed method and the details of its algorithmic implementation. Section 5 presents experimental evaluations of the proposed method on both a benchmark computer vision data set and a clinical imaging database for computer-aided diagnosis task. Section 6 summarizes our paper by outlining its main contributions and giving the possible future perspectives of this work.

2 Related Works

As classical feature selection methods [10] are not designed to work well under the assumption of distribution's shift, several methods were specifically proposed in the literature for feature selection in the context of domain adaptation. For instance, in [14], the authors search a latent low-dimensional subspace for two domains by jointly preserving the data structure and by selecting a subset of the latent features through a row-sparsity inducing regularization. While being quite effective in terms of classification results, this method, however, has two important drawbacks. First, it does not identify the original features that contribute to efficient adaptation but rather learns their embedding where the distributions' discrepancy is minimized. Second, its optimization procedure makes use of eigenvalue decomposition which has a high computational cost in large-scale applications. Another example of feature selection methods in domain adaptation are [17,25]. The contribution of the former paper consists in learning a least squares SVM in order to further remove the features that incur the smallest loss of the classification margin between the classes. The method described in the latter paper proposes to solve an optimization problem with two terms: the first term maximizes the relevance between source features and labels using the Hilbert-Schmidt Independence Criterion while the second term minimizes the shift between the domains using kernel embeddings. Contrary to our algorithm, the above mentioned methods are supervised as they both use annotations in the target domain. Finally, the method that is the most similar to ours is the feature selection algorithm for transfer learning presented in [23]. In this paper, the authors use a parametric maximum mean discrepancy distance in order to find a weight matrix that allows to identify invariant and shifting features in the original space. As we show in Sect. 4.1, this method and our contribution are closely related from the theoretical point of view, even though our method remains much more computationally attractive.

3   Preliminary Knowledge

In this section we give a brief overview of the basic elements related to the optimal transportation theory that are used later.

3.1   Optimal Transport

The theory of optimal transport has been introduced by Gaspard Monge in the 18th century and was recently revisited in [24]. In essence, this theory gives a mathematically founded tool that allows to align arbitrary probability distributions in an optimal way.

In the discrete case, it can be formalized as follows. Let $\hat{\mu}_S = \frac{1}{N_S} \sum_{i=1}^{N_S} \delta_{x_i^S}$ and $\hat{\mu}_T = \frac{1}{N_T} \sum_{i=1}^{N_T} \delta_{x_i^T}$ be two empirical probability measures defined as uniformly weighted sums of Diracs with mass at locations defined on two point sets $S = \{x_i^S \in \mathbb{R}^d\}_{i=1}^{N_S}$ and $T = \{x_i^T \in \mathbb{R}^d\}_{i=1}^{N_T}$ drawn from arbitrary probability distributions μ_S and μ_T. The Monge Kantorovich problem consists in finding a probabilistic coupling γ defined as a joint probability distribution over $S \times T$ that minimizes the cost of transport w.r.t. a metric $c : S \times T \to \mathbb{R}_+$:

$$\gamma^* = \arg\min_{\gamma \in \Pi(\hat{\mu}_S, \hat{\mu}_T)} \langle \gamma, C \rangle_F, \tag{1}$$

where $\langle \cdot, \cdot \rangle_F$ is the Frobenius dot product, $\Pi(\hat{\mu}_S, \hat{\mu}_T) = \{\gamma \in \mathbb{R}_+^{N_S \times N_T} | \gamma \mathbf{1} = \hat{\mu}_S, \gamma^T \mathbf{1} = \hat{\mu}_T\}$ is a set of doubly stochastic matrices and C is a dissimilarity matrix, i.e., for $x_i^S \in S$ and $x_j^T \in T$, we have $C_{ij} = c(x_i^S, x_j^T)$ which defines the energy needed to move a probability mass from x_i^S to x_j^T. This problem admits a unique solution γ^* and defines a metric on the space of probability measures (called the Wasserstein distance) as follows:

$$W(\hat{\mu}_S, \hat{\mu}_T) = \min_{\gamma \in \Pi(\hat{\mu}_S, \hat{\mu}_T)} \langle \gamma, C \rangle_F. \tag{2}$$

Despite its elegance and simplicity, the formulation of optimal transport given in Eq. 1 (abbreviated **OT**) is a Linear Programming problem that does not scale well because of its computational complexity.

3.2   Entropy-Regularized Optimal Transport

In order to solve this issue, [6] proposed to add the entropic regularization of γ to the Eq. 1 leading to the following optimization problem:

$$\gamma^* = \arg\min_{\gamma \in \Pi(\hat{\mu}_S, \hat{\mu}_T)} \langle \gamma, C \rangle_F - \frac{1}{\lambda} E(\gamma), \tag{3}$$

where $E(\gamma) := -\sum_{ij} \gamma_{ij} \log \gamma_{ij}$. The regularized optimal transport (abbreviated **OT2**) allows the source instances to be transported more or less uniformly to the target instances based on a hyper-parameter λ and can be optimized efficiently with the linear time Sinkhorn-Knopp algorithm [12].

3.3   Optimal Transport and Domain Adaptation

The use of optimal transport for domain adaptation has been studied for the first time in [5]. In this work, the authors present a new variant of optimal transport (abbreviated **OT3**) based on Eq. 3 by adding a class regularization $\ell_{\frac{1}{2},1}$:

$$\gamma^* = \underset{\gamma \in \Pi(\hat{\mu}_S, \hat{\mu}_T)}{\arg\min} \; \langle \gamma, C \rangle_F - \frac{1}{\lambda} E(\gamma) + \eta \Omega(\gamma), \tag{4}$$

where the $\Omega(\gamma) = \sum_j \sum_{\mathcal{L}} \|\gamma(I_{\mathcal{L}}, j)\|_1^{1/2}$ term prevents the source instances with different labels to be transported to the same target instance. $I_{\mathcal{L}}$ represents the list of sample indexes in S with label $\mathcal{L}$, and j goes through the sample indexes in T.

Using the optimal coupling matrix γ^* found with Eq. 1, 3 or 4, the authors propose to transport the source samples by solving for each of them:

$$\hat{x}_i^S = \underset{x \in \mathbb{R}^d}{\arg\min} \sum_j \gamma_{ij}^* c(x, x_j^T). \tag{5}$$

In case of the squared Euclidean distance, the closed form solution of this problem can be written as:

$$S_a = \mathrm{diag}\left((\gamma^* \mathbf{1})^{-1}\right) \gamma^* T. \tag{6}$$

When the marginals $\hat{\mu}_S$ and $\hat{\mu}_T$ are uniform (in practice, this is always the case for us), the Eq. 6 is simplified to

$$S_a = N_s \gamma^* T. \tag{7}$$

With this computation, each source instance is represented as the weighted barycenter of the target instances with which it has the highest values in γ^*.

For a graphical comparison of the **OT**, **OT2** and **OT3** algorithms, we refer the reader to the Supplementary material.

4   Proposed Approach

In this section we present our main contribution. We start by formally introducing a theoretical result that we use to derive our algorithm.

4.1   Theoretical Insight

From a theoretical point of view, domain adaptation problem is often formalized as follows: we define a domain as a pair consisting of a distribution μ_D on $\mathcal{X}$ and a labeling function $f_D : \mathcal{X} \rightarrow [0, 1]$. A hypothesis class H is a set of functions so that $\forall h \in H, h : \mathcal{X} \rightarrow \{0, 1\}$. Using the proposed notations, the definition of an error function can be given as follows.

Definition 1. *Given a convex loss-function l, the probability according to the distribution μ_D that a hypothesis $h \in H$ disagrees with a labeling function f_D (which can also be a hypothesis) is defined as*

$$\epsilon_D(h, f_D) = \mathbb{E}_{x \sim \mu_D} [l(h(x), f_D(x))].$$

When the source and target error functions are defined w.r.t. h and f_S or f_T, we use the shorthand $\epsilon_S(h, f_S) = \epsilon_S(h)$ and $\epsilon_T(h, f_T) = \epsilon_T(h)$.

The use of optimal transport in domain adaptation was first theoretically analyzed in [18]. In this paper, the authors proved that under some mild assumptions imposed on the form of the transport cost function, the source and target error function can be related through the following inequality

$$\epsilon_T(h) \leq \epsilon_S(h) + W(\mu_S, \mu_T) + \lambda, \tag{8}$$

where λ is the combined error of the ideal hypothesis h^* that minimizes $\epsilon_S(h) + \epsilon_T(h)$. This result shows that in order to upper bound the error of a classifier in the target domain, one has to minimize the source error function and the discrepancy between the source and target distributions given by the Wasserstein distance.

Below, we use this result as a starting point in order to develop our approach. To this end, we first notice that the source and target domains can be equivalently seen as 2-dimensional product spaces $\mathcal{X}_S \times \mathcal{F}_S$ and $\mathcal{X}_T \times \mathcal{F}_T$, where $\mathcal{X}_S$ (resp. $\mathcal{X}_T$) and $\mathcal{F}_S$ (resp. $\mathcal{F}_T$) denote the source (resp. target) instance and feature spaces. In this case, the probability distributions μ_S and μ_T are also product measures supported on $\mathcal{X}_S \times \mathcal{F}_S$ and $\mathcal{X}_T \times \mathcal{F}_T$ and can be written as $\mu_S^X \times \mu_S^f$ and $\mu_T^X \times \mu_T^f$, respectively. Using the results proved in [22] for concentration of measures in product spaces, we can upper bound the Wasserstein distance between μ_S and μ_T as follows:

$$W(\mu_S, \mu_T) \leq W(\mu_S^f, \mu_T^f) + \int_{\mathcal{F}_S} W(\mu_S^X | \mu_S^f, \mu_T^X) d\mu_S^f.$$

Note that in this inequality, measures μ_S^f (resp. μ_T^f) and μ_S^X (resp. μ_T^X) can be used interchangeably. We can see that the first term in the right-hand side stands for the Wasserstein distance between the measures defined on the feature spaces while the second term is the expectation of the Wasserstein distance between the source instances measure conditionally on the source features measure $\mu_S^X | \mu_S^f$

and the target instance measure μ_T^X. Now, by plugging it into the learning bound proposed in Eq. 8, we obtain

$$\epsilon_T(h) \le \epsilon_S(h) + W(\mu_S^f, \mu_T^f) + \int_{\mathcal{F}_S} W(\mu_S^X | \mu_S^f, \mu_T^X) d\mu_S^f + \lambda.$$

This inequality shows that when one considers probability measures over a product space of instances and features spaces, successful adaptation necessitates the minimization of the discrepancy between features distributions $\hat{\mu}_S^f$, $\hat{\mu}_T^f$ as well as that of the instances distributions μ_S^X, μ_T^X conditionally on the source features measure μ_S^f. Thus, it naturally leads to a two-stage procedure where the first goal is to reduce the discrepancy between the features sets of the two domains while the second is to apply an appropriate domain adaptation algorithm between their instances described by an optimal set of features obtained at the first stage.

In what follows, we introduce our method based on the idea of finding a coupling that aligns the distributions of features across the source and target domains. As suggested by the obtained bound, the selected features minimizing the $W(\mu_S^f, \mu_T^f)$ can be used then by a domain adaptation algorithm applied to the source and target samples of a reduced dimensionality. We also note that the Wasserstein distance here can be replaced, in practice, by the popular maximum mean discrepancy distance [9] often used in domain adaptation as both of them belong to a larger class of integral probability metrics defined over different functional classes. In this case, the feature selection algorithm proposed in [23][1] also indirectly minimizes the discrepancy between the marginals μ_S^f and μ_T^f. Nevertheless, the computational complexity of the proposed optimization procedure is polynomial thus making its use prohibitive in real-world applications.

4.2 Problem Setup

Until now the optimal transport was used in order to align empirical measures $\hat{\mu}_S$ and $\hat{\mu}_T$ defined based on the observable samples $S \in \mathbb{R}^{N_S \times d}$ and $T \in \mathbb{R}^{N_T \times d}$. The interpolation step performed using Eq. 7 aims at re-weighting the source instances so that their distribution matches the one of the target samples. The geometric interpretation is that, to minimize the divergence between μ_S and μ_T, we can associate the source samples with the target samples with which they have the highest coupling values.

As mentioned in the previous section, the idea of our method is to go from the sample space to the feature space. To this end, we now consider that S and T are drawn from 2-dimensional product spaces $\mathcal{X}_S \times \mathcal{F}_S$ and $\mathcal{X}_T \times \mathcal{F}_T$, where $\mathcal{X}_S, \mathcal{X}_T \subseteq \mathbb{R}^d$ while $\mathcal{F}_S \subseteq \mathbb{R}^{N_S}$ and $\mathcal{F}_T \subseteq \mathbb{R}^{N_T}$. In this case, we can define two empirical probability measures

$$\hat{\mu}_S^f = \frac{1}{d} \sum_{i=1}^d \delta_{f_i^S} \text{ and } \hat{\mu}_T^f = \frac{1}{d} \sum_{i=1}^d \delta_{f_i^T}$$

---

[1] Unfortunately, we were unable to use this method as a baseline in our experiments due to the lack of implementation details in their paper and the absence of a publicly available code.

based on the source and target features $\{f^S\}_{i=1}^d \in \mathcal{F}_S$, $\{f^T\}_{i=1}^d \in \mathcal{F}_T$, respectively. Our goal now would be to transport $\hat{\mu}_S^f$ to $\hat{\mu}_T^f$ by solving the entropic regularized optimal transportation problem given as follows:

$$\gamma^{*f} = \underset{\gamma^f \in \Pi(\hat{\mu}_S^f, \hat{\mu}_T^f)}{\arg\min} \; \langle \gamma^f, C^f \rangle_F - \frac{1}{\lambda} E\left(\gamma^f\right), \tag{9}$$

where $C_{ij}^f = \|f_i^S - f_j^T\|_2^2$.

In what follows, we show that the solution of this problem can lead to a principally different domain adaptation method that is based on feature selection approach rather than on the original instance re-weighting one.

4.3   Finding a Shared Feature Representation

At this point one may notice that in order to apply optimal transport between $\hat{\mu}_S^f$ and $\hat{\mu}_T^f$, it is necessary to calculate the cost matrix C^f which is possible only if the numbers of source and target instances are equal. Furthermore, as source and target features are described by supposedly shifted distributions, aligning them directly using any arbitrary sets of instances may not be appropriate due to the differences in the representation spaces that may exist across the two domains. In order to tackle both of these problems, we propose to find a matching between the sample number $i, \forall i = \{1, \dots, N_S\}$ describing the source features and the sample number $j, \forall j = \{1, \dots, N_T\}$ describing the target features based on the original optimal transportation problem. More formally, based on the solution γ^* of the optimization problem given by Eq. 1, we define the optimal subset of target instances $\boldsymbol{T_u}$ as:

$$\boldsymbol{T_u} := \{x_j \in T | j = \arg\max \gamma_{ij}^*, i \in \{1, \dots, N_S\}\}. \tag{10}$$

This particular choice of the algorithm **OT** rather than its regularized versions (**OT2** and **OT3**) is explained by the fact that we are interested in a sparse matching between the two sets, i.e., the one limiting the spread of mass[2].

This process, summarized in Algorithm 1[3], is a required preliminary step consisting in finding which examples will be used to describe the features in the source and target domains. The selection stage used to obtain $\boldsymbol{T_u}$ relies on the intrinsic capacity of the coupling matrix to describe the probability of associating each source instance with each target instance based on their similarity.

---

[2] The empirical justification of using the **OT** algorithm for sample selection is given in the Supplementary material.

[3] `zscore(X)`: for each column of X, subtract its mean and divide by its standard deviation.

Algorithm 1: Sample selection in target domain	**Algorithm 2:** Feature ranking for domain adaptation
Input : $S \in \mathbb{R}^{N_S \times d}$, $\quad\quad T \in \mathbb{R}^{N_T \times d}$, **Output:** $T_u \in \mathbb{R}^{N_S \times d}$ - optimal $\quad\quad$ subset of target instances $\mathsf{S} = \mathtt{zscore}(\mathtt{S}); \mathtt{T} = \mathtt{zscore}(\mathtt{T})$ $\gamma^* \leftarrow \mathtt{OT}(\mathtt{S},\mathtt{T})$ $T_u \leftarrow \{x_j \in \mathtt{T} \mid j = \underset{i=1,\dots,N_S}{\arg\max}\, \gamma^*_{ij}\}$	**Input** : $S \in \mathbb{R}^{N_S \times d}$, $\quad\quad T \in \mathbb{R}^{N_T \times d}$, **Output:** List F of d most similar $\quad\quad$ features from S and T $\mathtt{T_u} \leftarrow \mathtt{Algorithm1}(S,T)$ $\mathtt{S^T} = \mathtt{zscore}(\mathtt{S^T}); \mathtt{T_u^T} = \mathtt{zscore}(\mathtt{T_u^T})$ $\gamma^{*f} = \mathtt{OT2}(\mathtt{S^T}, \mathtt{T_u^T}, \lambda = 1)$ $\mathtt{F} = \mathtt{argSortDesc}(\{\{\gamma^{*f}\}_{ii} \mid i \in [1, d]\})$

4.4 Feature Selection

Now, we let $T := T_u$ meaning that in Eq. 9 the target features are described by the set T_u of the sample instances. Note that if $N_S > N_T$, we invert the roles of S and T in Algorithm 1 and instead let $S := S_u$. Furthermore, in a highly imbalanced classification setting, or in the presence of a large number of instances, we advise to first select a subset of source instances by balancing the samples according to their classes before applying Algorithm 1. This selection allows to capture a class information from the source domain without needing labeled samples from the target domain, and thus is still unsupervised w.r.t. the target domain.

We now solve the problem given in Eq. 9 and obtain the optimal coupling $\gamma^{*f} \in \mathbb{R}^{d \times d}$. Similar to what we have done at the sample selection step, we analyze the values of the coupling matrix in order to determine the less shifted features across the two domains. The important difference, however, is that we sort the features by analyzing only the diagonal of the coupling matrix. This peculiarity is explained by the fact that the values on the diagonal correspond to the similarities between the same features in the shared source and target representation space. By transporting the features with the **OT2** algorithm, each source feature is transported to its nearest target features. Because of this, if a given feature is shifted across the two domains, then its mass will be uniformly spread on the target features so that its mass on the corresponding target feature will be rather small. Similarly, if a feature is similar between the source and target domains, then the majority of the mass of this source feature should be found on its corresponding target feature.

Based on this idea, we propose to construct the ordered list of features F, where the feature number i in F is the one having the ith highest coupling value on the diagonal of the coupling matrix, i.e.:

$$F = \arg \text{sort}(\{\{\gamma^{*f}\}_{ii} \mid i \in \{1, \dots, d\}). \tag{11}$$

By varying the parameter λ in **OT2**, we can spread the mass of a source feature more or less uniformly when transporting it to the target features. Even though one may obtain different coupling values for different values of λ, it does not affect the order of features returned in F allowing us to fix $\lambda = 1$ in all empirical evaluations to avoid hyper-parameter tuning.

The pseudo-code given in Algorithm 2 summarizes our feature selection method. After having obtained the ordered list of features F, we can use its $d^* < d$ first features for the classification problem at hand. It is worth noting that the proposed method can be applied as a pre-processing before using any domain adaptation algorithm to discard the features that are completely different across the two domains. On the other hand, it can also be applied in "no adaptation setting" to select the common features between training and test data.

5   Experimental Evaluations

In this section, we provide an empirical study of the proposed algorithm based on the benchmark computer vision Office/Caltech data set and on clinical imaging database for computer-aided diagnostic task. Note that the optimal transport algorithms **OT** from Algorithm 1, **OT2** from Algorithm 2 and **OT3** are available in the Python POT library[4], making our method straightforward to implement. Nevertheless, we make the Python implementation and the data used in our experiments (except the medical data set) publicly available[5] for the sake of reproducibility.

5.1   Experiments on Visual Domain Adaptation Data

The main assumption of our method is that not all features are equally useful for adapting a classifier from source domain to the target one. This is especially the case for data sets described by features calculated using the Bag-of-Words (BoW) methods, such as, for instance, the features of the Office [19]/Caltech [8] data set.

Office/Caltech Data Set. For this data set, the classification task is to assign an image to a class based on its content. It is composed of 4 domains A, C, W and D containing 958, 1123, 295 and 157 images, respectively belonging each to one of 10 different classes. These domains form 12 domain adaptation pairs.

In what follows, we use three different types of features: (1) SURF features [2] of size 800 constructed using the BoW method; (2) CaffeNet features [11] that are obtained by feeding the images to a pre-trained neural network based on the prominent AlexNet [13]; (3) GoogleNet features [21] obtained in the way identical to CaffeNet features using GoogleNet network. In order to obtain CaffeNet and GoogleNet features, these two neural networks were first trained on ImageNet, a large data set containing millions of images distributed across 1000 different classes. We removed their classification layer of size 1000 to use the output of the previous layer, giving 4096 features for CaffeNet and 1024 features for GoogleNet. Note that we downloaded the pre-trained networks from the Caffe website [11] before using them to extract the features on our images,

---

[4] https://github.com/rflamary/POT.
[5] https://leogautheron.github.io.

Table 1. Results for CaffeNet features. This table presents mean ± standard deviation of recognition accuracy with no adaptation. Here, $\searchnothing X$ (resp. $\nearrow$) indicates the use of the X first features sorted by decreasing (resp. ascending) similarity computed with Algorithm 2.

DA pairs	$\searrow$ 512	$\nearrow$ 512	4096
A→C	**74.9 ± 2.0**	29.8 ± 2.4	71.7 ± 3.5
A→D	**78.8 ± 3.5**	20.4 ± 2.8	76.0 ± 3.5
A→W	**77.6 ± 1.9**	20.2 ± 3.5	66.0 ± 4.6
C→A	**83.7 ± 1.8**	38.7 ± 4.5	82.1 ± 2.2
C→D	**76.2 ± 3.6**	24.1 ± 3.4	74.2 ± 4.9
C→W	**75.4 ± 3.5**	20.3 ± 3.2	70.3 ± 5.3
D→A	**75.4 ± 2.1**	20.8 ± 3.8	68.7 ± 2.9
D→C	65.0 ± 2.6	21.5 ± 2.5	**66.6 ± 1.8**
D→W	**92.6 ± 2.0**	32.8 ± 5.1	91.9 ± 1.9
W→A	**81.5 ± 1.2**	18.8 ± 2.4	68.3 ± 3.0
W→C	**72.2 ± 1.1**	23.4 ± 2.1	61.2 ± 2.1
W→D	**96.5 ± 1.5**	49.7 ± 3.2	96.3 ± 1.0
Mean	**79.2 ± 2.2**	26.7 ± 3.3	74.4 ± 3.0

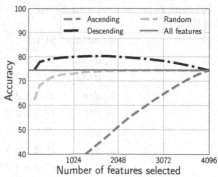

Fig. 1. Mean accuracies from Table 1. Our method corresponds to the 'Descending' curve consisting in selecting the features ordered by decreasing similarity between source and target domains.

and this without doing any fine-tuning or any other modification of the networks apart from removing their last layer.

The experimental protocol used to evaluate the proposed method is based on the one presented in [5]. For each adaptation pair $S \rightarrow T$, we randomly sample 20 images per class (8 if S is D). This gives us 200 images (resp. 80) for S. All images from T are considered. We then apply Algorithm 2 with S and T to obtain the ordered list of features F. For an increasing number of features d, we use the d first features of F to, first adapt S to T, and then use a 1-nearest neighbor classifier with the source adapted data as training set to compute the classification accuracy on the target data. We repeat this 19 times and report mean accuracies for each pair.

Classification Results. The classification results for CaffeNet features[6] are given in Table 1. From this table, we see that by selecting 512 features having the highest similarity between the source and target domains, we obtain a mean accuracy of 79.2% across the 12 adaptation pairs compared to 74.4% accuracy obtained using all 4096 features. This behavior is further confirmed by Fig. 1 that illustrates the obtained classification results for a number of features varying between 128 and 4096.

The general comparison for CaffeNet features, SURF and GoogleNet features is given in Table 2. As before, we observe an important difference between taking

---

[6] Due to the space limitations, we present the same detailed results for GoogleNet and SURF features in the Supplementary material.

Table 2. Mean accuracies over the 12 DA pairs without applying adaptation using 3 different type of features: SURF (d = 800), CaffeNet (d = 4096) and GoogleNet (d = 1024).

#features	SURF	CaffeNet	GoogleNet
$\searrow d/32$	21.3 ± 2.4	74.4 ± 2.9	80.0 ± 2.6
$\nearrow d/32$	12.7 ± 2.0	20.6 ± 3.0	24.2 ± 3.3
$\searrow d/8$	25.7 ± 2.6	79.2 ± 2.2	86.9 ± 1.8
$\nearrow d/8$	14.0 ± 2.2	26.7 ± 3.3	48.1 ± 3.9
$\searrow d/2$	29.9 ± 2.5	80.0 ± 2.2	88.1 ± 1.8
$\nearrow d/2$	16.2 ± 2.5	51.3 ± 4.4	77.2 ± 2.6
d	27.9 ± 2.2	74.4 ± 3.0	86.8 ± 1.8

the first most similar and dissimilar features across the two domains and note that better performances are obtained by taking a reduced number of features. Another noticeable point is that the performances of SURF features are far behind CaffeNet features, itself slightly worse than GoogleNet features. Even by taking a small number of 1024/32 = 32 GoogleNet features, we obtain a mean accuracy of 80.0% which is at least as good as all the other configurations using SURF and CaffeNet features. To summarize, the presented results clearly show that the order of features returned by our method is directly correlated with their adaptation capacities.

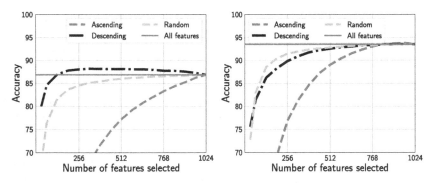

Fig. 2. Mean accuracies over the 12 DA pairs with GoogleNet features using no adaptation (**left**) and the **OT3** adaptation algorithm (**right**). We note that the classification performances are better with all features when we apply an adaptation algorithm: 86.8% without adaptation compared to 93.5% with.

We saw in the previous experiment that our method works for different types of features with the best performances obtained using GoogleNet descriptors. However, these performances were achieved without applying any adaptation algorithm. To this end, we present in Fig. 2 the impact of using an adaptation algorithm that takes as input a reduced set of GoogleNet features returned by our method. Several important conclusions can be made based on these results.

First, we notice that our algorithm does not improve the classification results compared to the performance of the **OT3** algorithm with a randomly selected subset of features. As explained in the introduction, **OT3** algorithm finds a new latent projection of source data in order to leverage the shift between the two domains. In this case, eliminating shifted features does not directly contribute to an improved classification performance as **OT3** algorithm can handle the reduction of shift between the two domains pretty well on its own. However, we can also observe the performance of **OT3** algorithm with a reduced "Ascending" set of features reaches its maximal value sooner than when no adaptation is performed. This is explained by the fact that the **OT3** algorithm successfully adapts the most shifted features. It is quite intuitive to assume that by selecting a subset of features, we decrease the computational complexity of the adaptation and classification algorithms that are used later. To support this claim, we present an additional study of the impact of reducing the number of features on both computational time and classification performance for several adaptation algorithms below.

Running Time Speed-Up. For this experiment, we evaluated the gain in computational time of different adaptation algorithms as a function of the number of features selected by our method. To this end, we compared the "no adaptation" setting with four state-of-the-art adaptation algorithms: **CORAL** [20], **SA** [7], **TCA** [16] and **OT3** [5]. We fixed the subspace dimensions of **SA** and **TCA** to 80 (or to the number of feature selected when smaller than 80) while for **OT3** we set $\lambda = 2$ and $\eta = 1$. Even if from Table 2 we obtained the best performances with GoogleNet features, we select for this experiment the CaffeNet features to better see the computational gain because they have the largest dimensionality (4096).

Table 3. Mean recognition accuracies in %, standard deviation and sum of total computational time (over the 12 DA pairs and 19 iterations) in seconds for different adaptation algorithms using the CaffeNet features.

Method	\512		\1024		\2048		4096	
No adapt.	79.2 ± 2.2	0.00 s	79.9 ± 2.3	0.00 s	80.0 ± 2.2	0.00 s	74.4 ± 3.0	0.00 s
CORAL	80.5 ± 1.8	110.43 s	80.8 ± 1.9	587.69 s	80.4 ± 1.7	3996.20 s	80.1 ± 1.7	29930.39 s
SA	81.8 ± 2.0	13.25 s	82.5 ± 1.8	32.09 s	82.9 ± 1.7	66.71 s	83.0 ± 1.7	169.71 s
TCA	83.5 ± 2.2	221.08 s	85.0 ± 1.9	223.62 s	85.8 ± 1.8	229.48 s	85.9 ± 1.7	242.71 s
OT3	84.2 ± 2.4	19.50 s	86.7 ± 1.9	31.76 s	88.8 ± 1.5	54.07 s	88.8 ± 1.4	97.47 s

The results of this evaluation are presented in Table 3. From these results, we see that by selecting 2048 out of 4096 most similar features, we are able to obtain slightly better classification performances for all adaptation methods compared to the case when all features are used. What's more, the computation time required by the algorithms greatly decrease. When only 512 features are used, an even more impressive speed up is obtained with a very slight drop in performance for the last three methods. These results confirm that our method is capable of

Table 4. Repartition of the MRI voxels between the Cancer and Non Cancer classes in source and target domains.

Class	#voxels 1.5T	#voxels 3T
Non cancer	363,222	846,556
Cancer	56,126	140,840
Total	419,348	987,396

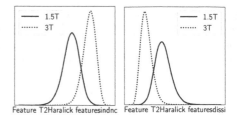

Feature T2Haralick featuresindnc   Feature T2Haralick featuresdissi

Fig. 3. Example distribution of 2 features illustrating the shift between the source and target domains.

finding subsets of similar features between source and target domains that can give comparable and sometimes even improved classification performances while decreasing considerably the computation time required for adaptation methods to converge.

5.2 Experiments on Medical Imaging Data Set

We now proceed to the evaluation of our method on a clinical data set of multi-parametric magnetic resonance images (mp-MRI) collected to train a computer-aided diagnosis system for prostate cancer mapping [1,15]. This system learns a binary decision model in a multidimensional feature space based on training samples (voxels) from different classes of interest. This model is then used to generate cancer probability maps.

Data Description. The considered database consists of 90 mp-MRI exams acquired with different imaging protocols on two different scanners (49 patients on a 1.5T scanner and 41 on a 3T scanner), thus producing heterogeneous data sets. Each individual voxel is described by a binary label (Cancer, Non Cancer) and a set of 95 handcrafted features consisting of image descriptors, texture coefficients, gradients and other visual characteristics (more details in [15]). Some of these 95 features have a clear shift between the two domains, as illustrated in Fig. 3. The number of available instances in both domains is shown in Table 4. Our goal is to learn a classifier on annotated 1.5T voxels, representing the source domain, performing well on 3T voxels, considered as the target domain, without using labels from the latter one.

Evaluation Protocol. We first randomly sample a set S of 1500 voxels equipro-portionally from the 49 1.5T exams and both classes of interest. Then, we use Algorithm 1 on S and on T as 20000 randomly sampled voxels from the 41 3T exams to obtain T_u. This step is followed by the adaptation of S to T_u, training a linear SVM on S_a and testing it on all voxels from the 3T target domain.

We used the area under the ROC Curve (AUC) as the diagnostic performance measure. This is due to the fact that both the source and the target domains data

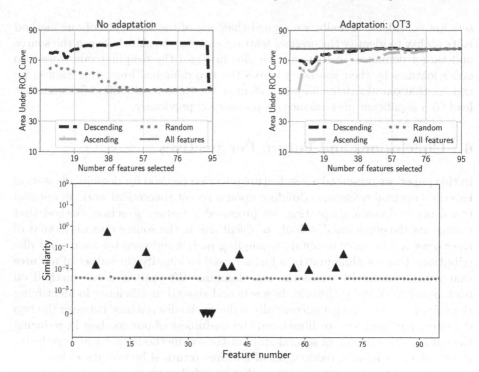

Fig. 4. Performance of our method on the clinical MRI database with no adaptation (top row, left) and using the **OT3** algorithm (top row, right). The log-scaled similarity of features across the two domains estimated by our algorithm is given in the bottom row. We observe that our method correctly identifies the three most shifted features that lead to an important drop in classifier's performance.

exhibit an important class imbalance with 86% of non-cancer voxels. In this case, the classification accuracy used in the previous experiments does not provide a truthful picture of the classifier's performance. Our feature selection method is used as a standalone method and in combination with the **OT3** adaptation algorithm. As before, we repeat this process 20 times, and we report the mean AUC over the 20 iterations.

Obtained Results. The results for this data set are shown in Fig. 4. When all the 95 features are used, we obtain an AUC of 50% without adaptation, corresponding to the worst possible performance with no distinction between Cancer and Non cancer classes. By applying our feature selection algorithm (the "Descending" curve) in a standalone manner, we are able to reach an AUC of 80% with a significant drop in performance when the 3 more dissimilar features are added. On the other hand and similar to the Office/Caltech data set, using our feature selection algorithm before applying an adaptation algorithm reduces greatly the number of features needed to achieve comparable performance. This benefit presents an important computational gain when high-dimensional data sets are considered. Finally, we argued that one of the strengths of our method is its ability to identify the original features causing the shift between the source and target domains. To this end, we plot in Fig. 4 the coupling values used to order features by their similarity across the two domains. From this Figure, we can see that our algorithm allows to identify the three most shifted features that lead to a significant performance drop observed previously.

6   Conclusions and Future Perspectives

In this paper, we presented a new feature selection method for domain adaptation based on optimal transport. Building upon a recent theoretical work on optimal transport in domain adaptation, we proposed a feature selection method that transports the empirical distribution of features in the source domain to that of the target one in order to obtain a coupling matrix representing their joint distribution. This coupling matrix is further used to identify the subset of features that remain unshifted across the two domains. We evaluated our method on both benchmark and real-world data sets and showed its efficiency in identifying the subset of features that successfully reduces the discrepancy between the two domains. Furthermore, we illustrated the usefulness of our method in reducing the computational time of several state-of-the-art methods that converge faster when taking as input a reduced set of features returned by our algorithm.

The possible future investigations that may follow up the presented work are many. First of all, we would like to combine our feature selection algorithm with a feature-transformation domain adaptation algorithm in a way such that the projection of data and the selection of features would be performed simultaneously. The potential interest of this joint approach would be to reduce the computational complexity of the adaptation methods and to improve their performance while maintaining the ease of interpretability of the obtained results. On the other hand, it would be also very interesting to extend the proposed framework to the general transfer learning scenario where the source and target domains tasks are not necessarily the same. In this case, the feature selection algorithm would have to take into account the discriminative power of each source feature in the target domain. Solving this problem in an unsupervised setting is a very challenging task that would require an efficient feature expressiveness measure to be introduced. We believe that this future perspective would be of a great interest in many real-world applications, notably the health-care one, where the manual labeling of the produced MRI scans represents an important bottleneck due to its highly time-consuming nature.

References

1. Aljundi, R., Lehaire, J., Prost-Boucle, F., Rouvière, O., Lartizien, C.: Transfer learning for prostate cancer mapping based on multicentric MR imaging databases. In: Bhatia, K.K., Lombaert, H. (eds.) MLMMI 2015. LNCS, vol. 9487, pp. 74–82. Springer, Cham (2015). https://doi.org/10.1007/978-3-319-27929-9_8
2. Bay, H., Tuytelaars, T., Van Gool, L.: SURF: speeded up robust features. In: Leonardis, A., Bischof, H., Pinz, A. (eds.) ECCV 2006. LNCS, vol. 3951, pp. 404–417. Springer, Heidelberg (2006). https://doi.org/10.1007/11744023_32
3. Ben-David, S., Blitzer, J., Crammer, K., Kulesza, A., Pereira, F., Vaughan, J.: A theory of learning from different domains. Mach. Learn. **79**, 151–175 (2010)
4. Ben-David, S., Blitzer, J., Crammer, K., Pereira, F.: Analysis of representations for domain adaptation. In: NIPS, pp. 137–144 (2007)
5. Courty, N., Flamary, R., Tuia, D.: Domain adaptation with regularized optimal transport. In: Calders, T., Esposito, F., Hüllermeier, E., Meo, R. (eds.) ECML PKDD 2014. LNCS (LNAI), vol. 8724, pp. 274–289. Springer, Heidelberg (2014). https://doi.org/10.1007/978-3-662-44848-9_18
6. Cuturi, M.: Sinkhorn distances: lightspeed computation of optimal transport. In: NIPS, pp. 2292–2300 (2013)
7. Fernando, B., Habrard, A., Sebban, M., Tuytelaars, T.: Unsupervised visual domain adaptation using subspace alignment. In: ICCV, pp. 2960–2967 (2013)
8. Gopalan, R., Li, R., Chellappa, R.: Domain adaptation for object recognition: an unsupervised approach. In: ICCV, pp. 999–1006 (2011)
9. Gretton, A., Borgwardt, K.M., Rasch, M.J., Schölkopf, B., Smola, A.: A kernel two-sample test. J. Mach. Learn. Res. **13**, 723–773 (2012)
10. Guyon, I., Elisseeff, A.: An introduction to variable and feature selection. J. Mach. Learn. Res. **3**, 1157–1182 (2003)
11. Jia, Y., et al.: Caffe: convolutional architecture for fast feature embedding. In: International Conference on Multimedia, pp. 675–678 (2014)
12. Knight, P.A.: The sinkhorn-knopp algorithm: convergence and applications. SIAM J. Matrix Anal. Appl. **30**(1), 261–275 (2008)
13. Krizhevsky, A., Sutskever, I., Hinton, G.: Imagenet classification with deep convolutional neural networks. In: NIPS, pp. 1097–1105 (2012)
14. Li, J., Zhao, J., Lu, K.: Joint feature selection and structure preservation for domain adaptation. In: IJCAI, pp. 1697–1703 (2016)
15. Niaf, E., Rouvière, O., Mège-Lechevallier, F., Bratan, F., Lartizien, C.: Computer-aided diagnosis of prostate cancer in the peripheral zone using multiparametric MRI. Phys. Med. Biol. **57**(12), 3833–51 (2012)
16. Pan, S.J., Tsang, I.W., Kwok, J.T., Yang, Q.: Domain adaptation via transfer component analysis. IEEE Trans. Neural Networks **22**(2), 199–210 (2011)
17. Persello, C., Bruzzone, L.: Kernel-based domain-invariant feature selection in hyperspectral images for transfer learning. IEEE Trans. Geosci. Remote Sens. **54**(5), 2615–2626 (2016)
18. Redko, I., Habrard, A., Sebban, M.: Theoretical analysis of domain adaptation with optimal transport. In: Ceci, M., Hollmén, J., Todorovski, L., Vens, C., Džeroski, S. (eds.) ECML PKDD 2017. LNCS (LNAI), vol. 10535, pp. 737–753. Springer, Cham (2017). https://doi.org/10.1007/978-3-319-71246-8_45
19. Saenko, K., Kulis, B., Fritz, M., Darrell, T.: Adapting visual category models to new domains. In: Daniilidis, K., Maragos, P., Paragios, N. (eds.) ECCV 2010. LNCS, vol. 6314, pp. 213–226. Springer, Heidelberg (2010). https://doi.org/10.1007/978-3-642-15561-1_16

20. Sun, B., Feng, J., Saenko, K.: Return of frustratingly easy domain adaptation. In: AAAI, p. 8 (2016)
21. Szegedy, C., et al.: Going deeper with convolutions. In: CVPR, pp. 1–9 (2015)
22. Talagrand, M.: Concentration of measure and isoperimetric inequalities in product spaces. Publications Mathématiques de l' I.H.E.S. **81**, 73–205 (1995)
23. Uguroglu, S., Carbonell, J.: Feature selection for transfer learning. In: Gunopulos, D., Hofmann, T., Malerba, D., Vazirgiannis, M. (eds.) ECML PKDD 2011. LNCS (LNAI), vol. 6913, pp. 430–442. Springer, Heidelberg (2011). https://doi.org/10.1007/978-3-642-23808-6_28
24. Villani, C.: Optimal Transport: Old and New, vol. 338. Springer Science & Business Media, Heidelberg (2008)
25. Yin, Z., Wang, Y., Liu, L., Zhang, W., Zhang, J.: Cross-subject EEG feature selection for emotion recognition using transfer recursive feature elimination. Frontiers Neurorobotics **11**, 19 (2017)

Web-Induced Heterogeneous Transfer Learning with Sample Selection

Sanatan Sukhija[✉] and Narayanan C. Krishnan

Indian Institute of Technology Ropar, Punjab 140001, India
{sanatan,ckn}@iitrpr.ac.in

Abstract. Transfer learning algorithms utilize knowledge from a data-rich source domain to learn a model in the target domain where labeled data is scarce. This paper presents a novel solution for the challenging and interesting problem of Heterogeneous Transfer Learning (HTL) where the source and target task have heterogeneous feature and label spaces. Contrary to common space based HTL algorithms, the proposed HTL algorithm adapts source data for the target task. The correspondence required for aligning the heterogeneous features of the source and target domain is obtained through labels across two domains that are semantically aligned using web-induced knowledge. The experimental results suggest that the proposed algorithm performs significantly better than state-of-the-art transfer approaches on three diverse real-world transfer tasks.

Keywords: Heterogeneous Transfer Learning · Sample Selection

1 Introduction

Traditional supervised algorithms require sufficient labeled data to learn a computational model with a reasonable generalization to unseen examples. However, for many real-world problems, collecting labeled data is often very expensive and cumbersome. Transfer learning approaches utilise knowledge from an auxiliary domain with abundant labeled data (source domain) to perform tasks in domains with scarce labeled data (target domain). HTL [35] algorithms transfer knowledge from one domain to the other when the two domains have different features. Due to the heterogeneous feature spaces, the first task of any HTL algorithm is to decide a "common" space for adaptation. The second task is to bridge the gap between the data differences that arise when the data from both the domains is projected onto the common space. This is generally achieved by leveraging some pivotal information that is shared among the domains. These pivots could be in the form of instance correspondences [39], overlapping features [13], shared label space [16,28,33], common meta-features/latent space [10,18,20,38] or any task specific/independent information [5,6].

Latent Space Transformation (LST) approaches to HTL project the data from both the domains onto a shared subspace for adaptation, thus learning

© Springer Nature Switzerland AG 2019
M. Berlingerio et al. (Eds.): ECML PKDD 2018, LNAI 11052, pp. 777–793, 2019.
https://doi.org/10.1007/978-3-030-10928-8_46

two transformations, one each for the source and target domain. On the other hand, Feature Space Remapping (FSR) approaches consider the common space as either of the two domains and determine a single transformation to transform data from the source domain to the target domain or vice-versa. The recent state-of-the-art HTL approaches leverage the common label space either to determine the cross-domain correspondences for learning the transformation(s) [16,28,33] or formulate a solution for obtaining the transformations as a minimization objective [17,26,27,31,37]. However, these approaches are not directly applicable for knowledge transfer between domains with heterogeneous label spaces.

We propose a novel FSR algorithm (refer [1,23] for our preliminary work) that works even when there are no shared features and instance correspondences between the source and target domain. It utilises the label space dependencies estimated through Normalised Google Distance to co-align the data from the two domains in the target space while preserving the original structure of the source data. Being a FSR framework, the proposed approach overcomes the need to determine an optimal shared subspace as compared to LST approaches and unlike [17,27], does not suffer from out-of-sample extension problem. The approach also utilises source instances whose labels are absent in the target domain, by encoding the absent labels using the inter-label relationships to the target labels. Along with inter-label dependencies across the heterogeneous label spaces, the approach also utilises intra-label relationships among target labels to learn a robust target model.

1.1   Problem Definition

Let $S \in \mathbb{R}^{n_S \times d_S}$ and $T \in \mathbb{R}^{n_T \times d_T}$ be the source and target domain data respectively where n_S and n_T represent the number of labeled data points in each domain respectively and $n_S \gg n_T$. The number of features in the source and target domain are denoted by d_S and d_T respectively. The features in the two domains are different and $d_S \neq d_T$. x^S denotes a labeled source instance with y^S as the associated label. Similarly, x^T is a labeled target instance with y^T as its label. The source and the target label space may or may not be overlapping. However, we assume that there exists semantic relationships within and across the label spaces. Let the number of unique labels in the source and target domain be L_S and L_T respectively. The goal of the proposed approach is to learn relevant source data points $B_S \in \mathbb{R}^{n_S \times d_T}$ that adapt well to the target task. The set of relevant source data is used along with the limited target data $\{x_i^T, y_i^T\}_{i=1}^{n_T}$ to learn the model for the target task.

2   Related Work

There have been many approaches for transfer learning, some of which have been extended for heterogenerous transfer learning. Manifold alignment based LST approaches [21,24,31,34] can be viewed as constrained dimensionality reduction frameworks that intend to find a low-dimensional embedding for multiple domains where the geometric structure of the original domains is preserved. These appraoches assume that the heterogeneous source and target domain share a smooth low-dimensional manifold (subspace). However, such a strong manifold assumption may not hold good for real-world heterogeneous transfer tasks, especially for datasets with high-dimensional features [37]. Supervised Heterogeneous Feature Augmentation (HFA) [17] is a SVM-based LST optimisation framework that uses a common augmented feature space for adaptation. Since HFA does not return the transformation matrices explicitly, it suffers from the out-of-sample extension problem. Subspace Co-Projection (SCP) [37] is a semi-supervised LST optimisation framework that learns the model weights in the projected subspace simultaneously with the transformations. The closed form solution of SCP requires large matrix inversions for high-dimensional datasets. Co-regularised Heterogeneous Transfer Learning (Co-HTL) [8] is a supervised LST approach that jointly aligns the data from the domains in the shared subspace. A common limitation of these LST approaches is that they require determining the optimal subspace by performing a grid-search on the dimension of the shared subspace (d).

In contrast, the FSR approaches directly map the features across the domains. However, learning the direct transformation involves estimation of a larger set of parameters in comparison with LST approaches. Sparse Heterogeneous Feature Remapping (SHFR) [16] leverages the common labels across the domains encoded using error correcting output codes (ECOC) as pivots to generate cross-domain correspondences. Supervised Heterogeneous Domain Adaptation using Random Forests (SHDA-RF) [33] relies on common label distributions that are obtained from leaf nodes of decision trees trained on labeled source and target domain data as the pivots. Both SHFR and SHDA-RF rely on a common set of labels between source and target domain to estimate correspondences and hence, cannot be directly applied to bridge two domains with heterogeneous label spaces.

The proposed FSR optimization framework overcomes the limitations of the above-mentioned approaches. It bridges the domains with heterogeneous feature and label spaces in a generic setting without relying on instance or feature correspondences. Even if there exists very few labeled instances in the target domain, the proposed algorithm is effective for transferring knowledge as asserted by the experimental results.

3   Proposed Methodology

Unlike LST approaches that have an inherent limitation of determining the optimal subspace for transfer, the proposed HTL algorithm, Web-Induced Heterogeneous Transfer Learning with Sample Selection (WIHTLSS), is conceived as a FSR minimisation objective with the goal of constructively utilising data from the source and target domains to learn a robust target model. Given the source domain data S and target domain data T, the proposed objective (Eq. 1) iteratively minimises the overall loss incurred by jointly aligning the data of the source and target domain in the target space while learning the transformed source data B_S and the transformation $P \in \mathbb{R}^{d_S \times d_T}$ that links the heterogeneous features of the source and target domain.

$$\min_{B_S,P} L(S, B_S, P) + \beta D(B_S, T) + \kappa G(B_S, T, m) + \lambda R(P, B_S) \qquad (1)$$

The first term, L, in Eq. 1 preserves the original structure in the transformed source data, while the second and the third term, D and G, align the transformed source data closer to the target data distribution. These two terms together determine the extent of alignment between transformed source and target instances. Excluding D and G will not adapt the transformed source to the target, and excluding L will overly bias the transformed source instances towards the limited labeled target instances, thus not generalising across the target. There is a trade-off between leveraging the relatedness to the source domain and the extent to which we want to adapt the transformed source data to the target task. This tradeoff is regulated by the variables β and κ. The last term, R, is the regulariser that prevents overfitting.

We define L in terms of the reconstruction error (Eq. 2), which measures the extent to which the structure of the original source data is preserved in the target domain. The reconstruction error computes the loss incurred due to projection of source data, i.e., the difference between the original domain data and the transformed data being remapped to the original space.

$$L(S, B_S, P) = \| S - B_S P' \|^2 \qquad (2)$$

Here, $P' \in \mathbb{R}^{d_T \times d_S}$ denotes the transpose of the transformation P. The reconstruction error takes advantage of the relatedness of the source and target domains to fill in the void in the target space with missing label data. An added benefit of using reconstruction error is that unlabeled data can be used to induce regularization which in turn helps to learn a more robust transformation.

The second and third term (D and G) that measure the mis-alignment between transformed source and target instances are defined in terms of weighted pairwise distances between the transformed source and target instances (presented in Eq. 3).

$$D(B_S, T) = \sum_{i=1}^{n_S} \sum_{j=1}^{n_T} \| B_{S_i} - x_j^T \|^2 W_{ij}$$

$$G(B_S, T, m) = \sum_{i=1}^{n_S} \sum_{j=1}^{n_T} \max(0, m - \| B_{S_i} - x_j^T \|^2)(1 - W_{ij}) \qquad (3)$$

Inspired by the contrastive loss [29], the second term D constrains the transformed source instances to be closer to the target instances with the same or related labels whereas the third term G ensures that dissimilar transformed source instances are pushed apart and kept at a minimum distance m. The similarity between the labels of the ith transformed source instance and the jth labeled target instance (denoted as W_{ij}) is assigned as the weight for aligning them.

As the domains have heterogeneous labels, weighting the distance based on the relatedness of the labels allows the use of source data tagged with related labels, when the exact target label is absent in the source domain. We define the encoding of the source labels in terms of target labels using the Normalised Google Distance (NGD) [7]. Equation 4 defines the NGD between two search keywords y_1 and y_2, where $f()$ denotes the number of web page hits returned by the Google search engine and N is the number of pages indexed by Google multiplied by the average number of singleton search keywords on those pages.

$$NGD(y_1, y_2) = \frac{\max\{\log f(y_1), \log f(y_2)\} - \log f(y_1, y_2)}{\log N - \min\{\log f(y_1), \log f(y_2)\}} \tag{4}$$

Since NGD is a dissimilarity measure that returns the value between $[0, \infty)$, we standardize it to a similarity measure W that outputs a value between $[0,1]$ (Eq. 5).

$$W_{ij} = \left(1 - \frac{NGD_{ij}}{Z}\right) \tag{5}$$

where Z denotes the maximum NGD score obtained from the labels across the domains. Using the similarity matrix W, we induce semantic co-alignment in our framework by exploiting the inter-label space similarities.

While minimising the inter-domain differences using the label information, there is a significant risk of over-fitting on the limited target training data. Hence, we adopt an explicit regulariser $R(.,.)$ in the objective function to penalise overfitting as depicted in Eq. 6.

$$R(P, B_S) = \| B_S \|^2 + \| P \|^2 \tag{6}$$

The overall objective $J()$ is given in Eq. 7.

$$J(.) = \min_{B_S, P} \| S - B_S P' \|^2 + \beta(\sum_{i=1}^{n_S} \sum_{j=1}^{n_T} W_{ij} \| B_{S_i} - x_j^T \|^2) +$$
$$\kappa(\sum_{i=1}^{n_S} \sum_{j=1}^{n_T} (1 - W_{ij}) \max(0, m - \| B_{S_i} - x_j^T \|^2)) + \lambda(\| B_S \|^2 + \| P \|^2) \tag{7}$$

The proposed optimisation problem is not jointly convex with respect to the two variables B_S and P. However, it is convex with respect to any one of them while the other has been fixed. Consequently, we utilise an alternating algorithm for solving the unconstrained optimisation (similar to the E-M process), by iteratively fixing one variable to estimate the remaining one until convergence. The proposed unconstrained optimization problem can be easily solved by using methods that are gradient-based or hessian-based. In contrast to gradient-based methods, hessian-based methods such as quasi-newton have a faster rate of convergence but demand extensive memory ($O(d_S.d_T)^2$) when dealing with high-dimensional datasets. The cross-lingual text transfer tasks involve high dimensional datasets where as the cross-domain activity recognition tasks have a significantly smaller dimension size. Hence, we use Conjugate Gradient Descent (CGD) method [9] (gradient-based) to solve the cross-lingual transfer tasks and quasi-newton method (hessian-based) for the cross-domain activity recognition tasks. The optimisation routine comprises of two alternating steps.

Step 1: Fix B_S, and solve for P, using the gradient update for P.

$$\nabla_P(J) = 2(-S'B_S + PB'_S B_S + \lambda P) \tag{8}$$

After updating P, the next step involves solving for B_S.

Step 2: Use P from Step 1 and update the value of B_S.

$$\nabla_{B_S}(J) = 2(-SP - B_S P'P + \beta(W_S B_S - WT) \\ -\kappa(W_D B_S - (1-W)T) + \lambda B_S) \tag{9}$$

Here, $(W_S)_{ii} = \sum_{j=1}^{n_T} W_{ij}$ and $(W_D)_{ii} = \sum_{j=1}^{n_T}(1 - W_{ij})$ where $W_S, W_D \in [0,1]^{n_S \times n_S}$. The two alternating steps of the proposed algorithm are repeated iteratively until convergence. These steps are summarized in Algorithm 1.

Algorithm 1. Web-Induced Heterogeneous Transfer Learning with Sample Selection (WIHTLSS)

Inputs: $S \in \mathbb{R}^{n_S \times d_S}$ and $T \in \mathbb{R}^{n_T \times d_T}$
Output: $B_S \in \mathbb{R}^{n_S \times d_T}, P \in \mathbb{R}^{d_S \times d_T}$
  (1) Randomly initialize P and B_S.
  repeat
     (2) Fix B_S. Find optimal P using quasi-newton or Conjugate Gradient Descent (CGD).
     (3) Use P obtained from the previous step and then find optimal B_S by using CGD or quasi-newton.
  until P and B_S are convergent

The existing HTL algorithms assume that the knowledge is transferred to a different but related target domain [25]. As the feature spaces are disparate, quantifying the extent of this relatedness apriori is a hard problem. In practice, the source and the target domain data can be very distant for real-world tasks and furthermore, only a handful of source data might be relevant to the target task. Since the proposed optimization objective tries to bring the similar instances closer while keeping the dissimilar ones away, we define a transformed source instance to be relevant if it is in the close proximity of at least one target instance. We select only the k-nearest transformed source instances for every target instance to train the final model. This helps to avoid negative transfer to a certain extent at the instance-level by selecting the relevant source instances for the target task. The optimal value of k is obtained through experimentation on a validation set. The extra computation cost of finding the nearest neighbors $O(n_S.n_T.d_T)$ is negligible as compared to the overall cost of the optimization routine.

3.1   Merging Heterogeneous Label Spaces

As the label spaces are heterogeneous, we have to link the label spaces first before using the relevant transformed source data and the target data T to train the final model. Given L_T unique labels in the target domain, we encode the label y of a training instance with a vector of size L_T. The ith entry of the encoded vector is the semantic similarity, computed via NGD, between y and the ith target label. This encoding encapsulates the inter-label space similarities of the source and target domain along with intra-label space similarities in the target domain. The associations between the labels within and across the domains enrich the transformation by providing additional discriminating information to learn a robust model. Replacing the label y with its encoding for every labeled instance $(x, y) \in (B_S \cup T)$ results in a multi-output regression problem. Independent single output regressors ignore relationships between the outputs. These are captured through stacked regressors [19]. Stacked regression is a 2-stage process that feeds the predictions made in the first stage as an augmented input for the second stage. For an unseen target instance, the target label encoding that is closest to the predicted values (computed via Euclidean distance) is chosen as the predicted label.

4   Experiments

The performance of WIHTLSS is compared against the following baseline and state-of-the-art transfer approaches:

- **Random Forest** (BRF) [22]: We train a random forest using 100 decision trees on only target training data where each tree is learned using $\sqrt{d_T}$ features [32]. Since the labeled target data is very limited, instance bagging is not used for tree construction.
- **Support Vector Machine with Error Correcting Output Codes** (SVM-ECOC): In our experiments, we use linear kernel and Error Correcting Output Codes (ECOC) on the target data to get the SVM baseline results. We obtain the optimal value of the box-penalty parameter by experimentation on the validation set.
- **Sparse Heterogeneous Feature Remapping** (SHFR) [16]: SHFR-ECOC is a FSR approach based on SVM-ECOC that utilizes common labels as pivots to generate cross-domain correspondences in the form of SVM weight vectors. A linear and sparse mapping is learned by employing Least Absolute Shrinkage and Selection Operator (LASSO) [30] on these correspondences. We perform a grid search to get the optimal box constraint parameter and the optimal length of the Error Correcting Output Codes using a validation set (for every fold). Since SHFR-ECOC reuses the source SVM model to evaluate the transfer performance, to ensure a fair comparison against WIHTLSS, we modify the original approach to SHFR-RF, where we train a random forest model on the transformed source data along with the target data.

- **Supervised Heterogeneous Domain Adaptation using Random Forests** (SHDA-RF) [33]: Similar to SHFR, SHDA-RF is another supervised FSR transfer approach that leverages the shared label space to generate cross-domain correspondences. After estimating the cross-domain correspondences from the random forest models trained on the source and target data, we fine tune the regulariser λ and report the best results over 16-folds. Both SHDA and SHFR estimate corresponding data by leveraging the common labels across the domains. For tasks with heterogeneous outputs, we vary the label-similarity threshold ($W_{ij} \in [0.3, 1]$) to generate the correspondences.
- **Domain Adaptation using Manifold Alignment** (DAMA) [31]: DAMA is a LST approach that learns a low-dimensional embedding where the projected data from the source and target domain is aligned using labels while preserving the original structure. We vary the hyper-parameter β to capture the trade-off between preserving the original topology and label-induced alignment in the shared subspace to get the best results. NGD is used to calculate the similarity of the labels across the domains (instead of 0/1 value used in the paper). The optimal size of the shared subspace was determined through experimentation on the validation set.
- **Co-regularised Heterogeneous Transfer Learning** (Co-HTL) [8]: Similar to DAMA, Co-HTL is another LST approach that also leverages label relationships to co-align source and target data in a common space. Here too, NGD is used to calculate the similarity between the labels of the domains (instead of the using the divergence between the label-embeddings obtained from word2vec [15] model or the divergence between topic distribution vectors obtained via Latent Dirichlet Allocation [14]). The optimal size of the shared subspace was determined through experimentation on the validation set.

We consider three heterogeneous transfer tasks to compare the performance of WIHTLSS against the aforementioned algorithms.

- **Cross-lingual Text Transfer**: We evaluate the cross-lingual transfer performance on two benchmark text datasets namely, (1) Amazon Cross-Lingual Sentiment (Amazon CLS) dataset[1] [11] and (2) Reuters Multi-Lingual dataset[2] [36]. We utilise the same experimental setting as mentioned by Zhou et al. [16] to test the performance of the proposed algorithm on cross-lingual sentiment and text classification tasks.
- **Cross-domain Activity Recognition**: We picked three single-resident CASAS[3] [12] datasets for this transfer task. We follow the same procedure as mentioned by Sukhija et al. [33] to construct the smart home datasets from raw sensor records. The labeled data from one smart home is treated as the source domain and limited labeled sensor data in another serves as the target domain. We employ the same experimental setting as mentioned by Sukhija et al. to evaluate six cross-domain activity recognition tasks.

---

[1] http://www.uni-weimar.de/en/media/chairs/webis/corpora/corpus-webis-cls-10/.
[2] https://archive.ics.uci.edu/ml/machine-learning-databases/00259/.
[3] http://ailab.wsu.edu/casas/datasets/.

– **Deep Representation Transfer**: In this experiment, the CIFAR-100[4] [3] image representations obtained from the last fully connected layer (4096 features) of the VGG-19 model [2] (pretrained on the Imagenet dataset[5]) act as the target domain whereas the related Imagenet[6] [4] image representations acquired from the last fully connected layer (4096 features) of the pre-trained VGG-16 model[7] act as the source domain. We perform dimensionality reduction using Principal Component Analysis (PCA) [39] to preserve 60% variance on the obtained source and target image representations. In every fold, for a source-target pair, we pick 100 images per label to form the source domain data and 5 images per label as the target training data. The optimal value of the hyper-parameters is determined using a disjoint validation set (100 images per label). We report the classification error on the test set[8] consisting of 100 images per label. We repeat this process over 16 different folds and report the mean error and standard deviation on the following (Source→Target) transfer tasks:

1. Imagenet(321–326, 118–121, 125) → CIFAR-100(Butterfly-14, Crab-26)
2. Imagenet(52–68, 33–37) → CIFAR-100(Snake-78, Turtle-93)

The numbers within the brackets for the Imagenet(.) dataset are the label identities[9] corresponding to the human readable labels.

5   Results and Discussion

We report and analyse the performance of several state-of-the-art transfer algorithms against WIHTLSS in three HTL scenarios, namely (1) when the source and target domain are characterised by heterogeneous feature spaces but share the same label space, (2) with heterogeneous feature spaces but an overlapping label space and (3) when the labels across the domains are also heterogeneous.

The cross-lingual transfer experiments belong to the first scenario where the vocabulary differences in the source and target documents lead to heterogeneous feature spaces. However, the same set of labels are shared between the domains. The cross-domain activity recognition experiments correspond to the second HTL scenario due to an overlapping set of activity labels across the source and target smart homes. However, the daily activities of different smart home residents lead to semantically related label spaces. The deep representation transfer tasks belong to the third scenario. With respect to a binary target task (distinguishing 2 classes from the CIFAR-100 dataset), we pick images with distinct but related labels as the source (related images from the Imagenet) to analyse the impact of leveraging semantic label relationships determined through NGD to align heterogeneous representations obtained from deep models.

---

[4] https://www.cs.toronto.edu/~kriz/cifar.html.
[5] https://keras.io/applications/#vgg19.
[6] http://www.image-net.org/.
[7] https://keras.io/applications/#vgg16.
[8] https://www.cs.toronto.edu/~kriz/cifar-100-matlab.tar.gz.
[9] https://gist.github.com/yrevar/942d3a0ac09ec9e5eb3a.

Table 1. Performance comparison of state-of-the-art algorithms on cross-lingual tasks, cross-domain activity recognition tasks and deep representation transfer tasks is shown in terms of mean error and standard deviation (%) over 16 folds. The best performance has been highlighted in bold and statistically significant WIHTLSS results against the second best performing algorithm are indicated by * respectively.

Amazon CLS dataset							
S→T	BRF	SVM-ECOC	DAMA	SHDA-RF	SHFR-RF	Co-HTL	WIHTLSS
English→French	43.90±2.34	51.83 ± 3.59	43.28 ± 1.39	37.01 ± 3.41	38.59 ± 3.57	34.18 ± 3.65	**32.05 ± 2.67***
English→German	44.21 ± 2.71	50.96 ± 4.01	42.82 ± 1.26	33.09 ± 3.93	37.51 ± 2.25	34.45 ± 2.48	**30.55 ± 2.27***
English→Japanese	49.0 ± 3.47	52.91 ± 4.60	47.75 ± 1.49	34.58 ± 4.89	37.79 ± 3.7	35.08 ± 2.71	**31.65 ± 2.98***
Reuters Multilingual Dataset							
S→T	BRF	SVM-ECOC	DAMA	SHDA-RF	SHFR-RF	Co-HTL	WIHTLSS
English→Spanish	32.13 ± 3.59	36.44 ± 4.30	31.84 ± 1.36	25.62 ± 1.15	27.93 ± 1.03	25.34 ± 1.19	**22.96 ± 1.10***
French→Spanish			30.48 ± 1.48	24.61 ± 0.85	26.52 ± 0.90	25.20 ± 1.0	**22.47 ± 1.23***
German→Spanish			30.73 ± 1.4	25.1±0.95	27.01 ± 1.23	24.54 ± 1.03	**22.32 ± 1.08***
Italian→Spanish			31.59 ± 1.82	23.57 ± 1.35	24.86 ± 1.47	22.82 ± 1.28	**21.03 ± 1.46***
CASAS horizon house (hh) datasets							
S→T	BRF	SVM-ECOC	DAMA	SHDA-RF	SHFR-RF	Co-HTL	WIHTLSS
hh102→hh113	29.67 ± 2.51	34.58 ± 2.20	31.98 ± 2.93	26.37 ± 2.23	27.98 ± 2.51	27.75 ± 2.55	**23.71 ± 2.36***
hh102→hh118	36.41 ± 2.42	43.51 ± 3.01	37.72 ± 2.52	31.50 ± 2.01	32.01 ± 2.09	30.54 ± 2.76	**26.85 ± 2.40***
hh113→hh102	36.70 ± 1.95	41.23 ± 2.93	36.21 ± 2.28	29.88 ± 1.76	32.81 ± 2.38	31.98 ± 2.44	**27.02 ± 2.11***
hh113→hh118	32.35 ± 2.56	39.41 ± 1.89	34.49 ± 2.42	28.0 ± 2.02	30.15 ± 2.44	29.33 ± 2.01	**25.75 ± 1.97***
hh118→hh102	38.95 ± 2.57	41.80 ± 2.21	39.55 ± 2.59	33.02 ± 2.13	35.40 ± 2.35	36.0 ± 2.45	**30.97 ± 2.18***
hh118→hh113	31.01 ± 1.80	34.73 ± 3.39	31.81 ± 2.09	27.73 ± 1.08	30.46 ± 2.95	30.81 ± 2.39	**26.45 ± 2.2***
Imagenet (VGG-16)→CIFAR-100(VGG-19)							
Target Task	BRF	SVM-ECOC	DAMA	SHDA-RF	SHFR-RF	Co-HTL	WIHTLSS
Butterfly v/s Crab	31.16 ± 5.61	27.78 ± 6.61	30.05 ± 2.95	24.71 ± 3.40	25.20 ± 3.28	26.09 ± 3.84	**22.05 ± 2.26***
Snake v/s Turtle	21.81 ± 8.27	18.53 ± 6.14	20.62 ± 4.34	15.92 ± 3.67	16.48 ± 3.53	16.80 ± 4.23	**13.24 ± 2.15***

The reason for learning a linear transformation stems from the characteristics of these transfer tasks. As we are trying to map bag-of-words representation for cross-lingual transfer tasks, the relationship between words (that are synonymous) within a language and across different languages tends to be linear [16]. Similarly for cross-domain activity recognition tasks, the feature values of the sensors (in a functional area such as kitchen, bedroom etc.) across different smart homes appear to be linearly related.

The baseline and transfer results on all transfer tasks are shown in Table 1. Since the baseline random forest (BRF) performed significantly better than classical linear SVM on most transfer tasks, we kept random forest as the final model for all transfer approaches. For cross-lingual transfer tasks on Amazon CLS dataset and Reuters Multilingual dataset, the notable performance improvement of the transfer approaches over the baseline BRF indicates the feasibility of cross-lingual knowledge transfer. It can be observed that the proposed framework WIHTLSS significantly outperforms (p-value < 0.05) SHFR-RF by 3.5–7%, SHDA-RF by 2.5–3%, DAMA by 7–15% and Co-HTL by 1.5–3.5% in every cross-lingual transfer setting.

The experimental results on CASAS datasets are also depicted in Table 1. It can be seen that WIHTLSS outperforms all the other approaches on the CASAS

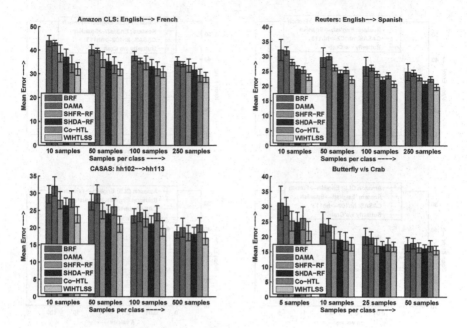

Fig. 1. Impact of the quantity of available labeled data in the target domain on the performance of baseline and transfer approaches.

datasets. Also, the performance of feature space remapping (FSR) approaches namely SHDA-RF and SHFR-RF is significantly better than DAMA. Among the FSR approaches, WIHTLSS outperforms the best performing algorithms, SHDA-RF and Co-HTL, by 3% on average (p-value < 0.05). Even for the deep representation transfer tasks, the proposed algorithm significantly outperforms (p-value < 0.05) all the baseline and transfer approaches.

We investigate the impact of having more labeled data in the target domain on the transfer performance of WIHTLSS. Due to lack of space, we show the results on a subset of the transfer tasks. The experimental results in Fig. 1 illustrate that having more labeled data in the target domain yields better transfer performance for all approaches. It can be observed that the transfer improvement is significant even when there are just 10 or 50 samples per label in the target domain. However, the utility of transfer reduces beyond certain point at which there are sufficient labeled instances in the target domain to directly learn a model. Similar trend was observed for other tasks as well.

5.1   Impact of the Hyper-parameters

We first perform a grid search on the hyper-parameter β (while keeping $\lambda = 0, \kappa = 0$) to assess the effectiveness of label-similarity induced alignment on the transfer performance of WIHTLSS. From Fig. 2(a), it is evident that aligning data of the domains in the target space by leveraging semantic relationships

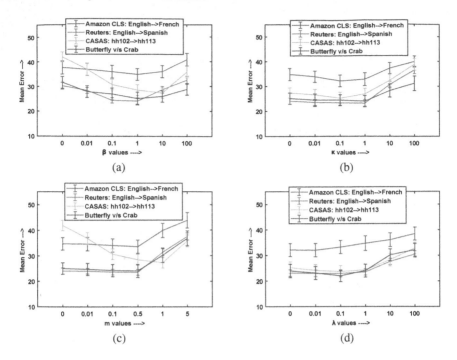

Fig. 2. (a) Impact of the hyper-parameters β (label-similarity induced semantic alignment), (b) κ (label-dissimilarity induced alignment), (c) λ (regulariser term) and (d) m (margin) on the performance of WIHTLSS.

between heterogeneous labels is beneficial. When $\beta = 0$, the performance is significantly poor in comparison to the baseline for all the transfer tasks. This suggests that only preserving the original structure of data in the target space does not guarantee a reasonable alignment where instances of related classes are closer than unrelated classes. We hypothesised to bring the instances with same or related labels closer to each other for learning a better transformation. It can be observed that there is substantial improvement in the transfer performance with increase in the importance of β. This indicates that leveraging the label similarities in the proposed formulation has a positive influence on the transfer performance of WIHTLSS. However, overemphasising the importance of label-space induced alignment has a detrimental effect on the transfer performance as the learned transformation becomes biased towards the limited labeled target data (which does not necessarily represent the true target distribution). Further, replacing NGD based similarity measure with divergence from the word2vec embeddings [15] did not result in significant change in the performance.

After obtaining optimal β, we vary κ to investigate the importance of the 3rd term G. For this experiment, we fix the margin $m = 1$, keep $\lambda = 0$ and use the optimal β for D. It can be observed from Fig. 2(b) that the transfer performance improves with increase in importance of G. This suggests that pushing dissimilar instances apart helps in learning a better transformation. However, over-regularization of this label-dissimilarity induced alignment worsens the transfer performance as it negatively affects the original structure. By fixing the optimal value of κ, we then determine the optimal margin (refer Fig. 2(c)). Our hypothesis here is that with low margin, dissimilar source instances will continue to remain close to the target instances resulting in poorer performance when compared to larger values of the margin. This can be observed in Fig. 2(c). The decrease in the performance for smaller values of margin is low for the cross-lingual and cross-domain activity recognition tasks, but significant in the deep representation transfer task. Similarly, beyond the optimal value of the margin, pushing away the dissimilar instances affects the distribution of the instances in the target space significantly, resulting in performance degradation. Finally, we fine tuned the regulariser term by varying the hyper-parameter λ. It can be observed from Fig. 2(d) that WIHTLSS performs the best, on average, when λ is set to 0.1.

Figure 3 depicts the impact of varying the number of nearest neighbors (k) on the transfer performance of WIHTLSS. It can be observed that the transfer performance improves with increase in the number of nearest neighbors. This result validates our hypothesis that the transformed source instances that were the nearest neighbors of any target instance are actually relevant for the target task. However, when we add more neighbors beyond a certain point, transformed source data points that are quite far also contribute towards learning the model. Apart from the deep transfer task (butterfly v/s crab), in the extreme case, when all the source instances are considered, it can be observed that the transfer performance suffers for all other tasks.

In our deep representation transfer experiment, for the binary target task (butterfly v/s crab), we used only related image representations as the source i.e. the source image representations only contain the encodings of different types of butterfly and crab. Hence, all the image representations of the source are actually relevant for the binary target task. This is in agreement with our observations as well (refer Fig. 3).

5.2 Convergence Analysis

The convergence results for the optimisation routines are shown in Fig. 4. With quasi-newton method for cross-domain activity recognition experiments, the optimisation error decreases gradually over iterations and convergence is achieved within 15–20 iterations (on average). However, with CGD algorithm on cross-lingual transfer experiments, the error drops sharply in the first few iterations and convergence is reached inside 5–10 iterations.

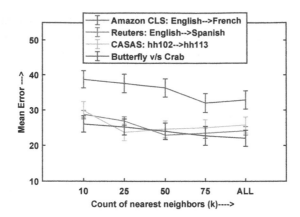

Fig. 3. Depicts the variation in the transfer performance (mean error and standard deviation) with increasing the nearest neighbors.

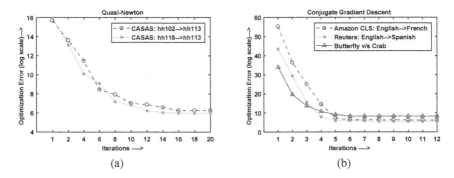

Fig. 4. Convergence with (a) quasi-newton algorithm on two cross-domain activity recognition tasks and (b) CGD on two cross-lingual transfer tasks and a deep representation transfer task

6   Summary and Future Work

WIHTLSS is an integration of collective inference and FSR based transfer learning approach that bridges domains with heterogeneous feature and label spaces without relying on instance or feature correspondences. Assuming some semantic relationships within and across the label spaces, it utilizes web knowledge to align the source data in the target space while preserving the original structure. It can be viewed as a simple version of dictionary learning without orthogonality or sparsity constraints but using L2 norm constraints on the dictionary and the new representation. The label-induced alignment terms are the modified Laplacian regularisation terms where the adjacency matrix is calculated based on the NGD. The experimental results on real-world HTL tasks with identical and different output spaces indicate the superiority of WIHTLSS over state-of-the-art supervised transfer approaches.

A limitation of the proposed approach is that it requires some amount of labeled data in the target domain. Consequently, if labeled data is absent in the target domain, annotating relatively small number of unlabeled data becomes an inescapable task. Besides leveraging the semantic label relationships to learn the transformation, one can investigate the possibility of leveraging unlabeled domain knowledge or the task associated characteristic properties to determine a quantifiable measure of the relatedness W. Further, the proposed algorithm utilises a single source domain for knowledge transfer, as part of future work, we wish to explore how to effectively combine data from multiple related modalities simultaneously to make an improved final prediction on the target.

Acknowledgment. This research is supported by the Department of Science and Technology, India under grant YSS/2015/001206, and by the Indian Institute of Technology Ropar under the ISIRD grant.

References

1. Sukhija, S.: Label space driven heterogeneous transfer learning with web induced alignment. In: Proceedings of the 32nd AAAI Conference on Artificial Intelligence (2018)
2. Simonyan, K., Zisserman, A.: Very deep convolutional networks for large-scale image recognition, arXiv preprint arXiv:1409.1556 (2014)
3. Krizhevsky, A., Hinton, G.: Learning multiple layers of features from tiny images, Citeseer (2009)
4. Deng, J., Dong, W., Socher, R., Li, L.-J., Li, K., Li, F.-F.: Imagenet: a large-scale hierarchical image database. In: IEEE Conference on Computer Vision and Pattern Recognition (CVPR) (2009)
5. Zhu,Y., et al.: Heterogeneous transfer learning for image classification. In: Proceedings of the AAAI National Conference on Artificial Intelligence (2011)
6. Prettenhofer, P., Stein, B.: Cross-language text classification using structural correspondence learning. In: Proceedings of the 48th Annual Meeting of the Association for Computational Linguistics (ACL) (2010)
7. Cilibrasi, R.L., Vitanyi, P.M.B.: The google similarity distance. In: IEEE Transactions on Knowledge and Data Engineering (2007)
8. Wei, Y., Zhu, Y., Leung, C.W.-k., Song, Y., Yang, Q.: Instilling social to physical: co-regularized heterogeneous transfer learning. In: Proceedings of the AAAI National Conference on Artificial Intelligence (2016)
9. Branch, M.A., Coleman, T.F., Li, Y.: A subspace, interior, and conjugate gradient method for large-scale bound-constrained minimization problems. SIAM J. Sci. Comput. **21**(1), 1–23 (1999)
10. Hu, D.H., Yang, Q.: Transfer learning for activity recognition via sensor mapping. In: Proceedings of the Twenty-Second International Joint Conference on Artificial Intelligence (IJCAI) (2011)
11. Lichman, M.: UCI Machine Learning Repository. University of California, Irvine, School of Information and Computer Sciences (2013). http://archive.ics.uci.edu/ml
12. Cook, D.J., Crandall, A.S., Thomas, B.L., Krishnan, N.C.: CASAS: a smart home in a box. Computer **46**(7), 62–69 (2013)

13. Pan, S.J., Yang, Q.: A survey on transfer learning. IEEE Trans. Knowl. Data Eng. **22**(10), 1345–1359 (2010)
14. Blei, D.M., Ng, A.Y., Jordan, M.I.: Latent dirichlet allocation. J. Mach. Learn. Res. **3**(Jan), 993–1022 (2003)
15. Mikolov, T., Sutskever, I., Chen, K., Corrado, G.S., Dean, J.: Distributed representations of words and phrases and their compositionality. In: Advances in Neural Information Processing Systems (2013)
16. Zhou, J.T., Tsang, I.W., Pan, S.J., Tan, M.: Heterogeneous domain adaptation for multiple classes. In: Proceedings of the International Conference on Artificial Intelligence and Statistics (2014)
17. Li, W., Duan, L., Xu, D., Tsang, I.W.: Learning with augmented features for supervised and semi-supervised heterogeneous domain adaptation. IEEE Trans. Pattern Anal. Mach. Intell. **36**(6), 1134–1148 (2014)
18. van Kasteren, T.L.M., Englebienne, G., Kröse, B.J.A.: Transferring knowledge of activity recognition across sensor networks. In: Floréen, P., Krüger, A., Spasojevic, M. (eds.) Pervasive 2010. LNCS, vol. 6030, pp. 283–300. Springer, Heidelberg (2010). https://doi.org/10.1007/978-3-642-12654-3_17
19. Borchani, H., Varando, G., Bielza, C., Larrañaga, P.: A survey on multi-output regression. Wiley Interdisc. Rev. Data Min. Knowl. Discovery **5**(5), 216–233 (2015)
20. He, J., Liu, Y., Yang, Q.: Linking heterogeneous input spaces with pivots for multitask learning. In: Proceedings of the SIAM International Conference on Data Mining (2014)
21. Wang, C., Mahadevan, S.: Manifold alignment using procrustes analysis. In: Proceedings of the International Conference on Machine learning (2008)
22. Breiman, L.: Random forests. Mach. Learn. **45**(1), 5–32 (2001)
23. Sukhija, S.: Label space driven feature space remapping. In: Proceedings of the ACM India Joint International Conference on Data Science and Management of Data (2018)
24. Wang, C., Mahadevan, S.: Manifold alignment without correspondence. In: Proceedings of the International Joint Conference on Artificial Intelligence (2009)
25. Rosenstein, M.T., Marx, Z., Kaelbling, L.P., Dietterich, T.G.: To transfer or not to transfer. In: NIPS 2005 Workshop on Transfer Learning (2005)
26. Hubert Tsai, Y.-H., Yeh, Y.R., Frank Wang, Y.-C.: Learning cross-domain landmarks for heterogeneous domain adaptation. In: Proceedings of the IEEE Conference on Computer Vision and Pattern Recognition (2016)
27. Yan, Y., et al.: Learning discriminative correlation subspace for heterogeneous domain adaptation. In: Proceedings of the 26th International Joint Conference on Artificial Intelligence (2017)
28. Sukhija, S., Krishnan, N.C., Kumar, D.: Supervised heterogeneous transfer learning using random forests. In: Proceedings of the ACM India Joint International Conference on Data Science and Management of Data (2018)
29. Hadsell, R., Chopra, S., LeCun, Y.: Dimensionality reduction by learning an invariant mapping. In: IEEE Computer Society Conference on Computer Vision and Pattern Recognition (CVPR) (2006)
30. Efron, B., Hastie, T., Johnstone, I., Tibshirani, R.: Least angle regression. Ann. Stat. **32**(2), 407–499 (2004)
31. Wang, C., Mahadevan, S.: Heterogeneous domain adaptation using manifold alignment. In: Proceedings of the International Joint Conference on Artificial Intelligence (2011)

32. Kyrillidis, A., Zouzias, A.: Non-uniform feature sampling for decision tree ensembles. In: Proceedings of the IEEE International Conference on Acoustics Speech and Signal Processing (2014)
33. Sukhija, S., Krishnan, N.C., Singh, G.: Supervised heterogeneous domain adaptation via random forests. In: Proceedings of the International Joint Conference on Artificial Intelligence (2016)
34. Wang, C., Mahadevan, S.: A general framework for manifold alignment. In: Proceedings of the AAAI Fall Symposium: Manifold Learning and Its Applications (2009)
35. Yang, Q., Chen, Y., Xue, G.-R., Dai, W., Yu, Y.: Heterogeneous transfer learning for image clustering via the social web. In: Proceedings of the International Joint Conference on Natural Language Processing (2009)
36. Amini, M., Usunier, N., Goutte, C.: Learning from multiple partially observed views-an application to multilingual text categorization. In: Advances in Neural Information Processing Systems (2009)
37. Xiao, M., Guo, Y.: Semi-supervised subspace co-projection for multi-class heterogeneous domain adaptation. In: Appice, A., Rodrigues, P.P., Santos Costa, V., Gama, J., Jorge, A., Soares, C. (eds.) ECML PKDD 2015. LNCS (LNAI), vol. 9285, pp. 525–540. Springer, Cham (2015). https://doi.org/10.1007/978-3-319-23525-7_32
38. Zhou, G., He, T., Wu, W., Hu, X.T.: Linking heterogeneous input features with pivots for domain adaptation. Proceedings of the 24th International Conference on Artificial Intelligence (IJCAI) (2015)
39. Zhou, J.T., Pan, S.J., Tsang, I.W., Yan, Y.: Hybrid heterogeneous transfer learning through deep learning. In: Proceedings of the AAAI National Conference on Artificial Intelligence (2014)

Towards More Reliable Transfer Learning

Zirui Wang[✉] and Jaime Carbonell

Language Technologies Institute, Carnegie Mellon University,
Pittsburgh, PA, USA
{ziruiw,jgc}@cs.cmu.edu

Abstract. Multi-source transfer learning has been proven effective when within-target labeled data is scarce. Previous work focuses primarily on exploiting domain similarities and assumes that source domains are richly or at least comparably labeled. While this strong assumption is never true in practice, this paper relaxes it and addresses challenges related to sources with diverse labeling volume and diverse reliability. The first challenge is combining domain similarity and source reliability by proposing a new transfer learning method that utilizes both source-target similarities and inter-source relationships. The second challenge involves pool-based active learning where the oracle is only available in source domains, resulting in an integrated active transfer learning framework that incorporates distribution matching and uncertainty sampling. Extensive experiments on synthetic and two real-world datasets clearly demonstrate the superiority of our proposed methods over several baselines including state-of-the-art transfer learning methods. Code related to this paper is available at: https://github.com/iedwardwangi/ReliableMSTL.

1 Introduction

Traditional supervised machine learning methods share the common assumption that training data and test data are drawn from the same underlying distribution. Typically, they also require sufficient labeled training instances to construct accurate models. In practice, however, one often can only obtain limited labeled training instances. Inspired by human beings' ability to transfer previously learned knowledge to a related task, *transfer learning* [23] addresses the challenge of data scarcity in the *target* domain by utilizing labeled data from other related *source* domain(s). Plenty of research has been done on the single-source setting [22,25,35], and it has been shown that transfer learning has a wide range of applications in areas such as sentiment analysis [3,21], computer vision [15], cross-lingual natural language processing [17], and urban computing [32].

Electronic supplementary material The online version of this chapter (https://doi.org/10.1007/978-3-030-10928-8_47) contains supplementary material, which is available to authorized users.

M. Berlingerio et al. (Eds.): ECML PKDD 2018, LNAI 11052, pp. 794–810, 2019.
https://doi.org/10.1007/978-3-030-10928-8_47

In recent years, new studies are contributing to a more realistic transfer learning setting with multiple source domains, both for classification problems [7,16,28,33] and regression problems [24,31]. These approaches try to capture the diverse *source-target* similarities, either by measuring the distribution differences between each source and the target or by re-weighting source instances, so that only the source knowledge likely to be relevant to the target domain is transferred. However, these works assume that all sources are equally reliable; in other words, all sources have labeled data of the same or comparable quantity and quality. Nonetheless, in real-world applications, no such assumption holds. For instance, recent sources typically contain less labeled data than long-established ones, and narrower sources contain less data than broader ones. Also some sources may contain more labeling noise than others. Therefore, it is common for source tasks to exhibit diverse reliabilities. However, to the best of our knowledge, not much work has been reported in the literature to compensate for source reliability divergences. Ideally, we thrive to find sources that are relevant and reliable at the same time, as a compromise in either aspect can hurt the performance. When this is infeasible, one has to include some less reliable sources and carefully weigh the trade-off. There are two reasons why including a source that is **not** richly labeled is desired. First, such a source may be very informative since it is closely related to the target domain. For example, in low-resource language processing, a language that is very similar to the target language is usually also relatively low-resource but it would still be used as an important source due to its proximity. In addition, while labeled data may be scarce, unlabeled data is often easier to obtain and one can acquire labels for unlabeled data from domain experts, assuming a budget is available via *active learning* [26].

Active learning addresses the problem of data scarcity differently than transfer learning by querying new labels from an oracle for the most informative unlabeled data. A fruitful line of prior work has developed various active learning strategies [14,18,20]. Recently, the idea of combining transfer learning and active learning has attracted increasing attention [6,13,30]. Most of existing work assumes that there is plenty of labeled data in source domain(s) and one can query labels for the target domain from an oracle. However, this assumption is not always true. Since transfer learning is mainly applied in target domains with restrictions on access to data, such as sensitive private domains or novel domains that lack human knowledge to generate sufficient labeled data, often there is little or **no** readily-available oracle in the target domain. For instance, in the influenza diagnosis task, transfer learning can be applied to assist diagnosing a new flu strain (target domain) by exploiting historical data of related flu strains (source domains), but there may be little expertise regarding the new strain. It is, however, often possible to query labels for a selected set of unlabeled data in source domains, as these may represent previous better-studied flu strains. Task similarity, corresponding to flu-strain diagnosis or treatment in this instance, may be established via commonalities in symptoms exhibited by patients with the new or old flu strains, or via analysis of the various influenza genomes. To the best of our knowledge, active learning for multi-source transfer learning has not been deeply studied.

In this paper, we focus on a novel research problem of transfer learning with multiple sources exhibiting diverse reliabilities and different relations to the target task. We study two related tasks. The first one is how to construct a robust model to effectively transfer knowledge from unevenly reliable sources, including those with less labeled data. The second task is how to apply active learning on multiple sources to stimulate better transfer, especially in such a scenario that sources have diverse quantities and qualities. Notice that these two tasks are related but can also be independent. For instance, in the low-resource language problem mentioned above, the first task is applicable but not the second one. This is because we may have an oracle for neither the source nor the target as they may both be of extremely limited resources. On the other hand, both tasks are relevant to the influenza example.

For the first task, inspired by [18], we propose a *peer-weighted multi-source transfer learning* method (*PW-MSTL*) that jointly measures source proximity and reliability. The algorithm utilizes *inter-source relationships*, which have been less studied previously, in two ways: (i) it associates the proximity coefficients based on distribution difference with the reliability coefficients measured using *peers*[1] to learn their target model relevance, and (ii) when a source is not confident in the prediction for a testing instance, it is allowed to query its peers. By doing so, the proposed method allows knowledge transfer among sources and effectively utilizes label-scarce sources. For the second task, we propose an active learning framework, called *adaptive multi-source active transfer* (*AMSAT*), that builds on the idea of Kernel Mean Matching (KMM) [8,11] and uncertainty sampling [29] to select unlabeled instances that are the most representative and avoid information redundancy. Experimental results, shown later, demonstrate that the combination of *PW-MSTL* and *AMSAT* significantly outperforms other competing methods. The proposed methods are generic and thus can be generalized to other learning tasks.

2   Related Work

Transfer Learning: Transfer learning aims at utilizing source domain knowledge to construct a model with low generalization error in the target domain. Transfer learning with multiple sources is challenging because of distribution discrepancies and complex inter-source dependencies. Most existing work focuses on how to capture the diverse domain similarities. Due to stability and capability of measuring fine-grained similarities explicitly, ensemble approaches are widely used. In [33], adaptive SVM (A-SVM) is proposed to adapt several trained source SVMs and learn model importance from meta-level features based on distributions. In contrast, [7] used Maximum Mean Discrepancy (MMD) [8] as source

---

[1] We define the peers of a source as the other sources weighted by inter-source relationships.

weights, while adding an additional manifold regularization based on a smoothness assumption of the target classifier. A more sophisticated two-stage weighting methodology based on distribution matching and conditional probability differences is presented in [28]. However, the valuable unlabeled data is not utilized in their method. Moreover, these methods assume that all sources are equally reliable and they have the limitation that inter-source relationships are ignored.

Multi-task Learning: Unlike multi-source transfer learning which focuses on finding a *single* hypothesis for the target task, multi-task learning [36] tries to improve performance on *all* tasks and finds a hypothesis for each task by adaptively leveraging related tasks. It is crucial to learn task relationships in multi-task learning problems, either used as priors [5] or learned adaptively [19]. While this is similar to the inter-source relationships we utilized in this paper, multi-task learning does not involve measuring the proximity between a source and the target nor the trade-off between proximity and reliability.

Active Learning: At the other end of the spectrum, active learning has been proven to be effective both empirically and theoretically. A new definition of sample complexity is proposed in [1] to show that active learning is strictly better than passive learning asymptotically. In recent years, there has been increasing interests in combining transfer learning and active learning to jointly solve the issue of insufficient labeled data. A popular representative of this family is the work by Chattopadhyay et al. [6] where the JO-TAL method is proposed. It is a unified framework that jointly performs active learning and transfer learning by matching Hilbert space embeddings of distributions. A key benefit of these methods is to address the cold start problem of active learning by introducing a version space prior learned from transfer learning [34]. However, unlike our approach, these methods all assume that source data are richly labeled and only perform active learning in the target domain.

3   Problem Formulation

For simplicity, we consider a binary classification problem for each domain, but the methods generalize to multi-class problems and are also applicable to regression tasks. We consider the following multi-source transfer learning setting that is common in real world when one tries to study a novel target domain with extremely limited resources. We denote by [N] consecutive integers ranging from 1 to N. Suppose we are given K auxiliary source domains, each contains both labeled and unlabeled data. Moreover, these sources typically have varying amount of labeled data (and thus with diverse reliabilities). Specifically, we denote by $S_k = S_k^L \cup S_k^U$ the dataset associated with the kth source domain. Here, $S_k^L = \{(x_i^{S_k^L}, y_i^{S_k^L})\}_{i=1}^{n_k^L}$ is the labeled set of size n_k^L, where $x_i^{S_k^L} \in \mathbb{R}^d$ is the ith labeled instance of the kth source and $y_i^{S_k^L}$ is its corresponding true label. Similarly, $S_k^U = \{x_i^{S_k^U}\}_{i=1}^{n_k^U}$ is the set of unlabeled data of size n_k^U. For the target domain, we assume that there is no labeled data available but we have sufficient unlabeled data, denoted as $T = \{x_i^T\}_{i=1}^{n_T}$ where instances are in the same feature space as sources. Furthermore, for the purpose of active learning, we assume that there is a uniform-cost oracle available only in the source domains, and the conditional probabilities $P(Y|X)$ are at least somewhat similar among both source and target domains.

4   Proposed Approach

4.1   Motivation

We first present the theoretical motivation for our methods. We analyze the expected target domain risk in the multi-source setting, making use of the theory of domain adaptation proved in [2,4].

Definition 1. *For a hypothesis space $\mathcal{H}$ for instance space $\mathcal{X}$, the symmetric difference hypothesis space $\mathcal{H}\Delta\mathcal{H}$ is the set of hypotheses defined as*

$$\mathcal{H}\Delta\mathcal{H} = \{h(x) \oplus h'(x) : h, h' \in \mathcal{H}\}, \tag{1}$$

where $\oplus$ is the XOR function. Then the $\mathcal{H}\Delta\mathcal{H}$-divergence between any two distributions D and D' is defined as

$$d_{\mathcal{H}\Delta\mathcal{H}}(D, D') \triangleq 2 \sup_{A \in \mathcal{A}_{\mathcal{H}\Delta\mathcal{H}}} \left| Pr_D[A] - Pr_{D'}[A] \right|, \tag{2}$$

where $\mathcal{A}_{\mathcal{H}\Delta\mathcal{H}}$ is the measurable set of subsets of $\mathcal{X}$ that are the support of some hypothesis in $\mathcal{H}\Delta\mathcal{H}$.

Now if we assume that all domains have the same amount of unlabeled data (which is a reasonable assumption since unlabeled data are usually cheap to obtain), i.e. $n_k^U = n_T = m'$, then we can show the following risk bound on the target domain.

Theorem 1. *Let $\mathcal{H}$ be a hypothesis space of VC-dimension d and $\hat{d}_{\mathcal{H}\Delta\mathcal{H}}(S_i^U, T)$ be the empirical distributional distance between the ith source and the target domain, induced by the symmetric difference hypothesis space. Then, for any $\mu \in (0, 1)$ and any $\hat{h} = \sum_i \alpha_i \hat{h}_i$ where $\hat{h}_i \in \mathcal{H}$ and $\sum \alpha_i = 1$, the following holds with probability at least $1 - \delta$,*

$$
\begin{aligned}
\epsilon_T(\hat{h}) \leq &\left[\sum_i^K \alpha_i \left[\mu \left[\hat{\epsilon}_{S_i}(\hat{h}_i) + \frac{1}{2}\hat{d}_{\mathcal{H}\Delta\mathcal{H}}(S_i^U, T) \right] \right.\right. \\
&\left.\left. + \frac{1-\mu}{K-1} \sum_{j \neq i}^K \left[\hat{\epsilon}_{S_j}(\hat{h}_i) + \frac{1}{2}\hat{d}_{\mathcal{H}\Delta\mathcal{H}}(S_j^U, T) \right] \right] \right. \\
&\left. + \left(\sum_i^K \alpha_i \sqrt{\frac{\mu^2}{\beta_i} + (\frac{1-\mu}{K-1})^2 \sum_{j \neq i} \frac{1}{\beta_j}} \right) \sqrt{\frac{d \log(2m) - \log(\delta)}{2m}} \right] \\
&+ 4\sqrt{\frac{2d \log(2m') + \log(\frac{4}{\delta})}{m'}} + \lambda_{\alpha,\mu},
\end{aligned}
\tag{3}
$$

where $\epsilon_*(h)$ is the expected risk of h in the corresponding domain, $m = \sum_i^K n_i^L$ is the sum of labeled sizes in all sources, $\beta_i = n_i^L/m$ is the ratio of labeled data in the ith source, and $\lambda_{\alpha,\mu}$ is the risk of the ideal multi-source hypothesis weighted by α and μ.

The proof can be found in the supplementary material. By introducing a concentration factor μ, we replace the size of labeled data for each source n_i^L with the total size m in the third line in Eq. (3), resulting in a tighter bound. Suppose that the hypothesis $\hat{h}_i$ is learned using data in the ith source, then the bound suggests to use peers to evaluate the reliability of $\hat{h}_i$ while the optimal weights should consider both proximity and reliability of the ith source. This inspired us to propose the following method.

4.2   Peer-Weighted Multi-source Transfer Learning

In this paper, we propose the idea of formulating all source domains jointly in a framework similar to the multi-task learning. Similar to the task relationships in multi-task learning, we learn the inter-source relationships for the multi-source transfer learning problem by training a source relationship matrix $\mathbf{R} \in \mathbb{R}^{K \times K}$ as follows:

$$\mathbf{R}_{i,j} = \begin{cases} \dfrac{\exp(\beta_1 \hat{e}_{S_i}(\hat{h}_j))}{\sum\limits_{j' \in [K], j' \neq i} \exp(\beta_1 \hat{e}_{S_i}(\hat{h}'_j))}, & i \neq j \\ 0, & \text{otherwise} \end{cases} \qquad (4)$$

Algorithm 1. PW-MSTL

1: **Input:** $S = S^L \cup S^U$: source data; T: target data; μ: concentration factor; b_1: confidence tolerance; T: test data size;
2: **for** $k = 1, ..., K$ **do**
3:     Compute α^k by solving (6).
4:     Train a classifier $\hat{h}_k$ on the α^k weighted S_k^L.
5: **end for**
6: Compute δ and $\mathbf{R}$ as explained in Section 4.2.
7: Compute ω as (5).
8: **for** $t = 1, ..., T$ **do**
9:     Observe testing example $x^{(t)}$.
10:     **for** $k = 1, ..., K$ **do**
11:         **if** $|\hat{h}_k(x^{(t)})| < b_1$ **then**
12:             Compute $\hat{p}_k^{(t)} = \sum\limits_{m \in [K], m \neq k} \mathbf{R}_{km}|\hat{h}_m(x^{(t)})|$.
13:         **else**
14:             Compute $\hat{p}_k^{(t)} = |\hat{h}_k(x^{(t)})|$.
15:         **end if**
16:     **end for**
17:     Predict $\hat{y}^{(t)} = sign(\sum_{k \in [K]} \omega_k \hat{p}_k^{(t)})$.
18: **end for**

---

where $\hat{h}_j$ is the classifier trained on the jth source, $\hat{\epsilon}_{S_i}(\hat{h}_j)$ is the empirical error of $\hat{h}_j$ measured on the ith source and β_1 is a parameter to control the spread of the error. In other words, $\mathbf{R}$ is a matrix such that all entries on the diagonal are 0 and the kth row, namely $\mathbf{R}_k \in \Delta^{K-1}$, is a distribution over all sources where Δ^{K-1} denotes the probability simplex. Note that we do not require $\mathbf{R}$ to be symmetric due to the asymmetric nature of information transferability. While classifiers trained on a more reliable source can be well transferred to a less reliable source, the reverse is not guaranteed to be true.

A key benefit of the matrix $\mathbf{R}$ is that it measures source reliabilities directly. The bound in Eq. (3) suggests that this direct measurement of reliability gives the algorithm extra confidence in lower generalization error. Intuitively, if a classifier $\hat{h}_i$ trained on the ith source has a low empirical error on the jth source and the distributional divergence (such as the $\mathcal{H}\Delta\mathcal{H}$-divergence) between the jth source and the target is small, then $\hat{h}_i$ should have a low error on the target domain as well. This is shown by the second line in Eq. (3).

We therefore parametrize the source importance weights, considering both source proximity and reliability, as follows:

$$\omega = \delta \cdot [\mu \mathbf{I_K} + (1 - \mu)\mathbf{R}] \tag{5}$$

where $\mathbf{I_K} \in \mathbb{R}^{K \times K}$ is the identity matrix, $\delta \in \mathbb{R}^K$ is a vector measuring pairwise source-target proximities, $\mu \in [0, 1]$ is a scalar and $\mathbf{R}$ is the source relationship matrix. The concentration factor μ introduces an additional degree of freedom in quantifying the trade-off between proximity and reliability. Setting $\mu = 1$ amounts to weight sources based on proximity only. In the next section, we obtain the effective heuristic of specifying $\mu = \frac{1}{K}$ in our experiments. The proximity vector δ may be manually specified according to domain knowledge or estimated from data. In this paper, we adapt the Maximum Mean Discrepancy (MMD) statistic [8,11] to estimate the distribution discrepancy. It measures the difference between the means of two distributions after mapping onto a Reproducing Kernel Hilbert Space (RKHS), avoiding the complex density estimation. Specifically, for the kth source, its re-weighted MMD can be written as:

$$\min_{\alpha^k} \left\| \frac{1}{n_k^L + n_k^U} \sum_{i=1}^{n_k^L + n_k^U} \alpha_i^k \Phi(x_i^{S_k}) - \frac{1}{n_T} \sum_{i=1}^{n_T} \Phi(x_i^T) \right\|_H^2 \tag{6}$$

where $\alpha_i^k \geq 0$ are the weights of the kth source aggregate data and $\Phi(x)$ is a feature map onto the RKHS H. By applying the kernel trick, the optimization in (6) can be solved efficiently as a quadratic programming problem using interior point methods.

For label-scarce sources, the ensemble approach often fails to utilize their information effectively. This unsatisfactory performance leads us to propose using confidence and source relationships to transfer knowledge. For many standard classification methods such as SVMs and perceptrons, we can use the distance to the boundary $|\hat{h}_k(x)|$ to measure the confidence of the kth classifier on the example x, as in previous works [18]. If the confidence is low, we allow the classifier to query its peers on this specific example, exploiting the source relationship matrix $\mathbf{R}$. Algorithm 1 summarizes our proposed method.

4.3   Adaptive Multi-source Active Learning

When an oracle is available in the source domain, we can acquire more labels for source data, especially for less reliable but target-relevant sources. This is different from traditional active learning, which tries to improve the accuracy within the target domain. The TLAS algorithm proposed in [12] performs the active learning on the source domain in the single-source transfer learning, by solving a biconvex optimization problem. However, their solution requires solving two quadratic programming problems alternatively at each iteration and is therefore computationally expensive, especially for multi-source problems of large scale. In this paper, we propose an efficient two-stage active learning framework. The pseudo-code is in Algorithm 2.

In the first stage, the algorithm selects the source domain for query in an adaptive manner. The idea is that while we want to explore label-scarce sources for high marginal gain in performance, we also want to exploit label-rich sources for reliable queries. Equation (3) suggests that the learner should prefer a uniform source labeled ratio. Therefore, we draw a Bernoulli random variable $P^{(t)}$ with probability $D_{KL}(\beta||uniform)$, i.e. the Kullback-Leibler (KL) divergence between the current ratio of source labeled data and the uniform distribution. If $P^{(t)} = 1$, then the algorithm explores sources that contain less labeled data.

Algorithm 2. AMSAT

1: **Input:** $S = S^L \cup S^U$: source data; T: target data; μ: concentration factor; B: budget;
2: **for** $k = 1, ..., K$ **do**
3:     Compute α^k by solving (6).
4:     Train a classifier $\hat{h}_k$ on the α^k weighted S_k^L.
5: **end for**
6: **for** $t = 1, ..., B$ **do**
7:     Compute $\beta_i^{(t)} = \frac{n_i^L}{\sum_i n_i^L}$.
8:     Draw a Bernoulli random variable $P^{(t)}$ with probability $D_{KL}(\beta^{(t)}||uniform)$.
9:     **if** $P^{(t)} = 1$ **then**
10:         Set $Q^{(t)} = \frac{1}{\beta^{(t)}}$.
11:     **else**
12:         Compute $\omega^{(t)}$ as (5) and set $Q^{(t)} = \omega^{(t)}$.
13:     **end if**
14:     Draw $k^{(t)}$ from [K] with distribution $Q^{(t)}$.
15:     Select $x^{(t)}$ according to (8) and query the label for it.
16:     Update $S_{k^{(t)}}^L \leftarrow S_{k^{(t)}}^L \cup \{x^{(t)}\}$.
17:     Update $S_{k^{(t)}}^U \leftarrow S_{k^{(t)}}^U \setminus \{x^{(t)}\}$.
18:     Update classifier $\hat{h}_{k^{(t)}}$.
19: **end for**

If $P^{(t)} = 0$, the algorithm exploits sources as demonstrated in line 12 of Algorithm 2. Notice that the combined weights proposed in Eq. (5) play a crucial role here by providing a measurement of sources with unequal reliabilities.

In the second stage, the algorithm queries the most informative instance within the selected source domain. Nguyen and Smeulders [20] propose a Density Weighted Uncertainty Sampling criterion as:

$$x = \arg\max_{x_i \in S_{k(t)}^U} E[(\hat{y}_i - y_i)^2 | x_i] p(x_i) \tag{7}$$

which picks the instance that is close to the boundary and relies in a denser neighborhood. In our setting, we propose to combine the distribution matching weights α in Eq. (6) and the uncertainty sampling to form the following selection criterion:

$$x = \arg\max_{x_i \in S_{k(t)}^U} E[(\hat{y}_i - y_i)^2 | x_i] \alpha_i^{k(t)} \tag{8}$$

This criterion selects instances that are representative in both source and target domains while it also takes uncertainty into consideration. Notice that we solve Eq. (6) once only and store the value of α. As shown in [27], such approach is as efficient as the base informativeness measure, i.e. uncertainty sampling. Therefore, the proposed method achieves efficiency, representativeness and minimum information overlap.

5   Empirical Evaluation

In the following experimental study we aim at two goals: (i) to evaluate the performance of *PW-MSTL* with sources with diverse reliabilities, and (ii) to evaluate the effectiveness of *AMSAT* in constructing more reliable classifiers via active learning. Due to space limitation, not all results are presented, but they show similar patterns. Unless otherwise specified, all model parameters are chosen via 5-fold cross validation and all results are averaged over experiments repeated randomly 30 times.

5.1   Datasets

Synthetic Dataset. We generate a synthetic data for 5 source domains and 1 target domain. The samples $x \in \mathbb{R}^{10}$ are drawn from Gaussian distributions $\mathcal{N}(\mu_T + \mathbf{p}\Delta\mu, \sigma)$, where μ_T is the mean of the target domain, $\Delta\mu$ is a random fluctuation vector and $\mathbf{p}$ is the variable controlling the proximity between each source and the target (higher $\mathbf{p}$ indicates lower proximity). We then consider a labeling function $f(x) = sign((w_0^T + \delta\Delta w)x + \epsilon)$, where w_0 is a fixed base vector, Δw is a random fluctuation vector and ϵ is a zero-mean Gaussian noise term. We set δ to small values as we assume labeling functions are similar. Using different $\mathbf{p}$ values, We generate 50 positive points and 50 negative points as training data for each source domain and additional 100 balanced testing samples for the target domain.

Spam Detection.[2] We use the task B challenge of the spam detection dataset from ECML PAKDD 2006 Discovery challenge. It consists of data from inboxes of 15 users and each user forms a single domain with 200 spam $(+)$ examples and 200 non-spam $(-)$ examples. Each example consists of approximately $150K$ features representing word frequencies and we reduce the dimension to 200 using the latent semantic analysis (LSA) method [10]. Since some spam types are shared among users while others are not, these domains form a multi-source transfer learning problem if we try to utilize data from other users to build a personalized spam filter for a target user.

Sentiment Analysis.[3] The dataset contains product reviews from 24 domains on Amazon. We consider each domain as a binary classification task by setting reviews with rating > 3 as positive $(+)$ and reviews with rating < 3 as negative $(-)$, while reviews with rating $= 3$ were discarded as they are ambiguous. We pick the 10 domains that each contains more than 2000 samples and for each round of experiment, we randomly draw 1000 positive reviews and 1000 negative reviews for each domain. Similar to the previous dataset, each review contains of approximately $350K$ features and we reduce the dimension to 200 using the same method.

For each set of experiments, we have 600 examples (100 examples per domain) for *synthetic*, 6000 emails (400 emails per user) for *spam* and 20000 reviews (2000 reviews per domain) for *sentiment*. We set one domain as the target and the rest as sources. For each source domain, we randomly divide the data into the labeled set and the unlabeled set, using a labeled-data fraction. For the target domain, we randomly divide it into two parts: 40% for testing and 60% as unlabeled training data (with the exception of *synthetic* as we have generated extra testing data). Note that we can set different labeled-data fractions for source domains to model diverse reliabilities.

5.2   Reliable Multi-source Transfer Learning

Competing Methods. To evaluate the performance of our approach, we compare the proposed PW-MSTL with five different methods: one single-source method, one aggregate method, and three ensemble methods. Specifically, Kernel Mean Matching (KMM) [11] is a conventional single-source method by distribution matching. We perform KMM for each single source and report the best performance (note that this is impractical in general as it requires an oracle to know which classifier is the best). Kernel Mean Matching-Aggregate (KMM-A) is the aggregate method which performs KMM on all sources' training data combined. For the ensemble approach, Adaptive SVM (A-SVM) [33] and Domain Adaptation Machine (DAM) [7] are two widely-adopted multi-source methods.

---

[2] http://ecmlpkdd2006.org/challenge.html.
[3] http://www.cs.jhu.edu/~mdredze/datasets/sentiment.

Finally, we also compare with our own baseline PW-MSTL$_b$, which is similar to DAM but directly uses Eq. (5) as model weights. We also compared with Transfer Component Analysis (TCA) [22] but the result is omitted due to its similar performance to KMM.

Setup. For all methods, we use SVM as the base model and fix $b_1 = 1$. For competing methods, we mainly follow the standard procedures for model selection as explained in their respective papers. For fair comparisons, we use the same Gaussion kernel $k(x_i, x_j) = e^{-\|x_i - x_j\|^2/\gamma}$ with the bandwidth γ set to the median pairwise distance of training data according to the *median heuristic* [9]. Using other kernels yields similar results. For ensemble methods, we set the proximity weight as $\delta_k = \frac{exp(-\beta_2 MMD^\rho(S_k,T))}{\sum_k exp(-\beta_2 MMD^\rho(S_k,T))}$, where β_2 and ρ are tuned for each dataset to control the spread of MMD as described in Sect. 4.2. For both PW-MSTL and PW-MSTL$_b$, we set $\mu = 0.2$ for all experiments and use the 0-1 loss to compute the relation matrix $\mathbf{R}$ (See Fig. 1 for results on varying μ).

Table 1. Classification accuracy (%) on the target domain, given that source domains contain diverse {1%, 5%, 15%, 30%} labeled data.

Method	Synthetic		Spam			Sentiment									
	case1	case2	user7	user8	user3	electronics	toys	music	apparel	dvd	kitchen	video	sports	book	health
KMM	82.7	88.8	92.0	91.8	89.7	77.6	77.4	71.0	78.3	72.4	78.4	72.1	79.1	71.2	77.4
KMM-A	87.3	91.4	92.0	92.0	91.8	74.6	76.3	70.3	75.8	72.4	75.2	70.5	76.7	69.7	74.9
A-SVM	70.8	89.4	84.5	87.8	86.8	70.8	73.7	67.7	73.6	62.6	72.8	62.5	73.7	66.9	71.4
DAM	75.8	91.0	83.8	85.4	86.8	71.3	73.7	68.0	75.1	62.5	72.1	62.0	73.0	68.0	72.5
PW-MSTL$_b$	85.5	90.8	91.5	92.6	90.3	78.0	78.7	70.7	79,5	73.2	78.3	72.5	79.5	71.5	77.7
PW-MSTL	**88.4**	**92.6**	**93.8**	**95.6**	**92.8**	**79.3**	**81.9**	**74.6**	**82.7**	**76.7**	**80.7**	**76.2**	**82.7**	**74.8**	**80.9**

Table 2. Classification accuracy (%) on the target domain, given that source domains contain the same fraction (%$_L$) of labeled data.

%$_L$	Method	Synthetic	Spam			Sentiment									
			user7	user8	user3	electronics	toys	music	apparel	dvd	kitchen	video	sports	book	health
10%	KMM	87.0	89.1	91.2	90.3	75.0	74.6	68.3	75.6	70.2	75.9	69.9	75.6	68.9	74.3
	KMM-A	91.1	91.3	90.7	91.0	74.8	76.5	70.2	76.8	71.3	77.6	71.6	77.7	71.3	75.4
	A-SVM	89.4	88.4	91.9	89.2	77.1	78.1	69.9	78.2	68.9	79.1	69.2	78.1	70.5	77.1
	DAM	89.7	89.6	90.4	91.3	77.5	79.0	69.9	79.8	69.0	79.5	68.9	78.4	71.9	77.7
	PW-MSTL$_b$	90.2	89.7	92.4	92.1	77.7	78.7	69.7	78.9	73.5	79.8	72.3	78.8	70.4	77.9
	PW-MSTL	91.2	**92.5**	**94.9**	**93.1**	**79.8**	**81.5**	**73.3**	**81.3**	**76.4**	**82.3**	**75.4**	**81.2**	**74.4**	**80.7**
50%	KMM	95.6	92.6	94.0	91.8	81.6	81.7	75.0	82.2	76.9	83.0	77.5	82.8	75.3	81.2
	KMM-A	97.2	91.4	93.8	**94.7**	80.4	82.4	74.5	82.7	77.1	83.8	76.5	82.8	76.0	79.6
	A-SVM	96.4	91.5	95.2	93.4	81.7	83.4	74.7	84.3	76.0	85.4	75.3	83.3	76.0	82.1
	DAM	96.6	92.7	93.1	93.2	83.5	84.5	73.4	84.4	77.3	86.7	76.5	84.8	76.8	83.6
	PW-MSTL$_b$	96.6	92.9	95.2	93.5	83.6	84.7	74.4	85.0	80.4	85.9	79.4	85.7	77.0	84.1
	PW-MSTL	97.2	**94.5**	**95.7**	93.7	**84.8**	**86.4**	**76.9**	**87.2**	**82.0**	**87.6**	**81.3**	**87.3**	**79.8**	**86.4**

Results and Discussion. Table 1^4 shows the classification accuracy of different methods on various datasets when sources contain different labeled fractions randomly chosen from the set $\{1\%, 5\%, 15\%, 30\%\}$, while Table 2 shows the results when sources have the same amount of labeled data. The method that outperforms all other methods within the column with $p < 0.001$ significance is highlighted and underlined (and thus we omit standard deviation due to space). We can observe that our proposed PW-MSTL method outperforms other methods in most cases. When source reliabilities are uneven, we get a significant improvement in the test set accuracy. By comparing results across the two tables, we can also observe that gap between our methods and other methods diminishes as source domains acquire more labeled data (more reliable).

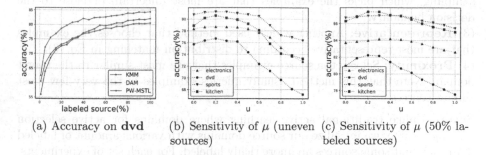

(a) Accuracy on **dvd**   (b) Sensitivity of μ (uneven sources)   (c) Sensitivity of μ (50% labeled sources)

Fig. 1. Empirical analysis: (a) incremental accuracy on **dvd** of KMM, DAM & PW-MSTL, when sources have same amount of labeled data; (b) sensitivity of μ when sources have different amount of labeled data; (c) sensitivity of μ when sources all have 50% of labeled data.

Note that we have built two scenarios for the *Synthetic* dataset in Table 1. We control the proximity variable **p** and the labeled fraction such that similar sources also contain more labeled data in case 1 while this is reversed in case 2. As a result, we observe that prior methods focusing on proximity measure obtain an accuracy boost under case 2 compared to case 1 while our methods are consistent. We also note that A-SVM and DAM perform poorly on **dvd** and **video** even when sources are balanced. This is because the distribution divergence between these two domains are very small, causing prior methods failure to utilize other less related sources effectively. We plot an incremental performance comparison on **dvd** in Fig. 1(a), showing that DAM slowly catches up with the single-source method as labeled fractions increase.

Finally, we study the sensitivity of μ. Figure 1(b) and (c) show the variation in accuracy for some cases in Tables 1 and 2 with varying μ over [0,1]. We can observe that similar patterns that accuracy values first increase and then decrease as μ increases in both cases, only that the drop in the performance is

---

[4] Due to space limits, we randomly show results of only three cases for *Spam* since their overall results are similar.

smoother in Fig. 1(c). This confirms the motivation of combining proximity and reliability and the theoretical result established in this paper. Results on other datasets are similar but we omit them in the graph due to different scales.

5.3 Multi-source Active Learning

Baselines. To the best of our knowledge, there is no existing study that can be directly applied to our setting. Therefore, to evaluate the performance of our proposed AMSAT algorithm, we compare to the following standard baselines:

(1) **Random**: Selects the examples to query randomly from all source domains. This is equivalent to passive learning.

(2) **Uncertainty**: This is a popular active learning strategy called uncertainty sampling, which picks the examples that are most uncertain or close to the decision boundary.

(3) **Representative**: Another widely used method that chooses the examples that are most representative according to distribution matching in Eq. (6).

(4) **Proximity**: Selects the source examples that are most similar to the target domain. Note that this method usually queries examples in the very few most similar sources.

Setup. Unlike traditional active learning where domains for active selection are label-scarce, we initialize our source domains with various amount of labeled data such that some sources are more richly labeled. For each set of experiments, we initialize each source domain with a labeled fraction randomly chosen from a fixed set. This same data partition is used for all comparing methods. After each query, we perform a multi-source transfer learning algorithm and record the accuracy on the test target data. The data partition is repeated randomly for 30 times and the average results are reported. For all of our experiments, we set the query budget to 10% of the total number of source examples in the dataset. For fair comparison, the selected examples are evaluated using the same transfer learning method.

Results and Discussion. The performance curve evaluated using PW-MSTL with increasing queries are plotted in Fig. 2. Due to space limitation, only a small fraction of results are presented and we only show results on *Sentiment* here because other datasets are less interesting due to diminishing gains. However, it should be noted that our proposed AMSAT method outperforms other methods with $p < 0.0001$ significance in a two-tailed test in all datasets. Figure 2 shows results with initial source labeled fractions randomly chosen from $\{1\%, 5\%, 15\%, 30\%\}$, same as experiments in Table 1. We can observe that AMSAT consistently outperforms baselines at almost all stages. The only exception is that Proximity sampling performs better at the early stage when there exists a source domain that is particularly close to the target domain, as in the case of **dvd** in Fig. 2(b). Figure 3(a) and (b) show examples of method performance using a different data initialization on the **kitchen** domain. While they show similar patterns compared to Fig. 2(d) (AMSAT consistently outperforming other baselines), we observe that AMSAT has comparable ending points in two plots due to diminishing gains. This suggests that performing active learning when sources have very uneven reliabilities is hard.

For the baselines, we observe that they are quite unstable when evaluated using PW-MSTL. For example, in some domains Uncertainty sampling performs better than Random sampling as in Fig. 2(d) while it is not the case in other domains such as in Fig. 2(a). When evaluated using other transfer learning methods such as DAM, often other baselines outperform Random sampling as expected. To show that combining PW-MSTL and AMSAT is superior, we compare several combinations of transfer learning and active learning methods. A typical example is shown in Fig. 3(c). Notice that the curves with the same active learning method are evaluated using different transfer learning methods on the same queried data. We observe that while AMSAT still performs better than baselines when evaluated using DAM, the performance gap is larger when evaluated using PW-MSTL. The combination of PW-MSTL and AMSAT significantly outperforms the rest, showing a synergy between the proposed methods.

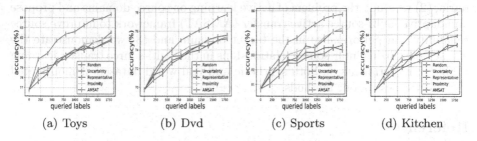

(a) Toys (b) Dvd (c) Sports (d) Kitchen

Fig. 2. Performance comparison of active learning methods on *Sentiment*: initial labeled fractions randomly selected from $\{1\%, 5\%, 15\%, 30\%\}$.

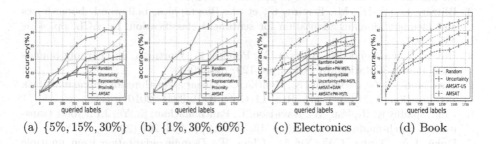

(a) $\{5\%, 15\%, 30\%\}$ (b) $\{1\%, 30\%, 60\%\}$ (c) Electronics (d) Book

Fig. 3. Performance comparison of different labeled data initialization for Kitchen in (a) and (b); (c) Performance comparison of different combinations of transfer learning methods and active learning methods; (d) Evaluation of the proposed source picking and example picking strategies by comparing AMSAT against Uncertainty and AMSAT-US.

Finally, we evaluate the effectiveness of both picking the source as line 14 in Algorithm 2 and picking the example according to Eq. (8). We add a baseline AMSAT-US which performs exactly the same as Algorithm 2 except that it picks the example according to uncertainty sampling in line 15. Therefore, the added baseline utilizes the proposed source picking strategy but not the example picking strategy. Figure 3(d) showcases the comparison among the four methods. We can observe that: (i) AMSAT-US performs better than Uncertainty, indicating the effectiveness of source picking strategy proposed; (ii) AMSAT outperforms AMSAT-US, showing the effectiveness of example picking strategy.

6   Conclusion

We study a new research problem of transfer learning with multiple sources exhibiting reliability divergences, and explore two related tasks. The contributions of this paper are: (1) we propose a novel peer-weighted multi-source transfer learning (PW-MSTL) method that makes robust predictions in the described scenario, (2) we study the problem of active learning on source domains and propose an adaptive multi-source active transfer (AMSAT) framework to improve source reliability and performance in the target domain, and (3) we demonstrate the efficacy of utilizing inter-source relationships in the multi-source transfer learning problem. Experiments on both synthetic and real world datasets demonstrated the effectiveness of our methods.

References

1. Balcan, M., Hanneke, S., Vaughan, J.W.: The true sample complexity of active learning. Mach. Learn. **80**(2), 111–139 (2010)
2. Ben-David, S., Blitzer, J., Crammer, K., Pereira, F.: Analysis of representations for domain adaptation. In: NIPS, pp. 137–144 (2007)
3. Blitzer, J., Dredze, M., Pereira, F., et al.: Biographies, bollywood, boom-boxes and blenders: domain adaptation for sentiment classification. In: ACL, pp. 440–447 (2007)
4. Blitzer, J., Crammer, K., Kulesza, A., Pereira, F., Wortman, J.: Learning bounds for domain adaptation. In: NIPS, pp. 129–136 (2008)
5. Cavallanti, G., Cesa-Bianchi, N., Gentile, C.: Linear algorithms for online multitask classification. J. Mach. Learn. Res. **11**(Oct), 2901–2934 (2010)
6. Chattopadhyay, R., Fan, W., Davidson, I., Panchanathan, S., Ye, J.P.: Joint transfer and batch-mode active learning. In: ICML, pp. 253–261 (2013)
7. Duan, L.X., Tsang, I.W., Xu, D., Chua, T.: Domain adaptation from multiple sources via auxiliary classifiers. In: ICML, pp. 289–296 (2009)
8. Gretton, A., Borgwardt, K.M., Rasch, M., Scholkopf, B., Smola, A.J.: A kernel method for the two-sample-problem. In: NIPS, pp. 513–520 (2007)
9. Gretton, A., et al.: Optimal kernel choice for large-scale two-sample tests. In: NIPS, pp. 1205–1213 (2012)
10. Halko, N., Martinsson, P.G., Tropp, J.A.: Finding structure with randomness: probabilistic algorithms for constructing approximate matrix decompositions. SIAM Rev. **53**(2), 217–288 (2011)

11. Huang, J.Y., Gretton, A., Borgwardt, K.M., Scholkopf, B., Smola, A.J.: Correcting sample selection bias by unlabeled data. In: NIPS, pp. 601–608 (2007)
12. Huang, S.J., Chen, S.: Transfer learning with active queries from source domain. In: IJCAI, pp. 1592–1598 (2016)
13. Kale, D., Ghazvininejad, M., Ramakrishna, A., He, J.R., Liu, Y.: Hierarchical active transfer learning. In: SIAM International Conference on Data Mining, pp. 514–522 (2015)
14. Konyushkova, K., Sznitman, R., Fua, P.: Learning active learning from data. In: NIPS, pp. 4228–4238 (2017)
15. Long, M.S., Wang, J.M., Jordan, M.I.: Deep transfer learning with joint adaptation networks. In: ICML, pp. 2208–2217 (2017)
16. Luo, P., Zhuang, F.Z., Xiong, H., Xiong, Y.H., He, Q.: Transfer learning from multiple source domains via consensus regularization. In: ACM Conference on Information and Knowledge Management, pp. 103–112 (2008)
17. Moon, S., Carbonell, J.: Completely heterogeneous transfer learning with attention-what and what not to transfer. In: IJCAI, pp. 2508–2514 (2017)
18. Murugesan, K., Carbonell, J.: Active learning from peers. In: NIPS, pp. 7011–7020 (2017)
19. Murugesan, K., Liu, H.X., Carbonell, J., Yang, Y.M.: Adaptive smoothed online multi-task learning. In: NIPS, pp. 4296–4304 (2016)
20. Nguyen, H.T., Smeulders, A.: Active learning using pre-clustering. In: ICML, p. 79 (2004)
21. Pan, S.J., Ni, X., Sun, J., Yang, Q., Chen, Z.: Cross-domain sentiment classification via spectral feature alignment. ACM Trans. Knowl. Discov. Data (TKDD) **8**(3), 12 (2014)
22. Pan, S.J., Tsang, I.W., Kwok, J.T., Yang, Q.: Domain adaptation via transfer component analysis. IEEE Trans. Neural Networks **22**(2), 199–210 (2011)
23. Pan, S.J., Yang, Q.: A survey on transfer learning. IEEE Trans. Knowl. Data Eng. **22**(10), 1345–1359 (2010)
24. Pardoe, D., Stone, P.: Boosting for regression transfer. In: ICML, pp. 863–870 (2010)
25. Raina, R., Battle, A., Lee, H., Packer, B., Ng, A.Y.: Self-taught learning: transfer learning from unlabeled data. In: ICML, pp. 759–766 (2007)
26. Settles, B.: Active learning literature survey. Technical report, University of Wisconsin, Madison (2010)
27. Settles, B., Craven, M.: An analysis of active learning strategies for sequence labeling tasks. In: EMNLP, pp. 1070–1079 (2008)
28. Sun, Q., Chattopadhyay, R., Panchanathan, S., Ye, J.P.: A two-stage weighting framework for multi-source domain adaptation. In: NIPS, pp. 505–513 (2011)
29. Tong, S., Koller, D.: Support vector machine active learning with applications to text classification. J. Mach. Learn. Res. **2**(Nov), 45–66 (2001)
30. Wang, X.Z., Huang, T.K., Schneider, J.: Active transfer learning under model shift. In: ICML, pp. 1305–1313 (2014)
31. Wei, P.F., Sagarna, R., Ke, Y.P., Ong, Y.S., Goh, C.K.: Source-target similarity modelings for multi-source transfer gaussian process regression. In: ICML, pp. 3722–3731 (2017)
32. Wei, Y., Zheng, Y., Yang, Q.: Transfer knowledge between cities. In: KDD, pp. 1905–1914 (2016)
33. Yang, J., Yan, R., Hauptmann, A.G.: Cross domain video concept detection using adaptive SVMs. In: ACM International Conference on Multimedia, pp. 188–197 (2007)

34. Yang, L., Hanneke, S., Carbonell, J.: A theory of transfer learning with applications to active learning. Mach. Learn. **90**(2), 161–189 (2013)
35. Zhang, L., Zuo, W.M., Zhang, D.: LSDT: latent sparse domain transfer learning for visual adaptation. IEEE Trans. Image Process. **25**(3), 1177–1191 (2016)
36. Zhang, Y., Yeung, D.Y.: A regularization approach to learning task relationships in multitask learning. In: WWW, pp. 751–760 (2010)

Differentially Private Hypothesis Transfer Learning

Yang Wang[1]([✉]), Quanquan Gu[2], and Donald Brown[1]

[1] Department of Systems and Information Engineering, University of Virginia,
Charlottesville, VA, USA
{yw3xs,deb}@virginia.edu
[2] Department of Computer Science, University of California, Los Angeles, CA, USA
qgu@cs.ucla.edu

Abstract. In recent years, the focus of machine learning has been shifting to the paradigm of transfer learning where the data distribution in the target domain differs from that in the source domain. This is a prevalent setting in real-world classification problems and numerous well-established theoretical results in the classical supervised learning paradigm will break down under this setting. In addition, the increasing privacy protection awareness restricts access to source domain samples and poses new challenges for the development of privacy-preserving transfer learning algorithms. In this paper, we propose a novel differentially private multiple-source hypothesis transfer learning method for logistic regression. The target learner operates on differentially private hypotheses and importance weighting information from the sources to construct informative Gaussian priors for its logistic regression model. By leveraging a publicly available auxiliary data set, the importance weighting information can be used to determine the relationship between the source domain and the target domain without leaking source data privacy. Our approach provides a robust performance boost even when high quality labeled samples are extremely scarce in the target data set. The extensive experiments on two real-world data sets confirm the performance improvement of our approach over several baselines. Data related to this paper is available at: http://qwone.com/~jason/20Newsgroups/ and https://www.cs.jhu.edu/~mdredze/datasets/sentiment/index2.html.

Keywords: Differential privacy · Transfer learning

1 Introduction

In the era of big data, an abundant supply of high quality labeled data is crucial for the success of modern machine learning. But it is both impractical and unnecessary for a single entity, whether it is a tech giant or a powerful government agency, to collect the massive amounts of data single-handedly and

© Springer Nature Switzerland AG 2019
M. Berlingerio et al. (Eds.): ECML PKDD 2018, LNAI 11052, pp. 811–826, 2019.
https://doi.org/10.1007/978-3-030-10928-8_48

perform the analysis in isolation. Collaboration among data centers or crowd-sourced data sets can be truly beneficial in numerous data-driven research areas and industries nowadays. As modern data sets get bigger and more complex, the traditional practice of centralizing data from multiple data owners turns out to be extremely inefficient. Increasing concerns over data privacy also create barriers to data sharing, making it difficult to coordinate large-scale collaborative studies especially when sensitive human subject data (e.g., medical records, financial information) are involved. Instead of moving raw data, a more efficient approach is to exchange the intermediate computation results obtained from distributed training data sets (e.g., gradients [15], likelihood values in MLE [2]) during the machine learning process. However, even the exchange of intermediate results or summary statistics is potentially privacy-breaching as revealed by recent evidence. There is a large body of research work addressing this problem in a rigorous privacy-preserving manner, especially under the notion of differential privacy [23,24].

In this paper, we consider the differentially private distributed machine learning problem under a more realistic assumption – the multiple training data sets are drawn from different distributions and the learned machine learning hypothesis (i.e. classifier or predictive model) will be tested and employed on a data set drawn from another different yet related distribution. In this setting, the training sets are referred to as the "source domains" and the test set is referred to as the "target domain". High quality labeled data in the target domain is usually costly, if not impossible, to collect, making it necessary and desirable to take advantage of the sources in order to build a reliable hypothesis on the target. The task of "transferring" source knowledge to improve target "learning" is known as transfer learning, one of the fundamental problems in statistical learning that is attracting increasing attention from both academia and industry. This phenomenon is prevalent in real-world applications and solving the problem is important because numerous solid theoretical justification, especially the generalization capability, in traditional supervised learning paradigm will break down under transfer learning. Plenty of research work has been done in this area, as surveyed in [16]. Meanwhile, the unprecedented emphasis on data privacy today brings up a natural question for the researchers:

How to improve the learning of the hypothesis on the target using knowledge of the sources while protecting the data privacy of the sources?

To our best knowledge, limited research work exists on differentially private multiple-source transfer learning. Two challenges emerge when addressing this problem – what to transfer and how to ensure differential privacy. In the transfer learning literature, there are four major approaches – instance re-weighting [6], feature representation transfer [20], hypothesis transfer [10] and relational knowledge transfer [14]. In particular, we focus on hypothesis transfer learning, where hypotheses trained on the source domains are utilized to improve the learning of target hypothesis and no access to source data is allowed. To better incorporate the source hypotheses, we also adapt an importance weighting mechanism in [9] to measure the relationship between sources and target by leveraging a public auxiliary unlabeled data set. Based on the source-target relationship, we can determine how much weight to assign to each source hypothesis in the process of constructing informative Gaussian priors for the parameters of the target hypothesis. Furthermore, the enforcement of differential privacy requires an additional layer of privacy protection because unperturbed source domain knowledge may incur privacy leakage risk. In the following sections, we explicitly apply our methodology to binary logistic regression. Our approach will provide end-to-end differential privacy guarantee by perturbing the importance weighting information and the source hypotheses before transferring them to the target. Extensive empirical evaluations on real-world data demonstrate its utility and significant performance lift over several important baselines.

The main contributions of this work can be summarized as follows.

- We propose a novel multiple-source differentially private hypothesis transfer learning algorithm and focus specifically on its application in binary logistic regression. It addresses the notorious problem of insufficient labeled target data by making use of unlabeled data and provides rigorous differential privacy guarantee.
- Compared with previous work, only one round of direct communication between sources and target is required in our computationally efficient one-shot model aggregation, therefore overcoming some known drawbacks (e.g. avoiding the complicated iterative hypothesis training process and privacy budget composition in [7,25], eliminating the need for a trusted third party in [8]).
- The negative impact of unrelated source domains (i.e. negative transfer) is alleviated as the weight assigned to each source hypothesis will be determined by its relationship with the target.
- The non-private version of our method achieves performance comparable to that of the target hypothesis trained with an abundant supply of labeled data. It provides a new perspective for multiple-source transfer learning when privacy is not a concern.

The rest of the paper is organized as follows: background and related work are introduced in Sect. 2. We then present the methods and algorithms in Sect. 3. It is followed by empirical evaluation of the method on two real-world data sets in Sect. 4. Lastly, we conclude the work in Sect. 5.

2   Background and Related Work

The privacy notion we use is differential privacy, which is a mathematically rigorous definition of data privacy [4].

Definition 1 (Differential Privacy). *A randomized algorithm M (with output space Ω and well-defined probability density $\mathcal{D}$) is (ϵ, δ)-differentially private if for all adjacent data sets S, S' that differ in a single data point s and for all measurable sets $\omega \in \Omega$:*

$$Pr[M(S) \in \omega] \leq \exp(\epsilon)Pr[M(S') \in \omega] + \delta. \tag{1}$$

Differential privacy essentially implies that the existence of any particular data point s in a private data set S cannot be determined by analyzing the output of a differentially private algorithm M applied on S. The parameter ϵ is typically known as the privacy budget which quantifies the privacy loss whenever an algorithm is applied on the private data. A differentially private algorithm with smaller ϵ means that it is more private. The other privacy parameter δ represents the probability of the algorithm failing to satisfy differential privacy and is usually set to zero.

Differential privacy can be applied to address the problem of private machine learning. The final outputs or the intermediate computation results of machine learning algorithms can potentially reveal sensitive information of individuals who contribute data to the training set. A popular approach to achieving differential privacy is output perturbation, derived from the sensitivity method in [4]. It works by adding carefully calibrated noises to the parameters of the learned hypothesis before releasing it. In particular, Chaudhuri et al. [3] adapted the sensitivity method into a differentially private regularized empirical risk minimization (ERM) algorithm of which logistic regression is a special case. They proved that adding Laplace noise with scale inversely proportional to the privacy parameter ϵ and the regularization parameter λ to the learned hypothesis provides $(\epsilon, 0)$-differential privacy.

More recently, research efforts have been focused on distributed privacy-preserving machine learning where private data sets are collected by multiple parties. One line of research involves exchanging differentially private information (e.g. gradients) among multiple parties during the iterative hypothesis training process [23,24]. An alternative line of work focuses on privacy-preserving model aggregation techniques. Pathak et al. [19] addressed the hypothesis ensemble problem by simple parameter averaging and publishing the aggregated hypothesis in a differentially private manner. Papernot et al. [18] proposed a different perspective – Private Aggregation of Teacher Ensembles (PATE), where private data set is split into disjoint subsets and different "teacher" hypotheses are trained on all subsets. The private teacher ensemble is used to produce labels on auxiliary unlabeled data by noisy voting and a "student" hypothesis is trained on the auxiliary data and released. Hamm et al. [8] explored a similar strategy of transferring local hypotheses and focused on convex loss functions. Their work involves a "trusted entity" who collects unperturbed local hypotheses trained on multiple private data sets and uses them to generate "soft" labels on an auxiliary unlabeled data set. A global hypothesis is trained on the soft-labeled auxiliary data and output perturbation is used to ensure its differential privacy.

In the limited literature of differentially private transfer learning, the most recent and relevant work to ours are [7,25] which focus on multi-task learning , one of the variants of transfer learning. Gupta et al. [7] proposed a differentially private mutli-task learning algorithm using noisy task relationship matrix and developed a novel attribute-wise noise addition scheme. Xie et al. [25] introduced a privacy-preserving proximal gradient algorithm to asynchronously update the hypothesis parameters of the learning tasks. Both work are iterative solutions and require multiple rounds of information exchange between tasks. However, one of the key drawbacks of iterative differentially private methods is that privacy risks accumulate with each iteration. The composition theorem of differential privacy [5] provides a framework to quantify the privacy budget consumption. Given a certain level of total privacy budget, there is a limit on how many iterations can be performed on a specific private data set, which severely affects the utility-privacy trade-off of iterative solutions.

3   The Proposed Methods

Let us assume that there are K sources in the transfer learning system (as shown in Fig. 1), indexed as $k = 1, ..., K$. For the k-th source, we denote the labeled samples as $S_l^k = \{(x_i^k, y_i^k) : 1 \leq i \leq n_l^k\}$ and the unlabeled samples as $S_{ul}^k = \{x_j^k : 1 \leq j \leq n_{ul}^k\}$, where $x_i^k, x_j^k \in \mathbb{R}^d$ are the feature vectors, y_i^k is the corresponding label, and n_l^k, n_{ul}^k are the sizes of S_l^k and S_{ul}^k respectively. The samples in the k-th source are drawn i.i.d. according to a probability distribution $\mathcal{D}^k$. There is also a target data set T with abundant unlabeled samples and very few labeled samples drawn i.i.d. from a different yet related distribution $\mathcal{D}^T$. Following the setting of [12], $\mathcal{D}^T$ is assumed to be a mixture of the source distributions $\mathcal{D}^k$'s. A public data set P of size n^P (not necessarily labeled) is accessible to both the sources and the target serving as an information intermediary.

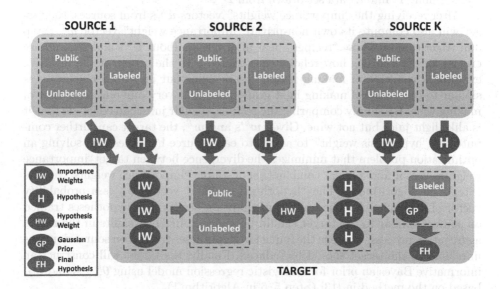

Fig. 1. Diagram of the multiple-source transfer learning system.

We present the full details of our method (Algorithm 1) here. Two pieces of knowledge need to be transferred to the target from each source in a differentially private manner. The first piece is the hypothesis locally trained on the labeled source samples S_l^k. In order to protect the data privacy, we apply the output perturbation method in [3] to perturb the source hypotheses into θ_{priv}^k ($k = 1, ..., K$) before transferring them to the target (Step 1 in Algorithm 1).

The second piece of knowledge is the differentially private "importance weights" vector with respect to samples in the public data set P, which is used to measure the relationship between source distribution and target distribution. It is a very difficult task to estimate high-dimensional probability densities and releasing the unperturbed distribution or the histogram can be privacy-breaching in itself. No existing work addresses the differentially private source-target relationship problem explicitly. We leverage a public data set P and propose a novel method based on the importance weighting mechanism in [9] (see Algorithm 2) which was originally developed for the purpose of differentially private data publishing. We take advantage of the "importance weights" vector and successfully quantify the source-target relationship without violating the data privacy of sources. To be more specific, every source k will compute the differentially private "importance weight" for each sample in P using its unlabeled samples S_{ul}^k (Step 2 in Algorithm 1). The "importance weight" of a data point in P is large if it is similar to the samples in S_{ul}^k, and small otherwise. The "importance weights" vector w^k is therefore an n^P-dimensional vector with non-negative entries that add up to n^P. By making a multiplicative adjustment to the public data set P using w^k, P will look similar to S_{ul}^k in proportion. In other words, w^k is the "recipe" of transforming P into a data set drawn from $\mathcal{D}^k$.

After receiving the "importance weights" vectors w^k's from sources, the target will also compute its own non-private "importance weight" vector w^T (Step 3 in Algorithm 1). These "recipes" bear information about $\mathcal{D}^k$'s and $\mathcal{D}^T$ and are crucial for determining how related each source is to the target. As an intuitive example, the recipe of making regular grape juice out of grapes and water is similar to the recipe of making light grape juice, but very different from that of making wine. By simply comparing the recipes, regular juice can conclude that it is alike light juice but not wine. Given w^k's and w^T, the target can further compute the "hypothesis weight" to assign to each source hypothesis by solving an optimization problem that minimizes the divergence between target "importance weights" vector and a linear combination of source "importance weights" vectors (Step 4 in Algorithm 1). The "hypothesis weights" vector w_H is a probability vector of dimension K. The implied assumption here is that hypotheses trained on sources similar to the target should be assigned higher weights in the model aggregation process, whereas the impact of hypotheses trained on source domains unrelated to the target should be reduced. Finally, the target will construct an informative Bayesian prior for its logistic regression model using θ_{priv}^k's and w_H based on the method in [13] (Step 5–6 in Algorithm 1).

To ensure differential privacy, both pieces of source knowledge transferred to the target need to be perturbed carefully. Note that the private hypothesis is trained on labeled source samples and the private "importance weights" vector is trained on the disjoint unlabeled source samples. Therefore no composition or splitting of privacy budget is involved. The privacy analysis is presented in Theorem 2 below.

Theorem 2. *Algorithm 1 is $(\epsilon, 0)$-differentially private with respect to both the labeled and the unlabeled samples in each source.*

Proof. We sketch the proof here for brevity. First, the perturbed source hypotheses θ^k_{priv}'s are proved to be differentially private with respect to the labeled source samples by [3]. Additionally, the "importance weights" vector w^k is $(\epsilon, 0)$-differentially private with respect to the unlabeled source samples by [9]. Therefore, the final target hypothesis θ^T built upon θ^k_{priv}'s and w^k's is also $(\epsilon, 0)$-differentially private with respect to the sources by the post-processing guarantee of differential privacy [5].

Algorithm 1. Differentially Private Hypothesis Transfer Learning (DPHTL)

Input: K private labeled source data sets S^k_l and unlabeled source data sets S^k_{ul}, a public data set P, a target data set T with limited labels, privacy parameter ϵ, regularization parameter for importance weighting mechanism λ_{IW}, regularization parameter for logistic regression model λ_{LR},

Output: A final target hypothesis $\theta^T \in \mathbb{R}^d$

1: Each source uses its labeled samples S^k_l to train a differentially private logistic regression model θ^k_{priv} under parameters λ_{LR} and ϵ. All the hypotheses θ^k_{priv} are sent to the target.

2: Each source fetches the public data set P, compute the differentially private "importance weights" vector $w^k \leftarrow \mathbf{DPIW}(S^k_{ul}, P, \epsilon, \lambda_{IW})$ and sends w^k to the target.

3: The target fetches the public data set P and compute the non-private "importance weights" vector $w^T \leftarrow \mathbf{DPIW}(T, P, \infty, \lambda_{IW})$.

4: The target calculates the "hypothesis weights" vector $w_H \in \mathbb{R}^K$ such that the Kullback-Leibler (KL) divergence between w^T and the linear combination of w^k weighted by w_H is minimized:

$$w_H = \underset{w \in \mathbb{R}^K_{\geq 0}, \sum w(k)=1}{\operatorname{argmin}} \mathbf{KL}\left(w^T, \sum_{k=1}^K w(k)w^k\right) \tag{2}$$

5: In order to construct an informative Gaussian prior using w_H and θ^k_{priv} from the sources, the target first calculates the mean $\mu^T(j)$ and standard deviation $\sigma^T(j)$ of each parameter $\theta^T(j)$ $(j = 1, 2, 3..., d)$ in the target logistic regression hypothesis:

$$\mu^T(j) = \sum_{k=1}^K w_H(k)\theta^k_{priv}(j), \quad \sigma^T(j) = \sqrt{\frac{K}{K-1}\sum_{k=1}^K w_H(k)(\theta^k_{priv}(j) - \mu^T(j))^2} \tag{3}$$

6: The target trains the Bayesian logistic regression model with the limited labeled target data set and the informative Gaussian prior following the method in [13] and return the posterior parameters θ^T.

Algorithm 2. Differentially Private Importance Weights (DPIW) [9]

Input: Private data set X of size N_X, public data set P of size N_P, privacy parameter ϵ, regularization parameter λ
Output: Differentially private importance weights vector $w \in \mathbb{R}^{N_P}$.
1: Label each data point in X as 1 and each data point in P as 0.
2: Fit the regularized logistic regression model on the combination of X and P with the new binary labels:

$$\beta^* = \underset{\beta}{\operatorname{argmin}} - \sum_{x \in P} \frac{\log(p(x \in P | x \in X \cup P))}{N_P}$$
$$- \sum_{x \in X} \frac{\log(p(x \in X | x \in X \cup P))}{N_X} + \frac{\lambda}{2} \|\beta\|^2, \qquad (4)$$

where $p(x \in X | x \in X \cup P) = 1 - p(x \in P | x \in X \cup P) = 1/(1 + \exp(-\beta^\mathsf{T}x))$.
3: Add Laplace noise to β^* to get the differentially private β_{priv}: $\beta_{priv} = \beta^* + \delta$, where $Pr(\delta) \sim \exp(-\epsilon\|\delta\|_2 N_X \lambda/\sqrt{d})$.
4: Output differentially private importance weight $w(x) = \exp(\beta_{priv}^\mathsf{T}x)N_P/Z$ for each x in P, where $Z = \sum_{x \in P} \exp(\beta_{priv}^\mathsf{T}x)$.

4   Experiments

In this section we empirically evaluate the effectiveness of our differentially private hypothesis transfer learning method (**DPHTL**) using publicly available real-world data sets. The experiment results show that the hypothesis obtained by our method provides a performance boost over the hypothesis trained on the limited labeled target data while guaranteeing differential privacy of sources and is also better than the hypotheses obtained by several baselines under various level of privacy requirements.

The first data set is the text classification data set 20NewsGroup (**20NG**)[1] [11], a popular benchmark data set for transfer learning and domain adaptation [17, 21]. It is a collection of approximately 20,000 newsgroup documents with stop words removed and partitioned evenly across 20 different topics. Some topics can be further grouped into a broader subject, for example, computer-related, recreation-related. To construct a binary classification problem, we randomly paired each of the 5 computer-related topics with 2 non-computer-related topics in each source domain. Our purpose is to build a logistic regression model to classify each document as either computer-related or non-computer-related. Specifically, there are 5 source domains with documents sampled from different topic groups as listed in Table 1. The target data set and the public data set are assumed to be mixtures of the source domains.

The second data set is the Amazon review sentiment data (**AMAZON**)[2], a famous benchmark multi-domain sentiment data set collected by Blitzer et al [1].

---

[1] http://qwone.com/~jason/20Newsgroups/.
[2] https://www.cs.jhu.edu/~mdredze/datasets/sentiment/index2.html.

Table 1. Topics of documents in each source domain for **20NG**.

Domain	Topic group
Source 1	comp.graphics
	rec.autos
	sci.space
Source 2	comp.os.ms-windows.misc
	talk.politics.guns
	sci.med
Source 3	comp.sys.ibm.pc.hardware
	soc.religion.christian
	talk.politics.mideast
Source 4	comp.sys.mac.hardware
	misc.forsale
	rec.sport.baseball
Source 5	comp.windows.x
	talk.religion.misc
	sci.crypt

It contains product reviews from 4 different types – *Book, DVD, Kitchen* and *Electronics*. Reviews with star rating > 3 are labeled positive, and those with star rating < 3 are labeled negative. Each source domain contains reviews from one type and the target domain and public data set are mixtures of the source domains. After initial preprocessing, each product review is represented by its count of unigrams and bigrams in the document and the amounts of positive reviews and negative reviews are balanced. Note that among the four product types, *DVD* and *Book* are similar domains sharing many common sentiment words, whereas *Kitchen* and *Electronics* are more correlated with each other because most of kitchen appliances belong to electronics.

4.1   Data Preprocessing

For the **20NG** data set, we removed the "headers", "footers" and "quotes". When building the vocabulary for each source domain, we ignored terms that have a document frequency strictly higher than 0.5 or strictly lower than 0.005. The total number of terms in the combination of source vocabularies is 5,588. To simulate the transfer learning setup, we sampled 1,000 documents from each source domain as our unlabeled source data set and another 800 documents with labels from each source domain as the labeled source data set. The public data set is an even mixture of the 5 sources with 1,500 documents in total. The target data set is also a mixture of all the source domains with insufficient labels. The final logistic regression model will be tested on a hold-out data set of size 320 sampled from the target domain. For the **AMAZON** data set, we selected the 1,500 most-frequently used unigrams/bigrams from each source domain and combined them to construct the feature set used in the logistic regression model. The total number of features is 2,969. We sampled 1,000 unlabeled reviews from each source domain as the unlabeled source data sets and another 1,000 labeled reviews as the labeled source data sets. The public data set is an even mixture of the 4 product types with 2,000 reviews in total. The counts of features were also transformed to a normalized tf-idf representation. The size of the target data set is 1,500 and the performance will be evaluated on a hold-out test set of size 600 sampled from the target domain. For both data sets, validation sets were reserved for the grid search of regularization parameters. We explored the performance of our method (**DPHTL**) across varying mixtures of sources, varying percentage of target labels and varying privacy requirements.

4.2   Differentially Private Importance Weighting

For the first set of experiments, we study the utility of the differentially private importance weighting mechanism (Step 2–4 in Algorithm 1) on both data sets. As shown by the results, our method can indeed reveal the proportion of each source domain in the target domain even under stringent privacy requirements. We set the regularization parameter λ_{IW} to be 0.001 after grid search among several candidates and plot the mean squared error (MSE) between the original proportion vector and the "hypothesis weights" vector determined by our method across varying privacy requirements ϵ. The experiments were repeated 20 times with different samples and the the error bars in the figures represent the standard errors.

The results (see Figs. 2 and 3) clearly show that the "hypothesis weights" vector is a good privacy-preserving estimation of the proportion of source domains in the target domain for both **20NG** and **AMAZON**. As the privacy requirement is relaxed (ϵ increases), the MSE will approach zero, reflecting the fact that the weight assigned to each source hypothesis is directly affected by the relationship between the source and the target. It is crucial in avoiding negative transfer as brute-force transfer may actually hurt performance if the sources are too dissimilar to the target [22].

4.3   Differentially Private Hypothesis Transfer Learning

For the second set of experiments, we investigate the privacy-utility trade-off of our **DPHTL** method and compare its performance with those of several baselines. The first baseline **SOURCE** is the best performer on the target test set among all the differentially private source hypotheses. The second baseline **AVERAGE** is the posterior Bayesian logistic regression model which assigns equal weights to all the source hypotheses in constructing the Gaussian prior. The third baseline is referred to as **SOFT**, which represents building the logistic regression model using soft labels. It is an adaptation of the work proposed by Hamm et al. in [8] under our transfer learning setting. More specifically, all

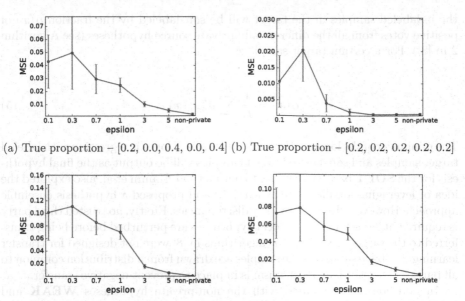

(a) True proportion – [0.2, 0.0, 0.4, 0.0, 0.4] (b) True proportion – [0.2, 0.2, 0.2, 0.2, 0.2]

(c) True proportion – [0.2, 0.0, 0.0, 0.8, 0.0] (d) True proportion – [0.0, 0.0, 0.6, 0.0, 0.4]

Fig. 2. MSE between differentially private hypothesis weight vector and true proportion (**20NG**).

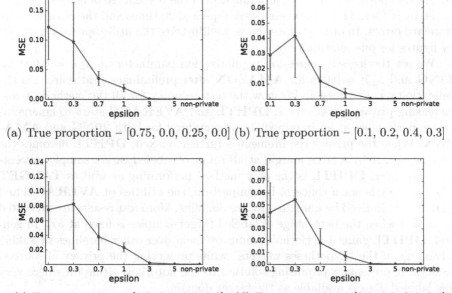

(a) True proportion – [0.75, 0.0, 0.25, 0.0] (b) True proportion – [0.1, 0.2, 0.4, 0.3]

(c) True proportion – [0.0, 0.5, 0.5, 0.0] (d) True proportion – [0.0, 0.2, 0.4, 0.4]

Fig. 3. MSE between differentially private hypothesis weight vector and true proportion (**AMAZON**).

the unlabeled samples in the target will be soft-labeled by the fraction $\alpha(x)$ of positive votes from all the differentially private source hypotheses (see Algorithm 2 in [8]). For a certain target sample x,

$$\alpha(x) = \frac{1}{K} \sum_{k=1}^{K} \theta_{priv}^{k}(x) \tag{5}$$

The minimizer of the loss function of the regularized logistic regression with labeled target samples and soft-labeled target samples will be output as the final hypothesis for the **SOFT** baseline. Similar to our method, Hamm et al. also explored the idea of leveraging auxiliary unlabeled data and proposed a hypothesis ensemble approach. However, there are two key discrepancies. Firstly, no trusted third party is required in our setting so the local hypotheses are perturbed before being transferred to the target. Secondly, the algorithms in [8] were not designed for transfer learning and by assumption the samples are drawn from a distribution common to all parties. Therefore, no mechanism is in place to prevent negative transfer.

In addition, we compared with the non-private hypotheses **WEAK** and **TARGET** trained on the target. The hypothesis **WEAK** is trained using the limited labeled samples in the target data set. The hypothesis **TARGET** is trained using the true labels of all the samples (both labeled and unlabeled) in the target data set and can be considered as the best possible performer since it has access to all the true labels and ignores privacy. The test accuracy of the obtained hypotheses is plotted as a function of the percentage of labeled samples in the target set. The experiments were repeated 20 times and the error bars are standard errors. In order to save space, we illustrate the utility-privacy trade-off by figures for one mixture only.

We set the logistic regression regularization parameter at $\lambda_{LR} = 0.003$ for **20NG** and $\lambda_{LR} = 0.005$ for **AMAZON** after preliminary grid search on the validation set. Figures 4 and 5 show the test accuracy for all the methods across increasing privacy parameter ϵ. **DPHTL** and **AVERAGE** start to emerge as better performers at an intermediate privacy level for both **20NG** and **AMA-ZON**. When the privacy requirement is further relaxed, **DPHTL** becomes the best performer by a large margin at all range of labeled target sample percentage. Moreover, **DPHTL** is the only method performing as well as **TARGET** when privacy is not a concern. In comparison, the utilities of **AVERAGE** and **SOFT** are hindered by unrelated source domains. More test results are presented in Table 2 when the percentage of labeled target samples is fixed at 5%. In general, **DPHTL** gains a performance improvement over other baselines by taking advantage of the "hypotheses weights" while preserving the privacy of sources. Besides, it presents a promising solution to the notorious situation where very few labeled data is available at the target domain.

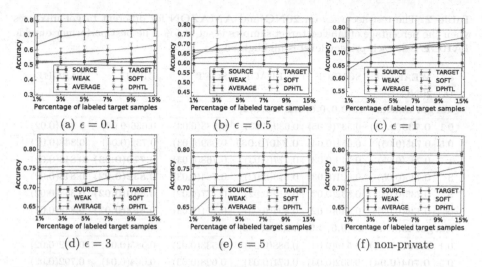

Fig. 4. The test accuracy comparison of **DPHTL** and baselines as a function of percentage of labeled target samples for **20NG**. In the target set, 60% are sampled from source domain 1 and 40% are sampled from source domain 4.

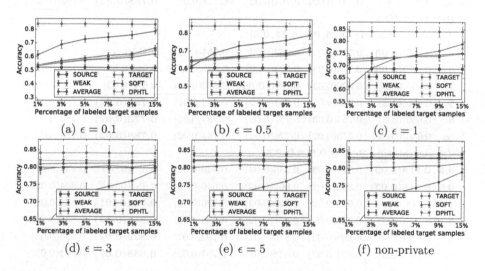

Fig. 5. The test accuracy comparison of **DPHTL** and baselines as a function of percentage of labeled target samples for **AMAZON**. In the target set, 25% are sampled from source domain 2 and 75% are sampled from source domain 4.

Table 2. The test accuracy on **20NG** and **AMAZON** for different target mixtures when the percentage of labeled target samples is set at 5%. The best performer outside **TARGET** at each privacy level is highlighted.

ϵ	WEAK	SOURCE	AVERAGE	SOFT	DPHTL	TARGET
(a) 20NG						
Target mixture 1: $[0.0, 0.6, 0.2, 0.0, 0.2]$						
0.1	**0.735(0.05)**	0.537(0.05)	0.639(0.03)	0.533(0.03)	0.636(0.03)	0.848(0.02)
0.5	0.735(0.05)	0.651(0.04)	**0.740(0.03)**	0.659(0.03)	0.732(0.04)	0.848(0.02)
1	0.735(0.05)	0.740(0.03)	0.777(0.03)	0.768(0.03)	**0.779(0.03)**	0.848(0.02)
3	0.735(0.05)	0.821(0.03)	0.813(0.02)	0.784(0.03)	**0.832(0.02)**	0.848(0.02)
5	0.735(0.05)	0.832(0.02)	0.819(0.02)	0.777(0.02)	**0.840(0.02)**	0.848(0.02)
∞	0.735(0.05)	0.840(0.02)	0.824(0.02)	0.771(0.03)	**0.853(0.02)**	0.848(0.02)
Target mixture 2: $[0.5, 0.0, 0.0, 0.5, 0.0]$						
0.1	**0.704(0.04)**	0.540(0.04)	0.588(0.04)	0.533(0.02)	0.585(0.04)	0.792(0.02)
0.5	**0.704(0.04)**	0.602(0.04)	0.671(0.03)	0.628(0.03)	0.664(0.04)	0.792(0.02)
1	0.704(0.04)	0.654(0.04)	0.714(0.02)	**0.725(0.03)**	0.716(0.03)	0.792(0.02)
3	0.704(0.04)	0.745(0.03)	0.747(0.02)	0.734(0.02)	**0.765(0.02)**	0.792(0.02)
5	0.704(0.04)	0.771(0.02)	0.751(0.02)	0.730(0.02)	**0.778(0.01)**	0.792(0.02)
∞	0.704(0.04)	0.781(0.02)	0.764(0.02)	0.728(0.02)	**0.788(0.02)**	0.792(0.02)
(b) AMAZON						
Target mixture 1: $[0.25, 0.0, 0.75, 0.0]$						
0.1	**0.650(0.03)**	0.523(0.03)	0.563(0.03)	0.554(0.02)	0.555(0.03)	0.780(0.02)
0.5	**0.650(0.03)**	0.574(0.03)	0.623(0.02)	0.625(0.03)	0.614(0.03)	0.780(0.02)
1	0.650(0.03)	0.629(0.03)	**0.681(0.02)**	0.676(0.02)	0.670(0.02)	0.780(0.02)
3	0.650(0.03)	0.723(0.02)	0.751(0.02)	0.744(0.02)	**0.755(0.02)**	0.780(0.02)
5	0.650(0.03)	0.753(0.03)	0.767(0.02)	0.754(0.02)	**0.773(0.02)**	0.780(0.02)
∞	0.650(0.03)	0.770(0.02)	0.776(0.02)	0.764(0.02)	**0.788(0.02)**	0.780(0.02)
Target mixture 2: $[0.0, 0.5, 0.0, 0.5]$						
0.1	**0.718(0.03)**	0.521(0.03)	0.586(0.03)	0.549(0.02)	0.575(0.03)	0.834(0.02)
0.5	**0.718(0.03)**	0.598(0.02)	0.662(0.02)	0.646(0.03)	0.657(0.03)	0.834(0.02)
1	0.718(0.03)	0.672(0.02)	0.729(0.02)	0.726(0.02)	**0.734(0.01)**	0.834(0.02)
3	0.718(0.03)	0.786(0.02)	0.801(0.02)	0.786(0.02)	**0.814(0.02)**	0.834(0.02)
5	0.718(0.03)	0.808(0.02)	0.815(0.01)	0.792(0.02)	**0.829(0.02)**	0.834(0.02)
∞	0.718(0.03)	0.823(0.02)	0.818(0.01)	0.794(0.02)	**0.838(0.02)**	0.834(0.02)

5   Conclusion

This paper proposes a multiple-source hypothesis transfer learning system that protects the differential privacy of sources. By leveraging the relatively abundant supply of unlabeled samples and an auxiliary public data set, we derive the relationship between sources and target in a privacy-preserving manner. Our hypothesis ensemble approach incorporates this relationship information to avoid negative transfer when constructing the Gaussian prior for the target logistic regression model. Moreover, our approach provides a promising and effective solution when the labeled target samples are scarce. Experimental results on benchmark data sets confirm our performance improvement over several baselines from recent work.

Acknowledgements. We would like to thank the anonymous reviewers for their helpful comments. This work is partially supported by a grant from the Army Research Laboratory W911NF-17-2-0110. Quanquan Gu is partly supported by the National Science Foundation SaTC CNS-1717206. The views and conclusions contained in this paper are those of the authors and should not be interpreted as representing any funding agencies.

References

1. Blitzer, J., Dredze, M., Pereira, F.: Biographies, bollywood, boom-boxes and blenders: domain adaptation for sentiment classification. In: Proceedings of the 45th Annual Meeting of the Association of Computational Linguistics, pp. 440–447 (2007)
2. Boker, S.M., et al.: Maintained individual data distributed likelihood estimation (middle). Multivar. Behav. Res. **50**(6), 706–720 (2015)
3. Chaudhuri, K., Monteleoni, C., Sarwate, A.D.: Differentially private empirical risk minimization. J. Mach. Learn. Res. **12**, 1069–1109 (2011)
4. Dwork, C., McSherry, F., Nissim, K., Smith, A.: Calibrating noise to sensitivity in private data analysis. In: Halevi, S., Rabin, T. (eds.) TCC 2006. LNCS, vol. 3876, pp. 265–284. Springer, Heidelberg (2006). https://doi.org/10.1007/11681878_14
5. Dwork, C., Roth, A.: The algorithmic foundations of differential privacy. Found. Trends Theoret. Comput. Sci. **9**(3–4), 211–407 (2014)
6. Garcke, J., Vanck, T.: Importance weighted inductive transfer learning for regression. In: Calders, T., Esposito, F., Hüllermeier, E., Meo, R. (eds.) ECML PKDD 2014. LNCS (LNAI), vol. 8724, pp. 466–481. Springer, Heidelberg (2014). https://doi.org/10.1007/978-3-662-44848-9_30
7. Gupta, S.K., Rana, S., Venkatesh, S.: Differentially private multi-task learning. In: Chau, M., Wang, G.A., Chen, H. (eds.) PAISI 2016. LNCS, vol. 9650, pp. 101–113. Springer, Cham (2016). https://doi.org/10.1007/978-3-319-31863-9_8
8. Hamm, J., Cao, Y., Belkin, M.: Learning privately from multiparty data. In: International Conference on Machine Learning, pp. 555–563 (2016)
9. Ji, Z., Elkan, C.: Differential privacy based on importance weighting. Mach. Learn. **93**(1), 163–183 (2013)
10. Kuzborskij, I., Orabona, F.: Stability and hypothesis transfer learning. In: Proceedings of The 30th International Conference on Machine Learning, pp. 942–950. ACM (2013)
11. Lang, K.: Newsweeder: learning to filter netnews. In: Machine Learning Proceedings 1995, pp. 331–339. Elsevier (1995)
12. Mansour, Y., Mohri, M., Rostamizadeh, A.: Domain adaptation with multiple sources. In: Advances in Neural Information Processing Systems, pp. 1041–1048 (2009)
13. Marx, Z., Rosenstein, M.T., Dietterich, T.G., Kaelbling, L.P.: Two algorithms for transfer learning. Inductive transfer: 10 years later (2008)

14. Mihalkova, L., Mooney, R.J.: Transfer learning by mapping with minimal target data. In: Proceedings of the Association for the Advancement of Artificial Intelligence (AAAI) Workshop on Transfer Learning for Complex Tasks (2008)
15. Nedic, A., Ozdaglar, A.: Distributed subgradient methods for multi-agent optimization. IEEE Trans. Autom. Control **54**(1), 48–61 (2009)
16. Pan, S.J., Yang, Q.: A survey on transfer learning. IEEE Trans. Knowl. Data Eng. **22**(10), 1345–1359 (2010)
17. Pan, W., Zhong, E., Yang, Q.: Transfer learning for text mining. In: Aggarwal, C., Zhai, C. (eds.) Mining Text Data, pp. 223–257. Springer, Boston (2012)
18. Papernot, N., Song, S., Mironov, I., Raghunathan, A., Talwar, K., Erlingsson, U.: Scalable private learning with PATE. In: International Conference on Learning Representations (2018). https://openreview.net/forum?id=rkZB1XbRZ
19. Pathak, M., Rane, S., Raj, B.: Multiparty differential privacy via aggregation of locally trained classifiers. In: Advances in Neural Information Processing Systems, pp. 1876–1884 (2010)
20. Raina, R., Battle, A., Lee, H., Packer, B., Ng, A.Y.: Self-taught learning: transfer learning from unlabeled data. In: Proceedings of the 24th International Conference on Machine Learning, pp. 759–766. ACM (2007)
21. Raina, R., Ng, A.Y., Koller, D.: Constructing informative priors using transfer learning. In: Proceedings of the 23rd International Conference on Machine Learning, pp. 713–720. ACM (2006)
22. Rosenstein, M.T., Marx, Z., Kaelbling, L.P., Dietterich, T.G.: To transfer or not to transfer. In: NIPS 2005 Workshop on Transfer Learning, vol. 898, pp. 1–4 (2005)
23. Shokri, R., Shmatikov, V.: Privacy-preserving deep learning. In: Proceedings of the 22nd ACM SIGSAC Conference on Computer and Communications Security, pp. 1310–1321. ACM (2015)
24. Wang, Y., et al.: Privacy preserving distributed deep learning and its application in credit card fraud detection. In: 2018 17th IEEE International Conference on Trust, Security and Privacy In Computing And Communications/12th IEEE International Conference on Big Data Science and Engineering (TrustCom/BigDataSE), pp. 1070–1078. IEEE (2018)
25. Xie, L., Baytas, I.M., Lin, K., Zhou, J.: Privacy-preserving distributed multi-task learning with asynchronous updates. In: Proceedings of the 23rd ACM SIGKDD International Conference on Knowledge Discovery and Data Mining, pp. 1195–1204. ACM (2017)

Information-Theoretic Transfer Learning Framework for Bayesian Optimisation

Anil Ramachandran[✉], Sunil Gupta, Santu Rana, and Svetha Venkatesh

Centre for Pattern Recognition and Data Analytics (PRaDA),
Deakin University, Geelong 3216, Australia
{aramac,sunil.gupta,santu.rana,svetha.venkatesh}@deakin.edu.au

Abstract. Transfer learning in Bayesian optimisation is a popular way to alleviate "cold start" issue. However, most of the existing transfer learning algorithms use overall function space similarity, not a more aligned similarity measure for Bayesian optimisation based on the location of the optima. That makes these algorithms fragile to noisy perturbations, and even simple scaling of function values. In this paper, we propose a robust transfer learning based approach that transfer knowledge of the optima using a consistent probabilistic framework. From the finite samples for both source and target, a distribution on the optima is computed and then divergence between these distributions are used to compute similarities. Based on the similarities a mixture distribution is constructed, which is then used to build a new information-theoretic acquisition function in a manner similar to Predictive Entropy Search (PES). The proposed approach also offers desirable "no bias" transfer learning in the limit. Experiments on both synthetic functions and a set of hyperparameter tuning tests clearly demonstrate the effectiveness of our approach compared to the existing transfer learning methods. Code related to this paper is available at: https://github.com/AnilRamachandran/ITTLBO.git and Data related to this paper is available at: https://doi.org/10.7910/DVN/LRNLZV.

1 Introduction

Experimental optimisation is a widely used technique in scientific studies and engineering design to find optimal solutions to a variety of problems via experimentation. In its most common form it is used to seek globally optimal solutions of unknown black-box functions, which are often expensive to evaluate. Bayesian optimisation offers a sample efficient solution to these kinds of problems. For example, it has been used in the scientific community for synthetic gene design [1], alloy optimisation [2] and fiber yield optimisation [3]. In machine learning it is often used to find the optimal hyperparameters for the learning algorithms and the optimisation routines [4]. Bayesian optimisation requires a probabilistic model of the function that is being optimised. Gaussian process is often a popular choice as the prior for the function model. Posterior computed based on the existing observations is then used to build a computationally cheap *acquisition*

© Springer Nature Switzerland AG 2019
M. Berlingerio et al. (Eds.): ECML PKDD 2018, LNAI 11052, pp. 827–842, 2019.
https://doi.org/10.1007/978-3-030-10928-8_49

function to seek the next evaluation point. There are a variety of acquisition functions such as Probability of Improvement [5], Expected Improvement [6], GP-Upper Confidence Bound (UCB) [7], Predictive Entropy Search (PES) [8] etc. Nearly, all the acquisition functions address the trade off between sampling the regions where the posterior mean is high (*exploitation*) and sampling the regions where uncertainty is high (*exploration*) [9]. The maximiser of the acquisition function offers the best chance of being the optima. Since the acquisition functions are computationally cheap they admit standard global optimisation routines such as DIRECT [10]. Bayesian optimisation runs in a loop with experiments sequentially being performed at the locations of the respective acquisition functions optima until an acceptable solution is found or the iteration budget is exhausted. However, the generic Bayesian optimisation approach is susceptible to the "cold start" problem where it may recommend several points with low function values before reaching a high function value region. For experiments which are highly expensive, for example, hyperparameter tuning of a large deep network on a massive training data that takes weeks to get trained on a many clusters of GPUs, the cost due to "cold start" can be quite substantial, and we like to avoid that.

A principled approach to alleviate the "cold start" issue is to utilize the knowledge acquired in any previous function (source) optimisations to achieve faster convergence in the optimisation of a new related function (target) via transfer learning. State of the art work include, [11], in which the authors assume high similarity in the rank-space of function values and develop a model that transfers function ranking between source and target. Similarly, [12] assumes high similarity between all the functions when their mean functions are subtracted out. Both these methods do not model task to task relatedness, and are hence susceptible to negative transfer in the presence of a very different source function. In [13], the authors assume that the task relationships are already known and utilize the knowledge as per source target relationship in a multi-task setting. In [14,15], the authors propose to compute overall similarity between functions based on all the observations and [16,17] uses meta-features to compute similarities between functions. Whilst the former can get tricked by scaling of the functions, the latter depends crucially on the availability of right kind of meta-features. Most of them assume function space similarity, instead of a more aligned similarity measure in terms of the location of the optima, and hence may fail to find the global optima if the functions have different scaling or relatedness between them. For this reason, a robust transfer learning framework for Bayesian optimisation which transfer knowledge based only on the proximity of the function optima is still an open problem.

In this paper, we propose a transfer learning framework for Bayesian optimisation that performs knowledge transfer based on the location of the optima in the input space and thus invariant to scaling and robust to noise in function. Since the location of global optima is not known perfectly with finite observations (be it source or target), we use probabilistic knowledge of the optima. Thompson sampling can be used to obtain the optima distribution for both sources and the target. Following that we propose a novel measure of relatedness by computing the divergence between these distributions. These distributions are then merged into a mixture distribution based on the similarities and is used to build the acquisition function. We use Predictive Entropy Search (PES) as a vehicle to build our new acquisition function. Our method offers a "no bias" algorithm, in a sense that in the limit ($T \to \infty$) the similarity to any random source tends to zero with probability 1, making Bayesian optimisation for the target free from any source induced bias in the limit. We validate our framework through application to optimisation of both synthetic functions and real world experiments. We compare our method with three well known transfer learning methods as well as with the generic Bayesian optimisation algorithm and demonstrate the superiority of our method.

2   Background

2.1   Gaussian Process

Gaussian process is an effective method for regression and classification problems and has received considerable attention in machine learning community. It serves as a probabilistic prior over smooth functions and is a generalization of infinite collection of normally distributed random variables. A Gaussian process can be completely specified by a mean function, $\mu(\mathbf{x})$ and covariance function, $k(\mathbf{x}, \mathbf{x}')$. Due to these properties, we can model a smooth function as a draw from a Gaussian process. Formally,

$$f(\mathbf{x}) \sim \mathrm{GP}(\mu(\mathbf{x}), k(\mathbf{x}, \mathbf{x}')) \tag{1}$$

where the function value at a point $\mathbf{x}$, i.e. $f(\mathbf{x})$ is a normal distribution and the relation between the function values at any two points $\mathbf{x}$ and $\mathbf{x}'$ is modeled by covariance function $k(\mathbf{x}, \mathbf{x}')$. Without loss in generality, the mean function can be defined to be zero thus making the Gaussian process fully specified by the covariance function [18,19]. Popular choices of covariance functions include squared exponential kernel, Matérn kernel, linear kernel, etc. We assume the function measurements are noisy, i.e. observations $y_i = f(\mathbf{x}_i) + \epsilon_i$, where $\epsilon_i \sim \mathcal{N}(0, \sigma^2)$ is the measurement noise. Collectively, we denote the observations as $\mathcal{D}_n = \{\mathbf{x}_{1:n}, \mathbf{y}_{1:n}\}$. The function values $[f(\mathbf{x}_1), \ldots, f(\mathbf{x}_n)]$ follow a multivariate Gaussian distribution $\mathcal{N}(0, \mathbf{K})$, where

$$\mathbf{K} = \begin{bmatrix} k(\mathbf{x}_1, \mathbf{x}_1) & \ldots & k(\mathbf{x}_1, \mathbf{x}_n) \\ \vdots & \ddots & \vdots \\ k(\mathbf{x}_n, \mathbf{x}_1) & \ldots & k(\mathbf{x}_n, \mathbf{x}_n) \end{bmatrix} \tag{2}$$

Given a new point $\mathbf{x}$, $\mathbf{y}_{1:n}$ and $f(\mathbf{x})$ are jointly Gaussian, then by the properties of Gaussian process we can write

$$\begin{bmatrix} \mathbf{y}_{1:n} \\ f(\mathbf{x}) \end{bmatrix} \sim \mathcal{N} \left(0, \begin{bmatrix} \mathbf{K} + \sigma^2 \mathbf{I} & \mathbf{k} \\ \mathbf{k}^T & k(\mathbf{x}, \mathbf{x}) \end{bmatrix} \right) \tag{3}$$

where $\mathbf{k} = [k(\mathbf{x}, \mathbf{x}_1)\, k(\mathbf{x}, \mathbf{x}_2)\ \ldots\ k(\mathbf{x}, \mathbf{x}_n))]$. Using Sherman-Morrison-Woodbury formula [18] we can write the predictive distribution at any $\mathbf{x}$ as

$$p(y \mid \mathcal{D}_n, \mathbf{x}) = \mathcal{N}(\mu_n(\mathbf{x}), \sigma_n^2(\mathbf{x})) \tag{4}$$

where the predictive mean $\mu_n(\mathbf{x})$ is given as

$$\mu_n(\mathbf{x}) = \mathbf{k}^{\mathbf{T}} \left[\mathbf{K} + \sigma^{\mathbf{2}}\mathbf{I}\right]^{-1} \mathbf{y}_{1:n} \tag{5}$$

and the predictive variance $\sigma_n^2(\mathbf{x})$ is given as

$$\sigma_n^2(\mathbf{x}) = k(\mathbf{x}, \mathbf{x}) - \mathbf{k}^{\mathbf{T}} \left[\mathbf{K} + \sigma^{\mathbf{2}}\mathbf{I}\right]^{-1} \mathbf{k}. \tag{6}$$

2.2 Bayesian Optimisation

Bayesian optimisation is an elegant approach for the global optimisation of expensive, black-box functions. Given a small set of observations from previous function evaluations, Bayesian optimisation proceeds by building a probabilistic model of the function generally using a Gaussian process (GP). However, other methods are also used for function modeling, e.g. random forests and Bayesian neural networks [20,21]. The model of the function is then combined with the existing observations to derive a posterior distribution. This distribution is then used to construct a surrogate utility function called *acquisition function*, which is cheap to evaluate and finally optimised to recommend the next function evaluation point while keeping a trade-off between exploitation and exploration [9,22].

Several popular acquisition functions are available in literature: Probability of improvement (PI), which takes into account the improvement in the probability over the current best function value [5], Expected improvement (EI), which considers the expected improvement over the current best [6] and GP-UCB, which selects the evaluation point based on the upper confidence bound [7]. These functions are based on predictive mean and variance of the model posterior. An alternative acquisition function that maximizes the expected posterior information gain about the location of the global optima over an input space grid is proposed in [23,24]. Another information-based acquisition function called Predictive Entropy Search (PES) extended this approach to continuous search spaces [8].

2.3 Transfer Learning for Bayesian Optimisation

Transfer learning methods in Bayesian optimisation utilize ancillary information acquired from previous function (source) optimisations to achieve faster optimisation for a new related function (target). The crucial requirement is to determine the source function which is highly similar to the target. Limited work exist for transfer learning in Bayesian optimisation and most of them have made assumption regarding similarity between the source and the target. For example, Bardenet et al. [11] proposed the first work on transfer learning for

Bayesian optimisation by transferring the source knowledge via a ranking function, which was assumed to be applicable for the target function as well. Another similar approach proposed in [12] assumes that the deviations of a function from its mean scaled through the standard deviation are transferable from source to target. Both these methods have strong assumption regarding source-target similarity and hence experience difficulty to find the global optima if among many sources, some have different function shapes or different optima compared to the target. To handle any potential differences between the source and the target, an alternate transfer learning framework was proposed by Joy et al. [14] modeling source data as noisy observations of the target function. The noise envelope is estimated by taking difference between any available source/target data and can be used to distinguish a related source from an unrelated one. Using the previous method as a base, Ramachandran et al. [25] proposed another transfer learning method for Bayesian optimisation that select sources which are highly related to the target. The authors use multi-armed bandit formulation for the selection of optimal sources. However, these methods would not be able to leverage from a related source having its output scale different from that of the target even though both source and target have their optima located at the same place. Meta-learning approaches proposed in [16,17] can estimate source/target similarities, however it requires meta features for source and target functions, which may not be available in general.

3   Proposed Method

We propose a new information-theoretic transfer learning framework for Bayesian optimisation that utilizes data from source and target functions to maximize the information about the global minima of the target function. We first discuss an information-theoretic framework for Bayesian optimisation known as Predictive Entropy Search (PES) and then present our proposed transfer learning model.

3.1   Optimisation of Black-Box Functions Using Predictive Entropy Search (PES)

Let $f(\mathbf{x})$ be an expensive black-box function and we need to find its global minimizer $\mathbf{x}_* = \underset{\mathbf{x} \in \mathcal{X}}{\mathrm{argmax}} f(\mathbf{x})$ over some domain $\mathcal{X} \subset \mathbb{R}^d$. Let $\mathcal{D}_n = \{(\mathbf{x}_i, y_i)\}_{i=1}^n$ denote the noisy observations from the function $f(\mathbf{x})$ under the observation model $y_i = f(\mathbf{x}_i) + \epsilon_i$ where $\epsilon_i \sim \mathcal{N}(0, \sigma^2)$ is the measurement noise. Predictive entropy search is an acquisition function for Bayesian optimisation that recommends the next function evaluation point with an aim to maximize the information about $\mathbf{x}_*$, whose posterior distribution is $p(\mathbf{x}_* \mid \mathcal{D}_n)$. The posterior distribution represents the likelihood of a location being the function global minimum after observing $\mathcal{D}_n$ observations. The information about $\mathbf{x}_*$ is measured in terms of the differential entropy between $p(\mathbf{x}_* \mid \mathcal{D}_n)$ and the expected value of $p(\mathbf{x}_* \mid \mathcal{D}_n \cup \{(\mathbf{x}, y)\})$. Formally, the PES acquisition function selects a point $\mathbf{x}_{n+1}$ that maximizes the information about $\mathbf{x}_*$ as

$$\mathbf{x}_{n+1} = \operatorname*{argmax}_{\mathbf{x} \in \mathcal{X}} \alpha_n(\mathbf{x}) = \mathbb{H}\left[p(\mathbf{x}_* \mid \mathcal{D}_n)\right] - \mathbb{E}_{p(y \mid \mathcal{D}_n, \mathbf{x})}\left[\mathbb{H}\left[p\left(\mathbf{x}_* \mid \mathcal{D}_n \cup \{(\mathbf{x}, y)\}\right)\right]\right] \quad (7)$$

where $\mathbb{H}[p(\mathbf{x})] = -\int p(\mathbf{x})\log p(\mathbf{x})d\mathbf{x}$ is the differential entropy and the expectation is with respect to the posterior predictive distribution of y given $\mathbf{x}$. Evaluation of the acquisition function in (7) is intractable and requires discretisation [24]. Noting that mutual information is a symmetric function, an easier yet equivalent formulation is as below:

$$\mathbf{x}_{n+1} = \operatorname*{argmax}_{\mathbf{x} \in \mathcal{X}} \alpha_n(\mathbf{x}) = \mathbb{H}\left[p(y \mid \mathcal{D}_n, \mathbf{x})\right] - \mathbb{E}_{p(\mathbf{x}_* \mid \mathcal{D}_n)}\left[\mathbb{H}\left[p\left(y \mid \mathcal{D}_n, \mathbf{x}, \mathbf{x}_*\right)\right]\right] \quad (8)$$

The first term in (8) involves the predictive distribution $p(y \mid \mathcal{D}_n, \mathbf{x})$, which is Gaussian under Gaussian process modeling of $f(\mathbf{x})$. Therefore, we have

$$\mathbb{H}\left[p(y \mid \mathcal{D}_n, \mathbf{x})\right] = 0.5 \log \left[2\pi e \left(v_n(\mathbf{x}) + \sigma^2\right)\right]$$

where $v_n(\mathbf{x})$ is the variance of $p(y \mid \mathcal{D}_n, \mathbf{x})$ and σ^2 is the variance due to measurement noise. The second term involves $p(y \mid \mathcal{D}_n, \mathbf{x}, \mathbf{x}_*)$, which is the posterior distribution for y given the observed data $\mathcal{D}_n$ and the location of the global minimizer of f. An exact form for the distribution $p(y \mid \mathcal{D}_n, \mathbf{x}, \mathbf{x}_*)$ is intractable and its entropy $\mathbb{H}[p(y \mid \mathcal{D}_n, \mathbf{x}, \mathbf{x}_*)]$ for a given $\mathbf{x}_*$ sample is usually computed using expectation propagation [26]. The expected value of the entropy with respect to the distribution $p(\mathbf{x}_* \mid \mathcal{D}_n)$ is approximated by averaging the entropy for Monte Carlo samples of $\mathbf{x}_*$. A well known approach called Thompson sampling is typically used to draw the $\mathbf{x}_*$ samples [27]. Using the estimated acquisition function, PES recommends the next evaluation point by the following maximization

$$\mathbf{x}_{n+1} = \operatorname*{argmax}_{\mathbf{x} \in \mathcal{X}} \alpha_n(\mathbf{x}) = 0.5 \log \left[v_n(\mathbf{x}) + \sigma^2\right] - \frac{1}{M} \sum_{i=1}^{M} 0.5 \log \left[v_n^{(i)}\left(\mathbf{x} \mid \mathbf{x}_*^{(i)}\right) + \sigma^2\right] \quad (9)$$

where M is the number of $\mathbf{x}_*$ samples drawn, $v_n(\mathbf{x})$ and $v_n^{(i)}\left(\mathbf{x} \mid \mathbf{x}_*^{(i)}\right)$ are the predictive variances given i-th sample of $\mathbf{x}_*$. A pseudo-code for the Bayesian optimisation using PES acquisition function is presented in **Algorithm** 1. For further details on PES, we refer the reader to [8].

3.2   The Proposed Transfer Learning Method

Since the goal of the Bayesian optimisation is to find the optima (or minima) of a function, we develop a transfer learning method that directly transfers knowledge about the location of global minima from source functions to the target. Let $\left\{\{\mathbf{x}_i^s, y_i^s\}_{i=1}^{N_s}\right\}_{s=1}^{S}$ be the observations from S sources under the observation model $y_i^s = f^s(\mathbf{x}_i^s) + \epsilon_i^s$ where $\epsilon_i^s \sim \mathcal{N}(0, \sigma^2)$ and $p^s(\mathbf{x}_*)$ be the global minima distribution of the global minimizer $\mathbf{x}_*$ from each source s. Similarly, let $\{\mathbf{x}_j, y_j\}_{j=1}^{n_0}$

Algorithm 1. Bayesian optimisation using PES as acquisition function

1. **Input:** Initial observations $\{(\mathbf{x}_i, y_i)\}_{i=1}^{n_0}$
2. **Output:** $\{\mathbf{x}_n, y_n\}_{n=1}^{T}$
3. **for** $n = n_0, \ldots, T$ **do**
   (a) Draw M samples of $\mathbf{x}_*$ from the posterior Gaussian process of the target function.
   (b) Use $\mathbf{x}_*$ samples to compute $\alpha_n(\mathbf{x})$ and maximize it as in (9) to recommend a new point $\mathbf{x}_n$.
   (c) Evaluate the target function at $\mathbf{x}_n$: $y_n = f(\mathbf{x}_n) + \epsilon_n$ where $\epsilon_n \sim \mathcal{N}(0, \sigma^2)$.
   (d) Augment $(\mathbf{x}_n, y_n)$ to the target observations and update the posterior Gaussian process.
4. **end for**

---

be the target observations under the observation model $y_j = f(\mathbf{x}_j) + \epsilon_j$ up to iteration n_0, where $\epsilon_j \sim \mathcal{N}(0, \sigma^2)$ and $p(\mathbf{x}_*)$ be the global minima distribution of its global minimizer $\mathbf{x}_*$. Our proposed transfer learning scheme intervenes into the distribution of $\mathbf{x}_*$ and modifies it to become a mixture distribution of $p(\mathbf{x}_*)$ from the target and $p^s(\mathbf{x}_*)$ from each source s. The proposed mixture distribution can be formally written as

$$p^{\mathrm{TL}}(\mathbf{x}_*) = \pi_0 p(\mathbf{x}_*) + \pi_1 p^1(\mathbf{x}_*) + \ldots + \pi_S p^S(\mathbf{x}_*) \tag{10}$$

where $\pi_0, \pi_1, \ldots, \pi_S$ are the mixture coefficients such that $\sum_{s=0}^{S} \pi_s = 1$. Our model sets these mixture coefficients in proportion to the similarity between the target and a source. We first define a similarity measure ψ_s between the target $p(\mathbf{x}_*)$ and a source $p^s(\mathbf{x}_*)$. ψ_0 is assumed to be the similarity of $p(\mathbf{x}_*)$ with itself and is set to 1. We define $\pi_0, \pi_1, \ldots, \pi_S$ as

$$\pi_s = \frac{\psi_s}{\sum_{s=0}^{S} \psi_s}. \tag{11}$$

Given the proposed mixture distribution $p^{\mathrm{TL}}(\mathbf{x}_*)$, our proposed information-theoretic transfer learning maximizes the following acquisition function

$$\mathbf{x}_{n+1} = \mathrm{argmax}_{\mathbf{x} \in \mathcal{X}} \alpha_n(\mathbf{x}) = \mathbb{H}\left[p(y \mid \mathcal{D}_n, \mathbf{x})\right] - \mathbb{E}_{p^{\mathrm{TL}}(\mathbf{x}_* \mid \mathcal{D}_n)}\left[\mathbb{H}\left[p(y \mid \mathcal{D}_n, \mathbf{x}, \mathbf{x}_*)\right]\right] \tag{12}$$

The entropies in the above equation are computed as in (9).

Source/Target Similarity Measure. Given two probability distributions $p(\mathbf{x}_*)$ and $p^s(\mathbf{x}_*)$, we measure a divergence between p and p^s using their Kullback-Leibler (KL) divergence. KL-divergence takes non negative values and is unbounded. After measuring the divergence (denoted as $D(p^s\|p)$), we map it to a similarity measure as $\psi_s = \exp(-\frac{D(p^s\|p)}{\eta})$, where $\eta > 0$ is a model hyperparameter. Any other divergence measures such Hellinger distance, total variation distance or χ^2-divergence could also be used [28].

Since we have access to only samples of $\mathbf{x}_*$ and no direct access to the closed form of the probability distributions $p(\mathbf{x}_*)$ and $p^s(\mathbf{x}_*)$, we need to estimate these density functions before computing the KL-divergence. A naive way to estimate probability density is via histograms with uniform binning, however, this method quickly becomes inefficient in the number of samples. To avoid this inefficiency, we estimate the KL-divergence based on nearest neighbor (NN) distances [29, 30], an approach that relies on non-parametric density estimation and then uses it to compute the KL-divergence. We refer to this estimate as NN-divergence. Let $\left\{ \mathbf{x}_*^{s,1}, \ldots, \mathbf{x}_*^{s,n} \right\}$ and $\left\{ \mathbf{x}_*^1, \ldots, \mathbf{x}_*^m \right\}$ be the d-dimensional samples drawn from $p^s(\mathbf{x}_*)$ and $p(\mathbf{x}_*)$ respectively. The NN divergence is estimated as

$$D_{n,m} \left(p^s(\mathbf{x}_*) \parallel p(\mathbf{x}_*) \right) = \frac{d}{n} \sum_{i=1}^{n} \log \frac{\tau_m(i)}{\rho_n(i)} + \log \frac{m}{n-1} \tag{13}$$

where

$$\tau_m(i) = \min_{j=1,\ldots,m} \| \mathbf{x}_*^{s,i} - \mathbf{x}_*^j \|$$

is the distance of $\mathbf{x}_*^{s,i}$ to its nearest neighbor in the target sample set $\left\{ \mathbf{x}_*^1, \ldots, \mathbf{x}_*^m \right\}$ and

$$\rho_n(i) = \min_{j=1,\ldots,n, j \neq i} \| \mathbf{x}_*^{s,i} - \mathbf{x}_*^{s,j} \|$$

is the distance of $\mathbf{x}_*^{s,i}$ to its nearest neighbor within the source sample set $\left\{ \mathbf{x}_*^{s,1}, \ldots, \mathbf{x}_*^{s,n} \right\}$.

Discussion. Since the target does not have many observations available in the initial iterations, the global minima samples from target, $p(\mathbf{x}_*)$ are distributed widely over the domain $\mathcal{X}$. Therefore the NN divergence of each source with the target will have similar values and this in turn causes our transfer learning framework to choose almost equal number of samples from each source in the initial iterations. As the iterations increase, the NN divergence estimate improves as more target observations are made. This increases the contribution of related sources and decreases the contribution of unrelated sources in the mixture distribution $p^{\mathrm{TL}}(\mathbf{x}_*)$. Asymptotically, the contribution of the $p^s(\mathbf{x}_*)$ distribution from an unrelated source becomes zero. This makes our transfer learning algorithm capable of preventing negative transfer from unrelated sources. In the limit when our algorithm has densely sampled the target function, $p(\mathbf{x}_*)$ becomes nearly an impulse function and its KL-divergence with any source becomes extremely large implying that $p^{\mathrm{TL}}(\mathbf{x}_*) \rightarrow p(\mathbf{x}_*)$ and becomes free of any bias from the sources. Figure 1 provides an illustration of the evolution of the mixture distribution $p^{\mathrm{TL}}(\mathbf{x}_*)$. We can see that the contribution of unrelated source reduces quickly whilst that of the related source increases. We considered two source functions as:

$$f(\mathbf{x}) = 1 - a * \exp\left(-\frac{1}{2}(\mathbf{x} - \mu)(\mathbf{x} - \mu)^T \right)$$

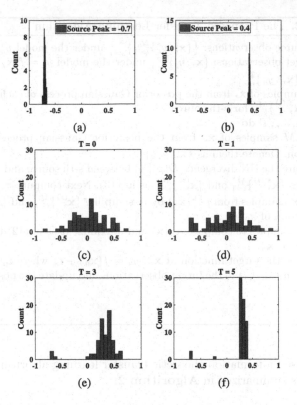

Fig. 1. Evolution of $p^{\mathrm{TL}}(\mathbf{x}_*)$ samples with increasing iterations for a one-dimensional target function (minimum at 0.3); (a)–(b) show histogram representation of $p^s(\mathbf{x}_*)$ for two one-dimensional source functions with minima at -0.7 and 0.4 respectively; (c) shows the histogram representation of $p(\mathbf{x}_*)$ at the start of optimisation; (d)–(f) show the histogram representation of $p^{\mathrm{TL}}(\mathbf{x}_*)$ at $T = 1, 3$ and 5 respectively. The contribution of the source farther from the target reduces quickly while that of the related source increases.

For source 1, $\mu = -0.7$, $a = 2$ and for source 2, $\mu = 0.4$, $a = 2$. The target function has a similar form with $\mu = 0.3$, $a = 1$. For each source, 20 data points are sampled randomly from the support $[-1.5, 1.5]$ and $p^s(\mathbf{x}_*)$ samples are drawn using their posterior Gaussian process models. Figures (1a) and (b) show the histogram counts of source $p^s(\mathbf{x}_*)$ samples. We also show the histogram count of $p(\mathbf{x}_*)$ samples drawn using the posterior Gaussian process models of initial target data (see Figure (1c)). Figures (1d)–(f) show the evolution of $p^{\mathrm{TL}}(\mathbf{x}_*)$ with increasing iterations. Initially $p^{\mathrm{TL}}(\mathbf{x}_*)$ samples are widely distributed. With increasing iterations, the mass of $p^{\mathrm{TL}}(\mathbf{x}_*)$ samples near the global minima location increases as it selects more samples from the closer source (minimum at 0.4). This example illustrates the typical behavior of our transfer learning algorithm in relying more on related sources.

Algorithm 2. The Proposed Transfer Learning Algorithm

---

1. **Input:** Source observations: $\left\{\{\mathbf{x}_i^s, y_i^s\}_{i=1}^{N_s}\right\}_{s=1}^{S}$ under the model $y_i^s = f^s(\mathbf{x}_i^s) + \epsilon_i^s$,
   Initial target observations: $\{\mathbf{x}_j, y_j\}_{j=1}^{n_0}$ under the model $y_j = f(\mathbf{x}_j) + \epsilon_j$.
2. **Output:** $\{\mathbf{x}_n, y_n\}_{n=1}^{T}$.
3. Draw M samples of $\mathbf{x}_*$ from the posterior Gaussian process of each source. Denote them as $\{\mathbf{x}_*^{s,(i)}\}_{i=1}^{M}$ for s-th source.
4. **for** $n = n_0, \ldots, T$ **do**
   (a) Draw M samples of $\mathbf{x}_*$ from the posterior Gaussian process of the target function. Denote them as $\{\mathbf{x}_*^{(i)}\}_{i=1}^{M}$.
   (b) Compute the NN-divergence $D(p^s\|p)$ between s-th source and the target using samples $\{\mathbf{x}_*^{s,(i)}\}_{i=1}^{M}$ and $\{\mathbf{x}_*^{(i)}\}_{i=1}^{M}$ as in (13). Next compute π_s using (11).
   (c) Draw $\mathbf{x}_*$ samples from $p^{\mathrm{TL}}(\mathbf{x}_*)$ by re-sampling $\{\mathbf{x}_*^{(i)}\}_{i=1}^{M}$ and $\{\mathbf{x}_*^{s,(i)}\}_{i=1}^{M}$ in the proportion of $\pi_0, \pi_1, \ldots, \pi_S$.
   (d) Use $\mathbf{x}_*$ samples to compute $\alpha_n(\mathbf{x})$ and maximize it as in (12) to recommend a new point $\mathbf{x}_n$.
   (e) Evaluate the target function at $\mathbf{x}_n$: $y_n = f(\mathbf{x}_n) + \epsilon_n$ where $\epsilon_n \sim \mathcal{N}\left(0, \sigma^2\right)$.
   (f) Augment $(\mathbf{x}_n, y_n)$ to the target observations and update the posterior Gaussian process.
5. **end for**

Our proposed information-theoretic transfer learning algorithm for Bayesian optimisation is summarized in **Algorithm 2**.

4   Experiments

We perform experiments using both synthetic and real optimisation tasks. Through synthetic experiments, we analyze the behavior of our proposed transfer learning method in a controlled setting. Through real data experiments, we show that our method can tune the hyperparameter for support vector machine and elastic net efficiently. For both synthetic and real experiments, we compare our proposed method with three other well known transfer learning methods and with Bayesian optimisation that does not use any knowledge from sources. The following baselines are used:

- **Env-GP:** This algorithm [14] models a source function as noisy measurements of the target function. The noise for each source is estimated separately and then the observations from each source are merged.
- **SMBO:** This algorithm [12] transfers deviation of a function from its mean scaled through the standard deviation. The observations from each source are first standardized and then data from all sources are merged.
- **SCoT:** This algorithm [11] transfers function ranking from a source to the target in latent space. The observations from each source are first adjusted and then data from all sources are merged.

– **Generic-BO (No Transfer):** This baseline is a PES based Bayesian optimisation algorithm (see Algorithm 1) and does not use any information from
source functions. This is used to assess the gain in optimisation performance
due to transfer learning.

4.1   Experimental Setting

We use the square-exponential kernel as the covariance function in Gaussian
process (GP) modeling. All GP hyperparameters are estimated using maximum
a posteriori (MAP) estimation. In all our experiments, the hyperparameter η
(used in similarity measure) is set to 10. All the results are averaged over 10
runs with random initializations.

4.2   Synthetic Experiments

We consider 4 source functions as:

$$f(\mathbf{x}) = 1 - a * \exp\left(-\frac{1}{2}(\mathbf{x} - \boldsymbol{\mu})(\mathbf{x} - \boldsymbol{\mu})^T\right)$$

For source 1, $\boldsymbol{\mu} = [-1.8, -1.8, -1.8]$, $a = 2$. For source 2, $\boldsymbol{\mu} = [-0.7, -0.7, -0.7]$,
$a = 2$. For source 3, $\boldsymbol{\mu} = [0.4, 0.4, 0.4]$, $a = 2$. For source 4, $\boldsymbol{\mu} = [1.5, 1.5, 1.5]$,
$a = 2$. The target function has a similar form with $\boldsymbol{\mu} = [0.3, 0.3, 0.3]$, $a = 1$. For each source, 50 data points were sampled randomly from $[-2, 2]$ along
each dimension. Our transfer learning method selects $p^{\text{TL}}(\mathbf{x}_*)$ samples from any
of the four sources based on the estimated source/target similarity. Figure 2a
shows the minimum function value obtained with respect to iterations for the
proposed method and the baselines. Each method starts with the same four
random observations. Our method outperforms the baselines by achieving 95%
of the minimum value in the 8th iteration. The performance of Env-GP method is
poor because it estimates source and target similarity by measuring the difference
between the source and the target functions, not their optima location. The other
two transfer learning baselines also show poor performance as some of the sources
are not similar to the target. Figure 2b shows the mixture proportions of sources
and the target distribution with respect to increasing iterations. As seen from
the Figure, the source with minima at $[\mathbf{0.4, 0.4, 0.4}]$ contributes maximally in
the mixture distribution. As the iterations increase, the contribution of sources
with distant minimum (from target's minimum) reduces to small values.

4.3   Hyperparameter Tuning

We tune hyperparameters of two classification algorithms: Support Vector
Machine (SVM) and Elastic net. We consider 5 binary classification datasets -
'banana', 'breast cancer', 'colon cancer', 'german numer' and 'diabetes' (LibSVM
repository [31]). A brief summary about these datasets is provided in **Table 1**.

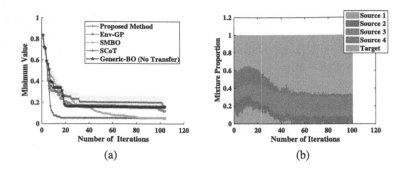

Fig. 2. Synthetic Experiments: (a) Minimum value vs optimisation iterations (b) Proportions of various sources and the target in the mixture distribution with respect to optimisation iterations.

Table 1. Binary datasets used in our experiments.

Dataset	Number of data points	Number of features
Diabetes	768	8
Banana	5300	2
Breast cancer	683	10
Colon cancer	62	2000
German numer	1000	24

We train a classifier for each dataset. For each dataset, 70% data is chosen randomly for training and the rest 30% used for validation. Hyperparameter tuning involves optimising validation performance (measured via AUC) as a *function* of hyperparameter values. We consider the first 4 hyperparameter tuning functions ('banana', 'breast cancer', 'colon cancer' and 'german numer') as source functions and the hyperparameter tuning function of the 'diabetes' as the target function. We assume that the several samples from the source functions are already available.

SVM with RBF kernel has two hyperparameters to tune: cost parameter (C) and kernel parameter (γ). The range for γ is set as $\left[10^{-3}, 10^{3}\right]$ and the same for C is $\left[2^{-3}, 2^{3}\right]$. We run our proposed method and other baseline methods and report the results in Fig. 3. Figure 3a shows the AUC performance on the held-out validation set. The baseline, Generic-BO (No Transfer)shows better performance than other three transfer learning baselines. In contrast, the proposed method is able to outperform Generic-BO (No Transfer) converging faster than all the baselines. Figure 3b shows the proportion of contributions from different sources versus iterations.

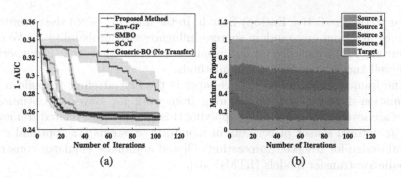

Fig. 3. Hyperparameter tuning (SVM): (a) 1-AUC vs optimisation iterations (b) Proportions of sources and the target in the mixture distribution with respect to optimisation iterations.

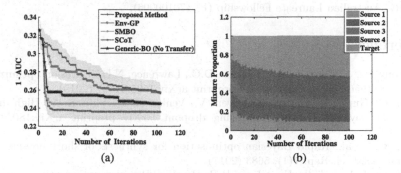

Fig. 4. Hyperparameter tuning (Elastic net): (a) 1-AUC vs optimisation iterations (b) Proportions of sources and the target in the mixture distribution with respect to optimisation iterations.

Elastic net has two hyperparameters to tune: L_1 and L_2 penalty weights. The ranges for both these hyperparameters are set as $[10^{-2}, 10^0]$. The performance in terms of AUC on the held-out validation set is shown in Fig. 4a. Our method performs significantly better than all the baselines. A plot depicting the proportions of contributions from different sources versus iterations is shown in Fig. 4b. The source code used for all these experiments are available at https://github. com/AnilRamachandran/ITTLBO.git and the datasets are available at https:// doi.org/10.7910/DVN/LRNLZV.

5   Conclusion

We propose a novel information-theoretic transfer learning algorithm for Bayesian optimisation. Our algorithm is based on constructing a mixture distribution of optima from both sources and the target combining them by the divergence between the optima distributions. This biased distributions is then used to formulate a new information theoretic acquisition function in a manner similar to the Predictive Entropy Search. In the limit $(T \to \infty)$ the optimisation becomes free from any random sources influence with probability 1. We evaluate our algorithm with diverse optimisation tasks and show that it outperforms other well known transfer learning methods.

The framework proposed in this paper is the first attempt to build a novel information-theoretic transfer learning framework for Bayesian optimisation. There are several possibilities for applying this idea in other related frameworks such as optimal sensor placements in monitoring systems [32], optimal experimental design for reservoir forecasting [33] and automatic emulator constructor for radiative transfer models (RTMs) [34].

Acknowledgment. This research was partially funded by the Australian Government through the Australian Research Council (ARC) and the Telstra-Deakin Centre of Excellence in Big Data and Machine Learning. Professor Venkatesh is the recipient of an ARC Australian Laureate Fellowship (FL170100006).

References

1. González, J., Longworth, J., James, D.C., Lawrence, N.D.: Bayesian optimization for synthetic gene design. arXiv preprint arXiv:1505.01627 (2015)
2. Li, C., Gupta, S., Rana, S., Nguyen, V., Venkatesh, S., Shilton, A.: High dimensional Bayesian optimization using dropout. arXiv preprint arXiv:1802.05400 (2018)
3. Li, C., et al.: Rapid Bayesian optimisation for synthesis of short polymer fiber materials. Sci. Rep. **7**(1), 5683 (2017)
4. Snoek, J., Larochelle, H., Adams, R.P.: Practical Bayesian optimization of machine learning algorithms. In: Pereira, F., Burges, C.J.C., Bottou, L., Weinberger, K.Q. (eds.) Advances in Neural Information Processing Systems 25, pp. 2951–2959. Curran Associates, Inc. (2012)
5. Kushner, H.J.: A new method of locating the maximum point of an arbitrary multipeak curve in the presence of noise. J. Basic Eng. **86**(1), 97–106 (1964)
6. Močkus, J., Tiesis, V., Žilinskas, A.: The application of bayesian methods for seeking the extremum. In: Toward Global Optimization, vol. 2, pp. 117–128. Elsevier (1978)
7. Srinivas, N., Krause, A., Kakade, S.M., Seeger, M.W.: Information-theoretic regret bounds for Gaussian process optimization in the bandit setting. IEEE Trans. Inf. Theory **58**(5), 3250–3265 (2012)
8. Hernández-Lobato, J.M., Hoffman, M.W., Ghahramani, Z.: Predictive entropy search for efficient global optimization of black-box functions. In: Ghahramani, Z., Welling, M., Cortes, C., Lawrence, N.D., Weinberger, K.Q. (eds.) Advances in Neural Information Processing Systems 27, pp. 918–926. Curran Associates, Inc. (2014)
9. Brochu, E., Cora, V.M., De Freitas, N.: A tutorial on Bayesian optimization of expensive cost functions, with application to active user modeling and hierarchical reinforcement learning. arXiv preprint arXiv:1012.2599 (2010)
10. Jones, D.R., Perttunen, C.D., Stuckman, B.E.: Lipschitzian optimization without the Lipschitz constant. J. Optim. Theory Appl. **79**(1), 157–181 (1993)
11. Bardenet, R., Brendel, M., Kégl, B., Sebag, M.: Collaborative hyperparameter tuning. In: ICML (2), pp. 199–207 (2013)

12. Yogatama, D., Mann, G.: Efficient transfer learning method for automatic hyper-parameter tuning. Transfer **1**, 1 (2014)
13. Swersky, K., Snoek, J., Adams, R.P.: Multi-task Bayesian optimization. In: Advances in Neural Information Processing Systems, pp. 2004–2012 (2013)
14. Joy, T.T., Rana, S., Gupta, S.K., Venkatesh, S.: Flexible transfer learning framework for bayesian optimisation. In: Bailey, J., Khan, L., Washio, T., Dobbie, G., Huang, J.Z., Wang, R. (eds.) PAKDD 2016. LNCS (LNAI), vol. 9651, pp. 102–114. Springer, Cham (2016). https://doi.org/10.1007/978-3-319-31753-3_9
15. Shilton, A., Gupta, S., Rana, S., Venkatesh, S.: Regret bounds for transfer learning in Bayesian optimisation. In: Artificial Intelligence and Statistics, pp. 307–315 (2017)
16. Feurer, M., Springenberg, J.T., Hutter, F.: Initializing Bayesian hyperparameter optimization via meta-learning. In: AAAI, pp. 1128–1135 (2015)
17. Feurer, M., Letham, B., Bakshy, E.: Scalable meta-learning for Bayesian optimization. arXiv preprint arXiv:1802.02219 (2018)
18. Rasmussen, C., Williams, C.: Gaussian processes for machine learning. Gaussian Processes for Machine Learning (2006)
19. Williams, C.K.I., Barber, D.: Bayesian classification with Gaussian processes. IEEE Trans. Pattern Anal. Mach. Intell. **20**(12), 1342–1351 (1998)
20. Hutter, F., Hoos, H.H., Leyton-Brown, K.: Sequential model-based optimization for general algorithm configuration. In: International Conference on Learning and Intelligent Optimization, pp. 507–523. Springer (2011)
21. Snoek, J., et al.: Scalable Bayesian optimization using deep neural networks. In: International Conference on Machine Learning, pp. 2171–2180 (2015)
22. Shahriari, B., Swersky, K., Wang, Z., Adams, R.P., de Freitas, N.: Taking the human out of the loop: a review of Bayesian optimization. Proc. IEEE **104**(1), 148–175 (2016)
23. Villemonteix, J., Vazquez, E., Walter, E.: An informational approach to the global optimization of expensive-to-evaluate functions. J. Glob. Optim. **44**(4), 509 (2009)
24. Hennig, P., Schuler, C.J.: Entropy search for information-efficient global optimization. J. Mach. Learn. Res. **13**, 1809–1837 (2012)
25. Ramachandran, A., Gupta, S., Rana, S., Venkatesh, S.: Selecting optimal source for transfer learning in bayesian optimisation. In: Geng, X., Kang, B.-H. (eds.) PRICAI 2018. LNCS (LNAI), vol. 11012, pp. 42–56. Springer, Cham (2018). https://doi.org/10.1007/978-3-319-97304-3_4
26. Minka, T.P.: A family of algorithms for approximate Bayesian inference. Ph.D. thesis, Massachusetts Institute of Technology (2001)
27. Chapelle, O., Li, L.: An empirical evaluation of Thompson sampling. In: Advances in Neural Information Processing Systems, pp. 2249–2257 (2011)
28. Liese, F., Vajda, I.: On divergences and informations in statistics and information theory. IEEE Trans. Inf. Theory **52**(10), 4394–4412 (2006)
29. Wang, Q., Kulkarni, S.R., Verdú, S.: A nearest-neighbor approach to estimating divergence between continuous random vectors. In: 2006 IEEE International Symposium on Information Theory, pp. 242–246. IEEE (2006)
30. Pérez-Cruz, F.: Kullback-Leibler divergence estimation of continuous distributions. In: IEEE International Symposium on Information Theory, 2008, ISIT 2008, pp. 1666–1670. IEEE (2008)
31. Chang, C.C., Lin, C.J.: LIBSVM: a library for support vector machines. ACM Trans. Intell. Syst. Technol. (TIST) **2**(3), 27 (2011)

32. Krause, A., Singh, A., Guestrin, C.: Near-optimal sensor placements in Gaussian processes: theory, efficient algorithms and empirical studies. J. Mach. Learn. Res. **9**, 235–284 (2008)
33. Busby, D.: Hierarchical adaptive experimental design for Gaussian process emulators. Reliab. Eng. Syst. Saf. **94**(7), 1183–1193 (2009)
34. Martino, L., Vicent, J., Camps-Valls, G.: Automatic emulator and optimized lookup table generation for radiative transfer models. In: Proceedings of IEEE International Geoscience and Remote Sensing Symposium (IGARSS) (2017)

A Unified Framework for Domain Adaptation Using Metric Learning on Manifolds

Sridhar Mahadevan[1](✉), Bamdev Mishra[2], and Shalini Ghosh[3]

[1] University of Massachusetts, Amherst, MA 01003, USA
mahadeva@cs.umass.edu
[2] Microsoft, Hyderabad 500032, Telangana, India
bamdevm@microsoft.com
[3] Samsung Research America, Mountain View, CA 94043, USA
shalini.ghosh@samsung.com

Abstract. We present a novel framework for domain adaptation, whereby both *geometric* and *statistical* differences between a labeled source domain and unlabeled target domain can be reconciled using a unified mathematical framework that exploits the curved Riemannian geometry of statistical manifolds. We exploit a simple but important observation that as the space of covariance matrices is both a Riemannian space as well as a homogeneous space, the shortest path geodesic between two covariances on the manifold can be computed analytically. Statistics on the SPD matrix manifold, such as the geometric mean of two SPD matries can be reduced to solving the well-known Riccati equation. We show how the Ricatti-based solution can be constrained to not only reduce the statistical differences between the source and target domains, such as aligning second order covariances and minimizing the maximum mean discrepancy, but also the underlying geometry of the source and target domains using diffusions on the underlying source and target manifolds. Our solution also emerges as a consequence of optimal transport theory, which shows that the optimal transport mapping between source and target distributions that are multivariate Gaussians is a function of the geometric mean of the source and target covariances, a quantity that also minimizes the Wasserstein distance. A key strength of our proposed approach is that it enables integrating multiple sources of variation between source and target in a unified way, by reducing the combined objective function to a nested set of Ricatti equations where the solution can be represented by a cascaded series of geometric mean computations. In addition to showing the theoretical optimality of our solution, we present detailed experiments using standard transfer learning testbeds from computer vision comparing our proposed algorithms to past work in domain adaptation, showing improved results over a large variety of previous methods. Code related to this paper is available at: https://github.com/sridharmahadevan/Geodesic-Covariance-Alignment.

© Springer Nature Switzerland AG 2019
M. Berlingerio et al. (Eds.): ECML PKDD 2018, LNAI 11052, pp. 843–860, 2019.
https://doi.org/10.1007/978-3-030-10928-8_50

1   Introduction

When we apply machine learning [19] to real-world problems, e.g., in image recognition [18] or speech recognition [13], a significant challenge is the need for having large amounts of (labeled) training data, which may not always be available. Consequently, there has been longstanding interest in developing machine learning techniques that can transfer knowledge across domains, thereby alleviating to some extent the need for training data as well as the time required to train the machine learning system. A detailed survey of transfer learning is given in [22].

Traditional machine learning assumes that the distribution of test examples follows that of the training examples [8], whereas in transfer learning, this assumption is usually violated. *Domain adaptation* (DA) is a well-studied formulation of transfer learning that is based on developing methods that deal with the change of distribution in test instances as compared with training instances [5,11]. In this paper, we propose a new framework for domain adaptation, based on formulating transfer from source to target as a problem of geometric mean metric learning on manifolds. Our proposed approach enables integrating multiple sources of variation between source and target in a unified framework with a theoretically optimal solution. We also present detailed experiments using standard transfer learning testbeds from computer vision, showing how our proposed algorithms give improved results compared to existing methods.

A general way to model domain adaptation is using the framework of optimal transport (OT) [9,23]. The OT problem, originally studied by Monge in 1781, is defined as the effort needed to move a given quantity of dirt (formally modeled by some source distribution) to a new location with no loss in volume (modeled formally by a target distribution), minimizing some cost of transportation per unit volume of dirt. A revised formulation by Kantorovich in 1942 allowed the transport map to be (in the discrete case) a bistochastic matrix, whose rows sum to the source distribution, and whose columns sum to the target distribution. When the cost of transportation is given by a distance metric, the optimal transport distance is defined as the *Wasserstein distance*. The solution we propose in this paper, based on the geometric mean of the source and target covariances, can be derived from optimal transport theory, under the assumption that the source and target distributions correspond to multivariate Gaussian distributions. Our approach goes beyond the simple solution proposed by optimal transport theory in that we take into account not only the cost of transportation, but also other factors, such as the geometry of source and target domains. In addition, we use the *weighted* geometric mean, giving us additional flexibility in tuning the solution.

Background: One standard approach of domain adaptation is based on modeling the *covariate shift* [1]. Unlike traditional machine learning, in DA, the training and test examples are assumed to have different distributions. It is usual in DA to categorize the problem into different types: (i) semi-supervised domain adaptation (ii) unsupervised domain adaptation (iii) multi-source domain adaptation (iv) heterogeneous domain adaptation.

Another popular approach to domain adaptation is based on aligning the distributions between source and target domains. A common strategy is based on the maximum mean discrepancy (MMD) metric [7], which is a nonparametric technique for measuring the dissimilarity of empirical distributions between source and target domains. Domain-invariant projection is one method that seeks to minimize the MMD measure using optimization on the Grassmannian manifold of fixed-dimensional subspaces of n-dimensional Euclidean space [2].

Linear approaches to domain adaptation involve the use of alignment of lower-dimensional subspaces or covariances from a data source domain $D_s = \{x_i^s\}$ with labels $L_s = \{y_i\}$ to a target data domain $D_t = \{x_i^t\}$. We assume both x_i^s and x_i^t are n-dimensional Euclidean vectors, representing the values of n features of each training example. One popular approach to domain adaptation relies on first projecting the data from the source and target domains onto a low-dimensional subspace, and then finding correspondences between the source and target subspaces. Of these approaches, the most widely used one is Canonical Correlation Analysis (CCA) [17], a standard statistical technique used in many applications of machine learning and bioinformatics. Several nonlinear versions [15] and deep learning variants [27] of CCA have been proposed. These methods often require explicit correspondences between the source and target domains to learn a common subspace. Because CCA finds a linear subspace, a family of manifold alignment methods have been developed that extend CCA [10,25] to exploit the nonlinear structure present in many datasets.

In contrast to using a single shared subspace across source and target domains, *subspace alignment* finds a linear mapping that transforms the source data subspace into the target data subspace [14]. To explain the basic algorithm, let $P_S, P_T \in \mathbb{R}^{n \times d}$ denote the two sets of basis vectors that span the subspaces for the "source" and "target" domains. Subspace alignment attempts to find a linear mapping M that minimizes

$$F(M) = \|P_S M - P_T\|_F^2.$$

It can be shown that the solution to the above optimization problem is simply the dot product between P_S and P_T, i.e.,:

$$M^* = \operatorname{argmin}_M F(M) = P_S^T P_T.$$

Another approach exploits the property that the set of k-dimensional subspaces in n-dimensional Euclidean space forms a curved manifold called the *Grassmannian* [12], a type of matrix manifold. The domain adaptation method called geodesic flow kernels (GFK) [16] is based on constructing a distance function between source and target subspaces that is based on the geodesic or shortest path between these two elements on the Grassmannian.

Rather than aligning subspaces, a popular technique called CORAL [24] aligns correlations between source and target domains. Let μ_s, μ_t and A_s, A_t represent the mean and covariance of the source and target domains, respectively.

CORAL finds a linear transformation A that minimizes the distance between the second-order statistics of the source and target features (which can be assumed as normalized with zero means). Using the Frobenius (Euclidean) norm as the matrix distance metric, CORAL is based on solving the following optimization problem:

$$\min_{A} \left\| A^T A_s A - A_t \right\|_F^2 . \tag{1}$$

where A_s, A_t are of size $n \times n$. Using the singular value decomposition of A_s and A_t, CORAL [24] computes a particular closed-form solution[1] to find the desired linear transformation A.

Novelty of Our Approach: Our proposed solution differs from the above previous approaches in several fundamental ways: one, we explicitly model the space of covariance matrices as a curved Riemannian manifold of symmetric positive definite (SPD) matrices. Note the difference of two SPD matrices is not an SPD matrix, and hence they do not form a vector space. Second, our approach can be shown to be both unique and globally optimal, unlike some of the above approaches. Uniqueness and optimality derive from the fact that we reduce all domain adaptation computations to nested equations involving solving the well-known *Riccati* equation [6].

The organization of the paper is as follows. In Sect. 2, we show the connection between the domain adaptation problem to the metric learning problem. In particular, we discuss the Riccati point of view for the domain adaptation problem. Section 3 discusses briefly the Riemannian geometry of the space of SPD matrices. Sections 4 and 5 discuss additional domain adaptation formulations. Our proposed algorithms are presented in Sects. 6 and 7. Finally, in Sect. 8 we show the experimental results on the standard Office and the extended Office-Caltech10 datasets, where our algorithms show clear improvements over CORAL.

2   Domain Adaptation Using Metric Learning

In this section, we will describe the central idea of this paper: modeling the problem of domain adaptation as a geometric mean metric learning problem. Before explaining the specific approach, it will be useful to introduce some background. The metric learning problem [4] involves taking input data in $\mathbb{R}^n$ and constructing a (non)linear mapping $\Phi : \mathbb{R}^n \to \mathbb{R}^m$, so that the distance between two points

---

[1] The solution characterization in [24] is non unique. [24, Theorem 1] shows that the optimal A, for full-rank A_s and A_t, is characterized as $A = U_s \Sigma_s^{-1} U_s^T U_t \Sigma_t^{1/2} U_t^T$, where $U_s \Sigma_s U_s^T$ and $U_t \Sigma_t U_t^T$ are the eigenvalue decompositions of A_s and A_t, respectively. However, it can be readily checked that there exists a continuous set of optimal solutions characterized as $A = U_s \Sigma_s^{-1/2} U_s^T O U_t \Sigma_t^{1/2} U_t^T$, where O is any orthogonal matrix, i.e., $OO^T = O^T O = I$ of size $n \times n$. A similar construction for non-uniqueness of the CORAL solution also holds for rank deficient A_s and A_t.

x and y in $\mathbb{R}^n$ can be measured using the distance $\|\varPhi(x) - \varPhi(y)\|$. A simple app-roach is to learn a squared *Mahalanobis* distance: $\delta_A^2(x, y) = (x - y)^T A(x - y)$, where $x, y \in \mathbb{R}^n$ and A is an $n \times n$ symmetric positive definite (SPD) matrix. If we represent $A = W^T W$, for some linear transformation matrix W, then it is easy to see that $\delta_A^2(x, y) = \|Wx - Wy\|_F^2$, thereby showing that the Mahalanobis dis-tance is tantamount to projecting the data into a potentially lower-dimensional space, and measuring distances using Euclidean (Frobenius) norm. Typically, the matrix A is learned using some weak supervision, given two sets of training examples of the form:

$$S = \{(x_i, x_j) | x_i \text{ and } x_j \text{ are in the same class}\},$$

$$D = \{(x_i, x_j) | x_i \text{ and } x_j \text{ are in different classes}\}.$$

A large variety of metric learning methods can be designed based on formulating different optimization objectives based on functions over the S and D sets to extract information about the distance matrix A.

For our purposes, the method that will provide the closest inspiration to our goal of designing a domain adaptation method based on metric learning is the recently proposed *geometric mean metric learning* (GMML) algorithm [28]. GMML models the distance between points in the S set by the Mahalanobis distance $\delta_A(x_i, x_j), x_i, x_j \in S$ by exploiting the geometry of the SPD matrices, and crucially, also models the distance between points in the disagreement set D by the *inverse* metric $\delta_{A^{-1}}(x_i, x_j), x_i, x_j \in D$. GMML is based on solving the objective function over all SPD matrices A:

$$\min_{A \succ 0} \sum_{(x_i, x_j) \in S} \delta_A^2(x_i, x_j) + \sum_{(x_i, x_j) \in D} \delta_{A^{-1}}^2(x_i, x_j),$$

where $A \succ 0$ refers to the set of all SPD matrices.

Several researchers have previously explored the connection between domain adaptation and metric learning. One recent approach is based on constructing a transformation matrix A that both minimizes the difference between the source and target distributions based on the previously noted MMD metric, but also captures the manifold geometry of source and target domains, and attempts to preserve the discriminative power of the label information in the source domain [26]. Our approach builds on these ideas, with some significant differences. One, we use an objective function that is based on finding a solution that lies on the geodesic between source and target (estimated) covariance matrices (which are modeled as symmetric positive definite matrices). Second, we use a cascaded series of geometric mean computations to balance multiple factors. We describe these ideas in more detail in this and the next section.

We now describe how the problem of domain adaptation can be considered as a type of metric learning problem, called geometric mean metric learning (GMML) [28]. Recall that in domain adaptation, we are given a source dataset D_s (usually with a set of training labels) and a target dataset D_t (unlabeled). The aim of domain adaptation, as reviewed above, is to construct an intermediate representation that combines some of the features of both the source and target domains, with the rationale being that the distribution of target features differs from that of the source. Relying purely on either the source or the target features is therefore suboptimal, and the challenge is to determine what intermediate representation will provide optimal transfer between the domains.

To connect metric learning to domain adaptation, note that we can define the two sets S and D in the metric learning problem as associated with the source and target domains respectively, whereby[2]

$$S \subseteq D_s \times D_s = \{(x_i, x_j)|x_i \in D_s, x_j \in D_s\}$$
$$D \subseteq D_t \times D_t = \{(x_i, x_j)|x_i \in D_t, x_j \in D_t\}.$$

Our approach seeks to exploit the nonlinear geometry of covariance matrices to find a Mahalanobis distance matrix A, such that we can represent distances in the source domain using A, but crucially we measure distances in the target domain using the inverse A^{-1}.

$$\min_{A \succ 0} \sum_{(x_i,x_j) \in S} (x_i - x_j)^T A (x_i - x_j) + \sum_{(x_i,x_j) \in D} (x_i - x_j)^T A^{-1} (x_i - x_j).$$

To provide some intuition here, we observe that as we vary A to reduce the distance $\sum_{(x_i,x_j) \in S}(x_i - x_j)^T A(x_i - x_j)$ in the source domain, we simultaneously increase the distance in the target domain by minimizing $\sum_{(x_i,x_j) \in D}(x_i - x_j)^T A^{-1}(x_i - x_j)$, and vice versa. Consequently, by appropriately choosing A, we can seek to minimize the above sum. We can now use the matrix trace to reformulate the Mahalanobis distances:

$$\min_{A \succ 0} \sum_{(x_i,x_j) \in S} \text{tr}(A(x_i - x_j)(x_i - x_j)^T) + \sum_{(x_i,x_j) \in D} \text{tr}(A^{-1}(x_i - x_j)(x_i - x_j)^T).$$

Denoting the source and target covariance matrices A_s and A_t as:

$$A_s := \sum_{(x_i,x_j) \in S} (x_i - x_j)(x_i - x_j)^T \tag{2}$$

$$A_t := \sum_{(x_i,x_j) \in D} (x_i - x_j)(x_i - x_j)^T, \tag{3}$$

we can finally write a new formulation of the domain adaptation problem as minimizing the following objective function to find the SPD matrix A such that:

$$\min_{A \succ 0} \omega(A) := \text{tr}(AA_s) + \text{tr}(A^{-1}A_t). \tag{4}$$

---

[2] We note that while there are alternative ways to define the S and D sets, the essence of our approach remains similar.

3   Riemannian Geometry of SPD Matrices

In this section, we outline some other formulations of domain adaptation that will be useful to discuss for presenting our overall approach.

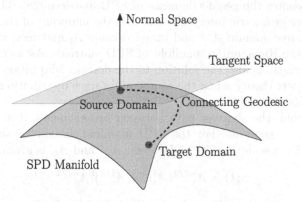

Fig. 1. The space of symmetric positive definite matrices forms a Riemannian manifold, as illustrated here. The methods we propose are based on computing geodesics (the shortest distance), shown in the dotted line, between source domain information and target domain information.

As Fig. 1 shows, our proposed approach to domain adaptation builds on the nonlinear geometry of the space of SPD (or covariance) matrices, we review some of this material first [6]. Taking a simple example of a 2×2 SPD matrix M, where:

$$M = \begin{bmatrix} a & b \\ b & c \end{bmatrix},$$

where $a > 0$, and the SPD requirement implies the positivity of the determinant $ac - b^2 > 0$. Thus, the set of all SPD matrices of size 2×2 forms the interior of a cone in $\mathbb{R}^3$. More generally, the space of all $n \times n$ SPD matrices forms a manifold of non-positive curvature in $\mathbb{R}^{n^2}$ [6]. In the CORAL objective function in Eq. (1), the goal is to find a transformation A that makes the source covariance resemble the target as closely as possible. Our approach simplifies Eq. (1) by restricting the transformation matrix A to be a SPD matrix, i.e, $A \succ 0$, and furthermore, we solve the resulting nonlinear equation *exactly* on the manifold of SPD matrices. More formally, we solve the *Riccati equation* [6]:

$$AA_sA = A_t, \text{ for } A \succ 0, \tag{5}$$

where A_s and A_t are source and target covariances SPD matrices, respectively. Note that in comparison with the CORAL approach in Eq. (1), the A matrix is symmetric (and positive definite), so A and A^T are the same. The solution to the above Riccati equation is the well-known *geometric mean* or *sharp mean*, of the two SPD matrices, A_s^{-1} and A_t.

$$A = A_s^{-1} \natural_{\frac{1}{2}} A_t = A_s^{-\frac{1}{2}} (A_s^{\frac{1}{2}} A_t A_s^{\frac{1}{2}})^{\frac{1}{2}} A_s^{-\frac{1}{2}},$$

where $\natural_{\frac{1}{2}}$ is denotes the geometric mean of SPD matrices [28]. The sharp mean has an intuitive geometric interpretation: it is the midpoint of the geodesic connecting the source domain A_s^{-1} and target domain A_t matrices, where length is measured on the Riemannian manifold of SPD matrices. As mentioned earlier, the geometric mean is also the solution to the domain adaptation problem from optimal transport theory when the source and target distributions are given by multivariate Gaussians [23].

In a manifold, the shortest path between two elements, if it exists, can be represented by a geodesic. For the SPD manifold, it can be shown that the geodesic $\gamma(t)$ for a scalar $0 \leq t \leq 1$ between A_s^{-1} and A_t, is given by [6]:

$$\gamma(t) = A_s^{-1/2} (A_s^{1/2} A_t A_s^{1/2})^t A_s^{-1/2}.$$

It is common to denote $\gamma(t)$ as the so-called "weighted" sharp mean $A_s^{-1} \natural_t A_t$. It is easy to see that for $t = 0$, $\gamma(0) = A_s^{-1}$, and for $t = 1$, we have $\gamma(1) = A_t$. For the distinguished value of $t = 1/2$, it turns out that $\gamma(1/2)$ is the *geometric mean* of A_s^{-1} and A_t, respectively, and satisfies all the properties of a geometric mean [6].

$$\gamma(1/2) = A_s^{-1/2} \natural_{1/2} A_t = A_s^{-1/2} (A_s^{1/2} A_t A_s^{1/2})^{1/2} A_s^{-1/2}.$$

The following theorem summarizes some of the properties of the objective function given by Eq. (4).

Theorem 1. *[28, Theorem 3] The cost function $w(A)$ in Eq. (4) is both strictly convex as well as strictly geodesically convex on the SPD manifold.*

Theorem 1 ensures the uniqueness of the solution to Eq. (4).

4   Statistical Alignment Across Domains

A key strength of our approach is that it can exploit both geometric and statistical information, and multiple sources of alignment are integrated by solving nested sets of Ricatti equations. To illustrate this point, in this section we explicitly introduce a secondary criterion of aligning the source and target domains so that the underlying (marginal) distributions are similar. As our results show later, we obtain a significant improvement over CORAL on a standard computer vision dataset (Office/Caltech/Amazon problem). The reason our approach outperforms CORAL is that not only are we able to solve the Riccati equation uniquely, whereas the CORAL solution proposed is [24] is only a particular solution due to non-uniqueness[3], we can exploit multiple sources of information.

---

[3] (see footnote 1).

A common way to incorporate the statistical alignment constraint is based on minimizing the maximum mean discrepancy metric (MMD) [7], a nonparametric measure of the difference between two distributions.

$$\left\| \frac{1}{n} \sum_{i=1}^{n} W x_i^s - \frac{1}{m} \sum_{i=1}^{m} W x_i^t \right\|^2 = \text{tr}(AXLX^T), \tag{6}$$

where $X = \{x_1^s, \ldots, x_n^s, x_1^1, \ldots, x_t^n\} \in \mathbb{R}^{(n+m)}$ and $L \in \mathbb{R}^{(n+m)(n+m)}$, where $L(i,j) = \frac{1}{n^2}$ if $x_i, x_j \in X_s$, $L(i,j) = \frac{1}{m^2}$ if $x_i, x_j \in X_t$, and $L(i,j) = -\frac{1}{mn}$ otherwise. It is straightforward to show that $L \succeq 0$, a symmetric positive-semidefinite matrix [26]. We can now combine the MMD objective in Eq. (6) with the previous geometric mean objective in Eq. (4) to give rise to the following modified objective function:

$$\min_{A \succ 0} \xi(A) := \text{tr}(AA_s) + \text{tr}(A^{-1}A_t) + \text{tr}(AXLX^T). \tag{7}$$

We can once again find a closed-form solution to the modified objective in Eq. (7) by taking gradients:

$$\nabla \xi(A) = A_s - A^{-1}A_t A^{-1} + XLX^T = 0,$$

whose solution is now given $A = A_m^{-1} \sharp_{\frac{1}{2}} A_t$, where $A_m = A_s + XLX^T$.

5   Geometrical Diffusion on Manifolds

So far we have shown how the solution to the domain adaptation problem can be shown to involve finding the geometric mean of two terms, one involving the source covariance information and the Maximum Mean Discrepancy (MMD) of source and target training instances, and the second involving the target covariance matrix. In this section, we impose additional geometrical constraints on the solution that involve modeling the nonlinear manifold geometry of the source and target domains.

The usual approach is to model the source and target domains as a nonlinear manifold and set up a diffusion on a discrete graph approximation of the continuous manifold [20], using a random walk on a nearest neighbor graph connecting nearby points. Standard results have been established showing asymptotic convergence of the graph Laplacian to the underlying manifold Laplacian [3]. We can use the above algorithm to find two graph kernels K_s and K_t that are based on the eigenvectors of the random walk on the source and target domain manifold, respectively.

$$K_s = \sum_{i=1}^{m} e^{-\frac{\sigma_s^2}{2\lambda_i^s}} v_i^s (v_i^s)^T$$

$$K_t = \sum_{i=1}^{n} e^{-\frac{\sigma_t^2}{2\lambda_i^t}} v_i^t (v_i^t)^T.$$

Here, v_i^s and v_i^t refer to the eigenvectors of the random walk diffusion matrix on the source and target manifolds, respectively, and λ_i^s and λ_i^t refer to the corresponding eigenvalues.

We can now introduce a new objective function that incorporates the source and target domain manifold geometry:

$$\min_{A \succ 0} \ \eta(A) := \text{tr}(AX(K + \mu L)X^T) + \text{tr}(AA_s) + \text{tr}(A^{-1}A_t), \qquad (8)$$

where $K \succ 0$ and $K = \begin{pmatrix} K_s^{-1} & 0 \\ 0 & K_t^{-1} \end{pmatrix}$, and μ is a weighting term that combines the geometric and statistical constraints over A.

Once again, we can exploit the SPD nature of the matrices involved, the closed-form solution to Eq. (8) is $A = A_{gs} \sharp_{\frac{1}{2}} A_t$, where $A_{gs} = A_s + X(K + \mu L)X^T$.

6   Cascaded Weighted Geometric Mean

One additional refinement that we use is the notion of a *weighted* geometric mean. To explain this idea, we introduce the following Riemannian distance metric on the nonlinear manifold of SPD matrices:

$$\delta_R^2(X, Y) \equiv \left\| \log(Y^{-\frac{1}{2}} X Y^{-\frac{1}{2}}) \right\|_F^2$$

for two SPD matrices X and Y. Using this metric, we now introduce a *weighted* version of the previous objective functions in Eqs. (4), (7), and (8).

For the first objective function in Eq. (4), we get:

$$\min_{A \succ 0} \ \omega_t(A) := (1 - t)\,\delta_R^2(A, A_s^{-1}) + t\,\delta_R^2(A, A_t), \qquad (9)$$

where $0 \le t \le 1$ is the weight parameter. The unique solution to (9) is given by the *weighted geometric mean* $A = A_s^{-1} \sharp_t A_t$ [28]. Note that the weighted metric mean is no longer strictly convex (in the Euclidean sense), but remains geodesically strictly convex [28, 6, Chap. 6].

Similarly, we introduce the weighted variant of the objective function given by Eq. (7):

$$\min_{A \succ 0} \ \xi_t(A) := (1 - t)\,\delta_R^2(A, (A_s + XLX^T)^{-1}) + t\,\delta_R^2(A, A_t), \qquad (10)$$

whose unique solution is given by $A = A_m^{-1} \sharp_t A_t$, where $A_m = A_s + XLX^T$ as before. A cascaded variant is obtained when we further exploit the SPD structure of A_s and XLX^T, i.e., $A_m = A_s \sharp_\gamma (XLX^T)$ (weighted geometric mean of A_s and (XLX^T) instead of $A_m = A_s + XLX^T$ (which is akin to the Euclidean mean of A_s and (XLX^T). Here, $0 \le \gamma \le 1$ is the weight parameter.

Finally, we obtain the weighted variant of the third objective function in Eq. (8):

$$\min_{A \succ 0} \eta_t(A) := (1 - t)\delta_R^2(A, (A_s + X(K + \mu I)X^T)^{-1} + t\, \delta_R^2(A, A_t), \qquad (11)$$

whose unique solution is given by $A = A_{gs}^{-1} \sharp_t A_t$, where $A_{gs} = A_s + X(K + \mu L)X^T$ as previously noted. Additionally, the cascaded variant is obtained when $A_s \sharp_\gamma (X(K + \mu L)X^T)$ instead of $A_{gs} = A_s + X(K + \mu L)X^T$.

7 Domain Adaptation Algorithms

We now describe the proposed domain adaptation algorithms, based on the above development of approaches reflecting geometric and statistical constraints on the inferred solution. All the proposed algorithms are summarized in Algorithm 1. The algorithms are based on finding a Mahalanobis distance matrix A interpolating source and target covariances (GCA1), incorporating an additional MMD metric (GCA2) and finally, incorporating the source and target manifold geometry (GCA3). It is noteworthy that all the variants rely on computation of the sharp mean, a unifying motif that ties together the various proposed methods. Modeling the Riemannian manifold underlying SDP matrices ensures the optimality and uniqueness of our proposed methods.

Algorithm 1. Algorithms for Domain Adaptation using Metric Learning

---

Given: A source dataset of labeled points $x_i^s \in D_s$ with labels $L_s = \{y_i\}$, and an unlabeled target dataset $x_i^t \in D_t$, with hyperparameters t, μ, and γ.

1. Define $X = \{x_1^s, \ldots, x_n^s, x_1^t, \ldots, x_m^t\}$.
2. Compute the source and target matrices A_s and A_t using Eqs. (2) and (3).
3. **Algorithm GCA1:** Compute the weighted geometric mean $A = A_s^{-1} \sharp_t A_t$ (see Eq. (9)).
4. **Algorithm GCA2:** Compute the weighted geometric mean taking additionally into account the MMD metric $A = A_m^{-1} \sharp_t A_t$ (see Eq. (10)), where $A_m = A_s + XLX^T$.
5. **Algorithm Cascaded-GCA2:** Compute the *cascaded* weighted geometric mean taking additionally into account the MMD metric $A = A_m^{-1} \sharp_t A_t$ (see Eq. (10)), $A_m = A_s \sharp_\gamma XLX^T$.
6. **Algorithm GCA3:** Compute the weighted geometric mean taking additionally into account the source and target manifold geometry $A = A_{gs}^{-1} \sharp_t A_t$ (see Eq. (11)), where $A_{gs} = A_s + X(K + \mu L)X^T$.
7. **Algorithm Cascaded-GCA3:** Compute the *cascaded* weighted geometric mean taking additionally into account the source and target manifold geometry $A = A_{gs}^{-1} \sharp_t A_t$ (see Eq. (11)), where $A_{gs} = A_s \sharp_\gamma (X(K + \mu L)X^T)$.
8. Use the learned A matrix to adapt source features to the target domain, and perform classification (e.g., using support vector machines).

---

8   Experimental Results

We present experimental results using the standard computer vision testbed used in prior work: the Office [16] and extended Office-Caltech10 [14] benchmark datasets. The Office-Caltech10 dataset contains 10 object categories from an office environment (e.g., keyboard, laptop, and so on) in four image domains: Amazon (**A**), Caltech256 (**C**), DSLR (**D**), and Webcam (**W**). The Office dataset has 31 categories (the previous 10 categories and 21 additional ones).

An exhaustive comparison of the three proposed methods with a variety of previous methods is summarized by the table in Table 1. The previous methods compared in the table refer to the unsupervised domain adaptation approach where a support vector machine (SVM) classifier is used. The experiments follow the standard protocol established by previous works in domain adaptation using this dataset. The features used (SURF) are encoded with 800-bin bag-of-words histograms and normalized to have zero mean and unit standard deviation in each dimension. As there are four domains, there are 12 ensuing transfer learning problems, denoted in Table 1 below as $\mathbf{A} \to \mathbf{D}$ (for Amazon to DSLR, etc.). For each of the 12 transfer learning tasks, the best performing method is indicated in boldface. We used 30 randomized trials for each experiment, and randomly sample the same number of labeled images in the source domain as training set, and use all the unlabeled data in the target domain as the test set. All experiments used a support vector machine (SVM) method to measure classifier accuracy, using a standard `libsvm` package. The methods compared against in Table 1 include the following alternatives:

- **Baseline-S:** This approach uses the projection defined by using PCA in the source domain to project both source and target data.
- **Baseline-T:** Here, PCA is used in the target domain to extract a low-dimensional subspace.
- **NA:** No adaptation is used and the original subspace is used for both source and target domains.
- **GFK:** This approach refers to the *geodesic flow kernel* [16], which computes the geodesic on the Grassmannian between the PCA-derived source and target subspaces computed from the source and target domains.
- **TCA:** This approach refers to the transfer component analysis method [21].
- **SA:** This approach refers to the subspace alignment method [14].
- **CORAL:** This approach refers to the correlational alignment method [24].
- **GCA1:** This is a new proposed method, based on finding the weighted geometric mean of the inverse of the source matrix A_s and the target matrix A_t.
- **GCA2:** This is a new proposed method, based on finding the (non-cascaded) weighted geometric mean of the inverse of the source matrix A_m and the target matrix A_t.
- **GCA3:** This is a new proposed method, based on finding the (non-cascaded) weighted geometric mean of the inverse of the source matrix A_{gs} and the target matrix A_t.

- **Cascaded-GCA2:** This is a new proposed method, based on finding the cascaded geometric mean of the inverse revised source matrix A_m and target matrix A_t
- **Cascaded-GCA3:** This is a new proposed method, based on finding the cascaded geometric mean of the inverse revised source matrix A_{gs} and target matrix A_t

Table 1. Recognition accuracy with unsupervised domain adaptation using SVM classifier (Office dataset + Caltech10). Our proposed methods (labeled GCAXX and Cascaded-GCAXX) perform better than all other methods in a majority of the 12 transfer learning tasks.

Method	$C \rightarrow A$	$D \rightarrow A$	$W \rightarrow A$	$A \rightarrow C$	$D \rightarrow C$	$W \rightarrow C$
NA	44.0	34.6	30.7	35.7	30.6	23.4
B-S	44.3	36.8	32.9	36.8	29.6	24.9
B-T	44.5	38.6	34.2	37.3	31.6	28.4
GFK	44.8	37.9	37.1	38.3	31.4	29.1
TCA	47.2	38.8	34.8	40.8	33.8	30.9
SA	46.1	**42.0**	**39.3**	39.9	35.0	31.8
CORAL	47.1	38.0	37.7	40.5	33.9	34.4
GCA1	48.4	40.1	38.6	**41.4**	**36.0**	**35.0**
GCA2	**48.9**	40.0	38.6	41.0	35.9	**35.0**
GCA3	48.4	40.9	37.6	41.1	36	33.6
Cascaded-GCA2	**49.5**	41.1	38.6	41.1	**36**	**35.1**
Cascaded-GCA3	**49.4**	41.0	38.5	41.0	35.9	35
Method	$A \rightarrow D$	$C \rightarrow D$	$W \rightarrow D$	$A \rightarrow W$	$C \rightarrow W$	$D \rightarrow W$
NA	34.5	36.0	67.4	26.1	29.1	70.9
B-S	36.1	38.9	73.6	**42.5**	34.6	75.4
B-T	32.5	35.3	73.6	37.3	34.2	80.5
GFK	32.5	36.1	74.6	39.8	34.9	79.1
TCA	36.4	39.2	72.1	38.1	36.5	80.3
SA	**44.5**	38.6	34.2	37.3	31.6	28.4
CORAL	38.1	39.2	84.4	38.2	39.7	85.4
GCA1	39.2	40.9	85.1	40.9	41.1	**87.2**
GCA2	39.9	41.4	85.1	40.1	41.1	**87.2**
GCA3	38.9	41.6	84.9	39.9	**41.4**	86.8
Cascaded-GCA2	40.4	40.7	**85.5**	41.3	40.3	**87.2**
Cascaded-GCA3	38.8	**42.4**	85.3	40.1	40.9	86.9

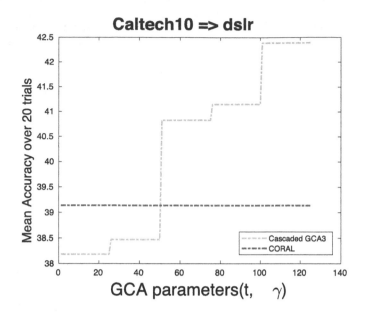

Fig. 2. Comparison of the cascaded GCA3 method with correlational alignment (CORAL). The parameters t and γ were varied between 0.1 and 0.9.

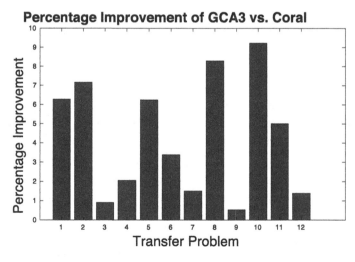

Fig. 3. Comparison of the cascaded GCA3 method with correlational alignment (CORAL) in terms of percentage improvement.

One question that arises in proposed algorithms is how to choose the value of t in computing the weighted sharp mean. Figure 2 illustrates the variation in performance of the cascaded GCA3 method over CORAL over the range $t, \gamma \in (0.1, 0.9)$, and μ fixed for simplicity. Repeating such experiments over all 12 transfer learning tasks, Fig. 3 shows the percentage improvement of the

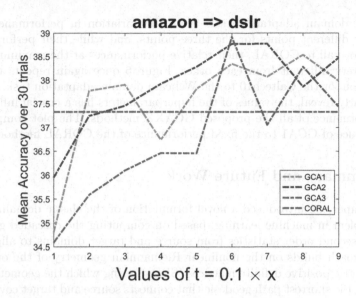

Fig. 4. Comparison of the three proposed GCA methods (GCA1, GCA2, and GCA3) with correlational alignment (CORAL).

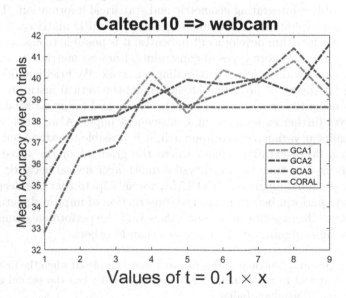

Fig. 5. Comparison of the three proposed GCA methods (GCA1, GCA2, and GCA3) with correlational alignment (CORAL).

cascaded GCA3 method over correlational alignment (CORAL), using the best discovered value of all three hyperparameters using cross-validation. Figure 4 compares the performance of the proposed GCA1, GCA2, and GCA3 methods where just the t hyperparameter was varied between 0.1 and 0.9 for the Amazon to DSLR domain adaptation task. Note the variation in performance with t occurs at different points for the three points, and while their performance is superior overall to CORAL, their relative performances at the maximum values are not very different from each other. Figure 5 once again repeats the same comparison for the Caltech10 to the Webcam domain adaptation task. As these plots clearly reveal, the values of the hyperparameters has a crucial influence on the performance of all the proposed GCAXX methods. The plot compares the performance of GCA1 to the fixed performance of the CORAL method.

9   Summary and Future Work

In this paper, we introduced a novel formulation of the classic domain adaptation problem in machine learning, based on computing the cascaded geometric mean of second order statistics from source and target domains to align them. Our approach builds on the nonlinear Riemannian geometry of the open cone of symmetric positive definite matrices (SPDs), using which the geometric mean lies along the shortest path geodesic that connects source and target covariances. Our approach has three key advantages over previous work: (a) Simplicity: The Riccati equation is a mathematically elegant solution to the domain adaptation problem, enabling integrating geometric and statistical information. (b) Theory: Our approach exploits the Riemannian geometry of SPD matrices. (c) Extensibility: As our algorithm development indicates, it is possible to easily extend our approach to capture more types of constraints, from geometrical to statistical.

There are many directions for extending our work. We briefly alluded to optimal transport theory as providing an additional theoretical justification for our solution of using the geometric mean of the source and target covariances, a link that deserves further exploration in a subsequent paper. Also, while we did not explore nonlinear variants of our approach, it is possible to extend our approach to develop a deep learning version where the gradient of the three objective functions is used to tune the weights of a multi-layer neural network. As in the case of correlational alignment (CORAL), we anticipate that the deep learning variants may perform better due to the construction of improved features of the training data. The experimental results show that the performance improvement tends to be more significant in some cases than in others.

Acknowledgments. Portions of this research were completed when the first and third authors were at SRI International, Menlo Park, CA and when the second author was at Amazon.com, Bangalore, India.

References

1. Adel, T., Zhao, H., Wong, A.: Unsupervised domain adaptation with a relaxed covariate shift assumption. In: AAAI, pp. 1691–1697 (2017)
2. Baktashmotlagh, M., Harandi, M.T., Lovell, B.C., Salzmann, M.: Unsupervised domain adaptation by domain invariant projection. In: ICCV, pp. 769–776 (2013)

3. Belkin, M., Niyogi, P.: Convergence of Laplacian eigenmaps. In: NIPS, pp. 129–136 (2006)
4. Bellet, A., Habrard, A., Sebban, M.: Metric Learning. Synthesis Lectures on Artificial Intelligence and Machine Learning. Morgan and Claypool Publishers, San Rafael (2015)
5. Ben-David, S., Blitzer, J., Crammer, K., Pereira, F.: Analysis of representations for domain adaptation. In: NIPS (2006)
6. Bhatia, R.: Positive Definite Matrices. Princeton Series in Applied Mathematics. Princeton University Press, Princeton (2007)
7. Borgwardt, K.M., Gretton, A., Rasch, M.J., Kriegel, H.P., Schölkopf, B., Smola, A.J.: Integrating structured biological data by kernel maximum mean discrepancy. In: ISMB (Supplement of Bioinformatics), pp. 49–57 (2006)
8. Cesa-Bianchi, N.: Learning the distribution in the extended PAC model. In: ALT, pp. 236–246 (1990)
9. Courty, N., Flamary, R., Tuia, D., Rakotomamonjy, A.: Optimal transport for domain adaptation. IEEE Trans. Pattern Anal. Mach. Intell. **39**(9), 1853–1865 (2017)
10. Cui, Z., Chang, H., Shan, S., Chen, X.: Generalized unsupervised manifold alignment. In: NIPS, pp. 2429–2437 (2014)
11. Daume, H.: Frustratingly easy domain adaptation. In: ACL (2007)
12. Edelman, A., Arias, T.A., Smith, S.T.: The geometry of algorithms with orthogonality constraints. SIAM J. Matrix Anal. Appl. **20**(2), 303–353 (1998)
13. Fayek, H.M., Lech, M., Cavedon, L.: Evaluating deep learning architectures for speech emotion recognition. Neural Netw. **92**, 60–68 (2017)
14. Fernando, B., Habrard, A., Sebban, M., Tuytelaars, T.: Subspace alignment for domain adaptation. Technical report, arXiv preprint arXiv:1409.5241 (2014)
15. Fukumizu, K., Bach, F.R., Gretton, A.: Statistical convergence of kernel CCA. In: NIPS, pp. 387–394 (2005)
16. Gong, B., Shi, Y., Sha, F., Grauman, K.: Geodesic flow kernel for unsupervised domain adaptation. In: CVPR, pp. 2066–2073 (2012)
17. Hotelling, H.: Relations between two sets of variates. Biometrika **28**, 321–377 (1936)
18. Krizhevsky, A., Sutskever, I., Hinton, G.E.: Imagenet classification with deep convolutional neural networks. Commun. ACM **60**(6), 84–90 (2017)
19. Murphy, K.P.: Machine Learning : A Probabilistic Perspective. MIT Press, Cambridge (2013)
20. Nadler, B., Lafon, S., Coifman, R.R., Kevrekidis, I.G.: Diffusion maps, spectral clustering and eigenfunctions of fokker-planck operators. In: NIPS, pp. 955–962 (2005)
21. Pan, S.J., Tsang, I.W., Kwok, J.T., Yang, Q.: Domain adaptation via transfer component analysis. IEEE Trans. Neural Netw. **22**(2), 199–210 (2011). https://doi.org/10.1109/TNN.2010.2091281
22. Pan, S., Yang, Q.: A survey on transfer learning. IEEE Trans. Knowl. Data Eng. **22**(10), 1345–1359 (2010)
23. Peyré, G., Cuturi, M.: Computational Optimal Transport. ArXiv e-prints, March 2018
24. Sun, B., Feng, J., Saenko, K.: Return of frustratingly easy domain adaptation. In: AAAI, pp. 2058–2065 (2016)
25. Wang, C., Mahadevan, S.: Manifold alignment without correspondence. In: IJCAI, pp. 1273–1278 (2009)

26. Wang, H., Wang, W., Zhang, C., Xu, F.: Cross-domain metric learning based on information theory. In: AAAI, pp. 2099–2105 (2014)
27. Wang, W., Arora, R., Livescu, K., Srebro, N.: Stochastic optimization for deep CCA via nonlinear orthogonal iterations. In: ALLERTON, pp. 688–695 (2015)
28. Zadeh, P., Hosseini, R., Sra, S.: Geometric mean metric learning. In: ICML, pp. 2464–2471 (2016)

Author Index

Printed in the United States
By Bookmasters